Photonic Quantum Technologies

Photonic Quantum Technologies

Science and Applications

Volume 1

Edited by Mohamed Benyoucef

WILEY-VCH

Editor

Mohamed Benyoucef
University of Kassel
Institute of Physics, Center for Interdisciplinary Nanostructure Science and Technology (CINSaT)
Heinrich-Plett-Str. 40
34132 Kassel
Germany

Cover Image: © Jurik Peter/Shutterstock

Library of Congress Card No.: applied for

British Library Cataloguing-in-Publication Data
A catalog record for this book is available from the British Library.

Bibliographic information published by the Deutsche Nationalbibliothek
The Deutsche Nationalbibliothek lists this publication in the Deutsche Nationalbibliografie; detailed bibliographic data are available on the Internet at <http://dnb.d-nb.de>.

Print ISBN: 978-3-527-35260-9
ePDF ISBN: 978-3-527-83741-0
ePub ISBN: 978-3-527-83743-4
oBook ISBN: 978-3-527-83742-7

Typesetting Straive, Chennai, India
Printing and Binding CPI Group (UK) Ltd, Croydon, CR0 4YY

C9783527352609_140524

I dedicate this book to my parents, for inspiring me throughout my life;
to my wife, and my children for their patience and support;
to my wonderful readers, thank you!

Contents

Contents to Volume 1

Preface *xv*
Abstracts and Keywords *xvii*

1 Introduction to Quantum Photonics *1*
Richard J. Warburton
1.1 The Photon *1*
1.2 The Light–Matter Interaction *2*
1.3 Single-Photon Sources *3*
1.4 Single-Photon Detectors *4*
1.5 Applications of the Photon in Quantum Technology *5*
1.5.1 Quantum Communication *5*
1.5.2 Quantum Simulation *5*
1.5.3 Quantum Computing *5*
References *6*

Part I Fundamentals of Quantum Technologies *7*

2 The Second Quantum Revolution: From Basic Concepts to Quantum Technologies *9*
Alain Aspect
2.1 Two Quantum Revolutions *9*
2.2 The First Quantum Revolution *11*
2.2.1 The Electronics and Information Age: Quantum Mechanics Applied *12*
2.3 Entanglement and Bell's Theorem *13*
2.3.1 The Bohr–Einstein Debate *13*
2.3.2 Bell's Theorem *14*
2.4 Quantum Mechanics and Single Objects *17*
2.4.1 From the Ensemble to the Single Quantum System *17*

2.4.2 Quantum Jumps in Action: New Clocks and New Theoretical Methods *19*
2.4.3 From Microscopic to Mesoscopic *19*
2.4.4 From Mesoscopic to Macroscopic: Decoherence *20*
2.5 The Second Quantum Revolution in Action: Quantum Information, Quantum Technologies *21*
2.5.1 Quantum Cryptography *21*
2.5.2 Quantum Computing [51, 52] *22*
2.5.3 Quantum Computing with Real, Imperfect Systems *23*
2.5.4 Other Quantum Technologies *24*
2.6 Conclusion: Questioning Quantum Mechanics Is Fruitful *25*
References *26*

3 Solid-State Quantum Emitters *31*
A. Mark Fox
3.1 Introduction *31*
3.2 Photon Correlation Functions *32*
3.3 Quantum Emitters *35*
3.4 Single-Photon Sources *36*
3.5 Purcell Enhancement *39*
3.6 Photon Coherence *41*
3.7 Deterministic Excitation of Quantum Emitters *44*
3.8 Conclusions *46*
References *48*

4 Single-Photon Sources for Multi-Photon Applications *53*
Stefan Frick, Robert Keil, Vikas Remesh, and Gregor Weihs
4.1 Motivation and Applications *53*
4.2 Nonlinear Optical Sources *54*
4.2.1 Principles *54*
4.2.1.1 Spontaneous Parametric Down-Conversion *54*
4.2.1.2 Spontaneous Four-Wave Mixing *55*
4.2.1.3 Phase Matching *55*
4.2.2 Photon-Number Purity and Heralding *56*
4.2.3 Indistinguishability *56*
4.2.4 Spectral Purity *57*
4.2.4.1 Methods for High-Purity Down-Conversion Sources *59*
4.2.5 Photon-Number Purity and Brightness *60*
4.2.6 Multiplexing Schemes *63*
4.3 Quantum Dots *63*
4.3.1 Advanced Excitation Schemes *65*
4.3.2 Engineering the Quantum Dot Environment *69*
4.3.3 Demultiplexing *73*
References *75*

5 Quantum Key Distribution Protocols *85*
Álvaro Navarrete, Víctor Zapatero, and Marcos Curty
5.1 Introduction *85*
5.2 Fundamentals of QKD *86*
5.3 Security Framework *89*
5.3.1 Security Definition *89*
5.3.2 Security Proof Techniques *91*
5.4 Practicalities *92*
5.4.1 Decoy-State QKD *92*
5.4.2 Measurement-Device-Independent QKD *93*
5.4.3 Other Practical QKD Protocols *94*
5.4.4 The Problem of Implementation Security *95*
5.5 Components of a QKD System *96*
5.5.1 Light Sources *96*
5.5.2 Single-Photon Detectors *98*
5.5.3 Linear Optical Elements *100*
5.6 Performance *100*
5.7 Conclusions and Future Challenges *101*
Acknowledgments *101*
References *102*

6 From Basic Science to Technological Development: The Case for Two Avenues *105*
Matteo Carlesso and Mauro Paternostro
6.1 Introduction *105*
6.2 Thermodynamics for an Energetically Efficient Quantum Information Processing *106*
6.2.1 On Step 2 *116*
6.2.2 On Step 4 *119*
6.3 Data Intensive Tools for Quantum Computing Science *120*
6.4 Conclusions *123*
Acknowledgments *123*
References *124*

7 Quantum Networks in Space *129*
Lisa Wörner
7.1 Global Communication *129*
7.2 Challenges in Global Secure Quantum Networks *132*
7.3 Untrusted Nodes *136*
7.3.1 Mirror Systems and Coupled Systems *137*
7.3.2 Sources *138*
7.3.3 Repeater *138*
7.4 Currently Available Technology *140*
7.4.1 Required Technology *140*
7.4.2 Technology Readiness Level (TRL) *140*

7.4.3 Optical Link Technologies *140*
7.4.4 Sources *141*
7.4.5 Repeater *142*
7.4.5.1 Overview *142*
7.4.5.2 Optical Fibers *142*
7.4.5.3 Memories *142*
7.4.5.4 Sum of Its Parts *144*
7.5 Summary *144*
7.6 Acronyms *144*
References *145*

Part II Atoms, Ions, and Molecules: From Experimental Techniques to Recent Progress *151*

8 Fluorescence Spectroscopy in Planar Dielectric and Metallic Systems *153*
Alexey I. Chizhik, Daja Ruhlandt, and Jörg Enderlein
8.1 Introduction *153*
8.2 Theory *154*
8.2.1 Ideal Electric Dipole Emitter *154*
8.2.2 Interaction of Plane Waves with Planar Layered Structures *155*
8.2.3 Electric Dipole Emission *158*
8.2.4 Angular Distribution of Radiation and Detection Efficiency *161*
8.2.5 Excitation Intensity Distribution *163*
8.2.6 Synthesis *165*
8.3 Applications *165*
8.3.1 Absolute Quantum Yield Measurements *165*
8.3.2 MIET/GIET imaging *171*
8.4 Conclusion *178*
References *178*

9 Single Trapped Neutral Atoms in Optical Lattices *183*
Andrea Alberti and Dieter Meschede
9.1 Introduction *183*
9.2 Tools for Trapping Single Neutral Atoms *184*
9.2.1 Trapping Neutral Atoms *184*
9.2.1.1 Magnetic Traps *184*
9.2.1.2 Optical Dipole Traps *185*
9.2.2 Preparing and Observing Individual Neutral Atoms *185*
9.2.2.1 Single Neutral Atoms in a MOT *186*
9.2.2.2 Single Neutral Atoms in Optical Dipole Traps *186*
9.2.2.3 Detecting Single Atoms in MOTs and ODTs with High Efficiency *187*
9.2.2.4 Detecting Single Atoms Using Optical Cavities *187*
9.2.2.5 Detecting the Qubit State of Trapped Atoms in Free Space *187*

9.2.2.6 Detecting the Qubit State of Atoms Trapped in Cavities *188*
9.2.2.7 Detecting the Position of Single Atoms with High Fidelity *189*
9.2.2.8 Simultaneous Detection of Quantum State and Position *190*
9.2.3 Precision Transport of Atoms *191*
9.2.3.1 Optical Conveyor Belt *191*
9.2.3.2 Spin-Dependent Optical Lattice *191*
9.2.3.3 Two Overlapped Optical Conveyor Belts with Orthogonal Polarizations *192*
9.2.3.4 Optical Conveyor Belt for Two Dimensions *192*
9.2.4 Controlling the Motional Atomic State *192*
9.2.5 Addressing and Controlling the Atom Position *193*
9.3 Quantum Control of Single Trapped Atoms *196*
9.3.1 Quantum State Transport *197*
9.3.2 Quantum Speed Limit of Atomic Motion *197*
9.3.3 Quantum Delocalization *198*
9.3.3.1 Spin-Dependent Transport *198*
9.3.3.2 Single-Atom Interferometer *198*
9.3.4 Quantum Walks of Single Atoms *198*
9.3.5 Applications of Quantum Walks *201*
9.3.5.1 Electric and Magnetic Quantum Walks *201*
9.3.5.2 Test of the Leggett–Garg Inequality *201*
9.3.6 Single Atoms as Sensors *202*
9.3.6.1 Single Atoms as Localized Quantum Probes *202*
9.3.6.2 Single Atoms for Environmental Applications *202*
9.4 Short Conclusions *203*
Acknowledgments *203*
References *204*

10 Long-Distance Entanglement of Atomic Qubits *209*
Kai Redeker, Wenjamin Rosenfeld, and Harald Weinfurter
10.1 Qubit Encoding in ^{87}Rb *210*
10.2 Trapping Single Atoms *212*
10.3 State Preparation, Control, and Measurement of Single Atoms *215*
10.3.1 Creation of Atom–Photon Entanglement *215*
10.3.2 Measurement of the Atomic State *216*
10.4 Coherence of the Atomic States *219*
10.4.1 Interaction with Magnetic and Optical Fields *219*
10.4.2 Control of the Magnetic Fields *220*
10.4.3 Decoherence Caused by Polarization Dependent AC-Stark Shift and Motion of the Atom *221*
10.5 Creation of Long-Distance Atom–Atom Entanglement *221*
10.5.1 Heralded Entanglement via Entanglement Swapping *222*
10.5.2 Operation of the Two-Trap Setup *223*
10.5.3 Creation and Verification of the Atom–Atom Entanglement *225*
10.6 Employing Distributed Entanglement *226*

10.6.1 Bell Test Simultaneously Closing Detection and Locality Loopholes *226*
10.6.2 Advanced Protocols Based on Distributed Entanglement *226*
10.6.3 Toward Long-Distance Quantum Networks *227*
Acknowledgments *227*
References *228*

11 Collective Light Emission of Ion Crystals in Correlated Dicke States *233*
Ferdinand Schmidt-Kaler and Joachim von Zanthier
11.1 Introduction *233*
11.2 Structure of the Article *235*
11.3 Fundamentals of Trapping and Laser Cooling of Ions in a Paul Trap *236*
11.4 Light Emission of Uncorrelated Ion Crystals *238*
11.5 Theory of Correlated Dicke States Among Trapped Ions via Projective Measurements of Scattered Photons *243*
11.6 Theory of Collective Light Emission of Ion Crystals in Correlated Dicke States *246*
11.7 Theory of Measuring Collective Light Emission from Linear Ion Crystals in Dicke-Correlated States *254*
11.8 Measurement of the Collective Light Emission of a Two Ion Crystal *259*
11.9 Conclusion and Outlook *263*
Acknowledgments *264*
Data availability *264*
References *264*

12 Single-Molecule Magnets Spin Devices *269*
Eufemio Moreno-Pineda and Wolfgang Wernsdorfer
12.1 Introduction *269*
12.2 SMMs and Quantum Effects *270*
12.2.1 Quantization *270*
12.2.2 Quantum Coherence *273*
12.2.3 Quantum Tunneling of the Magnetization *274*
12.2.4 Quantum Bits and Multilevel Systems *275*
12.3 SMMs for Single-Molecule Devices *277*
12.3.1 Spin Transistors *277*
12.3.2 Spin Valve *279*
12.4 SMMs for Quantum Technologies *282*
12.4.1 Quantum Sensing *282*
12.4.2 Quantum Simulations *285*
12.4.3 Quantum Computing *288*
12.4.4 Quantum Communication *292*
12.5 Conclusions *295*
References *295*

13 **Molecular-Ion Quantum Technologies** *305*
Mudit Sinhal and Stefan Willitsch
13.1 Introduction *305*
13.2 Experimental Techniques *306*
13.2.1 Ion Trapping *306*
13.2.2 Generation of Molecular Ions and Their State Initialization *309*
13.2.3 Cooling of Trapped Ions *311*
13.3 Destructive State-Readout Techniques *312*
13.4 Quantum-Logic Experiments on Single Trapped Molecular Ions *315*
13.4.1 Quantum-Nondemolition Molecular State Readout by State-Dependent Coherent Motional Excitation *315*
13.4.2 Molecular State Detection and Spectroscopy via a Motional Qubit *321*
13.4.3 Molecular Quantum Logic Spectroscopy Using Resolved-Sideband Raman Transitions *321*
13.5 Outlook on Future Developments and Conclusions *325*
Acknowledgments *326*
References *326*

14 **Optical Atomic Clocks** *333*
Ekkehard Peik
14.1 Introduction *333*
14.2 Optical Atomic Clocks *334*
14.3 Optical Clocks with a Single Trapped $^{171}Yb^{+}$ Ion *342*
14.4 Outlook on Future Developments *344*
Acknowledgments *346*
References *346*

Contents to Volume 2

Preface *xv*
Abstracts and Keywords *xvii*

Part III Spin Qubits and Quantum Memories: From Spin Properties to Physical Realizations *349*

15 **Coherent Spin Dynamics of Colloidal Nanocrystals** *351*
Dmitri R. Yakovlev, Anna V. Rodina, Elena V. Shornikova, Aleksandr A. Golovatenko, and Manfred Bayer

16 **Relaxation of Electron and Hole Spin Qubits in III–V Quantum Dots** *377*
Alexander Bechtold, Tobias Simmet, Friedrich Sbrezny, Nikolai A. Sinitsyn, Kai Müller, and Jonathan J. Finley

17 **Ensemble-Based Quantum Memory: Principle, Advance, and Application** *433*
Bo Jing and Xiao-Hui Bao

Part IV Solid-State and van der Waals Material Platforms: From Single Quantum Emitters to Hybrid Integration *463*

18 **Telecom Wavelengths InP-Based Quantum Dots for Quantum Communication** *465*
Mohamed Benyoucef and Anna Musiał

19 **Quantum Optics with Solid-State Color Centers** *509*
Joseph H.D. Munns, Laura Orphal-Kobin, Gregor Pieplow, and Tim Schröder

20 **Quantum Photonics with 2D Semiconductors** *563*
Mauro Brotons-Gisbert and Brian D. Gerardot

21 **Nano-Opto-Electro-Mechanical Systems for Integrated Quantum Photonics** *581*
Leonardo Midolo and Celeste Qvotrup

22 **Silicon Quantum Photonics - Platform and Applications** *599*
Cagin Ekici, Elizaveta Semenova, Davide Bacco, and Yunhong Ding

Part V Emerging Quantum Technologies: Challenges and Potential Applications *651*

23 **Photonic Realization of Qudit Quantum Computing** *653*
Yuchen Wang, Zixuan Hu, and Sabre Kais

24 **Fiber-Based Quantum Repeaters** *675*
Mohsen Razavi

25 **Long-Distance Satellite-Based Quantum Communication** *693*
Yuan Cao and Juan Yin

26 **Quantum Communication Networks for 6G** *715*
Riccardo Bassoli, Frank H.P. Fitzek, and Holger Boche

Index *737*

Preface

This book has been inspired by the growing interest in emerging quantum technologies. The ongoing miniaturization of photonic structures due to the availability of sophisticated nanofabrication has provided huge opportunities for physical research of novel phenomena in photonic quantum systems and quantum technological applications. These include quantum information processing such as quantum computing, quantum communication, quantum metrology, and quantum sensing.

The aim here is to bring together scientists with different expertise from across different disciplines. The manuscript provides selected topics on photonic quantum systems from research to applications. The chapters will represent a flavor of the cutting-edge research that is going on in this area. The studied material systems will range from ions, atoms, and molecules through color centers and other solid-state quantum emitters. The intention here is to provide an important guide for the students and researchers of the community. It is dedicated to one of today's most active and rapidly expanding fields of research and development in photonic quantum technologies, in particular the growing number of experimental implementations and practical applications of quantum systems. This book will be unique because it will provide researchers and scientists with various aspects of material development on different platforms and their applications in quantum technologies *in one place*.

Progress in the realization of quantum devices and circuits enables breakthroughs in experimental tests of quantum physics and in practical applications of quantum technologies in imaging, secure communication, ultrasensitive metrology, and quantum computing. Based on the current worldwide interdisciplinary research activities in these emerging quantum technologies, I strongly believe that the realization of a practical quantum device (e.g. quantum repeaters) will most probably be based on hybrid quantum systems. Therefore, the intention of this publication is to provide an overview of the latest cutting-edge research on photonic quantum technologies and to provide readers (academia and industry), researchers, and scientists with several ingredients *in one location*. This will help to bridge the gap between scientists working in academia and industry and guide them to understand the current development, the remaining challenges toward the realization of practical quantum systems, and the necessary requirements to overcome such challenges.

This book consists of 26 chapters, written by internationally prominent scientists in the field of quantum technologies from different countries: Austria, China, Denmark, France, Germany, Panama, Poland, Russia, Spain, Switzerland, the United Kingdom, and the United States. It is structured into five parts, covering various topics ranging from fundamentals to applications. I anticipate that the book will be helpful for students and researchers involved in this field and that it will encourage and guide engineers in their efforts for industrial implementations.

Though the cover page of the book was chosen before the announcement of the 2022 Nobel Prize in Physics, I am delighted that it matches the award's topic.

I would like to express my sincere gratitude to ***Alain Aspect***, the 2022 Nobel Prize Laureate in Physics, for taking time from his busy schedule to contribute a chapter. My sincere thanks go out to all my colleagues for their valuable contributions and their excellent cooperation throughout the writing and editing process.

Furthermore, I would like to thank M. Fox, S. Kais, P. van Loock, D. Oi, and M. Razavi for accepting my invitation to review some chapters and for providing valuable feedback and suggestions.

Special thanks to the editorial staff of Wiley-VCH and in particular to M. Preuss, D. Bez, H. Noethe, M. Abinaya, V. Shyamala, and C. Nussbeck for their support and patience in awaiting the final version and keeping this book on track for publication.

I would also like to thank the German Research Foundation (DFG) and the Federal Ministry of Education and Research (BMBF) for their continued support. I am also grateful for the financial support provided by the state of Hesse through the LOEWE program. My thanks also go out to the University of Kassel for its internal financial support.

Finally, I want to extend my gratitude to my parents for supporting me in pursuing my studies. My wife and children deserve a special thank you for their unconditional support and gratitude for my absence for several weekends.

Mohamed Benyoucef
Kassel, November 2022

Abstracts and Keywords

Chapter 1: Introduction to Quantum Photonics

*Richard J. Warburton**

*Corresponding author: richard.warburton@unibas.ch

A brief and non-technical introduction to contemporary quantum photonics is presented.

Keywords: photon; quantum optics; quantum technology.

Chapter 2: The Second Quantum Revolution: From Basic Concepts to Quantum Technologies

Alain Aspect,†*

*Corresponding author: alain.aspect@institutoptique.fr

†The 2022 Nobel Prize Laureate in Physics

The second quantum revolution was first conceptual, with the understanding and the experimental demonstration of the extraordinary character of entanglement, as evidenced by Bell inequalities violations. These experiments were permitted by the possibility to observe and manipulate individual quantum objects. The second quantum revolution is now at the stage of using these conceptual advances to develop quantum technologies that are able to go beyond those based on the first quantum revolution.

Keywords: second quantum revolution; entanglement; bell inequalities; quantum computing; quantum cryptography.

Chapter 3: Solid-State Quantum Emitters

*A. Mark Fox**

*Corresponding author: mark.fox@sheffield.ac.uk

This chapter gives a tutorial overview of the key parameters that are used to assess solid-state quantum emitters for applications in quantum photonics. Specifically,

it covers single-photon purity, coherence and photon indistinguishability, and the methods that are used to quantify them. The chapter concludes with a brief comparison of several of the more common material systems that are used in solid-state quantum optics.
Keywords: quantum emitter; single-photon source; photon statistics; photon anti-bunching; Purcell effect; Rabi rotations; Hanbury Brown-Twiss experiment; Hong-Ou-Mandel experiment; photon indistinguishability.

Chapter 4: Single-Photon Sources for Multi-photon Applications

*Stefan Frick, Robert Keil, Vikas Remesh, and Gregor Weihs**

*Corresponding author: gregor.weihs@uibk.ac.at

Much of the recent development of single-photon sources is driven by the desire to apply them to protocols and technologies that use the interference of two or more photons, such as quantum repeaters or boson sampling. In all of these cases, the indistinguishability of the produced photons is a key requirement. For those applications that want to scale to larger photon numbers, source efficiency is equally important. In this chapter, we will discuss two competing solutions, sources based on nonlinear optics and quantum dots, as the most-used single-quantum emitters.
Keywords: photon-pair sources; quantum dots; multi-particle interference; indistinguishability; number purity; spectral purity; excitation schemes.

Chapter 5: Quantum Key Distribution Protocols

Álvaro Navarrete, Víctor Zapatero, and Marcos Curty*

*Corresponding author: vzapatero@com.uvigo.es

In contrast to public-key cryptography, whose security relies on computational assumptions, quantum key distribution (QKD) provides an information-theoretically secure solution for the task of distributing cryptographic keys between distant parties through an insecure channel. In this chapter, we provide a general introduction to QKD protocols and their security, starting from the theoretical foundations of QKD – based on fundamental properties of quantum mechanics, such as the no-cloning theorem or entanglement – and moving to state-of-the-art QKD solutions and the latest technological advances.
Keywords: quantum key distribution; quantum communication; quantum cryptography; quantum technologies.

Chapter 6: From Basic Science to Technological Development: The Case for Two Avenues

*Matteo Carlesso and Mauro Paternostro**
*Corresponding author: m.paternostro@qub.ac.uk

We argue that, in the quest for the translation of fundamental research into actual quantum technologies, two avenues that have – so far – only partly explored should be pursued vigorously. On first entails that the study of energetics at the fundamental quantum level holds the promises for the design of a generation of more energy-efficient quantum devices. On second route to pursue implies a more structural hybridization of quantum dynamics with data science techniques and tools, for a more powerful framework for quantum information processing.
Keywords: quantum technologies; quantum thermodynamics; quantum machine learning.

Chapter 7: Quantum Networks in Space

*Lisa Wörner**
*Corresponding author: lisa.woerner@dlr.de

The inherent properties of quantum mechanics render it an important subject for fundamental research and applications alike. Entanglement stands out among those, allowing for novel secure communication and more rapid computation. To enable global secure communication, space based networks are required. A space network with trusted nodes for every user is an unviable feat, and could lead, given the amount of potential civil, governmental, and military users, to a cluttering of space. Additionally, global networks of trusted nodes exclude smaller to medium players from profiting from the quantum revolution and the benefits of secure communication. To remain inclusive to small businesses and enable broad usage of the technology, the operation of untrusted nodes in space is inevitable. Consequently, untrusted nodes need to be capable of receiving and re-transmitting the information with the possibility of storage of the information in-between. Hence, space-based quantum repeater, consisting of a receiver, a transmitter, and a coherent memory, could operate as untrusted nodes.

In this chapter, the challenges, necessities, and opportunities for global quantum networks will be discussed. This includes novel techniques for quantum repeater and their current technology readiness level.
Keywords: quantum network; untrusted node; quantum repeater.

Chapter 8: Fluorescence Spectroscopy in Planar Dielectric and Metallic Systems

*Alexey I. Chizhik, Daja Ruhlandt, and Jörg Enderlein**

*Corresponding author: jenderl@gwdg.de

In this chapter, we present an overview of our recent theoretical and experimental studies of the fluorescence modulation of quantum emitters by planar nanostructures. We will mainly focus on two methods that are based on this phenomenon: (i) axial localization of a fluorophore using energy transfer from the fluorophore to a planar metal film or graphene and (ii) absolute quantum yield measurements using a plasmonic nanocavity. Both methods have been used in a large number of recent studies. We present the full theoretical modeling of these experiments and give an overview of recent experimental results.

Keywords: single-molecule fluorescence; nanocavity electrodynamics; fluorescence lifetime spectroscopy; defocused single-molecule imaging; metal- and graphene-induced energy transfer.

Chapter 9: Single Trapped Neutral Atoms in Optical Lattices

*Andrea Alberti and Dieter Meschede**

*Corresponding author: meschede@uni-bonn.de

Controlling neutral atoms one by one has opened the path to create new experimental platforms for investigating their individual as well as their collective dynamics in a bottom-up approach. We discuss the tools for precisely detecting and preparing all degrees of freedom of trapped atoms at the ultimate quantum level, including their internal spin state, their motional state, and also their position, which has relevant quantum character. We summarize applications including single-atom interferometry and quantum walks.

Keywords: cold atoms; trapped neutral atoms; quantum information; quantum walks; single atoms; atom interferometry.

Chapter 10: Long-Distance Entanglement of Atomic Qubits

*Kai Redeker, Wenjamin Rosenfeld, and Harald Weinfurter**

*Corresponding author: h.w@lmu.de

Quantum networks connecting future quantum computers and enabling efficient long-distance quantum communication require as most important resource the distribution of entanglement. This chapter details how this can be achieved for the particular example of atomic quantum memories implemented with single, trapped, neutral rubidium atoms.

Keywords: quantum network; quantum communication quantum memory; entanglement.

Chapter 11: Collective Light Emission of Ion Crystals in Correlated Dicke States

Ferdinand Schmidt-Kaler and Joachim von Zanthier*

*Corresponding author: fsk@uni-mainz.de

We give an overview of collective light scattering of trapped ion crystals serving as arrays of correlated emitters of single, indistinguishable photons into free space. The collectively emitted light is recorded in the far field by detectors, which feature high spatial and high temporal resolution. We observe interference in the first- and second-order photon correlation functions as well as, bunched and antibunched photon statistics. We analyze theoretically how the observed features of collective light scattering off arrays of correlated single photon emitters result from projective quantum measurements into Dicke states.

Keywords: photon statistics and coherence theory; quantum description of interaction of light and matter; cooperative phenomena in quantum optical systems; single photon emitter; correlated emission of photons; single ion trapping and cooling.

Chapter 12: Single-Molecule Magnets Spin Devices

*Eufemio Moreno-Pineda and Wolfgang Wernsdorfer**

*Corresponding author: wolfgang.wernsdorfer@kit.edu

Since the first observation of quantum effects, scientists have striven to gain a deeper knowledge and understating of these characteristics with the ultimate goal of implementing them in practical applications. Today, due to important contributions of many pioneers, we are at the point in which quantum effects can be exploited, prompting the proposal of several futuristic technological applications such as quantum sensing, quantum simulation, quantum computing, and quantum communication. These technologies share some common ground: they rely on the manipulation and read-out of the electronic or nuclear states; hence, their quantum properties must fulfill certain characteristics. To accomplish these tasks, several systems have been proposed; however, among the several prospects, molecular systems have been shown to possess several advantageous characteristics, as compared to common platforms. In this chapter, key aspects of single-molecule magnets allowing their proposal in applications in quantum sensing, quantum simulation, quantum computing, and quantum communications will be revised.

Keywords: quantum sensing; quantum simulations; quantum computing; quantum communications; spin transistor; spin valve; quantum bit; qudit; quantum error correction; quantum tunneling of the magnetization.

Chapter 13: Molecular-Ion Quantum Technologies

*Mudit Sinhal and Stefan Willitsch**
*Corresponding author: Stefan.Willitsch@unibas.ch

Quantum-logic techniques for state preparation, manipulation, and nondestructive interrogation are increasingly being adopted for experiments on single molecular ions confined in traps. The ability to control molecular ions on the quantum level via a co-trapped atomic ion offers intriguing possibilities for new experiments in the realms of precision spectroscopy, quantum information processing, cold chemistry, and quantum technologies with molecules. The present article gives an overview of the basic experimental methods, recent developments, and prospects in this field.
Keywords: molecular ions; quantum-logic spectroscopy; precision measurements; quantum technologies.

Chapter 14: Optical Atomic Clocks

*Ekkehard Peik**
*Corresponding author: ekkehard.peik@ptb.de

Based on the development of atomic clocks and the definition of the unit of time via an atomic resonance frequency, time and frequencies are the most precisely measurable physical quantities. Over the past 30 years, the development of methods of cooling and trapping of atoms and ions and the development of low-noise laser oscillators and the invention of the optical frequency comb that establishes a link between microwave and optical frequencies with nearly perfect fidelity have led to impressive progress in the field. Optical clocks based on laser-cooled and trapped atoms and ions reach a systematic uncertainty in the low 10^{-18} range and below. This chapter presents the principles and methods that have enabled these performances.
Keywords: atomic clock; laser cooling and trapping; optical frequency comb; unit of time.

Chapter 15: Coherent Spin Dynamics of Colloidal Nanocrystals

Dmitri R. Yakovlev, Anna V. Rodina, Elena V. Shornikova, Aleksandr A. Golovatenko, and Manfred Bayer*
*Corresponding author: dmitri.yakovlev@tu-dortmund.de

This chapter gives an overview of the spin properties of semiconductor colloidal nanocrystals. We consider the energy and spin-level fine structures in colloidal quantum dots and nanoplatelets and their modification in an external magnetic field. We show how spectroscopic, time-resolved, and magneto-optical experimental approaches can be used for colloidal nanocrystals in order to get information on exciton states as well as the spin dynamics of excitons and charge carriers. We demonstrate that spin-dependent phenomena give access to further properties of colloidal nanocrystals, like the surface states and their spins, the photocharging dynamics, the carrier localization within a nanocrystal, etc.

Keywords: colloidal nanocrystals; colloidal quantum dots; semiconductor nanoplatelets; coherent spin dynamics; time-resolved Faraday rotation; spin-flip Raman scattering; exciton spin structure.

Chapter 16: Relaxation of Electron and Hole Spin Qubits in III–V Quantum Dots

*Alexander Bechtold, Tobias Simmet, Friedrich Sbrezny, Nikolai A. Sinitsyn, Kai Müller, and Jonathan J. Finley**

*Corresponding author: jonathan.finley@wsi.tum.de

The control of solid-state qubits for quantum information processing requires a detailed understanding of the mechanisms responsible for decoherence. During the past decade, considerable progress has been achieved for describing the qubit dynamics in relatively strong external magnetic fields. However, testing theoretical predictions at very low magnetic fields has proven difficult in optically active dots. Here, we describe our studies of electron and hole spin qubit dephasing in single InGaAs quantum dots using spin memory devices. The results show that without applied magnetic fields, the initially orientated electron spin rapidly loses its polarization due to precession around the fluctuating Overhauser field with an effective magnetic field amplitude of 10.5 mT. The inhomogeneous dephasing time associated with these hyperfine mediated dynamics is $T_2^* \sim 2$ ns. Over longer timescales, an unexpected stage of central spin relaxation is observed, namely the appearance of a second feature in the relaxation curve around $T_Q = 750$ ns arising from quadrupolar coupling. In comparison, hole spin qubits are shown couple significantly more weakly to the nuclear spin bath. We measure a $\sim 100\times$ times longer dephasing time $T_2^* \sim 210$ ns for hole spin qubits compared with the electron spin. We also obtain evidence for the impact of anisotropic hyperfine coupling on the spin polarization decay, allowing us to quantify the degree of anisotropy $\alpha = 0.19$ which is fundamental to the character of the confined hole spin wave function. By modeling this behavior, we derive the degree of light-hole heavy-hole mixing, which is an essential mechanism for enabling hole spin dephasing and thus refining the description of hole hyperfine coupling beyond the initially suggested pure Ising form.

Keywords: optically active quantum dots; electron and hole spin qubits; central spin problem; spin photon interfaces

Chapter 17: Ensemble-Based Quantum Memory: Principle, Advance, and Application

Bo Jing and Xiao-Hui Bao*

*Corresponding author: bjing@uestc.edu.cn

Quantum memory for photons plays an important role in quantum information science, such as quantum repeater, quantum network, quantum communica-

tion, and quantum computing. The realization of a quantum memory with high efficiency, long lifetime, high fidelity, large bandwidth, and multimode capacity is of great significance for its applications. Atomic ensemble, serving as one of the most appealing physical platforms for quantum memories, has been widely explored toward improving memory performance. Over the past decades, significant progress based on atomic ensembles has been made in developing new schemes and improving memory performances experimentally through various approaches. In this chapter, we first introduce the working principle of quantum memory by reviewing the widely used schemes. Next, we present several key figures of merit for judging memory performance. Afterward, we review the experimental progress made in several different systems. Finally, we give discussions on its application in a number of different directions. The scope of this chapter is limited to single-photon storage with atomic ensembles.
Keywords: quantum memory; atomic ensemble; rare-earth ions; electromagnetically induced transparency; atomic frequency comb; DLCZ.

Chapter 18: Telecom Wavelengths InP-Based Quantum Dots for Quantum Communication

Mohamed Benyoucef and Anna Musiał*
*Corresponding author: m.benyoucef@physik.uni-kassel.de

Quantum dots (QDs), also known as artificial atoms, are among the recently explored hardware platforms for photonic quantum technologies. They offer a range of advantages that make them particularly well-suited for building quantum communication networks over optical fiber channels. One of the most prominent features of QDs is the ability to be easily and deterministically integrated into photonic devices and photonic circuits. They also exhibit near-unity internal quantum efficiency and emit nearly Fourier-transform-limited single photons of high purity and entangled photons. QDs emitting at optical fiber communication wavelengths, especially in the telecom C-band, which corresponds to the spectral range of the lowest attenuation in silica fibers, are very promising for long-distance quantum communication. This makes the development of QD-based quantum light sources for these wavelength ranges of particular interest. This chapter gives an overview of recent developments in InP-based QDs, which are one of the candidates providing emission in this spectral window. These advancements include lowering the spatial density of nanostructures in this material system, increasing their in-plane symmetry, developing high-reflectivity distributed Bragg reflectors, and incorporating them in microcavities and other photonic structures to increase photon extraction efficiency to generate the highest-purity single photons at 1550 nm.
Keywords: quantum dots; III-V semiconductor nanostructures; molecular beam epitaxy growth; telecom wavelengths; single-photon sources; magneto-optics; photonic crystal cavities; fiber quantum network; extraction efficiency; symmetric InAs/InP quantum dots; internal quantum efficiency.

Chapter 19: Quantum Optics with Solid-State Color Centers

*Joseph H.D. Munns, Laura Orphal-Kobin, Gregor Pieplow, and Tim Schröder**
*Corresponding author: tim.schroeder@physik.hu-berlin.de

With the chapter "Quantum Optics with Solid-State Color Centers," we provide a comprehensive overview of color centers in wide-bandgap materials: We describe their generation and energy level schemes, how they are controlled both optically and via microwaves, and how they are used in quantum optics and technology. The chapter summarizes landmark experiments including applications ranging from single photon generation, over quantum memories, quantum gates, to entanglement generation. We structured the chapter such that the reader has access to a bottom-up introduction but can also directly jump to sections of interest. This work is directed at graduate students up to the principal investigator level.
Keywords: quantum optics; quantum technology; color centers; wide-bandgap materials; diamond; silicon carbide; review; single photon emitters; optically active solid-state spin defects.

Chapter 20: Quantum photonics with 2D semiconductors

*Mauro Brotons-Gisbert and Brian D. Gerardot**
*Corresponding author: b.d.gerardot@hw.ac.uk

In this chapter, we introduce the main properties of the most popular van der Waals materials for single-photon generation: hexagonal boron nitride(hBN), group-VI transition metal dichalcogenide semiconductors (TMDs), and 2D quantum devices consisting of different 2D crystals assembled into van der Waals heterostructures. We then focus primarily on quantum emitters (QEs) hosted by TMD semiconductors, including a description of their typical spin and optical properties. Finally, we consider the different means to engineer the QEs and their coupling to cavities and integrated photonic chips.
Keywords: 2D materials; nanophotonics; single photon; van der Waals heterostructure

Chapter 21: Nano-Opto-Electro-Mechanical Systems for Integrated Quantum Photonics

Leonardo Midolo and Celeste Qvotrup*
*Corresponding author: midolo@nbi.ku.dk

This chapter covers the topic of nano-opto-electro-mechanical systems (NOEMS) and their application in quantum photonic integrated circuits. When reducing the size of optical devices down to the nano-scale, as in nanophotonics and plasmonics, the properties of confined optical waves become increasingly affected by small geometrical variations of the materials. These variations produce a wide range

of electro- and opto-mechanical interactions that do not depend on any intrinsic material property and therefore can be operated at cryogenic temperatures and seamlessly integrated in quantum photonic circuits with embedded quantum emitters and single-photon detectors. We discuss the basic principles of operation of NOEMS, the techniques involved in their fabrication, and the functionalities that such devices can provide toward the realization of scalable quantum photonic integrated technologies.
Keywords: photonic integrated circuits; electro- and opto-mechanics; quantum emitters; micro and nanofabrication.

Chapter 22: Silicon Quantum Photonics - Platform and Applications

*Cagin Ekici, Elizaveta Semenova, Davide Bacco, and Yunhong Ding**

*Corresponding author: yudin@fotonik.dtu.dk

Silicon quantum photonics is the leading platform of integrated quantum photonics thanks to its advantages of ultradense integration, complementary metal–oxide semiconductor-compatible fabrication, etc. However, to enter the fault-tolerant, breakthroughs on the photon sources, optical loss, and quantum process fidelity have to be achieved before silicon-integrated quantum photonics can really achieve a useful scale. In this chapter, we will present the basics of silicon quantum photonics and give an overview of the state of the art. The chapter is structured as follows. In Section 22.2, we will discuss the typical fabrication process and necessary passive and active optical components. In Section 22.3, we will further present an overview of different quantum light sources on silicon, including quantum dot sources on silicon and nonlinear sources. In Section 22.4, we will discuss applications in processing of quantum information on silicon and quantum communication. Finally, we will give an outlook of silicon quantum photonics for further scaling.
Keywords: silicon photonics; quantum information technologies; single- photon source; nonlinear light source; quantum information processing; quantum communication.

Chapter 23: Photonic Realization of Qudit Quantum Computing

*Yuchen Wang, Zixuan Hu, and Sabre Kais**

*Corresponding author: kais@purdue.edu

In quantum computing, qudit is a multilevel computational unit. Qudit, as an alternative to qubit, has a larger state space for information storage and processing. Quantum circuit built with qudits can enhance algorithm efficiency, reduce the circuit complexity, and simplify the experiment. Photonic systems with their comparatively long decoherence time provide good platforms for quantum computing.

Some photonic systems also provide suitable working platforms for qudit quantum computing because qudit systems can be encoded naturally and conveniently with photons given the multidimensional nature in many photon properties. This chapter first introduces the basic knowledge and concepts of qudit quantum computing that include qudit gates, universality, and important qudit algorithms such as the quantum Fourier transform and the Phase Estimation Algorithm (PEA). It then reviews two implementations of qudit computing with photonic systems. The first system is an electro-optic platform that integrates on-chip modulators, pulse shapers, and other well-established fiber-optic components. This platform is capable of performing high-fidelity quantum Fourier transform pulse shaping and the two-qudit SUM gate via the time and frequency degrees of freedom, which paves the way to demonstrate a proof-of-principle qutrit PEA. The second system consists of microwave cavities and transmon qubits to implement a two-mode superconducting bosonic processor that enables the scalable simulation of molecular vibronic spectra.
Keywords: quantum information; quantum computing; qudit space; quantum phase estimation algorithm; electro-optic platform; superconducting bosonic processor and open quantum dynamics.

Chapter 24: Fiber-Based Quantum Repeaters

*Mohsen Razavi**

*Corresponding author: m.razavi@leeds.ac.uk

Quantum repeaters are the primary platform for the reliable exchange of quantum data in future quantum communications networks. In this chapter, I provide a brief overview of how different classes of quantum repeaters work, and, in particular, how they manage the two key challenges of loss and noise in quantum channels and systems. Along the way, the required resources for such repeaters and their expected performance will be discussed.
Keywords: quantum communications; quantum repeaters; quantum networks; quantum error correction codes; quantum data transfer; entanglement distribution; quantum memories.

Chapter 25: Long-Distance Satellite-Based Quantum Communication

*Yuan Cao and Juan Yin**

*Corresponding author: yinjuan@ustc.edu.cn

Long-distance quantum communication with satellites opens a promising avenue for the global secure quantum network and the large-scale test of quantum foundations. Free-space links could provide a better approach because photon loss and decoherence are almost negligible in the atmosphere. But there are still many new

challenges. The satellite moves fast and also has rotation and vibration. Besides, the environment of space is also complex. Many scientific research teams from different countries have spent the past decade performing systematic ground tests for satellite-based quantum communications to overcome all these difficulties. The world's first quantum science satellite Micius was launched on 16th August 2016, demonstrating a series of scientific experiments. Since then, many satellite plans have been proposed, making this field into focus. In this chapter, we briefly review the progress in long-distance satellite-based quantum communication, including step-by-step ground-based feasibility studies and developments of the critical technologies required, a series of satellite-ground quantum communication experiments with Micius, and ending with an outlook on the future work that needs to be done to build global-scale practical quantum networks eventually.
Keywords: quantum information; quantum communication; quantum key distribution; quantum entanglement.

Chapter 26: Quantum Communication Networks for 6G

Riccardo Bassoli, Frank H.P. Fitzek, and Holger Boche*
*Corresponding author: riccardo.bassoli@tu-dresden.de

The research and industrial communities are concentrating their efforts in two main trends in communications, which are seen as the new drivers not only in telecommunications but also in all the aspects of our lives: sixth generation (6G) and the Quantum Internet. While their research and standardization efforts are still independently faced, the limitations and challenges of these two systems can experience significant benefits from their integration. This chapter aims at making the reader familiar with current status of 6G and the Quantum Internet architectural design and how they can be integrated to gain a mutual advantage.
Keywords: 6G; quantum communication networks; entanglement-assisted communication; quantum campus network; quantum three-dimensional network.

1

Introduction to Quantum Photonics

Richard J. Warburton

Department of Physics, University of Basel, Klingelbergstrasse 82, CH4056 Basel, Switzerland

1.1 The Photon

The photon is the quantum particle of light. The concept of a photon was first introduced by Max Planck and Albert Einstein. The "birth" of a photon was described by Niels Bohr: in the shell-model of the atom, a photon is created when an electron jumps from a higher-lying shell to a lower-lying shell. All this work predated the development of formal quantum mechanics. In the meantime, quantum optics has developed into a highly coherent and well-tested branch of quantum physics. That light consists of quantum particles, photons, and that these photons exhibit both wave-like and particle-like behavior is uncontested. A new theme is the extent to which these quantum properties make the photon useful in quantum technology.

In classical physics, the energy associated with an electromagnetic field in a single cavity mode is divided equally between the electric and magnetic components. In quantum optics, the electric and magnetic fields constitute conjugate variables, corresponding to position and momentum, and are replaced with operators [1, 2]. In terms of the photon annihilation and creation operators, $\hat{a}$ and $\hat{a}^\dagger$, which satisfy the commutation relation $[\hat{a}, \hat{a}^\dagger] = 1$, the single-mode Hamiltonian becomes $\hat{H} = \hbar\omega(\hat{a}^\dagger\hat{a} + \frac{1}{2})$ where $\hbar\omega$ is the photon energy. The eigenstates of this Hamiltonian are the Fock states (equivalently, the number states), $|n\rangle$ with $n = 0,1,2,3,\ldots$. The Fock state with $n = 0$ represents the vacuum state. The fluctuations in the vacuum state constitute an important source of noise in quantum optics. The Fock state with $n = 1$ represents the single-photon state.

All the number states are highly quantum in the sense that their properties are very different to those of a classical electromagnetic wave. One example concerns the intensity autocorrelation function, $g^{(2)}(\tau)$. For the Fock state $|1\rangle$, $g^{(2)}(\tau = 0) = 0$. This reflects the particle-like nature of the photon: on detecting one photon at time $\tau = 0$, there is zero probability of detecting a second if the input state consists of just one photon. By contrast, for an ideal laser output, $g^{(2)}(\tau) = 1$ for all τ. A measurement of $g^{(2)}(\tau)$, for instance with a Hanbury Brown–Twiss interferometer, is widely used to characterize quantum light.

Photonic Quantum Technologies: Science and Applications, First Edition. Edited by Mohamed Benyoucef.

A second example concerns the behavior at a beam splitter. Assuming that two identical Fock states $|1\rangle$ arrive simultaneously at a 50:50 beam splitter, the output state is $\frac{i}{\sqrt{2}}(|0\rangle_A|2\rangle_B + |2\rangle_A|0\rangle_B)$ where A and B describe the two output modes [3]. This reflects the wave-like nature of the photon: the $|1\rangle_A|1\rangle_B$ state is not part of the output as a consequence of an interference phenomenon called two-photon interference. Although the two input states may be produced by completely different atoms, a highly entangled state is created at the output. The two-photon interference is only complete if the two input states are identical, "indistinguishable" in the language of quantum optics. A measurement of the $|1\rangle_A|1\rangle_B$ fraction is used as a metric to quantify the extent to which the two input states are indistinguishable.

The theory of quantum optics applies at all frequencies. However, the optical band covering approximately 200–600 THz is particularly important. On the one hand, highly coherent laser sources exist in this band. These sources are essential to create quantum states of light via the light–matter interaction. On the other hand, single-photon detectors also exist in this frequency range with both a high quantum efficiency and low timing jitter. The detectors are essential to probe the quantum states of light. These detectors also exploit the light–matter interaction.

A photon can be used as a quantum bit. One degree of freedom, for instance the polarization or the arrival time, is used to define the "north" and "south" poles on the Bloch sphere; superpositions cover all points on the sphere. By making a good choice, the quantum state of the photon can be preserved as it propagates over large distances. This feature renders the photon crucial in quantum communication and in interconnecting quantum computers in a future quantum internet. Here too photons at optical frequencies play a crucial role as optical fiber technology is highly developed: there are two windows with particularly low transmission loss, the O-band at wavelength 1300 nm and the C-band at 1550 nm. Photons at these wavelengths can be used to transport quantum information over large distances via standard optical fibers.

1.2 The Light–Matter Interaction

In the absence of matter, photons do not interact with each other. Photons do of course interact with matter. The process of spontaneous emission, the creation of a photon as an electron hops from one energy level to another, can be described by Fermi's golden rule: the lifetime depends on the product of a matrix element, an immutable property of the atom, and a photonic density of states factor. Edward Purcell's crucial insight is that, unlike the dipole matrix element, the density of states is not immutable – it can be controlled by engineering the atom's environment [4]. If the photonic engineering is carried out precisely enough, spontaneous emission from a single atom puts a single photon into one particular optical mode.

In a resonant cavity, the atom–photon interaction becomes more involved: the atom creates a photon in the cavity, and this photon can be reabsorbed by the atom such that the process repeats. In the limit that only two quantum states of the atom are relevant, a two-level atom, the interaction Hamiltonian is simply $\hat{H}_{\text{int}} = \hbar g(\sigma_+ a + \sigma_- a^\dagger)$ where σ_+ and σ_- are the atomic transition operators.

This Hamiltonian, the Jaynes–Cummings Hamiltonian, describes how a quantum of energy can be exchanged between the atom and the light in the cavity [1]. Equivalently, the eigenstates become dressed states, admixtures of photon- and atom-states. The energetic separation between the eigenstates "+" and "−" is $2\hbar g\sqrt{N+1}$ where N is the number of excitations. This N dependence uncovers a nonlinearity in the atom–photon interaction as a function of photon number. The nonlinearity arises as a consequence of the disparity between atom and photon: the photon number is arbitrary, yet the atom can be excited only once.

1.3 Single-Photon Sources

A single-photon source should produce single, indistinguishable photons on demand at a high repetition rate at a convenient wavelength. The metrics are the purity, $1 - g^{(2)}(\tau = 0)$, a measure of the extent to which the source produces single photons and not multiple photons ($g^{(2)}(\tau = 0) = 0$ for a perfect single-photon source); the two-photon interference visibility; the repetition rate; and the end-to-end efficiency.

The workhorse source of single photons is based on spontaneous parametric downconversion in a nonlinear crystal [1, 2]. Via a nonlinear process, a laser pulse creates two photons, the process satisfying energy and momentum conservation. This scheme is simple to implement. Furthermore, although the process is probabilistic – spontaneous parametric downconversion does not form the basis of an "on demand" source – photons are created pairwise: detection of one photon signals the creation of its twin. This heralding feature is powerful and overcomes in many experiments the weakness implied in the probabilistic nature of the source. On increasing the laser power, the probability of creating single- and two-photon states both increases such that an increase in efficiency results in a reduced purity. Typically, the application determines a maximum tolerable purity, and this in turn sets a limit to the efficiency.

In an attempt to create a truly on-demand source, alternative schemes to spontaneous parametric downconversion have been explored. The simplest scheme uses a two-level atom. The atom is excited to its excited state. One and only one photon is created via spontaneous emission as the atom returns to its ground state. A subsequent photon can only be created by re-exciting the atom. As such, at least in principle, the scheme can produce single photons on demand. In practice, two challenges need to be met. First, the photonic environment needs to be engineered such that the atom creates photons in one specific optical mode. Example optical modes are the confined mode in a cavity and the propagating mode of a waveguide. Second, to ensure the indistinguishability of the photons, environmental noise must be suppressed: noise in the environment results in random fluctuations in the photons' properties and hence low two-photon interference visibilities.

Much progress has been made in the quest of creating an on-demand single-photon source. One key line of development is to replace a real atom with an artificial atom, a semiconductor quantum dot [5, 6]. The tools in the semiconductor toolbox, layer-by-layer growth of semiconductor heterostructures and post-growth device fabrication, are exploited.

The temporal wave packet of single photons created via spontaneous emission from a two-level system consists of an abrupt turn-on and subsequent exponential decay. Single photons with arbitrary wave packets can be created using a three-level atom in which both ground states couple to a common excited state, a Λ-system. The ground-state splitting can be engineered by using different spin states of the atom. In fact, a Λ-system together with photonic engineering is a very powerful system. It can be used to create spin–photon entangled pairs in which the spin of the atom and one degree-of-freedom of the photon are entangled. Furthermore, by combining single-photon emission with spin control, cluster states can be created, a linear chain of photons in which each photon is entangled with its nearest neighbor [7].

Color centers with highly coherent spin states enable a Λ-system to be created in a solid-state host. For instance, the NV^- color center in diamond possesses a very coherent spin [8]. A Λ-system can be engineered in a semiconductor quantum dot by adding one electron (or hole) and using the spin state. The spin tends to dephase rather quickly on account of magnetic noise in the host material. However, schemes are now in place to suppress this noise by preparing the spin's environment [9].

1.4 Single-Photon Detectors

Quantum photonics in the optical domain benefits enormously from the availability of excellent laser sources and, significantly, excellent single-photon detectors [10]. The laser sources, both continuous-wave and pulsed, excite quantum processes. For instance, a laser pulse tuned in energy to satisfy the π-condition excites a two-level system from its ground state to its excited state. The created photons can be detected using a single-photon detector: demolition of a single photon results in an easy-to-measure voltage pulse at the output of the detector. The arrival time of the photon with respect to a trigger can be recorded using photon-counting hardware; correlation functions, notably $g^{(2)}$, can thereby be determined. In almost all cases, single-photon detectors are not photon-number resolving: they distinguish between no photons and one-or-more photons.

In the optical domain, a single-photon detector operates by converting the energy of a photon into an electronic excitation; detection proceeds by amplifying this electronic excitation. One way to implement this concept is to use silicon in a reverse-biased diode. A photon excites an electron across the band gap of the semiconductor; an avalanche process in the diode represents the amplification mechanism. Such silicon avalanche photodiodes work well for photon energies above the bandgap of silicon (wavelengths below 1100 nm). The quantum efficiency peaks at a photon energy of 1.6 eV, i.e. at a wavelength of 780 nm in the red part of the spectrum. At longer wavelengths, cooled InP-based single-photon detectors are available, but the higher defect concentration of InP with respect to silicon compromises the performance – after-pulsing, the creation of more than one detector pulse for each photon, is a common problem. These detectors work best in Geiger mode in which the detector is not free-running but is turned on occasionally with timing chosen to match the particular application. Silicon detectors suffer from a trade-off between quantum efficiency and timing jitter: high quantum-efficiency detectors have a thick absorption layer and consequently large timing jitter, hundreds of picoseconds; low timing jitter, tens of picoseconds, is achieved with a thinner absorption layer and hence a lower quantum efficiency. This may or may not represent a problem depending on the application.

An alternative to the semiconductor avalanche photodiode is a superconducting nanowire single-photon detector, SNSPD [11, 12]. In this case, a current is applied to a superconducting nanowire. The current is close to the critical current. On absorption of a photon, the material makes a local and temporary transition from the superconducting state to the normal state resulting in an easy-to-measure bias pulse. SNSPDs have the huge advantage of very high quantum efficiencies, above 90%, even at the telecoms wavelength of 1550 nm at which silicon detectors are completely blind. Furthermore, the quantum efficiency–timing jitter trade-off is broken: high quantum efficiency, low timing jitter (a few tens of picoseconds or less) single-photon detectors are now available. SNSPDs operate at cryogenic temperatures, typically a few Kelvin in a closed-cycle cryostat.

1.5 Applications of the Photon in Quantum Technology

1.5.1 Quantum Communication

Single photons can be used to send quantum information over large distances. The first protocol, BB84 [13], rests on the no-cloning theorem, the impossibility of creating a copy of an unknown quantum state. In practice, many implementations use a highly attenuated laser pulse instead of a single-photon source. Many of the output pulses are empty, a few populated by single photons. This compromises the security. A close-to-ideal single-photon source would remove this limitation, but insecurities still exist. Maximum security is guaranteed by so-called device-independent quantum key distribution in which full quantum-ness (for instance, violation of a Bell inequality) is demonstrably maintained at all times [14]. However, the hardware requirements are challenging to meet in practice. The photons should be almost perfectly indistinguishable and should be generated with very high efficiencies.

1.5.2 Quantum Simulation

Quantum states of light can be used to carry out tasks in quantum simulation. An example is boson sampling, the behavior of photons in a complex optical circuit for which quantum advantage has been demonstrated [15].

1.5.3 Quantum Computing

On the one hand, a photon represents a robust qubit. On the other hand, it is challenging to create entangled photonic states deterministically. Linear optical computing applies a circuit approach to quantum computing, using single-photon inputs and relying on two-photon interference to create entangled states [16]. It is a probabilistic method. An alternative is measurement-based quantum computing [17]: measurements and individual qubit operations are performed on a highly entangled multi-qubit state. This approach is well suited to photonics: potentially, one quantum source can create many qubits with the necessary entanglements;

measurements can be carried out quickly and reliably. It remains to be seen to what extent these ideas are practicable and to what extent they are competitive with respect to other approaches to quantum computing.

Some suggestions for further reading are included in the (highly incomplete) bibliography.

References

1 Gerry, C.C. and Knight, P.L. (2005). *Introductory Quantum Optics*. Cambridge University Press.

2 Fox, M. (2006). *Quantum Optics An Introduction*. Oxford University Press.

3 Hong, C.K., Ou, Z.Y., and Mandel, L. (1987). Measurement of subpicosecond time intervals between two photons by interference. *Physical Review Letters* 59: 2044–2046.

4 Purcell, E.M., Torrey, H.C., and Pound, R.V. (1946). Resonance absorption by nuclear magnetic moments in a solid. *Physical Review* 69: 37–38.

5 Lodahl, P., Mahmoodian, S., and Stobbe, S. (2015). Interfacing single photons and single quantum dots with photonic nanostructures. *Reviews of Modern Physics* 87: 347–400.

6 Senellart, P., Solomon, G., and White, A. (2017). High-performance semiconductor quantum-dot single-photon sources. *Nature Nanotechnology* 12: 1026–1039.

7 Lindner, N.H. and Rudolph, T. (2009). Proposal for pulsed on-demand sources of photonic cluster state strings. *Physical Review Letters* 103: 113602.

8 Balasubramanian, G., Neumann, P., Twitchen, D. et al. (2009). Ultralong spin coherence time in isotopically engineered diamond. *Nature Materials* 8: 383–387.

9 Gangloff, D.A., Ethier-Majcher, G., Lang, C. et al. (2019). Quantum interface of an electron and a nuclear ensemble. *Science* 364: 62–66.

10 Hadfield, R.H. (2009). Single-photon detectors for optical quantum information applications. *Nature Photonics* 3: 696–705.

11 Gol'tsman, G.N., Okunev, O., Chulkova, G. et al. (2001). Picosecond superconducting single-photon optical detector. *Applied Physics Letters* 79: 705.

12 Marsili, F., Verma, V.B., Stern, J.A. et al. (2013). Detecting single infrared photons with 93% system efficiency. *Nature Photonics* 7: 210–214.

13 Bennett, C.H. and Brassard, G. (1984). Quantum cryptography: public key distribution and coin tossing. *Proceedings of IEEE International Conference on Computers, Systems and Signal Processing* 175: 8.

14 Acín, A., Brunner, N., Gisin, N. et al. (2007). Device-independent security of quantum cryptography against collective attacks. *Physical Review Letters* 98: 230501.

15 Zhong, H.-S., Wang, H., Deng, Y.-H. et al. (2020). Quantum computational advantage using photons. *Science* 370: 1460–1463.

16 Knill, E., Laflamme, R., and Milburn, G.J. (2001). A scheme for efficient quantum computation with linear optics. *Nature* 409: 46–52.

17 Raussendorf, R. and Briegel, H.J. (2001). A one-way quantum computer. *Physical Review Letters* 86: 5188–5191.

Part I

Fundamentals of Quantum Technologies

2

The Second Quantum Revolution: From Basic Concepts to Quantum Technologies[1] [1]

Alain Aspect†

Université Paris-Saclay, Institut d'Optique, 2 Avenue Augustin Fresnel, 91120 Palaiseau, France

† *The 2022 Nobel Prize Laureate in Physics*

2.1 Two Quantum Revolutions

The development of quantum mechanics in the beginning of the twentieth century was a unique intellectual adventure, which obliged scientists and philosophers to change radically the concepts they used to describe the world [2]. After these heroic efforts, it became possible to understand the stability of matter, the mechanical and thermal properties of materials, the interaction between radiation and matter, and many other properties of the microscopic world that had been impossible to fully understand with classical physics. A few decades later, that *conceptual revolution* enabled a *technological revolution*, at the root of our information-based society. It is indeed with the quantum mechanical understanding of the structure and properties of matter that physicists and engineers were able to invent and develop the transistor and the laser – two key technologies that now permit the high bandwidth circulation of information, as well as many other scientific and commercial applications.

After such an accumulation of conceptual – and eventually technological – successes, one might think that by 1960, all the interesting questions about quantum mechanics had been raised and answered. However, in his now-famous paper of 1964 [3] – one of the most remarkable papers in the history of physics, John Bell drew the attention of physicists to the extraordinary features of entanglement: Quantum Mechanics describes a pair of entangled objects as a single global quantum system, impossible to be thought of as two individual objects, even if the two components are far apart. John Bell demonstrated that there is no way to understand entanglement in the framework of the notion of a physical reality localized in space time and obeying causality. This result was opposite to the expectations of Einstein,

1 The expression "Second quantum revolution" was used independently in "Dowling, J.P. and Milburn, G. J. (2003). Quantum technology: the second quantum revolution. *Philosophical Transactions of the Royal Society of London. Series A: Mathematical, Physical and Engineering Sciences* 361 (1809): 1655–1674.

Photonic Quantum Technologies: Science and Applications, First Edition. Edited by Mohamed Benyoucef.

who had first pointed out, with his collaborators Podolsky and Rosen, the possibility of entangled particles with strong correlations, and analyzed these correlations in the framework of local realism. The most remarkable feature of Bell's work was undoubtedly the possibility it offered to settle experimentally the question and decide whether or not Einstein's ideas could be kept. The experimental tests of *Bell inequalities* gave an unambiguous answer: entanglement *cannot* be understood as usual correlations, whose interpretation relies on the existence of common properties, originating in a common preparation, and remaining attached to each individual object after separation, as components of their physical reality.[2] More than five decades after the 1964 paper, the physics of entanglement is flourishing, and thousands of papers, theoretical or experimental, are found when one types "Bell inequalities" on a search engine.

Starting in the 1970s, another concept has progressively become more and more important in quantum physics: the description of *single objects*, in contrast to the statistical use of quantum mechanics to describe only properties of large ensembles (for instance, the fluorescence of an atomic vapor). That question had, like the EPR problem, been an object of debate between Bohr and Einstein [4], but it is the development of experimental abilities to isolate and observe single microscopic objects like electrons, ions, and atoms that prompted physicists to take quantum mechanical dynamics of single objects, including "quantum jumps," seriously. The experimental observation of quantum jumps (in the fluorescence light from a single ion) inspired new theoretical approaches, the so-called "Quantum Monte Carlo Wave Function" simulations, primarily used to describe "elementary" microscopic objects like ions, atoms, and small molecules. More recently, progress in nanofabrication, as well as experimental breakthroughs, has allowed physicists to create mesoscopic systems (e.g. electric and magnetic devices, gaseous Bose–Einstein condensates), which push the border of the quantum world to larger and larger systems that still need to be described as single quantum objects.

Before the realization of the importance of Bell's theorem, which happened only after the clear experimental violation of Bell inequalities, the conventional wisdom among physicists was that the "founding fathers" of quantum mechanics had settled all conceptual questions. Bell inequalities violations did not cast any doubt on the validity of quantum mechanics as a predictive tool. To the contrary, experiments found that nature definitely follows the quantum mechanical predictions even in those weird situations. But there was a lesson to be drawn: questioning the "orthodox" views, including the famous "Copenhagen interpretation," might lead to an improved understanding of the quantum mechanics formalism, even though that formalism remained impeccably accurate. In other words, physicists could question the belief that the conceptual understanding that had been achieved by the 1940s was the end of the story.

I think it is not an exaggeration to say that the realization of the importance of entanglement and the clarification of the quantum description of single

---

2 An example of usual correlations is the identity of the eye colors of twin brothers, linked to their identical chromosome sets. Correlations in entangled twin photons are different in nature, as we explain later.

objects have been at the root of a *second quantum revolution*. And it may well be that this once purely intellectual pursuit will also lead to *a new technological revolution*. Indeed, we should have no doubt that the advances in the quantum concepts used to describe single objects will certainly rejoin and play a key role in the ongoing *revolution of nanotechnology*. Even more amazing, physicists have endeavored to apply entanglement to "*quantum computation,*" and most of the systems that are being experimentally tested as elementary quantum processors are isolated individual quantum systems, such as a few interacting ions, atoms of superconducting nanodevices. Whether or not that second revolution will have a strong impact on our societies is not yet decided. But who would have imagined the ubiquitous presence of integrated circuits when the first transistor was invented?

2.2 The First Quantum Revolution

Searching for a consistent explanation of the black body radiation spectrum at both high and low frequencies, M. Planck introduced in 1900 the quantization of energy exchange between light and matter [5]. Albert Einstein took a step further by proposing the quantization of light itself to understand the photoelectric effect [6]. The properties he deduced were then tested by Robert Andrews Millikan in 1914 [7]. Convincing evidence of the existence of molecules – doubted until the beginning of the twentieth century – was provided by various observations, including Einstein's explanation of Brownian motion [8]. Together with many other experiments, these observations convinced physicists and philosophers to accept the granularity of matter and quantization of energy in the microscopic world.

In addition to rendering an account of experimental data on radiation, the new quantum mechanics resolved basic problems on matter. For instance, Niels Bohr's 1913 model of the atom explained both the absorption spectra of atomic gases and the stability of matter: without quantum mechanics, the Rutherford atom, composed of orbiting particles with opposite (i.e. attractive) charges, should radiate and collapse.

The first comprehensive paradigm of quantum mechanics centered about the Heisenberg and Schrödinger formalisms of 1925. The latter was a wave equation for matter, completing a beautiful duality: like light, matter can behave as either a particle or a wave. The wave-particle duality, first introduced by Einstein [9] about radiation, was proposed by L. de Broglie's about matter [10] and remains incomprehensible to the classical way of thinking. Within 20 years of its birth, the quantum mechanical formalism could describe consistently chemical bonds, electrical properties, and thermal properties of matter at a microscopic level. Continuing progress in physics was pushing along different directions: toward the incredibly small, with particle physics, or into the domain of more exotic properties of matter, such as superconductivity (the absence of resistance in some conductors at low temperature), or superfluidity (the absence of viscosity of liquid Helium at low temperature). Studies on light–matter interaction were refined by orders of magnitudes, thanks to experimental breakthroughs made possible by advances in microwave technology [11]. All this progress took place perfectly within the quantum mechanical framework, which had been refined to be applied both in the elementary phenomenon (Quantum Electrodynamics) and in complex situations encountered in condensed matter. But in the early 1950s, quantum mechanics still appeared as a game to be played by physicists only for the sake of progress in knowledge, without any impact on everyday life.

2.2.1 The Electronics and Information Age: Quantum Mechanics Applied

The applications of quantum physics are all around us in electronics and photonics. In technologies today, quantum mechanics is required to understand material properties (electrical, mechanical, optical, etc.) and the behavior of devices at the root of many technological achievements.

The transistor was invented in 1948 by a brilliant group of solid-state physicists after fundamental reflection about the quantum nature of electrical conduction [12]. This invention and its descendants, microfabricated integrated circuits [13], clearly had a monumental impact. Like the steam engine over a century earlier, the transistor changed our lives and gave birth to a new era, the information age.

The second technological progeny of quantum mechanics is the laser, developed in the late 1950s [14]. Some of its applications are present in everyday life: bar code readers, CD readers and players, medical tools, etc. Less visible but perhaps more important is the use of laser light in telecommunications, where it dramatically boosts the flow of information: terabits (millions of millions of information units) per second can be transmitted across the oceans through a single optical fiber. These information highways connect us to the stored knowledge and active computation distributed around the world. Starting from the few bits per second of the first telegraph operators, we have come a long way.

The quantum mechanical understanding of atom–photon interactions has also continued to develop and eventually led to applications. For example, in 1997, a Nobel prize was given to Steven Chu, Claude Cohen-Tannoudji, and William D. Phillips, for the development of laser cooling and trapping of atoms. Here as well, fundamental research soon led to a spectacular application, cold atom clocks, which have already allowed the accuracy of time measurement to reach a level of 10^{-18} (1 second accuracy in the age of the universe!). More is to come with cold atom or ion *optical* clocks (Chapter 14 provides a comprehensive description of optical atomic clocks). Atomic clocks are used in the global positioning system (GPS), and their replacement with cold atom clocks will eventually permit an improved positioning precision. Coming full circle, that improved technology of clocks can be applied to fundamental questions, such as tests of general relativity, or the search for slow variation of fundamental physical constants. The first quantum revolution, with its interplay between basic questions and applications, is still at work…

2.3 Entanglement and Bell's Theorem

2.3.1 The Bohr–Einstein Debate

Quantum mechanics was constructed at the price of several radical – and sometimes painful – revisions of classical concepts. For instance, to take into account particle-wave duality, quantum mechanics had to renounce the idea of a classical trajectory. This renunciation is best stated in the Heisenberg dispersion relations, which describe quantitatively the impossibility of defining precisely and simultaneously the position and velocity of a particle. One can also illustrate this renunciation of classical trajectories by remarking that in an interference experiment the particle "follows many paths at once."

In fact, such renunciations were so radical that several, including Einstein and de Broglie, could not admit their inevitability and differed with Bohr who had carved the Rosetta stone of interpretation of the new theory under the name "the Copenhagen interpretation." Einstein did not challenge the formalism and its provisions directly, but seemed to think that the renunciations put forward by Bohr could only signify the incompleteness of the quantum formalism. This position led to epic debates with Bohr, in particular that which started in 1935 with the publication of the article of Einstein, Podolsky, and Rosen (EPR), whose title posed the question, "Can Quantum-Mechanical description of physical reality be considered complete?" [15]. In this article, Einstein and his coauthors showed that the quantum formalism permitted the existence of certain two-particle states for which one can predict strong correlations both in velocity and in position even when the two particles are widely separated and no longer interact. They showed that measurements of positions would always give values symmetric about the origin, so that a measurement on one particle would allow one to know with certainty the value of the position of the other one. Similarly, measurements of the velocities of the two particles would always yield two opposite values so that a measurement on the first one would be enough to know with certainty the velocity of the other one. Of course, one has to choose between an accurate position and velocity measurement on the first particle, because of the Heisenberg relations. But the measurement on the first particle does not disturb the (distant) second particle, so that EPR concluded that the second particle already had, before measurement, well-determined values of position and velocity. And since the quantum formalism cannot give a simultaneous and precise value to these quantities, Einstein and his coauthors concluded that quantum mechanics was incomplete, and that physicists ought to devote themselves to try to complete it.

Niels Bohr was apparently bowled over by this argument [16], which rests on quantum mechanics itself to show its provisionary character. His writings show his profound conviction that if the EPR reasoning were correct, it would not be just a matter of completing the formalism of quantum mechanics, but it would be all of quantum physics that would collapse. He thus immediately contested the EPR reasoning [17], claiming that in such a quantum state, one could not speak about the individual properties of each particle, even if they are far from one another. In contrast to Bohr, Schrödinger acknowledged the importance of the situation described in the EPR paper and coined the term "entanglement," to characterize the lack of factorability of an EPR state [18].

In fact, when the EPR paper appeared in 1935, quantum mechanics was being crowned with one success after another, so apart from Bohr and Schrödinger, most physicists ignored this debate, which seemed rather academic. It seemed that adhering to one or the other position was a matter of personal taste (or epistemological position), but did not have any practical consequence on how quantum mechanics was used. Einstein himself apparently didn't contest this attitude, and we had to wait 30 years to see a resounding counter-argument to this relatively universal position.

2.3.2 Bell's Theorem

In 1964, a now famous short article [3] changed the situation dramatically. In this paper, John Bell takes seriously the EPR argument and introduces common supplementary parameters (also called "hidden variables" [19]) shared by the two particles at their initial preparation in an entangled state and carried along by each particle after separation. Reasoning on entangled states of two spin-1/2 particles – a simpler version of the EPR situation introduced by Bohm [20], Bell shows that one can easily explain the existence of correlations between the results of measurements on the two particles by writing that the result of a measurement on one particle depends only on the parameters carried by that particle and by the setting of the apparatus making that measurement. But then, a few lines of calculation suffice to show a quantitative contradiction with the predictions of quantum mechanics. More precisely, even if such a "hidden variable" theory can reproduce some of the predicted quantum correlations, it cannot mimic the quantum mechanical predictions for *all* possible settings of the measuring apparatus. Thus it is not possible, in general, to understand EPR-type correlations by "complementing" the quantum theory along the lines proposed by Einstein. This result, known as *Bell's theorem*, continues to surprise us even today, since we are used to explaining all types of correlations by a scheme akin to hidden variables.[3] For instance, if we have a pair of identical twins, we do not know what their blood type is before testing them, but if we determine the type of one, we know for sure that the other is the same type. We easily explain this by the fact that they were born with the same specific chromosomes that determined their blood type. What Bell's paper shows us is that if we try to describe the correlations between the entangled particles as we understand the correlations between the twins, we will be making a serious error.

A crucial hypothesis in Bell's reasoning is the "*locality hypothesis*," which needs to be fulfilled by the supplementary parameter models to lead to a conflict with quantum mechanics. This very natural assumption states that there is no direct non-local

3 A famous example is "Bertlmann's socks" (paper 16 in the book cited in Ref. [1]). Physicists are surprised as much as nonphysicists, may be more : see Mermin, D. (1985). Is the moon there? *Physics Today* 38: 11.

interaction between the two measuring apparatuses far from each other. In other words, the conflict arises only if the result of a measurement on the first particle does not depend on the setting of the second measuring apparatus.[4] As stressed by Bell in his 1964 paper [3], this very natural hypothesis would become a direct consequence of Einstein's views that no influence can travel faster than light, in an experimental scheme where the settings of the measuring apparatus are rapidly changed while the particles are flying between the source and the measuring apparatus [21].

To establish the incompatibility between quantum mechanics and the local hidden variable theories, Bell showed that the correlations predicted by any local hidden variable model are limited by inequalities – today called "Bell inequalities" – that are violated by certain quantum predictions. The choice between the positions of Einstein and Bohr was then no longer a question of personal taste. Instead, it became possible to settle the question experimentally, by looking carefully at measurements of correlations between entangled particles. Surprisingly, in 1964, there was no experimental result that permitted such a quantitative test. Experimentalists began then to think about how to construct an experiment to create and measure states for which quantum mechanics would predict a violation of Bell inequalities. In 1969 was published a version of inequalities well adapted to real experiments, where apparatus have some inefficiencies [22]: it became clear that conclusive experiments were possible, provided that the experimental imperfections remain small enough. Some experiments were carried out with γ-ray photons emitted in positronium annihilation, or with protons, but the most convincing were realized with photon pairs. After a first series of pioneering experiments [23], a new generation of experiments in the early 1980s [24, 25] obtained a collection of clear-cut results in agreement with quantum theory and violating Bell inequalities even with rapidly switched polarizers. A third generation of experiments, undertaken since the beginning of the 1990s, has definitely confirmed these results [26–29]. There is no doubt,[5] then, that two entangled photons are *not* two distinct systems carrying two identical copies of the same parameters. A pair of entangled photons must

4 The importance of a locality hypothesis had already been suggested in Bell's discussion of the "impossibility proofs" of hidden variables [19], and the locality assumption was clearly stated in the first paper presenting inequalities [3]. Actually, the locality (or separability) question appears in most of the papers of the book cited in Ref. [1].

5 Until 2015, there remained a loophole open for advocates of local hidden variable theories, when the efficiency of the detectors used in real experiments is small compared to unity, so that many photons remain undetected [23]. However, as stressed by John Bell [in: Atomic-cascade photons and quantum-mechanical nonlocality, *Comments on atomic and Molecular Physics* 9 (1980): 121–126], paper reproduced in Ref. [1] "it is difficult for me to believe that quantum mechanics, working very well for currently practical set-ups, will nevertheless fail badly with improvements in counter efficiency…" A first experiment with detection efficiency close to one [Rowe, M.A. et al. (2001). Experimental violation of a Bell's Inequality with efficient detection. *Nature* 409: 791–794] confirmed a clear violation of Bell's inequalities. In that experiment, however, the measurements were not space like separated. A series of more sophisticated experiments has followed, culminating in 2015 in so-called "loophole free experiments" where the sensitivity loophole is closed AND space-like separation is enforced between detection at one instrument and the setting of the other, distant instrument [29]. More recent loophole-free experiments confirm that Bell's inequalities are violated in a timing experiment with the sensitivity loophole closed.

instead be considered as a single, inseparable system, described by a global wave function that cannot be factorized into single-photon states.

The inseparability of an entangled photon state has been shown to hold even if the photons are far apart – including a "space-like" separation in the relativistic sense, that is a separation such that no signal traveling at a velocity less than or equal to the velocity of light can connect the two measurements. This was already the case in the 1982 experiments, where the photons were separated by 12 m at the time of the measurements [24]. In addition, it was possible to change the setting of the measuring polarizers during the 20 nanosecond flight of the photons between the source and the detector [25], in order to implement Bell's ideal scheme. In more recent experiments, where new sources have permitted the injection of entangled photons into two optical fibers, a violation of Bell Inequalities was observed at separations of hundreds of meters [27], and even more [28], and it has been possible to change *randomly* the setting of the polarizers during the propagation of the photons in the fibers [27]. "Timing experiments" with variable polarizers emphasize that everything happens as if the two entangled photons were still in contact, and as if the measurement of one photon would affect the other instantaneously. This seems to contradict the principle of relativistic causality that specifies that no interaction can propagate faster than the speed of light. It is referred to in the literature as "quantum non-locality" [30].

It should be stressed, however, that there is no violation of causality in the operational sense, and that one cannot use this non-separability to send a signal or a usable information faster than the speed of light [31]. To illustrate this impossibility, let us consider whether the observer at polarizer II, traditionally named Bob, can know immediately the orientation of the polarizer of the operator at polarizer I, traditionally named Alice. Since only the correlation depends on the relative angle, he should be able to compare his results to the results obtained by Alice. But the transmission of these results can be done only via a classical channel, which does not communicate at super-luminal speed [32]. It should be emphasized that a fundamental reason for that impossibility is the randomness of the outcome of the measurement at Alice's station. It is remarkable that randomness, a fundamental feature of quantum mechanics, is essential to enforce no signaling, in spite of quantum non-locality.

Another hypothetical way of signaling faster than light is forbidden by another fundamental quantum property, the quantum no-cloning theorem [33]. Indeed, if Bob could make many copies of one photon he receives, he could make many measurements with polarizers at various angles. He could then determine the direction at which one of the two orthogonal outcomes (+ or −) is maximum and instantaneously deduce the direction of Alice's polarizer (+ case) or of its orthogonal direction (−case) when analyzing the partner photon of the one analyzed by Bob.

Thus, as troubling as it may be, quantum entanglement does not allow us to build the super-luminal telegraph at work in science-fiction novels. Nevertheless, as we will describe in Section 2.5, fascinating applications of entanglement are developed in a new field: "quantum information."

2.4 Quantum Mechanics and Single Objects

The experimental evidence supporting quantum mechanics typically comes from large ensembles. For instance, atomic spectra are taken from clouds of myriads of atoms; semiconductors are bulk materials; and laser beams contain a tremendous number of photons, produced by an optical amplifier that contains a huge number of atoms. In these situations, we can apply without any problem the formalism of quantum mechanics, which usually yields probabilistic predictions. Since our observations bear on large ensembles, we realize statistical measurements, to which the quantum probabilities can be directly compared. These concepts are fully at work in the density matrix formalism of quantum mechanics, easily interpreted in the framework of large ensembles, even though the Copenhagen school has always claimed that the standard formalism could also be applied to individual objects. That issue was also deeply discussed between Einstein and Bohr [4], but it remained a matter of principle until experimentalists were able to deal with single microscopic objects.

2.4.1 From the Ensemble to the Single Quantum System

Beginning in the 1970s, physicists invented methods to manipulate and observe single elementary objects – such as a single electron or a single ion – and trap it for hours (or even days and months) with electric and magnetic fields that held the particle in a vacuum chamber, far from any material wall. In the decade that followed, scanning microscopy techniques allowed for the observation and manipulation of single atoms deposited on a surface [34]. These experimental advances, crowned by several Nobel prizes [35], had major consequences in fundamental physics. The achievement of trapping individual elementary objects led to remarkable improvements in the knowledge of certain microscopic quantities, whose values provide tests of fundamental theories. For instance, spectroscopy of a single trapped electron yields a measurement of the "g-factor" of the electron to 13 significant figures – a precision equivalent to measuring the distance between the earth and the moon to better than the diameter of a human hair! The g-factor of the electron is a fundamental quantity that can also be calculated using quantum electrodynamics, the refined theory of quantum mechanics applied to elementary electric charges and photons. The essentially perfect agreement between experiment and theory shows the incredible accuracy of the predictions of this theory. The trapping of elementary objects has also permitted crucial tests of the symmetry between matter and antimatter: one can verify with a spectacular precision proton–antiproton or electron–positron symmetry [36]. It is also possible to verify that two electrons, or two atoms of the same chemical element, have exactly the same properties. Indiscernibility has little sense in classical physics, where two beads, however identical they may seem, can always be distinguished by small defects or marks. On the other hand, strict identity of elementary natural quantum objects is in fact at the core of quantum physics [37].

In parallel with experimental achievements, the trapping of individual microscopic objects obliged physicists to think more carefully about the significance of quantum mechanics when applied to an individual object. We know that in general quantum mechanics gives probabilistic predictions. For example, one may calculate that an atom illuminated by a certain laser beam has a certain probability P_B to be in a "bright" state; and the complementary probability $P_D = 1 - P_B$ to be in a "dark" state [38]. By dark or bright, we mean that when illuminated by an auxiliary probe laser, an atom in a dark state radiates no photon, while in the bright state it emits many fluorescence photons, easy to observe with a photo-detector, or even the naked eye. If we have a vapor with a large number of atoms in this situation, the interpretation of the probabilistic quantum prediction is undemanding: it suffices to admit that a fraction P_B of the atoms are in the bright state (and thus scatter photons when probed with the auxiliary laser), while the remaining fraction are in the dark state (do not scatter photons). But what would happen for a single atom placed in the same situation? Asked to respond to this question, the "Copenhagen School" would respond that the atom is in a "superposition" of the dark and bright states. In such a superposition, the atom is simultaneously in both the bright and the dark states, and it is impossible to know in advance what will happen when we apply the probe laser: the atom may be found in the bright state or may be found in the dark state. Of course, the Copenhagen spokesperson would add, after repeating the measurement many times, one would observe a bright state for a fraction P_B of the cases, and the dark state for a fraction P_D.

In fact, this answer is not complete, since it tells us only about averaged results of repeated measurements. But how does the state of a *single* atom evolve with time if we observe it continuously? Or, more precisely, what would we observe if we would leave the probe laser on all the time? This question was academic in the 1930s, when experimentalists could not even imagine observing individual, isolated particles. However, the Copenhagen physicists had an answer that invoked the postulate of "wave packet reduction." When first illuminated by the probe laser, the atom in a superposition of the dark and bright states would collapse into one of the two basic states, say, for instance, the bright state, where fluorescence photons can be seen. A further evolution can lead the atom to collapse into the dark state, and the fluorescence then suddenly stops. Thus, one would predict that the atom would, at random moments, switch from the dark state to the bright state.

The existence of such "quantum jumps," implying a discontinuous evolution of the system, was strongly opposed by a number of physicists – including Schrödinger – who saw a convenient trick with a pedagogical value, but who contended that quantum mechanics inherently applies only to large ensembles and not to single quantum objects [39]. The experimental progress discussed above allowed the debate to be resolved experimentally in 1986 by the direct observation of quantum jumps in the fluorescence of a single trapped ion. In this type of experiment, one indeed observes that the ion evolves randomly between periods where it is invisible, and periods where it fluoresces intensely! This striking result resolved beyond all doubt that quantum jumps do exist, and that quantum theory can describe the behavior of a single object.

2.4.2 Quantum Jumps in Action: New Clocks and New Theoretical Methods

The – experimentally enforced – conceptual acceptance of quantum jumps led to surprising developments in both the theoretical and experimental arena. Experimentally, one can use the phenomenon of alternation between dark and bright states for the most precise ever spectroscopic measurements of ion spectral lines [40]. These weak lines can be used for new atom clocks, always more precise.

Quantum jumps have also inspired a new theoretical method called "Quantum Monte Carlo Wave Function" in which a possible history of the system is simulated by drawing random quantum jumps, whose probability is governed by quantum mechanical laws [41]. By generating a large number of these "possible histories," one can build the probability distribution of the results. In the large number limit, these distributions coincide with the density matrix predictions. There are some situations in which this calculation method is remarkably more efficient than traditional methods. Furthermore, quantum Monte Carlo methods have allowed the discovery that certain quantum processes obey unusual statistics – Lévy statistics, which are also encountered in domains as disparate as biology or the stock market. The use of these "exotic" statistical methods provides new and efficient methods to address quantum problems [42].

2.4.3 From Microscopic to Mesoscopic

Having established that quantum mechanics can describe the dynamics of a single system, one naturally asks how big that system can be. Certainly, we do not need quantum mechanics for macroscopic objects, which are well described by classical physics – this is probably a good reason why quantum mechanics seems so foreign to our everyday existence. Of course, we need quantum mechanics to understand properties of the bulk material of which the macroscopic object is made, but not for the behavior of the object as a whole. But between the scale of a single atom and the macroscopic world, there is a regime referred to as "mesoscopic," where though the object is not an elementary microscopic object, one cannot use classical physics to describe its behavior. In the mesoscopic regime, it is the object itself, and not only the material of which it is made, that needs be described by quantum mechanics.

For instance, nanofabricated conducting rings have demonstrated effects that can only be understood by treating their electrons with a global wave function [43]. Another famous example, for which the 2001 Nobel prize was awarded to Eric A. Cornell, Wolfgang Ketterle and Carl E. Wieman, is a gaseous Bose–Einstein condensate, which must also be treated as a single "large quantum" object, with a number of atoms usually ranging from a few thousands [44] to tens of millions [45] or more.

Even if these mesoscopic objects were for a long time curiosities in research labs, the uninterrupted miniaturization of microelectronics already obliges engineers to understand some of their circuits using the laws of quantum mechanics.

In Section 2.5, we will give the example of a would-be "quantum computer," but even ordinary transistors will exhibit new quantum properties if downscaled to the size of a few thousands of atoms

2.4.4 From Mesoscopic to Macroscopic: Decoherence

What then separates the microscopic and mesoscopic quantum world from the macroscopic classical world? John Bell was deeply concerned by that question, which was a major reason of his discomfort with the standard interpretation of quantum mechanics where this frontier plays a crucial role [46].

One of the most important features of quantum physics is to allow the existence of superpositions of states: if a system has several possible quantum states, it can not only be in any one of these states, but it can also be in a hybrid state, or "coherent superposition," composed of several basic states. Entanglement is a sophisticated case of state superposition for two or more quantum objects, but even single systems can be put in superposition states, and we have discussed above the case of an atom in a superposition of a dark state and a bright state. The situation becomes quite troubling when the two states involved are obviously incompatible. For instance, consider an atom arriving at an atomic beam splitter. The atom can either be transmitted or reflected, two options that can lead to well-separated paths. But the atom can also come out in a superposition of the reflected and transmitted states, i.e. simultaneously present in two clearly separated regions of space. One can show experimentally that this superposition state does exist, by recombining the two paths and observing interference fringes, which can only be explained by admitting that both paths were followed simultaneously. Such behavior has been observed with microscopic objects (electrons, photons, neutrons, atoms, molecules as large as C_{60} fullerenes), or mesoscopic objects (electric currents in nanocircuits), but never with macroscopic objects, even though this is not a priori forbidden by the quantum formalism. That problem has attracted the attention of many physicists, starting with Schrödinger, who gave an amusing ("burlesque" in his own words) illustration involving his famous cat [47]. In the proposed scenario, the cat's life rests upon a quantum event that could be in a superposition of states. Why, then, don't we find Schrödinger's cat in a coherent superposition of dead and alive?

To explain the nonexistence of superposition states of macroscopic objects, quantum physicists invoke quantum decoherence [48]. Decoherence comes from the interaction of the quantum system with the "outside world." For instance, in our example of the atom following two simultaneous paths in an interferometer, one can illuminate [49] the atomic trajectory by laser light, which allows one to see the atom's position and reveal which path is taken: this measurement reduces the superposition to the classical situation where the atoms have followed either one path or the other, and destroys the interference. As objects become larger and larger, they become more sensitive to external perturbations, which can destroy (partially or completely) coherent superpositions. This argument gives a plausible explanation for the difference in behaviors of the classical and the quantum world. Nobody knows, however, whether there is a hypothetical limit beyond which decoherence would absolutely be inevitable, or whether we always can, at least in principle, take sufficient precautions to protect the system against perturbations, no matter how large it is. A clear answer to that question would have immense consequences, both conceptually and for future quantum technologies.

2.5 The Second Quantum Revolution in Action: Quantum Information, Quantum Technologies

The existence of Bell inequalities, which establish a clear frontier between classical and quantum behavior, and their experimental violation are important conceptual results, which force us to recognize the extraordinary character of quantum entanglement. But in an unexpected way, it has been discovered that entanglement also offers completely new possibilities in the domain of information treatment and transmission. A new field has emerged, broadly called Quantum Information, which aims to implement radically new concepts that promise surprising applications. So far, there are two primary examples: quantum cryptography [50], already operational, and quantum computing [51, 52], still a nascent enterprise.

2.5.1 Quantum Cryptography

Cryptography is the science of encoding and/or transmitting a secret message without its being read/understood by a third party. Much of classical cryptography involves secure transmission on a public channel. As the field has progressed, methods have become more and more refined, involving sophisticated algorithms and driven by equally clever methods of "code breaking." Both encoding and code-breaking have progressed due to advances in mathematics and to the ever-increasing power of computers. When contemplating this continuing progress of coding and code breaking, it seems clear that the security of a transmission can be assured only on the hypothesis that the adversary (who is trying to break the code) has neither more advanced mathematics nor more powerful computers than the sender and intended receiver. In classical cryptography, the only absolutely sure transmission scheme uses the one-time pad method, where the emitter and the receiver share two identical coding keys, not shorter than the secret message to be transmitted, and that is used only once. The (preliminary to secure communication) distribution of the two copies of the key then becomes the critical stage, that involves secret channels, which might be intercepted by an "eavesdropper" with technologies more advanced than the one of the sender and intended receiver.

By contrast, in quantum cryptography, the security of a transmission rests on the fundamental physical laws at work in quantum mechanics. There, it is possible to detect an eavesdropper by using the trace that is *necessarily* left by such efforts [53], since in quantum physics all measurements perturb the system in some way. In the absence of such a trace, one can be certain that the message has passed without having been read by a spy.

A particular topic in quantum cryptography is especially spectacular: the use of EPR pairs to distribute securely the two copies of the random key that two distant partners will use later, for a one-time pad-coded transmission. How can they be sure that no one has read either copy of the key during the transmission? The use of pairs of entangled particles offers an elegant solution: the two partners (Alice and Bob) effecting measurements on the two particles of a single entangled pair will find random but perfectly correlated results. By repetition of such measurements, they generate two identical copies of a random sequence. And what we have learned from the violation of Bell inequalities is that, as long as the measurements have not been made, their results are not predictable, which means that the key does not yet exist. A nonexistent key cannot be read by any eavesdropper (Eve)! It is only at the moment of the measurement that the two identical keys appear in the apparatuses of the two partners. Bell inequalities play a crucial role in that scheme: their violation permits one to be sure that the particles received by Alice and Bob have not been fraudulently prepared by Eve in a state known by him, such that he would know in advance the sequences found by Alice and Bob. Demonstrations of this principle have been carried out [50].

2.5.2 Quantum Computing [51, 52]

In the early 1980s, the fundamental assumptions of classical information theory started to be challenged by several physicists who suggested that if one had a quantum computer, one could implement radically new algorithms to perform certain tasks. The names of Landauer, Feynman [54], Deutsch [55], and others can be quoted, and an important breakthrough happened in 1994 when P. Shor [56] showed that a quantum computer should allow one to factor large numbers, in times much shorter than with conventional methods. Factorization belongs to a class of problems (complexity class) whose solution (with classical computers) requires a time super-polynomial in the size of the problem (that is, the time needed grows faster than any power of the number of digits in the number to be factored). With a quantum computer, on the other hand, the computation time would only grow as a power of the size of the number.[6] This discovery had considerable conceptual implications, since it showed that contrary to what had been thought previously, the complexity class of a problem was not independent of the type of machine used. Besides that conceptual revolution, a quantum computer would certainly have applications beyond present imagination.

Several groups have started to develop the basic elements of a quantum computer: quantum bits and quantum gates. A quantum logic gate performs basic operations on quantum bits – or "qubits" – just as an electronic logic gate manipulates ordinary bits. However, in contrast to normal bits, which can take only one of two values, 0 or 1, quantum bits can be put in a superposition of two states. A quantum logic gate must thus be capable of combining two quantum bits to produce an entangled state.

6 It may make a "significant" difference: see, for instance, the example in Ref. [51], where the factorization time of a 400 digit number can be reduced from the universe age to a few years.

It is the possibility to work with such entangled states that opens new and powerful possibilities as compared to the classical algorithms.

Will the universal quantum computer of the theorists exist someday? It would be presumptuous to answer, but experimental research on various kinds of quantum bits and quantum gates is extremely active and has already obtained important results. Many approaches are being explored, with a diversity of physical realizations of qubits, including atoms, ions, photons, nuclear spins, and Josephson junctions [57].

For all these systems there are large unknowns. Quantum calculation relies on the ability to entangle dozens or even hundreds and thousands of quantum bits and perform thousands of operations before decoherence disrupts the quantum register. Decoherence results from the interaction with the outside world (see Section 2.4), and its effect is to wash out entanglement, putting previously entangled objects into a state where they behave as separated objects. The scalability to a large number of entangled qubits may turn out to be overwhelmingly difficult, since it is generally observed that decoherence dramatically increases when the number of the entangled particles increases. Here again, nobody knows whether there is a maximum size beyond which entanglement destruction by decoherence is definitely unavoidable, or whether it is only a matter of increasing experimental difficulty (or of finding special situations where the problem might not be as dramatic). An entire community of experimentalists and theorists are engaged in that quest. Understanding and reducing the effects of decoherence may well be the key question facing quantum computation, as a technological revolution. But even in the absence of an efficient quantum computer, the idea of quantum computation is certainly a milestone in computation science.

2.5.3 Quantum Computing with Real, Imperfect Systems

Experimental physics is not mathematics. Real systems follow approximatively only their basic theoretical description, and moreover, some perturbations may act upon the system. As a result, real qubits may be in a state somewhat different of what was intended, that is to say, the quantum information may have errors. Remarkably, error correction schemes do exist in quantum informatics. As in classical computing, it is based on redundancy, i.e. the quantum information is recorded on more qubits than would be necessary with ideal qubits. It turns out, however, that with the error correction schemes known in 2022, the overload is quite high, and is typically of 10^3 or more real qubits, of the best quality available today, to represent a single ideal qubit. It means that to have a quantum computer with 100 ideal qubits one needs at least 10^5 real qubits: we do not know yet any way to fulfill that demand.

It turns out, however, that an ensemble of a few hundred non-ideal qubits with controlled interactions may be used as quantum simulators of interesting physical systems, as suggested by Feynman in his founding paper [54]. For instance, a group at Institut d'Optique has an ensemble of Rydberg atoms that can be controlled individually and made to interact in such a way as to simulate an Ising problem of interacting spins one-half [58]. This problem has no exact solution, and numerical solutions are difficult, meaning that the size of the exact numerical calculation increases exponentially with the number of particles. Theorists must then develop approximated numerical methods to obtain a result in a reasonable amount of time, and it is remarkable that up to now the results of the physical simulation and of the approximate calculations are identical. It seems nevertheless that we have here an example of the celebrated "quantum advantage," i.e. the fact that a quantum machine can obtain useful results faster than numerical calculations for a large enough number of interacting systems. To speak of "quantum advantage," it is necessary, in my opinion, that the calculation bears on a true problem that was existing before addressing it with the quantum machine. It is the case for the example here. Such non-ideal entangled systems may also constitute what is called Noisy Intermediate Scale Quantum computers, to solve some difficult problems such as optimization problems, for instance, the traveling salesman problem, whose solution is the ground state of an Hamiltonian that can be implemented on a NISQ machine. This kind of system is promising enough that spin-off start-up companies propose such systems to customers who want to try to use them for solving their own problem. Although the quantum advantage has not yet been obtained for a real, useful problem, such as optimizing an electric grid, it does not seem out of reach.

In the same vein, such quantum simulators might be used to find the structure of complicated molecules envisaged as new drugs or of molecules of biological interest.

2.5.4 Other Quantum Technologies

The concepts of the second quantum revolution have many other applications. For instance, a single photon impinging on a beam splitter constitutes a remarkably simple random number generator, where randomness is guaranteed by the very laws of quantum physics. If that single photon belongs to an entangled pair, one can check that Bell's inequalities are violated, so one is sure that there is no hidden variable that would allow one to know in advance whether the photon will be reflected or transmitted. This sounds an ideal random number generator.

Entangled quantum objects may be used for metrology purposes, with fundamental uncertainty below what is called the standard quantum limit, which is in fact the classical limit associated to the fact that the measurement uncertainty for an ensemble of N identical particles scales as $N^{1/2}$, so that the signal-to-noise ratio increases as $N^{-1/2}$. Using entangled ensembles of particles, it may be possible [59] to make measurements where the signal-to-noise ratio decreases faster than $N^{-1/2}$.

Single quantum objects, e.g. an NV center at the tip of a scanning atomic force microscope, can be used as nanoprobes, e.g. of magnetic structures [60].

The range of applications is immense and limited only by the imagination of physicists or engineers, application oriented. The many unexpected applications of lasers constitute an interesting example. Who would have thought that they would be used for reading bar codes?

2.6 Conclusion: Questioning Quantum Mechanics Is Fruitful

Quantum mechanics was, and continues to be, revolutionary, primarily because it demands the introduction of radically new concepts to better describe the world. While the possibility to observe and control the quantum behavior of single quantum objects may seem natural to the young generation physicists, one should remember that it was not the case for many physicists in the middle of the twentieth century [39]. Moreover, quantum non-locality [30] of large-scale entangled systems is still hard to swallow, but it is a fact amply demonstrated by experiments on various different systems.

I started working in that domain at a time when the first quantum revolution had been so successful that nobody would "waste time" to consider questions about the very basic concepts at work in quantum mechanics. It took more than a decade to have these questions taken seriously.[7] For somebody who observed reactions to Bell's work on the EPR situation and entanglement, in the early 1970s, it is certainly amusing to see that an entry of the Physics and Astronomy Classification Scheme was later assigned to "Bell's inequalities" [61]. With his questions about entanglement, John Stuart Bell was able to clarify the Einstein–Bohr debate in an unanticipated manner, by offering the opportunity of experimentally settling the question. The experimental observation of the violation of Bell's inequalities contributed to trigger the second quantum revolution, primarily based on the recognition of the extraordinary features of entanglement [62], and pursued with efforts to use entanglement for quantum information. In fact, it not only triggered the domain, but Bell's theorem is an important tool, for instance, used to show that a quantum cryptography scheme is fundamentally secure, or that quantum computing is definitely different from classical computing [63].

Will the second quantum revolution turn out to be not only a conceptual revolution but also a technological revolution? Can it change our society as much as the first quantum revolution did it, or will its fruits remain in small niches? I think that it too early to give a fully qualified answer, but there is no doubt that it is an exciting topics, and the exploding eco-system of start-up companies based on its concepts, closely linked to academic research, will certainly produce unexpected new results that I am not smart enough to anticipate.

7 In his famous Lectures of the end of the 1950s, Feynman wrote, about wave-particle duality: "[it] *has in it the heart of quantum mechanics. In reality it contains the only mystery.*" But two decades later, in the paper considered as a founding paper for quantum information [54], he acknowledged that entanglement was different in nature:"*I've entertained myself always by squeezing the difficulty of quantum mechanics into a smaller and smaller place, so as to get more and more worried about this particular item.*" He then continued, referring to experimental violations of Bell's like inequalities: "*It seems to be almost ridiculous that you can squeeze it to a numerical question that one thing is bigger than another. But there you are – it is bigger than any logical argument can produce.*"

References

1 The present chapter is an updated version of Aspect, A. (2004, 2008). Introduction: John Bell and the second quantum revolution. In: *Speakable and Unspeakable in Quantum Mechanics: Collected Papers on Quantum Philosophy*, xvii–xl. Cambridge: Cambridge University Press.

2 Jammer, M. (1974). *The Philosophy of Quantum Mechanics*, 303. Wiley: this book is too early to render full justice to Bell's contribution, but it is a very valuable source of historical details and references for the first quantum revolution.

3 Bell, J.S. (1964). On the Einstein Einstein Podolsky Rosen paradox. *Physics* 1: 195–200.

4 See Chapter 10 in footnote 3.

5 Planck, M. (1901). On the law of the energy distribution in the normal spectrum. *Annals of Physics* 4: 553.

6 (a) Einstein, A. (1905). Uber einen die Erzeugung und Verwandlung des Lichtes bettrefenden heuristischen Gesichtpunkt. *Annalen der Physik* 17: 132. (b) Einstein, A. and Zur Theorie der Lichterzeugung und Lichtabsorption (1906). *Annalen der Physik* 20: 199.

7 Millikan, R.A. (1914). A direct determination of "h". *Physics Review* 4: 73; 6: 55 (1915); 7: 362 (1916).

8 Einstein, A. (1906). Zur Theorie des Brownschen Bewegung. *Annalen der Physik* 19: 371.

9 Einstein, A. (1909). On the development of our views concerning the nature and constitution of radiation. *Physikalishce Zeitschrift* 10 (22): 817–825.

10 de Broglie, L. (1925). *Recherches sur la théorie des quanta, Thèse de doctorat, Paris (1924)*, Annales de Physique, 10è série, vol. III, 22–128.

11 Lamb, W.E. and Retherford, R.C. (1947). Fine structure of the hydrogen atom by a microwave method. *Physics Review* 72: 241.

12 Bardeen, J. and Brattain, W.H. (1948). The transistor, a semi-conductor triode. *Physics Review* 74: 230–231.

13 Kilby, J.S., *Turning Potential into Reality: The Invention of the Integrated Circuit*. Alferov, Z.I., *The Double Heterostructure Concept and its applications in Physics, Electronics and Technology*. Herbert Kroemer, *Quasi-Electric Fields and Band Offsets: Teaching Electrons New Tricks*: Nobel Lectures (2000), Physics 1996–2000 (World Scientific), also available at http://www.nobel.se/physics/laureates/2000/.

14 Townes, C.H., *Production of coherent radiation by atoms and molecules*. Basov, N.G., *Semiconductor lasers*, Prokhorov, A.M., *Quantum electronics*, (1964), *in* Nobel Lectures, Physics 1963–1970 (Elsevier), also available at http://www.nobel.se/physics/.

15 Einstein, A., Podolsky, B., and Rosen, N. (1935). Can quantum-mechanical description of physical reality be considered complete? *Physical Reviews* 47: 777.

16 Bohr, T., *Presentation of the 2013 Niels Bohr Medal of the Danish Academy of Engineers –In association with the Niels Bohr Institute and the Royal Danish Society of Sciences and Letters.*

17 Bohr, N. (1935). Can quantum-mechanical description of physical reality be considered complete? *Physical Reviews* 48: 696.
18 Schrödinger, E. (1935). *Proceedings. Cambridge Philological Society* 31: 555–563.
19 John Bell had first proved wrong the famous von Neumann's "impossibility proof" of hidden variables in: Bell, J.S. (1966). On the problem of hidden variables in quantum theory. *Reviews of Modern Physics* 38: 447. Note that in contrast to the date of publication, this paper was written before the 1964 paper [3] where inequalities are shown for the first time (see footnote 3, p. 303).
20 Bohm, D. (1989). *Quantum Theory*. Dover: Prentice Hall, 1951.
21 Bohm, D. and Aharonov, Y. (1957). Discussion of experimental proofs for the paradox of Einstein, Rosen and Podolsky. *Physics Review* 108: 1070.
22 Clauser, J.F., Horne, M.A., Shimony, A., and Holt, R.A. (1969). Proposed experiments to test local hidden-variable theories. *Physical Review Letters* 23: 880.
23 See references in Clauser, J.F. and Shimony, A. (1978). Bell's theorem: experimental tests and implications. *Reports on Progress in Physics* 41: 1881.
24 Aspect, A., Grangier, P., and Roger, G. (1982). Experimental realization of Einstein-Podolsky-Rosen-Bohm Gedanken Experiment: a new violation of Bell's inequalities. *Physical Review Letters* 49: 91.
25 Aspect, A., Dalibard, J., and Roger, G. (1982). Experimental tests of Bell's inequalities using variable analysers. *Physical Review Letters* 49: 1804.
26 See references in several contributions to Bertlmann, R.A. and Zeilinger, A. (ed.) (2002). *Quantum [Un]speakables*. Springer, and in particular in Aspect, A. *Bell's Theorem: The Naïve View of an Experimentalist.*
27 Weihs, G., Jennewein, T., Simon, C. et al. (1999). Violation of Bell's inequalitiy under strict Einstein locality condition. *Physical Review Letters* 81: 5039.
28 (a) Tittel, W., Brendel, J., Zbinden, H., and Gisin, N. (1999). Violation of Bell inequalities by photons more than 10 km apart. *Physical Review Letters* 81: 3563. (b) Tapster, P.R., Rarity, J.G., and Owens, P.C.M. (1994). Violation of Bell's inequality over 4 km of optical fiber. *Physical Review Letters* 73: 1923.
29 Aspect, A. (2015). Closing the door on Einstein and Bohr's quantum debate. *Physics* 8: 123.
30 Brunner, N., Cavalcanti, D., Pironio, S. et al. (2014). Bell nonlocality. *Reviews of Modern Physics* 86 (2): 419.
31 Peres, A. and Terno, D. (2004). Quantum information and relativity theory. *Reviews of Modern Physics* 76 (1): 93.
32 Aspect, A. (1981). *Expériences sur les inégalités de Bell. J. Physique* (Paris), Colloque C2, p. C2–63. Note that this simple argument is the same as the one showing that quantum teleportation–another quantum effect based on entanglement–cannot be effected faster than light: Bennett, C.H., Brassard, G., Crépeau, C. et al. (1993, 1895). Teleporting an unknown quantum state via dual classical and Einstein-Podolsky-Rosen channels. *Physical Review Letters* 70.
33 Wooters, W.K. and Zurek, W.K. (1982). Quantum no-cloning theorem. *Nature* 299 (802): 16–23.
34 Eigler, D.M. and Schweizer, E.K. (1990). Positioning single atoms with a scanning tunnelling microscope. *Nature* 344 (6266): 524–526.

35 (a) Dehmelt, H. (1990). Experiments with an isolated subatomic particle at rest. *Reviews of Modern Physics* 62: 525–530. (b) Paul, W. (1990). Electromagnetic traps for charged and neutral particles. *Reviews of Modern Physics* 62: 531–540. (c) Binnig, G. and Rohrer, H. (1987). Scanning tunneling microscopy from birth to adolescence. *Reviews of Modern Physics* 59: 615–625.

36 (a) Gabrielse, G. (2001). Comparing the antiproton and proton, and opening the way to cold antihydrogen. In: *Advances in Atomic, Molecular, and Optical Physics*, vol. 45 (ed. B. Bederson and H. Walther), 1–39. New York: Academic Press. (b) Dehmelt, H., Mittelman, R., Van-Dyck, R.S. Jr., and Schwinger, P. (1999). Past electron-positron g-2 experiments yielded sharpest bound on CPT violation for point particles. *Physical Review Letters* 83: 4694.

37 Ramsey, N.F. (1995). Quantum-mechanics and precision-measurements. *Physica Scripta* 59: 26–28.

38 (a) Nagourney, W., Sandberg, J., and Dehmelt, H. (1986). Shelved optical electron amplifier: observation of quantum jumps. *Physical Review Letters* 56: 2797. (b) Bergquist, J.C., Hulet, R.G., Itano, W.M., and Wineland, D.J. (1986). Observation of quantum jumps in a single atom. *Physical Review Letters* 57: 1699.

39 Schrödinger, E. (1952). Are there quantum jumps? Part I. *The British Journal for the Philosophy of Science* 3 (10): 109–123.

40 Raffac, R.J., Young, B.C., Young, J.A. et al. (2000). Sub-dekahertz ultraviolet spectroscopy of $^{199}Hg^{+}$. *Physical Review Letters* 85: 2462.

41 (a) Dalibard, J., Castin, Y., and Moelmer, K. (1992). Wave function approach to dissipative processes in quantum optics. *Physical Review Letters* 68: 580. (b) Zoller, P. and Gardiner, C.W. (1997). Quantum noise in quantum optics : the stochastic Schrödinger equation. In: *Lecture notes for Les Houches Summer School LXIII (1995)* (ed. E. Giacobino and S. Reynaud). Elsevier, and references therein.

42 Bardou, F., Bouchaud, J.-P., Aspect, A., and Cohen-Tannoudji, C. (2002). *Lévy Statistics and Laser Cooling: When Rare Events Bring Atoms to Rest*. Cambridge University Press.

43 Lévy, L.P., Dolan, G., Dunsmuir, J., and Bouchiat, H. (1990). Magnetization of mesoscopic copper rings: evidence for persistent currents. *Physical Review Letters* 64: 2074.

44 Fried, D.G., Killian, T.C., Willmann, L. et al. (1998). Bose-Einstein condensation of atomic hydrogen. *Physical Review Letters* 81: 3811.

45 (a) Cornell, E.A. and Wieman, C.E. (2002). Bose-Einstein condensation in a dilute gas, the first 70 years and some recent experiments. *Reviews of Modern Physics* 74: 875. (b) Ketterle, W. (2002). Nobel lecture: when atoms behave as waves: Bose-Einstein condensation and the atom laser. *Reviews of Modern Physics* 74: 1131.

46 See Bell's preface to the first edition of the book *Speakable and Unspeakable in Quantum Mechanics: Collected Papers on Quantum Philosophy*. More than half of the papers of that volume address that question.

47 Schrödinger, E. (1935). *Naturwissenshaften* 23: 807, 823, 844.

48 Raimond, J.-M., Brune, M., and Haroche, S. (2001). Manipulating quantum entanglement with atoms and photons in a cavity. *Reviews of Modern Physics* 73: 565.

49 (a) Pfau, T., Spalter, S., Kurtsiefer, C. et al. (1994). *Loss of spatial coherence by a single spontaneous emission. Physical Review Letters* 73. (b) Chapman, M.S., Hammond, T.D., Lenef, A. et al. (1995). Photon scattering from atoms in an atom interferometer: coherence lost and regained. *Physical Review Letters* 75: 3783.

50 Gisin, N., Ribordy, G., Tittel, W., and Zbinden, H. (2002). Quantum cryptography. *Reviews of Modern Physics* 74: 145–195.

51 Preskill, J. (1999). Lecture notes for Physics 219: Quantum computation. *Caltech Lecture Notes*, 7.

52 (a) Nielsen, M.A., Chuang-Isaac, I., and Grover-Lov, K. (2002). Quantum computation and quantum information. *American Journal of Physics* 70: 558–559. (b) Nielsen, M.A. and Chuang-Isaac, I. (2000). *Quantum Computation and Quantum Information*. Cambridge University Press.

53 (a) Ekert, A.K. (1991). Quantum cryptography based on Bell's theorem. *Physical Review Letters* 67: 661. (b) Bennett, C.H., Brassard, G., and Mermin, N.D. (1992). Quantum cryptography without Bell's theorem. *Physical Review Letters* 68: 557. and references therein.

54 Feynman, R.P. (2018). Simulating physics with computers. In: *Feynman and Computation*, 133–153. CRC Press.

55 Deutsch, D. and Jozsa, R. (1992). Rapid solution of problems by quantum computation. *Proceedings of the Royal Society of London. Series A: Mathematical and Physical Sciences* 439 (1907): 553–558.

56 Shor, P.W. (1994). Algorithms for quantum computation: discrete logarithms and factoring. In: *Proceedings. 35th Annual Symposium on Foundations of Computer Science* (ed. S. Goldwasser), 124–134. Los Alamitos, CA: IEEE Comput. Soc. Press.

57 (a) Vandersypen, L.M.K., Steffen, M., Breyta, G. et al. (2001). Experimental realization of Shor's quantum factoring algorithm using nuclear magnetic resonance. *Nature* 414: 883. (b) Guide, S., Riebe, M., Lancaster, G.P.T. et al. (2003). Implementation of the Deutsch-Jozsa algorithm on an ion-trap quantum computer. *Nature* 421: 48. (c) Rauschenbeutel, A., Nogues, G., Osnaghi, S. et al. (1999). Coherent operation of a tunable quantum phase gate in cavity QED. *Physical Review Letters* 83: 5166. (d) Vion, D., Aassime, A., Cottet, A. et al. (2002). Manipulating the quantum state of an electrical circuit. *Science* 296: 886. and references therein. (e) Monroe, C., Meekhof, D.M., King, B.E. et al. (1995). Demonstration of a fundamental quantum logic gate. *Physical Review Letters* 75: 4714. (f) Blatt, R. and Wineland, D. (2008). Entangled states of trapped atomic ions. *Nature* 453 (7198): 1008–1015.

58 Browaeys, A. and Lahaye, T. (2020). Many-body physics with individually controlled Rydberg atoms. *Nature Physics* 16 (2): 132–142.

59 (a) Bouyer, P. and Kasevich, M.A. (1997). Heisenberg-limited spectroscopy with degenerate Bose-Einstein gases. *Physical Review A* 56 (2): R1083. (b) Wineland,

D.J., Bollinger, J.J., Itano, W.M., and Heinzen, D.J. (1994). Squeezed atomic states and projection noise in spectroscopy. *Physical Review A* 50 (1): 67.

60 Gross, I., Akhtar, W., Hrabec, A. et al. (2018). Skyrmion morphology in ultrathin magnetic films. *Physical Review Materials* 2 (2): 024406.

61 Physics and Astronomy Classification Scheme number 03.65.Ud. Now replaced by the Physics Subject Headings where "Bell's nequalities" is a subsection of "Nonlocality".

62 Freire, O. Jr., (2006). Philosophy enters the optics laboratory: Bell's theorem and its first experimental tests (1965–1982). *Studies in History and Philosophy of Science Part B: Studies in History and Philosophy of Modern Physics* 37 (4): 577–616.

63 Ref. [51], Sections 1.2 and 1.4.

3

Solid-State Quantum Emitters

A. Mark Fox

Department of Physics & Astronomy, University of Sheffield, Hicks Building, Hounsfield Road, Sheffield S3 7RH, United Kingdom

3.1 Introduction

A "quantum emitter" is a source of light that must be treated as a stream of photons rather than as an electromagnetic wave. The light must therefore be understood within the subject of *quantum* optics, rather than *classical* optics. A key characteristic of the light is that it has properties that are impossible to explain within the classical theory of light.

The photon model of light has been around since the early years of quantum theory, and it might therefore be thought that the observation of quantum-optical effects might be straightforward. However, this is not the case. Consider, for example, an experiment with a highly sensitive detector such as a photomultiplier tube under weak illumination. By attenuating the light beam very strongly, we can arrange that the average time between the photons impinging on the detector is longer than the detector response time. In these conditions, the output of the detector is a series of electrical pulses (often called "clicks") that are produced by the photoelectric effect when the photon strikes the photocathode. This might seem like clear proof that the light beam is a stream of individual quanta (i.e. photons) that release single electrons that get amplified to generate the pulses at the output. However, it can be shown that the individual clicks generated by the detector do not give clear evidence for the quantum nature of the light. Specifically, in the early 1960s, Mandel et al. developed a *semi-classical* theory of the photoelectric effect in which the clicks are explained by treating the atoms in the detector as quantized systems but the light as an electromagnetic wave rather than a stream of photons [1].

The first direct demonstration of quantum effects for light came from more sophisticated experiments performed in the 1970s. These involved making photon correlation experiments on the light emitted by *individual* atoms rather than from the multitude of atoms that make up a gas [2]. The key effect that was observed was called photon anti-bunching, as will be explained Section 3.2. The first generation of experiments was mainly carried out on gaseous systems, and it took about

Photonic Quantum Technologies: Science and Applications, First Edition. Edited by Mohamed Benyoucef.

15 years for the ideas to be transferred to the solid state. The strategy for observing quantum effects from solid-state systems involved isolating single emitters, which effectively act like individual "solid-state atoms." This was first observed for single molecules embedded within a crystal in 1992 [3] and then for single color centers and semiconductor quantum dots in 2000 [4, 5].

In the years that followed these first experiments, the subject has advanced very rapidly. In this chapter I will review the basic principles of quantum emission from solid-state systems and then give an overview of the main parameters by which they are characterized. The principles will usually be illustrated with examples from InAs quantum dot emitters, since this is the author's own research field, and is also the material system that is most developed. The chapter concludes with a brief survey of the main material systems that are commonly used. More detailed accounts of the latest state-of-the-art for telecom-wavelength quantum dots, color centers, and 2D materials may be found in chapters 18, 19, and 20, respectively, of the book.

3.2 Photon Correlation Functions

A beam of light can be classified by the time gap between the photons that form the beam. There are three general classifications: anti-bunched, random (also called coherent), and bunched. These three types of light are depicted schematically in Figure 3.1. The benchmark is the case in which the timing is random, shown in the middle panel. The bottom panel shows bunched light. As the name suggests, this corresponds to the case in which the photons come in bunches, so that the probability to detect a second photon after having detected a first one is higher at small time differences than at longer ones. The top panel shows the opposite possibility: anti-bunched light. Here, the photons are spread out so that the probability of detecting a second photon is very small at short time differences.

Consider an attenuated laser beam with constant power P impinging on a single-photon detector (i.e. one that is sensitive enough to give a click in response to a single photon[1]). The average number of photons $\overline{n}$ in time T is $PT/h\nu$, where h is Planck's constant and ν is the photon frequency. With strong attenuation, we can reduce P to the point where $\overline{n} < 1$ when T is the detector response time. In these conditions, the detector can register the individual photons and the output clicks reproduce their arrival times.

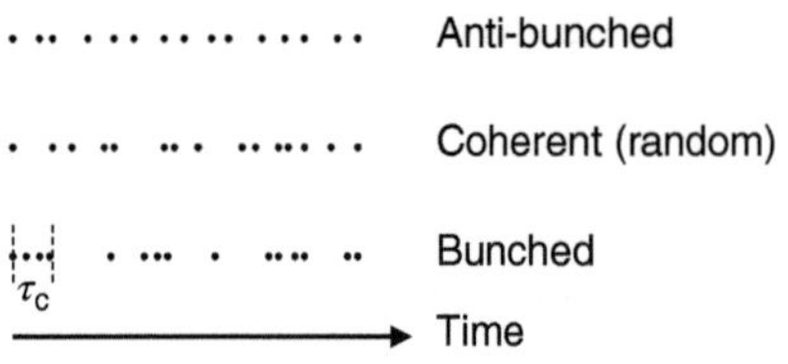

Figure 3.1 Schematic illustration of photon anti-bunching compared to coherent and bunched light. τ_c is the coherence time of the bunched light.

1 There are several different types of single-photon detectors available, including photomultiplier tubes, avalanche photodiodes, and superconducting nanowire detectors.

It might be thought that the clicks coming out of the detector generated by the laser would be regular, but this is not, in fact, the case. The *average* number of photons hitting the detector within time T is known, but not the precise timing. This is because the constant flow of energy has to be chopped into discrete pieces (i.e. into photons) and the only requirement is to make sure that the average number of energy quanta per unit time is constant. This means that the ideal laser beam with constant power has random timing between the photons, as illustrated by the middle panel in Figure 3.1. Since a highly stable single-frequency laser beam is the best approximation to a perfectly coherent electromagnetic wave,[2] this explains why the case with random time spacing is also called coherent.

The timing between the photons can be investigated in more detail by measuring the photon correlations in a Hanbury Brown–Twiss (HBT) experiment, where we split the photon stream with a 50:50 beam splitter, and register the two separated streams with two different single-photon detectors, as shown in Figure 3.2. The detectors are connected to a timer that records the time between the clicks. After registering many events, a histogram can be plotted of the number of events vs. the time difference between the clicks.

For the case of the ideal laser, the time between photons is random. Once a photon has generated a click on the start detector D1, we wait until another photon from the stream is directed toward D2 and generates the stop click. Since the timing between the photons is random, all time separations are equally likely. Hence the histogram will be flat, as shown in Figure 3.2c. On the other hand, if the photons come in bunches, the probability of measuring two clicks with small time difference is higher than for longer times. Hence the histogram would have a peak near time zero, as in Figure 3.2d. Finally, consider an anti-bunched source. The photons are spread out, and so the probability of getting events with small time spacing is lower

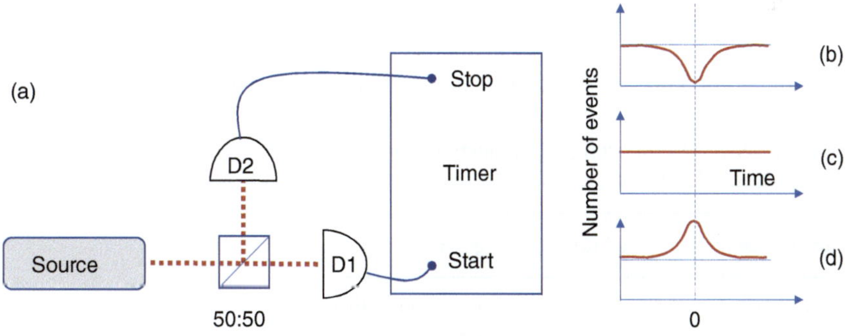

Figure 3.2 (a) Schematic of a Hanbury Brown–Twiss (HBT) experiment. D1 and D2 are single-photon detectors. (b–d) Idealized histograms of the number of events recorded vs. the time between the start and stop clicks for anti-bunched, coherent, and bunched light, respectively.

2 A perfectly coherent electromagnetic wave would have an electric field that varies sinusoidally in time with constant amplitude extending from $t = -\infty$ to $+\infty$.

than for longer times. This gives a histogram with a dip near time zero as shown in Figure 3.2b.

The results of an HBT experiment can be quantified in terms of the second-order correlation function $g^{(2)}(\tau)$ defined by[3]

$$g^{(2)}(\tau) = \frac{\langle I(t)I(t+\tau)\rangle}{\langle I(t)\rangle^2} \tag{3.1}$$

where $I(t)$ is the light intensity at time t, and $\langle \cdots \rangle$ indicates the time average. In a photon counting experiment, we can obtain $g^{(2)}(\tau)$ from

$$g^{(2)}(\tau) = CN(\tau) \tag{3.2}$$

where C is a normalization constant, and $N(\tau)$ is the number of events where a click on detector D2 occurs at a time $t + \tau$ after a click occurred on D1 at time t. Negative τ implies that the click on D2 occurred before the one on D1. The normalization constant is found by setting the value of $g^{(2)}(\tau)$ equal to 1 for $\tau \to \infty$. This assumes that there are no correlations between the arrival times of the photons at very long times, making all time separations equally likely. In practice, this requires $\tau \gg \tau_c$, where τ_c is the timescale of the source that determines the correlations between the photons, i.e. the coherence time (see Section 3.6).

Table 3.1 gives a summary of the key features of the $g^{(2)}(\tau)$ function for the three different types of light. For perfectly coherent light, the intensity does not change with time, and so it is obvious that $g^{(2)}(\tau) = 1$ for all values of τ, as shown in Figure 3.2c. With $g^{(2)}(\infty)$ normalized to 1 for all sources, we see that bunched light must have $g^{(2)}(0) > 1$, while anti-bunched light has $g^{(2)}(0) < 1$. Filtered thermal light (i.e. black-body radiation) and the light from standard atomic discharge lamps both exhibit photon bunching, with the width of the peak near $\tau = 0$ determined by the coherence time of the light. Quantum emitters, by contrast, exhibit photon anti-bunching, with the width of the dip determined by the radiative lifetime of the source.

It is shown in quantum-optics textbooks (see e.g. [6]) that the results of a HBT experiment for both coherent and bunched light can be explained within the

Table 3.1 Second-order correlation functions for different types of light.

Light source	$g^{(2)}(0)$	$g^{(2)}(\tau)$
Anti-bunched	<1	Increases with τ
Coherent	1	1 for all τ
Bunched	>1	Decreases with τ

3 The *second*-order correlation function contrasts with the first-order correlation function $g^{(1)}(\tau)$. Both consider the time dependence of the electric field $\mathcal{E}(t)$ of the light. $g^{(1)}(\tau)$ considers correlations in the first power of $\mathcal{E}(t)$, while $g^{(2)}(\tau)$ considers correlations of the second power of $\mathcal{E}(t)$, i.e. the intensity, which is proportional to $|\mathcal{E}(t)|^2$.

semi-classical model, in which the light is treated classically. On the other hand, photon anti-bunching cannot be explained in the semi-classical model, requiring a full quantum model in which the light is quantized as well as the atoms in the detector. The observation of photon anti-bunching is thus the tell-tale sign of a quantum emitter. A source will therefore be classified as being a quantum emitter if we observe a dip dropping below 1 at $\tau = 0$ in an HBT experiment.

3.3 Quantum Emitters

The original experiment performed by Hanbury Brown and Twiss in 1956 explored the second-order correlation function of light emitted from a mercury lamp [7]. Photon bunching was observed, consistent with the semi-classical model with the light beam considered as an electromagnetic wave with a fluctuating intensity. As explained in Section 3.2, it is photon anti-bunching rather than photon bunching that is the key signature of a quantum emitter. The first experiment demonstrating anti-bunching was performed by Kimble et al. in 1977 [2]. In their experiment, the quantum emitter was a single sodium atom within an atomic beam. The isolation of the light emitted by just one atom is non-trivial for a gaseous system and required very careful experimentation. The key to the success was to use a microscope objective lens to collect the light from a small volume of the beam and to arrange the flux of atoms within the beam so that there was a very small probability of there being more than one atom within the field of view of the objective lens during the detection time. Their experiment established the recipe for observing anti-bunched light, namely to collect the light from a single atom.

We can see why the isolation of the light from a single atom is key to observing anti-bunched light by thinking about the way the atom emits light. Consider an atom with a series of quantized electron states above its ground state. In order to emit a photon, the atom must be promoted to an excited state, from which it decays back to the ground state by a series of optical transitions of different frequencies, as shown schematically in Figure 3.3. If we isolate a single atom, there can only be one photon emitted for each transition, and we therefore obtain only one photon of a particular frequency each time the atom is excited. This photon can be selected by placing a filter in the collection path. In order to emit another photon of the same wavelength, the atom must be excited again, and follow the same decay path. Since the excitation and emission processes take a finite time, the probability of emitting two photons of the same wavelength is very small at short time differences. This gives rise to an anti-bunching dip as shown in Figure 3.2b, with the width of the dip determined by the radiative emission time.

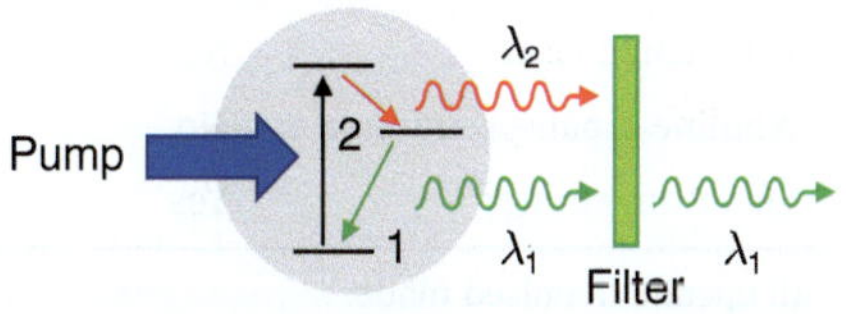

Figure 3.3 Excitation of a single atom producing a single photon on a specific transition. In this case, the photon emitted by the 2 → 1 transition is selected after filtering.

The emission from single atoms presented in the previous paragraph can be contrasted with the emission of light from a discharge lamp containing millions of atoms. Each individual atom emits only one photon of a particular wavelength each time it is excited. However, this process is going on randomly for a large number of emitters, with each atom emitting independently of the others. This produces chaotic light, which is bunched at short time delays determined by the coherence time of the light. Anti-bunching is only observed by isolating the light from single atoms, which is not routinely done.

3.4 Single-Photon Sources

A single-photon source is a pulsed light source in which the pulses contain just one photon. There are a number of different ways to make single-photon sources, and Table 3.2 considers three different approaches that are widely used.

Consider first a pulsed laser. If the laser has a repetition frequency of f and the average power of the beam is P, then the average energy per pulse is P/f. The *average* number of photons per pulse is then $\overline{n} = P/fh\nu$, where ν is the laser frequency. A normal pulsed laser emits billions of photons per pulse, but we can reduce the average power P to the point where $\overline{n} \sim 1$ by strong attenuation. It might seem that an attenuated pulsed laser with $\overline{n} = 1$ would make a good single-photon source. However, this is not the case. As mentioned in Section 3.2, the timing between the photons in a laser beam is random, and so the probability $\mathcal{P}(n)$ that a particular pulse contains n photons is given by a Poisson distribution:

$$\mathcal{P}(n) = \frac{\overline{n}^n}{n!}\mathrm{e}^{-\overline{n}} \tag{3.3}$$

For $\overline{n} = 1$, we find $\mathcal{P}(0) = \mathcal{P}(1) = 0.368$, and $\mathcal{P}(n \geq 2) = 0.264$. This means that about 37% of the pulses contain no photons, which reduces the single-photon rate. Of more concern is the fact that about 26% of the pulses contain more than one photon. For applications such as quantum cryptography, it is essential that the pulse does not contain more than one photon. For $\overline{n} = 1$, we find $\mathcal{P}(n \geq 2)/\mathcal{P}(1) = 0.72$, which is unacceptably high.

A work-around for the pulsed laser is to attenuate it more strongly so that $\overline{n} \ll 1$. A typical value might be $\overline{n} = 0.1$, for which we have $\mathcal{P}(0) = 0.9048$, $\mathcal{P}(1) = 0.0905$, and $\mathcal{P}(n \geq 2) = 0.0047$. The ratio of photon-containing pulses with more than one

Table 3.2 Different types of single-photon sources.

Source	On demand	Notes
Attenuated laser	No	Typically operates with $\overline{n} \leq 0.1$
Nonlinear pair production	No	Probabilistic, but heralded
Quantum emitter	Yes	Best results with resonant excitation

All operate in pulsed mode.

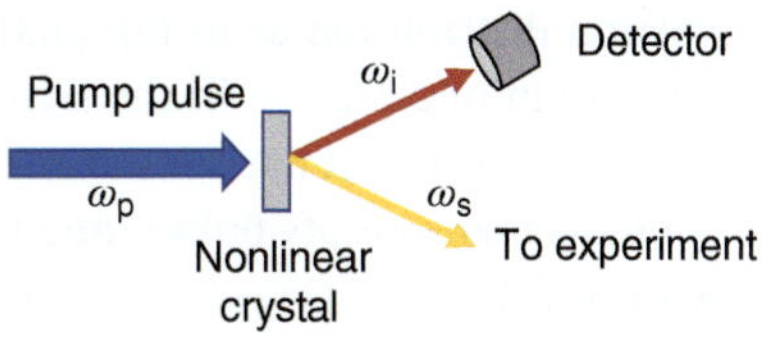

Figure 3.4 Heralded single-photon source based on photon pair production in a nonlinear crystal pumped by a laser pulse at frequency ω_p. The detection of the idler photon at frequency ω_i heralds the presence of the signal photon with frequency ω_s.

photon to those with just one is now reduced to 5%.[4] This ratio might be acceptable in practice, but the data rate has been severely limited, with more than 90% of the pulses containing no photons. Despite these less than ideal properties, attenuated pulsed lasers are in fact extensively used in tests of quantum cryptography on account of their simplicity of use. However, in the long run, we can do better using the other two types of single-photon source listed in Table 3.2.

Consider now single photons produced by selecting one photon from a photon pair. The photons are generated by nonlinear processes in which a crystal emits a pair of photons in response to a "pump" laser pulse, as shown in Figure 3.4. In *second-order* nonlinear processes, a single photon from the pump laser at frequency ω_p is converted to two photons at frequency ω_s and ω_i while conserving energy, such that:

$$\omega_p = \omega_s + \omega_i \tag{3.4}$$

where the subscripts *s* and *i* stand for *signal* and *idler*, respectively. The process is generally called *down conversion* as the frequency of the photons is split symmetrically about $\omega_p/2$, namely half the pump laser frequency. By contrast, in *third-order* nonlinear processes we use two photons from the pump laser to generate the signal and idler photons according to:

$$2\omega_p = \omega_s + \omega_i \tag{3.5}$$

This is generally called *four-wave mixing* as four photons are involved: two in and two out. The signal and idler frequencies split symmetrically about ω_p.

In both types of nonlinear process, the signal and idler photons are separated by some method and one of them is detected, indicating the presence of the other. This separation could be spatial, with the two photons going in different directions as shown in Figure 3.4, or spectral, with the photon pair split by dichroic filters. The detected photon acts as a *herald* that indicates the presence of the other one and allows post-selection of time bins to sift out events when a single photon was present. Hence these types of sources are called *heralded* single-photon sources.

The heralding of the single photons is important because the probability of the photon pair production has to be kept low, which means that many time bins are empty. The reason for this is to prevent the emission of two or more pairs at the same time. There is nothing to restrict the nonlinear processes to producing just a single pair of photons,[5] and the actual probability for pair production follows a

4 The general result for $\overline{n} \ll 1$ is $\mathcal{P}(n \geq 2)/\mathcal{P}(1) = \overline{n}/2$.

5 The nonlinear processes under consideration here are routinely used to convert the frequency of laser pulses, for example. in optical parametric amplifiers and oscillators, leading to signal and idler pulses containing very large numbers of photons.

Poisson distribution as in Eq. (3.3). In order to reduce the possibility of obtaining more than one photon in the signal or idler pulse, the pump intensity is reduced so that $\overline{n} \sim 0.1$, just like for the attenuated laser. Hence the heralded photon source contains many empty pulses, and the ones that do contain photons come randomly with probability $\mathcal{P}(1) \sim 0.1$. The heralded source is, however, an improvement on the attenuated laser because it is known which of the time bins contains a photon, but the process is still probabilistic as it cannot be predicted in advance when the photon will arrive.

The final type of single-photon source to consider is the quantum emitter shown schematically in Figure 3.3. As discussed in Section 3.3, it is essential to isolate a single quantum emitter, e.g. a single atom, molecule, quantum dot, color center, etc. The difference to Section 3.3 is that we are now considering pulsed rather than continuous excitation. If the pulse duration is short compared to the emission time, the emitter will be excited just once during each pulse cycle, and will therefore emit just one photon per transition. Hence by using short pulse excitation with a filter to select the photons on a specific transition, we obtain one photon per pulse, as required.

An ideal quantum emitter behaves as a *deterministic* single-photon source in the sense that we know when the photons are coming. In other words, we have $\mathcal{P}(1) = 1$, and $\mathcal{P}(n \neq 1) = 0$. This contrasts with attenuated laser and heralded single-photon sources that are probabilistic ($\mathcal{P}(1) \sim 0.1$), so that we can only say that a particular time bin might contain a photon with a certain probability. An alternative way to say this is to describe the quantum emitter as being an *on demand* source: we obtain a single photon with near unity probability in response to a trigger pulse.

A question remains whether we really can really obtain truly deterministic sources with $\mathcal{P}(1) = 1$. The reasons why we might obtain $\mathcal{P}(1) < 1$ for the quantum emitter could be either internal or external to the source. Consider first internal issues that could prevent the emitter from producing photons in response to the excitation pulse. One possibility is that the quantum emitter might have several decay paths from the excited state, so that a particular transition would only be obtained according to a branching probability. Another possibility is that the emitter jumps randomly between bright states that emit photons and dark states that do not. This produces an intermittent stream of photons, an effect known as *blinking*. As regards external factors, the challenge is to capture the photons that are emitted. The emitter could be very efficient, but the photons will be emitted over the full 4π solid angle, and many will be lost unless something is done to channel them efficiently toward the collection optics.

The goal of producing single, on-demand photons with high efficiency motivates much current research. At present, the best results have been obtained using epitaxial semiconductor quantum dots. (See [8] for a review.) The branching problem has been overcome by using π-pulse resonant excitation (see Section 3.7) in which the quantum dot (QD) is promoted to the upper level of the desired transition with near unity efficiency and there is only one decay path. The blinking problem has been overcome by embedding the emitters in diode structures with controlled gating, so that the free carriers that cause jumps between bright and dark states are swept away, leading to stable charge states. The photon collection problem has been overcome by embedding the QDs inside nano-cavities that funnel all the photons to a specific optical channel with high efficiency. In this way it has been demonstrated that QDs can approximate very well to ideal deterministic single-photon source in the right conditions [9].

Despite the fact that QDs have shown the best results, practical considerations make researchers resort to heralded or attenuated sources for many applications in the field. One such consideration is the fact that the QD sources only work at their best when they are cooled to cryogenic temperatures. Nevertheless, the results show that the potential is there, and the challenge is to optimize the source for practical deployment in the field.

3.5 Purcell Enhancement

The speed at which a quantum emitter can emit photons is governed by its radiative lifetime τ_R, giving a maximum photon emission rate of $\sim\tau_R^{-1}$. The lifetime is determined by Fermi's golden rule for the transition rate W_{21}:

$$W_{21} = \frac{2\pi}{\hbar}|M_{21}|^2 g(E) \tag{3.6}$$

where M_{21} is the transition matrix element and $g(E)$ is the density of final states. The subscripts indicate that we are making a transition from the upper level (2) to the lower one (1), which lies at an energy E below level 2, as shown schematically in Figure 3.3. The lifetimes are typically calculated for the case in which the emitter is embedded in a uniform dielectric medium (e.g. air) and emits uniformly over the full 4π solid-angle. This gives typical lifetimes of ~10 ns for strong radiative transitions in atoms at optical frequencies,[6] implying maximum photon emission rates of $\sim10^8$ s^{-1}. The radiative lifetimes of some solid-state quantum emitters can be faster on account of their larger optical dipole moment (and hence matrix element), which opens the possibility for higher single-photon rates. For example, the radiative lifetime of an InAs QD emitting at ~900 nm and embedded within a uniform semiconductor crystal is around 1 ns, and some molecules embedded within solid-state materials can be faster still.

It might be thought that the radiative lifetime is a fixed parameter for each emitter, and that there is not much scope for its manipulation. This is not, in fact, the case, as it is possible to speed up the emission by using the Purcell effect in an optical cavity. Purcell's original work considered emission at microwave frequencies [10], but the field has now developed into cavity quantum electrodynamics, which considers the interaction of a quantum emitter in resonance with a cavity mode.

The principle of the Purcell effect is to use the cavity to alter the transition rate given by Fermi's golden rule. We consider an atom in resonance with an optical cavity as illustrated schematically in Figure 3.5. The density of states factor $g(E)$ in Eq. (3.6) is determined by the density of electronic states in the lower level, and the

6 For example, the sodium D-Line transition at 589 nm has a radiative lifetimes of 16 ns.

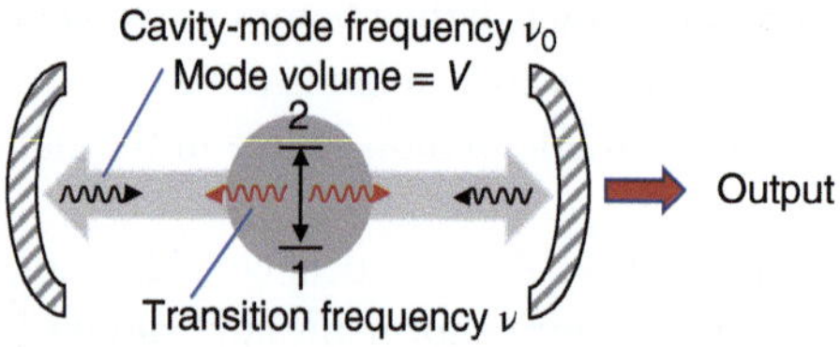

Figure 3.5 Two-level atom in an optical cavity with mode volume V. Resonance occurs when the atom's transition frequency ν matches the cavity-mode frequency ν_0. The cavity is designed so that one of the end mirrors has a lower reflectivity than the other one, so that photons are coupled out preferentially at one side.

density of photon states into which the photon is emitted. For an atom emitting into a uniform dielectric material with refractive index n, the density of photon states per unit volume at frequency ν is given by:

$$g(\nu) = \frac{8\pi}{(c/n)^3}\, \upsilon^2 \tag{3.7}$$

On the other hand, if the atom is enclosed in a cavity that supports only one mode at frequency ν_0 with volume V, the density of states factor changes to[7]:

$$g(\nu)^{\text{cavity}} = \frac{1}{V}\,\frac{2}{\pi\Delta\nu}\,\frac{\Delta\upsilon^2}{4(\nu-\nu_0)^2+\Delta\nu^2} \tag{3.8}$$

This is a Lorentzian function with full width at half maximum of $\Delta\nu$ and is normalized so that $V\int_0^\infty g(\nu)^{\text{cavity}}\, d\nu = 1$, as appropriate for a single mode of volume V. The cavity density of states is largest at ν_0, where we have $g(\nu_0)^{\text{cavity}} = 2/V\pi\Delta\nu = 2Q/V\pi\nu_0$, where Q is the *quality factor* of the cavity given by $Q = \nu_0/\Delta\nu$. The ratio of the density of state when the atom is in resonance with the cavity mode to that in a uniform medium is then given by:

$$\frac{g(\nu_0)^{\text{cavity}}}{g(\nu_0)} = \frac{2Q}{V\pi\nu_0} \div \frac{8\pi n^3\nu_0^2}{c^3} = \frac{1}{4\pi^2}\left(\frac{\lambda}{n}\right)^3\frac{Q}{V} \tag{3.9}$$

where $\lambda = c/\nu_0$ is the vacuum wavelength of the photon. We then obtain the Purcell factor F_P for the ratio of the transition rates:

$$F_\text{P} = \frac{W_{21}^{\text{cavity}}}{W_{21}^{\text{free space}}} = \frac{\tau_\text{R}^{\text{free space}}}{\tau_\text{R}^{\text{cavity}}} = \frac{3}{4\pi^2}\left(\frac{\lambda}{n}\right)^3\frac{Q}{V} \tag{3.10}$$

where the extra factor of three accounts for the random orientation of the dipoles in free space compared to having the dipole aligned with the cavity mode. The conclusion is that we can obtain enhanced emission (i.e. shorter radiative lifetime) for high-Q cavities with small modal volumes.

The prospect of enhancing the emission rate in the cavity encourages intense research efforts to develop micro- and nano-cavities with very small modal volumes

7 The factor of $1/V$ in Eq. (3.8) is required to maintain dimensional consistency with the density of states per unit volume for a uniform medium given in Eq. (3.7).

and high-quality factors, building on pioneering work demonstrating a shortened lifetime for quantum dots coupled to a micro-pillar cavity in 1998 [11]. The cavities are designed with asymmetric end-mirror reflectivities, so that photons can be coupled out preferentially on one side, as shown schematically in Figure 3.5. Maximum Purcell enhancement is achieved when the emitter is tuned to resonance with the cavity and positioned with maximal overlap with the cavity field. The latter point is not so straightforward for solid-state emitters that are located at random positions within a crystal, and in practice, it is usually necessary to fabricate many devices and search for one with an emitter at the right position. In the long run, it is more efficient to determine the position of the emitter first and then to fabricate the cavity around it. (See e.g. [12].)

An additional benefit of using a cavity is that the photons are preferentially emitted into the cavity mode rather than into 4π solid angle. The parameter of interest is the β-factor that quantifies the fraction of photons emitted into the mode. This is clearly related to the Purcell factor since the latter determines the emission rate to the cavity mode. In fact, the two parameters are related to each other, with:

$$\beta = \frac{F_{\mathrm{P}}}{1 + F_{\mathrm{P}}} \tag{3.11}$$

Hence, a large Purcell factor gives a large β factor approaching the optimal value of unity. By collecting the photons coupled out of the cavity, the collection efficiency can be greatly enhanced, thus overcoming the photon collection problem mentioned at the end of Section 3.4. The cavity can thus benefit in two ways – speeding up the emission and improving the collection efficiency – which explains why many of the best single-photon sources now routinely employ cavities.

3.6 Photon Coherence

Some applications in quantum photonics place additional constraints on the single-photon source in addition to the obvious one of excluding multiple photons within the pulse. In particular, applications in linear-optics quantum information processing require that the photons should be indistinguishable from each other. The photon indistinguishability is closely related to the coherence of the source, and hence, it is necessary to understand how the coherence is quantified and what determines it.

In classical optics, the coherence of light determines the visibility of the interference fringes observed when the light beam is split and then recombined after having traveled by different paths. An intuitive way to do this is to use a Michelson interferometer, in which the light is split into two paths and then recombined with a path difference d between them. This effectively interferes the electric field of the light at time t with that at time $t + \tau$, where $\tau = d/c$. Interference fringes can be observed if τ is shorter than the coherence time τ_{c} of the light, but not when $\tau \gg \tau_{\mathrm{c}}$.

The coherence time of a quantum emitter is determined by two main parameters: the radiative lifetime of the source, and the dephasing time. This relationship is traditionally written using notation that was borrowed from the Bloch sphere model of nuclear magnetic resonance (NMR):

$$\frac{1}{T_2} = \frac{1}{2T_1} + \frac{1}{T_2^*} \tag{3.12}$$

Here, $T_2 \equiv \tau_c$ is the coherence time, T_1 is the population decay time (i.e. τ_R if the only decay channel is by radiative emission), and T_2^* is the dephasing time.

The Bloch sphere model applies to a two-state system with wave functions ψ_1 and ψ_2. A coherent superposition can be formed where the system wave function ψ is given

$$\psi = c_1\psi_1 + c_2\psi_2 \tag{3.13}$$

The values of the coefficients (c_1, c_2) can be represented diagrammatically by the Bloch vector **r**. Since wave-function normalization requires that

$$|c_1|^2 + |c_2|^2 = 1 \tag{3.14}$$

the length of the vector is unity, and we can determine the spherical-polar angles (θ, ϕ) of **r** from:

$$\begin{aligned} c_1 &= \cos(\theta/2), \\ c_2 &= e^{i\phi}\sin(\theta/2) \end{aligned} \tag{3.15}$$

This is illustrated in Figure 3.6, where we have mapped the state with $(c_1, c_2) = (1, 0)$ to quantum state $|0\rangle$ at the top of the sphere and $(c_1, c_2) = (0, 1)$ to $|1\rangle$ at the bottom, following the notation generally used in quantum information. In the original work by Bloch, the two quantum states were magnetic spin states split by the Zeeman energy, but in quantum optics, they represent the lower and upper levels of the atomic transition.

We can understand the importance of the T_1 and T_2 times in Eq. (3.12) by considering the difference between coherent and mixed states. A coherent state has a well-defined phase relationship between c_1 and c_2 and is represented by points on the surface of the Bloch sphere. Mixed states, by contrast, do not contain phase information. All we know is the relative probability that the system is in state 1 or 2, which is governed by the ratio $|c_1|^2/|c_2|^2$. A completely incoherent mixed state would lie on the z axis of the Bloch sphere, and processes that destroy coherence move the state horizontally toward the z axis, as illustrated in Figure 3.6. The rate at which these dephasing processes occur is governed by T_2^*, which gives the timescales for

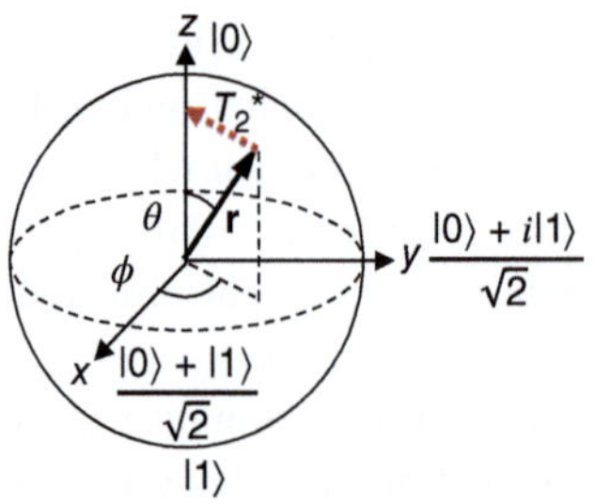

Figure 3.6 Schematic depiction of the Bloch sphere. The state of the system is represented by the Bloch vector **r**. The horizontal arrow labeled T_2^* illustrates a dephasing process that destroys coherence while maintaining the population.

elastic scattering events that preserve the populations but destroy phase information. The T_1 time gives the population decay time, which is the time for the system to return spontaneously to the lower level. T_1 is ultimately limited by the radiative lifetime, but could be shorter than τ_R is there are non-radiative decay paths between the two levels. A decay back to the ground state at the north pole of the Bloch sphere clearly destroys the coherent superposition state and hence limits the coherence time. Hence the total dephasing rate is the sum of the pure dephasing rate $1/T_2^*$ and the population dephasing rate $1/2T_1$, as indicated in Eq. (3.12). The factor of 2 that is associated with the population decay arises from the fact that the light emission decays as e^{-t/T_1} and is proportional to $|\mathcal{E}|^2$, where $\mathcal{E}$ is the electric field amplitude. Hence $\mathcal{E}(t) \propto e^{-t/2T_1}$, and it is the decay of the electric field that determines the coherence, as measured, for example, by the fringe visibility in interference experiments.

The ideal scenario for dephasing is where the only decay channel between the upper and lower states is radiative, so that $T_1 = \tau_R$ and the dephasing rate $1/T_2^*$ is negligible compared to the radiative decay. In this situation, the overall coherence time T_2 is equal to $2\tau_R$ and the spectral width of the emission is determined only by τ_R. This is called the Fourier limit.

The challenge in solid-state systems is to control the environment of the emitter so that dephasing processes occur on a much slower timescale than the radiative emission. For quantum-dot sources, the uncontrolled charges associated either with non-resonant excitation or with defects cause dephasing by generating random electric fields that perturb the quantum dot via the Stark effect. These effects can be eliminated by embedding the QD in a diode that controls its electrical environment and using resonant excitation to prevent generation of unwanted free carriers in the device [13]. Phonon scattering, however, is always an issue, and the sources generally only have high coherence when operated at low temperatures.

An alternative approach to mitigating against dephasing is to speed up the radiative emission using the Purcell effect to the point that $1/T_2^*$ is negligible compared to $1/2\tau_R$. Very large Purcell factors are possible for small nano-cavities with high-quality factors, and this has been shown to improve the coherence of a QD source to the Fourier limit even in devices with average electrical properties [14].

In some application in quantum photonics, it is not sufficient just to have one photon in each pulse; it is also essential that the photons in separate pulses are indistinguishable from each other. The degree of indistinguishability can be measured by performing Hong–Ou–Mandel (HOM) experiments [15]. In this experiment, two single photons impinge on separate input ports of a 50:50 beam splitter, with detectors D1 and D2 positioned at the two output ports, as shown in Figure 3.7. There are three possible outcomes: both photons go to detector D1, both go to D2, or one goes to D1 and the other to D2. The table in Figure 3.7 gives the probabilities for these three outcomes for two extreme cases. If the photons are completely distinguishable (e.g. if they arrive at different times), then both exit the beam splitter with 50% probability of going either to D1 or to D2. This gives a 50% probability of coincidence events where both detectors click. On the other hand, if the photons are completely indistinguishable, then the path amplitudes interfere, and the probability of observing coincidence events drops to zero. The degree of photon indistinguishability can

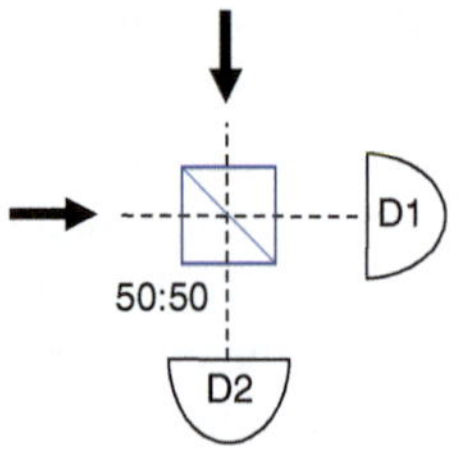

Outcomes	Distinguishable	Indistinguishable
Both to D1	25%	50%
One to D1, other to D2	50%	0
Both to D2	25%	50%

Figure 3.7 Schematic depiction of a Hong–Ou–Mandel experiment. The table gives the probabilities for the possible outcomes when single photons enter the beam splitter via the two input ports.

therefore be determined by measuring the drop in coincidence events in a HOM experiment. One way to do this is to compare the number of coincidence events when the photons arrive at the same time to the number when they arrive at different times. This ratio can be normalized to give the HOM visibility $\mathcal{V}$, where $\mathcal{V} = 1$ implies that the photons are totally indistinguishable.

High photon indistinguishability is only obtained when pure dephasing processes are eliminated, since the interaction between the emitter and the environment provides, in principle, a memory in the system that time-tags a specific photon. Hence, the HOM visibility is given by:

$$\mathcal{V} = \frac{T_2}{2T_1} \tag{3.16}$$

where T_2 and T_1 are related to each other through Eq. (3.12). For an ideal Fourier-transform-limited source, pure dephasing processes are negligible (i.e. $T_2^* \gg 2T_1$), and $T_2 = 2T_1$, implying $\mathcal{V} = 1$. By using high-quality quantum dots with controlled environments under resonant excitation, values of $\mathcal{V}$ approaching unity are obtainable [16]. The indistinguishability remains high for photons separated by long times compared to the 13 ns pulse separation, but eventually drops off due to dephasing occurring on microsecond timescales caused by spectral diffusion due to charge fluctuations in the vicinity of the dot [17]. More recent data using Purcell enhancement and a very high-quality dot have demonstrated negligible drop-off in the indistinguishability for photons emitted with ~ 1 μs time separation [18].

3.7 Deterministic Excitation of Quantum Emitters

Quantum emitters must be excited before they emit a photon. One way to do this is to embed the quantum emitter within a diode structure and to inject electrons and holes via an electrical pulse, with subsequent photon emission when the carriers recombine. A single-photon LED was demonstrated in this way in 2002 [19]. However, electrical excitation is not ideal, as timing jitter is introduced by the process of capturing the electron and hole into the quantum emitter from the bands of the semiconductor. Moreover, large numbers of free charge carriers are present in the device that can introduce noise and dephasing processes. The same is true for *non-resonant* optical excitation in which electrons and holes are created in the energy states above the transition levels of the quantum emitter after absorption of a laser pulse and must undergo relaxation before the emission occurs.

A way to avoid these timing jitter problems and to ensure that the emitter is excited with 100% probability for each pulse cycle is to use coherent control methods with a *resonant* laser. These methods can be understood by considering the Bloch-vector of the quantum emitter as in Figure 3.6 and borrowing techniques routinely used in NMR.

Consider a system that is prepared in the $|0\rangle$ state with the Bloch vector pointing up at the North Pole. In NMR experiments, this could be done by applying a large field so that all the spins are pointing along the field. In quantum optics, it would be done by ensuring that the emitter is in the ground state. The direction of the Bloch vector can be rotated by applying an electromagnetic pulse that is *resonant* with the transition frequency ν_0. In NMR experiments, the angle Θ through which the Bloch vector rotates is determined by the Rabi frequency Ω_R, and the pulse duration T, with $\Theta = \Omega_R T$. The Rabi frequency is given by:

$$\Omega_R = \left| \frac{\mu_{12}\mathcal{E}_0}{\hbar} \right| \tag{3.17}$$

where μ_{12} is the transition dipole moment and $\mathcal{E}_0$ is the electric-field amplitude, which is assumed to be constant during the pulse. This only works when the $T \ll T_2$, so that the system is coherent. For NMR experiments, this is not such an issue, as the coherence time can be in the millisecond range or longer. However, for a quantum emitter with $T_1 \sim 1$ ns or faster, we need short laser pulses with duration in the ps range. In this case, the amplitude of the electric field varies with time, and we therefore have a time-varying Rabi frequency. The rotation angle is then given by:

$$\Theta = \int_{-\infty}^{+\infty} \Omega_R(t)\mathrm{d}t \tag{3.18}$$

where $\Omega_R(t) = |\mu_{12}\mathcal{E}_0(t)/\hbar|$, with $\mathcal{E}_0(t)$ following the pulse envelope.

The case of a π-pulse where $\Theta = \pi$ is of special interest. If the system starts in the $|0\rangle$ state with the Bloch vector pointing up, then a π-pulse causes a rotation by π about an axis in the x–y plane so that the Bloch vector ends up pointing down with the system in the $|1\rangle$ state. We therefore have a *deterministic* method for preparing the system in the upper level of the transition, from which it can then emit the photon. Deterministic state preparation is the preferred method for *on-demand* single-photon sources in which the system produces a photon in response to a trigger pulse with probability approaching 100%.

Despite the benefits, the use of resonant π-pulse excitation methods does present some technical challenges that makes it difficult to implement in the field. The problem is that the photon that is emitted has the same frequency as the laser pulse and hence has to be carefully separated. This problem is in fact common to all the resonance fluorescence experiments that have been mentioned so far in this chapter. The standard way to separate the pump laser and the single photon is to use polarization optics. We pump using one polarization (e.g. X) and only collect light with the opposite polarization (i.e. Y) with a polarizer in the collection path to block the scattered laser photons. This works if the emitter dipole is oriented at 45° to the X and Y axes, so that the driving field has a component parallel to the dipole, and the emitting dipole has a component along the collection polarization. The need to resolve the X-polarized pump laser along the 45° direction is not a problem, as we can just double the power to get the same rotation angle, but the 45° orientation of the emitter dipole relative to the Y-polarized collection path is an issue, as it means that we lose half the photons that are emitted.

An ingenious method to get around this problem is to use an elliptical cavity with two oppositely polarized cavity modes with slightly different frequencies [20]. The QD is in resonance with the Y-polarized cavity, which enhances the emission probability for Y-polarized photons through the Purcell effect. The dot is pumped resonantly via the X-polarized cavity mode, which is off-resonance with the dot. The coupling of the laser though the off-resonance cavity mode is therefore relatively weak, but this is not a problem, as the laser power can simply be turned up. The key point in that we can pump the dot through the X mode, but all the photons are emitted into the Y mode due to the Purcell effect, giving very high emission probability.

3.8 Conclusions

We have seen in this chapter that solid-state quantum emitters can serve as excellent sources of single photons for applications in quantum photonics. The key parameters that quantify their performance are listed below.

Wavelength: The wavelength that is required is determined by the application. Emitters optimized for optical fiber systems are of particular interest. Short-range local area networks tend to operate around 900 nm, while long-range systems work in the telecom O-band (1260–1360 nm) and C-band (1530–1565 nm) wavelength range. Emitters that are optimized for spectral regions where atmospheric losses are low (e.g. the visible spectral region) are also of particular interest for free-space quantum channels.

Brightness: The brightness of the source is determined by the speed at which photons can be emitted and by the efficiency with which they can be collected. The speed is governed by the radiative lifetime, which can be enhanced by using the Purcell effect in a micro- or nano-cavity. (See Section 3.5.) The cavity can also help in funneling the photons into the collection optics.

Single-photon purity: The single-photon purity is quantified by the value of $g^{(2)}(0)$ measured in a HBT experiment. (See Section 3.2.) The value of $g^{(2)}(0)$ for a pulse containing n photons is $n(n-1)/n^2$ (see e.g. [6].) An ideal single-photon source thus has $g^{(2)}(0) = 0$, while a pulse that contains two photons gives $g^{(2)}(0) = 0.5$. The value of $g^{(2)}(0)$ gradually approaches the classical value of unity for larger values of n. A key requirement for a source to be considered as a quantum emitter is $g^{(2)}(0) < 1$, but demanding applications require that $g^{(2)}(0)$ should be as close to zero as possible.

Photon indistinguishability: The photon indistinguishability is measured by the visibility obtained in a HOM experiment. (See Section 3.6.) An ideal source would give $\mathcal{V} = 1$ over a wide range of photon separation times. Some applications (e.g. single-photon based quantum cryptography) do not require that the photons should be indistinguishable, but the value of $\mathcal{V}$ is critical for more demanding applications such as linear-optics quantum information processing.

Reviews of the present state of research in solid-state single-photon sources may be found in [21] and [22]. The types of materials that are used broadly fall into four main categories: quantum dots, color centers, strain centers in 2D materials, and individual emissive dopants (e.g. single molecules and rare-earth ions.) Table 3.3 gives a list of some of the materials that have been used along with the wavelength of the single photons that they generate. It is apparent that a wide range of wavelengths is available that opens up a great variety of applications.

As noted at the start of the chapter, the first demonstration of a solid-state quantum emitter was achieved for a single molecule system [3]. However, it is the

Table 3.3 Incomplete list of solid-state emitters.

Material system	Material details	Wavelength range	References
Epitaxial quantum dots	(InGa)As/GaAs	~900–1300 nm	[8, 9, 23, 24]
	GaAs/AlGaAs	~780 nm	[25]
	InAs/InP	C-band	[26, 27]
	InAs/GaAs on metamorphic buffer	C-band	[28, 29]
Colloidal quantum dots	II–VI semiconductor	Visible, NIR	[30–32]
	Perovskite	Visible	[33]
Diamond color centers	NV	637 nm	[34, 35]
	SiV	738 nm	[36]
	GeV	602 nm	[37–39]
	SnV	619 nm	[40, 41]
Silicon	Optically-active point defects	1100–1550 nm	[42]
Silicon carbide	Color centers	540–1400 nm	[43, 44]
Silicon nitride	Defect centers	550–750 nm	[45]
2D materials	TMDCs: WS_2, WSe_2, $MoSe_2$	NIR	[46–48]
	TMDCs: $MoTe_2$	1080–1550 nm	[49]
	hBN	Visible	[50]
Molecular	Polyaromatic hydrocarbons	Visible, NIR	[51]
Rare earth ions	$Nd^{3+}:YVO_4$	880 nm	[52]
	$Yb^{3+}:LiNbO_3$	980 nm	[53]

The works cited give a flavor of some of the relevant recent work, or are review articles. They should not be considered as being an exhaustive account of the latest state-of-the-art. NIR, near infrared; TMDC, transition-metal dichalcogenide; hBN, hexagonal boron nitride.

epitaxial quantum dots based on III–V semiconductors that have given the best results. This is because of the availability of advanced semiconductor technology that enables growth of samples with very high purity and fabrication of highly sophisticated nano-photonic devices. In this way, nearly ideal single-photon source performance has been demonstrated for InAs/GaAs QD devices that emit around 900 nm, and much progress has been made in developing QDs for the longer wavelength telecom bands. Shorter wavelength QDs based on the GaAs/AlGaAs system are also being developed for coupling to Rb atomic quantum memories at 780 nm. See Table 3.3 for references.

In parallel with the development of QD single-photon sources, there has also been a great deal of work done on color centers. Initial interest focused on color centers in diamond (see [54] for a review), but now single-photon emission has been observed for a large number of color centers in a great variety of crystals. Some of these materials are listed in Table 3.3.

In general, all of the solid-state quantum emitters work best at low temperatures. Indeed, the nearly ideal results for the InAs/GaAs QDs are only obtained with the sample held at liquid helium temperatures ($\sim$4 K), as phonon scattering degrades the performance at the temperature increases. In this context, some of the color-center sources show special promise, since they are relatively robust to temperature and can give good results even at room temperature. (see e.g. [34, 43, 45]). Hexagonal boron nitride (hBN) and colloidal QDs can also give good results at room temperature (see e.g. [55] and [30]), although none of these room-temperature results match the performance demonstrated for the InAs QDs at cryogenic temperatures.

The purpose of this chapter has been to give an overview of the basic principles that underpin solid-state quantum emitters, without going into details of the latest state of the art. More detailed information about telecom-wavelength quantum dots, color centers, and 2D materials may be found in Part IV of the book.

References

1 Mandel, L., Sudarshan, E.C.G., and Wolf, E. (1964). Theory of photoelectric detection of light fluctuations. *Proceedings of the Physical Society* 84: 435–444.

2 Kimble, J., Dagenais, M., and Mandel, L. (1977). Photon antibunching in resonance fluorescence. *Physical Review Letters* 36: 691–695.

3 Basché, T., Moerner, W.E., Orrit, M., and Talon, H. (1992). Photon antibunching in the fluorescence of a single dye molecule trapped in a solid. *Physical Review Letters* 69: 1516–1519.

4 Brouri, R., Beveratos, A., Poizat, J.-P., and Grangier, P. (2000). Photon antibunching in the fluorescence of individual color centers in diamond. *Optics Letters* 25: 1294–1296.

5 Michler, P., Imamoglu, A., Mason, M.D. et al. (2000). Quantum correlation among photons from a single quantum dot at room temperature. *Nature* 406: 968–970.

6 Fox, M. (2006). *Quantum Optics: An Introduction.* Oxford University Press.

7 Hanbury Brown, R. and Twiss, R.Q. (1956). Correlation between photons in two coherent beams of light. *Nature* 177: 27–29.

8 Arakawa, Y. and Holmes, M.J. (2020). Progress in quantum-dot single photon sources for quantum information technologies: a broad spectrum overview. *Applied Physics Reviews* 7: 021309.

9 Senellart, P., Solomon, G., and White, A. (2017). High-performance semiconductor quantum-dot single-photon sources. *Nature Nanotechnology* 12: 1026–1039.

10 Purcell, E.M. (1946). Spontaneous emission probabilities at radio frequencies. *Physical Review* 69: 681.

11 Gérard, J.-M., Sermage, B., Gayral, B. et al. (1998). Enhanced spontaneous emission by quantum boxes in a monolithic optical microcavity. *Physical Review Letters* 81: 1110–1113.

12 Shao, H., Ying, G., Lennon, S.A. et al. (2020). Purcell enhancement of a deterministically coupled quantum dot in an SU-8 laser patterned photonic crystal heterostructure. *Applied Physics Letters* 117: 043103.

13 Kuhlmann, A.V., Prechtel, J.H., Houel, J. et al. (2015). Transform-limited single photons from a single quantum dot. *Nature Communications* 6: 8204.

14 Liu, F., Brash, A.J., O'Hara, J. et al. (2018). High Purcell factor generation of indistinguishable on-chip single photons. *Nature Nanotechnology* 13: 835–840.

15 Hong, C.K., Ou, Z.Y., and Mandel, L. (1987). Measurement of subpicosecond time intervals between two photons by interference. *Physical Review Letters* 59: 2044–2046.

16 Dusanowski, L., Kwon, S.H., Schneider, C., and Höfling, S. (2019). Near-unity indistinguishability single photon source for large-scale integrated quantum optics. *Physical Review Letters* 122: 173602.

17 Wang, H., Duan, Z.C., Li, Y.H. et al. (2016). Near-transform-limited single photons from an efficient solid-state quantum emitter. *Physical Review Letters* 116: 213601.

18 Tomm, N., Javadi, A., Antoniadis, N.O. et al. (2021). A bright and fast source of coherent single photons. *Nature Nanotechnology* 16: 399–403.

19 Yuan, Z., Kardynal, B.E., Stevenson, R.M. et al. (2002). Electrically driven single-photon source. *Science* 295: 102–105.

20 Wang, H., He, Y.-M., Chung, T.H. et al. (2019). Towards optimal single-photon sources from polarized microcavities. *Nature Photonics* 13: 770–775.

21 Zhang, G., Cheng, Y., Chou, J.-P., and Gali, A. (2020). Material platforms for defect qubits and single-photon emitters. *Applied Physics Reviews* 7: 031308.

22 Lee, J., Leong, V., Kalashnikov, D. et al. (2020). Integrated single photon emitters. *AVS Quantum Science* 2: 031701.

23 Uppu, R., Pedersen, F.T., Wang, Y. et al. (2020). Scalable integrated single-photon source. *Science Advances* 6: eabc8268.

24 Ollivier, H., de Buy Wenniger, I.M., Thomas, S. et al. (2020). Reproducibility of high-performance quantum dot single-photon sources. *ACS Photonics* 7: 1050–1059.

25 Zhai, L., Lobl, M.C., Nguyen, G.N. et al. (2020). Low-noise GaAs quantum dots for quantum photonics. *Nature Communications* 11: 4745.

26 Musiał, A., Holewa, P., Wyborski, P. et al. (2019). High-purity triggered single-photon emission from symmetric single InAs/InP quantum dots around the telecom C-band window. *Advanced Quantum Technologies* 3: 1900082.

27 Anderson, M., Müller, T., Skiba-Szymanska, J. et al. (2021). Coherence in single photon emission from droplet epitaxy and Stranski–Krastanov quantum dots in the telecom C-band. *Applied Physics Letters* 118: 014003.

28 Portalupi, S.L., Jetter, M., and Michler, P. (2019). InAs quantum dots grown on metamorphic buffers as non-classical light sources at telecom C-band: a review. *Semiconductor Science and Technology* 34: 053001.

29 Nawrath, C., Vural, H., Fischer, J. et al. (2021). Resonance fluorescence of single In(Ga)As quantum dots emitting in the telecom C-band. *Applied Physics Letters* 118: 244002.

30 Lin, X., Dai, X., Pu, C. et al. (2017). Electrically-driven single-photon sources based on colloidal quantum dots with near-optimal antibunching at room temperature. *Nature Communications* 8: 1132.

31 Feng, S.W., Cheng, C.Y., Wei, C.Y. et al. (2017). Purification of single photons from room-temperature quantum dots. *Physical Review Letters* 119: 143601.

32 Eich, A., Spiekermann, T.C., Gehring, H. et al. (2022). Single-photon emission from individual nanophotonic-integrated colloidal quantum dots. *ACS Photonics* 9: 551–558.

33 Zhu, C., Marczak, M., Feld, L. et al. (2022). Room-temperature, highly pure single-photon sources from all-inorganic lead halide perovskite quantum dots. *Nano Letters* 22: 3751–3760.

34 Bogdanov, S.I., Shalaginov, M.Y., Lagutchev, A.S. et al. (2018). Ultrabright room-temperature sub-nanosecond emission from single nitrogen-vacancy centers coupled to nanopatch antennas. *Nano Letters* 18: 4837–4844.

35 Riedel, D., Söllner, I., Shields, B.J. et al. (2017). Deterministic enhancement of coherent photon generation from a nitrogen-vacancy center in ultrapure diamond. *Physical Review X* 7: 031040.

36 Marseglia, L., Saha, K., Ajoy, A. et al. (2018). Bright nanowire single photon source based on SiV centers in diamond. *Opt Express* 26: 80–89.

37 Iwasaki, T., Ishibashi, F., Miyamoto, Y. et al. (2015). Germanium-vacancy single color centers in diamond. *Scientific Reports* 5: 12882.

38 Chen, D., Mu, Z., Zhou, Y. et al. (2019). Optical gating of resonance fluorescence from a single germanium vacancy color center in diamond. *Physical Review Letters* 123: 033602.

39 Siampour, H., Kumar, S., Davydov, V.A. et al. (2018). On-chip excitation of single germanium vacancies in nanodiamonds embedded in plasmonic waveguides. *Light Science & Applications* 7: 61.

40 Rugar, A.E., Aghaeimeibodi, S., Riedel, D. et al. (2021). Quantum photonic interface for tin-vacancy centers in diamond. *Physical Review X* 11: 031021.

41 Trusheim, M.E., Pingault, B., Wan, N.H. et al. (2020). Transform-limited photons from a coherent tin-vacancy spin in diamond. *Physical Review Letters* 124: 023602.

42 Durand, A., Baron, Y., Redjem, W. et al. (2021). Broad diversity of near-infrared single-photon emitters in silicon. *Physical Review Letters* 126: 083602.

43 Wang, J., Zhou, Y., Wang, Z. et al. (2018). Bright room temperature single photon source at telecom range in cubic silicon carbide. *Nature Communications* 9: 4106.

44 Castelletto, S. (2021). Silicon carbide single-photon sources: challenges and prospects. *Materials for Quantum Technology* 1: 023001.

45 Senichev, A., Martin Zachariah, O., Peana, S. et al. (2021). Room-temperature single-photon emitters in silicon nitride. *Science Advances* 7: eabj0627.

46 Kianinia, M., Xu, Z.-Q., Toth, M., and Aharonovich, I. (2022). Quantum emitters in 2D materials: emitter engineering, photophysics, and integration in photonic nanostructures. *Applied Physics Reviews* 9: 011306.

47 Turunen, M., Brotons-Gisbert, M., Dai, Y. et al. (2022). Quantum photonics with layered 2D materials. *Nature Reviews Physics* 4: 219–236.

48 Azzam, S.I., Parto, K., and Moody, G. (2021). Prospects and challenges of quantum emitters in 2D materials. *Applied Physics Letters* 118: 240502.

49 Zhao, H., Pettes, M.T., Zheng, Y., and Htoon, H. (2021). Site-controlled telecom-wavelength single-photon emitters in atomically-thin $MoTe_2$. *Nature Communications* 12: 6753.

50 Shaik, A.B.D. and Palla, P. (2021). Optical quantum technologies with hexagonal boron nitride single photon sources. *Scientific Reports* 11: 12285.

51 Toninelli, C., Gerhardt, I., Clark, A.S. et al. (2021). Single organic molecules for photonic quantum technologies. *Nature Materials* 20: 1615–1628.

52 Zhong, T., Kindem, J.M., Bartholomew, J.G. et al. (2018). Optically addressing single rare-earth ions in a nanophotonic cavity. *Physical Review Letters* 121: 183603.

53 Xia, K., Sardi, F., Sauerzapf, C. et al. (2022). Tunable microcavities coupled to rare-earth quantum emitters. *Optica* 9: 445–450.

54 Janitz, E., Bhaskar, M.K., and Childress, L. (2020). Cavity quantum electrodynamics with color centers in diamond. *Optica* 7: 1232–1252.

55 Grosso, G., Moon, H., Lienhard, B. et al. (2017). Tunable and high-purity room temperature single-photon emission from atomic defects in hexagonal boron nitride. *Nature Communications* 8: 705.

4

Single-Photon Sources for Multi-Photon Applications

Stefan Frick, Robert Keil, Vikas Remesh, and Gregor Weihs

Institut für Experimentalphysik, Universität Innsbruck, Technikerstraße 25, 6020 Innsbruck, Austria

4.1 Motivation and Applications

Linear optical quantum information processing relies on the interference of photons, often of multi-photon states. This is true for gate-based [1, 2], measurement-based [3, 4] and fusion-based [5, 6] quantum computing implementations, in quantum communication [7, 8], as well as in (Gaussian) boson sampling [9–11], which can be utilized for quantum simulation [12]. Whether one wants to build multi-photon entangled states, implement a two-qubit optical quantum gate, or perform Bell-state analysis, interference of two or more photons is at the heart of the process.

The quality of such a process is then directly connected to the quality (visibility, contrast) of the interference. Both at the single- and multi-particle levels, interference of quantum waves (amplitudes, evolutions) is intimately connected to the fundamental indistinguishability of two or more evolutions that connect the same initial and final states. The interferometer must thus couple two or more modes that we usually call external modes. Distinguishability can originate from any degrees of freedom of the interfering quanta that are not touched by the couplings of the external modes and are thus called internal [13]. Which degrees of freedom are internal and which are external is not always straightforward to identify, though. A simple situation is given when photons in a number of spatial modes, be it optical beams or waveguide modes, are subjected to a unitary operation on these modes in the form of a network of beam splitters and phase shifters. Then, polarization, frequency, and temporal wavefunction are possible internal degrees of freedom. These will introduce distinguishability, if they are not identical for all the photons that are involved in a particular interference scenario. If not all photons can be created with mutual indistinguishability, particular scenarios still work [14]. In the past, authors considered almost exclusively the pairwise indistinguishability, usually defined via the visibility of Hong–Ou–Mandel (HOM) interference [15], see Section 4.2.3, but recently it was realized [16, 17] that there are nontrivial multi-particle phases and associated (in-)distinguishability, such that pairwise interference tests are insufficient to predict the rich dynamics of the whole multi-photon state.

Photonic Quantum Technologies: Science and Applications, First Edition. Edited by Mohamed Benyoucef.

In this book chapter, we cover two widely used technologies for the generation of single- and multi-photon states, namely nonlinear optical sources, see Section 4.2, and semiconductor quantum dots, see Section 4.3. These types of sources operate on fundamentally different principles, have their own advantages and challenges, and vary substantially in the properties of the emitted photons. For any source, though, we do care about the mutual indistinguishability, either of photons emitted by multiple copies of the source or of subsequent photons emitted by the same source, see e.g. Sections 4.2.4 and 4.3.3.

4.2 Nonlinear Optical Sources

The generation of photon pairs via nonlinear optical processes is a long-established method. Several techniques exist, the most prominent ones, which we will cover in this section, are based on either spontaneous parametric down-conversion (SPDC, see Section 4.2.1.1) or spontaneous four-wave mixing (SFWM, Section 4.2.1.2). Compared to single-quantum emitters, such as quantum dots, they have the fundamental disadvantage of a random emission process, with detrimental consequences to their number purity (see Section 4.2.2). Nevertheless, they are still widely used due to their relatively straightforward implementation, their operation at room temperature, their spectral flexibility and the possibility of integration in optical chips.

4.2.1 Principles

4.2.1.1 Spontaneous Parametric Down-Conversion

SPDC or parametric fluorescence is a nonlinear optical process, which converts one high-energetic pump photon (p) into a pair of photons with lower energy, termed signal (s) and idler (i) [18, 19]. SPDC is a second-order ($\chi^{(2)}$) nonlinear process, requiring non-centrosymmetric (hence anisotropic) materials, and it needs to satisfy energy and (approximate) momentum conservation, imposing

$$\omega_\mathrm{p} = \omega_\mathrm{s} + \omega_\mathrm{i} \tag{4.1}$$

$$\vec{k}_\mathrm{p} = \vec{k}_\mathrm{s} + \vec{k}_\mathrm{i} \tag{4.2}$$

The energy conservation condition (4.1) is illustrated in Figure 4.1a. For $\omega_\mathrm{s} = \omega_\mathrm{i}$, the SPDC is said to be in the degenerate regime, otherwise it is termed non-degenerate. The wave vectors in the momentum conservation condition (4.2) depend on the involved frequencies as well as the refractive properties of the conversion medium, which leads to a variety of possible phase-matching configurations discussed in Section 4.2.1.3.

In resonators with round-trip losses being smaller than the round-trip gain, the stimulated version of the process takes over, leading to optical parametric oscillation (OPO) [20], which is widely applied in classical nonlinear optics, but can also be used in Gaussian boson sampling [21]. SPDC of a single photon pair relates to OPO as spontaneous emission relates to stimulated emission in a laser.

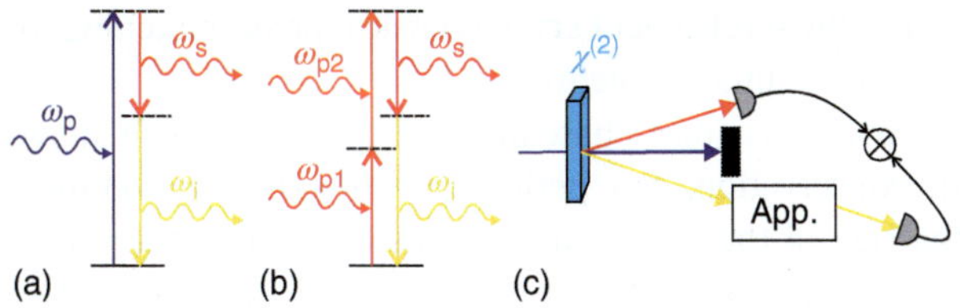

Figure 4.1 Energy level schemes of SPDC (a) and SFWM (b). All levels except the ground state are virtual, such that no absorption and re-emission occurs physically. (c) Heralding (here illustrated for SPDC, works analogously for SFWM): Instead of recording all photons transmitted through some application (App.), only coincidences between the signal (here acting as the herald, roles can be exchanged) and the idler photons are recorded. This ensures single-photon-like number statistics of the idler photons.

4.2.1.2 Spontaneous Four-Wave Mixing

Another frequently used mechanism for the generation of photon pairs is SFWM. In this process, two pump photons are converted to a pair of signal and idler photons, again satisfying energy conservation $\omega_{p1} + \omega_{p2} = \omega_s + \omega_i$, see Figure 4.1b, and momentum conservation $\vec{k}_{p1} + \vec{k}_{p2} = \vec{k}_s + \vec{k}_i$. Relying on a third-order ($\chi^{(3)}$) nonlinear effect, SFWM works also in centrosymmetric materials. Therefore, it is a popular method for photon-pair generation in optical waveguides and fibers, where the lower pair generation efficiency of a $\chi^{(3)}$-process compared to a $\chi^{(2)}$-process in suitable media can be compensated via the transverse confinement of light over extensive lengths [22–25]. If either $\omega_s = \omega_i$ or $\omega_{p1} = \omega_{p2}$ the process is termed degenerate, otherwise it is called non-degenerate.

4.2.1.3 Phase Matching

The momentum conservation of Eq. (4.2) (or its SFWM-counterpart) is equivalent to requiring that the field oscillations of the pump laser(s) on the one hand and the signal and idler on the other stay in phase when propagating through the conversion medium. Then, the signal(idler) fields generated in different sections of the converter interfere constructively with each other, producing a bright output. Otherwise, destructive interference will occur, preventing efficient emission. The dispersion relation of the involved materials $k_{p,s,i} = \frac{\omega_{p,s,i}}{c} n(\omega_{p,s,i})$ implies that phase matching can be either achieved by a judicious combination of frequencies, refractive indices, and propagation directions. There are three types of phase matching, which are defined by the polarization of the involved fields:

- *Type-0*: Pump, signal, and idler share the same linear polarization,
- *Type-1*: Signal and idler share the same linear polarization, which is orthogonal to the polarization of the pump,
- *Type-2*: Signal and idler are orthogonally polarized, the pump polarization is parallel to one of them.

As the simultaneous solution of Eqs. (4.1) and (4.2) is only possible for polarization-dependent refractive indices, phase matching in a homogeneous material must be of type-1 or type-2 and is also termed birefringent phase matching. One distinguishes further whether this matching of refractive indices is achieved via adjusting the orientation of the birefringent crystal (critical phase matching) or tuning its temperature (non-critical phase matching).

An often used alternative method is quasi-phase matching [26, 27], which allows for imperfect momentum conservation, but periodically inverts the nonlinear susceptibility (and thus the sign of the phase mismatch) via poling of the crystal domains during fabrication (see Section 4.2.4.1 for more details). A comprehensive discussion of phase matching techniques for SPDC can be found in [28].

4.2.2 Photon-Number Purity and Heralding

Several protocols in quantum computing, simulation, communication, or metrology entirely rely on or at least benefit from a precisely defined number of quanta. For example, a Bell experiment requires entangled photon pairs, i.e. exactly two photons. A non-vanishing probability for two or more emitted photons per mode leads to ambiguities in the detection patterns and biases the violation of the Bell inequality [29]. For an ideal single-photon source, the photon number statistics should consist only of zero photons or one photon in a given time interval, whereas all higher photon numbers should never occur. The absence of two or more photons is quantified by the second-order intensity autocorrelation $g^{(2)}(0)$, which can be measured in a Hanbury Brown & Twiss experiment [30]. One obtains $g^{(2)}(0) = 0$ if and only if one has at most one photon at a time in a given mode and $g^{(2)}(0) = 1$ for the Poissonian photon number distribution of a single-mode laser.

As the nonlinear optical photon generation processes of SPDC and SFWM are intrinsically probabilistic, the number of emitted photons per time window of interest (either the duration of the pump pulse or some application-dependent detection time window) cannot be controlled. The number statistics of either signal or idler alone follows a super-Poissonian distribution (with details depending on the spectral purity, see Section 4.2.4), i.e. one has $1 \leq g^{(2)}(0) \leq 2$. This is undesired in all applications, where true single-photon character is needed. The situation can be remedied by sacrificing one of the two output modes of the source as an auxiliary "herald" and using only the other in the application. Only when an auxiliary "heralding" photon is detected, events from the other channel will be registered (see Figure 4.1c). One can show that with this heralding the photon-number statistics in the so "heralded" channel approaches the statistics of a perfect single-photon emitter in the limit of low count rates (see Section 4.2.5 for more details).

4.2.3 Indistinguishability

If photons in a two- or multi-photon interference scenario match in all their internal degrees of freedom, they are said to be indistinguishable. The paradigmatic example for a two-photon (TP) interference is given by the HOM effect [15]: Two photons impinge on the input modes of a balanced 50/50 beam splitter and the coincidences between photon detections in the two output modes are recorded, see Figure 4.2. For indistinguishable input photons, no coincidences will be detected due to the destructive interference between the involved amplitudes, while distinguishable photons

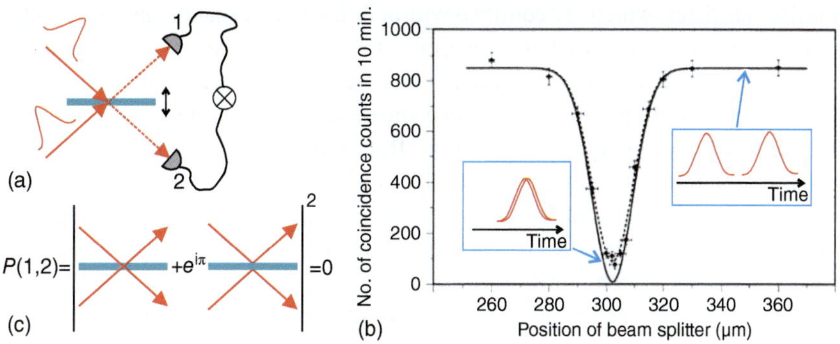

Figure 4.2 Hong–Ou–Mandel interference. a) Photons impinge on a beam splitter with an adjustable position, tuning the photon's temporal overlap. (b) For maximal overlap the detected two-photon coincidences vanish in what is known as the HOM dip. (c) This phenomenon arises from a complete destructive interference of the two-photon amplitudes of double transmission and double reflection at the beam splitter. Source: Reproduced with permission from Hong et al. [15]/American Physical Society.

(i.e. orthogonal in at least one internal mode) pass the beam splitter independently from one another.

In reality, the indistinguishability of the photons can be compromised by a variety of effects, such as polarization rotations, time jitter, imperfect spectral overlap, or spectral correlations (see Section 4.2.4). All these effects reduce the overlap of at least one internal mode. For two photons, the indistinguishability can be measured with a beam splitter and is given by the visibility of the HOM dip [Figure 4.2b]: $V = \left(C_\infty - C_0\right)/C_\infty$, with C_0 as the coincidence rate at zero delay (maximal temporal overlap) and C_∞ as the coincidence rate at large delay (zero overlap).

For multiple photons, there is no unique measure of indistinguishability. A plausible, application oriented, definition is to link indistinguishability with the contrast of a multi-particle interference in the interferometer of interest [13, 31, 32]. A zero or perfect visibility of all pairwise HOM-dips then guarantees the complete absence or the maximal contrast of the interference, respectively. In the important intermediate regime of partial distinguishability, however, an involved dependence of the interference on the modulus [33, 34] or phase of the internal state overlaps can arise [16] and pairwise dip visibilities in general provide only upper and lower bounds on the visibility of the combined interference [35]. Indeed, there can be even scenarios where some pairs of particles are fully distinguishable (no HOM dip occurs), yet multi-particle interference can arise due to collective phases [36] or symmetries [14, 37], in the latter case even with full contrast. An unambiguous measure of multi-particle indistinguishability can be defined in cyclic multi-port interferometers, where the interference contrast is equally sensitive to the internal states of all particles and all their mutual overlaps [17, 38].

4.2.4 Spectral Purity

Photonic quantum information processing relies on interactions between photons being exclusively mediated by interference. Most of the time, interference between multiple particles, which are confined within different external modes, is necessary to solve computational problems. Generally, the more complex these problems get, more photons and modes are required to map the problem to a photonic circuit.

For effective multi-photon interference, the indistinguishability of photons arising from different pair-creation events, either from separate sources or from the same source at different points in time, needs to be ensured. This is often trivially satisfied in some internal degrees of freedom, such as polarization. The spectral correlation imposed on each signal-idler pair by the energy conservation (4.1), however, can cause a substantial complication: Even if all the signal(idler) photons from the various conversion events have identical *average* spectra, their individual spectra can still get "marked" by the wavelength of its co-produced idler(signal) photon. Consequently, the indistinguishability between photons from different pairs is reduced. This effect can be avoided by ensuring the photon pairs to be created in a single well-defined spectral mode. This property, termed *spectral purity*, leaves the photon pairs no room for correlations between signal and idler spectra. Given no fluctuations or mismatches in other internal degrees of freedom, spectral purity then implies indistinguishability between photons from separate emission events.

The de-facto measure for the spectral purity of a down-conversion process, becomes the Schmidt number (or rank) of the joint spectral amplitude (JSA) of the produced down-converted state. The Schmidt number K is defined as

$$K = 1 \Big/ \sum_i \lambda_i^4 \tag{4.3}$$

where $\{\lambda_i\}_i$ are the singular values of the Schmidt decomposition of the JSA. K is an estimate for the number of spectral modes involved in the down-conversion process [39]. As can be seen in Figure 4.3, the higher the spectral (anti-)correlation within the JSA is, the higher the Schmidt number becomes. Since a good state for quantum computing tasks only occupies one single spectral mode, to guarantee *indistinguishability*, JSAs with Schmidt number $K = 1$ are favorable. This becomes true when the JSA becomes factorizable and no correlations between ω_s and ω_i exist.

Spectral purity is also intrinsically linked to the photon-number statistics of a down-conversion process. For $K = 1$ (i.e. perfect spectral purity), the number statistics is equivalent to a thermal state ($g^{(2)}(0) = 2$) [40], whereas for a maximally

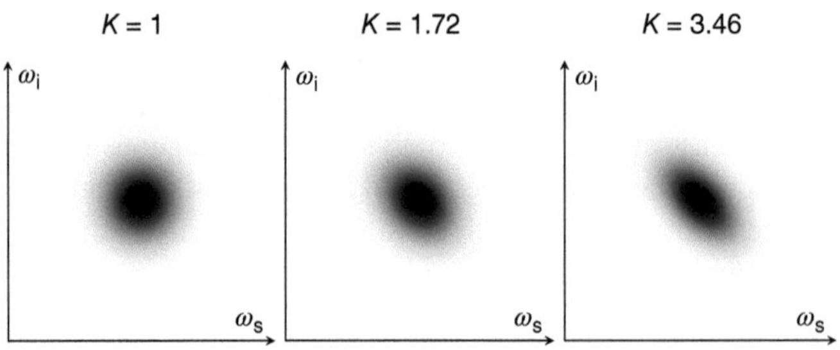

Figure 4.3 Joint spectral amplitudes with increasing Schmidt number. The higher the spectral correlation between the signal and idler modes becomes, the more modes are within the support of the joint spectral amplitudes.

correlated JSA ($K \to \infty$), one gets a Poissonian photon-number distribution ($g^{(2)}(0) = 1$). Generally, one gets $g^{(2)}(0) \approx 1 + 1/K$ in the low-gain regime [39, 41].

4.2.4.1 Methods for High-Purity Down-Conversion Sources

Effective heralded single-photon sources for quantum information processing need to be spectrally pure to enable their indistinguishability. However, when signal and idler photons are spectrally correlated, they will occupy more than a single spectral mode, and thus indistinguishability cannot be guaranteed. Two main techniques are used to counteract the effect of a correlated JSA: filtering and source engineering.

Appropriate spectral filtering is often employed to eliminate correlations between signal and idler photons. Yet, this technique always has to introduce additional loss. Besides reducing overall count rates, loss has also a detrimental effect on number purity (see Section 4.2.2). This becomes immediately obvious in an easy Gedanken-experiment. If a state of exactly N photons, i.e. a Fock state with photon number purity 1 (that is $g^{(2)}(0) = 1 - 1/N$), is sent through a medium with absorption probability > 0, every photon will be absorbed with that probability. This transforms the original Fock state to a Binomial distribution where all photon numbers between N and 0 occur. The photon number purity has thus been decreased. While filtering increases the spectral purity of a single-photon source, it will inevitably reduce the number purity.

Additionally, the same is true for symmetric heralding efficiency. Tighter spectral filtering allows for higher spectral purity but, at the same time, reduces the symmetric heralding efficiency of the photon pair source [42]. This is especially problematic in cases where the heralding efficiency needs to be symmetric or where both photons of a pair are utilized within an experiment (see Figure 4.4).

Another, more promising approach is to tailor the phase matching function of the down-conversion process. If the group velocity dispersion is chosen carefully, such that the respective phase-matching angle, i.e. the angle of the phase-matching function in the JSA (θ in Figure 4.4), is close to 45°, a high spectral purity can be achieved without the need of filtering [43]. The spectral power curve of the pump used in the SPDC process $a(\omega_p) = a(\omega_s + \omega_i)$ will always exhibit a −45° angle within the coordinate system of the JSA. Thus, if the group velocity of signal, idler, and pump field create a phase-matching curve orthogonal to the spectral power curve, the spectral width of the pump can be chosen such that a JSA with high spectral purity is ensured.

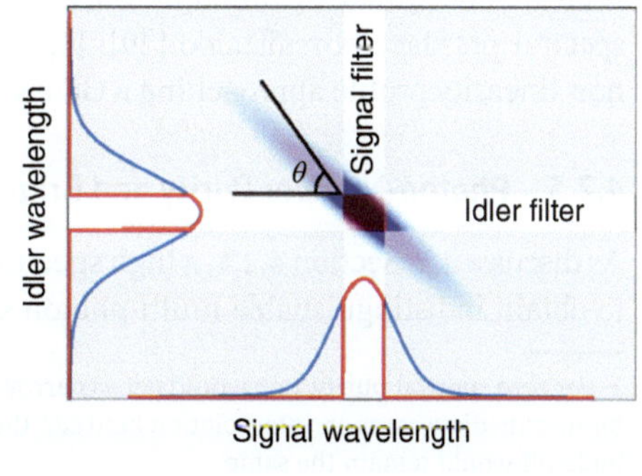

Figure 4.4 Filtering of a correlated joint spectral amplitude. Because the filter always introduces additional loss the photon number purity is reduced. While the red part of the JSA remains correlated, the remaining transmitted spectra of signal and idler photons (blue) diminish the heralding efficiency. Figure reprinted with permission from [42].

In principle, high purity can be achieved with any phase-matching angle between 0° and 90°. In this case, however, not with circular but with elliptical JSAs [44].

Unfortunately, this method of engineering JSAs is relatively restrictive and can only be applied for certain wavelength ranges and in certain nonlinear crystal materials. Another, more versatile method of engineering leverages on the flexibility of a phase-matching mechanism called quasi-phase-matching (QPM) [26, 27]. QPM allows to phase-match wavelengths in a co-linear fashion, where usually no down-converted light would be generated. This is achieved by domain engineering. Nonlinear crystals such as ppLN and ppKTP, can be manufactured in a way such that the lattice orientation changes ever so often throughout the crystal. The change in lattice orientation creates an effective π phase-shift of the nonlinear polarization inside the crystal, thus reversing the sign of $\chi^{(2)}$. A volume of same lattice orientation is called a domain and two domains, each poled in opposite directions make up a poling period Λ. If Λ is chosen correctly, Eq. (4.2) can be rewritten as

$$k_{\mathrm{p}} - k_{\mathrm{s}} - k_{\mathrm{i}} = \frac{2\,\pi}{\Lambda} \tag{4.4}$$

QPM is a versatile mechanism, which enables co-linear phase-matching in situations where usually this could not be achieved, virtually only limited by the manufacturing tolerances on small poling periods [45] and the overall crystal length. Not only extends QPM the wavelength range of nonlinear materials, it also enables more advanced domain-engineering techniques. Apodized poling structures can be used to implement a variation of the effective nonlinearity over the length of the crystal. One tuning knob is the order of the poling: if an odd multiple m of Λ in (4.4) is chosen as the poling period a down-converted electric field is still generated throughout the nonlinear crystal, however, at a lower rate. Additionally, the duty cycle D of the poling can be varied. Giving the two domains of one poling period different lengths, allows to finely tune the effective non-linearity d_m within each poling period,

$$d_m = 2\,d\,\frac{\sin(\pi\,D)}{m\,\pi} \tag{4.5}$$

with d being the bulk nonlinearity. Approaching a Gaussian profile with this technique is especially interesting, since the phase-matching function (given by the Fourier transform of the nonlinearity profile) will consequently also be of Gaussian shape, allowing the JSA to become separable and exhibit high spectral purity [46]. Using domain engineering to achieve high purity down-conversion sources has been demonstrated in different wavelength ranges [47, 48], but many different spectral profiles are realizable [49]. Figure 4.5 shows an example for an effective non-linearity profile approaching a Gaussian shape.

4.2.5 Photon-Number Purity and Brightness

As discussed in Section 4.2.4, a high spectral purity of the photon pairs is paramount to obtain indistinguishable multi-photon states. Given perfect spectral purity,[1] the

---

1 For zero spectral purity, one would get a (narrower) Poissonian distribution instead of (4.6) and for intermediate cases an interpolation between the two [50, 51]. The nature of the here discussed trade-off would remain the same.

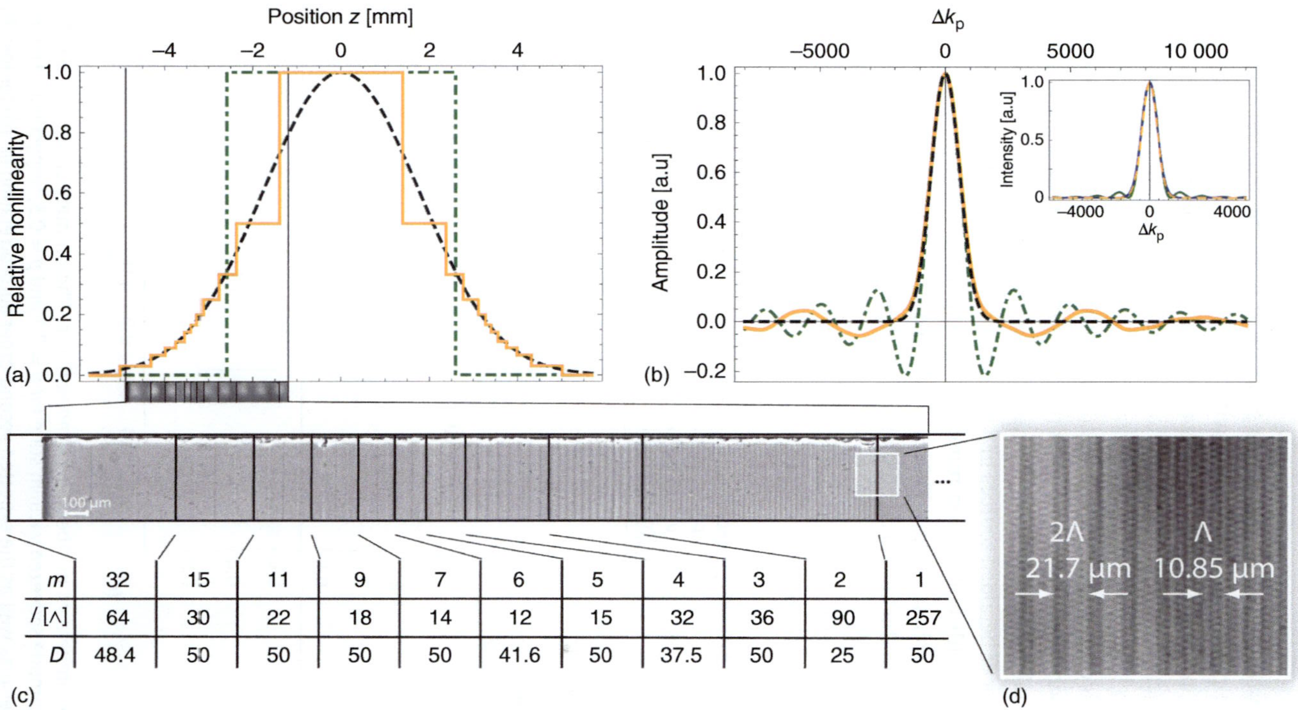

m	32	15	11	9	7	6	5	4	3	2	1
l [Λ]	64	30	22	18	14	12	15	32	36	90	257
D	48.4	50	50	50	50	41.6	50	37.5	50	25	50

Figure 4.5 Example of a nonlinearity profile approaching a Gaussian shape. (a) Shows the nonlinear profile in orange with the usual top-hat shape in green and the ideal Gaussian shape in black. (b) Demonstrates the corresponding spectral responses. (c) Depicts the poling structure of the crystal with poling order *m* and duty cycle *D*. (d) Zoom-in view of the poling structure from order $m = 1$ to $m = 2$. Figure reprinted with permission from [47].

resulting two-mode squeezed vacuum state has a probability distribution of the photon number n of the signal or idler per pump pulse of

$$p(n) = (1 - p)p^n \tag{4.6}$$

with $p = (\tanh r)^2$ as the probability to generate at least one pair per pulse and r the squeezing parameter, which scale with the effective nonlinearity of the material, interaction length, and pump power. Figure 4.6a,b illustrate two exemplary cases. With increasing p, the average photon number per pulse $\mu = \langle n \rangle = p/(1 - p)$ [53], and with it the brightness of the source are increasing. At the same time, the multi-photon contributions with $n > 1$ (from multi-pair emissions) get more pronounced, which increases the spread of the distribution and with it the uncertainty in photon number. In many schemes, heralding is used to produce a single-photon output state (or N heralded sources in parallel to produce an N-photon state). However, heralding with non-photon number resolving detectors cannot rule out the presence of multi-photon terms. This can be formally evaluated by the second-order autocorrelation in the heralded channel, which is given by [54]:

$$g_{\mathrm{h}}^{(2)}(0) = \frac{\langle n^2 \rangle_{\mathrm{h}} - \langle n \rangle_{\mathrm{h}}}{\langle n \rangle_{\mathrm{h}}^2} \tag{4.7}$$

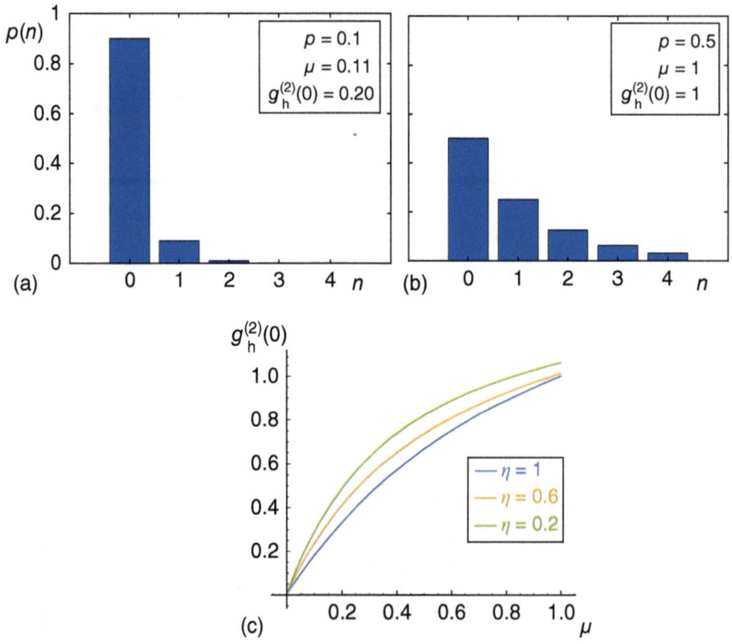

Figure 4.6 Brightness vs. number purity trade-off. (a,b) Photon-number distribution of one mode of a two-mode squeezed vacuum state for $p = 0.1$ and $p = 0.5$, respectively. The mean photon number μ as well as the expected second-order autocorrelation $g_{\mathrm{h}}^{(2)}(0)$ for loss-free heralding is also shown for each case. (c) Heralded second-order autocorrelation as a function of mean photon number and heralding efficiency η, evaluated according to Eqs. (7) and (8) in [52].

with $\langle\rangle_h$ denoting averaging with respect to the photon-number distribution after heralding, which is given in the loss-free case as

$$p_h(n) = \frac{1-p}{p} p^n \tag{4.8}$$

for $n > 0$ and $p_h(0) = 0$ [52]. As evident from the blue curve in Figure 4.6c the photon-number purity deviates increasingly from the ideal case of a single-photon emitter with $g_h^{(2)}(0) = 0$ as the brightness of the source is ramped up. For finite heralding efficiency η, the situation deteriorates further (orange and green curves). This illustrates the fundamental trade-off between the brightness of a nonlinear optical source and its photon-number purity. In other words, a nonlinear optical source cannot have a high probability of photon emission and a precisely defined number of emitted photons at the same time. To overcome this intrinsic limitation, one can either resort to multiplexing (see Section 4.2.6) or true single-photon emitters, such as quantum dots (see Section 4.3).

4.2.6 Multiplexing Schemes

As discussed in Section 4.2.5, nonlinear optical sources suffer from an intrinsic trade-off between brightness and number-purity. As a consequence, the probability p to produce at least one photon pair per pump pulse (or coherence length in cw-pumped schemes) in a target signal/idler mode must be kept small, in order to also keep the multi-pair emission rate small. Then it seems logical to increase the total rate of photon emission by keeping p small, but increasing the number of modes, into which a conversion can occur. This is the domain of multiplexing.

The fundamental idea is to have several possible signal/idler modes, heralding which of the modes are populated and active optical switching to route the heralded photon(s) into the target mode(s) for further applications. Depending on which degree of freedom is used for this enterprise, these schemes are termed spatial multiplexing (several sources emitting in parallel) [55–58], temporal multiplexing (one source emitting at multiple possible times) [59–61], spectral multiplexing (one source emitting at multiple possible spectral channels) [62] or combinations thereof [63]. Clearly, temporal and spectral multiplexing are less resource-demanding than spatial multiplexing when scaling to large numbers of modes, but require fast and precisely timed switches or nonlinear optical frequency conversion, respectively. As of today, only temporal multiplexing schemes have managed to beat the theoretical limits on simultaneous number-purity and brightness of single-source SPDC [52, 60, 64]. If only the rates of (multi-)photon events are of interest, but not their particular timing, then also passive temporal multiplexing without active switching can provide an advantage [65]. For a comprehensive review of multiplexing techniques, see [52].

4.3 Quantum Dots

For multi-photon applications, the major disadvantage of SPDC-based sources is that they produce photon pairs probabilistically. The resulting statistics impose an intrinsic trade-off between brightness and photon-number purity. Semiconductor quantum dots offer great potential to overcome this limitation, boasting their high brightness, photostability, and fairly easily scaleable growth processes. If the quantum dot can be isolated and addressed individually, it will usually absorb or emit only a single photon with specific properties but not a second, identical one, ensuring non-classical light emission. Without intrinsic limits on brightness or purity, quantum dots are therefore among the best sources to deterministically deliver single photons and entangled photon pairs with high rates.

This section considers only epitaxial quantum dots focusing on advanced excitation methods, which overcome the limitations of the basic excitation schemes (see Chapter 2), compares the typical photonic nanostructures where quantum dots are embedded to ensure better photon properties, and finally discusses the time-to-space demultiplexing of single photon streams. These are the three most active areas that researchers consolidate toward the goal of quantum-dot-based single photon sources for multi-photon applications.

As for any single quantum emitter, an important consideration for the practical use of quantum dots is the availability and efficacy of excitation methods. In many cases, incoherent and even direct electrical excitation pathways exist.[2] For cases where no specific properties of the emitted single photons are required, e. g. in some quantum key distribution protocols, these are sufficient and practical. However, more demanding applications, such as advanced quantum communication protocols (entanglement swapping, etc.), in optical quantum information processing, or in quantum sensing with multi-photon states, one often requires a certain level of coherence between the subsequently emitted photons or photons emitted by different sources, as will be discussed in Section 4.3.1. Depending on the requirements in photon properties in contrast to the complexity involved, the choice of excitation method varies.

Another basic feature of light–matter interaction is that it depends both on the internal composition and structure of the emitter as well as on the photonic environment surrounding it. Often it is easier to design and control the latter rather than the former, for example, by hosting the emitter in a purposefully designed optical cavity or waveguide structure. The optical modes supported or suppressed by the electromagnetic environment will help select individual emitters that happen to be resonant, favor or suppress certain transitions, modify the lifetime of an excited state, and/or change the spectral properties of the emitted light compared to the one of a bare emitter. In Section 4.3.2, we discuss this in detail.

While the interference of single photons emitted from different quantum dots has been realized, for a realistic and easier yet high-fidelity demonstration, the only

---

2 Because direct electrical excitation of a quantum dot would, of course, be ideal for applications it has seen much interest [66–70]. A critical advantage over optical excitation is that the emitted photon rate is no longer limited by the repetition rate of a pump laser, and pump rates as high as 0.5 GHz have been achieved [67]. Controlling the energy of charge carriers via electrical method is challenging however, and coherent electrical excitation of quantum dot exciton has not been achieved directly yet. A way out may be to electrically drive an on-chip nanolaser to resonantly excite quantum dots [71, 72].

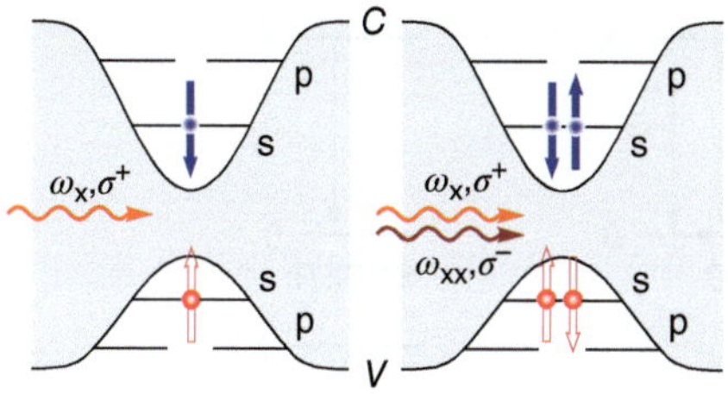

Figure 4.7 Most quantum dots can be described by finite potential wells in the conduction and valence bands, which lead to quantized levels for electrons and holes. Many electron and hole configurations are possible, but the two predominant, optically accessible, neutral states are the exciton (X, 1e−1h) and the biexciton (XX, 2e−2h). The exciton has two possible spin configurations with the electron spin ($\pm 1/2$) antiparallel to the (heavy) hole spin ($\pm 3/2$). The biexciton has a singlet electron pair and a singlet hole pair.

way forward appears to be to use subsequently emitted single photons from a single quantum dot. Demultiplexing re-routes these photons so that they simultaneously enter the multi-photon experiment. Approaches to this technique are discussed in Section 4.3.3.

Quantum dots are epitaxially grown semiconductor nanocrystals that are surrounded by higher bandgap materials with the appropriate band alignment, so that electrons and holes are confined in all three dimensions. Figure 4.7 illustrates the energy levels and confined carrier states of the so-called neutral exciton and biexciton states. These are, in most general cases, the target states to generate single photons or entangled photon pairs respectively, through spontaneous decay into the ground state. In addition to these neutral states, there are charged states with an unequal occupation of electrons and holes. For some purposes, it may be advantageous to operate with these charged states, especially when using a single electron or hole to store quantum information in its spin. Another group of states is optically dark, because these states have optically inaccessible angular momentum configurations. For example, the dark neutral exciton has spin ± 2. Generally the lifetime of dark states in quantum dots is orders of magnitude longer than that of bright ones, therefore they may be suitable for storing quantum information, or to act as metastable states for more elaborate state preparation schemes [73].

4.3.1 Advanced Excitation Schemes

To ensure high-efficiency generation of single photons with optimal properties – such as spectral purity and indistinguishability, it turns out that the excited state of the quantum dot must be prepared with near-unity efficiency to avoid any dissipation which inevitably leads to noise. Figure 4.8 sketches an overview of various excitation methods commonly in use. Here we emphasize the advanced methods that bypass the inherent challenges or limitations of the basic schemes. For instance, the incoherent above-band excitation schemes are less-selective and offer poor photon properties, while resonant excitation of exciton demands sensitive polarization filtering.

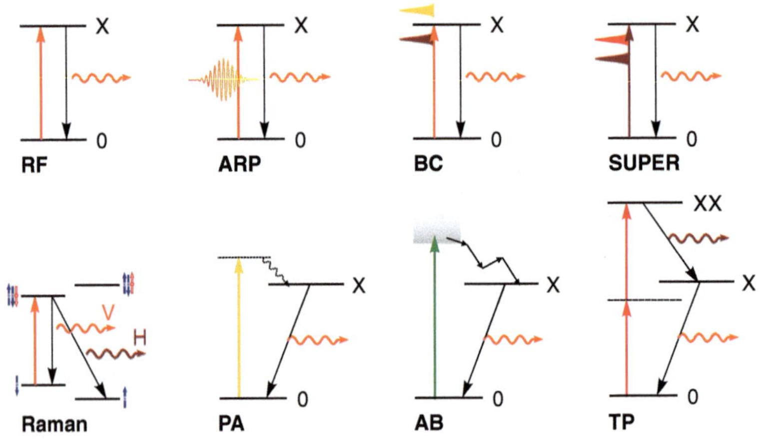

Figure 4.8 Except for the Raman scheme, the most prominent (RF, resonance fluorescence; ARP, adiabatic rapid passage; BC, bichromatic; SUPER, swing up of quantum emitter population; PA, phonon assisted; AB, above band; TP, two-photon resonant) optical excitation schemes for single photon generation from quantum dots all start from the ground state (0) and target the exciton (X) or biexciton (XX) states. The Raman scheme on the other hand couples charged states, here a single electron state with a trion state.

Because quantum dots are typically grown in a solid state environment, the coupling of confined carriers to the surrounding lattice atoms offers the possibility of off-resonant, **phonon-assisted** excitation of excitonic [74] and biexcitonic states [75]. Generally speaking, phonon-mediated interactions in quantum dots can occur in two ways. Firstly, there are acoustic phonons that couple to the occupation of quantum dot states (and not transitions) resulting in pure dephasing processes, for example, damping of Rabi oscillations. Longitudinal phonons, in contrast, provide efficient relaxation channels to the fundamental resonances in quantum dots. Here, one employs blue-detuned laser pulses (typically 2–3 meV) to populate the excitonic states following phonon emission, which is therefore incoherent. This has the advantage that the targeted excited state population is only weakly dependent on the detuning, provided it is in a range of sufficient phonon spectral density. The phonon-assisted excitation scheme has achieved remarkably high exciton ($\approx 80\%$ for pulse areas up to 15π) [74] and biexciton preparation fidelities ($\approx 95\%$ for pulse areas up to 3π) [75] for carefully chosen pulse durations and detunings.

In a special version of coherent excitation targeted at single-photon generation, quantum state transfer via spin-flip **Raman** transitions is an interesting choice [76, 77]. Consider a quantum dot charged with an excess electron or hole. In the absence of magnetic field, ground and excited (trion) states are degenerate, but under an external magnetic field they undergo a Zeeman splitting, depending on the electron–hole g factors. If this splitting is considerable large, each trion state forms an independent Λ system with the ground state, resulting in a double-Λ system. Following a two-pulse excitation addressing the trion spin states (first pulse, red detuned, initializes the electron spin-up, while the second pulse, blue detuned, recycles to spin-down), spin-flip Raman transitions can be induced. Here, the spin state of the electron that stores the quantum information in the quantum dot is mapped onto the polarization of the emitted photon, governed by optical selection rules; diagonal transitions are orthogonally polarized with respect to vertical transitions and have equal probabilities to occur. Consequently, one can opt for a clever polarization filtering of the spontaneous emission. In the decay, a photon is emitted whose polarization is entangled with the decay channel and thus the final state, if the path information can be erased by fast-timed detection. The scheme has the potential to achieve 100% brightness, as the emission can be spectrally filtered and is less sensitive to quantum dot size variations and excited state dephasing.

As mentioned earlier, the resonant excitation scheme is extremely sensitive to the quantum dot parameters and excitation conditions. To circumvent this, one resorts to robust schemes that are well known in atomic and molecular physics and magnetic resonance spectroscopy, like adiabatic rapid passage (ARP) [78]. In this technique, the instantaneous frequency of the excitation laser pulse is swept across (i.e. chirped) the quantum dot resonance to achieve a complete population inversion. If the frequency sweep rate is much slower than the transition (Rabi) frequency, the system adiabatically evolves from ground state to the target excited state via an avoided energy-level crossing. The electric field of a chirped laser pulses is described by

$$E(t) = \frac{1}{2}E_\mathrm{p}(t)\exp\left[-\mathrm{i}\left(\omega_\mathrm{l}t + \alpha t^2\right)\right]$$

where $E_\mathrm{p}(t)$, ω_l, and α denote the pulse envelope, the central laser frequency and the chirp parameter respectively. The introduction of α results not only in temporal stretching of the pulse from its Fourier limit, but also the frequency order in time: for $\alpha > 0$, red frequencies arrive before blue ones and vice versa if $\alpha < 0$. By virtue of the relation

$$\alpha = 2\phi'' / \left[\tau_0^4/(2\ln(2))^2 + \left(2\phi''\right)^2\right]$$

rendering α is often done via frequency domain (Fourier) shaping using a 4f pulse shaper, which manipulates the group delay dispersion (GDD) ϕ''. The frequency ordering governed by the sign of ϕ'' (or α) results in an interesting effect: it determines the inversion efficiency of the target state in the quantum dot. For positive GDD, the system traverses the lower-energy adiabatic branch, and hence is disengaged from the phonon influence, obtaining near-unity efficiency, while for negative GDD, preparation efficiencies suffer from phonon influences. ARP has been used to demonstrate high fidelity excitation of the exciton [79, 80] and biexciton states [81, 82]. Albeit requiring higher pulse energy than resonant excitation, ARP is practically immune to laser power fluctuations and offers a wider resonance energy range. The latter feature implies that one could afford large detuning and thus excite multiple quantum dots using the same laser pulse, for example, for producing frequency-multiplexed single photons.

Attempts to go beyond the coherent monochromatic excitation methods with their difficult filtering led to the use of **bichromatic** laser fields to achieve the resonant Rabi frequency in average. It relies on the argument that the combined effect of two pulses (red- and blue-) detuned from the resonant frequency becomes equivalent to a single resonant pulse with a modified envelope, as in

$$\begin{aligned} f(t) &= E_{\mathrm{p}}(t)\cos\left[\left(\omega_0 - \Delta\right)t)\right] + E_{\mathrm{p}}(t)\cos\left[\left(\omega_0 + \Delta\right)t\right] \\ &= 2E_{\mathrm{p}}(t)\cos(\Delta t)\cos\left(\omega_0 t\right) = E'_{\mathrm{p}}(t)\cos\left(\omega_0 t\right) \end{aligned}$$

where $E_{\mathrm{p}}(t)$ is the envelope of detuned pulses (chosen identical here) and $E'_{\mathrm{p}}(t)$ is the modified envelope, with ω_0 and Δ denoting the resonant frequency and the detunings (chosen identical here) respectively. In the nonlinear regime, coherent driving of a two-level system is possible without actually resonant light, as long as there exists a non-zero time-integrated Rabi frequency associated with the transition. Consequently, challenging polarization filtering can be averted. The required bichromatic laser pulses can be generated from two separate laser sources [83], or with a mechanical slit in a 4f pulse shaper [84, 85] or can be amplitude-shaped out of a broadband laser source [86]. These works demonstrated exciton state population close to 60% [84, 85], with single photon characteristics on par with the resonant schemes. This scheme requires higher pulse power and might possibly excite unwanted higher excited states, however. In a recent creative development, it was shown [86] that one could realize a "spectral hole" excitation as a combination of the bichromatic and ARP schemes to mitigate the sensitivity to excitation pulse parameters and enhance the preparation efficiency in bichromatic implementations. The process combines an excitation frequency spectrum given by

$$A(\omega) = 1 - \mathrm{e}^{-\ln(2)\left(\omega - \omega_0\right)^2/\delta^2}$$

with the GDD (described earlier) phase as in conventional ARP. Here, with positive GDD, one can suppress the phonon influence in the excitation and can outweigh the limitations in inversion efficiency.

In a surprising further development, it was shown that a coherent two level driving can also be realized through a gradual Swing-UP of the quantum EmitteR population (**SUPER**). The scheme relies on two coherently coupled, red detuned pulses (i.e. below the quantum dot absorption edge), which achieves a gradual rise in the exciton state population by modulating the Rabi frequency through the beating of the two interactions [87]. In the frequency domain, the excitation spectral response is given by

$$E(\omega) = A_1 \exp\left(-\frac{\left(\omega - \omega_1\right)^2}{2\sigma_1^2}\right) + A_2 \exp\left(-\frac{\left(\omega - \omega_2\right)^2}{2\sigma_2^2}\right)$$

while the two detunings are defined as $\Delta_i = \hbar(\omega_i - \omega_0)$. If the difference between the two detunings coincides with the Rabi frequency, i.e. $|\Delta_2 - \Delta_1| = \hbar\Omega_1^{\mathrm{Rabi}}$, implying the condition $|\Delta_2| > 2|\Delta_1|$, the SUPER mechanism results in a complete population inversion of a quantum emitter. The proof-of-principle implementations of this scheme were demonstrated in [88], relying on amplitude shaping of a broadband laser source with spatial light modulator [88], or using multiple pulse shapers [89]. The authors demonstrated that the population fidelity matches the one achieved by TP excitation (see below) and the calculated single-photon purity matches with that under resonant excitation. Because it works below the absorption edge of the quantum emitter, no spurious resonant excitation of higher-lying states can occur, if it is applied to the fundamental (exciton) resonance. The SUPER scheme, therefore guarantees truly off-resonant, polarization-filter free, high-fidelity state preparation.

For the direct generation of entangled photon pairs one usually uses the biexciton–exciton cascade. Due to optical selection rules, coherent excitation of the biexciton proceeds via a two-photon excitation (TPE) process with the associated Rabi oscillation [90]. Due to its binding energy, the exciton-biexciton transition is energetically separated from the ground state to exciton one, which can therefore be spectrally selected. The photon pair emitted in the cascade can be entangled in polarization [91, 92] or in time-bins [93]. For polarization entanglement, one needs to eliminate or mitigate the so-called fine-structure splitting of the exciton, the methods for which are beyond the scope of this chapter.

The biexciton–exciton cascade results in a time jitter, which fundamentally limits the indistinguishability of the emitted photons [94]. To circumvent this problem, following the TPE of the biexciton, a subsequent **timed stimulating** pulse can selectively and coherently depopulate it to the desired exciton state as demonstrated by multiple research groups [95–97]. In addition to restoring the high indistinguishability [97], this technique also ensures a definite polarization of the emitted exciton photon [98]. Furthermore, the scheme also achieves a lower multiphoton error rate compared to TP excitation and the indistinguishability is on par with resonant excitation. Although the power requirements for the TP laser pulse scale quadratically, the stimulation pulse power scales only linearly [95].

In summary, while resonant excitation of exciton state in a quantum dot is conceptually simple, it comes with the requirement of polarization filtering of the scattered pump laser light, thus sacrificing 50% of the brightness. The alternative schemes that only require the simpler spectral filtering including phonon-assisted excitation [99] and bichromatic schemes [84, 85, 88] offer high preparation efficiency and single photon quality matching that of resonant excitation, albeit at much higher pulse powers. Regarding multiphoton experiments with their low coincidence count rates, an important consideration is stable long-term operation of a single-photon source. Thus the source needs to be resilient against power and/or spectral fluctuations of the exciting laser. Then one would opt for ARP [82], or further innovations that combine ARP with bichromatic schemes [86]. The same helps address the intrinsic, growth-induced variability in quantum dot properties. For example, in Ref. [100], the authors demonstrated the potential of pulse shaping techniques to achieve simultaneous rotation of different qubits. Similarly, one could, in principle, extend the ARP scheme toward simultaneous entanglement generation from a number of non-degenerate quantum emitters. In Table 4.1 we present an overview on the performance characteristics of various excitation schemes.

4.3.2 Engineering the Quantum Dot Environment

Once a suitable trade-off on the excited state preparation is fixed, en route to realizing high brightness quantum dot single- or multiphoton sources, one faces the

Table 4.1 Overview of different excitation schemes.

References	Excitation method	Purity	Indistinguish-ability	Preparation fidelity (%)
Wang et al. [101]	Resonant TPE XX	0.013(1)	0.71(1)	70
Wei et al. [97]	TPE + stimulation	0.002(1)	0.926(4)	
Koong et al. [84] and He et al. [85]	Bichromatic	0.012(1)	0.962(6)	60
Karli et al. [88]	SUPER	0.06(1)		
Thomas et al. [99]	Phonon assisted X	0.011(1)	0.909(4)	85
Reindl et al. [102]	Phonon assisted XX	0.023(7)	0.65(7)	80
Wei et al. [80]	ARP X	0.003(2)	0.979	
He et al. [77]	Raman X	0.01(1)	0.95(3)	

NB: For ARP-XX, one does not expect the preparation fidelity to be higher than in the resonant TPE of XX, due to the intrinsic limit [94]. There is no intrinsic limit for the preparation fidelity in resonant TPE of XX, ARP-XX, TPE+ timed stimulation, ARP, SUPER and bichromatic + ARP.

uphill task of ensuring the low-loss deployment of generated photons to the communication channel. The challenges here are two-fold: enhancing the photon rate and engineering the photon extraction efficiency.

The semiconductor material around a quantum dots causes the emitted light to undergo total internal reflection at the semiconductor–air interface. The large refractive index at this interface prevents photon extraction from the air side. For a critical angle of θ, the extraction efficiency to the air side is given by

$$(1 - \cos\theta)/2$$

which, for a quantum dot in bulk GaAs, with a critical angle of ca. 16° turns out to be about 2% [103]. Consequently, to increase the light extraction efficiency, one would tune the emission mode by controlling its angular distribution to match the external collection optics.

The traditional approach is to couple the quantum dot to a nanoantenna, that converts the incident field to the localized field; providing an effective route to channel photons to and from the quantum dot. When one also desires to modify the light-matter interaction of the quantum dot one embeds it in a nanocavity. In this case, the spontaneous lifetime of a quantum dot photon source is shortened and thus the emission rate and **brightness** are enhanced by the Purcell factor F_P, given by

$$F_P = \frac{\gamma_C}{\gamma_0} = \frac{3}{4\pi^2}\left(\frac{\lambda_C}{n_{\text{eff}}}\right)^3\left(\frac{Q}{V}\right)$$

where γ_C and γ_0 are the cavity and quantum dot radiative rates, respectively, λ_C is the free space wavelength of the emitter, n_{eff} is the refractive index of the surroundings, Q denotes the nanocavity Q-factor and V its mode volume. Therefore, one seeks a design with higher Q and lower V. Plasmonic metal nanoantennas offer low Q (≈ 20) but also very small V theoretically resulting in huge F_P. However, metallic nanocavities suffer from ohmic losses, often only work for a particular polarization and are limited in terms of on-chip functionality. High-refractive-index dielectric nanoantennas are more attractive; they have higher Q due to lower material losses in the near infrared regime, and lower mode volumes are achieved by clever cavity designs.

Usually, one also makes a choice between narrowband and broadband cavity designs, with corresponding complexity level and fabrication tolerances. A **broadband** cavity affords more spectroscopic flexibility as, for example, the simultaneous enhancement of both photons of the biexciton-exciton cascade. Furthermore it reduces the requirement of precise tuning and stabilization of the cavity resonance to the quantum dot emission wavelength. In turn, it will not be as effective in suppressing phonon sidebands. The simplest and most straightforward designs for **broadband** operation include **planar dielectric microcavities** [104], **circular Bragg gratings** (**CBGs**) [105, 106], and **photonic crystal waveguides** [107, 108] (see Figure 4.9, which provide a modest F_P ($\approx$4, with theoretical limit 25), and estimated Q-factors of $\approx$150. Such cavity designs also leave the choice of excitation method flexible; however, large cavity sizes ($\sim$10 μm) reduce the coupling efficiency to the external optics.

A **narrowband** cavity, on the other hand focuses strictly on engineering high brightness for single photons at a particular wavelength, thereby effectively suppressing the phonon sidebands. Common design strategies in this direction include **micropillar** cavities (Q-factors $\approx$ 6000; F_P between 4 and 6, see Figure 4.9) [109], or open Fabry–Perot cavities (Q-factors $\approx$ 12 000; F_P between 10 and 12) [110]. The strict requirement of cavity mode - quantum dot spectral overlap is a major drawback when attempting advanced excitation methods that involve excitation laser detuning (for example SUPER and bichromatic) or multiphoton processes

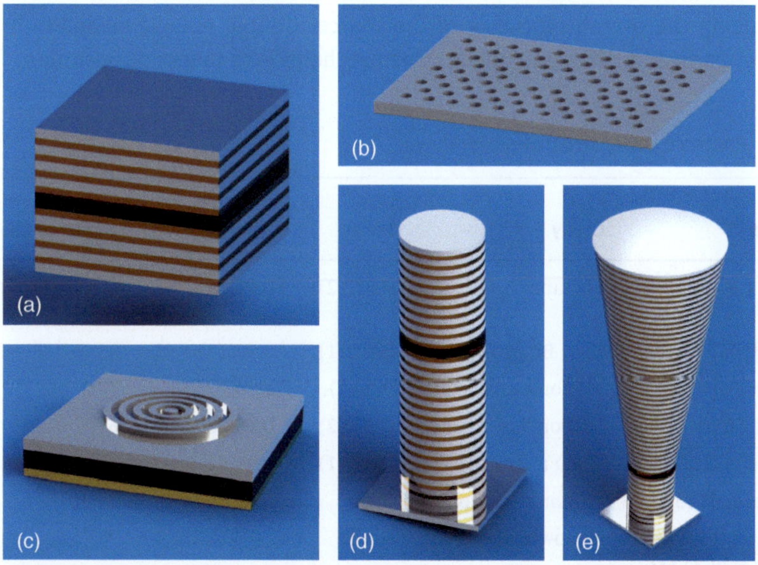

Figure 4.9 Visualizations of the most prominent monolithic microcavities used for quantum dots: (a) planar, (b) photonic crystal, (c) bulls eye (circular Bragg grating), (d) micro-pillar, and (e) photonic trumpet.

(for example TPE, ARP, and phonon-assisted schemes that target the biexciton state). For a detailed overview on micropillar cavities, we refer the reader to [111].

A further challenge is to couple the emitted single photons to single mode fiber by engineering the **extraction efficiency** of the photons. Quantum dot samples are often operated with a solid immersion lens (SIL) to modify the angular divergence. Several other promising design strategies to tailor the divergence angle include nanowires and nanoposts [112, 113], microlenses [114], the photonic trumpet [115], "hourglass" [116], nanorings [117], and many more. Such designs boost the extraction efficiency and often improve the free-space collection, too. The ultimate goal, however, would be a compact, robust fiber-coupled single photon source that can be operated without any optical alignment. Solutions based on tapered microfibers [118, 119] and gradient index lenses [120] are under development. In Table 4.2 we present an overview of the performance characteristics of few selected photonic cavities.

For high-multiphoton coincidence rates, one would like to start with the brightest single-photon source possible. Compared to many other quantum emitters, quantum dots exhibit some of the fastest radiative decay times, where, for example, lifetimes of less than 200ps have been observed for the GaAs/AlGaAs system [121]. By coupling these quantum dots to CBG cavities [106], the lifetimes can be further reduced, thereby promising photon rates well above few tens of GHz. Another issue regarding the rate is the so-called blinking. For quantum dots embedded in a semiconductor matrix, the diffusion of carriers from the vicinity (and vice versa) results in the formation of undesired (e.g. charged) states, effectively rendering the quantum dot dark for the chosen excitation scheme. This results in a random on-off modulation of the emitted light on relatively long time scales of micro- to milliseconds, reducing the overall efficiency of the source. Charge tunable samples, for example, by embedding in a p-i-n diode structure, helps long timescale operation.

Table 4.2 Photonic cavity structures overview.

References	Cavity	Purity	Indistinguish-ability	First lens efficiency (%)
He et al. [104] and Chen et al. [121]	Planar cavity	0.012(2)	0.91(2)	6
Liu et al. [106]	Circ. Bragg Grating	0.001(1)	0.901(3)	85
Madsen et al. [122]	Photonic crystal	0.03(4)	0.70	44.3
Ding et al. [109]	Micropillar	0.009(2)	0.964(30)	66
Tomm et al. [110]	Open microcavity	0.021(1)	0.913(1)	
Munsch et al. [115]	Photonic trumpet	0.31		75(10)
Dalacu et al. [112] and Laferrière et al. [123]	Nanowires	0.005		23
Schöll et al. [124]	SIL on planar	0.0025(2)	0.9	20(3)
Gschrey et al. [114]	Microlenses	0.01	0.76(4)	19(2)

4.3.3 Demultiplexing

There are two possible routes to produce multi-photon states from quantum dots: Source multiplexing, i.e. several quantum dots emitting in parallel, or temporal-to-spatial demultiplexing of the photons from a single quantum dot, i.e. routing the subsequently emitted photons into separate spatial modes. From a technology perspective, these two routes hold very similar challenges as spatial and temporal multiplexing of nonlinear-optical sources (see Section 4.2.6).

Source multiplexing does not require active switching components, thus avoiding additional loss, but has to overcome the substantial challenges of inter-dot variability as well as of fluctuating emission properties: Not only do the central wavelength and natural linewidth of the different emitters have to be matched, but also noise processes (dephasing and spectral diffusion) have to be suppressed. Substantial progress has been made in this direction by optimized material purities as well as spectral matching and noise suppression via electric gating, reaching an indistinguishability of 93% between two photons from quantum dots in separate cryostats [125]. Other methods include strain tuning via micrometer-sized piezoelectric films [126] or laser-machined piezoelectric substrates [127]. Upscaling these method to multiple photons would require either enormous resources (one cryostat per photon) or very advanced sample and optics designs to address and control multiple quantum dots in a single cryostat.

In contrast, the demultiplexing strategy requires only a single quantum dot at the cost of additional loss due to the switching. The basic idea is to send a stream of subsequently emitted photons through a routing setup, which feeds the incoming photons into separate optical delays, whose length is matched to the pulse spacing of the emitter, see Figure 4.10a. It benefits from the fact that the properties of a single emitter typically vary a lot less over the comparatively short time scale of a few emission cycles than the properties of different emitters vary within an ensemble. The conceptually most simple form is passive demultiplexing: A cascade of $N-1$ beam splitters is employed to parallelize N subsequently emitted photons into N spatial modes [128]. This is clearly unscalable as the success probability of routing each photon into the correct channel drops as $(1/N)^N$. Active demultiplexing, on the other hand, employs active optical components in a switching tree to deterministically route individual photons to their dedicated target mode. If the probability to route a photon successfully[3] into the correct target mode is η_r, the overall N-photon rate will scale as η_r^N/N. The two main challenges here are to keep losses and routing errors low (i.e. η_r high), while at the same time being able to switch fast enough to route subsequent photons into separate modes (one-by-one routing). The latter constraint can be relaxed by burst routing, i.e. sending packets of several photons into the same channel (with correspondingly adjusted delays), such that photons from distant emission events form the multi-photon state. This allows using slower active components, but necessitates better long-term stability of the emitter, as the temporal separation between interfering photon increases [Figure 4.10b].

3 Due to the different delay lengths, routing errors, i.e. photons routed to wrong output modes, are equivalent to loss but do not compromise the output state.

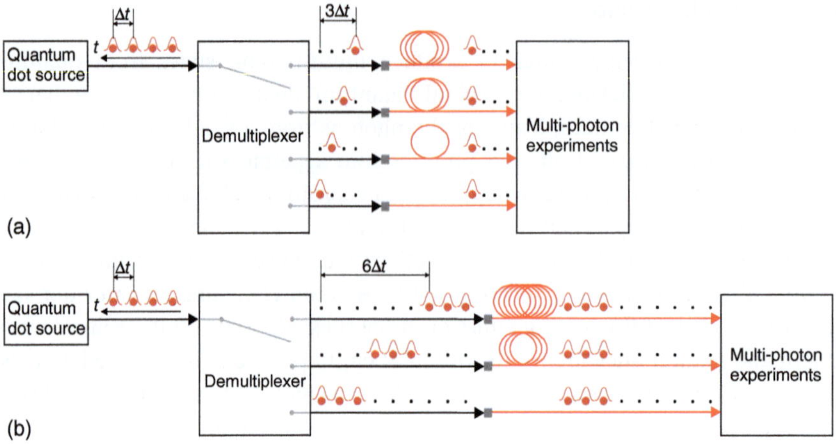

Figure 4.10 Active temporal-to-spatial demultiplexing of photons from a quantum dot emitted with pulse spacing Δt. (a) One-by-one routing: Each subsequent photon gets routed to a different channel, necessitating fast switching (up to rate $(2\Delta t)^{-1}$ in current implementations). The maximum emission time separation of interfering photons is $(N-1)\Delta t$ for N demultiplexing modes. (b) Burst routing: Photons get routed in packages of size b (here $b = 3$ is shown). Slower switches suffice, but the emission time difference increases to $b(N-1)\Delta t$.

Fast switching (at 40MHz switching rate, which allows one-by-one routing of photons with 80MHz repetition rate) has been achieved in waveguide-integrated lithium niobate electro-optic modulators (EOMs), but only with high losses ($\eta_r = 23\%$) [129]. Low-loss active burst routing (1MHz switching rate) has been implemented with free-space EOMs [130, 131] as well as with acousto-optic modulators (AOM) [17]. A burst-routing experiment with EOMs driven in an intermediate frequency regime (10MHz) has also been reported [132]. Only recently, a low-loss one-by-one routing has been demonstrated [133]. In another context, active demultiplexing has been used to circumvent the slow recovery time of superconducting transition-edge photon detectors in a temporal-mode Gaussian boson sampling experiment [21].

Table 4.3 Active demultiplexing schemes in comparison.

References	Technology	Type	N	Switch rate (MHz)	η_r (%)
Lenzini et al. [129]	Integrated EOM	1 by 1	4	40	23
Hummel et al. [130]	Free-space broadband EOM	Burst	4	0.95	77
Wang et al. [131]	Free-space broadband EOM	Burst	20	0.76	84
Pont et al. [17]	AOM	Burst	4	1.4	65
Antón et al. [132]	Free-space resonant EOM	Burst	3	10	50
Münzberg et al. [133]	Free-space resonant EOM	1 by 1	4	38	79
Madsen et al. [21]	Free-space resonant EOM	1 by 1	16	3	85

Source: [21, 133].

Table 4.3 summarizes the performance parameters of these active demultiplexing schemes.

Quantum dots have now been developed as single-photon sources for over two decades and yet the worldwide research activity in the field appears to increase ever more. Quantum dots are arguably the best single-photon sources for many purposes. Virtually all applications be it in multi-photon state generation, optical quantum computing, quantum communication, or quantum sensing benefit from high photon emission rates, which is one of the strongest arguments in favor of quantum dots. Those properties that are not yet perfect are being overcome as discussed above, so that we see a very bright future for quantum dots.

References

1 Knill, E., Laflamme, R., and Milburn, G. (2001). A scheme for efficient quantum computation with linear optics. *Nature* 409: 46. https://doi.org/10.1038/35051009.

2 Lanyon, B.P., Weinhold, T.J., Langford, N.K. et al. (2007). Experimental demonstration of a compiled version of Shor's algorithm with quantum entanglement. *Physical Review Letters* 99: 250505. https://doi.org/10.1103/PhysRevLett.99.250505.

3 Raussendorf, R. and Briegel, H.J. (2001). A one-way quantum computer. *Physical Review Letters* 86: 5188. https://doi.org/10.1103/PhysRevLett.86.5188.

4 Walther, P., Resch, K.J., Rudolph, T. et al. (2005). Experimental one-way quantum computing. *Nature* 434: 169. https://doi.org/10.1038/nature03347.

5 Browne, D.E. and Rudolph, T. (2005). Resource-efficient linear optical quantum computation. *Physical Review Letters* 95: 010501. https://doi.org/10.1103/PhysRevLett.95.010501.

6 Bartolucci, S., Birchall, P., Bombin, H. et al. (2021). Fusion-based quantum computation. *arXiv:2101.09310* https://doi.org/10.48550/arXiv.2101.09310.

7 Briegel, H.-J., Dür, W., Cirac, J.I. et al. (1998). Quantum repeaters: the role of imperfect local operations in quantum communication. *Physical Review Letters* 81: 5932. https://doi.org/10.1103/PhysRevLett.81.5932.

8 Jennewein, T., Weihs, G., Pan, J.-W. et al. (2002). Experimental nonlocality proof of quantum teleportation and entanglement swapping. *Physical Review Letters* 88: 017903. https://doi.org/10.1103/PhysRevLett.88.017903.

9 Aaronson, S. and Arkhipov, A. (2011). The computational complexity of linear optics. *Proceedings of the 43rd Annual ACM Symposium on Theory of Computing*, pp. 333–342.

10 Hamilton, C.S., Kruse, R., Sansoni, L. et al. (2017). Gaussian boson sampling. *Physical Review Letters* 119: 170501. https://doi.org/10.1103/PhysRevLett.119.170501.

11 Zhong, H.-S., Wang, H., Deng, Y.-H. et al. (2020). Quantum computational advantage using photons. *Science* 370: 1460. https://doi.org/10.1126/science.abe8770.

12 Arrazola, J.M., Bergholm, V., Brádler, K. et al. (2021). Quantum circuits with many photons on a programmable nanophotonic chip. *Nature* 591: 54. https://doi.org/10.1038/s41586-021-03202-1.

13 Dittel, C., Dufour, G., Weihs, G. et al. (2021). Wave-particle duality of many-body quantum states. *Physical Review X* 11: 031041. https://doi.org/10.1103/PhysRevX.11.031041.

14 Münzberg, J., Dittel, C., Lebugle, M. et al. (2021). Symmetry allows for distinguishability in totally destructive many-particle interference. *PRX Quantum* 2: 020326. https://doi.org/10.1103/PRXQuantum.2.020326.

15 Hong, C.K., Ou, Z.Y., and Mandel, L. (1987). Measurement of subpicosecond time intervals between two photons by interference. *Physical Review Letters* 59: 2044. https://doi.org/10.1103/PhysRevLett.59.2044.

16 Menssen, A.J., Jones, A.E., Metcalf, B.J. et al. (2017). Distinguishability and many-particle interference. *Physical Review Letters* 118: 153603. https://doi.org/10.1103/PhysRevLett.118.153603.

17 Pont, M., Albiero, R., Thomas, S.E. et al. (2022). Quantifying n-photon indistinguishability with a cyclic integrated interferometer. https://doi.org/10.1103/PhysRevX.12.031033.

18 Harris, S.E., Oshman, M.K., and Byer, R.L. (1967). Observation of tunable optical parametric fluorescence. *Physical Review Letters* 18: 732. https://doi.org/10.1103/PhysRevLett.18.732.

19 Burnham, D.C. and Weinberg, D.L. (1970). Observation of simultaneity in parametric production of optical photon pairs. *Physical Review Letters* 25: 84. https://doi.org/10.1103/PhysRevLett.25.84.

20 Giordmaine, J.A. and Miller, R.C. (1965). Tunable coherent parametric oscillation in $linbo_3$ at optical frequencies. *Physical Review Letters* 14: 973. https://doi.org/10.1103/PhysRevLett.14.973.

21 Madsen, L.S., Laudenbach, F., Askarani, M.F. et al. (2022). Quantum computational advantage with a programmable photonic processor. *Nature* 606: 75. https://doi.org/10.1038/s41586-022-04725-x.

22 Sharping, J.E., Lee, K.F., Foster, M.A. et al. (2006). Generation of correlated photons in nanoscale silicon waveguides. *Optics Express* 14: 12388. https://doi.org/10.1364/OE.14.012388.

23 Azzini, S., Grassani, D., Strain, M.J. et al. (2012). Ultra-low power generation of twin photons in a compact silicon ring resonator. *Optics Express* 20: 23100. https://doi.org/10.1364/OE.20.023100.

24 Silverstone, J.W., Bonneau, D., Ohira, K. et al. (2014). On-chip quantum interference between silicon photon-pair sources. *Nature Photonics* 8: 104. https://doi.org/10.1038/nphoton.2013.339.

25 Spring, J.B., Mennea, P.L., Metcalf, B.J. et al. (2017). Chip-based array of near-identical, pure, heralded single-photon sources. *Optica* 4: 90. https://doi.org/10.1364/OPTICA.4.000090.

26 Armstrong, J.A., Bloembergen, N., Ducuing, J. et al. (1962). Interactions between light waves in a nonlinear dielectric. *Physical Review* 127: 1918. https://doi.org/10.1103/PhysRev.127.1918.

27 Hum, D.S. and Fejer, M.M. (2007). Quasi-phasematching. *Comptes Rendus Physique* 8: 180. https://doi.org/10.1016/j.crhy.2006.10.022.

28 Christ, A., Fedrizzi, A., Hübel, H. et al. (2013). *Parametric Down-Conversion*, Chapter 11, vol. 45, 351–410. Academic Press. ISBN 1079-4042, https://doi.org/10.1016/B978-0-12-387695-9.00011-1.

29 Meraner, S., Chapman, R.J., Frick, S. et al. (2021). Approaching the Tsirelson bound with a Sagnac source of polarization-entangled photons. *SciPost Physics* 10: 17. https://doi.org/10.21468/SciPostPhys.10.1.017.

30 Brown, R.H. and Twiss, R.Q. (1956). Correlation between photons in two coherent beams of light. *Nature* 177: 27. https://doi.org/10.1038/177027a0.

31 Tichy, M.C. (2014). Interference of identical particles from entanglement to boson-sampling. *Journal of Physics B* 47: 103001. https://doi.org/10.1088/0953-4075/47/10/103001.

32 Dittel, C., Dufour, G., Walschaers, M. et al. (2018). Totally destructive many-particle interference. *Physical Review Letters* 120: 240404. https://doi.org/10.1103/PhysRevLett.120.240404.

33 Tichy, M.C., Lim, H.-T., Ra, Y.-S. et al. (2011). Four-photon indistinguishability transition. *Physical Review A* 83: 062111. https://doi.org/10.1103/PhysRevA.83.062111.

34 Ra, Y.-S., Tichy, M.C., Lim, H.-T. et al. (2013). Nonmonotonic quantum-to-classical transition in multiparticle interference. *Proceedings of the National Academy of Sciences of the United States of America* 110: 1227. https://doi.org/10.1073/pnas.1206910110.

35 Brod, D.J., Galv ao, E.F., Viggianiello, N. et al. (2019). Witnessing genuine multiphoton indistinguishability. *Physical Review Letters* 122: 063602. https://doi.org/10.1103/PhysRevLett.122.063602.

36 Jones, A.E., Menssen, A.J., Chrzanowski, H.M. et al. (2020). Multiparticle interference of pairwise distinguishable photons. *Physical Review Letters* 125: 123603. https://doi.org/10.1103/PhysRevLett.125.123603.

37 Dittel, C., Dufour, G., Walschaers, M. et al. (2018). Totally destructive interference for permutation-symmetric many-particle states. *Physical Review A* 97. https://doi.org/10.1103/PhysRevA.97.062116.

38 Tichy, M.C., Tiersch, M., Mintert, F. et al. (2012). Many-particle interference beyond many-boson and many-fermion statistics. *New Journal of Physics* 14: 093015. https://doi.org/10.1088/1367-2630/14/9/093015.

39 Christ, A., Laiho, K., Eckstein, A. et al. (2011). Probing multimode squeezing with correlation functions. *New Journal of Physics* 13: 033027. https://doi.org/10.1088/1367-2630/13/3/033027.

40 Blauensteiner, B., Herbauts, I., Bettelli, S. et al. (2009). Photon bunching in parametric down-conversion with continuous-wave excitation. *Physical Review A* 79. https://doi.org/10.1103/PhysRevA.79.063846.

41 Ou, Z.Y., Rhee, J.-K., and Wang, L.J. (1999). Photon bunching and multiphoton interference in parametric down-conversion. *Physical Review A* 60: 593. https://doi.org/10.1103/PhysRevA.60.593.

42 Meyer-Scott, E., Montaut, N., Tiedau, J. et al. (2017). Limits on the heralding efficiencies and spectral purities of spectrally filtered single photons from photon-pair sources. *Physical Review A* 95: 061803. https://doi.org/10.1103/PhysRevA.95.061803.

43 U'Ren, A.B., Silberhorn, C., Erdmann, R. et al. (2005). Generation of pure single photon wavepackets by conditional preparation based on spontaneous parametric downconversion. *Laser Physics* 15: 146. https://doi.org/10.1117/12.616191.

44 Mosley, P.J., Lundeen, J.S., Smith, B.J. et al. (2008). Heralded generation of ultrafast single photons in pure quantum states. *Physical Review Letters* 100: 133601. https://doi.org/10.1103/PhysRevLett.100.133601.

45 Fejer, M., Magel, G., Jundt, D. et al. (1992). Quasi-phase-matched second harmonic generation: tuning and tolerances. *IEEE Journal of Quantum Electronics* 28: 2631. https://doi.org/10.1109/3.161322.

46 Graffitti, F., Kelly-Massicotte, J., Fedrizzi, A. et al. (2018). Design considerations for high-purity heralded single-photon sources. *Physical Review A* 98: 053811. https://doi.org/10.1103/PhysRevA.98.053811.

47 Brańczyk, A.M., Fedrizzi, A., Stace, T.M. et al. (2011). Engineered optical nonlinearity for quantum light sources. *Optics Express* 19: 55. https://doi.org/10.1364/OE.19.000055.

48 Pickston, A., Graffitti, F., Barrow, P. et al. (2021). Optimised domain-engineered crystals for pure telecom photon sources. *Optics Express* 29: 6991. https://doi.org/10.1364/oe.416843.

49 Dosseva, A., Cincio, L., and Brańczyk, A.M. (2016). Shaping the joint spectrum of down-converted photons through optimized custom poling. *Physical Review A* 93: 013801. https://doi.org/10.1103/PhysRevA.93.013801.

50 Riedmatten, H.D., Scarani, V., Marcikic, I. et al. (2004). Two independent photon pairs versus four-photon entangled states in parametric down conversion. *Journal of Modern Optics* 51: 1637. https://doi.org/10.1080/09500340408232478.

51 Ou, Z.Y. (2006). Temporal distinguishability of an n-photon state and its characterization by quantum interference. *Physical Review A* 74: 063808. https://doi.org/10.1103/PhysRevA.74.063808.

52 Meyer-Scott, E., Silberhorn, C., and Migdall, A. (2020). Single-photon sources: approaching the ideal through multiplexing. *Review of Scientific Instruments* 91: 041101. https://doi.org/10.1063/5.0003320.

53 Gerry, C.C. and Knight, P.L. (2005). *Introductory Quantum Optics*. Cambridge University Press. ISBN 9780511791239. https://doi.org/10.1017/CBO9780511791239.

54 Stevens, M.J. (2013). Photon statistics, measurements, and measurements tools. In: *Single-Photon Generation and Detection, Experimental Methods in the Physical Sciences*, Chapter 2, vol. 45 (ed. A. Migdall, S.V. Polyakov, J. Fan et al.), 25–68. Academic Press. https://doi.org/10.1016/B978-0-12-387695-9.00002-0.

55 Migdall, A.L., Branning, D., and Castelletto, S. (2002). Tailoring single-photon and multiphoton probabilities of a single-photon on-demand source. *Physical Review A* 66: 053805. https://doi.org/10.1103/PhysRevA.66.053805.

56 Ma, X.-s., Zotter, S., Kofler, J. et al. (2011). Experimental generation of single photons via active multiplexing. *Physical Review A* 83: 043814. https://doi.org/10.1103/PhysRevA.83.043814.

57 Collins, M.J., Xiong, C., Rey, I.H. et al. (2013). Integrated spatial multiplexing of heralded single-photon sources. *Nature Communications* 4: 2582. https://doi.org/10.1038/ncomms3582.

58 Francis-Jones, R.J.A., Hoggarth, R.A., and Mosley, P.J. (2016). All-fiber multiplexed source of high-purity single photons. *Optica* 3: 1270. https://doi.org/10.1364/OPTICA.3.001270.

59 Pittman, T.B., Jacobs, B.C., and Franson, J.D. (2002). Single photons on pseudo-demand from stored parametric down-conversion. *Physical Review A* 66: 042303. https://doi.org/10.1103/PhysRevA.66.042303.

60 Kaneda, F., Christensen, B.G., Wong, J.J. et al. (2015). Time-multiplexed heralded single-photon source. *Optica* 2: 1010. https://doi.org/10.1364/OPTICA.2.001010.

61 Xiong, C., Zhang, X., Liu, Z. et al. (2016). Active temporal multiplexing of indistinguishable heralded single photons. *Nature Communications* 7: 10853. https://doi.org/10.1038/ncomms10853.

62 Joshi, C., Farsi, A., Clemmen, S. et al. (2018). Frequency multiplexing for quasi-deterministic heralded single-photon sources. *Nature Communications* 9: 847. https://doi.org/10.1038/s41467-018-03254-4.

63 Mendoza, G.J., Santagati, R., Munns, J. et al. (2016). Active temporal and spatial multiplexing of photons. *Optica* 3: 127. https://doi.org/10.1364/OPTICA.3.000127.

64 Kaneda, F. and Kwiat, P.G. (2019). High-efficiency single-photon generation via large-scale active time multiplexing. *Science Advances* 5: eaaw8586. https://doi.org/10.1126/sciadv.aaw8586.

65 Broome, M.A., Almeida, M.P., Fedrizzi, A. et al. (2011). Reducing multi-photon rates in pulsed down-conversion by temporal multiplexing. *Optics Express* 19: 22698. https://doi.org/10.1364/OE.19.022698.

66 Yuan, Z., Kardynal, B.E., Stevenson, R.M. et al. (2002). Electrically driven single-photon source. *Science* 295: 102. https://doi.org/10.1126/science.1066790.

67 Ellis, D.J., Bennett, A.J., Dewhurst, S.J. et al. (2008). Cavity-enhanced radiative emission rate in a single-photon-emitting diode operating at 0.5 GHz. *New Journal of Physics* 10: 043035. https://doi.org/10.1088/1367-2630/10/4/043035.

68 Heindel, T., Schneider, C., Lermer, M. et al. (2010). Electrically driven quantum dot-micropillar single photon source with 34% overall efficiency. *Applied Physics Letters* 96: 011107. https://doi.org/10.1063/1.3284514.

69 Patel, R.B., Bennett, A.J., Farrer, I. et al. (2010). Two-photon interference of the emission from electrically tunable remote quantum dots. *Nature Photonics* 4: 632. https://doi.org/10.1038/nphoton.2010.161.

70 Zrenner, A., Beham, E., Stufler, S. et al. (2002). Coherent properties of a two-level system based on a quantum-dot photodiode. *Nature* 418: 612. https://doi.org/10.1038/nature00912.

71 Munnelly, P., Heindel, T., Thoma, A. et al. (2017). Electrically tunable single-photon source triggered by a monolithically integrated quantum dot microlaser. *ACS Photonics* 4: 790. https://doi.org/10.1021/acsphotonics.7b00119.

72 Kreinberg, S., Grbesˇ ić, T., Strauß, M. et al. (2018). Quantum-optical spectroscopy of a two-level system using an electrically driven micropillar laser as a resonant excitation source. *Light: Science & Applications* 7: 1. https://doi.org/10.1038/s41377-018-0045-6.

73 Lüker, S., Kuhn, T., and Reiter, D. (2015). Direct optical state preparation of the dark exciton in a quantum dot. *Physical Review B* 92: 201305. https://doi.org/10.1103/physrevb.92.201305.

74 Ardelt, P.-L., Hanschke, L., Fischer, K.A. et al. (2014). Dissipative preparation of the exciton and biexciton in self-assembled quantum dots on picosecond time scales. *Physical Review B* 90: 241404. https://doi.org/10.1103/physrevb.90.241404.

75 Bounouar, S., Müller, M., Barth, A.M. et al. (2015). Phonon-assisted robust and deterministic two-photon biexciton preparation in a quantum dot. *Physical Review B* 91: 161302. https://doi.org/10.1103/physrevb.91.161302.

76 Fernandez, G., Volz, T., Desbuquois, R. et al. (2009). Optically tunable spontaneous Raman fluorescence from a single self-assembled InGaAs quantum dot. *Physical Review Letters* 103: 087406. https://doi.org/10.1103/physrevlett.103.087406.

77 He, Y., He, Y.-M., Wei, Y.-J. et al. (2013). Indistinguishable tunable single photons emitted by spin-flip Raman transitions in InGaAs quantum dots. *Physical Review Letters* 111: 237403. https://doi.org/10.1103/PhysRevLett.111.237403.

78 Loy, M.M. (1974). Observation of population inversion by optical adiabatic rapid passage. *Physical Review Letters* 32: 814. https://doi.org/10.1103/physrevlett.32.814.

79 Wu, Y., Piper, I.M., Ediger, M. et al. (2011). Population inversion in a single InGaAs quantum dot using the method of adiabatic rapid passage. *Physical Review Letters* 106: 67401. https://doi.org/10.1103/PhysRevLett.106.067401.

80 Wei, Y.-J., He, Y.-M., Chen, M.-C. et al. (2014). Deterministic and robust generation of single photons from a single quantum dot with 99.5% indistinguishability using adiabatic rapid passage. *Nano Letters* 14: 6515. https://doi.org/10.1021/nl503081n.

81 Debnath, A., Meier, C., Chatel, B. et al. (2013). High-fidelity biexciton generation in quantum dots by chirped laser pulses. *Physical Review B* 88: 201305. https://doi.org/10.1103/PhysRevB.88.201305.

82 Kaldewey, T., Lüker, S., Kuhlmann, A.V. et al. (2017). Coherent and robust high-fidelity generation of a biexciton in a quantum dot by rapid adiabatic passage. *Physical Review B* 95: 161302(R). https://doi.org/10.1103/PhysRevB.95.161302.

83 Peiris, M., Konthasinghe, K., Yu, Y. et al. (2014). Bichromatic resonant light scattering from a quantum dot. *Physical Review B* 89: 155305. https://doi.org/10.1103/PhysRevB.89.155305.

84 Koong, Z.X., Scerri, E., Rambach, M. et al. (2021). Coherent dynamics in quantum emitters under dichromatic excitation. *Physical Review Letters* 126: 47403. https://doi.org/10.1103/PhysRevLett.126.047403.

85 He, Y.M., Wang, H., Wang, C. et al. (2019). Coherently driving a single quantum two-level system with dichromatic laser pulses. *Nature Physics* 15: 941. https://doi.org/10.1038/s41567-019-0585-6.

86 Wilbur, G.R., Binai-Motlagh, A., Clarke, A. et al. (2022). Notch-filtered adiabatic rapid passage for optically driven quantum light sources. *APL Photonics* 7 (11). https://aip.scitation.org/doi/full/10.1063/5.0090048.

87 Bracht, T.K., Cosacchi, M., Seidelmann, T. et al. (2021). Swing-up of quantum emitter population using detuned pulses. *PRX Quantum* 2: 40354. https://doi.org/10.1103/prxquantum.2.040354.

88 Karli, Y., Kappe, F., Remesh, V. et al. (2022). Super scheme in action: experimental demonstration of red-detuned excitation of a quantum dot. *Nano Letters* **22**: 6567 https://doi.org/10.1021/acs.nanolett.2c01783.

89 Boos, K., Sbresny, F., Kim, S.K. et al. (2022). Coherent dynamics of the swing-up excitation technique. Preprint arXiv:2211.14289.

90 Stufler, S., Machnikowski, P., Ester, P. et al. (2006). Two-photon Rabi oscillations in a single $\text{In}_x\ \text{Ga}_1 - x$ As / GaAs quantum dot. *Physical Review B* 73: 125304. URL https://doi.org/10.1103/physrevb.73.125304.

91 Benson, O., Santori, C., Pelton, M. et al. (2000). Regulated and entangled photons from a single quantum dot. *Physical Review Letters* 84: 2513. https://doi.org/10.1103/physrevlett.84.2513.

92 Akopian, N., Lindner, N., Poem, E. et al. (2006). Entangled photon pairs from semiconductor quantum dots. *Physical Review Letters* 96: 130501. https://doi.org/10.1103/physrevlett.96.130501.

93 Jayakumar, H., Predojević, A., Kauten, T. et al. (2014). Time-bin entangled photons from a quantum dot. *Nature Communications* 5: 1. https://doi.org/10.1038/ncomms5251.

94 Schöll, E., Schweickert, L., Hanschke, L. et al. (2020). Crux of using the cascaded emission of a three-level quantum ladder system to generate indistinguishable photons. *Physical Review Letters* 125: 233605. https://doi.org/10.1103/physrevlett.125.233605.

95 Sbresny, F., Hanschke, L., Schöll, E. et al. (2022). Stimulated generation of indistinguishable single photons from a quantum ladder system. *Physical Review Letters* 128: 093603. https://doi.org/10.1103/physrevlett.128.093603.

96 Yan, J., Liu, S., Lin, X. et al. (2022). Double-pulse generation of indistinguishable single photons with optically controlled polarization. *Nano Letters* 22: 1483. https://doi.org/10.1021/acs.nanolett.1c03543.

97 Wei, Y., Liu, S., Li, X. et al. (2022). Tailoring solid-state single-photon sources with stimulated emissions. *Nature Nanotechnology* 17: 470. https://doi.org/10.1038/s41565-022-01092-6.

98 Akimov, I., Andrews, J., and Henneberger, F. (2006). Stimulated emission from the biexciton in a single self-assembled II-VI quantum dot. *Physical Review Letters* 96: 067401. https://doi.org/10.1103/physrevlett.96.067401.

99 Thomas, S., Billard, M., Coste, N. et al. (2021). Bright polarized single-photon source based on a linear dipole. *Physical Review Letters* 126: 233601. https://doi.org/10.1103/physrevlett.126.233601.

100 Gamouras, A., Mathew, R., Freisem, S. et al. (2013). Simultaneous deterministic control of distant qubits in two semiconductor quantum dots. *Nano Letters* 13: 4666. https://doi.org/10.1021/nl4018176.

101 Wang, H., Hu, H., Chung, T.-H. et al. (2019). On-demand semiconductor source of entangled photons which simultaneously has high fidelity, efficiency, and indistinguishability. *Physical Review Letters* 122: 113602. https://doi.org/10.1103/physrevlett.122.113602.

102 Reindl, M., Jöns, K.D., Huber, D. et al. (2017). Phonon-assisted two-photon interference from remote quantum emitters. *Nano Letters* 17: 4090. https://doi.org/10.1021/acs.nanolett.7b00777.

103 Ma, Y., Kremer, P.E., and Gerardot, B.D. (2014). Efficient photon extraction from a quantum dot in a broad-band planar cavity antenna. *Journal of Applied Physics* 115: 023106. andhttps://doi.org/10.1063/1.4861723.

104 He, Y.-M., He, Y., Wei, Y.-J. et al. (2013). On-demand semiconductor single-photon source with near-unity indistinguishability. *Nature Nanotechnology* 8: 213. https://doi.org/10.1038/nnano.2012.262.

105 Davanco, M., Rakher, M.T., Schuh, D. et al. (2011). A circular dielectric grating for vertical extraction of single quantum dot emission. *Applied Physics Letters* 99: 041102. https://doi.org/10.1364/qels.2011.qwc5.

106 Liu, J., Su, R., Wei, Y. et al. (2019). A solid-state source of strongly entangled photon pairs with high brightness and indistinguishability. *Nature Nanotechnology* 14: 586. https://doi.org/10.1038/s41565-019-0435-9.

107 Englund, D., Fattal, D., Waks, E. et al. (2005). Controlling the spontaneous emission rate of single quantum dots in a two-dimensional photonic crystal. *Physical Review Letters* 95: 013904. https://doi.org/10.1103/physrevlett.95.013904.

108 Lund-Hansen, T., Stobbe, S., Julsgaard, B. et al. (2008). Experimental realization of highly efficient broadband coupling of single quantum dots to a photonic crystal waveguide. *Physical Review Letters* 101: 113903. https://doi.org/10.1103/physrevlett.101.113903.

109 Ding, X., He, Y., Duan, Z.-C. et al. (2016). On-demand single photons with high extraction efficiency and near-unity indistinguishability from a resonantly driven quantum dot in a micropillar. *Physical Review Letters* 116: 020401. https://doi.org/10.1103/physrevlett.116.020401.

110 Tomm, N., Javadi, A., Antoniadis, N.O. et al. (2021). A bright and fast source of coherent single photons. *Nature Nanotechnology* 16: 399. https://doi.org/10.1038/s41565-020-00831-x.

111 Reitzenstein, S. and Forchel, A. (2010). Quantum dot micropillars. *Journal of Physics D* 43: 033001. https://doi.org/10.1088/0022-3727/43/3/033001.

112 Dalacu, D., Mnaymneh, K., Lapointe, J. et al. (2012). Ultraclean emission from InAsP quantum dots in defect-free wurtzite InP nanowires. *Nano Letters* 12: 5919. https://doi.org/10.1021/nl303327h.

113 Michler, P., Kiraz, A., Becher, C. et al. (2000). A quantum dot single-photon turnstile device. *Science* 290: 2282. https://doi.org/10.1126/science.290.5500.2282.

114 Gschrey, M., Thoma, A., Schnauber, P. et al. (2015). Highly indistinguishable photons from deterministic quantum-dot microlenses utilizing three-dimensional in situ electron-beam lithography. *Nature Communications* 6: 1. https://doi.org/10.1038/ncomms8662.

115 Munsch, M., Malik, N.S., Dupuy, E. et al. (2013). Dielectric GaAs antenna ensuring an efficient broadband coupling between an InAs quantum dot and a gaussian optical beam. *Physical Review Letters* 110: 177402. https://doi.org/10.1103/physrevlett.110.177402.

116 Osterkryger, A.D., Claudon, J., Gérard, J.-M. et al. (2019). Photonic hourglass design for efficient quantum light emission. *Optics Letters* 44: 2617. https://doi.org/10.1364/ol.44.002617.

117 Haws, C., Perez, E., Davanco, M. et al. (2022). Broadband, efficient extraction of quantum light by a photonic device comprised of a metallic nano-ring and a gold back reflector. *Applied Physics Letters* 120: 081103.

118 Cadeddu, D., Teissier, J., Braakman, F.R. et al. (2016). A fiber-coupled quantum-dot on a photonic tip. *Applied Physics Letters* 108: 011112. https://doi.org/10.1063/1.4939264.

119 Schlehahn, A., Fischbach, S., Schmidt, R. et al. (2018). A stand-alone fiber-coupled single-photon source. *Scientific Reports* 8: 1. https://doi.org/10.1038/s41598-017-19049-4.

120 Northeast, D.B., Dalacu, D., Weber, J.F. et al. (2021). Optical fibre-based single photon source using InAsP quantum dot nanowires and gradient-index lens collection. *Scientific Reports* 11 (1). https://doi.org/10.1038/s41598-021-02287-y.

121 Chen, Y., Zopf, M., Keil, R. et al. (2018). Highly-efficient extraction of entangled photons from quantum dots using a broadband optical antenna. *Nature Communications* 9: 1. https://doi.org/10.1038/s41467-018-05456-2.

122 Madsen, K.H., Ates, S., Liu, J. et al. (2014). Efficient out-coupling of high-purity single photons from a coherent quantum dot in a photonic-crystal cavity. *Physical Review B* 90: 155303. https://doi.org/10.1103/physrevb.90.155303.

123 Laferrière, P., Yeung, E., Miron, I. et al. (2022). Unity yield of deterministically positioned quantum dot single photon sources. *Scientific Reports* 12: 1. https://doi.org/10.1038/s41598-022-10451-1.

124 Schöll, E., Hanschke, L., Schweickert, L. et al. (2019). Resonance fluorescence of GaAs quantum dots with near-unity photon indistinguishability. *Nano Letters* 19: 2404. https://doi.org/10.1117/12.2568638.

125 Zhai, L., Nguyen, G.N., Spinnler, C. et al. (2022). Quantum interference of identical photons from remote GaAs quantum dots. *Nature Nanotechnology* https://doi.org/10.1038/s41565-022-01131-2.

126 Chen, Y., Zhang, J., Zopf, M. et al. (2016). Wavelength-tunable entangled photons from silicon-integrated III-V quantum dots. *Nature Communications* 7: 1. https://doi.org/10.1038/ncomms10387.

127 Trotta, R., Martín-Sánchez, J., Wildmann, J.S. et al. (2016). Wavelength-tunable sources of entangled photons interfaced with atomic vapours. *Nature Communications* 7: 1. https://doi.org/10.1038/ncomms10375.

128 Loredo, J.C., Broome, M.A., Hilaire, P. et al. (2017). Bosonsampling with single-photon fock states from a bright solid-state source. *Physical Review Letters* 118: 130503. https://doi.org/10.1103/PhysRevLett.118.130503.

129 Lenzini, F., Haylock, B., Loredo, J.C. et al. (2017). Active demultiplexing of single-photons from a solid-state source. *Laser & Photonics Reviews* 11: 1600297. https://doi.org/10.1002/lpor.201600297.

130 Hummel, T., Ouellet-Plamondon, C., Ugur, E. et al. (2019). Efficient demultiplexed single-photon source with a quantum dot coupled to a nanophotonic waveguide. *Applied Physics Letters* 115: 021102. https://doi.org/10.1063/1.5096979.

131 Wang, H., Qin, J., Ding, X. et al. (2019). Boson sampling with 20 input photons and a 60-mode interferometer in a 10^{14}-dimensional hilbert space. *Physical Review Letters* 123: 250503. https://doi.org/10.1103/PhysRevLett.123.250503.

132 Antón, C., Loredo, J.C., Coppola, G. et al. (2019). Interfacing scalable photonic platforms: solid-state based multi-photon interference in a reconfigurable glass chip. *Optica* 6: 1471. https://doi.org/10.1364/OPTICA.6.001471.

133 Münzberg, J., Draxl, F., Covre da Silva, S.F. et al. (2022). Fast and efficient demultiplexing of single photons from a quantum dot with resonantly enhanced electro-optic modulators. *APL Photonics* 7: 070802. https://doi.org/10.1063/5.0091867.

5

Quantum Key Distribution Protocols

Álvaro Navarrete[1,2,3], Víctor Zapatero[1,2,3], and Marcos Curty[1,2,3]

[1] *Vigo Quantum Communication Center, University of Vigo, Estrada de Marcosende 89, E-36310, Vigo, Spain*
[2] *EI Telecomunicación, Dept. of Signal Theory and Communications, University of Vigo, Rúa Maxwell s/n, E-36310, Vigo, Spain*
[3] *atlanTTic Research Center, University of Vigo, Rúa Maxwell s/n, E-36310, Vigo, Spain*

5.1 Introduction

The ability to transmit confidential information between distant parties is a fundamental resource for multiple applications of our digital society. Open networks like Internet are typically protected with public-key cryptography, whose security relies on the difficulty of solving certain mathematical problems, such as, for instance, the problem of factoring large integers. However, with the advent of quantum computers, the security of most widely used public-key cryptographic schemes will fall apart.

It is necessary to develop and deploy quantum-safe cryptosystems able to protect our communications against the threat of quantum computers. For this purpose, there are two main approaches. The first one – called post-quantum cryptography – belongs to the family of public-key cryptography. The key idea is to employ mathematical problems that are believed to be resistant to quantum algorithms. This is essentially a software solution, and thus, it facilitates a smooth transition from current methods. The main drawback of post-quantum cryptography is, however, that its security has not been fully established yet, which makes it unsuitable for applications that require long-term security. The second approach consists of combining quantum key distribution (QKD) with conventional private-key cryptosystems like e.g. the one-time pad encryption scheme, whose security can be proven to be information-theoretic, i.e. it holds independently of the computational capabilities of a potential eavesdropper (typically called Eve).

The role of QKD is to securely distribute secret cryptographic keys between the legitimate users of the system (typically called Alice and Bob), which constitute the essential resource required by private-key cryptography. Importantly, the security of QKD relies on the laws of quantum physics. In particular, the quantum no-cloning theorem states that any attempt by Eve to copy (or access) information encoded in

Photonic Quantum Technologies: Science and Applications, First Edition. Edited by Mohamed Benyoucef.

non-orthogonal quantum states inevitably introduces detectable errors. Most importantly, based on the observed disturbance, quantum mechanics allows to quantify the amount of information that Eve could have learnt about the key, and this leaked information can be removed from the final key by simply using privacy amplification techniques.

Since its first theoretical proposal in the 1980s, much theoretical and experimental progress has been made in QKD. This includes the development of QKD systems able to operate at 10 GHz repetition rates, to deliver secret key rates above 10 Mb/s, and to achieve transmission distances beyond 800 km using optical fiber and around 7600 km using satellite-based quantum communication. Moreover, quantum and classical signals can be multiplexed in the same fiber nowadays, chip-based QKD is being developed, and quantum networks based on a trusted node architecture are being deployed worldwide.

In this chapter, we introduce the fundamentals of QKD in Section 5.2. Then, in Section 5.3, we discuss the security framework considered. This includes the assumptions usually employed in security proofs, the relevant security definition to guarantee universal composability, and a brief discussion of the main methodologies to prove the security of a QKD protocol. Section 5.4 addresses the main limitations of practical QKD due to device imperfections and introduces two key techniques – decoy-state QKD and measurement-device-independent QKD – to mitigate them. In this section, we also discuss the problem of implementation security. Then, in Section 5.5 we present the main components included in the optical layer of a QKD setup, making special emphasis on the light-sources and photo-detectors. Finally, in Section 5.6, we review the performance of state-of-the-art QKD demonstrations and summarize the chapter in Section 5.7.

5.2 Fundamentals of QKD

A QKD protocol has two phases: a quantum communication phase and a classical post-processing phase. In the former, Alice and Bob exchange and measure quantum signals by using a quantum channel potentially controlled by Eve. As a result, they obtain correlated classical data a and b, respectively, which follow a certain joint probability distribution $p_{a,b}$. The key advantage of QKD, when compared to classical key distribution methods, is that quantum mechanics allows Alice and Bob to upper bound Eve's information on a and b, based only on their knowledge about the signals sent and measurements performed.

Ideally, Alice and Bob would like to share Bell states of the form, say, $|\Phi^+\rangle_{AB} = \frac{1}{\sqrt{2}}(|00\rangle_{AB} + |11\rangle_{AB})$, where A and B denote the quantum systems they measure to obtain the respective outcomes a and b. Indeed, by measuring A and B in the Z basis $\{|0\rangle, |1\rangle\}$, a and b will be fully random but equal. Moreover, the monogamy of entanglement forbids Eve to be correlated with them. That is, $a = b$ represents a perfect secret key.

In practice, however, channel loss and noise, as well as Eve's possible intervention, might prevent Alice and Bob from sharing perfect Bell states $|\Phi^+\rangle_{AB}$. Importantly,

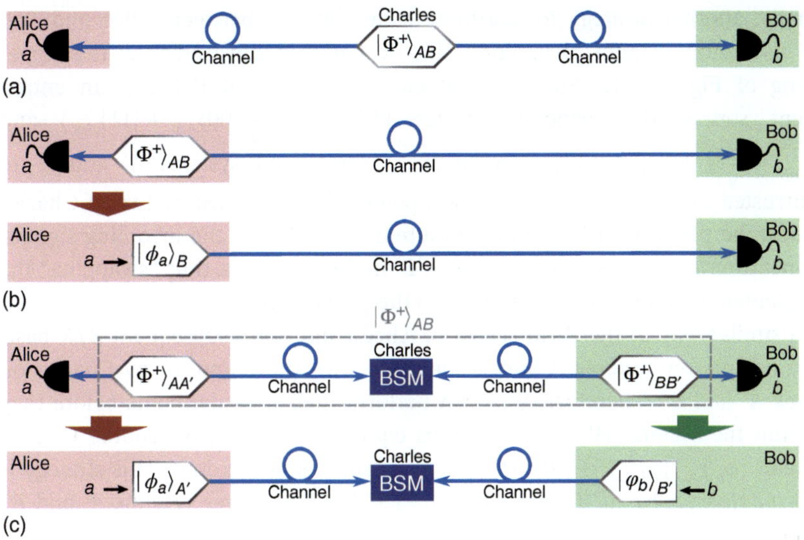

Figure 5.1 Strategies to establish quantum correlations between Alice and Bob. (a) Entanglement-based QKD scheme, (b) prepare-and-measure QKD scheme, and (c) interference-based QKD scheme.

they can verify if this is the case by measuring a random subset of the systems A and B in the X basis defined as $\{|+\rangle, |-\rangle\}$, with $|\pm\rangle = \frac{1}{\sqrt{2}}(|0\rangle \pm |1\rangle)$. If, according to the Z/X-basis measurement statistics, it turns out that the shared states are sufficiently close to $|\Phi^+\rangle_{AB}$, Alice and Bob can use local operations and public classical communication to distill a smaller fraction of perfect states $|\Phi^+\rangle_{AB}$, which essentially specifies the achievable secret key rate. This latter process is performed in the classical post-processing phase of the protocol.

There are various possible strategies to distribute states $|\Phi^+\rangle_{AB}$ between Alice and Bob. In *entanglement-based* (EB) schemes, it is an untrusted third party Charles – which could even be Eve – who does it. This is illustrated in Figure 5.1a.

Alternatively, say, Alice could prepare and share the states $|\Phi^+\rangle_{AB}$ with Bob. This is illustrated in the upper drawing of Figure 5.1b. One technical difference with respect to the EB scenario is that now, since Alice is trusted, it is guaranteed that she prepares the correct states. This means that the reduced density matrix $\rho_A = \text{Tr}_B(|\Phi^+\rangle\langle\Phi^+|_{AB})$ is fixed and cannot be modified by Eve. Most importantly, Alice's subsequent Z/X-basis measurements on the systems A commute with those applied to the systems B. Therefore, Alice could measure the systems A before she actually sends the systems B to Bob. This is equivalent to Alice randomly sending Bob the states $\{|0\rangle, |1\rangle, |+\rangle, |-\rangle\}$, a strategy called *prepare-and-measure* (P&M), which is illustrated in the lower drawing of Figure 5.1b.

In general, in a P&M scheme, Alice sends Bob states from a certain set $S_A = \{|\phi_a\rangle\}_a$ selected with a priori probabilities p_a. As explained, this is equivalent to preparing entangled states of the form $|\Psi\rangle_{AB} = \sum_a \sqrt{p_a}|a\rangle_A|\phi_a\rangle_B$ followed by measuring the systems A in the orthonormal basis $\{|a\rangle\}_a$.

Finally, another strategy to distribute states $|\Phi^+\rangle_{AB}$ between Alice and Bob is to use entanglement swapping. This scenario is illustrated in the upper drawing of Figure 5.1c. Suppose that each of Alice and Bob has an entanglement source that generates states $|\Phi^+\rangle_{AA'} = \frac{1}{\sqrt{2}}(|00\rangle_{AA'} + |11\rangle_{AA'})$ and $|\Phi^+\rangle_{BB'} = \frac{1}{\sqrt{2}}(|00\rangle_{BB'} + |11\rangle_{BB'})$, respectively. They send systems A' and B' to an untrusted intermediate node Charles, potentially controlled by Eve. If Charles is honest, he performs a Bell-state measurement (BSM) on the incoming signals. This measurement projects the systems $A'B'$ in the Bell basis, thus "swapping" the entanglement contained in AA' and BB' to the systems AB.[1]

In a similar way to the P&M scenario, Alice's and Bob's subsequent Z/X-basis measurements on systems A and B commute with those applied by Charles on systems A' and B'. Therefore, they can perform their measurements before they send him the systems A' and B'. This is equivalent to directly sending Charles the states $|\phi_a\rangle_{A'}, |\varphi_b\rangle_{B'} \in \{|0\rangle, |1\rangle, |+\rangle, |-\rangle\}$ selected at random. This strategy is called *interference-based* (IB) because Charles' BSM interferes systems A' and B', and it is illustrated in the lower drawing of Figure 5.1c.

Once the quantum communication phase ends, Alice and Bob post-process their observed classical data a and b to distill a secret key. As already mentioned, in an equivalent EB scenario, this essentially requires to first distill perfect states $|\Phi^+\rangle_{AB}$, and then measure systems A and B in the Z basis. Fortunately, this process can be implemented classically, by directly manipulating the raw bit strings a and b. It requires though to have access to a public *authenticated classical channel*, which can be achieved with information-theoretic security given that Alice and Bob pre-share a small initial secret key. After each QKD session, a small portion of the generated fresh secret key is used to replace the key consumed for authentication purposes.

In a nutshell, the classical post-processing phase runs as follows. First, Alice and Bob publicly announce certain information about their setting choices (e.g. their measurement basis selection for each system), and often also reveal a random subset of their raw bit strings a and b. This allows them to identify the presumably correlated data contributing to the secret keys – in a step called *sifting* – and to lower bound the extractable secret-key length – in a step called *parameter estimation* –. Next, they run an *error correction* protocol to reconcile their sifted keys, often followed by an *error verification* step to check the consistency of the resulting strings. Lastly, the parties execute a *privacy amplification* step (typically based on two-universal hashing) that "erases" Eve's knowledge about the keys by shrinking them into shorter final bit strings.

As an example, Box 1 includes a description of the famous BB84 protocol introduced by Bennett and Brassard in 1984 [1].

---

1 After performing a BSM on systems $A'B'$, one obtains $|\Omega\rangle_{AB} = \frac{{}_{A'B'}\langle\Omega|\Phi^+,\Phi^+\rangle_{AA'BB'}}{\sqrt{p_\Omega}}$, where $p_\Omega = \left\| {}_{A'B'}\langle\Omega|\Phi^+,\Phi^+\rangle_{AA'BB'} \right\|^2$ is the probability of finding a Bell state $|\Omega\rangle \in \{|\Phi^+\rangle, |\Phi^-\rangle, |\Psi^+\rangle, |\Psi^-\rangle\}$ and $|\Phi^+,\Phi^+\rangle \equiv |\Phi^+\rangle \otimes |\Phi^+\rangle$. Given the Bell state $|\Omega\rangle_{AB}$, it can be transformed into $|\Phi^+\rangle_{AB}$ by performing local operations.

BB84 Protocol 5.1

1. *Quantum communication*: In each round, Alice randomly sends Bob a state $\{|0\rangle, |1\rangle, |+\rangle, |-\rangle\}$. The states $|0\rangle$ and $|+\rangle$ ($|1\rangle$ and $|-\rangle$) encode the bit value 0 (1). Bob measures each received signal in the Z or X basis, which he also selects at random. This process is repeated N times.
2. *Sifting*: They broadcast their basis choices, and discard the data associated to basis mismatched events. The remaining bits form their sifted keys.
3. *Parameter estimation*: They reveal a random subset of their sifted keys to estimate the quantum bit-error rate in both bases. If it is too high, they abort the protocol.
4. *Information reconciliation*: They perform error correction (and possibly error verification) to obtain two identical sifted keys with high probability.
5. *Privacy amplification*: They apply privacy amplification to their corrected sifted keys to obtain two identical secret keys.

5.3 Security Framework

Before we introduce the formal definition of security in QKD, it is important to emphasize that a security proof of a QKD protocol relies on certain assumptions, besides the natural requirement that quantum mechanics is both correct and complete [2]. Note that the former refers to the correctness of its predictions, while the later excludes the possibility that an underlying theory more predictive than quantum mechanics exists, possibly allowing Eve to access more information on the quantum information carriers than we estimate. Notably though, it is known that theories of this kind necessarily exhibit very unphysical properties.

Importantly, there are a few further assumptions that typically must hold. Precisely, Alice and Bob must be able to faithfully generate random numbers – an assumption that is ubiquitous in all of cryptography – and have access to an authenticated classical channel, which has already been discussed in Section 5.2. In addition, since security proofs of QKD unavoidably apply to mathematical models of the physical devices involved, they are only fully significant if the behavior of the real devices sticks to these models. Therefore, it is often necessary to characterize the functioning of the QKD devices precisely. Moreover, one usually demands that no unwanted information leakage occurs through the boundaries of Alice's and Bob's labs. Keeping up with these requirements in practical systems is a challenging task mainly due to engineering limitations, a problem that is termed *implementation security*. We will come back to this in Section 5.4.4.

5.3.1 Security Definition

The security notion lies in the so-called real-world vs. ideal-world paradigm. In this formalism, the security of a real protocol is quantified by measuring its distinguishability to an ideal execution of the cryptographic task at hand [2]. In the context of

QKD, conditioned on the event of not aborting the protocol, Alice's and Bob's keys together with Eve's correlated quantum register – which contains all her information about the key – can be abstractly described by a classical-classical-quantum state of the form

$$\rho^{\top}_{ABE} = \frac{1}{1-p^{\perp}} \sum_{k_A,k_B} p_{k_A,k_B} |k_A, k_B\rangle \langle k_A, k_B|_{AB} \otimes \rho_E^{k_A,k_B} \tag{5.1}$$

where $p^{\perp}$ is the abortion probability, $k_A, k_B \in \{0,1\}^l$ (l being the finite length of the keys) and p_{k_A,k_B} is the joint probability mass function of k_A and k_B. In this context, the ideal classical-classical-quantum state pursued by QKD reads

$$\tilde{\rho}^{\top}_{ABE} = \frac{1}{2^l} \sum_k |k, k\rangle \langle k, k|_{AB} \otimes \rho^{\top}_E \tag{5.2}$$

where $k \in \{0,1\}^l$ and $\rho^{\top}_E$ is an arbitrary quantum state. That is, Alice's and Bob's keys are equal, uniformly random, and uncorrelated to the state of Eve's register.

Given $\rho^{\top}_{ABE}$ and $\tilde{\rho}^{\top}_{ABE}$, a QKD protocol is ϵ-secure according to the real-world vs. ideal-world paradigm if

$$\left(1-p^{\perp}\right) D\left(\rho^{\top}_{ABE}, \tilde{\rho}^{\top}_{ABE}\right) \leq \epsilon \tag{5.3}$$

where $D(\tau,\sigma) = \frac{1}{2}\,||\tau - \sigma||_1$ is the trace distance between the states τ and σ, with $||\rho||_1 = \sqrt{\rho\rho^{\dagger}}$. Notably, the trace distance admits the simple interpretation of a distinguishing advantage compared to a random guess: if a "distinguisher" is given either τ or σ with equal priors, its probability to correctly guess the delivered state with a quantum measurement is upper bounded by $(1 + D(\tau,\sigma))/2$.

Remarkably, the *correctness* criterion on the key – namely, that Alice's and Bob's keys be identical – and the *secrecy* criterion – namely, that Alice's and Bob's keys be uniformly distributed and uncorrelated to Eve's register – can be decoupled from each other, as we explain next.

Precisely, a QKD protocol is said to be ϵ_{cor}-correct if the final state $\rho^{\top}_{ABE}$ satisfies

$$(1-p^{\perp})\,\Pr\left(k^{\top}_A \neq k^{\top}_B\right) \leq \epsilon_{\text{cor}} \tag{5.4}$$

where $k^{\top}_A$ and $k^{\top}_B$ are Alice's and Bob's keys conditioned on not aborting the protocol.

Similarly, a protocol is ϵ_{sec}-secret if the reduced states $\rho^{\top}_{AE} = \text{Tr}_B\left[\rho^{\top}_{ABE}\right]$ and $\tilde{\rho}^{\top}_{AE} = \text{Tr}_B\left[\tilde{\rho}^{\top}_{ABE}\right] = 2^{-l}\sum_k |k\rangle\langle k|_A \otimes \rho^{\top}_E$ satisfy

$$(1-p^{\perp}) D\left(\rho^{\top}_{AE}, \tilde{\rho}^{\top}_{AE}\right) \leq \epsilon_{\text{sec}} \tag{5.5}$$

The feature that justifies the division of the security claim into two parts is that, if a QKD protocol is ϵ_{cor}-correct and ϵ_{sec}-secret, it is ϵ-secure for all $\epsilon \geq \epsilon_{\text{cor}} + \epsilon_{\text{sec}}$.

A crucial implication of the real-world vs. ideal-world paradigm is that it guarantees *universal composability* [3, 4], according to which the security parameters are additive under the composition of protocols. For instance, using an ϵ-secure key in an ϵ'-secure protocol results in an $(\epsilon + \epsilon')$-secure combination.

In practice, the attainability of certain security parameters ϵ_{cor} and ϵ_{sec} relies on the classical post-processing of the raw keys generated in the quantum communication phase of the protocol. In particular, upon success of the error verification step, a correctness claim of the form of Eq. (5.4) follows, and a given value of ϵ_{sec} can be reached by suitably choosing the output length of the privacy amplification step.

5.3.2 Security Proof Techniques

Among QKD security proofs, there exist two main distinct methodologies, which have recently been proven equivalent: the entropic approach [5] and the phase-error correction (PEC) approach [6].

The entropic approach relies on a fundamental theorem known as the quantum leftover hash lemma (LHL) [7], which is a generalization of a former LHL in classical cryptography. The quantum LHL establishes the amount of privacy amplification necessary to achieve ϵ_{sec}-secrecy against a quantum adversary. To be precise, it states that the extractable secret key length is determined by the *conditional smooth min-entropy* of the key given Eve's available information, an entropic measure of Eve's uncertainty. As a consequence, the goal of the security proofs in this approach is to lower bound the conditional smooth min-entropy of the key on the basis of the observables of the QKD protocol.

Alternatively, the second approach is based on the notion of PEC, an algorithm to correct the *phase errors* between Alice's and Bob's measurement outcomes in an EB formulation of the protocol. The phase errors are defined as the bit errors that would have arisen from measuring Alice's and Bob's systems associated to the key generation rounds in the complementary basis – e.g. the X basis – of the key generation basis – e.g. the Z basis –.

To be precise, in the PEC approach one constructs a virtual QKD protocol with a quantum PEC algorithm embedded inside, and the analysis of the security is reduced to setting an upper bound on the failure probability of the quantum PEC (by means, for instance, of a coding theorem). In this regard, a significant improvement of this approach was proposed by Koashi [8], and the resulting technique is sometimes called the *complementarity* approach. Notably, this simplified framework totally decouples privacy amplification from error correction, a main complication in the previous PEC-based security proofs.

When proving the security of a QKD protocol, it is also common to consider asymptotic analyses where an infinite number of signals is exchanged, or to restrict Eve to perform limited types of attacks. These assumptions are often invoked for mathematical simplicity. In particular, the consideration of an asymptotic regime allows to neglect the effect that statistical fluctuations have on the estimation of the extractable secret-key length. In a similar fashion, limiting Eve's attacking power allows to simplify certain technicalities in the security proofs. In this context, Eve's attacks are usually classified in three main groups: coherent, collective, and individual attacks. Coherent attacks correspond to the most general scenario, where Eve is presumed to interact arbitrarily and coherently with all the signals in the quantum channel. In the weaker setting of collective attacks, Eve's interaction factors between rounds as a product of identical and independent maps, each of them acting on the joint state of the quantum signal and a single-round probe system controlled by Eve. In spite of this tensor product structure, Eve can store all of her probes and perform a joint quantum measurement on them, possibly a posteriori of the public discussion between Alice and Bob. Lastly, a further relaxation of collective attacks are individual attacks, where not only Eve's interaction but also her measurements on the quantum probes factor from round to round, all of them being identical as well.

5.4 Practicalities

In the discussion of Section 5.2, we assume that Alice and Bob can distribute and measure qubits, which could be encoded in the state of single photons. However, high-rate on-demand single-photon sources at telecom wavelengths are not available yet. Therefore, most current implementations of QKD rely on laser sources instead, which might emit multi-photon pulses. Similarly, it has been shown that single-photon detectors (SPDs) are the Achilles' heel of QKD because their functioning could be controlled by Eve, for instance, by illuminating them with bright light. Here, we introduce two main solutions – decoy-state QKD and measurement-device-independent QKD – to overcome these limitations.

5.4.1 Decoy-State QKD

The state emitted by a phase-randomized laser source of intensity μ can be written as a classical mixture of Fock states $|n\rangle$, i.e.

$$\rho_\mu = \sum_{n=0}^{\infty} \mathrm{e}^{-\mu} \frac{\mu^n}{n!} |n\rangle \langle n| \tag{5.6}$$

These sources emit multi-photon pulses with a non-negligible probability $1 - \mathrm{e}^{-\mu} - \mu\mathrm{e}^{-\mu}$.

It is well known that multi-photon pulses may reveal Eve complete information about their encoded bits without introducing errors, thanks to the photon-number splitting (PNS) attack. In this attack, Eve first measures the number of photons contained in each pulse emitted by Alice with a quantum non-demolition measurement. If a pulse contains two or more photons, she extracts one of them and stores it in a quantum memory, while she resends Bob the remaining photons through a lossless quantum channel to guarantee that he detects them. On the other hand, Eve only sends him those single-photon pulses that are necessary to reproduce the expected detection statistics, while she blocks the rest. Once Alice reveals the encoding basis, Eve measures the stored photons and learns the encoded bits.

As a result of the PNS attack, the performance of QKD with laser sources is strongly limited, as only the bits encoded in single-photon pulses might be secure. Moreover, to estimate their detection probabilities and error rates one must consider the worst-case scenario in which all the multi-photon pulses are detected, and all the observed errors originate from the single photon pulses. This provides rather poor estimations for the quantities of interest.

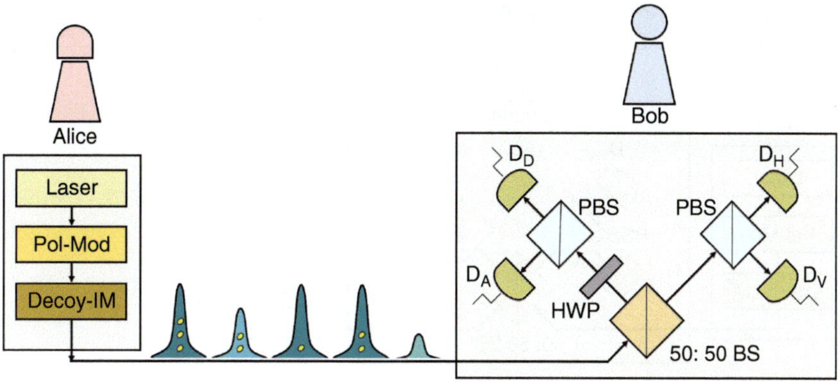

Figure 5.2 Illustration of a decoy-state QKD setup. Pol-Mod, polarization modulator; Decoy-IM, decoy intensity modulator; BS, beamsplitter; HWP, half-wave plate; PBS, polarizing beamsplitter.

However, these quantities can be estimated precisely with the decoy-state method [9–11], which can provide a performance similar to that of single-photon sources. The idea is rather simple. Suppose that Alice wants to estimate a probability $p_{b|n}$ that an emitted n-photon pulse produces a particular measurement outcome b at Bob's side. Then, she simply prepares phase-randomized laser pulses of various intensities, which she selects at random for each pulse from a predefined set $\{\mu_i\}_i$. In doing so, we have that the conditional probability that Bob observes the measurement outcome b given that Alice selected the intensity μ_i can be written as

$$Q_{b|\mu_i} = \sum_{n=0}^{\infty} \mathrm{e}^{-\mu_i} \frac{\mu_i^n}{n!} p_{b|n} \tag{5.7}$$

Importantly, this defines a set of linear equations for the different intensity values μ_i, where the quantities $Q_{b|\mu_i}$ are directly observed in the experiment and the probabilities $\mathrm{e}^{-\mu_i}\frac{\mu_i^n}{n!}$ are known. By solving it, one can obtain tight bounds on the desired quantities, say the probabilities $p_{b|1}$ associated to the single photon pulses. A typical decoy-state QKD setup is illustrated in Figure 5.2.

5.4.2 Measurement-Device-Independent QKD

Recent years have witnessed several quantum hacking attacks that modify the behavior of the SPDs at the QKD receiver to compromise the security of the secret key. For instance, in a *blinding attack* [12], Eve changes the operating mode of the SPDs from Geiger mode to linear mode by illuminating them with bright light, which renders them insensitive to weak light pulses. By sending Bob additional tailored bright pulses, Eve can decide which detector clicks each given time. If this strategy is combined with an intercept-and-resend attack in which Eve first measures out Alice's pulses with a setup similar to Bob's, Eve can force him to obtain exactly the same results as she does without introducing errors, and thus, no secure key can be distilled.

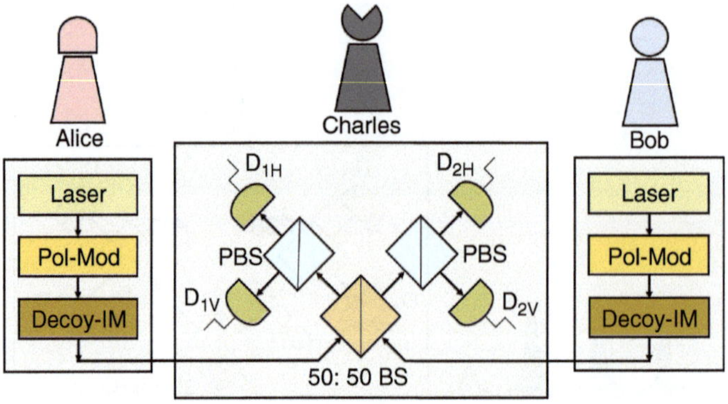

Figure 5.3 Illustration of an MDI-QKD setup. Pol-Mod, polarization modulator; Decoy-IM, decoy intensity modulator; BS, beamsplitter; PBS, polarizing beamsplitter.

An efficient solution to counter any possible hacking attack targeting the SPDs is measurement-device-independent (MDI) QKD [13]. Essentially, MDI-QKD is an IB type of scheme that exploits two-photon interference. As already discussed, in IB schemes all the measurements are performed by an untrusted third party Charles. So, its security can be guaranteed without characterizing the measurement unit. Precisely, in MDI-QKD, Alice and Bob send BB84 states to Charles, who performs a BSM on the received signals and then announces the results. Importantly, by varying the intensities of their lasers, Alice and Bob can use the decoy-state method to tightly estimate the detection and error rates associated to those rounds in which both of them emitted single-photon signals, from which a secret key can be distilled. A schematic of MDI-QKD is shown in Figure 5.3.

IB type of schemes can also be based on single-photon interference. A prominent example is *twin-field* (TF) QKD [14], which requires that a single photon from either Alice or Bob arrives at Charles' untrusted node. As a result, the transmission distance basically doubles the one attainable with MDI-QKD. Like all IB schemes, TF-QKD is also immune to any hacking attack against the measurement unit.

5.4.3 Other Practical QKD Protocols

There exist many other practical QKD schemes with their own merits. Differential phase reference protocols – such as the differential phase shift (DPS) [15] or the coherent one-way (COW) [16] – have been proposed as an alternative to the decoy-state method to counter PNS attacks. The idea is to encode the information coherently between adjacent pulses, since any measurement of the number of photons contained in a pulse breaks the coherence in a detectable manner. For instance, in the DPS protocol, the information is encoded in the relative phase between consecutive coherent states, which is revealed at Bob's site by means of an unbalanced interferometer. The performance of DPS can be further improved if Bob measures the coherence between arbitrarily separated signals instead of consecutive ones, using a more complex receiver.

The case of COW is more tricky. Alice sends Bob coherent pulses arranged in pairs of time-bins. In particular, for key generation Alice sends him the states $|\alpha\rangle\,|0\rangle$ and $|0\rangle\,|\alpha\rangle$, where $|\alpha\rangle$ ($|0\rangle$) denotes a coherent (vacuum) pulse. These states can be distinguished by Bob via "early" and "late" detection events. In addition, Alice also emits states $|\alpha\rangle\,|\alpha\rangle$ to test the channel. However, since vacuum states naturally break the coherence between pulses, this scheme is not as efficient as DPS, and its performance turns out to be comparable to that of the BB84 protocol without decoy states.

The secret-key rate of QKD can be increased by enlarging the dimension of the information carriers, in an alternative called high-dimensional (HD) QKD. One can use, for instance, spatial degrees of freedom of the photons or time-energy entanglement. Also, HD-QKD schemes may provide larger tolerance to noise than their qubit-based counterparts, although they typically require more complex experimental setups.

All QKD protocols discussed so far use SPDs at the receiver, but approaches based on standard photo-detectors also exist. In these schemes, called continuous-variable (CV) QKD, the secret keys are encoded in the quadratures of the quantized electric field, and then measured with coherent detection techniques. These measurements project the phase and amplitude of the field onto the quadrature axes, thus yielding a CV outcome. Many CV-QKD protocols use Gaussian-modulated coherent states for the encoding [17]. By sending overlapping Gaussian distributions in the optical phase-space of the quadratures, any attempt by Eve to distinguish them will necessarily introduce detectable errors. Notably, CV-QKD can usually provide higher key generation rates than those schemes using SPDs, but they are also more sensitive to channel loss, which often renders them more convenient for metropolitan communications.

5.4.4 The Problem of Implementation Security

IB schemes close all security loopholes from the measurement unit. The missing step to guarantee implementation security is, therefore, to protect the source. Besides multi-photon emissions, there are other device imperfections within the transmitter that need to be taken into account. We briefly address this issue in this section.

A particularly relevant type of imperfection is the leakage of sensitive information from the QKD devices, which may occur actively – provoked by a *Trojan-horse attack* (THA) – or passively – in the form of electromagnetic radiation, power consumption, the encoding of information in undesired degrees of freedom of the optical signals, and pulse correlations –. For instance, in a THA Eve injects bright light pulses into the QKD transmitter and then measures the back-reflected light, aiming to extract information about the internal setting choices (e.g. the bit values, the bases or the decoy settings). Fortunately, given an upper bound on the amount of information leaked (whether the leakage is active or passive), this information can be essentially removed via privacy amplification. The practical challenge is how to obtain the necessary upper bound.

Besides information leakage, other source imperfections are: imperfect phase randomization of the laser pulses, intensity fluctuations or encoding flaws. Remarkably, efficient techniques to incorporate these imperfections in the security analysis of

QKD have been developed in recent years. Certainly, a sharpening of the transmitter model leads to a larger tolerance against imperfections, although an exhaustive characterization of the functioning of all the devices might be a cumbersome task in practice.

Remarkably, this "precise characterization problem" can be avoided in a variant of EB schemes termed *device-independent* (DI) QKD. In this approach, no device at all needs to be internally characterized.[2] Its security requires that the measurement outcomes of the parties largely violate a Bell inequality in a loophole-free fashion. Roughly speaking, the observation of such violation certifies that Alice and Bob actually share entangled states in a fully DI fashion. From an experimental point of view, implementing a Bell test is a challenging task with current technology. Notwithstanding, several conclusive Bell tests have been realized in recent years, and the first successful implementations of DI-QKD experiments have just been reported. One must note that the performance of DI-QKD today is far from being practical for applications, both in terms of secret key generation rate and attainable distances. But technology is improving, and DI-QKD may become a practical technology in the future.

5.5 Components of a QKD System

The following components are typically included in the optical layer of a QKD system.

5.5.1 Light Sources

Most P&M and IB schemes use laser diodes. They can operate at telecom wavelengths and permit clock rates of the order of GHz. In decoy-state based QKD, phase randomization is typically achieved by operating the laser under *gain-switching* conditions, i.e. by driving it well above and below threshold. In doing so, the generation of each optical pulse is triggered by an independent process of spontaneous emission, and thus its phase is completely random. This technique can produce very narrow pulses, of the order of tens of picoseconds. Alternatively, an external phase modulator can also imprint a random phase to each pulse.

As a downside, the optical pulses generated by fast-driven semiconductor laser diodes may suffer from *frequency chirp* – time dependence of the instantaneous frequency – and *timing jitter* – deviations of their temporal positions from those expected in a perfectly periodic pulse train –. The *optical injection locking* technique – see Figure 5.4a – can be used to mitigate these effects, in addition to enabling precise control of the phase between pulses and increasing the laser modulation bandwidth. For this, a low-noise master laser injects light into a slave laser, whose wavelength varies until it equals that of the master, locking both its frequency and

2 Crucially though, in the DI setting one must rule out certain types of information leakage in the QKD receivers, which possibly requires partial characterization of the measurement devices.

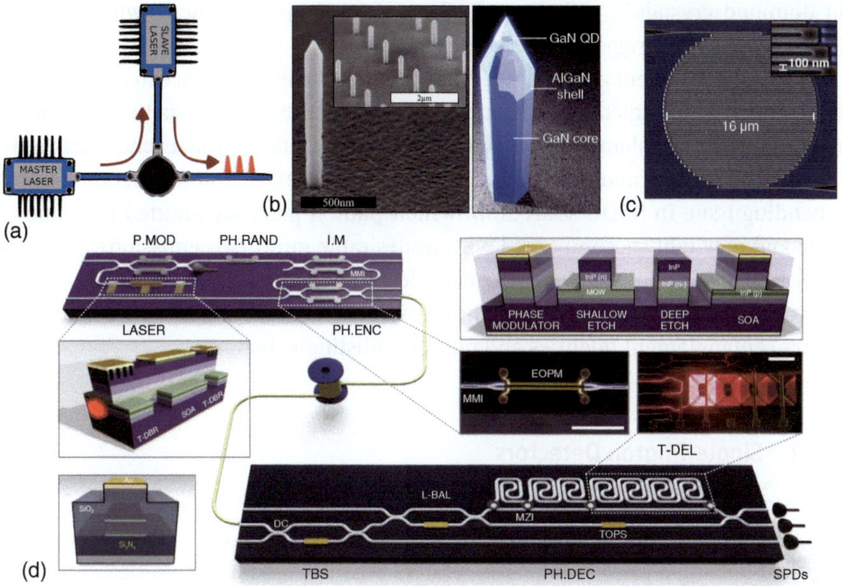

Figure 5.4 (a) Optical injection locking technique. (b) Nanowire-QDs. (c) Single-pixel SNSPD. (d) Integrated photonic devices for QKD. Source: (b) Holmes et al. [18], with permission of American Chemical Society. (c) Wollman et al. [19]/Society of Photo-Optical Instrumentation Engineers (SPIE)/CC BY 4.0. (d) Sibson et al. [20]/Springer Nature/CC BY 4.0.

phase. Furthermore, protocols like TF-QKD need to maintain long-term phase stability between the signals emitted by Alice and Bob. This can be achieved by using either a Sagnac-loop configuration or an *optical phase-locked loop* architecture together with compensation techniques.

Recent years have witnessed a great effort to develop single-photon sources. There are various methods, the one based on semiconductor quantum dots (QDs) being among the most promising ones. A semiconductor QD consists of a nanoscale semiconductor that possesses atomic-like quantized energy levels of the confined charge – fact for which they are often considered as a kind of artificial atoms – and hence a discrete emission spectrum. These structures are often placed in optical cavities – e.g. cylindrical Fabry–Pérot microcavities or nanowires – to efficiently collect the emitted photons, see Figure 5.4b. Most QD-based single-photon emitters require cryogenic cooling but room temperature operation is also possible at the cost of reducing the coherence of the emission. This technology has been used in satellite-based QKD to produce on-demand single-photon pulses with low multi-photon emission probabilities. Moreover, by improving the design of the cavities and photonic structures it is possible to enhance the emission directionality and reduce losses, pointing to photon extraction efficiencies up to ~90%. Alternative methods for single-photon generation include the trapping and subsequent optical excitation of individual atoms/ions, the use of discrete electronic states in crystallographic point defects – e.g. nitrogen-vacancies or silicon-vacancies in diamond crystals –, and the use of spontaneous parametric down-conversion (SPDC) sources as triggered single-photon sources.

As for entanglement generation, SPDC is undoubtedly the most extended method for producing entangled-photon pairs, having been used in various Bell-test experiments and QKD implementations. Unfortunately, despite having reached near-unity indistinguishability and heralding efficiency, on-demand photon generation is still a pending issue in SPDC sources, in which photon pairs are emitted probabilistically and inevitably accompanied with undesirable multi-pair emissions. Recently, deterministic QD-based sources of entangled photons have been developed, achieving pair generation rates and extraction efficiencies of ~60%, apart from near unity (~90%) entanglement fidelities and photon indistinguishabilities.

5.5.2 Single-Photon Detectors

The most relevant components of a QKD receiver certainly are the SPDs, which dictate the limits of the achievable performance. Indeed, the extreme sensitivity required for single-photon detection renders existing standard photo-detectors inadequate for this task. The workhorse for single-photon detection in practical applications is the single-photon avalanche diode (SPAD). The most widely used are silicon SPADs and Indium Gallium Arsenide (InGaAs) SPADs, the former being suitable for visible to near-infrared wavelengths typically used in free-space QKD, and the latter being adequate for telecom wavelengths (1310 and 1550 nm) used in fiber-based QKD. Nowadays, both alternatives allow for detection efficiencies of up to 30–50%, or even larger.

The basis of an SPAD is a semi-conductor p-n junction that exploits the photoelectric effect to generate a photo-current operating in a Geiger mode. To be precise, in an SPAD, a high reverse bias voltage exceeding the breakdown voltage is used, which can result in a self-sustained avalanche photo-current in response to the absorption of a single photon. In addition, SPADs rely on a so-called *quenching circuit* to stop the avalanche and reset the device for a new detection event. Notably, the electronics of this quenching circuit are determinant for the performance of the SPAD, in particular for the timing jitter that usually dominates the timing resolution. The timing jitter here refers to the variation from event to event in the delay between the absorption of a photon and the generation of an output electrical signal, and active quenching may allow to reach jitters below 100 ps.

A limitation of SPADs is that the strong photo-current can provoke electrons to get trapped in defects and be spontaneously released later on, triggering a so-called *afterpulse*. Afterpulses can be mitigated by setting a time delay between bias voltage gates for trapped electrons to relax, which unfortunately typically limits the maximum gating frequency to about 10 MHz. The gating frequencies can be pushed to the realm of GHz by means of a *self-differencing* technique that allows to detect much weaker avalanches. Importantly as well, SPADs (particularly, InGaAs SPADs) often exhibit large dark-count rates that degrade the signal-to-noise ratio and become prohibitive for long-distance QKD. Typical values of the dark-count rate in SPADs operating at telecom wavelengths are around 10^{-4} or 10^{-5}. This can be slightly reduced by cooling the detectors to temperatures of 210–250 K using thermoelectric coolers.

To further reduce the dark-count rate and be able to achieve longer distances with QKD, one may use superconducting nanowire single-photon detectors (SNSPDs). They consist of a thin (~5 nm) and narrow (~100 nm) superconducting wire that is typically hundreds of μm long, patterned in a compact meander shape to create a square or circular pixel that provides a practical sensitive area to enhance the detection efficiency. This is illustrated in Figure 5.4c. The wire is cooled well below its critical temperature – e.g. from 1 to 4 K in a typical nanowire made of niobium nitride – and biased with a DC current at a level just below the critical current density. In this state, the absorption of a single photon locally triggers the appearance of a normal resistance region or *hotspot*, which provokes the current density to increase in the adjacent regions, thus exceeding the critical current density. As a consequence, a normal resistance region is formed throughout the width of the wire, yielding a voltage peak that indicates the detection of a single photon. Importantly, different techniques allow to boost the efficiency of the SNSPD, such as integration within an optical cavity or a photonic waveguide, or the addition of nanoantenna structures. As an example, recently, SNSPDs with detection efficiencies of 98% at 1550 nm and extremely low dark-count rates (of the order of 10^{-8}) have been demonstrated. Moreover, SNSPDs exhibit an extremely low timing jitter, as low as 3 ps in the last-generation devices.

The SPDs discussed above are threshold SPDs, which means that they do not provide photon-number resolution, but only output "click" (at least one photon is detected) or "no click" (no photon is detected). In certain applications, however, it might be advantageous to be able to determine the number of photons contained in an optical pulse; for instance, to employ an SPDC source as a triggered single-photon source. For this, one usually operates the detectors in linear mode, such that the output signal is proportional to the number of incident photons. A popular candidate is the superconducting transition-edge sensor (TES). In a TES, a thin layer of superconducting material is forced to operate at the cusp temperature of its superconducting transition by means of a constant voltage bias. In this situation, any change in temperature causes an abrupt change in resistance. The absorption of an incident photon heats the sensor increasing its resistance, which results in a reduction of the current flow recorded as a photo-detection event. TESs have a natural ability for photon-counting because the signal is proportional to the energy of a photon, possibly resolving up to 8 photons or more. Moreover, when embedded in an optical cavity structure, they can reach detection efficiencies of ~95% at 1550 nm and of ~98% at 850 nm, and they exhibit negligible dark-count rates as well. Current limitations include the need to operate at temperatures below 0.1 K, and the limited performance both in terms of recovery time – with maximum counting rates of the order of MHz – and timing jitter (e.g. ~5 ns for tungsten-based TES).

A certain degree of photon-number resolution is also achievable in SNSPD arrays through the use of time-binning, multiplexing, or advanced readout schemes. Although this solution does not match the intrinsic photon-number resolution of a superconducting TES, it enables higher temperatures and counting rates.

5.5.3 Linear Optical Elements

Several optical elements that are common in standard optical communication systems are also used in QKD transmitters and receivers. This includes e.g. filters, attenuators, circulators or isolators, beamsplitters, polarization beamsplitters, Faraday mirrors, and amplitude and phase modulators.

Notably, current approaches tend to embed all the optical elements into integrated photonic circuits, as shown in Figure 5.4d.

5.6 Performance

The most limiting factor to the performance of QKD is channel loss – typically around 0.2 dB/km in telecom fibers –. Indeed, fundamental limits for point-to-point QKD configurations prohibit the secret-key rate to scale better than linearly with the transmittance of the channel, which decreases exponentially with the distance. To overcome these limits, one needs to place intermediate nodes along the quantum channel. This is the approach followed by quantum networks based on trusted nodes – which locate several trusted nodes where the key is available between the end users –, some IB schemes such as TF-QKD – for which the secret-key rate scales with the square root of the transmittance of the channel thanks to the use of single-photon interference –, or architectures based on quantum repeaters – which still require for an important technological development to be practically implementable, but will remove the need to trust the intermediate nodes –.

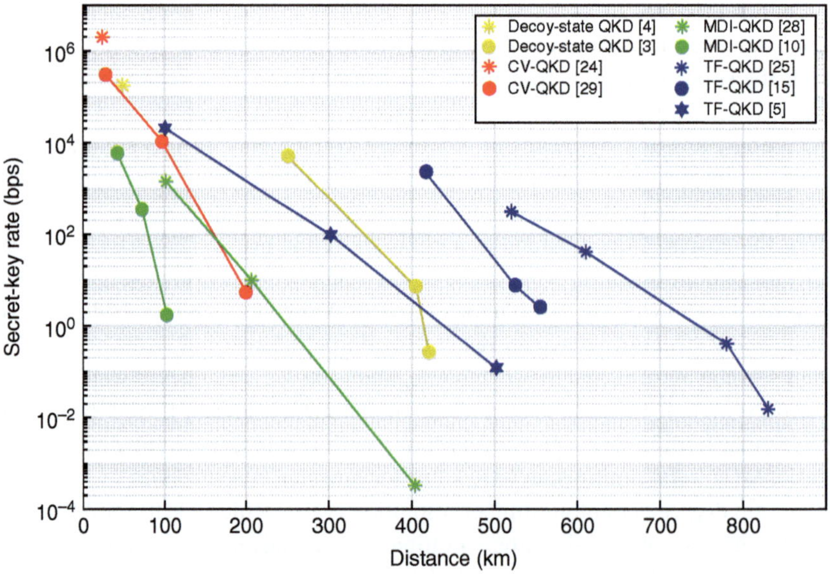

Figure 5.5 Secret-key rate vs. distance in some recent QKD experiments.

Figure 5.5 summarizes the secret-key rate vs. distance attained by some recent fiber-based QKD experiments. In particular, the experiments reported in [21–24] are proven to be secure against collective attacks, whereas those reported in [25–29] are secure against general coherent attacks. The figure highlights the important advantage of TF-QKD over the other competing QKD protocols in point-to-point links for long-distance communication.

5.7 Conclusions and Future Challenges

In the present chapter, we have presented an overview of QKD protocols and their technology. The motivation of QKD as an information-theoretically secure solution to distribute secret keys for private-key cryptography, its fundamental working principles and its security paradigm were addressed in the first place. Secondly, we dove into the practical issues of QKD implementations and the challenge of achieving implementation security, and we discussed the state of the art of QKD equipment with a special emphasis on the light sources and the photo-detection technologies. To finish with, we ended presenting the rate-vs-distance performance of several recent QKD experiments. The take-home-message from all of it is that QKD has evolved from a theoretical curiosity to a consolidated technology, whose worldwide deployment has just begun.

Certainly, however, there are still plenty of challenges to address. Among other breakthroughs, on-demand single photon sources, photonic integration for chip-based QKD and practical DI-QKD implementations are under the radar. Also, much effort is being dedicated to the spread of trusted node and satellite based QKD networks, and the development of practical quantum repeaters is a must in the near future aiming to cover arbitrary distances in fiber-based QKD without the need of trusted nodes. Importantly as well, faster QKD transmitters are being designed, which combined with multiplexing techniques and high time-resolution SPDs will translate in an enhancement of the secret-key generation rates.

Putting it all together, QKD is expected to become a prolific industry in the years to come.

Acknowledgments

We thank Marco Lucamarini for discussions about the contents of this chapter. This work was supported by MICINN with funding from the European Union NextGenerationEU (PRTR-C17.I1) through the "Planes Complementarios de I+D+I con las Comunidades Autónomas" in Quantum Communication, by the Galician Regional Government (consolidation of Research Units: AtlantTIC), the Spanish Ministry of Economy and Competitiveness (MINECO), and the Fondo Europeo de Desarrollo Regional (FEDER) through Grant No. PID2020-118178RB-C21.

References

1 Bennett, C.H. and Brassard, G. (1984). *Proceedings of the IEEE International Conference on Computers, Systems and Signal Processing*, Bangalore, India, pp. 175–179.

2 Portmann, C. and Renner, R. (2021). Security in quantum cryptography. *arXiv preprint arXiv:2102.00021.*

3 Ben-Or, M., Horodecki, M., Leung, D.W. et al. (2005). The universal composable security of quantum key distribution. In: *Theory of Cryptography Conference, TCC 2005*, Lecture Notes in Computer Science, vol. 3378 (ed. J. Kilian), 386–406. Springer-Verlag.

4 Renner, R. and König, R. (2005). Universally composable privacy amplification against quantum adversaries. *Theory of Cryptography Conference*, pp. 407–425. Springer.

5 Renner, R. (2008). Security of quantum key distribution. *International Journal of Quantum Information* 6 (01): 1–127.

6 Shor, P.W. and Preskill, J. (2000). Simple proof of security of the BB84 quantum key distribution protocol. *Physical Review Letters* 85 (2): 441.

7 Tomamichel, M., Schaffner, C., Smith, A., and Renner, R. (2011). Leftover hashing against quantum side information. *IEEE Transactions on Information Theory* 57 (8): 5524–5535.

8 Koashi, M. (2007). Complementarity, distillable secret key, and distillable entanglement. *arXiv preprint arXiv:0704.3661.*

9 Hwang, W.-Y. (2003). Quantum key distribution with high loss: toward global secure communication. *Physical Review Letters* 91 (5): 057901.

10 Lo, H.-K., Ma, X., and Chen, K. (2005). Decoy state quantum key distribution. *Physical Review Letters* 94 (23): 230504.

11 Wang, X.-B. (2005). Beating the photon-number-splitting attack in practical quantum cryptography. *Physical Review Letters* 94 (23): 230503.

12 Makarov, V. (2009). Controlling passively quenched single photon detectors by bright light. *New Journal of Physics* 11 (6): 065003.

13 Lo, H.-K., Curty, M., and Qi, B. (2012). Measurement-device-independent quantum key distribution. *Physical Review Letters* 108 (13): 130503.

14 Lucamarini, M., Yuan, Z.L., Dynes, J.F., and Shields, A.J. (2018). Overcoming the rate–distance limit of quantum key distribution without quantum repeaters. *Nature* 557 (7705): 400–403.

15 Inoue, K., Waks, E., and Yamamoto, Y. (2002). Differential phase shift quantum key distribution. *Physical Review Letters* 89 (3): 037902.

16 Stucki, D., Brunner, N., Gisin, N. et al. (2005). Fast and simple one-way quantum key distribution. *Applied Physics Letters* 87 (19): 194108.

17 Grosshans, F. and Grangier, P. (2002). Continuous variable quantum cryptography using coherent states. *Physical Review Letters* 88 (5): 057902.

18 Holmes, M.J., Choi, K., Kako, S. et al. (2014). Room-temperature triggered single photon emission from a III-nitride site-controlled nanowire quantum dot. *Nano Letters* 14 (2): 982–986.

19 Wollman, E.E., Verma, V.B., Walter, A.B. et al. (2021). Recent advances in superconducting nanowire single-photon detector technology for exoplanet transit spectroscopy in the mid-infrared. *Journal of Astronomical Telescopes, Instruments, and Systems* 7 (1): 011004.

20 Sibson, P., Erven, C., Godfrey, M. et al. (2017). Chip-based quantum key distribution. *Nature Communications* 8 (1): 1–6.

21 Boaron, A., Boso, G., Rusca, D. et al. (2018). Secure quantum key distribution over 421 km of optical fiber. *Physical Review Letters* 121 (19): 190502.

22 Bunandar, D., Lentine, A., Lee, C. et al. (2018). Metropolitan quantum key distribution with silicon photonics. *Physical Review X* 8 (2): 021009.

23 Wang, H., Pi, Y., Huang, W. et al. (2020). High-speed gaussian-modulated continuous-variable quantum key distribution with a local local oscillator based on pilot-tone-assisted phase compensation. *Optics Express* 28 (22): 32882–32893.

24 Zhang, Y., Chen, Z., Pirandola, S. et al. (2020). Long-distance continuous-variable quantum key distribution over 202.81 km of fiber. *Physical Review Letters* 125 (1): 010502.

25 Yin, H.-L., Chen, T.-Y., Yu, Z.-W. et al. (2016). Measurement-device-independent quantum key distribution over a 404 km optical fiber. *Physical Review Letters* 117 (19): 190501.

26 Liu, H., Wang, W., Wei, K. et al. (2019). Experimental demonstration of high-rate measurement-device-independent quantum key distribution over asymmetric channels. *Physical Review Letters* 122 (16): 160501.

27 Fang, X.-T., Zeng, P., Liu, H. et al. (2020). Implementation of quantum key distribution surpassing the linear rate-transmittance bound. *Nature Photonics* 14 (7): 422–425.

28 Pittaluga, M., Minder, M., Lucamarini, M. et al. (2021). 600-km repeater-like quantum communications with dual-band stabilization. *Nature Photonics* 15 (7): 530–535.

29 Wang, S., Yin, Z.-Q., He, D.-Y. et al. (2022). Twin-field quantum key distribution over 830-km fibre. *Nature Photonics* 16: 154–161.

6

From Basic Science to Technological Development: The Case for Two Avenues

Matteo Carlesso and Mauro Paternostro

Centre for Quantum Materials and Technologies, School of Mathematics and Physics, Queen's University Belfast, BT7 1NN Belfast, UK

6.1 Introduction

The promises of quantum technologies seem limitless: from ultra-secure communication to more accurate weather prediction [1], from the design of target-specific drug [2, 3] to new and revolutionary materials [4]. The ambitions for quantum technologies encompass innovations spread broadly across quantum information science, materials, and sensing technologies.

There is little doubt that the development of a successful framework for quantum technologies *should* build on the very substantial body of work that has been produced, over the course of the past 30 years, on basic science. The best approach is always to put science first. Areas of particular significance in this context include quantum information theory, the foundations of quantum mechanics, material science, computer engineering, and the framework of system identification and process characterization. However, it also includes research that goes beyond the field of pure quantum technologies, such as the study of quantum effects in thermodynamic or biological processes. While the community has identified four main strands for research and development that will enable the construction of a comprehensive work infrastructure for quantum technologies, the gain of new knowledge underpinning the activities of such *pillar*, and thus contributing to technological advances and new applications in the long term, has to stem from basic science.

Solving basic science problems in quantum science across all its complexities will enable transformative scientific and industrial progress that, in time, will transition into major drivers for scientific advancement, economy, and even national security.

The provision of a comprehensive overview of the various directions along which a study of fundamental physics could lead to technological advancement goes well beyond the scopes of this work. Instead, we shall aim at arguing in favor of two specific areas of investigation, thermodynamics of quantum processes, and machine learning for quantum problems, which have recently gained considerable attention from the community interested in the control of quantum systems and dynamics,

Photonic Quantum Technologies: Science and Applications, First Edition. Edited by Mohamed Benyoucef.

and comment on how they are significantly contributing to the process of *virtuous* two-way influence between fundamental science and technological developments.

The thermodynamic interpretation of fundamental energy-exchange process occurring among the elements of a quantum device allows for the establishment of fundamental constraint to the efficiency of certain information-processing tasks, most remarkably quantum computation. This will contribute to the technological step change entailed by the Second Quantum Revolution by addressing the energetic and entropic footprints of quantum devices. It thus addresses a blind spot in the work program that followed the publication of the "Quantum Manifesto" upon which the EU-funded Quantum Flagship initiative has been built that must be addressed to achieve the paradigm shift promised by quantum technologies, and thus bypass the energy-consumption issues in information technologies such as those based on CMOS technology. In doing so, the development of a thermodynamic approach to the energetics of quantum information processing will allow:

(1) The assessment of the energy cost of creating and consuming quantum coherence in the experimental platforms that embody the building blocks of upcoming quantum processors. Their thermodynamic assessment will allow for the quantification of the energetic cost of processing quantum information, including computational tasks;
(2) The establishment of a framework for the exploration and understanding of the interplay between logical and thermodynamic irreversibility, which is the fundamental source of heat dissipation in computing systems.

From a different, yet related perspective, the interplay between data-intensive research and *quantum* is leading to the development of new hybrid methods for control and performance optimization of quantum processes, while opening up the possibility to design less expensive methods for system identification, performance validation, and quantum-property reconstruction. The use of artificial intelligence in the management of quantum devices widens the already significant pool of hardware problems benefiting of machine learning, providing a boost to the development of near-term quantum devices enhanced by artificial intelligence accelerators. In time, it will also allow for the use of quantum computing devices as neural networks for the systematic adaptation of the controls of quantum processors, the classification of datasets resulting from quantum processes, and the inexpensive reconstruction of states or properties of quantum devices.

6.2 Thermodynamics for an Energetically Efficient Quantum Information Processing

The application of quantum mechanics to the mesoscopic and macroscopic scales faces the limitations imposed by the surrounding environment, which can have a strong influence on the quantum system that one wants to study or employ in quantum technology applications. The comprehension of the interaction of such an open quantum system with its environment becomes pivotal for the development and enhancement of quantum technologies as quantum sensing and communication.

To be eventually able to suppress the action of the environment in open quantum systems, one needs to identify all – or at least the main – contributions. Clearly, depending on the quantum system under scrutiny, some environmental mechanisms will contribute more than other to the total influence. These effects can be studied with the mathematical tools provided by the open quantum system theory. In such a framework, the system S is considered in interaction with a surrounding environment E with a total evolution defined by a unitary dynamics $\hat{U}_t$ with respect to the total Hamiltonian of the system + environment. Namely, the total statistical operator evolves as $\hat{\rho}_{\mathrm{SE}}(t) = \hat{U}_t \hat{\rho}_{\mathrm{SE}}(0)\hat{U}_t^\dagger$. Then, one averages over the degrees of freedom of the environment to get the effective (reduced) dynamics of the system alone. This is usually described by the dynamical equation known as master equation $\frac{\mathrm{d}\hat{\rho}_{\mathrm{S}}(t)}{\mathrm{d}t} = \mathbb{L}_t[\hat{\rho}_{\mathrm{S}}]$, whose form is defined by the generator $\mathbb{L}_t$. The latter can encode decoherence (loss of quantum coherence, and decay of quantum superpositions of the system, namely suppresses the quantum features of the system) and dissipation (energy exchange between system and environment) due to the environment. Typically, the decoherence is described through the master equation (in position representation) [5]

$$\frac{\mathrm{d}\left\langle x|\hat{\rho}_{\mathrm{S}}(t)|x'\right\rangle}{\mathrm{d}t} = -\frac{i}{\hbar}\left\langle x|[\hat{H}, \hat{\rho}_{\mathrm{S}}(t)]|x'\right\rangle - \Gamma(x-x')\left\langle x|\hat{\rho}_{\mathrm{S}}(t)|x'\right\rangle, \tag{6.1}$$

where the rate $\Gamma(x-x')$ quantifies the deviations from the unitary dynamics defined by the system's Hamiltonian $\hat{H}$.

Such a rate Γ scales quadratically with $\Delta_x = |x - x'|$ for small spatial separations with respect to the characteristic length scale of the specific decoherence process (namely, the long-wavelength limit), while it saturates to a constant value for large separations (short-wavelength limit). A way to connect these two limits is to approximate the decoherence rate to $\Gamma(\Delta_x) \sim \gamma[1 - \exp(-\Lambda\Delta_x^2/\gamma)]$, which introduces the rate of events γ and the diffusion constant Λ [cf. Figure 6.1]. The typical example of such a separation of length scales is provided by the collisional decoherence [6]. This is the case of a mesoscopic particle, having the role of the system of interest,

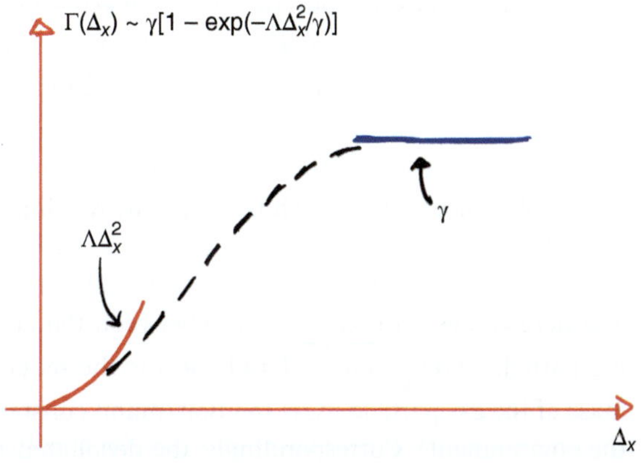

Figure 6.1 The decoherence function Γ as a function of the superposition distance Δ_x in the two relevant limits. In the long-wavelength limit $\Gamma(\Delta_x) \sim \Lambda\Delta_x^2$ (red line), while in the short- wavelength limit $\Gamma(\Delta_x) \sim \gamma$ (blue line). The dashed gray line represent a possible parametrization of $\Gamma(\Delta_x)$ that connects the two limiting cases. Source: Adapted from Gasbarri et al. [14].

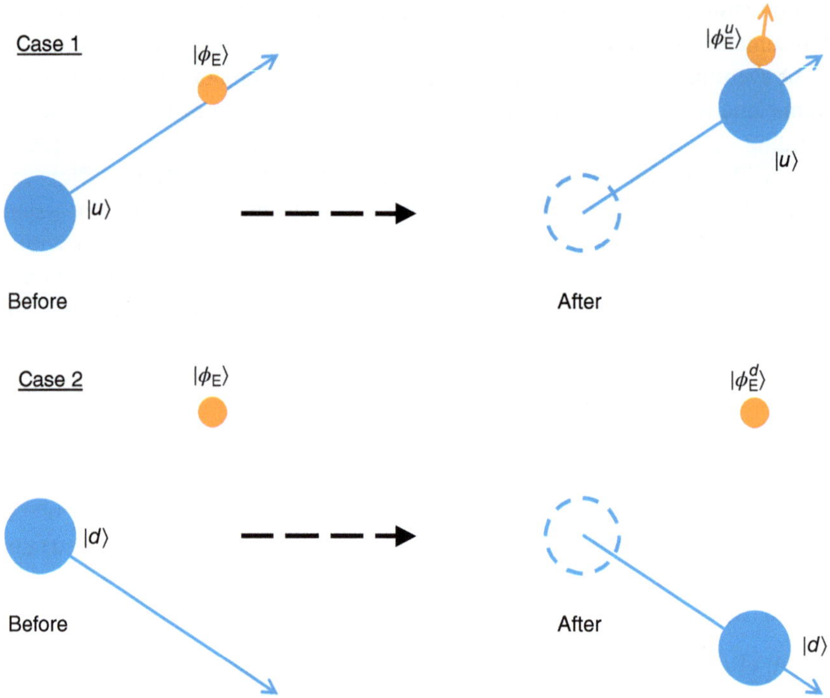

Figure 6.2 Scheme of the collisional decoherence. Case 1: the system is initially prepared in the state $|u\rangle$, then it will collide with the gas particle which is in the state $|\phi_E\rangle$, eventually ending in the state $|\phi_E^u\rangle$. Case 2: conversely, if the system is initially prepared in $|d\rangle$, it will leave the state $|\phi_E\rangle = |\phi_E^d\rangle$ untouched.

under the action of the collisions of the particles constituting the residual gas in the chamber where the experiment is performed. Let us assume that the initial state of the system $|\psi\rangle$ is in a superposition of the particle going up $|u\rangle$ and going down $|d\rangle$: $|\psi\rangle \propto |u\rangle + |d\rangle$, and that the single collision with the gas particle takes place only if the systems is in the state $|u\rangle$ (see Figure 6.2). In particular, one can describe the collision as $|u\rangle \otimes |\phi_E\rangle \to |u\rangle \otimes |\phi_E^u\rangle$.

Now, the collision on the full superposition reads $(|u\rangle + |d\rangle) \otimes |\phi_E\rangle \to |u\rangle \otimes |\phi_E^u\rangle + |d\rangle \otimes |\phi_E^d\rangle$, where $|\phi_E^d\rangle = |\phi_E\rangle$. Correspondingly, the reduced density matrix (obtained after averaging over the environmental degrees of freedom) in the $\{|u\rangle, |d\rangle\}$ basis will have off-diagonal terms being proportional to $\langle \phi_E^u | \phi_E^d \rangle$. Now, it is clear that more the states $|\phi_E^u\rangle$ and $|\phi_E^d\rangle$ differ each other stronger becomes the decoherence effect, which is eventually maximized when the two environmental states are orthogonal. The difference between such states is imposed by the momentum transfer between the system and the gas particle during the collision, and here comes into play the ratio between the characteristic length scale of the gas particle $2\pi\hbar/\sqrt{2\pi m_p k_B T}$ (where $\hbar$ is the reduced Planck constant, m_p is the mass of the gas particle, k_B is the Boltzmann constant, and T is the temperature of the environment). Correspondingly, the decoherence effects scale with $m_p\bar{v}PR^2\Delta_x^2$

Table 6.1 Decoherence rates $\Gamma(\Delta_x)$ in the long ($\Gamma(\Delta_x) \sim \Lambda\Delta_x^2$) and short ($\Gamma(\Delta_x) \sim \gamma$) wavelength limits with respect to the superposition distance Δ_x [8].

	Λ (long-wavelength limit)	γ (short-wavelength limit)
Collisions	$\dfrac{8\sqrt{2\pi m_{\mathrm{p}}}\overline{v}PR^2}{3\sqrt{3}\hbar^2}$	$\dfrac{16\pi\sqrt{2\pi}PR^2}{\sqrt{3}\overline{v}m_{\mathrm{p}}}$
Scattering	$8!\dfrac{8\zeta(9)}{9\pi}R^6c\,\Re\left(\dfrac{\epsilon-1}{\epsilon+2}\right)^2\left(\dfrac{k_{\mathrm{B}}T}{\hbar c}\right)^9$	$8!\dfrac{8\zeta(9)\pi^{1/3}}{9}R^6c\,\Re\left(\dfrac{\epsilon-1}{\epsilon+2}\right)^2\left(\dfrac{k_{\mathrm{B}}T}{\hbar c}\right)^7$
Abs.&Emis.	$\dfrac{16\pi^5}{189}R^3c\,\Im\left(\dfrac{\epsilon-1}{\epsilon+2}\right)\left(\dfrac{k_{\mathrm{B}}T_{(\mathrm{i,e})}}{\hbar c}\right)^6$	$\dfrac{16\pi^6\pi^{1/3}}{189}R^3c\,\Im\left(\dfrac{\epsilon-1}{\epsilon+2}\right)\left(\dfrac{k_{\mathrm{B}}T_{(\mathrm{i,e})}}{\hbar c}\right)^4$

The collisions with the residual gas are characterized by the length scale $2\pi\hbar/\sqrt{2\pi m_{\mathrm{p}}k_{\mathrm{B}}T}$, while scattering, absorption, and emission of blackbody radiation by $\pi^{2/3}\hbar c/k_{\mathrm{B}}T$. Here, we distinguish the internal temperature T_{i} (relevant for the emission process) from the external temperature T_{e} (relevant for the absorption process).

in the long-wavelength limit (depending on the pressure P in the experiment, the mean velocity $\overline{v}$ of the gas, and the linear dimension $2R$ of the system). Conversely, in the short-wavelength limit go as $PR^2/\overline{v}m_{\mathrm{p}}$. The effects of collisional decoherence are particularly important in matter-wave interferometry experiments [7], which have been key to establish the emergence of quantum effects at the mesoscopic scale by dealing with system of increasing size and complexity (Table 6.1).

Another example of decoherence is provided by three processes involving blackbody radiation, which are scattering, absorption, and emission, and are characterized by the length scale $\pi^{2/3}\hbar c/k_{\mathrm{B}}T$. The first among these three processes is the equivalent of the collisions with the residual gas particle: the blackbody radiation elastically scatters on the system, and there is a momentum exchange between the two. Conversely, during the absorption, the scattering is completely inelastic, the photon is absorbed, and the system gains completely the photon's momentum. Finally, the emission process can be understood as the exactly the time-inverted process of the absorption, where a photon is emitted and the particle losses momentum. Depending on the process, they are characterized by a different scaling in the long-wavelength limit: $R^6T^9\Re(\epsilon-1/\epsilon+2)^2$ for the scattering and $R^3T^6\Im(\epsilon-1/\epsilon+2)$ for emission and absorption where ϵ is the dielectric constant of the system. Similarly, in the short-wavelength limit, one has $R^6T^7\Re(\epsilon-1/\epsilon+2)$ for the scattering and $R^3T^4\Im(\epsilon-1/\epsilon+2)$ for emission and absorption.

When dealing with much smaller systems, where the abovementioned decoherence mechanisms can be negligible, other mechanisms kick in. For example, in Bose–Einstein condensates (BEC), the main concern in loosing quantum properties is the reduction of the number of particles in the condensate rather than decoherence understood as suppression of quantum coherences. Here, two processes are relevant. The first concerns the collisions of the atoms in the condensates with those of the background thermal could, which is made of the atoms of the cloud that do not appertain to the condensate. The second process is that of the three-body recombination, which accounts for the inelastic collisions between atoms in the BEC that lead to the molecule formation. These two processes can be described together through [9]

$$\frac{dN_t}{dt} = -K_1 N_t - \tilde{K} N_t^3, \tag{6.2}$$

where K_1 is the one-body decay rate due to the interaction with the thermal cloud, and $\tilde{K} = K_3/((2\pi)^3 3^{3/2} \sigma_t^6)$ depends on the position variance σ_t of the BEC and the three-body loss coefficient K_3. For large BECs, the stronger reduction effect comes from the three-body recombination process, thus indicating that to generate and maintain a large BEC for long times becomes pivotal to reduce $\tilde{K}$ as much as possible. To be quantitative, for keeping the reduction below the 10% over 10 s when assuming $K_1 \sim 10^{-3}\ \mathrm{s}^{-1}$, one requires $\tilde{K} \sim 10^{-12}\ \mathrm{s}^{-1}$ for an initial value of atoms of $N_0 \sim 10^5$, while $\tilde{K}$ needs to drop below $4 \times 10^{-14}\ \mathrm{s}^{-1}$ when considering $N_0 \sim 5 \times 10^5$. This corresponds to almost two orders of magnitude reductions of $\tilde{K}$ when one considers *only* an increase of a factor 5 in the number of atoms in the condensate.

The progress of open quantum system's theory has provided the grounds to quantify the limiting effects imposed by the environment surrounding a system and thus, carefully characterize the way the former might affect quantum technologies. Typically, the process of interaction between a quantum system and its environment implies the exchange of energy among the parties. The washing out of the degrees of freedom of the environment results in such process to be interpreted as an exchange of heat, which might eventually lead it to equilibration and, in the case of an environment already at thermal equilibrium, thermalization of the system.

Such an interaction moves the state of a system away from those delicate conditions necessary for the emergence of fragile genuine quantum effects. Usually, one tries to depart from such a hindering situation and keep the system as isolated as possible. This implies lowering the temperature of the system, thus avoiding high-temperature thermal states, reducing the pressure of the surrounding residual gas – thus minimizing their collisions with the system – and isolates the system from all possible noises that could perturb it. Other than the noises from the surrounding environment meant as residual gas and blackbody radiation, one needs to account for the total noise budget also those imposed by the specific experimental setup that is used. Among the different sources, we can mention the vibrational noises that can have various sources as the Earth seismic activity that is particularly relevant at low frequencies, the urban traffic, and the refrigerator that can be predominant at the corresponding characteristic frequencies. Some experiments are so sensitive that can feel variations of the local gravitational field when, for example, a train passes through the nearby train station. To mitigate the vibrational noise, one typically detaches as much as possible the system from ground by employing a system of springs that attenuate such noises at the relevant frequencies. Figure 6.3 shows a possible implementation of these spring systems when applied to the mirrors of the gravitational wave detector LIGO [10]. Alternatively, one can perform experiments in free-fall where several components of vibrational noises are avoided completely. One option is constituted by experiments performed in drop towers

Figure 6.3 Suspensions of the Advanced LIGO's mirrors. Top figure: the Hydraulic External Pre-Isolator (HEPI) in blue provides the first stage of isolation outside the vacuum chamber where the LIGO's mirror are placed. Bottom figure: the Seismic Attenuation System (SAS) within the vacuum chamber constitutes the second seismic isolation. Figures from [10].

as that of ZARM in Bremen [11] or the Einstein Elevator [12], which can provide up to around 9 s and 4 s of free fall, respectively. However, this option undergoes strong limitations in the time for the run of the experiment, which is determined by the height of the drop tower, and it would be still subject to vibrations due to the friction between the residual gas in the drop tower and the capsule containing the actual experiment. Another option is to perform experiments in space [13–17], which would provide much longer free-evolution times, and it would avoid the problem of the friction. To be quantitative, LISA Pathfinder – which is the prototype for the planned space-based gravitational wave detector LISA – has demonstrated an acceleration noise floor of $10^{-15}\mathrm{m\,s^{-2}}/\sqrt{\mathrm{Hz}}$ that constitutes an improvement of around 10 orders of magnitude with respect to the value of $10^{-5}\mathrm{m\,s^{-2}}/\sqrt{\mathrm{Hz}}$ provided by the ZARM drop tower [18].

Being able to control the environment is pivotal to possible quantum technological applications such as quantum sensing or quantum communication, but it is also the basis to perform test of basic science as the detection of gravitational waves or the generation of quantum superposition with massive objects. Science-driven experiments demand and, with time, generate a significant progress, at both the fundamental and technological level, in the design and the achievement of the conditions able to meet the requirements for a very low-noise, controlled operational regime [14, 15]. Examples are provided by the 30-year endeavor of gravitational wave detector such as LIGO (Laser Interferometer Gravitational-wave Observatory) or – more recently – by its space counterpart Laser Interferometer Space Antenna (LISA) [19] that is planned to be launched in 2037. Here, the noise's reduction is fundamental to being able to detect the fainted signals of the passage of gravitational waves. Other science-driven experiments are focused on the detection of non-standard effects appearing as noises conversely to a signal as in the case of gravitational waves. In such a case, the challenge is to distinguish one (non-standard) noise within a background of other noises. Examples of such effects run from deviations of quantum mechanics [20, 21], gravitational decoherence [22], quantum gravity effects [23], dark matter/energy [24], and others.

Clearly, in some situations, changing the environmental parameters, such as temperature and pressure, is not sufficient to achieve the desired conditions. The paradigmatic example is the achievement of the ground state in optomechanical systems [25], which is something hardly possible by simply reducing the temperature of the environment, and one needs to find alternative routes toward this aim. Indeed, satisfying the condition of $\hbar\omega_\mathrm{m} \gg k_\mathrm{B}T$, where ω_m is the mechanical frequency of the system harmonically trapped, is challenging also for dilution refrigerators providing a 20 mK environment as one would require ω_m to be larger than 10 GHz. Active cooling techniques were thus developed, such as, among others, resolved sideband or feedback cooling in optomechanical system.

In the case of sideband cooling (top panel of Figure 6.4), the mechanical resonator, which plays the role of the main system to be cooled and characterized by the frequency ω_m, is placed in mutual interaction with an optical field of frequency Ω. Both the systems are coupled to the same thermal bath at temperature T. However, since the frequency of the optical field Ω is assumed to be much larger than that of the resonator ω_m and such that $\hbar\Omega \gg k_\mathrm{B}T$, from the perspective of the optical field, the bath is in the vacuum state, i.e. in the ground state. This means that the optical field can only lose energy, at rate κ toward the bath without gaining any. Then, by employing an external laser at frequency ω_L, one can stimulate the transition between the $|m,0\rangle$ state, which indicates m excitations in the mechanical resonator and 0 in the optical field, and the $|m-1,1\rangle$; then, in the limit of $\kappa \gg \omega_\mathrm{m}$, the optical field will decay from $|1\rangle$ to $|0\rangle$, emitting the photon in the thermal bath, before the excitation goes back to the mechanical resonator. To summarize, the process leads to the following transition

$$|m,0\rangle \rightarrow |m-1,1\rangle \rightarrow |m-1,0\rangle, \tag{6.3}$$

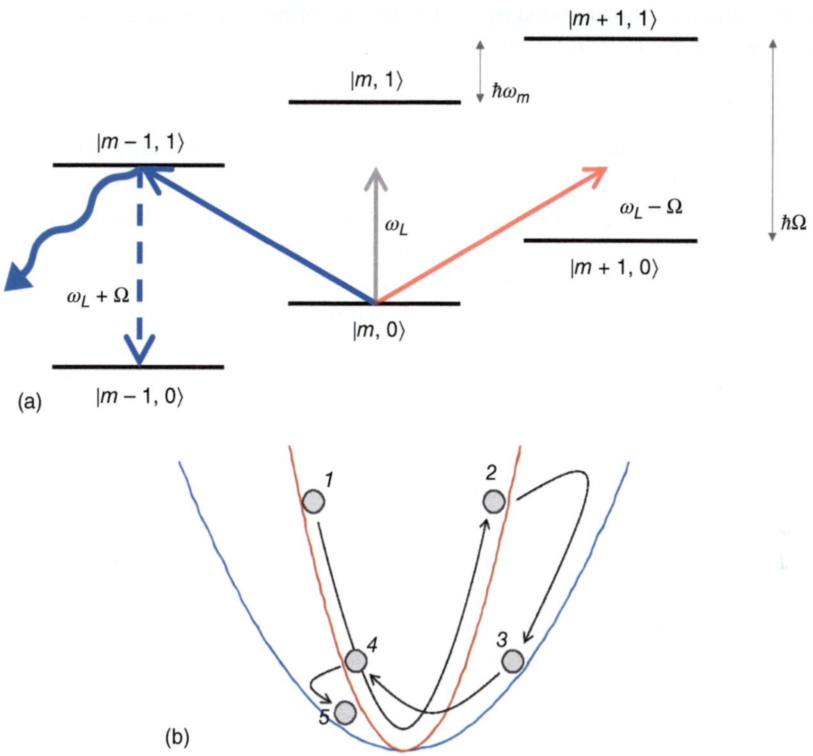

Figure 6.4 Conceptual scheme of the working process behind the resolved sideband cooling (a) and the parametric feedback cooling (b).

meaning that effectively the mechanical resonator has lost an excitation. If the rate of this transition process, being $g^2/(\kappa/2)$ where g is the coupling between the mechanical resonator and the optical field, is larger than the effective heating $\gamma\overline{n}(\omega_{\mathrm{m}})/2$ of the mechanical resonator, where $\overline{n}$ is the thermal occupation number, one can cool the system's occupation number to the value of

$$n = \overline{n}(\omega_{\mathrm{m}})\frac{\kappa\gamma}{4g^2} + \left(\frac{\kappa}{4\omega_{\mathrm{m}}}\right)^2, \tag{6.4}$$

which can go below the unity, indicating that the ground state cooling regime has been reached.

The concept behind the feedback cooling is instead completely different. In such a case, one measures one among the observables of the system – for example a quadrature or the energy – and it acts on it depending on the measurement outcome. In particular, the feedback process can be implemented directly through the use of the measurement apparatus, since it already acts on the system inducing back-action effects. Such effects can be however modified to cool the system toward the ground state. An example is given by the case of an harmonic oscillator whose momentum is continuously monitored [26]. The outcome of such a position $\langle\hat{x}\rangle$ is then used in the feedback loop to generate the feedback force, which will act on the harmonic oscillator, and it is implemented by modifying the stiffness of the harmonic trap. In particular, such a force is proportional to the measurement outcome

$$F_{\mathrm{bf}} = -k_{\mathrm{fb}}(t)\langle\hat{x}\rangle, \tag{6.5}$$

where $k_{\mathrm{bf}}(t)$ is the change of the stiffness due to the feedback. To reduce the amplitude of motion of the particle, that we assume can be parametrized harmonically as $x(t) = \langle\hat{x}(t)\rangle = x_0 \sin(\omega_{\mathrm{m}}t)$, the trap stiffness is increased as the particle climbs the potential well so that its kinetic energy is reduced. When instead the particle goes toward the center, the trap stiffness is reduced. In particular, one needs to modify the stiffness at twice the frequency of the harmonic oscillator to achieve the energy reduction, which leads to an effective extra damping $\delta\Gamma$ with respect to that of the case without feedback Γ. Eventually, one can reach the center-of-mass temperature

$$T_{\mathrm{c.m.}} = \frac{\Gamma}{\Gamma + \delta\Gamma}\, T, \tag{6.6}$$

which is lower than the environmental temperature T.

Cooling techniques were developed also for other quantum systems as for Bose–Einstein condensates (BEC) as, for example the evaporative cooling, the Doppler laser cooling, or the delta-kick cooling (see Figure 6.5). In the case of the evaporative cooling [27], the atomic cloud is initially trapped and let thermalize. Then, the potential that traps the cloud is ramped down so that the more energetic

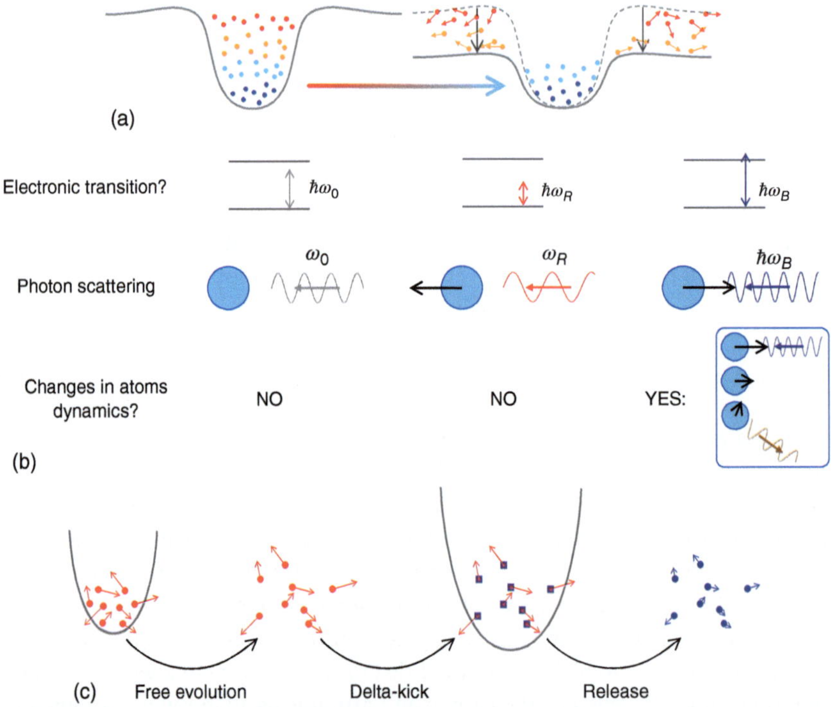

Figure 6.5 Conceptual scheme of the working process behind the evaporative cooling (a), Doppler cooling (b), and delta-kick cooling (c).

atoms can escape the trap boundaries. Those atoms that remain in the trap are those with less kinetic energy, meaning that the corresponding energy distribution can be associated to a lower temperature.

The second cooling technique for atomic systems is the Doppler laser cooling, which is based on the Doppler effect [28, 29]. Here, one shines the atoms with a laser having a frequency slightly below the electron transition, namely it is a red detuned laser. This means that the electrons of a laser will not absorb the corresponding photon, as its energy is not sufficient to lead to an energetic transition. Then, one here exploits the Doppler effect: the atom moving away from the laser source will see the photons being even more red detuned (i.e. at lower frequency), while if it moves toward the source, it will see it blue detuned (i.e. at higher frequency). Then, when moving away, the atom's electron will not get excited; on the contrary, when moving toward the laser source, the gap between the transition energy and the initially red detuned laser is filled, and they can absorb the photon. Due to the conservation of momentum, the atom will then slower and, eventually, its electron will spontaneously emit the absorbed photon. Since the direction of the emission will be random, there will be no net increase of energy when taking the average over the ensemble of atoms.

The third cooling technique we mention is the delta-kick cooling [30]. Here, one neatly traps the ensemble of atoms and lets them thermalize. Then, one leaves the atoms to evolve freely for some limited time before switching on a more broad harmonic trap. During the free evolution, the atoms having more kinetic energy will move further away from the center of the trap. This is then exploited by the harmonic trap, as it will imprint a force on each atom that is proportional to the distance just covered $F \propto -x$. Therefore, more energetic atoms will get a stronger "kick" back toward the center of the trap, thus strongly cooling the entire atom cloud.

Despite the efforts along the lines illustrated above, and the enormous progress achieved in the construction of devices that are only mildly affected by environmental effects, any piece of quantum technology will have to be considered as intrinsically and inherently noisy. As such, it is only meaningful to address the challenge posed by the coupling between a quantum system and its environment from a thermodynamic perspective that treats the energetics of quantum processes from a fundamental perspective but aims directly to tangible technological development [31].

Thermodynamics was developed in the nineteenth century to improve the efficiency of steam engines. Its impact has been enormous and affected fundamental science, technology, and everyday life alike. In the third millennium, the challenge is to address quantum thermodynamics to design efficient quantum machines. Quantum thermodynamics (QT) describes stochastic processes of a quantum system out of its state of equilibrium. The intrinsic stochastic nature of the process can be described in terms of probability density functions, which are characterized by quantum fluctuation relations. These regulate the amount of work and heat that can be respectively performed or exchanged by the system in an arbitrary non-equilibrium process. Such an amount is bounded by the thermodynamic irreversibility of the process (equivalently to the second principle of thermodynamics) and can be characterized in terms of the entropy production (EP) and the entropy production rate (EPR) [32]. Under this perspective, QT provides the means to identify among different protocols for state preparation, which is the most efficient. Indeed, it has been shown that, in a heat engine, the efficiency of a cycle is given by the Carnot efficiency reduced by a term proportional to the EPR [32]. Thus, at minimum EPR corresponds maximum efficiency, and this can be exploited to determine the most efficient protocol. The irreversibility characterization provided by quantum thermodynamics can be used to quantify the performance cost of quantum protocols and thus, find the optimal one with respect to suitable figures of merit, as for example the energy required to perform a specific protocol.

The community working in quantum technology is progressively developing awareness of this challenge [33], although a workplan designed to establish the necessary knowledge baseline for advancing an energetics-based approach to the design and realization of quantum devices is still lacking [34]. Such workplan should be built through the following concrete steps:

Step 1 Design and demonstrate processes that enable the manipulation of energy (work extraction and heat flux steering) by means of quantum systems – both elementary and complex ones – against (classical and quantum) noise;

Step 2 Identify the best schemes for the enhancement of the performance of such processes through genuine quantum resources (including quantum measurements) and sophisticated quantum control schemes;

Step 3 Benchmark the points above in noisy intermediate-scale quantum systems embodying important test beds for a future generation of larger-scale energy effective devices;

Step 4 Demonstrate new functionalities for the efficient control of quantal energy-exchange processes, including novel cooling techniques, and strategies to harness heat fluxes resulting from such processes;

Step 5 The identification of thermodynamics-inspired strategies for the minimization of the energetic cost of quantum information processing.

In what follows, we will address some of these intended steps in some quantitative details.

6.2.1 On Step 2

The fundamental connections between information and thermodynamics date back to the seminal contributions by Maxwell, Szilard, and Landauer. The process of acquiring information can impact the entropic balance of a given physical process. Such information must thus be accounted for and considered on equal footing to other thermodynamic quantities, such as heat and work. This is particularly relevant for processes involving microscopic systems, which are fundamentally dominated by fluctuations: the acquisition of information through measurements introduces additional stochasticity and makes the overall process strongly dependent on the monitoring methodology. When assessing a monitored system, one must distinguish between conditioned and unconditional dynamics, the former being affected by the measurement records [35].

To this end, let us consider the dynamics of a continuously measured open quantum system subjected to a Markovian evolution can be described by a stochastic master equation (SME) describes the evolution of the quantum state of the system *conditioned* on the outcomes of the continuous measurement [36–38]. Upon averaging over all stochastic trajectories, weighted by the outcomes probabilities, the stochastic part vanishes leaving a deterministic Lindblad master equation for the open system dynamics, whose dynamics will be called *unconditional* throughout the paper.

Solving SMEs is in general a complex goal, made simpler when dealing with Gaussian systems, for which the problem is translated into solving a simpler system of stochastic equations. In order to fix the ideas, let us consider a simple system consisting of a single harmonic oscillator described by a pair of quadrature operators $(\hat{q}, \hat{p})$ with $[\hat{q}, \hat{p}] = i\hbar$, and define the vector of operators $\hat{\mathbf{x}} = (\hat{q}, \hat{p})$. The generalization to an n-oscillator system has been reported in Ref. [39]. In the Gaussian framework, the Hamiltonian is bilinear in the quadratures and can be written as $\hat{H} = \frac{1}{2}\hat{\mathbf{x}}^T H_s \hat{\mathbf{x}} + \mathbf{b}^T \Omega \hat{\mathbf{x}}$, where H_s is a 2×2 matrix, $\mathbf{b}$ is a two-dimensional vector accounting for a (possibly time-dependent) linear driving, and $\Omega = i\sigma_{y,j}$ is the symplectic matrix ($\sigma_{y,j}$ is the y-Pauli matrix of subsystem j). By modeling the monitoring process through Gaussian measurements, the SME preserves the Gaussianity of any initial state. In this case, the vector of average momenta $\overline{\mathbf{x}} = \langle \hat{\mathbf{x}} \rangle$ and the Covariance Matrix (CM) $\sigma_{ij} = \langle \{\hat{\mathbf{x}}_i, \hat{\mathbf{x}}_j\} \rangle / 2 - \langle \hat{\mathbf{x}}_i \rangle \langle \hat{\mathbf{x}}_j \rangle$ of the system describe completely the dynamics,through the equations

$$\begin{aligned} \dot{\sigma} &= A\sigma + \sigma A^T + D - \chi(\sigma), \\ \mathrm{d}\overline{\mathbf{x}} &= (A\overline{\mathbf{x}} + \mathbf{b})\mathrm{d}t + (\sigma C^T + \Gamma^T)\mathrm{d}\mathbf{w}, \end{aligned} \tag{6.7}$$

where $\mathrm{d}\mathbf{w}$ is a vector of Wiener increments, A (D) is the drift (diffusion) matrix characterizing the unconditional open dynamics of the system, and $\chi(\sigma)$ is a σ-dependent term that accounts fully for the measurement process. Notwithstanding the stochasticity of the overall dynamics, the equation for the covariance matrix is deterministic.

This is translated in the following expression for the Wigner entropy $S = -\int W \ln W \, \mathrm{d}^{2n}\mathbf{x}$ of the state of the system, which we adopt as our entropic measure [40]

$$S = -\ln(\mathcal{P}) + k = \frac{1}{2}\ln(\det \sigma) + k, \tag{6.8}$$

with k an inessential constant and $\mathcal{P}$ the purity of the state of the system, which for a Gaussian state reads $\mathcal{P} = (\det \sigma)^{-1/2}$. Therefore, S coincides (modulo the constant k_n) with the Rényi-2 entropy, tends to the von-Neumann entropy in the classical limit [41], and is a fully deterministic quantity despite the continuous-monitoring process.

As the Wigner entropy only depends on the CM of the system, its evolution is deterministic even for continuously measured system. The same then holds true for the *entropy rate*

$$\frac{\mathrm{d}S}{\mathrm{d}t} = \frac{1}{2}\frac{\mathrm{d}}{\mathrm{d}t}(\mathrm{Tr}[\log \sigma]) = \frac{1}{2}\,\mathrm{Tr}[2A + \sigma^{-1}(D - \chi(\sigma))], \tag{6.9}$$

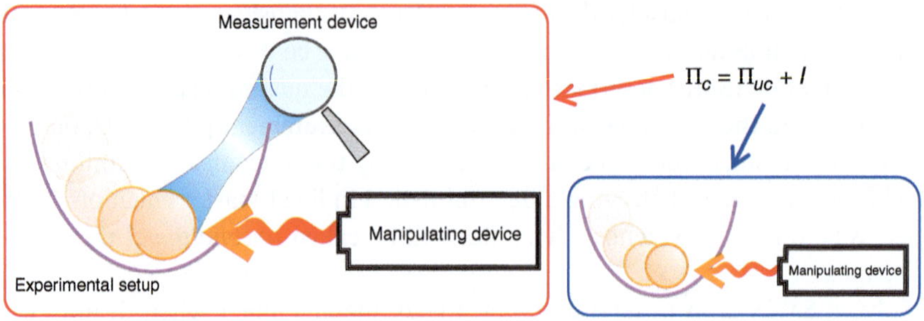

Figure 6.6 Schematic representation describing the difference between the process of unconditional (uc) and conditional (c) dynamics, which, respectively, provide the unconditional Π_{uc} and conditional Π_c entropy production rate (EPR). Their connection is given by Eq. (6.11).

which, from the study of unconditional non-equilibrium thermodynamics [32] of quantum processes, can be split in two contributions as $\dot{S}_{\rm uc} = \Phi_{\rm uc} + \Pi_{\rm uc}$ with $\Phi_{\rm uc}$ ($\Pi_{\rm uc}$) the unconditional entropy flux (production) rate. In the conditional case, while a similar splitting is indeed possible [39] both the entropy flux and production rate are inherently stochastic, as they depend on the first moments of the quadrature operators. Upon taking the average over the outcomes of the measurement, the expression for $\dot{S}$ can be recast into the very elegant form

$$\dot{S} = \dot{S}_{uc} + I, \tag{6.10}$$

where the term $I = \frac{1}{2}\,\mathrm{Tr}[\sigma^{-1}\tilde{D} - \sigma_{\rm uc}^{-1}D]$ accounts for the excess entropy production resulting from the measurement process, and it is thus information theoretical in nature. In turn, this enables a similar splitting for the EPR as

$$\Pi_c = \Pi_{\rm uc} + I. \tag{6.11}$$

This result – which holds regardless of the Gaussian nature of the system at hand [42] – connects the entropy production rate of the unmonitored open system to the akin quantity for the monitored one via the informational term I (see Figure 6.6). The second law for the un-monitored system, which reads $\Pi_{\rm uc} \geq 0$ [32], is now refined as $\Pi \geq I$, which shows the connection between non-equilibrium thermodynamics and information theory. The I term depends explicitly on the measurement strategy implemented to acquire information, which in turn affects the dynamics of the measured system, driving it to different final states. We thus have two different yet related mechanisms that we can exploit here: on the one hand, we have the dynamics itself, which drives the evolution of a system with the aim, potentially, to optimize thermodynamic performance. On the other hand, we have the well-known possibility to condition the dynamics of a quantum system through the back-action induced by a measurement process. This clearly offers an exploitable mechanism to effectively drive the open dynamics of a system toward a thermodynamics-based criterion for the choice of the specific protocol to be implemented. Indeed, EPR embodies a cost function whose value one aims to control when optimizing the conditional dynamics of the system at hand.

6.2.2 On Step 4

We have previously addressed a scheme for feedback cooling of a mechanical system that relied on a closed-loop approach. This is not the only possibility at hand, and one can exploit a different method that, building on the capabilities offered by conditional dynamics ensuing quantum measurements, achieves efficient cooling performances.

Let us consider a harmonic oscillator of frequency ω, characterized by annihilation and creation operators a and $a^\dagger$, respectively, such that $[a, a^\dagger] = 1$. Such bosonic mode or harmonic oscillator comprises a stiffening Duffing-like deformation with strength $\epsilon > 0$, such that $\epsilon \ll \omega$, as found in different experimental platforms. In addition, the bosonic mode is coupled to a (spin-like) two-level system via a Jaynes–Cummings interaction. The Hamiltonian of the system reads

$$H_s = \hbar\frac{\omega_A}{2}\sigma_z + \omega a^\dagger a + \lambda(a\sigma^+ + a^\dagger\sigma^-) \tag{6.12}$$

where ω_A and λ denote the Bohr frequency and coupling strength of the two-level system, respectively. We have introduced the spin Pauli matrices, $\sigma_{x,y,z}$ such that $[\sigma_i, \sigma_j] = 2i\delta_{ijk}\sigma_k$ and $\sigma_z = |e\rangle\langle e| - \langle g|\langle g|$ with $|e\rangle$ ($|g\rangle$) the excited (ground) state of the two-level system. Finally, $\sigma^+ = (\sigma^-)^\dagger = |e\rangle\langle g|$ is the spin raising operator. At the resonant condition $\omega_A = \omega$, the dynamics governed by the Jaynes–Cummings interaction takes a state $|g, n+1\rangle$ and transforms it into $|e, n\rangle$ in a time $T_n = \pi/(2\lambda\sqrt{n+1})$ with $n \geq 0$.

The goal is to cool an initial thermal state $\rho_r^{\text{th}} = \sum_{k=0} p_k |k\rangle\langle k|$ (with $p_k = n_{\text{th}}^k/(1+n_{\text{th}})^{k+1}$ and $n_{\text{th}} = \text{Tr}[a^\dagger a \rho_r^{\text{th}}]$) of the oscillator down to its ground state by performing measurements on the spin degree of freedom. That is,

$$\rho_r^{\text{th}} \to |\psi_{\text{gs}}\rangle\langle\psi_{\text{gs}}| \approx |0\rangle\langle 0|. \tag{6.13}$$

The model in Eq. (6.12) can be realized in a number of different platforms. Among them, levitated nanoparticles [43, 44], trapped ions [45], circuit quantum electrodynamics [46], and optomechanical systems. In the latter context, a spin defect of frequency $\omega_A \in [0.5, 1.5]\omega$ coupled to a mechanical resonator, $\omega \approx 200$ MHz, and $Q \approx 10^6$, can achieve coupling strengths as large as $\lambda \approx 0.05\omega$ at negligible spin damping rates [47].

The cooling schemes put forward in Refs. [48, 49] consider the concatenation of free evolution following Eq. (6.12) and projective measurements on the spin degrees of freedom. While Ref. [48] uses random detection times, the approach of Ref. [49] implies measuring the spin at regular intervals. When N_{rep} cycles are implemented, starting from the initial state

$$\rho_s(0) = |g\rangle\langle g| \otimes \rho_r^{\text{th}}, \tag{6.14}$$

and when the free evolution in each cycle takes place for a time $T_n = \pi/(2\lambda\sqrt{n+1})$, we bring in populations from $|g, n+1\rangle$ to $|e, n\rangle$ states with $n \geq 0$ by sweeping each of the subspaces at a time. This is achieved by evolving $\rho_s(0)$ during a time, i.e. $\rho_s(T_n) = U(T_n)\rho_s(0)U^\dagger(T_n)$ with $U(t) = e^{-itH_s}$ the evolution operator. In this manner, we remove excitations and entropy and thus, cool down the oscillator. After a total time

$$T_f = \frac{\pi}{2\lambda} \sum_{n=0}^{N_{\text{rep}}-1} \frac{1}{\sqrt{n+1}}, \tag{6.15}$$

where a negligible detection time was assumed. The probability of a successful detection of the spin in its ground state $|g\rangle$ upon the evolution $U(T_n)$ is given by $p_{g;n} = \text{Tr}[|g\rangle\langle g|\,\rho(T_n)]$, which is lower-bounded by the probability $p_0 = (1 + n_{\text{th}})^{-1}$ to find the oscillator in its ground state. Upon N_{rep} repetitions, success is achieved for $\Pi_{n=0}^{N_{\text{rep}}-1} p_{g;n} \approx p_0$, which turns out to be achievable for $N_{\text{rep}} \gg 1$. Hence, one can already notice that this method can be favorable to cool down states of a resonator containing few excitations. In particular, if $n_{\text{th}} \lesssim 10$, we have $p_0 \gtrsim 1/10$ with $p_k \lesssim 10^{-3}$ for $k \gtrsim 50$, so that $N_{\text{rep}} \lesssim 50$ would be sufficient to achieve a significant reduction on the occupation number. However, it is thermodynamically impossible to achieve the ground state of a quantum system in a finite time in light of the third law of thermodynamics and the unattainability principle [50], while being still possible to get very close to the actual ground state.

Remarkably, this approach is effective in cooling also oscillators that showcase a small degree of nonlinearity [49]. For instance, for a quartic potential of the Duffing type $\epsilon(a + a^{\dagger})^4$, despite the coupling between states in the Jaynes–Cummings spectrum that do not have the same energy [something impossible to achieve through Eq. (6.12) only], similar cooling performances can be achieved, although oscillators with large values of ϵ require longer concatenation cycles.

6.3 Data Intensive Tools for Quantum Computing Science

Quantum computing and artificial intelligence are both transformational technologies in need of each other to achieve significant progress. Although artificial intelligence produces functional applications with classical computers, it is limited by the computational capabilities of the latter. Quantum computing can provide a computation boost to artificial intelligence, enabling it to tackle more complex problems. The visionary goal from the computer science perspective is the achievement of artificial general intelligence, namely the engineering of a system capable of human-level thinking, continuously improving its performances and reaching far beyond current (classical) computational capabilities. For quantum computing, instead, the interplay with artificial intelligence, specifically through machine learning methods, holds the promises to exploit the computational advantages of quantum processors so as to achieve results that are not possible with classical ones. This embodies a new and exciting way to combine a special form of fundamental science, this time deeply rooted into the mathematical and statistics-based efforts aimed at the development of sophisticated algorithmic methods for the processing of large dataset, and quantum technologies for computation and information processing at large [51].

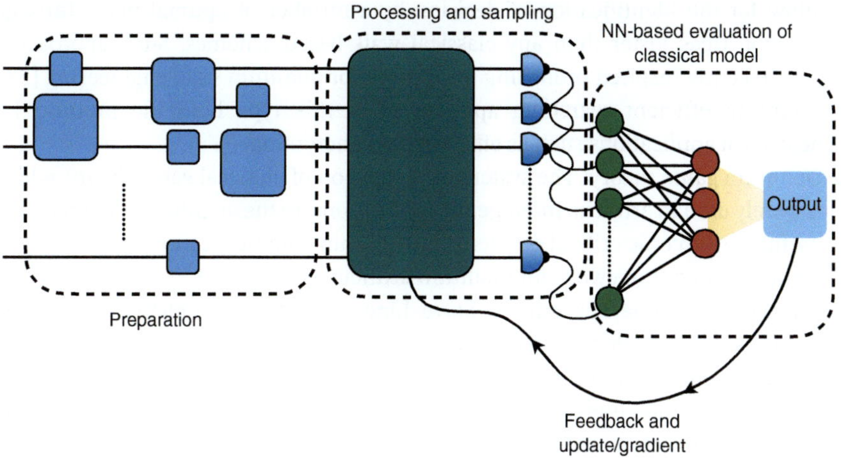

Figure 6.7 General architecture for a machine learning-assisted quantum computational process: the outputs of the quantum processing stage is harvested through a sampling and fed into an NN-based architecture for the evaluation of a classical cost function, whose value informs a feedback loop that changes the parameters characterizing the quantum circuit, aiming at optimization.

In a typical implementation of a machine-learning-assisted quantum computational problem, quantum data obtained from a suitable preparation stage are processed and manipulated by a quantum circuit, implementing the quantum computational stage of the process. Classical information is harvested through a sampling stage where the output state of the circuit is probed by an array of quantum measurements. Such classical information is then fed into a classical processing architecture aimed at evaluating a classical model – for instance through a neural network (NN)-like implementation – that results in the evaluation of a cost function whose value informs a feedback/updating stage of the quantum circuit, aimed at optimizing the quantum processing steps till convergence is reached [cf. Figure 6.7]. The typical complexity of quantum data (stemming from quantum features such as superposition or entanglement) and their fragility to classical and quantum noise make it useful to deploy machine-learning-based methods for their interpretation and manipulation.

Crudely, the following embodies a short list of applications of machine learning to quantum information processing problems that are currently being pursued to drive the design of more process efficient quantum technology [52]

- Development of quantum algorithms for quantum generalization of classical learning models. The scope is the achievement of quantum-induced speed-ups in the deep learning training process, for instance through the fast identification of the optimal values of the weights and links in a NN architecture.
- Quantum algorithms for optimal classical decision problems. The formulation of optimal decision trees is in general complex and cannot be efficiently address by dichotomic approaches leading to classical random walks. Quantum walks, through the coherence-induced fast exploration of configuration space, could

allow for the identification of decision by a number of optimal paths through decision trees faster than any classical walk-based schemes. Such approaches have recently inspired promising techniques for quantum state engineering [53]. Quantum efficient searching approaches of this type hold the promise for near-term applications such as efficient data encryption.

- Quantum game theory: The systematic extension of classical game theory, which is widely used in artificial intelligence applications, to the quantum domain will be useful to overcome critical problems in quantum communication and the implementation of a framework for quantum artificial intelligence.
- Quantum Machine Learning to Solve Linear Algebraic Problems: Many Data Analysis and machine learning problems are solved by performing matrix operation on vectors in a high-dimensional vector space. Quantum computers can solve common linear algebraic problems such as the Fourier Transformation, finding eigenvectors and eigenvalues, and solving linear sets of equations over vector spaces in time that is polynomial in the dimension of the space (and exponentially faster than classical computers due to the quantum speedup). One of the examples is the Harrow, Hassidim, and Lloyd (HHL) algorithm [54].
- Quantum Principal Component Analysis: Principal Component Analysis is a dimensionality reduction technique that is used when managing large datasets. Dimensionality reduction comes at the cost of accuracy, as one needs to decide which variables to eliminate without losing important information. Classically, dealing with such request efficiently is hard: over large input datasets with many features and attributes, classical methods of principal component analysis will fail because it will be hard for us to visualize the importance of each variable.
  Another issue with classical computers is the calculation of eigenvectors and eigenvalues, whose number grows with the dimensionality of the input. Quantum computers can solve this problem very efficiently and at a very high speed by using Quantum Random Access Memory (QRAM) [55] to choose a data vector at random. It maps that vector into a quantum state. The summarized vector that we get after Quantum Principal Component Analysis has logarithmic size in the number of qubits involved. By repeatedly sampling the data and using a trick called density matrix exponentiation, combined with the quantum phase estimation algorithm (which calculates the eigenvectors and eigenvalues of the matrices), we can take the quantum version of any data vector and decompose it into its principal components. Both the computational complexity and time complexity are thus reduced exponentially.
- Quantum Support Vector Machines: Support Vector Machine is a classical machine learning algorithm used both for classification and regression. For classification tasks it is used to classify linearly separable datasets into their respective classes. Suppose, if the data are not linearly separable, then its dimensions are increased till it is linearly separable. Quantum computers can perform Support Vector Algorithm at an exponentially faster rate owing to superposition and entanglement.
- Quantum Optimization: Optimization is used in a machine learning model to improve the learning process so that it can provide the most adequate and accurate

estimations. The main aim of optimization is to minimize a loss function. A more considerable loss function means there will be more unreliable and less accurate outputs, which can be costly and lead to wrong estimations. Most methods in machine learning require iterative optimization of their performance. Quantum optimization algorithms suggest improvement in solving optimization problems in machine learning through the use of multi-party superpositions.

- Deep Quantum Learning: Quantum computing can be combined with deep learning to reduce the time required to train a neural network. By this method, we can introduce a new framework for deep learning and performing underlying optimization. We can mimic classical deep learning algorithms on an actual, real-world quantum computer. When multilayer perceptron architectures are implemented, the computational complexity increases as the number of neurons increases. Dedicated GPU clusters can be used to improve the performance, significantly reducing training time. However, even this will increase when compared with quantum computers. Quantum computers are designed in such a way that the hardware can mimic the neural network instead of the software used in classical computers. Here, a qubit acts as a neuron that constitutes the basic unit of an NN. Thus, a quantum network can act as an NN and can be used for deep learning applications at a rate that surpasses any classical machine learning algorithm.

On the other hand, a number of applications of classical machine learning are driving the development of significantly disruptive methods aimed at demonstrating prototype quantum sensors able to optimally extract information about its environment, achieve fully automatic calibration and operation of multi-qubit circuits, and improve the performance of quantum algorithms for quantum chemistry.

6.4 Conclusions

The transfer of research results from the field of basic research to the development of quantum devices may be the only challenge facing the community working in quantum technologies, which will be considered as we enter a period of domain and territorial integration. Its successful implementation requires a multidisciplinary approach. While this has only been partially addressed so far, such transfer is critical to the success of our endeavors in building a quantum technological framework and infrastructure. This contribution has focused on two such aspects, namely energetics at the fundamental quantum level and the interplay between artificial intelligence and quantum, as two tantalizing pathways to explore while pursuing a such tantalizing ultimate goal.

Acknowledgments

We acknowledge support from the European Union's Horizon 2020 FET-Open project TEQ (766900), the Leverhulme Trust Research Project Grant UltraQuTe (grant RGP-2018-266), the Royal Society Wolfson Fellowship (RSWF/R3/183013), the UK EPSRC (grant EP/T028424/1), and the Department for the Economy Northern Ireland under the US-Ireland R&D Partnership Programme, the Horizon Europe EIC Pathfinder project QuCoM (Grant Agreement No.101046973).

References

1 Cardoso, M.C., Silva, M., Vellasco, M.M.B.R., and Cataldo, E. (2015). Quantum-inspired features and parameter optimization of spiking neural networks for a case study from atmospheric. *Procedia Computer Science* 53: 74.

2 Jackson, M. and McAdams, S. (2022). The future of quantum drug discovery. https://medium.com/cambridge-quantum-computing/the-future-of-quantum-drug-discovery-909aa5140bff (accessed 12 May 2022).

3 Elfving, V.E., Broer, B.W., Webber, M. et al. (2020). How will quantum computers provide an industrially relevant computational advantage in quantum chemistry? arXiv:2009.12472.

4 Giustino, F., Lee, J.H., Trier, F. et al. (2020). The 2021 quantum materials roadmap. *Journal of Physics: Materials* 3: 042006.

5 Joos, E. and Zeh, H.D. (1985). The emergence of classical properties through interaction with the environment. *Zeitschrift für Physik B Condensed Matter* 59: 223–243.

6 Schlossauer, M. (2019). Quantum decoherence. *Physics Reports* 831: 1–57.

7 Hornberger, K., Uttenthaler, S., Brezger, B. et al. (2003). Collisional decoherence observed in matter wave interferometry. *Physical Review Letters* 90: 160401.

8 Romero-Isart, O. (2011). Quantum superposition of massive objects and collapse models. *Physical Review A* 84: 052121.

9 Tolra, B.L., O'Hara, K.M., Huckans, J.H. et al. (2003). QObservation of reduced three-body recombination in a fermionized 1D Bose gas. arXiv e-prints cond-mat/0312003.

10 LIGO Scientific collaboration. https://www.ligo.org/index.php (accessed 28 June 2022).

11 Könemann, T., Kaczmarczik, U., Gierse, A. et al. (2015). Concept for a next-generation drop tower system. *Advances in Space Research* 55: 1728–1733.

12 Lotz, C., Froböse, T., Wanner, A. et al. (2020). Einstein-Elevator: a new facility for research from μg to 5 g. *Gravitational and Space Research* 5: 11–27.

13 Belenchia, A., Carlesso, M., Donadi, S. et al. (2021). Test quantum mechanics in space–invest US $1 billion. *Nature* 596: 32–34.

14 Gasbarri, G., Belenchia, A., Carlesso, M. et al. (2021). Testing the foundations of quantum physics in space interferometric and non-interferometric tests with large particles. *Communications Physics* 4: 155.

15 Belenchia, A., Carlesso, M., Bayraktar, O. et al. (2022). Quantum physics in space. *Physics Reports* 951: 1.

16 Kaltenbaek, R., Aspelmeyer, M., Barker, P.F. et al. (2016). Macroscopic quantum resonators (MAQRO): 2015 update. *EPJ Quantum Technology* 3: 1–47.

17 Sidhu, J.S., Joshi, S.K., Gündo?an, M. et al. (2021). Advances in space quantum communications. *IET Quantum Communication* 2: 182–217.
18 Selig, H., Dittus, H., and Lämmerzahl, C. (2010). Drop tower microgravity improvement towards the Nano-g level for the MICROSCOPE payload tests. *Microgravity Science and Technology* 22: 539–549.
19 Danzmann, K. and the LISA Study Team (1996). LISA: laser interferometer space antenna for gravitational wave measurements. *Classical and Quantum Gravity* 13: A247.
20 Arndt, M. and Hornberger, K. (2014). Testing the limits of quantum mechanical superpositions. *Nature Physics* 10: 271–277.
21 Carlesso, M., Donadi, S., Ferialdi, L. et al. (2022). Present status and future challenges of non-interferometric tests of collapse models. *Nature Physics* 18: 243–250.
22 Bassi, A., Großardt, A., and Ulbricht, H. (2017). Gravitational decoherence. *Classical and Quantum Gravity* 34: 193002.
23 Carney, D., Stamp, P.C.E., and Taylor, J.M. (2019). Tabletop experiments for quantum gravity: a user's manual. *Classical and Quantum Gravity* 36: 034001.
24 Monteiro, F., Afek, G., Carney, D. et al. (2020). Search for composite dark matter with optically levitated sensors. *Physical Review Letters* 125: 181102.
25 Barzanjeh, S., Xuereb, A., Gröblacher, S. et al. (2022). Optomechanics for quantum technologies. *Nature Physics* 18: 15.
26 Vovrosh, J., Rashid, M., Hempston, D. et al. (2017). Parametric feedback cooling of levitated optomechanics in a parabolic mirror trap. *Journal of the Optical Society of America B* 34: 1421–1428.
27 Petrich, W., Anderson, M.H., Ensher, J.R., and Cornell, E.A. (1995). Stable, tightly confining magnetic trap for evaporative cooling of neutral atoms. *Physical Review Letters* 74: 3352.
28 Wineland, D.J. and Dehmelt, H. (1975). Proposed $10^{14}\Delta\nu < \nu$ Laser Fluorescence Spectroscopy on Tl^+ Mono-Ion Oscillator III. *Bulletin of the American Physical Society* 20: 637.
29 Hänsch, T.W. and Schawlow, A.L. (1975). Cooling of gases by laser radiation. *Optics Communications* 13: 68–69.
30 Ammann, H. and Christensen, N. (1997). Delta kick cooling: a new method for cooling atoms. *Physical Review Letters* 78: 2088.
31 Deffner, S. and Campbell, S. (2019). *Quantum Thermodynamics*. Morgan & Claypool Publishers.
32 Landi, G.T. and Paternostro, M. (2021). Irreversible entropy production, from quantum to classical. *Reviews of Modern Physics* 93: 035008.
33 (a) Cimini, V., Gherardini, S., Barbieri, M. et al. (2020). Experimental characterization of the energetics of quantum logic gates. *npj Quantum Information* 6: 96; (b) Deffner, S. (2021). Energetic cost of Hamiltonian quantum gates. *EPL* 134: 40002; (c) Abah, O., Puebla, R., Kiely, A. et al. (2019). Energetic cost of quantum control protocols. *New Journal of Physics* 21: 103048; (d) Landi, G.T., de Oliveira, A.L.F. and Buksman, E. (2020). Thermodynamic analysis of quantum error-correcting engines. *Physical Review A* 101: 042106; (e) Buffoni, L. and

Campisi, M. (2020). Thermodynamics of a quantum annealer. *Quantum Science and Technology* 5: 035013; (f) Karimi, B., Brange, F., Samuelsson, P., and Pekola, J.P. (2020). Reaching the ultimate energy resolution of a quantum detector. *Nature Communications* 11: 1; (g) Linpeng, X., Bresque, L., Maffei, M. et al. (2022). Energetic cost of measurements using quantum, coherent, and thermal light. *Physical Review Letters* 128: 220506; (h) Stevens, J. (2021). Energetics of a Single Qubit Gate. arXiv:2109.09648.

34 Auffèves, A. (2022). Quantum technologies need a quantum energy initiative. *PRX Quantum* 3: 020101.

35 Manzano, G. and Zambrini, R. (2022). Quantum thermodynamics under continuous monitoring: a general framework. *AVS Quantum Science* 4: 025302.

36 Doherty, A.C. and Jacobs, K. (1999). Feedback control of quantum systems using continuous state estimation. *Physical Review A* 60: 2700.

37 Serafini, A. (2017). *Quantum Continuous Variables: A Primer of The- oretical Methods*. CRC Press.

38 Genoni, M.G., Lami, L., and Serafini, A. (2016). *Contemporary Physics* 57: 331.

39 Belenchia, A., Mancino, L., Landi, G.T., and Paternostro, M. (2020). Entropy production in continuously measured gaussian quantum systems. *npj Quantum Information* 6: 97.

40 Adesso, G., Girolami, D., and Serafini, A. (2012). *Physical Review Letters* 109: 190502.

41 Santos, J.P., Landi, G.T., and Paternostro, M. (2017). *Physical Review Letters* 118: 220601.

42 Landi, G.T., Paternostro, M., and Belenchia, A. (2022). Informational steady-states and conditional entropy production in continuously monitored systems. *PRX Quantum* 3: 010303.

43 Fonseca, P.Z.G., Aranas, E.B., Millen, J. et al. (2016). Nonlinear dynamics and strong cavity cooling of levitated nanoparticles. *Physical Review Letters* 117: 173602.

44 Setter, A., Vovrosh, J., and Ulbricht, H. (2019). Characterization of non-linearities through mechanical squeezing in levitated optomechanics. *Applied Physics Letters* 115: 153106.

45 Home, J.P., Hanneke, D., Jost, J.D. et al. (2011). Normal modes of trapped ions in the presence of anharmonic trap potentials. *New Journal of Physics* 13: 073026.

46 Ong, F.R., Boissonneault, M., Mallet, F. et al. (2011). Circuit QED with a nonlinear resonator: ac-Stark shift and dephasing. *Physical Review Letters* 106: 167002.

47 Tian, L. (2011). Cavity cooling of a mechanical resonator in the presence of a two-level-system defect. *Physical Review B* 84: 035417.

48 Li, Y., Wu, L.-A., Wang, Y.-D., and Yang, L.-P. (2011). Nondeterministic ultrafast ground-state cooling of a mechanical resonator. *Physical Review B* 84: 094502.

49 Puebla, R., Abah, O., and Paternostro, M. (2020). Measurement-based cooling of a nonlinear mechanical resonator. *Physical Review B* 101: 245410.

50 Masanes, L. and Oppenheim, J. (2017). A general derivation and quantification of the third law of thermodynamics. *Nature Communications* 8: 14538.

51 Carleo, G., Cirac, I., Cranmer, K. et al. (2019). Machine learning and the physical sciences. *Reviews of Modern Physics* 91: 045002.

52 (a) Bromley, T.R., Arrazola, J.M., Jahangiri, S. et al. (2020). Applications of near-term photonic quantum computers: software and algorithms. *Quantum Science and Technology* 5: 034010; (b) Schuld, M., Sinayskiy, I., and Petruccione, F. (2015). An introduction to quantum machine learning. *Contemporary Physics* 56: 172; (c) Dunjko, V. and Briegel, H.J. (2018). Machine learning & artificial intelligence in the quantum domain. *Reports on Progress in Physics* 81: 074001.

53 Innocenti, L., Majury, H., Giordani, T. et al. (2017). Quantum state engineering using one-dimensional discrete-time quantum walks. *Physical Review A* 96: 062326.

54 Harrow, A.W., Hassidim, A., and Lloyd, S. (2009). Quantum algorithm for solving linear systems of equations. *Physical Review Letters* 15: 150502.

55 Giovannetti, V., Lloyd, S., and Maccone, L. (2008). Quantum random access memory. *Physical Review Letters* 100: 160501.

7

Quantum Networks in Space

Lisa Wörner

German Aerospace Center, Institute for Quantum Technologies, DLR-QT, Wilhelm-Runge Strasse 11, 89081 Ulm, Baden Württemberg, Germany

7.1 Global Communication

Communication of information is of crucial importance to human nature and global communication has proven to be indispensable for dissemination of information and impacts all aspects of society. Several studies have investigated the necessity and impact of instant availability of information and efficient global communication [1–4]. With increasing amount and faster accessibility of available information, the necessity for information security is rising [5]. The protection of data becomes more prevalent in case of sensitive data, such as medical [6], economic [7], and military information [8], but also spans personal exchange and messaging [9]. As an example, stock market exchange, which relies on fast transmission of exchange requires high standards of transmission security [10, 11].

The necessity of long range secure communication is not a new invention. Most of warfare relied on gaining an advantage over the other party by either intercepting transmissions or spying. The enigma machine [12], deployed during world war two has reached infamous fame, not least by the 2014 movie "The Imitation Game" starring Benedict Cumberbatch.

With quantum mechanics, a novel form of secure communication is available to humanity [13]. Quantum mechanics present a unique branch of physics, in which phenomena are described that are un-intuitive at first sight. One of such phenomena is entanglement [14], in which a system is described by its overall properties without having knowledge about the properties of the individual constituents. Such a system is the hydrogen atom, in which the overall spin has to amount to a bosonic particle, without knowledge about the spin positions of the involved proton and electron being available. Entanglement is the property enabling secure quantum communication [15]. Here, two entangled photons are prepared with one being send to the partner. The photons are transferred via an optical fiber or free space with the setup details being transmitted classically. The longest terrestrial free space exchange of entangled photons has been established on the Canary islands and spanned 144 km [16]. With longer distances being

Photonic Quantum Technologies: Science and Applications, First Edition. Edited by Mohamed Benyoucef.

achieved by fiber-based systems [17]. Space based experiments are on the rise with recent experiments demonstrating entanglement over a distance of 1.200 km [18]. Deploying a combination of fiber bound and free space links including necessary relay stations, quantum key distribution (QKD) over networks of total lengths of 4.600 km [19] have been achieved.

As long range entanglement transmission has been demonstrated, it is important to focus on the advantages of the novel technology. While a portion of the information has to be transmitted classically, the inherent security is not compromised by any leakage of the classical channel. In the contrary, due to quantum entanglement, this enables the two parties to know if the information transfer has been intercepted. This possibility renders the deployment of quantum entanglement ideal for the distribution of keys instead of full messages, QKD [20]. If only a key is exchanged, the knowledge of interception allows for the key to be discarded and to start a new sequence. The details on QKD between two partners including sources, detectors, protocols, experimental needs, and open challenges are described in reviews, such as reference [21–23]. Here, the focus shall lie on the challenges of enabling global secure communication through quantum technologies.

As it has been established above, global communication impacts all parts of society with most of humankind relying on the availability of information. Some applications require higher standards of security, while others focus on fast transmission of information. There are mainly three sources of information transfer available at this time:

- *Cables and fibers*: For this method of information transfer, cables connecting different users are deployed. Mostly these cables are installed under ground or, as in the example of the internet glass fibers, on the oceans floor. Over ground cables are more cost effective in installation but have the disadvantage of higher probability of cable damage, for instance by environmental occurrences.
- *Terrestrial free space*: Broadcasting signals, especially radio and mobile phone signals are transferred in free space over long distances. For this method, transmission masts are installed and connected to existing networks to enable global coverage.
- *Satellite-based free space*: With satellite-based information transfer, broader coverage is possible. An important example are the global positioning service (GPS) or the European Galileo satellites, providing position, navigation, and timing information worldwide. These satellites usually communicate with ground using emissions in the radio-frequency range.

Let us examine these options for viability for QKD: Photonic quantum entanglement, as it is currently deployed in QKD, requires the transmission of photons. The deployed photons are usually in the infrared range, with 1550 nm due to availability of sources and equipment. The reason for the wide availability is due to the transmission windows in glass around 850, 1310, and 1550 nm. With transmission profiles strongly dependent on the material, adaptations, and tailoring are possible but costly. In consequence telecom applications rely on 1550 nm, leading to the development of various components, increased availability, and in consequence, reduction in component costs.

Other wavelengths are studied and, in the following, we will discuss, why other wavelengths might be interesting and necessary to advance the technology to establish global networks. All of the above discussed methods of global networks operate on electromagnetic emissions in the radio-frequency range. Here, now the usability of these methods in the optical regime shall be discussed.

Obviously, the first option for establishing a global network appears to be the usage of optical fibers. Those yield the absence of precise pointing and easy circumvention of obstacles, such as buildings or landmarks. However, the loss rates in optical fibers are high if compared to loss rates of free space links. Loss rates in free space propagation are dominated by diffractive spreading and absorption by the surrounding medium. The resulting intensity reduction in the far field regime is best described by an inverse square law [24] and consequently trumps over the exponential loss rate in optical fibers, which has dropped to be in the order of 0.2 dB/km at 1550 nm [25].

This leaves the discussion of free-space transmission. As pointed out above, free-space transmission requires optimal pointing. In a terrestrial environment, long-range free-space transmission faces three main challenges:

1. *Earths curvature*:
   For short-range transmissions, the curvature of the Earth is negligible. As such, the curvature could be countered by a sufficient number of intermediate transmission masts. However, for truly global coverage, long-distances, such as oceanic bodies, have to be overcome, where relay stations are challenging to implement and to maintain.
2. *Atmospheric turbulences and losses*:
   As mentioned above, absorption and random scattering are the two major sources of signal loss of a free space beam. This becomes more prevalent the denser the gas, the photons have to traverse, and the larger the cross section of absorption. Here, infrared optical beams trump over ultraviolet optical beams as the cross section usually increases with decreasing wavelength. Most molecules and atoms present in the air will, however, absorb photons in the visible spectrum at a sufficient rate to be non-negligible for horizontal free space laser links over long distances.

   Additionally, the atmosphere is prone to pollution, turbulences caused, for instance, by thermal fluctuations or storms, and the occurrence of other weather phenomena [26]. This increases the optical density of the atmosphere and thereby losses in the transmitted photonic signal. Furthermore, widening of the optical beam and deflections of the beam are a result of varying atmospheric densities.

   As a consequence, it is advisable to reduce the distance any optical signal has to travel through the atmosphere. It is to be remarked, that the longest free space entanglement experiment was performed at an high altitude in optimal conditions [16].

3. *Landmarks and buildings*:
   Finally, free space optical beams in the infrared range can be blocked by landmarks, such as mountains, and tall buildings. As the demand for secure information is especially high in populated areas, this presents a challenge to the distribution.
   Applications in less densely populated areas, which might require temporary optical communication, could be installed on masts and would be prone to blockage by trees or dust storms.

In conclusion, a global secure quantum network can not be established based solely on terrestrial transmitters. To reduce the impact of atmospheric turbulences, depending on space based systems is the solution that appears the most reasonable. This system requires optical downlinks to the Earth to enable the secure transmission and make use of quantum mechanics in form of photon entanglement. Satellite-based quantum networks allow global coverage and need to be supplemented by terrestrial systems to transfer the information between the two intended partners. As such, it seems advisable to build ground stations with the highest level of security and transfer information from there with classical protocols, which do not include all of the sensitive information or for which a lower level of security is understood.

7.2 Challenges in Global Secure Quantum Networks

As established in Section 7.1, communication over long distances or on a global scale is essential to civilization as we know it with impacts in all sectors. It was also discussed, that the most promising solution for a global secure quantum network is a satellite-based system to reduce the impact of atmospheric or other disturbances.

Let us now take a look at what a global secure quantum network based on a satellite network entails and what is required to set it up.

To avoid confusion, the following notation will follow the general rules, in which Alice is usually understood as the sender of information and Bob receives them. In this chapter, the interception by a third party, usually called Eve, will not be discussed. The inherent security of quantum communication is based on entanglement. Entanglement is a quantum mechanical phenomena in which solely the state of the overall system is known but not that of the individual constituents, see for instance Ref. [16]. The phenomena is again touched on in the following section headed "Untrusted Nodes" 7.3.

In the following, first, the different orbital constellations and involved challenges will be discussed. The constellations are schematically depicted in Figure 7.1. Here, it is assumed that two partners, Alice and Bob want to exchange information that requires a high level of security. The other depicted buildings and vehicles that are shown are representations of other possible receivers or transmitters of secure communication. To keep the readability of the figure, the required classical channel for the exchange of secure quantum information between Alice and Bob is inserted only once as a landline. The classical channel, necessary to the generation of the transmitted key, needs to be authenticated to preserve security.

Other channels means of traditional communication between the two are, of course, possible.

In addition to the secure transmission, it is possible to envisage communication through non-quantum channels to end-users as depicted on the right hand side of the figure. Here, it has to be understood, that the level of security is no longer available. However, if only parts of the message are relayed and not the key and encrypted message, the key is not corrupted and could still be used by Alice and Bob. In such a scenario, probably only parts of the information will be forwarded to third parties and secure classical channels deployed.

The first idea that comes to mind when discussing global secure quantum communication is that of having a satellite with a reflector on board. In this scenario, the information is directly relayed without any storage in-between. While such a scenario is not encumbered by landmarks or buildings, the distance between Alice and Bob is still limited. Without any storage, the two stations have to be within the field of view of the satellite. The length of the transferred information is limited as the optical line of sight has to be established for both stations at the same time. Finally, the radar equation, describes that the losses in ideal transmission conditions are proportional to the distance to the power of four, limiting the maximum achievable range. The described scenario is marked **scenario one** in Figure 7.1.

To enlarge the distance between the two ground stations and increase the length of the transferable information in this scenario, it is possible to store the information on board of a satellite before sending it to the receiver. This scenario is depicted in Figure 7.1 as **scenario two**. Here, the signal is transmitted to a satellite by the ground station and then stored on-board the satellite. The satellite travels a given

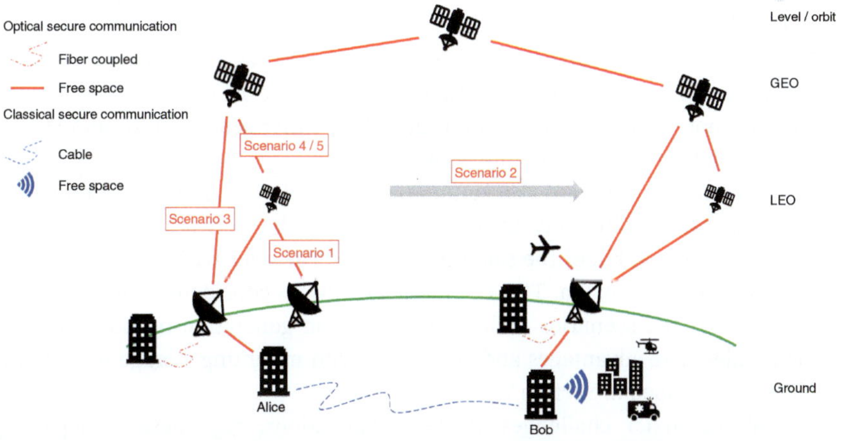

Figure 7.1 Depiction of different scenarios for a space-based quantum network. Exemplary, the transmitter (Alice) and the receiver (Bob) are chosen in this image. Also, only for those two the required additional classical connection is shown. The only exception is scenario one, in which the receiver is not depicted.

time before the information is relayed to the second ground station. With this scenario it is theoretically possible to connect two ground stations on opposite sides of the Earth. The main limitation of the achievable distance between Alice and Bob is the travel time of the satellite and with that its coherent storage capability. In 2017, the idea for such constellations has been outlined [27]. This scenario sets requirements towards the deployed quantum repeater and their memory types. Those are outlined in Table 7.1.

Another option to establish a secure quantum network is by stationing several individual satellites connected by optical links. This option requires short term storage on board the satellites before sending the information on to the next relay station and finally to the receiver Bob. This scenario is depicted as **scenario three** in Figure 7.1.

This scenario considers geostationary satellites that are then linked to a ground station. While geostationary satellites yield the advantage of being permanently available, their distance to the ground station, shorter time of sight from the ground station, and increased radiation levels. present difficult challenges. With the increased distance, the single photon transmission might get difficult to discriminate against the background radiation and light pollution.

Similar challenges are posed in navigation, where the current radio-frequency-based technology is targeted by more precise optical clocks that also require optical links between satellites and ground stations. For this scenario, the Kepler constellation was proposed [28, 29]. In this constellation, two sets of satellites are foreseen with one subset operating in a low Earth orbit (LEO) and the other operating in geostationary positions.

To transfer the concept, a set of LEO and geostationary satellites could be envisaged which transmit the information between Alice and Bob. This way, the closer satellites could act as relay stations to enhance the signal before distributing it globally through the interconnected geostationary satellites. This scenario is depicted as **scenario four** in Figure 7.1.

Another option in this constellation is to generate the two entangled photons not on ground at Alice but in orbit (**scenario five** in Figure 7.1). In this scenario, the entangled photons are created in orbit and then either send to a ground station directly or transferred via a satellite network.

The different scenarios are also summarized in Table 7.1. Here, a differentiation between scenarios with only mirrors establishing the network and those also including storage on board the satellites is made. The ones with storage on board are denoted an additional a. Those storages are always considered untrusted node satellites. Only the scenarios including the photon generation on board require trusted nodes. The advantages and challenges with mounting storage units to the satellites is described in Section 7.3.

One of the major challenges of the constellations, regardless of deploying untrusted or trusted nodes, is range limitation. Range limitations arise from the loss of photon signal. This is influenced by the desired protocol, atmospheric effects, the quality of the repeater, and required photon transmission. Exact numbers are not given in this chapter, as not all of the factors can be included satisfactory. Further information on repeaters can be found in Section 7.3.

Table 7.1 Overview over the scenarios depicted in Figure 7.1.[a)]

Scenario #	Orbit	Storage	Storage time	Photon generation
1	LEO	None	n/a	On ground
1a	LEO	Yes	Short term (order of milliseconds)	On ground
2	LEO	Yes	Up to orbit time (LEO typically 90 minutes)	On ground
3	GEO	None	n/a	On ground
3a	GEO	Yes	Short term (order of milliseconds)	On ground
4	LEO and GEO	None	n/a	On ground
4a	LEO and GEO	Yes	Short term (order of milliseconds)	On ground
5	LEO and GEO	None	n/a	In orbit
5a	LEO and GEO	Yes	Short term (order of milliseconds)	In orbit

a) The scenarios 1a, 3a, 4a, and 5a differ from the respective scenarios by usage of additional storage on-board the satellites. The necessity and advantages of which will be discussed in Section 7.3.

Regardless of the chosen orbital constellation, With systems in orbit, the hardware can no longer be accessed or updated if a security leak occurs or a component has been found not to be functioning as envisaged. The latter could be targeted with sufficient redundancy on board the satellite, allowing for components to fail with the satellite relay or storage station remaining utilizable. As reliability, resilience, and availability are crucial features in secure information transfer, it appears advisable to instate redundant satellites and expect updates or exchange of satellites in case of errors.

The system most prone to requiring updates appears to be scenario five. As in this case the entangled photons are generated in orbit, novel technologies will have the highest impact on the constellation hardware.

But, even with implemented means to increase resilience and the entangled photon generation set on ground, attacks on quantum networks can not be completely dismissed [23, 30]. Once hardware is send to space, it is vulnerable. Due to the long operation in orbit and the sensitive nature of the transmitted information, developing target protocols becomes both profitable and achievable. With ongoing technology development and resulting new protocols, sources, and transmission options, hardware updates are required to ensure the desired security level. Such updates are easily incorporated if the system is ground based but increasingly difficult if a satellite system, especially one in higher orbits, is involved. Not only does this require an update of the hardware but also preparation of the setup for operation in space. Reaching the technology readiness level (TRL) necessary for robust operation in orbit, usually takes several years. Within this time frame, the best ground technology advances and the prepared systems security is overhauled. To mitigate this risk, the operation of untrusted nodes at the relay stations appears the sensible option. In conclusion, regular updates following novel developments and technologies, render exchange of satellites in a quantum network very likely.

The final challenge discussed in this chapter is the operation of the satellite network. To guarantee security in the transmission of information, the operation of the network appears to be the obvious solution. While this might be feasible for scenario 1 and 1a, this poses increasingly severe challenges for the other scenarios. Especially a full network, as proposed in scenarios four and five, can only be established and operated by entities with extensive budgets and require the involvement of several people to build, qualify, and operate the hardware. It can easily be seen that, with even only two separate entities desiring to instate a likewise network, the relevant orbits become busy with collision probabilities increasing with every additional relay station in orbit. Even a network, which is completely operated by the relevant users, does not prevent malicious parties to attempt and intercept the established optical links or infiltrate the assembly or operation of the hardware.

If, however, a network could enable safe usage of quantum entanglement by different users without compromising the security level of the transmitted information, the system could become much more cost effective and lead to the realization of a quantum internet [31]. Such a system needs to rely on untrusted nodes and will make use of all forms of distribution of information: space-based free space links, terrestrial free space links, and fiber bound connections. It would be complemented by small, isolated networks in relevant areas. Such a complex system enables small to medium businesses to participate in establishing the technology and benefit from the global revenues. Of course, this system requires ongoing error correction and synchronization between the different systems and compensation for atmospheric disturbances.

7.3 Untrusted Nodes

In Section 7.2, the opportunities and challenges involved in setting up a global secure quantum network have been discussed. Two key messages can be drawn from the content:

1. A complete global secure quantum network can not be operated by one player alone and will be a complex system of hardware build and operated by different parties.
2. To ensure the security of the transmitted information, untrusted nodes are required that do not compromise the security of the information.

Let us now take a look at the meaning of untrusted nodes, the advantage of deployment of such systems, and the challenges involved.

An untrusted node is a relay station that is not operated by either of the two parties that want to transmit information. The key property of an untrusted node is that these can not access or temper with the transmitted information. In classical physics, such untrusted nodes are difficult to implement, as any storage, amplification, or relay station has to read the transmitted information and any tempering or copying is not visible to the receiver. Quantum entanglement, as described above, differs in this important feature.

To generate a quantum key, different protocols are implemented. One of the most common protocols, not dependent on entanglement, is the BB84 protocol, named after the inventors (Bennett and Brassard) and the year they first proposed it (1984) [32, 33]. This protocol allows for a generation of a key between two partners without either of them having any prior knowledge about the final product. Thus arises the necessity of the classical exchange between the two parties on the experimental setup. Hence, if during the generation of the key, a photon goes amiss it, provided the time synchronization between the receiver and sender is sufficient, is disregarded from the key generation. This leads to a situation in which a relay station tempering with the code or trying to generate the key with the original sender, will compromise the result and giving away their intrusion, even if they have gained access to the classically transferred supplementary information. A more comprehensive discussion on the working principle of untrusted nodes is given in Ref. [34]. Based on this principle, the security level can be conserved with an untrusted node relaying the information.

As the key is generated between the two parties as a result of a measurement, the proposed scenario 5, including the source on a satellite to reduce the impact of atmospheric turbulences, results in secure communication between the two receivers (still denoted Alice and Bob) without any other party being able to impinge on the information without rendering the key unusable.

Having established that untrusted nodes in QKD are a viable option, let us take a quick look at what untrusted nodes could entail and how the distribution of information potentially benefits form them. There are several stages of untrusted nodes.

7.3.1 Mirror Systems and Coupled Systems

As described above, the simplest form of relay stations are mirror systems that simply deflect the incoming beam and transmit it to another site. Such systems could for instance be envisaged for small systems with a single satellite connecting two defined sites or usage in terrestrial applications between moving partners which could use a beacon in-between to connect.

In the context of a broader network, mirror systems could facilitate the transmission between a ground station and the end-user. Similar to mirror systems, in ground based dissemination of the information, coupling of the information into or out of a fiber are feasible.

One of the major challenges in this system is the loss in signal, which scales with r^4. As such, the loss in signal in this constellation limits the possibility of large distance transfer. The loss in signal is further increased by outside effects, such as atmospheric disturbances.

7.3.2 Sources

As discussed above, entangled photon sources could be operated by third parties without impacting the security of the generated key. Consequently, a satellite with a source transmitting entangled photons to two independent sites is a possibility for an untrusted node, even though it is not, in the strict sense, a relay station. In the spirit of a complex secure space-based quantum network, however, it seems correct to list sources on third-party operated source satellites as potential nodes.

7.3.3 Repeater

Repeater are systems that receive a signal and emit it again. In the process, the signal is amplified to pass over larger distances or to improve on the signal quality. In general, this could lead to an amplification of noise and or (unintended) alteration of the original signal. It is important to note, that a quantum repeater, in difference to a classical amplifier, increases the probability of successful transmission but cannot increase the signal. In QKD, the first becomes an issue with noise levels being the main source for incorrect measurements, while the second issue is a lesser concern.

In these terms, repeater for quantum based systems usually comprise in-coupling optics, possibly a storage matter, and then out-coupling optics for the signal to be relayed. In this chapter, two types of repeaters will be discussed, those based on fibers and those based on other memories. A schematic overview over the two systems is given in Figure 7.2.

Fiber based repeater are repeaters in which a fiber is set as storage matter. This concept is depicted as the upper schematic in Figure 7.2. As discussed above, optical fiber losses are high, especially compared to those encountered in free space propagation in vacuum. As such, a fiber based repeater could not act as an amplifier and

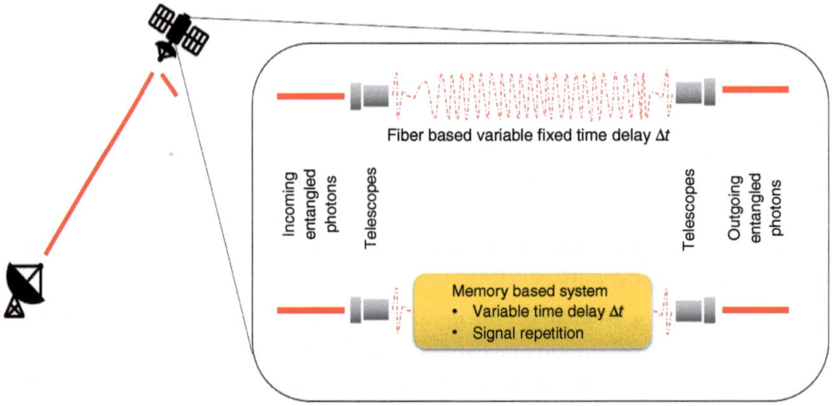

Figure 7.2 Depiction of the two different repeater schemes mounted to a possible satellite system. The upper position depicts a fiber based repeater in which the length of the fiber determines the time delay in receiving and emitting the signal. Below that the repeater concept of a memory based system is depicted. Other necessary systems, such as satellite control, repeater control, and thermal control, are not depicted, see also Section 7.4.5.4.

quite to the contrary could lead to additional signal loss or alteration. However, especially in scenarios 1a and 2, fiber based repeaters could trump over memory based repeaters. The fiber length determines the residence time of any given photon. With different length fibers, photons can be stored for precise times before being send to ground. Due to loss rates and fiber lengths, these types of repeaters are only useful for short storage times, for instance, to increase the quantum key length in scenario 1. It is important to remark, that on sufficiently short time scales, losses can be accepted and do not hinder the quantum key generation.

Memory based repeater are repeater in which the storage matter is a quantum memory. This concept is depicted as the lower schematic in Figure 7.2. Quantum memories are systems with the capability of storing photons. Quantum memories can be categorized according to what kind of information can be stored, scalability, the level of read-out fidelity, on-demand readout, and maximum storage times.

Different applications can make use of different memory types. For QKD, most of the above set qualities are required, but most importantly, the quantum mechanical properties of the entangled system should not be altered. This is directly linked to the quality of coherence of the memory.

There are different memory types that are best deployed in different scenarios. Options range from systems with lower internal coherence, such as warm vapor cells to systems of condensed atomic gases, rare earth ion doped (REID) crystals, solid state systems, and nitrogen vacancy (NV) centers in diamond. These can be divided into two classes: Memories that are based on the ensemble, such as Bose–Einstein condensates (BECs), and those based individual emitters, such as the NV centers in diamond. Overviews over these systems and their readiness for space based operation can be found in references [34, 35].

The most important quality of quantum memories is, of course, the possibility to store and transmit the information. For this purpose, different protocols can be implemented, which are chosen according to the platform they are used for. Such protocols are, for example, electromagnetically induced transparency (EIT) [36], Raman type schemes [37], gradient echo memories (GEMs) [38], controlled reversible in-homogeneous broadening (CRIB) [39], and atomic frequency comb (AFC) protocols [40]. Based on the memory type, these protocols promise different storage efficiencies, fidelities, and possibility to store single photons. While those protocols can be utilized with different memory types, usually, EIT and Raman type schemes are usually deployed for atomic systems, such as warm vapor cells and BECs, with AFC protocols usually applied to REID crystal memories.

Memory based repeater also suffer from loss rates based on absorption cross sections and emission probabilities. However, these systems trump over fiber based systems in this regard and could be envisioned as amplifier in long range experiments for fundamental physics, such as the deep space quantum link proposed by NASA [41]. In this experiment, transmitting entangled photons over long distances, is proposed to be exploited for purposes of testing Bells inequality [42].

If quantum networks, as in the case of the deep space quantum link, need to be extended to other bodies in the solar system, repeater including amplifier, such as quantum memories, become a necessity.

Additionally, such systems allow for exploration of the limits of quantum mechanics and photonic collapse models [43]. Understanding the underlying quantum mechanical boundaries allows to deploy the systems most efficiently and push technology to the limits.

7.4 Currently Available Technology

7.4.1 Required Technology

In Sections 7.1, 7.2, the necessity for global communication networks, secure information transfer, and untrusted nodes has been described. Alongside, different scenarios for establishing a global secure quantum network have been outlined in Figure 7.1 and Table 7.1. With those in mind, the requirements and resulting hardware proposals have been detailed. This includes a description of untrusted nodes and the various options for those systems. This section shall now demonstrate the current level of technology readiness and open challenges.

7.4.2 Technology Readiness Level (TRL)

Within the scope of this chapter, the necessary technology shall be rated with respect to the TRL. TRL is a measure set and agreed upon by the space agencies. It ranges from 1, where the basic principles of a technology have been proven, to 9, where a system has been operated in space. It covers the different stages a technology goes through when being qualified for a space mission. These include miniaturization as well as environmental testing. They can include operation on sounding rockets, where the environmental requirements are similar to those of a rocket launch. It is an important measure to explain the required effort to advance an item and can weigh the development cost against those of a qualified commercial of the shelf (COTS) item.

Regardless of previous experiments, however, it is custom in space operation that a TRL 9 item, that has to be adapted to a new experiment or is foreseen to be operated in another orbit, can no longer be considered with such a high TRL number and has to undergo testing, both with respect to functionality and environmental robustness again. Consequently, no TRL number can be given to the necessary technology without understanding the requirements for a specific mission. Still, efforts will be made to judge the current levels of maturity and lay out open challenges and required developments on ground and in space.

Please note, that the TRL is also used by other institutions and industry with the different levels adapted to their respective situation.

7.4.3 Optical Link Technologies

Optical links in space are required for various applications, ranging from navigation [28, 29], to Earth observation [44, 45], gravitational field measurements of other bodies, such as the moon [46], gravitational wave detection [47, 48], fundamental science, and, obviously, photon transfer [49, 50]. Depending on the application, optical links are deployed to transmit frequency information, as in the case of connection clock networks, measure distances precisely, for the measurement of gravitational gradients impacting the satellites, and to transmit classical and quantum information. With this high interest in optical link technology for deployment in space, it is no surprise, that it is readily available. Most of the technology focuses on the telecom wavelengths in the order of 1550 nm. A prominent example is the OSIRIS terminal [51]. The hardware is developed in close collaboration between the German Aerospace Center and TESAT and prepared for commercialization.

The OSIRIS terminal is prepared with data transfer in mind requiring adaptations to fit the needs of entangled photon experiments and the described global quantum networks. With experiments having demonstrated ground-to-satellite entanglement, the optical link technology has proven its feasibility and applicability for quantum communication. Further technological advancements with respect to collection efficiency, field of view, adaptability to various wavelengths, and radiation hardness required for geostationary orbits or deep space operation are emerge in the upcoming years.

As mentioned above, loss rates in optical links increase with the fourth power of distance. To collect as much signal as possible, the natural optimization are large apertures of the telescopes. Consequently, a trade-off has to be done to accommodate systems on a satellite, which is limited in the available size, mass, and power budget while having sufficient aperture to collect and re-send photons.

7.4.4 Sources

As described above, sources for entangled photons can be regarded as individual nodes in a global quantum network. They can be deployed in space, on ground, and on board of vehicles. For these different applications, varying source requirements exist. They are most relaxed when it comes to stationary systems in laboratory buildings and transmission sites and are increasingly restrictive when it comes to operation on vehicles or spacecrafts. Especially the latter imposes limitations on the size, mass, and power (SWAP) budgets. According sources have been operated successfully on cube satellites, such as Qube [52] and SpooQy-1 [53], efficiently demonstrating the feasibility of compact sources in space.

In the scenarios displayed in Figure 7.1, the scenario (scenario 5) with sources as space-borne nodes foresees the sources solely in LEO. As such, the technology for building such nodes in a secure quantum network is very advanced.

The argument might be made that, if quantum entanglement experiments shall be performed over larger distances or secure communication shall reach further targets, further investigations into resilience and robustness against cosmic radiation are necessary. Similarly, networks with medium Earth orbits (MEOs) as opposed to LEO operation could be preferable. In combination with increased efficiency, adaptation to specific missions, closing of loopholes, enhanced resilience against attacks, and possibly choosing different wavelengths, the current technology can be judged as very mature with adaptations and modifications surely arising as research advances.

7.4.5 Repeater

7.4.5.1 Overview

As described above, repeaters exist in different configurations. This sections is dedicated to describing the necessary technologies, currently available hardware, proposed concepts, and open challenges of the individual constituents. The exception to these are the sources and optical link technologies described above.

7.4.5.2 Optical Fibers

First on the list are optical fibers. Space qualified optical fibers exist. Still, the material is prone to defects caused by cosmic radiation and consequential, operation in high- level radiation environments, such as geostationary orbits, has to be carefully prepared. In recent years efforts have been made to increase robustness in the fibers for deployment in Earths orbit [54, 55]. Such measures include housing and shielding of the fibers. This comes at the added benefit of being able to control the thermal gradient over the fibers length. However, the latest experiments demonstrate, that, even with cosmic radiation shielding in place, fading of the fibers over time can not be completely eradicated.

Coming back to the scenarios in Figure 7.1, long optical fibers are required to make a difference in scenario 1a with respect to scenario 1. While those nodes are operated predominantly in a LEO or MEO orbit, the impact of radiation on long fibers can not be neglected. As cosmic radiation effects are statistically distributed in place of incident and severity of impact, redundant systems with automatic interchangeability appear appropriate even though it is possible that the redundant system is hit harder by cosmic radiation than the original one. The second requirement is in line with the requirement on on-demand readout, in which the time spend on the satellite is chosen to fit the distance between the two parties exchanging the information.

Still, scenarios 2–5 require optical preparation and connection. For the nodes in these scenarios, however, shorter fibers could be considered, reducing the probability of impacts fatal to the utility of the node.

As it was mentioned previously, the satellite-based nodes are not accessible during operation and, consequently, need to be robust against environmental impacts as well as outside attacks. The only updates and repairs that can be executed from ground during operation are software updates or exchange of nodes. With this in mind, the challenge involving the optical fibers becomes both, more critical and less severe.

Fiber failure, is a critical and very costly failure, as the fibers can not be repaired or exchanged without replacing the affected node. On the other hand, if exchanges of satellites due to outdated or vulnerable technology are planned activities in the operation of a global secure quantum network, the deterioration of the fibers could be kept to a minimum by implementing shields and exchange the systems latest when the losses in the nodes become too severe.

7.4.5.3 Memories

Nodes incorporating quantum memories can act as untrusted nodes with lower losses than those with optical fibers. As discussed before different types of memories are currently envisaged. Looking back at Table 7.1, different scenarios require different memories. While in scenario two, the storage of several single photons over prolonged times is required the other scenarios do not need such lengthy storage times. With recent developments having demonstrated storage times of several hours in nuclear spins [56, 57], single photon storage efficiencies remain challenging. The currently most promising systems for single photon memories appear to be warm vapor cells [37] and cold atomic ensembles [58], having demonstrated efficiencies over 70%.

With having discussed the available efficiencies and storage time requirements, it is important to take a look at the space readiness of the systems and the demonstration in space:

Warm vapor cells have been deployed in several different systems. Especially in frequency standard applications, rubidium cells have been deployed. One prominent example is the rubidium microwave clock on board of the Galileo satellites [59]. This technology has been further developed [60], leading to smaller and more robust systems. Additionally, optical frequency references based on various atomic and molecular species have demonstrated the working principal and have been tested on ground [61] and in sounding rocket missions [62]. The adaptation for those systems for quantum memories is on its way with experiments on rubidium [63] and cesium vapor cells [64] having been performed. With warm vapor cells having demonstrated their suitability both, for operation as memories and in space, the next step is to demonstrate that warm vapor cell based memories are capable to cope with the environmental requirements for space-based operation.

The situation of cold and condensed atomic ensembles is similar to the one of warm vapor cells. However, one key feature of condensed atomic ensembles is condensation into one overarching wavefunction, leading to higher coherence than exhibited by warm vapor cells. At this point, the efficiencies and storage times observed in BECs are not on par with the ones observed in warm vapor cells [34]. The leading cause for this lag is the technological and experimental accessibility of cold and condensed atom ensembles, which first arose in the late 1990s [65]. It is therefore to be expected that the performance level of deploying BECs as quantum memories will increase quickly.

Most of the research on cold atoms in space is currently focused on fundamental research [66, 67], Earth observation, testing of the equivalence principle [68], and gravitational wave detection [69]. Additional steps are undertaken to deploy this exotic state of matter in ground based applications, such as inertial measurement units (IMUs) [70].

Regardless, these research and application interests lead to the development and improvement of space-qualified setups. Especially with the campaigns in the drop tower [71, 72], on board of planes [73], on board of sounding rockets [74], and in orbit [66], the readiness of the technology for space-borne applications has been demonstrated.

REID crystals require high magnetic fields and cryogenic temperatures. This increases the setup SWAP budgets. While those systems have demonstrated the longest storage times so far, additional efforts will have to be undertaken in miniaturization and environmental compliance. This is aided by developments from other communities, where cryogenic systems are also key technologies. Recently, cryogenic systems have been developed for space deployment [75], paving the road for REID crystals as quantum memories in orbit.

In terms of NV centers, their application as magnetometers has rendered them interesting for industrial usage. Unsurprisingly, the requirements for commercial use and for deployment in space are similar when it comes to SWAP budgets, autonomous operation, and reliability. Additionally, current projects are developing space-qualified systems based on the NV-centers in diamond. Similar to the REIDs, BECs, and warm vapor cells, additional tests with systems optimized for single photon storage are required before those can be deployed in orbit.

7.4.5.4 Sum of Its Parts

With the individual components discussed above, it is worth mentioning, that a repeater is more than just the sum of its parts. An important quality of quantum memories and quantum repeaters is their possibility of deploying entanglement swapping [76]. In addition to the fiber or quantum memory at the center of the repeater, items, such as coupling optics, and operation software are required. Those are depicted schematically in Figure 7.2. Especially the coupling optics present an open challenge based on the desired transmission wavelength and center item of the repeater. Space operation with any of the above described configurations or scenarios, also requires general systems, such as alignment of the space segment, thermal and power management, and radiation shielding. Additionally, the environmental requirements during launch and operation in orbit need to be fulfilled by an assembled and integrated quantum repeater as opposed to its constituents.

7.5 Summary

Within this chapter, the necessity, opportunities, and challenges of a global secure quantum network have been described. It has been established, that it is infeasible to operate a global quantum secure network by an individual party and that the operation of untrusted nodes by third parties does not undermine the security of the exchanged information. Untrusted nodes are nodes in a network, that act as relay stations. In case of a global secure quantum network, such relay stations would be based on satellites and could constitute sources, memories, or fiber based storage. Those global satellite based systems could potentially be supported by smaller, local networks, transferring information between individual sites or moving targets.

To conclude, a global secure quantum network is only feasible if several players are involved and has therefore to be regarded as more than just the sum of its parts. Combining different players, technologies, ground stations, and transmission protocols to build a cohesive global secure quantum network is going to be an important challenge in its success.

7.6 Acronyms

In this section, the acronyms used in the text are summarized (Table 7.2).

Table 7.2 Acronyms used in this chapter.

Acronym	Description
AFC	Atomic frequency comb
BB84	QKD Protocol named after the inventors and the year it was proposed
BEC	Bose–Einstein condensate
COTS	Commercial of the shelf
CRIB	Controlled reversible in-homogeneous broadening
DLR	German Aerospace Center
EIT	Electromagnetically induced transparency
GEM	Gradient echo memory
GEO	Geo stationary orbit
GPS	Global positioning system
IMU	Inertial measurement unit
LEO	Low Earth orbit
MEO	Medium Earth orbit
NASA	National Aeronautics and Space Administration
NV	Nitrogen vacancy (center)
QKD	Quantum key distribution
QT	Quantum Technologies
REID	Rare Earth ion doped (crystal)
SWAP	Size, mass, and power
TRL	Technology readiness level

References

1 Wilkin, P. (2001). *The Political Economy of Global Communication: An Introduction*. London: Pluto Press.

2 Gilboa, E. (2006). Global communication and foreign policy. *Journal of Communication* 52 (4): 731–748. https://doi.org/10.1111/j.1460-2466.2002.tb02571.x.

3 Chakrabarti, S., Gupta, M., and Choi, J.-D. (1996). Global communication analysis and optimization. *SIGPLAN Notices* 31 (5): 68–78, https://doi.org/10.1145/249069.231391

4 Marusich, L.R., Bakdash, J.Z., Onal, E. et al. (2016). Effects of information availability on command-and-control decision making: performance, trust, and situation awareness. *Human Factors* 58 (2): 301–321.

5 Lin, H. (2012). A virtual necessity: some modest steps toward greater cybersecurity. *Bulletin of the Atomic Scientists* 68 (5): 75–87. https://doi.org/10.1177/0096340212459039.

6 Roberts, S.J. (2014). The necessity of information security in the vulnerable pharmaceutical industry. *Journal of Information Security* 5 (04): 147.

7 Lábodi, C. and Michelberger, P. (2010). Necessity or challenge-information security for small and medium enterprises. *Annals of the University of Petrosani Economics* 10 (3): 207–216.

8 Kim, S., Yu, H., Gang, S., and Kim, J. (2017). Matter-wave beam splitter on an atom chip for a portable atom interferometer. *Applied Physics B* 123 (5): 154. https://doi.org/10.1007/s00340-017-6719-6.

9 Dowling, B. and Hale, B. (2020). There can be no compromise: the necessity of ratcheted authentication in secure messaging. *IACR Cryptology ePrint Archive*, 2020/541.

10 Campbell, K., Gordon, L.A., Loeb, M.P., and Zhou, L. (2003). The economic cost of publicly announced information security breaches: empirical evidence from the stock market. *Journal of Computer security* 11 (3): 431–448.

11 Bansal, G., Hasija, V., Chamola, V. et al. (2019). Smart stock exchange market: a secure predictive decentralized model. *2019 IEEE Global Communications Conference (GLOBECOM)*, pp. 1–6. IEEE.

12 Gagajewski, W.A. (1999). *British Cryptographic Efforts in World War II: the struggle against the German ENIGMA machine*. California State University, Dominguez Hills.

13 Barrett, J., Hardy, L., and Kent, A. (2005). No signaling and quantum key distribution. *Physical Review Letters* 95 (1): 010503.

14 Vedral, V. (2014). Quantum entanglement. *Nature Physics* 10 (4): 256–258.

15 Hughes, R.J., Buttler, W.T., Kwiat, P.G. et al. (1999). Quantum cryptography for secure free-space communications. *Proceedings SPIE 3615, Free-Space Laser Communication Technologies XI*, Volume 3615, pp. 98–103. https://doi.org/10.1117/12.346170.

16 Ursin, R., Tiefenbacher, F., Schmitt-Manderbach, T. et al. (2007). Entanglement-based quantum communication over 144 km. *Nature Physics* 3 (7): 481–486. https://doi.org/10.1038/nphys629.

17 Zhang, Y., Chen, Z., Pirandola, S. et al. (2020). Long-distance continuous-variable quantum key distribution over 202.81 km of fiber. *Physical Review Letters* 125 (1): 010502. https://doi.org/10.1103/PhysRevLett.125.010502.

18 Bedington, R., Arrazola, J.M., and Ling, A. (2017). Progress in satellite quantum key distribution. *npj Quantum Information* 3 (1): 30. https://doi.org/10.1038/s41534-017-0031-5.

19 Chen, Y.-A., Zhang, Q., Chen, T.-Y. et al. (2021). An integrated space-to-ground quantum communication network over 4,600 kilometres. *Nature* 589 (7841): 214–219. https://doi.org/10.1038/s41586-020-03093-8.

20 Muralidharan, S., Kim, J., Lütkenhaus, N. et al. (2014). Ultrafast and fault-tolerant quantum communication across long distances. *Physical Review Letters* 112: 250501. https://doi.org/10.1103/PhysRevLett.112.250501.

21 Gisin, N., Ribordy, G., Tittel, W., and Zbinden, H. (2002). Quantum cryptography. *Reviews of Modern Physics* 74 (1): 145. https://doi.org/10.1103/RevModPhys.74.145.

22 Diamanti, E., Lo, H.-K., Qi, B., and Yuan, Z. (2016). Practical challenges in quantum key distribution. *npj Quantum Information* 2 (1): 1–12. https://doi.org/10.1038/npjqi.2016.25.

23 Belenchia, A., Carlesso, M., Ömer, B. et al. (2022). Quantum physics in space. *Physics Reports* 9510: 1–70. https://doi.org/10.1016/j.physrep.2021.11.004.

24 Feynman, R.P., Leighton, R.B., Sands, M., and Treiman, S.B. (1964). *The Feynman Lectures on Physics*, vol. I. Addison–Wesley.

25 Li, M.-J. and Hayashi, T. (2020). Advances in low-loss, large-area, and multicore fibers. In: *Optical Fiber Telecommunications VII*, Chapter 1 (ed. A.E. Willner), 3–50. Academic Press. https://www.sciencedirect.com/science/article/pii/B9780128165027000014.

26 Moll, F., Botter, T., Marquardt, C. et al. (2019). Stratospheric QKD: feasibility analysis and free-space optics system concept. *Quantum Technologies and Quantum Information Science V*, Volume 11167, pp. 34–42. SPIE. https://doi.org/10.1117/12.2539076.

27 Wittig, S. (2017). Concept for single-satellite global quantum key distribution using a solid state quantum memory. *IAC Proceeding* 17: x36863. http://iafastro.directory/iac/paper/id/36863/summary/

28 Giorgi, G., Schmidt, T.D., Trainotti, C. et al. (2019). Advanced technologies for satellite navigation and geodesy. *Advances in Space Research* 64 (6): 1256–1273.

29 Glaser, S., Michalak, G., Männel, B. et al. (2020). Reference system origin and scale realization within the future gnss constellation "Kepler". *Journal of Geodesy* 94 (12): 1–13.

30 Jain, N., Anisimova, E., Khan, I. et al. (2014). Trojan-horse attacks threaten the security of practical quantum cryptography. *New Journal of Physics* 16 (12): 123030. https://doi.org/10.1088/1367-2630/16/12/123030.

31 Wehner, S., Elkouss, D., and Hanson, R. (2018). Quantum internet: a vision for the road ahead. *Science* 362 (6412). https://doi.org/10.1126/science.aam9288.

32 Bennett, C. and Brassard, G. (1984). Bb84. *IEEE International Conference on Computers, Systems, and Signal Processing*.

33 Bennett, C. (2014). Quantum cryptography: public key distributions and coin tossing. *Theoretical Computer Science* 560: 7–11. https://doi.org/10.1016/j.tcs.2014.05.025.

34 Mol, J.-M., Esguerra, L., Meister, M. et al. (2021). Quantum memories for fundamental physics in space. Decadal Survey on Biological and Physical Sciences Research in Space 2023–2032.

35 Sidhu, J.S., Joshi, S.K., Gundogan, M. et al. (2021). Advances in space quantum communications. arXiv e-prints, 2103–12749.

36 Ma, L., Slattery, O., and Tang, X. (2017). Optical quantum memory based on electromagnetically induced transparency. *Journal of Optics* 19 (4): 043001.

37 Guo, J., Feng, X., Yang, P. et al. (2019). High-performance raman quantum memory with optimal control in room temperature atoms. *Nature Communications* 10 (1): 1–6.

38 Hétet, G., Longdell, J., Sellars, M. et al. (2008). Multimodal properties and dynamics of gradient echo quantum memory. *Physical Review Letters* 101 (20): 203601.

39 Lauritzen, B., Minář, J., De Riedmatten, H. et al. (2011). Approaches for a quantum memory at telecommunication wavelengths. *Physical Review A* 83 (1): 012318.

40 Sabooni, M., Li, Q., Kröll, S., and Rippe, L. (2013). Efficient quantum memory using a weakly absorbing sample. *Physical Review Letters* 110 (13): 133604.

41 Mohageg, M., Strekalov, D., Dolinar, S., Shaw, M., Yu, N. (2018). Deep space quantum link. *Deep Space Gateway Concept Science Workshop*, Proceedings of the workshop held February 27-March 1, 2018 in Denver, Colorado. LPI Contribution No. 2063, id.3039, Volume 2063, p. 3039.

42 Bell, J.S. (1964). On the Einstein Podolsky Rosen paradox. *Physics Physique Fizika* 1 (3): 195. https://doi.org/10.1103/PhysicsPhysiqueFizika.1.195.

43 Shadbolt, P., Mathews, J.C., Laing, A., and O'brien, J.L. (2014). Testing foundations of quantum mechanics with photons. *Nature Physics* 10 (4): 278–286.

44 Tapley, B.D., Bettadpur, S., Ries, J.C. et al. (2004). GRACE measurements of mass variability in the earth system. *Science* 305 (5683): 503–505. https://doi.org/10.1126/science.1099192.

45 Velicogna, I., Mohajerani, Y., Geruo, A. et al. (2020). Continuity of ice sheet mass loss in Greenland and Antarctica from the GRACE and GRACE Follow-On missions. *Geophysical Research Letters* 47 (8): e2020GL087291. https://doi.org/10.1029/2020GL087291.

46 Zuber, M.T., Smith, D.E., Watkins, M.M. et al. (2013). Gravity field of the moon from the gravity recovery and interior laboratory (GRAIL) mission. *Science* 339 (6120): 668–671.

47 Abou El-Neaj, Y., Alpigiani, C., Amairi-Pyka, S. et al. (2020). AEDGE: Atomic experiment for dark matter and gravity exploration in space. *EPJ Quantum Technology* 7: 6. https://doi.org/10.1140/epjqt/s40507-020-0080-0.

48 Bartolo, N., Caprini, C., Domcke, V. et al. (2016). Science with the space-based interferometer LISA. IV: Probing inflation with gravitational waves. *Journal of Cosmology and Astroparticle Physics* 2016 (12): 026. https://doi.org/10.1088/1475-7516/2016/12/026.

49 Yin, J., Cao, Y., Li, Y.-H. et al. (2017). Satellite-to-ground entanglement-based quantum key distribution. *Physical Review Letters* 119 (20): 200501. https://doi.org/10.1103/PhysRevLett.119.200501.

50 Jennewein, T., Grant, C., Choi, E. et al. (2014). The NanoQEY mission: ground to space quantum key and entanglement distribution using a nanosatellite. *SPIE Proceedings*, Volume 9254.

51 Fuchs, C. and Schmidt, C. (2019). Update on DLR's OSIRIS program. *International Conference on Space Optics–ICSO 2018*, Volume 11180, p. 111800. International Society for Optics and Photonics.

52 Haber, R., Garbe, D., Busch, S. et al. (2018). Qube - a cubesat for quantum key distribution experiments. *32th Annual AAIA/USU Conference on Small Satellites (SSC18-III-05).*

53 Villar, A., Lohrmann, A., Bai, X. et al. (2020). Entanglement demonstration on board a nano-satellite. *Optica* 7 (7): 734–737.
54 Girard, S., Kuhnhenn, J., Gusarov, A. et al. (2013). Radiation effects on silica-based optical fibers: recent advances and future challenges. *IEEE Transactions on Nuclear Science* 60 (3): 2015–2036. https://doi.org/10.1109/TNS.2012.2235464.
55 Girard, S., Morana, A., Ladaci, A. et al. (2018). Recent advances in radiation-hardened fiber-based technologies for space applications. *Journal of Optics* 20 (9): 093001.
56 Zhong, M., Hedges, M.P., Ahlefeldt, R.L. et al. (2015). Optically addressable nuclear spins in a solid with a six-hour coherence time. *Nature* 517 (7533): 177–180.
57 Hedges, M.P., Longdell, J.J., Li, Y., and Sellars, M.J. (2010). Efficient quantum memory for light. *Nature* 465 (7301): 1052–1056.
58 Yang, S.-J., Wang, X.-J., Bao, X.-H., and Pan, J.-W. (2016). An efficient quantum light–matter interface with sub-second lifetime. *Nature Photonics* 10 (6): 381–384.
59 Jeanmaire, A., Rochat, P., and Emma, F. (1999). Rubidium atomic clock for Galileo. *Proceedings of the 31st Annual Precise Time and Time Interval Systems and Applications Meeting*, pp. 627–636.
60 Bandi, T., Affolderbach, C., Calosso, C.E., and Mileti, G. (2011). High-performance laser-pumped rubidium frequency standard for satellite navigation. *Electronics Letters* 47 (12): 698–699.
61 Martin, K.W., Phelps, G., Lemke, N.D. et al. (2018). Compact optical atomic clock based on a two-photon transition in rubidium. *Physical Review Applied* 9 (1): 014019.
62 Döringshoff, K., Gutsch, F.B., Schkolnik, V. et al. (2019). Iodine frequency reference on a sounding rocket. *Physical Review Applied* 11: 054068. https://doi.org/10.1103/PhysRevApplied.11.054068.
63 Wolters, J., Buser, G., Horsley, A. et al. (2017). Simple atomic quantum memory suitable for semiconductor quantum dot single photons. *Physical Review Letters* 119 (6): 060502.
64 Katz, O. and Firstenberg, O. (2018). Light storage for one second in room-temperature alkali vapor. *Nature Communications* 9 (1): 1–6.
65 Ketterle, W. and Druten, N.J.V. (1996). Evaporative cooling of trapped atoms. *Advances in Atomic, Molecular, and Optical Physics* 37: 181–236. https://doi.org/10.1016/S1049-250X(08)60101-9.
66 Aveline, D., Elliott, E., Williams, J., Thompson, R. et al. (2020). Observation of Bose–Einstein condensates in an Earth-orbiting research lab. *Nature* 582 (7811): 193–197.
67 Frye, K., Abend, S., Bartosch, W. et al. (2021). The Bose-Einstein condensate and cold atom laboratory. *EPJ Quantum Technology* 8 (1): 1–38. https://doi.org/10.1140/epjqt/s40507-020-00090-8.
68 Wolf, P., Blanchet, L., Bouyer, P. et al. (2019). Exploring the foundations of the physical universe with space tests of the equivalence principle. arXiv e-prints, 1908–117853.

69 El-Neaj, Y.A., Alpigiani, C., Amairi-Pyka, S. et al. (2020). AEDGE: Atomic experiment for dark matter and gravity exploration in space. *EPJ Quantum Technology* 7 (1): 1–27. https://doi.org/10.1140/epjqt/s40507-020-0080-0.

70 Bongs, K., Holynski, M., Vovrosh, J. et al. (2019). Taking atom interferometric quantum sensors from the laboratory to real-world applications. *Nature Reviews Physics* 1 (12): 731–739. https://doi.org/10.1038/s42254-019-0117-4.

71 Müntinga, H., Ahlers, H., Krutzik, M. et al. (2013). Interferometry with Bose-Einstein condensates in microgravity. *Physical Review Letters* 110: 093602. https://doi.org/10.1103/PhysRevLett.110.093602.

72 van Zoest, T., Gaaloul, N., Singh, Y. et al. (2010). Bose-Einstein condensation in microgravity. *Science* 328 (5985): 1540–1543. https://doi.org/10.1126/science.1189164.

73 Geiger, R., Ménoret, V., Stern, G. et al. (2011). Detecting inertial effects with airborne matter-wave interferometry. *Nature Communications* 2: 474. https://doi.org/10.1038/ncomms1479.

74 Lachmann, M.D., Ahlers, H., Becker, D. et al. (2021). Ultracold atom interferometry in space. *Nature Communications* 12 (1317): 1–6. https://doi.org/10.1038/s41467-021-21628-z.

75 You, L., Quan, J., Wang, Y. et al. (2018). Superconducting nanowire single photon detection system for space applications. *Optics Express* 26 (3): 2965–2971. https://doi.org/10.1364/OE.26.002965.

76 Pan, J.-W., Bouwmeester, D., Weinfurter, H., and Zeilinger, A. (1998). Experimental entanglement swapping: entangling photons that never interacted. *Physical Review Letters* 80 (18): 3891.

Part II

Atoms, Ions, and Molecules: From Experimental Techniques to Recent Progress

Part 4

8

Fluorescence Spectroscopy in Planar Dielectric and Metallic Systems

Alexey I. Chizhik[1], *Daja Ruhlandt*[1], *and Jörg Enderlein*[1,2]

[1] *Third Institute of Physics – Biophysics, Georg August University, 37077, Göttingen, Friedrich-Hund-Platz 1, Germany*
[2] *Cluster of Excellence "Multiscale Bioimaging: from Molecular Machines to Networks of Excitable Cells" (MBExC), Georg August University, 37077, Göttingen, Friedrich-Hund-Platz 1, Germany*

8.1 Introduction

Modification of a fluorophore's photoluminescence properties by placing it in a close proximity to planar interfaces has been the focus of numerous theoretical and experimental studies in recent decades [1–5]. Major examples of fundamental research in this field are cavity quantum electrodynamics studies that brought the Nobel Prize to one of its pioneers, Serge Haroche and coworkers [6], or Karl–Heinz Drexhage's fundamental experiments concerning the excited state lifetime modulation of a molecule in front of a mirror [7]. Despite a great variety of currently existing highly complex nanostructures that are used to tailor fluorescence properties, planar nanostructures possess several important advantages that still make them attractive candidates for many state-of-the-art applications. First, the relative technical simplicity of their manufacture enhances the reproducibility and precision in their application. Second, the broad availability of instrumentation for their manufacture and for the verification of their properties makes planar nanostructures accessible to a wide research community. Finally, robust theoretical models that have been developed and verified in recent decades make the modeling of arbitrary planar metal/dielectric structures straightforward and fast, in contrast to numerically complex theoretical models that are required for modeling highly complex nanostructures.

In this chapter, we present an overview of two fluorescence spectroscopy methods that we developed in our research group and that are based on the modulation of the excited state lifetime of fluorophores by planar metal/dielectric structures. The first method is based on the excited state lifetime modulation of a fluorophore placed in an optical resonator for measuring its absolute quantum yield. The method does not require any calibration or reference samples, is applicable to any type of quantum emitters, and allows for measuring quantum yields even for single fluorophores.

Photonic Quantum Technologies: Science and Applications, First Edition. Edited by Mohamed Benyoucef.

The second technique is metal- or graphene-induced energy transfer (MIET or GIET, respectively), a new fluorescence-based method, which uses the energy transfer from an optically excited donor molecule to a thin metal film or a single sheet of graphene, and which allows for achieving an axial localization of a fluorophore with down to 1 nanometer accuracy. This goes far beyond the diffraction limit of light microscopy and surpasses in accuracy almost all light-based techniques used for enhancing the axial resolution.

8.2 Theory

In this section, we develop the full theoretical framework for describing the fluorescence excitation and emission in arbitrary planar metal/dielectric structures, such as gold layers on a glass surface or a metal nanocavity. We start with the electromagnetic field of an electric dipole emitter in free space, investigate then its interaction with an arbitrary nanocavity, derive its emission power, i.e. its excited state lifetime, consider next its angular distribution of radiation, which allows us to model the detection efficiency of observing such an emitter with a microscope objective, and finally derive the nonuniform excitation intensity distribution in a nanocavity when focusing light into it with a microscope objective. This all provides a full theoretical description of any fluorescence spectroscopy/imaging experiment in planar metal/dielectric structures.

8.2.1 Ideal Electric Dipole Emitter

The electromagnetic emission of basically all fluorescent emitters (organic dyes, fluorescent proteins, luminescent nano-crystals aka quantum dots) can be perfectly described by that of an oscillating electric dipole. For finding the electromagnetic field of an oscillating electric dipole with dipole moment amplitude $\mathbf{p}$ and oscillation frequency ω at position $\mathbf{r}_d = \{x_d, y_d, z_d\}$ in a sample medium with refractive index n_d, we have to solve Maxwell's equations in the presence of such a dipole. In what follows, we assume that all fields have the same time dependence $\exp(-i\omega t)$ as the dipole moment oscillations, and we consider only the amplitudes of the electric and magnetic fields. Thus, we have jointly to solve the following Maxwell's equations

$$\begin{aligned} \mathbf{rot\ E} &= \frac{i\omega}{c}\mathbf{B} \\ \mathbf{rot\ B} &= -\frac{i\omega\epsilon_d}{c}\mathbf{E} + \frac{4\pi}{c}\mathbf{j} \end{aligned} \tag{8.1}$$

where $\epsilon_d = n_d^2$ is the dielectric constant of the sample solution and $\mathbf{j} = -i\omega\mathbf{p}\delta(\mathbf{r} - \mathbf{r}_d)$ is the electric current generated by the oscillating dipole. Thus, we find the following equation for the electric field $\mathbf{E}_d$ of the dipole emitter

$$\mathbf{rot\ rot\ E}_d - k_d^2\mathbf{E}_d = 4\pi k_0^2\mathbf{p}\delta\left(\mathbf{r} - \mathbf{r}_d\right) \tag{8.2}$$

where $k_0 = \omega/c$ and $k_d = n_d k_0$. Passing to Fourier space yields the following equation for the Fourier amplitude $\tilde{\mathbf{E}}_d$

$$(k'^2 - k_d^2)\tilde{\mathbf{E}}_d - \mathbf{k}'(\mathbf{k}' \cdot \tilde{\mathbf{E}}_d) = 4\pi k_0^2\mathbf{p}\exp\left(-i\mathbf{k}' \cdot \mathbf{r}_d\right) \tag{8.3}$$

where $\mathbf{k}'$ is the Fourier space coordinate. Multiplying the last equation by $\mathbf{k}'$ yields

$$\mathbf{k}' \cdot \tilde{\mathbf{E}}_d = -\frac{4\pi}{\epsilon}\mathbf{k}' \cdot \mathbf{p} \exp\left(-i\mathbf{k}' \cdot \mathbf{r}_d\right) \tag{8.4}$$

so that we finally find

$$\tilde{\mathbf{E}}_d = \frac{4\pi \exp(-i\mathbf{k}' \cdot \mathbf{r}_d)}{\epsilon(k'^2 - k_d^2)} \left[k^2\mathbf{p} - \mathbf{k}'(\mathbf{k}' \cdot \mathbf{p})\right] \tag{8.5}$$

Back-transforming this result to real space yields then the solution

$$\mathbf{E}_d = \int \frac{d^3\mathbf{k}'}{2\pi^2\epsilon} \left[k_d^2\mathbf{p} - \mathbf{k}'(\mathbf{k}' \cdot \mathbf{p})\right] \frac{\exp\left[i\mathbf{k}' \cdot (\mathbf{r} - \mathbf{r}_d)\right]}{k'^2 - k_d^2} \tag{8.6}$$

Performing the integration over the variable k'_z results in

$$\mathbf{E}_d = \frac{i}{2\pi\epsilon} \int \frac{d^2\mathbf{q}}{w_d} \left[k_d^2\mathbf{p} - \mathbf{k}_d(\mathbf{k}_d \cdot \mathbf{p})\right] \exp\left[i\mathbf{q} \cdot (\rho - \rho_d) + iw_d\left|z - z_d\right|\right] \tag{8.7}$$

where we used the abbreviations $\mathbf{k}_d = \{\mathbf{q}, w\}$, and $w_d = \sqrt{k_d^2 - q^2}$, and the two-dimensional integration over $\mathbf{q}$ extends over an infinite (Fourier) plane oriented perpendicular to the optical axis. The z-component w of $\mathbf{k}_d$ is assumed to have always zero or positive imaginary part. Eq. (8.7) is the so-called Weyl representation, a plane wave representation of the electric field of a free oscillating dipole, which is particularly suited when modeling the interaction of a dipole emitter with planar dielectric/metallic structures.

8.2.2 Interaction of Plane Waves with Planar Layered Structures

Next, we will consider the interaction of the dipole emitter with a planar structure consisting of a stack of layers of dielectric and metallic materials with generally complex-valued refractive indices n_j and thickness values d_j, where j enumerates the layer. For that purpose, we will first consider the interaction of a planar wave with one interface dividing layer j and $j + 1$ of such a stack. We first study the case of a p-polarized plane wave where the electric field vector lies within the incident plane formed by the wave vector of the plane wave and the normal to the interface. Furthermore, for simplifying all subsequent derivations, we will work in a unit system where the vacuum wave vector has length 1 (vacuum wavelength equal to 2π), so that the wave vector amplitude in the jth layer is n_j. Now, the electric field in layers j and $j + 1$ is given by

$$\begin{aligned} \mathbf{E}_{p,j} &= \mathbf{E}^+_{p,j} \exp\left(i\mathbf{q} \cdot \rho + iw_j z\right) + \mathbf{E}^-_{p,j} \exp\left(i\mathbf{q} \cdot \rho - iw_j z\right) \\ \mathbf{E}_{p,j+1} &= \mathbf{E}^+_{p,j+1} \exp\left(i\mathbf{q} \cdot \rho + iw_{j+1} z\right) + \mathbf{E}^-_{p,j+1} \exp\left(i\mathbf{q} \cdot \rho - iw_{j+1} z\right) \end{aligned} \tag{8.8}$$

where $\mathbf{q}$ is the lateral part of the wave vector (parallel to the interfaces), which has to be the same in all layers due to the boundary conditions, which requires same periodicity of the fields on both sides of an interface, $w_j = \sqrt{n_j^2 - q^2}$ is the (generally complex-valued) perpendicular component of the wave vector in the jth

layer, the $\mathbf{E}^{\pm}_{p,j}$ are the electric field amplitude vectors, $\boldsymbol{\rho}$ is the horizontal (parallel to interfaces) part of the position vector, and coordinate z is perpendicular to the interfaces. The incidence angle of the plane waves in the jth layer is defined by $\sin\theta_j = q/n_j$. Because the electric field amplitude vectors have to be perpendicular to the direction of the wave vector, one has the relations

$$\begin{aligned} E^{\pm}_{p,j\parallel} &= \pm\frac{w_j}{n_j}E^{\pm}_{p,j} \\ E^{\pm}_{p,j\perp} &= \frac{q}{n_j}E^{\pm}_{p,j} \end{aligned} \tag{8.9}$$

where $\perp$ refers to the components perpendicular to the interface and $\parallel$ to the components parallel to the interface. The magnetic fields are found via Maxwell's equations to be $\mathbf{B}^{\pm}_{p,j} = \mathbf{n}^{\pm}_j \times \mathbf{E}^{\pm}_{p,j}$, where $\mathbf{n}^{\pm}_j$ are vectors parallel to the wave vectors $\mathbf{k}^{\pm}_j$ but having amplitude n_j.

The boundary conditions for the electric and magnetic fields at an interface require that the lateral components of these fields have to be continuous. Mathematically, this leads to the matrix equation

$$\begin{pmatrix} \frac{w_j}{n_j} & -\frac{w_j}{n_j} \\ n_j & n_j \end{pmatrix} \cdot \begin{pmatrix} E^{+}_{p,j} \\ E^{-}_{p,j} \end{pmatrix} = \begin{pmatrix} \frac{w_{j+1}}{n_{j+1}} & -\frac{w_{j+1}}{n_{j+1}} \\ n_{j+1} & n_{j+1} \end{pmatrix} \cdot \begin{pmatrix} E^{+}_{p,j+1} \\ E^{-}_{p,j+1} \end{pmatrix} \tag{8.10}$$

where the first line describes the continuity of the lateral electric field components, and the second line that of the lateral magnetic field components. This equation leads to the following relation connecting the electric field components in adjacent layers:

$$\begin{pmatrix} E^{+}_{p,j} \\ E^{-}_{p,j} \end{pmatrix} = \frac{1}{2}\begin{pmatrix} n+\frac{w}{n} & n-\frac{w}{n} \\ n-\frac{w}{n} & n+\frac{w}{n} \end{pmatrix} \cdot \begin{pmatrix} E^{+}_{p,j+1} \\ E^{-}_{p,j+1} \end{pmatrix} \tag{8.11}$$

where we have used the abbreviations $n = n_{j+1}/n_j$ and $w = w_{j+1}/w_j$.

Similarly, the boundary conditions for a s-wave (electric field amplitude vectors are parallel to the interfaces) read

$$\begin{pmatrix} 1 & 1 \\ w_j & -w_j \end{pmatrix} \cdot \begin{pmatrix} E^{+}_{s,j} \\ E^{-}_{s,j} \end{pmatrix} = \begin{pmatrix} 1 & 1 \\ w_{j+1} & -w_{j+1} \end{pmatrix} \cdot \begin{pmatrix} E^{+}_{s,j+1} \\ E^{-}_{s,j+1} \end{pmatrix} \tag{8.12}$$

leading to the following relation between field amplitudes in adjacent layers

$$\begin{pmatrix} E^{+}_{s,j} \\ E^{-}_{s,j} \end{pmatrix} = \frac{1}{2}\begin{pmatrix} 1+w & 1-w \\ 1-w & 1+w \end{pmatrix} \cdot \begin{pmatrix} E^{+}_{s,j+1} \\ E^{-}_{s,j+1} \end{pmatrix} \tag{8.13}$$

This kind of matrix formulation is extremely useful for deriving reflection and transmission coefficients for arbitrary stacks of dielectric/metallic layers. For example, for a plane wave incident on a single interface ($j = 1, 2$), one can set $E^{+}_2 = 1$ and $E^{-}_2 = 0$, which leads to

$$\begin{pmatrix} E^{+}_{p,1} \\ E^{-}_{p,1} \end{pmatrix} = \frac{1}{2}\begin{pmatrix} n+\frac{w}{n} & n-\frac{w}{n} \\ n-\frac{w}{n} & n+\frac{w}{n} \end{pmatrix} \cdot \begin{pmatrix} 1 \\ 0 \end{pmatrix} \tag{8.14}$$

so that the reflection and transmission coefficients for a p-wave are given by

$$\begin{aligned} r_p &= \frac{E^-_{p,1}}{E^+_{p,1}} = \frac{n^2 - w}{n^2 + w} \\ t_p &= \frac{1}{E^+_{p,1}} = \frac{2n}{n^2 + w} \end{aligned} \tag{8.15}$$

and similarly for an s-wave by

$$\begin{aligned} r_s &= \frac{1 - w}{1 + w} \\ t_s &= \frac{2}{1 + w} \end{aligned} \tag{8.16}$$

Here again, n stands for n_2/n_1 and w for $w_2/w_1 = \sqrt{n_2^2 - q^2}\Big/\sqrt{n_1^2 - q^2}$, where the propagation angles on both sides are given by $\sin\theta_{1,2} = q/n_{1,2}$. Remember that we work in a unit system where the vacuum wavelength is 2π and the wave vector amplitudes are thus $k_{1,2} = n_{1,2}$.

Next, let us consider the propagation of plane waves through a stack of many layers divided by interfaces $j \to j + 1$. In each layer j, we have now to consider the electric field in layer j at the $j - 1 \to j$ interface, which we denote by $E^{L\pm}_{ps,j}$, and at the $j \to j + 1$ interface, which we denote by $E^{R\pm}_{ps,j}$ (L and R as the first letters of "left" and "right" when visualizing the stack oriented from left to right). Then, for the $j \to j + 1$ interface, we have (see above)

$$\begin{pmatrix} E^{R+}_{ps,j} \\ E^{R-}_{ps,j} \end{pmatrix} = \hat{M}^{ps}_{j,j+1} \cdot \begin{pmatrix} E^{L+}_{ps,j+1} \\ E^{L-}_{ps,j+1} \end{pmatrix} \tag{8.17}$$

where we have defined the matrices

$$\hat{M}^{p}_{j,j+1} = \frac{1}{2}\begin{pmatrix} \frac{n_{j+1}}{n_j} + \frac{w_{j+1}n_j}{w_j n_{j+1}} & \frac{n_{j+1}}{n_j} - \frac{w_{j+1}n_j}{w_j n_{j+1}} \\ \frac{n_{j+1}}{n_j} - \frac{w_{j+1}n_j}{w_j n_{j+1}} & \frac{n_{j+1}}{n_j} + \frac{w_{j+1}n_j}{w_j n_{j+1}} \end{pmatrix} \tag{8.18}$$

and

$$\hat{M}^{s}_{j,j+1} = \frac{1}{2}\begin{pmatrix} 1 + \frac{w_{j+1}}{w_j} & 1 - \frac{w_{j+1}}{w_j} \\ 1 - \frac{w_{j+1}}{w_j} & 1 + \frac{w_{j+1}}{w_j} \end{pmatrix}. \tag{8.19}$$

For the propagation of the field from the $j - 1 \to j$ interface toward the $j \to j + 1$ interface in layer j, we have now an additional equation connecting the $E^{L\pm}_{ps,j}$ with the $E^{R\pm}_{ps,j}$:

$$\begin{pmatrix} E^{L+}_{ps,j} \\ E^{L-}_{ps,j} \end{pmatrix} = \hat{M}^{d}_{j} \cdot \begin{pmatrix} E^{R+}_{ps,j} \\ E^{R-}_{ps,j} \end{pmatrix} \tag{8.20}$$

with the propagation matrix

$$\hat{M}^{d}_{j} = \begin{pmatrix} \exp\left(-iw_j d_j\right) & 0 \\ 0 & \exp\left(iw_j d_j\right) \end{pmatrix}. \tag{8.21}$$

Now, we can write the full equation for a plane wave propagation from an infinite half space (index $j = 0$) to another infinite half space (index $j = N + 1$) through a stack of layers with refractive indices n_j and thickness values d_j:

$$\begin{pmatrix} E^{R+}_{ps,0} \\ E^{R-}_{ps,0} \end{pmatrix} = \hat{M}^{ps}_{0,1} \cdot \prod_{j=1}^{N} \hat{M}^{d}_{j} \hat{M}^{ps}_{j,j+1} \begin{pmatrix} E^{L+}_{ps,N+1} \\ E^{L-}_{ps,N+1} \end{pmatrix} \tag{8.22}$$

And as in the case of a single interface, by setting $E^{L+}_{ps,N+1} = 1$ and $E^{L-}_{ps,N+1} = 0$, one can calculate the compound reflection and transmission coefficients via

$$\begin{aligned} R_{ps} &= \frac{E^{R-}_{ps,0}}{E^{R+}_{ps,0}} \\ T_{ps} &= \frac{1}{E^{R+}_{ps,0}} \end{aligned} \tag{8.23}$$

Please note that these coefficients depend on the lateral wave vector component q (which is constant through all interfaces) and thus on the propagation angles θ_j that are given by $\sin\theta_j = q/n_j$. The just found result is quite general and will be also valid for metallic layers where the refractive indices become complex-valued.

8.2.3 Electric Dipole Emission

Now, we have everything in place for modeling the interaction of the dipole emitter with an arbitrary planar nanocavity. We assume that the emitter is located in a dielectric layer with refractive index n_d and thickness h, which is bounded on both sides by different stacks of dielectric/metallic layers with compound reflection and transmission coefficients $R^{\pm}_{p,s}$ and $T^{\pm}_{p,s}$, where we imagine all layers oriented horizontally so that the index + refers to the stack above the emitter and the index – for the stack below the emitter, see Figure 8.1.

To model the interaction of the electric field of the dipole emitter, eq. (8.7), with such a cavity, we split the electric field amplitude vector in eq. (8.7) into its p- and s-polarized component:

$$k_d^2\mathbf{p} - \mathbf{k}_d(\mathbf{k}_d \cdot \mathbf{p}) = k_d^2 \left[\left(\mathbf{p} \cdot \hat{\mathbf{e}}_{\parallel}\right) \hat{\mathbf{e}}_{\parallel} + \left(\mathbf{p} \cdot \hat{\mathbf{e}}^{\pm}_{\perp}\right) \hat{\mathbf{e}}^{\pm}_{\perp} \right] \tag{8.24}$$

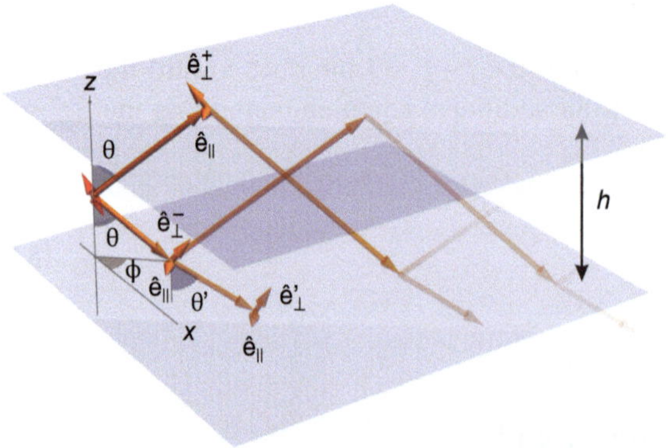

Figure 8.1 Dipole emission in a nanocavity.

where the unit vectors $\hat{\mathbf{e}}_\parallel$ and $\hat{\mathbf{e}}_\perp^\pm$ are given by $\hat{\mathbf{e}}_\parallel = \{-\sin\phi, \cos\phi, 0\}$ and $\hat{\mathbf{e}}_\perp^\pm = \hat{\mathbf{e}}_\parallel \times \hat{\mathbf{k}}_d^\pm$, where the plus sign + refers to plane waves traveling upward ($z > z_d$) and the minus sign – for plane waves traveling downward. Then, the full electric field inside the cavity is given by the dipole's source field plus its infinite forward and backward reflections at the top and bottom interfaces of the cavity. As we will see, of greatest interest are the fields directly on the top and bottom interfaces of the cavity. The field at the top interface is given by

$$\mathbf{E}(z = h) = \frac{ik_d^2}{2\pi\epsilon_d} \int \frac{d^2\mathbf{q}}{w_d} \sum_{\alpha,\beta=\pm} \left[L_s^{\alpha\beta} \left(\mathbf{p} \cdot \hat{\mathbf{e}}_\parallel\right) \hat{\mathbf{e}}_\parallel + L_p^{\alpha\beta} \left(\mathbf{p} \cdot \hat{\mathbf{e}}_\perp^\alpha\right) \hat{\mathbf{e}}_\perp^\beta\right] \exp\left[i\mathbf{q} \cdot \left(\rho - \rho_d\right) + iw_d \left(h - z_d\right)\right] \tag{8.25}$$

where the $L_{p,s}^{\alpha\beta}$ are defined by

$$\begin{aligned}
L_{p,s}^{++} &= \frac{1}{1 - R_{p,s}^+ R_{p,s}^- \exp(2iw_d h)} \\
L_{p,s}^{+-} &= \frac{R_{p,s}^+}{1 - R_{p,s}^+ R_{p,s}^- \exp(2iw_d h)} \\
L_{p,s}^{-+} &= \frac{R_{p,s}^- \exp(2iw_d z_d)}{1 - R_{p,s}^+ R_{p,s}^- \exp(2iw_d h)} \\
L_{p,s}^{--} &= \frac{R_{p,s}^+ R_{p,s}^- \exp(2iw_d z_d)}{1 - R_{p,s}^+ R_{p,s}^- \exp(2iw_d h)}.
\end{aligned} \tag{8.26}$$

These coefficients take into account the multiple back and forth reflections of the plane waves within the cavity, see Figure 8.1. Similarly for the field at the bottom of the cavity ($z = 0$), one finds

$$\mathbf{E}(z = 0) = \frac{ik_d^2}{2\pi\epsilon_d} \int \frac{d^2\mathbf{q}}{w_d} \sum_{\alpha,\beta=\pm} \left[L_s^{\alpha\beta} \left(\mathbf{p} \cdot \hat{\mathbf{e}}_\parallel\right) \hat{\mathbf{e}}_\parallel + L_p^{\alpha\beta} \left(\mathbf{p} \cdot \hat{\mathbf{e}}_\perp^\alpha\right) \hat{\mathbf{e}}_\perp^\beta\right] \exp\left[i\mathbf{q} \cdot \left(\rho - \rho_d\right) + iw_d z_d\right] \tag{8.27}$$

with the $L_{p,s}^{\alpha\beta}$ now defined by

$$\begin{aligned}
L_{p,s}^{++} &= \frac{R_{p,s}^+ R_{p,s}^- \exp[2iw_d(h - z_d)]}{1 - R_{p,s}^+ R_{p,s}^- \exp(2iw_d h)} \\
L_{p,s}^{+-} &= \frac{R_{p,s}^+ \exp[2iw_d(h - z_d)]}{1 - R_{p,s}^+ R_{p,s}^- \exp(2iw_d h)} \\
L_{p,s}^{-+} &= \frac{R_{p,s}^-}{1 - R_{p,s}^+ R_{p,s}^- \exp(2iw_d h)} \\
L_{p,s}^{--} &= \frac{1}{1 - R_{p,s}^+ R_{p,s}^- \exp(2iw_d h)}.
\end{aligned} \tag{8.28}$$

The magnetic fields are found by remembering that for each plane wave component we have $\mathbf{B} = \mathbf{n}_d \times \mathbf{E}$ where the vector $\mathbf{n}_d$ is parallel to the wave vector $\mathbf{k}$ of the plane wave component but having length n_d. Thus, the magnetic fields are given by the same expressions as the electric fields, but by doing the following replacements: For the magnetic field **B**, replace in the equation for **E**

$$\begin{aligned} L_s^{\alpha\beta}\left(\mathbf{p}\cdot\hat{\mathbf{e}}_\parallel\right)\hat{\mathbf{e}}_\parallel &\to -n_d L_s^{\alpha\beta}\left(\mathbf{p}\cdot\hat{\mathbf{e}}_\parallel\right)\hat{\mathbf{e}}_\perp^\beta \\ L_p^{\alpha\beta}\left(\mathbf{p}\cdot\hat{\mathbf{e}}_\perp^\alpha\right)\hat{\mathbf{e}}_\perp^\beta &\to n_d L_p^{\alpha\beta}\left(\mathbf{p}\cdot\hat{\mathbf{e}}_\perp^\alpha\right)\hat{\mathbf{e}}_\parallel \end{aligned} \tag{8.29}$$

Knowing now both the electric and magnetic fields at both interfaces, we can calculate the *full emission power* of the dipole via the Poynting vector through the two-dimensional integrals

$$S_{\text{total}} = \frac{c}{8\pi}\int d^2\rho\ \hat{\mathbf{z}}\cdot\text{Re}\left[(\mathbf{E}\times\mathbf{B}^*)_{z=h} - (\mathbf{E}\times\mathbf{B}^*)_{z=0}\right] \tag{8.30}$$

where $\hat{\mathbf{z}}$ is a unit vector along the positive z-direction, c is the vacuum speed of light, and a star index denotes complex conjugation.

For an emitter in a homogeneous environment with refractive index n_d, the emission power is equal to

$$S_0 = \frac{c n_d p^2 k^4}{3} \tag{8.31}$$

where $k = 2\pi/\lambda$ is the vacuum wave vector amplitude. Knowing both S_{total} and S_0 allows us now to calculate the change in observable fluorescence lifetime. The core idea is that the ratio S_{total}/S_0 tells us how much the *radiative* transition of a fluorescent molecule from its excited to its ground state is accelerated. Thus, for a fluorescing molecule with quantum yield Φ, the observable fluorescence lifetime τ will be given by

$$\frac{\tau}{\tau_0} = \frac{S_0}{\Phi S_{\text{total}} + (1-\Phi)S_0} \tag{8.32}$$

One important aspect of the emission power of a dipole emitter close to a surface or in a nanocavity is that this power is orientation-dependent. This can be seen when inspecting eqs. (8.25) and (8.27): the integrands contain the scalar products $\mathbf{p}\cdot\hat{\mathbf{e}}_\perp^\pm$ and $\mathbf{p}\cdot\hat{\mathbf{e}}_\parallel$, which are formed with the dipole moment vector $\mathbf{p}$. However, for an arbitrary dipole vector $\mathbf{p} = p_x\hat{\mathbf{x}} + p_y\hat{\mathbf{y}} + p_z\hat{\mathbf{z}}$, one has

$$\begin{aligned} \mathbf{p}\cdot\hat{\mathbf{e}}_\perp^\pm &= \mp p_x\sin\theta\cos\phi \mp p_y\sin\theta\sin\phi + p_z\cos\theta \\ \mathbf{p}\cdot\hat{\mathbf{e}}_\parallel &= -p_x\sin\phi + p_y\cos\phi \end{aligned} \tag{8.33}$$

which shows that the in-plane components $p_{x,y}$ of the dipole moment vector lead to ϕ-dependencies of $\cos\phi$ and $\sin\phi$, while the vertical component p_z leads to an ϕ-independent term. Thus, when inserting all into the integral on the r.h.s. of eq. (8.30) and integrating over ϕ (see also eqs. (8.25) and (8.27)), only expression that vary as $\cos^2\phi$, $\sin^2\phi$, or do not vary in ϕ at all, will survive. This shows that the total emission rate S_{total} factors into

$$S_{\text{total}} = S_\parallel \sin^2\beta + S_\perp \cos^2\beta \tag{8.34}$$

where β is the angle of the dipole moment vector with the vertical z-axis, and $S_\parallel$ and $S_\perp$ are the emission rates for a purely horizontally or purely vertically

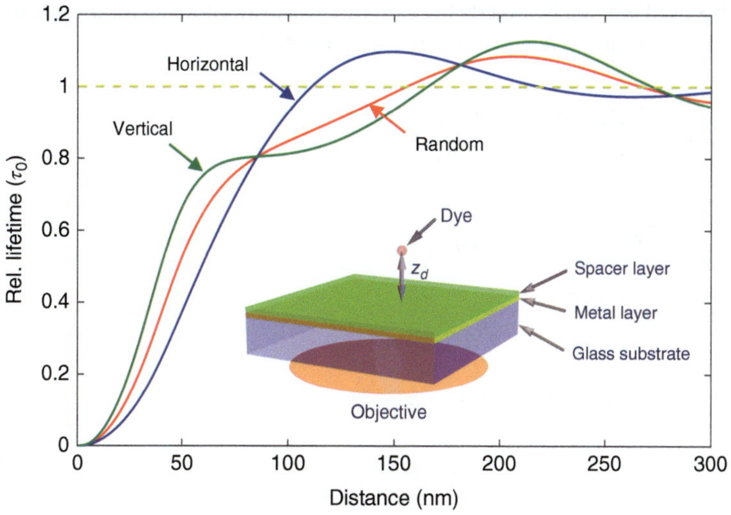

Figure 8.2 Dependence of the excited state lifetime of an emitter in water as a function of its distance from a gold-coated glass surface.

oriented dipole, respectively. Thus, in practice, one has only to calculate these two extreme cases and has then immediately the emission rate for any arbitrary dipole orientation.

As an example, we consider an emitter in water (refractive index 1.33) above a thin 20 nm gold layer on a glass surface (refractive index 1.5) emitting at 670 nm. In that case, the equations eqs. (8.25) and (8.27) simplify because there is no reflection from a stack of layers above the dipole emitter (i.e. $R^{+}_{p,s} = 0$). However, our general formalism remains fully valid and can be used also for this simpler problem. The complex-valued refractive index for gold at 670 nm was assumed to be $n_{\mathrm{Au}} = 0.1739 + 3.6125i$ [8]. The resulting fluorescence lifetime curve is shown in Figure 8.2, for two different emitter dipole orientations (horizontal and vertical), and assuming a quantum yield of fluorescence $\Phi = 1$. As can be seen, the presence of the gold layer on glass dramatically changes the fluorescence lifetime when the emitter is approaching the surface. This strong lifetime dependence on distance is the foundation of Metal-Induced Energy Transfer (MIET) imaging, see section Applications below.

8.2.4 Angular Distribution of Radiation and Detection Efficiency

In Section 8.2.3, we have derived all equations required for calculating the total emission power of a dipole emitter in a cavity. When observing such an emitter from outside the cavity, normally by collecting its emission with a microscope objective, one needs also to know its *angular distribution of radiation* (ADR) in the infinite half space from where one observes. Thus, we need to consider the electric field outside the cavity. This field is given by exactly the same expression as derived for the electric field inside the cavity at $z = 0$, eq. (8.27), but with the $L^{\alpha\beta}_{\parallel,\perp}$ now defined as

$$
\begin{aligned}
L^{++}_{p,s} &= \frac{T^{-}_{p,s}R^{+}_{p,s}R^{-}_{p,s}\exp[2iw_d(h-z_d)]}{1-R^{+}_{p,s}R^{-}_{p,s}\exp(2iw_d h)} \\
L^{+-}_{p,s} &= \frac{T^{-}_{p,s}R^{+}_{p,s}\exp[2iw_d(h-z_d)]}{1-R^{+}_{p,s}R^{-}_{p,s}\exp(2iw_d h)} \\
L^{-+}_{p,s} &= \frac{T^{-}_{p,s}R^{-}_{p,s}}{1-R^{+}_{p,s}R^{-}_{p,s}\exp(2iw_d h)} \\
L^{--}_{p,s} &= \frac{T^{-}_{p,s}}{1-R^{+}_{p,s}R^{-}_{p,s}\exp(2iw_d h)}.
\end{aligned}
\tag{8.35}
$$

where the $T^{-}_{p,s}$ are the compound transmission coefficient for the bottom stack of dielectric/metal layers. In the lower half space, each plane wave component in eq. (8.27) has to be additionally multiplied by an additional factor $\exp(-iw_0 z)$ with $w_0 = \sqrt{n_0^2 - q^2}$ describing the plane wave propagation in the lower half space with refractive index n_0 (immersion medium of objective). Each plane wave component with electric field amplitude $\mathbf{E}_A$ is connected with an energy flux density $cn_0 \left|\mathbf{E}_A\right|^2/8\pi$ along the direction of its wave vector. This allows us to use eq. (8.27) together with the coefficients eq. (8.35) for writing down the emission into solid angle $d\Omega^2 = \sin\theta_0 d\theta_0 d\phi = (q/w_0)dqd\phi$ as

$$
\frac{d^2S}{d\Omega^2} = \frac{cn_0}{8\pi}\left|\frac{k_d^2 w_0}{\epsilon_d w_d}\right|^2 \left|\sum_{\alpha,\beta=\pm}\left[L^{\alpha\beta}_s\left(\mathbf{p}\cdot\hat{\mathbf{e}}_{\parallel}\right)\hat{\mathbf{e}}_{\parallel} + L^{\alpha\beta}_p\left(\mathbf{p}\cdot\hat{\mathbf{e}}^{\alpha}_{\perp}\right)\hat{\mathbf{e}}^{\beta}_{\perp}\right]\right|^2 \tag{8.36}
$$

where the exact factor in front of the squared absolute values on the right-hand side is found by demanding that $d^2S/d\Omega^2$ in free space (no layers) has to be equal to the known exact analytical result $d^2S/d\Omega^2 = cn_0 p^2 k^4 \sin^2\theta/8\pi$ (where again $k = 2\pi/\lambda$ is the vacuum wave vector amplitude).

Knowing this ADR, one can then directly calculate how much light is collected by an objective with a half angle of light collection Θ by integrating $\sin\theta d^2S/d\Omega^2$ from $\phi = 0$ to $\phi = 2\pi$ and from $\theta = 0$ to $\theta = \Theta$. As an example, Figure 8.3 shows the compound ADR for a homogeneous distribution of emitters with emission wavelength of 500 nm inside a metallic nanocavity formed by two gold layers, where the bottom layer has a thickness of 10 nm and the top layer of 1000 nm. Both layers are deposited on glass (refractive index 1.5), and the complex-valued refractive index of gold at 500 nm was set to $n_{\mathrm{Au}} = 0.8651 + 1.8488i$ [8]. The figure shows a comparison of the ADR for a cavity formed only by glass without gold cover with that of the gold coated cavity. Calculations where done for randomly oriented dipoles, i.e. by integrating the ADR over all possible orientations of the dipole moment vector $\mathbf{p}$. The density plots inside the cavity show the position-dependent detection efficiency, which was obtained by integrating the ADR over the solid angle of light collection of a water immersion objective with numerical aperture NA = 1.2, which has a half angle of

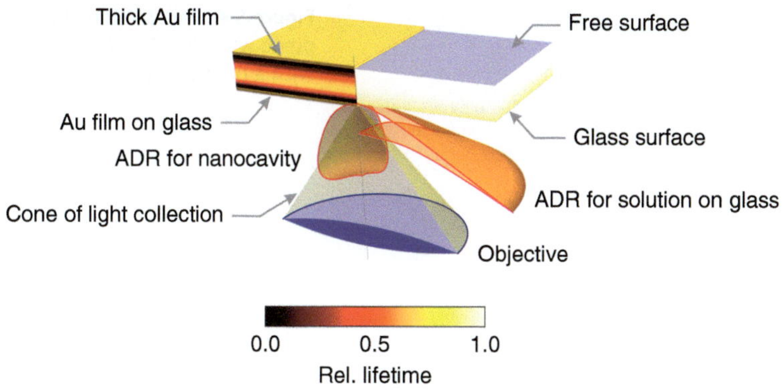

Figure 8.3 Comparison of the emission coming from a thin layer of aqueous solution on glass (right side) with that coming from a gold covered nanocavity (left side). Shown are the ADRs as three-dimensional polar plots, and as a density plot the relative fluorescence lifetime (relative to its free space value in solution) for emitters with unity quantum yield.

light collection $\Theta = \arcsin \mathrm{NA}/n_0 \approx 64^\circ$, where $n_0 = 1.33$ is the refractive index of water.

8.2.5 Excitation Intensity Distribution

Last but not least, we have to consider how well emitters are excited within a nanocavity. The presence of the cavity will strongly alter the excitation intensity distribution as compared to the no-cavity situation. For simplicity, we consider here the focusing of a linearly polarized plane wave through an objective (diffraction-limited focus) with numerical aperture NA into the cavity, see Figure 8.5. In free space, without the cavity, the electric field distribution around the focus is given by [9, 10]

$$\mathbf{E}_{ex}(\mathbf{r}) \propto \int_0^{\Theta} d\theta \sin\theta \sqrt{\cos\theta} \int_0^{2\pi} d\phi \left(-\sin\phi \hat{\mathbf{e}}_{\parallel} + \cos\phi \hat{\mathbf{e}}_{\perp}^{0}\right) \exp\left(i\mathbf{k}_0 \cdot \mathbf{r}\right) \tag{8.37}$$

where we have assumed that the incoming electric plane wave is polarized along the $\phi = 0$ direction in Figure 8.5, $\mathbf{k}_0 = k_0\{-\sin\theta\cos\phi, -\sin\theta\sin\phi, \cos\phi\}$ is the wave vector in the objective's immersion medium with refractive index n_0 ($k_0 = 2\pi/n_0\lambda$), and the focus position is at $\mathbf{r} = \mathbf{0}$. The unit vectors in eq. (8.37) are defined (in Cartesian coordinates) by $\hat{\mathbf{e}}_{\parallel} = \{-\sin\phi, \cos\phi, 0\}$ and $\hat{\mathbf{e}}_{\perp}^{0} = \{\cos\theta\cos\phi, \cos\theta\sin\phi, \sin\theta\}$. Thus, the electric field is again given by a plane wave representation, and similarly to what we did for calculating the electric field generated by a dipole emitter inside the cavity, we can use the compound transmission and reflection coefficients of the two stacks of layers forming the cavity for calculating the electric field inside the cavity. This is now given by

$$\begin{aligned}\mathbf{E}_{ex}(\mathbf{r}) \propto & \int_0^{\Theta} d\theta \sin\theta \sqrt{\cos\theta} \int_0^{2\pi} d\phi \\ & \sum_{\alpha=\pm} \left(-L_s^{\alpha} \sin\phi \hat{\mathbf{e}}_{\parallel} + L_p^{\alpha} \cos\phi \hat{\mathbf{e}}_{\perp}^{\alpha}\right) e^{i(\mathbf{q}\cdot\rho + w_d z - w_0 z_f)}\end{aligned} \tag{8.38}$$

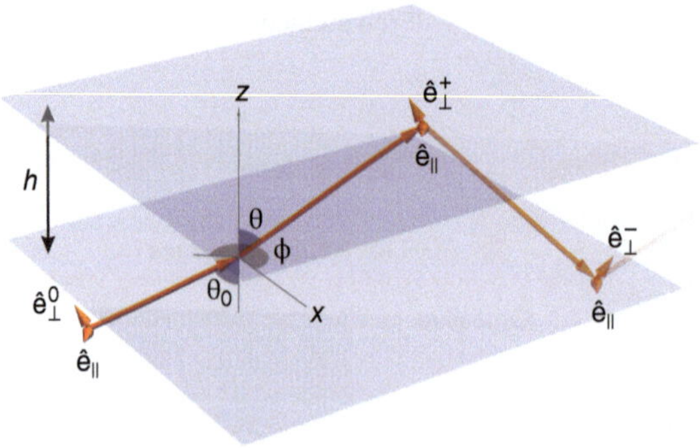

Figure 8.4 Interaction of a plane wave incident from below with a nanocavity.

where z_f is the axial position to which the objective focuses the light, $w_0 = \sqrt{n_0^2 - q^2}$ is the axial wave vector component in the objective's immersion medium, and the $L_{p,s}^{\alpha}$ are defined by (see also Figure 8.4)

$$\begin{aligned} L_{p,s}^{+} &= \frac{T_{p,s}}{1 - R_{p,s}^{+} R_{p,s}^{-} \exp(2iw_d h)} \\ L_{p,s}^{-} &= \frac{T_{p,s} R_{p,s}^{+} \exp(iw_d h)}{1 - R_{p,s}^{+} R_{p,s}^{-} \exp(2iw_d h)} \end{aligned} \tag{8.39}$$

where $T_{p,s}$ are the compound transmission coefficients for plane p- and s-waves through the bottom of the cavity, and the $R_{p,s}^{\pm}$ are the compound reflection coefficients inside the cavity for the bottom (−) and top (+) cavity walls (same coefficients as in eqs. (8.26), (8.28) and (8.35)).

As an example, Figure 8.5 shows a comparison of the excitation intensity distribution inside and outside a metal nanocavity. All parameter values (gold layer

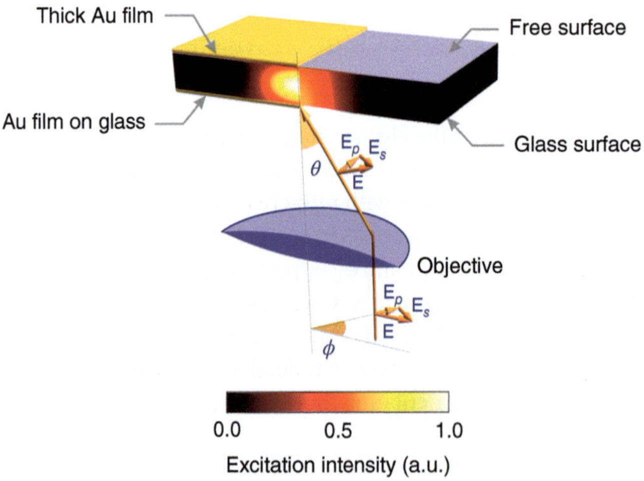

Figure 8.5 Comparison of the excitation intensity generated by focusing a plane wave of light through an objective. Right-hand side shows a density plot of the intensity distribution in aqueous solution above a glass surface, and left-hand side a density plot of the intensity distribution inside a gold covered nanocavity.

thickness values, refractive indices, wavelength, numerical aperture) are the same as used for the calculation of Figure 8.3. As can be seen, inside the cavity, the excitation intensity is strongly nonuniform along the optical axis and enhanced in the cavity mid-plane.

8.2.6 Synthesis

With the knowledge of the excitation intensity distribution, one has the last required information for fully describing a fluorescence spectroscopy experiment in an arbitrary planar system. For example, when interested in calculating the observable mean fluorescence lifetime in a cavity experiment, one first calculates the position-dependent lifetime using the equations presented in subsection 8.2.3 and then calculates the mean lifetime as a weighted average using the product of (position-dependent) excitation intensity (last subsection) times detection efficiency (Section 8.2.3) as weight function. However, the situation is complicated by reorientation effects (rotational diffusion) of the emitters. Two limiting cases are of biggest interest: The rotational diffusion of the emitters is much faster than the fluorescence lifetime (this is typically the case when studying small organic fluorescent molecules in aqueous solution), or the rotational diffusion is much slower than the fluorescence lifetime (large molecules such as fluorescent proteins or molecules embedded into a polymer). In the first case, one has to calculate the total emission rate eq. (8.34) as an average over all dipole orientation angles before inserting the result into eq. (8.32). In the second case, on has to average eq. (8.32) over all orientation angles, which gives often a significantly different result, see, e.g. [11]. For the intermediate case, one has to apply a more complicated approach that involves solving the rotational diffusion equation, and we refer the reader to Ref. [12]. However, in almost all cases of practical interest, one usually has to deal only with the limiting cases of very fast or very slow rotational diffusion.

8.3 Applications

8.3.1 Absolute Quantum Yield Measurements

The fluorescence quantum yield (QY) is one of the photophysical parameters of an emitter that are most difficult to determine. The QY is a measure of the conversion efficiency of absorbed light into emitted photons and is thus a key parameter of its photophysics [13]. The era of quantitative QY measurements began in 1924 with Vavilov's invention of the first reliable method for measuring the absolute luminescence efficiency in solution by comparing the luminescent emission with scattering intensities [14]. Nowadays, the QY is most often determined in a comparative manner by measuring the total luminescence emission intensity against that of a fluorescent standard of known QY [15–18] or against the scattering intensity of a non-luminescent scatterer [19]. Another way to determine the QY is to measure the thermal power released after excitation in a so-called thermal lensing measurement [19, 20]. This method provides one of the most accurate ways to determine the QY, but it is technically challenging and requires large sample concentrations. Finally, the photoluminescence QY value can be absolutely obtained by measuring the number of emitted photons per number of absorbed photons using an integrating sphere setup [21, 22]. While this technique avoids inaccuracies that typically occur when performing referential QY measurements, its precision relies on the correct radiometric characterization of the integration sphere setup and on the proper account of re-absorption effects [22].

The fluorescence QY is defined as the ratio of the radiative (k_r) to the non-radiative (k_{nr}) transition rate from the excited to the ground state of an emitter

$$\Phi = \frac{k_r}{k_r + k_{nr}} \tag{8.40}$$

Thus, if one is able to specifically vary either the radiative or non-radiative transition rate while measuring the total de-excitation rate $k_r + k_{nr}$ (inverse excited state lifetime), one can deduce the value of the QY of an emitter. In recent years, several publications have dealt with the controlled modification of the radiative transition rate of an emitter by tuning the local density of states (LDOS) of the electromagnetic field at its position. As was pointed out by Purcell, by changing the LDOS of the electromagnetic field one changes the coupling of an emitter's excited-to-ground-state transition to that field, thus leading to a modification of the radiative transition rate [23]. This so-called Purcell effect has been measured for fluorophores placed between two gold nanoparticles [24], close to a dielectric interface [25, 26], to a metallic mirror [7, 27, 28], or to a sharp tip of a scanning probe microscope [29, 30]. A very efficient way of changing the LDOS of the electromagnetic field is to embed an emitter into a plane-parallel optical resonator [11, 31–35], which allows for precise control of the LDOS by varying the cavity length. Moreover, this length can be precisely monitored by measuring the cavity's transmission spectrum, which eliminates potential systematic errors caused by mechanical instabilities of the system. We have recently employed such a tunable cavity to measure the QY of single molecules at fixed positions inside a resonator [34], and for investigating the impact of rotational diffusion on the electrodynamic coupling of dipole emitters to the cavity modes [11].

We apply the tunable nanocavity for the absolute, calibration-free measurement of QY values of fluorophores in solution. The core idea of the nanocavity-based method is to measure the change of the fluorescence lifetime of a solution of fluorophores inside a planar metal nanocavity as a function of cavity length. Changing the cavity length changes the LDOS inside the cavity and thus the radiative transition rate of the enclosed fluorophores [36]. Employing a complete and quantitative theoretical model of the expected change of k_r as a function of cavity length, and knowing the full excited-to-ground state transition rate $k_r + k_{nr}$ from the lifetime measurement, one can use Eq. (8.40) for fitting an absolute value of the QY. As will be shown below, this kind of measurement is easy to set up, yields an absolute and reference-free value of the QY, and needs only minute amounts of sample (microliters of micromolar or less concentration). Thus, the method will be of considerable interest for all applications where rapid and absolute QY measurements on small sample quantities are required.

A schematic of the experimental setup is shown in Figure 8.6. A homemade nanocavity consists of two silver mirrors placed at sub-wavelength distance one from each other. The bottom silver mirror (35 nm thick) was prepared by vapor deposition onto a cleaned glass cover slide (thickness 170 μm) using an electron beam source (Univex 350, Laybold) under high-vacuum conditions ($\sim 10^{-6}$ mbar). The top silver layer (75 nm thick) was prepared by vapor deposition of silver on the surface of a plan-convex lens (focal length of 150 mm) under the same conditions. During vapor deposition, film thickness was monitored using an oscillating quartz unit and afterward verified by atomic force microscopy measurements. The complex-valued wavelength-dependent dielectric constants of the silver films were determined by ellipsometry (Nanofilm ep3, Accurion GmbH, Göttingen).

Transmitting light from a halogen lamp through the cavity results in a white light transmission pattern shown in inset of figure 8.6 (bottom-right corner). The dark spot in the center of the pattern corresponds to the center of the cavity where the mirrors are in direct mechanical contact with each other. As a result, no constructive interference of the visible light occurs in that area. The inner color ring around the dark spot corresponds to the first interference region of the cavity (so-called $\lambda/2$ region), where the cavity-induced modulation of fluorescence properties of a fluorophore is maximized. Larger color rings correspond to higher orders of interference of the light going through the cavity. The local cavity length was determined by measuring the white light transmission spectrum [32, 34] using a spectrograph (SR 303i, Andor) and a CCD camera (iXon DU897 BV, Andor).

For the QY measurements, a droplet of a micromolar solution of dye molecules in water (either rhodamine 6G, OregonGreen 488, Alexa 488, or Atto 488) was placed between the cavity mirrors. For fluorescein, we used a 0.1 N aqueous sodium hydroxide solution. Fluorescence lifetime measurements were performed with a home-built confocal microscope equipped with a high numerical aperture objective (UPLSAPO, 60×, N.A.=1.2 water immersion, Olympus). A white-light laser system (SC400-4-80, Fianium) with an acousto-optical tunable filter (AOTFnC-400.650-TN, AA Optic) served as the excitation source ($\lambda_{exc} = 488$ nm). The excitation light was reflected by a dichroic mirror (BrightLine FF484-FDi01, Semrock) toward the objective. Back-scattered excitation light was blocked with a long-pass filter (EdgeBasic BLP01-488R, Semrock). Collected fluorescence was focused onto the active area of a single-photon detection module (PDM series, MPD). Data acquisition was accomplished with a multichannel picosecond event timer (HydraHarp 400, PicoQuant GmbH). Photon arrival times were histogrammed (bin width 50 ps) for obtaining fluorescence decay curves, and all curves were recorded until reaching at least 10^4 counts at maximum. The fluorescence decay curves were fitted with a multi-exponential decay model, from which the average excited state lifetime was calculated. Positioning of the focal spot inside the cavity was done by moving the sample with a piezo nano-positioning stage PI P-562.

After placing a 10 μl droplet of a micromolar solution of dye between the cavity mirrors, we recorded fluorescence decay curves for a series of values of cavity length. The spherical shape of the upper mirror allowed us to reversibly tune the cavity length by moving the cavity laterally with respect to the excitation focus with

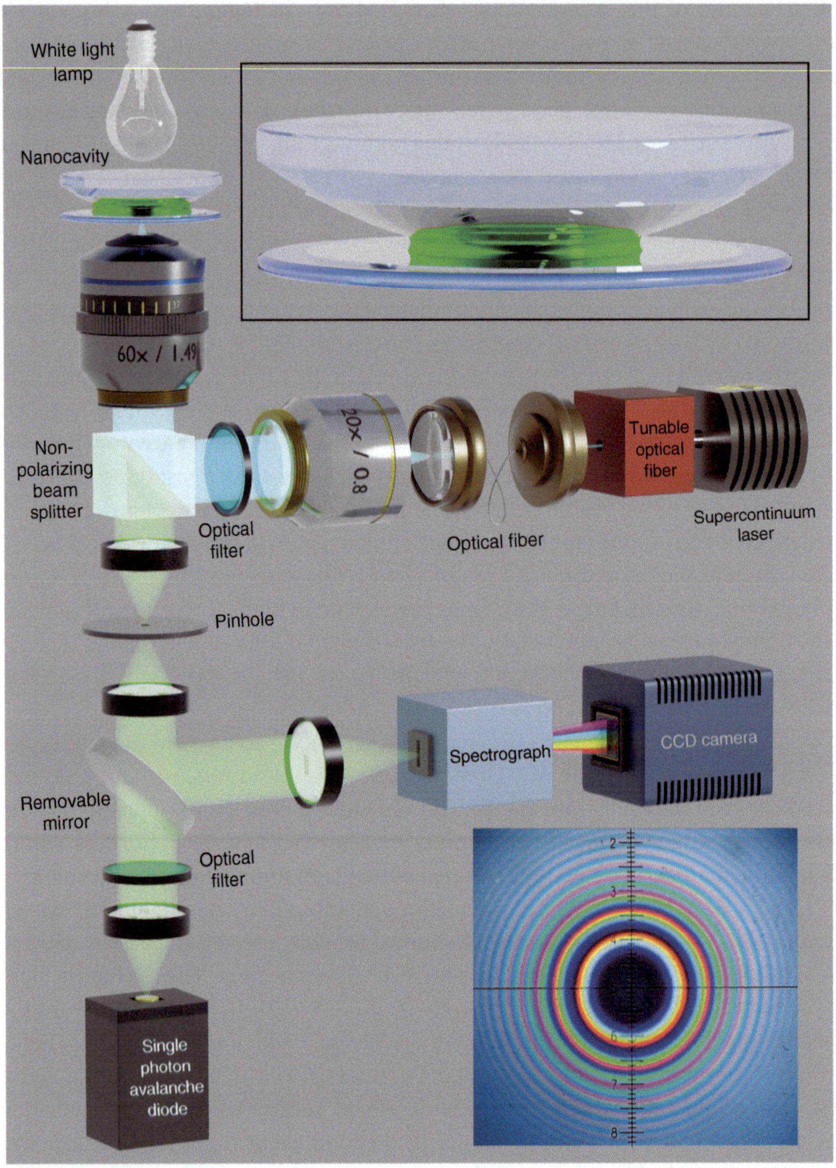

Figure 8.6 Experimental setup. The nanocavity consists of two silver layers, deposited on the glass surface. The upper silver layer is sputtered on the surface of a plano-convex lens, which allows one to tune the cavity length by moving the cavity in a horizontal plane. It should be noted that within the focal spot of a diffraction-limited objective lens focal spot, the cavity acts as a plane-parallel resonator. Top-right corner: the nanocavity with a droplet of solution placed between the mirrors. Bottom-right corner: a white light transmission pattern of the cavity.

the piezo nano-positioning stage. It should be noted that across the excitation focus, the cavity can be considered as a plane-parallel resonator [32]. To exclude any influence of potential mechanical instabilities of the cavity on the obtained results, we recorded white light transmission spectra of the cavity before and after acquiring each fluorescence decay curve. By fitting the white light transmission curves, we determined exact value of the cavity geometry. For doing that we had to know the exact values of the wavelength-dependent complex-valued refractive index of the silver layers, which we determined by ellipsometry measurements.

Acquisition of fluorescence decay curves was done in a range of the white light transmission maxima between 500 and 650 nm, corresponding to the fluorescence emission bands of the selected dyes. Figure 8.7 shows the results of the measurements for rhodamine 6G, Alexa 488, Oregon Green 488, Atto 488, and fluorescein (red circles) as a function of maximum transmission wavelength, which is linearly proportional to the cavity length. The curves show a strong, dye-specific decrease of the lifetime values with increasing cavity length. The solid lines show fits of the theoretical model to the experimental data, where the only free fit parameters were the free space lifetime τ_0, and the fluorescence QY value Φ. In the model calculations, we assumed that the rotational diffusion time is much faster than the fluorescence decay time, which is justified for small dyes in aqueous solutions [11]. For determining the errors of our nanocavity measurement method, we employ Efron's bootstrap algorithm [37]. The errors typically do not exceed 2-3% for high QYs and stay within 5% for QYs below 10-20%.

One of the key advantages of the nanocavity-based methods is that it allows one to measure QY of complex systems that are impossible to measure using standard methods. Figure 8.8 summarizes recent QY measurements of various types of samples. First, the exceptional sensitivity of the method allowed us to use it even at the single molecule level [34]. It has been shown that single molecules have different QY when placed inside a solid polymer matrix. Placing SiO_2 nanoparticles between the cavity mirrors on a dielectric spacer showed that their intrinsically disordered structure leads to a particle-to-particle variation of the QY [35, 38].

The insensitivity of the nanocavity-based method to non-luminescent species allowed us to measure QY of graphene nanoparticles [39]. Conventional measurements of QY are prone to errors when the sample contains also absorbing but non-luminescent species, which is commonly encountered in complex systems such as graphene quantum dots that are either optically inactive themselves or contain impurities formed during nanoparticle synthesis. Their presence can lead to a gross underestimation of the quantum yield of the luminescent species. By comparing these values with those obtained with a comparative method, the average size and relative concentration of the non-luminescent particles were determined.

The nanocavity-based method allows one to measure several QY values simultaneously in a multi-chromo-phoric sample. It has been used for measuring QY in a mixture of three different sizes of semiconductor quantum dots [40] or even in a mixture of fluorophores emitting and absorbing in the same spectral range both in liquid and solid phases [41]. Recently, the method was used for measuring QY of molecules

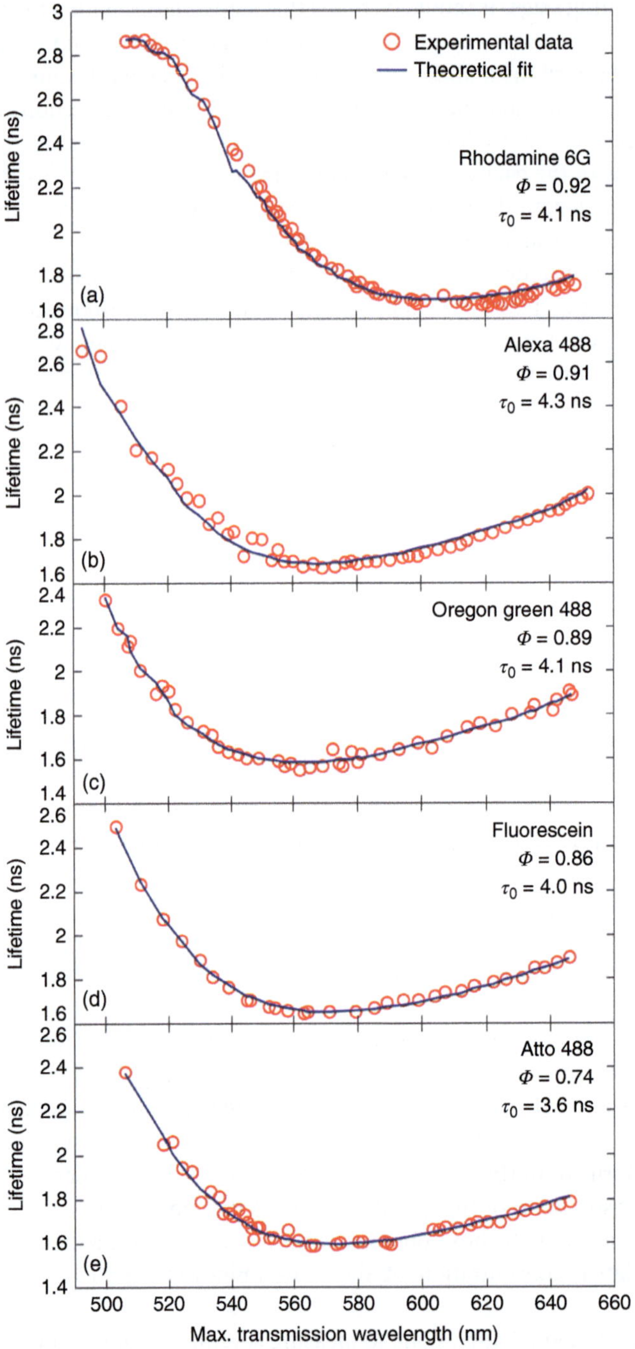

Figure 8.7 Cavity-controlled fluorescence lifetime of a rhodamine 6G, Alexa 488, OregonGreen 488 and Atto 488 in water and fluorescein in 0.1 M NaOH aqueous solution as a function of the maximum transmission wavelength of the resonator, which is linearly proportional to the cavity length. Red circles are the experimental data, blue curves are fits of the model. Φ and τ_0 indicate the values of the QY and free space fluorescence lifetime, respectively. Source: Chizhik et al. [12]/With permission of John Wiley & Sons.

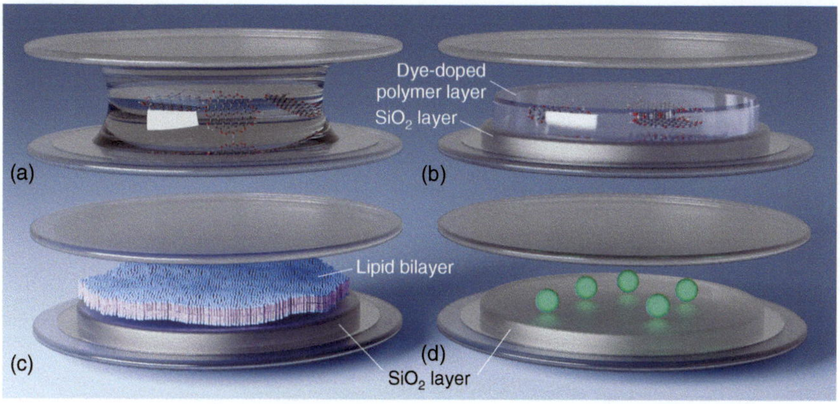

Figure 8.8 Different samples inside the nanocavity: (a) fluorophores in a solvent; (b) fluorophores immobilized in a polymer film on a SiO_2 spacer; (c) a lipid bilayer on a SiO_2 spacer; (d) fluorophores placed on the surface of a SiO_2 spacer.

placed inside a lipid bilayer [42]. Finally, QYs of 23 fluorescent proteins (including photo-switchable ones) were determined using the nanocavity-based method [43].

8.3.2 MIET/GIET imaging

It was predicted by Edward Purcell in 1946 [23] that placing a fluorescent molecule in the vicinity of a metal quenches its fluorescence emission and decreases its excited state lifetime. From a physics point of view, the mechanism behind this phenomenon is similar to that of Förster resonance energy transfer (FRET) [44]: energy from the excited molecule is transferred, via electromagnetic coupling, into plasmons of the metal, where energy is either dissipated or re-radiated as light. This fluorophore–metal interaction was extensively studied in the 1970s and 1980s [4], and a quantitative theory developed on the basis of semi-classical quantum optics [5, 45], see section Theory at the beginning of this chapter. The achieved quantitative agreement between experimental measurement and theoretical prediction was excellent.

Owing to the fact that the energy transfer rate is dependent on the distance of a molecule from the metal layer, the fluorescence lifetime can be directly converted into a distance value (Figure 8.9). The theoretical basis for the success of this conversion is the perfect quantitative understanding of MIET [46]. It is important to emphasize that the energy transfer from the molecule to the metal is dominated by the interaction of the molecule's near-field with the metal and is thus a thoroughly near-field effect, similar to FRET. However, due to the planar geometry of the metal film, which acts as the acceptor, the distance dependency of the energy transfer efficiency is much weaker than the sixth power of the distance, which leads to a monotonous relation between lifetime and distance over a size range between zero and ~250 nm above the surface.

Experimentally, one needs a standard scanning confocal microscope capable of fluorescence lifetime imaging (Figure 8.10), that is, equipped with a pulsed

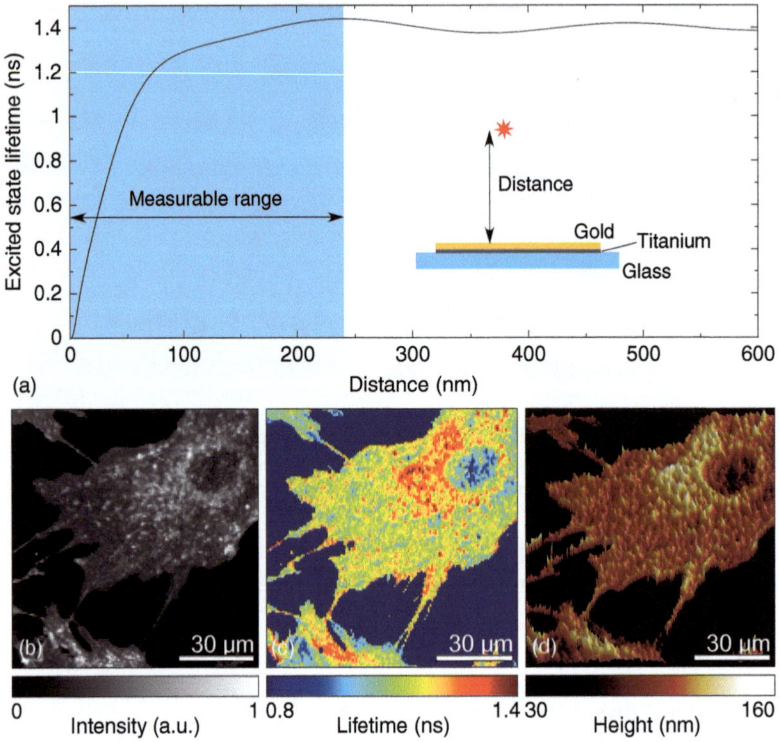

Figure 8.9 (a) Calculated dependence of molecule lifetime on its axial position over the metal film. Curves are calculated for an emission wavelength of 660 nm and a gold film thickness of 15 nm deposited on an intermediate titanium layer of 2 nm on the glass cover slide. Simultaneously acquired fluorescence intensity (b) and lifetime (c) images of the basal membrane of a living NMuMG cell grown on a gold-covered glass substrate, acquired with a standard confocal microscope. (d) Three-dimensional profile of the basal cell membrane reconstructed from lifetime image (c) using the curve MIET calibration curve (a). Source: Baronsky et al. [48], with permission of American Chemical Society.

excitation laser and a single photon avalanche diode. The only addition that is required for MIET imaging is coating the substrate with a semitransparent metal film, typically of 10–15 nm thickness. Gold as a coating material combines such crucial properties as non-toxicity for living cells, absence of oxidation, and high transparency compared to other metals.

The applicability of MIET for live-cell imaging has been first shown by mapping the basal membrane of living cells with nanometer accuracy [47]. Knowledge of the precise cell-substrate distance as a function of time and location with unprecedented resolution provides a new means to quantify cellular adhesion and dynamics, as is required for a deeper understanding of fundamental biological processes such as cell differentiation, tumor metastasis, and cell migration.

As a biological model system, three adherent cell lines were chosen: MDA-MB-231 human mammary gland adenocarcinoma cells, A549 human lung carcinoma cells, which are able to form metastasis *in vivo* models, as well as MDCK II from canine kidney tissue as a benign epithelial cell line. Interestingly, significant differences in

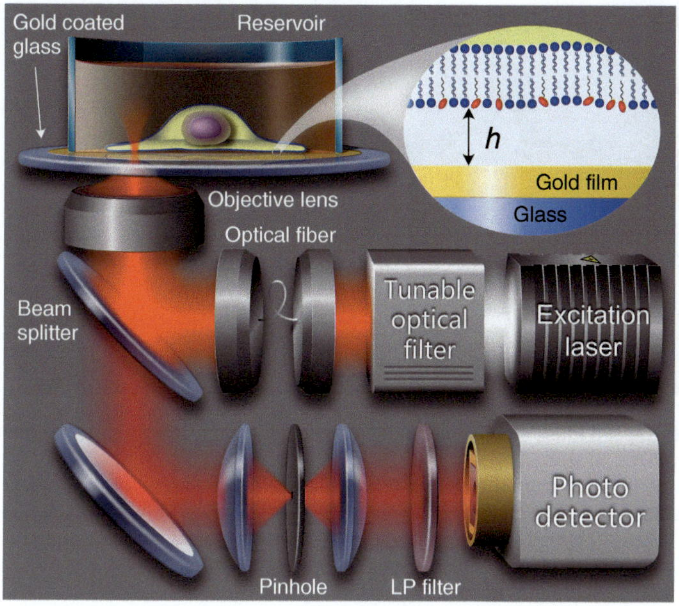

Figure 8.10 Schematic of the experimental setup for MIET imaging. LP filter = long-pass filter. Source: Baronsky et al. [48], with permission of American Chemical Society.

the cell–interface distance between a normal epithelial cell and cancerous cell lines were observed.

Figures 8.9(a) and (b) show the measured intensity and lifetime images that were used to obtain the 3D reconstruction of the basal cell membrane. Because the variation of the fluorescence intensity is not only dependent on the metal-induced quenching, but also on the homogeneity of labeling, exclusively the lifetime information was used for reconstructing a three-dimensional map of the basal membrane. On the other hand, the intensity distribution was used to discriminate the membrane fluorescence against the background. Regions with no cells are difficult to identify from the lifetime images alone, as the lifetime values can become exceedingly scattered at low signal-to-noise ratios. Figure 8.9(c) shows the result of recalculation of the lifetime image into the 3d height profile.

A relatively fast scanning speed of a confocal microscope that is used for MIET imaging allows to monitor dynamic processes. Figure 8.11 shows the spreading behavior of MDCK-II cells. Generally, the spreading process of adherent cells can be divided into three distinct temporal phases. The first phase is characterized by the formation of initial bonds between adhesion molecules and molecules of the extracellular matrix. This process of tethering is followed by the second phase, which comprises the initial cell spreading and that is driven by actin polymerization. The latter forces the cell surface area to increase by drawing membrane from a reservoir of folded regions. The third phase encompasses recruitment of additional plasma membrane from the internally stored membrane buffer and extension of lamellipodia to occupy a larger area.

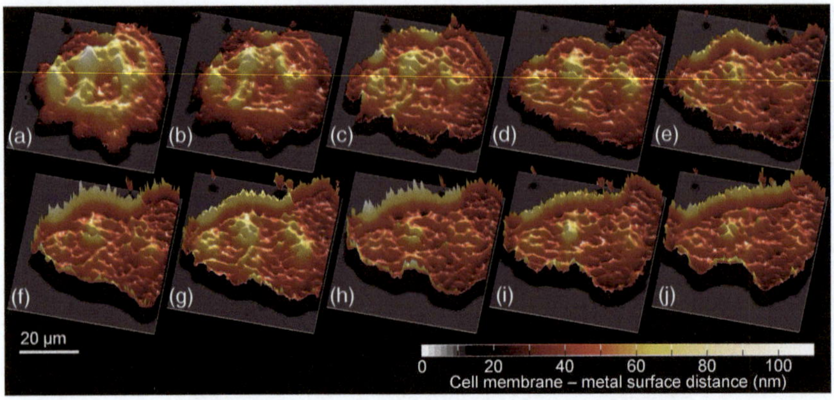

Figure 8.11 Time-elapsed MIET images recorded in 5-minute time intervals showing the late stages of cell (MDCK II) spreading on gold. The cell forms tightly attached protrusions/lamellipodia away from the center of the cell. The cell occupies a larger area with time and presses down more closely. A darker color refers to lower cell–substrate distance. At later stages (k-n), first lamellipodia are formed that exhibit a low cell–substrate distance. Source: Chizhik et al. [47], with permission of Springer Nature.

The axial resolution of the recorded images can be determined by calculating the standard deviation of cell–substrate distance. The resolution depends on the photon rate and varies between 2 nm and 4 nm for typically measured fluorescence intensities (ca. 5,000–10,000 detected photons per scan position) and can be further enhanced to 1 nm by increasing the number of detected photons.

The unprecedented axial resolution of MIET allowed us to monitor the cell-substrate distance of epithelial NMuMG cells during the biological process of the epithelial-to-mesenchymal transition (Figure 8.12) [48]. (EMT) allows epithelial cells to enhance their migratory and invasive behavior and plays a key role in embryogenesis, fibrosis, wound healing, and metastasis. Among the multiple biochemical changes from an epithelial to a mesenchymal phenotype, the alteration of cellular dynamics in cell–cell as well as cell–substrate contacts is crucial. It was shown that, in the very first hours of the transition, the cell–substrate distance increases by several tens on nanometers, but later in the process after reaching the mesenchymal state, this distance is reduced again to the level of untreated cells.

Dual-color MIET allowed for reconstructing the 3D profile of the nuclear envelope over the whole basal area of HeLa cells [49]. The profilometry was done by measuring the axial distance between the proteins Lap2β and Nup358 as components of the nuclear envelope and the nuclear pore complex, with defined localizations at the inner nuclear membrane and the cytoplasmic side of the protein complex, respectively (Figure 8.13). The obtained thickness of the nuclear envelope of 30–35 nm is in very good agreement with the values that were obtained using electron microscopy. This study has shown that optical microscopy allows one not only to measure the distance between the outer and inner nuclear membrane but also to reconstruct its 3D profile over the whole basal area.

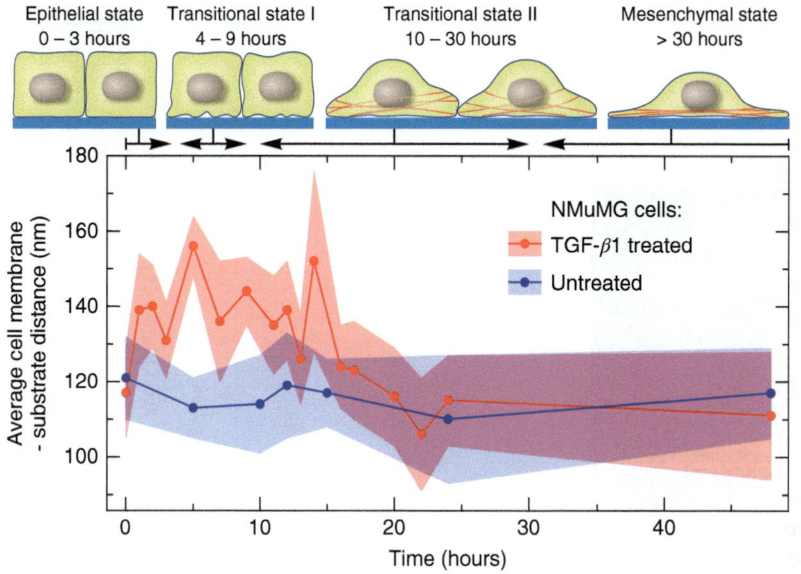

Figure 8.12 Average cell membrane–substrate distance of untreated (blue) and TGF-β1 treated NMuMG cells (red) over time. NMuMG cells detach from the surface by more than 20 nm on average in response to TGF-β1 administration. After 20 h, the initial cell–substrate distance is restored. The standard error of mean (SEM) is illustrated as colored area around the data points.

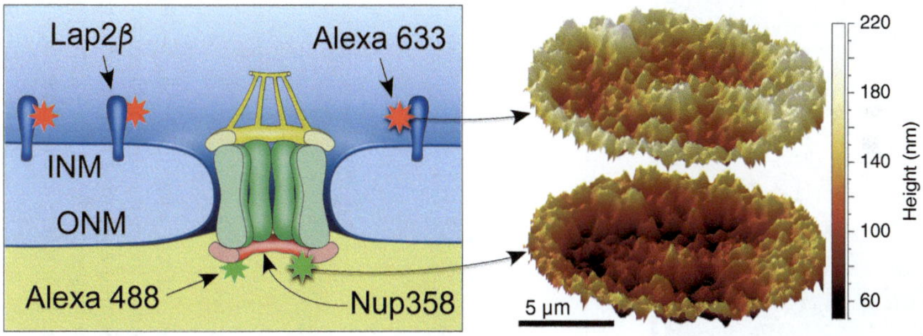

Figure 8.13 Schematic of the positions of Lap2β and Nup358 in the inner nuclear membrane and the nuclear pore complex, respectively. HeLa cells were fixed and subjected to indirect immunofluorescence using goat anti-Nup358. Three-dimensional height profiles of the inner (top) and outer (bottom) nuclear membrane of a typical HeLa cell nucleus, as determined by MIET imaging. The outer nuclear membrane roughly follows the profile of the inner nuclear membrane.

Recently, dual-color MIET was combined with FRET for studies of cytoskeletal elements and adhesions in human mesenchymal stem cells [50]. In addition to resolving nanometric structural details along the z-axis using MIET, FRET was used to measure the distance between actin and vinculin at focal adhesions. The analysis of the temporal evolution of actin heights shows that the actin filaments move closer to the surface while the cell is spreading and firmly adhering (Figure 8.14).

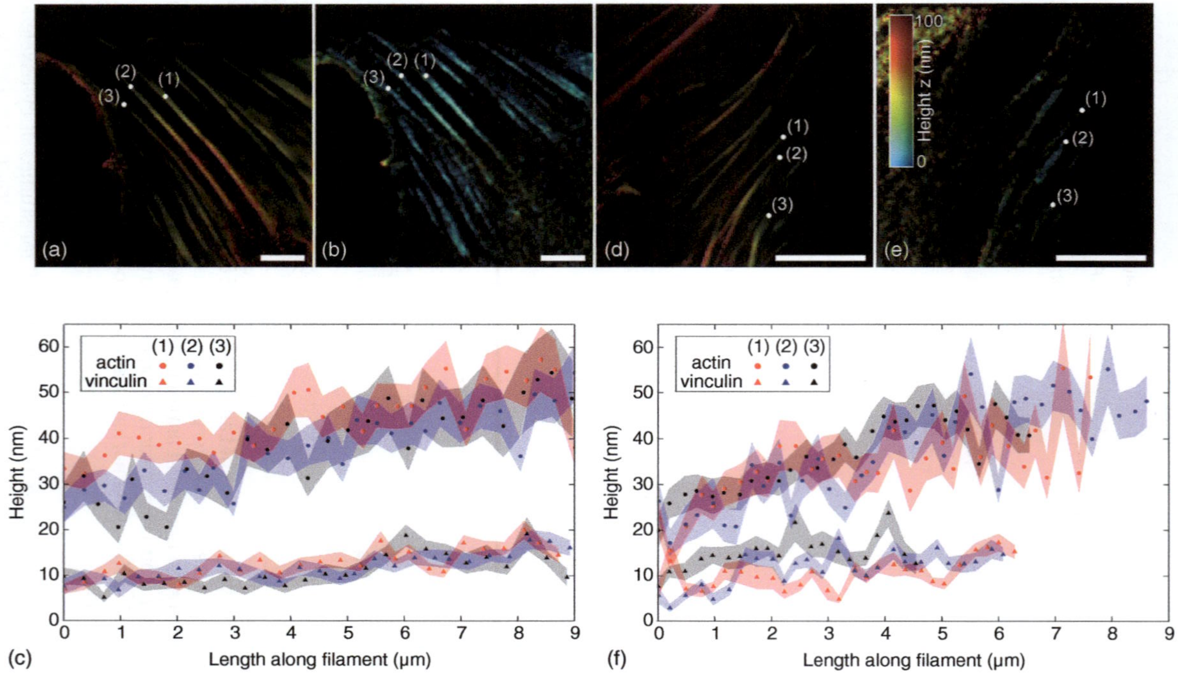

Figure 8.14 3D architecture of stress fibers at focal adhesions changes from 12 to 24 h. Height profiles along actin filaments and vinculin complexes after 12 and 24 h. Images a and b correspond to intensity-weighted ensemble heights of actin and vinculin, respectively, for a cell fixed 12 h after seeding. Images d and e correspond to intensity-weighted ensemble heights of actin and vinculin, respectively, for a cell fixed 24 h after seeding. White points (1), (2), and (3) on the intensity-weighted height images indicate the starting points of the height profiles shown in images (c) and (f). They show the height of actin filaments (circles) and vinculin clusters (triangles) at the same focal adhesion. The shaded areas mark the 1σ-regions of the height values. Scale bar is 10 μm.

Although the fibers are distributed over a broad height range during an early phase (1–6 h), their distance to the surface reduces around 12h and later time points to 40 nm. On the other hand, during maturation of focal adhesion complexes, vinculin aggregates grow larger as indicated by an increase in height, and the mean height of the actin bundles above the surface is decreasing. The nanometer-precise height information along the fibers and of the vinculin clusters (figure 8.14) gives a detailed picture of stress fibers anchoring at focal adhesions and spanning the cell at a slight inclination of below 1°.

Use of single-photon counting detectors for MIET measurements allows one not only to achieve nanometer resolution of sub-cellular structures with high labeling density, but also to do nanometer axial localization of single molecules. The proof-of-principle study was done by Karedla et al., where the authors determined the height of dye molecules deposited on a dielectric spacer of a known thickness [51]. By varying the thickness of the spacer, the authors showed that the axial position of molecules can be determined with accuracy better than 2.5 nm. The excellent agreement between the known thickness and the height values that were obtained using MIET showed its applicability for single-molecule studies with accuracy that is unachievable with conventional microscopy techniques.

Isbaner et al. used MIET for co-localizing two single fluorescent emitters along the optical axis with nanometer accuracy [52]. For this purpose, the authors used step-wise photobleaching to find the fluorescence lifetime values of each emitter on one DNA origami pillar, which allowed them to determine their individual heights from the surface and thus their mutual axial distance. The determined distance of 32±11 nm is in excellent agreement with the design value of 32 nm.

Despite the advantages of gold over other metals for MIET microscopy, other electrically conducting materials can be interesting candidates for being used in this method. It has been recently shown that by replacing a gold layer with graphene, one can make the MIET calibration curve significantly steeper (see Figure 8.15) [53, 54].

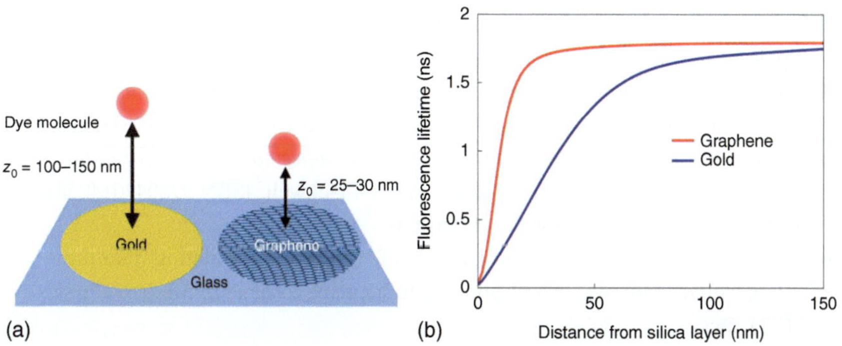

Figure 8.15 (a), The substrate consists of a titan/gold/titan multilayer or a single graphene sheet on a glass coverslip, covered by a protecting silica layer if desired (not shown). A fluorescent molecule within a range of up to 100–150 nm (for metal) or 25–30 nm (for graphene) can electromagnetically couple to the layers, which leads to a distance-dependent quenching of its fluorescence in the visible spectral range. (b), Fluorescence lifetime as a function of distance of a molecule from the surface of a silica spacer of 5 nm thickness deposited on gold (blue) or graphene (red) layer. Calculations were performed for Atto655 for a single emission wavelength of 680 nm. The refractive index of the upper medium was set to 1.33 (water), and the refractive index of glass was set to 1.52.

This, on the one hand, shortens the measurable range to just a few tens of nanometers, but, on the other hand, increases the resolution of the method. Whereas the qualitative dependence of the MIET resolution on photon number remains valid for graphene, the steeper curve results in a higher localization accuracy as compared to the use of a gold layer. It has been shown that the axial localization accuracy is enhanced approximately by an order of magnitude when the gold layer is replaced with graphene [53, 54]. In particular, sub-nanometer accuracy of GIET is sufficient to resolve the distance between the leaflets of a lipid bilayer.

8.4 Conclusion

We have presented the full theory describing fluorescence spectroscopy experiments in arbitrary planar metal/dielectric systems, such as interfaces or nanocavities. We gave an overview of important applications of the interaction of fluorophores with planar metal/dielectrics structures, such as the nanocavity-based method for QY measurement or MIET microscopy. These methods require only a conventional confocal microscope capable of fluorescence lifetime imaging, thus making them accessible to a wide community interested in fluorescence spectroscopy or super-resolution microscopy.

References

1 Barnes, W.L. (1998). Fluorescence near interfaces: the role of photonic mode density. *Journal of Modern Optics* 45 (4): 661–699.
2 Chance, R.R., Miller, A.H., Prock, A., and Silbey, R. (1975). Fluorescence and energy transfer near interfaces: the complete and quantitative description of the $Eu^{+}3$/mirror systems. *Journal of Chemical Physics* 63 (4): 1589–1595.
3 Mertz, J. (2000). Radiative absorption, fluorescence, and scattering of a classical dipole near a lossless interface: a unified description. *JOSA B* 17 (11): 1906–1913.
4 Drexhage, K.H. (1974). Iv interaction of light with monomolecular dye layers. *Progress in Optics* 12: 163–232. https://doi.org/10.1016/S0079-6638(08)70266-X.
5 Chance, R.R., Prock, A., and Silbey, R. (2007). *Molecular Fluorescence and Energy Transfer Near Interfaces*, 1–65. Wiley-Blackwell. ISBN 9780470142561. https://doi.org/10.1002/9780470142561.ch1.
6 Raimond, J.-M., Brune, M., and Haroche, S. (2001). Manipulating quantum entanglement with atoms and photons in a cavity. *Reviews of Modern Physics* 73 (3): 565.
7 Drexhage, K.H., Kuhn, H., and Schäfer, F.P. (1968). Variation of the fluorescence decay time of a molecule in front of a mirror. *Berichte der Bunsengesellschaft für Physikalische Chemie* 72 (2): 329–329.
8 Rakić, A.D., Djurisˇ ić, A.B., Elazar, J.M., and Majewski, M.L. (1998). Optical properties of metallic films for vertical-cavity optoelectronic devices. *Applied Optics* 37 (22): 5271–5283.

9 Wolf, E. (1959). Electromagnetic diffraction in optical systems-I. An integral representation of the image field. *Proceedings of the Royal Society of London. Series A. Mathematical and Physical Sciences* 253 (1274): 349–357.

10 Richards, B. and Wolf, E. (1959). Electromagnetic diffraction in optical systems, II. Structure of the image field in an aplanatic system. *Proceedings of the Royal Society of London. Series A. Mathematical and Physical Sciences* 253 (1274): 358–379.

11 Chizhik, A.I., Gregor, I., Schleifenbaum, F. et al. (2012). Electrodynamic coupling of electric dipole emitters to a fluctuating mode density within a nanocavity. *Physical Review Letters* 108 (16): 163002.

12 Chizhik, A.I., Gregor, I., Ernst, B., and Enderlein, J. (2013). Nanocavity-based determination of absolute values of photoluminescence quantum yields. *ChemPhysChem* 14 (3): 505–513.

13 Lakowicz, J.R. (2013). *Principles of Fluorescence Spectroscopy*. Springer Science & Business Media.

14 Vavilov, S.I. (1924). Die fluoreszenzausbeute von farbstofflösungen. *Zeitschrift für Physik* 22: 266–272.

15 Crosby, G.A. and Demas, J.N. (1971). Measurement of photoluminescence quantum yields. Review. *Journal of Physical Chemistry* 75 (8): 991–1024.

16 Demas, J.N. (1982). Measurement of photon yields. In: *Measurement of Photoluminescence*, (ed. K.D. Mielenz), 195–247. New York: Academic Press.

17 Resch-Genger, U., Hoffmann, K., Nietfeld, W. et al. (2005). How to improve quality assurance in fluorometry: fluorescence-inherent sources of error and suited fluorescence standards. *Journal of Fluorescence* 15 (3): 337–362.

18 Würth, C., Grabolle, M., Pauli, J. et al. (2011). Comparison of methods and achievable uncertainties for the relative and absolute measurement of photoluminescence quantum yields. *Analytical Chemistry* 83 (9): 3431–3439.

19 Brannon, J.H. and Magde, D. (1978). Absolute quantum yield determination by thermal blooming. Fluorescein. *Journal of Physical Chemistry* 82 (6): 705–709.

20 Gordon, J.P., Leite, R.C.C., Moore, R.S. et al. (1965). Long-transient effects in lasers with inserted liquid samples. *Journal of Applied Physics* 36 (1): 3–8.

21 de Mello, J.C., Wittmann, H.F., and Friend, R.H. (1997). An improved experimental determination of external photoluminescence quantum efficiency. *Advanced Materials* 9 (3): 230–232.

22 Würth, C., Lochmann, C., Spieles, M. et al. (2010). Evaluation of a commercial integrating sphere setup for the determination of absolute photoluminescence quantum yields of dilute dye solutions. *Applied Spectroscopy* 64 (7): 733–741.

23 Purcell, E.M. (1946). Spontaneous emission probabilities at radio frequencies. *Physical Review* 69: 681.

24 Ringler, M., Schwemer, A., Wunderlich, M. et al. (2008). Shaping emission spectra of fluorescent molecules with single plasmonic nanoresonators. *Physical Review Letters* 100 (20): 203002.

25 Brokmann, X., Coolen, L., Dahan, M., and Hermier, J.P. (2004). Measurement of the radiative and nonradiative decay rates of single CdSe nanocrystals through a controlled modification of their spontaneous emission. *Physical Review Letters* 93 (10): 107403.

26 Macklin, J.J., Trautman, J.K., Harris, T.D., and Brus, L.E. (1996). Imaging and time-resolved spectroscopy of single molecules at an interface. *Science* 272 (5259): 255–258.

27 Buchler, B.C., Kalkbrenner, T., Hettich, C., and Sandoghdar, V. (2005). Measuring the quantum efficiency of the optical emission of single radiating dipoles using a scanning mirror. *Physical Review Letters* 95 (6): 063003.

28 Cesa, Y., Blum, C., van den Broek, J.M. et al. (2009). Manipulation of the local density of photonic states to elucidate fluorescent protein emission rates. *Physical Chemistry Chemical Physics* 11 (14): 2525–2531.

29 Ambrose, W.P., Goodwin, P.M., Keller, R.A., and Martin, J.C. (1994). Alterations of single molecule fluorescence lifetimes in near-field optical microscopy. *Science* 265 (5170): 364–367.

30 Trabesinger, W., Kramer, A., Kreiter, M. et al. (2002). Single-molecule near-field optical energy transfer microscopy. *Applied Physics Letters* 81 (11): 2118–2120.

31 Vahala, K.J. (2003). Optical microcavities. *Nature* 424 (6950): 839–846.

32 Meixner, A.J., Steiner, M., Schleifenbaum, F. et al. (2005). Microcavity-controlled single molecule fluorescence. *Laser Science*, p. LTuC2.

33 Chizhik, A., Schleifenbaum, F., Gutbrod, R. et al. (2009). Tuning the fluorescence emission spectra of a single molecule with a variable optical subwavelength metal microcavity. *Physical Review Letters* 102 (7): 073002.

34 Chizhik, A.I., Chizhik, A.M., Khoptyar, D. et al. (2011). Probing the radiative transition of single molecules with a tunable microresonator. *Nano Letters* 11 (4): 1700–1703.

35 Chizhik, A.I., Chizhik, A.M., Kern, A.M. et al. (2012). Measurement of vibrational modes in single SiO_2 nanoparticles using a tunable metal resonator with optical subwavelength dimensions. *Physical Review Letters* 109 (22): 223902.

36 Fermi, E. (1932). Quantum theory of radiation. *Reviews of Modern Physics* 4 (1): 87.

37 Efron, B. (1979). The 1977 RIETZ lecture. *Annals of Statistics* 7: 1–26.

38 Tarpani, L., Ruhlandt, D., Latterini, L. et al. (2016). Photoactivation of luminescent centers in single SiO_2 nanoparticles. *Nano Letters* 16 (7): 4312–4316.

39 Ghosh, S., Oleksiievets, N., Enderlein, J., and Chizhik, A.I. (2021). Radiative rate modulation reveals near-unity quantum yield of graphene quantum dots. *Advanced Optical Materials* 9 (19): 2100314.

40 Chizhik, A.I., Gregor, I., and Enderlein, J. (2013). Quantum yield measurement in a multicolor chromophore solution using a nanocavity. *Nano Letters* 13 (3): 1348–1351.

41 Karedla, N., Enderlein, J., Gregor, I., and Chizhik, A.I. (2014). Absolute photoluminescence quantum yield measurement in a complex nanoscopic system with multiple overlapping states. *Journal of Physical Chemistry Letters* 5 (7): 1198–1202.

42 Schneider, F., Ruhlandt, D., Gregor, I. et al. (2017). Quantum yield measurements of fluorophores in lipid bilayers using a plasmonic nanocavity. *Journal of Physical Chemistry Letters* 8 (7): 1472–1475.

43 Ruhlandt, D., Andresen, M., Jensen, N. et al. (2020). Absolute quantum yield measurements of fluorescent proteins using a plasmonic nanocavity. *Communications Biology* 3 (1): 1–7.

44 Ha, T., Enderle, T., Ogletree, D.F. et al. (1996). Probing the interaction between two single molecules: fluorescence resonance energy transfer between a single donor and a single acceptor. *Proceedings of the National Academy of Sciences of the United States of America* 93 (13): 6264–6268.

45 Lukosz, W. and Kunz, R.E. (1977). Light emission by magnetic and electric dipoles close to a plane interface. I. Total radiated power. *Journal of the Optical Society of America* 67 (12): 1607–1615. https://doi.org/10.1364/JOSA.67.001607.

46 Enderlein, J. (2002). Theoretical study of single molecule fluorescence in a metallic nanocavity. *Applied Physics Letters* 80 (2): 315–317. https://doi.org/10.1063/1.1434314.

47 Chizhik, A.I., Rother, J., Gregor, I. et al. (2014). Metal-induced energy transfer for live cell nanoscopy. *Nature Photonics* 8 (2): 124.

48 Baronsky, T., Ruhlandt, D., Brückner, B.R. et al. (2017). Cell-substrate dynamics of the epithelial-to-mesenchymal transition. *Nano Letters* 17 (5): 3320–3326.

49 Chizhik, A.M., Ruhlandt, D., Pfaff, J. et al. (2017). Three-dimensional reconstruction of nuclear envelope architecture using dual-color metal-induced energy transfer imaging. *ACS Nano* 11 (12): 11839–11846. https://doi.org/10.1021/acsnano.7b04671. PMID: 28921961.

50 Chizhik, A.M., Wollnik, C., Ruhlandt, D. et al. (2018). Dual-color metal-induced and Förster resonance energy transfer for cell nanoscopy. *Molecular Biology of the Cell* 29 (7): 846–851.

51 Karedla, N., Chizhik, A.M., Stein, S.C. et al. (2018). Three-dimensional single-molecule localization with nanometer accuracy using metal-induced energy transfer (MIET) imaging. *Journal of Chemical Physics* 148 (20): 204201.

52 Isbaner, S., Karedla, N., Kaminska, I. et al. (2018). Axial colocalization of single molecules with nanometer accuracy using metal-induced energy transfer. *Nano Letters* 18 (4): 2616–2622.

53 Ghosh, A., Chizhik, A.I., Karedla, N., and Enderlein, J. (2021). Graphene- and metal-induced energy transfer for single-molecule imaging and live-cell nanoscopy with (sub)-nanometer axial resolution. *Nature Protocols* 16: 3695–3715.

54 Ghosh, A., Sharma, A., Chizhik, A.I. et al. (2019). Graphene-based metal-induced energy transfer for sub-nanometre optical localization. *Nature Photonics* 13: 860–865.

9

Single Trapped Neutral Atoms in Optical Lattices

Andrea Alberti and Dieter Meschede

Institut für Angewandte Physik, Universität Bonn, Wegelerstr. 8, 53115 Bonn, Germany

9.1 Introduction

In recent years, the application of quantum mechanical principles has reached the point of full control of all microscopic degrees of freedom governing the dynamics of isolated quantum systems. An early and groundbreaking milestone was the preparation and observation of a single trapped ion more than 40 years ago [1]. This breakthrough was accomplished by using electromagnetic traps to localize single particles almost to a point in space and laser cooling to reduce their motion to near the motional quantum ground state.

The first observations of microscopic particles can be traced back to Wilson's cloud chamber [2] more than a century ago. The invention of the Penning trap [3] for confining charged particles is another important milestone on the journey toward single particle control. It was once stated by Dehmelt that *an isolated atom at rest in free space is a spectroscopist's dream* [4]. Indeed, the possibility of slowing the motion of charged particles with lasers [5] opened the door toward Dehmelt's dream, enabling new experiments aiming not only to observe but also to precisely control the individual ions and atoms.

Quantum systems (technical realizations based on certain physical components are frequently called "platforms") used for such applications now include photons, neutral atoms, ions, artificial atoms such as color center defects and quantum dots, and more, both at the few particle and at the many particle level. A common trait of these quantum platforms is the usage of electromagnetic waves mostly at optical wavelengths to exert control of the relevant degrees of freedom. The laser is the pivotal instrument rendering the advances in this photonic technology possible. We may furthermore distinguish the "bottom up"-approach discussed here from an alternative path ("top down") starting with many body systems of ultracold atoms such as a Bose Einstein condensate [6].

In this article, we collect information and resources on the state of art of preparing, controlling, and applying neutral atoms with focus at the single and few-particle levels trapped by optical lattices carried out in our research group at the University of Bonn.

Photonic Quantum Technologies: Science and Applications, First Edition. Edited by Mohamed Benyoucef.

9.2 Tools for Trapping Single Neutral Atoms

To confine atoms in space over an extended period of time, optical pumping and laser cooling [7] play a pivotal role since only feeble electromagnetic forces based on magnetic and – most prominently – optical dipole interactions are available. Furthermore, for many applications, clean two-level systems in up and down (pseudo-) spin states are required, which, however, are not naturally available for most atoms. Optical pumping methods were established long ago to prepare pseudo spin 1/2 system with excellent two-level character. With the advent of quantum information science, we call such two-level system now qubits. Trapped neutral atoms offer an excellent platform for realizing qubits and applications in quantum information processing.

9.2.1 Trapping Neutral Atoms

The development of laser cooling methods in the 1970s and 1980s opened the route to reducing the speed of atoms in a gas moving at thermal speeds of several 100 m/s down to the mm/s scale. Especially the unexpected observation of so-called sub-Doppler temperatures in optical molasses allowed to slow atomic velocities indeed by more than 5 or 6 orders of magnitude below room temperature taking advantage of a fairly robust device: the magneto-optical trap (MOT) [7]. The MOT, which relies on dissipative friction forces, has become the workhorse for initializing experiments with cold atoms, both at the few-particle and the many-particle levels. Conservative trapping forces, however, are needed to control both the internal and the motional quantum state of trapped atoms.

9.2.1.1 Magnetic Traps

Magnetic traps were the technology of choice in the early stages of the quest for trapping and controlling single neutral atoms [8]. The energy scale of an atomic magnetic moment of order μ_B in a magnetic field B can be estimated from $\Delta E_B = \mu_B B$. For laboratory fields of order $B = 1\text{T}$, we find that the magnetic energy is approximately

$$\Delta E_B = \mu_B B\big|_{B=1\text{T}} \approx 0.93 \times 10^{-23}\text{J} \approx k_B \times 1\text{ K}$$

Magnetic traps are constructed with quadrupole or hexapole configurations and have their potential minimum located where B vanishes. At the zero-field point of the trap, however, where the most interesting low energy states are confined, spin flip transitions (known as Majorana flops) cause the mixing of trapping and non-trapping quantum states, yielding strong losses unless more complex arrangements such as that of an orbiting potential is made [9]. Furthermore, the internal degree of freedom, most importantly the orientation of the magnetic moment, of the atoms is no longer available in a magnetic trap, neither for quantum control of the motional state of the trapped atom nor as a qubit system for quantum information processing.

9.2.1.2 Optical Dipole Traps

Optical dipole traps (ODTs) are based on light–matter interaction where an atom with an induced electric dipole moment interacts with a spatially varying light field and experiences a trapping force toward the high (low) field intensities for red (blue) detunings from atomic resonance. Such optical traps have been realized with standing wave light fields in 1, 2, and 3 dimensions (optical lattices), or with focused traveling laser beams (optical tweezers). Here the relevant energy is given by the spatially varying AC Stark shift energy $\Delta E_{AC} = (ea_0E_0)^2/(4\,\hbar\delta)$ with Bohr radius a_0 and detuning $\delta = \omega - \omega_0$ of the laser frequency ω from an atomic resonance transition ω_0. Assuming that the total laser power $P = \sigma_\lambda \frac{c\epsilon_0}{2} E_0^2$ is focused to $\sigma_\lambda = 3\lambda^2/2\pi$, the classical resonant atomic absorption cross section, and using the convenient relationship $\epsilon_0 = e^2/(2hc\alpha)$, we replace $e^2E_0^2 \to 4h\alpha P/\sigma_\lambda$ with $\alpha \approx 1/137$ the fine structure constant:

$$\begin{aligned}\Delta E_{AC} &= \alpha \frac{a_0^2}{\sigma_\lambda} \frac{\lambda}{\delta\lambda} \frac{\lambda P}{c}\bigg|_{P=1\ \mathrm{W};\lambda=1\ \mu\mathrm{m}} \approx 10^{-23} \frac{\lambda}{\delta\lambda}\ \mathrm{J} \\ &\approx \frac{\omega}{\delta}\ k_B \times 1\mathrm{K}\end{aligned} \tag{9.1}$$

The choice of the detuning factor, ω/δ in Eq. (9.1), is a compromise between obtaining deep trap depths at laser powers available (calling for small detunings) and reducing the residual spontaneous scattering rate, which scales as $(\omega/\delta)^2$ for a *far-off-resonance dipole trap*, when the detuning is much larger than the linewidth of the relevant transitions [10]. Importantly, the different scaling of the two terms implies that spontaneous scattering can be strongly suppressed by increasing the detuning to produce light potentials that behave like nearly perfect conservative potentials [11]. The processing time is then limited by residual scattering rates experimentally adjustable to below 1 s^{-1} even for deeper optical potentials [12].

Let us also estimate the properties of the motional atomic quantum states, i.e. their energy spacing and the extension of its ground state wave packet in dipole traps. Assuming a Gaussian laser beam focused on an ultimate waist size of order $w_0 \approx 1\ \mu\mathrm{m}$ we have

$$\nu_{vib} \approx \frac{1}{2\pi}\sqrt{\frac{4\Delta E_{AC}}{m_{Atom} w_0^2}} \tag{9.2}$$

which for $\Delta E_{AC} \approx k_B \cdot 1$ K makes typical vibration frequencies of order 0.1–10 MHz even for heavier particles such as the widely used Rb and Cs atoms experimentally accessible (see Figure 9.9).

9.2.2 Preparing and Observing Individual Neutral Atoms

Ions can be confined in space using Coulomb or Lorentz forces acting on their charge with radiofrequency control fields, while leaving their internal state almost unaffected. The first observation of a single trapped fluorescing ion in 1980 [1] opened the door for experimentalists to move from pure observation of the single quantum particles to a new era where these particles can be individually controlled. In contrast to the strong confining potential for trapping ions, dipole forces acting of neutral

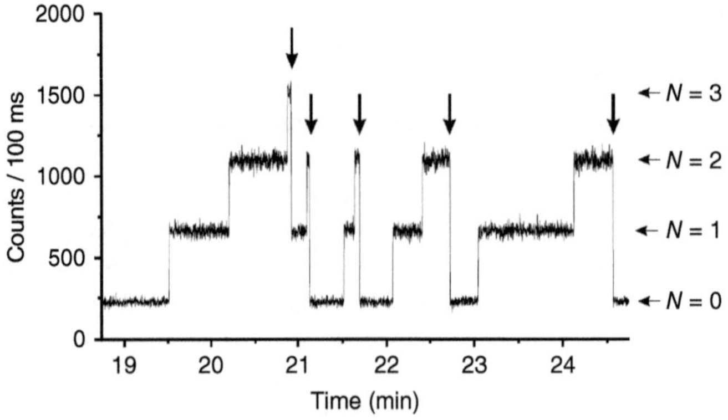

Figure 9.1 Fluorescence of a fluctuating small integer number of Cs atoms in a MOT. Two atom losses (arrows) indicate radiatively assisted collisions causing losses. Source: Ueberholz et al. [16]/IOP Publishing/Public Domain.

atoms make use of the internal degrees of freedom – the atom polarizability – which limits their strengths according to Eq. (9.1) much below typical ion trap depths. It is thus owed to the advent of efficient laser cooling techniques in the 1980s including the MOT that neutral atom control became an alternative to trapping ions where the Coulomb charge has advantages as outlined for trapping but also disadvantages once interactions of atoms are of relevance.

9.2.2.1 Single Neutral Atoms in a MOT

First experiments with trapped individual atoms in MOTs confining the atoms to a typical volume of 20 μm diameter were reported beginning in 1994 [8, 13, 14]. Already with the MOT setup it was possible to identify phenomena only observable at the single particle level, including anti-bunching and fluctuations of atomic orientation via polarization correlations [15].

Figure 9.1 shows an early example of dynamically evolving fluorescence record from a MOT showing discrete intensity levels associated with an integer number of trapped Cs atoms [16]. The traces also show clear two atom loss events due to light assisted two body collisions. In fact, the rate of two body losses can be experimentally adjusted such that only a single atom at a time can reside in the trap (collisional blockade [17, 18]).

Even near the center of the MOT both magnetic field strength and polarizations vary rapidly in space leading to continued optical pumping of the trapped atom among magnetic sublevels [15] at all relevant timescales. Thus, non-dissipative traps offering nearly conservative potentials are required for controlled applications of single trapped atoms.

9.2.2.2 Single Neutral Atoms in Optical Dipole Traps

To eliminate the rapidly fluctuating character of the internal atomic quantum state imposed by the MOT environment atoms are transferred into ODTs, which in terms of optical lattices [12] or optical tweezers [18] are now dominating all experiments with cold atoms both at the single- and many-particle level. In contrast to magnetic traps, the magnetic moment is available in ODTs and has proven to be an independent and hence invaluable degree of freedom for controlling both internal and external quantum states of trapped atoms. A detailed account for the dynamics of single atoms in ODTs is found in Ref. [19].

9.2.2.3 Detecting Single Atoms in MOTs and ODTs with High Efficiency

In Figure 9.1, we show the fluorescence of a MOT with individual atoms entering and leaving the trap, where the integration time of 100 ms lets us extract the integer number of atoms with visibly good fidelity. For continuous imaging, trapped atoms are irradiated with slightly red detuned laser light providing scattering rates up to $\sim 10^6$ photons/s with no preferred direction. Using off-the-shelf lenses, imaging systems with numerical apertures (NAs) of order 0.3–0.5 can be built [20], which allow collection of about 1% of the fluorescence yield. Thereby, count rates of order 10 kHz per atom or more are readily achieved. The signal-to-noise ratio improves when atoms are trapped in ODTs, since the tighter localization enables spatial filtering for background suppression. AC Stark shifts of the fluorescence transition can be straightforwardly compensated by adjusting laser frequencies.

9.2.2.4 Detecting Single Atoms Using Optical Cavities

Optical cavities can strongly enhance the interaction of an atom with a light field already at the single photon level. Consider an atom with classical resonant absorption cross section $\sigma = 3\lambda^2/2\pi$ coupled to a cavity field with waist size πw_0^2. In a simplified picture a cavity with highly reflecting mirrors circulates the light field with $n \sim \mathcal{F}$ roundtrips ($\mathcal{F}$: cavity finesse) and hence enhances the absorption probability of the photon up to the cooperativity factor $C \sim \mathcal{F}\cdot(\sigma/\pi w_0^2)$, which is a measure of enhancement by the so-called Purcell effect [21]. For $C \geq 1$ atomic resonance transition and cavity photon field become strongly and coherently coupled and the common energy diagram is split (Figure 9.2a).

While in early experiments atoms transiently passing a cavity were detected by this method [23], the conveyor belt technique (see below) allowed to precisely position trapped atoms inside the optical cavity field. In Figure 9.2b an atom entering the cavity suppresses transmission of a probe laser field on resonance with the empty cavity. While in past experiments, bulk cavities were used for this purpose in more recent times fiber Fabry–Perot cavities integrated with conventional optical fibers have become almost common place in such applications, representing the trend toward photonic miniaturization. Excellent fidelity of qubit state detection of the atom–cavity system can be achieved with this scheme [24]; for a summary see [25].

9.2.2.5 Detecting the Qubit State of Trapped Atoms in Free Space

The spin-up and -down states constituting the qubit states of typical trapped atoms (alkalis, earth alkalis, ...) are typically selected hyperfine states separated by energy splittings at microwave frequencies. Projective measurements onto the internal quantum states are accomplished in destructive and non-destructive ways:

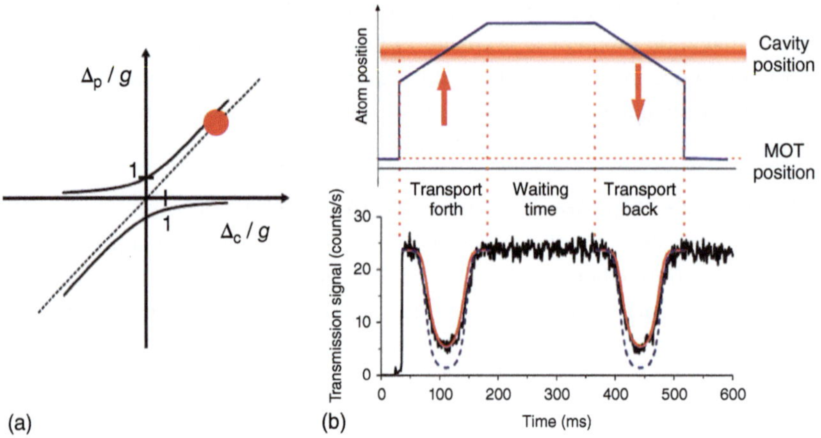

Figure 9.2 (a) Schematic illustration of the coupled atom-cavity state showing a splitting at perfect resonance ($\Delta_{p,c}$: detunings of cavity (*c*) and probe laser (*p*) from atom resonance, respectively. Normalized to the atom–cavity coupling rate *g*). The red dot indicates the chosen parameters. (b) Transmission of the probe laser transmission as a single atom is slowly swept back and forth across the cavity mode over a distance of 100 μm within 150 ms using a conveyor belt. Khudaverdyan et al. [22]/IOP Publishing/Public Domain.

- (a) The *push-out* method ejects atoms in one of the spin states by state selective resonant light pressure form the trap. Subsequently, all remaining atoms are detected through resonance fluorescence at their initial position and thus projected to the other, opposite spin state. The method is destructive since atoms in the "wrong" quantum state are lost and hence after each step reloading of atoms is necessary [26, 27].
- (b) Scattering of about 10^6 photons requires closing even weak leaky transitions, which would remove atoms from the driven closed two-level system. This is experimentally achieved by means of additional repumping lasers, which, however, also eliminate the discrimination of spin-up and -down states. It was shown that the state selective fluorescence yield terminated before the loss described above occurs (*photon burst* method) is sufficient to selectively detect the quantum state of the trapped atom [28].

An example demonstrating the excellent quality of the push-out method is given in Figure 9.3 where many Rabi oscillation cycles are monitored. With a period of about 16 μs, the measurements show processing times for the qubit state orders of magnitude shorter than storage times (>10 s) and scattering time of off-resonant photons (>100 ms).

9.2.2.6 Detecting the Qubit State of Atoms Trapped in Cavities

As outlined above, the presence of an atom inside a cavity can be very efficiently detected by monitoring transmission of (or reflection off) the cavity [30]. In addition, the atom–cavity system distinguishes the two qubit states since only one of them is strongly coupled to the cavity, while the other one is "dark". Quantum jumps between the two states therefore cause the reflection (or transmission) signal of a

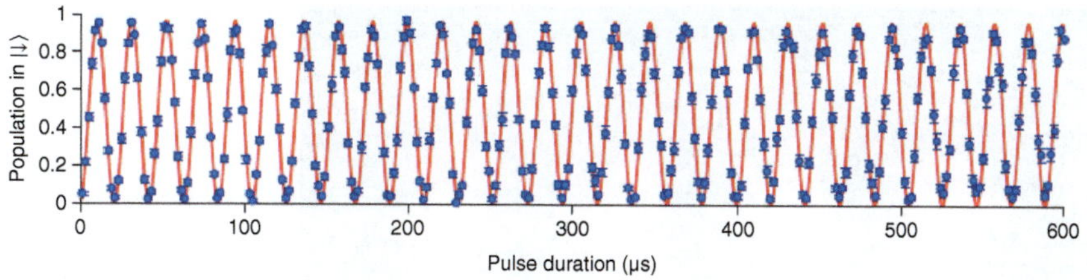

Figure 9.3 Rapid Rabi oscillation of single trapped Cs atoms with 60 kHz frequency. No significant decoherence is observed during the interrogation time of almost 1 millisecond. Source: Robens [29].

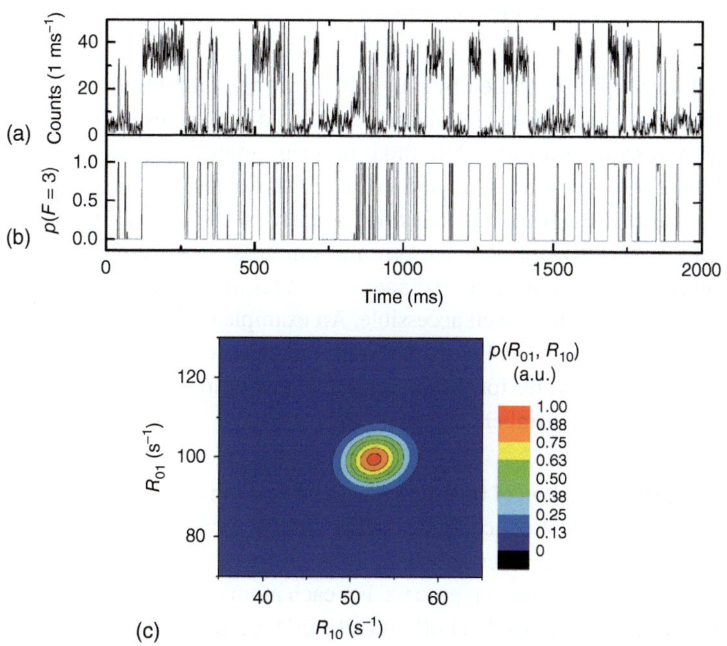

Figure 9.4 (a) Random telegraph signal of a strongly coupled atom–cavity system. (b) Bayes reconstruction of the qubit state (c) Probability distribution for the rates. Source: Kampschulte [31].

probe laser interrogating the atom–cavity system to show a random telegraph signal, Figure 9.4a. Using Bayes inference, the qubit state of the system can be dynamically assigned to the signal and, e.g. the transition rates of the system can be extracted, as shown in Figure 9.4b,c [31].

9.2.2.7 Detecting the Position of Single Atoms with High Fidelity

In optical lattices as well as in arrays of optical tweezers, it is essential to also detect the position, i.e. identifying the exact lattice site where the atoms are trapped. While for tweezer arrays with spacings of order 10 μm, the resolution with NAs ~ 0.5 is sufficient to discriminate adjacent micro potentials, in optical lattices neighboring

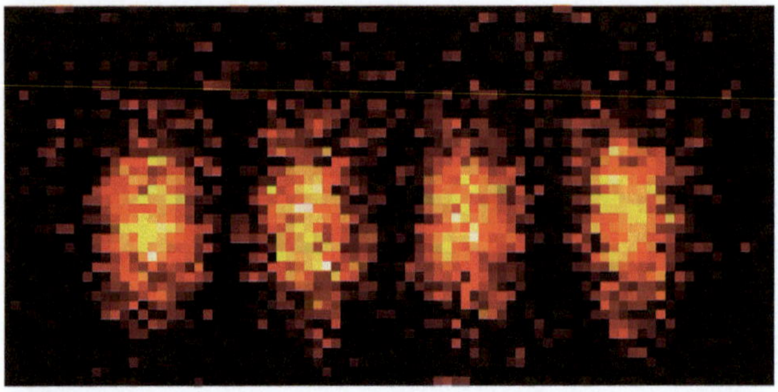

Figure 9.5 Four atoms trapped in a one dimensional optical lattice. Integration time 100 ms, atom separation 10 lattice sites; adapted from [29].

micro potentials are typically spaced by only $\lambda/2$ of the trapping laser field wavelength, i.e. distances below 0.5 μm. Resolving atoms in adjacent sites is thus a technical challenge and requires advanced methods of image analysis [32, 33] as well as the construction of objective lenses adapted to the needs of observing trapped atoms confined by optical lattices ("quantum gas microscopes," see Refs. [34, 35]). Today, EMCCD cameras with still improving background suppression as well as high performance numerical image analysis have made position resolved detection of neutral atoms trapped in optical lattices well accessible. An example of the fluorescence of four trapped Cs atoms is shown in Figure 9.5. With the precision transport methods described below sorting of atoms for the generation of low-entropy states is possible and seems scalable to large numbers of atoms [36].

9.2.2.8 Simultaneous Detection of Quantum State and Position

Even when experiments deal conceptually with "single-" or "few-atom physics," they are often carried out with an ensemble of up to several 100 atoms in parallel in order to speed up the data acquisition process. For each atom of the ensemble, quantum information is indeed encoded in both its (integer!) position and spin state. It is therefore essential to simultaneously detect both the quantum state and the position simultaneously. Methods used include:

- Application of the *push out* method (see above) and registration of fluorescence by a high resolution imaging system and a camera immediately realizes position detection conditioned on one of the two spin quantum states. The disadvantage of the method is its destructiveness.
- The state selective *photon burst* method can be extended to detect the quantum state and position of atoms in parallel [37].
- In case the initial positions of an ensemble of single atoms are known, the spin-dependent conveyor belt (see below) allows to map the internal spin state of atoms onto position information through state conditioned transport operations, e.g. shifting or not shifting atoms depending on their spin by a given number of lattice sites. Readout of atom positions thus contains position as well as internal state information [38].

9.2.3 Precision Transport of Atoms

Loading an optical lattice from a MOT results in statistical distribution of the atom across lattice sites. Since it remains difficult to obtain site occupancy in excess of 50%, [39], it is important to obtain full control of the position as well as the motional quantum state of the atoms, i.e. single atoms need to be delivered deterministically to given positions. Transport of atoms within a given plane can be achieved for tweezer array by acousto-optic beam deflectors [40], and for optical lattices by moving the intensity pattern of the associated standing wave. Here we concentrate on the latter.

9.2.3.1 Optical Conveyor Belt

The idea of the conveyor belt [41, 42] is very simple: counter propagating two laser beams at exactly the same frequency creates a standing wave with a periodic series of micro potentials with spacing $\lambda/2 = \pi c/\nu$. Using acousto-optical modulators for both arms, the relative frequency between the two arms can be controlled with radiofrequency precision. For instance, a detuning between the two arms of 1 Hz causes the standing wave to shift by 1 lattice site/s.

9.2.3.2 Spin-Dependent Optical Lattice

As outlined above, it is important to have the magnetic moment degree of freedom of the atoms available as a control degree of freedom. Following a suggestion by Cirac and Zoller [43] it was shown by Mandel et al. [44] that a wise choice of wavelengths (at the so called *magic* wavelength) and lattice polarization state enabled the realization of a spin-dependent optical lattice. In a simplified view, a linearly polarized standing wave is composed of two lattices with orthogonal circular polarization, Figure 9.6. In a spin-dependent optical lattice, spin-up and spin-down states of the atom are energy shifted by only one of the two sub lattices, respectively. Transport of an atom by means of the optical lattice (see next paragraph) can hence be conditioned on its spin state!

In early experiments, the range of moving lattices was accomplished by active polarization using electro-optical modulators, which was limited by the finite dynamic range of the polarization rotation [44]. The radio frequency method

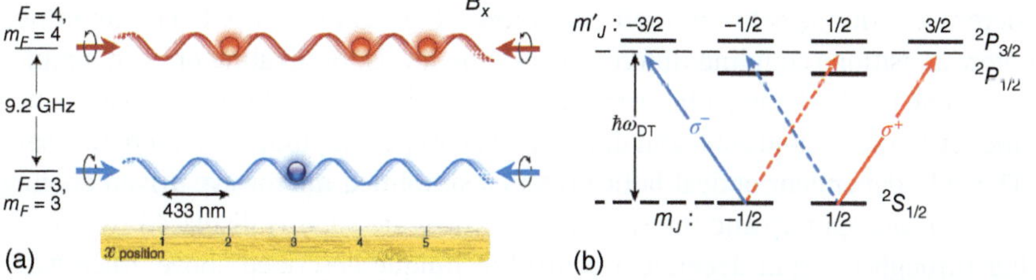

Figure 9.6 (a) State-dependent optical lattices acting selectively on either one of two long-lived hyperfine states of a cesium atom. Upper and lower lattices originate from orthogonal circularly polarized standing wave light fields, respectively. (b) By choosing a so-called magic wavelength. Source: Robens [29].

overcomes this problem and allows to travel arbitrarily long distances. In fact, transport distances of order 1 m should be possible today.

9.2.3.3 Two Overlapped Optical Conveyor Belts with Orthogonal Polarizations

As indicated in Figure 9.6 arbitrary polarization states of a traveling light field can be constructed from a superposition of a right-handed and a left-handed optical wave. Implementing this concept with optical components and counter-propagating this light field with a single linearly polarized reference wave amounts to the creation of two conveyor belts, which can be fully independently steered with Å-precision and almost MHz-bandwidth [45]. The purity of the polarization state indeed is only limited by the inhomogeneity of polarization components.

In conclusion, the spin-dependent optical conveyor belt allows not only full control on atomic positions – we can position the two lattices at any relative position and intensity with each other –, but also provides an essential tool to access all relevant internal (pseudo spin) and external (motional) quantum states.

9.2.3.4 Optical Conveyor Belt for Two Dimensions

While for a one dimensional standing wave the concept is straightforwardly understood and implemented, a more subtle arrangement is required in two dimensions since the two orthogonal directions of space interfere with each other. A good solution was found by Robens et al. [45], who realized that two orthogonal and independently controlled conveyor belts could be realized by using a single linearly polarized reference wave and interfering it with two control waves at opposing right angles (Figure 9.7b). Sideband spectra (c) document the 95% ground state occupation in this lattice. Imaging resolves atoms in adjacent lattice sites (a), and the action of two independent conveyor belts oriented along orthogonal directions is shown in (d).

9.2.4 Controlling the Motional Atomic State

Within each lattice site (or tweezer site), it is essential to render the atom as point-like as possible, i.e. to prepare the atoms in the motional ground state. Since the vibrational states of trapped atoms have large separations of order 100 kHz and more sideband cooling methods can be applied, where coherent transitions remove a quantum $h\nu$ of vibrational energy and a repumper closes the optical cycle. With neutral atoms, coherent transitions between motional states are typically induced via Raman two photon transitions coupling different hyperfine levels for alkaline or even simpler for alkaline earth atoms with narrow transitions ${}^1S_0 \rightarrow {}^3P_1$. This state decays automatically to the ground state and does not even require a pumper, nor a repumper.

The spin-dependent optical lattice offers a simplified microwave driven cooling method: If the spin-up and the spin-down lattices are slightly displaced from each other through the spin-dependent control technique described above, microwave transitions can conveniently drive side bands of spin flip transitions, i.e. $|\uparrow, n\rangle \rightarrow |\downarrow, n-1\rangle$ and vice versa. Experiments investigating microwave cooling properties (Figure 9.8) have also shown that heating processes caused by lattice fluctuations can be kept at levels not impairing coherent quantum processes.

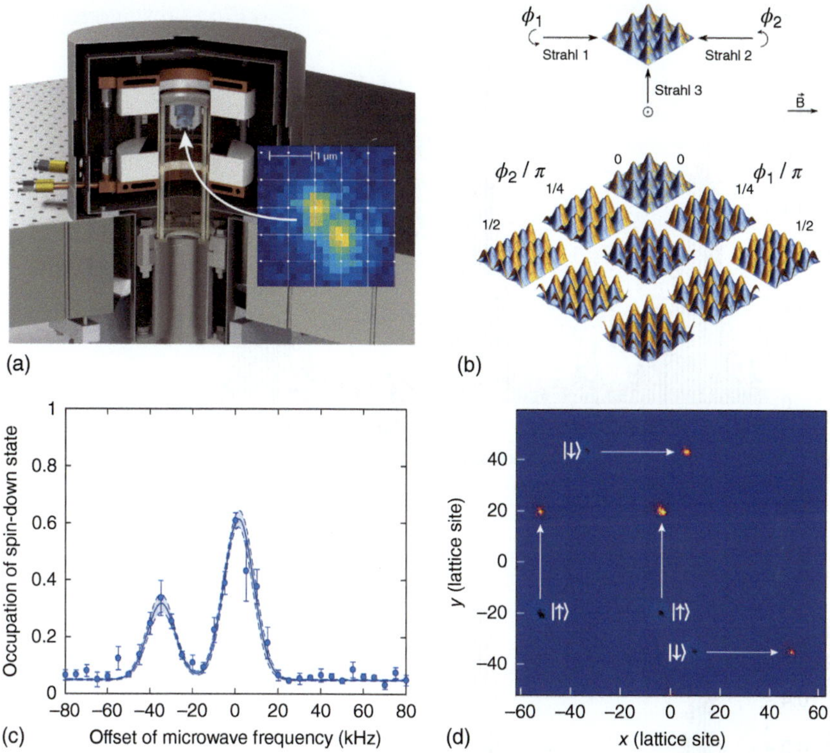

Figure 9.7 The most advanced version of the 2D optical conveyor belt. (a) Apparatus with high NA microscope objective and a fluorescence image of two atoms spaced by 0.7 μm. Source: Dr. Stefan Brakhane (b) Interference arrangement of the spin-dependent 2D conveyor belt. (c) Sideband spectra showing 95% 3d ground state occupation. (d) Camera image indicating 2D spin-dependent transport of four atoms.

It is clear that once atoms are cooled to a single ground state coherent excitation of any n-vibrational state can be achieved by selecting a suitable displacement of the σ^+/σ^--lattices and driving microwave π-pulses. This is documented by the spectrum shown in Figure 9.9: It displays all motional states provided by a sinusoidal optical lattice. It clearly shows the equidistant spacing deep inside the lattice and a series of (unresolved) levels towards the "ionization threshold" [46].

9.2.5 Addressing and Controlling the Atom Position

Initially atoms are randomly distributed across the lattice sites. Using movable optical tweezers steered by acousto-optic deflector, it is now straightforward to arrange 100 or more atoms in to ordered patterns [48, 49]. For our optical lattices, we employ the spin-dependent conveyor belt, which allows us to transport atoms conditioned on their spin state, with Å-precision and at almost MHz bandwidth. All transport protocols need to initially find and address atoms at specific sites (e.g. where they had been detected) and apply subsequent deterministic transport operations. There

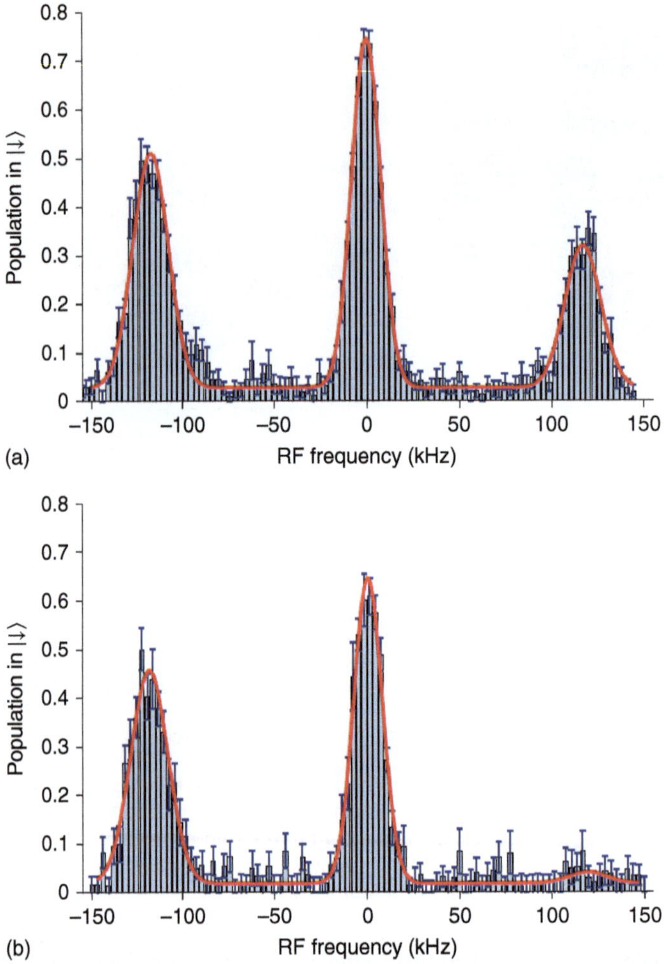

Figure 9.8 Sideband spectra of neutral atoms subject to microwave cooling in a spin dependent optical lattice. (a) Before cooling; (b) after cooling. The effective temperature can be inferred from the ratio of the red to the blue sideband and indicates about 95% ground state population in this case. Source: Robens [29].

are numerous different protocols for atom position control available depending on the specific application. Here we only give few examples.

For addressing a trapped atom following position detection, active discrimination of its position can be achieved through e.g. focused interaction with a laser beam, a method widely used with ion traps. Here we instead concentrate on magnetic field gradients encoding the atomic position by means of the spin flip transition frequencies. An example of such a procedure is shown in Figure 9.10. The experiments showed deterministic transport of atoms with high fidelity [41].

Specific transport protocols depend on the desired application, where we restrict ourselves to two simple examples:

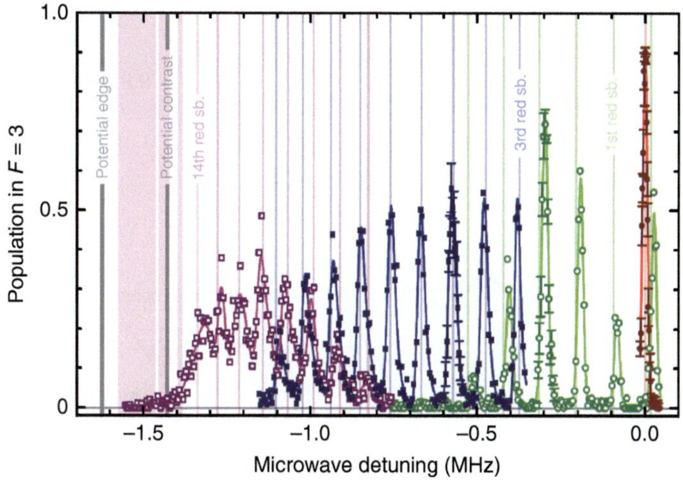

Figure 9.9 Spectrum of vibrational quantum states in a deep sinusoidal optical dipole trap (optical lattice) by monitoring transitions. The transitions were monitored by initializing an atom in a spin-down motional ground state with $n_{\downarrow} = 0$. By slightly displacing the spin-up lattice from perfect overlap with the spin-down lattice, all carrier and sideband transitions $n_{\downarrow} \rightarrow n'_{\uparrow}$ become accessible [47].

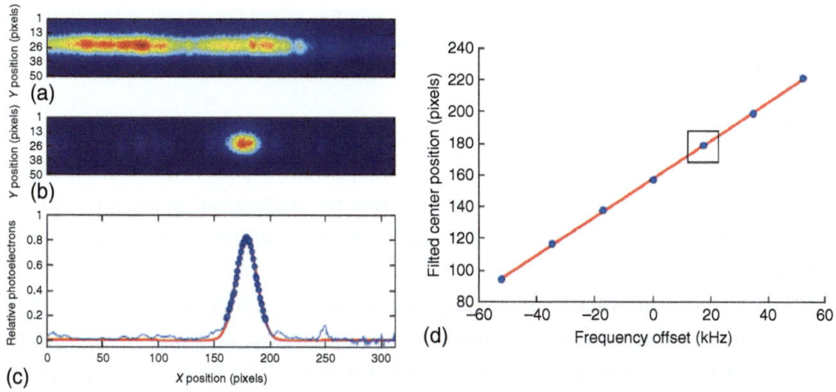

Figure 9.10 Position-dependent microwave addressing of atoms using a magnetic field gradient. (a) Averaged image of 20 initial distributions of atoms, which are trapped in the optical lattice. (b) Averaged image of the 20 distributions after employing a position-dependent Gaussian-shaped microwave π-pulse and the push out beam. (c) Vertically integrated image, normalized to the initial distribution. The red line represents a Gaussian fit. (d) Resulting magnetic field gradient calibration after repeating several position dependent microwave spectra for different transition frequencies, each represented by a blue dot. The black square represents the data shown in (a)–(c). Error bars are smaller than the dots [29].

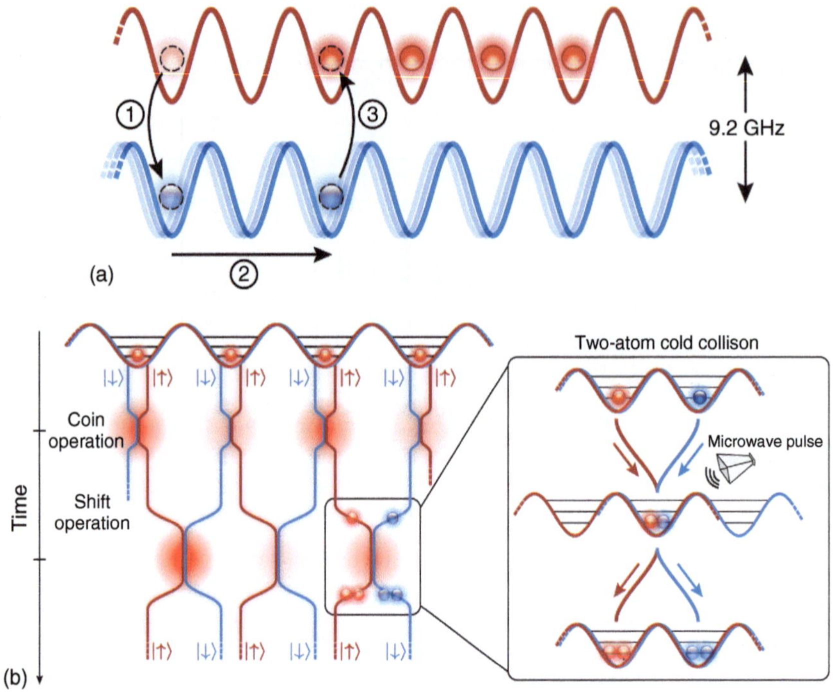

Figure 9.11 Operations to induce a quantum collision of two particles. (a) Sorting algorithm to generate low entropy states [36]. The lattice for the spin-up atoms serves as a storage register, while the other lattice for the spin-down atoms serves as a shift register. (b) Illustration of the multi-particle quantum walk scheme. When two atoms meet on the same site, they can interact through ultracold collisions, potentially assisted by microwave-induced spin rotations.

- Transporting a single atom to a predefined location while leaving all others at rest. Here all atoms are initialized in e.g. the spin down state. Afterwards, a specific atom is selectively addressed via a microwave pulse rotating the atom to the spin-up conveyor belt. Then the spin-up transport operation will act only onto this atom.
- For a collisional interaction of exactly two atoms, the procedure in Figure 9.11 can be used. Arbitrary atomic positions are initialized, and one of the atoms is transported to a site adjacent with a second atom. The final step superposes the two atoms in space where microwave pulses may induce simultaneous spin rotations.

9.3 Quantum Control of Single Trapped Atoms

The quantum states of the trapped atoms are product states given by

$$|\psi\rangle = |\pm\rangle|i_x\rangle|n_{\text{vib}}\rangle \tag{9.3}$$

where $|\pm\rangle$ is the internal spin state, $|i_x\rangle$ the lattice site (or tweezer index in a tweezer array), and $|n_{\mathrm{vib}}\rangle$ the motional state. We have shown in Sections 1.2 that we can determine and initialize all quantum numbers of a trapped single atom with high fidelity. The atom exhibits quantum character when it evolves into a superposition of the states in Eq. (9.3). Superpositions of the internal spin state are indeed common place, and in fact, experiments with trapped single atom qubit states closely resemble operation of single atom clocks, cf. Figure 9.3. Similarly coherent preparation of vibrational levels following initialization of atom in their motional ground state follows well known routes from, e.g. ion trapping experiments.

A more specific characteristic of single trapped neutral atoms is the possibility of delocalizing them over different positional states. It is hence most interesting to explore spatial quantum state superpositions of different lattice sites $|i_x\rangle$.

9.3.1 Quantum State Transport

The first question to ask is: if an initial superposition of internal quantum states is created, e.g. by coherent microwave π-pulses – does it survive transport over significant distances, i.e. across at least several of lattice sites? Or does the transport cause rapid decoherence leaving atoms after transport in a mixture of states? The latter would render the trapped atoms tiny but effectively classical objects with respect to transport operations. The experiments, however, confirm that atoms are true "quantum marbles" and may e.g. be used for interesting protocols involving the quantum information stored in the spatial degree of freedom.

Experiments have used Ramsey spectroscopy techniques where an internal quantum coherence is induced at an initial site and which is phase sensitively interrogated after transport to another site. Experiments [26, 50–52] give a clear answer: internal quantum state superpositions do survive transport operations! Excellent fidelity can furthermore be obtained by applying optimal control strategies to the transport process, as described below.

9.3.2 Quantum Speed Limit of Atomic Motion

A question closely related to quantum transport is: how fast can we transport an atom by accelerating and decelerating the optical lattice without loosing quantum coherence? Ramsey interferometry provides a sensitive tool to investigate this question: We create a (spin-up) copy of a (spin-down) atomic ground state wave packet (which amounts to a Gaussian coherent state), which we leave as a reference wave at rest. We now study the fidelity after deterministic transport over a single lattice site and back. The transport operations are optimized using optimal control strategies in order to eliminate motional excitations, which would otherwise occur for fast transport operations, i.e. when the adiabatic conditions are not fulfilled. After the return the interference of the incoming with the reference wave packet is monitored using Ramsey spectroscopy. The fidelity is directly reflected in the contrast of the recorded Ramsey interference fringe. The measurements show again that quantum state transport is robust up to the so-called quantum speed limit [53]. For even faster transport, i.e. exceeding the quantum speed limit, the transport fidelity significantly drops because, fundamentally, motional excitations cannot be longer controlled, unless the trap depth is further increased.

9.3.3 Quantum Delocalization

In Born's interpretation, the spatial wave function of quantum objects is associated with a probability distribution to find the particle at any point in space after a projective measurement is carried out. In the case of an optical lattice, this measurement amounts to detecting the position of an atom as shown in Figure 9.5. In fact, finding an atom at a discrete position yields an integer quantum number, i_x, associated with its lattice site.

9.3.3.1 Spin-Dependent Transport

An atom initially prepared in a single lattice site and a pure spin state will be transported by "its" optical conveyor belt either to right or to the left, depending on its spin state, see Figure 9.6. After multiple repetitions, the distance traveled by the atom will grow linearly. For the other spin state, the same situation exists for the conveyor belt associated with the orthogonal polarization state of light [45].

The most interesting situation occurs, when initially an internal spin superposition state is created. For a one-dimensional conveyor belt, an atom prepared in a superposition of both spin states is shifted into opposite directions at the same time: The internal spin superposition is thus mapped onto a spatial superposition state, which delocalizes (here: separates) the components of the spatial wave function by two or even very many lattice sites (see below, quantum walks) with a potentially large spacing in between. For instance, the two trajectories of the single atom interferometer in Figure 9.12c are easily resolved with a microscope. Note that in this case we deterministically create (single particle) entanglement between internal and external degrees of freedom, e.g. $|+, i_x\rangle \rightarrow (|+, i_x\rangle + |-, j_x\rangle)/\sqrt{2}$.

9.3.3.2 Single-Atom Interferometer

It is clear that the spin-dependent transport protocol described above amounts to the realization of a spatial atom "splitter." This is in close analogy with an optical polarizing beam splitter, where impinging photons in a superposition of the two polarizations exit the beam splitter in a superposition of both output ports. We can extend this analogy to optical interferometers to describe the splitting and recombination of quantum trajectories of an atom. An example of a Mach–Zehnder-like single-atom interferometer is shown in Figure 9.12 [54].

9.3.4 Quantum Walks of Single Atoms

Quantum walks are the multiple trajectory extension of the two trajectory interferometer discussed above.

The idea is illustrated in Figure 9.13a. Each step of a quantum walk consists of two operations acting on the "walker":

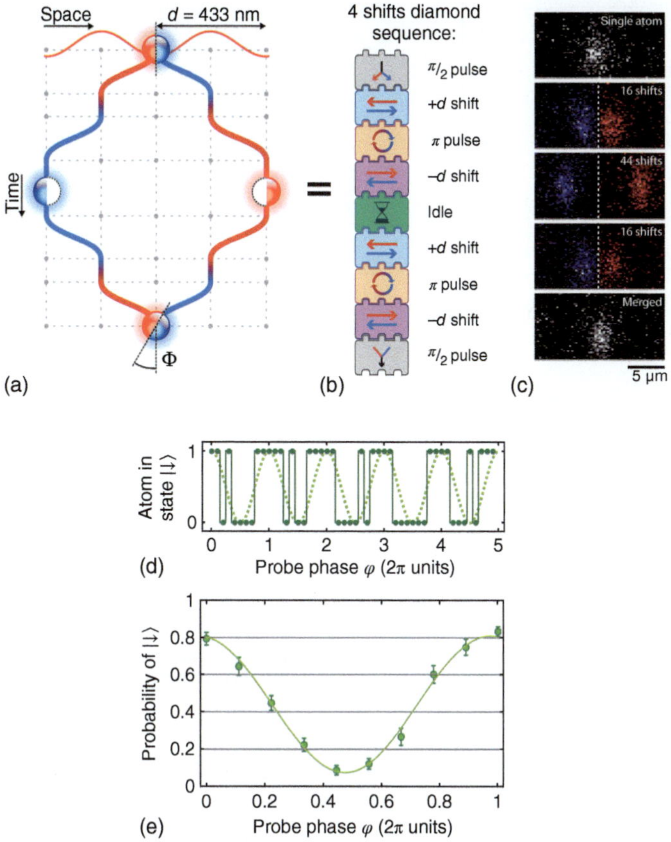

Figure 9.12 (a) Schematic of a single-atom four-step Mach–Zehnder interferometer. A $\pi/2$-pulse brings the atom into a superposition of two spin states, which are then shifted to a separation of four lattice sites. After remaining at this distance for a given time, they are recombined and a final $\pi/2$-pulse with control phase ϕ maps the phase difference into state populations. (b) The experimental sequence represented in elementary operations. With the spin-flip operations, it is sufficient to shift the conveyor belt back and forth by a single site in order to achieve large spatial separations of the atom trajectories. (c) Here the interferometer separation reaches $\sim$ 10 μm using 44 shifts. The images were made by recording two times after half the sequence, finding the atom right or left. (d) Discrete signal of individual atom events while scanning the interferometer phase ϕ. (e) Interferometer fringe averaged over many single-atom events. Source: Steffen et al. [54]/with permission of National Academy of Science.

- A "coin operation" causes a spin rotation. Most often, the rotation angle is chosen equal to $\pi/2$ pulse, meaning that the coin operation realizes a Hadamard gate.
- A "shift operation" causes a step to the left or the right conditioned on the spin-up or spin-down state.

This protocol is called a "discrete-time" quantum walk since the "time" can be thought to evolve in discrete and deterministic steps. The duration of operations is only technically limited; for instance, transport times of order of

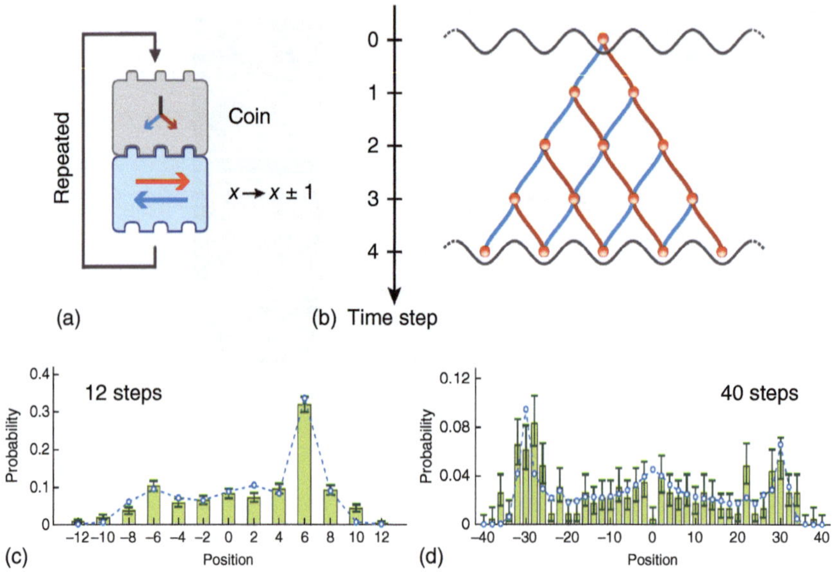

Figure 9.13 (a) Quantum walks: multiple repetitions of the coin-shift operation. (b) Trajectories (e.g. trapped atoms) with multiple interferences. (c) Twelve step quantum walk with initial state in spin up. (d) Forty step quantum walk with initial state in a superposition of spin-up and spin-down states. Alberti et al. [55]/IOP Publishing/Public Domain.

10 μs can be achieved, limited by the maximum trap depth of the optical lattice potential. In contrast, "continuous-time" quantum walks rely on tunneling processes studied, e.g. with quantum gas microscopes in shallow optical lattices, where the "step" duration is intrinsically limited to longer time scales of order 1 ms [56, 57].

An N-step quantum walk, which consists in a N-fold application of the step operation, will then cause the interference of multiple trajectories in every lattice site accessible by the walker, as conceptually shown in Figure 9.13b. In Figure 9.13c,d, we show the experimentally recorded probability distribution of the quantum walk process: the spread of the probability distribution grows $\sim N/\sqrt{2}$. The linear growth in the number of steps N stands in striking contrast to the classical equivalent of random walks. Note that the quantum walk can be reduced to its classical equivalent by allowing for decoherence in between the steps, which experimentally is simply done by inserting a long idle time between steps. Then the quantum walk turns into a classical random walk, where we expect the spread to grow diffusively $\sim \sqrt{N}$. Unitarity of the quantum walk operations in contrast also allows to reverse the quantum evolution. In Ref. [58], it was shown that indeed reversal of the quantum walk refocuses the distribution back to the initial site, where decoherence also causes a diffusive pedestal to appear. In Figure 9.13, the influence of decoherence can be seen with the slight cusp of the distribution near the initial position "0." By theoretical analysis, this deviation from a perfect quantum walk can be traced to the result of spin decoherence [55].

9.3.5 Applications of Quantum Walks

There are several applications of quantum walks as a tool for quantum information science [59], quantum simulations, and fundamental tests. In relation to the latter two, let us present a couple of examples:

9.3.5.1 Electric and Magnetic Quantum Walks

An atom in an optical lattice performs a quantum walk over a periodic array of lattice sites. It is hence interesting to investigate the motion of a single walker in this periodic potential and apply a constant force. This situation resembles electric forces acting on electrons moving in a periodic crystal potential and gives rise to the notion "electric quantum walk." With an optical lattice, a constant force is simulated through the inertial force of a constant (spin-independent) acceleration of the optical lattice imposed by the conveyor belt. Note that in experiments, there is enough time left to study quantum walks under this condition of an accelerated reference frame, before the system moves out of the field of view of the microscope.

For discrete time quantum walks the extra force is accounted for by an extra phase $e^{i\Phi x}$ imprinted on the atom, depending on the (integer) lattice site x [60]. This phase (note the gauge-field character) is experimentally tuned by varying the acceleration (or the duration of the acceleration operation). For "rational" electric fields, $\Phi = 2\pi n/m$, where n and m have no common factor, we observe spreading oscillations of the walker distribution in space with signatures of refocusing after m steps, reminiscent of Bloch oscillations of electrons in a periodic crystal. For "irrational" electric fields, e.g. $\Phi = 2\pi/\phi$ with $\phi = (1 + \sqrt{5})/2 = 1.618\ldots$ the golden ratio, in contrast, the spatial distribution of the walker experiences dynamical localization for all times.

Following the realization of "electric quantum walks," the interest in the magnetic analog is readily justified. While for the electric force simulation, a one-dimensional lattice is sufficient, magnetic forces need an advanced concept to simulate two-dimensional gauge fields. In [61], protocols for neutral atoms walking in a two-dimensional optical lattice are proposed: An atom walking in a few steps around a closed loop acquires a so called Peierls phase. Here experiments remain one of the future challenges.

9.3.5.2 Test of the Leggett–Garg Inequality

The superposition of multiple trajectories in quantum walks is an essential aspect, which makes the motion of the "walker" in space truly quantum. There exist, however, alternative theories to standard quantum mechanics, known as macro-realistic theories, which postulate the presence of spatial decoherence terms modifying the coherent Schrödinger equation [62]. In this framework, the walker can take well-defined (classical) trajectories with a certain probability, which depends on the relative magnitude of coherent vs. decohering terms in the equations that describe the motion of massive particles such as Cs atoms. Leggett and Garg haven constructed an inequality [63], which allows to distinguish (more precisely falsify à la Popper) whether the walker has evolved along well-defined trajectories. One prerequisite for these experiments is the ability to perform ideal negative measurements, i.e. being able to detect the absence of a particle in places where the particle is not, without – from a macro-realistic point of view – disturbing the subsequent evolution of the particle. Such ideal negative measurements can be performed with the atom quantum walk platform, which offers the possibility of direct manipulation of individual trajectories involved in the walker propagation. The experiments [64] show a strong violation (6σ) of the Leggett–Garg inequality, thus rejecting macro-realistic interpretations. At the same time, the experimental data are in excellent agreement with the expectation from standard quantum theory. Thus, we can justifiably call these experiments quantum walks.

9.3.6 Single Atoms as Sensors

9.3.6.1 Single Atoms as Localized Quantum Probes

In all experiments presented so far, the atoms were isolated (i.e. embedded into an electromagnetic vacuum). One interesting line of experiments follows the question what properties can be expected if single atoms are immersed into another quantum system made of potentially many particles. The larger quantum system effectively constitutes a controllable environment for the single particles. The scheme of such experiments based on the single atom methods described above is shown in Figure 9.14 [65]. Studies of the dynamic behavior of single atoms in cold quantum gases show applications [66] of the single atoms as very localized probes for extracting local properties of the larger quantum system, including thermodynamical quantities such as the local temperature, which is relevant when the larger system is not at equilibrium.

9.3.6.2 Single Atoms for Environmental Applications

Trace elements are important indicators of dynamical environmental evolutions or also hazards. The methods developed for detecting and controlling for basic

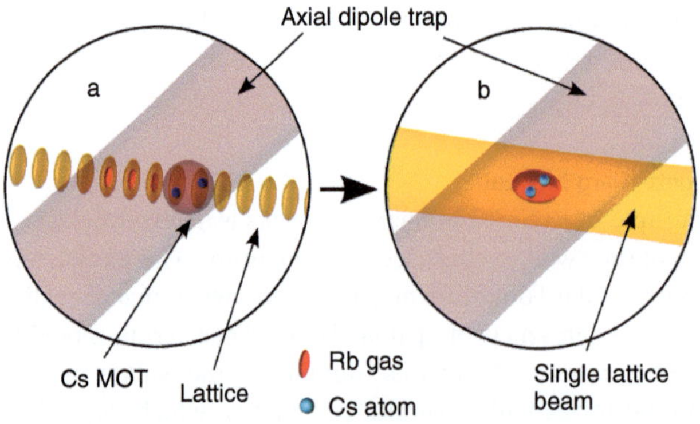

Figure 9.14 Schematic example of immersing single Cs atoms by means of a conveyor belt into an ultracold gas of Rb atoms. Source: Spethmann et al. [65]/American Association for the Advancement of Science.

physical research have indeed also been adapted to detect very rare elements such as noble gases $^{38,39}Ar$ and ^{91}Kr (ATTA-method, Atom Trap Trace Analysis) [67, 68]. The ATTA method draws its advantage from an enormous selectivity for the isotope under observation and is able to provide count rates of order 1 atom/h and below.

9.4 Short Conclusions

We have presented an overview on methods and applications to gain quantum control of single neutral atoms. Many experimental techniques are now mature and available to address new questions in more complex quantum systems. The "bottom-up" approach meets with the "top-down" scheme to gain insight into many-body phenomena at the constituent particle level. At the few-particle level, enhanced atom–atom interactions induced by Rydberg-blockade-type schemes have opened new routes to control, e.g. atom–photon systems [69]. A merger of the recent techniques for controlling Rydberg excitations with the experimental approaches presented here, in particular the possibility of moving and delocalizing single atoms in space with high fidelity, may offer improved quantum control in the future.

Acknowledgments

It is a pleasure to express our gratitude to a large number of students obtaining their master's or PhD degree or working with us at the postdoctoral level – there are too many to name them all individually. They have shared our enthusiasm to experiment with new techniques for single-atom control. Together with them, we have been witnessing a transition to an era of quantum technologies in which passion for precision physics has become essential for future advances in quantum science. It is an occasion to offer special thanks to our close collaborator, senior scientist Wolfgang Alt, for his invaluable scientific and technical contributions in ensuring the quality of our work.

We certainly learned (and had to learn) a lot from our colleagues, be it as visitors or direct collaborators. Let us mention Reinhard Werner of the Leibniz University Hannover, Artur Widera of the University of Kaiserslautern, Arno Rauschenbeutel of the Humboldt University Berlin, Jean Michel Raimond of ENS Paris, Yoav Sagi and Gal Ness of the Technion in Haifa, Janos Asboth of the Wigner Research Center Budapest, Klaus Moelmer of the Niels Bohr Institute Kopenhagen, Nir Davidson of the Weizmann Institute in Rehovot, Clive Emary of Newcastle University, and more.

We also thank the European Research Council (ERC), the European Commission (EC), the Deutsche Forschungsgemeinschaft (DFG), the Alexander von Humboldt Foundation, and the Bundesministerium für Bildung und Forschung (BMBF) for continued support over more than two decades in terms of programs including an Advanced Grant, a Humboldt fellowship, a Forscherguppe, an SFB, a Verbundprojekt, and more.

References

1 Neuhauser, W., Hohenstatt, M., Toschek, P.E., and Dehmelt, H. (1980). Localized visible Ba^+ mono-ion oscillator. *Physical Review A* 22: 1137.

2 Wilson, C.T.R. (1897). Condensation of water vapour in the presence of dust-free air and other gases. *Philosophical Transactions of the Royal Society of London Series A* 189: 265.

3 Dehmelt, H.G. (1968). Radiofrequency spectroscopy of stored ions I: storage. *Advances in Atomic and Molecular Physics* 3: 53.

4 Dehmelt, H. (1983). Stored-ion spectroscopy. In: *Advances in Laser Spectroscopy: NATO Advanced Study Institute, 1981, San Miniato*, vol. 6 (eds F. Strumia, F.T. Arecchi, and H. Walther), 153. Plenum Press.

5 Neuhauser, W., Hohenstatt, M., Toschek, P., and Dehmelt, H. (1978). Optical-sideband cooling of visible atom cloud confined in parabolic well. *Physical Review Letters* 41: 233.

6 Bloch, I., Dalibard, J., and Nascimbène, S. (2012). Quantum simulations with ultracold quantum gases. *Nature Physics* 8: 267.

7 Harold, P.S. and Metcalf, J. (1999). *Laser Cooling and Trapping*. Springer-Verlag.

8 Haubrich, D., Schadwinkel, H., Strauch, F. et al. (1996). Observation of individual neutral atoms in magnetic and magneto-optical traps. *Europhysics Letters* 34: 663.

9 Anderson, M.H., Ensher, J.R., Matthews, M.R. et al. (1995). Observation of Bose-Einstein condensation in a dilute atomic vapor. *Science* 269: 198.

10 Grimm, R., Weidemüller, M., and Ovchinnikov, Y.B. (2000). Optical dipole traps for neutral atoms. *Advances In Atomic, Molecular, and Optical Physics* 42: 95.

11 Miller, J.D., Cline, R.A., and Heinzen, D.J. (1993). Far-off-resonance optical trapping of atoms. *Physical Review A* 47: R4567.

12 Frese, D., Ueberholz, B., Kuhr, S. et al. (2000). Single atoms in an optical dipole trap: towards a deterministic source of cold atoms. *Physical Review Letters* 85: 3777.

13 Hu, Z. and Kimble, H.J. (1994). Observation of a single atom in a magneto-optical trap. *Optics Letters* 19: 1888.

14 Ruschewitz, F., Bettermann, D., Peng, J.L., and Ertmer, W. (1996). Statistical investigations on single trapped neutral atoms. *Europhysics Letters* 34: 651.

15 Gomer, V., Strauch, F., Ueberholz, B. et al. (1998). Single-atom dynamics revealed by photon correlations. *Physical Review A* 58: 1657.

16 Ueberholz, B., Kuhr, S., Frese, D. et al. (2000). Counting cold collisions. *Journal of Physics B: Atomic and Molecular Physics* 33: L135.

17 Willems, P.A., Boyd, R.A., Bliss, J.L., and Libbrecht, K.G. (1997). Stability of magneto-optical traps with large field gradients: limits on the tight confinement of single atoms. *Physical Review Letters* 78: 1660.

18 Schlosser, N., Reymond, G., and Grangier, P. (2002). Collisional blockade in microscopic optical dipole traps. *Physical Review Letters* 89: 023005.

19 Alt, W., Schrader, D., Kuhr, S. et al. (2003). Single atoms in a standing-wave dipole trap. *Physical Review A* 67: 033403.

20 Alt, W. (2002). An objective lens for efficient fluorescence detection of single atoms. *Optik* 113: 142.

21 Gallego, J., Alt, W., Macha, T. et al. (2018). Strong purcell effect on a neutral atom trapped in an open fiber cavity. *Physical Review Letters* 121: 173603.

22 Khudaverdyan, M., Alt, W., Dotsenko, I. et al. (2008). Controlled insertion and retrieval of atoms coupled to a high-finesse optical resonator. *New Journal of Physics* 10: 073023.

23 Boca, A., Miller, R., Birnbaum, K.M. et al. (2004). Observation of the vacuum Rabi spectrum for one trapped atom. *Physical Review Letters* 93: 233603.

24 Gehr, R., Volz, J., Dubois, G. et al. (2010). Cavity-based single atom preparation and high-fidelity hyperfine state readout. *Physical Review Letters* 104: 203602.

25 Pfeifer, H., Ratschbacher, L., Gallego, J., et al. (2022). Achievements and perspectives of optical fiber Fabry-Perot cavities. *Applied Physics B* 128: 29.

26 Kuhr, S., Alt, W., Schrader, D. et al. (2003). Coherence properties and quantum state transportation in an optical conveyor belt. *Physical Review Letters* 91: 213002.

27 Schrader, D., Dotsenko, I., Khudaverdyan, M. et al. (2004). Neutral atom quantum register. *Physical Review Letters* 93: 150501.

28 Fuhrmanek, A., Bourgain, R., Sortais, Y.R.P., and Browaeys, A. (2011). Free-space lossless state detection of a single trapped atom. *Physical Review Letters* 106: 133003.

29 Robens, C. (2017). Microwave control of atomic motion in a spin dependent optical lattice. PhD thesis. University of Bonn.

30 Khudaverdyan, M., Alt, W., Kampschulte, T. et al. (2009). Quantum jumps and spin dynamics of interacting atoms in a strongly coupled atom-cavity system. *Physical Review Letters* 103: 123006.

31 Kampschulte, T. (2011). Coherently driven three-level atoms in an optical cavity. PhD thesis. University of Bonn.

32 Karski, M., Förster, L., Choi, J.M. et al. (2009). Nearest-neighbor detection of atoms in a 1D optical lattice by fluorescence imaging. *Physical Review Letters* 102: 053001.

33 Alberti, A., Robens, C., Alt, W. et al. (2016). Super-resolution microscopy of single atoms in optical lattices. *New Journal of Physics* 18: 053010.

34 Bakr, W.S., Gillen, J.I., Peng, A. et al. (2009). A quantum gas microscope for detecting single atoms in a Hubbard-regime optical lattice. *Nature* 462: 74.

35 Robens, C., Brakhane, S., Alt, W. et al. (2017). High numerical aperture (NA = 092) objective lens for imaging and addressing of cold atoms. *Optics Letters* 42: 1043.

36 Robens, C., Zopes, J., Alt, W. et al. (2017). Low-entropy states of neutral atoms in polarization-synthesized optical lattices. *Physical Review Letters* 118: 065302.

37 Martinez-Dorantes, M., Alt, W., Gallego, J. et al. (2017). Fast nondestructive parallel readout of neutral atom registers in optical potentials. *Physical Review Letters* 119: 180503.

38 Robens, C., Alt, W., Emary, C. et al. (2017). Atomic "bomb testing": the Elitzur-Vaidman experiment violates the Leggett-Garg inequality. *Applied Physics B-Lasers and Optics* 123: 12.

39 Fung, Y., Sompet, P., and Andersen, M. (2016). Single atoms preparation using light-assisted collisions. *Technologies* 4: 4.

40 Kaufman, A.M. and Ni, K.-K. (2021). Quantum science with optical tweezer arrays of ultracold atoms and molecules. *Nature Physics* 17: 1324.

41 Kuhr, S., Alt, W., Schrader, D. et al. (2001). Deterministic delivery of a single atom. *Science* 293: 278.

42 Schrader, D., Kuhr, S., Alt, W. et al. (2001). An optical conveyor belt for single neutral atoms. *Applied Physics B-Lasers and Optics* 73: 819.

43 Cirac, J.I. and Zoller, P. (2003). How to manipulate cold atoms. *Science* 301: 176.

44 Mandel, O., Greiner, M., Widera, A. et al. (2003). Coherent transport of neutral atoms in spin-dependent optical lattice potentials. *Physical Review Letters* 91: 010407.

45 Robens, C., Brakhane, S., Alt, W. et al. (2018). Fast, high-precision optical polarization synthesizer for ultracold-atom experiments. *Physical Review Applied* 9: 034016.

46 Förster, L., Karski, M., Choi, J.-M. et al. (2009). Microwave control of atomic motion in optical lattices. *Physical Review Letters* 103: 233001.

47 Förster, L. (2010). Testing the quantumness of atom trajectories. PhD thesis. University of Bonn.

48 Barredo, D., Lienhard, V., de Léséleuc, S. et al. (2018). Synthetic three-dimensional atomic structures assembled atom by atom. *Nature* 561: 79.

49 Ebadi, S., Wang, T.T., Levine, H. et al. (2021). Quantum phases of matter on a 256-atom programmable quantum simulator. *Nature* 595: 227.

50 Beugnon, J., Tuchendler, C., Marion, H. et al. (2007). Two-dimensional transport and transfer of a single atomic qubit in optical tweezers. *Nature Physics* 3: 696.

51 Kuhr, S., Alt, W., Schrader, D. et al. (2005). Analysis of dephasing mechanisms in a standing-wave dipole trap. *Physical Review A* 72: 023406.

52 Bluvstein, D., Levine, H., Semeghini, G. et al. (2022). A quantum processor based on coherent transport of entangled atom arrays. *Nature* 604: 451.

53 Lam, M.R., Peter, N., Groh, T. et al. (2021). Demonstration of quantum brachistochrones between distant states of an atom. *Physical Review X* 11: 011035.

54 Steffen, A., Alberti, A., Alt, W. et al. (2012). Digital atom interferometer with single particle control on a discretized space-time geometry. *Proceedings of the National Academy of Sciences of the United States of America* 109: 9770.

55 Alberti, A., Alt, W., Werner, R., and Meschede, D. (2014). Decoherence models for discrete-time quantum walks and their application to neutral atom experiments. *New Journal of Physics* 16: 123052.

56 Weitenberg, C., Endres, M., Sherson, J.F. et al. (2011). Single-spin addressing in an atomic Mott insulator. *Nature* 471: 319.

57 Preiss, P.M., Ma, R., Tai, M.E. et al. (2015). Strongly correlated quantum walks in optical lattices. *Science* 347: 1229.

58 Karski, M., Förster, L., Choi, J.-M. et al. (2009). Quantum walk in position space with single optically trapped atoms. *Science* 325: 174.

59 Kempe, J. (2003). Quantum random walks: an introductory overview. *Contemporary Physics* 44: 307.

60 Genske, M., Alt, W., Steffen, A. et al. (2013). Electric quantum walks with individual atoms. *Physical Review Letters* 110: 190601.

61 Sajid, M., Asbóth, J.K., Meschede, D. et al. (2019). Creating anomalous Floquet Chern insulators with magnetic quantum walks. *Physical Review B* 99: 214303.

62 Arndt, M. and Hornberger, K. (2014). Testing the limits of quantum mechanical superpositions. *Nature Physics* 10: 271.

63 Emary, C., Lambert, N., and Nori, F. (2014). Leggett-Garg inequalities. *Reports on Progress in Physics* 77: 016001.

64 Robens, C., Alt, W., Meschede, D. et al. (2015). Ideal negative measurements in quantum walks disprove theories based on classical trajectories. *Physical Review X* 5: 011003.

65 Spethmann, N., Kindermann, F., John, S. et al. (2012). Dynamics of single neutral impurity atoms immersed in an ultracold gas. *Physical Review Letters* 109: 235301.

66 Bouton, Q., Nettersheim, J., Adam, D. et al. (2020). Single-atom quantum probes for ultracold gases boosted by nonequilibrium spin dynamics. *Physical Review X* 10: 011018.

67 Ebser, S., Kersting, A., Stöven, T. et al. (2018). ^{39}Ar dating with small samples provides new key constraints on ocean ventilation. *Nature Communications* 9: 5046.

68 Tong, A.L., Gu, J.-Q., Yang, G.-M. et al. (2021). An atom trap system for ^{39}Ar dating with improved precision. *Review of Scientific Instruments* 92: 063204.

69 Morgado, M. and Whitlock, S. (2021). Quantum simulation and computing with Rydberg-interacting qubits. *AVS Quantum Science* 3: 023501.

10

Long-Distance Entanglement of Atomic Qubits

Kai Redeker[1,2], Wenjamin Rosenfeld[1,3], and Harald Weinfurter[1,4]

[1] *Faculty of Physics, LMU Munich, Schellingstr. 4, 80799, Munich, Germany*
[2] *Federal Office for Information Security, P.O. Box 200363, 53133, Bonn, Germany*
[3] *Blickfeld GmbH, Barthstr. 12, 80399, Munich, Germany*
[4] *Max-Planck-Institute for Quantum Optics, Hans-Kopfermann-Str. 1, 85748, Garching, Germany*

Entanglement of separated stationary qubits serving as quantum memories is the key ingredient for future large-scale quantum networks [1] enabled by using quantum repeater protocols for the efficient communication of quantum information over large distances [2]. Such networks are key to perform, e.g. distributed quantum computation, secure communication over long distances, or precise clock synchronization [3]. In this context, the so-called device-independent (DI) protocols for self-testing of connections between quantum nodes arose [4, 5] and acted, besides tests of extensions of quantum physics based on Bell's inequality, as strong driver for experiments on the distribution of entanglement.

This chapter shows, how entanglement between two single, neutral ^{87}Rb atoms can be achieved. For this, single atoms are trapped [6] in two setups 398 m apart. To create entanglement between the atoms, the spin of each atom is entangled with the polarization of a single photon employing excitation and spontaneous photon emission [7, 8]. The emitted photons are coupled into single-mode optical fibers and guided to a photon measurement setup. The successful detection of the two photons in entangled states, i.e. a Bell state projection measurement of the two photons, swaps the entanglement to the atoms [9–11]. It also heralds the atom–atom entanglement and allows for an event-ready measurement of the atomic state. The measurement of the atomic states is based on a state selective ionization and a subsequent detection of the ionization fragments [12–14] and makes an efficient measurement of the atomic state possible in less than 1.1 μs [15], enabling space like separation of parallel atomic measurements [16]. The short time this measurement requires is essential especially for so-called loophole-free tests of Bell's inequality and the application of device-independent protocols as it is sufficient to separate the laboratories by approximately 400 m (Figure 10.1).

In the following, we will introduce both the experimental setup required to trap atoms and to generate atom–photon entanglement and the main experimental

Photonic Quantum Technologies: Science and Applications, First Edition. Edited by Mohamed Benyoucef.

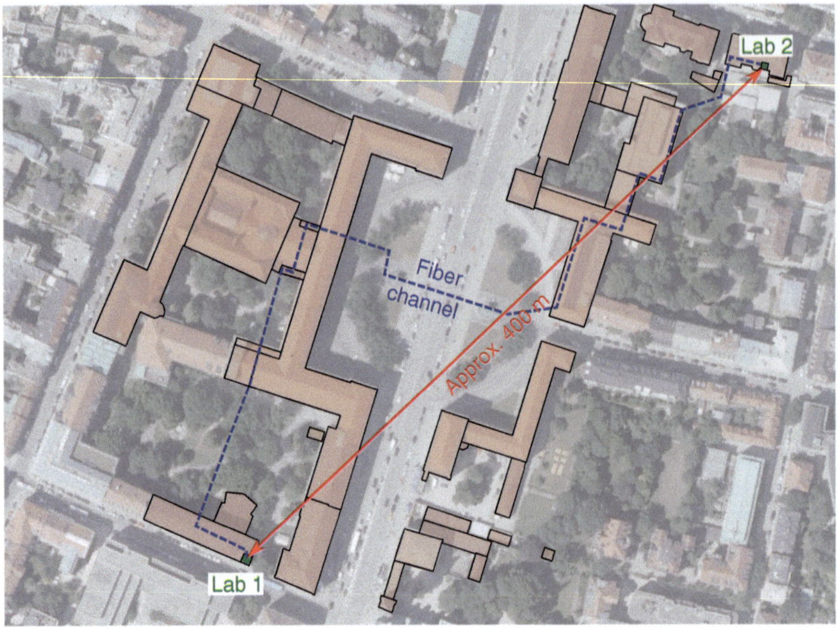

Figure 10.1 The location of the two laboratories at the main campus of Ludwig-Maximilians-Universität in Munich. Laboratory 1 (Lab 1) is located in the basement of the faculty of physics at Schellingstraße 4. Laboratory 2 (Lab 2) is located in the basement of the department of economics at Schackstraße 4. Both labs are connected with glass fibers for communication and for guiding the fluorescence photons emitted by the atoms. Source: Map data provided by State Office for Digitization, Broadband and Surveying.

methods. First, the main features of rubidium are discussed, which will be used for the experiments discussed herein. This is followed by a description of the setup used to trap single neutral atoms, along with the procedure to control and measure the atomic state. Finally, the process of creating entanglement of two single atoms in separated traps is described. At the end, we briefly summarize the first applications already enabled by such entanglement distribution.

10.1 Qubit Encoding in ^{87}Rb

To realize an atomic qubit, an atom with suitable quantum states is needed. For that purpose, the element needs to fulfill certain requirements. The first is the possibility for entanglement of the stationary atomic qubit with a "flying" qubit to distribute the entanglement over long distances. Next, in view of applications based on tests of a Bell inequality, the qubit should allow for a high-fidelity readout within a time short relative to the distance between atomic quantum memories, and finally, the atomic coherence time must be long enough to measure the atomic states with a high state fidelity also after entanglement generation. The alkali metal rubidium is chosen as the quantum states of its single valence electron are particularly well suited for this

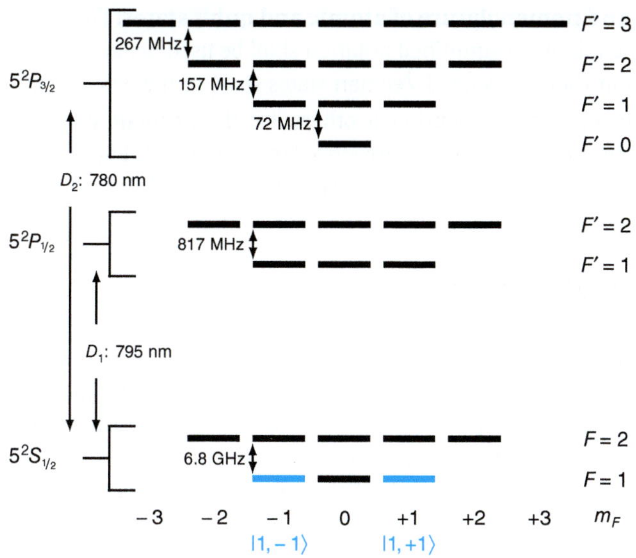

Figure 10.2 Level scheme for the valence electron of ^{87}Rb [17]: the ground state $5^2S_{\frac{1}{2}}$ is split in two hyper-fine levels $F = 1$ and $F = 2$. The excited state $^2P_{\frac{1}{2}}$ is split also in two hyper-fine levels $F' = 1$ and $F' = 2$ while the excited state $^2P_{\frac{3}{2}}$ is split into four hyper-fine levels $F' = 0$, $F' = 1$, $F' = 2$ and $F' = 3$. All hyper-fine levels are additionally split into Zeeman sublevels m_F. The transition wavelength for $^2S_{\frac{1}{2}}$ to $^2P_{\frac{1}{2}}(\mathrm{D}_1)$ is 795 nm and for $^2S_{\frac{1}{2}}$ to $^2P_{\frac{3}{2}}(\mathrm{D}_2)$ is 780 nm. The atomic qubit is encoded in the $5^2S_{\frac{1}{2}}$, $F = 1$, $m_F = \pm 1$ ground states with $m_F = -1 = |1,-1\rangle = |\downarrow\rangle_z$ and $m_F = +1 = |1,+1\rangle = |\uparrow\rangle_z$ (blue). Source: Adapted from Steck [17].

kind of experiment. Of the two naturally occurring isotopes ^{85}Rb and ^{87}Rb, the latter with nuclear spin of $I = \frac{3}{2}$ is used due to its better fitting hyper-fine level structure.

The qubit is encoded in the degenerate Zeeman substates $F = 1, m_F = \pm 1$ of the $5^2S_{\frac{1}{2}}$ ground state (Figure 10.2), satisfying all listed requirements. The spontaneous decay of the $F' = 0$ hyper-fine level of the $5^2P_{\frac{3}{2}}$ excited state to the ground state generates a photon whose polarization state is entangled with the Zeeman state of the $F = 1$ of the $5^2S_{\frac{1}{2}}$ ground state. Dipole selection rules allow to individually address the Zeeman states with polarized light. By this method and by additionally employing particle detectors, a fast and efficient readout scheme can be implemented. As a down side, the employed Zeeman states are highly susceptible to magnetic and light fields, which need to be controlled to enable long state coherence (Section 10.4).

For using single ^{87}Rb-atoms as stationary qubits, or as a so-called quantum memory, the atoms need to be trapped and well localized. The methods of optical dipole trapping in combination with laser cooling are well suited for this purpose. For an optical dipole trap (ODT) [18] as employed here, it is necessary to first slow down the atoms before trapping by cooling them [19]. ^{87}Rb can be conveniently laser cooled via Doppler cooling [20]. Here, radiation pressure and the Doppler effect are employed to slow down atoms.

Formal definition and nomenclature of atomic and qubit states: For convenience of the reader, the following simplified notation shall be used henceforth: the hyper-fine states (F) and their associated Zeeman states (m_F) of the $5^2S_{\frac{1}{2}}$ ground state are written as $|F, m_F\rangle$. If not mentioned otherwise, the Zeeman states are defined with respect to the quantization axis introduced in section 10.2 (Figure 10.4). With this, the qubit states are defined as $|1,-1\rangle = |\downarrow\rangle_z$ and $|1,+1\rangle = |\uparrow\rangle_z$.

10.2 Trapping Single Atoms

Central to the experiment are the two traps for single ^{87}Rb-atoms, which are set up in two laboratories 400 m apart on different sides of the campus (Figure 10.1). For trapping of atoms and obtaining long coherence times, it is necessary to isolate them from the environment. Collisions with other atoms and molecules can only be suppressed in an ultrahigh-vacuum (UHV) environment. In addition to an UHV-vacuum chamber (pressure below 10^{-9} mBar), a good optical access for the many laser beams is required and also particle detectors for the fast and efficient atomic-state measurement of ionized atoms. All this can be achieved using a specially designed glass cell, which is attached to the steel part of standard UHV components. This glass cell is specially coated for allowing to control electric fields and designed such that both particle detectors can be placed inside, all without impeding optical access to the trap region (Figure 10.3) [15, 21].

To actually trap atoms inside the vacuum chamber, a two-step scheme is implemented: first, a magneto-optical trap (MOT) [22] is used to obtain a cloud of cooled, trapped atoms with a diameter of about 1 mm containing on the order of 10^4 atoms [8]. Doppler cooling and especially so-called polarization gradient cooling [23] result in temperatures of only 30–40 μK [16, 24].

Figure 10.3 The glass cell connected to the main UHV set-up: the two particle detectors for the fast atomic state readout are placed close to the region of the trap.

At this point, the atoms are slow enough to be captured in an ODT [18, 25]. It relies on the interaction between an induced atomic dipole and a light field. The latter is far detuned from any transition such that other potential effects, such as optical excitation and photon scattering, can be neglected. The light field thus acts like a conservative potential, depending on the detuning and local intensity of the light field. For Gaussian light beams, one obtains with good approximation an harmonic potential with the deepest point at the point of highest intensity. It is typically shallow and allows only the trapping of very slow atoms with an energy equivalent of a few mK.

To implement the ODT in our experiment, a laser with a wavelength of 852 nm is focused at the position of the cloud of cold ^{87}Rb-atoms trapped by the MOT. The focal waist of the laser is $w_0 = 1.92$ μm with a Rayleigh length of $z_R = 13.6$ μm [15] where only one single atom can be trapped due to collisional blockade effects [6, 26]. This waist size together with an optical power of 60 mW results in an trapping potential $U_{\mathrm{ODT}} = k_B \cdot 3.2$ mK. The transverse and longitudinal oscillation frequencies of atoms in the trap are $\omega_T = 2\pi \cdot 92$ kHz and $\omega_L = 2\pi \cdot 13$ kHz [16].

To enable the atom being used as qubit, it is not only necessary to trap a single atom, but also to be able to collect and detect light emitted by it. Here a confocal configuration with a single high numerical aperture (NA) microscope objective[1] is used to focus the ODT laser and to collect the fluorescence light (Figure 10.4). The microscope objective is designed for both wavelengths 852 nm and 780 nm and thanks to its long working distance, it can be mounted outside the vacuum setup to collect the fluorescence light from inside the vacuum glass cell. A dichroic beam splitter separates the light emitted by the atom with a wavelength of 780 nm from the 852 nm laser used for the ODT. The fluorescence light is then coupled into a single-mode optical fiber and guided to the single photon detection setup (Figure 10.11). The combined collection and detection efficiency of the fluorescence light, η_{col}, is ca 0.215%. This low efficiency is due to the small fraction of light that can be coupled into the optical fiber by the microscope objective, which is limited by the collection and aberration effects and transmission losses in the fiber τ_{fib}.

The procedure to trap a single atom runs as follows: first, the cooling beams, the current in the quadrupole coils, and the dipole trap laser are switched on. By this a cloud of cold atoms is trapped by the MOT. One of these cold atoms eventually enters the dipole trap potential and is trapped there provided it scatters light inside the potential. The ongoing scattering of cooling light from this atom enables one to observe the trapped atom due to an increase in the photon detection rate. If the photon detection rate is above a certain threshold, the current in the quadrupole coils is turned off. The cloud of cold atoms thus dissipates, while the single atom in the ODT remains trapped, and is now ready for experiments. If the atom is lost after some time the photon detection rate drops back to the background level. As soon as the photon detection rate is below a certain threshold, typically set lower than the first one (Figure 10.5), the current in the quadrupole coils is turned on to restart the trapping sequence.

---

1 Mitutoyo, G Plan Apo 50, NA = 0.5 corrected for a 3.5 mm glass plate, working distance 13.89 mm, wavelength range 435–655 nm, reduced performance in the near infrared.

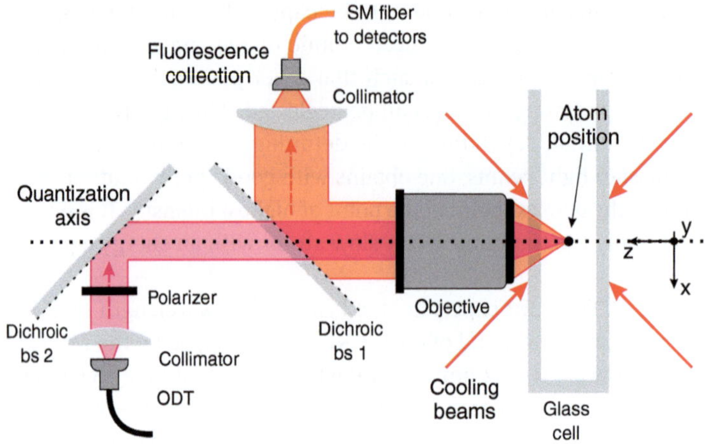

Figure 10.4 Confocal ODT and fluorescence collection setup: the linearly polarized ODT laser (magenta) is focused with a microscope objective at the position of the atom cloud trapped by the MOT (red arrows: cooling beams) inside the vacuum glass cell. A single atom is trapped a the point of highest intensity (focus) of the ODT laser (black dot: position of the trapped single atom). The fluorescence light emitted by the trapped single atom (red) is collected with the same objective. The collection of the fluorescence light via the microscope objective into the single-mode fiber defines the quantization axis of the system. The coordinate system of the experiment is defined so that the z-axis coincides with the quantization axis.

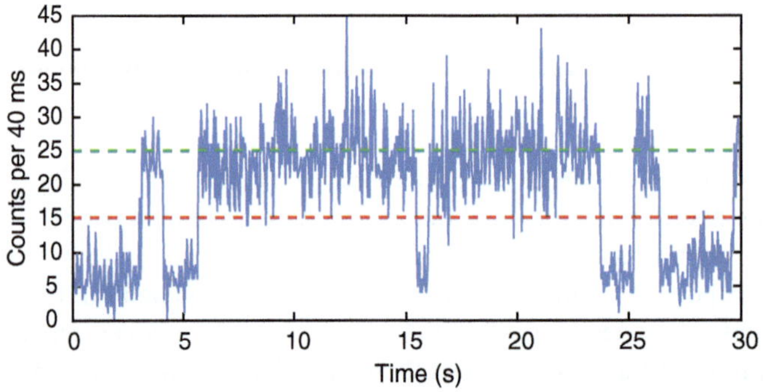

Figure 10.5 Trace of photon counts: photons collected in the confocal setup are registered by single-photon detector with the count rate integrated for 40 ms (blue). The background count level is caused by the cooling beams and has an average of 6.75 counts in 40 ms. If an atom is trapped the average counts rise to 24.75. The threshold above which an atom can be considered trapped is 25 counts (green). If fewer than 15 counts (red) are recorded, the atom is considered lost. This hysteresis is needed to identify correctly whether an atom is trapped or lost.

10.3 State Preparation, Control, and Measurement of Single Atoms

After trapping the single atoms, they are ready to be used as atomic qubits. For this it is necessary to prepare, control, and measure atomic states with high fidelity. For the creation and detection of long-distance entanglement via entanglement swapping, it is necessary to entangle the atomic qubit with a photon, perform measurements of the qubit state, and control the coherence of these states.

10.3.1 Creation of Atom–Photon Entanglement

Spontaneous emission of an excited atom generally leads to entanglement of the emitted photon state with the atomic state after the decay. The atom is first prepared in the $5^2S_{\frac{1}{2}},\ F=1,\ m_F=0$ state and then excited to the $5^2P_{\frac{3}{2}},\ F'=0$ state via a short laser pulse. The preparation of $5^2S_{\frac{1}{2}},\ F=1,\ m_F=0$ is performed via optical pumping [27] using multiple, polarized light beams resonant to the transition between $5^2S_{\frac{1}{2}},\ F=2$ and $5^2P_{\frac{3}{2}},\ F'=1$ and π-light resonant to the transition between $5^2S_{\frac{1}{2}},\ F=1$ and $5^2P_{\frac{3}{2}},\ F'=1$.

Next, the atom, prepared $5^2S_{\frac{1}{2}},\ F=1,\ m_F=0$ ground state, is excited with a short laser pulse that is π polarized to the $5^2P_{\frac{3}{2}},\ F'=0,\ m_F=0$ state. The excited state decays with a short lifetime of 26.24 ns [17].

After preparing a sufficient population in the $F=1, m_F=0$ state, the atom is excited with a short laser pulse that is π polarized to the $5^2P_{\frac{3}{2}},\ F'=0$ state (lifetime 26.24 ns) from where it decays back to the ground state. From this state the atom only decays to the $5^2S_{\frac{1}{2}},\ F=1$ ground state via three possible decay channels (Figure 10.6) to $m_F=+1$ while emitting a σ^- (left hand circular) polarized photon, to $m_F=0$ while emitting a π (parallel to the quantization axis) polarized photon, and to $m_F=-1$ while emitting a σ^+ (right hand circular) polarized photon. These processes occur with equal probability of $\frac{1}{3}$. Considering quantization along the z-axis, which is defined by the collection optics of the experiment (Figure 10.4), the entangled atom–photon state has the form

$$|\Psi\rangle_{\text{AP}} = \frac{1}{\sqrt{3}}\left(\left|\sigma^+\right\rangle|1,-1\rangle + |\pi\rangle\,|1,0\rangle + \left|\sigma^-\right\rangle|1,+1\rangle\right)$$

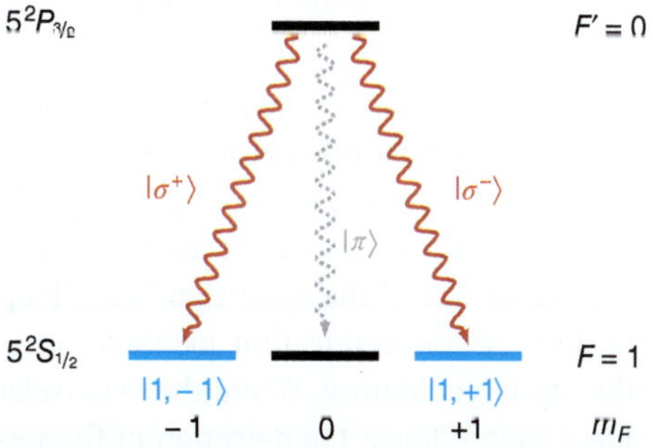

Figure 10.6 Creation of atom–photon entanglement via decay from the $5^2P_{\frac{3}{2}},\ F=0,\ m_F=0$ state resulting in the entangled atom–photon state $|\Psi_{\text{AP}}\rangle = \frac{1}{\sqrt{2}}(|\sigma^+\rangle|1,-1\rangle + |\sigma^-\rangle|1,+1\rangle)$.

As π-light is not coupled to the single-mode fiber, we are left with only two decay channels, resulting in the atom–photon state

$$\begin{aligned}|\Psi\rangle_{\mathrm{AP}} &= \frac{1}{\sqrt{2}}\left(|\sigma^+\rangle|1,-1\rangle + |\sigma^-\rangle|1,+1\rangle\right)\\ &= \frac{1}{\sqrt{2}}(|\mathrm{L}\rangle|1,-1\rangle + |\mathrm{R}\rangle|1,+1\rangle)\\ &= \frac{1}{\sqrt{2}}\left(|\mathrm{L}\rangle\,|\downarrow\rangle_z + |\mathrm{R}\rangle\,|\uparrow\rangle_z\right)\end{aligned} \tag{10.1}$$

with $|\mathrm{L}\rangle \equiv |\sigma^+\rangle$, $|\mathrm{R}\rangle \equiv |\sigma^-\rangle$). $|\mathrm{L}\rangle$ and $|\mathrm{R}\rangle$ represent the left and right circular polarization states of the photon in the reference frame of the laboratory, respectively. The emitted photons are collected with the confocal microscope set-up, effectively filtering out the π polarized photons. The remaining photons are then guided to the photon detection setup.

The fidelity of the atom–photon state (10.1) is limited by imperfections of the pumping and excitation process, leading to unwanted effects, such as off-resonant excitation resulting in two photon emission during one excitation process. These effects strongly depend on the polarization, temporal shape, and intensity of the excitation pulse. For optimizing these parameters, it is necessary to consider the photon collection efficiency, the photon detection efficiency, as well as the detector dark counts. A rigorous analysis is given in [11]. For an approximately Gaussian-shaped excitation pulse with a full width at half maximum (FWHM) duration of 20.35 ns and with an intensity yielding approximately 80% of the maximal excitation efficiency to reduce two-photon emissions, and a time window of 208 ns, one obtains a total photon detection probability of $\eta = 0.17\%$ per excitation pulse.

This low detection probability makes a fast repetition of the pumping and excitation process necessary to achieve a reasonable event rate. Yet, here again heating of the atom by repeated pumping and excitation has to be considered. The heating does not only lead to a possible loss of the trapped atom, but also to decoherence of the atomic state caused by stronger motion of the trapped atom (Section 10.4).

10.3.2 Measurement of the Atomic State

To read out the qubit encoded in the atomic state, a projection measurement with a freely chosen measurement direction has to be performed. For this purpose, state-selective ionization is used. It is composed of two steps: first, a selected superposition of Zeeman states is excited from the $5^2S_{\frac{1}{2}}$, $F = 1$ ground state to the $5^2P_{\frac{1}{2}}$, $F' = 1$ excited state. Then the excited atoms are ionized by an additional laser beam providing sufficient energy to release the valence electron (see Figure 1.7). The ionized atoms are no longer trapped, and the cooling light and detection of fluorescence photons can be used to measure if an atom is still trapped or not.

To close one of the most significant loopholes in Bell experiments, the so-called locality or communication loophole, a very fast state measurement with a high fidelity is mandatory. While the state-selective ionization process is very fast and has a high fidelity, the detection of fluorescence light in this setup has also a high

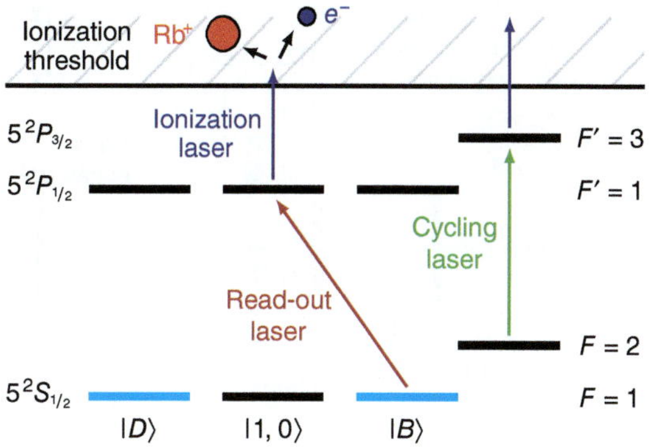

Figure 10.7 Fast readout scheme based on Zeeman-state selective ionization. The polarization χ_{ro} (10.2) of the readout leaser (red) selects which superposition of the $|1, \pm 1\rangle$ Zeeman states, called *bright state* $|B\rangle$ (10.3), is excited to the $5^2P_{\frac{1}{2}}$, F′ = 1 state, while the orthogonal superimposed *dark state* $|D\rangle$ is not affected by the readout laser and stays in the ground state. The third Zeeman state $|1, 0\rangle$ can be excited by any polarization χ_{ro} of the readout laser. The excited atoms are ionized by the ionization laser (blue). To overcome the detrimental effect of the decay to $F = 2$ ground state, this state is excited by the additional "cycling" laser (green) to the $5^2P_{\frac{3}{2}}$, F′ = 3 excited state. Additionally, atoms that decay back to the *bright state* $|B\rangle$ are re-excited by the readout laser (red) (decay channels not shown).

fidelity but needs long measurement times of more than 30 ms. To enable a faster decision, particle detectors, which detect the $^{87}Rb^+$-ion and the electron directly, are employed to confirm the ionization.

To excite only a specific superposition of the Zeeman levels for the $F = 1$, $m_F = \pm 1$, laser light (readout laser) resonant to the $5^2S_{\frac{1}{2}}$, $F = 1$ to $F' = 1$ transition (D1 line) is used (Figure 10.7). The incidence direction of the readout laser coincides with the quantization axis, counter propagating to the dipole trap (Figure 10.8). Selection rules resulting from transition dipole moments of the atom determine which polarization of the readout laser χ_{ro} (10.2) excites which superposition of the Zeeman-levels. The superposition transferred $|B\rangle_{\chi_{ro}}$ will be called *bright state* (10.3) and the orthogonal superposition, which is not transferred $|D\rangle_{\chi_{ro}}$ *dark state* (10.4). The polarization of the readout laser is set via a quarter and a half wave plate, thus any measurement direction can be chosen.

$$\chi_{ro} = \cos(\alpha) \cdot V + e^{-i\phi} \sin(\alpha) \cdot H \tag{10.2}$$

$$|B\rangle_{\chi_{ro}} = \cos(\alpha) \frac{-1}{\sqrt{2}} (|1,-1\rangle - |1,+1\rangle) + \sin(\alpha) e^{i\phi} \frac{i}{\sqrt{2}} (|1,-1\rangle + |1,+1\rangle) \tag{10.3}$$

$$|D\rangle_{\chi_{ro}} = \sin(\alpha) \frac{1}{\sqrt{2}} (|1,-1\rangle - |1,+1\rangle) + \cos(\alpha) e^{i\phi} \frac{i}{\sqrt{2}} (|1,-1\rangle + |1,+1\rangle) \tag{10.4}$$

The ionization threshold for the excited $5^2P_{\frac{1}{2}}$, $F' = 1$ state is 473.67 nm and the ionization threshold for the ground state $5^2S_{\frac{1}{2}}$, $F = 1$ is 296.82 nm [28]. To ionize only the excited atoms, a laser with a wavelength below 473 nm is used, which is focused on the atom with the same microscope objective used for fluorescence collection and for focusing the dipole trap (Figure 10.8). By integrating the fluorescence

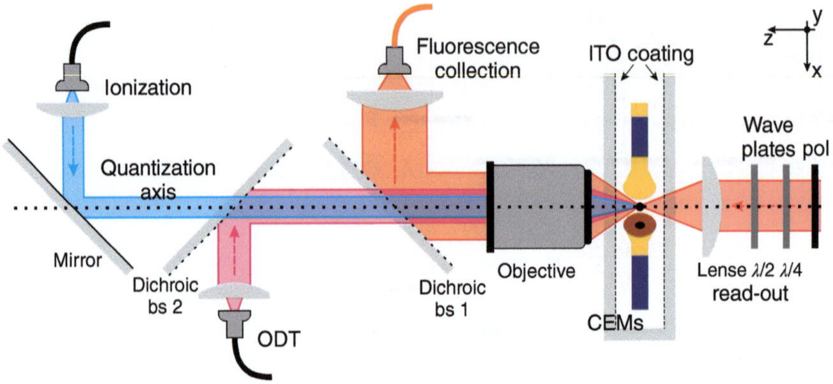

Figure 10.8 Experimental setup for the fast and efficient state readout. The readout laser (dark red) propagating in positive z-direction (counter propagating to the ODT laser) is focused on the atom (black dot). A linear polarizer followed a $\frac{\lambda}{2}$ and a $\frac{\lambda}{4}$ wave plate allow to set any desired polarization χ_{ro}. The ionization laser (blue) is overlapped with the ODT laser by a dichroic beam splitter and focused with the objective on the atom. Two channel electron multipliers (CEMs) are located near the atom inside the vacuum glass cell. The two inner surfaces of the glass cell are coated with transparent and conducting indium tin oxide (ITO) allowing for applying of electric fields.

counts measured at the detectors, it is possible to determine whether the atom was ionized and lost from the trap or not. The probability to identify a state as a bright state with this method, meaning the atom is ionized, for a given atomic state ρ is $Pr_B = \langle B| \rho |B\rangle + \langle 1,0| \rho |1,0\rangle$ while the probability to identify the state as dark state, in this case the atom is still trapped, is $Pr_D = \langle D| \rho |D\rangle$ (disregarding imperfections).

The fidelity of this measurement is limited by the lifetime of the excited state and by off-resonant excitation of the dark state to the state $5^2P_{\frac{1}{2}}$, $F' = 2$. The short lifetime of 27.7 ns leads to spontaneous decay to the ground states before the excited atom is ionized (Figure 10.7). With a probability of $\frac{5}{6}$, the excited state decays to the $F = 2$ ground state. This can be overcome by exciting this state with an additional laser to the $5^2P_{\frac{3}{2}}$, $F' = 3$ state, allowing for the ionization of the atom. With a probability of $\frac{1}{6}$, the excited state decays to the $F = 1$ ground state and in equal parts to *bright state* $|B\rangle_{ro}$ and to *dark state* $|D\rangle_{ro}$. While the part in $|B\rangle_{ro}$ can be excited again, the decay into $|D\rangle_{ro}$ leads to a reduction of ionization probability of the initial *bright state* resulting in additional infidelity. With a pulse length of 140 ns and optical power of 1.24 μW, ionization probabilities of $Pr_{\text{ionized}}\left(|B\rangle_{\chi_{ro}}\right) \simeq 0.98$ for the bright state and $Pr_{\text{ionized}}\left(|D\rangle_{\chi_{ro}}\right) \simeq 0.04$ for the dark state can be obtained, resulting in a contrast of $C_{ro} \simeq 0.94$ [15]. The total time for selection of the measurement basis together with the state-selective ionization is less than 400 ns.

In order to close the locality loophole of a Bell experiment, it is necessary to obtain the result of the ionization as fast as possible in order to keep the required spatial separation low, i.e. much faster than with fluorescence detection. This can be achieved by a direct detection of the ionization fragments [14] using channel electron multipliers (Figure 10.8).

In conclusion, state-selective ionization employing the polarization of an excitation pulse gives an easy option to full state analysis without additional manipulation of the atom. Using fluorescence detection to confirm the ionization, the overall measurement takes > 30 ms and the measurement fidelity is ca 0.97. However, using particle Detectors, the result is obtained within only 570 ± 3 ns after the start of the readout laser pulse [16, 29]. Still, the measurement fidelity of ca 0.965 [15] is very high and allows for both a fast and efficient atomic state readout.

10.4 Coherence of the Atomic States

In the experiments presented here, the atomic state is measured only if the signal heralding atom–atom entanglement is received. This is called an *event-ready* scheme. For the experiments to be successful, the initially prepared state needs to be conserved until the state is measured after receiving the heralding signal. A long temporal coherence of the atomic state is thus necessary. As the heralding signal has to be sent back to the trap setups after the photon detection, the time from excitation to receiving the heralding signal should be significantly shorter than the coherence time of the atomic states. Considering the 700 m long fiber channel this time is about 7 μs.

The atomic qubit is encoded in the $|1, \pm 1\rangle$ Zeeman states of the $S_{\frac{1}{2}}$ ground state. As these are the energetically lowest states, there is no spontaneous decay possible to other atomic states. Still, the atomic qubit state is subject to external influence, which can result in dephasing. For single atoms trapped inside UHV, these external influences are limited to laser light of the setup and the external magnetic field.

10.4.1 Interaction with Magnetic and Optical Fields

Due to the residual motion of the atom in the QDT, the interaction with spatially varying magnetic and optical fields is the main reason for dephasing and loss of fidelity of the quantum states. The interaction between the atomic spin and a magnetic field $\vec{B}$ is described by the interaction Hamiltonian $\hat{H}_B$. For a quantization axis in z-direction and the basis vectors $\vec{e}_1 = |1, +1\rangle_Z$, $\vec{e}_2 = |1, 0\rangle_Z$, and $\vec{e}_3 = |1, -1\rangle_Z$ and a magnetic field $\vec{B}$ the Hamiltonian is $\hat{H}_B = \frac{\mu_B g_F}{\hbar}\vec{B}\vec{\hat{F}}$ with the vector of the angular momentum operators for a spin 1 system the Larmor frequency $\omega_L = \frac{1}{\hbar} g_F \mu_B ||\vec{B}||$, the Bohr magneton μ_B, the Landé factor g_F for the hyperfine state.

During the time between excitation and state readout, the only laser interacting with the atom is the ODT laser. Since the ODT laser is far red off-resonant, scattering of photons is very rare and can be neglected. However, the ac-stark shift needs to be considered for state evolution. It depends on the intensity I of the ODT laser, the decay rate, and transition frequency of the central D-line Γ and ω_0, the detuning $\Delta_{1,F}$ and $\Delta_{2,F}$ with respect to the transition of the D_1 and D_2 line, on the magnetic sublevel m_F, and the ellipticity P of the trap light polarization ($P = 0$ for linear and $P = \pm 1$ for right/left circular polarization, respectively). The energy shift thus is equivalent to an energy shift $\Delta E_{bz} = g_F \mu_B m_F \, P \cdot I$ caused by an additional magnetic field in

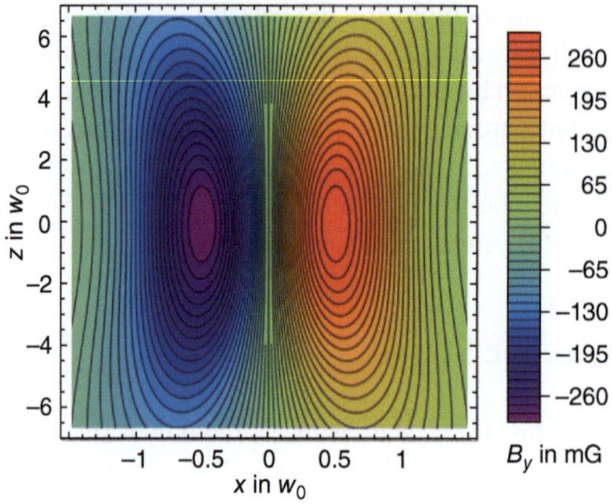

Figure 10.9 Magnetic field component $B_y(\vec{x})$ in the x–z plane at $y = 0$ caused by the longitudinal polarization components E_z that arise near the focus.

z-direction ($b_x = b_y = 0$, $b_z = 1$). Since the laser beam profile is a Gaussian TEM_{00} mode, one has a position-dependent intensity. In conclusion, the evolution can be seen as due to an effective magnetic field, which depends on the ellipticity of the light field and on the position of the atom in the trapping potential.

Ideally, for a linearly polarized light field with $P = 0$, the energy shift ΔE_{m_F} vanishes. However, when strongly focusing a linearly polarized Gaussian beam longitudinal polarization near the focus emerges off-axis resulting in additional, spatially dependent effective magnetic fields [30, 31]. As the magnitude of the QDT-laser intensity is considerably high, there is a strong contribution to the effective magnetic field. Figure 10.9 exemplary shows this for the x–z plane reaching values as high as 250 mG.

10.4.2 Control of the Magnetic Fields

External magnetic fields originate from multiple sources with a broad variety of strengths and temporal behavior. The strongest contribution is caused by the ion getter pump of the vacuum setup, which has a large permanent magnet leading to a static field up to 1000 mG at the position of the atom. Another static field is caused of course by the earth magnetic field of about 480 mG [32]. Time-varying fields are caused by the Munich subway, located only 60 m from the trap position, which changes its amplitude on a timescale of below one minute, and sources inside the lab like multiple power supplies and other electronics contributing to the magnetic field with timescales below one second. Compensation coils are used to control the magnetic field present at the position of the atom, allowing to create any desired magnetic field strength and direction in the range of ±5000 mG based on an active feedback loop referenced to a 3D magnetic field sensor placed less than 2 cm from the atom. Two ways of achieving long coherence times of the atomic state become possible: first, employing a magnetic guiding field for a controlled oscillation between the Zeeman states, which also allows to suppress the influence od external field fluctuations, or second, setting the magnetic field to (almost) zero with (ideally) no state evolution at all.

10.4.3 Decoherence Caused by Polarization Dependent AC-Stark Shift and Motion of the Atom

The circular polarization components of the ODT laser influence the atomic state evolution similarly to a magnetic field. But in contrast to the external magnetic field, which is homogeneous over the trapping region of a few μm^3, the effect of the ODT laser strongly depends on the position of the atom.

For a hypothetical case in which the atom is not moving, it is possible to compensate all effects of the ODT on the atomic state evolution perfectly. Yet in practice, the trapped atom is only cooled to a temperature of about 40 μK and thus oscillates in the trap, with a transverse trap oscillation period of typically $T_{\mathrm{ODT,t}} = 10.9$ μs and a longitudinal period of $T_{\mathrm{ODT,l}} = 108.9$ μs [16]. As the ODT corresponds in a good approximation to a harmonic potential, the oscillation frequency is independent of the actual energy. The evolution of the Zeeman states, however, depends on the actual trajectory of the atom during the time between excitation and state measurement. This trajectory is different for each trial, and averaging over multiple trials of the experiment yields the average over all those trajectories. This results in a rapid dephasing and loss of fidelity of the atomic state [30].

Fortunately, the spatial distribution of the optically induced magnetic field is antisymmetric across the transverse dimension of the trap (Figure 10.9). Thus, after one transverse oscillation period, the state evolution on both sides of the optical axis cancels, leaving the atom in the starting state. This effect is independent of the actual starting position of transverse motion of the atom in the trapping light field. This "rephasing" of the different trajectories is slightly reduced by two effects: first, the additional motion of the atom along the longitudinal axis of the trap leads to a variation of the effective magnetic field, and this prevents a perfect rephasing after one period. Second, the Gaussian-shaped potential of the ODT is not exactly harmonic; thus, the trap frequencies for atoms with different energies are slightly different. This leads to a temperature dependence of the rephasing; the colder the atoms, the better the fidelity of the state [16].

The measurements shown in Figure 10.10 demonstrate that the atomic qubit can be readout with a very high fidelity more than 10 μs after its preparation. Yet, one has to consider the rephasing and decide on a suited rephasing point before the experiment.

10.5 Creation of Long-Distance Atom–Atom Entanglement

Once a long-distance atom–atom entanglementhigh level of control of the trapped atoms and a high fidelity state readout are obtained, the experiment is ready for the next step: creation of entanglement between two single atoms trapped in independent setups separated by 398 m. Here, the two atoms are entangled via the entanglement swapping process [9, 33]. This enables the transfer of entanglement of two atom–photon pairs to the atoms by performing a joint measurement on the

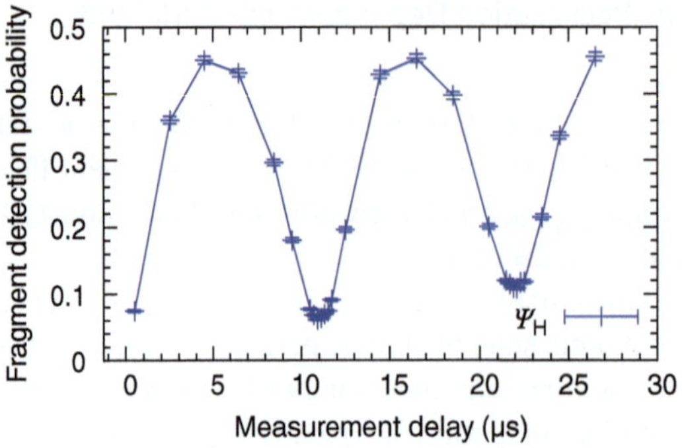

Figure 10.10 Measurement of the time evolution of atoms prepared in the states $|\Psi(0)\rangle = \frac{i}{\sqrt{2}}(|1,+1\rangle - |1,-1\rangle)$, which is most affected by the additional effective field of the ODT. The ionization fragment detection probability on the y-axis shows the reduction of the probability to find the atom in the initial state after a certain time. Rephasing occurs after one transverse oscillation of the atom in the trap (10.9 μs). Note the first measurement point at 0.5 μs is already influenced by the effect of the longitudinal polarization.

photons [10, 11]. In order to achieve a high fidelity of the photon measurement, the photons emitted by independent atoms should be indistinguishable in all degrees of freedom except polarization where the quantum state of interest is encoded. In particular, the control of the temporal degree of freedom (temporal shape) requires synchronization of the processes in the two setups.

10.5.1 Heralded Entanglement via Entanglement Swapping

For the entanglement swapping, we start with two entangled pairs of particles. Here, each pair consists of an atom and a photon. The entangled atom–photon pairs initially are in the states (for $i\epsilon 1, 2$)

$$\left|\Psi_i\right\rangle = \frac{1}{\sqrt{2}}\left(|\uparrow\rangle_{\mathrm{A},i}|\downarrow\rangle_{\mathrm{P},i} + |\downarrow\rangle_{\mathrm{A},i}|\uparrow\rangle_{\mathrm{P},i}\right)$$

with the subscripts indicating atom A and photon P. The four-particle state can be written as $\left|\Psi_{\mathrm{all}}\right\rangle = \left|\Psi_1\right\rangle \otimes \left|\Psi_2\right\rangle$, where a projection of the photons onto one of the four Bell states $\left|\Psi_{\mathrm{Bell}}\right\rangle \in \left\{|\Psi^-\rangle, |\Psi^+\rangle, |\Phi^-\rangle, |\Phi^+\rangle\right\}$ results in an entangled atom–atom state of the same type. Thus, a measurement of the photonic Bell state will transfer the entanglement to the atoms. The outcome of this measurement will herald, which entangled atom–atom state is prepared.

Ideally, for projecting on a Bell state, one would employ a quantum logic operation like a *controlled Not*. As this is not easily available for photons, interferometric Bell state analysis is the tool of choice. There, the two photons are overlapped at a beam splitter and propagate, depending on the symmetry of their spatial component of the wave function either both into the same output port or into different output ports [34, 35]. Together with polarization analysis in the output arms of the beam splitter projection onto the states $\left|\Psi^{\pm}\right\rangle_{\mathrm{PP}} = \frac{1}{\sqrt{2}}(|H\rangle|V\rangle \pm |V\rangle|H\rangle)$ is achieved

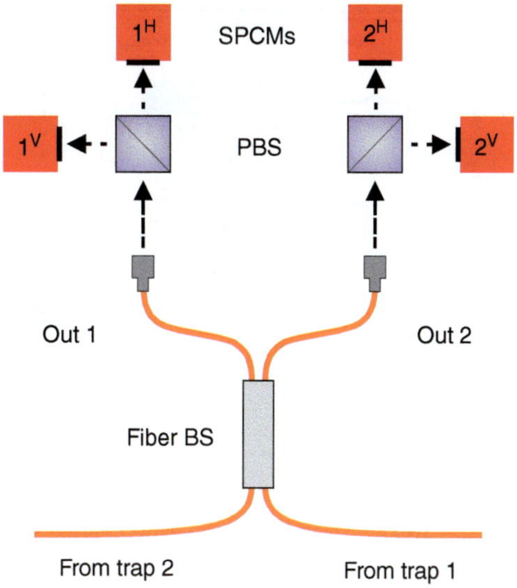

Figure 10.11 Bell-state measurement (BSM) setup with a fiber beam splitter. The fiber beam splitter (gray and orange) allows for perfect spatial overlap. The subsequent polarization analysis is done free space with polarizing beam splitters (PBS, purple) and single- photon detectors (SPCMs, red) yielding a high fidelity of the polarization measurement.

preparing and heralding the atomic state

$$|\Psi^{\pm}\rangle_{AA} = \frac{1}{\sqrt{2}}\left(|\downarrow\rangle_x|\uparrow\rangle_x \pm |\uparrow\rangle_x|\downarrow\rangle_x\right) \tag{10.5}$$

For the implementation of such an interferometric Bell-state measurement, there are two important points. First, the beam splitter should be polarization-independent, and, second, as mentioned above, the photons need to be indistinguishable in every degree of freedom. Spectral and temporal properties are defined by the excitation and emission process (Figure 10.11). The spatial degree of freedom is best defined in single-mode fiber beam splitters as this ensures a perfect overlap of the spatial modes of the two photons. Polarization analysis is performed with free space optics on the light coupled out of the fiber ends and registered with single-photon counting modules.

10.5.2 Operation of the Two-Trap Setup

To create entanglement between two atoms in independent and separated traps (and buildings), the fibers guiding the fluorescence photons are connected to the Bell state measurement (BSM) setup. In the experiments described here, the BSM was located in lab 1 (Figure 10.12). Trap 1 is connected via a 5 m long fiber, while trap 2 is connected via a 700 m fiber across the main campus of the LMU. The polarization encoding of the photon state makes compensation of the birefringence in the fibers necessary. The birefringence in the shorter fiber to trap 1 and in the fiber beam splitter, which are both inside Lab 1, changes only very slowly and is manually compensated every couple of days. In contrast, polarization drifts in the 700 m fiber are on a timescale below 1 hour and a automatized compensation of the polarization is employed [36] there.

Furthermore, the excitation process needs to be synchronized in a way that the emitted photons overlap temporally at the BSM setup. This means, that while both

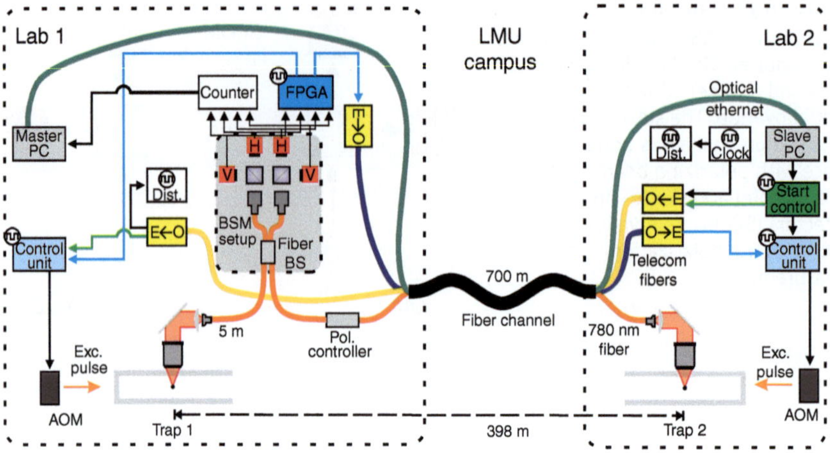

Figure 10.12 Overview of the experimental setup with two independent traps: The trap setups are located 400 m apart in different buildings of the LMU main campus (see Figure 10.1). They are connected with a 700 m fiber link cable, which contains fibers for guiding the fluorescence photons at 780 nm and for communication at telecom wavelengths. The Bell-state measurement setup for entanglement swapping is located in lab 1. It is connected with trap 1 via a 5 m fiber and to trap 2 via the fiber link (orange). For maintaining the polarization of the fluorescence photons, the long fiber is compensated for polarization drifts using an automatized polarization controller. Time critical components of both traps and the BSM setup are synchronized with a clock located in lab 2. The signal is sent via a communication fiber (yellow) to lab 1 and distributed via two clock distribution systems in each laboratory. The experiment is controlled by a master computer in lab 1 that is responsible for the loading process in lab 1 and monitoring the photon detection counts from the single-photon detectors. For the loading procedure in lab 2, the master computer sends commands to a slave computer in lab 2. When two atoms are trapped, the start control (green) in lab 2 is triggered and starts the synchronized excitation process (see Figure 10.13) by the control units (light blue). Its signal is transmitted via an asynchronous communication channel to lab 1. A so-called field-programmable-gate-array (FPGA) processor monitors the single-photon detection signals after the excitation pulses for a coincidence that heralds the creation of atom–atom entanglement. In case of a valid coincidence, the FPGA (blue) sends a signal to the control units of both labs, triggering the readout of the atomic states.

trap setups are able to work completely independently from each other, the systems need to communicate and to be very precisely synchronized on a sub-ns scale. For this the signal from a 100 MHz master clock, located in lab 2, is distributed to all synchronized components via an analog optical communication channel. This allows for an overall synchronization of the control units with a jitter below 150 ps rms [16].

Figure 10.12 shows an overview of the two-trap setup. Each trap is operated via a local computer that sets most experimental parameters and controls the atom loading procedure. In the two-trap experiment, the computer in lab 1 serves as master that controls the experiment. It sends commands via the fiber-based Ethernet connection to the computer in lab 2 serving as slave. The master monitors the trapping procedure and when two atoms are trapped, it sends a command to the slave that triggers a synchronized start control unit. Upon receiving the start signal, the control units take over. They execute the experimental sequence until one of the atoms is lost and the loading procedure starts again.

10.5.3 Creation and Verification of the Atom–Atom Entanglement

The first of three steps is trapping a single atom in each of the setups. Second, both trapped atoms are excited and each emits a photon entangled with the emitting atom. These photons are collected and guided to the Bell-state measurement setup, where the measurement on both photons swaps the entanglement onto the atoms. And third, this measurement outcome heralding the entanglement is sent back to the trap setup triggering the last step, the measurement of the atomic states.

Due to the low combined photon collection and detection probabilities of $\eta_1 \approx 1.7 \times 10^{-3}$ for trap 1 and $\eta_2 \approx 0.85 \times 10^{-3}$ for trap 2, the probability for a two-photon coincidence heralding atom–atom entanglement is only $Pr_{AA} = \frac{1}{2}\eta_1\eta_2 \approx 7 \times 10^{-7}$ [16]. Such a low success probability makes a high repetition rate necessary. Its inverse includes the time needed for the excitation scheme, starting with state preparation together with a waiting time needed to receive the heralding signal. Due to the asymmetric location of the BSM in lab 1 next to trap 1, this time is much larger for trap 2 than for trap 1. Altogether, this time is 10.36 μs. After 40 excitations, cooling for 350 μs is required to maintain a low atomic temperature resulting in an average excitation repetition rate of $52.2 \times 10^3 \frac{1}{s}$ (see Figure 10.13) [16]. Finally, the two-photon coincidence signal from the BSM heralds the presence of an entangled state and "readiness" of the system. Now entanglement is ready for further usage at the two nodes.

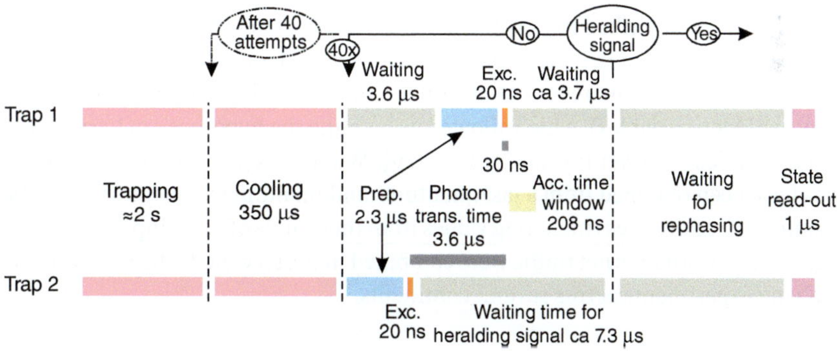

Figure 10.13 Timing scheme for creation of atom–atom entanglement: after trapping of an atom in each trap (red), the synchronized excitation process starts. Preparation (blue) and excitation (orange) processes in both traps are timed taking into account the photon transmission times in the optical fibers (gray) such that the two photons overlap in time at the fiber beam splitter and then are detected within the acceptance time window (yellow). In the case of a two-photon coincidence detection heralding entanglement, the atomic state readout process (magenta) is triggered, otherwise the preparation and excitation processes are repeated. After 40 unsuccessful attempts, the repetition is interrupted with an additional cooling period (red) for both atoms to assure a low atom temperature before the scheme continues until a valid coincidence occurs or one of the atoms is lost.

10.6 Employing Distributed Entanglement

The field of quantum communication envisages quantum networks, where entanglement is shared between the nodes of the network, thereby enabling the application of a number of different quantum protocols. Foremost, quantum teleportation allows one to transfer quantum information between the nodes making it the core tool for other protocols like storing quantum information in the nodes or connecting quantum computers to perform tasks like blind or distributed quantum computing. Here we briefly describe the first steps already achieved with the two-atom system.

10.6.1 Bell Test Simultaneously Closing Detection and Locality Loopholes

The correlation between measurement results from two distant, entangled systems cannot be described by classical statistics and thus violate a Bell inequality. This allows an experimental test ruling out any local-realistic description of nature. While such tests are conceptually simple, there are strict requirements concerning the detection efficiency of the involved measurements, as well as the enforcement of space-like separation between the measurement events.

The experiment described here has a number of advantages for performing a test of local hidden variable theories using the Bell inequality. As a heralding signal is delivered once entanglement is distributed, the remaining task can focus on the fast measurement on the two atoms. Provided the atoms are distant enough in relation to the required readout time, space-like separation can be achieved. For the distance of 398 m in our experiment, one has more than 1 μs to obtain a fresh random number for selecting the measurement basis, to set it and then to perform the read out. The experiments are described in [29, 37] and analyzed in detail in [16] and [38].

A clear violation of a Bell inequality closing essential loopholes was obtained with $S = 2.221 \pm 0.033$ (compared to the maximal value of 2 achievable with models based on local hidden variables), which allowed us to refute the hypothesis of local realism with a significance level $P < 2.57 \times 10^{-9}$ [29]. While here the measurement directions have been obtained from a fast quantum random number generator, a number of sources of randomness including values from thousands of users input in a mobile phone app or from internet traffic were combined in the so-called "Big Bell Test" [37] involving experiments across several continents.

10.6.2 Advanced Protocols Based on Distributed Entanglement

Moreover, the violation of a Bell inequality was seen to enable one to identify the quantumness of a system or to make a quantum communication protocol device independent. The violation essentially gives a measure on whether the given task could have been performed also with classical means only. One such task, for example is to certify that the generation of the entangled states between the two labs indeed is quantum, i.e. that it is indeed a quantum link enabling protocols impossible with any classical connection [39]. After improving the optics for higher collection efficiency [40] and increasing the coherence of the system by further cooling the violation of a Bell inequality could be increased. This finally made it possible to generate a secure secret bit string using entanglement-based quantum key distribution, where the security is no longer threatened by any malfunction introduced by the producer of the devices [41].

10.6.3 Toward Long-Distance Quantum Networks

Quantum repeaters will allow for scalable quantum networks, where photon-mediated entanglement is distributed among quantum memories at stationary nodes. The entanglement between stationary quantum memories and photonic channels is the essential resource but useful for practical applications only if shared over large distances. When using optical fiber connections for distributing the photons, wavelength conversion to telecom wavelengths is necessary to reduce transmission losses. We used polarization-preserving quantum frequency conversion to transform the wavelength of a photon and could thus increase the distance significantly [42]. In a first experiment we could increase the distance to observe atom–photon entanglement to 20 km. Thanks to the high external device conversion efficiency of 57%, an entanglement fidelity between the atom and the telecom photon of $\geq 78.5 \pm 0.9\%$ could be observed after transmission through 20 km of optical fiber, mainly limited by decoherence of the atomic state.

Combining two such setups finally enables to distribute entanglement over sizeable distances. Although the two atoms physically are still separated by the same 398 m, we could introduce two fiber spools with about 15 km of fiber, such that the effective distance between the two atoms was increased to 33 km. Clearly, the fidelity decreases with the distance and the resulting long waiting times for the heralding signal. But, given the much longer coherence times of about 50 μs, the fidelity of the state with respect to a maximally entangled state was still as high as 0.622 ± 0.015, which proves the distribution of entanglement even for the full distance [43].

There is still a long way to go in order to really provide a quantum network. Most important is the realization of quantum memories with very long coherence times, possibly supported by error correction. Another task is the development of local quantum logic operations between several quantum memories. This is necessary for locally connecting quantum memories entangled with other nodes, as well as for implementing entanglement purification, enabling to improve the quality of the transmission to finally bridge large distances in an efficient way.

Acknowledgments

This experiment is a long running enterprise with essential contributions from so many colleagues. Let us acknowledge Christian Kurtsiefer, Markus Weber, Jürgen

Volz, Florian Henkel, Julian Hofmann, Norbert Ortegel, Michael Krug, Daniel Burchardt, Kai Redeker, Robert Garthoff, Tim van Leent, Florian Fertig, Pooja Malik, and Yiru Zhou for their effort in the different stages of the experiment. The project was supported by DFG (Cluster of Excellence MCQST), by BMBF (quorope, qlink.X, QR.X).

References

1 Kimble, H.J. (2008). The quantum internet. *Nature* 453: 1023.

2 Briegel, H.-J., Dür, W., Cirac, J.I., and Zoller, P. (1998). Quantum repeaters: the role of imperfect local operations in quantum communication. *Physical Review Letters* 81 (26): 5932–5935.

3 Jozsa, R., Abrams, D.S., Dowling, J.P., and Williams, C.P. (2000). Quantum clock synchronization based on shared prior entanglement. *Physical Review Letters* 85 (9): 2010–2013.

4 Mayers, D. and Yao, A. (2004). Self testing quantum apparatus. *Quantum Information and Computation* 4 (4): 273–286.

5 Šupić, I. and Bowles, J. (2020). Self-testing of quantum systems: a review. *Quantum* 4: 337.

6 Weber, M. (2005). Quantum optical experiments towards atom-photon entanglement. PhD thesis. Ludwig-Maximilians-Universität München. http://xqp.physik.uni-muenchen.de/publications/theses_phd/phd_weber.html (accessed 5 December 2022).

7 Volz, J., Weber, M., Schlenk, D. et al. (2006). Observation of entanglement of a single photon with a trapped atom. *Physical Review Letters* 96 (3): 030404.

8 Volz, J. (2006). Atom-photon entanglement. PhD thesis. Ludwig-Maximilians-Universität München. http://xqp.physik.uni-muenchen.de/publications/theses_phd/phd_volz.html (accessed 5 December 2022).

9 Zukowski, M., Zeilinger, A., Horne, M.A., and Ekert, A.K. (1993). Event-ready-detectors" bell experiment via entanglement swapping. *Physical Review Letters* 71 (26): 4287–4290.

10 Hofmann, J., Krug, M., Ortegel, N. et al. (2012). Heralded entanglement between widely separated atoms. *Science* 337 (6090): 72.

11 Hofmann, J. (2014). Heralded atom-atom entanglement. PhD thesis. Ludwig-Maximilians-Universität München. http://xqp.physik.uni-muenchen.de/publications/theses_phd/phd_hofmann.html (accessed 5 December 2022).

12 Krug, M. (2018). Ionization based state read out of a single Rb-87 atom. PhD thesis. Ludwig-Maximilians-Universität München. http://xqp.physik.uni-muenchen.de/publications/theses_phd/phd_krug.html (accessed 5 December 2022).

13 Henkel, F., Krug, M., Hofmann, J. et al. (2010). Highly efficient state-selective submicrosecond photoionization detection of single atoms. *Physical Review Letters* 105: 253001.

14 Henkel, F. (2011). Photoionisation detection of single Rb-87 atoms using channel electron multipliers. PhD thesis. Ludwig-Maximilians-Universität München. http://xqp.physik.uni-muenchen.de/publications/theses_phd/phd_henkel.html (accessed 5 December 2022).

15 Ortegel, N. (2016). State readout of single Rubidium-87 atoms for a loophole-free test of Bell's inequality. PhD thesis. Ludwig-Maximilians-Universität München. http://xqp.physik.uni-muenchen.de/publications/theses_phd/phd_ortegel.html (accessed 5 December 2022).

16 Burchardt, D. (2018). A Rigorous Test of Bell's Inequality and Quantum Teleportation Employing Single Atoms. PhD thesis. Ludwig-Maximilians-Universität München. http://xqp.physik.uni-muenchen.de/publications/theses_phd/phd_burchardt.html (accessed 5 December 2022).

17 Steck, A.D. (2003). Rubidium 87 D Line Data. https://steck.us/alkalidata/rubidium87numbers.1.6.pdf (accessed 5 December 2022).

18 Chu, S., Bjorkholm, J.E., Ashkin, A., and Cable, A. (1986). Experimental observation of optically trapped atoms. *Physical Review Letters* 57: 314–317.

19 Chu, S., Hollberg, L., Bjorkholm, J.E. et al. (1985). Three-dimensional viscous confinement and cooling of atoms by resonance radiation pressure. *Physical Review Letters* 55: 48–51.

20 Hänsch, T.W. and Schawlow, A.L. (1975). Cooling of gases by laser radiation. *Optics Communications* 13 (1): 68–69.

21 Redeker, K.S.M. (2012). Detektion von Ionisationsfragmenten von Rb-87 Atomen. Diploma thesis. Ludwig-Maximilians-Universität München. http://xqp.physik.uni-muenchen.de/publications/theses_diplom/diplom_redeker.html (accessed 5 December 2022).

22 Raab, E.L., Prentiss, M., Cable, A. et al. (1987). Trapping of neutral sodium atoms with radiation pressure. *Physical Review Letters* 59: 2631–2634.

23 Dalibard, J. and Cohen-Tannoudji, C. (1989). Laser cooling below the Doppler limit by polarization gradients: simple theoretical models. *Journal of the Optical Society of America B* 6 (11): 2023–2045.

24 Koschmieder, P. (2019). Efficient control of magnetic fields in single atom experiments. Master's thesis. Ludwig-Maximilians-Universität München. https://xqp.physik.uni-muenchen.de/publications/files/theses_master/master_koschmieder.pdf (accessed 5 December 2022).

25 Grimm, R., Weidemüller, M., and Ovchinnikov, Y.B. (1999). Optical dipole traps for neutral atoms. *arXiv:physics/9902072*. https://arxiv.org/abs/physics/9902072.

26 Schlosser, N., Reymond, G., and Grangier, P. (2002). Collisional blockade in microscopic optical dipole traps. *Physical Review Letters* 89: 023005.

27 Redeker, K. (2020). Entanglement of single rubidium atoms: from a bell test towards applications. PhD thesis. Ludwig-Maximilians-Universität München. https://xqp.physik.lmu.de/publications/theses_phd/phd_redeker.html (accessed 5 December 2022).

28 Lee, S.A., Helmcke, J., Hall, J.L., and Stoicheff, B.P. (1978). Doppler-free two-photon transitions to Rydberg levels: convenient, useful, and precise reference wavelengths for dye lasers. *Optics Letters* 3 (4): 141–143.

29 Rosenfeld, W., Burchardt, D., Garthoff, R. et al. (2017). Event-ready bell test using entangled atoms simultaneously closing detection and locality loopholes. *Physical Review Letters* 119: 010402.

30 Rosenfeld, W. (2008). Experiments with an entangled system of a single atom and a single photon. PhD thesis. Ludwig-Maximilians-Universität München. http://xqp.physik.uni-muenchen.de/publications/theses_phd/phd_rosenfeld.html (accessed 5 December 2022).

31 Dorn, R., Quabis, S., and Leuchs, G. (2003). The focus of light – linear polarization breaks the rotational symmetry of the focal spot. *Journal of Modern Optics* 50 (12): 1917–1926.

32 Thébault, E., Finlay, C.C., Beggan, C.D. et al. (2015). International geomagnetic reference field: the 12th generation. *Earth, Planets and Space* 67 (1): 79.

33 Pan, J.-W., Bouwmeester, D., Weinfurter, H., and Zeilinger, A. (1998). Experimental entanglement swapping: entangling photons that never interacted. *Physical Review Letters* 80 (18): 3891–3894.

34 Weinfurter, H. (1994). Experimental bell-state analysis. *Europhysics Letters* 25: 559.

35 Pan, J.-W., Chen, Z.-B., Lu, C.-Y. et al. (2012). Multiphoton entanglement and interferometry. *Reviews of Modern Physics* 84: 777–838.

36 Rosenfeld, W., Hocke, F., Henkel, F. et al. (2008). Towards long-distance atom-photon entanglement. *Physical Review Letters* 101 (26): 260403.

37 Abellán, C. and The BIG Bell Test Collaboration (2018). Challenging local realism with human choices. *Nature* 557 (7704): 212–216.

38 Redeker, K. (2020). Entanglement of single rubidium atoms: from a bell test towards applications. PhD thesis. Ludwig-Maximilians-Universität München. http://xqp.physik.uni-muenchen.de/publications/theses_phd/phd_redeker.html (accessed 5 December 2022).

39 Bancal, J.-D., Redeker, K., Sekatski, P. et al. (2021). Self-testing with finite statistics enabling the certification of a quantum network link. *Quantum* 5: 401.

40 Garthoff, R. (2021). Efficient single photon collection for single atom quantum nodes. PhD thesis. Ludwig-Maximilians-Universität München. http://xqp.physik.uni-muenchen.de/publications/theses_phd/phd_garthoff.html (accessed 5 December 2022).

41 Zhang, W., van Leent, T., Redeker, K. et al. (2022). A device-independent quantum key distribution system for distant users. *Nature* 607: 687.

42 van Leent, T., Bock, M., Garthoff, R. et al. (2020). Long-distance distribution of atom-photon entanglement at telecom wavelength. *Physical Review Letters* 124: 010510.

43 van Leent, T., Bock, M., Fertig, F. et al. (2022). Entangling single atoms over 33 km telecom fibre. Nature 607, 69–73. https://doi.org/10.1038/s41586-022-04764-4.

11

Collective Light Emission of Ion Crystals in Correlated Dicke States

Ferdinand Schmidt-Kaler[1] and Joachim von Zanthier[2]

[1] *QUANTUM, Institut für Physik, Johannes Gutenberg-Universität Mainz, Staudingerweg 7, 55128 Mainz, Germany*
[2] *AG Quantum Optics and Quantum Information, Department Physik, Friedrich-Alexander-Universität Erlangen-Nürnberg, Staudtstr. 1, 91058 Erlangen, Germany*

11.1 Introduction

Even if it might appear on first sight that the spontaneously emitted radiation by an atom is an immutable property, it can be modified. To this end, one can alter the boundary conditions of the electromagnetic field modes of the vacuum, as exploited in cavity-quantum-electrodynamics (cavity-QED). Pioneering work has been done using superconducting mirrors and microwave transitions in Rydberg atoms [1, 2] or employing ultrahigh-reflective mirrors with dielectric coatings in combination with cold atoms [3, 4]. More recently, the modification of the boundary conditions of the electromagnetic field within nano-optical devices has been in the focus [5, 6], with potential applications in quantum communication [7].

A completely different pathway to change the spatiotemporal characteristics of spontaneous decay exploits correlations among atoms of an ensemble in free space. Several methods have been explored to generate such interatomic correlations. On the one hand, one may use direct interactions, e.g., the exchange of virtual photons caused by the dipole–dipole interaction between the atoms. This occurs at interatomic separations much smaller than the transition wavelength, in which case modified energy levels and collective decay rates are produced. The latter have also been observed experimentally [8–16]. Recently, the cooperative response resulting from such an interaction has been measured in the directional reflection of light by a two-dimensional array of atoms trapped in an optical lattice, in combination with subradiance [17]. For larger interatomic separations, atom–atom correlations can be induced via the dynamical exchange of photons, mediated, e.g., by the coupling to nano-optical devices [6, 18] such as photonic wave-guides [19–24]. A further scheme for producing correlations among atoms makes use of long-range interactions, such as the Rydberg dipolar blockade mechanism or the Coulomb

Photonic Quantum Technologies: Science and Applications, First Edition. Edited by Mohamed Benyoucef.

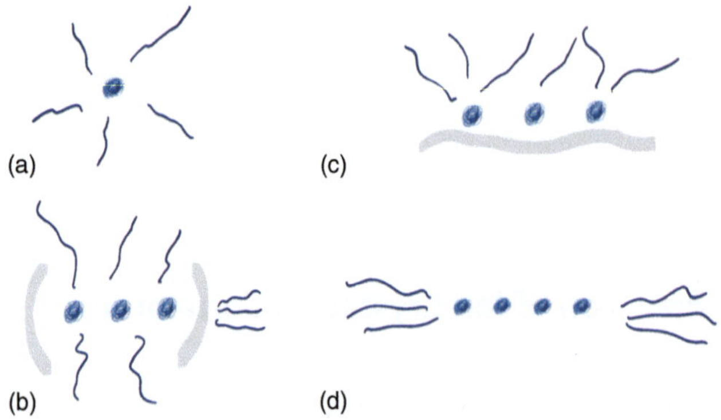

Figure 11.1 Single atom emission in free space (a), modified via boundary conditions in a cavity (b), due to a near field nano-optical structure (c), or as a result of atom-atom correlations (d).

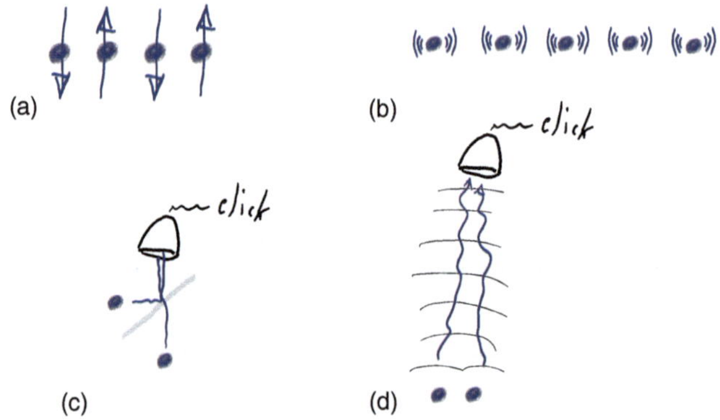

Figure 11.2 Entanglement can be generated by long-range Rydberg interactions (a) or via Coulomb-mediated spin interactions (b). If the emission of a photon from two identical atoms is superimposed onto one optical mode behind a beam splitter, a photon-click detector cannot extract any *welcher-Weg* information resulting in entanglement of the atoms (c); an alternative is to measure the photon in the far field, in which case also no *welcher-Weg* information is obtained (d). In all cases (a)–(d), the spontaneous emission into free space may be altered as a result of the collective emission of light by the entangled emitters. We concentrate in this paper on the projectively generated correlations of emitters as in (d), constituted by a photon-click detector in the far field.

interaction in case of trapped ions [25–27]. Coulomb crystals in ion traps are an established platform to generate, manipulate, and analyze such correlated states, either for fundamental quantum optical experiments or in the context of quantum computation.

But quantum mechanics provides us with yet an alternative method for generating correlations among emitters in free space. The method relies on the detection of photons scattered from indistinguishable emitters by carefully excluding *Welcher-Weg* information so that from a detection event one is unable to identify the photon source [28–43]. Two approaches have been identified so far that implement this scheme. Either the photons scattered from different atoms are superposed in a common mode by mixing them in a beam splitter [30, 33, 35–37, 41, 42] or, alternatively, the photons are recorded by detectors in the far field of the emitters [28, 29, 31, 32, 34, 38–40, 43]. Note that in both schemes, initially uncorrelated atoms are transferred into correlated and potentially entangled states via projective measurements of photons, even in the absence of any direct interaction between the emitters. Note further that the method is particularly useful for the experimental platform of trapped ions [33, 36, 42–44] because the mutual Coulomb repulsion prohibits separations below the optical wavelength, thus excluding efficient interactions based on the exchange of virtual photons [45]. Yet besides trapped ions, the method has been implemented also in other important platforms, such as neutral atoms [35, 40], NV-centers [37], or quantum dot emitters [41, 46].

The scheme has been also used to propose theoretically various complex multi-partite entangled states, including GHZ- and Dicke states [28, 29, 32, 34, 47]. Such states are predicted to feature interesting collective emission characteristics including superradiant and subradiant spontaneous decay [48–50]. However, in almost all experiments performed so far, single spatial modes have been picked out for measuring the scattered photons. This measuring scheme inhibits the observation of the genuine collective spatial emission patterns produced by the correlated atomic ensemble. In other words, in order to observe the full spatiotemporal emission characteristics of entangled atomic arrays, including super- and subradiant photon emission, a far-field free-space detector setting is indispensable.

11.2 Structure of the Article

We start in Section 11.2 by a short recapitulation of the fundamentals of trapping and laser cooling of ions in a Paul trap. In Section 11.3, we investigate the emission of light of uncorrelated continuously laser excited ion crystals, in order to highlight the difference to the collective light emission of ions in correlated states investigated in the remainder of the chapter. In Section 11.4, we present the theory for generating entangled Dicke states among trapped ions via projective measurements of photons scattered by the particles. Trapped ions in free space prepared in entangled Dicke states of different excitations are the starting point for the collective emission of spontaneous radiation, investigated in Section 11.5. In Section 11.6, we explore how the collective light emission of a ion crystal in an entangled Dicke state can be observed experimentally. To this aim we introduce the tool of higher-order correlation functions giving access to the entire spectrum of the spatiotemporal collective spontaneous emission characteristics of correlated ion crystals. Finally, in Section 11.7, we present a measurement of the collective light emission by a two ion crystal, the most elementary building block of an atomic ensemble. Finally, we conclude with potential future extensions and applications in Section 11.8.

11.3 Fundamentals of Trapping and Laser Cooling of Ions in a Paul Trap

The trapping and cooling of ion crystals have been developed since many decades to realize ultrahigh-precision frequency standards, optical clocks [51], quantum computers, and quantum simulators [52]. Here, we are interested in trapped and laser-cooled ions for observing the collective emission of light. For these investigations, several technological advantages of the trapped ion platform can be readily used:

- The combination of DC and AC electric fields in the trap leads to a (dynamical) harmonic trapping potential with secular frequencies that can be tuned between a few 100 kHz and a few MHz. Trapping of the particles is determined only by the charge and the mass of the ions, independent of the electronic internal degrees of freedom of the ions. In turn, optical transitions are not affected by the trap and long coherence times can be reached for transitions between long-lived atomic levels [53–55].
- A number of N ions assemble in the three-dimensional harmonic trap potential to either form a linear crystal or crystals in planar or even three-dimensional shape [56–59] (see Figure 11.4 for some examples). Here, the dimensionality and distances between the ions are set by the number N in combination with the applied trap control voltages. We find typically inter-ion distances between 20 μm and 2 μm. Using modern micro-segmented ion traps [60–63], the potentials can be even varied to form double wells [64, 65] or multiple wells for grouping of ions or to form any other non-harmonic shape.
- Ions are trapped under ultrahigh-vacuum conditions and in deep electrical potentials, thus continuous and stable trapping of ion crystals is possible for days and even weeks. This allows for excessively long data acquisition times.
- The matter wave function of a single trapped and laser-cooled ion is localized to a fraction of the optical wavelength, for ground state cooling to a few 10 nm, for higher temperatures, e.g., after Doppler cooling, to $\sim$ 100 nm. If the optical wavelength is exceeding the size of the matter wave packet, the ion will interact with light both, in absorption and in emission, in the Lamb Dicke regime. In this case, the photon recoil is transferred to the trap as a whole with high probability and is not inducing additional motion to the trapped ion. Using a narrow optical transition of the trapped ion, the laser excitation frequency can be chosen such that the ion absorbs light without any change in motional quantum number [66]. Such sort of control and decoupling of motional degrees of freedom from internal excitation are instrumental for generating indistinguishable photons. It is also the basis of a highly efficient collection of photons.
- Several optical cooling methods have been developed for larger ion crystals to generate low motional excitation close to the quantum ground state $|n{=}0\rangle$ [58, 67, 68]. Under this condition, an ion crystal, which is acting as an array of single photon emitters, becomes a near-to-ideal model system for studying the collective emission of light.

After discussing the general advantages of using trapped ions for quantum optical light scattering experiment, we focus in the following on the trapping of $^{40}Ca^+$ ions. We employ a segmented Paul trap [70], allowing to position each ion in a linear crystal at will. In a typical setting, i.e., with trap frequencies of $\omega_{r_1,r_2,z}/(2\pi)$=(1.9, 2.6, 1.0) MHz, the ions form linear crystals, which align along the weakest trap axis $\mathbf{e}_z$. The electric dipole transition $4^2S_{1/2} \rightarrow 4^2P_{1/2}$ of $^{40}Ca^+$ near 397 nm is used for Doppler cooling and light scattering, where the $4^2P_{1/2}$ state decays with a probability of 7% to the metastable $3^2D_{3/2}$ level [71]. In order to maintain continuous Doppler cooling, we employ a laser near 866 nm for repumping the ion out of the $3^2D_{3/2}$ level (see Figure 11.3(a)). The radial modes ω_{r_1,r_2} are aligned along the $\mathbf{e}_{\pm x+y}$ direction, respectively, whereas the cooling and repumping laser illuminate the ion crystals along the (x,y,z)=$(\pm 1,0,-1)/\sqrt{2}$ direction, respectively, so that the **k**-vectors of the laser beams have a projection on all vibrational axes of the ion crystal (see Figure 11.3(b)). A quantization axis is defined by a magnetic field of $B = 0.62$ mT, implemented via three pairs of Helmholtz coils, oriented along $\mathbf{e}_y$, i.e., vertical to the detection plane. The laser beam near 397 nm, with a waist of about 600 μm at the ions' positions, is linearly polarized along this axis and thus excites the $\Delta m = 0$ transitions (see Figure 11.3(a)). The light scattered by the ions is collected by an f/1.6 objective L1 (focal length 67 mm) at a working distance of 48.5 mm and focused at a distance of about 770 mm, after being sent through a polarization beam splitter

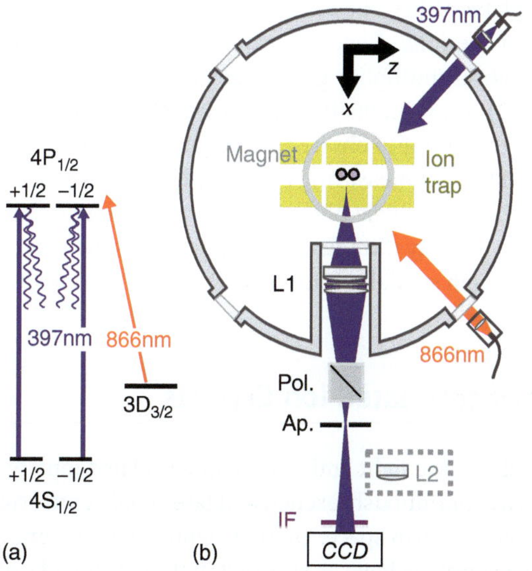

Figure 11.3 (a) Level scheme and relevant transitions of the $^{40}Ca^+$ ion including the metastable $3^2D_{3/2}$ state. (b) Sketch of the experimental setup used to study the emission of light by ion crystals; the ions are held in a segmented micro trap (at center of the vaccum vessel), forming linear crystals along the z-axis and are illuminated by laser light near 397 nm and 866 nm. The light scattered by the ion crystal is collected by lens L1 and observed in the far field (or imaged by use of lens L2) via a charge-coupled-device (CCD) camera, for details see text. Source: Adapted from Wolf et al. [69]/with permission of American Physical Society.

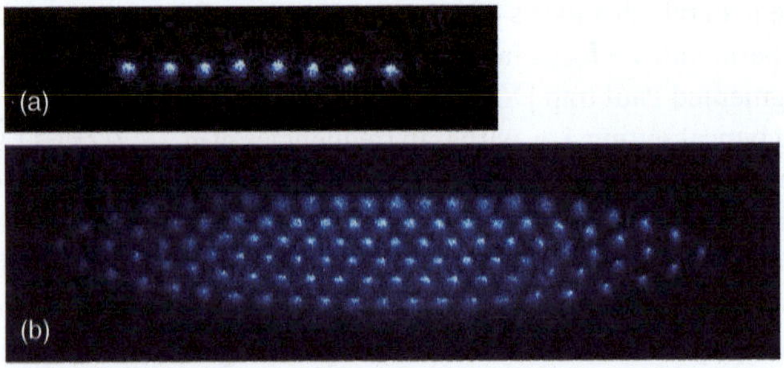

Figure 11.4 While the fluorescence light emitted on the $4^2P_{1/2} \rightarrow 4^2S_{1/2}$ transiton near 397 nm of the trapped $^{40}Ca^+$ crystals is typically detected in the far field (see Figure 11.3), we can also take images of the ion crystal. To this aim, lens L2 in Figure 11.3 is placed into the scattered light beam to focus the ion crystal on the EMCCD camera; (a) linear crystal with N= 8, (b) planar ion crystal with N=100.

(Pol.), also oriented along $\mathbf{e}_y$, i.e., the same axis as the cooling laser polarization to ensure indistinguishably of the scattered photons (see Figure 11.3b). An aperture (Ap.) (diameter ~ 400 μm) is placed at the back focal plane of the objective to suppress unwanted stray light in combination with an infrared filter (IF, center wavelenght $\lambda = 394 \pm 10$ nm).

The scattered light is finally recorded by diverse sort of detectors, positioned ~ 100 mm behind the back focal plane of the objective to observe the light in the far field, i.e., the Fourier plane of the ions. Depending on the specific measurement task, we used either an electron multiplier gain intensifier enhanced CCD camera (EMCCD, Andor iXon 860) or an intensified CCD camera (ICCD, Andor iStar 334T), which allows for fast triggering to observe non-equilibrium processes, respectively, as studied in Section 11.3. For measuring photon–photon correlations, investigated in Section 11.7, we employed single photon counting modules (Count blue, Laser Components) and more recently ultra-fast single-photon resolving MCP (microchannel plate) cameras with high spatial resolution (LINCam, Photonscore), see Figure 11.15.

11.4 Light Emission of Uncorrelated Ion Crystals

We start our investigations by studying the light emission of uncorrelated ion crystals, i.e., the scattering of photons by continuously excited and laser-cooled uncorrelated trapped ions, localized at fixed points in space. On first sight, one might argue that in this configuration the ions simply behave as slits of a grating or double slit as originally devised by Young [72]. However, we will see that replacing the slits by atoms involves nonlinear effects and thus leads to a more intricate scattering behavior.

In a seminal experiment by D. Wineland and coworkers, Young-type interferences have been observed with two mercury atoms trapped in a linear ion trap, weakly excited by near-resonant laser light [73], as shown in Figure 11.5 (see also [29, 74–78]).

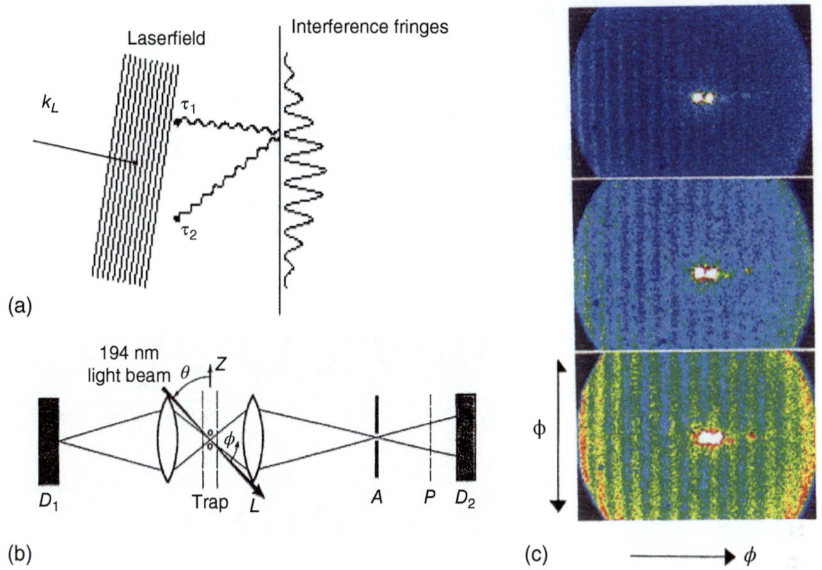

Figure 11.5 (a) Scheme for weak coherent driving of two ions trapped in a linear Paul trap; (b) setup of the experiment. Source: Eichmann et al. [73]; (c) interference pattern obtained for three different separations of the two $^{198}Hg^+$ ions, namely (from top to bottom) 5.4 μm, 4.3 μm, and 3.7 μm; the two white spots are caused by stray reflections of the laser beam. Source: Adapted from Eichmann et al. [73]/with permission of American Physical Society.

We extended this work recently by trapping longer linear crystals and obtained similar interference patterns for two, three, and four ions, observed on an EMCCD (see Figure 11.6). The fits to the interference patterns in Figure 11.6 are obtained from the source distribution via Fourier transformation, taking into account the resolution of the imaging device. From the fit parameters, we are able to determine the distance d between the ions, the width w of the point spread function (PSF), and the visibility $\mathcal{V}$ of the interference fringes. In this way, we obtained from Figure 11.6(a) a width of the PSF of $w = 3.6$ μm and an inter-ion distance in the two-ion crystal of $d = 6.4$ μm. Taking experimental errors into account, this is in good agreement with the independently deduced two-ion distance $d_{theo.} = 5.8$ μm, based on a spectroscopic determination of the COM-mode frequency of the crystal [79].

Modern trap technology [62, 80], where the DC trap potential is shaped by multiple control segments, allows one to modify the trap potential along $\mathbf{e}_z$ and thus the inter-ion distances. This becomes particularly relevant for crystals with ≥ 4 ions. If a crystal with four ions is kept in a harmonic trap, the equilibrium positions of the ions are non-equidistant [79], e.g., for trap frequencies $\omega_{r_1,r_2,z}/(2\pi)$=(1.978, 2.180, 0.429) MHz, the distance between the innermost ions is 7.2 μm and between the outer and the inner ions 7.6 μm, respectively. This results in an interference fringe signal with two spatial Fourier frequencies as shown in Figure 11.6(c). By adjusting the trap control electrode voltages, we can generate a non-harmonic potential [65], which features equal ion separations of 9.1 μm (see Figure 11.6(d)).

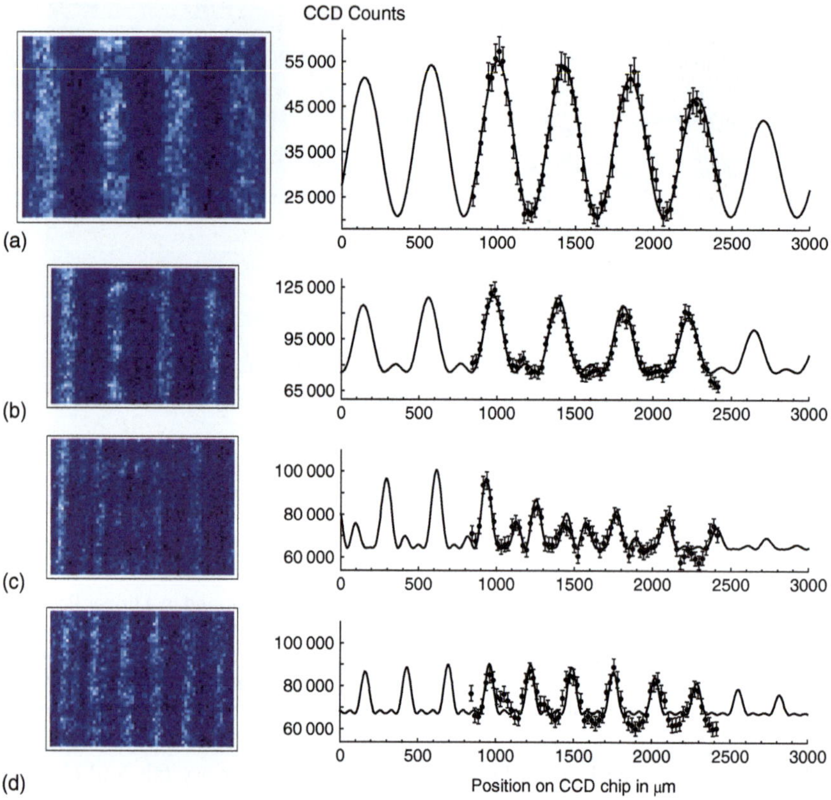

Figure 11.6 Images of the EMCCD camera (left) and interference fringe patterns (right) for (a) two (b) three (c) four ions in a harmonic trap potential. In (d) the data stem from a crystal with four equidistant ions. The EMCCD images have been rotated, distortion-corrected, and the background has been subtracted. Note that the data of the EMCCD camera are integrated over an exposure time of 60 s and include the internal avalanche gain. The fringe patterns (right) are obtained from the corrected EMCCD images by integration over the vertical axis. Errors on each data point correspond to photon shot noise, dark noise, and read-out noise. From a fit of the experimental curves, we obtain a visibility $\mathcal{V}$ of the fringe patterns of 45.2(6)%, 22.7(6)%, 22(1)%, and 15(1)% for the two-, three-, four-ion crystal, and the equidistant four-ion array, respectively; all errors represent the root mean square deviation of each fit. Source: Adapted from Wolf et al. [69]/with permission of American Physical Society.

The corresponding fringe pattern matches indeed the intensity distribution of a coherently illuminated 4-slit grating.

However, in comparison to classical slits, laser-driven trapped ions involve the internal atomic dynamics resulting in the emission of more complex light fields, ranging from fully coherent to partially coherent and fully incoherent light. This transition arises from the different process of photon scattering by ions with respect to slits: According to the quantum theory of light [81–83], in the case of atoms or ions, the scattering event involves the destruction of an incoming photon and the creation of an outgoing photon. For low intensities, elastic coherent scattering dominates this process, so that the outgoing photon has the same frequency and a fixed phase relationship with respect to the incoming one [9, 84]. As a consequence, the scattered light fields of *all* ions have a fixed phase relation, resulting in coherent light emission and a stationary interference pattern in the far field.

Yet, when increasing the intensity of the driving laser, the atomic emitters undergo an internal dynamics, which alters the frequency and phase of the scattered photons with respect to the incoming photons. Such inelastic scattering processes lead to a reduced mutual coherence of the light fields of the ions and thus to the emission of merely partially coherent light, resulting in a decrease of the visibility of the interference patterns. In the case of very intense laser driving, the atoms emit fully incoherent fluorescence light [85]; in this case, the visibility of the fringe pattern vanishes.

In the paraxial approximation, and for identical polarization of excitation and detection, the intensity produced by a two-ion crystal is given by [86]

$$\begin{aligned} I(\mathbf{r},t) =& I_1(\mathbf{r},t) + I_2(\mathbf{r},t) \\ &+ 2\sqrt{I_1(\mathbf{r},t)}\sqrt{I_2(\mathbf{r},t)}\mathrm{Re}\left\{\gamma(\mathbf{r}_1,\mathbf{r}_2,\tau)\right\}, \end{aligned} \tag{11.1}$$

where $I_1(\mathbf{r},t)$ $(I_2(\mathbf{r},t))$ is the intensity at $\mathbf{r}$ if ion 2 (ion 1) is absent, and $\mathrm{Re}\,\{.\}$ denotes the real part. In Eq. (11.1), $\gamma(\mathbf{r}_1,\mathbf{r}_2,\tau) = \langle E_1(\mathbf{r}_1,t-\tau)E_2^*(\mathbf{r}_2,t)\rangle / \sqrt{\langle|E_1(\mathbf{r}_1)|^2\rangle\langle|E_2(\mathbf{r}_2)|^2\rangle}$ corresponds to the complex degree of coherence, which describes the mutual coherence of the two light fields $E_1(\mathbf{r}_1,t)$ and $E_2(\mathbf{r}_2,t)$, generated by ion 1 at $\mathbf{r}_1$ and ion 2 at $\mathbf{r}_2$, respectively. Assuming identical excitation strength $I_1(\mathbf{r},t) = I_2(\mathbf{r},t) \equiv I_0$ of the two ions, the visibility of the interference fringes becomes equal to the modulus of the complex degree of coherence and the fringe modulation is determined by the phase φ, where $\varphi = k\,c\,\tau = k\,(|\mathbf{r}-\mathbf{r}_1| - |\mathbf{r}-\mathbf{r}_2|)$ is the relative phase accumulated by the two light fields at $\mathbf{r}$.

In a three-level model and with the ions at fixed positions, the intensity distribution is [69]

$$I(\mathbf{r}) = 2\,I_0\,(1 + |\rho_{sp}|^2/\rho_{pp}\cos\varphi), \tag{11.2}$$

where ρ_{sp} denotes the single atom coherence between states $\mathrm{s} = \mathrm{S}_{1/2}$ and $\mathrm{p} = \mathrm{P}_{1/2}$, and ρ_{pp} is the population of the excited state decaying either to level s or level $\mathrm{d} = \mathrm{D}_{3/2}$. According to Eq. (11.2), the visibility of the interference pattern is given by

$$\mathcal{V} = |\gamma| = |\rho_{sp}|^2/\rho_{pp}. \tag{11.3}$$

A reduction of $\mathcal{V}$ is thus predicted for growing population ρ_{pp} and reduced coherence ρ_{sp}. If we model the ions as two-level atoms coherently excited on the transition $\mathrm{S}_{1/2} \rightarrow \mathrm{P}_{1/2}$ of $^{40}\mathrm{Ca}^+$ near 397 nm (for which Eqs. (11.2) and (11.3) equally hold), this occurs for increased laser saturation s_{397}. However, the two-level model does not take into account the modification of ρ_{sp} and ρ_{pp} due to the additional decay channel toward the $^{40}\mathrm{Ca}^+$ $\mathrm{D}_{3/2}$-state. In this case, ρ_{sp} and ρ_{pp}, and thus Eq. (11.3), become more involved functions of the laser parameters.

The measured visibility $\mathcal{V}$ produced by two-ion crystals as a function of s_{397} is shown in Figure 11.7. A reduction of $\mathcal{V}$, corresponding to the emission of *partially coherent light*, is clearly visible when increasing s_{397}, what agrees well with the two-level model. When the saturation of the repumping laser is increased by a

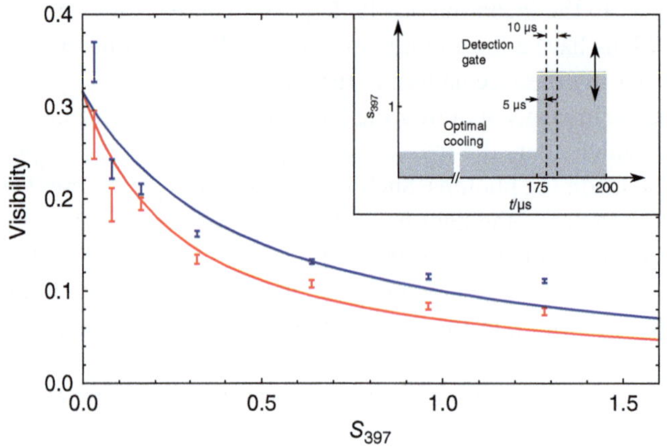

Figure 11.7 Interference fringe visibility at the crossover of elastic to inelastic scattering for a two-ion crystal as a function of laser saturation s_{397}. The repumping laser saturation corresponds to s_{866}= 0.032 (red) and s_{866}= 0.15 (blue). (Adapted from Wolf et al. [69])

factor of ∼ 4 we observe, however, again an improved visibility. This behavior is well described by the three-level model fit curves in Figure 11.7 [69].

Apart from these fundamental aspects of light scattering, also other practical implications have to be taken into account: Since the same laser is used for fluorescence emission by the ions as well as for cooling of the ions, the intensity of the driving laser affects the temperature of the ions, i.e., their external degrees of freedom. The ion temperature plays an important role for the fringe visibility as it determines the localization of the scatterers. We investigated the fringe visibility of the interference patterns produced by the scattered light off up to four trapped ions employing a gated detection method to clearly separate the effect of inelastic scattering from that of reduced atom cooling and localization [69]. To this aim, we introduced a gated cooling probe detection (GCPD) scheme for the used intensifier enhanced CCD camera. The GCPD scheme works as follows (see Figure 11.7): The ion crystals are initialized during 175 µs via Doppler cooling under optimum conditions for the saturation s_{397} and s_{866} of the cooling and repumping lasers at 397 nm and 866 nm, i.e., well below the respective saturation intensities, and with a cooling and repumping laser detuning of $\Delta_{397} = -10$ MHz and $\Delta_{866} = +60$ MHz, respectively. Thereafter the saturation of the cooling laser s_{397} is switched to a different value using an acousto-optical modulator. After a delay of 5 µs to allow for proper switching of the laser, the CCD is gated for 10 µs to observe the scattered light at 397 nm for the selected cooling laser saturation. As the motional states of the ion crystals evolve over much longer timescales [69], they are unable to adapt to the modified cooling laser saturation within this short detection time. In this way, the mutual coherence of the scattered light fields is determined uniquely by the internal degrees of freedom of the ions. Employing the GCPD scheme, we could thus determine the dependency of the visibility of the interference pattern as a function of the laser saturation as in Figure 11.7 without being affected by the ion temperature.

11.5 Theory of Correlated Dicke States Among Trapped Ions via Projective Measurements of Scattered Photons

In the previous Section 11.3 we discussed the appearance of interference patterns in the light scattered by weakly laser-excited ions, i.e. for the case of coherent light emission, where the dipole moments of all atoms are synchronized and oscillating in phase [69, 73, 74]. We learned that this configuration is close to the one where transversally coherent light passes a multi-slit grating. For this reason, we coin it the regime of *classical coherence* [87]. However, we also learned that in this regime the visibly of the interference pattern disappears if the intensity of the driving laser field is increased. This is due to the fact that spontaneous emission, i.e., inelastic and incoherent processes where the phase of the photons emitted by each atom fluctuates from emission process to emission process, starts to dominate the scattering. The question arises whether it is nevertheless possible to prepare the ion array in such a way that the incoherent light emitted by the array via spontaneous decay forms a stationary interference pattern. The theory addressing this question was presented almost 70 years ago by Robert Dicke in a seminal paper entitled *Coherence in Spontaneous Radiation Processes*, introducing the phenomenon of collective light emission dubbed *super- and subradiance* [88].

As it turns out, a prerequisite for the observation of super- and subradiance is that the atoms occupy so-called Dicke states [88]. These are highly entangled states, which are difficult to produce. But how could they be generated? As discussed in Section 11.1, there are principally two ways: On the one hand, one can exploit the long-range Coulomb interaction in combination with a series of appropriate laser pulses [26, 27], a well-established method to generate entangled states, extensively used in the context of quantum computation with trapped ions. Another option exploits the measurement of photons such that the detection is unable to identify the individual photon source [28–44], i.e., either by mixing the photons in a beam splitter [30, 33, 35–37, 41, 42] or recording the photons in the far field [28, 29, 31, 32, 34, 38–40, 43]. In the following, we focus on the latter scheme (see Figure 11.2(d)).

As it turns out, the latter technique allows for generating any symmetric Dicke state among N 2-level atoms with upper level $|e\rangle$ and lower level $|g\rangle$. In the case of Ca^+, we identify the two levels $|g\rangle = S_{1/2}$ and $|e\rangle = P_{1/2}$. Note that for N spin-$\frac{1}{2}$ systems, the Dicke states are denoted by $|S, M\rangle$ [88]. They are defined as the simultaneous eigenstates of both, the square of the total spin operator $\boldsymbol{s}^2$ and its z-component s^z, where $S(S+1)\hbar^2$ and $M\hbar$ are the corresponding eigenvalues, with $S = N/2, N/2-1, N/2-2, \ldots, > 0$ the cooperation number and M the inversion number $-S \leq M \leq +S$ [88]. The $2N+1$ states with the highest value of the cooperation number $S = N/2$ form a special subset of all 2^N Dicke states, called symmetric Dicke states. These states are the only ones that are totally symmetric under permutation of the particles. They can be written in the compact form

$$|\frac{N}{2}, M\rangle = \begin{pmatrix} N \\ \frac{N}{2}+M \end{pmatrix}^{-\frac{1}{2}} \sum_k P_k(|e_1, e_2, \ldots, e_{\frac{N}{2}+M}, g_1, g_2, \ldots, g_{\frac{N}{2}-M}\rangle), \tag{11.4}$$

where $\{P_k\}$ denotes the complete set of all possible distinct permutations of the atoms.

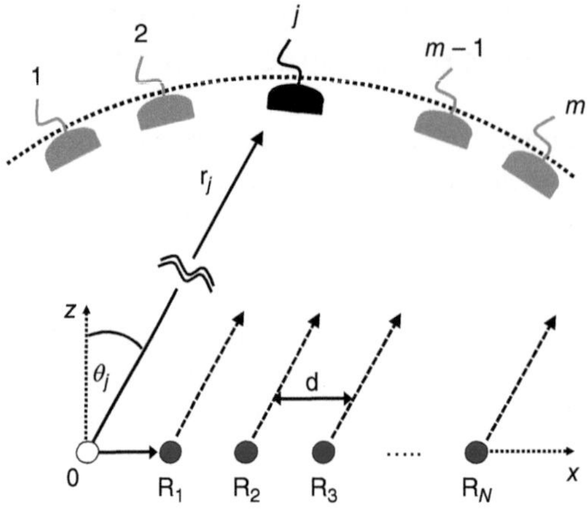

Figure 11.8 Considered setup: N identical 2-level ions are aligned in a row with equal spacing $d \gg \lambda$ at positions $\mathbf{R}_l$, $l = 1, \ldots, N$; the light scattered by the ions is measured by m detectors at positions $\mathbf{r}_j$, $j = 1, \ldots, m$, in the far field; the jth detector is placed at $\mathbf{r}_j$, where it sees all ions under an angle θ_j with respect to a line perpendicular to the symmetry axis of the ion crystal. Source: Wiegner et al. [39]/with permission of American Physical Society.

In order to produce any symmetric Dicke state in the ion array via projective measurements of spontaneously emitted photons, we assume all ions to be initially excited into the upper state $|e\rangle$ by a laser π-pulse. To simplify the calculations, but without loss of generality, we suppose the N ions to be regularly arranged along one axis with equal spacing d (see Figure 11.8). A number of $m \leq N$ detectors are placed at distinct positions $\mathbf{r}_n$ ($n = 1, \ldots, m$) in the far field of the ions. Again without loss of generality, we assume that the emitters and the detectors are in one plane and that the atomic dipole moments of all ions of the transition $|e\rangle \to |g\rangle$ are oriented perpendicular to this plane. In addition, we assume that via post-selection each of the detectors registers one and only one photon. In the far field of the ions, the detectors are unable to distinguish which particular ion has emitted a registered photon. Therefore, after the detection of the first photon by one of the detectors, all atoms will form a correlated symmetric Dicke state with one atom in the ground state [29–32].

The production of entanglement among the ions, even if separated by large distances $d \gg \lambda$, is a consequence of two ingredients: the impossibility of the detectors to determine which ion emitted a particular photon together with the projection postulate which states that the state of the ions after the detection of a photon is transferred into a state compatible with the outcome of the measurement [29]. In the following, we introduce a third ingredient to this scheme. It exploits the geometrical phase differences of all m-photon quantum paths resulting from the $\binom{N}{m} \cdot m!$ possibilities that m of the N ions emit a photon, which are subsequently registered by m detectors. As will be shown below, these geometrical phase differences allow to prepare symmetric Dicke states of arbitrary inversion M.

In order to study this in more detail, let us introduce the coordinate system shown in Figure 11.8. The position of the lth ion, $l = 1, \ldots, N$, is given by $\mathbf{R}_l = ld\,\mathbf{u}$, where $\mathbf{u}$ is a unit vector along the array axis. Denoting the unit vector along the direction of the jth detector by $\mathbf{e}_j := \mathbf{r}_j/r_j$, we introduce the angle θ_j shown in Figure 11.8 so that

$\mathbf{R}_l \cdot \mathbf{e}_j = l\, d\, \sin\theta_j$. The phase difference δ_j between two photons of wavenumber k, emitted by adjacent ions and both detected at $\mathbf{r}_j$, can then be written as

$$\delta_j := k\,(\mathbf{R}_{l+1} \cdot \mathbf{e}_j - \mathbf{R}_l \cdot \mathbf{e}_j) = k\, d\, \sin\theta_j. \tag{11.5}$$

Assuming initially all N ions in the upper state $|e\rangle$, the initial state $|\Psi^i_N\rangle$ of the atoms is given by

$$|\Psi^i_N\rangle = |e, e, \ \dots\ , e\rangle_N, \tag{11.6}$$

where the dimension of the state is indicated by the subscript N. The m photons, subsequently emitted by the N ions, are detected by m detectors at $\mathbf{r}_j$, $n = 1, \ \dots, \ m$. Hereby, each detection event has to take into account that one (unknown) ion out of N possible scatterers has emitted the photon. Using the coordinate system of Figure 11.8, the (unnormalized) operator describing the detection event of the jth photon at $\mathbf{r}_j$ can thus be written as [29, 32]

$$\hat{D}_j := \hat{D}(\delta_j) = \sum_{l=1}^{N} e^{-il\delta_j} |g\rangle_l\langle e|, \tag{11.7}$$

where the operator $|g\rangle_l\langle e|$ projects the lth ion from the state $|e\rangle$ to the ground state $|g\rangle$ and δ_j is the phase introduced in Eq. (11.5).

With the detector operator of Eq. (11.7), we can describe the detection processes of all m photons emitted by the N ions. As an example, let us consider the case of $N = 3$ ions. After a first photon is detected at $\mathbf{r}_1$, we obtain from Eqs. (11.6) and (11.7):

$$\hat{D}_1\ |\Psi^i_3\rangle = e^{-i\delta_1}|g, e, e\rangle + e^{-i2\delta_1}|e, g, e\rangle + e^{-i3\delta_1}|e, e, g\rangle. \tag{11.8}$$

The detection of the second photon may occur at $\mathbf{r}_2$, and we describe this event by applying successively the two detector operators $\hat{D}_1$ and $\hat{D}_2$ on the initial state $|\Psi^i_3\rangle$. The (not normalized) final state $|\Psi^f_3\rangle$ of the three atoms can then be written as:

$$\begin{aligned} |\Psi^f_3\rangle &= \hat{D}_2\ \hat{D}_1\ |\Psi^i_3\rangle \\ &= e^{-i\delta_1 - i2\delta_2}|g, g, e\rangle + e^{-i\delta_1 - i3\delta_2}|g, e, g\rangle \\ &\quad + e^{-i\delta_2 - i2\delta_1}|g, g, e\rangle + e^{-i2\delta_1 - i3\delta_2}|e, g, g\rangle \\ &\quad + e^{-i\delta_2 - i3\delta_1}|g, e, g\rangle + e^{-i2\delta_2 - i3\delta_1}|e, g, g\rangle. \end{aligned} \tag{11.9}$$

As can be seen from Eq. (11.9), the geometrical phase differences δ_j, with $j = 1, 2$, determine the symmetry of the state. In particular, to generate the symmetric Dicke states $|\frac{3}{2}, M\rangle$, with $M = -3/2, -1/2, +1/2, +3/2$, the phases δ_j should adopt multiples of 2π. The latter can be realized by a suitable localization of the detectors according to Eq. (11.5). In particular, this means that we can generate all four symmetric Dicke states

$$\begin{aligned} |\tfrac{3}{2}, +\tfrac{3}{2}\rangle &= |e, e, e\rangle \\ |\tfrac{3}{2}, +\tfrac{1}{2}\rangle &= 3^{-\frac{1}{2}}\,(|e, e, g\rangle + |e, g, e\rangle + |g, e, e\rangle) \\ |\tfrac{3}{2}, -\tfrac{1}{2}\rangle &= 3^{-\frac{1}{2}}\,(|e, g, g\rangle + |g, e, g\rangle + |g, g, e\rangle) \\ |\tfrac{3}{2}, -\tfrac{3}{2}\rangle &= |g, g, g\rangle. \end{aligned} \tag{11.10}$$

Note that the product states $|\frac{3}{2},+\frac{3}{2}\rangle$ and $|\frac{3}{2},-\frac{3}{2}\rangle$ can be obtained by simply measuring no photons and $m = N = 3$ photons, respectively. It is, however, also possible to generate the entangled states $|\frac{3}{2},+\frac{1}{2}\rangle$ and $|\frac{3}{2},-\frac{1}{2}\rangle$. In this case, one or two detectors have to be used, respectively, located with the phases δ_j in such a way as to adopt multiples of 2π.

The generalization to arbitrary Dicke state $|\frac{N}{2},M\rangle$ with N ions is straightforward. To this aim, we have to place again all m detectors at positions $\mathbf{r}_1, \ldots, \mathbf{r}_m$ such that the phases δ_j adopt multiples of 2π. The unnormalized state of the N ions after a first photon has been detected at $\mathbf{r}_1$ is then obtained by applying the operator $\hat{D}_1$ on the initial state Eq. (11.6). From this procedure we get

$$\hat{D}_1|\Psi_N^i\rangle = \sum_k P_k(|g,e,\ldots,e\rangle_N), \tag{11.11}$$

where $\{P_k\}$ denotes the set of all possible permutations of the N qubits. In analogy to the case discussed above, we assume that the $m-1$ remaining photons (with $m \le N$) are detected at positions $\mathbf{r}_2, \mathbf{r}_3, \ldots, \mathbf{r}_m$, respectively. We can calculate the unnormalized final state of the atoms, i.e., the state after all m photons have been recorded by the m detectors, by applying the detector operators $\hat{D}_2, \hat{D}_3, \ldots, \hat{D}_m$ on the intermediate state (11.11). With this scheme, we obtain the state of Eq. (11.4) with $M = N/2 - m$.

11.6 Theory of Collective Light Emission of Ion Crystals in Correlated Dicke States

So far we derived that an arbitrary symmetric Dicke state $|\frac{N}{2},M\rangle$ among N ions can be generated from the fully excited state Eq. (11.6) upon detection of $m = N/2 - M$ photons, if the m detectors are placed at positions $\mathbf{r}_1, \ldots, \mathbf{r}_m$ such that the phases δ_j adopt multiples of 2π. In this section, we study that once the ions occupy the symmetric Dicke state $|\frac{N}{2},M\rangle$, they emit the subsequent photon in a collective manner dubbed *superradiance* [88]. This means in particular that for $M \approx 0$ the radiated intensity scales with N^2 even though the emission is based entirely on spontaneous decay and thus takes place incoherently. This surprising result led Dicke to entitle his seminal paper *Coherence in Spontaneous Radiation Processes* [88].

That the radiated intensity of a symmetric Dicke state $|\frac{N}{2},M\approx 0\rangle$ scales quadratically with N can be proven rather simply in the small sample limit, i.e., if the emitters are separated by distances much smaller than the transition wavelength λ. To show this, we start by noting that the Hamiltonian for N non-interacting 2-level atoms is given by

$$H_S = \hbar\omega_0 \sum_{l=1}^{N} s_l^z \tag{11.12}$$

where $s_l^z = \frac{1}{2}(s_l^+ s_l^- - s_l^- s_l^+)$ is the z-component of the pseudo-spin operator $\boldsymbol{s}_l$, and $s_l^\pm$ are the corresponding pseudo-spin raising and lowering operators of the

lth ion, $l \in \{1, ..., N\}$, respectively. We further define the collective pseudo-spin operators

$$\mathbf{s} = \sum_{l=1}^{N} \mathbf{s}_l, \; s^z = \sum_{l=1}^{N} s_l^z,$$
$$s^+ = \sum_{l=1}^{N} s_l^+, \; s^- = \sum_{l=1}^{N} s_l^-. \tag{11.13}$$

The symmetric Dicke states $|\frac{N}{2}, M\rangle$ are eigenstates of H_S as well as of s^2 and s_z and obey the eigenvalue equations [88]

$$s^2|N/2, M\rangle = \hbar^2(N/2)(N/2+1)|N/2, M\rangle, \tag{11.14}$$

$$s_z|N/2, M\rangle = \hbar M|N/2, M\rangle. \tag{11.15}$$

Moreover, the collective pseudospin raising and lowering operators act on the symmetric Dicke states in the following way [88]

$$s^+|N/2, M\rangle = \hbar\sqrt{(N/2-M)(N/2+M+1)}|N/2, M+1\rangle, \tag{11.16}$$

$$s^-|N/2, M\rangle = \hbar\sqrt{(N/2+M)(N/2-M+1)}|N/2, M-1\rangle. \tag{11.17}$$

Recalling that the rate of photon emission for a single atom is [48, 88, 89]

$$W_1 = 2\gamma \left\langle s_l^+ s_l^- \right\rangle, \tag{11.18}$$

we obtain for the radiation of N ions in the small sample limit (where we can neglect the phase factors of the electromagnetic field in the interaction Hamiltonian) [48, 88, 89]

$$W_N = 2\gamma \left\langle s^+ s^- \right\rangle = 2\gamma(N/2+M)(N/2-M+1) \tag{11.19}$$

Eq. (11.19) shows that the rate of photon emission by the ion array starts from a value $2\gamma N$ for $M = N/2$, i.e., the fully excited state, to a value equal to $2\gamma N/2(N/2+1)$ for a state with $M \approx 0$, i.e., half way down the ladder of symmetric Dicke states, ending up eventually in no photon emission for $M = -N/2$, i.e., when the ion crystal has reached the collective ground state. We stress that this behavior is very different from the exponential decay obtained in the case of N uncorrelated ions. In particular, we realize that the radiated intensity initially *increases* when the ions pass down the ladder of symmetric Dicke states, displaying altogether a burst with a peak intensity proportional to N^2 for $M \approx 0$, and a width inversely proportional to N [48, 88, 89] (see Figure 11.9).

This behavior is commonly attributed to a synchronization of the dipole moments of the ions, oscillating in phase once the first photons have been emitted, equivalent to coherently oscillating antennas as discussed in Section 11.3, [48, 50, 86, 90]. However, this classical interpretation picture ignores the fact that the Dicke states do not carry any dipole moment [39, 87]. Hence the physical picture of synchronized dipoles coherently oscillating in phase appears not appropriate to describe the emission of N emitters prepared in entangled Dicke states.

In the following, we demonstrate that a different type of coherence is at the basis of the N^2-scaling law of Dicke superradiance, not linked to the classical picture of

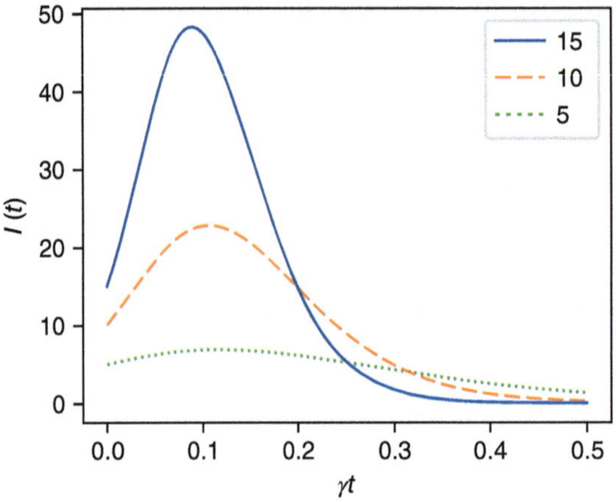

Figure 11.9 Intensity $I(t)$ against γt in the case of a small sample of initially fully excited ions in the state $|N/2, N/2\rangle$ for three different numbers of ions N. Due to the collective emission in that regime, a short burst of radiation is obtained instead of an exponential decay. Source: Adapted from Bojer and von Zanthier [49]/ArXiv/CC BY-4.0.

coherently oscillating and synchronized dipole moments but stemming from the quantum correlations due to the entanglement of the Dicke states [39, 49, 87]. In fact, we will show how this quantum nature of the Dicke states dictates their radiative characteristics, by employing a description that makes use of quantum paths and quantum interferences appearing in the emission process [91]. We emphasize that the quantum path interpretation holds also in the case of inter-ion separations $d \gg \lambda$, particularly important for the case of trapped ions.

For the proof, we assume for simplicity again a linear array of N ions with upper level $|e\rangle$ and ground state $|g\rangle$, localized at positions $\mathbf{R}_1, \dots, \mathbf{R}_N$ with equal spacing $d \gg \lambda$ such that any dipole–dipole interaction between the particles can be neglected (see Figure 11.8). The initial state of the ion array is taken to be the symmetric Dicke state $|\frac{N}{2}, M\rangle$ (see Eq. (11.4)). Denoting with m the number of detected photons to prepare the state $|\frac{N}{2}, M\rangle$ from the initial state $|\Psi_N^i\rangle = |e, e, \dots, e\rangle_N$ according to scheme presented in Section 11.4, we find that $m = N/2 - M = N - n_e$, where n_e (n_g) is the number of excited (ground state) ions in the Dicke state, with $N = n_e + n_g$. The Dicke state Eq. (11.4) can thus be also written in the form

$$|D_{n_e,N-n_e}\rangle \equiv \binom{N}{n_e}^{-\frac{1}{2}} \sum_k \mathbf{P}_k |S_{n_e,N-n_e}\rangle. \tag{11.20}$$

with $\{\mathbf{P}_k\}$ referring to the complete set of all distinct possible permutations of the ions and $|S_{n_e,N-n_e}\rangle$ denoting the separable state defined by

$$|S_{n_e,N-n_e}\rangle \equiv \prod_{\alpha=1}^{n_e} |e_\alpha\rangle \prod_{\beta=n_e+1}^{N} |g_\beta\rangle. \tag{11.21}$$

In the following, we will compare the radiative properties of the separable state $|S_{n_e,N-n_e}\rangle$ to the radiative properties of the entangled Dicke state $|D_{n_e,N-n_e}\rangle$.

To this aim, let us calculate the intensity at position $\mathbf{r}$ radiated by the state $|S_{n_e,N-n_e}\rangle$ and the state $|D_{n_e,N-n_e}\rangle$, respectively. The intensity derives from the expectation value

$$\mathrm{I}(\mathbf{r}) = \langle \hat{E}^{(-)}(\mathbf{r})\, \hat{E}^{(+)}(\mathbf{r}) \rangle, \tag{11.22}$$

where the positive frequency part of the electric field operator coupling the two levels of the ions is given by

$$\hat{E}^{(+)} \sim \sum_l e^{-i\, l\, \delta_j}\, \hat{s}_l^- = \sum_l e^{-i\varphi_l}\, \hat{s}_l^-, \tag{11.23}$$

and the negative frequency part of the electric field operator $\hat{E}^{(-)}$ is obtained by Hermitian conjugation $\hat{E}^{(-)} = \hat{E}^{(+)^\dagger}$. Since $k\, d \gg 1$, in contrast to Eq. (11.19), we have to take into account the different phase factors stemming from the optical paths accumulated by a photon emitted by ion l at $\mathbf{R}_l$ toward the detector at $\mathbf{r}$. This leads to the following expression for the radiated intensity

$$\begin{aligned} \mathrm{I}(\mathbf{r}) &= \sum_{i,j} \langle \hat{s}_i^+ \hat{s}_j^- \rangle\, e^{i\,(\varphi_i - \varphi_j)} \\ &= \sum_i \langle \hat{s}_i^+ \hat{s}_i^- \rangle + \left(\sum_{i \neq j} \langle \hat{s}_i^+ \rangle \langle \hat{s}_j^- \rangle \right. \\ &\quad \left. + \sum_{i \neq j} (\langle \hat{s}_i^+ \hat{s}_j^- \rangle - \langle \hat{s}_i^+ \rangle \langle \hat{s}_j^- \rangle) \right) e^{i\,(\varphi_i - \varphi_j)}. \end{aligned} \tag{11.24}$$

From Eq. (11.24) we see that the intensity depends on the correlations $\langle \hat{s}_i^+ \hat{s}_j^- \rangle$ as well as the dipole moments $\langle \hat{s}_i^+ \rangle$ of the ion array.

In the case that the ion array is in the separable state $|S_{n_e,N-n_e}\rangle$, the intensity calculates to

$$\mathrm{I}_{|S_{n_e,N-n_e}\rangle} = \sum_{i,j=1}^{n_e} \langle \hat{s}_i^+ \hat{s}_j^- \rangle\, e^{i\,(\varphi_i - \varphi_j)} = \sum_{i=1}^{n_e} \langle \hat{s}_i^+ \hat{s}_i^- \rangle = n_e. \tag{11.25}$$

Here, we have explicitly used the fact that for a separable state we have $\langle \hat{s}_i^+ \hat{s}_j^- \rangle = \langle \hat{s}_i^+ \rangle \langle \hat{s}_j^- \rangle = 0$ for $i \neq j$, since the dipole moment $\langle \hat{s}_j^+ \rangle$, as well as the correlations $\langle \hat{s}_i^+ \hat{s}_j^- \rangle$ for $i \neq j$, vanishes. According to Eq. (11.25), the intensity distribution of a separable state is thus a constant, corresponding to the number of initially excited ions n_e, i.e., $\mathrm{I}_{|S_{n_e,N-n_e}\rangle} = \mathrm{I}_{|S_{n_e,0}\rangle}$, independent of N. The interpretation is that from an uncorrelated state every ion radiates independently.

Before deriving the intensity of the Dicke state $|D_{n_e,N-n_e}\rangle$, we illustrate the key idea of the difference to the separable state $|S_{n_e,N-n_e}\rangle$ by a simple example. Consider a system of three ions prepared in the single-excited symmetric Dicke state

$$|D_{1,2}\rangle = \frac{1}{\sqrt{3}} (|e\; g\; g\rangle + |g\; e\; g\rangle + |g\; g\; e\rangle). \tag{11.26}$$

The intensity of this state calculates to

$$\begin{aligned} \mathrm{I}_{|D_{1,2}\rangle} &= \sum_{i,j=1}^{3} \langle \hat{s}_i^+ \hat{s}_j^- \rangle\, e^{i\,(\varphi_i - \varphi_j)} \\ &= \sum_{i=1}^{3} \langle \hat{s}_i^+ \hat{s}_i^- \rangle + \sum_{\substack{i,j-1 \\ i \neq j}}^{3} \langle \hat{s}_i^+ \hat{s}_j^- \rangle\, e^{i\,(\varphi_i - \varphi_j)}, \end{aligned} \tag{11.27}$$

with the dipole moments $\langle \hat{s}_j^+ \rangle$ being again zero. The first sum of the last line of Eq. (11.27) corresponds to the intensity of the separable state $|S_{1,2}\rangle$ (see Eq. (11.25)), if we keep in mind the normalization factor of the state $|D_{1,2}\rangle$. In the second sum, the inter-ion correlations $\langle \hat{s}_i^+ \hat{s}_j^- \rangle$ of $|D_{1,2}\rangle$ calculate for $i = 1$ and $j = 2$ to

$$\begin{aligned}\langle \hat{s}_1^+ \hat{s}_2^- \rangle &= \frac{1}{\sqrt{3}} (\langle e\ g\ g| + \langle g\ e\ g| + \langle g\ g\ e|)\, \hat{s}_1^+ \hat{s}_2^- |D_{1,2}\rangle \\ &= \frac{1}{3} \langle g\ g\ g|\ \hat{s}_2^- (|e\ g\ g\rangle + |g\ e\ g\rangle + |g\ g\ e\rangle) \\ &= \frac{1}{3} \langle g\ g\ g | g\ g\ g\rangle = \frac{1}{3},\end{aligned} \tag{11.28}$$

and equally for the other permutations of i and j. Taking into account in Eq. (11.27) the phase factors from the different optical paths accumulated by a photon at the detector at $\mathbf{r}$ after being emitted by an ion at $\mathbf{R}_l$, we obtain altogether for the radiated intensity

$$\begin{aligned}\mathrm{I}_{|D_{1,2}\rangle} &= \frac{1}{3} \left(3 + \sum_{\substack{i,j=1 \\ i \neq j}}^{3} e^{i\,(\varphi_i - \varphi_j)} \right) \\ &= 1 + \frac{2}{3} \sum_{i<j=1}^{3} \cos(\varphi_i - \varphi_j),\end{aligned} \tag{11.29}$$

displaying an angular dependency and a maximum of $\left[\mathrm{I}_{|D_{1,2}\rangle}\right]^{\mathrm{Max}} = 3$ at $\varphi_1 = \varphi_2 = \varphi_3$ (fulfilled for $\theta = 0, \pm\pi$), i.e., thrice as large as $\mathrm{I}_{|S_{1,2}\rangle}$. This is, however, so far a purely mathematical derivation. In order to gain physical insight into the enhanced emission of radiation by the entangled Dicke states $|D_{n_e,N-n_e}\rangle$ compared to the separable states $|S_{n_e,N-n_e}\rangle$, let us examine the transition amplitude (quantum path) for each photon detection event and coherently sum over all possible quantum paths contributing to the total intensity $\mathrm{I}(\mathbf{r})$ [91].

In the following, we focus on the symmetric Dicke state $|D_{2,1}\rangle$ with two excited ions and one ion in the ground state. We start by investigating the quantum paths occurring for the initially separable state $|S_{2,0}\rangle = |e\ e\rangle$ (see Figure 11.10). There are

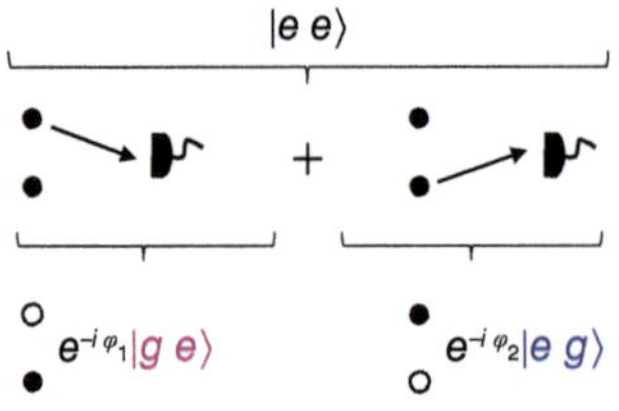

Figure 11.10 Possible quantum paths of the initially separable state $|S_{2,0}\rangle$. Black (white) circles denote ions in the excited (ground) state. The middle row depicts the different quantum paths and the lower row the final states of the ions and the phases accumulated by the photon along the different quantum paths. Source: Wiegner et al. [91]/with permission of American Physical Society.

two distinct possibilities: either the photon (black arrow) is scattered by the first ion (black circle) transferring the ion into the ground state (white circle) where a phase $e^{-i\,\varphi_1}$ is accumulated by the photon or the photon is emitted by the second ion accumuling the phase $e^{-i\,\varphi_2}$. Each quantum path leads to an orthogonal final state so that no interference terms show up when taking the modulus square of the coherent sum of the two different quantum paths to compute the intensity. Explicitly, from Figure 11.10, we obtain for the intensity produced by the state $|S_{2,0}\rangle$

$$\mathrm{I}_{|S_{2,0}\rangle} = ||e^{-i\,\varphi_1}|g\;e\rangle||^2 + ||e^{-i\,\varphi_2}|e\;g\rangle||^2 = 2, \tag{11.30}$$

where the norm of the state vector $|\Psi\rangle$ is given by $||\Psi||^2 = \langle\Psi|\Psi\rangle$, in agreement with Eq. (11.25).

Let us compare this result to the intensity obtained for the Dicke state $|D_{2,1}\rangle = (|e\;e\;g\rangle + |e\;g\;e\rangle + |g\;e\;e\rangle)/\sqrt{3}$, which also has only two excitations. The different quantum paths leading to a successful photon detection event are sketched in Figure 11.11. Considering only the first term, the state $|e\;e\;g\rangle$ basically leads to the same quantum paths as the state $|S_{2,0}\rangle$: either the first ion emits a photon leading to the final state $|g\;e\;g\rangle$ by accumulating the phase $e^{-i\,\varphi_1}$, or the second ion scatters the photon leading to the final state $|e\;g\;g\rangle$ with a phase corresponding to $e^{-i\,\varphi_2}$. However, for the state $|D_{2,1}\rangle$, in difference to the separable state $|S_{2,0}\rangle$, we have to consider the two other terms as well, i.e., $|e\;g\;e\rangle$ and $|g\;e\;e\rangle$. The coherent superposition of all possible quantum paths from the three terms leads altogether to six quantum paths, resulting in a different outcome than Eq. (11.30) (see Figure 11.11). Considering all six quantum paths, we obtain for the intensity of the state $|D_{2,1}\rangle$

$$\begin{aligned}\mathrm{I}_{|D_{2,1}\rangle} &= \frac{1}{3}\left|\left|(e^{i\,\varphi_1} + e^{-i\,\varphi_2})|g\;g\;e\rangle\right|\right|^2\\ &+ \frac{1}{3}\left|\left|(e^{-i\,\varphi_1} + e^{-i\,\varphi_3})|g\;e\;g\rangle\right|\right|^2\\ &+ \frac{1}{3}\left|\left|(e^{-i\,\varphi_2} + e^{-i\,\varphi_3})|e\;g\;g\rangle\right|\right|^2.\end{aligned} \tag{11.31}$$

Focusing only on the maximum of the intensity distribution, we get from Eq. (11.31)

$$\begin{aligned}\left[\mathrm{I}_{|D_{2,1}\rangle}\right]^{\mathrm{Max}} &= \frac{6}{3} + \frac{6}{3}\\ &\equiv \left[\mathrm{I}_{|S_{2,0}\rangle}\right]^{\mathrm{Max}} + \text{interference terms},\end{aligned} \tag{11.32}$$

what is twice the result obtained for $|S_{2,0}\rangle$ in Eq. (11.30).

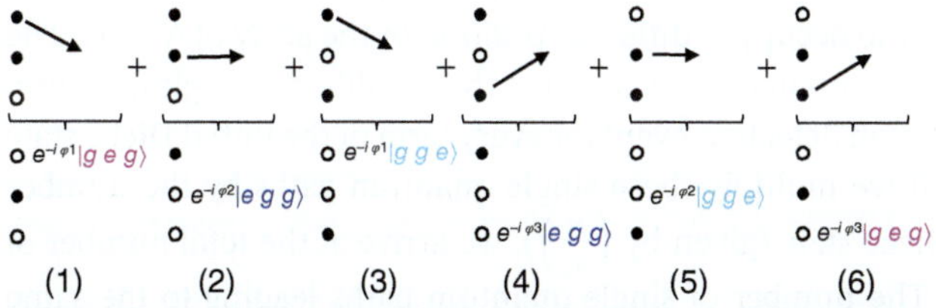

Figure 11.11 Possible quantum paths of the initial Dicke state $|D_{2,1}\rangle$. Source: Wiegner et al. [91]/with permission of American Physical Society.

Casting the foregoing argument into a general formula for the maximum intensity of a symmetric Dicke state, we obtain [91]

$$\left[\mathrm{I}_{|D\rangle}\right]^{\mathrm{Max}} = \left[\mathrm{I}_{|S\rangle}\right]^{\mathrm{Max}} + (\#QP) \times (\#|f\rangle) \times \left(\mathcal{N}\right). \tag{11.33}$$

Hereby, $(\#QP)$ stands for the number of interfering quantum path leading to the same final state. Multiplied by the number of final states $(\#|f\rangle)$, we arrive at the total number of interfering quantum paths (interference terms) contributing to the maximum of the intensity. Together with the squared normalization constant $\left(\mathcal{N}\right)$ of the corresponding Dicke state, the expression $(\#QP) \times (\#|f\rangle) \times \left(\mathcal{N}\right)$ equals for $\theta = 0, \pm\pi$ the constructive contribution of all interference terms to the maximal intensity.

Let us apply Eq. (11.33) to rederive the maximum of the intensity $\mathrm{I}_{|D_{2,1}\rangle}$. In Eq. (11.30), we already calculated the maximum intensity of the corresponding separable state to be 2. The number of interfering quantum path leading to the same final state can be obtained by counting (see Figure 11.11): we could either pick out the two quantum paths $(1, 6)$ and $(6, 1)$ *or* the quantum paths $(2, 4)$ and $(4, 2)$ *or* the quantum paths $(3, 5)$ and $(5, 3)$, i.e., $(\#QP) = 2$. The number of different final states is $(\#|f\rangle) = 3$ and the squared normalization of the state $|W_{2,1}\rangle$ is $\left(\mathcal{N}\right) = \frac{1}{3}$. Thus, we obtain for the maximum intensity (cf. Eq. (11.31))

$$\left[\mathrm{I}_{|D_{2,1}\rangle}\right]^{\mathrm{Max}} = 2 + 2 \times 3 \times \frac{1}{3} = 4, \tag{11.34}$$

in agreement with Eq. (11.32).

Adopting the foregoing argument to an arbitrary symmetric Dicke state $|D_{n_e,N-n_e}\rangle$ with n_e excited ions and $N - n_e$ ions in the ground state as in Eq. (11.20), the general formula for the maximum intensity can be derived using combinatorial considerations and the maximum of the intensity of the separable state $\mathrm{I}_{|S_{n_e,N-n_e}\rangle}$ as given in Eq. (11.25). In this way we obtain [91]

$$\begin{aligned}\left[\mathrm{I}_{|D_\star\rangle}\right]^{\mathrm{Max}} &= n_e + \left[(\#QP) \times (\#|f\rangle) \times \left(\mathcal{N}\right)\right]_{|D_\star\rangle} \\ &= n_e + n_g(n_g + 1) \times \binom{N}{n_e - 1} \times \binom{N}{n_e}^{-1} \\ &= n_e(n_g + 1),\end{aligned} \tag{11.35}$$

coinciding with Eq. (11.19) when noting that $M = n_e - N/2 = N/2 - n_g$. Here, $\left(\mathcal{N}\right)$ is the squared normalization constant of the generalized symmetric Dicke state (see Eq. (11.20)), whereas the number of final states $(\#|f\rangle)$ is given by $\binom{N}{n_e-1}$. The latter can be seen by noting that after the detection of a photon, there are $n_e - 1$ excited atoms left, which are able to occupy N different position in the array of N ions. The crucial term $(\#QP)$ can be computed by considering that n_e different single quantum paths are leading to a photon detection event for every term of the initial Dicke state (see, e.g., Eq. (11.31)). If we multiply these single quantum paths by the number of terms of the initial Dicke state (given by $\binom{N}{n_e}$), we arrive at the total number of single quantum paths. The number of single quantum paths leading to the same final states - abbreviated by $(\#sqp)$ – is then obtained by dividing the total number of single quantum paths by the number of final states [91]

$$(\#sqp) = \frac{n_e \binom{N}{n_e}}{\binom{N}{n_e-1}} = n_g + 1. \tag{11.36}$$

These $n_g + 1$ single quantum paths, which lead to the same final state, now interfere among each other producing in total $(\#QP) = n_g(n_g + 1)$ interfering quantum path pairs. Clearly $(\#QP)$, i.e., the enhancement of the intensity is zero, if $n_g = 0$. A 3D-plot of the maximum intensity of the generalized Dicke state $|D_{n_e,N-n_e}\rangle$ as a function of n_e and N is shown in Figure 11.12. Here, the global maximum intensity, namely $N/2(N/2+1)$, obtained for the symmetric Dicke state $|D_{N/2,N/2}\rangle$, is highlighted by the solid line in Figure 11.12.

Note that we have focused so far our discussion on a physical explanation for the maxima of the intensity radiated by symmetric Dicke states. However, besides studying the maximum intensity, we can also derive the angular dependency of the intensity. Using Eq. (11.24), the intensity distribution radiated by the ion array in a symmetric Dicke state calculates to [91]

$$\mathrm{I}_{|D_{n_e,n_g}\rangle}(\theta) = \frac{n_e\,(n_e - 1)}{N-1} + \frac{n_e\,(N - n_e)}{N(N-1)} \frac{\sin^2(\frac{\varphi_N}{2})}{\sin^2(\frac{\varphi_1}{2})}. \tag{11.37}$$

With $\delta\varphi_N = N\,k\,d\,\cos\theta\,\delta\theta$ (see Eq. (11.5)), the width of the distribution Eq. (11.37) is given by

$$\delta\theta = \frac{2\pi}{N\,k\,d}, \tag{11.38}$$

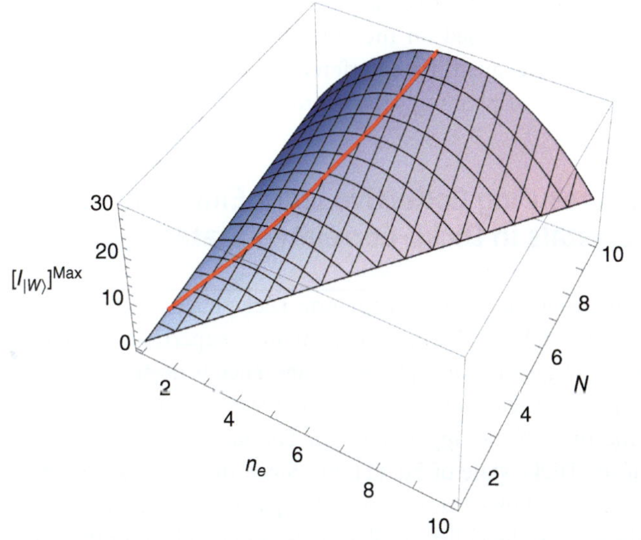

Figure 11.12 Plot of the maximum intensity $\left[\mathrm{I}_{|D_{n_e,N-n_e}\rangle}\right]^{\mathrm{Max}}$ of the Dicke state $|D_{n_e,N-n_e}\rangle$ as a function of n_e and $N = n_e + n_g$. $\left[\mathrm{I}_{|D_{N/2,N/2}\rangle}\right]^{\mathrm{Max}}$ is highlighted by the solid line. Source: Wiegner et al. [91]/with permission of American Physical Society.

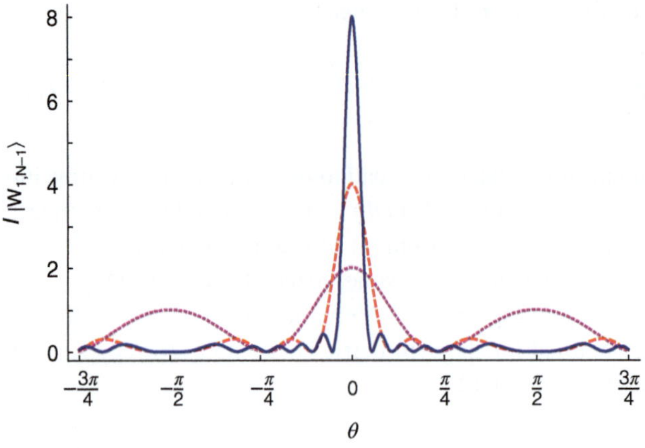

Figure 11.13 Intensity distribution $I_{|D_{1,N-1}\rangle}(\theta)$ as a function of the position θ of the detector recording the photon from the single excited symmetric Dicke state $|D_{1,N-1}\rangle$ for different numbers of ions N ($N = 2$ dotted, $N = 4$ striped, $N = 8$ solid, with $k\,d = \frac{3}{2}\pi$). A strong directionality in the emission of spontaneous radiation by the symmetric Dicke state $|D_{1,N-1}\rangle$ can be seen. Source: Wiegner et al. [91]/with permission of American Physical Society.

which depends on the wavelength λ, the distance d between adjacent atoms and the total number of atoms N.

We close by noting that in the regime of classical coherence an enhanced radiation rate is related to a large dipole moment. Yet, the Dicke states exhibit no dipole moment. By contrast, we could show by use of a quantum path interpretation that the Dicke states display enhanced collective radiation, i.e., a superradiant emission behavior. We call the coherence based on the entanglement of the Dicke states responsible for the constructive quantum interference phenomena displayed in Eq. (11.33) and Figure 11.13 correspondingly the regime of *quantum coherence* [87].

11.7 Theory of Measuring Collective Light Emission from Linear Ion Crystals in Dicke-Correlated States

In the previous section we deduced how symmetric Dicke states of the form of Eq. (11.4), or equivalently of Eq. (11.20), emit light in a superradiant manner. However, these states are fragile as their quantum coherence is destroyed as soon as the ions start to decay from the upper state $|e\rangle$ to the lower state $|g\rangle$, what occurs on average after the life time of the upper state $|e\rangle$. The latter is therefore called the coherence time of the Dicke state of Eq. (11.20). Since this time is very short, on the order of ns for dipole allowed transitions, the question can be raised how after preparation of these state the collective light emission can be observed. This is discussed in the present section where we study an appropriate measurement scheme [38, 39].

According to the scheme presented in Section 11.4, for the production of the symmetric Dicke state $|D_{n_e,N-n_e}\rangle$ of Eq. (11.20), we can measure $m - 1 = N - n_e$ photons by $m-1$ detectors at fixed positions $\mathbf{r}_j, j = 1, \ldots, m-1$, such that the corresponding phases take the values $\delta_j = \alpha \cdot 2\pi$, $\alpha \in \mathbb{N}$. The measurement of a further photon at an arbitrary position $\mathbf{r}_m$ then could be used as a probe to observe the collective light emission resulting from the produced symmetric Dicke state $|D_{n_e,N-n_e}\rangle$, as discussed in Section 11.5. This setting corresponds to measuring the m-th order photon correlation function. It allows (a) to produce any symmetric Dicke state $|D_{n_e,N-n_e}\rangle$ of the form of Eq. (11.20) from statistically independent and initially uncorrelated ions in the fully excited state $|\Psi_N^i\rangle$ (see Eq. (11.6)), and (b) to observe subsequently the corresponding superradiant emission characteristics of the corresponding Dicke state. Note that this scheme can be regarded also as a generalized Hanbury Brown and Twiss setup where the m-th-order photon correlation function is measured at m different positions $\mathbf{r}_j, j = 1, \ldots, m$ [38, 39].

To see this in more detail, let us calculate the m-th-order photon correlation function for N fully excited independent ions initially in the state $|\Psi_N^i\rangle$. For arbitrary photo-detector positions, this function is defined as

$$G^{(m)}(\mathbf{r}_1, \ldots, \mathbf{r}_m) \equiv : \prod_{j=1}^{m} E^{(-)}(\mathbf{r}_j) E^{(+)}(\mathbf{r}_j) :, \tag{11.39}$$

where $: \ldots :$ denotes the (normally ordered) quantum mechanical expectation value and the electric field operators $E^{(+)}(\mathbf{r}_j)$ and $E^{(-)}(\mathbf{r}_j)$ are given by Eq. (11.23). Note that for simplicity, we define as in Eq. (11.23) the field and hence all correlation functions of m-th order dimensionless; the actual values can be obtained by multiplying $G^{(m)}$ with m times the intensity of a single source.

Starting with all ions in the state $|\Psi_N^i\rangle$, we find from Eq. (11.39)

$$\begin{aligned} &\mathrm{G}^{(m)}_{\Psi_N^i}(\mathbf{r}_1, \ldots, \mathbf{r}_m) \\ &\sim \| \sum_{\sigma_1=1}^{N} \sum_{\substack{\sigma_2=1 \\ \sigma_1 \neq \sigma_2}}^{N} \cdots \sum_{\substack{\sigma_m=1 \\ \sigma_i \neq \sigma_m;\ i<m}}^{N} \prod_{j=1}^{m} e^{-i\,\varphi_{\sigma_j j}} |g_{\sigma_j}\rangle \|^2 \\ &= \sum_{\sigma_1=1}^{N} \sum_{\sigma_2=\sigma_1+1}^{N} \cdots \sum_{\sigma_m=\sigma_{m-1}+1}^{N} | \sum_{\substack{\sigma_1, \ldots, \sigma_m \\ \in S_m}} \prod_{j=1}^{m} e^{-i\,\varphi_{\sigma_j j}} |^2 . \end{aligned} \tag{11.40}$$

Here, $\||\psi\rangle\|^2 = \langle\psi|\psi\rangle$ defines the norm of the state vector $|\psi\rangle$, $|\ldots|$ abbreviates absolute values, and the expression $\sum_{\substack{\sigma_1, \ldots, \sigma_m \\ \in S_m}}$ denotes the sum over the symmetric group S_m with elements $\sigma_1, \ldots, \sigma_m$. In Eq. (11.40), the products $\prod_{j=1}^{m} e^{-i\,\varphi_{\sigma_j j}}$ represent m-photon quantum paths with phases $\sum_{j=1}^{m} \varphi_{\sigma_j j}$, accumulated by m photons emitted from m ions at $\mathbf{R}_{\sigma_j}$ and recorded by m detectors at $\mathbf{r}_j$. Since the particular source of a recorded photon is unknown, we have to sum over all possible combinations of m-photon quantum paths, which is expressed by the sum $\sum_{\sigma_1=1}^{N} \sum_{\sigma_2=1}^{N} \cdots \sum_{\sigma_m=1}^{N}$ in the second line of Eq. (11.40). Hereby, the additional condition $\sigma_i \neq \sigma_j$ $(i \neq j)$ ensures that each ion emits at most one photon. Considering that several combinations of m-photon quantum paths lead to the *same* final state and thus have to be added coherently, we end up with the modulus square in the third line of Eq. (11.40).

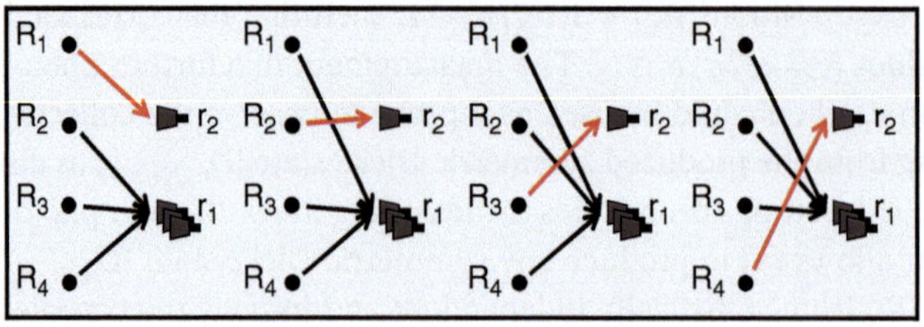

Figure 11.14 (Color online) N-photon quantum paths of N statistically independent ions where $(m-1)$ photons are detected at $\mathbf{r}_1$ and the mth photon at $\mathbf{r}_2$ (see Eq. (11.41)). The interference signal $G^{(M)}_{\Phi_N}(\mathbf{r}_1, \dots, \mathbf{r}_1, \mathbf{r}_2)$ can be reduced to a superposition of N indistinguishable yet different m-photon quantum paths. In the figure, the explicit case of $N = m = 4$ sources and detectors is shown. Source: Wiegner et al. [39]/with permission of American Physical Society.

Hereby, for the $\binom{N}{m}$ *different* final states, the corresponding transition probabilities $|\dots|^2$ have to be summed incoherently, which results in the combination of sums $\sum_{\sigma_1=1}^{N} \sum_{\sigma_2=\sigma_1+1}^{N} \cdots \sum_{\sigma_m=\sigma_{m-1}+1}^{N}$ in the third line of Eq. (11.40).

Assuming again an array of N identical 2-level ions, regularly aligned in a row with equal spacing $d \gg \lambda$ at positions $\mathbf{R}_l$, $l = 1, \dots, N$, we choose this time that $(m-1)$ detectors are placed in the far field at the *same* position $\mathbf{r}_1$, whereas the m-th detector is placed at an arbitrary position $\mathbf{r}_2$ (see Figure 11.14). Under this condition, we can make use of the fact that out of the $m!$ different m-photon quantum paths, denoted by $\sum_{\sigma_1,\dots,\sigma_m \in S_m} \prod_{j=1}^{m} e^{-i\,\varphi_{\sigma_j j}}$ in Eq. (11.40), $(m-1)!$ quantum paths are equal. In this case, Eq. (11.40) takes the form

$$
\begin{aligned}
&G^{(m)}_{\Phi_N^i}(\mathbf{r}_1, \dots, \mathbf{r}_1, \mathbf{r}_2) \\
&= \sum_{\substack{\sigma_1,\dots,\sigma_m = 1 \\ \sigma_1 < \dots < \sigma_m}}^{N} |(m-1)! \sum_{j=1}^{m} e^{-i\,(\varphi_{\sigma_j 2} - \varphi_{\sigma_j 1})}|^2 \\
&= ((m-1)!)^2 \sum_{\substack{\sigma_1,\dots,\sigma_m = 1 \\ \sigma_1 < \dots < \sigma_m}}^{N} |\sum_{j=1}^{m} e^{-i\,(\varphi_{\sigma_j 2} - \varphi_{\sigma_j 1})}|^2 \\
&= ((m-1)!)^2 \binom{N}{m} \left(\frac{m\,(N-m)}{N-1} + \frac{M\,(m-1)}{N\,(N-1)} \frac{\sin^2(N\frac{\varphi_{11}-\varphi_{12}}{2})}{\sin^2(\frac{\varphi_{11}-\varphi_{12}}{2})} \right),
\end{aligned}
\tag{11.41}
$$

with $\sum_{\substack{\sigma_1,\dots,\sigma_m = 1 \\ \sigma_1 < \dots < \sigma_m}}^{N} = \sum_{\sigma_1=1}^{N} \sum_{\sigma_2=\sigma_1+1}^{N} \cdots \sum_{\sigma_m=\sigma_{m-1}+1}^{N}$. We thus finally obtain for the m-th order correlation function [38, 39]

$$
\begin{aligned}
&G^{(m)}_{\Phi_N^i}(\mathbf{r}_1, \dots, \mathbf{r}_1, \mathbf{r}_2) \\
&= \frac{N!(m-1)!}{(N-m)!} \left(\frac{N-m}{N-1} + \frac{m-1}{N(N-1)} \frac{\sin^2(N\frac{\varphi_{11}-\varphi_{12}}{2})}{\sin^2(\frac{\varphi_{11}-\varphi_{12}}{2})} \right).
\end{aligned}
\tag{11.42}
$$

According to Eq. (11.42), $G^{(m)}_{\Phi^i_N}(\mathbf{r}_1, ..., \mathbf{r}_1, \mathbf{r}_2)$ as a function of $\mathbf{r}_2$ displays maxima and minima with a visibility of

$$\mathcal{V}_{SPE} = \frac{m-1}{m+1-\frac{2m}{N}}, \tag{11.43}$$

and an angular width of the central maximum

$$\delta\theta_2 \approx \frac{2\pi}{N\,k\,d}, \tag{11.44}$$

where we have used the relation $\delta\varphi_{12} = kd\,\cos\theta_2\,\delta\theta_2 \cong kd\,\delta\theta_2$. For growing numbers of ions, an increased focusing of the m-th-order photon correlation function in the direction of $\mathbf{r}_2 = \mathbf{r}_1$ is thus observed.

Note that for $m = 1$, the visibility $\mathcal{V}_{SPE}$ vanishes, which illustrates the fact that there is no preferred direction of emission of the first photon. Indeed, for $m = 1$, all ions are initially in the excited state $|\Psi^i_N\rangle$, which is well known to not show any focused emission behavior. For $m > 1$, however, superradiance and collective emission of the m-th photon set in, preconditioned by the measurement of $(m-1)$ photons in a particular fixed direction. Finally, for $m = N$, a maximum visibility of $\mathcal{V}_{SPE} = 100\%$ is obtained. According to Eq. (11.42), the height of the central maximum grows as $(N!)^2$. Moreover, recalling that

$$\begin{aligned}&\frac{1}{2\pi}\int G^{(m)}_{\Psi^i_N}(\mathbf{r}_1, \dots, \mathbf{r}_1, \varphi_{12})\,\mathrm{d}\varphi_{12}\\ &= \left\langle G^{(m)}_{\Psi^i_N}\right\rangle_{\varphi_{12}} = ((m-1)!)^2\binom{N}{m-1}(N-m+1),\end{aligned} \tag{11.45}$$

we find that the normalized m-th-order correlation function $G^{(m)}_{\Psi^i_N}/\left\langle G^{(m)}_{\Psi^i_N}\right\rangle_{\varphi_{12}}$ has a central maximum, which is proportional to N for $m = N$. Note that the scaling of the maximum as N and the angular width as $1/N$ are the typical features of superradiance from a Dicke state with *one* excitation (see Eq. (11.35)). This already suggests a close connection between the m-th-order correlation function for the initially fully excited state $G^{(m)}_{\Phi^i_N}(\mathbf{r}_1, ..., \mathbf{r}_1, \mathbf{r}_2)$ with the superradiant behavior of a suitably prepared Dicke state $|D_{n_e,N-n_e}\rangle$ with $(m-1)$ deexcitations.

In order to demonstrate the identity between superradiance and the m-th-order intensity correlation function starting from the fully excited state (apart from a proportionality factor), let us write $G^{(m)}_{\Psi^i_N}(\mathbf{r}_1, \dots, \mathbf{r}_1, \mathbf{r}_2)$ as [39]

$$\begin{aligned}&G^{(m)}_{\Psi^i_N}(\mathbf{r}_1, \dots, \mathbf{r}_1, \mathbf{r}_2)\\ &= \mathrm{Tr}[\tilde{\rho}^{(m-1)}E^{(-)}(\mathbf{r}_2)E^{(+)}(\mathbf{r}_2)]G^{(m-1)}_{\Psi^i_N}(\mathbf{r}_1, \dots, \mathbf{r}_1),\end{aligned} \tag{11.46}$$

where we defined

$$\begin{aligned}\tilde{\rho}^{(m-1)} &= \frac{\left(E^{(+)}(\mathbf{r}_1)\right)^{m-1}|\Psi^i_N\rangle\langle\Psi^i_N|\left(E^{(-)}(\mathbf{r}_1)\right)^{m-1}}{G^{(m-1)}_{\Psi^i_N}(\mathbf{r}_1, \dots, \mathbf{r}_1)}\\ &= |\Phi_{m-1}\rangle\langle\Phi_{m-1}|,\end{aligned} \tag{11.47}$$

i.e.,

$$
\begin{aligned}
&G^{(m)}_{\Psi^i_N}(\mathbf{r}_1,\dots,\mathbf{r}_1,\mathbf{r}_2) \\
&\quad = G^{(1)}_{\Phi_{m-1}}(\mathbf{r}_2)\, G^{(m-1)}_{\Psi^i_N}(\mathbf{r}_1,\dots,\mathbf{r}_1).
\end{aligned}
\tag{11.48}
$$

This shows already that the variation of $G^{(m)}_{\Psi^i_N}(\mathbf{r}_1,\dots,\mathbf{r}_1,\mathbf{r}_2)$ with respect to $\mathbf{r}_2$, i.e., θ_2, is indeed a function of the state $|\Phi_{m-1}\rangle$. For simplicity, we can set $\theta_1 = 0$. In this case we find

$$
E^{(+)}(\mathbf{r}_1) = \sum_l s_l^- = s^-,
\tag{11.49}
$$

where s^- is the collective lowering operator of Eq. (11.13). Using Eq. (11.49) in Eq. (11.47) we obtain for $|\Phi_{m-1}\rangle$

$$
\begin{aligned}
|\Phi_{m-1}\rangle &= \frac{(s^-)^{m-1}|\Psi^i_N\rangle}{\sqrt{\langle\Psi^i_N|(s^+)^{m-1}(s^-)^{m-1}|\Psi^i_N\rangle}} \\
&= |\frac{N}{2},\frac{N}{2}-(m-1)\rangle \\
&= |D_{N-(m-1),m-1}\rangle,
\end{aligned}
\tag{11.50}
$$

where $|\frac{N}{2},M\rangle = |D_{N-(m-1),m-1}\rangle$ is the symmetric Dicke state from Eq. (11.4) and Eq. (11.20), respectively, with $(m-1)$ ions in the ground state and $N-(m-1)$ ions in the excited state, for $m > 1$. The normalization factor $G^{(m-1)}_{\Psi^i_N}(\mathbf{r}_1,\dots,\mathbf{r}_1)$ in Eq. (11.47) for $\theta_1 = 0$ is equal to $\langle\frac{N}{2},\frac{N}{2}|(s^+)^{m-1}(s^-)^{m-1}|\frac{N}{2},\frac{N}{2}\rangle = \binom{N}{m-1}((m-1)!)^2$, so that Eq. (11.46) can also be written as [39]

$$
\begin{aligned}
&G^{(m)}_{\Psi^i_N}(\mathbf{r}_1,\dots,\mathbf{r}_1,\mathbf{r}_2) \\
&\quad = \langle\Phi_{m-1}|E^{(-)}(\mathbf{r}_2)E^{(+)}(\mathbf{r}_2)|\Phi_{m-1}\rangle \binom{N}{m-1}((m-1)!)^2 \\
&\quad = I_{\Phi_{m-1}}(\mathbf{r}_2)\binom{N}{m-1}((m-1)!)^2,
\end{aligned}
\tag{11.51}
$$

with

$$
I_{\Phi_{m-1}}(\mathbf{r}_2) = \langle\Phi_{m-1}|E^{(-)}(\mathbf{r}_2)E^{(+)}(\mathbf{r}_2)|\Phi_{m-1}\rangle.
\tag{11.52}
$$

Here, $I_{\Phi_{m-1}}$ is the intensity of the radiation produced by N ions prepared in the Dicke state $\Phi_{m-1} = |\frac{N}{2},\frac{N}{2}-(m-1)\rangle = |D_{N-(m-1),m-1}\rangle$, i.e. [91],

$$
\begin{aligned}
&I_{\Phi_{m-1}}(\mathbf{r}_2) \\
&\quad = (N-m+1)\left(\frac{N-m}{N-1} + \frac{m-1}{N(N-1)}\frac{\sin^2(N\frac{\varphi_{12}}{2})}{\sin^2(\frac{\varphi_{12}}{2})}\right),
\end{aligned}
\tag{11.53}
$$

what coincides with Eq. (11.37) in view of the fact that $m-1 = n_g$ and $N-m+1 = n_e$. In particular, for $m = N$, we find for Eq. (11.53) the intensity distribution $\mathrm{I}_{|D_{1,N-1}\rangle}(\theta)$ displayed in Figure 11.13.

We thus proved that the superradiant character of the Dicke state $|\frac{N}{2}, \frac{N}{2} - (m-1)\rangle$ can be observed by measuring the m-th-order photon correlation function $G^{(m)}_{\Psi^i_N}(\mathbf{r}_1, \ldots, \mathbf{r}_1, \mathbf{r}_2)$ for $\theta_1 = 0$ of the fully excited system [39]

$$G^{(m)}_{\Psi^i_N}(\mathbf{r}_1, \ldots, \mathbf{r}_1, \mathbf{r}_2) = \frac{N!(m-1)!}{(N-m)!} \left(\frac{N-m}{N-1} + \frac{m-1}{N(N-1)} \frac{\sin^2(N\frac{\varphi_{12}}{2})}{\sin^2(\frac{\varphi_{12}}{2})} \right). \tag{11.54}$$

The first $(m-1)$ detection events prepare the system conditionally in the Dicke state $|D_{N-(m-1),m-1}\rangle = |\frac{N}{2}, \frac{N}{2} - (m-1)\rangle$ whose behavior is then probed by the detection of the m-th photon.

11.8 Measurement of the Collective Light Emission of a Two Ion Crystal

Having established a measurement scheme to produce any desired symmetric Dicke state $|\frac{N}{2}, \frac{N}{2} - (m-1)\rangle = |D_{N-(m-1),m-1}\rangle$ from the fully excited state $|\Psi^i_N\rangle$ and subsequently observe the corresponding superradiant emission behavior, we present in this section a concrete example with $N = 2$ ions, the most elementary building block of a potentially collective ensemble. This simple system already exhibits a great wealth of collective spatiotemporal emission characteristics, which can be observed by measuring the second-order photon correlation function $G^{(2)}_{|e,e\rangle}(\boldsymbol{r}_1, \boldsymbol{r}_2, \tau)$ according to the scheme presented in Section 11.6. In particular, we will show that the collective spontaneous emission behavior of the two-ion crystal displays spatial super- and subradiance in combination with photon antibunching and bunching, respectively. To observe these particular features, we employ ultra-fast single photon resolving cameras with high spatial resolution in a Hanbury–Twiss setup, a prerequisite to disclose these very particular collective spatiotemporal emission phenomena (see Figure 11.15).

For the measurement, we employ again a pair of $^{40}Ca^+$ ions, stored in a linear Paul trap. However, instead of using a π-pulse excitation as discussed in the previous sections, the ions are continuously excited and laser-cooled near the Ca^+ $S_{1/2} - P_{1/2}$ transition by a laser at 397 nm (see Figure 11.15). To model the dynamical evolution of the ions including dissipation via spontaneous decay, we have to solve the master equation for the density matrix ρ of two identical 2-level ions with upper states $|e\rangle_i$ and ground states $|g\rangle_i$, $i = 1, 2$, with dipole moment $\mathbf{d}_{eg}$, located at fixed positions $\mathbf{R}_1$ and $\mathbf{R}_2$ [89]. In the rotating-wave, Markov and Born approximations, the master equation reads [29]

$$\dot{\rho} = -i\Omega \sum_{l=1}^{2} \left[exp\left[i\left(\mathbf{k}_L \cdot \mathbf{R}_l - \omega_L t \right) \right] s_l^+ + \text{h.c.}, \rho \right]$$
$$- \sum_{l=1}^{2} \gamma \left(s_l^+ s_l^- \rho + \rho s_l^+ s_l^- - 2 s_l^- \rho s_l^+ \right), \tag{11.55}$$

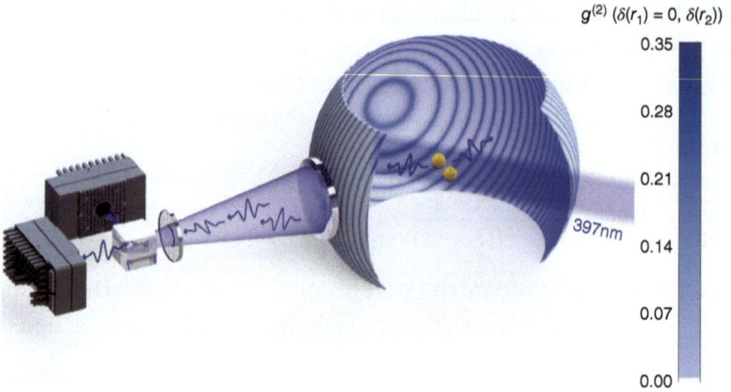

Figure 11.15 To measure $g^{(2)}(\boldsymbol{r}_1, \boldsymbol{r}_2, \tau)$, two $^{40}Ca^+$ ions are trapped and continuously excited on the $S_{1/2} - P_{1/2}$ transition while the scattered light is collected by a lens and, after passing a polarization filter, fed into a Hanbury-Brown and Twiss setup using two MCP cameras. The sphere around the two ions sketches the spatial pattern of $g^{(2)}(\boldsymbol{r}_1, \boldsymbol{r}_2, \tau = 0)$ as a function of $\boldsymbol{r}_2$ for $\delta(\mathbf{r}_1) = 0.34\pi$.

where 2Ω is the Rabi frequency of the ion-laser interaction, $\mathbf{k}_L$ is the laser wave vector, ω_L the laser angular frequency, $s_l^{\pm}$ are the raising and lowering operators for ion l at $\mathbf{R}_l$, and $2\gamma = \frac{4}{3}|\mathbf{d}_{eg}|^2\frac{\omega^3}{\hbar c^3}$ is the Einstein A coefficient for a single ion.

Taking into account the dynamics of the two-ion system, the two-time second-order correlation function in the Heisenberg picture is defined as

$$G^{(2)}(\mathbf{r}_1, t; \mathbf{r}_2, t+\tau) = \sum_{\mu,\nu,\lambda,\rho=1}^{2} \left[ik(\mathbf{R}_{\lambda,\nu} \cdot \hat{\mathbf{r}}_1 + \mathbf{R}_{\rho,\mu} \cdot \hat{\mathbf{r}}_2)\right] \times \langle s_\lambda^+(t) s_\rho^+(t+\tau) s_\mu^-(t+\tau) s_\nu^-(t)\rangle, \tag{11.56}$$

with $\mathbf{R}_{\mu,\nu} := \mathbf{R}_\mu - \mathbf{R}_\nu$ and $\hat{\mathbf{r}} = \frac{\mathbf{r}}{|\mathbf{r}|}$. $G^{(2)}(\mathbf{r}_1, t; \mathbf{r}_2, t+\tau)$ can be calculated using the quantum regression theorem and solving the master equation for the density matrix Eq. (11.55) [29].

In steady state and for an ion separation $d \gg \lambda$, the normalized second-order correlation function of Eq. (11.56) at $\tau = 0$ takes the form [29]

$$g^{(2)}(\boldsymbol{r}_1, \boldsymbol{r}_2, \tau = 0) = \frac{(1+s)^2 \cos^2\left(\left(\delta(\boldsymbol{r}_1) - \delta(\boldsymbol{r}_2)\right)/2\right)}{\left(1 + s + \cos\delta(\boldsymbol{r}_1)\right)\left(1 + s + \cos\delta(\boldsymbol{r}_2)\right)} \tag{11.57}$$

were s denotes the saturation parameter of the atomic transition $|g\rangle \leftrightarrow |e\rangle$ and $\delta(\boldsymbol{r}) = (\boldsymbol{k}_L - k_L\hat{\boldsymbol{r}})\cdot\boldsymbol{d}$ is the optical phase accumulated by a photon recorded at position $\boldsymbol{r}$ if scattered by ion 1 with respect to a photon scattered by ion 2, with $\boldsymbol{d}$ the distance vector between the ions. From Eq. (11.57), we infer that $g^{(2)}(\boldsymbol{r}_1, \boldsymbol{r}_2, \tau = 0)$ features a minimum detection probability for the second photon at $\delta(\boldsymbol{r}_2) = 0$ if the first photon has been recorded at $\delta(\boldsymbol{r}_1) = \pi$, corresponding to spatial subradiance, see Figure 11.16(a). Oppositely, $g^{(2)}(\boldsymbol{r}_1, \boldsymbol{r}_2, \tau = 0)$ exhibits a maximum detection probability for the second photon at $\delta(\boldsymbol{r}_2) = 0$ if the first photon has been recorded also at $\delta(\boldsymbol{r}_1) = 0$, which corresponds to spatial superradiance [38, 91], see Figure 11.16(b).

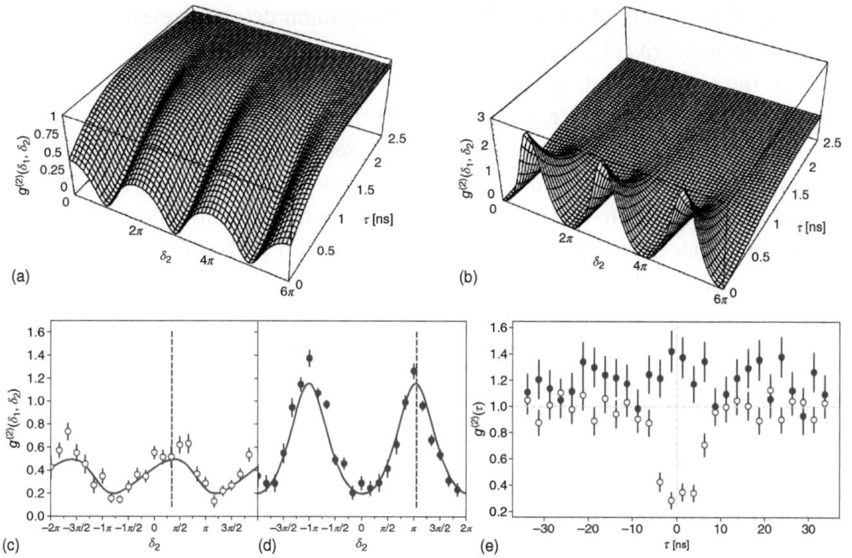

Figure 11.16 Theoretically derived spatiotemporal two-photon correlation function $g^{(2)}(\boldsymbol{r}_1, \boldsymbol{r}_2, \tau)$ for (a) $\delta(\mathbf{r}_1) = 0$ and (b) $\delta(\mathbf{r}_1) = \pi$; either spatial superradiance (a) or spatial subradiance (b) is predicted Source: Adapted from Skornia et al. [29]/with permission of American Physical Society. The experimental data shown in (c) for $\delta(\mathbf{r}_1) = 0.34\pi$ and (d) for $\delta(\mathbf{r}_1) = 1.05\pi$ confirm the theoretical predictions to a high degree; for the autocorrelation function displayed in (e) we find $g^{(2)}(\boldsymbol{r}_1, \boldsymbol{r}_1, \tau = 0) < 1$ for $\delta(\mathbf{r}_1) = 0.34\pi$, corresponding to photon antibunching, and $g^{(2)}(\boldsymbol{r}_1, \boldsymbol{r}_1, \tau = 0) > 1$ for $\delta(\mathbf{r}_1) = 1.05\pi$, corresponding to photon bunching; the related positions for $\delta(\mathbf{r}_2)$ are indicated as dotted lines in (c) and (d); error bars correspond to one statistical standard deviation for all subplots.

In order to measure $g^{(2)}(\boldsymbol{r}_1, \boldsymbol{r}_2, \tau)$, the two $^{40}Ca^+$ ions are stored again in a segmented linear Paul trap with trap frequencies $\omega_{(\mathrm{ax,\,rad1,\,rad2})}/2\pi = (0.76, 1.275, 1.568)$ MHz for the axial and the two radial modes, respectively, see Figure 11.3. An $f/1.6$ lens system in the far field collects 2.5% of the photons scattered by the ions. After passing a polarization filter, the photons are fed into a Hanbury-Brown and Twiss setup consisting of a non-polarizing 50/50 beam splitter and two synchronized MCP LinCam detectors, see Figure 11.15. With 1000×1000 virtual spatial bins and a timing resolution of 50 ps, the MCPs combine both high spatial and ultrahigh temporal resolution. As each of them has a dead time of 600 ns, we have to use two of such cameras to resolve $g^{(2)}(\boldsymbol{r}_1, \boldsymbol{r}_2, \tau)$ to better than the ion's coherence time given by $\tau_{P_{1/2}} = 6.9$ns [71].

With a single photon count rate of about ~ 7 kHz and a coincidence rate of ~ 68 mHz, we record data for about 205 hours. To increase the signal-to-noise ratio per bin, the two-dimensional 1000×1000 spatial data set of each frame of each MCP camera is reduced to 1×96 bins, taking advantage of the fact that the relevant spatial modulation of the photon correlations occurs parallel to the ion distance vector $\boldsymbol{d}$, similar to the first-order photon correlation function displayed in Figures 11.5(c) and 11.6. Choosing time bins of 2.5 ns instead of 50ps, the two-photon correlations are stored in a data structure with $96 \times 96 \times 38$ bins, encoding the position of the first (the second) photon detection event at $\boldsymbol{r}_1$ (at $\boldsymbol{r}_2$) and the photon arrival time difference τ.

From the total data set, we select at $\tau = 0$ two different values for the phase of the first photon, namely $\delta(\mathbf{r}_1) = 0.34\pi$ and 1.05π, and display $g^{(2)}(\boldsymbol{r}_1, \boldsymbol{r}_2, \tau = 0)$ as a function of $\delta(\mathbf{r}_2)$ in Figures 11.16(c) and (d), respectively. In the two figures, the two-photon correlation functions are binned to only two periods $\delta(\mathbf{r}_2) \in [-2\pi, +2\pi]$, to increase the signal-to-noise ratio per bin even further. From Figure 11.16(d) we can see that indeed a pattern with minimum emission probability in the direction $\delta(\mathbf{r}_2) \approx 0$ is obtained if the first photon has been recorded at $\delta(\mathbf{r}_1) = 1.05\ \pi$. In this case, the two-ion system has been projected approximately into the anti-symmetric Dicke state $|0, 0\rangle$, corresponding to the regime of spatial subradiance [29, 91]. In contrast, in Figure 11.16(c), we observe a pattern with maximum emission probability for the second photon in the direction $\delta(\mathbf{r}_2) \approx 0$ in case that the first photon has been recorded at $\delta(\mathbf{r}_1) = 0.34\ \pi$. For this setting, the two-ion system has been projected approximately into the symmetric Dicke state $|D_{1,1}\rangle = |1, 0\rangle$, i.e., we observe the regime of spatial superradiance [29, 38, 39, 91].

To fully model the experimentally measured data, we extend Eq. (11.57) by including the Debye-Waller factor $f_{DWF}^{\boldsymbol{r}}$, which describes the residual motion of the trapped ions as well as the momentum transfer induced onto the two-ion system by the absorbed and emitted photons. Additionally, we take an offset of $\Delta \approx 0.2$ into account resulting from dark counts and events when the ions decay into the metastable $D_{3/2}$ state and have to be pumped back into the fluorescence cycle by the laser near 866 nm [92] (~ 6.5 % of the deexcitations). The offset includes also a loss of contrast due to the convolutions in the temporal and spatial binning process outlined above. From Eq. (11.57), we thus derive the theoretical model function

$$g^{(2)}(\boldsymbol{r}_1, \boldsymbol{r}_2, \tau = 0) = \frac{(1+s)^2 \cos^2\left(\left(\delta(\boldsymbol{r}_1) - \delta(\boldsymbol{r}_2)\right)/2\right) f_{DWF}^{\boldsymbol{r}_1} f_{DWF}^{\boldsymbol{r}_2}}{\left(1 + s + \cos\delta(\boldsymbol{r}_1) f_{DWF}^{\boldsymbol{r}_1}\right)\left(1 + s + \cos\delta(\boldsymbol{r}_2) f_{DWF}^{\boldsymbol{r}_2}\right)} + \Delta, \tag{11.58}$$

where $f_{DWF}^{\boldsymbol{r}_i}$ is given by [74]

$$f_{DWF}^{\boldsymbol{r}_i} = \exp\left[-\frac{\hbar q_x^2(\boldsymbol{r}_i)}{m\omega_{roc_x}}\left(\langle N_{roc_x}\rangle + \frac{1}{2}\right) - \frac{\hbar q_y^2(\boldsymbol{r}_i)}{m\omega_{roc_y}}\left(\langle N_{roc_y}\rangle + \frac{1}{2}\right) - \frac{\hbar q_z^2(\boldsymbol{r}_i)}{m\omega_b}\left(\langle N_b\rangle + \frac{1}{2}\right)\right]. \tag{11.59}$$

Here $q_\alpha(\boldsymbol{r}_i)$, $i = 1, 2$, describes the projection of the momentum transfer vector $\boldsymbol{q}(\boldsymbol{r}_i) = \boldsymbol{k}_L - k_L\hat{\boldsymbol{r}}_i$ onto the basis axis $\alpha = x, y, z$ of the underlying vibrational modes ω_κ occupied by $\langle N_\kappa\rangle$ phonons, where ω_κ are the frequencies of the relevant vibrational modes $\kappa = roc_x, roc_y, b$. Note that for a two-ion system, out of the six possible modes only the contrast-reducing modes have to be taken into account, namely the breathing mode ω_b, affecting the inter-ion distance, and the two rocking-modes ω_{roc_x} and ω_{roc_y}, acting perpendicular to the ion distance vector $\boldsymbol{d}$ and describing the shear movement of the ions along two orthogonal axes. For the investigated two-ion system, the Debye-Waller factors calculate to $f_{DWF}^{\boldsymbol{r}_1} \approx f_{DWF}^{\boldsymbol{r}_2} = 0.51$, taking into account the chosen angle of $\alpha \approx 135°$ between the laser wave vector $\boldsymbol{k}_L$ and the detector positions $\boldsymbol{r}_i$, $i = 1, 2$, as well as the residual phonon numbers $\langle N_b \rangle \approx 15$, $\langle N_{roc_x} \rangle \approx 9$, and $\langle N_{roc_y} \rangle \approx 7$ after cooling the two-ion system to the Doppler-limit (see Figure 11.3). The saturation parameter of the cooling laser is determined to $s = 0.65$, derived experimentally independently from the Rabi oscillations of the auto-correlation function $g^{(2)}(\tau)$ using a single trapped ion for different laser powers [84].

Turning to the discussion of the temporal features in the measured photon correlation function $g^{(2)}$, we observe that the photon auto-correlation function $g^{(2)}(\boldsymbol{r}_1, \boldsymbol{r}_1, \tau = 0)$ displays antibunching for $\delta(\mathbf{r}_1) = \delta(\mathbf{r}_2) = 0.34\ \pi$ and bunching for $\delta(\mathbf{r}_1) = \delta(\mathbf{r}_2) = 1.05\ \pi$, see Figure 11.16(e) [43]. Combining both the spatial and the temporal aspects of the collective light emision of the two ion system, we recognize that the maximum of superradiance observed at a position $\delta(\mathbf{r}_2) = \delta(\mathbf{r}_1) \approx 0$ is accompagnied by photon antibunching, whereas the maximum of spatial subradiance appearing at $\delta(\mathbf{r}_2) = \delta(\mathbf{r}_1) \approx \pi$ comes along with bunching of the photon stream (see also figures 11.16(a) and (b)).

11.9 Conclusion and Outlook

In this article, we have given a comprehensive overview on the collective scattering of light by trapped ion crystals serving as arrays of correlated emitters of indistinguishable photons into free space. Since for trapped ions the mutual Coulomb repulsion prohibits separations below the optical transition wavelength, we have focussed the analysis on the particular case of atoms with separations d >> λ. The interatomic correlations are produced in this case in free space via the detection of photons in the far field, i.e., by excluding Welcher-Weg information. The collectively emitted light is subsequently recorded in the far field by pixelated cameras, which feature high spatial and high temporal resolution. We observe classical interference based on classical coherence of phase-coherently oscillating synchronized dipoles if we aim for measuring the first-order photon correlation function. However, we observe quantum interferences based on quantum coherence [49] of entangled ion arrays when measuring higher-order photon correlation functions $G^{(m)}_{\Psi^i_N}(\mathbf{r}_1, \ldots, \mathbf{r}_1, \mathbf{r}_2)$ with $m > 1$. For a two-ion crystal, we recorded spatial superradiance in combination with anti-bunching and spatial subradiance in combination with bunching, depending on the position of detection of the first photon when starting from the fully excited state. We analyzed theoretically how the observed features of collective light emission from arrays of correlated ions result from measurements projecting the initially non-entangled ions in the fully excited state into entangled Dicke states. The experiments with a two-ion crystal fully confirmed these predictions.

In the future, we aim for scaling up the number of correlated ions, extending the linear crystal structures toward two-dimensional ion crystals as emitters of indistinguishable photons. For the detection we will employ high numerical aperture objectives in combination with sets of single photon resolving cameras with high quantum efficiency. The increased detection efficiency will boost the photon correlation data acquisition rates, eventually enabling us to conduct photon correlation measurements with $m > 2$. We further aim for quantifying the state fidelity of the projectively prepared multi-ion Dicke states and the corresponding entanglement from the measured multi-photon correlation interference patterns.

Acknowledgments

We thank the scientists with whom we have collaborated over many years on the subject of collective light emission from trapped ions, in particular T. Bastin, D. Bhatti, M. Bojer, A. Classen, S. Mährlein, S. Oppel, S. Richter, U. Schilling, C. Skornia, E. Solano, C. Thiel, J. Wechs, S. Wolf, R. Wiegner, and specifically G. S. Agarwal for a long-standing fruitful collaboration. We thank Photonscore GmbH for providing the synchronized MCP camera system for the experiment outlined in Section 11.7. We acknowledge funding by the Deutsche Forschungsgemeinschaft (DFG, German Research Foundation) within the TRR 306 QuCoLiMa ("Quantum Cooperativity of Light and Matter") – Project-ID 429529648 and acknowledge support from the Graduate School of Advanced Optical Technologies (SAOT) as well as the International Max-Planck Research School – Physics of Light, Erlangen.

Data availability

Source data for the plots shown in Section 11.7 are provided under https://dx.doi.org/10.22000/560. The experimental data that support the findings of this study are available from the corresponding author upon reasonable request.

References

1 Raimond, J.M., Brune, M., and Haroche, S. (2001). *Reviews of Modern Physics* 73: 565.
2 Rempe, G., Schmidt-Kaler, F., and Walther, H. (1990). *Physical Review Letters* 64: 2783.
3 Thompson, R.J., Rempe, G., and Kimble, H.J. (1992). *Physical Review Letters* 68: 1132.
4 Boca, A., Miller, R., Birnbaum, K.M. et al. (2004). *Physical Review Letters* 93: 233603.
5 Haroche, S. and Raimond, J. (1985). *Advances in Atomic and Molecular Physics* 20: 347–411.
6 Chang, D.E., Douglas, J.S., González-Tudela, A. et al. (2018). *Reviews of Modern Physics* 90: 031002.
7 Reiserer, A. and Rempe, G. (2015). *Reviews of Modern Physics* 87: 1379.
8 DeVoe, R. and Brewer, R. (1996). *Physical Review Letters* 76: 2049.
9 Eschner, J., Raab, C., Schmidt-Kaler, F., and Blatt, R. (2001). *Nature* 413: 495.

10 González-Tudela, A., Hung, C.-L., Chang, D.E. et al. (2015). *Nature Photonics* 9: 320.

11 Facchinetti, G., Jenkins, S.D., and Ruostekoski, J. (2016). *Physical Review Letters* 117: 243601.

12 Bettles, R.J., Gardiner, S.A., and Adams, C.S. (2016). *Physical Review Letters* 116: 103602.

13 Shahmoon, E., Wild, D.S., Lukin, M.D., and Yelin, S.F. (2017). *Physical Review Letters* 118: 113601.

14 Asenjo-Garcia, A., Moreno-Cardoner, M., Albrecht, A. et al. (2017). *Physical Review X* 7: 031024.

15 Bradac, C., Johnsson, M.T., van Breugel, M. et al. (2017). *Nature Communications* 8: 1205.

16 Ferioli, G., Glicenstein, A., Robicheaux, F. et al. (2021). *Physical Review Letters* 127: 243602.

17 Rui, J., Wei, D., Rubio-Abadal, A. et al. (2020). *Nature* 583: 369.

18 Lodahl, P., Mahmoodian, S., Stobbe, S. et al. (2017). *Nature* 541: 473.

19 van Loo, A.F., Fedorov, A., Lalumière, K. et al. (2013). *Science* 342: 1494.

20 Sipahigil, A., Evans, R.E., Sukachev, D.D. et al. (2016). *Science* 354: 847.

21 Evans, R.E., Bhaskar, M.K., Sukachev, D.D. et al. (2018). *Science* 362: 662.

22 Kim, J.-H., Aghaeimeibodi, S., Richardson, C.J.K. et al. (2018). *Nature* 18: 4734.

23 Grim, J.Q., Bracker, A.S., Zalalutdinov, M. et al. (2019). *Nature Materials* 18: 963.

24 Prasad, A.S., Hinney, J., Mahmoodian, S. et al. (2020). *Nature Photonics* 14: 719.

25 Ebadi, S., Wang, T.T., Levine, H. et al. (2021). *Nature* 595: 227–232.

26 Monroe, C., Meekhof, D.M., King, B.E. et al. (1995). *Physical Review Letters* 75: 4714.

27 Monz, T., Schindler, P., Barreiro, J.T. et al. (2011). *Physical Review Letters* 106: 130506.

28 Cabrillo, C., Cirac, J.I., Garcia-Fernandez, P., and Zoller, P. (1999). *Physical Review A* 59: 1025.

29 Skornia, C., von Zanthier, J., Agarwal, G.S. et al. (2001). *Physical Review A* 64: 063801.

30 Browne, D.E., Plenio, M.B., and Huelga, S.F. (2003). *Physical Review Letters* 91: 067901.

31 Simon, C. and Irvine, W.T.M. (2003). *Physical Review Letters* 91: 110405.

32 Thiel, C., von Zanthier, J., Bastin, T. et al. (2007). *Physical Review Letters* 99: 193602.

33 Mochring, D.L., Maunz, P., Olmschenk, S. et al. (2007). *Nature* 449: 68.

34 Bastin, T., Thiel, C., von Zanthier, J. et al. (2009). *Physical Review Letters* 102: 053601.

35 Hofmann, J., Krug, M., Ortegel, N. et al. (2012). *Science* 337: 72.

36 Slodička, L., Hétet, G., Röck, N. et al. (2013). *Physical Review Letters* 110: 083603.

37 Bernien, H., Hensen, B., Pfaff, W. et al. (2013). *Nature* 497: 86.

38 Oppel, S., Wiegner, R., Agarwal, G., and von Zanthier, J. (2014). *Physical Review Letters* 113: 263606.

39 Wiegner, R., Oppel, S., Bhatti, D. et al. (2015). *Physical Review A* 92: 033832.

40 McConnell, R., Zhang, H., Hu, J. et al. (2015). *Nature* 519: 439.
41 Delteil, A., Sun, Z., Gao, W.-b. et al. (2016). *Nature Physics* 12: 218.
42 Araneda, G., Higginbottom, D.B., Slodička, L. et al. (2018). *Physical Review Letters* 120: 193603.
43 Wolf, S., Richter, S., von Zanthier, J., and Schmidt-Kaler, F. (2020). *Physical Review Letters* 124: 063603.
44 Richter, S., Wolf, S., von Zanthier, J., and Schmidt-Kaler, F. (2022). *Physical Review Research* 5: 013163.
45 Meir, Z., Schwartz, O., Shahmoon, E. et al. (2014). *Physical Review Letters* 113: 193002.
46 Stockill, R., Stanley, M.J., Huthmacher, L. et al. (2017). *Physical Review Letters* 119: 010503.
47 Maser, A., Schilling, U., Bastin, T. et al. (2009). *Physical Review A* 79: 033833.
48 Gross, M. and Haroche, S. (1982). *Physics Reports* 93: 301.
49 Bojer, M. and von Zanthier, J. (2022). *Superradiance of non-interacting atoms* arXiv:2206.14646.
50 Bonifacio, R. and Lugiato, L.A. (1975). *Physical Review A* 11: 1507.
51 Ludlow, A.D., Boyd, M.M., Ye, J. et al. (2015). *Reviews of Modern Physics* 87: 637.
52 Blatt, R. and Roos, C. (2012). *Nature Physics* 8: 277.
53 Schmidt-Kaler, F., Gulde, S., Riebe, M. et al. (2003). *Journal of Physics B: Atomic, Molecular and Optical Physics* 36: 623.
54 Ruster, T., Schmiegelow, C., Kaufmann, H. et al. (2016). *Applied Physics B* 122: 254.
55 Wang, P., Luan, C., Qiao, M. et al. (2021). *Nature Communications* 12: 233.
56 Mortensen, A., Nielsen, E., Matthey, T., and Drewsen, M. (2006). *Physical Review Letters* 96: 103001.
57 Kaufmann, H., Ulm, S., Jacob, G. et al. (2012). *Physical Review Letters* 109: 263003.
58 Joshi, M.K., Fabre, A., Maier, C. et al. (2021). *New Journal of Physics* 22: 103013.
59 Qiao, M., Cai, Z., Wang, Y. et al. (2022). arXiv:2204.07283.
60 Moehring, D., Highstrete, C., Stick, D. et al. (2011). *New Journal of Physics* 13: 075018.
61 Ragg, S., Decaroli, C., Lutz, T., and Home, J. (2019). *Review of Scientific Instruments* 90: 103203. https://doi.org/10.1063/1.5119785.
62 Kaushal, V., Lekitsch, B., Stahl, A. et al. (2020). *AVS Quantum Science* 2: 014101. https://doi.org/10.1116/1.5126186.
63 Revelle, M. (2020). arXiv:2009.02398.
64 Bowler, R., Gaebler, J., Lin, Y. et al. (2012). *Physical Review Letters* 109: 080502.
65 Ruster, T., Warschburger, C., Kaufmann, H. et al. (2014). *Physical Review A* 90: 033410.
66 Leibfried, D., Blatt, R., Monroe, C., and Wineland, D. (2003). *Reviews of Modern Physics* 75: 281.
67 Morigi, G., Eschner, J., and Keitel, C.H. (2000). *Physical Review Letters* 85: 4458.
68 Lechner, R., Maier, C., Hempel, C. et al. (2016). *Physical Review A* 93: 053401.

69 Wolf, S., Wechs, J., von Zanthier, J., and Schmidt-Kaler, F. (2016). *Physical Review Letters* 116: 183002.

70 Jacob, G., Groot-Berning, K., Wolf, S. et al. (2016). *Physical Review Letters* 117: 043001.

71 Hettrich, M., Ruster, T., Kaufmann, H. et al. (2015). *Physical Review Letters* 115: 143003.

72 Shamos, M.H. (1959). *Great Experiments in Physics: Firsthand Accounts from Galileo to Einstein*. Courier Corporation.

73 Eichmann, U., Bergquist, J.C., Bollinger, J.J. et al. (1993). *Physical Review Letters* 70: 2359.

74 Itano, W.M., Bergquist, J.C., Bollinger, J.J. et al. (1998). *Physical Review A* 57: 4176.

75 Ficek, Z. and Swain, S. (2005). *Quantum Interference and Coherence: Theory and Experiments*, vol. 100. Springer Science & Business Media.

76 DeVoe, R.G. and Brewer, R.G. (1996). *Physical Review Letters* 76: 2049.

77 Wong, T., Tan, S.M., Collett, M.J., and Walls, D.F. (1997). *Physical Review A* 55: 1288.

78 Schön, C. and Beige, A. (2001). *Physical Review A* 64: 023806.

79 James, D.F.V. (1998). *Applied Physics B* 66: 181.

80 Jacob, G., Groot-Berning, K., Wolf, S. et al. (2014). arxiv.org:1405.6480.

81 Loudon, R. (2000). *The Quantum Theory of Light*. Oxford University Press.

82 Cohen-Tannoudji, C., Dupont-Roc, J., and Grynberg, G. (2004). *Photons and Atoms: Introduction to Quantum Electrodynamics*. Wiley-VCH.

83 Scully, M.O. and Zubairy, M.S. (1997). *Quantum Optics*. Cambridge University Press.

84 Diedrich, F. and Walther, H. (1987). *Physical Review Letters* 58: 203.

85 Mollow, B.R. (1969). *Physical Review* 188: 1969.

86 Mandel, L. and Wolf, E. (1995). *Optical Coherence and Quantum Optics*. Cambridge University Press.

87 Bhatti, D., Bojer, M., and von Zanthier, J. (2021). *Physical Review A* 104: 052401.

88 Dicke, R.H. (1954). *Physical Review* 93: 99.

89 Agarwal, G.S. (1974). *Quantum Optics*, 1–128. Berlin: Springer-Verlag.

90 Scully, M.O. and Svidzinsky, A.A. (2009). *Science* 325: 1510.

91 Wiegner, R., von Zanthier, J., and Agarwal, G. (2011). *Physical Review A* 84: 023805.

92 Ramm, M., Pruttivarasin, T., Kokish, M. et al. (2013). *Physical Review Letters* 111: 023004.

12

Single-Molecule Magnets Spin Devices

Eufemio Moreno-Pineda[1,2] *and Wolfgang Wernsdorfer*[3,4]

[1] *Facultad de Ciencias Naturales, Exactas y Tecnología, Depto. de Química-Física, Universidad de Panamá, Campus Universitario Octavio Méndez Pereira, 3366, Panamá*
[2] *Sistema Nacional de Investigadores (SNI), Secretaría Nacional de Ciencias y Tecnología (SENACYT), Clayton Ciudad del Saber Edif.205, 0816-02852, Panamá*
[3] *Karlsruhe Institute of Technology (KIT), Institute for Quantum Materials and Technology (IQMT), Hermann-von-Helmholtz-Platz 1, D-76344 Eggenstein-Leopoldshafen, Germany*
[4] *Physikalisches Institut, Karlsruhe Institute of Technology (KIT), Wolfgang-Gaede-Str. 1, D-76131 Karlsruhe, Germany*

12.1 Introduction

The impressive and unique characteristics of single-molecule magnets (SMMs), systems possessing an energy barrier to the relaxation of the magnetization at single-molecule level, stimulated intensive investigation of these systems for their implementation in technological applications. The quantum properties exhibited by SMMs could allow the long-sought control of quantum effects and their implementation in practical applications. This goal has been in the mind of scientists since the quantum mechanical description by Erwin Schrödinger [1], with scientists continuously exploring new quantum effects and developing the understanding of the quantum properties. The acquired knowledge of the quantum nature during the so-called *first quantum revolution* allowed researchers to understand the chemical properties and physics of systems, e.g. semiconductors, ultimately leading to inventions such as the laser and transistors. As a subsequent step, a so-called *second quantum revolution* attempts to utilize the acquired knowledge of the quantum principles to engineering quantum systems to implement them in technological applications [2, 3]. The continuous miniaturization of technological devices, as described by Moore's law, and the quantum nature taking over at the molecular and atomic scale are the main driving forces.

The foreseen advantages offered by the implementation of the quantum effects in practical applications are numerous. For instance, by exploiting quantum coherence and entanglement of electronic states, it is possible to measure small fluctuations surpassing classical sensing devices, offering new sensing tools with applicability ranging from physics to biology [4–12]. Furthermore, quantum effects can allow the

Photonic Quantum Technologies: Science and Applications, First Edition. Edited by Mohamed Benyoucef.

simulation of specific quantum mechanical problems, as proposed by Feynman [13] and Lloyd [14]. One of the most attractive implementations is related to the development of a quantum computer, which would enable the realization of extremely difficult and even unachievable tasks, such as the factorization of large numbers, fast computation of extremely large problems, to the true universal simulation of large quantum mechanical problems, among others. The development of a quantum computer triggered the investment of several companies and consortiums eventually resulting in a 16-qubit processor developed by the IBM Q-experience [15], a 72-qubit quantum chip developed by Google [16], a 19-qubits computer tested by Rigetti [17], a silicon spin-based 2-qubits processor achieved by QuTech [18], and even a NASA–Google partnership has claimed quantum supremacy [19].

Currently, there are several platforms explored for the development and implementation in technological applications. Among the several systems explored toward this end, SMMs are very appealing due to the displayed quantum effects and the tailored control over not just the structural aspects of the systems, but also over their electronic and nuclear properties. In this chapter, some key aspects of SMMs for the integration in novel technological applications such as quantum simulators, quantum sensing, and quantum computing are discussed. We will also revise the development of an emerging technology, known as quantum communications, employing magnetic molecules, in which the read-out and manipulation of the encoded information can be realized by light.

12.2 SMMs and Quantum Effects

The development of new technologies based on quantum mechanics is contingent upon the observed quantum effects, such as quantization, superposition, entanglement, tunneling, and coherence [2]. For the development of quantum sensors, quantum simulators, and quantum computers, a set of characteristics must be met, which are very similar to the requirements for quantum bits [20]. In this section, a brief description of certain key properties of SMMs, which makes the candidates to be considered in these novel technologies, is provided.

12.2.1 Quantization

One key aspect for molecular systems to be employed in technological applications is the quantization of their ground state. Quantization in a molecular system was first observed in the archetypal $[Mn_{12}(\mu_3\text{-}O_{12})(CH_3COO)_{16}(H_2O)_4]\cdot 2CH_3COOH]\cdot 2H_2O$ (or Mn_{12}) complex [21]. At low temperatures, an $S = 10$ ground state characterized the Mn_{12} cage, which arises from the strong interaction of an inner tetrahedron of Mn^{+4} and an external octagon of Mn^{+3}. For transition metals, although small, the spin orbit coupling (SOC) plays an important role since it induces magnetic anisotropy in the molecule. The latter can be described by a spin Hamiltonian of the following form: $\mathscr{H} = DS_z^2 + E\left(S_x^2 + S_y^2\right)$, where D is the zero-field splitting parameter and E is the rhombic term. When considering the first term in the spin Hamiltonian, the S states are quantized and labeled with the quantum numbers $m_S = -S \ldots S$ being

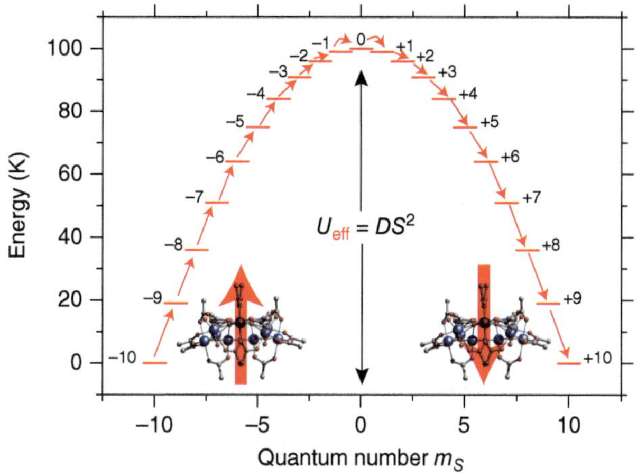

Figure 12.1 Quantum Effects in Single-Molecule Magnets: Schematic view of the energy diagram of Mn_{12} with $S = 10$ with $U_{eff} = DS^2$. The reversal of the magnetization occurs through the m_S states, represented by the red arrows between m_S states.

doubly degenerated: $E(\pm m_S) = -|D|\,m_S^2$, hence, following a parabolic behavior with an energy barrier to the relaxation with the form $U_{eff} = -|D|\,m_S^2$. In the parabola, each side corresponds to an opposed orientation of the magnetic moment; hence, D gives the approximate barrier for the spin to flip its orientation. The reversal of the magnetization occurs through the m_S states via phonon absorption (excitation of the spin system to the top of the barrier) and emission (de-excitation to the bottom of the other side of the barrier) (Figure 12.1). At the beginning of the SMMs era, due to the spin dependence of the energy barrier, transition metal SMMs with large spin multiplicity were the aim.

Although quantization was first observed in transition metal SMMs, undoubtedly, the most striking results in the field of molecular magnetism were observed for lanthanide-based SMMs [22]. The new era for the molecular magnetism field began with the observation of an energy barrier to the relaxation in a single lanthanide ion complex with the formula TBA[$TbPc_2$] [22]. The complex comprised a Tb^{3+} ion sandwiched between two phthalocianinato (Pc) anionic ligands and charged balanced by a tetrabutylammonium (TBA) cation (Figure 12.2).

In stark contrast to transition metal SMMs, where polymetallic cages were required, the SMM characteristics of the 4f-SMMs arose from the intrinsic strong SOC due to the large nuclear charge and the unquenched orbital momentum. For 4f-elements their magnetic moment is best described by the Russell–Saunders (RS) coupling scheme ($J = |L - S| \leq J \leq L + S$), given that neither S nor L is a good quantum number. For $[TbPc_2]^-$ a $S = 3$ and $L = 3$, hence, $J = L + S = 6$ and due to the large separation between the $J = 6$ and $J = 5$ (c. 2900 K), solely the $J = 6$ multiplet is populated at normal conditions. Taking into account the D_{4h} symmetry the Tb^{3+} ion, sandwiched between the Pc ligands, ligand field spin Hamiltonian takes the form [23]:

$$\mathcal{H}_{lf} = B_2^0 O_2^0 + B_4^0 O_4^0 + B_4^4 O_4^4 + B_6^0 O_6^0 + B_6^4 O_6^4 \quad (12.1)$$

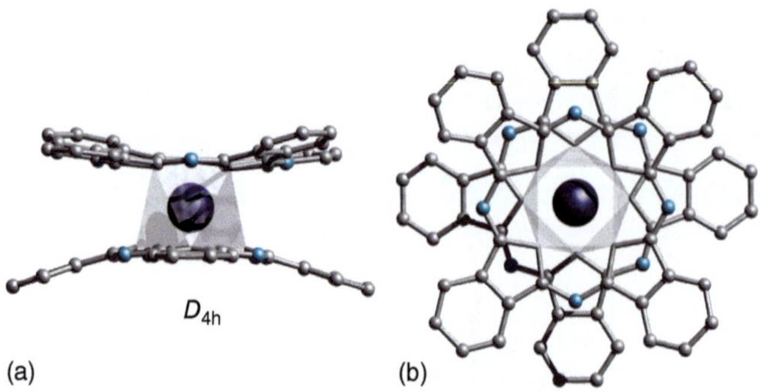

Figure 12.2 TBA[$TbPc_2$] First 4f-SMM: (a) front and (b) side view of [$TbPc_2$]$^-$ SMM composed of two phtalocianinato (Pc) ligands and a Tb^{3+} ion (Color code: Tb, blue; N, cyan; C, light gray), with local D_{4h} symmetry. TBA counter cation and hydrogen atoms omitted for clarity.

where B_q^k are experimentally determined parameters. Diagonalization of the spin Hamiltonian reveals that the ligand field lifts the degeneracy of the $2J+1$ states of the $J = 6$ multiplet, rendering a $m_J = \pm 6$ ground state with a separation of c. 600 K from the $m_J = \pm 5$ first excited states (Figure 12.3 red and blue traces). The crystal field acting upon the Tb^{3+} ion generates a quantization axis for the $m_J = \pm 6$ ground state perpendicular to the Pc plane, which is frozen below a certain threshold temperature.

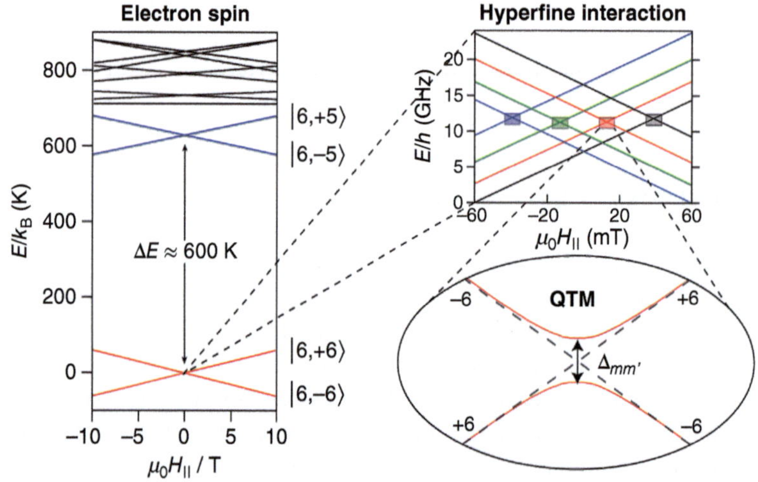

Figure 12.3 **Quantum Tunneling of the Magnetization:** Electronic energy level for TBA[$TbPc_2$] (left) with the first excited state (blue traces) separated by c. 600 K from the ground doublet state (red traces). Zoomed region shows the hyperfine split ground doublet state by nuclear spin $I = 3/2$ into $m_I = -3/2$ (black), $-1/2$ (red), $+1/2$ (green), and $+3/2$ (blue) square regions (right). The zoomed region of the hyperfine levels shows the avoided level crossing arising from transverse field. Reversal can occur via *hf*-QTM through the barrier when the states are in resonance. Source: Adapted from Moreno-Pineda et al. [24] with permission from The Royal Society of Chemistry.

Another remarkable aspect of 4f-SMMs is their strong hyperfine coupling, which is of utmost importance for the observation of other quantum effects as well as their implementation in quantum computing applications as described later [25]. The electronic spin couples strongly to the nuclear spin $I = 3/2$ of Tb^{3+} via the hyperfine and the quadrupolar interaction. The spin Hamiltonian hence takes the form [26, 27]:

$$\mathcal{H}_{\mathrm{TbPc_2}} = \mathcal{H}_{\mathrm{lf}} + \mu_{\mathrm{B}}\mu_0 \boldsymbol{H}\hat{g}\boldsymbol{J} + A_{\mathrm{hyp}}\boldsymbol{I}\cdot\boldsymbol{J} + \boldsymbol{I}\hat{P}_{\mathrm{quad}}\boldsymbol{I} \tag{12.2}$$

with the second term being the Zeeman interaction, and the third and fourth terms the hyperfine and quadrupolar interactions, respectively. The hyperfine interaction splits the $m_J = 6$ state further in $2I + 1$ levels, i.e. $m_I = +3/2, +1/2, -1/2$, and $-3/2$, while the quadrupolar term couples the electric field gradient of the magnetic moment, causing an uneven separation between m_I states (zoomed region in Figure 12.3 (right)); this playing an important role for the manipulation of the hyperfine levels.

12.2.2 Quantum Coherence

An important aspect for the implementation of SMMs in quantum technologies is to be able to carry out logical operations without losing the information before the operation has been completed. Two parameters are commonly employed to evaluate the performance of the systems: the spin–spin relaxation time, T_2, and the spin–lattice relaxation time, T_1. T_2 is the coherence time and dictates the superposition lifetime, i.e. the computational time. More precisely, T_2 is the time taken by the magnetization to decay on the xy-plane after being tipped 90° from the z axis. T_1 describes how long it takes for the magnetization to return to the thermal equilibrium along the z-axis after being tipped 90° (Figure 12.4). These two relaxations are intertwined, as the coherence time is limited by T_1, as $2T_1 \geq T_2$, especially at high temperatures. SMMs have been shown to possess relaxation times similar or even better than common platforms and due to this characteristic have been proposed as the basic unit of quantum computers, i.e. the quantum bit or qubit.

Although the electronic ground state is expected to remain frozen in a certain direction below certain temperatures, the temperature-dependent relaxation

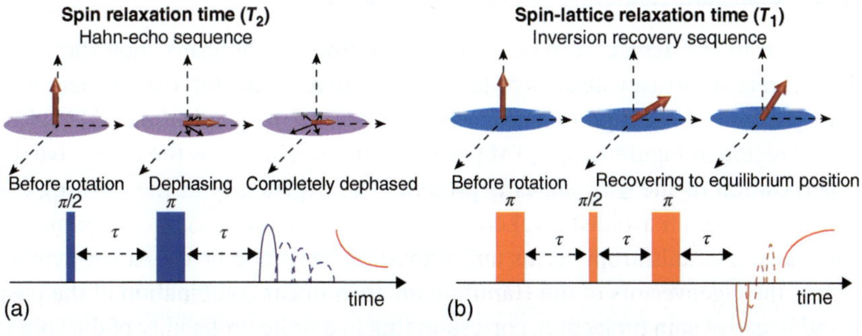

Figure 12.4 Quantum Coherence and Relaxation: Schematic representations of (a) T_2 and (b) T_1 and common sequences employed for their determination.

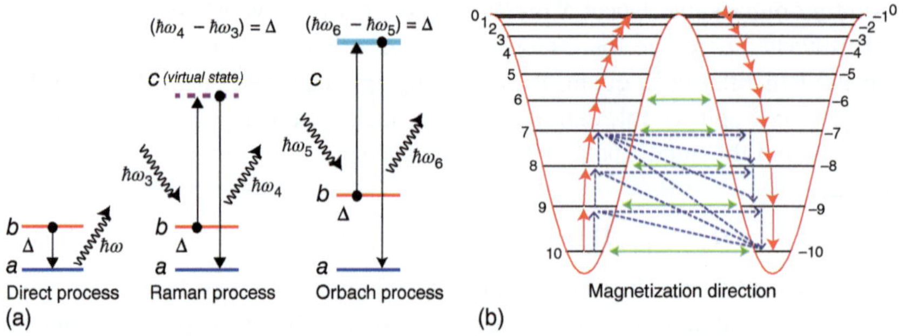

Figure 12.5 Quantum Coherence and Relaxation: (a) Common relaxation mechanism in SMMs. (b) Energy diagram for the relaxation occurring in SMMs. Red arrows between m_S and $m_S + 1$ states represent the Orbach relaxation, blue arrows are the Raman relaxation, while the green arrows represent QTM.

mechanisms become important when the thermal energy is comparable to the barrier; thus, the system can relax via several thermally activated processes. Spin–lattice relaxation or phonon-assisted mechanisms contribute to the relaxation of the magnetization greatly by modulating the ligand field via vibrations, causing a modulation on the crystal field, hence, inducing transitions between the different spin states. These processes can occur through absorption and emission of one or two phonons and are contingent upon the nature of the ion involved, i.e. Kramers or non-Kramers ions. For a Kramers ion, a minimum degeneracy of two is always retained, while in non-Kramers ions the degeneracy can be completely lifted. Overall, three processes are mainly involved in the relaxation dynamics of *f*-SMMs: (i) direct, (ii) Raman, and (iii) Orbach process (Figure 12.5). In the direct mechanism, the relaxation takes place through the emission of one phonon of energy. However, in Orbach and Raman, the relaxation involves the absorption and reemission of phonons. The difference is that while the first process goes through a real excited state, the second uses a virtual excited state to relax. The understanding of the relaxation pathways [28–33] has allowed the synthesis of SMMs with remarkably enhanced properties [34–36].

12.2.3 Quantum Tunneling of the Magnetization

In addition to the temperature-activated relaxation mechanisms, non-thermally activated relaxation can also provide an alternative route for the relaxation of the magnetization, through quantum tunneling of the magnetization (QTM) [37] (zoomed region in Figure 12.3). QTM permits spin reversal below the energy barrier if superposition of the $\pm m_J$ states is present. The degeneracy at the crossing can be converted into an avoided level crossing, with a separation $\Delta_{m,m'}$, consequence of the transverse anisotropy terms and transverse magnetic fields. At the avoided crossing, the eigenvectors of the Hamiltonian are a linear combination of the positive and negative spin projection, corresponding to a finite probability of the spin to be at both sides of the barrier. The resonance of the spin between both orientations is denoted as spin tunneling. The tunneling probability, P_{LS}, between states m and

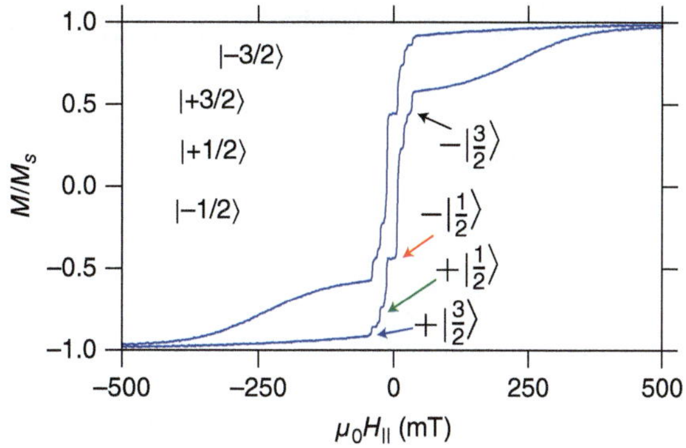

Figure 12.6 Quantum Tunneling of the Magnetization: μSQUID hysteresis loops and *hf*-QTM events. Source: Adapted from Moreno-Pineda et al. [24] with permission from The Royal Society of Chemistry.

m' upon sweeping the magnetic field at a rate, α, through the resonance is given by the Landau–Zener–Stückelberg (LZS) model:

$$P_{LS} = 1 - \exp\left(-\pi\omega_T \frac{\delta H_0}{\alpha}\right) \tag{12.3}$$

where ω_T is the angular frequency of oscillation between states m and m', which is related to the tunnel splitting by $\Delta_{m,m'} = 2\hbar\omega_T^{m,m'}$. The field interval where tunneling is predicted to occur is given by $\delta H_0 = \frac{\Delta_{m,m'}}{g\mu_B\mu_0|m-m'|}$, known as bare tunnel width.

Hyperfine-driven QTM (*hf*-QTM) has been observed in TBA[$TbPc_2$] (Figure 12.6) in μSQUID studies. Note that despite having a large separation between the ground and first excited state, *hf*-QTM acts as relaxation pathway for the magnetization [26]. Although QTM acts as temperature-independent relaxation pathways, considered a nuisance for high-density data storage applications, it plays an important role in the initialization, manipulation, and read-out of nuclear spins in the $TbPc_2$ complex, which ultimately lead to the realization of Grover's quantum algorithm [25, 38]. QTM has also allowed the read-out of the hyperfine-coupled nuclear spin states of a Tb_2Pc_3 SMM in a single crystal [39] and more recently in a spin transistor configuration [40].

12.2.4 Quantum Bits and Multilevel Systems

Due to the quantum characteristic exhibited by SMMs, these systems have been proposed to act as quantum bits or qubits. Unlike the classical bit, the quantum nature of the qubit allows for the superposition of the $|1\rangle$ or $|0\rangle$ states, i.e. $|\psi\rangle = a_0|0\rangle + a_1|1\rangle$ where the squares of a_0 and a_1 are the amplitude of the probability following $|a_0|^2 + |a_1|^2 = 1$ (Figure 12.7). The non-orthogonal configurations providing 2^N states, with N being the number of qubits, give QC the potential to perform immensely large and complex operations.

Many SMMs have been shown to fulfill the so-called DiVincenzo criteria [41], that is: (i) well-defined scalable levels; (ii) long coherence times in order to carry out the computational operation; (iii) to be able to initialize the system in well-defined states; (iv) to be able to perform universal quantum gates via entanglement and/or superposition of states; (v) successful read-out of the quantum states after

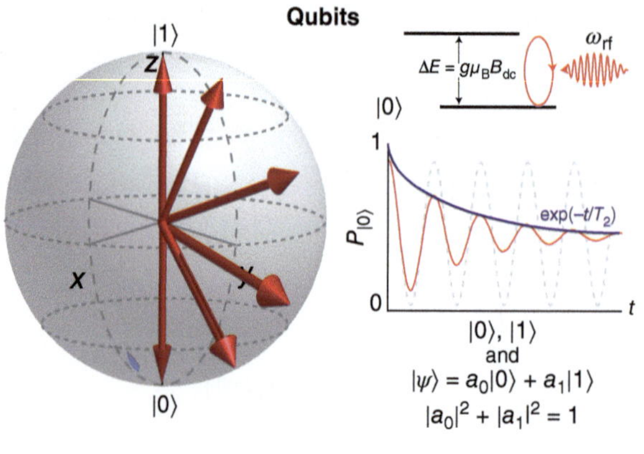

Figure 12.7 Quantum Computer Building Blocks: Schematic representation of qubit, with the four states comprising it. For a qubit, the coherence should be long enough to allow the manipulation, operation, and read-out before the information is lost.

the operation. For quantum communications, two other factors are of importance [41]: (vi) interconversion between dynamic and stationary qubits and (vii) exact transmission of dynamic qubits. These two prerequisites are essential for the effective transmission of information when employing the entanglement of photons and for non-local qubits [42].

All the necessary characteristics for qubits are gathered by SMMs. Due to the facile manipulation of the electronic spins, these have been proposed as electron spin qubits whereby through application of thermal stimuli, magnetic fields, or electromagnetic pulses [43–49] they can be manipulated. More recently, systems known as qu*d*its (Figure 12.8), where the *d* stands for the multilevel character of the system, have gained much attention due to the possibility to employ the multilevel nature of the systems to perform complex algorithms in a single physical unit [24, 40, 43, 44, 46, 50–58]. Additionally, qu*d*its offer several advantages over qubits, such as d^N orthogonal states, allowing parallelization in a single unit with lower error rates [45, 59, 60]. The higher energy states in magnetic molecules can be used for the realization of complex quantum gates [20, 24, 40, 52]. Qu*d*its are promising systems since they have been shown to possess numerous advantages over qubits, such as:

(i) do not require inter-qubit interaction to perform qugates [44, 46, 54];
(ii) the number of gates required to execute a computational task is reduced [59];

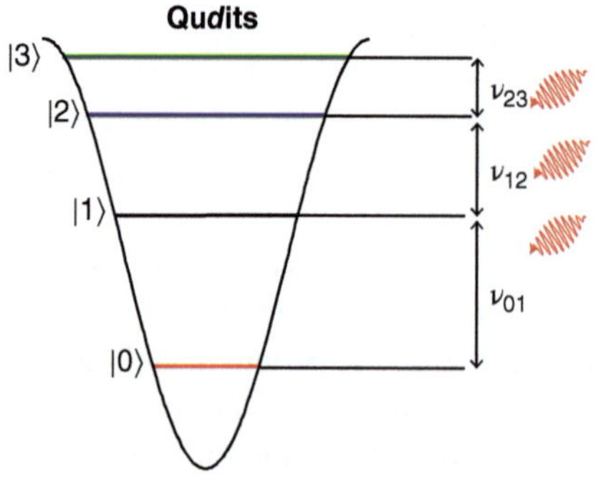

Figure 12.8 Quantum Computer Building Blocks: Multilevel representation of a qu*d*it ($d = 4$). The different accessible excited states can be employed to perform quantum gates in a single physical unit.

(iii) they can parallelize information in a smaller number of physical units [45];
(iv) possess lower error rates [60, 61];
(v) complex gates can be implemented in a single unit [43, 46, 53, 62];
(vi) entangled qu*d*its are more suitable for quantum cryptography [59];
(vii) simpler quantum simulations with similar Hilbert space [63];
(viii) large dimensions of operational state with a smaller number of processing units [61, 64].

12.3 SMMs for Single-Molecule Devices

The integration of SMMs in single-molecule devices has been shown to render extraordinary results, highlighting the quantum nature of the complexes and the possibility to not solely observe these effects, but also, implement them. Through this approach, it has been possible to create spintronic molecular devices, enabling the manipulation and read-out of the spin states of the SMM to perform quantum operations. In this section, we highlight some key results of single-molecule devices and their quantum characteristics.

12.3.1 Spin Transistors

A spin transistor is a three-terminal device, where electrons are passed through an SMM, bridging the gap of the junction, from the source to the drain electrode. This device configuration has allowed the manipulation and read-out of the electronic states of the SMM. The first experiment involving an SMM in such configuration was attempted with the archetypal Mn_{12}; however, the structural integrity of the system was compromised in such configuration [65]. Better results were obtained for an Fe_4 [66] and $TbPc_2$ [67–71] molecules. The most important results for such configuration employing SMMs have been achieved when a $TbPc_2$ complex is trapped into gold junctions through electro-migration [67–71] (Figure 12.9).

The extremely stable redox state of the Tb^{3+} ion and the π-radical delocalized over the Pc ligands allowing the conduction of electrons while maintaining the SMM intact [67–71]. An indirect coupling allows the read-out of the nuclear spins. The read-out dot is achieved due to the strong interaction between the π-radical and the Tb^{3+} spins, allowing the observation of change in conductance near zero field. A single charge-degeneracy point with a weak spin $S = ½$ Kondo effect, which is ascribed to the π-radical delocalized over the Pc rings, was observed in differential conductance (dI/dV) studies as a function of drain–source voltage (V_{ds}) and gate voltage (V_g). Due to the ferromagnetic coupling between the $S = ½$ of the π radical to the magnetic moment carried by the Tb^{3+} ion, and its hyperfine coupling to the nuclear spin states of the Tb^{3+} ion, the spin cascade $|S = 1/2\rangle||J = 6\rangle||I = 3/2\rangle$ is reflected by the transport properties (Figure 12.10). The spin cascade allows the read-out of the nuclear spin states by the spin-dot with experimental conditions close to a charge-degeneracy point. Four abrupt jumps in the differential conductance are obtained when sweeping the field, corresponding to the reversal

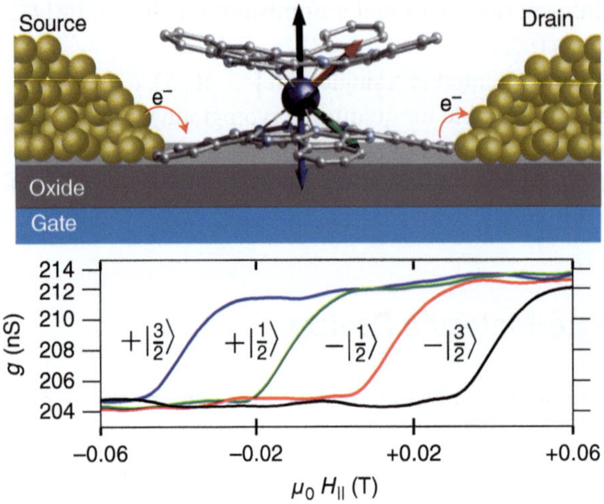

Figure 12.9 **$TbPc_2$ Quantum Spin Transistor:** Pictorial representation of molecular spin transistor, with a $TbPc_2$ molecule embedded between the source and drain (top). The conductance jumps correspond to the four level-crossings associated with the nuclear spin $I = 3/2$ of Tb^{3+} (bottom). Source: Thiele et al. [70] and Moreno-Pineda and Wernsdorfer [72] with permission from Springer Nature (copyrights © 2014) and The American Association for the Advancement of Science (copyrights © 2021), respectively.

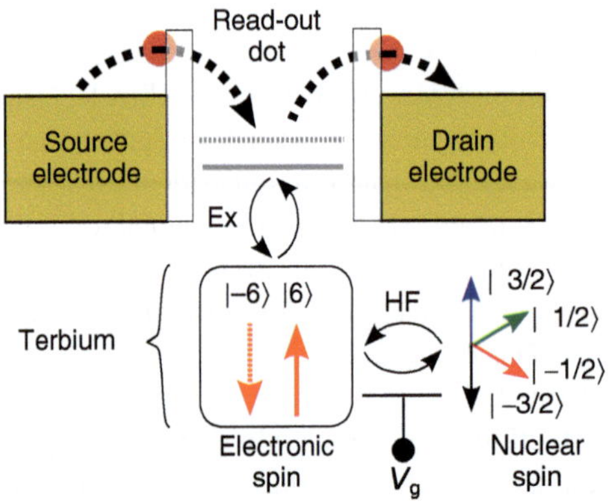

Figure 12.10 **$TbPc_2$ Quantum Spin Transistor:** Schematic view of the read-out cascade occurring in [$TbPc_2$] in the transistor device. Source: Thiele et al. [70] with permission from Springer Nature (copyrights © 2014).

of the Tb^{3+} electron spin through QTM at the nuclear spin crossing (Figure 12.9 [lower panel]).

The highly efficient relaxation through QTM at low fields ($\mu_0 H_z < \pm 50\,mT$) is of paramount importance for the detection of the spin reversals at these four level-crossings causing each time a change in the transport properties of the read-out dot [67, 70]. Moreover, T_1 and T_2 of 20 and 300 microseconds, respectively, have been found, leading to their implementation of Grover's algorithm using the four states of the nuclear spin [25] (vide infra) and the realization of an iSWAP gate [71]. Likewise, employing a spin transistor configuration, incorporating a Tb_2Pc_3 unit, it has been possible to the read-out of several nuclear spin states, generated by the indirect coupling of the electronic states of the Tb^{3+}, leading to seven

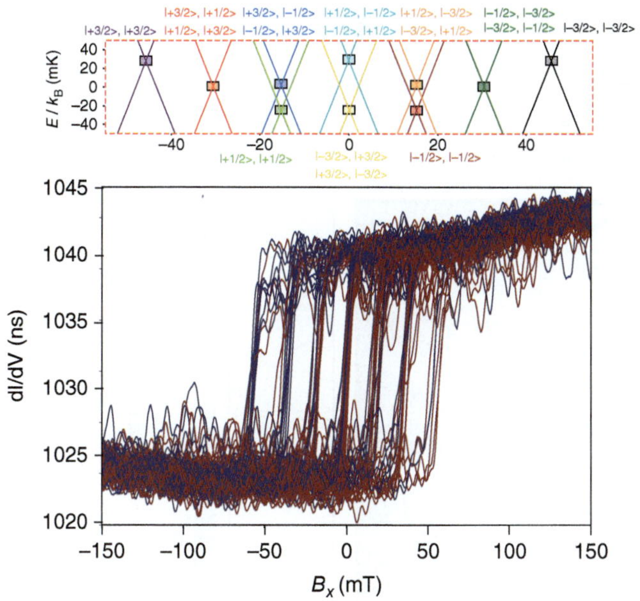

Figure 12.11 Tb_2Pc_3 Quantum Spin Transistors: dI/dV as a function of V_{ds} and V_g for a Tb_2Pc_3 spin transistor device. The conductance jumps correspond to spin reversal at the allowed QTM crossings. Source: Biard et al. [40] with permission from Springer Nature/CC BY-4.0.

non-degenerated nuclear states, opening the possibility of the implementation of more complex algorithms [40] (Figures 12.11 and 12.12).

Besides the direct coupling transistor configuration, it is possible to integrate SMMs into hybrid architectures by indirect coupling, in which the molecule is subjected to a weaker back-action. In this configuration, the spin dot (SMM) is connected to the electrodes by a non-magnetic molecular conductor (read-out dot), where the coupling between the spin dot and the read-out dot is regulated by the gate voltage.

12.3.2 Spin Valve

Another configuration in which SMMs have been integrated into devices is the supramolecular spin valve. Urdampilleta et al. [73–75]. studied such a system composed of $TbPc_2$, acting as a spin dot, and a carbon nanotube (CNT), acting as a read-out dot. The exceptional mechanical, structural, and electrical characteristics of CNTs make them suitable to act as read-out dot. The sensible coupling between the SMM and CNTs, possessing 1D conductor character, along with the Kondo effect and Coulomb blockade exhibited at low temperatures, makes them conductance-sensitive to charge fluctuations including the spin reversal process. In this configuration, the CNT-$TbPc_2$ device permitted the determination of the electronic and nuclear spin characteristic of the Tb^{3+} ion [73–75]. Read-out was accomplished, through magneto-transport measurements, due to the strong interaction

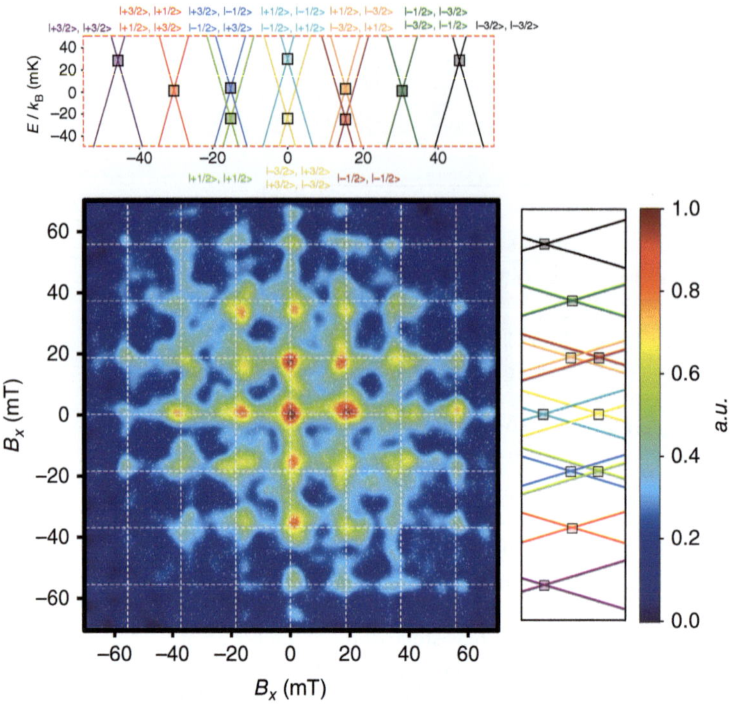

Figure 12.12 Tb_2Pc_3 Quantum Spin Transistors: Spins reversals correlation measurement. The *x*-axis corresponds to conductance jumps during a magnetic field sweep, and the *y*-axis represents the conductance jumps during subsequent sweeps. Source: Biard et al. [40] with permission from Springer Nature/CC BY-4.0.

between the $TbPc_2$ molecules suspended on CNTs [73–75]. A maximum in the conductance is observed when the electronic spin of the $TbPc_2$ molecules, suspended on the CNT, was aligned in a parallel configuration (ferromagnetic coupling). Conversely, a minimum conductance was obtained when the electron spin of the $TbPc_2$ was antiparallel (antiferromagnetic coupling) (Figure 12.13).

Conductance jumps were observed between ±500 mT corresponding to the direct relaxation process. Moreover, the highly sensitive measurements revealed differently oriented molecules attached to the CNT. Likewise for the spin transistor, the nuclear spin states for each $TbPc_2$ attached to the CNT were detected by measuring the tunneling probability as a function of sweep rate (Figure 12.14). Nevertheless, in contrast to the spin transistor configuration, where the four level-crossings of nuclear spin $I = 3/2$ are associated to relaxation through QTM, the relaxation in the spin vale occurs through the direct relaxation mechanism. In the spin valve configuration, QTM can be suppressed at low temperatures due to one-dimensional phonons associated to the mechanical motion of the CNT and its strong coupling with the $TbPc_2$ unit [33, 76]. One-dimensional phonons, associated with the mechanical motion of CNT, allow relaxation to occur solely via a single mode (direct relaxation), associated with the mechanical motion of the CNT. The so-called quantum Einstein–de Hass effect could in principle allow coherent spin

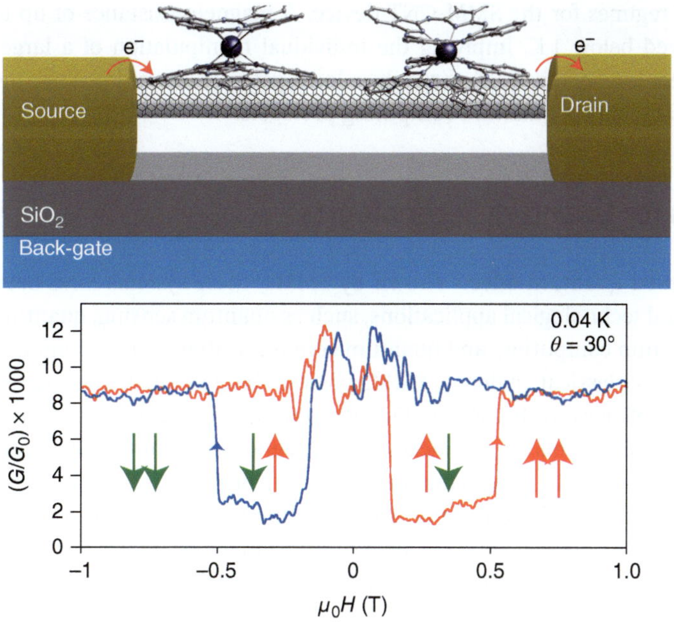

Figure 12.13 Supramolecular Spin Valves: Schematic view of the spin-valve device (top) and conductance measurements as a function of the magnetic field. The red curve corresponds to the conductance under increasing field (−1 to +1 T) and the blue curve under decreasing field. The conductance jumps around zero field are attributed to molecules experiencing direct relaxation. Source: Urdampilleta et al. [73].

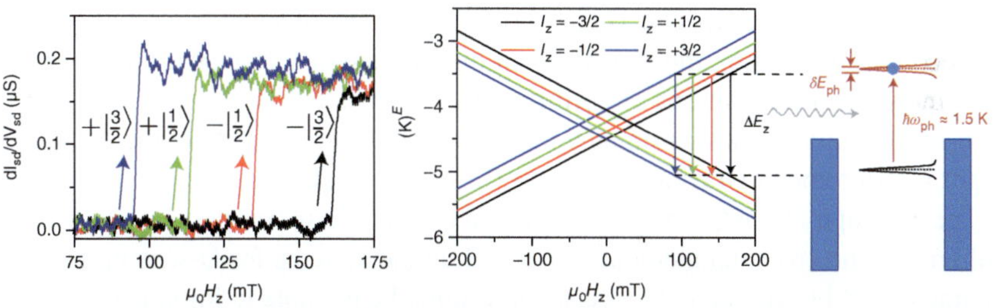

Figure 12.14 Supramolecular Spin Valves: Magnetization reversal of the Tb^{3+} when sweeping the magnetic field $\mu_0 H_z$ resulting in an abrupt increase in the differential conductance through the dot. The switching field is contingent on the nuclear spin state (bottom). Source: Ganzhorn et al. [76] with permission from Springer Nature (copyrights © 2011).

manipulation of the spin states opening the possibility of coherent manipulation of the spin and entanglement [33].

Furthermore, it was shown that when two $TbPc_2$ molecules were suspended on the CNT, a supramolecular spin-valve behavior was observed, acting as a spin polarizer-analyzer revealing a strong magnetoresistive effect. The fine-tuning and the electronic detection/manipulation of a single magnetic moment were achieved for two working regimes for the SMM-CNT device. A magnetoresistance of up to 300% was observed below 1 K, implying the individual manipulation of a larger number of molecules on the CNT through a local gate. This would allow the implementation of complex quantum computing protocols.

12.4 SMMs for Quantum Technologies

Due to the observed quantum effects shown by SMMs, these systems have been proposed in several technological applications, such as quantum sensing, quantum simulations, quantum computing, and quantum communications. The advantages foreseen surpass any classical analogue. SMMs possess all quantum characteristics to be at the heart of these technologies. Furthermore, in the previous section, it was shown that these systems can be successfully integrated in spin-transistor and spin-valve devices, paving the way to quantum device hybrid architectures. In this section, a brief description of the most important requirements of SMMs for each application is discussed. Furthermore, an emerging area in which light is employed for the read-out of the information for quantum communication technologies is also discussed.

12.4.1 Quantum Sensing

Quantum effects are at the heart of novel technologies, where the goal is the utilization of these effects for the realization of certain tasks not feasible with classical systems. Quantum sensing, or Q-sensing, is an example of the utilization and exploitation of quantum effects for the development of measuring and sensing protocols. By exploiting the sensitivity of quantum states, it is possible to explore extremely small fluctuations or perturbations, as demonstrated through the detection of gravitational waves employing squeezed states of light by the Advanced Light Interferometer Gravitational-Wave Observatory (aLIGO) [77].

The development of Q-sensing is based on two important properties: (i) quantum coherence and (ii) entanglement [78–80]. Suitable systems for Q-sensing must be characterized by well-defined levels, and it must be possible to initialize them and read the final state out. Also, in certain cases, the coherent manipulation of the states of the system might be required. In contrast to the basic unit of quantum computers, for a successful Q-sensing protocol, the Q-sensor must interact with the system to be sensed, i.e. the probe is prepared in a defined state, which is dependent on the parameter to be determined. The interaction between the Q-sensor and the systems to be sensed causes some change in the initial properties of the sensor, which is inferred in the final outcome of the measurement. Based on the interaction between the sensor and the system to be sensed, two possible configurations are devised: (i) in the first configuration, the sensor is prepared in an initial state ρ_0, which is allowed to interact with the system to be sensed or field under investigation (red area in Figure 12.15). The system under study is characterized by the unknown parameter γ (Figure 12.15a). After the interaction between the sensor and the system

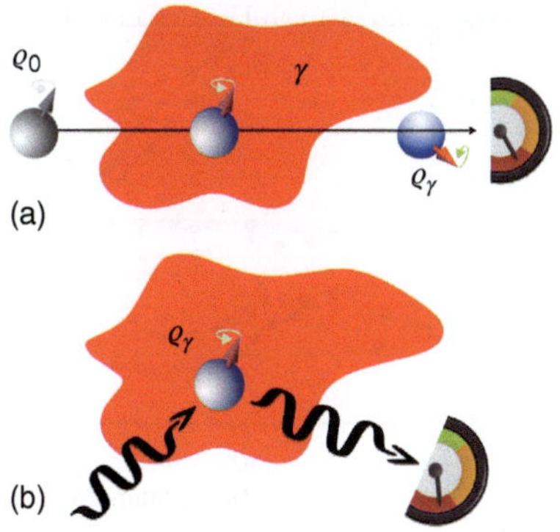

Figure 12.15 Quantum Sensing Configurations: (a) The sensor is prepared in an initial state and it's allowed to interact with the system under investigation. During the interaction, information of the unknown parameter is obtained. Measurements of the final state of the sensor allow the determination of the information of the system under study. (b) In the second configuration, the sensor is embedded in a structure, and its state contains information on some internal parameters of interest. Source: Troiani et al. [78] with permission from Elsevier B.V. All rights reserved (copyrights © 2019).

under investigation, the quantum probe obtains information about γ. Finally, measurements of the sensor allow the extraction of information. (ii) The second configuration consists of a quantum probe embedded in a larger structure, and its state contains information about the internal parameters of interest (Figure 12.15b).

The advantages of quantum sensors are based on the exploitation of quantum effects, such as the superposition of states and entanglement, hence allowing the sensing of effects not achievable with classical sensors. Several systems are actively considered as Q-sensors, ranging from NV centers [5–12, 81, 82], superconducting circuits [79], light [4, 77, 83], silicon vacancies [84, 85], and nuclear and electronic spins embedded in magnetic molecules [78, 80], among others. For sensing at a nanometer scale, the quantum characteristics of single atoms and molecules are better suited than any classical sensor, due to their spatial resolution [78, 80]. Moreover, the chemical control over the synthesis of magnetic molecules allows for the engineering of the structural, electronic, and nuclear characteristics. One main aspect for the implementation of such a system as Q-sensors is the coherence and the possibility of preserving such coherence in the presence of noise; thus, long coherence times (T_2) are essential for Q-sensors [78–80]. In this sense, the thorough understanding and the gained control of the synthetic methodologies for the rational design of magnetic molecules have permitted the observation of magnetic molecules with long coherence times [86–88] (Figure 12.16). Long coherence times have been observed in the well-known $\{Cr_7Ni\}$ wheels via pulsed EPR. Likewise, through chemical design, it has been possible to enhance the coherence times of these systems [89], which can remain robust when attached to one [90–92] or more units [93] ultimately leading to their proposal for logic gates. Additionally, other examples show that through a rational design, it has been possible to enhance coherence by eliminating the decoherence source in the environment of the molecule allowing the detection of long T_1 and T_2 times at room temperature and in bulk crystals [55, 94]. Moreover, it has been possible to observe long coherence times in vanadyl-based complexes [50, 51, 55–58, 95–98], yielding coherence times comparable to NV centers [99].

Another strategy to achieve long coherence times consists of the preparation of magnetic noise resilient systems, eliminating the necessity of magnetic dilution. Magnetic noise resilient systems are obtained due to an interaction that mixes the qubit states, e.g. hyperfine interaction, resulting in an avoided level-crossing,

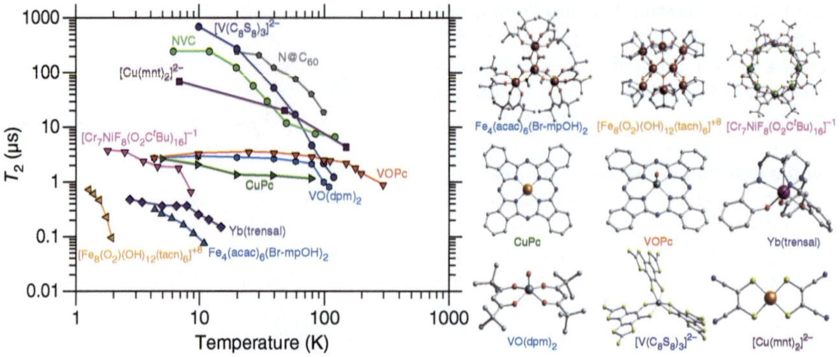

Figure 12.16 Coherence Times in SMMs: Temperature dependence of the electron spin–spin relaxation (T_2) for several prospect qubits (left) and crystal structure of the complexes (right). Color code: C, gray; S, yellow; N, blue; Fe, dark red; Cr, green; V, aqua; Cu, orange; Yb, purple; O, red; Ni, blue gray. Source: Adapted from Moreno-Pineda et al. [24] with permission from The Royal Society of Chemistry.

termed as "clock transitions" (CTs). As a result, the spin qubits become insensitive to external fluctuations of magnetic fields, yielding long T_2. A $[Ho(W_5O_{18})_2]^{9-}$ system (HoW_{10}) with D_{4d} pseudo-axial symmetry was shown to be magnetic noise resilient [98]. The combination of crystal field, hyperfine interactions, and Zeeman splitting originate yields avoided level-crossings between the m_J levels of the same m_I. Long T_2 at 5 K was found at the CT, providing an alternative strategy to achieve long T_2 values in SMMs.

SMMs have been proposed as Q-sensors for the detection of physical parameters, such as temperature, electrical or magnetic fields, currents, nuclear spins, spins, and magnons to dark matter [78, 80]. Magnetic molecules have been shown to possess long coherence times, while the quantum effects shown by magnetic molecules, as well as their nanometric scale, offering spatial resolution, and their tunability make these systems promising candidates for Q-sensing. Furthermore, through chemical design, it can be possible to position sensors on the desired systems by exploitation of interactions. An example of Q-sensing is given by a single $TbPc_2$ molecule in a transistor configuration, acting as a local thermometer [100]. In this system, the $TbPc_2$ molecule is coupled to a magnetic field and is in equilibrium with the thermal bath, hence acting as a temperature sensor. In the transistor configuration, the $TbPc_2$ molecule is embedded between two gold leads. Current is passed through the organic radical characterizing the $TbPc_2$ molecule, acting as a quantum dot for electrical read-out. Through electrical read-out and monitoring of the spin-flip of the electronic spin of the Tb ion, after application of non-resonant microwave pulses, it was found that the spin flip rate follows an exponential law, hence acting as a thermometer. Another example for local sensing employing a single spin is the spin-polarized scanning tunneling microscope (SP-STM) [101, 102] in combination with electromagnetic pulses [78]. The Q-sensing protocol consists of the application of a microwave pulse, which induces an electronic transition, hence, changing the population of the center, which is subsequently detected by the

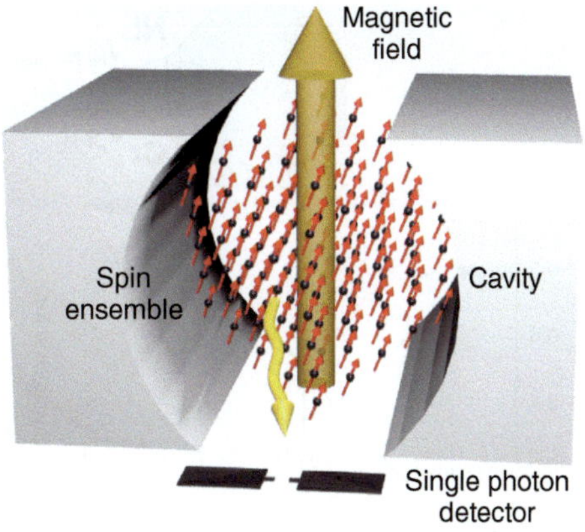

Figure 12.17 SMMs for Quantum Sensing: Scheme of a spin ensemble in a microwave cavity. The system is driven, at low temperatures, in the strong coupling regime for the efficient transfer of magnetic excitation to microwave photon. Source: Troiani et al. [78] with permission from Elsevier B.V. All rights reserved (copyrights © 2019).

SP-STM tip [72, 103–109]. Similarly, it has been shown that through an analogous protocol, it is possible to detect the electronic and nuclear spin properties of a $TbPc_2$ system. Likewise, spin assembles can allow the detection of magnetic fields in the strong coupling regime and for the detection of dark matter [78] (Figure 12.17).

12.4.2 Quantum Simulations

In the early 1980s, Richard Feynman hypothesized that quantum systems would be able to simulate quantum effects, highlighting the limitations of classical computers [13]. Quantum simulators are systems that can be prepared so that they can mimic the dynamics of complex quantum mechanical problems. Magnetic molecules are particularly appealing systems given that these have shown all the requirements to act as qubits [24, 86–88, 110, 111]. Furthermore, they can mimic the evolution of quantum mechanical problems due to their quantum properties, and their structural and electronic characteristics can be tuned through chemical means. An important aspect for the realization of quantum operation is the possibility of bringing two qubits into proximity to realize quantum gates. Chemically, it has been shown that the SMMs can be linked leading to the proposal of quantum gates [90, 91, 112–117]. Magnetic molecules with a switchable interaction between two units have been proposed for quantum simulations. Example of this is the $\{Cr_7Ni\}$ dimers to simulate quantum gates [91, 118–120]. Through ab initio many body calculations, and the experimentally determined magnetic couplings, it has been shown that the switchable interaction mediated by a metallic linker between $\{Cr_7Ni\}$ units allows the simulation of CNOT and $\sqrt{iSWAP}$ quantum gates with high fidelities [91, 118–121] (Figure 12.18).

Furthermore, the multilevel character of these systems can be exploited, utilizing the various two-level systems as qubit units [24, 122]. In this respect, multilevel systems, or qu*d*its, are especially appealing due to the possibility of implementing complex algorithms in a smaller number of physical units, with smaller error rates [24, 40, 43, 44, 46, 50–58, 114]. An important example of quantum simulation using a multilevel molecular system was shown by Atzori and coworkers [114]. The demonstration was carried out employing a dimeric

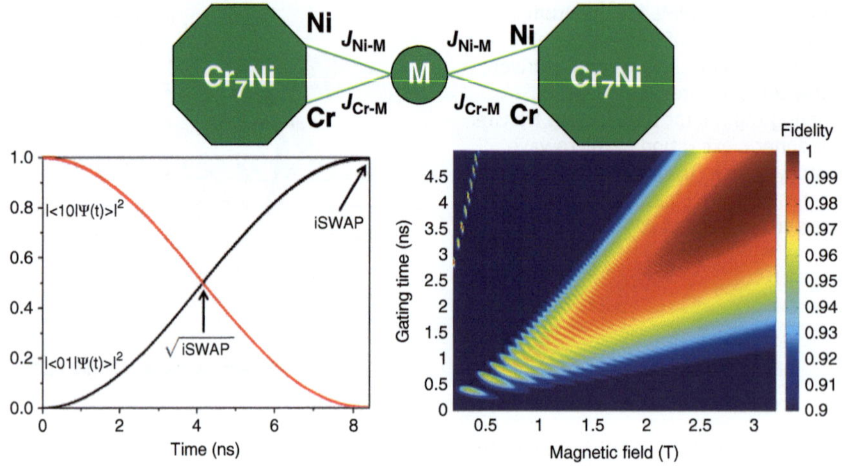

Figure 12.18 Quantum Simulations with Switchable {Cr$_7$Ni} wheels: Schematic representation of coupled {Cr$_7$Ni} qubits linked by a metallic switchable linker (top). Simulation of $\sqrt{iSWAP}$ employing a {Cr$_7$Ni}Co complex (left) and calculated fidelities (right). Source: Adapted with permission from Ferrando-Soria et al. [91] Springer Nature/CC BY 4.0.

vanadium(IV) complex with formula $[PPh_4]_4[(VO)_2(L_1)_2]$ (L_1 = tetraanion of $C_6H_3(OH)_2$–CONH–C_6H_4–CONH–$C_6H_3(OH)_2$), exploiting both the electronic and nuclear spins. In the dimer, the electronic spins of the V(IV) are coupled via an exchange interaction, while the electronic and nuclear states are coupled via a hyperfine interaction, as revealed by EPR spectroscopy. The generated states can be separated according to the total M_S component in the applied field. The generated logical states of the qubit, $|1\rangle$ and $|0\rangle$, are encoded in the $m_s = -1/2$ state of the electron spin of the $M_S = -1$ level, and the four states of the $(2I + 1)$ [2] nuclear spin manifold. The states correspond to the $m_I = 7/2$ and $m_I = 5/2$ states of each V nucleus. Single qubits rotations can therefore be carried out between the $m_I = 7/2$ and $m_I = 5/2$ states (Figure 12.19).

Due to the interaction, the rotation of the electronic states is contingent upon the nuclei; hence, the system can be implemented in two-qubits gates, where an indirect interaction between the electronic spins coupled to the nuclear states acts as a switch. The long coherence and the strong hyperfine interaction made the system very appealing [50, 55, 56, 96, 123, 124]. Exploiting the multilevel character of the complex, it was possible to perform a controlled-shift ($C\varphi$) two qubits gate, which generates entanglement between the qubits adding a phase to one of states, while leaving the remaining three states unaffected. Employing this system as a quantum simulator, it was possible to simulate the quantum tunneling of the magnetization for $S = 1$, with high fidelities (Figure 12.20).

More recently, a multilevel molecular system, qu*d*its unit [24, 40, 43, 44, 46, 50–58], comprising a dimer based on two spins, i.e. an $S = 1/2$ and $S \geq 3/2$ spins, which can be manipulated by microwave pulses, has been proposed as quantum simulators for light–matter interactions. The proposal is based on the exploitation of the spin

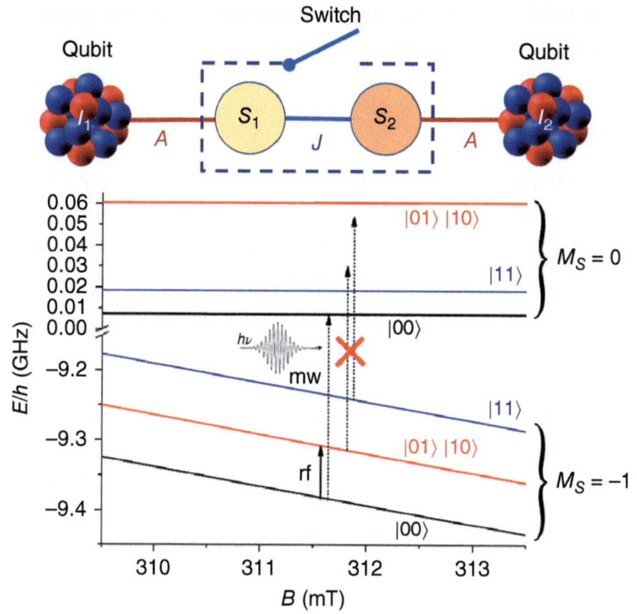

Figure 12.19 Simulation of QTM with a Vanadyl Electron-nuclear Spin Dimer: Pictorial view of vanadyl dimer (top). Zeeman diagram for the vanadyl complex based on experimental EPR data, and the rf pulse for a single-qubit rotation. The dashed arrows represent the implementation of a $C\varphi$ gate. Source: Atzori et al. [114] with permission from The Royal Society of Chemistry.

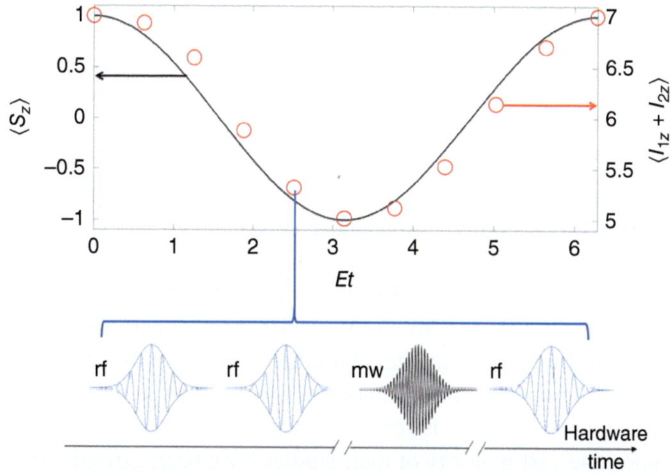

Figure 12.20 Simulation of QTM with a Vanadyl Electron-nuclear Spin Dimer: Quantum simulation of the oscillation of the magnetization for an $S = 1$ spin experiencing QTM. Source: Adapted from Atzori et al. [114] with permission from The Royal Society of Chemistry.

S ion to encode the photon field enabling the digital simulation of a wide range of spin–boson models with a higher efficiency than multi-qubit registers [125].

12.4.3 Quantum Computing

Quantum computing was first proposed by Richard Feynman, in the early 1980s, when he observed that certain properties of systems could not be simulated efficiently with classical computers [126]. Feynman's proposal [13] was later on backed by Shor [127], Grover [38], Lloyds [14], and others, when they showed that a quantum computer would offer unparalleled advantages over classical computers. Nowadays, the field of quantum computing is an active field, in which several companies and consortiums have invested staggering amounts of money in the quest for a functional quantum computer.

Due to the relatively facile manipulation of electronic spins, these systems have been proposed to act as qubits, whereby the manipulation is achieved via thermal stimuli, magnetic fields, or electromagnetic pulses [43–49]. One of the proposed methods for manipulation of the electronic spins of magnetic molecules is electron paramagnetic resonance (EPR) pulses. The $\{Cr_7Ni\}$ antiferromagnetic wheels are an example of electron spin qubits, which can be manipulated via EPR pulses. The strong antiferromagnetic coupling between the Cr^{3+} and Ni^{2+} in the wheel, leads to a well-defined spin $S = ½$ ground state, isolated from excited states below 10 K [128]. In addition to transition metal complexes, lanthanide-based magnetic molecules have also been proposed as qubits [98, 116, 129] owing to their inherent magnetic anisotropy and ground doublet state characteristics. A representative example is an asymmetric lanthanide dimer in which a small interaction between the Ce^{3+} and Er^{3+} lanthanide ions could in principle be addressed through manipulation of the resonance frequencies or fields, leading to the proposal of a CNOT gate [116].

Furthermore, the first manipulation of the electronic states embedded in a qu*d*it was shown employing a gadolinium polyoxometalate complex with the formula $K_{12}Gd(H_2O)P_5W_{30}O_{110}·27.5H_2O$ (GdW_{30}) [130, 131]. The large spin multiplicity, $S = 7/2$, and a small, yet sizable anisotropy ensure eight possible states for manipulation, as shown by EPR studies [110] (Figure 12.21).

T_2 ranging between 400 and 600 ns and T_1 values between 2.3 and 2.6 μs were determined for the GdW_{30} complex. Furthermore, Rabi oscillations and a controlled-controlled-NOT (CCNOT) or Toffoli gate were carried out employing the molecular qu*d*it, demonstrating the advantages of multilevel systems (Figure 12.22).

Although the electronic spin in magnetic molecules is actively explored for their applicability as qubits, the electron spin is very susceptible to interactions with the spin bath. Furthermore, the vast majority of their studies have been carried out in crystals on large ensembles, which would compromise the initialization step due to inhomogeneity effects. This is of high importance when the ultimate goal is their integration in hybrid spintronic devices, in which the magnetic molecules would be in direct contact with metallic leads [132]. The $TbPc_2$ molecule is an example of a nuclear spin qu*d*it, in which the four nuclear states have been employed to perform a quantum algorithm. Due to its remarkable stability, the $TbPc_2$ has been deposited

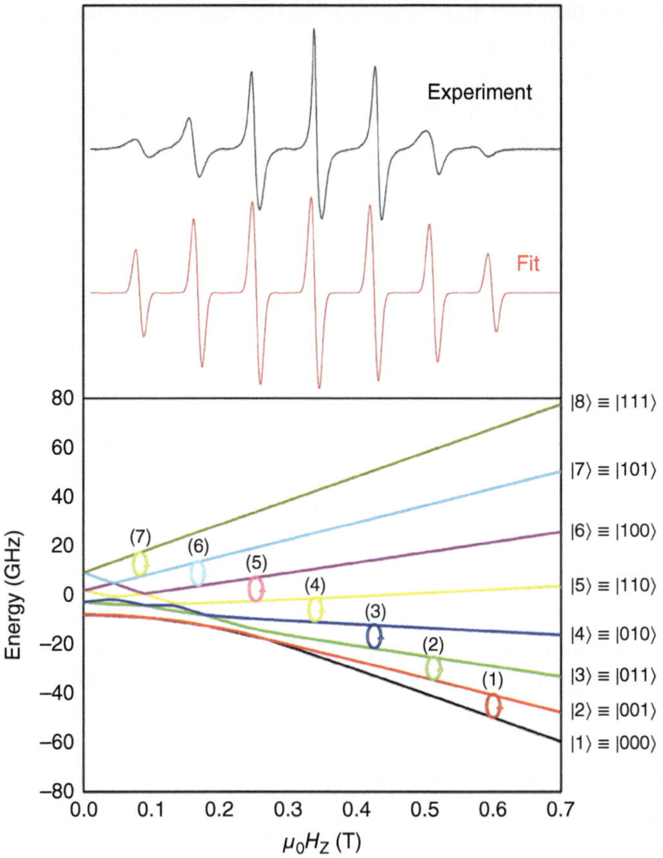

Figure 12.21 Three-qubits CNOT Gate in a GdW_{30} SMM: (A) X-band cw EPR spectrum on a single crystal of $Y_{0.99}Gd_{0.01}W_{30}$ (top) and Zeeman diagram with accessible qubits states (bottom). Source: Jenkins et al. [130] with permission from the American Physical Society.

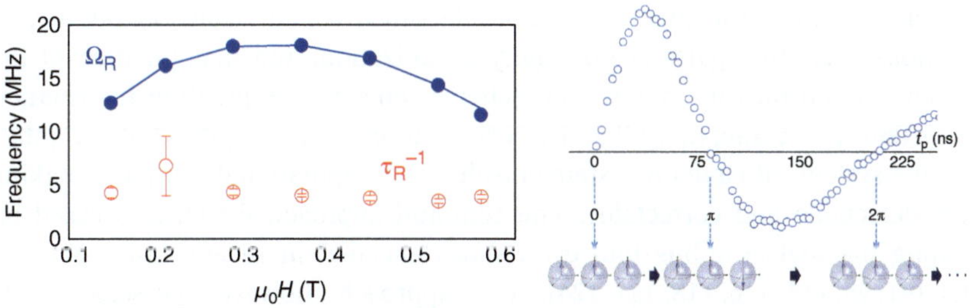

Figure 12.22 Three-qubits CNOT Gate in a GdW_{30} SMM: Rabi Frequencies $\Omega_{R,n}$ and damping rates $1/\tau_{R,n}$ for the oscillations (left). Rabi oscillations for transition (1) showing the coherent evolution between |000⟩ and |001⟩ three-qubit states (Right). Source: Jenkins et al. [130] with permission from the American Physical Society.

in a large range of substrates [132] and integrated transistors and spin valve configurations (vide supra). Initialization of the nuclear states is achieved at sub-kelvin temperatures and low applied fields, corresponding to four nuclear qu*d*its states, i.e. $|m_I = \pm {}^1\!/_2\rangle$ and $|m_I = \pm {}^3\!/_2\rangle$ [68, 70], while manipulation of these states is possible by employing pulse sequences corresponding to the uneven separation between the states arising from the quadrupolar interaction [70]. Long T_1 and T_2 values have been obtained for these systems, which are in combination with the ability to read out, initialize, and manipulate the nuclear states.

The algorithm is achieved by a succession of two gates: (i) a superposition of all qu*d*it states is generated with a Hadamard gate, and (ii) amplification of the amplitude of the sought state, initially labeled via its phase or energy, is obtained by the Grover gate. Employing the quantum amplitudes to determine the probabilities of an event, the sought state is found. The Hadamard gate plays an important role in the algorithm since it creates superposition of states, i.e. for a n number of qubits prepared at an initial state, a Hadamard gate applied to each qubit, creating a total of n-qubits superpositions. The superposed states embody all possible combinations of the n qubits from 0 to $2n - 1$, i.e. $H|0\rangle \otimes H|0\rangle \otimes \cdots \otimes H|0\rangle = \frac{1}{\sqrt{2^n}} \sum_{j=0}^{2^n-1} |j\rangle$, representing all possible solutions of a given problem, acting as shortcuts accelerating the computation process. The highly superposed state created by the Hadamard gate allows the Grover's algorithm to succeed in a quadratic speed up, compared to classical algorithms [38]. Through multifrequency pulses, the simultaneous manipulation of the m_I states allows the superposition of the four nuclear spin states creating a qu*d*it (with $d = 4$ for $I = {}^3/_2$) (Figure 12.23). Once the superposition of states has been achieved, pulses parameters are tuned to reach a resonance condition for the sought state. Consequently, the qu*d*it populations oscillate and the population of the labeled state increases. This experiment demonstrated for the first time the experimental Grover's algorithm on an SMM [25].

An important aspect for the realization of quantum computing is to be able to protect the quantum information from decoherence sources and quantum noise. To this end, quantum error correction (QEC) is essential for the development of fault-tolerant universal quantum computing, since it can reduce errors arising from noise, quantum gates errors, faulty initialization, and manipulation of the information and measurements. QEC schemes encode the quantum information into systems possessing more than two energy levels "logical qubits"; these qubits bring the system into a defined state outside the computational subspace, making errors detectable and correctable. The common approach for QEC is based on encoding the logical qubits into extra states encoded in several other physical units [91, 98, 114–116, 118, 121, 124]. This approach, however, possesses some difficulties, due to the requirement of non-local states in distinct objects to carry out QEC and quantum computation. An alternative approach to circumvent this problem is to encode the logical qubit into a single multilevel quantum object [24, 40, 50–52, 55–58].

In this regard, the multilevel characteristics, at the electronic and nuclear level, of SMMs make them viable architectures to encode and protect the qubit through QEC. An example of an SMM employed for QEC schemes is a heterometallic trinuclear

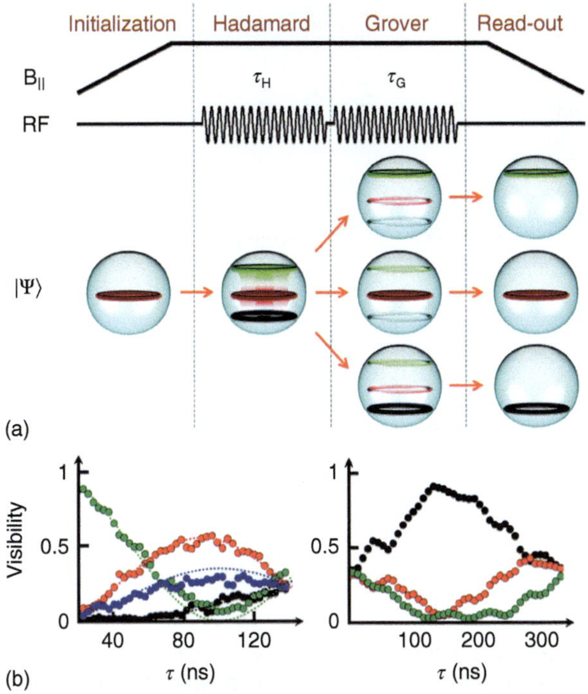

Figure 12.23 Grover Algorithm with $TbPc_2$ Spin Transistor: (a) The Grover algorithm is implemented using four different steps: initialization, Hadamard gate, Grover gate, and final read-out. (b) (left) Evolution of the population as function of the Hadamard gate pulse length. (Right) Evolution of the population in function of the Grover gate pulse. Starting from a superposed state (obtained by an Hadamard pulse sequence), the system oscillated between this superposed state and a desired state (black points). Source: Adapted from Moreno-Pineda et al. [24] with permission from The Royal Society of Chemistry.

[ErCeEr] complex, with an interaction connecting the Er—Ce pair, as determined by SQUID magnetometry, heat capacity, and EPR spectroscopy (Figure 12.24b,c) [133]. Through numerical simulation, it was shown that the complex can be exploited in a three-qubit repetition code to protect qubits from single bit or phase flip errors (Figure 12.24a) [134]. The correction scheme comprises *encoding*, *decoding*, and *correction* steps. The *encoding* step involves two controlled-NOT (cNOT) two-qubit gates inducing a flip on the target qubit, which is protected from error due to entanglement. A *memory* time stores the qubit, while the decoding (reverse encoding) is implemented after. Correction is finally implemented via a single controlled-controlled-NOT (ccNOT) gate applied to the qubit. Numerical simulations show that the SMMs can efficiently protect the encoded qubit from decoherence.

Another example of molecular magnetic systems for QEC schemes is $\{Cr_7Ni\}$-Cu system, in which the electronic state of the $\{Cr_7Ni\}$ and the nuclear state embedded in the Cu ion can be exploited for error correction [135]. Execution of complex quantum algorithms required idle times, in which the qubits remain idle between sets of

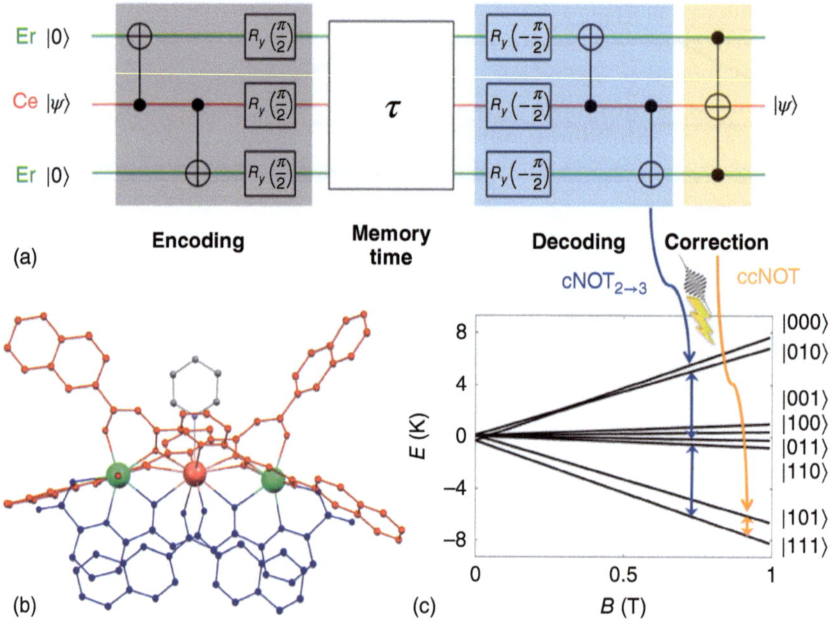

Figure 12.24 Quantum Error Correction with SMMs: (a) Quantum circuit for a three-qubit phase-flip repetition code. The central qubit carries the information, while the remaining are auxiliary qubits. (b) Molecular structure of *f*-SMM (Er, green, and Ce, light red) corresponding to the qubits of the circuit shown in (a). (c) Energy levels as a function of the external field *B* applied along *z* (the Er—Ce direction). Source: Macaluso et al. [133] with permission from The Royal Society of Chemistry.

quantum operations for long times. The nuclear spins embedded in the Cu^{2+}, can, therefore, be employed as quantum memory to store information during idle times. Simulation shows that by exploiting the nuclear states in the Cu^{2+}, acting as storage states with quantum error correction, information can be protected for times much longer than the processor coherence. Undoubtedly, these examples paved the way for SMMs for universal fault-tolerant quantum computer architectures [136].

12.4.4 Quantum Communication

The implementation of quantum effects in communication technologies could revolutionize specific areas such as secure communication and computing. For quantum communication or distributed quantum computing, optical connection between photons and qubits is essential [137]. For the creation of quantum networks nodes, the integration of local error corrections is necessary, since they can enable secure communication and distributed quantum computing, and it is the basic component of a future quantum internet. For such implementation, qubits with long-lived memory, that is, long coherence times, efficiently interfaced to photons is required.

Among the proposed building blocks for quantum communication and networking, lanthanides have been shown to have promising properties since they possess nuclear spin states with very long coherence times [138] in which quantum states can be stored. Moreover, lanthanide ions provide 4f-4f optical transitions with remarkable coherence times [139], which can be exploited for direct optical read-out and control of the spin states. Due to these characteristics, lanthanides systems are promising candidates for the creation of long-lasting quantum memories and qubits for the realization of quantum gates. Optical manipulation of individual ions has been shown [140–142], evidencing the potential for scalability lanthanide systems. Despite the remarkable results, the exceptional results are limited to a very small number of high-quality bulk single crystals and nanomaterials, such as Y_2SiO_5 [140, 143, 144]. Unfortunately, these inorganic materials lack the systematic control over their synthesis and pose some difficulties in their integration into photonic structures. Furthermore, these inorganic photonic crystals lack the tunability capability and scalable synthesis and are limited to near-infrared wavelengths, whereas promising species, e.g. for the realization of quantum gates (Eu^{3+}, Pr^{3+}) need to be addressed in the visible region.

On the molecular magnetism side, the field of optospintronic has emerged as method for the manipulation and read-out of the electronic properties employing photons. The rationale behind this is that even though remarkable results have been obtained for SMMs incorporated in devices, for practical applications the read-out and manipulation of the electronic and nuclear states is too slow. Optospintronic seeks to employ light as means to manipulate and control the spin states of the SMMs, leading to more efficient read-out methodologies. For optospintronic devices, high-quality quantum emitters are required, allowing the optical read-out of the spin states of the SMMs, without altering their quantum states. Optical techniques such as optical fiber scanning cavities [145], surface-enhanced fluorescence [146], and surface-enhanced Raman scattering [147, 148] could be employed to enhance the light–matter interaction leading to reliable and faster optical read-out. In this scheme, a quantum emitter is adapted to the size of the SMM, probing it locally. The quantum emitters can be linked at specific sites of molecules, allowing a read-out at specific positions. Furthermore, by employing fluorophores, with different fluorescence, could in principle enable the independent read-out of different spin centers on the same molecule. This is one of the main advantages of this approach, where several different fluorophores can be chemically attached to the same molecule, whereas attaching two electrodes to the same molecule is extremely demanding. Recently, Bayliss and coworkers have reported the optical read-out and manipulation of the electronic states of a transition metal complex with formula CrR_4 (with R = *o*-tolyl, 2,3-dime- thylphenyl, 2,4-dimethylpheny) [149] diluted in the diamagnetic hots. Employing a variation of our proposed approach, Bayliss et al. were able to determine the relaxation times and also induce Raby oscillations in the molecule, highlighting the novelty and richness of the growing field. Although these are important results in the field of optospintronic, the coherence time remains short for quantum communication applications.

Despite the short times, these results highlight the versatility of molecular systema for technological applications due to the tunable characteristics. The first demonstration of the importance of magnetic molecules for quantum communication

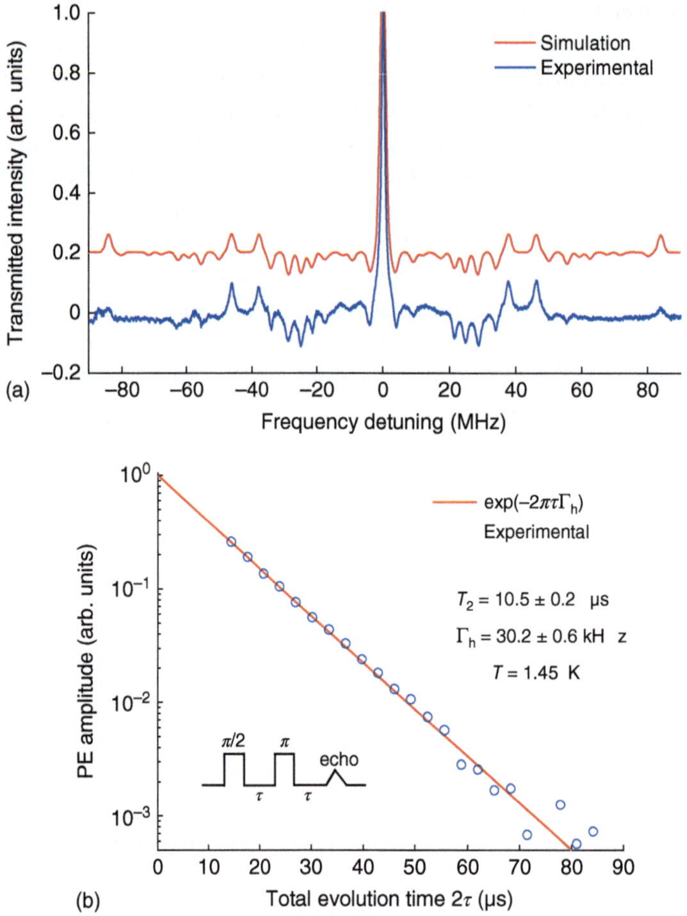

Figure 12.25 Long Optical Coherence for Quantum Communications: (a) Hole burning experimental spectrum of the $^{151}Eu^{3+}$ (blue) and simulation (red). (b) Two-pulse photon echo decay for the $^7F_0 \rightarrow {}^5D_0$ transition. Source: Serrano et al. [151] with permission from Springer Nature (copyrights © 2022).

applications was shown in a Eu^{3+} complex. Through spectral hole burning (SHB) in a dinuclear compound, lifetimes of a few seconds, long enough to allow efficient optical pumping, were observed [150]. Subsequently, longer optical coherence times, up to 10 μs, were determined in mono-nuclear Eu^{3+} complex [151] (Figure 12.25). The optical coherence times in the mononuclear Eu^{3+} complex were found to be four orders of magnitude narrower than those reported to date in molecular complexes. A suitable system would in principle allow coherent spin–phonon interactions with lifetimes limited by the linewidth of the photons. Additionally, the inhomogeneous linewidth is comparable systems such as Y_2O_3. For the Eu^{3+} complex, narrow-linewidth spectral hole burning with spin pumping over 90% was accomplished. Furthermore, optical storage, employing the atomic frequency comb quantum memory protocol, was achieved and optically controlled phase shifts, showing the basic mechanism of two qubit gates, were observed. The results highlight the importance and potential of lanthanide-based magnetic molecules. In addition, spectroscopic studies indicate that important improvements could be achieved at mK temperatures.

12.5 Conclusions

Since the discovery of magnetization relaxation in magnetic molecules in the early 1990s, SMMs have been at the heart of many studies. These systems have been shown to exhibit bewildering quantum effects, leading to their proposal of several futuristics applications. The systematic control gained over their syntheses makes them promising building-block architectures. This has been exemplified by the development of SMMs operating above liquid nitrogen temperatures [34–36], systems possessing remarkable long coherence times [86–88] and the manipulation of their quantum states to their integration of hybrid spin architectures [40, 67–71, 73–75], ultimately, leading to the implementation of a quantum algorithm [25]. The aforementioned examples highlight the results of the first quantum revolution, that is, the acquired knowledge gained of quantum effects.

Toward the true implementation of quantum effects in practical application, the second quantum revolution attempts to engineer these effects and implement them in hybrid devices. At this stage, due to the extraordinary control of SMMs, not solely over their structural synthetic aspect, but also over their electronic and nuclear characteristics, SMMs are at the point of being integrated into novel technologies. The coherence, along with the entanglement and spatial resolution, makes them prospect quantum sensors. Furthermore, they have been shown to possess all the properties to execute quantum simulations and fault-tolerant quantum computing architectures. More recently, to realize quantum networks, the characteristics of SMMs for quantum communications are being investigated, yielding promising results [151]. Clearly, SMMs have been shown to possess all the characteristics to be integrated into futuristic technologies, thus, bringing the research field to verge toward novel technological devices governed by quantum effects.

References

1 Schrödinger, E. (1926). Quantisierung als Eigenwertproblem. *Ann. Phys.* 385: 437–490.

2 Dowling, J.P. and Milburn, G.J. (2003). Quantum technology: the second quantum revolution. *Philosophical Transactions of the Royal Society A - Mathematical Physical and Engineering Sciences* 361: 1655–1674.

3 Atzori, M. and Sessoli, R. (2019). The second quantum revolution: role and challenges of molecular chemistry. *Journal of the American Chemical Society* 141: 11339–11352.

4 Lawrie, B.J., Lett, P.D., Marino, A.M., and Pooser, R.C. (2019). Quantum sensing with squeezed light. *ACS Photonics* 6: 1307–1318.

5 Radu, V., Price, J.C., Levett, S.J. et al. (2020). Dynamic quantum sensing of paramagnetic species using nitrogen-vacancy centers in diamond. *ACS Sensors* 5: 703–710.

6 Lee-Wong, E., Xue, R., Ye, F. et al. (2020). Nanoscale detection of magnon excitations with variable wavevectors through a quantum spin sensor. *Nano Letters* 20: 3284–3290.

7 Jenkins, A., Pelliccione, M., Yu, G. et al. (2019). Single-spin sensing of domain-wall structure and dynamics in a thin-film skyrmion host. *Physical Review Materials* 3: 83801.

8 Barry, J.F., Turner, M.J., Schloss, J.M. et al. (2016). Optical magnetic detection of single-neuron action potentials using quantum defects in diamond. *Proceedings of the National Academy of Sciences of the United States of America* 113: 14133–14138.

9 Dovzhenko, Y., Casola, F., Schlotter, S. et al. (2018). Magnetostatic twists in room-temperature skyrmions explored by nitrogen-vacancy center spin texture reconstruction. *Nature Communications* 9: 2712.

10 Boss, J.M., Cujia, K.S., Zopes, J., and Degen, C.L. (2017). Quantum sensing with arbitrary frequency resolution. *Science* 356: 837–840.

11 Shi, F., Zhang, Q., Wang, P. et al. (2015). Single-protein spin resonance spectroscopy under ambient conditions. *Science* 347: 1135–1138.

12 Lovchinsky, I., Sushkov, A.O., Urbach, E. et al. (2016). Nuclear magnetic resonance detection and spectroscopy of single proteins using quantum logic. *Science* 351: 836–841.

13 Feynman, R.P. (1982). Simulating physics with computers. *International Journal of Theoretical Physics* 21: 467–488.

14 Lloyd, S. (1993). A potentially realizable quantum computer. *Science* 261: 1569–1571.

15 García-Pérez, G., Rossi, M.A.C., and Maniscalco, S. (2020). IBM Q experience as a versatile experimental testbed for simulating open quantum systems. *npj Quantum Information* 6: 1.

16 Kelly, J., Barends, R., Fowler, A.G. et al. (2015). State preservation by repetitive error detection in a superconducting quantum circuit. *Nature* 519: 66–69.

17 Otterbach, J.S., Manenti, R., Alidoust, N. et al. (2017). Unsupervised machine learning on a hybrid quantum computer. arXiv:1712.05771 [quant-ph].

18 Watson, T.F., Philips, S.G.J., Kawakami, E. et al. (2018). A programmable two-qubit quantum processor in silicon. *Nature* 555: 633–637.

19 Arute, F., Arya, K., Babbush, R. et al. (2019). Quantum supremacy using a programmable superconducting processor. *Nature* 574: 505–510.

20 Leuenberger, M.N. and Loss, D. (2001). Quantum computing in molecular magnets. *Nature* 410: 789–793.

21 Sessoli, R., Gatteschi, D., Caneschi, A., and Novak, M.A. (1993). Magnetic bistability in a metal-ion cluster. *Nature* 365: 141–143.

22 Ishikawa, N., Sugita, M., Ishikawa, T. et al. (2003). Lanthanide double-decker complexes functioning as magnets at the single-molecular level. *Journal of the American Chemical Society* 125: 8694–8695.

23 Ishikawa, N., Sugita, M., Okubo, T. et al. (2003). Determination of ligand-field parameters and f-electronic structures of double-decker bis(phthalocyaninato)lanthanide complexes. *Inorganic Chemistry* 42: 2440–2446.

24 Moreno-Pineda, E., Godfrin, C., Balestro, F. et al. (2018). Molecular spin qudits for quantum algorithms. *Chemical Society Reviews* 47: 501–513.

25 Godfrin, C., Ferhat, A., Ballou, R. et al. (2017). Operating quantum states in single magnetic molecules: implementation of Grover's quantum algorithm. *Physical Review Letters* 119: 187702.

26 Ishikawa, N., Sugita, M., and Wernsdorfer, W. (2005). Quantum tunneling of magnetization in lanthanide single-molecule magnets: bis(phthalocyaninato)terbium and bis(phthalocyaninato)dysprosium anions. *Angewandte Chemie, International Edition* 44: 2931–2935.

27 Taran, G., Bonet, E., and Wernsdorfer, W. (2019). The role of the quadrupolar interaction in the tunneling dynamics of lanthanide molecular magnets. *Journal of Applied Physics* 125: 142903.

28 Briganti, M., Santanni, F., Tesi, L. et al. (2021). A complete ab initio view of Orbach and Raman spin–lattice relaxation in a dysprosium coordination compound. *Journal of the American Chemical Society* 143: 13633–13645.

29 Santanni, F., Albino, A., Atzori, M. et al. (2021). Probing vibrational symmetry effects and nuclear spin economy principles in molecular spin qubits. *Inorganic Chemistry* 60: 140–151.

30 Mirzoyan, R. and Hadt, R.G. (2020). The dynamic ligand field of a molecular qubit: decoherence through spin-phonon coupling. *Physical Chemistry Chemical Physics* 22: 11249–11265.

31 Albino, A., Benci, S., Tesi, L. et al. (2019). First-principles investigation of spin-phonon coupling in vanadium-based molecular spin quantum bits. *Inorganic Chemistry* 58: 10260–10268.

32 Kragskow, J.G.C., Marbey, J., Buch, C.D. et al. (2022). Analysis of vibronic coupling in a 4f molecular magnet with FIRMS. *Nature Communications* 13: 825.

33 Ganzhorn, M., Klyatskaya, S., Ruben, M., and Wernsdorfer, W. (2016). Quantum Einstein-de Haas effect. *Nature Communications* 7: 11443.

34 Goodwin, C.A.P., Ortu, F., Reta, D. et al. (2017). Molecular magnetic hysteresis at 60 kelvin in dysprosocenium. *Nature* 548: 439–442.

35 Guo, F.S., Day, B.M., Chen, Y.C. et al. (2018). Magnetic hysteresis up to 80 kelvin in a dysprosium metallocene single-molecule magnet. *Science* 362: 1400–1403.

36 Gould, C.A., McClain, K.R., Reta, D. et al. (2022). Ultrahard magnetism from mixed-valence dilanthanide complexes with metal-metal bonding. *Science* 375: 198–202.

37 Gatteschi, D. and Sessoli, R. (2003). Quantum tunneling of magnetization and related phenomena in molecular materials. *Angewandte Chemie, International Edition* 42: 268–297.

38 Grover, L.K. (1997). Quantum mechanics helps in searching for a needle in a haystack. *Physical Review Letters* 79: 325–328.

39 Moreno-Pineda, E., Klyatskaya, S., Du, P. et al. (2018). Observation of cooperative electronic quantum tunneling: increasing accessible nuclear states in a molecular qudit. *Inorganic Chemistry* 57: 9873–9879.

40 Biard, H., Moreno-Pineda, E., Ruben, M. et al. (2021). Increasing the Hilbert space dimension using a single coupled molecular spin. *Nature Communications* 12: 4443.

41 DiVincenzo, D.P. (2000). The physical implementation of quantum computation. *Fortschritte der Physik* 48: 771–783.

42 Milburn, G.J. (2009). Photons as qubits. *Physica Scripta, T.* T137: 014003.

43 Mohammadi, M., Niknafs, A., and Eshghi, M. (2011). Controlled gates for multi-level quantum computation. *Quantum Information Processing* 10: 241–256.

44 Kiktenko, E.O., Fedorov, A.K., Strakhov, A.A., and Man'Ko, V.I. (2015). Single qudit realization of the Deutsch algorithm using superconducting many-level quantum circuits. *Physics Letters, Section A: General, Atomic and Solid State Physics* 379: 1409–1413.

45 O'Leary, D.P., Brennen, G.K., and Bullock, S.S. (2006). Parallelism for quantum computation with qudits. *Physical Review A: Atomic, Molecular, and Optical Physics* 74: 032334.

46 Balakrishnan, S. (2014). Various constructions of qudit SWAP gate. *Physiotherapy Research International* 2014: 1–5.

47 Popov, A., Kiktenko, E., Fedorov, A., and Man'ko, V.I. (2016). Information processing using three-qubit and qubit–qutrit encodings of noncomposite quantum systems. *Journal of Russian Laser Research* 37: 581–590.

48 Kiktenko, E.O., Fedorov, A.K., Man'ko, O.V., and Man'ko, V.I. (2015). Multilevel superconducting circuits as two-qubit systems: operations, state preparation, and entropic inequalities. *Physical Review A* 91: 042312.

49 Rungta, P., Munro, W.J., Nemoto, K. et al. (2007). Qudit Entanglement. *Directions in Quantum Optics. Lecture Notes in Physics* 561: 149–164.

50 Yu, C.J., Graham, M.J., Zadrozny, J.M. et al. (2016). Long coherence times in nuclear spin-free vanadyl qubits. *Journal of the American Chemical Society* 138: 14678–14685.

51 Fataftah, M.S., Zadrozny, J.M., Coste, S.C. et al. (2016). Employing forbidden transitions as qubits in a nuclear spin-free chromium complex. *Journal of the American Chemical Society* 138: 1344–1348.

52 Moreno-Pineda, E., Damjanović, M., Fuhr, O. et al. (2017). Nuclear spin isomers: engineering a $Et_4N[DyPc_2]$ spin qudit. *Angewandte Chemie, International Edition* 56: 9915–9919.

53 Luo, M. and Wang, X. (2014). Universal quantum computation with qudits. *Science China - Physics Mechanics & Astronomy* 57: 1712–1717.

54 Gedik, Z., Silva, I.A., Çakmak, B. et al. (2015). Computational speed-up with a single qudit. *Scientific Reports* 5: 14671.

55 Atzori, M., Tesi, L., Morra, E. et al. (2016). Room-temperature quantum coherence and rabi oscillations in vanadyl phthalocyanine: toward multifunctional molecular spin qubits. *Journal of the American Chemical Society* 138: 2154–2157.

56 Atzori, M., Morra, E., Tesi, L. et al. (2016). Quantum coherence times enhancement in vanadium(IV)-based potential molecular qubits: the key role of the vanadyl moiety. *Journal of the American Chemical Society* 138: 11234–11244.

57 Atzori, M., Tesi, L., Benci, S. et al. (2017). Spin dynamics and low energy vibrations: insights from vanadyl-based potential molecular qubits. *Journal of the American Chemical Society* 139: 4338–4341.

58 Atzori, M., Benci, S., Morra, E. et al. (2018). Structural effects on the spin dynamics of potential molecular qubits. *Inorganic Chemistry* 57: 731–740.

59 Richart, D., Fischer, Y., and Weinfurter, H. (2012). Experimental implementation of higher dimensional time-energy entanglement. *Applied Physics B: Lasers and Optics* 106: 543–550.

60 Gottesman, D. Fault-tolerant quantum computation with higher-dimensional systems. in *Lecture Notes in Computer Science (Including Subseries Lecture Notes in Artificial Intelligence and Lecture Notes in Bioinformatics)* 1509, 302–313 (1999).

61 Kues, M., Reimer, C., Roztocki, P. et al. (2017). On-chip generation of high-dimensional entangled quantum states and their coherent control. *Nature* 546: 622–626.

62 Ralph, T.C., Resch, K.J., and Gilchrist, A. (2007). Efficient Toffoli gates using qudits. *Physical Review A: Atomic, Molecular, and Optical Physics* 75: 022313.

63 Neeley, M., Ansmann, M., Bialczak, R.C. et al. (2009). Emulation of a quantum spin with a superconducting phase qudit. *Science* 325: 722–725.

64 Neves, L., Lima, G., Gómez, J.G.A. et al. (2005). Generation of entangled states of qudits using twin photons. *Physical Review Letters* 94: 100501.

65 Heersche, H.B., De Groot, Z., Folk, J.A. et al. (2006). Electron transport through single Mn_{12} molecular magnets. *Physical Review Letters* 96: 206801.

66 Nossa, J.F., Islam, M.F., Canali, C.M., and Pederson, M.R. (2013). Electric control of a $\{Fe_4\}$ single-molecule magnet in a single-electron transistor. *Physical Review B: Condensed Matter and Materials Physics* 88: 224423.

67 Vincent, R., Klyatskaya, S., Ruben, M. et al. (2012). Electronic read-out of a single nuclear spin using a molecular spin transistor. *Nature* 488: 357–360.

68 Thiele, S., Vincent, R., Holzmann, M. et al. (2013). Electrical readout of individual nuclear spin trajectories in a single-molecule magnet spin transistor. *Physical Review Letters* 111: 037203.

69 Godfrin, C., Thiele, S., Ferhat, A. et al. (2017). Electrical read-out of a single spin using an exchange-coupled quantum dot. *ACS Nano* 11: 3984–3989.

70 Thiele, S., Balestro, F., Ballou, R. et al. (2014). Electrically driven nuclear spin resonance in single-molecule magnets. *Science* 344: 1135–1138.

71 Godfrin, C., Ballou, R., Bonet, E. et al. (2018). Generalized Ramsey interferometry explored with a single nuclear spin qudit. *npj Quantum Information* 4: 53.

72 Moreno-Pineda, E. and Wernsdorfer, W. (2021). Measuring molecular magnets for quantum technologies. *Nature Reviews Physics* 3: 645–659.

73 Urdampilleta, M., Klyatskaya, S., Cleuziou, J.P. et al. (2011). Supramolecular spin valves. *Nature Materials* 10: 502–506.

74 Urdampilleta, M., Klyatskaya, S., Ruben, M., and Wernsdorfer, W. (2013). Landau-Zener tunneling of a single Tb^{3+} magnetic moment allowing the electronic read-out of a nuclear spin. *Physical Review B* 87: 195412.

75 Urdampilleta, M., Klayatskaya, S., Ruben, M., and Wernsdorfer, W. (2015). Magnetic interaction between a radical spin and a single-molecule magnet in a molecular spin-valve. *ACS Nano* 9: 4458–4464.

76 Ganzhorn, M., Klyatskaya, S., Ruben, M., and Wernsdorfer, W. (2013). Strong spin-phonon coupling between a single-molecule magnet and a carbon nanotube nanoelectromechanical system. *Nature Nanotechnology* 8: 165–169.

77 Aasi, J., Abadie, J., Abbott, B.P. et al. (2013). Enhanced sensitivity of the LIGO gravitational wave detector by using squeezed states of light. *Nature Photonics* 7: 613–619.

78 Troiani, F., Ghirri, A., Paris, M.G.A. et al. (2019). Towards quantum sensing with molecular spins. *Journal of Magnetism and Magnetic Materials* 491: 165534.

79 Degen, C.L., Reinhard, F., and Cappellaro, P. (2017). Quantum sensing. *Reviews of Modern Physics* 89: 035002.

80 Yu, C.J., Von Kugelgen, S., Laorenza, D.W., and Freedman, D.E. (2021). A molecular approach to quantum sensing. *ACS Central Science* 7: 712–723.

81 Kim, D., Ibrahim, M.I., Foy, C. et al. (2019). A CMOS-integrated quantum sensor based on nitrogen–vacancy centres. *Nature Electronics* 2: 284–289.

82 Cooper, A., Sun, W.K.C., Jaskula, J.C., and Cappellaro, P. (2020). Identification and control of electron-nuclear spin defects in diamond. *Physical Review Letters* 124: 83602.

83 Panajotov, K.P., Arizaleta, M., Gomez, V. et al. (2004). Semiconductor lasers for quantum sensing. *Quantum Sensing and Nanophotonic Devices* 5359: 360.

84 Morley, G.W. (2014). Towards spintronic quantum technologies with dopants in silicon. In: *Electron Paramagnetic Resonance*, Chapter 3, 62–76.

85 Ohshima, T., Satoh, T., Kraus, H. et al. (2018). Creation of silicon vacancy in silicon carbide by proton beam writing toward quantum sensing applications. *Journal of Physics D: Applied Physics* 51: 62–76.

86 Coronado, E. (2020). Molecular magnetism: from chemical design to spin control in molecules, materials and devices. *Nature Reviews Materials* 5: 87–104.

87 Gaita-Ariño, A., Luis, F., Hill, S., and Coronado, E. (2019). Molecular spins for quantum computation. *Nature Chemistry* 11: 301–309.

88 Wasielewski, M.R., Forbes, M.D.E., Frank, N.L. et al. (2020). Exploiting chemistry and molecular systems for quantum information science. *Nature Reviews Chemistry* 4: 490–504.

89 Wedge, C.J., Timco, G.A., Spielberg, E.T. et al. (2012). Chemical engineering of molecular qubits. *Physical Review Letters* 108: 107204.

90 Timco, G.A., Carretta, S., Troiani, F. et al. (2009). Engineering the coupling between molecular spin qubits by coordination chemistry. *Nature Nanotechnology* 4: 173–178.

91 Ferrando-Soria, J., Moreno Pineda, E., Chiesa, A. et al. (2016). A modular design of molecular qubits to implement universal quantum gates. *Nature Communications* 7: 11377.

92 Fernandez, A., Moreno Pineda, E., Muryn, C.A. et al. (2015). G-engineering in hybrid rotaxanes to create AB and AB_2 electron spin systems: EPR spectroscopic studies of weak interactions between dissimilar electron spin qubits. *Angewandte Chemie, International Edition* 54: 10858–10861.

93 Ferrando-Soria, J., Fernandez, A., Moreno Pineda, E. et al. (2015). Controlled synthesis of Nanoscopic metal cages. *Journal of the American Chemical Society* 137: 7644–7647.

94 Bader, K., Dengler, D., Lenz, S. et al. (2014). Room temperature quantum coherence in a potential molecular qubit. *Nature Communications* 5: 5304.

95 Graham, M.J., Zadrozny, J.M., Shiddiq, M. et al. (2014). Influence of electronic spin and spin-orbit coupling on decoherence in mononuclear transition metal complexes. *Journal of the American Chemical Society* 136: 7623–7626.

96 Zadrozny, J.M., Niklas, J., Poluektov, O.G., and Freedman, D.E. (2014). Multiple quantum coherences from hyperfine transitions in a vanadium(IV) complex. *Journal of the American Chemical Society* 136: 15841–15844.

97 Tesi, L., Lucaccini, E., Cimatti, I. et al. (2016). Quantum coherence in a processable vanadyl complex: new tools for the search of molecular spin qubits. *Chemical Science* 7: 2074–2083.

98 Shiddiq, M., Komijani, D., Duan, Y. et al. (2016). Enhancing coherence in molecular spin qubits via atomic clock transitions. *Nature* 531: 348–351.

99 Graham, M.J., Zadrozny, J.M., Fataftah, M.S., and Freedman, D.E. (2017). Forging solid-state qubit design principles in a molecular furnace. *Chemistry of Materials* 29: 1885–1897.

100 Godfrin, C., Lumetti, S., Biard, H. et al. (2019). Microwave-assisted reversal of a single electron spin. *Journal of Applied Physics* 125: 142801.

101 Wiesendanger, R., Güntherodt, H.J., Güntherodt, G. et al. (1990). Observation of vacuum tunneling of spin-polarized electrons with the scanning tunneling microscope. *Physical Review Letters* 65: 247–250.

102 Wiesendanger, R. (2009). Spin mapping at the nanoscale and atomic scale. *Reviews of Modern Physics* 81: 1495–1550.

103 Willke, P., Yang, K., Bae, Y. et al. (2019). Magnetic resonance imaging of single atoms on a surface. *Nature Physics* 15: 1005–1010.

104 Willke, P., Bae, Y., Yang, K. et al. (2018). Hyperfine interaction of individual atoms on a surface. *Science* 362: 336–339.

105 Yang, K., Paul, W., Phark, S.H. et al. (2019). Coherent spin manipulation of individual atoms on a surface. *Science* 366: 509–512.

106 Willke, P., Singha, A., Zhang, X. et al. (2019). Tuning single-atom electron spin resonance in a vector magnetic field. *Nano Letters* 19: 8201–8206.

107 Willke, P., Paul, W., Natterer, F.D. et al. (2018). Probing quantum coherence in single-atom electron spin resonance. *Science Advances* 4: eaaq1543.

108 Yang, K., Bae, Y., Paul, W. et al. (2017). Engineering the eigenstates of coupled spin- 1/2 atoms on a surface. *Physical Review Letters* 119: 227206.

109 Bae, Y., Yang, K., Willke, P. et al. (2018). Enhanced quantum coherence in exchange coupled spins via singlet-triplet transitions. *Science Advances* 4: eaau4159.

110 Sugisaki, K., Nakazawa, S., Toyota, K. et al. (2019). Quantum chemistry on quantum computers: quantum simulations of the time evolution of wave functions under the S^2 operator and determination of the spin quantum number: S. *Physical Chemistry Chemical Physics* 21: 15356–15361.

111 Crippa, L., Tacchino, F., Chizzini, M. et al. (2021). Simulating static and dynamic properties of magnetic molecules with prototype quantum computers. *Magnetochemistry* 7: 117.

112 Montanaro, A. (2016). Quantum algorithms: an overview. *npj Quantum Information* 2: 15023.

113 Nakazawa, S., Nishida, S., Ise, T. et al. (2012). A synthetic two-spin quantum bit: G-engineered exchange-coupled biradical designed for controlled-NOT gate operations. *Angewandte Chemie, International Edition* 51: 9860–9864.

114 Atzori, M., Chiesa, A., Morra, E. et al. (2018). A two-qubit molecular architecture for electron-mediated nuclear quantum simulation. *Chemical Science* 9: 6183–6192.

115 Luis, F., Repollés, A., Martínez-Pérez, M.J. et al. (2011). Molecular prototypes for spin-based CNOT and SWAP quantum gates. *Physical Review Letters* 107: 117203.

116 Aguilà, D., Barrios, L.A., Velasco, V. et al. (2014). Heterodimetallic [LnLn'] lanthanide complexes: toward a chemical design of two-qubit molecular spin quantum gates. *Journal of the American Chemical Society* 136: 14215–14222.

117 Aromí, G., Aguilà, D., Gamez, P. et al. (2012). Design of magnetic coordination complexes for quantum computing. *Chemical Society Reviews* 41: 537–546.

118 Chiesa, A., Whitehead, G.F.S., Carretta, S. et al. (2014). Molecular nanomagnets with switchable coupling for quantum simulation. *Scientific Reports* 4: 7423.

119 Chiesa, A., Santini, P., and Carretta, S. (2016). Supramolecular complexes for quantum simulation. *Magnetochemistry* 2: 37.

120 Santini, P., Carretta, S., Troiani, F., and Amoretti, G. (2011). Molecular nanomagnets as quantum simulators. *Physical Review Letters* 107: 230502.

121 Ferrando-Soria, J., Magee, S.A., Chiesa, A. et al. (2016). Switchable interaction in molecular double qubits. *Chem* 1: 727–752.

122 Stepanenko, D., Trif, M., and Loss, D. (2008). Quantum computing with molecular magnets. *Inorganica Chimica Acta* 361: 3740–3745.

123 Graham, M.J., Krzyaniak, M.D., Wasielewski, M.R., and Freedman, D.E. (2017). Probing nuclear spin effects on electronic spin coherence via EPR measurements of vanadium(IV) complexes. *Inorganic Chemistry* 56: 8106–8113.

124 Zadrozny, J.M., Niklas, J., Poluektov, O.G., and Freedman, D.E. (2015). Millisecond coherence time in a tunable molecular electronic spin qubit. *ACS Central Science* 1: 488–492.

125 Tacchino, F., Chiesa, A., Sessoli, R. et al. (2021). A proposal for using molecular spin qudits as quantum simulators of light-matter interactions. *Journal of Materials Chemistry C* 9: 10266–10275.

126 Tacchino, F., Chiesa, A., Carretta, S., and Gerace, D. (2020). Quantum computers as universal quantum simulators: state-of-the-art and perspectives. *Advanced Quantum Technologies* 3: 1900052.

127 Shor, P.W. (1997). Polynomial-time algorithms for prime factorization and discrete logarithms on a quantum computer. *SIAM Journal on Computing* 26: 1484–1509.

128 Troiani, F., Ghirri, A., Affronte, M. et al. (2005). Molecular engineering of antiferromagnetic rings for quantum computation. *Physical Review Letters* 94: 207208.

129 Pedersen, K.S., Ariciu, A.M., McAdams, S. et al. (2016). Toward molecular 4f single-ion magnet qubits. *Journal of the American Chemical Society* 138: 5801–5804.

130 Jenkins, M.D., Duan, Y., Diosdado, B. et al. (2017). Coherent manipulation of three-qubit states in a molecular single-ion magnet. *Physical Review B* 95: 064423.

131 Martínez-Pérez, M.J., Cardona-Serra, S., Schlegel, C. et al. (2012). Gd-based single-ion magnets with tunable magnetic anisotropy: molecular design of spin qubits. *Physical Review Letters* 108: 247213.

132 Moreno Pineda, E., Komeda, T., Katoh, K. et al. (2016). Surface confinement of $TbPc_2$-SMMs: structural, electronic and magnetic properties. *Dalton Transactions* 45: 18417–18433.

133 Macaluso, E., Rubín, M., Aguilà, D. et al. (2020). A heterometallic [LnLn'Ln] lanthanide complex as a qubit with embedded quantum error correction. *Chemical Science* 11: 10337–10343.

134 Nielsen, M.A. and Chuang, I.L. (2012). *Quantum Computation and Quantum Information*. Cambridge University Press.

135 Lockyer, S.J., Chiesa, A., Timco, G.A. et al. (2021). Targeting molecular quantum memory with embedded error correction. *Chemical Science* 12: 9104–9113.

136 Chiesa, A., Macaluso, E., Petiziol, F. et al. (2020). Molecular Nanomagnets as qubits with embedded quantum-error correction. *Journal of Physical Chemistry Letters* 11: 8610–8615.

137 Awschalom, D.D., Hanson, R., Wrachtrup, J., and Zhou, B.B. (2018). Quantum technologies with optically interfaced solid-state spins. *Nature Photonics* 12: 516–527.

138 Zhong, M., Hedges, M.P., Ahlefeldt, R.L. et al. (2015). Optically addressable nuclear spins in a solid with a six-hour coherence time. *Nature* 517: 177–180.

139 Böttger, T., Thiel, C.W., Cone, R.L., and Sun, Y. (2009). Effects of magnetic field orientation on optical decoherence in Er^{3+}: Y_2SiO_5. *Physical Review B* 79: 115104.

140 Zhong, T. and Goldner, P. (2019). Emerging rare-earth doped material platforms for quantum nanophotonics. *Nanophotonics* 8: 2003–2015.

141 Kindem, J.M., Ruskuc, A., Bartholomew, J.G. et al. (2020). Control and single-shot readout of an ion embedded in a nanophotonic cavity. *Nature* 580: 201–204.

142 Chen, S., Raha, M., Phenicie, C.M. et al. (2020). Parallel single-shot measurement and coherent control of solid-state spins below the diffraction limit. *Science* 370: 592–595.

143 Ortu, A., Tiranov, A., Welinski, S. et al. (2018). Simultaneous coherence enhancement of optical and microwave transitions in solid-state electronic spins. *Nature Materials* 17: 671–675.

144 Serrano, D., Karlsson, J., Fossati, A. et al. (2018). All-optical control of long-lived nuclear spins in rare-earth doped nanoparticles. *Nature Communications* 9: 2127.

145 Vahala, K.J. (2003). Optical microcavities. *Nature* 424: 839–846.

146 Deng, W., Xie, F., Baltar, H.T.M.C.M., and Goldys, E.M. (2013). Metal-enhanced fluorescence in the life sciences: here, now and beyond. *Physical Chemistry Chemical Physics* 15: 15695–15708.

147 Etchegoin, P.G. and Le Ru, E.C. (2008). A perspective on single molecule SERS: current status and future challenges. *Physical Chemistry Chemical Physics* 10: 6079–6089.

148 Tian, J.H., Liu, B., Li, X. et al. (2006). Study of molecular junctions with a combined surface-enhanced Raman and mechanically controllable break junction method. *Journal of the American Chemical Society* 128: 14748–14749.

149 Bayliss, S.L., Laorenza, D.W., Mintun, P.J. et al. (2020). Optically addressable molecular spins for quantum information processing. *Science* 370: 1309–1312.

150 Kumar, K.S., Serrano, D., Nonat, A.M. et al. (2021). Optical spin-state polarization in a binuclear europium complex towards molecule-based coherent light-spin interfaces. *Nature Communications* 12: 2152.

151 Serrano, D., Kuppusamy, S.K., Heinrich, B. et al. (2022). Ultra-narrow optical linewidths in rare-earth molecular crystals. *Nature* 603: 241–246.

13

Molecular-Ion Quantum Technologies

Mudit Sinhal and Stefan Willitsch

Department of Chemistry, University of Basel, Klingelbergstrasse 80, 4056 Basel, Switzerland

13.1 Introduction

Over the past years, the development of experimental methods that enable the control of single isolated quantum systems has made impressive progress. While sophisticated techniques for the cooling and coherent manipulation of atomic systems are now well established and form the basis for their application in modern quantum science including quantum computation, quantum simulation, and precision measurements [1–5], their adaption to molecular systems has remained a persistent challenge. The complexity of the molecular energy-level structure reflecting electronic, vibrational, and rotational motions as well as electronic- and nuclear-spin degrees of freedom poses, in general, severe difficulties for their translational cooling as well as the preparation, manipulation, and readout of individual quantum states. While direct laser cooling of diatomic molecules like SrF [6] and CaF [7, 8] as well as polyatomic molecules like H_2CO [9], CaOH [10], and $CaOCH_3$ [11] has been demonstrated, the vast majority of molecules cannot be laser cooled because they lack closed optical-cycling transitions. Thus, alternative cooling schemes have to be implemented [12]. By the same token, sensitive optical-cycling methods that are used for state readout in atomic systems [13–15] cannot readily be applied.

While the additional degrees of freedom offered by molecules, in particular rotations and vibrations, impose challenges for their quantum control, they at the same time provide a rich resource for new applications. Spectroscopic transitions in molecular systems span a wide range of frequency bands from the kHz to the PHz, many of which are not easily accessible in atomic systems. Within the high density of interacting molecular levels, transitions with excellent coherence properties, which are scarce in atomic systems, can readily be identified. Such transitions are attractive candidates for qubits and for precision measurements [16]. Moreover, molecules offer prospects as novel platforms for tests of fundamental physical concepts such as the electric dipole moment of electron (eEDM) [17, 18] and possible parity violation effects at the molecular level [19–21], for precise determinations of fundamental constants [22–26], for precision spectroscopic measurements [27, 28],

Photonic Quantum Technologies: Science and Applications, First Edition. Edited by Mohamed Benyoucef.

for tests of ab initio calculations and molecular quantum theory [26, 29, 30], and for investigations of state- and energy-controlled atom–molecule [31, 32] and ultracold molecule–molecule [33] collisions. Molecules are also being increasingly considered for new frequency standards and clocks [16, 34–36]. Last but not least, molecules are the building blocks of chemistry giving access to a vast variety of chemical and biochemical phenomena, the study of which immensely benefits from precise experimental methods [37–39].

In this chapter, we discuss upcoming quantum technologies for the coherent control and manipulation of molecular ions in radiofrequency (RF) traps and their application in various domains of molecular science. Quantum-logic schemes in which single molecular ions are trapped together with single atomic ions in ion traps have recently emerged as general and flexible approaches to the nondestructive quantum-state manipulation of single molecules. In such experiments, operations that cannot directly be performed on the molecular ion due to its complex energy-level structure can be implemented via the co-trapped atomic ion. This new methodology constitutes a paradigm change in how molecules are interrogated: single molecules are used instead of large ensembles, investigations can be performed in a completely nondestructive and even quantum-non-demolition manner, and dramatic improvements in measurement sensitivity and precision can therefore be realized. As will be reviewed in this chapter, such protocols have so far successfully been employed for the cooling of molecular ions to the motional ground state of the trap, for the preparation and nondestructive detection of molecular quantum states, for extremely sensitive spectroscopic measurements, and for the implementation of coherent operations [40–46], while a variety of further applications has been proposed.

We begin by discussing the basic concepts and methods used in ion trapping and cooling. To contrast with recent developments, we briefly summarize destructive techniques that have been used in the past to probe trapped molecular ions. We then give an overview of the most recent developments in molecular-ion quantum technologies with a focus on advances in the nondestructive manipulation and coherent control of single molecular ions. Finally, we conclude with an outlook on future developments in the field.

13.2 Experimental Techniques

13.2.1 Ion Trapping

The experiments discussed here all rely on the confinement of ions in RF traps, also referred to as Paul traps. The theory of RF traps has been extensively treated in text books and review articles, see, e.g. Refs. [13, 47–52]. Here, we briefly discuss the principles of linear-quadrupole RF traps (LQTs), which are of particular relevance in the present context.

Figure 13.1 shows a schematic of a generic linear Paul trap consisting of four cylindrical electrodes arranged in a quadrupolar configuration. Note that LQTs

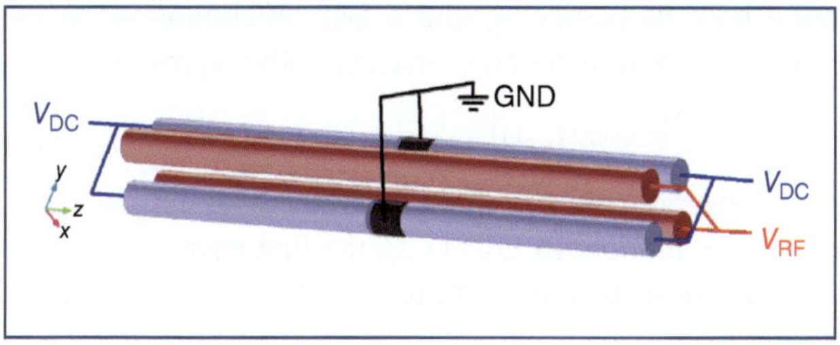

Figure 13.1 Schematic representation of a linear-quadrupole radiofrequency ion trap consisting of four cylindrical electrodes arranged in a quadrupolar configuration. Ions are confined in the center of the trap by applying radiofrequency voltages V_{RF} and static voltages V_{DC} to the electrodes as indicated.

can be configured in a variety of three- and two-dimensional electrode geometries including cylinders, wires, blades, wafers, and surface structures [50, 51]. Because Earnshaw's theorem forbids the generation of an electrostatic potential minimum, ion traps employ RF fields in order to confine ions dynamically. A dynamic trapping field in the radial (x, y) directions is generated by applying a time-varying voltage $V_{\mathrm{RF}} = V_0 \cos(\Omega_{\mathrm{RF}} t)$ to a pair of diagonally opposed electrodes (shown in red in Figure 13.1). Spatial confinement of the ions along the axial (z) direction is provided by applying static potentials, V_{DC}, to the endcaps of the trap (shown in blue in Figure 13.1).

An LQT is designed to generate an approximately quadrupolar electric potential close to the trap center. The equations of motion for a single ion in such a trap take the form of Mathieu equations [47, 49, 53],

$$\frac{d^2u}{d\tau^2} + \left[a_u - 2q_u \cos(2\tau)\right] u = 0 \tag{13.1}$$

where $u \in \{x, y, z\}$. The Mathieu parameters, a_u and q_u, are given by,

$$a_x = a_y = -\frac{1}{2} a_z = -\kappa \frac{4eV_{\mathrm{DC}}}{m z_0^2 \Omega_{\mathrm{RF}}^2} \tag{13.2a}$$

$$q_x = -q_y = \frac{2eV_{\mathrm{RF}}}{m r_0^2 \Omega_{\mathrm{RF}}^2}, \quad q_z = 0 \tag{13.2b}$$

and

$$\tau = \frac{\Omega_{\mathrm{RF}} t}{2} \tag{13.2c}$$

where m and e are the mass and the charge of the trapped ion, respectively, $2z_0$ is the distance between the endcaps, $2r_0$ is the distance between two diagonally opposed electrodes, and κ is a geometry parameter describing the stiffness of the static axial trapping potential that is considered to be harmonic [49]. Stable trapping of ions is possible for certain ranges of the a_u and q_u parameters, which is usually represented in terms of stability diagrams [49].

In the limit of $|a|$, $|q| \ll 1$, the motion of a trapped ion according to Eq. (13.1) can approximately be described in terms of two superimposed components, a slow, "secular" harmonic motion with frequency ω_u and a fast "micromotion" at the frequency Ω_{RF} of the RF drive. The trajectory of the ion can thus be expressed as [54],

$$u(t) \approx A_u \cos(\omega_u t + \phi)\left[1 - \frac{q_u}{2}\cos(\Omega_{RF}t)\right] \tag{13.3}$$

where ϕ is a phase describing the ion's initial position and velocity, and $A_u \equiv A_x, A_y$, and A_z are the amplitudes of the motion. In Eq. (13.3), the first term corresponds to the secular and the second term to the micromotion. The secular motional frequencies are given by

$$\omega_u = \frac{\Omega_{RF}}{2}\sqrt{a_u + \frac{q_u^2}{2}} \tag{13.4}$$

In the quantum limit, the secular motion of the ion can be represented by a quantum harmonic oscillator with discrete motional states $|n_u\rangle$ characterized by motional quantum numbers $n_u = 0, 1, 2, 3, \ldots$ and energies $E_{n_u} = (n_u + 1/2)\hbar\omega_u$. For an ion in the motional ground state $|n_u = 0\rangle$, the extent of its wavefunction is given by,

$$u_0 = \sqrt{\frac{\hbar}{2m\omega_u}} \tag{13.5}$$

In the interaction of the ion with light, the quantity $\eta = ku_0$ (the "Lamb–Dicke parameter") represents the ratio between the wavelength of the radiation with wave number $k = 2\pi/\lambda$ and u_0 [13]. In the Lamb–Dicke regime, defined as $\eta^2(2\,\overline{n}_u + 1) \ll 1$, where $\overline{n}_u$ represents the average motional quantum number of the ion in the trap, the extent of the ion's wavefunction is much smaller than the wavelength of the radiation coupling to the ion. In this regime, first-order Doppler shifts due to the motion of the ion vanish and transitions between different motional states appear as discrete sidebands modulated onto an ionic spectrum [13].

Figure 13.2a depicts the combined internal-motional energy levels of a hypothetical two-level ion, with internal ground ($|g\rangle$) and an excited ($|e\rangle$) state, trapped in a harmonic potential. As the ion's motional frequencies (of order MHz) are usually much smaller than its internal transition frequencies (typically hundreds of THz for electronic transitions), transitions involving motional energy levels appear as modulations on electronic excitations. Transitions of the form $|g, n_u\rangle \rightarrow |e, n_u\rangle$ or $|e, n_u\rangle \rightarrow |g, n_u\rangle$ in which the ion is excited or de-excited without a change in the motional quantum number are termed as carrier transitions (gray arrows in Figure 13.2a). Transitions in which the motional quantum number of the trapped ion changes are known as sideband transitions. When a sideband appears at a higher frequency than the carrier (blue arrows), i.e. $|g, n_u\rangle \rightarrow |e, n_u + 1\rangle$ or $|e, n_u\rangle \rightarrow |g, n_u - 1\rangle$, the transition is termed a blue sideband (BSB). By contrast, a transition of the form $|g, n_u\rangle \rightarrow |e, n_u - 1\rangle$ or $|e, n_u\rangle \rightarrow |g, n_u + 1\rangle$ is termed a red sideband (RSB) featuring a lower frequency than the carrier (red arrows). In Figure 13.2a, only first-order sidebands with $\Delta n_u = \pm 1$ are indicated, but it should be noted that also higher order sidebands with $|\Delta| n_u > 1$ exist [13].

In the context of experiments with multiple trapped ions, the Coulomb interaction between the charged particles couples the motion of the individual ions.

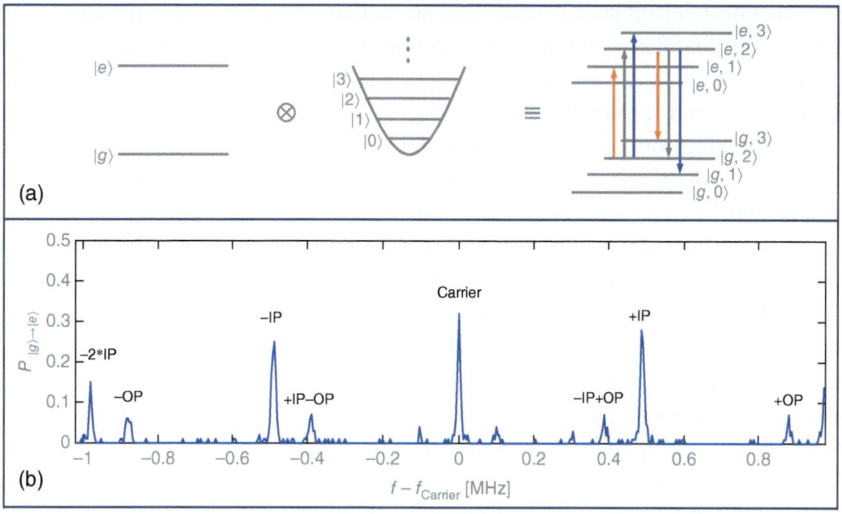

Figure 13.2 (a) Illustration of the combined internal-motional energy levels of a two-level system trapped in a harmonic potential. In the Lamb–Dicke regime, transitions between the quantized motional states of the particle appear as red and blue sidebands (red and blue arrows, respectively) modulated onto carrier transitions (gray arrows). (b) Probability ($P_{|g\rangle\to|e\rangle}$) for exciting the $|g\rangle = (4s)^2S_{1/2}(m = -1/2) \to |e\rangle = (3d)^2D_{5/2}(m = -5/2)$ transition in Ca^+ (also see Figure 13.4) for a $Ca^+-N_2^+$ two-ion Coulomb crystal as a function of the laser frequency, f, around the carrier frequency, $f_{Carrier}$. Red (−) and blue (+) sidebands of the in-phase (IP) and out-of-phase (OP) axial motional modes of the ion crystal are observed. Reproduced from Ref. [42] with permission from the Royal Society of Chemistry.

The collective motions of a string of ions are then described in terms of normal modes [55, 56]. For quantum-logic experiments with two ions of mass m_1 and m_2, the axial secular motion of the two-ion string can be described by an in-phase (IP) mode with a frequency ω_- and an out-of-phase (OP) mode with a frequency ω_+ given by [56],

$$\omega_\mp^2 = \frac{\omega_z}{m_1}\left(1 + \frac{1}{M_{12}} \mp \sqrt{1 + \frac{1}{M_{12}^2} - \frac{1}{M_{12}}}\right) \tag{13.6}$$

where ω_z is the axial secular frequency of the ion with the higher mass, m_1, and $M_{12} = m_1/m_2$ is the mass ratio of the two ions. Figure 13.2b shows a typical spectrum of a $Ca^+-N_2^+$ two-ion chain in an ion trap depicting the carrier and the red- and blue-sideband transitions (represented by "−" and "+" signs, respectively) corresponding to the IP and OP axial modes.

13.2.2 Generation of Molecular Ions and Their State Initialization

For quantum experiments with trapped molecular ions, the clean chemical preparation of the ions in the trap and their initialization in the required internal state is imperative. Molecular hydrides like CaH^+ and MgH^+, which are frequently used species in the present context, are usually produced by chemical reactions [43–46,

57–59] of the relevant laser-cooled alkaline-earth ions with hydrogen gas leaked into the vacuum chamber. Homonuclear ions like O_2^+, N_2^+, H_2^+ and their isotopomers can be generated by photoionization [60–64] or electron-impact ionization [65] of their neutral precursor molecules.

The initialization of the molecular ions in specific internal, i.e. electronic, vibrational, rotational, and even hyperfine-Zeeman, states (usually the absolute ground state) can be achieved by a variety of methods. Polar molecular ions whose rotational–vibrational (rovibrational) degrees of freedom couple to dipole radiation can be prepared in specific rovibrational levels by blackbody-radiation-assisted optical pumping techniques [43, 44, 66, 67], see Figure 13.3a for examples. Extensions of these schemes have also been used for hyperfine-state preparation [69]. As a further development, broadband optical-pumping schemes using femtosecond lasers that address multiple rotational transitions simultaneously have been shown to achieve state preparation on considerably reduced timescales [70].

By contrast, homonuclear diatomic ions like N_2^+ and O_2^+ do not feature a permanent electric dipole moment. Consequently, their rovibrational degrees of freedom

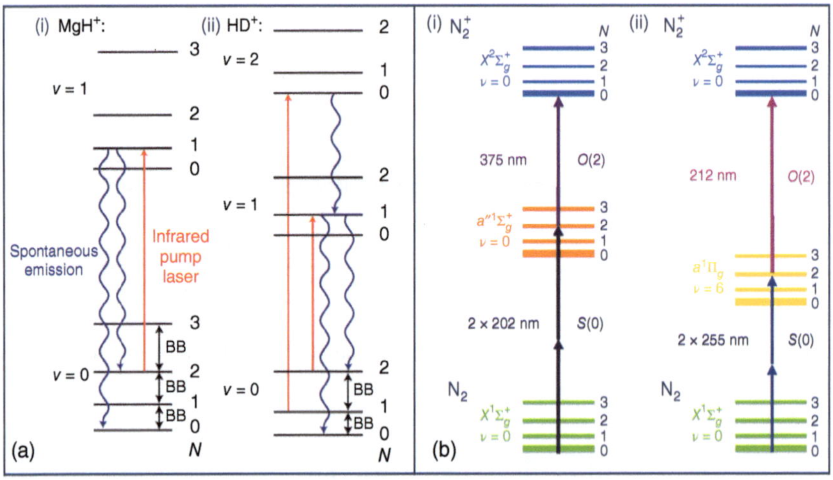

Figure 13.3 Rovibrational quantum-state preparation techniques for molecular ions: (a) Trapped (i) MgH^+ [66] and (ii) HD^+ [67] molecular ions are prepared in the rovibrational ground state ($N = 0, v = 0$) by continuously pumping population from selected rotational levels in the ground vibrational state ($v = 0$) to excited levels using infrared lasers (red arrows). The excited molecules can decay to the ground state ($N = 0, v = 0$) by spontaneous emission (blue wavy arrows). Coupling to the ambient blackbody-radiation field (BB; black arrows) leads to a redistribution of the level populations, which ultimately accumulate in the rotational ground state. Willitsch [68]/Springer Nature. (b) Two variants of state-selective photoionization schemes for the preparation of N_2^+ molecular ions in the rovibronic ground state [62, 64]. Both schemes start from rotational levels of the vibrationless ($v = 0$) $X^1\Sigma_g^+$ ground electronic state, but differ in the choice of intermediate electronic and vibrational states for the resonance-enhanced photoionization sequence. In (i), the $a''^1\Sigma_g^+, v = 0$ state is employed [62] while in (ii) the $a^1\Pi_g, v = 6$ state is chosen [64]. In both cases, the wavelength of the ionization laser is selected so as to only populate the rotational ground state of the cation.

do not couple to dipole radiation, which renders direct optical-pumping schemes inefficient. As an alternative, state-selective photoionization schemes can be implemented for producing the ions in well-defined rotational states. The technique relies on optical selection and propensity rules, which lead to the population of a limited number of rotational states in the ion upon photoionization of its neutral precursor molecule. By exciting the neutral molecule to a well-defined intermediate electronic state in a first step and subsequently setting the photoionization energy slightly above the lowest ionic state accessible due to selection rules (threshold photoionization), molecular ions could be generated in single rotational levels, see Figure 13.3b [62, 63].

Finally, a probabilistic state preparation of the molecular ions can also be achieved by the state-readout schemes discussed in Section 13.4 below.

13.2.3 Cooling of Trapped Ions

Ions loaded into a trap are typically hot with secular translational temperatures vastly exceeding room temperature. Thus, multistage cooling schemes are usually required to prepare the ions in their motional ground state, which is the starting point for the advanced experimental protocols discussed in this article.

In a first Doppler-cooling stage, photons are repeatedly scattered from a closed optical transition with a laser frequency slightly detuned to the red from resonance [71]. Alkaline-earth ions like Be^+, Mg^+, Ca^+, Ba^+, and Sr^+ exhibit simple energy-level structures that enable the implementation of the required closed optical cycles. As an example, Figure 13.4 depicts the Doppler-cooling scheme employed for Ca^+ ions. As most molecular ions do not possess optical-cycling transitions suitable for laser cooling, their motion has to be cooled sympathetically by the interaction with simultaneously trapped, laser-cooled atomic ions [59]. The limiting temperature achievable by Doppler laser cooling is given by [71],

$$T_D \approx \frac{\hbar\gamma}{2k_B} \tag{13.7}$$

where γ is the natural linewidth of the cooling transition, $\hbar$ is the reduced Planck constant, and k_B is the Boltzmann constant. T_D is typically on the order of 1 mK, which under typical trapping conditions is usually sufficient for the establishment of the Lamb–Dicke regime. Under these conditions, the ions localize in the trap forming "Coulomb crystals" in which individual particles can be observed, addressed, and manipulated, see Figure 13.5 [59, 74, 75].

Following Doppler cooling, a second cooling stage needs to be implemented that repeatedly addresses red-sideband transitions to optically pump the ions into their motional ground state. This can either be direct sideband cooling on a narrow optical transition, e.g. the $(4s)^2S_{1/2}(m = -1/2) \leftrightarrow (3d)^2D_{5/2}(m = -5/2)$ transition in Ca^+ (Figure 13.2) [76], Raman sideband cooling as is frequently used in Be^+ [77], or electromagnetically induced transparency (EIT)-based sideband cooling [78, 79]. In a mixed-species ion chain, sideband cooling in the Lamb–Dicke regime needs to address specific collective modes of the ion motion [80]. For many protocols, it is

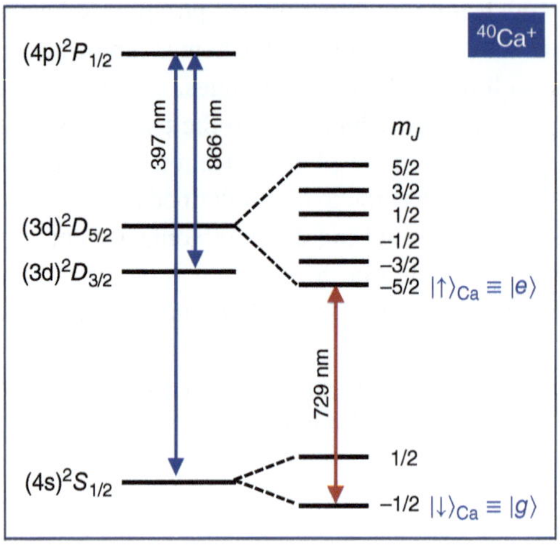

Figure 13.4 Energy-level scheme for the laser cooling of $^{40}Ca^+$ ions. Cooling to temperatures close to the Doppler limit is achieved by the repeated scattering of photons on the $(4s)^2S_{1/2} \rightarrow (4p)^2P_{1/2}$ transition at 397 nm. A second laser beam at 866 nm is required to repump population from the metastable $(3d)^2D_{3/2}$ state in order to close the laser cooling cycle. Red sidebands can be addressed repeatedly on, e.g. the narrow $(4s)^2S_{1/2}(m = -1/2) \leftrightarrow (3d)^2D_{5/2}(m = -5/2)$ electric-quadrupole transition in order to cool the ion to the motional ground state. In this case, an additional repumping laser at 854 nm is employed in order to drive the $(3d)^2D_{5/2} \rightarrow (4p)^2P_{3/2}$ transition (omitted for clarity in the above representation) from which the ion can relax to the ground state by spontaneous emission. With permission of American Association for the Advancement of Science, from Sinhal et al. [40]; permission conveyed through Copyright Clearance Center, Inc.

sufficient to cool only one mode of motion to its ground state, e.g. the IP or OP axial mode.

13.3 Destructive State-Readout Techniques

Prior to the advent of quantum-logic-spectroscopic readout schemes, various action-spectroscopy techniques have been used in order to probe Coulomb-crystallized molecular ions. As chemical changes of the trapped molecular ions can, in principle, be detected with unit probability using mass spectrometry [81], these methods are highly sensitive and have been widely used in various spectroscopic experiments and for the measurement of state populations.

As an example, laser-induced charge transfer (LICT) was implemented for performing electronic and vibrational spectroscopy of N_2^+ molecular ions co-trapped with atomic Ca^+ ions (Figure 13.5a) [62, 63, 72]. These experiments capitalized on a state-dependent charge-transfer reaction of N_2^+ with Ar atoms according to $N_2^+(v \geq 1) + Ar \rightarrow N_2 + Ar^+$ which only occurs in excited vibrational states of N_2^+ with vibrational quantum numbers $v \geq 1$ on energetic grounds [82]. Starting from the vibrational ground state, vibrationally excited states were populated either

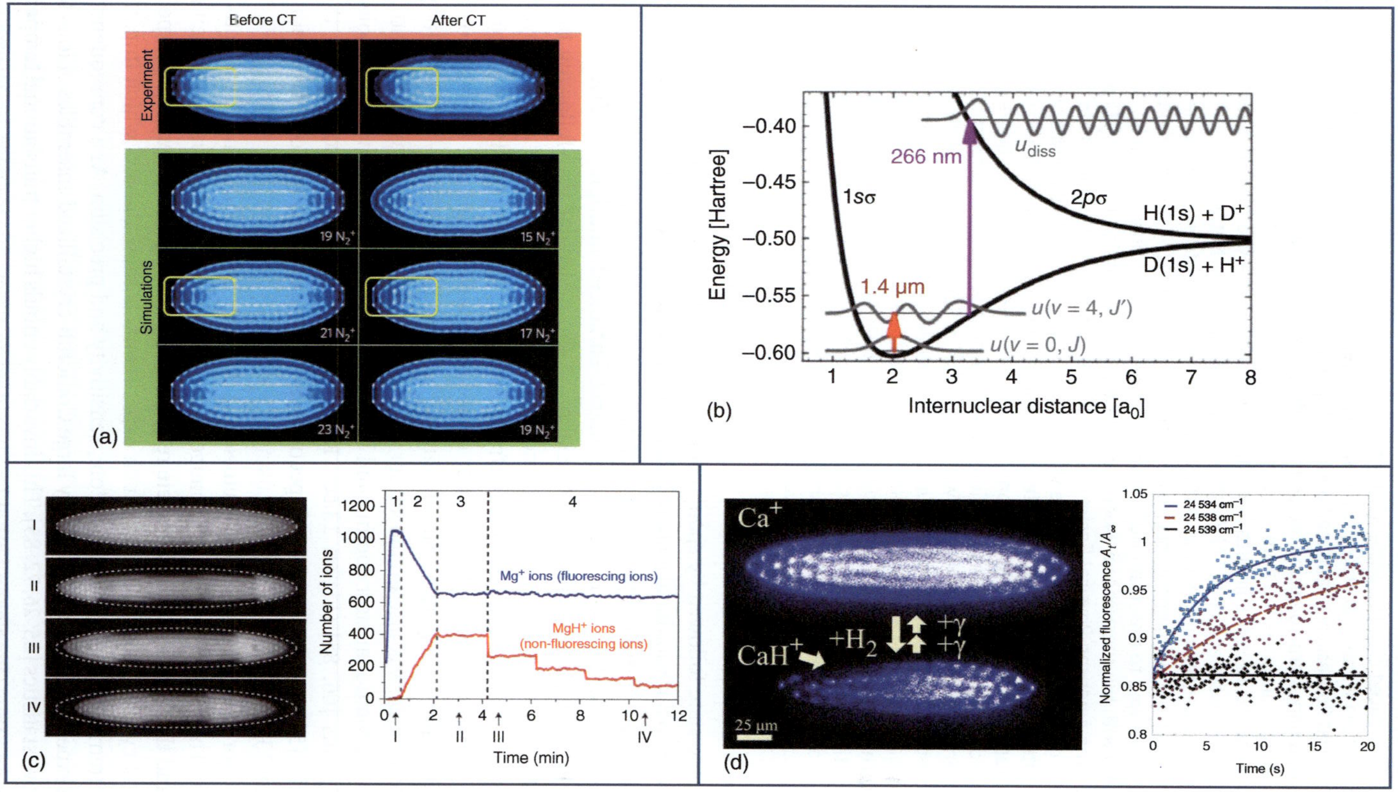

Before CT
After CT
Experiment
Simulations
19 N$_2^+$
15 N$_2^+$
21 N$_2^+$
17 N$_2^+$
23 N$_2^+$
19 N$_2^+$
(a)
Energy [Hartree]
−0.40
−0.45
−0.50
−0.55
−0.60
1sσ
2pσ
266 nm
1.4 μm
u_{diss}
H(1s) + D$^+$
D(1s) + H$^+$
$u(v = 4, J')$
$u(v = 0, J)$
Internuclear distance [a$_0$]
(b)
I
II
III
IV
Number of ions
Mg$^+$ ions (fluorescing ions)
MgH$^+$ ions
(non-fluorescing ions)
Time (min)
(c)
Ca$^+$
CaH$^+$
+H$_2$
+γ
25 μm
Normalized fluorescence A_t/A_∞
24 534 cm^{-1}
24 538 cm^{-1}
24 539 cm^{-1}
Time (s)
(d)

Figure 13.5 Destructive state detection and spectroscopic techniques for trapped molecular ions: (a) Fluorescence images of a Coulomb crystal of Ca^+ and N_2^+ ions before and after laser-induced charge transfer (CT) with Ar atoms following their infrared excitation to the first excited vibrational state [72]. The images were obtained by recording the laser-induced fluorescence produced during Doppler cooling of the Ca^+ ions. Molecular ions appear as a dark, nonfluorescing region in the center of the crystal. The number of molecular ions disappearing from the crystal due to CT was found by comparison of the experimental fluorescence images (red panel) with molecular-dynamics simulations (green panel). Germann et al. [72]/with permission of Springer Nature. (b) Resonance-enhanced multiphoton dissociation (REMPD) state-detection scheme implemented for rovibrational spectroscopy of HD^+ at 1.4 μm [73]. A single photon at 266 nm was employed to dissociate HD^+ ions from the $\upsilon = 4$ vibrational state following excitation with the spectroscopy laser. Reproduced with permission from [73]. Copyright 2007 by the American Physical Society. (c) REMPD of sympathetically cooled MgH^+ ions [66]. The images on the left-hand side show (I) the initial pure Mg^+ Coulomb crystal, (II) the crystal after loading MgH^+ ions (visible as the nonfluorescing region at the extremities of the crystal), and the crystal after (III) one, and (IV) four REMPD cycles. The figure on the right-hand side shows the number of ions deduced from the images as a function of the experiment time. Staanum et al. [66]/with permission of Springer Nature. (d) REMPD of Coulomb-crystallized CaH^+ ions produced by the chemical reaction of laser-cooled Ca^+ ions with H_2 gas [58]. The CaH^+ ions are photodissociated thus regenerating the original Ca^+ crystal. The figure on the right-hand side shows the total fluorescence yield of the Ca^+ ions as a function of the exposure time to the dissociation laser for various laser wavenumbers. Calvin et al. [58]/with permission of American Chemical Society.

in direct infrared excitation [72] or indirectly via excitation to the first excited electronic state followed by spontaneous emission [62, 63]. The latter scheme was used to map out rotational state populations following the generation of the ions by state-selective threshold photoionization.

A complementary technique is resonance-enhanced multiphoton dissociation (REMPD), which relies on the dissociation of the molecular ions following their excitation to a suitable intermediate molecular state. Such a scheme was employed, for instance, in the high-resolution infrared spectroscopy of trapped HD^+ molecular ions sympathetically cooled by laser-cooled Be^+ ions [25, 26, 73, 83], as shown schematically in Figure 13.5b, and in the rotational spectroscopy of HD^+ [84]. The technique was also employed for the read-out of rotational state populations following rotational laser cooling and buffer-gas cooling of sympathetically cooled MgH^+ ions [66, 85] (Figure 13.5c). REMPD techniques were also used for the vibronic [57] and rovibronic [58] spectroscopy of sympathetically cooled CaH^+ ions (Figure 13.5d).

A major drawback of action-spectroscopic methods is their inherently destructive nature. They rely on a chemical change of the molecular ion following photoexcitation and thus necessitate a complete reinitialization of the experiment every experimental cycle. The duty cycle of such experiments is therefore low which impairs measurement statistics and, therefore, sensitivity and precision. As a consequence, these experiments typically employ large Coulomb-crystallized ensembles of ions to improve statistics [25, 26, 72, 85]. This invariably entails higher translational temperatures and thus Doppler and other line broadening effects due to ensemble averaging (except in special cases in which spectroscopic experiments can be performed in the Lamb–Dicke regime under these conditions, see, e.g. Ref. [84]). However, in the context of advanced applications such as coherent and ultrahigh-resolution experiments on molecular ions, it is advantageous, and in many cases necessary, to work with single particles. It thus became expedient to develop novel techniques capable of interrogating single molecular ions that do not destroy the molecule and, ideally, also preserve its quantum state.

13.4 Quantum-Logic Experiments on Single Trapped Molecular Ions

Over the past few years, a range of coherent protocols have been developed for manipulating and probing trapped molecular ions. These schemes can be regarded as variants and extensions of the "quantum-logic spectroscopy," which was originally conceived in the framework of the Al^+ ion optical clock [86] and adapted to molecular species. These schemes are inherently nondestructive, thus mitigating the problems associated with the destructive techniques discussed in the previous section, and pave the way for a range of new applications of trapped molecular ions within the realms of quantum science and precision measurements.

In the context of experiments with single trapped molecular ions, we discuss the following three different implementations of quantum-logic protocols. As a first example, we consider the quantum-non-demolition state detection of a single N_2^+ ion performed via a co-trapped Ca^+ ion [40, 42]. The scheme relies on the coherent motional excitation of the ions using molecular-state-dependent optical-dipole forces (ODFs). This approach enabled the tracking of molecular quantum state with a high time resolution as well as rovibronic molecular spectroscopy. In an extension of this scheme [41], specific molecular states embedded within a dense energy-level structure could be identified. As a second example, we discuss the nondestructive state detection of MgH^+ ions [43]. In these experiments, an ODF was implemented in order to map the molecular state onto a motional qubit which was then read out on a co-trapped Mg^+ ion. As a final example, we highlight molecular state manipulation schemes implemented for CaH^+ ions via co-trapped Ca^+ ions [44, 45]. These experiments relied on exciting motional sidebands with Raman transitions on the molecule and allowed for coherent spectroscopy of selected molecular rotational states. Moreover, entanglement between the molecule and the atom could be demonstrated [46].

13.4.1 Quantum-Nondemolition Molecular State Readout by State-Dependent Coherent Motional Excitation

Figure 13.6 illustrates a quantum-non-demolition experiment for the measurement of the spin-rovibronic state of a single N_2^+ via a co-trapped Ca^+ ion [40, 42]. In the first step of the experimental protocol, the IP mode of the Ca^+–N_2^+ two-ion string was cooled to the ground state of the trap. An optical-dipole force (ODF), the strength of

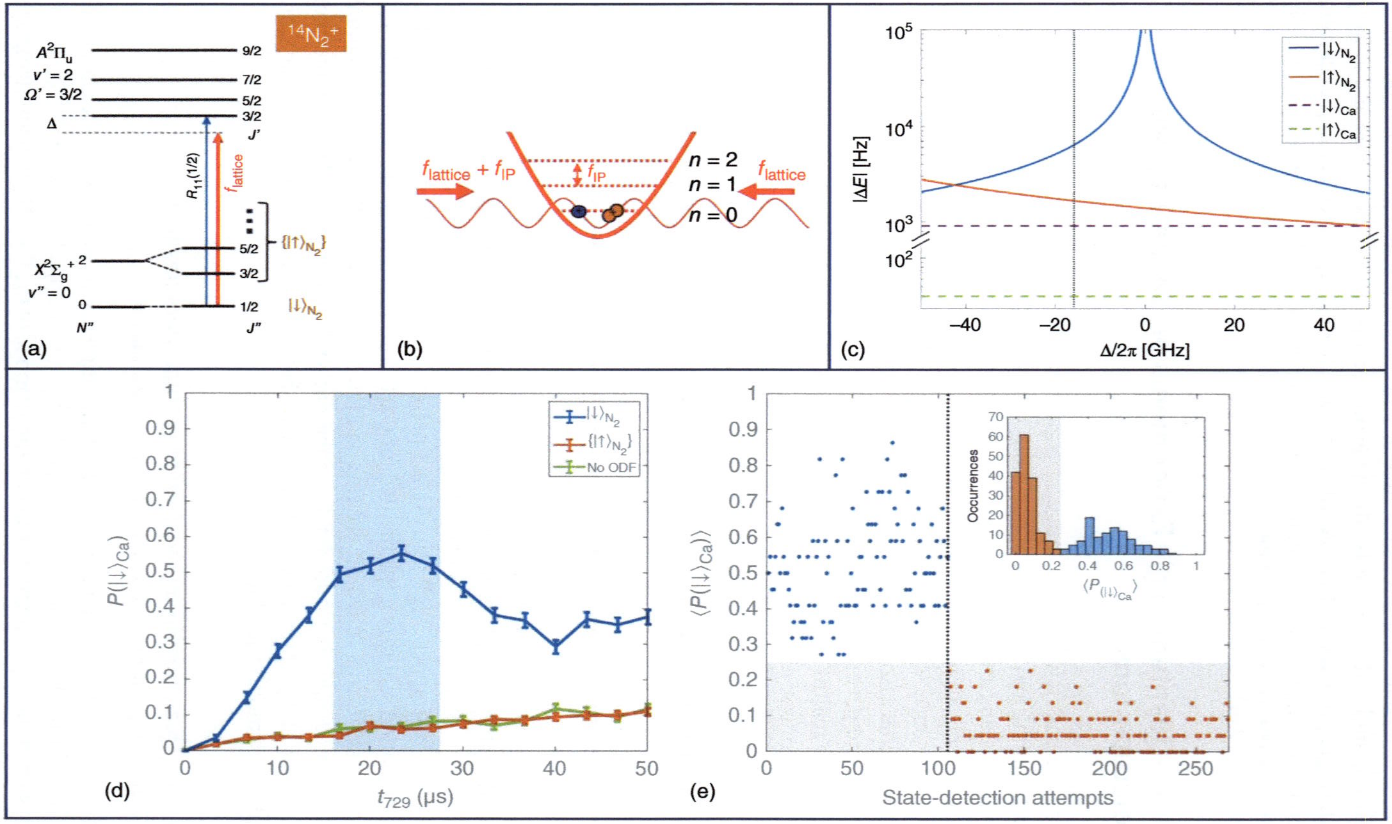

(a)
$^{14}N_2^+$
$A^2\Pi_u$
$v' = 2$
$\Omega' = 3/2$
Δ
9/2
7/2
5/2
3/2
J'
$R_{11}(1/2)$
$f_{lattice}$
$X^2\Sigma_g^+$
$v'' = 0$
2
0
N''
5/2
3/2
1/2
J''
$\{|\uparrow\rangle_{N_2}\}$
$|\downarrow\rangle_{N_2}$
(b)
$f_{lattice} + f_{IP}$
f_{IP}
$n = 2$
$n = 1$
$n = 0$
$f_{lattice}$
(c)
$|\Delta E|$ [Hz]
$\Delta/2\pi$ [GHz]
$|\downarrow\rangle_{N_2}$
$|\uparrow\rangle_{N_2}$
$|\downarrow\rangle_{Ca}$
$|\uparrow\rangle_{Ca}$
(d)
$P(|\downarrow\rangle_{Ca})$
t_{729} (μs)
$|\downarrow\rangle_{N_2}$
$\{|\uparrow\rangle_{N_2}\}$
No ODF
(e)
$\langle P(|\downarrow\rangle_{Ca})\rangle$
State-detection attempts
Occurrences
$\langle P_{(|\downarrow\rangle_{Ca}}\rangle$

Figure 13.6 Quantum-non-demolition state detection of molecular ions: (a) simplified energy-level scheme of the N_2^+ ion relevant for the quantum-logic-based state determination technique implemented in Refs. [40, 41]. (b) Schematic of the protocol implemented for the state readout of N_2^+. Two counter-propagating laser beams form a one-dimensional optical lattice interacting with a Ca^+–N_2^+ two-ion string cooled to the motional ground state of the in-phase (IP) mode in a harmonic trap. One of the lattice beams is detuned by the frequency of the IP mode, f_{IP}, to resonantly excite its motion by an optical dipole force (ODF) depending on the internal state of the N_2^+ ion. (c) Calculated magnitude of the ac-Stark shift, $|\Delta E|$, which gives rise to the ODF, experienced by the N_2^+ ion as a function of the laser-frequency detuning, Δ, from a strong spin-rovibronic transition in the molecule. At the lattice-laser wavelength indicated by the vertical black dotted line, the N_2^+ ion experienced a larger ac-Stark shift in its ground rovibronic state ($|\downarrow\rangle_{N_2}$) compared to the all other states ($\{|\uparrow\rangle_{N_2}\}$) leading to the state selectivity of the detection scheme. In the experiments, the Ca^+ ion was shelved in an excited state $|\uparrow\rangle_{Ca}$ in which it experienced a negligible ac-Stark shift (green-dashed line) in comparison to the N_2^+ ion. Conversely, a Ca^+ ion in its ground state would experience a considerably increased ac-Stark shift (purple-dashed line) leading to a spurious background signal in the experiment. (d) Rabi spectroscopy on the Ca^+ $|\uparrow\rangle_{Ca} \rightarrow |\downarrow\rangle_{Ca}$ blue-sideband (BSB) transition for N_2^+ ion in the $|\downarrow\rangle_{N_2}$ state (blue) and in one of the $\{|\uparrow\rangle_{N_2}\}$ states (red) as a function of the BSB pulse length t_{729}, following coherent motional excitation by the state-dependent ODF. $P(|\downarrow\rangle_{Ca})$ denotes the population in the $|\downarrow\rangle_{Ca}$ state. The green trace represents a background measurement of the Rabi oscillation signal without ODF. (e) Time trace of N_2^+ state-detection attempts (blue and orange dots) for measuring the molecular state. A threshold of 0.25 (gray-shaded area) was used to distinguish between N_2^+ in the $|\downarrow\rangle_{N_2}$ or $\{|\uparrow\rangle_{N_2}\}$ states. A histogram of the state detection attempts is shown in the inset. Sinhal et al. [40]/American Association for the Advancement of Science.

which depended on the molecular state, was then used to map the state information onto the motion of the two-ion string. In case the N_2^+ was in the state targeted by the ODF ($|\downarrow\rangle_{N_2}$ in Figure 13.6a), the Ca^+–N_2^+ string experienced a large ODF which was used to coherently excite the motion of the previously cooled IP mode. By contrast, for all other molecular states (summarily labeled $\{|\uparrow\rangle_{N_2}\}$ in Figure 13.6), the magnitude of the ODF was significantly reduced so that only negligible motional excitation occurred. Subsequently, the motional excitation of the two-ion string was detected by Rabi sideband thermometry on the Ca^+ ion thus revealing the information about the internal state of the N_2^+ ion.

The ODF for the state detection was implemented by two counter-propagating laser beams that formed a one-dimensional optical lattice. In order to coherently excite the IP mode of motion, the amplitude of the ODF was modulated by detuning one of the lattice beams by the mode frequency, f_{IP} (Figure 13.6b). The magnitude of the ODF experienced by the Ca^+–N_2^+ string was dependent on the ac-Stark shift experienced by the N_2^+ ion due to the lattice laser. Figure 13.6c depicts the calculated ac-Stark shift, ΔE, for $|\downarrow\rangle_{N_2}$ (here designated as the rovibronic ground state of the molecule) and the $\{|\uparrow\rangle_{N_2}\}$ states as a function of the lattice-laser detuning from a rovibronic transition (here the $X^2\Sigma_g^+(v'' = 0) - A^2\Pi_u(v' = 2)$ $R_{11}(1/2)$ line [87], where $v'(v'')$ represents the vibrational quantum number of the upper(lower) electronic state originating from the $|\downarrow\rangle_{N_2}$ state of N_2^+. The frequency of the lattice laser, $f_{lattice}$ (black-dotted line), was set such that a molecule in the $|\downarrow\rangle_{N_2}$ state experienced a large ac-Stark shift and hence a large ODF. Conversely, molecules in excited rotational or vibrational states experienced a much weaker ODF due to the lattice laser being farther detuned from any possible spectroscopic transition originating from these states thus resulting in a smaller motional excitation. For the lattice intensity and polarization chosen in these experiments, a Ca^+ ion in the ground $|\downarrow\rangle_{Ca}= (4s)^2S_{1/2}(m = -1/2)$ state (Figure 13.4b) experienced a large, nearly detuning-independent ac-Stark shift (purple-dashed line in Figure 13.6c) leading to a background ODF originating from Ca^+, which impaired the signal-to-noise (SNR) ratio for the detection of the state of the N_2^+ ion. Thus, in order to minimize the coupling of the atomic ion to the lattice laser, the Ca^+ ion was shelved in the $|\uparrow\rangle_{Ca}= (3d)^2D_{5/2}(m = -5/2)$ state that experienced a negligible ac-Stark shift (green- dashed line) resulting in no appreciable motional excitation due to the Ca^+ ion.

Figure 13.6d shows the results of Rabispectroscopy experiments performed on Ca^+ ion after the implementation of the ODF. For N_2^+ in the $|\downarrow\rangle_{N_2}$ state, significant motion was excited resulting in Rabi oscillations (blue trace) on the BSB of the IP mode of the Ca^+ $|\uparrow\rangle_{Ca} \rightarrow |\downarrow\rangle_{Ca}$ transition. By contrast, for molecules in the $\{|\uparrow\rangle_{N_2}\}$ states, no Rabi oscillations were observed (orange trace) as BSB transitions could not be excited when the ion string remained cooled to the ground motional state of the IP mode. Indeed, the signal obtained was comparable to a background measurement of the blue-sideband oscillations on a ground-state cooled $Ca^+-N_2^+$ chain without the application of any ODF (green trace). Each of the traces in Figure 13.6d resulted from thousands of measurements performed on the same molecular ion which was possible due to the nondestructive nature of the state-detection scheme.

Figure 13.6e depicts the results of the experiments as a function of molecular-state detection attempts. Each data point represents the average of 22 Rabi experiments for which the pulse time was set within the region of maximum contrast indicated by the light-blue area in Figure 13.6d. In this case, a single state-detection attempt took only few 100 ms, which allowed for tracking of the molecular state with a high time resolution. For the measurements recorded at the beginning of the experiment (blue dots), the molecule was determined to be in the rovibronic ground state 105 times with zero false events when setting the detection threshold above the gray-shaded area in Figure 13.6e. Subsequently, the molecule changed its state and was detected not to be in the rovibronic ground state 163 times with zero false detection attempts (orange dots). In this experiment, the fidelity for the faithful detection of the rovibronic ground state amounted to >99%.

An application of this state-detection protocol is shown in Figure 13.7. Since the magnitude of the excited motion is dependent on the ac-Stark shift giving rise to the ODF that in turn depends on the detuning of the lattice laser from a spectroscopic transition in the molecule, the scheme can be employed for noninvasive spectroscopy on the molecular ion. By extracting the magnitude of the ac-Stark shift from the observed Rabi oscillations as a function of the lattice-laser detuning, the spectral line shape of the resonance could be mapped out (Figure 13.7a). From the determined ac-Stark shifts, quantitative spectroscopic parameters like vibronic Einstein-A coefficients, A_{vibronic}, were obtained (Figure 13.7b). Both line positions

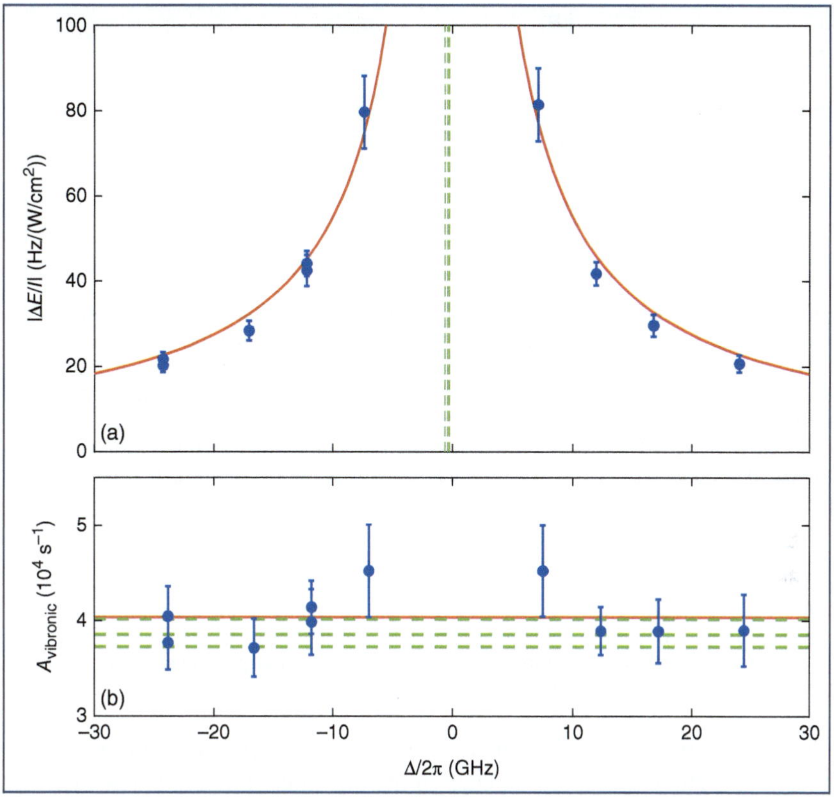

Figure 13.7 Noninvasive molecular spectroscopy of single molecular ions: (a) line shape of the $X^2\Sigma_g^+(v''=0) - A^2\Pi_u(v'=2)\ R_{11}(1/2)$ rovibronic transition in N_2^+ measured by determining the ac-Stark shift experienced by the ion at various lattice-laser detunings (blue data points) [40]. The line center was found by a fit (orange trace) to the measurements in very good agreement with the literature values (green-dashed lines). (b) Vibronic Einstein *A* coefficients, A_{vibronic}, extracted from the measured ac-Stark shifts in panel A at various detunings (blue data points). The mean of the measurements (orange line) was found to be in very good agreement with previous literature values (green-dashed lines). Sinhal et al., [40]/American Association for the Advancement of Science.

and Einstein coefficients were in very good agreement with literature values (green-dashed lines in Figure 13.7) thus validating the accuracy of the scheme.

In a variation of this scheme, not only the magnitude, but also the sign of the ac-Stark shift was retrieved [41]. As the ac-Stark shift changes sign across a resonance, its sign encodes the direction of the detuning from resonance (red or blue) and thus valuable additional information for molecular-state identification, which is particularly useful in molecules with a dense energy-level structure. Figure 13.8 depicts an experiment on a $Ca^+ - N_2^+$ chain in which in addition to the N_2^+ ion also the Ca^+ ion experienced a finite and constant detuning-independent ODF. An interferometric measurement as shown in Figure 13.8A was then performed in order to detect the molecular state. As before, the measurement started with preparing the $Ca^+ - N_2^+$ string in the ground state of the IP mode of motion. The Ca^+ ion was

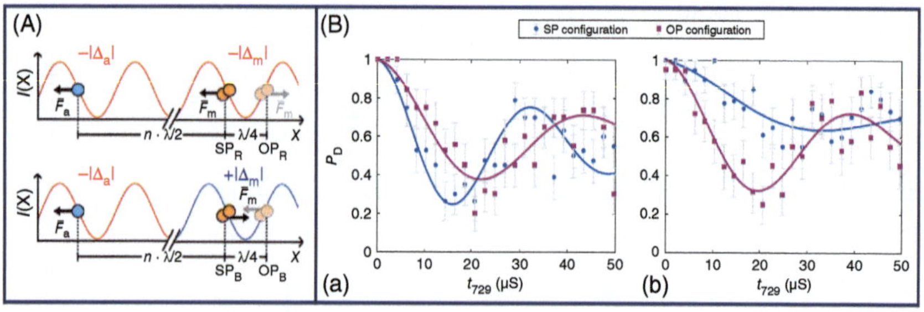

Figure 13.8 Phase-sensitive forces for molecular state identification. (A) Same-phase (SP) and opposite-phase (OP) configurations of the positions of a N_2^+ (orange dots) and Ca^+ (blue dot) ion in an optical lattice in order to measure the amplitude and phase of the ODF experienced by the N_2^+ ion. The Ca^+ ion was always red-detuned from its closest resonance and thus always experienced an approximately constant negative ac-Stark shift, $-|\Delta_a|$ and therefore a positive ODF. Depending on the molecular quantum state, the N_2^+ ion was either red ($-|\Delta_m|$) or blue ($+|\Delta_m|$) detuned from its nearest resonance resulting in attractive and repulsive optical forces, respectively. (B) If the ODF on N_2^+ and Ca^+ have the same sign (subpanel a), a stronger combined force and therefore stronger Rabi signal are measured on the Ca^+ ion after motional excitation in the SP configuration (blue trace) compared to the OP configuration (magenta trace). Conversely, if the ODF on N_2^+ and Ca^+ have opposite signs, the OP configuration gives a stronger Rabi signal (subpanel b). The relative sign of the ODF on both ions thus yields additional information on the molecular state that is particularly valuable for state identification within dense energy-level structures. Najafian et al. [41]/Springer Nature.

then prepared in the $|\uparrow\rangle_{Ca}$ state such that it was always red-detuned from its closest resonance and experienced an ODF in the direction of higher lattice intensity. The N_2^+ ion was generated in an arbitrary excited rotational state with the aim to identify this state within the dense Zeeman-spin-rotational energy-level structure in this region. Thus, for N_2^+ the amplitude and sign of the ac-Stark shift (and hence the ODF) depended on its specific rovibronic, hyperfine, and Zeeman state, while it remained constant and negative for Ca^+. In order to identify the molecular state, two combined measurements were then performed. In the first experiment, the two ions were placed in the trap such that their separation corresponded to an integer multiple of the lattice nodes. This lattice configuration was denoted as "same phase" (SP). In the SP configuration, the ODFs on the molecule and the atom acted in the same (opposite) direction when the molecular ion was red (blue) detuned from its closest resonance (SP_R and SP_B in Figure 13.8A). In the second experiment, the distance between the ions was changed by half a lattice spacing, such that the two ions always experienced an opposite gradient of the lattice field, i.e. the "opposite phase" (OP) of the lattice. In the OP configuration, the forces acted in the same (opposite) direction when the molecular ion was blue (red) detuned (OP_B and OP_R in Figure 13.8A). Figure 13.8B depicts Rabi oscillation signals on the $|\uparrow\rangle_{Ca} \rightarrow |\downarrow\rangle_{Ca}$ BSB transition for the SP (blue) and OP (purple) lattice configurations after an ODF pulse for two different molecular states (panels a and b). In panel a, the stronger SP signal (larger frequency and amplitude of the Rabi oscillation) suggested that the lattice laser was red detuned from the closest molecular transition. By comparing the measured amplitude and sign of the ac-Stark shift with theory, the spin-rotational state of the molecule could be identified as $N = 6, J = 11/2$, where in this case N and J are the rotational and spin-rotational quantum numbers, respectively. In panel b, the stronger OP signal suggested that the lattice laser was blue detuned from the closest molecular transition. In this case, all rotational levels $N \leq 4$ could be excluded.

13.4.2 Molecular State Detection and Spectroscopy via a Motional Qubit

Figure 13.9 shows a complementary approach to molecular-state identification in the presence of a spurious detuning-independent ODF experienced by the atomic ion. In these experiments [43], which represented the first quantum-logic spectroscopy performed on a molecular ion, a specific motional state was engineered on a Mg^+–MgH^+ two-ion chain. An ODF was then implemented in order to map the state of the molecular ion onto the motion that was subsequently detected by sideband measurements on the atomic ion. The experimental sequence is shown in Figure 13.9A. Here, both the IP and the OP motional modes of the two-ion chain were cooled to the ground state by resolved sideband cooling [89]. In a first step, the motion was initialized in a state $|\downarrow\rangle_{\text{motion}} = |1\rangle_{\text{IP}}|0\rangle_{\text{OP}}$ corresponding to one quantum of excitation in the IP mode. The atomic ion was prepared in an excited state $|\uparrow\rangle_{\text{atom}}$. An ODF was then applied using a 1D optical lattice with parameters chosen such that the motional state of the two-ion crystal was changed to $|\uparrow\rangle_{\text{motion}} = |0\rangle_{\text{IP}}|1\rangle_{\text{OP}}$ by the ODF acting solely on the atom in case the molecular ion was not in the $J = 1$ rotational state (Figure 13.9B(b)). By contrast, if the MgH^+ ion was in the targeted rotational state, the Mg^+–MgH^+ two-ion chain experienced an additional state-dependent ODF due to the molecular ion resulting in a different dynamics on the Bloch sphere of the motional qubit (Figure 13.9B(c)). For suitably chosen experimental parameters, the protocol corresponds to a CNOT logic gate performed on the motional qubit with the molecular state as the control allowing to detect the $J = 1$ rotational state in the ground vibronic manifold. In a subsequent step, a sideband measurement on the atomic ion was used to detect the motional state of the two-ion chain resulting in the determination of the molecular state. Figure 13.9C shows the probability of exciting the sideband on the Mg^+ ion after the ODF pulse as a function of the detuning, Δ_{MgH}, of the lattice laser frequency on both the red and blue sides of a molecular transition starting from the $J = 1$ rotational state. The theoretical estimates, including the measurement uncertainties, are given by the green areas and the experimental realizations are represented by the data points. By determining the strength of the ODF acting on the molecular ion, a spectroscopic measurement on the relevant molecular transition was performed as shown in Figure 13.9D.

13.4.3 Molecular Quantum Logic Spectroscopy Using Resolved-Sideband Raman Transitions

A different approach to molecular-state preparation and detection was implemented by Chou et al. [44] in CaH^+ (see Figure 13.10). As in the previously discussed protocols, the experiment was started by cooling the shared motion of a Ca^+–CaH^+

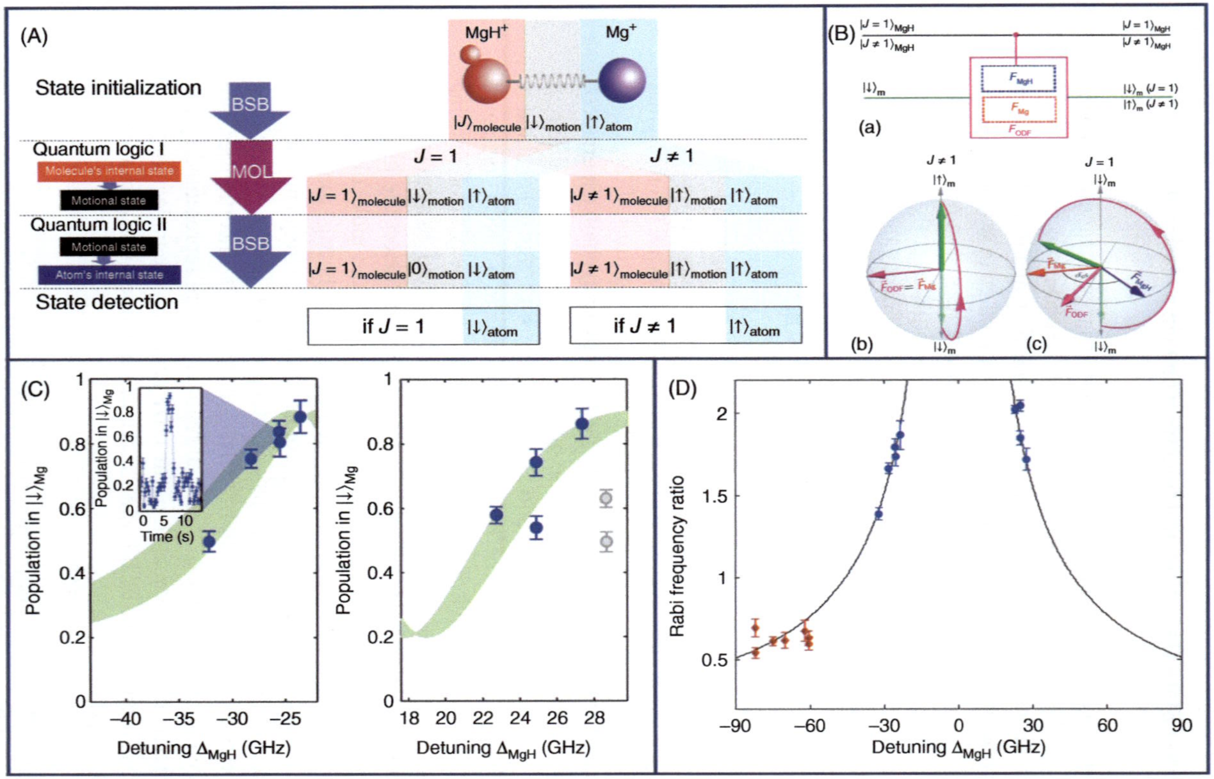

Figure 13.9 (A) Schematic representation of the quantum-logic spectroscopy implemented for the state detection of MgH^+ in Ref. [43]. See text for details. Reproduced with permission from F. Wolf for [88]. (B) Circuit diagram of the state-mapping process from the molecular to the motional state (panel a). A Bloch-sphere representation depicting the final motional qubit state for the $J \neq 1$ (panel b) and $J = 1$ (panel c) rotational states is also shown. (C) Theoretical estimation (green-shaded region) and experimental observations (data points) for the population measurement in the ground state $|\downarrow\rangle_{Mg}$ of the Mg^+ ion provided that the molecule is in the $J = 1$ rotational state, after the implementation of the state-dependent force. (D) Mapping of a rovibronic transition in the MgH^+ ion by application of the nondestructive state detection technique. Panels (B)–(D) are reproduced with permission from the Springer Nature Customer Service GmbH: Springer Nature for Wolf et al., [43], Copyright 2016.

two-ion chain and preparing the Ca^+ ion in an excited state. In the room-temperature environment of the experiment in Ref. [44], the molecules were prepared in their electronic and vibrational ground state and preparation of a specific rotational state relied on blackbody-assisted population redistribution. Raman laser beams were then applied in order to drive specific BSB transitions in the molecule such that a change in the molecular state was accompanied by an excitation of the shared motion of the two-ion chain (Figure 13.10a). The excited motion was subsequently detected by driving a RSB on the atomic ion followed by fluorescence detection leading to projection of the molecule into a specific quantum state by the measurement. By allowing sufficient time for the molecule to re-equilibrate with the blackbody environment, Raman spectra of several molecular transitions could be recorded by sideband measurements on the atomic ion.

Since the times for population redistribution of the rotational states by blackbody radiation (BBR) were long (about 100 ms to more than 2 seconds at room temperature for $J < 8$) compared to the time required to drive transitions (<5 ms), the population of the rotational states could be accumulated in specific magnetic sublevels by optical pumping and the targeted quantum states could then be coherently manipulated. The eigenstates of CaH^+ are labeled by the quantum numbers $|J, m, \xi\rangle$, where $J = 0, 1, 2, 3, \ldots$ is the rotational angular momentum quantum number in this case, $m \in \{-J - 1/2, -J + 1/2, J - 1/2, J + 1/2\}$ denotes the sum of the components of the rotational angular momentum and the proton spin along the magnetic field direction. $\xi \in \{+, -\}$ indicates the relative sign of a superposition of product states with the same m but opposite proton spin. Figure 13.10b depicts coherent spectroscopy and operations on the molecular ion on the $|J, -J - 1/2, -\rangle \leftrightarrow |J, -J + 1/2, -\rangle$ transitions for $J = 1$ (panels 1, 2, and 3) and $J = 2$ (panels 4, 5, and 6).

Based on these state detection and preparation techniques, a high-resolution rotational spectrum of a single CaH^+ molecule was recorded with a linewidth <1 kHz and an accuracy of ~1 part per billion [45]. The spectra were obtained by coherently driving stimulated-Raman rotational transitions using an optical frequency comb (Figure 13.10c). Here, the molecule was prepared in the $|\mathcal{J}''\rangle = |2, -5/2, -\rangle$ state followed by probing the $|\mathcal{J}'\rangle \leftarrow |\mathcal{J}''\rangle$ transition where $|\mathcal{J}'\rangle = |4, -7/2, -\rangle$. The probabilities of the molecule being in either states as a function of the Raman difference frequency from the resonance are shown in panels 1 and 3. When prepared in the $|\mathcal{J}'\rangle$ state, the coherent Rabi flops to and from the $|\mathcal{J}''\rangle$ state are shown in panels 2 and 4.

Moreover, by applying tailored pulse sequences on the Ca^+ and CaH^+ ion, a quantum entanglement was generated between the atom and the molecule [46]. Figure 13.10d shows parity measurements of entangled states of the atom and a low-frequency ($|\psi_l\rangle$) as well as a high-frequency ($|\psi_h\rangle$) molecular qubit. The state parities, Π, were measured as a function of the phase of an "analysis" pulse, ϕ_a, and oscillate as $C\cos(2\phi_a + \phi_0)$ where ϕ_0 is an offset in phase and $C \geq 0$ is the observed contrast. The fidelities for the generation of the low-frequency and high-frequency entangled states in this experiment were estimated to be $F_l = 0.87(3)$ and $F_h = 0.76(3)$ indicating bipartite entanglement for both cases [90].

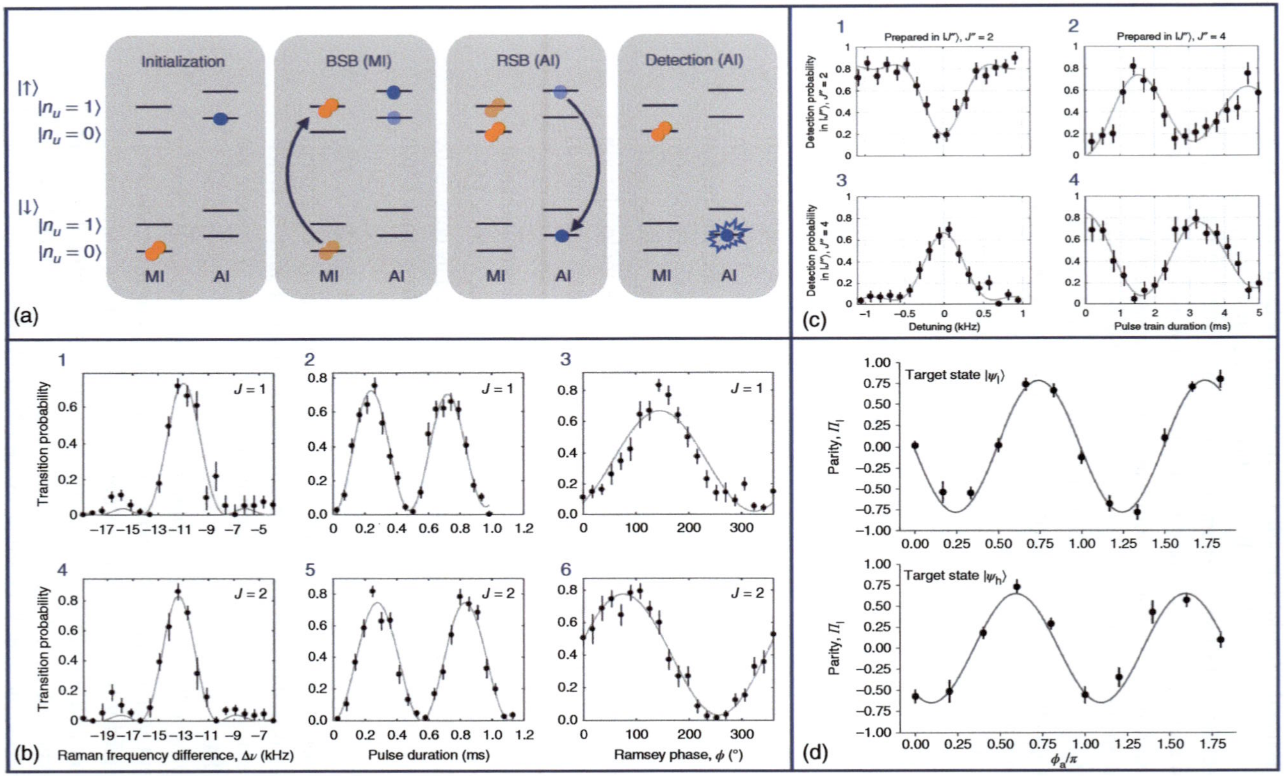

Figure 13.10 Nondestructive state determination and coherent manipulation of CaH^+ ions. (a) Schematic of the quantum-logic technique implemented for molecular-state detection and projective preparation in Ref. [44] (MI = molecular ion, AI = atomic ion). See text for details. Note that the definition of a red sideband as indicated in the third panel of Figure 13.10a adopted by the authors in Ref. [44] differs from the one used throughout this chapter (for example see Figure 13.2a where such a transition would be labeled as a blue sideband). (b) Coherent spectroscopy and manipulation on the $J = 1$ (panels 1, 2, and 3) and $J = 2$ (panels 4, 5, and 6) rotational states of the CaH^+ ion on transitions of the form $|J, -J-1/2, -\rangle \leftrightarrow |J, -J+1/2, -\rangle$. Frequency scans are shown in panels 1 and 4, Rabi flops are depicted in panels 2 and 5, Ramsey fringes as a function of the relative phase between two $\pi/2$ pulses with a wait time of 15 ms are shown in panels 3 and 6. Chou et al., [44]/American Association for the Advancement of Science. (c) High-resolution frequency spectra (panels 1 and 3) and Rabi flops (panels 2 and 4) driven by an optical frequency comb for $\Delta J = 2$ rotational transitions in a single CaH^+ ion. Chou et al., [45]/ American Association for the Advancement of Science. (d) Parity measurements of entangled states between an atom and a low-frequency ($|\psi_l\rangle$) as well as a high-frequency ($|\psi_h\rangle$) molecular qubit. Reproduced with permission from the Springer Nature Customer Service GmbH: Springer Nature for Lin et al., [46], Copyright 2020.

13.5 Outlook on Future Developments and Conclusions

In addition to the already realized experiments discussed earlier, alternative quantum-logic protocols have been proposed for molecular ions. Mur-Petit et al. [91] have suggested temperature-independent geometric-phase gates based on state-dependent forces for nondestructive molecular-state detection. These operations are insensitive to the motional state of a trapped atom-molecule two-ion crystal and can thus be applied to Doppler cooled ions possibly leading to simpler and faster experimental implementations.

Loh et al. [92] have proposed a state-detection technique relying on state-dependent magnetic g factors, which lead to distinct Zeeman splittings for electronic, rotational, and hyperfine quantum states of a molecule. Such a Zeeman-splitting-assisted quantum logic spectroscopy (ZSQLS) scheme would employ lasers far detuned from one-photon transitions in the molecule to drive Raman transitions between Zeeman sublevels without scattering photons and thus preserve the quantum state of the molecular ion. Apolar molecules like O_2^+ and polar molecules like SO^+, CO^+, and SiO^+ could be probed by the ZSQLS technique under suitable experimental conditions [92].

As an extension of the CaH^+ state-detection scheme implemented by Chou et al. [44], Wolf et al. [93] proposed a quantum-logic technique for nondestructive state detection of O_2^+ ions that would allow for amplification of the state detection signal, which is particularly important if single-shot readout of the atomic ion is not possible for technical reasons. The scheme employs bichromatic far-detuned Raman lasers to drive red and BSB transitions between selected Zeeman states simultaneously. For specific rotational states, the phase of the bichromatic drive can be chosen such that a Schrödinger cat state of the form $|\psi\rangle_{SC} = |+\rangle\,|\alpha\rangle \pm |-\rangle\,|-\alpha\rangle$ is produced, where $|\pm\rangle = (|1/2\rangle \pm |3/2\rangle)/\sqrt{2}$ and $|\pm\alpha\rangle$ denotes a coherent state with complex amplitude $\pm\alpha$. By detecting the depletion of the motional ground state due to the emergence of the Schrödinger cat state after application of the Raman pulses, the successful drive of the transition can be detected on the atomic ion making it possible to infer the molecular state.

Another series of proposals explored the potential of using the dipole moments of trapped polar molecular ions as a resource for quantum logic, either by capitalizing on direct dipole–dipole couplings between the molecules [94] or the coupling of the dipoles to the normal modes of a Coulomb crystal [95, 96]. The latter approach offers possibilities for implementing protocols which do not require a direct optical addressing of the molecules and their cooling to the motional ground state. Proposals for further developments of the dipole–phonon toolbox include the application of microwave fields to address both the internal states and the motions of the trapped polar ions [97].

A variant of quantum-logic spectroscopy makes use of the recoil imparted by the absorption or emission of photons to the motion of ground-state cooled ions. The resulting change of motional state can then be detected by the usual sideband-spectroscopy techniques. This approach has proven useful in the precise measurement of strong, dipole-allowed transitions in atomic ions [98]. Extensions of this method to infrared spectroscopy [99] and to probing the vibrational dynamics of trapped molecular ions [100] have recently been proposed.

The new coherent methods for molecular ions discussed are not only of interest for applications like quantum science and precision spectroscopy, but also have a considerable potential in completely different realms such as collision studies and chemistry. Proof-of-principle experiments were presented in Refs. [16, 40] in which changes of the internal state [40] and of the chemical identity [16] of single N_2^+ ions were tracked as a consequence of collisions demonstrating the utility of the present methods as noninvasive probes for investigating inelastic and reactive collisions of molecular ions.

In parallel to the methodological advances outlined earlier, new technological developments are addressing an ever-growing variety of molecular systems. Recently, laser cooling and optical-pumping schemes for molecular cations like BH^+ [101], SiO^+ [102, 103], TeH^+ [104], $AlCl^+$, and AlF^+ [105] and anions like C_2^- [106] have been explored theoretically. Apart from the systems already used in current experiments, i.e. MgH^+, CaH^+, and N_2^+, quantum-logic techniques for molecular ions like H_2^+ and its isotopologues [23, 35, 36], O_2^+ [93, 107] and polyatomics such as cyanoacetylene [100] and propanediol [108] have been explored in various proposals. Clearly, with the vast number of molecular systems available, the possibilities are virtually endless.

The coherent experimental tools available for trapped molecular ions have now reached a similar level of sophistication as the ones employed for atomic ions. Although the field of molecular-ion quantum technologies is still in its infancy, the presently realized experiments as well as the current proposals for future advances discussed in this review illustrate the immense potential for the quantum control and manipulation of molecular ions and the ensuing applications in precision spectroscopy, quantum science, and even chemistry.

Acknowledgments

Our work is funded by the Swiss National Science Foundation (SNSF) under project grants CRSII5_183579, TMAG-2_209193 and 200021_204123. It is also supported as a part of NCCR QSIT, a National Centre of Competence in Research, funded by the SNSF (Grant number 51NF40-185902).

References

1 Ladd, T.D., Jelezko, F., Laflamme, R. et al. (2010). Quantum computers. *Nature* 464: 45.

2 Georgescu, I.M., Ashhab, S., and Nori, F. (2014). Quantum simulation. *Reviews of Modern Physics* 86: 153.

3 Acín, A., Bloch, I., Buhrman, H. et al. (2018). The quantum technologies roadmap: a European community view. *New Journal of Physics* 20 (8): 080201.

4 Bruzewicz, C.D., Chiaverini, J., McConnell, R., and Sage, J.M. (2019). Trapped-ion quantum computing: progress and challenges. *Applied Physics Reviews* 6: 021314.

5 Ludlow, A.D., Boyd, M.M., Ye, J. et al. (2015). Optical atomic clocks. *Reviews of Modern Physics* 87: 637.

6 Shuman, E.S., Barry, J.F., and DeMille, D. (2010). Laser cooling of a diatomic molecule. *Nature* 467: 820.

7 Williams, H.J., Caldwell, L., Fitch, N.J. et al. (2018). Magnetic trapping and coherent control of laser-cooled molecules. *Physical Review Letters* 120: 163201.

8 Anderegg, L., Augenbraun, B.L., Bao, Y. et al. (2018). Laser cooling of optically trapped molecules. *Nature Physics* 14: 890.

9 Prehn, A., Ibrügger, M., Glöckner, R. et al. (2016). Optoelectrical cooling of polar molecules to submillikelvin temperatures. *Physical Review Letters* 116: 063005.

10 Baum, L., Vilas, N.B., Hallas, C. et al. (2021). Establishing a nearly closed cycling transition in a polyatomic molecule. *Physical Review A* 103 (4): 043111.

11 Mitra, D., Vilas, N.B., Hallas, C. et al. (2020). Direct laser cooling of a symmetric top molecule. *Science* 369 (6509): 1366–1369.

12 Krems, R.V., Stwalley, W.C., and Friedrich, B. (eds.) (2009). *Cold Molecules: Theory, Experiment, Applications*. Boca Raton, FL: CRC Press.

13 Leibfried, D., Blatt, R., Monroe, C., and Wineland, D. (2003). Quantum dynamics of single trapped ions. *Reviews of Modern Physics* 75: 281.

14 Häffner, H., Roos, C.F., and Blatt, R. (2008). Quantum computing with trapped ions. *Physics Reports* 469: 155.

15 Myerson, A.H., Szwer, D.J., Webster, S.C. et al. (2008). High-Fidelity Readout of a Trapped-Ion Qubit. *Physical Review Letters* 100: 200502.

16 Najafian, K., Meir, Z., and Willitsch, S. (2020). From megahertz to terahertz qubits encoded in molecular ions: theoretical analysis of dipole-forbidden spectroscopic transitions in N_2^+. *Physical Chemistry Chemical Physics* 22 (40): 23083–23098.

17 Ho, C.J., Devlin, J.A., Rabey, I.M. et al. (2020). New techniques for a measurement of the electron's electric dipole moment. *New Journal of Physics* 22 (5): 053031.

18 Baron, J., Campbell, W.C., DeMille, D. et al. (2014). Order of magnitude smaller limit on the electric dipole moment of the electron. *Science* 343: 269.

19 Altuntaş, E., Ammon, J., Cahn, S.B., and DeMille, D. (2018). Measuring nuclear-spin-dependent parity violation with molecules: experimental methods and analysis of systematic errors. *Physical Review A* 97 (4): 042101.

20 Crassous, J., Monier, F., Dutasta, J.P. et al. (2003). Search for resolution of chiral fluorohalogenomethanes and parity-violation effects at the molecular level. *ChemPhysChem* 4 (6): 541–548.

21 Quack, M., Stohner, J., and Willeke, M. (2008). High-resolution spectroscopic studies and theory of parity violation in chiral molecules. *Annual Review of Physical Chemistry* 59: 741.

22 Shelkovnikov, A., Butcher, R.J., Chardonnet, C., and Amy-Klein, A. (2008). Stability of the proton-to-electron mass ratio. *Physical Review Letters* 100 (15): 150801.

23 Schiller, S. and Korobov, V. (2005). Tests of time independence of the electron and nuclear masses with ultracold molecules. *Physical Review A* 71: 032505.

24 Kajita, M., Gopakumar, G., Abe, M. et al. (2014). Test of m_p/m_e changes using vibrational transitions in N_2^+. *Physical Review A* 89: 032509.

25 Kortunov, I.V., Alighanbari, S., Hansen, M.G. et al. (2021). Proton–electron mass ratio by high-resolution optical spectroscopy of ion ensembles in the resolved-carrier regime. *Nature Physics* 17 (5): 569–573.

26 Patra, S., Germann, M., Karr, J.-Ph. et al. (2020). Proton-electron mass ratio from laser spectroscopy of HD^+ at the part-per-trillion level. *Science* 369 (6508): 1238–1241.

27 Semeria, L., Jansen, P., Camenisch, G.M. et al. (2020). Precision measurements in few-electron molecules: the ionization energy of metastable $^4He_2$ and the first rotational interval of $^4He_2^+$. *Physical Review Letters* 124 (21): 213001.

28 Melosso, M., Diouf, M.L., Bizzocchi, L. et al. (2021). Hyperfine-resolved near-infrared spectra of $H_2^{17}O$. *Journal of Physical Chemistry A* 125 (36): 7884–7890.

29 Hölsch, N., Beyer, M., Salumbides, E.J. et al. (2019). Benchmarking theory with an improved measurement of the ionization and dissociation energies of H_2. *Physical Review Letters* 103002: 122.

30 Alighanbari, S., Giri, G.S., Constantin, F.L. et al. (2020). Precise test of quantum electrodynamics and determination of fundamental constants with HD^+ ions. *Nature* 581: 152.

31 Jurgilas, S., Chakraborty, A., Rich, C.J.H. et al. (2021). Collisions between ultracold molecules and atoms in a magnetic trap. *Physical Review Letters* 126 (15): 153401.

32 Dörfler, A.D., Eberle, P., Koner, D. et al. (2019). Long-range versus short-range effects in cold molecular ion-neutral collisions. *Nature Communications* 10: 5429.

33 Cheuk, L.W., Anderegg, L., Bao, Y. et al. (2020). Observation of collisions between two ultracold ground-state CaF molecules. *Physical Review Letters* 125 (4): 043401.

34 Kajita, M. (2015). N_2^+ quadrupole transitions with small Zeeman shift. *Physical Review A* 92: 043423.

35 Karr, J.-Ph. (2014). H_2^+ and HD^+: candidates for a molecular clock. *Journal of Molecular Spectroscopy* 300: 37.

36 Schiller, S., Bakalov, D., and Korobov, V.I. (2014). Simplest molecules as candidates for precise optical clocks. *Physical Review Letters* 113: 023004.

37 Willitsch, S. (2017). Chemistry with controlled ions. *Advances in Chemical Physics* 162: 307.

38 Heazlewood, B. and Softley, T.P. (2015). Low-temperature kinetics and dynamics with coulomb crystals. *Annual Review of Physical Chemistry* 66: 475–495.

39 Heazlewood, B. and Softley, T.P. (2021). Towards chemistry at absolute zero. *Nature Reviews Chemistry* 5: 125.

40 Sinhal, M., Meir, Z., Najafian, K. et al. (2020). Quantum-nondemolition state detection and spectroscopy of single trapped molecules. *Science* 367 (6483): 1213–1218.

41 Najafian, K., Meir, Z., Sinhal, M., and Willitsch, S. (2020). Identification of molecular quantum states using phase-sensitive forces. *Nature Communications* 11 (1): 1–10.

42 Meir, Z., Hegi, G., Najafian, K. et al. (2019). State-selective coherent motional excitation as a new approach for the manipulation, spectroscopy and state-to-state chemistry of single molecular ions. *Faraday Discussions* 217: 561–583.

43 Wolf, F., Wan, Y., Heip, J.C. et al. (2016). Non-destructive state detection for quantum logic spectroscopy of molecular ions. *Nature* 530: 457.

44 Chou, C.W., Kurz, C., Hume, D.B. et al. (2017). Preparation and coherent manipulation of pure quantum states of a single molecular ion. *Nature* 545: 203.

45 Chou, C.W., Collopy, A.L., Kurz, C. et al. (2020). Frequency-comb spectroscopy on pure quantum states of a single molecular ion. *Science* 367 (6485): 1458–1461.

46 Lin, Y., Leibrandt, D.R., Leibfried, D., and Chou, C.W. (2020). Quantum entanglement between an atom and a molecule. *Nature* 581 (7808): 273–277.

47 Major, F.G., Gheorghe, V.N., and Werth, G. (2005). *Charged Particle Traps*. Berlin and Heidelberg: Springer-Verlag.

48 March, R.E. and Todd, J.F. (2005). *Quadrupole Ion Trap Mass Spectrometry*, 2e. Hoboken, NJ: Wiley.

49 Drewsen, M. and Brøner, A. (2000). Harmonic linear Paul trap: stability diagram and effective potentials. *Physical Review A* 62: 045401.

50 Knoop, M., Madsen, N., and Thompson, R.C. (2014). *Physics with Trapped Charged Particles: Lectures from the Les Houches Winter School*. World Scientific.

51 Thompson, R.C., Madsen, N., and Knoop, M. (2016). *Trapped Charged Particles: A Graduate Textbook with Problems and Solutions*, vol. 1. World Scientific.

52 Ghosh, P.K. (1995). *Ion Traps*. Oxford: Clarendon Press.

53 March, R.E. (1997). An introduction to quadrupole ion trap mass spectrometry. *Journal of Mass Spectrometry* 32 (4): 351–369.

54 Berkeland, D.J., Miller, J.D., Bergquist, J.C. et al. (1998). Minimization of ion micromotion in a Paul trap. *Journal of Applied Physics* 83: 5025.

55 James, D.F.V. (1998). Quantum computation and quantum information theory. *Applied Physics B* 66: 181–190.

56 Morigi, G. and Walther, H. (2001). Two-species Coulomb chains for quantum information. *European Physical Journal D* 13: 261.

57 Rugango, R., Calvin, A.T., Janardan, S. et al. (2016). Vibronic spectroscopy of sympathetically cooled CaH^+. *ChemPhysChem* 17 (22): 3764–3768.

58 Calvin, A.T., Janardan, S., Condoluci, J. et al. (2018). Rovibronic spectroscopy of sympathetically cooled $^{40}CaH^+$. *Journal of Physical Chemistry A* 122 (12): 3177–3181.
59 Mølhave, K. and Drewsen, M. (2000). Formation of translationally cold MgH^+ and MgD^+ molecules in an ion trap. *Physical Review A* 62: 011401.
60 Dochain, A. and Urbain, X. (2015). Production of a rovibrationally selected O_2^+ beam for dissociative recombination studies. *EPJ Web Conference, Volume* 84, p. 05001. EDP Sciences.
61 Schmidt, J., Louvradoux, T., Heinrich, J. et al. (2020). Trapping, cooling, and photodissociation analysis of state-selected H_2^+ ions produced by (3 + 1) multiphoton ionization. *Physical Review Applied* 14: 024053.
62 Tong, X., Winney, A.H., and Willitsch, S. (2010). Sympathetic cooling of molecular ions in selected rotational and vibrational states produced by threshold photoionization. *Physical Review Letters* 105: 143001.
63 Tong, X., Wild, D., and Willitsch, S. (2011). Collisional and radiative effects in the state-selective preparation of translationally cold molecular ions in ion traps. *Physical Review A* 83: 023415.
64 Gardner, A., Softley, T., and Keller, M. (2019). Multi-photon ionisation spectroscopy for rotational state preparation of N_2^+. *Scientific Reports* 9: 506.
65 Blythe, P., Roth, B., Fröhlich, U. et al. (2005). Production of ultracold trapped molecular hydrogen ions. *Physical Review Letters* 95: 183002.
66 Staanum, P.F., Højbjerre, K., Skyt, P.S. et al. (2010). Rotational laser cooling of vibrationally and translationally cold molecular ions. *Nature Physics* 6: 271.
67 Schneider, T., Roth, B., Duncker, H. et al. (2010). All-optical preparation of molecular ions in the rovibrational ground state. *Nature Physics* 6: 275.
68 Willitsch, S. (2010). Very cool molecular ions. *Nature Physics* 6: 240.
69 Bressel, U., Borodin, A., Shen, J. et al. (2012). Manipulation of individual hyperfine states in cold trapped molecular ions and application to HD^+ frequency metrology. *Physical Review Letters* 108: 183003.
70 Lien, C.-Y., Seck, C.M., Lin, Y.-W. et al. (2014). Broadband optical cooling of molecular rotors from room temperature to the ground state. *Nature Communications* 5: 4783.
71 Metcalf, H.J. and van der Straten, P. (1999). *Laser Cooling and Trapping*. New York: Springer.
72 Germann, M., Tong, X., and Willitsch, S. (2014). Observation of electric-dipole-forbidden infrared transitions in cold molecular ions. *Nature Physics* 10: 820.
73 Koelemeij, J.C.J., Roth, B., and Schiller, S. (2007). Blackbody thermometry with cold molecular ions and application to ion-based frequency standards. *Physical Review A* 76: 023413.
74 Willitsch, S., Bell, M.T., Gingell, A.D. et al. (2008). Cold reactive collisions between laser-cooled ions and velocity-selected neutral molecules. *Physical Review Letters* 100: 043203.
75 Willitsch, S. (2012). Coulomb-crystallised molecular ions in traps: methods, applications, prospects. *International Reviews in Physical Chemistry* 31: 175.

76 Schmidt-Kaler, F., Häffner, H., Gulde, S. et al. (2003). How to realize a universal quantum gate with trapped ions. *Applied Physics B* 77: 789.
77 Monroe, C., Meekhof, D.M., King, B.E. et al. (1995). Resolved-sideband Raman cooling of a bound atom to the 3D zero-point energy. *Physical Review Letters* 75: 4011.
78 Morigi, G., Eschner, J., and Keitel, C.H. (2000). Ground state laser cooling using electromagnetically induced transparency. *Physical Review Letters* 85 (21): 4458.
79 Schmidt-Kaler, F., Eschner, J., Morigi, G. et al. (2001). Laser cooling with electromagnetically induced transparency: application to trapped samples of ions or neutral atoms. *Applied Physics B* 73 (8): 807–814.
80 Home, J.P. (2013). Quantum science and metrology with mixed-species ion chains. *Advances in Atomic and Molecular Physics* 62: 231.
81 Drewsen, M., Mortensen, A., Martinussen, R. et al. (2004). Nondestructive identification of cold and extremely localized single molecular ions. *Physical Review Letters* 93: 243201.
82 Schlemmer, S., Kuhn, T., Lescop, E., and Gerlich, D. (1999). Laser excited N_2^+ in a 22-pole ion trap: experimental studies of rotational relaxation processes. *International Journal of Mass Spectrometry* 185/186/187: 589.
83 Biesheuvel, J., Karr, J.P., Hilico, L. et al. (2016). Probing QED and fundamental constants through laser spectroscopy of vibrational transitions in HD^+. *Nature Communications* 7: 10385.
84 Alighanbari, S., Hansen, M.G., Korobov, V.I., and Schiller, S. (2018). Rotational spectroscopy of cold and trapped molecular ions in the Lamb–Dicke regime. *Nature Physics* 14: 555.
85 Hansen, A.K., Versolato, O.O., Klosowski, L. et al. (2014). Efficient rotational cooling of coulomb-crystallized molecular ions by a helium buffer gas. *Nature* 508: 76–79.
86 Schmidt, P.O., Rosenband, T., Langer, C. et al. (2005). Spectroscopy using quantum logic. *Science* 309: 749.
87 Wu, Y.D., Ben, J.W., Li, B. et al. (2007). Study of (2, 0) band of $A^2\Pi_u - X^2\Sigma_g^+$ system of N_2^+ by optical heterodyne detected velocity modulation spectroscopy. *Chinese Journal of Chemical Physics* 20: 285.
88 Wolf, F. (2019). Motional quantum state engineering for quantum logic spectroscopy of molecular ions. PhD thesis. Gottfried Wilhelm Leibniz Universität Hannover.
89 Wan, Y., Gebert, F., Wolf, F., and Schmidt, P.O. (2015). Efficient sympathetic motional-ground-state cooling of a molecular ion. *Physical Review A* 91 (4): 043425.
90 Leibfried, D., Knill, E., Seidelin, S. et al. (2005). Creation of a six-atom 'Schrödinger cat' state. *Nature* 438 (7068): 639–642.
91 Mur-Petit, J., García-Ripoll, J.J., Pérez-Ríos, J. et al. (2012). Temperature-independent quantum logic for molecular spectroscopy. *Physical Review A* 85: 022308.
92 Loh, H., Ding, S., Hablutzel, R. et al. (2014). Zeeman-splitting-assisted quantum-logic spectroscopy of trapped ions. *Physical Review A* 90 (6): 061402.

93 Wolf, F., Heip, J.C., Zawierucha, M.J. et al. (2020). Prospect for precision quantum logic spectroscopy of vibrational overtone transitions in molecular oxygen ions. *arXiv preprint arXiv:2002.05584*.

94 Hudson, E.R. and Campbell, W.C. (2018). Dipolar quantum logic for freely rotating trapped molecular ions. *Physical Review A* 98: 040302.

95 Campbell, W.C. and Hudson, E.R. (2020). Dipole-phonon quantum logic with trapped polar molecular ions. *Physical Review Letters* 125: 120501.

96 Mills, M., Wu, H., Reed, E.C. et al. (2020). Dipole-phonon quantum logic with alkaline-earth monoxide and monosulfide cations. *Physical Chemistry Chemical Physics* 22: 24964.

97 Hudson, E.R. and Campbell, W.C. (2021). Laserless quantum gates for electric dipoles in thermal motion. *Physical Review A* 104: 042605.

98 Wan, Y., Gebert, F., Wübbena, J.B. et al. (2014). Precision spectroscopy by photon-recoil signal amplification. *Nature Communications* 5: 3096.

99 Clausen, E.H., Jarlaud, V., Fisher, K. et al. (2022). Unresolved sideband photon recoil spectroscopy of molecular ions. *Physical Review A* 105: 063709.

100 Schindler, P. (2019). Ultrafast infrared spectroscopy with single molecular ions. *New Journal of Physics* 21: 083025.

101 Nguyen, J.H.V., Viteri, C.R., Hohenstein, E.G. et al. (2011). Prospects for laser cooling of molecular ions. *New Journal of Physics* 13: 063023.

102 Stollenwerk, P.R., Odom, B.C., Kokkin, D.L., and Steimle, T. (2017). Electronic spectroscopy of a cold SiO^+ sample: Implications for optical pumping. *Journal of Molecular Spectroscopy* 332: 26–32.

103 Li, R., Yuan, X., Liang, G. et al. (2019). Laser cooling of the SiO^+ molecular ion: a theoretical contribution. *Chemical Physics* 525: 110412.

104 Stollenwerk, P.R., Kokish, M.G., Oliveira-Filho, D. et al. (2018). Optical pumping of TeH^+: implications for the search for varying m_p/m_e. *Atoms* 6 (3): 53.

105 Kang, S.Y., Kuang, F.G., Jiang, G. et al. (2017). Ab initio study of laser cooling of AlF^+ and $AlCl^+$ molecular ions. *Journal of Physics B: Atomic, Molecular and Optical* 50 (10): 105103.

106 Yzombard, P., Hamamda, M., Gerber, S. et al. (2015). Laser cooling of molecular anions. *Physical Review Letters* 114 (21): 213001.

107 Hanneke, D., Carollo, R.A., and Lane, D.A. (2016). High sensitivity to variation in the proton-to-electron mass ratio in O_2^+. *Physical Review A* 94: 050101.

108 Patterson, D. (2018). Method for preparation and readout of polyatomic molecules in single quantum states. *Physical Review A* 97: 033403.

14

Optical Atomic Clocks

Ekkehard Peik

Department Time and Frequency, Physikalisch-Technische Bundesanstalt, Bundesallee 100, 38116 Braunschweig, Germany

14.1 Introduction

While the present development of quantum technology is intensely focused on what is called a "second generation" of devices that make use of quantum features like superposition and entanglement, the atomic clock can be seen as a prominent example of quantum technology of the "first generation." It makes use of the resonant frequencies of electromagnetic radiation that is absorbed by atoms in the transition between electronic energy levels. Provided that a transition is used that is experimentally well accessible and not easily perturbed by external interactions, this can be applied to define a universal reference frequency and to build a frequency standard of very high accuracy. The concept was devised by I. I. Rabi and his group at Columbia University in the early 1940s based on the successes of their experiments with radiofrequency spectroscopy of atomic beams. He proposed to use the ground-state hyperfine transition at 9.193 GHz in ^{133}Cs atoms [1]. The first atomic clock based on this principle became operative in 1955 in the British National Physical Laboratory [2], and since then, this technology has been progressively introduced in all fields of precision metrology of time and frequency. Atomic clocks have enabled the realization of global navigation satellite systems like GPS, a technology that is now applied ubiquitously in mass-market devices.

The development of atomic clocks has pushed the frontier of accuracy of time and frequency measurements since the 1950s by about 10 orders of magnitude into the 10^{-18} range. For the present definition of the SI second with Cs clocks, the lowest uncertainty is about 2×10^{-16}, making time intervals and frequencies the most precisely measurable quantities in physics. The relative uncertainties in state-of-the-art mechanical and electrical measurements are in the range from 10^{-8} to 10^{-10}. Just to give an impression of the scale factor 1×10^{-18}, in terms of time, it corresponds to 0.5 seconds relative to the age of the universe of 14 billion years. Given the fundamental role that time plays as one of the coordinates of space-time, it is not surprising that experiments with atomic clocks are used in research on the foundations of the

Photonic Quantum Technologies: Science and Applications, First Edition. Edited by Mohamed Benyoucef.

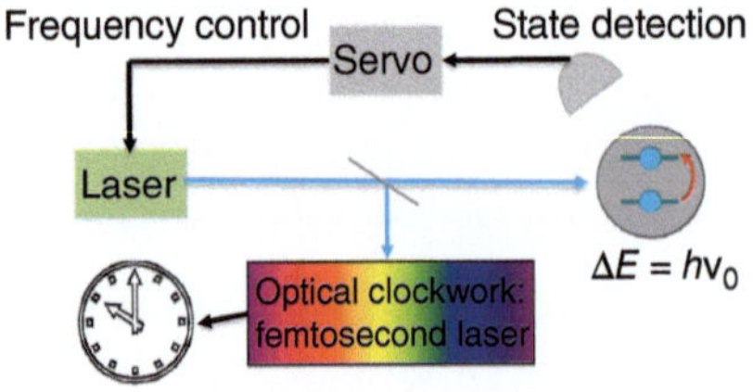

Figure 14.1 Schematic of an optical atomic clock: a laser oscillator is locked to an atomic transition frequency ν_0. An optical clockwork is used to provide a standard output frequency or a 1-pulse-per-second time signal.

physical theories, both in the quantum domain and in special and general relativity. The motivation of these studies is to find indications of "new physics" as deviations from the predictions of the established theoretical framework of general relativity and the standard model, and that such experimental results may serve as a guide or a test case in the development of a unified theoretical description of physics. The concept is to use the high precision of frequency comparisons to detect small effects in table-top atomic physics experiments as an alternative to high-energy experiments in particle physics [3].

Atomic clocks derive their stability from an ideally long-lived and unperturbed coherent superposition of two atomic energy levels E_1 and E_2 via the relation $\nu_0 = (E_2 - E_1)/h$ with the Planck constant h (see Figure 14.1). An oscillator is stabilized to the atomic resonance frequency and provides a macroscopic signal for distribution and comparison to other oscillators. In the case of microwave clocks, the oscillator is commonly derived from an oscillating quartz or another piezoelectric crystal, whereas for an optical clock, it is a prestabilized single-mode laser. The research on atomic clocks and frequency standards has been strongly influenced over the past 30 years by the application of methods of laser cooling and trapping of atoms. Being able to observe atoms at rest, for extended periods of time and well isolated from external perturbations, has led to advances in the accuracy of atomic clocks by several orders of magnitude. This chapter will focus on the ideas, concepts, and recent achievements in the field of optical atomic clocks, necessarily leaving some technical details and rigorous derivations aside. For a more in-depth coverage, I recommend the comprehensive review article Ref. [4] on optical atomic clocks.

14.2 Optical Atomic Clocks

Before we start to look in more detail at the development of *optical* atomic clocks, let us start by briefly summarizing the present status of microwave cesium clocks with laser-cooled atoms [5]. A modern primary cesium clock is a so-called atomic fountain. Cs atoms are collected from a dilute vapor or a slowed atomic beam, are laser-cooled to a temperature of a few μK in a region created by several counter-propagating laser beams (optical molasses) and then accelerated by an optical force and thrown upward against gravity to a launch height of about 0.5 m above the cooling zone, providing a free time of flight of about one second after the interaction with the laser light. During this time, the ground-state hyperfine splitting is probed while the atoms pass twice through a microwave cavity fed with a 9.19 GHz field, once during ascent and once during descent. This realizes Ramsey's method of separated oscillatory fields [6]. This method is essential for clocks based on atomic beams or atoms in free flight because it allows to probe the atoms phase-coherently even though they move by much more than one wavelength (of 3.3 cm in this case) during the interrogation cycle. The spectroscopic signal obtained in this way resembles a two-slit interference pattern in frequency space and therefore is called Ramsey fringes. The central fringe shows a linewidth of about 1 Hz determined by the free flight time of the atoms between the two passages through the cavity. The ratio of 1 Hz linewidth over 9.19 GHz transition frequency is nearly six orders of magnitude higher than the achievable systematic uncertainty in the low 10^{-16} range, showing that a good understanding of the lineshape and long averaging times of several hours is needed to achieve this low level of uncertainty in a frequency measurement. This observation shows the importance of the operation frequency for the stability of a clock signal, and it explains why progress in the development of clocks proceeds in parallel with increases of this frequency. This has already been the case in the past, when large steps were made from $\approx$1 Hz in mechanical clocks, to $\approx 10^4$ Hz in quartz clocks to $\approx 10^{10}$ Hz in microwave atomic clocks. The next step will lead to $\approx 10^{15}$ Hz in optical clocks.

The optical clock provides key advantages against the microwave clock in terms of stability and accuracy. The stability of an atomic clock is generally described by an Allan deviation $\sigma_y(\tau)$, a two-sample deviation calculated from a sequence of frequency measurements during intervals of duration τ. If technical noise of the oscillator or in the atom detection is minimized, the limit to the atomic signal-to-noise ratio and to the stability of the clock is set by quantum projection noise resulting from the state measurement at the end of the interrogation cycle [7]. Especially in the case of a clock with a *single* atom, the importance and character of quantum projection noise become obvious: while the atom can in principle be prepared in any superposition state $|\Psi\rangle = \alpha|0\rangle + \beta|1\rangle$, $(\alpha^2 + \beta^2 = 1)$, the outcome of a single-state measurement will only be one of the two states $|0\rangle$ or $|1\rangle$. For a clock with N atoms, the Allan deviation $\sigma_y(\tau)$ is given by:

$$\sigma_y(\tau) = C\frac{\Delta\nu}{\nu_0}\sqrt{\frac{t_c}{N\tau}} \tag{14.1}$$

where $\Delta\nu$ is the linewidth, ν_0 the frequency, t_c the time required for one interrogation cycle, and C a numerical constant of order unity [4]. It is implicit here via the factor t_c/τ that all clocks with laser-cooled atoms are operated in a cyclic fashion, where the atom is first prepared, then interrogated, and finally, its state is read out. In order to obtain maximum sensitivity of the atomic signal to frequency fluctuations of the oscillator, the cycle time should be used as much as possible for the interrogation, with the time used for laser cooling and state readout as short as possible. In order not to lose signal contrast, the interrogation time must not exceed the coherence time of the laser, i.e. the inverse linewidth of the prestabilized laser. A second limiting criterion is given by heating of the motion of the atom from coupling to the thermal environment during the interrogation when laser cooling is not active.

In practice, most optical clocks are operating with an interrogation time in the range from 0.2 to 2 seconds.

The accuracy of a primary atomic clock is ultimately determined by the uncertainty in the realization of the unperturbed atomic transition frequency to which it is referenced. The ideal clock would reproduce the atomic resonance as it would be in the absence of any interactions with the environment. An uncertainty evaluation has to take all the external perturbations and possible measurements errors into account that would lead to a shift of the clock output frequency. In many cases, the shift can be analyzed and corrected for, and only the uncertainty of the correction enters the final uncertainty budget of the clock. The atom is regarded as unperturbed when it is at rest in the reference frame of the clock and in the absence of external fields. These conditions can be well approximated by cooling the atoms as well as their environment close to $T = 0$. For practical purposes, however, in many cases, only the atoms are laser-cooled, while the environment remains at room temperature. The Zeeman and Stark shifts that atomic energy levels undergo in the interaction with external magnetic and electric fields depend on atomic parameters like the magnetic moment and polarizability in the specific electronic configuration. A given magnetic or electric field strength therefore induces a characteristic level energy shift, and the influence via the relative shift of a transition frequency is smaller when the frequency is higher. Consequently, optical clocks are generally less sensitive to field-induced frequency shifts than microwave clocks.

This reasoning that the perturbation is relatively less important at higher frequency ν_0 does not hold for all types of systematic frequency shifts. The relativistic Doppler shift resulting from motion of the atom is given to second order in v/c by:

$$\frac{\delta\nu}{\nu_0} = \frac{\langle v_\parallel \rangle}{c} - \frac{\langle v^2 \rangle}{2c^2} + \frac{\langle v_\parallel \rangle^2}{c^2} \tag{14.2}$$

where $\langle v_\parallel \rangle$ denotes the velocity of the atom parallel to the wave vector of the light field. The shift is proportional to ν_0, so going to higher frequency does not mitigate the problem. The first-order term can be made to vanish by interrogating trapped atoms, confined in space around a fixed average position. If the confinement is provided by a harmonic trap with an oscillation frequency that is higher than the linewidth of the clock signal, the frequency modulation from the linear Doppler effect results in sidebands that are separated from a "Doppler-free" carrier. If, in addition, the amplitude of motion is smaller than the transition wavelength $\lambda = c/\nu_0$, the sidebands are weaker than the carrier, so that most of the transition strength remains measurable in a single spectral line. This criterion is known as the Lamb–Dicke regime [8]. The second-order Doppler shift can be seen as a form of time dilation between the moving atom and the stationary oscillator of the clock. Since $\langle v^2 \rangle / c^2 = k_B T/(mc^2)$, the shift can be reduced by cooling the atoms: At $T = 1$ mK, as readily reached with laser cooling and for an atomic mass $m = 100$ amu, the second-order Doppler shift becomes $\delta\nu/\nu_0 = -5 \times 10^{-19}$.

The first detailed proposal of an optical clock was made by Hans Dehmelt in the 1970s, bringing together the techniques of high-resolution laser spectroscopy with laser cooling and trapping of ions in radiofrequency ion traps (Paul trap).

In a publication from 1981, Dehmelt formulated the target for this research: "Thus the current promise of an atomic line spectral resolution of 1 part in 10^{18} or 10^8 times better than achieved to date may be realized in the not too far future." [9]. Envisioning an improvement by eight orders of magnitude through the use of a single trapped and laser-cooled ion, the idea was regarded as visionary but was also considered as highly optimistic by many experts at the time. Nevertheless, it rapidly became established as the goal and challenge for scientists working on optical frequency standards. The level of the challenge can be seen by looking at simple proportions: Obtaining such an uncertainty necessitates the counting of optical oscillations at 1×10^{15} Hz over 1000 seconds without losing more than a single cycle. In terms of a systematic frequency shift, it corresponds to the relativistic time dilation $v^2/2c^2$ observed for a clock that moves at slow pedestrian speed of $v = 0.4$ m/s.

The method of ion trapping in radiofrequency (RF) Paul traps [10] (see Figure 14.2) provides the strong advantage that the Coulomb force that is used to localize the ion acts on the integral charge but exerts only a minimal influence on the internal level structure. This is in contrast to the trapping of neutral atoms where a force (other than gravity) can only be applied by displacing electronic energy levels in spatially inhomogeneous electric or magnetic fields. Coulomb repulsion between like charges on the other hand also introduces a disadvantage of trapped ion experiments: They work at very low density or particle number, in extremum just with a single ion, at the deficit of signal strength in comparison to an experiment with many atoms. This summarizes the dominant characteristics of the two main directions of research in terms of the preparation of the atomic systems for optical clocks: the ion trap and the optical lattice.

Ion traps used in modern optical clocks are geometrical variants of the classic three-electrode Paul trap [10] that have in common that they make use of an electric quadrupole potential that results in a time-averaged harmonic trap for the ion. For experiments with single ions, the so-called endcap trap [11] is a convenient choice because it provides open access for laser beams from different directions. If structured trapping potentials for several ions are required, more complex electrode patterns can be realized in 2D or 3D in the fashion of printed circuit boards [12, 13].

The optical lattice (see Figure 14.2 for schematics of 1D and 2D configurations) is an interference pattern of two (or more) intersecting laser beams that form a spatially periodic structure of intensity maxima (and minima) that produce an array of traps for atoms that are attracted by regions of higher (or lower) electric field strengths. Depending on the wavelength of the laser, i.e. on its detuning from the atomic resonances, each atomic level will be shifted with a specific sensitivity. It is therefore possible to tune the laser to a "magic" wavelength where the displacement of the lower and the upper level of the reference transitions is identical, so that the atom can be trapped *without* shifting the frequency of a selected transition [14–16], like for the ion in an ion trap. The optical lattice clock could ideally be regarded as a massive parallel version of a single-ion clock, with an important resulting gain in signal-to-noise ratio. The individual traps could be populated by a single or multiple atoms, leading to a trade-off between atom number and potential collisional frequency shifts.

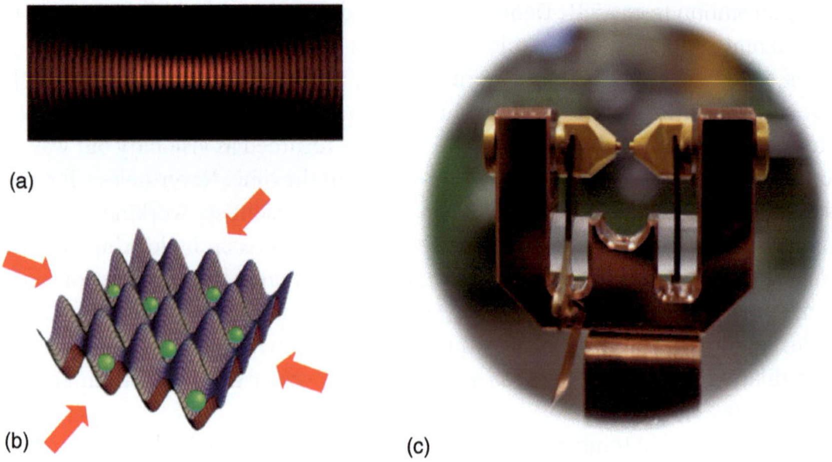

Figure 14.2 Types of traps used in optical atomic clocks: (a) the simplest atom trap for an optical lattice clock is created by a standing wave laser beam shown here schematically (not to scale). In the beam direction, the intensity maxima in which the atoms are trapped are 400 nm wide, the lateral extension is typically 150 μm. (b) Schematic of the trap potential in a two-dimensional optical lattice, created by interference of four laser beams. (c) Radiofrequency ion trap (endcap trap) used at PTB for experiments with Yb^+ ions: Ions are held in the center between the two conical gold-coated electrodes where an oscillating quadrupole potential is created. For loading of the trap, ions are created from an atomic beam (source not shown) by photo ionization or electron impact.

Most experiments so far have been performed with a one-dimensional optical lattice, resulting in a stack of traps with oblate atomic density distributions in each trap. In special experiments, a 3D configuration with single-atom occupancy has also been investigated [17, 18].

Dehmelt also made a far-sighted proposal for the type of reference transition to be studied. In general, one will employ an optical resonance line for laser cooling and fluorescence detection, and a "forbidden" transition as the reference of the clock, i.e. one that does not fulfill the angular momentum selection rules for electric dipole radiation. This criterion follows from the requirement of obtaining a narrow (ideally sub-Hz) linewidth that is not limited by the lifetime of the excited state for radiative decay. It could be fulfilled by transitions within a fine structure multiplet (same parity, $\Delta J = \pm 1$, magnetic dipole transition) or by excited states that differ by $\Delta |J| > 1$ from all lower levels (higher order multipole radiation). Dehmelt pointed out that using a transition between levels with vanishing electronic angular momentum presents advantages in avoiding or reducing frequency shifts induced by electric and magnetic fields because of the intrinsic symmetry of such levels. Such states are obtained when two valence electrons couple with oppositely oriented spins to a $^1S_0$ state or with aligned spins to $^3P_0$. In the absence of any J-mixing, symmetry-breaking interaction, a radiative single-photon transition between two such levels would be rigorously forbidden by the conservation of angular momentum. In the presence of a nuclear spin, or a symmetry-breaking external magnetic field, a small admixture of $^3P_1$ and $^3P_0$ is created and the transition becomes weekly dipole allowed. This type of configuration appears, for example, in the level schemes of the neutral alkaline earth elements and in the singly charged ions of group 13. The first optical clock that reported an uncertainty evaluation resulting in a value below 10^{-17} in 2010 has been built at NIST using this type of transition in Al^+ [19]. The predominantly scalar interaction of the $^1S_0$ and $^3P_0$ levels with far detuned light fields has also made this type of transition the best choice for the optical lattice clock [14], where Sr, Yb, and Hg are presently the elements that are investigated most.

Figure 14.3 shows the main elements of an optical clock (see also Figure 14.1): Besides the atomic reference, the laser oscillator and a femtosecond-laser frequency comb are required. The reference laser obtains its linewidth and short-term stability from locking to a stable passive Fabry–Perot cavity. Two highly reflective mirrors are optically contacted to a rigid spacer with vanishing or small coefficient of thermal expansion, mounted inside a vacuum tank on a vibration-isolation platform. Significant advances in the frequency stability of such laser systems have been obtained, reaching into the regime of $\sigma_y(1\text{s}) \approx 10^{-17}$. The frequency stability is fundamentally limited by the thermomechanical noise (Brownian motion) of the mirror spacing. In recent years, it has been possible to understand and model this noise quantitatively and to reduce it significantly by using materials of high mechanical stiffness and working at cryogenic temperature. In general, the effect of thermal noise can always be reduced by making the mirror spacing and the beam

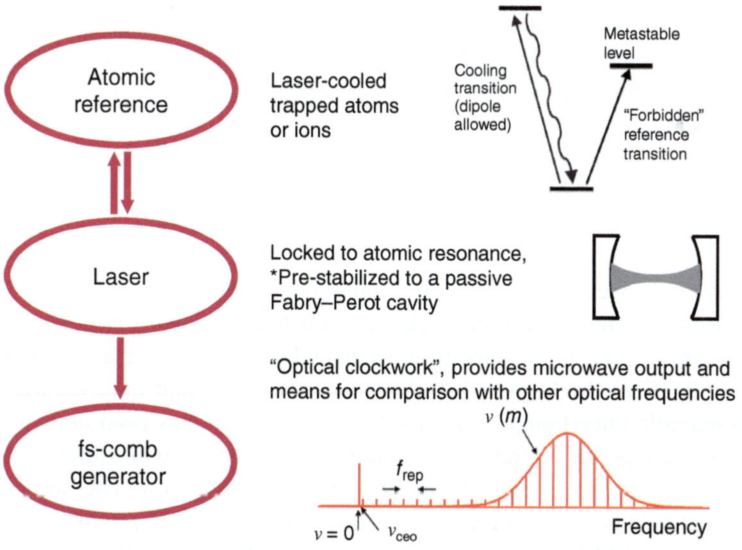

Figure 14.3 Main elements of an optical clock (cf. Figure 14.1). (i) Atomic reference with a *V*-level scheme for laser cooling and reference transition. (ii) Laser oscillator used for interrogation of the atom and as a frequency flywheel. (iii) Femtosecond laser frequency comb, acting like a ruler in frequency space and serves as an optical clockwork for division of optical frequencies into microwaves. The frequency of each optical mode of the femtosecond laser $\nu(m)$ can be related to two measurable microwave frequencies f_{rep} and ν_{ceo}: $\nu(m) = mf_{rep} + \nu_{ceo}$.

diameter on the resonator mirrors as large as possible. Good results have been achieved with resonators operated at 124 K whose spacer and mirror substrates are made of silicon monocrystals [20]. Further improvements are obtained by going to 4 K and by using low-noise monocrystalline mirror coatings [21].

The conversion of the optical reference frequency to the microwave domain is performed with a femtosecond laser frequency comb generator [22, 23]. This is an essential element of an optical clock for all applications that require an oscillator with electronically countable cycles, for example for establishing a timescale. The central element of the frequency comb is a mode-locked fs-laser that emits a very regular temporal sequence of pulses that each have a duration of a few optical cycles only. In the frequency domain, this produces a comb of equally spaced optical frequencies $f_m = mf_{\text{rep}} + f_{\text{ceo}}$. The pulse repetition rate f_{rep} and f_{ceo} can be measured and controlled in the microwave domain, and the mode number m is a large integer of order 10^5, which can be determined and fixed. In the first experiments, Ti:Sa lasers have been used with a spectrum around 800 nm that was further broadened over more than one octave by supercontinuum generation through self-phase modulation in a nonlinear photonic crystal fiber. For reasons of robustness and reliability in continuous operation, the erbium-doped fiber laser frequency comb with wavelengths around 1500 nm is today the most commonly used system. Harmonic generation and frequency mixing can be applied to reach different wavelength regions from the visible to the near infrared. The femtosecond laser frequency comb generator may work as an optical synthesizer or as an optical clockwork, for example by adjusting f_{ceo} to zero and by stabilizing one comb line f_m to the optical frequency f_0 so that f_{rep} is an exact subharmonic to order m of f_0. The precision of this transfer scheme has been investigated and was found to be so high that it will not limit the performance of optical clocks at a 10^{-19} systematic uncertainty.

Over the past two decades, and more than 40 years after the initial proposals, the field of optical clocks has seen huge progress in different systems, with laser-cooled atoms and ions of different elements and with different types of transitions [4]. Until the end of 2021, groups at four institutes in the USA, Japan, and Germany have reported systematic uncertainties in the 10^{-18} range (see Table 14.1).

In comparison to primary cesium clocks – cesium fountains with laser-cooled atoms – this represents an improvement in accuracy by two orders of magnitude. Table 14.1 summarizes the key specifications of experiments that have reported a relative systematic uncertainty below 1×10^{-17}. This status has been obtained with two different types of reference transitions so far: The $J = 0 \rightarrow 0$ transition mentioned above, studied in neutral Sr, Yb, and in the Al^+ ion. The second transition in this field is even more special: the electric octupole (E3) transition of Yb^+. The spectrum of Yb^+ is in some properties similar to those of the elements with a single valence electron. It possesses a $S_{1/2}$ ground state and strong $S-P$ resonance lines. But the lowest excited state arises from an excitation of the filled 4f shell, leading to a $F_{7/2}$ state that because of the large difference in electronic angular momentum $\Delta J = \Delta L = 3$ can decay to the ground only under the emission of octupole radiation, with a lifetime of about 1.58 years [35]. Since the electrons of the 4f shell are on average closer to the nucleus than the outer 5s or 6s electrons, the excited state is not easily polarizable by external fields, providing low

Table 14.1 List of optical atomic clock experiments with a reported systematic uncertainty below 1×10^{-17}.

Atom	Transition	Lab.	Syst. unc.	σ_y(1s)	Δ(2 systs.)	References
			(10^{-18})	(10^{-15})	(10^{-18})	
Cs	GS HFS		≈200	≈20	≈200	
Al^+	0–0	NIST	8.6	2.8	18 ± 7	[19]
Yb	0–0	NIST		0.32		[24]
Sr	0–0	JILA	6.4	0.34	28 ± 54	[25]
Sr	0–0	RIKEN	7.2	0.18	2 ± 10	[26]
Sr	0–0	JILA	2.1	0.22		[27]
Yb^+	E3	PTB	3.2	5		[28]
Yb	0–0	NIST		0.06		[29]
Yb	0–0	NIST	1.4	0.15	0.7 ± 1.0	[30]
Yb^+	E3	PTB	2.7	1	2.8 ± 4.2	[31]
Al^+	0–0	NIST	0.94	1.2		[32]
Sr	0–0	JILA	2.1	0.048		[33]
Sr	0–0	RIKEN	3.5		0.3 ± 4.7	[34]

sensitivity to systematic frequency shifts. At the same time, the motion of the 4f electrons is highly relativistic in the electric field of the nucleus, making this Yb^+ transition frequency sensitive in fundamental tests that search for variations in the value of the fine structure constant [36] or in violations of local Lorentz invariance [31].

The second column of Table 14.1 lists the instability σ_y obtained for $\tau = 1$ second of averaging. From this value the instability at longer averaging times can be estimated based on the scaling $\sigma_y(\tau) \propto 1/\sqrt{\tau}$. One can distinguish the single-ion clocks where this number is about 10^{-15}, limited by the quantum projection noise [7]: A single-state measurement on a single atom can provide a signal-to-noise ratio of only ≈1, and if the measurement is performed on a 1-Hz-wide line at $\approx 10^{15}$ Hz, this leads to an instability in the range of 10^{-15} in 1 second. For the lattice clock with the higher atom number N, the limit can be lower by a factor $1/\sqrt{N}$. However, in most of the practical realizations of lattice clocks so far, the short-term instability has been limited by the laser oscillator.

The column Δ(2 systems) indicates the best agreement that has been observed between two similar systems built by the same group. One may see that this crucial test of performance has not yet been performed for all systems. This, and the measurement of optical frequency ratios between the different transition frequencies, will likely be a key aspect of the work for the next few years, in order to establish a fully consistent evaluation of this new generation of clocks. The measurement of frequency ratios provides a highly valuable diagnostic because they can be determined as dimensionless numbers with a relative uncertainty that is not limited by the available primary frequency standards [37].

14.3 Optical Clocks with a Single Trapped $^{171}Yb^+$ Ion

As a specific example of an optical clock and for the kind of considerations that enter into an uncertainty evaluation in the 10^{-18} range, we discuss the Yb^+ clocks that are developed at PTB in Germany, NPL in the United Kingdom and in other laboratories. The work on this ion has shown several important, partly unexpected advantages and has made it possible to obtain one of the lowest systematic uncertainties at present. The isotope $^{171}Yb^+$ with nuclear spin $I = 1/2$ is used. Here, the electron and nuclear spins couple to a hyperfine level of the ground state with vanishing total angular momentum $F = 0$ (see Figure 14.4 for the level scheme). This level is nondegenerate and can therefore be prepared easily by hyperfine optical pumping, and its energy is insensitive to the magnetic field. The high atomic mass leads to small Doppler shift at a given temperature. In experiments with trapped Yb^+ ions, very long storage times exceeding several months have been observed [38], facilitating the long-term continuous operation of the clock. While in other ions, chemical reactions with background gas ultimately limit the storage time, and this loss process is prevented for Yb^+ by the photodissociation of the product ion YbH^+ with the cooling laser light. Two optical reference transitions in $^{171}Yb^+$ are being studied: the $^2S_{1/2} - {}^2D_{3/2}$ electric quadrupole (E2) transition [39] and the $^2S_{1/2} - {}^2F_{7/2}$ electric octupole (E3) transition [40]. The octupole transition between the $^2S_{1/2}$ ground state and the lowest excited $^2F_{7/2}$ state is unusual because of its extremely small natural linewidth in the nanohertz range [35]. While allowing for very high resolution, at the limit imposed by the linewidth of the interrogation laser, an associated disadvantage is a significant light shift of the transition frequency [41]. This shift is proportional to

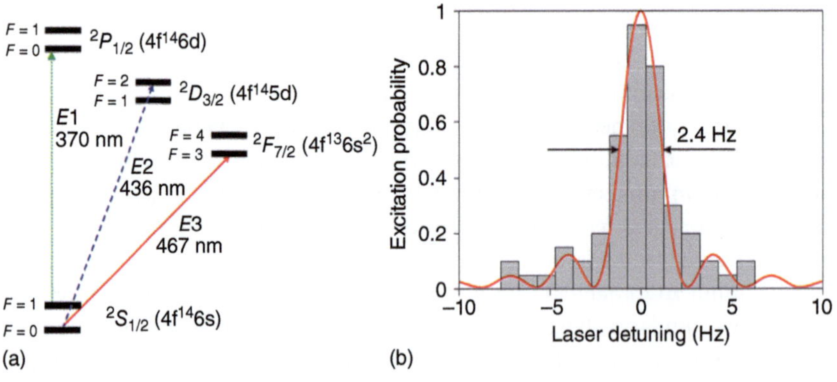

Figure 14.4 (a) Scheme of the lowest electronic energy levels of $^{171}Yb^+$, showing the cooling transition (E1) and the two reference transitions E2 and E3. (b) Excitation of a single $^{171}Yb^+$ on the E3 transition, obtained with pulses of 335 ms duration and 50 μW laser power, showing a Fourier-limited linewidth of 2.4 Hz.

the laser intensity so that a π-pulse with Fourier-limited spectral width Δf causes a shift proportional to $(\Delta f)^2$. The shift contains both scalar and tensorial contributions and scales like 0.65(3) $\text{Hz}^{-1}(\Delta f)^2$ if the polarization and magnetic field orientation are chosen to maximize the excitation probability [42]. To treat this problem, we use modified versions of Ramsey's method of separated oscillatory fields [43–45], optical excitation schemes that consist of a sequence of excitation pulses and dark periods during which the atomic coherence and the laser phase evolve freely and the atom is not subject to light shift. Pulse sequences can be applied to produce a resonance signal essentially from the dark period, which is immune to the light shift and other shifts of the transition frequency that are correlated with the probe pulses [45]. With the excellent control of laser intensity and pulse areas that is possible in a trapped ion experiment, such schemes are ideally suited to an optical clock based on the E3 transition of $^{171}Yb^+$.

Apart from the light shift, the sensitivities of the Yb^+ octupole transition frequency to static electric field–induced shifts are relatively low as has been pointed out in theoretical estimates [46] and measured in the frequency standards [42]. Qualitatively, this can be explained by the electronic configuration ($4f^{13}6s^2$) of the $^2F_{7/2}$ level that consists of a hole in the 4f shell that is surrounded by the filled 6s shell. Comparing the sensitivity factors for field-induced shifts of the E2 and E3 transitions, the quadratic Zeeman, quadratic Stark (scalar and tensor), and electric quadrupole shifts of the latter are smaller by factors of 25, 8, 62, and 68, respectively, for identical external fields. It is therefore very efficient to use the E2 transition in the same ion as an *in situ* diagnosis for perturbations through external fields and to deduce the much smaller corrections and uncertainties of the E3 frequency from these measurements. Table 14.2 presents the systematic shifts and related uncertainties of the $^{171}Yb^+$ E3 optical clock as evaluated at the PTB in 2016 [28].

Table 14.2 Fractional frequency shifts $\delta\nu/\nu_0$ (10^{-18}) and related relative uncertainties u/ν_0 (10^{-18}) in the realization of the unperturbed $^2S_{1/2}(F=0) \rightarrow {}^2F_{7/2}(F=3)$ transition frequency ν_0 of a single trapped $^{171}Yb^+$ ion.

Effect	$\delta\nu/\nu_0$ (10^{-18})	u/ν_0 (10^{-18})
Second-order Doppler shift	−3.7	2.1
Blackbody radiation shift	−70.5	1.8
Probe light related shift	0	1.1
Second-order Zeeman shift	−40.4	0.6
Quadratic dc Stark shift	−1.2	0.6
Background gas collisions	0	0.5
Servo error	0	0.5
Quadrupole shift	0	0.3
Total	−115.8	3.2

Source: From Huntemann et al. [28]/American Physical Society.

The remaining largest shift of the transition frequency is caused by the Stark shift induced by the thermal blackbody radiation (BBR) emitted by the ion's environment. It is an advantage of trapped ion optical clocks that they generally have smaller BBR shifts than those based on neutral atoms [4]. This is primarily due to the tighter binding of the remaining electrons after ionization, which shifts the atomic transitions to shorter wavelengths, further away from the infrared thermal radiation. A complication in the determination of the BBR shift for trapped ions is that the strong radiofrequency electric fields used to confine the ions will heat the trap structure via dielectric losses in the insulators and joule heating in the conductors. Heat dissipation from the trap is provided by conduction and radiation. A combination of finite element modeling with measurements made with an infrared camera and temperature sensors at test points of a copy of the operational trap has allowed us to determine the effective temperature rise of the radiation seen by the ion as 2.1(1.1) K [47]. An optimized trap design with an exclusive use of low-loss dielectrics and improved thermal contact from the trap electrodes to the vacuum feedthrough allows one to reduce the temperature rise and uncertainty further [48].

Correction of the BBR shift requires knowledge of the temperature and of the differential polarizability between the ground and excited state. While the polarizability of neutral atoms can be measured by placing the atoms in a static electric field, ions can only be held in alternating fields, for example in the focus of an infrared laser beam. For the Yb^+ E3 transition, the problem is facilitated by the fact that all electric dipole transitions that are relevant for the differential Stark shift are at wavelengths smaller than 380 nm so that the wavelength dependence of the polarizability in the infrared range is essentially flat. A precise measurement of the differential polarizability has been performed by measuring the light shift induced by several near-infrared lasers between 0.85 and 1.5 μm [28]. Combining the results of the thermal analysis of the trap [47] and the polarizability measurements, the BBR shift for the Yb^+ E3 transition is −45 mHz and the contribution to the relative uncertainty of the frequency standard 1.8×10^{-18} [28].

In total, the 2016 evaluation of the Yb^+ E3 optical clock has resulted in a systematic uncertainty of 3.2×10^{-18}, with the largest contribution caused by the relativistic Doppler shift from residual motion of the ion [28]. Since then, an improved ion trap with gold-coated electrodes (see Figure 14.2) has resulted in lower motional heating rates and a better thermal homogeneity. With additional measures like ground state cooling of the ion, it seems realistic to obtain a systematic uncertainty for this clock below 1×10^{-18}.

14.4 Outlook on Future Developments

Since the demonstrated accuracy and stability obtained with several different optical clocks have now surpassed the performances of primary cesium clocks, discussions and preparatory work have been started toward a new definition of the International system of units (SI) second [49–51], where the 9.19 GHz cesium reference frequency would be replaced by one of the more stable and accurate reference frequency signals in the optical range. Since until now cesium clocks and the infrastructure of time-keeping and frequency standards in the microwave range serve the vast majority of practical applications very well, the question has not obtained the same momentum as that for the redefinitions of the electrical units and the kilogram that have been redefined in terms of fundamental constants in 2019. In preparation for the use of optical frequency standards, a working group of the consultative committee for time and frequency CCTF is collecting and evaluating the results of frequency measurements of suitable transitions, in order to calculate and publish a set of recommended frequencies for so-called secondary realizations of the SI second [52]. Presently, nine different transitions are recommended: the hyperfine frequency of ^{87}Rb, three species of neutral atoms in optical lattices ^{87}Sr, ^{171}Yb, and ^{199}Hg, and five transitions in trapped ions: $^{27}Al^+$, $^{88}Sr^+$, $^{171}Yb^+$ (with two reference transitions), and $^{199}Hg^+$. The variety of promising candidates and the rapidly progressing development of techniques and methods do not allow us today to identify a front-runner as the successor of cesium. It should rather be seen as a sign of the scientific vitality of the field that the rankings of the "best" optical clocks are changing frequently.

In addition, novel systems that have been proposed and are in earlier phases of experimental realization may possibly reach the leading group within a few years. I would like to briefly discuss two new types of reference systems: highly charged ions (HCI) [53] and the nucleus ^{229}Th with a low-energy nuclear transition [54, 55]. In HCI, the remaining electrons are tightly bound and therefore less polarizable and less prone to level shifts through external fields. One would expect intuitively that with increasing charge state, relevant electronic transition wavelengths are moved away from the optical range and into the X-ray region. However, transitions of the fine or even hyperfine structure are now appearing in the optical range. In addition, there is a large class of investigated optical clock candidates in HCI that are based on so-called level crossings between different configurations like 5s and 4f that approach each other (or cross) as a function of charge state, leading to forbidden transitions with attractive properties in the visible spectral range. Some pioneering experiments in this field have been performed with a fine structure transition in $^{40}Ar^{13+}$ at a wavelength of 441 nm, making use of sympathetic laser cooling of the Ar ions and quantum logic readout of their laser excitation [56].

Based on the experimental finding that the ^{229}Th nucleus possesses a low-energy isomer, initially located at 3.5 eV transition energy, nuclear laser spectroscopy and a nuclear optical clock have been proposed [54]. A distinctive feature of the *nuclear* clock in comparison to present *atomic* clocks is that the electronic environment (charge state, electronic state, etc.) under which the nuclear reference transition is probed can be selected to provide almost complete immunity to field-induced systematic frequency shifts [54, 57]. Despite this strong motivation and various experimental approaches, a direct laser excitation of the nuclear transition, now expected at about 8.2 eV or 150 nm wavelength, has not been demonstrated so far (see [58, 59] for reviews of recent experimental work). Experiments are under way to develop the required laser systems in this vacuum ultraviolet (VUV) region, together with the methods to trap and cool Th ions or to prepare them in suitable wide-bandgap transparent host crystals. Besides the interest for metrology, these systems promise insights into so far inaccessible domains of atomic and nuclear physics, regarding the motion of electrons in the strong electric field close to the nucleus, and the collective dynamics of a heavy nucleus that will be measurable in unprecedented detail.

Acknowledgments

I would like to thank my colleagues from PTB for their contributions to the work that is summarized here and for many discussions: N. Huntemann, C. Sanner, B. Lipphardt, Chr. Tamm, R. Lange, M. Okhapkin, S. Weyers, A. Bauch, U. Sterr, and C. Lisdat.

References

1 Ramsey, N.F. (2005). *Metrologia* 42: S1.
2 Henderson, D. (2005). *Metrologia* 42: S4.
3 Safronova, M.S., Budker, D., DeMille, D. et al. (2018). *Reviews of Modern Physics* 90: 025008.
4 Ludlow, A.D., Boyd, M.M., Jun, Y. et al. (2015). *Reviews of Modern Physics* 87: 637.
5 Wynands, R. and Weyers, S. (2005). *Metrologia* 42: S64.
6 Ramsey, N.F. (1990). *Reviews of Modern Physics* 62: 541.
7 Itano, W.M., Bergquist, J.C., Bollinger, J.J. et al. (1995). *Physical Review A* 47: 3554.
8 Dicke, R.H. (1953). *Physical Review* 89: 472.
9 Dehmelt, H. (1981). *Journal of Physiology (Paris)* 42: C8299.
10 Paul, W. (1990). *Reviews of Modern Physics* 62: 531.
11 Schrama, C.A., Peik, E., Smith, W.W., and Walther, H. (1993). *Optics Communications* 101: 32.
12 Seidelin, S., Chiaverini, J., Reichle, R. et al. (2006). *Physical Review Letters* 96: 253003.
13 Herschbach, N., Pyka, K., and Keller, J. (2012). Mehlstäubler. *Applied Physics B* 107: 891.
14 Katori, H. (2002). Spectroscopy of Strontium atoms in the Lamb-Dicke confinement. *Proceedings of the 6th Symposium on Frequency Standards and Metrology*. Singapore: World Scientific.
15 Ye, J., Kimble, J., and Katori, H. (2008). *Science* 320: 1734.
16 Derevianko, A. and Katori, H. (2011). *Reviews of Modern Physics* 83: 331.
17 Akatsuka, T., Takamoto, M., and Katori, H. (2010). *Physical Review A* 81: 023402.
18 Campbell, S., Hutson, R.B., Marti, G.E. et al. (2017). *Science* 358: 90.
19 Chou, C.W., Hume, D.B., Koelemeij, J.C.J. et al. (2010). *Physical Review Letters* 104: 070802.
20 Matei, D.G., Legero, T., Häfner, S. et al. (2017). *Physical Review Letters* 118: 263202.

21 Robinson, J.M., Oelker, E., Milner, W.R. et al. (2019). *Optica* 6: 240.

22 Udem, Th., Holzwarth, R., and Hänsch, T.W. (2002). *Nature* 416: 233.

23 Hall, J.L. (2006). *Reviews of Modern Physics* 78: 1279.

24 Hinkley, N., Sherman, J.A., Phillips, N.B. et al. (2013). *Science* 341: 1215.

25 Bloom, B.J., Nicholson, T.L., Williams, J.R. et al. (2014). *Nature* 506: 71.

26 Ushijima, I., Takamoto, M., Das, M. et al. (2015). *Nature Photonics* 9: 185.

27 Nicholson, T.L., Campbell, S.L., Hutson, R.B. et al. (2015). *Nature Communications* 6: 6896.

28 Huntemann, N., Sanner, C., Lipphardt, B. et al. (2016). *Physical Review Letters* 116: 063001.

29 Schioppo, M., Brown, R.C., McGrew, W.F. et al. (2017). *Nature Photonics* 11: 48.

30 McGrew, W.F., Zhang, X., Fasano, R.J. et al. (2018). *Nature* 564: 87.

31 Sanner, C., Huntemann, N., Lange, R. et al. (2019). *Nature* 567: 204.

32 Brewer, S.M., Chen, J.S., Hankin, A.M. et al. (2019). *Physical Review Letters* 123: 033201.

33 Bothwell, T., Kedar, D., Oelker, E. et al. (2019). *Metrologia* 56: 065004.

34 Takamoto, M., Ushijima, I., Ohmae, N. et al. (2020). *Nature Photonics* 14: 411.

35 Lange, R., Peshkov, A.A., Huntemann, N. et al. (2021). *Physical Review Letters* 127: 213001.

36 Lange, R., Huntemann, N., Rahm, J.M. et al. (2021). *Physical Review Letters* 126: 011102.

37 Boulder Atomic Clock Optical Network (BACON) Collaboration (2021). *Nature* 591: 564.

38 Tamm, C., Weyers, S., Lipphardt, B., and Peik, E. (2009). *Physical Review A* 80: 043403.

39 Tamm, C., Engelke, D., and Bühner, V. (2000). *Physical Review A* 61: 053405.

40 Roberts, M., Taylor, P., Barwood, G.P. et al. (1997). *Physical Review Letters* 78: 1876.

41 Webster, S.A., Taylor, P., Roberts, M. et al. (2002). *Physical Review A* 65: 052501.

42 Huntemann, N., Okhapkin, M.V., Lipphardt, B. et al. (2012). *Physical Review Letters* 108: 090801.

43 Yudin, V.I., Taichenachev, A.V., Oates, C.W. et al. (2010). *Physical Review A* 82: 011804.

44 Huntemann, N., Lipphardt, B., Okhapkin, M.V. et al. (2012). *Physical Review Letters* 109: 213002.

45 Sanner, C., Huntemann, N., Lange, R. et al. (2018). *Physical Review Letters* 120: 053602.

46 Lea, S.N. (2007). *Reports on Progress in Physics* 70: 1473.

47 Dolezˇ al, M., Balling, P., Nisbet-Jones, P.B.R. et al. (2015). *Metrologia* 52: 842.

48 Nisbet-Jones, P.B.R., King, S.A., Jones, J.M. et al. (2016). *Applied Physics B* 122: 57.

49 Gill, P. (2011). *Philosophical Transactions of the Royal Society A: Mathematical, Physical and Engineering Sciences* 369: 4109.

50 Riehle F. (2015). *Comptes Rendus Physique* 16: 506.

51 Bize, S. (2019). *Comptes Rendus Physique* 20: 153.

52 Riehle, F., Gill, P., Arias, F., and Robertsson, L. (2018). *Metrologia* 55: 188.

53 Kozlov, M.G., Safronova, M.S., Lopez-Urrutia, J.R.C., and Schmidt, P.O. (2018). *Reviews of Modern Physics* 90: 045005.

54 Peik, E. and Tamm, C. (2003). *Europhysics Letters* 61: 181.

55 Peik, E. and Okhapkin, M. (2015). *Comptes Rendus Physique* 16: 516.

56 Micke, P., Leopold, T., King, S.A. et al. (2020). *Nature* 578: 60.

57 Campbell, C.J., Radnaev, A.G., Kuzmich, A. et al. (2012). *Physical Review Letters* 108: 120802.

58 Beeks, K., Sikorsky, T., Schumm, T. et al. (2021). *Nature Review Physics* 3: 238.

59 Peik, E., Schumm, T., Safronova, M.S. et al. (2021). *Quantum Science and Technology* 6: 034002.

Photonic Quantum Technologies

Photonic Quantum Technologies

Science and Applications

Volume 2

Edited by Mohamed Benyoucef

WILEY-VCH

Editor

Mohamed Benyoucef
University of Kassel
Institute of Physics, Center for Interdisciplinary Nanostructure Science and Technology (CINSaT)
Heinrich-Plett-Str. 40
34132 Kassel
Germany

Cover Image: © Jurik Peter/Shutterstock

Library of Congress Card No.: applied for

British Library Cataloguing-in-Publication Data
A catalog record for this book is available from the British Library.

Bibliographic information published by the Deutsche Nationalbibliothek
The Deutsche Nationalbibliothek lists this publication in the Deutsche Nationalbibliografie; detailed bibliographic data are available on the Internet at <http://dnb.d-nb.de>.

Print ISBN: 978-3-527-35261-6
ePDF ISBN: 978-3-527-83741-0
ePub ISBN: 978-3-527-83743-4
oBook ISBN: 978-3-527-83742-7

Typesetting Straive, Chennai, India
Printing and Binding CPI Group (UK) Ltd, Croydon, CR0 4YY

C9783527352616_140524

I dedicate this book to my parents, for inspiring me throughout my life;
to my wife, and my children for their patience and support;
to my wonderful readers, thank you!

Contents

Contents to Volume 1

Preface *xv*
Abstracts and Keywords *xvii*

1 **Introduction to Quantum Photonics** *1*
Richard J. Warburton

Part I Fundamentals of Quantum Technologies *7*

2 **The Second Quantum Revolution: From Basic Concepts to Quantum Technologies** *9*
Alain Aspect

3 **Solid-State Quantum Emitters** *31*
A. Mark Fox

4 **Single-Photon Sources for Multi-Photon Applications** *53*
Stefan Frick, Robert Keil, Vikas Remesh, and Gregor Weihs

5 **Quantum Key Distribution Protocols** *85*
Álvaro Navarrete, Víctor Zapatero, and Marcos Curty

6 **From Basic Science to Technological Development: The Case for Two Avenues** *105*
Matteo Carlesso and Mauro Paternostro

7 **Quantum Networks in Space** *129*
Lisa Wörner

Part II Atoms, Ions, and Molecules: From Experimental Techniques to Recent Progress *151*

8 **Fluorescence Spectroscopy in Planar Dielectric and Metallic Systems** *153*
Alexey I. Chizhik, Daja Ruhlandt, and Jörg Enderlein

9 **Single Trapped Neutral Atoms in Optical Lattices** *183*
Andrea Alberti and Dieter Meschede

10 **Long-Distance Entanglement of Atomic Qubits** *209*
Kai Redeker, Wenjamin Rosenfeld, and Harald Weinfurter

11 **Collective Light Emission of Ion Crystals in Correlated Dicke States** *233*
Ferdinand Schmidt-Kaler and Joachim von Zanthier

12 **Single-Molecule Magnets Spin Devices** *269*
Eufemio Moreno-Pineda and Wolfgang Wernsdorfer

13 **Molecular-Ion Quantum Technologies** *305*
Mudit Sinhal and Stefan Willitsch

14 **Optical Atomic Clocks** *333*
Ekkehard Peik

Contents to Volume 2

Preface *xv*
Abstracts and Keywords *xvii*

Part III Spin Qubits and Quantum Memories: From Spin Properties to Physical Realizations *349*

15 **Coherent Spin Dynamics of Colloidal Nanocrystals** *351*
Dmitri R. Yakovlev, Anna V. Rodina, Elena V. Shornikova, Aleksandr A. Golovatenko, and Manfred Bayer
15.1 Introduction *351*
15.2 Spin-Level Structure of Neutral and Charged Excitons *353*
15.3 Photoluminescence in Magnetic Field *355*
15.4 Time-Resolved Faraday Rotation *359*
15.5 Dynamics of Photocharging Visualized via Electron Spin Coherence *361*
15.6 Spin–Flip Raman Scattering *363*

15.7 Surface Magnetism *364*
15.8 Diluted Magnetic Semiconductor Colloidal NCs *367*
15.9 New Materials: Perovskite QDs *368*
15.10 Conclusions *369*
Acknowledgments *370*
References *370*

16 Relaxation of Electron and Hole Spin Qubits in III–V Quantum Dots *377*
Alexander Bechtold, Tobias Simmet, Friedrich Sbrezny, Nikolai A. Sinitsyn, Kai Müller, and Jonathan J. Finley
16.1 Introduction *377*
16.1.1 Solid-State Spin Qubits *377*
16.1.2 Spin-Qubits for Measurement-Based Quantum Information Processing (QIP) *378*
16.1.3 Contents of the Chapter *379*
16.2 Fundamental Properties and Devices Investigated *381*
16.2.1 Key Physical Electronic and Optical Properties of Self-Assembled Quantum Dots *381*
16.2.2 Optical Spin Storage Photodiodes *383*
16.2.3 Reset and Charge Readout *385*
16.2.4 Optical Manipulation: Geometric Phase Control and Spin Echo *389*
16.3 Relaxation Dynamics of Electron Spin Qubits *392*
16.3.1 Theoretical Background: A Historical Perspective *392*
16.3.2 Inhomogeneous Dephasing in a Fluctuating Overhauser Field *393*
16.3.3 Decoherence as a Result of Time Dependent Changes in the Nuclear Spin Bath *396*
16.3.4 Complete Depolarization Due to Dipolar Interactions in the Nuclear Spin Bath *398*
16.4 Electron Spin Relaxation Studied in Experiments *399*
16.4.1 Monitoring the Electron Spin Qubit Relaxation *400*
16.5 Hole Spin Relaxation in Single Quantum Dots *405*
16.5.1 Theory of the Anisotropic Hyperfine Coupling of Holes *405*
16.5.2 Nuclear Spin Dynamics *409*
16.5.3 Hole Spin Storage Devices *411*
16.5.4 Fast Dynamics of the Central Spin *413*
16.5.5 Slow Part of the Relaxation *415*
16.6 Summary *420*
Ackowledgments *421*
References *422*

17 Ensemble-Based Quantum Memory: Principle, Advance, and Application *433*
Bo Jing and Xiao-Hui Bao
17.1 Introduction *433*

17.2 Memory Schemes *434*
17.2.1 Optical Delay Line *434*
17.2.2 Electromagnetically Induced Transparency *435*
17.2.3 Atomic Frequency Comb *438*
17.2.4 DLCZ Scheme *440*
17.2.5 Other Schemes *442*
17.3 Performance Criteria *442*
17.3.1 Working Wavelength *443*
17.3.2 Efficiency *443*
17.3.3 Storage Time *444*
17.3.4 Fidelity *444*
17.3.5 Bandwidth *445*
17.3.6 Multimode Capacity *446*
17.3.7 Integratability *446*
17.4 Physical Realization *447*
17.4.1 Gas Atomic Ensemble *447*
17.4.2 Solid-State Atomic Ensemble *448*
17.5 Applications *450*
17.5.1 Linear-Optical Quantum Computing *450*
17.5.2 Quantum Repeater *451*
17.5.3 Quantum Key Distribution *452*
17.5.4 Detection of Single Photons *454*
17.6 Summary and Outlook *455*
References *455*

Part IV Solid-State and van der Waals Material Platforms: From Single Quantum Emitters to Hybrid Integration *463*

18 Telecom Wavelengths InP-Based Quantum Dots for Quantum Communication *465*
Mohamed Benyoucef and Anna Musiał
18.1 Introduction *465*
18.2 Basic Concepts *468*
18.2.1 Artificial and Real Atoms *468*
18.2.2 Formation of Quantum Dots *470*
18.2.3 Excitons in Quantum Dots *471*
18.3 Low-Density InP-Based Quantum Dots *475*
18.3.1 Low-Density InAs QDs on Distributed Bragg Reflectors *476*
18.3.2 Quantum Dots in Quaternary Barriers: Round-Shaped Dots *477*
18.3.2.1 Epitaxial Quantum Dot Growth *477*
18.3.2.2 Structural Properties of Quantum Dots *478*
18.3.2.3 Charged Quantum Dots *479*
18.3.3 Quantum Dots in Binary Barriers: Symmetric Dots *481*

18.3.3.1 Optical Properties *482*
18.3.3.2 Magneto-Optical Studies *483*
18.3.3.3 Photon Extraction Efficiency *486*
18.3.3.4 Quantum Dots in Photonic Crystal Cavities *491*
18.3.3.5 Radiative Lifetime *493*
18.4 Symmetric InP-Based Quantum Dots as Quantum Light Sources *494*
18.5 Challenges and Future Directions *496*
Acknowledgments *497*
References *498*

19 Quantum Optics with Solid-State Color Centers *509*
Joseph H.D. Munns, Laura Orphal-Kobin, Gregor Pieplow, and Tim Schröder
19.1 Introduction *509*
19.2 Color Centers *511*
19.2.1 Diamond *512*
19.2.1.1 Material Considerations *512*
19.2.1.2 Nitrogen-Vacancy Defect Center *513*
19.2.1.3 Group IV Defect Centers *514*
19.2.1.4 Other Color Centers in Diamond *516*
19.2.2 Silicon Carbide *516*
19.2.2.1 Material Properties *516*
19.2.2.2 Prominent Color Centers *517*
19.2.3 Other Host Materials *518*
19.2.4 Optical Micro- and Nanostructures for Controlled Light-Matter Interaction *519*
19.2.4.1 Fabrication and Nanostructures in Diamond *521*
19.2.4.2 Fabrication and Nanostructures in SiC *523*
19.3 Applications *524*
19.3.1 Color Centers as Quantum Emitters *524*
19.3.1.1 Quantum Emitters at Room Temperature *525*
19.3.1.2 Quantum Emitters at Cryogenic Temperatures *525*
19.3.1.3 State-of-the-art *528*
19.3.2 Color Centers as Quantum Memories *530*
19.3.2.1 Considerations and Figures of Merit *530*
19.3.2.2 Physical Considerations on Cycling Transitions: Spin Initialization and State Read-out of Qubits in Diamond *532*
19.3.2.3 State-of-the-Art *534*
19.3.3 Color Centers for 2-Qubit Quantum Gates *537*
19.3.3.1 Color Centers for Spin–Photon Entanglement *537*
19.3.3.2 Color Centers in Solid-State Multi-Qubit Registers *539*
19.3.3.3 Color Centers for Remote Spin–Spin Entanglement *541*
19.4 Proposals and Perspectives *543*
19.4.1 Applications *543*
19.4.2 Directions and Remaining Challenges *544*
References *545*

20 Quantum Photonics with 2D Semiconductors *563*
Mauro Brotons-Gisbert and Brian D. Gerardot
20.1 Introduction *563*
20.1.1 Hexagonal Boron Nitride *563*
20.1.2 Transition Metal Dichalcogenide Semiconductors *565*
20.1.3 Van der Waals Heterostructures *567*
20.2 Semiconductor Quantum Emitters *567*
20.3 Engineering 2D Semiconductor Quantum Emitters *567*
20.3.1 Dielectric and Electrostatic Engineering *568*
20.3.2 Strain-Tuning *569*
20.3.3 Site-Controlled Implantation *569*
20.3.4 Moiré Heterostructures *570*
20.3.5 Heterostructure Device Functionalization *570*
20.3.6 Integrated Photonics with 2D Quantum Emitters *572*
20.4 Outlook *573*
Acknowledgments *573*
References *573*

21 Nano-Opto-Electro-Mechanical Systems for Integrated Quantum Photonics *581*
Leonardo Midolo and Celeste Qvotrup
21.1 Introduction and Overview *581*
21.2 Device Principles *583*
21.2.1 Tunable Beam Splitters *584*
21.2.2 Device Speed *586*
21.2.3 Electro-mechanical Actuation *587*
21.3 NOEMS Fabrication *588*
21.4 Application of NOEMS to Quantum Photonics *590*
21.4.1 Routing and Switching Single Photons *590*
21.4.2 Controlling Light–Matter Interaction *592*
21.5 Challenges and Perspectives *593*
References *594*

22 Silicon Quantum Photonics - Platform and Applications *599*
Cagin Ekici, Elizaveta Semenova, Davide Bacco, and Yunhong Ding
22.1 Introduction *599*
22.1.1 Fabrication Process *601*
22.1.2 Passive Components *602*
22.1.3 Active Components *605*
22.2 Quantum State of Light on Silicon Photonics Platform *608*
22.2.1 Toward Deterministic and Efficient Quantum Light Sources on Silicon Platform *608*
22.2.2 Nonlinear Sources on Silicon *612*
22.2.2.1 Fundamentals of Nonlinear Photon Sources *612*
22.2.2.2 Multiplexing for Deterministic Photon Generation *617*

22.2.2.3 Parasitic Effects *618*
22.3 Applications *619*
22.3.1 Quantum Information Processing *619*
22.3.1.1 Measurement-Based Model of Quantum Computation *622*
22.3.1.2 Rudimentary Quantum Computers *622*
22.3.2 Quantum Communications *624*
22.3.2.1 Quantum Key Distribution *625*
22.3.2.2 Future Quantum Networks *630*
22.4 Outlook *630*
References *631*

Part V Emerging Quantum Technologies: Challenges and Potential Applications *651*

23 Photonic Realization of Qudit Quantum Computing *653*
Yuchen Wang, Zixuan Hu, and Sabre Kais
23.1 Introduction to Qudit Quantum Computing *653*
23.1.1 Universality and Examples of Qudit Gates *653*
23.1.2 Examples of Qudit Quantum Algorithms *655*
23.2 Qudit Implementation on Photonic Systems *658*
23.2.1 Qudits in Time and Frequency Degrees of Freedom *658*
23.2.2 Superconducting Bosonic Processor *665*
23.3 Summary and Future Outlooks *669*
Acknowledgments *670*
References *670*

24 Fiber-Based Quantum Repeaters *675*
Mohsen Razavi
24.1 Quantum Repeater Toolbox *677*
24.1.1 Entanglement Distribution *678*
24.1.2 Quantum Memories *678*
24.1.3 Bell-State Measurement *679*
24.1.4 Entanglement Distillation *679*
24.1.5 Quantum Error Correction *680*
24.2 Quantum Repeaters Based on Heralded Entanglement Distribution *680*
24.2.1 Probabilistic Quantum Repeaters *681*
24.2.2 Semi-Probabilistic Quantum Repeaters *682*
24.2.3 Encoded Quantum Repeaters *684*
24.3 Memory-Less Quantum Repeaters *685*
24.3.1 All-photonic Quantum Repeaters *685*
24.3.2 One-way Quantum Repeaters *686*
24.4 Summary and Discussion *687*
References *688*

25 Long-Distance Satellite-Based Quantum Communication *693*
Yuan Cao and Juan Yin
25.1 Introduction *693*
25.2 Ground-Based Feasibility Studies *694*
25.3 Satellite-Based Quantum Communication Experiments with Micius *699*
25.3.1 Satellite-to-Ground Quantum Key Distribution *699*
25.3.2 Satellite-Based Quantum Entanglement Distribution *702*
25.3.3 Ground-to-Satellite Quantum Teleportation *704*
25.4 Other Quantum Satellite Projects *706*
25.5 Outlook *707*
References *709*

26 Quantum Communication Networks for 6G *715*
Riccardo Bassoli, Frank H.P. Fitzek, and Holger Boche
26.1 Introduction *715*
26.2 What Is 6G? *717*
26.3 6G Intrinsic Limitations: Why Do We Need Other Technologies? *721*
26.4 The Vision of the Quantum Internet *725*
26.5 The Architectural Convergence of Quantum Technologies and 6G *731*
References *735*

Index *737*

Preface

This book has been inspired by the growing interest in emerging quantum technologies. The ongoing miniaturization of photonic structures due to the availability of sophisticated nanofabrication has provided huge opportunities for physical research of novel phenomena in photonic quantum systems and quantum technological applications. These include quantum information processing such as quantum computing, quantum communication, quantum metrology, and quantum sensing.

The aim here is to bring together scientists with different expertise from across different disciplines. The manuscript provides selected topics on photonic quantum systems from research to applications. The chapters will represent a flavor of the cutting-edge research that is going on in this area. The studied material systems will range from ions, atoms, and molecules through color centers and other solid-state quantum emitters. The intention here is to provide an important guide for the students and researchers of the community. It is dedicated to one of today's most active and rapidly expanding fields of research and development in photonic quantum technologies, in particular the growing number of experimental implementations and practical applications of quantum systems. This book will be unique because it will provide researchers and scientists with various aspects of material development on different platforms and their applications in quantum technologies *in one place*.

Progress in the realization of quantum devices and circuits enables breakthroughs in experimental tests of quantum physics and in practical applications of quantum technologies in imaging, secure communication, ultrasensitive metrology, and quantum computing. Based on the current worldwide interdisciplinary research activities in these emerging quantum technologies, I strongly believe that the realization of a practical quantum device (e.g. quantum repeaters) will most probably be based on hybrid quantum systems. Therefore, the intention of this publication is to provide an overview of the latest cutting-edge research on photonic quantum technologies and to provide readers (academia and industry), researchers, and scientists with several ingredients *in one location*. This will help to bridge the gap between scientists working in academia and industry and guide them to understand the current development, the remaining challenges toward the realization of practical quantum systems, and the necessary requirements to overcome such challenges.

This book consists of 26 chapters, written by internationally prominent scientists in the field of quantum technologies from different countries: Austria, China, Denmark, France, Germany, Panama, Poland, Russia, Spain, Switzerland, the United Kingdom, and the United States. It is structured into five parts, covering various topics ranging from fundamentals to applications. I anticipate that the book will be helpful for students and researchers involved in this field and that it will encourage and guide engineers in their efforts for industrial implementations.

Though the cover page of the book was chosen before the announcement of the 2022 Nobel Prize in Physics, I am delighted that it matches the award's topic.

I would like to express my sincere gratitude to ***Alain Aspect***, the 2022 Nobel Prize Laureate in Physics, for taking time from his busy schedule to contribute a chapter. My sincere thanks go out to all my colleagues for their valuable contributions and their excellent cooperation throughout the writing and editing process.

Furthermore, I would like to thank M. Fox, S. Kais, P. van Loock, D. Oi, and M. Razavi for accepting my invitation to review some chapters and for providing valuable feedback and suggestions.

Special thanks to the editorial staff of Wiley-VCH and in particular to M. Preuss, D. Bez, H. Noethe, M. Abinaya, V. Shyamala, and C. Nussbeck for their support and patience in awaiting the final version and keeping this book on track for publication.

I would also like to thank the German Research Foundation (DFG) and the Federal Ministry of Education and Research (BMBF) for their continued support. I am also grateful for the financial support provided by the state of Hesse through the LOEWE program. My thanks also go out to the University of Kassel for its internal financial support.

Finally, I want to extend my gratitude to my parents for supporting me in pursuing my studies. My wife and children deserve a special thank you for their unconditional support and gratitude for my absence for several weekends.

Mohamed Benyoucef
Kassel, November 2022

Abstracts and Keywords

Chapter 1: Introduction to Quantum Photonics

*Richard J. Warburton**

*Corresponding author: richard.warburton@unibas.ch

A brief and non-technical introduction to contemporary quantum photonics is presented.
Keywords: photon; quantum optics; quantum technology.

Chapter 2: The Second Quantum Revolution: From Basic Concepts to Quantum Technologies

Alain Aspect,†*

*Corresponding author: alain.aspect@institutoptique.fr

*†The 2022 Nobel Prize Laureate in Physics

The second quantum revolution was first conceptual, with the understanding and the experimental demonstration of the extraordinary character of entanglement, as evidenced by Bell inequalities violations. These experiments were permitted by the possibility to observe and manipulate individual quantum objects. The second quantum revolution is now at the stage of using these conceptual advances to develop quantum technologies that are able to go beyond those based on the first quantum revolution.
Keywords: second quantum revolution; entanglement; bell inequalities; quantum computing; quantum cryptography.

Chapter 3: Solid-State Quantum Emitters

*A. Mark Fox**

*Corresponding author: mark.fox@sheffield.ac.uk

This chapter gives a tutorial overview of the key parameters that are used to assess solid-state quantum emitters for applications in quantum photonics. Specifically,

it covers single-photon purity, coherence and photon indistinguishability, and the methods that are used to quantify them. The chapter concludes with a brief comparison of several of the more common material systems that are used in solid-state quantum optics.
Keywords: quantum emitter; single-photon source; photon statistics; photon anti-bunching; Purcell effect; Rabi rotations; Hanbury Brown-Twiss experiment; Hong-Ou-Mandel experiment; photon indistinguishability.

Chapter 4: Single-Photon Sources for Multi-photon Applications

*Stefan Frick, Robert Keil, Vikas Remesh, and Gregor Weihs**

*Corresponding author: gregor.weihs@uibk.ac.at

Much of the recent development of single-photon sources is driven by the desire to apply them to protocols and technologies that use the interference of two or more photons, such as quantum repeaters or boson sampling. In all of these cases, the indistinguishability of the produced photons is a key requirement. For those applications that want to scale to larger photon numbers, source efficiency is equally important. In this chapter, we will discuss two competing solutions, sources based on nonlinear optics and quantum dots, as the most-used single-quantum emitters.
Keywords: photon-pair sources; quantum dots; multi-particle interference; indistinguishability; number purity; spectral purity; excitation schemes.

Chapter 5: Quantum Key Distribution Protocols

Álvaro Navarrete, Víctor Zapatero, and Marcos Curty*

*Corresponding author: vzapatero@com.uvigo.es

In contrast to public-key cryptography, whose security relies on computational assumptions, quantum key distribution (QKD) provides an information-theoretically secure solution for the task of distributing cryptographic keys between distant parties through an insecure channel. In this chapter, we provide a general introduction to QKD protocols and their security, starting from the theoretical foundations of QKD – based on fundamental properties of quantum mechanics, such as the no-cloning theorem or entanglement – and moving to state-of-the-art QKD solutions and the latest technological advances.
Keywords: quantum key distribution; quantum communication; quantum cryptography; quantum technologies.

Chapter 6: From Basic Science to Technological Development: The Case for Two Avenues

*Matteo Carlesso and Mauro Paternostro**

*Corresponding author: m.paternostro@qub.ac.uk

We argue that, in the quest for the translation of fundamental research into actual quantum technologies, two avenues that have – so far – only partly explored should be pursued vigorously. On first entails that the study of energetics at the fundamental quantum level holds the promises for the design of a generation of more energy-efficient quantum devices. On second route to pursue implies a more structural hybridization of quantum dynamics with data science techniques and tools, for a more powerful framework for quantum information processing.
Keywords: quantum technologies; quantum thermodynamics; quantum machine learning.

Chapter 7: Quantum Networks in Space

*Lisa Wörner**

*Corresponding author: lisa.woerner@dlr.de

The inherent properties of quantum mechanics render it an important subject for fundamental research and applications alike. Entanglement stands out among those, allowing for novel secure communication and more rapid computation. To enable global secure communication, space based networks are required. A space network with trusted nodes for every user is an unviable feat, and could lead, given the amount of potential civil, governmental, and military users, to a cluttering of space. Additionally, global networks of trusted nodes exclude smaller to medium players from profiting from the quantum revolution and the benefits of secure communication. To remain inclusive to small businesses and enable broad usage of the technology, the operation of untrusted nodes in space is inevitable. Consequently, untrusted nodes need to be capable of receiving and re-transmitting the information with the possibility of storage of the information in-between. Hence, space-based quantum repeater, consisting of a receiver, a transmitter, and a coherent memory, could operate as untrusted nodes.

In this chapter, the challenges, necessities, and opportunities for global quantum networks will be discussed. This includes novel techniques for quantum repeater and their current technology readiness level.
Keywords: quantum network; untrusted node; quantum repeater.

Chapter 8: Fluorescence Spectroscopy in Planar Dielectric and Metallic Systems

*Alexey I. Chizhik, Daja Ruhlandt, and Jörg Enderlein**
*Corresponding author: jenderl@gwdg.de

In this chapter, we present an overview of our recent theoretical and experimental studies of the fluorescence modulation of quantum emitters by planar nanostructures. We will mainly focus on two methods that are based on this phenomenon: (i) axial localization of a fluorophore using energy transfer from the fluorophore to a planar metal film or graphene and (ii) absolute quantum yield measurements using a plasmonic nanocavity. Both methods have been used in a large number of recent studies. We present the full theoretical modeling of these experiments and give an overview of recent experimental results.
Keywords: single-molecule fluorescence; nanocavity electrodynamics; fluorescence lifetime spectroscopy; defocused single-molecule imaging; metal- and graphene-induced energy transfer.

Chapter 9: Single Trapped Neutral Atoms in Optical Lattices

*Andrea Alberti and Dieter Meschede**
*Corresponding author: meschede@uni-bonn.de

Controlling neutral atoms one by one has opened the path to create new experimental platforms for investigating their individual as well as their collective dynamics in a bottom-up approach. We discuss the tools for precisely detecting and preparing all degrees of freedom of trapped atoms at the ultimate quantum level, including their internal spin state, their motional state, and also their position, which has relevant quantum character. We summarize applications including single-atom interferometry and quantum walks.
Keywords: cold atoms; trapped neutral atoms; quantum information; quantum walks; single atoms; atom interferometry.

Chapter 10: Long-Distance Entanglement of Atomic Qubits

*Kai Redeker, Wenjamin Rosenfeld, and Harald Weinfurter**
*Corresponding author: h.w@lmu.de

Quantum networks connecting future quantum computers and enabling efficient long-distance quantum communication require as most important resource the distribution of entanglement. This chapter details how this can be achieved for the

particular example of atomic quantum memories implemented with single, trapped, neutral rubidium atoms.
Keywords: quantum network; quantum communication quantum memory; entanglement.

Chapter 11: Collective Light Emission of Ion Crystals in Correlated Dicke States

Ferdinand Schmidt-Kaler and Joachim von Zanthier*
*Corresponding author: fsk@uni-mainz.de

We give an overview of collective light scattering of trapped ion crystals serving as arrays of correlated emitters of single, indistinguishable photons into free space. The collectively emitted light is recorded in the far field by detectors, which feature high spatial and high temporal resolution. We observe interference in the first- and second-order photon correlation functions as well as, bunched and antibunched photon statistics. We analyze theoretically how the observed features of collective light scattering off arrays of correlated single photon emitters result from projective quantum measurements into Dicke states.
Keywords: photon statistics and coherence theory; quantum description of interaction of light and matter; cooperative phenomena in quantum optical systems; single photon emitter; correlated emission of photons; single ion trapping and cooling.

Chapter 12: Single-Molecule Magnets Spin Devices

*Eufemio Moreno-Pineda and Wolfgang Wernsdorfer**
*Corresponding author: wolfgang.wernsdorfer@kit.edu

Since the first observation of quantum effects, scientists have striven to gain a deeper knowledge and understating of these characteristics with the ultimate goal of implementing them in practical applications. Today, due to important contributions of many pioneers, we are at the point in which quantum effects can be exploited, prompting the proposal of several futuristic technological applications such as quantum sensing, quantum simulation, quantum computing, and quantum communication. These technologies share some common ground: they rely on the manipulation and read-out of the electronic or nuclear states; hence, their quantum properties must fulfill certain characteristics. To accomplish these tasks, several systems have been proposed; however, among the several prospects, molecular systems have been shown to possess several advantageous characteristics, as compared to common platforms. In this chapter, key aspects of single-molecule magnets allowing their proposal in applications in quantum sensing, quantum simulation, quantum computing, and quantum communications will be revised.
Keywords: quantum sensing; quantum simulations; quantum computing; quantum communications; spin transistor; spin valve; quantum bit; qudit; quantum error correction; quantum tunneling of the magnetization.

Chapter 13: Molecular-Ion Quantum Technologies

*Mudit Sinhal and Stefan Willitsch**
*Corresponding author: Stefan.Willitsch@unibas.ch

Quantum-logic techniques for state preparation, manipulation, and nondestructive interrogation are increasingly being adopted for experiments on single molecular ions confined in traps. The ability to control molecular ions on the quantum level via a co-trapped atomic ion offers intriguing possibilities for new experiments in the realms of precision spectroscopy, quantum information processing, cold chemistry, and quantum technologies with molecules. The present article gives an overview of the basic experimental methods, recent developments, and prospects in this field.
Keywords: molecular ions; quantum-logic spectroscopy; precision measurements; quantum technologies.

Chapter 14: Optical Atomic Clocks

*Ekkehard Peik**
*Corresponding author: ekkehard.peik@ptb.de

Based on the development of atomic clocks and the definition of the unit of time via an atomic resonance frequency, time and frequencies are the most precisely measurable physical quantities. Over the past 30 years, the development of methods of cooling and trapping of atoms and ions and the development of low-noise laser oscillators and the invention of the optical frequency comb that establishes a link between microwave and optical frequencies with nearly perfect fidelity have led to impressive progress in the field. Optical clocks based on laser-cooled and trapped atoms and ions reach a systematic uncertainty in the low 10^{-18} range and below. This chapter presents the principles and methods that have enabled these performances.
Keywords: atomic clock; laser cooling and trapping; optical frequency comb; unit of time.

Chapter 15: Coherent Spin Dynamics of Colloidal Nanocrystals

Dmitri R. Yakovlev, Anna V. Rodina, Elena V. Shornikova, Aleksandr A. Golovatenko, and Manfred Bayer*
*Corresponding author: dmitri.yakovlev@tu-dortmund.de

This chapter gives an overview of the spin properties of semiconductor colloidal nanocrystals. We consider the energy and spin-level fine structures in colloidal quantum dots and nanoplatelets and their modification in an external magnetic field. We show how spectroscopic, time-resolved, and magneto-optical experimental approaches can be used for colloidal nanocrystals in order to get information on exciton states as well as the spin dynamics of excitons and charge carriers. We demonstrate that spin-dependent phenomena give access to further properties of colloidal nanocrystals, like the surface states and their spins, the photocharging dynamics, the carrier localization within a nanocrystal, etc.

Keywords: colloidal nanocrystals; colloidal quantum dots; semiconductor nanoplatelets; coherent spin dynamics; time-resolved Faraday rotation; spin-flip Raman scattering; exciton spin structure.

Chapter 16: Relaxation of Electron and Hole Spin Qubits in III–V Quantum Dots

*Alexander Bechtold, Tobias Simmet, Friedrich Sbrezny, Nikolai A. Sinitsyn, Kai Müller, and Jonathan J. Finley**

*Corresponding author: jonathan.finley@wsi.tum.de

The control of solid-state qubits for quantum information processing requires a detailed understanding of the mechanisms responsible for decoherence. During the past decade, considerable progress has been achieved for describing the qubit dynamics in relatively strong external magnetic fields. However, testing theoretical predictions at very low magnetic fields has proven difficult in optically active dots. Here, we describe our studies of electron and hole spin qubit dephasing in single InGaAs quantum dots using spin memory devices. The results show that without applied magnetic fields, the initially orientated electron spin rapidly loses its polarization due to precession around the fluctuating Overhauser field with an effective magnetic field amplitude of 10.5 mT. The inhomogeneous dephasing time associated with these hyperfine mediated dynamics is $T_2^* \sim 2$ ns. Over longer timescales, an unexpected stage of central spin relaxation is observed, namely the appearance of a second feature in the relaxation curve around $T_Q = 750$ ns arising from quadrupolar coupling. In comparison, hole spin qubits are shown couple significantly more weakly to the nuclear spin bath. We measure a $\sim 100\times$ times longer dephasing time $T_2^* \sim 210$ ns for hole spin qubits compared with the electron spin. We also obtain evidence for the impact of anisotropic hyperfine coupling on the spin polarization decay, allowing us to quantify the degree of anisotropy $\alpha = 0.19$ which is fundamental to the character of the confined hole spin wave function. By modeling this behavior, we derive the degree of light-hole heavy-hole mixing, which is an essential mechanism for enabling hole spin dephasing and thus refining the description of hole hyperfine coupling beyond the initially suggested pure Ising form.
Keywords: optically active quantum dots; electron and hole spin qubits; central spin problem; spin photon interfaces

Chapter 17: Ensemble-Based Quantum Memory: Principle, Advance, and Application

Bo Jing and Xiao-Hui Bao*

*Corresponding author: bjing@uestc.edu.cn

Quantum memory for photons plays an important role in quantum information science, such as quantum repeater, quantum network, quantum communica-

tion, and quantum computing. The realization of a quantum memory with high efficiency, long lifetime, high fidelity, large bandwidth, and multimode capacity is of great significance for its applications. Atomic ensemble, serving as one of the most appealing physical platforms for quantum memories, has been widely explored toward improving memory performance. Over the past decades, significant progress based on atomic ensembles has been made in developing new schemes and improving memory performances experimentally through various approaches. In this chapter, we first introduce the working principle of quantum memory by reviewing the widely used schemes. Next, we present several key figures of merit for judging memory performance. Afterward, we review the experimental progress made in several different systems. Finally, we give discussions on its application in a number of different directions. The scope of this chapter is limited to single-photon storage with atomic ensembles.
Keywords: quantum memory; atomic ensemble; rare-earth ions; electromagnetically induced transparency; atomic frequency comb; DLCZ.

Chapter 18: Telecom Wavelengths InP-Based Quantum Dots for Quantum Communication

Mohamed Benyoucef and Anna Musiał*
*Corresponding author: m.benyoucef@physik.uni-kassel.de

Quantum dots (QDs), also known as artificial atoms, are among the recently explored hardware platforms for photonic quantum technologies. They offer a range of advantages that make them particularly well-suited for building quantum communication networks over optical fiber channels. One of the most prominent features of QDs is the ability to be easily and deterministically integrated into photonic devices and photonic circuits. They also exhibit near-unity internal quantum efficiency and emit nearly Fourier-transform-limited single photons of high purity and entangled photons. QDs emitting at optical fiber communication wavelengths, especially in the telecom C-band, which corresponds to the spectral range of the lowest attenuation in silica fibers, are very promising for long-distance quantum communication. This makes the development of QD-based quantum light sources for these wavelength ranges of particular interest. This chapter gives an overview of recent developments in InP-based QDs, which are one of the candidates providing emission in this spectral window. These advancements include lowering the spatial density of nanostructures in this material system, increasing their in-plane symmetry, developing high-reflectivity distributed Bragg reflectors, and incorporating them in microcavities and other photonic structures to increase photon extraction efficiency to generate the highest-purity single photons at 1550 nm.
Keywords: quantum dots; III-V semiconductor nanostructures; molecular beam epitaxy growth; telecom wavelengths; single-photon sources; magneto-optics; photonic crystal cavities; fiber quantum network; extraction efficiency; symmetric InAs/InP quantum dots; internal quantum efficiency.

Chapter 19: Quantum Optics with Solid-State Color Centers

*Joseph H.D. Munns, Laura Orphal-Kobin, Gregor Pieplow, and Tim Schröder**
*Corresponding author: tim.schroeder@physik.hu-berlin.de

With the chapter "Quantum Optics with Solid-State Color Centers," we provide a comprehensive overview of color centers in wide-bandgap materials: We describe their generation and energy level schemes, how they are controlled both optically and via microwaves, and how they are used in quantum optics and technology. The chapter summarizes landmark experiments including applications ranging from single photon generation, over quantum memories, quantum gates, to entanglement generation. We structured the chapter such that the reader has access to a bottom-up introduction but can also directly jump to sections of interest. This work is directed at graduate students up to the principal investigator level.
Keywords: quantum optics; quantum technology; color centers; wide-bandgap materials; diamond; silicon carbide; review; single photon emitters; optically active solid-state spin defects.

Chapter 20: Quantum photonics with 2D semiconductors

*Mauro Brotons-Gisbert and Brian D. Gerardot**
*Corresponding author: b.d.gerardot@hw.ac.uk

In this chapter, we introduce the main properties of the most popular van der Waals materials for single-photon generation: hexagonal boron nitride(hBN), group-VI transition metal dichalcogenide semiconductors (TMDs), and 2D quantum devices consisting of different 2D crystals assembled into van der Waals heterostructures. We then focus primarily on quantum emitters (QEs) hosted by TMD semiconductors, including a description of their typical spin and optical properties. Finally, we consider the different means to engineer the QEs and their coupling to cavities and integrated photonic chips.
Keywords: 2D materials; nanophotonics; single photon; van der Waals heterostructure

Chapter 21: Nano-Opto-Electro-Mechanical Systems for Integrated Quantum Photonics

Leonardo Midolo and Celeste Qvotrup*
*Corresponding author: midolo@nbi.ku.dk

This chapter covers the topic of nano-opto-electro-mechanical systems (NOEMS) and their application in quantum photonic integrated circuits. When reducing the size of optical devices down to the nano-scale, as in nanophotonics and plasmonics, the properties of confined optical waves become increasingly affected by small geometrical variations of the materials. These variations produce a wide range

of electro- and opto-mechanical interactions that do not depend on any intrinsic material property and therefore can be operated at cryogenic temperatures and seamlessly integrated in quantum photonic circuits with embedded quantum emitters and single-photon detectors. We discuss the basic principles of operation of NOEMS, the techniques involved in their fabrication, and the functionalities that such devices can provide toward the realization of scalable quantum photonic integrated technologies.
Keywords: photonic integrated circuits; electro- and opto-mechanics; quantum emitters; micro and nanofabrication.

Chapter 22: Silicon Quantum Photonics - Platform and Applications

*Cagin Ekici, Elizaveta Semenova, Davide Bacco, and Yunhong Ding**

*Corresponding author: yudin@fotonik.dtu.dk

Silicon quantum photonics is the leading platform of integrated quantum photonics thanks to its advantages of ultradense integration, complementary metal–oxide semiconductor-compatible fabrication, etc. However, to enter the fault-tolerant, breakthroughs on the photon sources, optical loss, and quantum process fidelity have to be achieved before silicon-integrated quantum photonics can really achieve a useful scale. In this chapter, we will present the basics of silicon quantum photonics and give an overview of the state of the art. The chapter is structured as follows. In Section 22.2, we will discuss the typical fabrication process and necessary passive and active optical components. In Section 22.3, we will further present an overview of different quantum light sources on silicon, including quantum dot sources on silicon and nonlinear sources. In Section 22.4, we will discuss applications in processing of quantum information on silicon and quantum communication. Finally, we will give an outlook of silicon quantum photonics for further scaling.
Keywords: silicon photonics; quantum information technologies; single- photon source; nonlinear light source; quantum information processing; quantum communication.

Chapter 23: Photonic Realization of Qudit Quantum Computing

*Yuchen Wang, Zixuan Hu, and Sabre Kais**

*Corresponding author: kais@purdue.edu

In quantum computing, qudit is a multilevel computational unit. Qudit, as an alternative to qubit, has a larger state space for information storage and processing. Quantum circuit built with qudits can enhance algorithm efficiency, reduce the circuit complexity, and simplify the experiment. Photonic systems with their comparatively long decoherence time provide good platforms for quantum computing.

Some photonic systems also provide suitable working platforms for qudit quantum computing because qudit systems can be encoded naturally and conveniently with photons given the multidimensional nature in many photon properties. This chapter first introduces the basic knowledge and concepts of qudit quantum computing that include qudit gates, universality, and important qudit algorithms such as the quantum Fourier transform and the Phase Estimation Algorithm (PEA). It then reviews two implementations of qudit computing with photonic systems. The first system is an electro-optic platform that integrates on-chip modulators, pulse shapers, and other well-established fiber-optic components. This platform is capable of performing high-fidelity quantum Fourier transform pulse shaping and the two-qudit SUM gate via the time and frequency degrees of freedom, which paves the way to demonstrate a proof-of-principle qutrit PEA. The second system consists of microwave cavities and transmon qubits to implement a two-mode superconducting bosonic processor that enables the scalable simulation of molecular vibronic spectra.
Keywords: quantum information; quantum computing; qudit space; quantum phase estimation algorithm; electro-optic platform; superconducting bosonic processor and open quantum dynamics.

Chapter 24: Fiber-Based Quantum Repeaters

*Mohsen Razavi**

*Corresponding author: m.razavi@leeds.ac.uk

Quantum repeaters are the primary platform for the reliable exchange of quantum data in future quantum communications networks. In this chapter, I provide a brief overview of how different classes of quantum repeaters work, and, in particular, how they manage the two key challenges of loss and noise in quantum channels and systems. Along the way, the required resources for such repeaters and their expected performance will be discussed.
Keywords: quantum communications; quantum repeaters; quantum networks; quantum error correction codes; quantum data transfer; entanglement distribution; quantum memories.

Chapter 25: Long-Distance Satellite-Based Quantum Communication

*Yuan Cao and Juan Yin**

*Corresponding author: yinjuan@ustc.edu.cn

Long-distance quantum communication with satellites opens a promising avenue for the global secure quantum network and the large-scale test of quantum foundations. Free-space links could provide a better approach because photon loss and decoherence are almost negligible in the atmosphere. But there are still many new

challenges. The satellite moves fast and also has rotation and vibration. Besides, the environment of space is also complex. Many scientific research teams from different countries have spent the past decade performing systematic ground tests for satellite-based quantum communications to overcome all these difficulties. The world's first quantum science satellite Micius was launched on 16th August 2016, demonstrating a series of scientific experiments. Since then, many satellite plans have been proposed, making this field into focus. In this chapter, we briefly review the progress in long-distance satellite-based quantum communication, including step-by-step ground-based feasibility studies and developments of the critical technologies required, a series of satellite-ground quantum communication experiments with Micius, and ending with an outlook on the future work that needs to be done to build global-scale practical quantum networks eventually.
Keywords: quantum information; quantum communication; quantum key distribution; quantum entanglement.

Chapter 26: Quantum Communication Networks for 6G

Riccardo Bassoli, Frank H.P. Fitzek, and Holger Boche*
*Corresponding author: riccardo.bassoli@tu-dresden.de

The research and industrial communities are concentrating their efforts in two main trends in communications, which are seen as the new drivers not only in telecommunications but also in all the aspects of our lives: sixth generation (6G) and the Quantum Internet. While their research and standardization efforts are still independently faced, the limitations and challenges of these two systems can experience significant benefits from their integration. This chapter aims at making the reader familiar with current status of 6G and the Quantum Internet architectural design and how they can be integrated to gain a mutual advantage.
Keywords: 6G; quantum communication networks; entanglement-assisted communication; quantum campus network; quantum three-dimensional network.

Part III

Spin Qubits and Quantum Memories: From Spin Properties to Physical Realizations

Part III

15

Coherent Spin Dynamics of Colloidal Nanocrystals

Dmitri R. Yakovlev[1,2], *Anna V. Rodina*[2], *Elena V. Shornikova*[1], *Aleksandr A. Golovatenko*[2], *and Manfred Bayer*[1,2]

[1] *Experimental Physics 2, Department of Physics, TU Dortmund University, Otto-Hahn Str. 4a, 44221 Dortmund, Germany*
[2] *Ioffe Institute, Russian Academy of Sciences, Polytekhnicheskaya 26, 194021 St. Petersburg, Russia*

15.1 Introduction

Semiconductor quantum dots (QDs) provide three-dimensional geometric confinement of charge carriers on the nanometer-scale and, in that sense, mimic atoms in nature. Electronic excitations in atoms or ions are prototypes for quantum bits due to their well-defined character and long-lived coherence, the latter of which is connected to the isolation from the environment. It was natural to study QDs as hardware for quantum information applications [1–3], as they promise also rather long coherence times due to the suppression of relaxation mechanisms related to free carrier motion. Further, semiconductor-based systems offer a variety of attractive features such as established fabrication tools providing reproducibility, possible integration with conventional electronics, controllable properties through external perturbations, etc. In addition, QDs possess large electric dipole moments for excitation across the band gap by which optical manipulation on picosecond timescale may be achieved.

On the other hand, excitations in QDs, despite the similarity to atoms, are nevertheless embedded in a potentially hostile environment. This environment, even if ideal, can host excitations coupling to the quantum bits such as lattice vibrations or nuclear spins. Deviations from ideality like charge fluctuations aggravate the situation. Due to the high quality of the material, the main corresponding activities have been focused on QDs based on epitaxially grown nanostructures. Remarkable achievements have been obtained such as the demonstration of high-quality sources of single and entangled photons. On the other hand, also challenges have been identified such as the limited coherence of quantum bits compared to atoms. Disregarding the nuclear spins of the host lattice, the longest coherence time has been demonstrated for confined carrier spins, for which many threats can be suppressed: vibrations can be frozen out by cooling the sample to cryogenic temperatures, charge fluctuations can be reduced by resonant excitation of optical transitions. Still, even

Photonic Quantum Technologies: Science and Applications, First Edition. Edited by Mohamed Benyoucef.

for these conditions, the coupling to the nuclear spins limits the spin coherence in semiconductor QDs to microsecond times.

Colloidal nanocrystals (NCs) are semiconductor particles of nanometer size with very strong confinement of charge carriers (electrons and holes). NC is the general term used for nanoparticles of different shapes: a spherical QD, an elongated nanorod (NR), or an atomically flat nanoplatelet (NPL). The confinement and shape determine the electronic and spin-level structure of NCs, making them attractive for basic research and applications [4–6]. Colloidal NCs can be fabricated by chemical synthesis in solution, which is a widely used technology, or by synthesis in a glass matrix. Different semiconductors can be used for that, among them II–VI materials, like CdSe and CdS, are very popular and mostly studied. Also various III–V materials, like InP [7], and recently lead halide perovskites ($CsPbBr_3$, $FAPbI_3$) with high quantum yield at room temperature [8] became available.

The possibility of using colloidal semiconductor NCs as sources of single photons in photonic quantum technologies has been studied for more than 20 years [9]. The following features of state-of-the-art colloidal NCs contribute to their application in this field: (i) the size-dependent emission color across a wide spectral range from the UV to near IR, (ii) the high photoluminescence (PL) quantum yield, (iii) the suppressed PL blinking. Depending on the charge state of a NC, its emission is governed either by neutral excitons or charged excitons. All those features make colloidal NCs also interesting for applications in spin-based electronics or quantum information.

A great amount of knowledge has been accumulated nowadays for the optical and spin properties of semiconductor QDs grown by molecular-beam epitaxy. Thus, it is attractive to use this knowledge for understanding colloidal NCs, but this has to be done carefully, with taking into account their specifics. Here we list important differences distinguishing colloidal NCs from epitaxially grown QDs.

- Strong carrier confinement provided by the small NC size down to 1 nm and high (practically infinite) barriers between semiconductor and organic ligands or glass matrix.
- Strong enhancement of the Coulomb interaction between electrons and holes due to the large dielectric contrast between the semiconductor and its ligand environment. For example, in CdSe NPLs the exciton binding energy reaches 315 meV, which is 20 times greater than in bulk CdSe [10].
- Large exchange splitting between bright and dark exciton states (ΔE_{AF} up to 20 meV) and important role of the ground dark exciton state, whose emission is prohibited [11–14].
- Important role of surface spins in NCs for the spin dynamics and radiative recombination of the dark excitons [15–17].
- Complexity of doping of NCs [18]. However, singly charged NCs as key elements of quantum technologies can be obtained by long-lived photocharging. For that a controllable surface modification to provide surface capture of either electrons or holes can be used [19, 20].
- Great variety of shapes of colloidal NCs and their arrangement in supercrystals.
- Variety (commonly random) of orientation of NCs in ensemble, which complicates theoretical treatment and interpretation of experimental data [21, 22].

- Carrier spin coherence observed up to room temperature [23, 24], evidencing that spin relaxation mechanisms in colloidal NCs is less efficient. One of the reasons could be the absence of the phonon reservoir of the surrounding barrier, which is unavoidable for epitaxially grown QDs.

A further comparison of colloidal and epitaxially grown QDs can be found in Ref. [25].

In this chapter, we give an overview of the spin properties of semiconductor colloidal NCs and of related spin dependent phenomena. We consider the energy and spin level fine structures in colloidal QDs and NPLs and their modification in an external magnetic field. We show how spectroscopic, time-resolved, and magneto-optical experimental approaches can be used for colloidal NCs in order to get information on exciton states as well as the spin dynamics of excitons and charge carriers. We demonstrate that spin-dependent phenomena give access to further properties of colloidal NCs, like the surface states and their spins, the photocharging dynamics, the carrier localization within a NC, etc.

15.2 Spin-Level Structure of Neutral and Charged Excitons

Colloidal NCs can be synthesized from different semiconductors and their combinations and in plenty of different shapes and sizes. This allows one to tune widely not only the absorption spectra of NCs, but also the fine energy structure of neutral and charged excitons, which controls their emission and spin-dependent optical properties. For instance, in colloidal quasi-two-dimensional NPLs, the neutral or charged excitons formed by the heavy-hole and light-hole states are strongly split (with the heavy-hole exciton being the ground state) similar to epitaxially grown quantum wells and self-organized QDs. In one-dimensional nanorods, on the contrary, the ground state is the light-hole exciton. In colloidal QDs, the total anisotropy splitting Δ between the heavy and light holes is determined by the non-spherical shape and the presence of the crystal field in wurtzite semiconductors like wz-CdSe [12].

The fine energy structure of the negatively charged exciton (negative trion comprising two electrons with opposite spins and one hole) for the case $\Delta > 0$ is shown in Figure 15.1a. Importantly, the electron–hole exchange interaction is compensated in the singlet state of the negative trion and both heavy-hole and light-hole trion states are bright. As a result, the temperature-induced redistribution of their relative populations, as well as the state mixing induced by any perturbation including an external magnetic field, does not affect the PL dynamics.

For neutral excitons, the electron–hole exchange interaction is drastically enhanced due to both the spatial and the dielectric confinement in colloidal NCs. It mixes the heavy-hole and light-hole exciton states. In spherical or "quasi-spherical" NCs with $\Delta = 0$ [26], the exciton states split by the electron–hole exchange energy Δ_{exch} are characterized by the total spin $\mathcal{F} = 1$ and $\mathcal{F} = 2$. The joint action of

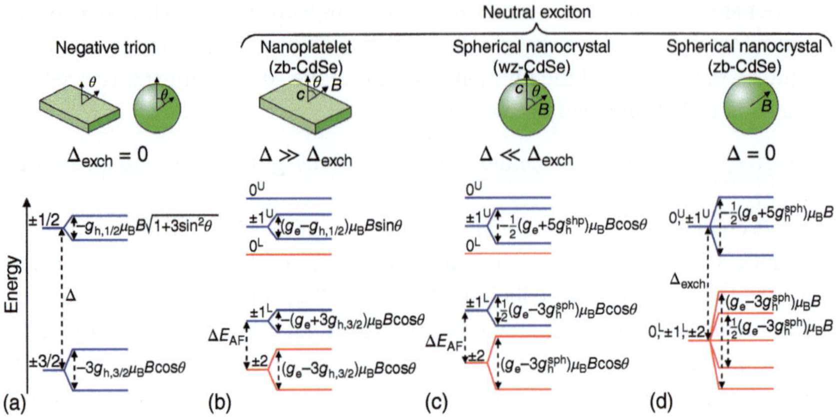

Figure 15.1 Negative trion (a) and neutral exciton spin-level structure in NCs of various symmetry for (b) $\Delta \gg \Delta_{exch}$, (c) $\Delta \ll \Delta_{exch}$, and (d) $\Delta = 0$. The states are marked by the hole spin projection for the negative trion and by the total momentum projection F for the neutral exciton on the anisotropic c-axis. The upper indexes L and U correspond to the lower and upper exciton states with the same F as was first introduced in Ref. [11]. Bright (dark) trion and exciton states are shown by blue (red) lines. Scheme of the Zeeman splitting of spin states in magnetic field is shown for $\mu_B B \ll \Delta_{exch}, \Delta$. θ is the angle between the anisotropy axis and the magnetic field direction. Source: Adapted from [11].

the exchange interaction and the anisotropic splitting results in formation of five exciton states labeled by the total spin projection $F \equiv \mathcal{F}_z$ on the anisotropy axis $c \parallel z$ shown in Figure 15.1 for the limits $\Delta \gg \Delta_{exch}$ (b), $\Delta \ll \Delta_{exch}$ (c), and $\Delta = 0$ (d). The values of Δ and Δ_{exch} scale differently with the diameter of spherical NCs, resulting in a size dependence of the exciton energy level splitting that is controlled by their ratio [11].

The energy splitting ΔE_{AF} between the lowest state of the exciton fine structure, which is a nominally dipole-forbidden dark exciton with projections $\mathcal{F}_z = \pm 2$, and the lowest dipole-allowed bright exciton state $\mathcal{F}_z = \pm 1^L$ can amount up to 20 meV. As a result, the dark exciton determines the low-temperature PL properties. In turn, it can be used as a probe of the interaction of charge carriers with magnetic ions and surface spins, which activates its recombination. The same exciton-level structure remains valid for quasi-two-dimensional NPLs, where the anisotropy splitting Δ caused by the strong quantization normal to the plane is much larger than the enhanced electron–hole interaction Δ_{exch} (Figure 15.1b). Importantly, the temperature-induced redistribution of the populations between the dark (red lines) and bright (blue lines) neutral exciton states or the bright-to-dark exciton state mixing induced by any perturbation including an external magnetic field directly affects the PL dynamics.

The key parameter in spin physics is the Landé factor (or g-factor) of charge carriers. It determines the response of electrons and holes, as well as of excitonic complexes that they form, to an external magnetic field including the Zeeman energy splitting between spin sublevels (Figures 15.1 and 15.3a), and the Larmor frequency of the spin precession (Figure 15.5a). Therefore, it is important to know how the effective g-factors are changed with respect to their known bulk values, depending on the shape and size of the NCs. The change of the electron g-factor, g_e, in colloidal NCs is mostly controlled by the electron quantization energy, which modifies the admixture of the valence band states to the conduction band and results in a corresponding g_e size dependence (see Figure 15.5b) [27, 28]. The renormalization of the hole g-factor, g_h, is mostly related to the modification of the heavy- and light-hole mixing caused by a strong spin-orbit interaction in the valence band and is strongly affected by the particular shape of a NC [27]. Its size dependence is weaker than for the electrons and is mainly caused by the admixture of the spin–orbit split valence band states. The Zeeman splitting of the heavy-hole states (angular momentum projection $M = \pm 3/2$ on the anisotropy axis) and the light-hole states ($M = \pm 1/2$) (and of the respective trion states) in the magnetic field **B** directed along the anisotropy *c*-axis is defined as [27]

$$\Delta E_{|M|} = E_{-M} - E_{+M} = 2Mg_{h,|M|}\mu_B B \tag{15.1}$$

Here μ_B is the Bohr magneton. In anisotropic structures with $\Delta \neq 0$, the splittings $\Delta E_{|M|}$ acquire a factor $\sqrt{\cos^2\theta + (9/2 - 2M^2)\sin^2\theta}$ describing their dependence on the angle θ between the magnetic field and the anisotropy *c*-axis. In spherical NCs, $g_{h,3/2} \approx g_{h,1/2} = g_h^{\text{sph}}$.

The exciton Zeeman splitting comprises the electron and hole g-factors in a complicated way depending on the ratio $\Delta/\Delta_{\text{exch}}$. The limiting cases are shown in Figure 15.1. In the case $\Delta \neq 0$, all exciton states with $F \neq 0$ split with the anisotropic Zeeman splittings $\Delta E_{|F|}$ proportional to the component of magnetic field parallel to the anisotropy *c*-axis as far as the Zeeman splitting is smaller than the fine structure splittings at zero magnetic field. In stronger magnetic fields, this component can also mix the exciton spin levels with the same projection F, while the perpendicular component of the magnetic field mixes the light-hole to the heavy-hole states with $\Delta M = \pm 1$, as well as bright to dark exciton states with $\Delta F = \pm 1$. Importantly, the anisotropy results in a dispersion of the exciton parameters in the ensemble of NCs with different orientations of the NC axis.

Experimentally, the exciton fine structure and its modification in magnetic field can be addressed by various optical and magneto-optical techniques, which also allow one to distinguish between neutral and charged excitons [29, 30]. Examples for CdSe NPLs are given in Refs. [10, 14, 31, 32].

15.3 Photoluminescence in Magnetic Field

The exciton emission line from a single CdSe QD at liquid helium temperature can be as narrow as 0.02 meV, if its spectral wandering related to charge fluctuations is suppressed [33]. In a QD ensemble, the PL spectrum has considerable inhomogeneous broadening of a few tens of meV caused by the size dispersion of the colloidal QDs. This broadening is larger than the typical fine structure splitting (e.g. ΔE_{AF}), Zeeman splitting (which is about 1 meV at $B = 10$ T for the spin states with $g = 2$), exciton diamagnetic shift, and binding energy of charged trions, which limits the use of spectral filtering for exciton spectroscopy. One of the experimental approaches to bypass this limitation is the fluorescence line narrowing (FLN) technique based on resonant excitation within the inhomogeneously broadened exciton line [34]. For that, a narrow laser line excites a small subensemble of QDs with the corresponding narrow spectral distribution, which results in narrowing of the Stokes shifted emission lines from the dark exciton and from LO photon-assisted replica. Alternative approaches for exciton spectroscopy are time-resolved measurements of recombination dynamics and magnetic-field-induced circular polarization (degree of circular polarization [DCP]) of the PL. Here, we demonstrate their potential to identify and distinguish between neutral and charge excitons and to give insight in their recombination and spin dynamics.

As representative examples we use two samples of core/shell CdSe/CdS QDs with 5 nm diameter of the CdSe core [35]. One sample has a 2-nm-thick CdS shell, which is too thin to provide efficient photocharging even at cryogenic temperatures, so the emission of these QDs is dominated by neutral excitons. In another sample with a 10-nm-thick shell, the photocharging is efficient, here the hole is captured by surface states and the electron is left in the CdSe core. As a result, its emission is dominated by negative trions. However, the PL spectra on both samples are quite similar showing a broad band with about 60 meV width (Figure 15.2b).

As we described in Section 15.2, the main difference between the trion and exciton fine structure is the presence of the lowest dark exciton state determining the PL at low temperatures, while all trion states are bright (Figure 15.1). However, the ± 2 dark exciton state (denoted as F in Figure 15.2a) is not completely dark in colloidal QDs and NPLs, as its radiative recombination becomes possible due to admixture of the bright exciton states by various perturbations [13]. At liquid helium temperatures, the long-living recombination from this state can be seen as a long tail in the PL decay with a characteristic lifetime of the order of hundreds of nanoseconds (Figure 15.2c). With increasing temperature, the lifetime shortens due to population of the next higher-lying bright exciton state $\pm 1^L$ (denoted as A), and the PL decay is accelerated. Also, a shortening of the lifetime can be induced by a magnetic field, which admixes the bright exciton state to the dark one (inset). In the case of the trion, the lowest fine structure state is a dipole-allowed state, so the trion lifetime τ_r is independent of magnetic field and temperature (Figure 15.2d).

Insight into the spin properties of excitons and trions, as well as identification of the trion type (negative – two electrons and one hole, or positive – two holes and one electron) can be obtained from DCP studies. The DCP method exploits the Zeeman splitting ΔE_Z of exciton and trion spin states in magnetic field. For the magnetic field being parallel to the k-vector of the emitted light (Faraday geometry) and to the anisotropy c-axis, the emission from different spin states has opposite circular polarization (Figure 15.3). At low temperatures, when the thermal energy $k_B T$ is comparable with ΔE_Z, the lowest spin state is predominantly populated and the emission is circularly polarized. It is remarkable that the inhomogeneous broadening, which is typically much larger than the Zeeman splitting, makes no effect on the circular polarization degree, so that NC ensembles with large inhomogeneous broadening can be studied.

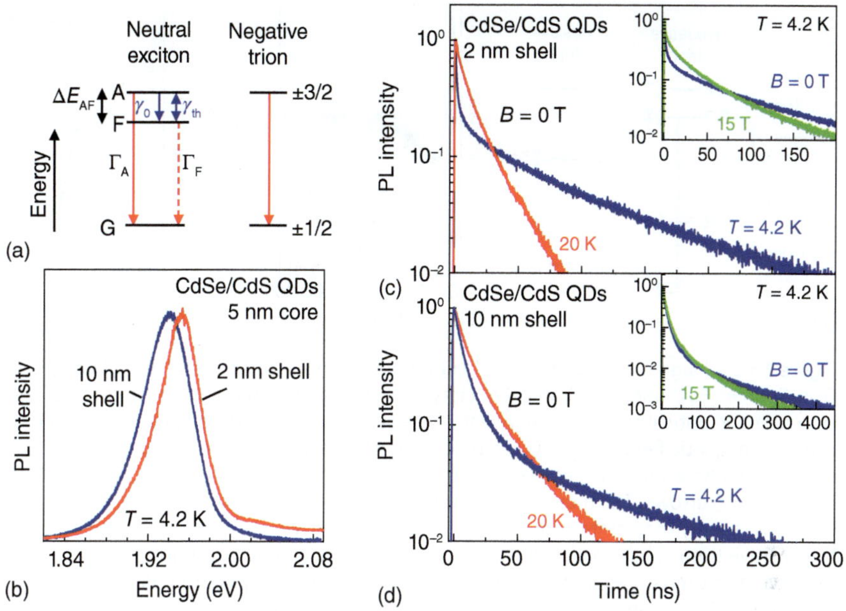

Figure 15.2 Photoluminescence of neutral and charged CdSe/CdS QDs [35]. (a) Energy-level scheme of neutral exciton and negative trion. A, F, and G denote bright exciton, dark exciton, and crystal ground state, respectively. ΔE_{AF} is the bright–dark splitting. γ_0 is the spin–relaxation rate from the bright to the dark exciton state. γ_{th} is the thermally activation rate from the dark to the bright state. Γ_A and Γ_F are recombination rates of bright and dark excitons. The spin of the negative trion state $\pm 3/2$ is contributed by the heavy hole. A resident electron with spin $\pm 1/2$ is left in the CdSe core after trion recombination. (b) PL spectra of thin-shell (2 nm) and thick-shell (10 nm) CdSe/CdS QDs under continuous-wave excitation. Diameter of CdSe core is 5 nm. (c,d) PL decay in QDs with different shell thickness measured at $B = 0$ T for two temperatures of 4.2 and 20 K. Insets: normalized PL decays at $B = 0$ and 15 T for $T = 4.2$ K. Source: Adapted from [35].

The DCP is defined by

$$P_c = \frac{I^+ - I^-}{I^+ + I^-} \tag{15.2}$$

Here, $I^{\pm}$ are the $\sigma^{\pm}$ circularly polarized PL intensities. They can be measured with pulsed excitation in the time-resolved regime (Figure 15.3b, so that intensities $I^{\pm}(t)$ give the $P_c(t)$ dynamics according to Eq. (15.2)), or with continuous-wave excitation (Figure 15.3c). The DCP sign allows one to distinguish between negative and positive trions if the sign and value of the electron and hole g-factors are known, i.e. are measured by other magneto-optical techniques. One can see from the scheme in Figure 15.3a, that in CdSe-based QDs with $\Delta > 0$, where the electron g-factor is positive and the hole g-factor is negative, negative and positive trions have opposite signs of DCP. The PL is stronger in σ^- polarization for the negative trion and in σ^+ for the positive trion. In thick-shell CdSe/CdS QDs, where the trion emission was identified through the recombination dynamics [35], σ^- emission is stronger in magnetic field (Figure 15.3b,c). This allows one to identify the emission as stemming

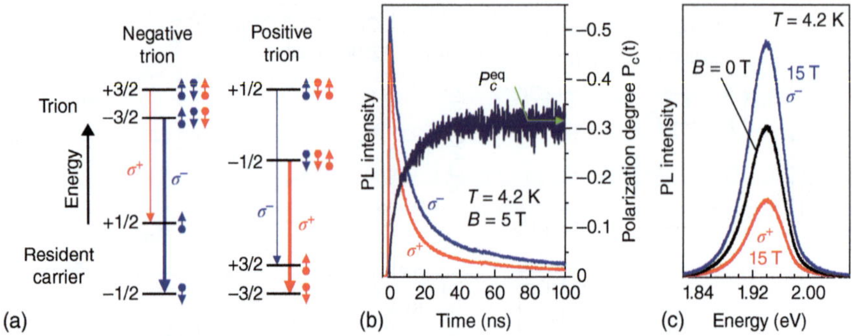

Figure 15.3 Magnetic-field-induced circular polarization of thick-shell (10 nm) CdSe/CdS QDs. Source: [35]/American Physical Society. (a) Schematic presentation of spin-level structure and optical transitions between these levels for negative and positive trions in an external magnetic field for $g_e > 0$ and $g_h < 0$. Short blue and red arrows indicate the electron and hole spins, respectively. Polarized optical transitions are shown by red (σ^+) and blue (σ^-) arrows. The more intense emission, shown by a thicker arrow, comes from the lowest-in-energy trion state with spin −3/2 for the negative trion and with spin −1/2 for the positive trion. (b) Dynamics of circularly polarized PL and polarization degree of negative trions measured under pulsed excitation. (c) Circularly polarized PL spectra measured at $B = 0$ and 15 T under continuous-wave excitation.

from the negative trion. The same behavior was reported for CdSe/CdS NPLs with thick shells [31].

The time evolution of $P_c(t)$ gives access to the spin relaxation time τ_s. One can see in Figure 15.3b that at $t \gg \tau_s$ the dependence $P_c(t)$ saturates at the equilibrium value P_c^{eq}, which is controlled by the $\Delta E_Z/k_B T$ ratio at the given magnetic field. The DCP values measured with continuous-wave excitation correspond to the time-integrated values P_c^{int}, obtained from $I^{\pm} = \int I^{\pm}(t)dt$. The magnetic field dependences of $P_c^{eq}(B)$ and $P_c^{int}(B)$ measured for negative trions in CdSe/CdS QDs are shown in Figure 15.4a

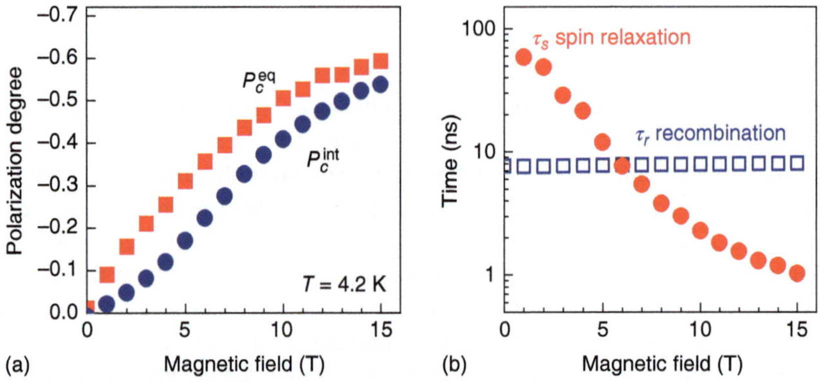

Figure 15.4 (a) Magnetic field dependences of equilibrium (P_c^{eq}) and time-integrated (P_c^{int}) DCP in thick-shell (10 nm) CdSe/CdS QDs. Source: Adapted from [35]. (b) Magnetic field dependences of trion spin relaxation time τ_s and PL decay time τ_r.

[35]. The difference between P_c^{int} and P_c^{eq} is caused by the so-called dynamical factor $d = \tau_r/(\tau_r + \tau_s)$, describing the competition between spin-relaxation and recombination:

$$P_c^{\text{int}}(B) = \frac{\tau_r}{\tau_r + \tau_s} P_c^{\text{eq}}(B) = d(B) \tanh \frac{\Delta E_Z(B)}{2k_B T} \tag{15.3}$$

In the case of the negative trion, the anisotropic splitting $\Delta E_Z = \Delta E_{3/2}$ is given by Eq. (15.1) with the factor $\cos\theta$. The trion recombination time $\tau_r = 9$ ns is independent of magnetic field, while its spin relaxation time τ_s drastically decreases with increasing field (Figure 15.4b). As a result, the nearly order-of-magnitude difference between P_c^{eq} and P_c^{int} at small magnetic fields becomes negligible at $B = 15$ T (Figure 15.4a).

As the spin relaxation of the negative trion is determined by the hole, this allows us to evaluate the longitudinal spin relaxation of the hole, $\tau_s = 60$ ns, at $B = 0$ T. Its magnetic field dependence is caused by the light-to-heavy hole admixture through the perpendicular to *c*-axis component of the applied magnetic field. Therefore, in the analysis of the trion DCP from the ensemble of randomly oriented NCs, one has to take into account that not only the Zeeman splitting $\Delta E_{3/2}(B, \theta)$, but also the dynamical factor $d(B, \theta)$ are functions of the angle θ [35].

For the DCP from the neutral excitons, one has to consider the anisotropic Zeeman splitting of the dark excitons $\Delta E_Z = \Delta E_2(B, \theta)$ in Eq. (15.3) (Figure 15.1b,c), the dependence of the dark exciton recombination time $\tau_r(B, \theta)$ (and thus the dynamical factor $d(B, \theta)$) on the magnetic field and NC orientation, and the mechanisms of the dark exciton radiative recombination [17, 22]. Importantly, the strong exchange interaction between the electron and the hole in a neutral exciton not only complicates its fine structure, but is responsible for the extremely fast spin relaxation in colloidal NCs, also in external magnetic fields [35, 36]. As a result, the spin relaxation time τ_s between the Zeeman split sublevels of the neutral exciton is of the order of 1 ns, even in weak magnetic fields, so that d is close to unity and $P_c^{\text{int}} \approx P_c^{\text{eq}}$. This additionally allows one to distinguish the emission from charged and neutral excitons [35]. Examples of the exciton and trion spin properties studied by means of polarized PL can be found for chemically synthesized [37–40] and glass-embedded [22] CdSe QDs, CdSe/CdS QDs [35, 41, 42], CdTe QDs [43], CdSe/CdS dot-in-rods [21], CdSe [17] and CdSe/CdS NPLs [31], and $CsPbBr_3$ perovskite QDs [44].

Spin-polarized excitons can be generated using polarized resonant excitation even at zero magnetic field [45]. Optical orientation (for circularly polarized excitation) and optical alignment (for linearly polarized excitation) of excitons were reported for $CsPbI_3$ QDs [46] and CdSe/CdS NPLs [47].

15.4 Time-Resolved Faraday Rotation

Time-resolved pump-probe Faraday rotation (FR) is a powerful technique to study carrier spin coherence and measure various spin parameters, such as g-factors, spin relaxation times, hyperfine interaction constants with nuclei, etc. [48, 49].

It was successfully used for investigating the coherent spin dynamics in colloidal NCs, both at cryogenic and room temperatures. Most of these studies were performed on chemically synthesized CdSe QDs [19, 20, 23, 26, 50–56], but also CdS QDs [24, 56–61], CdSe/ZnS QDs [19] and CdSe, CdSe/CdS as well as CdSe/ZnS NPLs [31, 62] were investigated. Coherent spin transfer between coupled CdSe QDs was demonstrated [63, 64]. For NCs in glass the reported data are limited to $CdS_{0.5}Se_{0.5}$ QDs [65] and CdSe QDs [66].

Time-resolved FR uses pulsed lasers with pulse durations of 100 fs to few ps. The laser beam is split into the pump and probe beams. The circularly polarized pump generates spin-oriented carriers. The spin dynamics of these carriers is monitored via the rotation of the polarization plane of the linearly polarized probe (FR) by varying the time delay between pump and probe pulses. In some cases, it is more convenient to measure the ellipticity instead of the FR [67]. At zero magnetic field and in Faraday geometry, when the magnetic field is parallel to the light propagation direction, the longitudinal spin relaxation time T_1 can be measured. In Voigt geometry, when the magnetic field is perpendicular to the orientation of the photogenerated carrier spins, the spins precess about the magnetic field. In this case, the FR signal oscillates with the Larmor frequency, $\omega_L = g\mu_B B/\hbar$. It decays with the ensemble spin dephasing time T_2^* due to various mechanisms: g-factor dispersion, spin relaxation, electron–nuclear hyperfine interaction, and electron–hole recombination. This allows one to evaluate the g-factor and identify the type of carriers (electrons or holes) contributing to the signal.

The time-resolved ellipticity dynamics of CdSe QDs measured at room temperature is shown in Figure 15.5a [19]. As-grown QDs in toluene solution demonstrate a single Larmor frequency in ellipticity corresponding to $g_2 = 1.56$ (blue line), see also the inset showing fast Fourier transform (FFT) spectra. When the hole acceptor

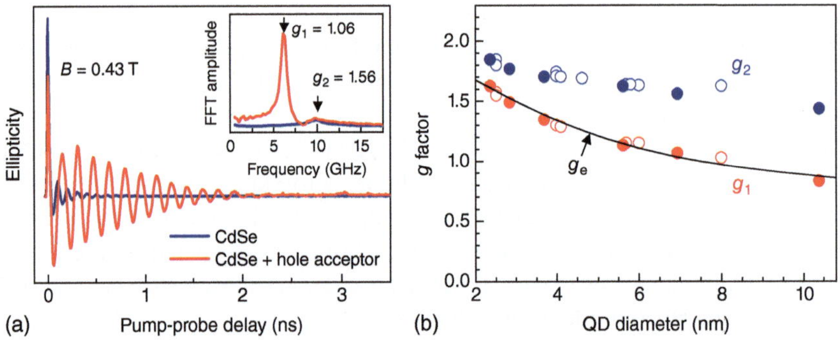

Figure 15.5 Coherent spin dynamics of CdSe QDs measured in solution at room temperature. (a) Time-resolved ellipticity of as-grown and n-type photodoped CdSe QDs (diameter 6.9 nm) using the hole acceptor Li[Et_3BH]. Source: [19]/American Chemical Society. Inset shows FFT spectrum for frequency $\omega_L/2\pi$. $B = 0.43$ T. (b) Electron g-factor as a function of QD diameter. Experimental data from [19] are shown by filled circles. Open circles show experimental results from Refs. [26, 50, 54]. Solid line shows calculated electron g-factor, g_e, for the lowest quantum confined state in CdSe QDs. Source: Adapted from Refs. [26, 27, 50, 54].

$Li[Et_3BH]$ is added in solution, the spin signal amplitude (red line) increases strongly and signal corresponding to a second Larmor frequency with $g_1 = 1.06$ appears. The hole acceptor captures photogenerated holes, which leads to QDs charged with electrons.

The dependences of the g-factors g_1 and g_2 on the diameter of wz-CdSe QDs reported in Ref. [19] are shown in Figure 15.5b by filled circles. They are in good agreement with the data from previous pump-probe studies [26, 50, 54] shown by open circles. Recently, we found the g_2-related signal in CdSe QDs in a glass matrix [66]. The g_1-value can be confidently assigned to the resident electron g-factor g_e [26, 27], while the origin of the g_2-related signal had remained unclear. In Ref. [26], it was suggested that this signal corresponds to the precession of the lowest level excitons in quasi-spherical QDs with the frequency corresponding to the splitting between +1 and −1 states, shown for $\Delta = 0$ in Figure 15.1d. However, an accurate analysis shows that the exciton precession should occur at half of this frequency, and with realistic values for the hole g-factor, g_h^{sph}, one cannot describe the g_2 size dependence within the exciton model [27, 66].

In Ref. [19], we show that a specific spin component can be enhanced by the choice of the acceptor type for core-only CdSe QDs, while in core/shell CdSe/ZnS QDs, the spin signals are significantly weaker. This evidences that surface states play an important role in the appearance of the spin signals, and both Larmor frequencies are related to the coherent spin precession of electrons. Namely, the lower frequency signal $g_1 = g_e$ can be indeed assigned to the electron confined inside the QDs, while the higher frequency signal g_2 is associated with the electron localized in the vicinity of the QDs surface.

It was found that two types of hole acceptors, $Li[Et_3BH]$ and 1-octanethiol, result in a distinctly different electron spin dynamics in CdSe QDs [55]. The differences include the electron g-factors, spin dephasing, and spin relaxation times as well as mechanisms. This reveals that the condition of the surface and the surrounding can strongly affect the electron spin dynamics in NCs.

15.5 Dynamics of Photocharging Visualized via Electron Spin Coherence

Photocharging of colloidal NCs results from spatial separation of photogenerated electron–hole pairs, after which the core keeps an electron or a hole (negative or positive photocharging, respectively), while the carrier with opposite charge is trapped at the NC surface or ejected from the NC into the surrounding matrix [68]. Because of the large surface-to-volume ratio in colloidal NCs, photocharging is a common phenomenon with significant consequences for the optical and spin properties.

Time-resolved FR is a powerful tool to study the photocharging dynamics in colloidal NCs, especially that it works for NCs in solution and at room temperature [24, 58]. The g-factor evaluated from the Larmor frequency allows one to identify the type of the carrier and also the carrier location, e.g. whether an electron is in the

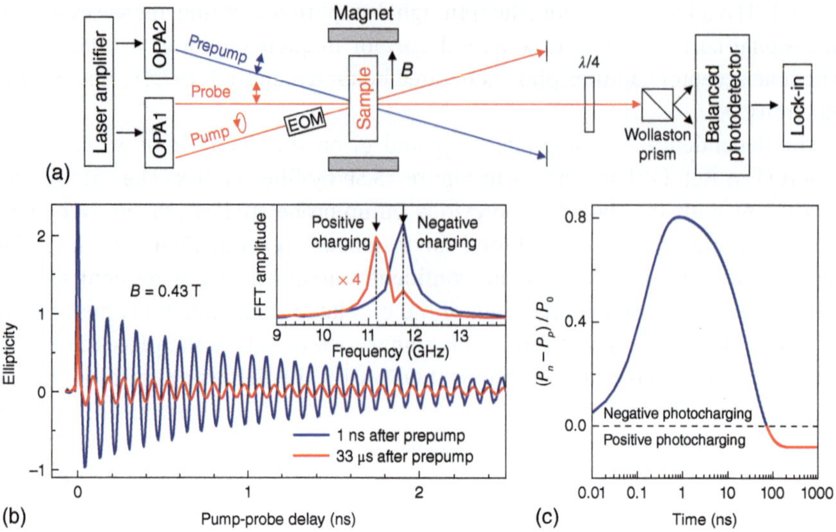

Figure 15.6 Photocharging dynamics in CdS QDs (5.5 nm diameter) measured by time-resolved ellipticity at room temperature. Source: Adapted from [24]/American Chemical Society. (a) Scheme of prepump (linearly polarized) – pump (circularly) – probe (linearly) experiment. (b) Ellipticity dynamics showing electron Larmor precession measured at 1 ns and 33 μs delays after prepump. Inset shows corresponding fast Fourier transform spectra. For short delays, the spectra show one mode at 11.77 GHz ($g_1 = 1.955$), while at long delays a second mode with 11.17 GHz ($g_2 = 1.856$) dominates. (c) Simulation of the dynamical evolution from negative to positive photocharging in CdS QDs on basis of the experimental data. Parameters: hole trapping time 0.2 ns, hole detrapping time 1 ns, electron trapping time 60 ns. The photocharging dynamics covers the temporal range from 100 ps to 1 μs. P_n and P_p are the probabilities for a QD to be charged by an electron or a hole, respectively. P_0 is the probability of QD photoexcitation.

center of the NC or in the vicinity of its surface. For that, the common two-pulse technique is modified for a prepump-pump-probe protocol (Figure 15.6a) by adding a third linearly polarized prepump, which photogenerates unpolarized carriers. By that the photocharging dynamics can be studied across a large temporal range from 1 ps up to 1 ms using a laser with a low repetition rate of 1–35 kHz [24].

Typical results for CdS QDs are shown in Figure 15.6b. In a magnetic field of 0.43 T and at room temperature, the electron spin dephasing time T_2^* extracted from the signal decay exceeds 1 ns. The Larmor precession frequency changes with delay time between the prepump and pump pulses, which is well seen in the FFT spectra (inset). It corresponds to $g_1 = 1.955$ at a short delay of 1 ns, which is associated with QDs each charged with an electron, i.e. the photogenerated hole is captured by the surface. At a long delay of 33 μs, the FFT spectrum has two components, and the dominating one with $g_2 = 1.856$ corresponds to precession of the electron in the positively charged exciton (trion), which is formed in QDs each charged with a hole. That means that the initially established negative photocharging changes for the positive charging at a timescale from 100 ns to 10 μs, which is a rather surprising behavior. The photocharging dynamics evaluated from experiment with a rate-equation model is shown in Figure 15.6c. It allows us to suggest the following scenario [24]. Upon laser irradiation, an electron–hole pair is generated in the QD core. Then, the QD became negatively photocharged due to fast surface trapping of the hole within 7–500 ps. Meanwhile, due to thermally activated hole detrapping, an equilibrium is established between core and surface hole states. The core electron depopulates owing to electron–hole recombination and electron trapping at the dot surface. During 100 ns, hole detrapping from the surface and electron trapping at the surface convert the QD to positive photocharging, which is maintained for hundreds of microseconds. These findings help to understand the photophysical processes in colloidal QDs, especially PL blinking. It is important to note that even at room temperature, the photocharging can be upheld for hours and even up to a month [20], which makes it a valuable approach for providing the necessary hardware for spin manipulation in singly charged NCs.

15.6 Spin–Flip Raman Scattering

Spin–flip Raman scattering (SFRS) is a coherent optical process involving spin excitation in an external magnetic field [69–71]. The SFRS signal intensity is strongly enhanced when the exciting laser photon energy is tuned to the exciton resonance, which serves as an intermediate state for the light scattering. The scattered light is shifted from the laser energy by the Zeeman energy $\Delta E_Z = g\mu_B B$, which is the splitting between the initial and final spin states of the exciton or resident charge carriers participating in the process (Figure 15.7a). The Raman shift can be used for evaluation of the g-factor and its anisotropy. The polarization properties of the scattered light measured in linear and circular polarizations deliver information about the origin of the involved electronic states, their symmetry and the underlying spin–flip mechanisms. Neutral excitons, negative trions, and positive trions can be identified by SFRS [71–74]. The SFRS signals are detected in close vicinity of the laser line with shifts as small as 0.1 meV. For that, high-end double or triple spectrometers are needed to provide large dispersion and high suppression of the laser light.

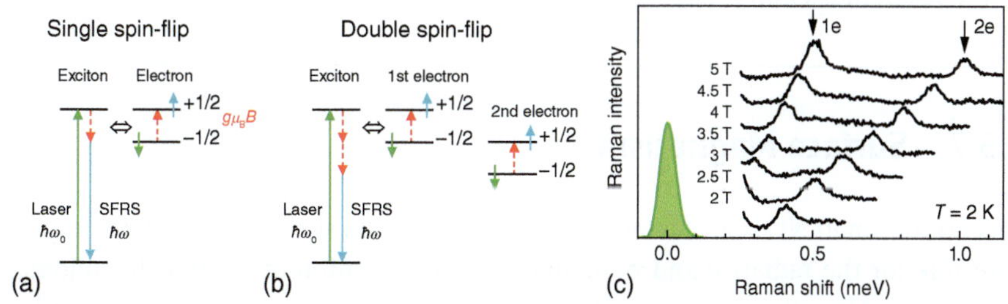

Figure 15.7 Spin-flip Raman scattering in CdSe NPLs [73]. (a,b) Mechanisms of single electron spin–flip resulting in the 1e line and double electron spin–flip for the 2e line. $\hbar\omega_0$ and $\hbar\omega$ are the energies of the incident and scattered photons, respectively. (c) SFRS spectra of four monolayer thick CdSe NPLs measured in parallel linear polarization (HH) in the Voigt geometry for various magnetic field strengths. Green line at zero Raman shift indicates the laser photon energy at 2.541 eV, resonant with the exciton.

In nanostructures, the SFRS intensity is drastically increased by the presence of resident carriers interacting with photogenerated excitons. As discussed above, resident carriers in colloidal NCs can be created by photocharging processes. In two-dimensional NPLs, the emission from neutral excitons and negative trions can be spectrally separated [32, 73]. A set of SFRS spectra measured on CdSe NPLs [73] with the excitation slightly above the exciton emission line is shown in Figure 15.7c. The two lines corresponding to a single electron spin–flip (1e) and a double electron spin–flip (2e) shift linearly away from the laser line with increasing magnetic field. Both signals become enhanced when additional illumination of the NPLs providing photocharging is applied. The corresponding electron *g*-factor is $g_e = 1.74$. Note that the double spin–flip is unusual and a rare phenomenon for diamagnetic semiconductors [75, 76]. The theoretical analysis of the polarization selection rules evidences that this process is provided by two resident electrons interacting with the same exciton (Figure 15.7b). Moreover, the developed theory for single and double electron spin flips in NPLs [74] demonstrates that the observation of the double spin flip signal is possible only when the exchange interaction between the electrons is much smaller than the electron–hole exchange interaction in the exciton. In the opposite case, when a negative trion state is created as intermediate state, the double electron spin flip is less probable. For example, only the single electron SFRS signal due to the presence of resident electrons is observed in thick-shell CdSe/CdS NPLs with efficient photocharging and dominant trion emission [31]. Here, the intensity of the spin-flip line in the Faraday and Voigt geometries provides information about the preferential orientation of the NPLs in the ensemble. We found it to be random in dense ensembles, while the NPLs are lying flat on the substrate in diluted ensembles.

A SFRS process for the spin flip of a neutral photogenerated exciton should be accompanied by emission or absorption of acoustic phonons with energy equal to ΔE_Z. It can be distinguished from the SFRS process on a resident electron by the selection rules for circular polarized light in the Faraday geometry. In addition, the spin flip of a neutral exciton can be observed via the resonance excitation of the biexciton state, also with the selection rules different from those for the spin flip of the resident exciton [77].

15.7 Surface Magnetism

Due to the small NC size of a few nanometers only, the surface plays a very important role for the radiative and nonradiative recombination of excitons in colloidal structures. Also, the spin properties and spin dynamics can be strongly influenced by surface spins, which are unavoidably present in nominally nonmagnetic (diamagnetic) NCs [16, 17]. The origin of the surface spins is associated with the spins of dangling bonds. The chemical bonds of ions at the NC surface are passivated by organic ligands. But the passivation is not perfect and the surface has a considerable concentration of dangling bonds, which can carry charge and spin.

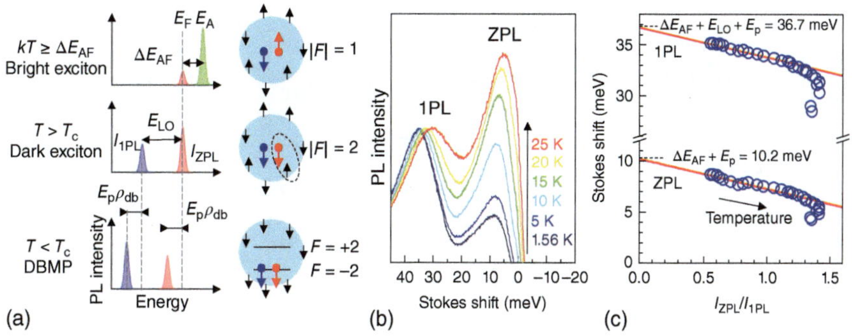

Figure 15.8 Magnetic polaron on surface spins in CdSe QDs [16]. (a) Schematics of the DBMP formation. Arrows are electron (red), hole (blue) and surface (black) spins. The bright exciton (E_A) is photogenerated, after a spin–flip of the electron it relaxes into a dark exciton (E_F), which recombination requires a flip-flop of the electron spin with a surface spin. After many acts of dark exciton recombination all surface spins are polarized and the DBMP is formed. (b) PL of zero-phonon line (ZPL) and one LO-phonon-assisted line (1PL) for selective excitation at various temperatures. (c) Parametric dependences of PL line energy vs. PL intensity ratio tuned by a temperature increase. Red lines are model calculations. Source: Adapted from [16].

Inspired by theoretical predictions on the mechanism providing radiative recombination of spin-forbidden dark excitons via interaction with surface spins and on the formation of a dangling bond magnetic polaron (DBMP) [15], we found both effects experimentally in CdSe QDs [16]. The experiment is schematically presented in Figure 15.8a. The laser excites resonantly the bright exciton state (E_A shown by the green line), which emits in a resonant decay at the same energy so that it cannot be detected. The bright exciton relaxes to the dark exciton (E_F), which either recombines without involvement of phonons (zero-phonon line [ZPL], red line) or assisted by emission of one LO phonon (1PL, blue line). The ZPL emission requires an electron spin flip, which is provided by a flip-flop process of the electron in the dark exciton with a surface spin (middle scheme). The relative intensity of the ZPL and 1PL lines is proportional to the number of surface spins with suitable orientation. After many recombination events of dark excitons in a QD, the surface spins become dynamically polarized (bottom scheme). As a result, they cannot assist anymore the dark exciton recombination, but form all together a DBMP.

The DBMP binding energy E_p can be measured from the additional spectral shift $E_p\rho_{db}$, where ρ_{db} is the polarization of the surface spins. With increasing temperature from 1.56 to 25 K, the polaron is destroyed, due to which the ZPL line exposes a high energy shift and increases its intensity (Figure 15.8b). The parametric dependences of the ZPL and 1PL emission shown in Figure 15.8c allow us to evaluate $E_p = 7$ meV (the total exciton-surface spins exchange energy) and a number of 60 surface spins in a QD [16, 78]. Note that the DBMP has similarity with the exciton magnetic polaron (EMP) in diluted magnetic semiconductors (DMSs) (Section 15.8), while they differ in their formation mechanisms. In the EMP the Mn^{2+} spins are polarized in the exchange field of the exciton, which is mainly contributed by the hole exchange. In the case of the DBMP, this mechanism is too

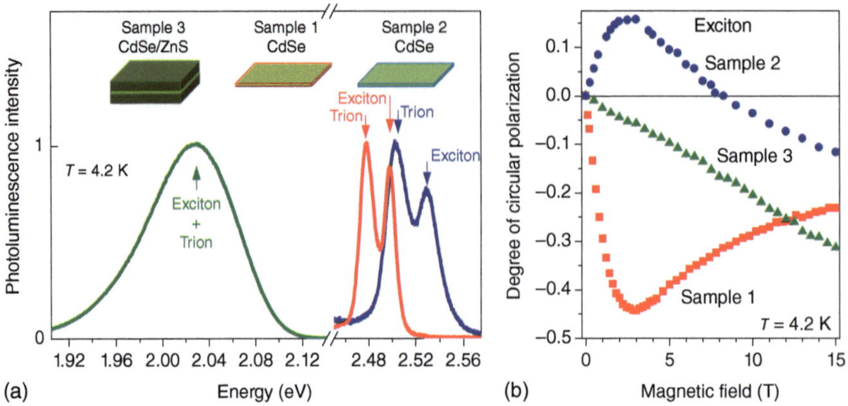

Figure 15.9 Surface spins in CdSe NPLs. Source: [17]/Springer Nature. (a) PL-spectra of CdSe NPLs (4 monolayers thickness) synthesized in argon (red) and in air (blue) and of core/shell CdSe/ZnS NPLs (green). (b) Magnetic field dependence of DCP in the NPLs from panel (a) with $E_p > 0$ in sample 1 (red) and $E_p < 0$ in sample 2 (blue). In sample 3 (green) $E_p = 0$.

weak to provide sufficient polarization, which is gained under illumination via dynamic polarization of surface spins [16, 78].

In CdSe NPLs, the surface spins are monitored by DCP measured in magnetic fields, namely by the strongly nonmonotonic dependence $P_c(B)$ [17]. The emission spectrum of CdSe NPLs at $T = 4.2$ K comprises the two narrow lines of excitons and negatively charged trions [32] (Figure 15.9a). Adding a ZnS shell shifts and broadens the emission lines, making excitons and trions spectrally unresolvable. In CdSe/ZnS NPLs, where excitons are well isolated from the surface by the shell, the $P_c(B)$ dependence increases linearly reaching −32% in a magnetic field of 15 T (green symbols in Figure 15.9b), which is the expected behavior for thermal spin polarization (see Section 15.3). But in CdSe NPLs grown in argon (red symbols), $P_c(B)$ increases much faster reaching −45% already at 3 T and then decreases its value down to −25% at 15 T. The behavior reminds to DMSs, while here the surface spins play the role of the magnetic ions. We did not find such unusual behavior for the negative trion, which polarization is controlled by the hole, as the two electrons are in a singlet state. From that we conclude that in the exciton, the electron interaction with the surface spins provides the main mechanism. It is interesting that the polarization sign depends on the synthesis conditions, as in weak fields P_c is positive for NPLs grown in air (blue symbols). The theoretical analysis shows that the exciton exchange interaction with the surface spins polarized by the external magnetic field modifies the intrinsic Zeeman splitting ΔE_2 of the dark exciton (Eq. (15.1)) by the value $2E_p\rho_{\text{db}}$, thus affecting the DCP depending on the sign of the exchange energy E_p (Figure 15.9). In addition, the DCP in CdSe NPLs is contributed by the spin-dependent recombination of the dark exciton, which is also caused by its interaction with polarized surface spins [17].

15.8 Diluted Magnetic Semiconductor Colloidal NCs

To extend the spin-dependent phenomena, one can use the approach of fabricating DMSs and implement magnetic Mn^{2+} ions with spin 5/2 in colloidal NCs. Quantum confinement greatly enhances the interaction of carriers (electrons and holes) with the magnetic ions, which results in a strong modification of the spin dynamics and giant Zeeman splitting. A strong decrease and even a sign reversal of the exchange integral of the electron interaction with localized Mn^{2+} spins in QDs were predicted in Ref. [79]. The importance of Mn^{2+} spin fluctuations increases as their effective field acting on the electrons in $Cd_{1-x}Mn_xSe$/CdS QDs reaches 15–30 T [80]. Also carrier and exciton magnetic polarons, both in the fluctuation and collective regimes, gain energy in confined systems [79, 81].

We give here a few examples of magneto-optical effects in DMS colloidal NCs. In CdSe/$Cd_{1-x}Mn_xS$ NPLs, the magnetic ions are located in the barriers, therefore, electrons and holes, being mostly confined in the CdSe core, interact with them through the tails of their wave functions [82]. The Mn^{2+} concentration of 0.009 (i.e. 0.9%) is very low. Despite of that, the exciton Zeeman splitting is controlled by the hole interaction with the magnetic ions, as we conclude from the sign of the DCP and its characteristic dependence on magnetic field with a saturation (red symbols in Figure 15.10a). Note that the DCP is negative and monotonously increasing in nonmagnetic CdSe/CdS NPLs (green symbols). The exchange interaction between holes and magnetic ions contributes to the Zeeman splitting and changes the sign of the DCP in DMS NPLs.

The *g*-factor of heavy holes in NPLs is strongly anisotropic (Figure 15.1) with the in-plane component close to zero. This allows us to induce multiple Mn^{2+} spin–flips

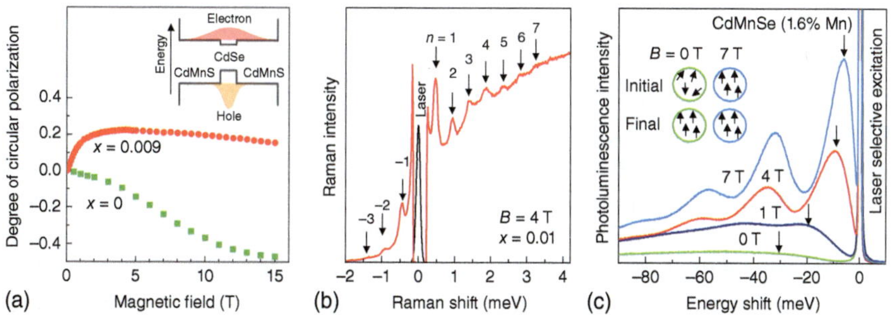

Figure 15.10 Diluted magnetic semiconductor colloidal NCs. (a) Magnetic field dependence of DCP in CdSe/$Cd_{1-x}Mn_xS$ NPLs with $x = 0$ and 0.009 [82]. $T = 4.2$ K. Inset shows scheme of electron (red) and hole (yellow) wave functions confined in NPL. (b) Mn^{2+} spin-flip Raman scattering spectra in CdSe/$Cd_{1-x}Mn_xS$ NPLs with $x = 0.01$ measured at $T = 1.6$ K [82]. Up to 7 (3) resonances are observed in the Stokes (anti-Stokes) spectral ranges. (c) Exciton magnetic polaron in $Cd_{1-x}Mn_xSe$ QDs ($x = 0.016$) [83]. PL-spectra measured under resonant excitation of excitons at 2.11 eV in various magnetic fields at $T = 1.8$ K. Arrows mark magnetic polaron maximum shifted from the laser line by EMP energy. Inset shows schematically orientation of Mn^{2+} spins within the exciton volume at the exciton photogeneration moment (initial) and after formation of magnetic polaron (final). Source: (a, b) Adapted from [82]. (c) Adapted from [85].

in a magnetic field applied in the Voigt geometry and detect them via Raman scattering (Figure 15.10b), similar to previous reports for $Cd_{1-x}Mn_xTe$-based quantum wells [79]. Also, resonant heating of the Mn^{2+} spin system in the condition of magnetic resonance [84], when the applied microwave radiation of 60 GHz matches the Mn^{2+} Zeeman splitting, was demonstrated in CdSe/$Cd_{1-x}Mn_xS$ NPLs [85]. The dynamics of the induced changes in DCP and luminescence intensity by the microwave radiation allow us to measure the spin–lattice relaxation time of the Mn^{2+} spin system and suggest a method for evaluation of the Mn concentration in colloidal NCs.

An EMP is formed in DMS due to polarization of Mn^{2+} spins within the exciton volume through the hole exchange field [81, 83]. For localized excitons in bulk $Cd_{1-x}Mn_xTe$ or confined in $Cd_{1-x}Mn_xTe$-based quantum wells, the EMP is formed for large Mn concentrations exceeding 0.1 (i.e. 10%). However, in $Cd_{1-x}Mn_xSe$ colloidal QDs even at very small $x = 0.004 - 0.016$ large polaron binding energies up to 26 meV are observed [83]. They result from a strong effective exchange field of about 10 T of the confined exciton on the Mn^{2+} spins. Resonant excitation of the exciton PL provides direct access to the EMP binding energy via its Stokes spectral shift from the laser photon energy [81]. The EMP part of this shift is suppressed in external magnetic fields polarizing the Mn^{2+} spins (Figure 15.10c).

The coherent spin dynamics of Mn^{2+} spins triggered by the exchange field of spin-polarized holes were studied in $Cd_{1-x}Mn_xSe$ QDs by means of pump-probe FR [80, 86]. A giant Zeeman splitting of electrons resulting in their Larmor precession at frequencies exceeding 2 THz in magnetic field of 7 T has been reported.

15.9 New Materials: Perovskite QDs

Lead halide perovskite semiconductors $APbX_3$ (A = Cs, MA, or FA and X = I, Br, or Cl) is a class of materials that attract recently great attention due to their high potential for photovoltaics, optoelectronics, sensorics, etc. Bulk perovskites show strong response to optical and magneto-optical techniques used in spin physics (optical orientation, spin polarization of carriers and excitons in magnetic field, time-resolved Faraday/Kerr rotation, spin–flip Raman scattering, optically detected nuclear magnetic resonance, etc.) [44, 87–89]. The coherent and incoherent spin dynamics of carriers are sufficiently long-lived and comparable with III–V and II–VI semiconductors. Also an efficient interaction of carriers with the nuclear spin system is found, where the hole interaction is considerably greater than the electron one. Perovskites offer a new testbed due to "inverted" band structure, which is highly attractive for basic research, as here the holes are dominating in many spin-dependent phenomena, namely due to the stronger interaction with the nuclear spins [88].

Spin phenomena have been also studied in perovskite QDs. Figure 15.11a shows the coherent spin dynamics contributed by negatively- and positively charged solution-grown $CsPbBr_3$ QDs, evidencing two Larmor frequencies [90]. The respective g-factors for the electrons and holes are $g_e = 1.73$ and $g_h = 0.83$ and the spin dephasing times reach 5 ns.

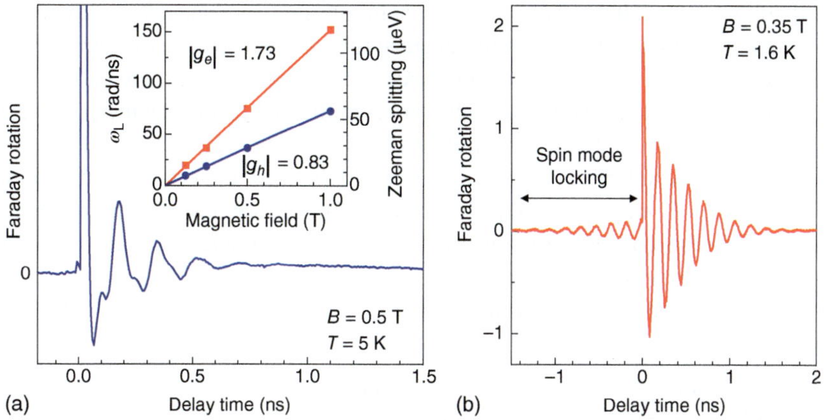

Figure 15.11 Carrier spin coherence in perovskite QDs. (a) Time-resolved Faraday rotation dynamics in solution-grown $CsPbBr_3$ QDs. Source: Adapted from [90]. Inset shows the magnetic field dependences of the Larmor precession frequencies of electrons (red) and holes (blue) and corresponding Zeeman splittings. (b) Spin mode locking in CsPb $(Cl_{0.56}Br_{0.44})_3$ QDs in glass. Source: Adapted from [91]. Laser photon energy is 2.737 eV.

For $CsPb(Cl_{0.56}Br_{0.44})_3$ QDs in glass we have found the spin mode locking effect for holes with $g_h = 1.20$ (Figure 15.11b) [91]. It occurs for periodic laser excitation (here with a repetition rate of 75.6 MHz and a repetition period $T_R = 13.2$ ns) when the spin coherence time of the resident hole, $T_{2,h} > T_R$, which allows for accumulation of spin coherence in spin synchronized Larmor precession modes [49, 92, 93]. Experimentally, it shows up as a spin coherent signal at negative time delays with increasing amplitude by approaching the pump pulse at zero delay time. This unusual spin phenomenon has been found previously only in singly charged (In,Ga)As/GaAs QDs grown by molecular-beam epitaxy [49, 92, 93]. It may be enhanced by the nuclear focusing effect to tune all QDs in an inhomogeneous ensemble to the same Larmor precession frequency and extend the ensemble spin dephasing time T_2^* (on the order of a nanosecond) up to the spin coherence time T_2 of a single QD of few microseconds [94, 95]. We evaluate for the studied perovskite QDs $T_{2,h}$ exceeding 13 ns.

15.10 Conclusions

The experimental methods of spin physics developed for semiconductor bulk materials and epitaxial nanostructures have been successfully used for colloidal NCs. New spin-dependent effects are found, among them the DBMP formation as most surprising one, and detailed information on the spin parameters of colloidal NCs and their properties has been obtained. Colloidal NCs differ in many respects from the epitaxially grown quantum wells and QDs. Therefore, the "common knowledge" about the spin properties of epitaxial structures can be transferred only with caution onto colloidal NCs.

Still one has to admit that coherent spin physics as required for quantum information technologies is at an initial stage for colloidal structures, compared to the status achieved for epitaxial systems. As indicated above, for them carrier spin coherence times of tens of nanoseconds were demonstrated, during which spins can be initialized and manipulated multiple times by ultrashort laser pulses. However, progress in colloidal technology, involving developing new materials and designing new structures, is exciting, so that it demands further studies of spin-dependent phenomena. The current situation as described here provides a solid basis for pursuing such activities.

A possible outlook for nearest future studies of the spin physics in colloidal NCs could be:

- Demonstration of implementation of a single Mn^{2+} ion in a colloidal CdSe QD [96], opening a way for optical addressing and manipulating localized spins.
- Interaction of localized spins of magnetic ions with surface spins and their cumulative effect on carriers and excitons.
- Spin functionalization of NC surfaces, e.g. for nanoscale magnetic sensors.
- Establishing further perovskites representing a new and promising platform for colloidal NCs, offering a great, tailorable variety of chemical compositions.
- Composing hybrid metal–semiconductor NCs, e.g. for plasmonic enhancement, as has been demonstrated for Au-CdSe QDs [52].
- Fabricating ordered ensembles of NCs, e.g. NPLs or nanorods with the same orientation.
- Testing and developing protocols for coherent spin manipulations at room temperature.

Acknowledgments

The results presented in this chapter were gained in collaboration of technologists, experimentalists, and theorists. The authors are thankful to all colleagues contributing to this research and acknowledge among them Al. L. Efros, E. L. Ivchenko, M. A. Semina, N. A. Gippius, L. Biadala, F. Liu, D. H. Feng, V. F. Sapega, I. V. Kalitukha, D. Kudlacik, A. Greilich, E. A. Zhukov, E. Kirstein, G. Qiang, D. O. Tolmachev, V. Yu. Ivanov, V. V. Belykh, Yu. G. Kusrayev, S. A. Crooker, P. C. M. Christianen, M. Nasilowski, B. Dubertret, I. Moreels, T. Aubert, Z. Hens, S. Delikani, H. V. Demir, E. Lhuillier, and E. V. Kolobkova. This work has been supported by the Deutsche Forschungsgemeinschaft and Russian Foundation for Basic Research (Project 19-52-12064 NNIO-a) in the frame of the ICRC TRR 160 (projects A1, B1, B2).

References

1 Michler, P. (ed.) (2017). *Quantum Dots for Quantum Information Technologies*. Springer International Publishing AG.

2 Awschalom, D.D., Loss, D., and Samarth, N. (eds) (2002). *Semiconductor Spintronics and Quantum Computation*. Berlin: Springer-Verlag.

3 de Arquer, F.P.G., Talapin, D.V., Klimov, V.I. et al. (2021). Semiconductor quantum dots: technological progress and future challenges. *Science* 373: eaaz8541.

4 Talapin, D.V., Lee, J.-S., Kovalenko, M.V., and Shevchenko, E.V. (2010). Prospects of colloidal nanocrystals for electronic and optoelectronic applications. *Chemical Review* 110: 389–458.

5 Pietryga, J.M., Park, Y.-S., Lim, J. et al. (2016). Spectroscopic and device aspects of nanocrystal quantum dots. *Chemical Review* 116: 10513–10622.

6 Efros, Al.L. and Brus, L.E. (2021). Nanocrystal quantum dots: from discovery to modern development. *ACS Nano* 15: 6192–6210.

7 Micic, O.I., Curtis, C.J., Jones, K.M. et al. (1994). Synthesis and characterization of InP quantum dots. *Journal of Physical Chemistry* 98: 4966–4969.

8 Kovalenko, M.V., Protesescu, L., and Bondarchuk, M.I. (2017). Properties and potential optoelectronic applications of lead halide perovskite nanocrystals. *Science* 358: 745–750.

9 Michler, P., Imamoglu, A., Mason, M.D. et al. (2000). Quantum correlation among photons from a single quantum dot at room temperature. *Nature* 406: 968.

10 Shornikova, E.V., Yakovlev, D.R., Gippius, N.A. et al. (2021). Exciton binding energy in CdSe nanoplatelets measured by one- and two-photon absorption. *Nano Letters* 21: 10525–10531.

11 Efros, Al.L., Rosen, M., Kuno, M. et al. (1996). Band-edge exciton in quantum dots of semiconductors with a degenerate valence band: dark and bright exciton states. *Physical Review B* 54: 4843–4856.

12 Efros, Al.L. (2003). Fine structure and polarization properties of band-edge excitons in semiconductor nanocrystals. In: *Semiconductor and Metal Nanocrystals: Synthesis and Electronic and Optical Properties*, Chapter 3 (ed. V.I. Klimov), 97–132. New York: Marcel Dekker.

13 Rodina, A.V. and Efros, Al.L. (2016). Radiative recombination from dark excitons: activation mechanisms and polarization properties. *Physical Review B* 93: 155427.

14 Shornikova, E.V., Biadala, L., Yakovlev, D.R. et al. (2018). Addressing the exciton fine structure in colloidal nanocrystals: the case of CdSe nanoplatelets. *Nanoscale* 10: 646–656.

15 Rodina, A.V. and Efros, Al.L. (2015). Magnetic properties of nonmagnetic nanostructures: dangling bond magnetic polaron in CdSe nanocrystals. *Nano Letters* 15: 4214–4222.

16 Biadala, L., Shornikova, E.V., Rodina, A.V. et al. (2017). Magnetic polaron on dangling bond spins in CdSe colloidal nanocrystals. *Nature Nanotechnology* 12: 569.

17 Shornikova, E.V., Golovatenko, A.A., Yakovlev, D.R. et al. (2020). Surface spin magnetism controls the polarized exciton emission from CdSe nanoplatelets. *Nature Nanotechnology* 15: 277–282.

18 Norris, D.J., Efros, Al.L., and Erwin, S.C. (2008). Doped nanocrystals. *Science* 319: 1776–1779.

19 Hu, R.R., Yakovlev, D.R., Liang, P. et al. (2019). Origin of two Larmor frequencies in the coherent spin dynamics of colloidal CdSe quantum dots revealed by controlled charging. *Journal of Physical Chemistry Letters* 10: 3681–3687.
20 Hu, R.R., Wu, Z., Zhang, Y.Y. et al. (2019). Long-lived negative photocharging in colloidal CdSe quantum dots revealed by coherent electron spin precession. *Journal of Physical Chemistry Letters* 10: 4994–4999.
21 Siebers, B., Biadala, L., Yakovlev, D.R. et al. (2015). Exciton spin dynamics and photoluminescence polarization of CdSe/CdS dot-in-rod nanocrystals in high magnetic fields. *Physical Review B* 91: 155304.
22 Qiang, G., Golovatenko, A.A., Shornikova, E.V. et al. (2021). Polarized emission of CdSe nanocrystals in magnetic field: the role of phonon-assisted recombination of the dark exciton. *Nanoscale* 13: 790–800.
23 Gupta, J.A., Awschalom, D.D., Peng, X., and Alivisatos, A.P. (1999). Spin coherence in semiconductor quantum dots. *Physical Review B* 59: R10421–R10424.
24 Feng, D.H., Yakovlev, D.R., Pavlov, V.V. et al. (2017). Dynamic evolution from negative to positive photocharging in colloidal CdS quantum dots. *Nano Letters* 17: 2844–2851.
25 Bayer, M. (2019). Bridging two worlds: colloidal versus epitaxial quantum dots. *Annalen der Physik (Berlin)* 531: 1900039.
26 Gupta, J.A., Awschalom, D.D., Efros, Al.L., and Rodina, A.V. (2002). Spin dynamics in semiconductor nanocrystals. *Physical Review B* 66: 125307.
27 Semina, M.A., Golovatenko, A.A., and Rodina, A.V. (2021). Influence of the spin-orbit split-off valence band on the hole g factor in semiconductor nanocrystals. *Physical Review B* 104: 205423.
28 Tadjine, A., Niquet, Y.M., and Delerue, C. (2017). Universal behavior of electron g-factors in semiconductor nanostructures. *Physical Review B* 95: 235437.
29 Yakovlev, D.R. and Rodina, A.V. (2018). Spin dynamics of charged and neutral excitons in colloidal nanocrystals. *Journal of Electronic Materials* 47: 4260–4271.
30 Rodina, A.V., Golovatenko, A.A., Shornikova, E.V., and Yakovlev, D.R. (2018). Spin physics of excitons in colloidal nanocrystals. *Physics of the Solid State* 60: 1537–1553.
31 Shornikova, E.V., Biadala, L., Yakovlev, D.R. et al. (2018). Electron and hole g-factors and spin dynamics of negatively charged excitons in CdSe/CdS colloidal nanoplatelets with thick shells. *Nano Letters* 18: 373–380.
32 Shornikova, E.V., Yakovlev, D.R., Biadala, L. et al. (2020). Negatively charged excitons in CdSe nanoplatelets. *Nano Letters* 20: 1370–1377.
33 Littleton, B., Fernée, M., Rubinsztein-Dunlop, H. et al. (2009). High-resolution line width measurement of single CdSe nanocrystals at long time scales. *Journal of Physical Chemistry C* 113: 5345.
34 Nirmal, M., Murray, C.B., and Bawendi, M.G. (1994). Fluorescence line-narrowing in CdSe quantum dots: surface localization of the photogenerated exciton. *Physical Review B* 50: 2293.
35 Liu, F., Biadala, L., Rodina, A.V. et al. (2013). Spin dynamics of negatively charged excitons in CdSe/CdS colloidal nanocrystals. *Physical Review B* 88: 035302.

36 He, J., Zhong, H., and Scholes, G.D. (2010). Electron-hole overlap dictates the hole spin relaxation rate in nanocrystal heterostructures. *Physical Review Letters* 105: 046601.

37 Johnston-Halperin, E., Awschalom, D.D., Crooker, S.A. et al. (2001). Spin spectroscopy of dark excitons in CdSe quantum dots to 60 T. *Physical Review B* 63: 205309.

38 Furis, M., Hollingsworth, J.A., Klimov, V.I., and Crooker, S.A. (2005). Time- and polarization-resolved optical spectroscopy of colloidal CdSe nanocrystal quantum dots in high magnetic fields. *Journal of Physical Chemistry B* 109: 15332–15338.

39 Granados del Águila, A., Pettinari, G., Groeneveld, E. et al. (2017). Optical spectroscopy of dark and bright excitons in CdSe nanocrystals in high magnetic fields. *Journal of Physical Chemistry C* 121: 23693–23704.

40 Htoon, H., Crooker, S.A., Furis, M. et al. (2009). Anomalous circular polarization of photoluminescence spectra of individual CdSe nanocrystals in an applied magnetic field. *Physical Review Letters* 102: 017402.

41 Javaux, C., Mahler, B., Dubertret, B. et al. (2013). Thermal activation of non-radiative Auger recombination in charged colloidal nanocrystals. *Nature Nanotechnology* 8: 206–212.

42 Pinchetti, V., Shornikova, E.V., Qiang, G. et al. (2019). Dual-emitting dot-in-bulk CdSe/CdS nanocrystals with highly emissive core- and shell-based trions sharing the same resident electron. *Nano Letters* 19: 8846–8854.

43 Liu, F., Rodina, A.V., Yakovlev, D.R. et al. (2014). Exciton spin dynamics of colloidal CdTe nanocrystals in magnetic fields. *Physical Review B* 89: 115306.

44 Canneson, D., Shornikova, E.V., Yakovlev, D.R. et al. (2017). Negatively charged and dark excitons in $CsPbBr_3$ perovskite nanocrystals revealed by high magnetic fields. *Nano Letters* 17: 6177–6183.

45 Kusrayev, Yu.G. (2008). Optical orientation of excitons and carriers in quantum dots. *Semiconductor Science and Technology* 23: 114013.

46 Nestoklon, M.O., Goupalov, S.V., Dzhioev, R.I. et al. (2018). Optical orientation and alignment of excitons in ensembles of inorganic perovskite nanocrystals. *Physical Review B* 97: 235304.

47 Smirnova, O.O., Kalitukha, I.V., Rodina, A.V. et al. (2022). Optical alignment and orientation of excitons in ensemble of colloidal CdSe/CdS core/shell nanoplatelets. ArXive CondMat 2022. https://doi.org/10.48550/arXiv.2212.06134.

48 Awschalom, D.D. and Samarth, N. (2002). Optical manipulation, transport and storage of spin coherence in semiconductors. In: *Semiconductor Spintronics and Quantum Computation* (ed. D.D. Awschalom, D. Loss, and N. Samarth), 147–193. Berlin: Springer-Verlag.

49 Yakovlev, D.R. and Bayer, M. (2017). Coherent spin dynamics of carriers. In: *Spin Physics in Semiconductors*, Chapter 6 (ed. M.I. Dyakonov), 155–206. Cham: Springer International Publishing.

50 Stern, N.P., Poggio, M., Bartl, M.H. et al. (2005). Spin dynamics in electrochemically charged CdSe quantum dots. *Physical Review B* 72: 161303.

51 Li, Y.Q., Steuerman, D.W., Berezovsky, J. et al. (2006). Cavity enhanced Faraday rotation of semiconductor quantum dots. *Applied Physics Letters* 88: 193126.

52 Zhang, J., Tang, Y., Lee, K., and Ouyang, M. (2010). Tailoring light-matter-spin interactions in colloidal hetero-nanostructures. *Nature* 466: 91.

53 Fumani, A.K. and Berezovsky, J. (2013). Magnetic-field-dependent spin decoherence and dephasing in room-temperature CdSe nanocrystal quantum dots. *Physical Review B* 88: 155316.

54 Zhang, Z.B., Jin, Z.M., Ma, H. et al. (2014). Room-temperature spin coherence in zinc blende CdSe quantum dots studied by time-resolved Faraday ellipticity. *Physica E: Low-dimensional Systems and Nanostructures* 56: 85–89.

55 Wu, Z., Zhang, Y.Y., Hu, R.R. et al. (2021). Hole-acceptor-manipulated electron spin dynamics in CdSe colloidal quantum dots. *Journal of Physical Chemistry Letters* 12: 2126–2132.

56 Zhang, Y.Y., Jiang, M.Z., Wu, Z. et al. (2021). Hyperfine-induced electron-spin dephasing in negatively charged colloidal quantum dots: a survey of size dependence. *Journal of Physical Chemistry Letters* 12: 9481–9487.

57 Feng, D.H., Li, X., Jia, T.Q. et al. (2012). Long-lived, room-temperature electron spin coherence in colloidal CdS quantum dots. *Applied Physics Letters* 100: 122406.

58 Feng, D.H., Shan, L.F., Jia, T.Q. et al. (2013). Optical manipulation of electron spin coherence in colloidal CdS quantum dots. *Applied Physics Letters* 102: 062408.

59 Tong, H.F., Feng, D.H., Li, X. et al. (2013). Room-temperature electron spin generation by femtosecond laser pulses in colloidal CdS quantum dots. *Materials* 6: 4523–4531.

60 Li, X., Feng, D.H., Tong, H.F. et al. (2014). Hole surface trapping dynamics directly monitored by electron spin manipulation in CdS nanocrystals. *Journal of Physical Chemistry Letters* 5: 4310–4316.

61 Masumoto, Y., Umino, H., Sun, J.H., and Suzumura, E. (2015). Enhanced electron spin rotation in CdS quantum dots. *Physical Chemistry Chemical Physics* 17: 25278–25282.

62 Feng, D.H., Yakovlev, D.R., Dubertret, B., and Bayer, M. (2020). Charge separation dynamics in CdSe/CdS core/shell nanoplatelets addressed by coherent electron spin precession. *ACS Nano* 14: 7237–7244.

63 Ouyang, M. and Awschalom, D.D. (2003). Coherent spin transfer between molecularly bridged quantum dots. *Science* 301: 1074–1078.

64 Meier, F., Cerletti, V., Gywat, O. et al. (2004). Molecular spintronics: coherent spin transfer in coupled quantum dots. *Physical Review B* 69: 195315.

65 Gupta, J.A. and Awschalom, D.D. (2001). Spin precission and the optical Stark effect in a semiconductor-doped glasses. *Physical Review B* 63: 085303.

66 Qiang, G., Zhukov, E.A., Evers, E. et al. (2022). Electron spin coherence in CdSe nanocrystals in glass matrix. *ACS Nano* 16: 18838.

67 Glazov, M.M., Yugova, I.A., Spatzek, S. et al. (2010). Effect of pump-probe detuning on the Faraday rotation and ellipticity signals of mode-locked spins in (In,Ga)As/GaAs quantum dots. *Physical Review B* 82: 155325.

68 Efros, Al.L. and Nesbitt, D.J. (2016). Origin and control of blinking in quantum dots. *Nature Nanotechnology* 11: 661–671.

69 Sirenko, A.A., Ruf, T., Cardona, M. et al. (1997). Electron and hole g factors measured by spin-flip Raman scattering in CdTe/$Cd_{1-x}Mg_xTe$ single quantum wells. *Physical Review B* 56: 2114.

70 Sirenko, A.A., Belitsky, V.I., Ruf, T. et al. (1998). Spin-flip and acoustic-phonon Raman scattering in CdS nanocrystals. *Physical Review B* 58: 2077–2087.

71 Debus, J., Dunker, D., Sapega, V.F. et al. (2013). Spin-flip Raman scattering of the neutral and charged excitons confined in a CdTe/(Cd,Mg)Te quantum well. *Physical Review B* 87: 205316.

72 Debus, J., Sapega, V.F., Dunker, D. et al. (2014). Spin-flip Raman scattering of the resident electron in singly charged (In,Ga)As/GaAs quantum dot ensembles. *Physical Review B* 90: 235404.

73 Kudlacik, D., Sapega, V.F., Yakovlev, D.R. et al. (2020). Single and double electron spin-flip Raman scattering in CdSe colloidal nanoplatelets. *Nano Letters* 20: 517–525.

74 Rodina, A.V. and Ivchenko, E.L. (2020). Theory of single and double electron spin-flip Raman scattering in semiconductor nanoplatelets. *Physical Review B* 102: 235432.

75 Scott, J.F. and Damen, T.C. (1972). Anomalous double spin-flip Raman scattering in CdS, and a visible spin-flip laser. *Physical Review Letters* 29: 107–109.

76 Economou, E.N., Ruvalds, J., and Ngai, K.L. (1972). Theory of multiple spin-flip Raman scattering in semiconductors. *Physical Review Letters* 29: 110–113.

77 Rodina, A.V., Ivchenko E.L. (2022). Theory of resonant Raman scattering due to spin flips of resident charge carriers and excitons in perovskite semiconductors. *Physical Review B* **106**: 245202.

78 Rodina, A.V., Golovatenko, A.A., Shornikova, E.V. et al. (2018). Effect of dangling bond spins on the dark exciton recombination and spin polarization in CdSe colloidal nanostructures. *Journal of Electronic Materials* 47: 4338–4344.

79 Merkulov, I.A. and Rodina, A.V. (2010). Exchange interaction between carriers and magnetic ions in quantum size heterostructures. In: *Introduction to the Physics of Diluted Magnetic Semiconductors*, Chapter 3 (ed. J. Kossut and J.A. Gaj), 65–101. Heidelberg: Springer-Verlag.

80 Rice, W.D., Liu, W.Y., Baker, T.A. et al. (2016). Revealing giant internal magnetic fields due to spin fluctuations in magnetically doped colloidal nanocrystals. *Nature Nanotechnology* 11: 137–142.

81 Yakovlev, D.R. and Ossau, W. (2010). Magnetic polarons. In: *Introduction to the Physics of Diluted Magnetic Semiconductors*, Chapter 7 (ed. J. Kossut and J.A. Gaj), 221–262. Heidelberg: Springer-Verlag.

82 Shornikova, E.V., Yakovlev, D.R., Tolmachev, D.O. et al. (2020). Magneto-optics of excitons interacting with magnetic ions in CdSe/CdMnS colloidal nanoplatelets. *ACS Nano* 14: 9032–9041.

83 Rice, W.D., Liu, W., Pinchetti, V. et al. (2017). Direct measurements of magnetic polarons in CdMnSe nanocrystals from resonant photoluminescence. *Nano Letters* 17: 3068–3075.

84 Yakovlev, D.R. and Merkulov, I.A. (2010). Spin and energy transfer between carriers, magnetic ions, and lattice. In: *Introduction to the Physics of Diluted Magnetic Semiconductors*, Chapter 8 (ed. J. Kossut and J.A. Gaj), 263–303. Heidelberg: Springer-Verlag.

85 Tolmachev, D.O., Ivanov, V.Yu., Yakovlev, D.R. et al. (2020). Optically detected magnetic resonance in CdSe/CdMnS nanoplatelets. *Nanoscale* 12: 21932–21939.

86 Nelson, H.D., Bradshaw, L.R., Barrows, C.J. et al. (2015). Picosecond dynamics of excitonic magnetic polarons in colloidal diffusion-doped $Cd_{1-x}Mn_xSe$ quantum dots. *ACS Nano* 11: 11177–11191.

87 Belykh, V.V., Yakovlev, D.R., Glazov, M.M. et al. (2019). Coherent spin dynamics of electrons and holes in $CsPbBr_3$ perovskite crystals. *Nature Communications* 10: 673.

88 Kirstein, E., Yakovlev, D.R., Glazov, M.M. et al. (2022). Lead-dominated hyperfine interaction impacting the carrier spin dynamics in halide perovskites. *Advanced Materials* 34: 2105263.

89 Kirstein, E., Yakovlev, D.R., Glazov, M.M. et al. (2022). The Landé factors of electrons and holes in lead halide perovskites: universal dependence on the band gap. *Nature communications* 13: 3062.

90 Grigoryev, P.S., Belykh, V.V., Yakovlev, D.R. et al. (2021). Coherent spin dynamics of electrons and holes in $CsPbBr_3$ colloidal nanocrystals. *Nano Letters* 21: 8481.

91 Kirstein, E., Kopteva, N.E., Yakovlev, D.R. et al. (2023). Mode locking of hole spin coherences in $CsPb(Cl,Br)_3$ perovskite nanocrystals in glass matrix. *Nature Communications*, 14: 699.

92 Greilich, A., Yakovlev, D.R., Shabaev, A. et al. (2006). Mode locking of electron spin coherences in singly charged quantum dots. *Science* 313: 341.

93 Greilich, A., Shabaev, A., Yakovlev, D.R. et al. (2007). Nuclei-induced frequency focusing of electron spin coherence. *Science* 317: 1896.

94 Greilich, A., Spatzek, S., Yugova, I.A. et al. (2009). Collective single-mode precession of electron spins in an ensemle of singly charged (In,Ga)As/GaAs quantum dots. *Physical Review B* 79: 201305(R).

95 Evers, E., Kopteva, N.E., Yugova, I.A. et al. (2021). Suppression of nuclear spin fluctuations in an InGaAs quantum dot ensemble by GHz-pulsed optical excitation. *npj Quantum Information* 7: 60.

96 Fainblat, R., Barrows, C.J., Hopmann, E. et al. (2016). Giant excitonic exchange splittings at zero field in single colloidal CdSe quantum dots doped with individual Mn^{2+} impurities. *Nano Letters* 16: 6371–6377.

16

Relaxation of Electron and Hole Spin Qubits in III–V Quantum Dots

Alexander Bechtold[1], Tobias Simmet[1], Friedrich Sbrezny[2], Nikolai A. Sinitsyn[3], Kai Müller[2], and Jonathan J. Finley[1]

[1] *Technical University of Munich, Walter Schottky Institute, Department of Physics, Am Coulombwall 4, 85748 Garching, Germany*
[2] *Technical University of Munich, ZEITLab, Department of Electronic and Computer Engineering, Hans-Piloty-Str. 1, 85748 Garching, Germany*
[3] *Los Alamos National Laboratory, Theoretical Division, Los Alamos, NM 87545, USA*

16.1 Introduction

We are currently entering a second technological revolution that aims to leverage the subtle concepts of quantum physics, such as those of coherence and entanglement, to realize new ideas from Quantum Science and Technology (QST). While many proof-of-concept demonstrations have already been performed, especially in the hardware domain, scaling remains the major technological challenge that must be overcome on the way to developing performant quantum technologies. The key building blocks needed by quantum machines are *qubits*; two-level quantum systems that can be initialized into a specific superposition state, manipulated via external control methods or using gatable qubit–qubit interactions, and read out with high efficiency. The temporal evolution of the quantum state of the qubits should be predictable, and the initialization, control, and readout steps should all occur over timescales that are short compared with the one over which decoherence occurs. It is only in this way that entanglement can be used as a *resource* to provide new algorithmic and functional capabilities.

16.1.1 Solid-State Spin Qubits

Localized and optically addressable spins in solid-state nanosystems are a highly attractive resource for quantum technologies due to their intrinsic robustness against decoherence and frequency selective coupling to photons [1]. Moreover, modular and scalable device engineering concepts [2, 3] exist that allow spins to be integrated into photonic integrated circuits (PICs), facilitating the realization of modular components that can be combined and scaled up to increase complexity and enhance computational power, the security of communication systems and

Photonic Quantum Technologies: Science and Applications, First Edition. Edited by Mohamed Benyoucef.

the sensitivity of detectors and sensors [4]. A variety of different optically active spin systems have been studied including rare-earth ions [5], single molecules [6], impurity centers in wide bandgap insulators such as the negatively charge nitrogen vacancy center in diamond (NV^- 1– [7], and point defects in hexagonal-BN [8]. The localized electron spin in a III–V semiconductor [9] subject to a real or effective magnetic field is a natural qubit choice since the *s*-like Bloch character of the conduction band states has two eigenstates separated by $g(\Delta)\mu_B B$, where g is the effective component of the g-tensor, and Δ is a parameter that can be optically switched via laser fields or electrically by exploiting exchange couplings. Trapped electron spins couple weakly to the main source of decoherence in bulk semiconductors, the lattice vibrations (phonons – [10], while being strong enough to interact with experimental probes at the quantum limit, unlike nuclear spins in solids. Other optically active spin systems such as the NV^- center in diamond have a number of apparent advantages over III–V quantum dots (QDs). For example, (i) the atomic scale of the wavefunction and low atomic number of the carbon atoms very effectively weaken the spin–orbit (SO)-coupling and (ii) spin dephasing arising from exchange and dipolar coupling to proximal nuclear spins can be suppressed by isotopically purifying the material to remove ^{13}C atoms and extend the coherence time beyond 1 ms at room temperature [11]. As described in this chapter, in comparison to NV^- the coherence (T_2^*) times for electron (hole) spins in III–V semiconductors extends only up to a few (hundred) nanoseconds at liquid-He temperatures giving the impression that the situation for III–V QDs seems hopeless. However, diamond is more challenging to process into large numbers of nanophotonic devices with a high yield [12], posing technological challenges for all-diamond-based spin-based functional quantum systems. Moreover, charges trapped in III–V quantum dots have large wavefunctions extending over several nanometers in the crystal ($\sim 10^4$–10^5 nuclei) resulting in very large dipole moments that allow spins to be initialized, manipulated, and read out using the toolbox of quantum optical techniques [13, 14]. Moreover, the coupling of a single central spin with large reservoir of nuclear spins provides a window into controlling quantum many body states [15].

16.1.2 Spin-Qubits for Measurement-Based Quantum Information Processing (QIP)

Optically active III–V QD nanostructures are likely to play a central role in the area of *measurement-based* quantum computation and communication. Measurement-based quantum technologies are powered by multi-particle entangled states, known as photonic cluster states when photonic qubits are used [16–20]. These quantum states consist of many photons connected by multiple entanglement links to provide a required entanglement structure. A particular advantage of the measurement-based approach is the built-in error correction provided by a redundancy of entanglement: If one or more photons are lost or not successfully detected, sufficient entangled photons remain to complete the quantum algorithm. This redundancy and the resulting fault-tolerant effects require that deterministic sources of large cluster states are available. Such measurement-based approaches to universal photonic quantum computation are fully equivalent to the better-known gate-based approaches [17–19]. Moreover, the main advantage of using photons over other technologies and the use of semiconductor nanofabrication technologies is their potential for massive upscaling using PICs [21]. These can achieve the level of component integration, complexity, and performance required for full-stack quantum processors that integrate millions of qubits and components on a single chip. PICs are very stable because they are typically constructed from solid-state platforms that minimize mismatches between adjacent optical components and reduce the effects of noise introduced by mechanical vibration and/or temperature variations. Moreover, by using CMOS-compatible nanofabrication methods and established concepts from integrated photonics [22], PICs can be reconfigured and interfaced with chip-external photon sources and control and signal processing electronics, making them available to remote end users. Quantum photonics is one of the most promising platforms for quantum computing because photons do not suffer from decoherence (the most important limitation for matter qubits) and therefore quantum computations can be performed at room temperature using compact, stable, and highly scalable PICs. The main technological problems are: (i) photon loss, (ii) the quality of currently available sources of highly entangled cluster states that produce large numbers of photons, and (iii) the efficiency and scalability of integrated single photon detectors with photon number resolving capabilities. The first problem, photon loss, is actually manageable, since up to 50% total photon loss can be tolerated in modern measurement-based quantum computing protocols [23]. The second problem is more critical, as the generation of quantum entanglement with linear photon-photon gates is inefficient. The direct availability of entangled cluster states with many photons would be a real breakthrough, as it would eliminate the need for such linear entanglement gates and completely circumvent this bottleneck. In advanced measurement-based quantum computing, [24] an entangled state with many photons is used as the initial state, and all quantum calculations are based solely on measurements. Therefore, measurement-based optical quantum computation combined with quasi-deterministic generation of photon cluster states is one of the most promising approaches for quantum computation [21, 25] due to the very favorable error thresholds and scaling behavior of advanced protocols [20, 23].

16.1.3 Contents of the Chapter

In this chapter, we focus exclusively on epitaxially grown, optically active III–V semiconductor quantum dots (QDs) due to their favorable optical and spin properties Such single dots have many properties that make them ideally suited for on-demand generation of photonic cluster states: (i) They are strongly optically active in the infrared spectral region, (ii) they can be tuned electrically or by strain to remove dot-to-dot inhomogeneities, (iii) they can be precisely stabilized to suppress noise, (iv) they host coherent electron and hole spin-states that can be addressed optically, and (v) they can be coupled to scale up the dimensionality of the quantum resource available to multiple qubits. At the same time, QD nanostructures can be monolithically embedded into state-of-the-art photonic nanostructures, such as Bulls-eye gratings [26] and optical nanocavities [27], to push the photon in- and out-coupling efficiencies into fiber-based quantum channels up toward unity and enhance the strength of light-matter couplings. Recent major advances have demonstrated unparalleled performance metrics in the photon generation [28, 29] and spin–photon entanglement rates [30], state-of-the-art purities and indistinguishability of the quantum photonic states developed and the potential to scale QD-based devices from single QDs to artificial QD-molecules [31, 32]. As described elsewhere in this issue single photon sources with photon extraction efficiencies $> 50\%$ have been demonstrated, with exceptionally high indistinguishability of $> 95\%$ and multi-photon probabilities $g^2(0) < 10^{-4}$. With such performance metrics, QD-based sources are now >50× brighter than any previously available photonic source of similar quality. Efficient coupling into a single optical mode, as well as the possibility to generate long trains of indistinguishable photons, was also reported [33, 34]. However, the coherence times of electron spins are typically limited to a few nanoseconds placing an upper limit of the attainable spin–photon entanglement fidelities and ultimately the size of the photonic cluster states that can be formed. Thus, while the vast majority of the properties of QD-based modular photonic modes make them highly attractive for realizing quantum repeater (QR) schemes, their comparatively short spin-coherence times appears, at a first glance, to be their Achilles heel.

We demonstrate that coherence for electron spin qubits in optically active QDs has the potential to extend beyond microsecond timescales limited only by interactions with the many body system formed by the $\sim 10^4$–10^5 nuclear spins within the wavefunction. Without application of external magnetic fields, the initially orientated electron spin rapidly loses its polarization due to coupling to an effective nuclear magnetic (Overhauser) field with a fluctuating amplitude of $\sim$10.5 mT. The inhomogeneous dephasing time associated with these hyperfine-mediated dynamics is shown to be of the order of $T_2^* \sim 2$ ns. Over longer timescales, an unexpected stage of central spin relaxation is observed, namely the appearance of a second feature in the relaxation curve around $T_Q = 750$ ns. By comparing results with theoretical simulations, this additional decoherence channel is shown to arise from coherent dynamics in the nuclear spin bath itself.

In contrast to electron spin qubits, the hole spin features a p-type wave function, which vanishes at the position of the nuclei. This strongly suppresses the hyperfine contact interaction with the nuclei and is expected to result in longer coherence times for hole spins compared to electrons. We explore mechanisms responsible for the depolarization of a single hole spin at finite and zero magnetic field. Thereby time-resolved and magnetic field resolved pump-probe measurements are applied to probe hole spin dynamics over timescales ranging from 1 ns to 40 μs and magnetic fields of $\pm$50 mT. This allows us to model the temporal decay of the hole spin projection along the optical axis, retrieve the coupling strength to the Overhauser field as well as new information about the anisotropy of hyperfine coupling. The signature of these quantum dynamics of the nuclear ensemble becomes visible in the hole spin decay transient which differs strongly from the decay behavior known from conduction band electrons [35–38]. We provide two phenomenological models, which aim at developing a more physical description of the quadrupolar broadened nuclear spin spectrum than the simple isotropic Gaussian distribution that is often used to model the nuclear environment of electron qubits. These models provide good agreement with experiments and allow the estimation of the average quadrupolar coupling strength. Hole spin relaxation dynamics are shown to be more than an order of magnitude slower than for electrons.

16.2 Fundamental Properties and Devices Investigated

This section summarizes some of the salient electronic and optical properties of III–V semiconductors than make them particularly suitable for forming optically active spin-qubits. Moreover, we explore the device geometry used in our work to deterministically prepare single spins (electrons *or* holes) in the QD using resonant optical excitation, store them for timescales longer than those over which spin dynamics occur, and then probe the quantum state of the spin using optical methods.

16.2.1 Key Physical Electronic and Optical Properties of Self-Assembled Quantum Dots

The QDs explored in this chapter are formed from the III–V semiconductor material system GaAs and the ternary alloy InGaAs. Within the tight-binding approximation, electrons in the conduction band have a *s*-like Bloch-wavefunction symmetry with the total angular momentum quantum number $J = 1/2$. In comparison, holes in the valence band have a *p*-like symmetry of the Bloch-wavefunction [39]. The resulting SO interaction for the valence band states leads to a splitting into two sets of states with a total momentum of $J = 3/2$ and $J = 1/2$. The states with $J = 3/2$ represent the heavy hole (hh) and light hole (lh) valence bands with components $|J_z| = 3/2$ and $|J_z| = 1/2$ along the magnetic field quantization axis, respectively. The $J = 1/2$ state produces a third valence band (the SO band) that lies at significantly lower energy with respect to both hh and lh bands. In the fully confined electronic structure the energetic separation between the bulk bands having hh-, lh-, and SO- character is further increased owing to different effective masses m^* of the holes and the impact of strain, such that the SO-band can safely be neglected and the highest energy orbital state formed from the valence band has predominant hh-character [14].

During the initial stages of epitaxial growth of InGaAs on GaAs using molecular beam epitaxy (MBE), the lattice constant of InGaAs initially adapts that of GaAs and growth proceeds via the usual Frank–van der Merwe (layer by layer) growth mode [40]. As a result of the $\sim$7% lattice mismatch between GaAs and InGaAs layers, strain accumulates as the growth front evolves [41] such that strain relaxes by spontaneous formation of nanoscale islands – the quantum dots [42] – after deposition of a few monolayers. This growth method is termed Stranski–Krastanow mode and generally occurs during the growth of lattice mismatched materials where island formation is induced by a strain relaxation. The QDs grown with this method typically arrange randomly over the surface during their formation with remarkably small fluctuations in size and shape. After island formation the QDs have to be capped with a GaAs layer to provide a three-dimensional confinement [43]. In addition to such a spatial confinement, the capping layer isolates the QDs from surface defects providing excellent optical properties. The QDs grown in this fashion are typically lens-shaped with a diameter of ~10–20 nm and a height of ~3–5 nm depending on precise growth conditions. These dimensions give rise to the strongest motional confinement along the z-axis (Figure 16.1a) and weaker confinement associated with the in-plane (x, y)-directions. As such, the discrete orbital structure of the dot is typically associated with the motion of the particle in the in-plane directions in the dot.

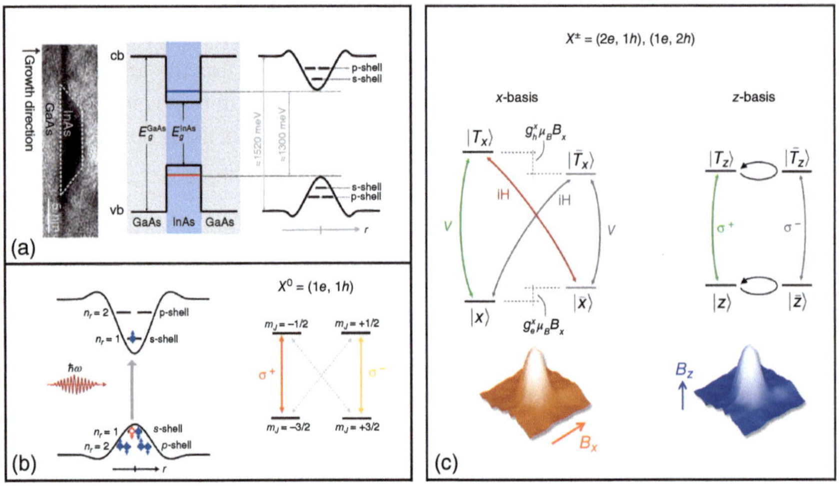

Figure 16.1 (a) Cross-sectional transmission electron microscopy (TEM) micrograph of a single InGaAs QD illustrating the strong motional quantization associated with motion along the z-axis and the formation of a series of orbital states associated with the radial motion. (lower) TEM image of vertically correlated quantum dots showing how strain in the GaAs capping layer of the lower dot causes the upper dot to be positioned directly above it. Tunnel coupling between orbital states in the QDM (QD-molecule) facilitates the formation of spatially direct excitons with strong optical activity and indirect excitons with weaker oscillator strength. (b) Schematic representation of the orbital states in a typical InGaAs QD. In the electron picture, the primarily hh-like valence orbitals are filled besides a single missing electron that has been excited into the conduction band via optical excitation. This corresponds to a single valence band hh and a single conduction band e in the dot. Note: The m_j values on this figure are labeled in the electron (i.e. non-hole) basis. The transition selection rules are such that optical polarization maps directly to the spin projection of the electron and hole along the optical axis (parallel to the growth axis). Hereby, circularly polarized photons result in the promotion of a single electron from the valence to the conduction band orbitals with anti-parallel electron and hole (parallel *electron*) spins. (c) Selection rules for electron – trion optical transitions in x-eigenbasis and z-basis representation. In-plane magnetic field symmetry couples the spin states of electron and trion, respectively, forming a double λ-system. A circularly polarized laser field selectively couples to one λ-subsystem (colored arrows), enabling optical spin control.

When Coulomb interactions between excited electrons (e) and holes (h) are taken into account, optically excited single and few-particle states have discrete transition frequencies that depend on the number and spin of optically excited charges within the dot [44–48]. In this case, the charge neutral exciton ($X^0 = (1e, 1h)$) behaves like a quasi 2-level system with a spin structure $(1e, 1h) = (\uparrow, \Downarrow)$ and $(\downarrow, \Uparrow)$ that is determined by the optical polarization of the driving field. Thus, $(1e, 1h) = (\uparrow, \Downarrow)$ having $\Delta J_z = -1$ couples to σ^- polarized light and $(\downarrow, \Uparrow)$ having $\Delta J_z = +1$ will couple to σ^+ polarized light (Figure 16.1b). The selection rules for trion transitions are presented in Figure 16.1c in the x- and z-basis representations, respectively. In-plane magnetic field (Voigt geometry) couples the spin states of electron and trion, respectively, forming a double λ-system. A circularly polarized laser field selectively couples to one λ-subsystem (colored arrows), enabling optical spin control.

16.2.2 Optical Spin Storage Photodiodes

As depicted schematically in Figure 16.2a, the QDs explored in this chapter are embedded within the intrinsic region of an n-type Schottky photodiode, separated by a 5 nm thick GaAs layer from a 20 nm thick $Al_x Ga_{1-x}As$ blocking barrier, the $x = 30\%$ Al-content inhibiting tunneling escape from photogenerated electrons from the QD. Note, holes can still readily tunnel out of the QD at a rate determined by the

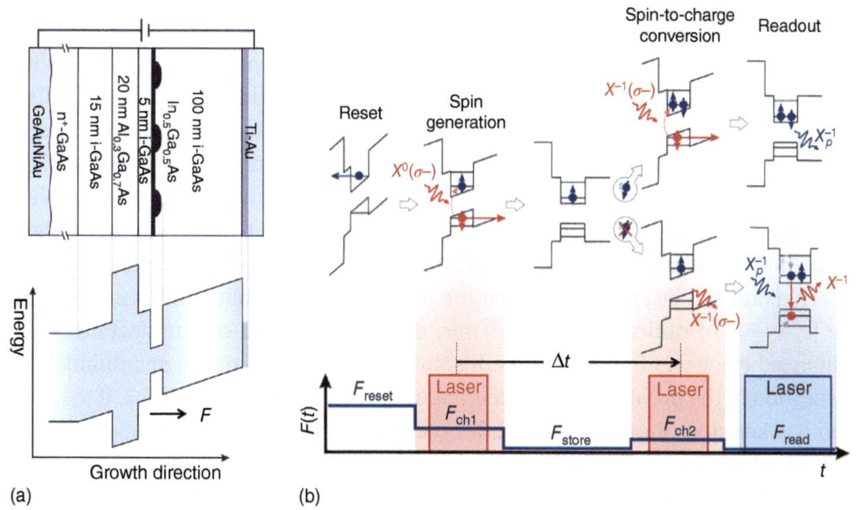

Figure 16.2 (a) Typical epitaxial layer sequence and schematic band profile for an optical charging devices (b) Schematic profile of a typical single measurement cycle used to probe spin-dynamics. After a reset phase of the measurement where all charges are removed from the QD, the electric field $F(t)$ is reduced to the level where holes are selectively removed from the dot over few picosecond timescales and a resonant charging laser is turned on. After storing the optically generated spin for a time Δt the spin-charge conversion step described in the body text is performed with a second resonant laser pulse, after which the electric field is reduced allowing driving of an optical cycling transition to produce the charge occupancy dependent readout signal.

axial electric field, itself governed by the applied gate voltage (V_g). As schematically depicted on the rightmost panel of Figure 16.2 the basic working principle of the spin storage device can be divided into five basic phases of operation. All experiments start with a reset operation (labeled Reset on Figure 16.2) by an application of a strong electric field (F_{reset}) in order to fully empty the QD from residual charges enabling a fresh start for every single measurement cycle. The electric field is related to the applied gate voltage via $F = (V_{bi} - V_g)/d$, where $d = 140$ nm is the total thickness of the intrinsic GaAs layer including the AlGaAs tunnel barrier. In a next step, denoted as charging in Figure 16.2, we switch the electric field strength to a lower value (F_{ch1}) and apply a laser pulse, that is energetically tuned into resonance with the $|\text{cgs}\rangle \rightarrow |X^0\rangle$ optical transition. In this phase of operation the optically generated excitonic spin state is defined by the polarization selection rules, such that a circularly polarized pump pulse resonantly creates the exciton with an electron spin up or down configuration with respect to the optical axis, as discussed in Section 16.2.1. After the generation of the neutral exciton with a predefined spin state, the hole tunnels out of the QD as a consequence of the local applied electric field leaving behind the single electron stored in the QD. The electric field is chosen such that the hole removal time is much faster than the timescale for exciton fine structure precession providing a high spin initialization fidelity [10, 32]. In contrast to the short hole lifetime, electron tunneling is strongly suppressed by the AlGaAs barrier and can achieve spin storage times in the order of seconds [49, 50]. Consequently, the electron can be stored in the QD for a controlled time Δt at selected electric fields (F_{store}) until the spin state is tested. In a next step, instead of directly probing the electron spin state, we perform a spin-to-charge conversion before readout operation takes place, as illustrated in Figure 16.2. The benefit of implementing such an intermediate step is the increased luminescence yield since the charge state of the QD represents a physically more stable quantity, whereas the central electron spin itself is subject to interactions with the environment [51]. During this step, we tune the electric field to an intermediate value F_{ch2} corresponding to an electric field regime where hole tunneling is possible. To map the spin state onto a charge occupancy, a second circularly polarized laser pulse, termed the probe pulse, is applied in resonance with the $|e^-\rangle \rightarrow |X^-\rangle$ optical transition. Thus, depending on the spin projection of the initialized electron after Δt, the Pauli spin blockade either allows or inhibits light absorption of the probe pulse. As schematically illustrated in Figure 16.2, if the electron has kept its spin polarization after the storage time, the QD becomes charged with $1e$, whereas in the case when the electron spin changes its state over time, Pauli spin blockade is lifted and X^- optical creation is possible which is followed by a rapid hole tunneling leaving behind the QD charged with $2e$.

Finally, the device is biased into the charge readout mode by reducing the electric field to F_{read} where optical recombination of the electron and hole dominates over tunneling. In order to read the charge occupancy of the QD we turn on a laser field that is energetically tuned into resonance with an excited state of the single, negatively charged exciton. As shown in Figure 16.2, in the case when the QD is charged with two electrons, the readout laser is out-of-resonance with the optical transition and no photoluminescence (PL) signal is produced in the readout phased of the experiment. In contrast, a QD charged with only one electron results in absorption of the read laser creating the excited state of the negatively charged trion, whereafter relaxation into the ground state and optical recombination a PL signal is produced, as schematically depicted in Figure 16.2 [36]. By monitoring the yield of the PL signal obtained from the $|X_T^-\rangle \rightarrow |e^-\rangle$ recombination, we extract the spin information of the electron. We continue to optimize and characterize the performance of the used operations of the spin storage and readout method.

16.2.3 Reset and Charge Readout

Owing to the tunnel barrier and the resulting asymmetry in the tunneling rates of the charge carriers, electrons tend to accumulate in the QD during laser illumination. To prevent an electron accumulation the sample has to be emptied periodically using the reset operation. To demonstrate the necessity of preventing a charge accumulation, we present in Figure 16.3 photoluminescence spectra with (a) and without (b) using a reset operation. As can be seen in the electro-optical pulse sequence in the upper part of the figure, we apply a laser pulse during the application of F_{read} and compare both PL spectra as a function of F_{read} (lower part of figure). Here, the laser

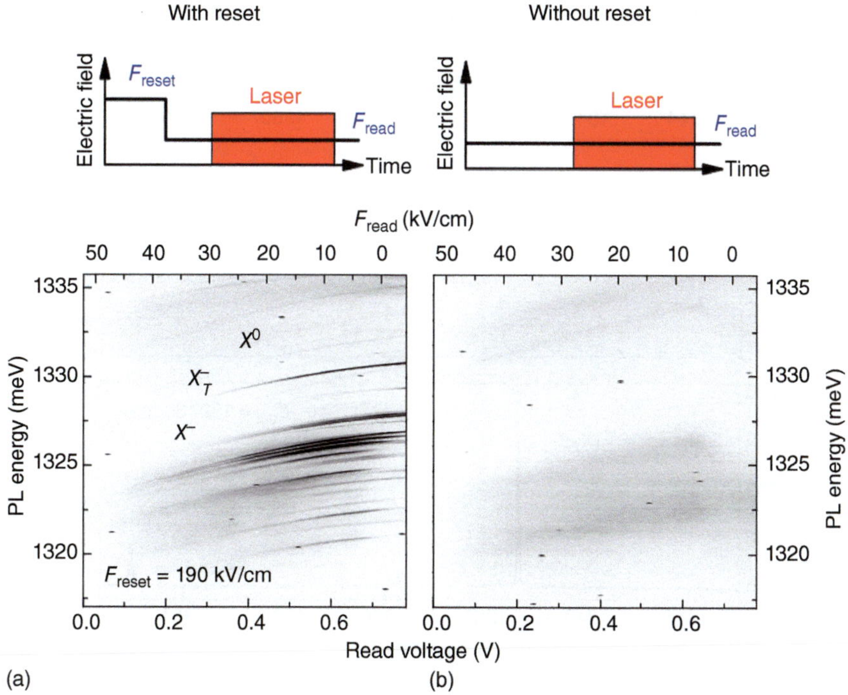

Figure 16.3 Electric field (F_{read}) dependent PL spectra with (a) and without (b) application of a reset operation prior to laser illumination. The corresponding electro-optical pulse sequence is schematically illustrated in the upper part. The laser energy is tuned into quasi-continuum of the wetting layer with 1 μs duration during the read-plateau with 1.5 μs duration. The duration of F_{reset} is set to 500 ns.

energy is tuned quasi-resonantly to the wetting layer of the QD. By comparing both spectra, PL signatures from the excitonic states can be observed only in the case when a strong reset pulse is applied at the device prior to the readout step, thus, demonstrating that accumulated electrons tunnel out the QD despite the presence of the tunnel barrier and, with this, resulting in a stabilization of the excitonic PL lines. The impact of the reset duration and F_{reset} on the discharging efficiency of the QD is discussed in detail in [52]. Here, we use the optimized values of $t_{reset} = 500$ ns and $F_{reset} = 190$ kV/cm.

Under quasi-resonant excitation in the wetting layer continuum, however, the strict requirement for a charge-sensitive readout laser absorption is not feasible. Therefore, an excitation of higher orbital excitonic states of the negative trion (X_T^-) is more suitable for the charge-read operation. In order to identify the interband transitions of excited orbital states, we performed photoluminescence excitation (PLE) spectroscopy at fixed $F_{read} = 9$ kV/cm. The basic principle of the PLE measurement is shown in the inset of Figure 16.4a. Here, the energy of the excitation laser E_{exc} is tuned over higher orbital states and once the laser energy coincides with an optical interband transition, an electron–hole pair is created in higher orbital shells, followed by a charge relaxation and ground state optical recombination with energy E_{det}. The result of such a PLE measurement is presented in Figure 16.4a showing excited orbital states of the negatively charged

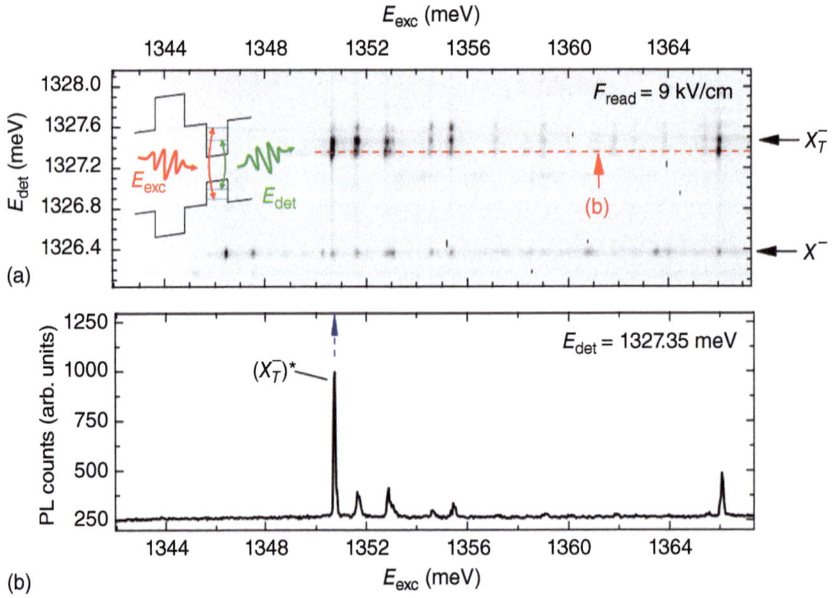

Figure 16.4 (a) PL excitation spectrum recorded for the negatively charged trions as a function of excitation laser energy with $F_{reset} = 190$ kV/cm for 500 ns and $F_{read} = 9$ kV/cm for 1.5 μs. (b) PL intensity of X_T^- ground state recombination by driving the excitation laser over excited orbital states. For the readout operation in our spin storage method the read laser is fixed to the excitation energy indicated by the blue arrow at 1350.5 meV, i.e. the energetically lowest excited state of X_T^-.

excitons, X^- and X_T^-, respectively. A typical signature of the trion triplet complex X_T^- is the threefold peak structure in the luminescence intensity, energetically separated by typically ~100 μeV arising from the isotropic electron–hole exchange interaction [53]. A further indication is that the trion triplet states are typically blue shifted by ~1 meV with respect to the negatively charged exciton X^-, as observed in Fig ure 16.4a. Figure 16.4b shows a cut of the PLE contour plot in (a) through the energetically lowest triplet state of the trion (red dashed line) for better visibility of the PL intensities. For the implementation of the X_T^- state into our charge readout method we set the read laser energy to optically excite the first excited trion state (X_T^- *), as indicated by the blue arrow in Figure 16.4b, since this excitonic transition shows the strongest optical dipole element improving the PL yield of the readout operation. Now that the conditions for the reset and read operations have been determined, we construct the electrical and optical pulse sequences for electron spin generation and spin-to-charge conversion, respectively.

Electron spin initialization and spin-to-charge conversion. For single electron spin preparation, we extend the pulse sequence by an additional 5 ps laser pulse, termed the pump pulse, as schematically depicted in the top part of Figure 16.5a. The spin state of the electron is only defined by the helicity of the pump pulse owing to strong optical selection rules in the QD [54], as discussed above in Section 16.2.1. For a σ^+ polarized pump pulse the electron is initiated in $|\downarrow\rangle$ z-spin orientation. The electric field during this charging mode is set to $F_{\text{ch}} = 59$ kV/cm such that a relatively fast hole tunneling takes place after X^0 generation to isolate the single electron. A detailed analysis of hole and electron tunneling rates as a function of the applied electric field is given in [52]. However, to find the X^0 resonance we tune the pump laser energy E_{pump} over the QDs transition frequencies and observe a peak in the readout signal corresponding to $|\text{cgs}\rangle \rightarrow |\downarrow\Uparrow\rangle$ optical transition, as presented in Figure 16.5a. Owing to the relatively fast hole tunneling after exciton generation, the energy bandwidth Γ of the excitonic transition is mainly determined by the hole tunneling time ($\tau_h = \tau_X$, from which the hole lifetime can be extracted. According to the estimation $2\Gamma = \hbar/\tau_X \sim 850$ μeV, we obtain a hole tunneling time of $\tau_h \sim 5\ ps$, a value which is comparable to the pulse duration of the pump laser as measured to be ~5 ps by using autocorrelation. As a result of comparable pulse duration and X^0-lifetime, we expect strongly damped Rabi oscillations where the state population is pushed toward complete population inversion [55], as measured and demonstrated in Figure 16.5a by monitoring the PL intensity as a function of laser power P_{pump}. Here, the pump laser energy is fixed to $E_{\text{pump}} = 1325.1$ meV driving the neutral exciton optical transition, as indicated by the red arrow in the figure. However, for the spin initialization method, we fix the pump laser energy corresponding to the peak maximum of $|\text{cgs}\rangle \rightarrow |X^0\rangle$ and choose a pump laser power corresponding to the first maximum in the power trace, as the red arrows in Figure 16.5 indicate.

For the spin sensitive readout in our method, we extend the measurement sequence by a spin-to-charge conversion step, as illustrated in Figure 16.5b. This is achieved by adding an additional 5 ps laser pulse to our measurement sequence, termed the probe laser, and keeping the electric field constant ($F_{\text{ch1}} = F_{\text{ch2}} = F_{\text{charge}}$). The probe laser has the same polarization as the pump laser in order to ensure

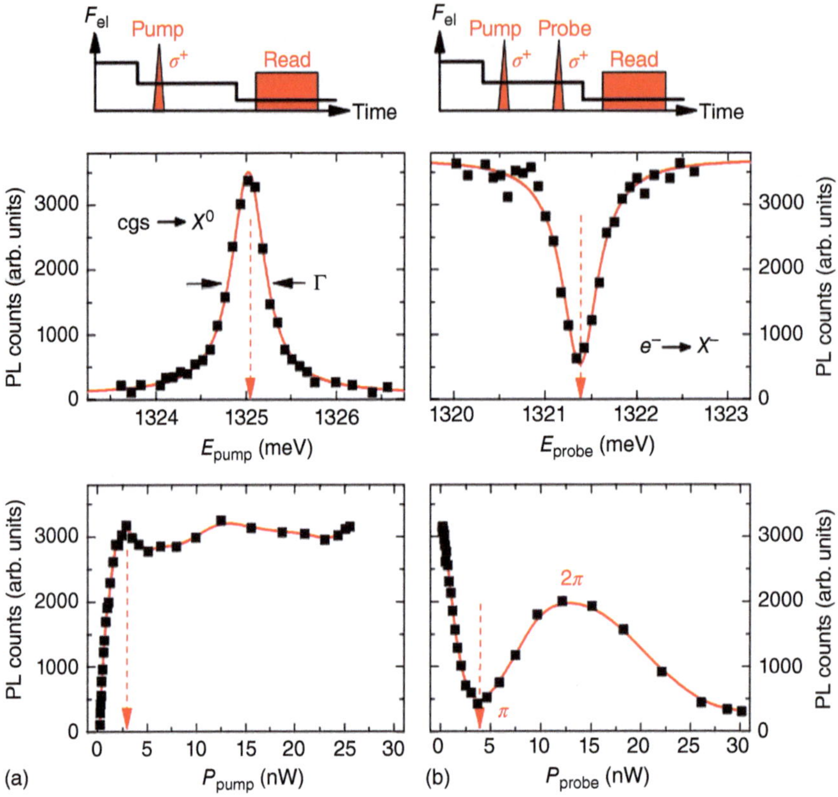

Figure 16.5 (a) Pump-read measurement cycle for electron spin generation and charge readout: $F_{reset} = 190$ kV/cm for 500 ns, $F_{charge} = 59$ kV/cm for 500 ns and $F_{read} = 9$ kV/cm for 1.5 μs. The lower parts show the resonance of the neutral exciton, fitted with a Lorentz function (red line) and Rabi oscillations (red line is guide to the eye). (b) Pump-probe-read measurement cycle for spin generation, spin-to-charge conversion, and charge readout. Electric field sequence as in (a). A weak in-plane magnetic field $B_x = 300$ mT is applied such that the probe pulse arrives when the electron spin is flipped after a half of a Larmor precession period lifting Pauli spin blockade. The lower parts show the resonance of the negatively charged trion, fitted with a Lorentz function (red line) and Rabi oscillations of this transition (red line is guide to the eye).

that in case of a spin flip, Pauli spin blockade is lifted, a second electron is created in the QD, and no PL signal is produced during the charge read step. However, to find the resonance condition of the $|e\rangle \rightarrow |X^-\rangle$ optical transition, we tune the probe laser energy E_{probe} and observe a dip in the PL readout intensity, as shown in Figure 16.5b. Here, we applied a weak in-plane magnetic field of $B_x = 300$ mT to force a spin flip and, thus, a lifting of the Pauli spin blockade inducing the breakdown of the readout PL intensity in resonance condition. Additionally, we set the time delay between pump and probe to obtain a half of a Larmor precession period, $\Delta t = \hbar/(2g_e\mu_B B_x) \sim 215$ ps, with $g_e = -0.5$. However, the position of the dip minimum, indicated by the red arrow in Figure 16.5b, is the probe laser energy of choice for the spin sensing method.

The PL readout intensity as a function of probe laser power P_{probe} during $|e^-\rangle \rightarrow |X^-\rangle$ excitation is depicted in the lower part of Figure 16.5b, demonstrating Rabi oscillations up to a Rabi area of π of this transition. Note the "up-side-down" behavior of the oscillation in the figure that arises from the pump-probe measurement technique. However, to convert the spin information into a charge occupancy, we choose the probe laser power having a Rabi-area of π, as indicated by the red arrow in the figure. A laser power of Rabi-area of π corresponds to a projective measurement, as intended for the spin sensing method. Moreover, for spin control using, e.g. geometric phase control, a laser power equal to a Rabi-area of 2π is needed.

16.2.4 Optical Manipulation: Geometric Phase Control and Spin Echo

In order to coherently control the state of the spin after the electron has been initialized, one can apply a laser pulse with a Rabi-area of $\theta = 2\pi$, energetically adjusted in resonance with the $|e^-\rangle \rightarrow |X^-\rangle$ optical transition [56–58]. An additional key requirement for spin control is the existence of a coupling between the spin states of both, $|e^-\rangle$ and $|X^-\rangle$. Such a coupling can be achieved by applications of in-plane magnetic fields, while the optical axis remains in the growth direction of the QD, defined as the z-direction, thus, perpendicular to the magnetic field direction. Such a configuration is termed the Voigt geometry and discussed in Section 16.2.4.

The energy levels and optical selection rules of the spin system in Voigt configuration are shown in Figure 16.1a in the x-basis. The magnetic field defines the spin eigenstates along the x-axis, energetically separated by the electron Zeeman splitting. Such electron eigenstates along the x-direction are symmetric and antisymmetric linear combinations of the spin states along the optical z-axis, $|x\rangle = |z\rangle + |z^-\rangle$ and $|x^-\rangle = |z\rangle - |z^-\rangle$ [59]. The eigenstates of the negatively charged trions, $|T_x\rangle$ and $|\overline{T}_x\rangle$, are defined by the quantization axis along x separated by the hole spin Zeeman energy. Optical transitions are coupled by linearly polarized light [60], H and V, introducing a double λ-system as presented in Figure 16.1c. Both electron spin states are coupled equally strongly to the trion states. In contrast, circularly polarized light can selectively couple to one λ-subsystem, as indicated by the colored arrows in Figure 16.1c. Even though the bandwidth of the control laser spans over all four transition frequencies, different phase evolution of the transitions with respect to each other enable the selectivity when using circularly polarized light. A much simpler representation of the coupled system is achieved in the optical basis along the z-direction, shown on the right of Figure 16.1c. The spin-z components of the electron ($|z\rangle$,$|\overline{z}\rangle$) and the trion ($|T\rangle$,$|\overline{T}\rangle$) are coupled by the in-plane magnetic field, where the temporal phase evolution is given by the Larmor frequency, as illustrated with black arrows in Figure 16.1c. According to optical selection rules in the QD only circularly polarized light couples to the electron-trion system in the z-basis. However, optical transitions in the z-basis are not part of the systems eigenbasis, but a linear combination of circularly polarized light according to $|H\rangle = |\sigma^+\rangle + i|\sigma^-\rangle$ and $|V\rangle = |\sigma^+\rangle - i|\sigma^-\rangle$. Therefore, optical selection rules enable the ability to optically address one of the two λ-subsystems without the need for frequency selectivity.

Without the application of a coherent optical field the eigenstates of both electron and the trion are coupled by the in-plane magnetic field resulting in a temporal evolution of the phases at the Larmor frequency. By applying a circularly polarized broadband laser with Rabi-area of 2π the temporal evolution of the phases are effectively frozen during time instants of the laser duration resulting in an optical phase gate [57]. In the Bloch sphere representation, the angle of z-rotations in the phase plane is then given by $\phi_z = 2\arctan(\beta/\Delta)$, where β is the laser bandwidth and Δ the detuning of the laser energy with respect to the optical transition frequencies.

In order to experimentally demonstrate the phase control over the electron spin, we first applied an in-plane magnetic field of $B_x = 1$ T and measured the electron spin Larmor precession using the pump-probe-read pulse sequence as discussed above. The result is shown in Figure 16.6a. Here, we used a σ^+-polarized pump pulse to generate the $|\downarrow\rangle_z$-electron, corresponding to a point in the equatorial plane in the Bloch sphere (see the figure at $t = 0$). The north and south poles of the Bloch sphere are defined by the magnetic field quantization axis, directed along x, and correspond to the states $|\downarrow\rangle_x$ and $|\uparrow\rangle_x$. After time delay t_1 we optically probe the probability of

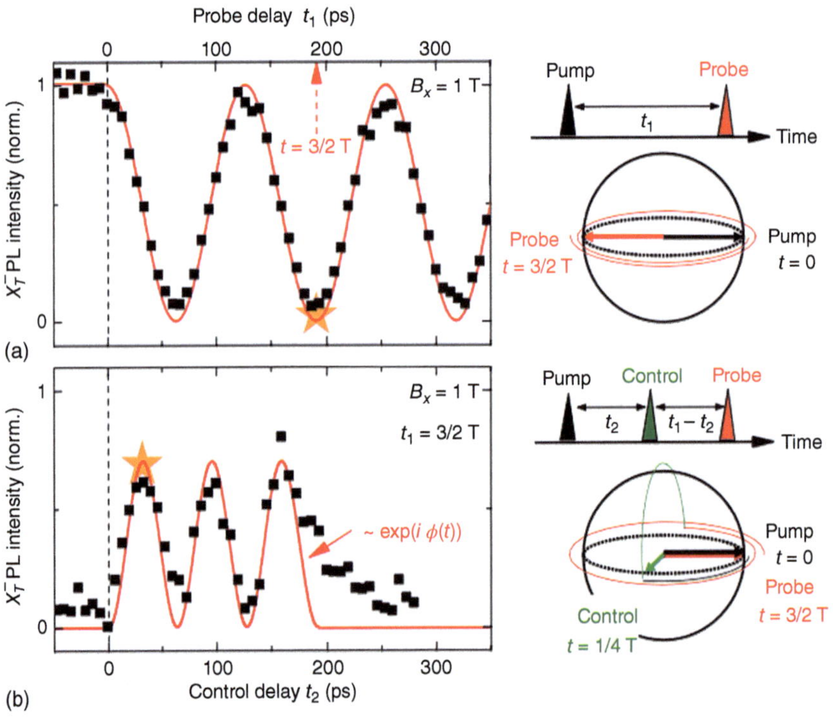

Figure 16.6 Demonstration of spin phase control using a control pulse with Rabi-area of 2π. (a) Pump-probe-read sequence to measure Larmor oscillations of the electron spin in an in-plane magnetic field of $B_x = 1$ T as a function of probe delay t_1. The red solid line is a cosinusoidal fit for $t_1 > 0$ with amplitude equal to "1." (b) Pump-control-probe-read measurement cycle with fixed t_1, corresponding to 3/2 Larmor periods, as a function of control delay t_2. Stars in (a) and (b) indicate time moments for which the spin evolution in the Bloch sphere is shown.

finding the electron in $|\downarrow\rangle_z$ using the probe pulse. As Figure 16.6a shows, by increasing the time delay $t_1 > 0$ we observe Larmor oscillations as expected for electron spin precession in in-plane magnetic fields. Here, the PL intensity is normalized such that an intensity equal to "1" ("0") corresponds to the case of finding the electron spin in $|\downarrow\rangle_z$ ($|\uparrow\rangle_z$) state. In a next step, we fix the time delay t_1 to 3/2 of a Larmor period, as indicated by the red arrow in Figure 16.6a. In such a configuration, we find the electron in a spin-$|\uparrow\rangle_z$ state, when no control pulse is used, thus, no PL signal is produced after the measurement sequence. Figure 16.6b shows the result of a pump-control-probe-read measurement cycle as a function of control delay time t_2, with t_1 fixed to 3/2 of a Larmor precession period. Most notably, at timescales of multiples n of half Larmor periods ($t_2 = nT/2 + T/4$) we observe an increased probability of finding the spin in its initialized state again, manifesting in an increase of the PL intensity after the measurement cycle. This behavior is illustrated in the Bloch sphere in Figure 16.6b and can be explained as follows: after the spin initialization at $t = 0$ the spin starts to precess with the Larmor frequency. At time instants $t = 1/4$, 3/4, and 5/4 the control pulse changes the phase of the spin by $\phi_z = \pi$ resulting in a phase flip to the opposite side of the Bloch sphere. After the remaining time (total time is set to 3/2 of Bloch rotations) the Bloch vector always points toward the initialized direction resulting in increased PL intensities in the measurement sequence. The phase control as demonstrated in Figure 16.6b exhibits a relatively poor fidelity of ~70% seen in reduced PL intensities of the peak amplitudes ($I_{PL} < 1$). This arises from the fact that the fidelity of the Rabi-2π control pulse only amounts to ~70%, as observed in Figure 16.6b. However, the spin component along the optical z-axis can be modeled with a function $\exp(i\phi(t))$, where the phase evolution $\phi(t)$ is defined as

$$\phi(t) = \omega \int_0^t \alpha(t')\mathrm{d}t' \tag{16.1}$$

where $\alpha(t') = -1$ for $0 \leq t' \leq t_0$, $\alpha(t') = +1$ for $t_0 \leq t' \leq t$ and $\alpha(t') = 0$ elsewhere. In addition, in Eq. (16.1) $\omega = g_{ex}\mu_B B_x/\hbar$ is the electron spin precession frequency, t_0 is the instant in time when the control pulse is applied, t is the total time, i.e. when the probe pulse arrives. The application of this model with fixed total time $t = t_1$ and variable control delay $t_0 = t_2$ is shown as the red solid line in Figure 16.6b and agrees very well with a rotation fidelity of 0.7. At timescales $t_1 \sim t_2$, both laser pulses, control and probe, arrive at approximately the same time at the sample which leads to a laser induced increase of PL readout intensity explaining the deviation between data and model at ~200 ps.

For a spin echo pulse sequence the condition $t_0 = t/2$ must be fulfilled. This corresponds to a complete inversion of the phases at time instant $t/2$ (half of total time) where fast phases are flipped behind slow phases using the 2π control pulse. At time t the phases catch up each with each other and, thereby, refocus again [56, 59, 61]. Due to the inversion of the phase evolution such a spin echo sequence is able to remove linear dephasing components enabling measurements of intrinsic decoherence times T_2, which are typically covered by the inhomogeneous dephasing T_2^*.

16.3 Relaxation Dynamics of Electron Spin Qubits

The control of solid-state qubits for quantum technologies requires a detailed understanding of the mechanisms responsible for decoherence. Considerable progress has been achieved in this field for qubit dynamics in strong external magnetic fields [62–65]. However, the processes causing decoherence at very low magnetic fields are less well understood, in particular the role of the quadrupolar coupling of nuclear spins. For spin qubits in semiconductor QDs, phenomenological models of decoherence, such as the Merkulov–Efros–Rosen (MER) model, currently recognize three basic types of spin relaxation [66–68]; fast ensemble dephasing in the nanosecond timescale due to the coherent precession of spin qubits around a nearly static but randomly distributed Overhauser field [69–72], and a much slower process of irreversible monotonic relaxation of the spin qubit polarization due to nuclear spin co-flips with the central spin, also known as the Knight field [73–76], or due to other complex many-body interaction effects [77] that occur in the microsecond timescale. Weak dipole–dipole interactions between neighboring nuclear spins lead to a final stage of decoherence, where the electron spin is completely depolarized at timescales of 100 ms [78, 79], such that the spin information is ultimately lost. In this section, we demonstrate that such a view on decoherence of electron spin qubits is somewhat oversimplified; the relaxation of a spin qubit state at low magnetic fields is determined by an additional stage corresponding to the effect of coherent precession processes that occur in the nuclear spin bath itself. This leads to a relatively fast but incomplete non-monotonic relaxation of the central spin polarization at intermediate ($\sim$750 ns) timescales.

16.3.1 Theoretical Background: A Historical Perspective

Among different sources of decoherence relevant for an electron spin in a QD, decoherence due to interactions with a spin bath is the most dominant mechanism at magnetic fields less than a few tesla and low temperatures [51]. Much research has been focused on this interesting field [37, 62, 67, 68, 80–84], the most important semi-classical description of such "central spin problem" is the MER model in which three distinct timescales are relevant for the electron spin relaxation [66]. The predictions of the MER model are schematically depicted in Figure 16.7 showing the characteristic relaxation timescales typical for confined electrons in self-assembled QDs, each reducing the spin polarization by a factor of $1/3$ until complete depolarization takes place over timescales of several hundred microseconds.

We continue to give a brief theoretical introduction to each relaxation channel in the framework of a semi-classical picture. Although the following models are based on a semi-classical picture, the large number of nuclear spins contributing to the hyperfine interaction ($N \sim 10^5$ in InGaAs QDs) justify their validity. Moreover, in [69] and in [81], it was shown that a full quantum mechanical treatment of this problem reproduces the semi-classical approach. This shows that considering the nuclear spin bath as a Markovian ensemble of classical magnetic moments is mostly sufficient to describe the basic relaxation stages of electron spin qubits.

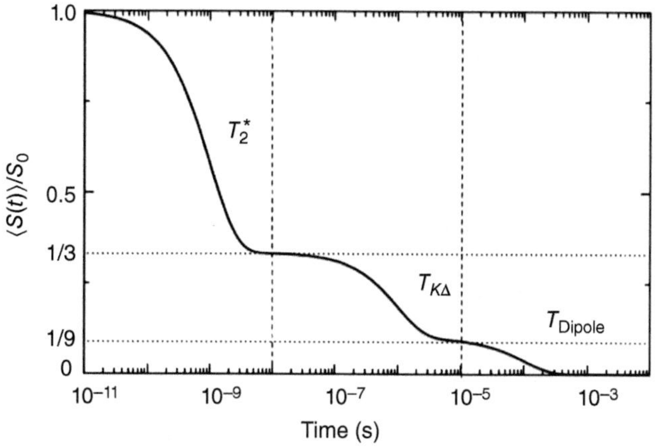

Figure 16.7 Predictions of the Merkulov–Efros–Rosen model: electron spin relaxation induced by hyperfine and dipolar interactions in the electron-nuclear spin system is characterized by three distinct stages. Figure adapted from Merkulov et al. [79].

16.3.2 Inhomogeneous Dephasing in a Fluctuating Overhauser Field

The first relaxation arises from a coupling of the central spin to the nuclear spin bath via the hyperfine Fermi contact interaction [67, 85]. This coupling gives rise to an effective magnetic field $\mathbf{B}_n$, the Overhauser field [66, 86], around which the central spin precesses typically at nanosecond timescales in optically active QDs [64, 87–89]. The Hamiltonian of this interaction can be written as [85]

$$\hat{H}^{\text{hf}}_{\text{cont}} = \frac{v_0}{2} \sum_j A^j |\psi(\mathbf{r}_j)|^2 (2\hat{I}^j_z \hat{\sigma}_z + [\hat{I}^j_+ \hat{\sigma}_- + \hat{I}^j_- \hat{\sigma}_+]) \tag{16.2}$$

where v_0 is the volume of the primitive unit cell, A^j the strength of the hyperfine coupling constant, $\mathbf{r}_j$ the position of nucleus j with spin $\mathbf{I}^j$, $|\psi(\mathbf{r}_j)|$ and $\hat{\sigma}$ are the electron wavefunction and the Pauli matrices, respectively. In the mean field approach, that is when the sum of contributions from a large number of nuclei is averaged over N nuclear spin wavefunctions, the effective nuclear magnetic field acting on the central spin is given by

$$\mathbf{B}_n = \frac{v_0}{\mu_B g_e} \left\langle \sum_j A^j |\psi(\mathbf{r}_j)|^2 \hat{\mathbf{I}}^j \right\rangle_N \tag{16.3}$$

Here, μ_B is the Bohr magneton and g_e is the electron Landé g-factor. The magnitude and direction of this effective magnetic field fluctuate with time in a random way. Moreover, unlike the situation for holes discussed later in this chapter, the fluctuations in the field are expected to be isotropically distributed due to the s-like nature of the electron central spin wavefunction. These fluctuations are, thereby, expected to follow a Gaussian probability density distribution function [66], i.e.

$$W(\mathbf{B}_n) = \frac{1}{(2\pi\sigma_n)^{3/2}} \exp\left(-\frac{\mathbf{B}_n^2}{2\sigma_n^2}\right) \tag{16.4}$$

where σ_n characterizes the dispersion of the Overhauser field distribution. As it will be shown later in this section, the magnitude σ_n is the only relevant value describing the timescale of the first relaxation mechanism. During the initial 100 ns, the Overhauser field can be considered to be time independent [51, 66], also known as quasi-static approximation (QSA). In this case, the electron spin will move around the frozen fluctuation of the hyperfine magnetic field $\mathbf{B}_n$. Even though the electron spin evolves coherently in the frozen field, the time averaging over many Larmor precession periods leads to a reduction of the electron spin polarization seen as inhomogeneous dephasing. The equation of motion of a spin $\mathbf{S}$ in a fixed magnetic field is given by

$$\mathbf{S}(t) = (\mathbf{S}_0 \cdot \mathbf{n})\mathbf{n} + \left(\mathbf{S}_0 - \left(\mathbf{S}_0 \cdot \mathbf{n}\right)\mathbf{n}\right)\cos(\omega t) + \left(\mathbf{S}_0 \times \mathbf{n}\right)\sin(\omega t) \tag{16.5}$$

where $\mathbf{S}_0$ is the initial electron spin polarization at time $t = 0$, $\mathbf{n} = \mathbf{B}_n/B_n$ is the unit vector orientated along the direction of the nuclear magnetic field and $\omega = \mu_B g_e B_n/\hbar$ is the Larmor frequency of the electron spin precession in this field. Usually, the validation of a spin polarization is based on measurement techniques where the measured spin state is obtained by averaging over many repeated measurement cycles meaning that Eq. (16.5) has to be averaged over the magnetic field distribution of Eq. (16.4). From this, we obtain the time dependence of the averaged electron spin polarization, namely

$$\langle \mathbf{S}(t) \rangle = \int \mathbf{S}(t).W(\mathbf{B}_n).B_n^2.\cos\theta.\mathrm{d}B_n.\mathrm{d}\phi.\mathrm{d}\theta \tag{16.6}$$

The integration over spherical coordinates results in an expression describing the time evolution of the electron spin at zero external magnetic fields:

$$\langle \mathbf{S}(t) \rangle = \frac{\mathbf{S}_0}{3}\left\{1 + 2\left[1 - \frac{t}{T_2^*}\right]^2 \exp\left[-\frac{1}{2}\left(\frac{t}{T_2^*}\right)^2\right]\right\} \tag{16.7}$$

Here, T_2^* is the inhomogeneous dephasing time of the electron spin resulting from the random electron spin precession frequencies in the quasi static Overhauser field. The inhomogeneous dephasing time is then defined as

$$T_2^* = \frac{\hbar}{\mu_B g_e \sigma_n} \tag{16.8}$$

and depends only on the amplitude of the Overhauser field dispersion σ_n. It can be shown that the Overhauser field dispersion itself depends mainly on characteristic parameters of the jth nucleus, such as the hyperfine coupling constant A^j and nuclear spin I^j, and is proportional to $\sigma_n \propto \frac{1}{\sqrt{N}}$ with N being the number of nuclei interacting with the electron. Thus, the fluctuation of the effective Overhauser field is related to N via

$$\sigma_n^2 = \frac{16}{3N}\cdot\frac{1}{(\mu_B g_e)^2}\cdot\sum_j I^j(I^j + 1)(A^j)^2 \tag{16.9}$$

For InGaAs quantum dots with $N \sim 10^5$ nuclei, the Overhauser field dispersion is in the range of tens of mT resulting in a typical dephasing time of $T_2^* \sim 1$ ns. Considering that the electron lifetime in the QD in our experiments (Δt – see Figure 16.2) is much longer than timescales described by the hyperfine interaction, Eq. (16.6) predicts that in the absence of external magnetic fields the coherent character of this precession leads to a characteristic dip in the central spin relaxation, i.e. the spin polarization reaches a minimum during a few nanoseconds from which it recovers before reaching a nearly steady level at 1/3 of the initial polarization $\mathbf{S}_0$ [66, 90]. In contrast, a confined electron having a finite lifetime τ within timescales of the hyperfine interaction, i.e. some nanoseconds, the average electron spin polarization obtains the form $\left(\frac{1}{\tau}\right) \int \langle S(t)\rangle . \exp(t/\tau) .dt$. This results in a "smeared out" dip as depicted in Figure 16.8 during the first phases of the coherent spin dynamics.

The application of an external magnetic field parallel to the initial spin direction ($\mathbf{B}_{ext} \parallel \mathbf{S}_0$) significantly changes the temporal evolution of $\langle \mathbf{S}(t)\rangle$. In such a situation the total magnetic field $\mathbf{B} = \mathbf{B}_{ext} + \mathbf{B}_n$ acting on the electron spin is effectively directed along the external magnetic field, providing $B_{ext} \gg B_n$. This results in a stabilization of the electron spin component along $\mathbf{B}_{ext}$ and no relaxation takes place in this regime, as can be seen in Figure 16.8a for magnetic fields $B_{ext} > 100$ mT. The

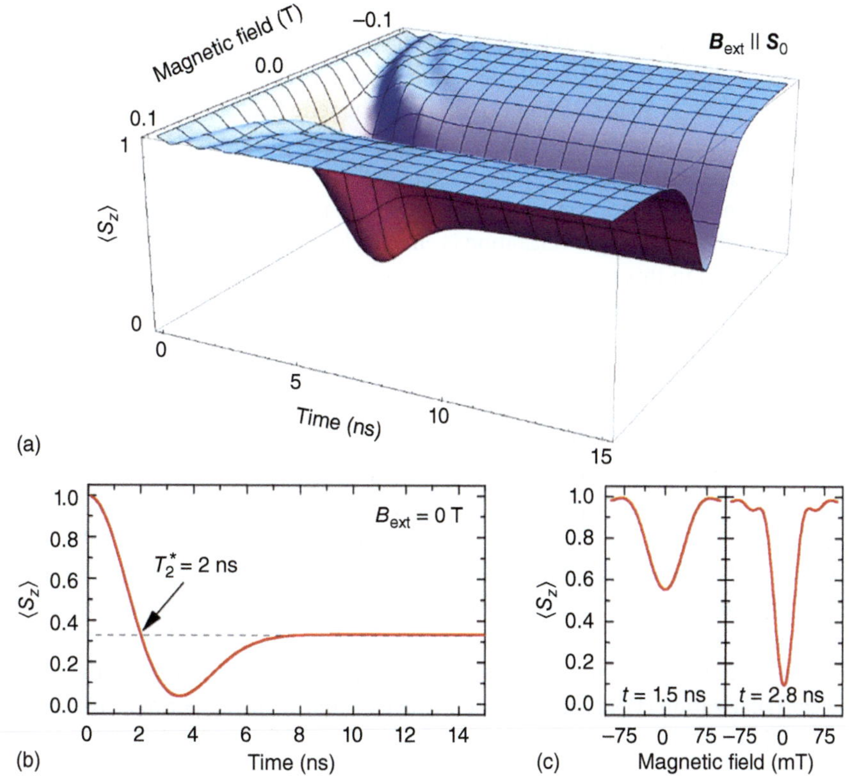

Figure 16.8 (a) Electron spin dynamics as a function of magnetic field applied along the optical axis ($\mathbf{B}_{ext} \parallel \mathbf{S}_0$) using Eq. (16.6) with $g_e = 0.55$ and $\sigma_n = 10$ mT. (b) and (c) are line-plots of (a) along constant external magnetic fields ($B_{ext} = 0$) and constant storage time ($t = 1.5$ and 2.8 ns). An application of external magnetic fields ($B_{ext} \gg B_n$) to inhibit out-of-plane spin dynamics and relaxation.

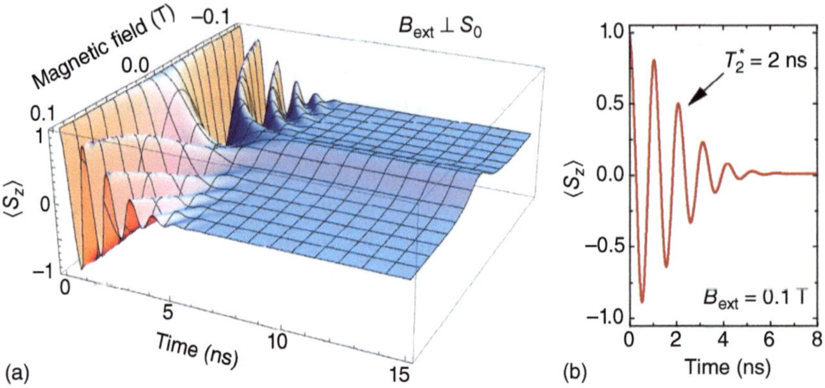

Figure 16.9 (a) Electron spin dynamics as a function of in-plane magnetic fields ($B_{ext} \perp S_0$) using Eq. (16.7) with $g_e = 0.55$ and $\sigma_n = 10$ mT. (b) shows a line-plot of (a) along $B_{ext} = 100$ mT. Despite the application of external magnetic fields ($B_{ext} \gg B_n$), the in-plane electron spin components are still affected by fluctuations of the nuclear field resulting in a dephasing at timescale T_2^*

figure shows the application of Eq. (16.6) by changing the expression of the magnetic field to $\mathbf{B}_n \rightarrow \mathbf{B}_{ext} + \mathbf{B}_n$ and using realistic parameters for InGaAs quantum dots, $|g_e| = 0.55$ and $\sigma_n = 10$ mT. When no external magnetic field is applied, the degree of electron spin polarization shows a characteristic dip during the initial nanoseconds within a dephasing time T_2^* and recovers to $S_0 = 1/3$ for longer times, as can be seen in Figure 16.8b. Once the magnitude of the external magnetic field exceeds the hyperfine field, electron spin relaxation is effectively suppressed and $\langle S_z(t) \rangle \simeq 1$ for all storage times in this regime, as can be seen in Figure 16.8c for $B_{ext} \geq 80$ mT. At this point it is important to note that the electron spin is still subject to a dephasing since the Overhauser field fluctuations are still present in the QD. Although comparatively strong external magnetic fields suppress the spin relaxation along the direction of this field, the spin components perpendicular to the external field are entirely affected by the nuclear field fluctuations such that these spin components still dephase within the timescale T_2^*, as demonstrated in Figure 16.9a. The figure shows the application of Eq. (16.4) with $B_{ext} \perp S_0$ using the same set of parameters as in Figure 16.7. The resulting Larmor oscillations decohere quickly, as can be seen in Figure 16.9b, and obtains the same value of T_2^* as in the case when no magnetic field is applied at the QD. This confirms that prolonging the macroscopic coherence time cannot simply be achieved through an application of a magnetic field. Instead, a narrowing of the hyperfine field dispersion is necessary, for example via dynamic nuclear polarization (DNP), as has recently been achieved using periodic driving of the nuclei to induce cooling and suppress fluctuations [15, 91, 92].

16.3.3 Decoherence as a Result of Time Dependent Changes in the Nuclear Spin Bath

The second stage of decoherence of an electron spin qubit in a QD is governed by the nuclear spin precession in the inhomogeneous hyperfine field of the localized electron [73–76, 78]. Due to the fast electron spin precession around the nuclear magnetic field $\mathbf{B}_n$, the nuclei only see the long time average of the electron spin directed along the Overhauser field, $\langle \mathbf{S} \rangle \parallel \mathbf{B}_n$. Since the electron wavefunction is inhomogeneously distributed over the nuclei, the direction of the total nuclear spin $\mathbf{I} = \sum_j \mathbf{I}^j$ differs from $\mathbf{B}_n \propto A^j.\left|\psi(\mathbf{r}_j)\right|^2.\mathbf{I}^j$, and consequently $\mathbf{B}_n \parallel \mathbf{I}^j$. This leads to a Larmor precession of $\mathbf{I}^j$ around $\mathbf{B}_n$ with different precession rates proportional to the square of the electron wavefunction at the position of $\mathbf{I}^j$ [66]. In a simplified classical picture the average electron spin polarization can be described as an effective magnetic field $\mathbf{B}_K$. This is the Knight field discussed earlier, around which the nuclei precess. The time averaged Knight field acting on one specific nucleus j in the presence of an electron is given by [51]

$$\sigma_n^2 = \frac{16}{3N} \cdot \frac{1}{(\mu_B g_e)^2} \cdot \sum_j I^j(I^j + 1)(A^j)^2 \tag{16.10}$$

with g_N and μ_N the nuclear g-factor and the nuclear magneton, respectively. As a result of the nuclear spin precession around the electron's Knight field, the nuclear spin system cannot be considered to be static anymore. Such a time-dependent nuclear spin bath changes the direction of the Overhauser field over time, which in turn affects the electron spin evolution $\langle \mathbf{S}(t) \rangle$. As a result of the temporal evolution of the Overhauser field, the central electron spin experiences a second relaxation channel within the timescale T_K^Δ, reducing the electron spin polarization further to $\sim S_0/9$ as schematically depicted in Figure 16.7. The precession of the electron in the macroscopic fluctuation of the Overhauser field is $\sqrt{N}$ times faster than the precession of a nucleus in the Knight field of an electron, from which it follows that the nuclear precession period can be approximated to $T_{K\Delta} \simeq T_2^* \sqrt{N}$ [66]. Here, the value N denotes the size of the nuclear spin bath consisting of N nuclei.

The fate of the remaining polarization of $S_0/9$ has been subject of considerable debates. Early studies predicted, e.g. that the spatial randomness of the hyperfine coupling can lead to an intermediate stage of a qubit relaxation that has qualitatively similar features to a dephasing stage [66]. Subsequent theoretical [67, 77, 81, 93] and numerical studies [35, 68] showed that the randomness of parameters leads only to a slow, $\langle \mathbf{S}_z \rangle \sim 1/\log(t)$, and incomplete monotonic relaxation toward a non-decaying fraction of electron spin polarization in a fashion controlled by the spatial distribution of the hyperfine coupling constants A^j, i.e. the precise form of the electron wavefunction in the QD.

Figure 16.10 shows examples of a result of a numerical calculation of the hyperfine interaction is for different hyperfine anisotropy factors β. For s-type wavefunctions, such as conduction band electrons, the hyperfine interaction is isotropic ($\beta = 1$) [86]. As can be seen in the figure, after the initial decay to $S_0/3$ following the deep local minimum as being due to the coherent character of initial precession around the fluctuating Overhauser field: For a spin initially orientated along the z-axis, the component of $\mathbf{B}_n$ along the z-axis (B_n^z) has no impact on the expectation value $\langle S_z(t) \rangle$, whereas B_n^x and B_n^y result in averaging of $\langle S_z(t) \rangle \to 0$ for $t \to \infty$. The electron spin additionally relaxes during longer timescales that is of the order of microseconds in real QDs. This additional relaxation is a direct consequence from co-flip effects

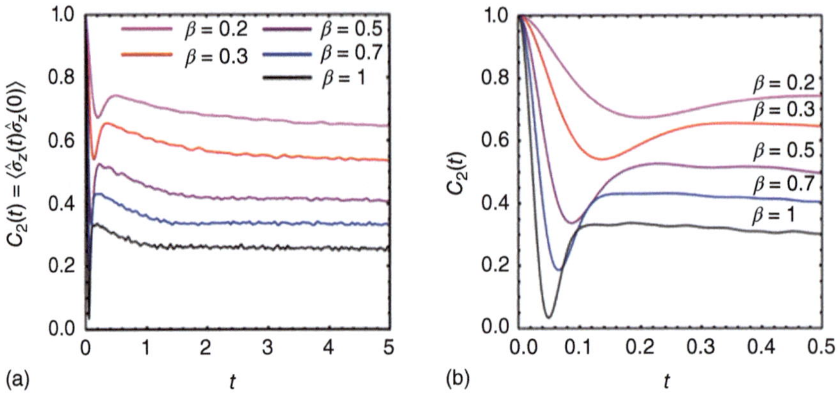

Figure 16.10 Numerical calculation of the spin correlator $C_2 \equiv \langle S_z \rangle$ at $B = 0$ for different values of the hyperfine anisotropy factor β and for $N = 700$ nuclear spins. (a) shows the spin relaxation at longer timescales arising from co-flips between central spin and nuclear spin bath (Knight field). (b) shows the central spin relaxation at short timescales owing to Overhauser field fluctuations. The time t is in units of the hyperfine coupling constant $1/A$. The hyperfine interaction for electrons is isotropic, therefore, $\beta = 1$. The impact of anisotropies in the hyperfine interaction (e.g. for hh spins) will be explored later in the chapter. Figure adapted from [35].

between the electron spin and the nuclear spin bath and saturates at values different from $S_0/9$ as predicted by the relatively simple MER-model.

16.3.4 Complete Depolarization Due to Dipolar Interactions in the Nuclear Spin Bath

The final stage of electron spin relaxation, important for confined electrons in QDs, is attributed to co-flips between nuclear spins via dipole–dipole interactions. The dipole–dipole interaction of a nucleus n with the other nuclei n separated by the translation vector $\mathbf{r}_{nn}$ can be written as [85]

$$\hat{H}_{\text{dd}} = \frac{\mu_N^2}{2} \sum_{n \neq n'} \frac{g_n \cdot g_{n'}}{r_{nn'}^3} \left(\hat{\mathbf{I}}^n \hat{\mathbf{I}}^{n'} - 3 \frac{(\hat{\mathbf{I}}^n \hat{\mathbf{r}}_{nn'})(\hat{\mathbf{I}}^{n'} \hat{\mathbf{r}}_{nn'})}{r_{nn'}^2} \right) \tag{16.11}$$

As a result of the dipole–dipole interaction, each nucleus experiences a local effective magnetic field B_{loc} created by the other nuclei and is estimated to be in the range of ~0.1 mT [75, 94]. The non-secular part of this interaction (second term in Eq. (16.11)) leads to spin flip-flop processes between the nuclei at different sites such that the total nuclear spin is not conserved within a volume due to spin diffusion. The co-flip processes cause a spatial reorientation of the nuclear spin distribution [95], which in combination with the hyperfine coupling leads to an electron spin relaxation through spectral diffusion. This non-spin-conserving process gives an upper bound for the intrinsic electron spin coherence time $T_2 \sim 100$ μs, as indicated with T_{Dipole} in Figure 16.7. In addition, such a reconfiguration of the nuclear spin bath affects the amplitude and direction of the Overhauser field and is responsible for the inhomogeneous dephasing T_2^* when averaging the electron spin polarization over times much larger than T_{Dipole}.

We continue to apply novel experimental techniques that not only clearly resolve the precession dip in the spin qubit relaxation but also provide new insights into the time-dependence of the central spin qubit for timescales four orders of magnitude longer. We demonstrate experimentally as well as theoretically the appearance of a hitherto unexpected stage of central spin relaxation, namely the appearance of a second dip in the relaxation curve after several hundred nanoseconds. We show that this feature reflects coherent dynamic processes in the nuclear spin bath itself induced by quadrupolar couplings of nuclear spins to the strain induced electric field gradients [96–98]. Eventually, the combined effect of quadrupolar coherent nuclear spin dynamics and incoherent co-flips of nuclear spins with the central spin induce a stage with monotonic relaxation that occurs over microsecond timescales and low magnetic fields with almost a complete loss of coherence. Our findings in the following discussions have been partly published in [36].

16.4 Electron Spin Relaxation Studied in Experiments

To probe the spin relaxation dynamics with and without externally applied magnetic fields, we use the devices introduced in Section 16.2 to optically prepare a single electron in the QD at time $t = 0$, orientate its spin via the optical polarization and probing the z-projection of the spin wavefunction at a time $t = \Delta t$ after generation. The time dependent electric field profile and optical pulse sequence applied is presented in Figure 16.11b.

The QD is first emptied by the application of $F_{\text{reset}} = 190\,\text{kV/cm}$ for 500 ns. During the charging mode ($F_{\text{charge}} = 70\,\text{kV/cm}$) a 5 ps duration $\sigma^{\pm}$ polarized laser pulse resonantly drives the cgs $\rightarrow X^0$ transition indicated with pump in Figure 16.11a,b. For the spin-to-charge conversion after a storage time T_{store} a second $\sigma^{\pm}$-polarized laser pulse with 5 ps duration is applied to resonantly excite the $1e \rightarrow X$ transition at $F_{\text{charge}} = 70\,\text{kV/cm}$, as indicated with probe in Figure 16.11a,b. For the charge read mode ($F_{\text{read}} = 13\,\text{kV/cm}$) a 1 μs duration laser pulse with 1350.6 meV laser energy resonantly drives an excited state of the hot trion transition $1e \rightarrow X^-$ [99], probing the charge occupancy of the QD and, therefore, the electron spin polarization after $\Delta t = T_{\text{store}}$ by measuring the PL yield from the X recombination, indicated with read in Figure 16.11a,b. To obtain the degree of spin polarization $\langle S_z(t) \rangle$ we perform two different measurement sequences; a reset-pump-read measurement cycle (red points in Figure 16.11c) to obtain the PL intensity as a reference when only one electron is present in the QD and a reset-pump-probe-read measurement sequence (black points in Figure 16.11c) from which we deduce the average charge occupation of the QD, $1e$ or $2e$, by comparing the PL intensities of the X ground state recombination, I_{1e} or I_{2e}. Note that both measurement sequences are only sensitive to the 1e charge state of the QD, the average $2e$ charge state is not measured. To determine the $2e$ charge occupancy we first measure the PL signal I_{pump} obtained using the reset-pump-read sequence. This signal reflects the PL intensity when

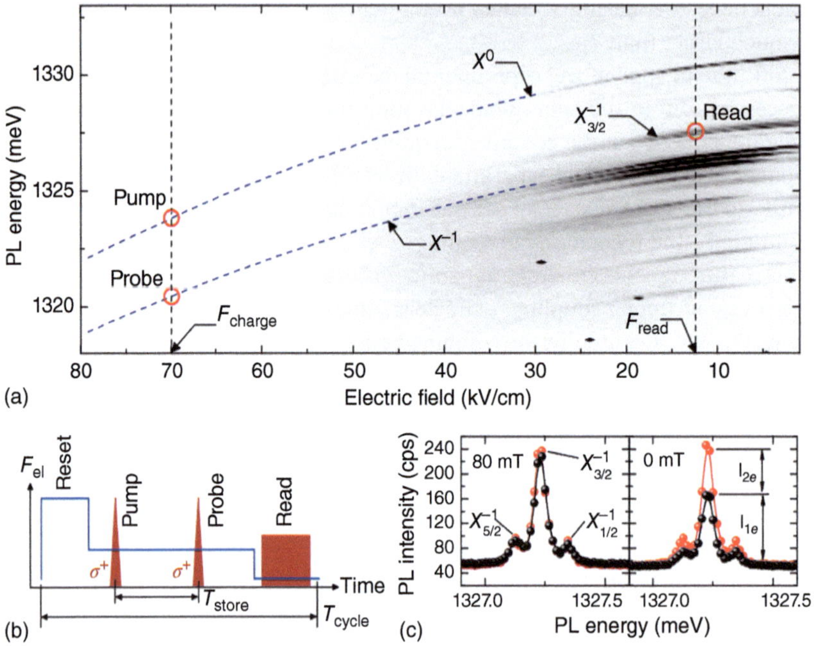

Figure 16.11 (a) Photoluminescence spectra as a function of electric field with optical excitation in the wetting layer. (b) Representation of the applied electric field and optical pulse sequence as a function of time. The measurement cycle consists of four phases; (i) discharging the QD (Reset), (ii) electron spin preparation (Pump), (iii) spin-to-charge conversion for spin measurement (Probe) and (iv) charge readout (Read). (c) The PL signature of the electron spin readout for storage times of $T_{store} = 2.8$ ns. The PL intensity of the negatively charged trion measured during the readout phase of the measurement reflects the charge state of the QD (1*e* or 2*e*) and, thereby the spin polarization.

only one electron is present in the QD and only one electron is created with the pump pulse. We need this value as a reference for deducing the upper bound of the PL signal obtained during the spin-sensitive reset-pump-probe-read sequence. The latter measurement sequence produces a PL signal I_{probe} which is smaller or equal to the reference reflecting the 1*e* charge occupation of the QD ($I_{probe} = I_{2e}$). The difference between both PL intensities reflects the 2*e* charge occupation of the QD ($I_{pump} - I_{probe} = I_{2e}$), as indicated in Figure 16.11c. The degree of spin polarization as a function of the storage time $\Delta t = T_{store}$ is then given by

$$\langle S_z(t) \rangle = \frac{I_{1e}(t) - I_{2e}(t)}{I_{1e}(t) + I_{2e}(t)} \tag{16.12}$$

As can be seen in Figure 16.11c, upon reducing the magnetic field from 80 to 0 mT, the probability of finding the dot charged with 2*e* rises ($I_{2e} > 0$) indicating that electron spin relaxation has occurred and consequently we find $\langle S_z(t \neq 0) \rangle$.

16.4.1 Monitoring the Electron Spin Qubit Relaxation

Using the experimental methods described in Section 16.2 we prepare the electron spin S_0 in the *z*-direction and monitor the temporal evolution of $\langle S_z(t) \rangle$

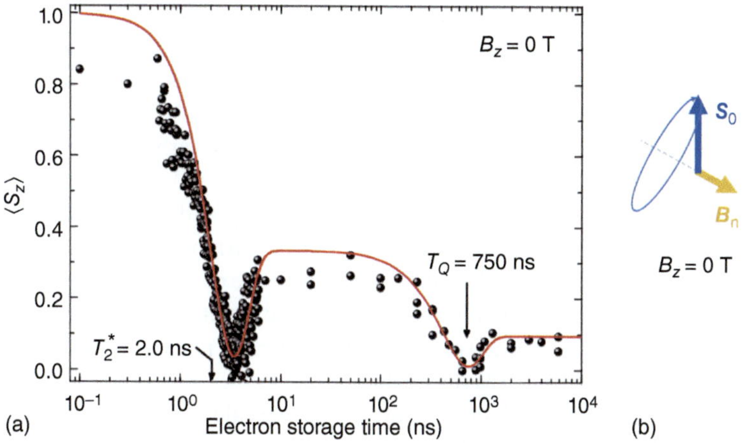

Figure 16.12 Dynamics of the electron spin relaxation at zero external magnetic field. (a) Experimental data reveal a fast inhomogeneous electron spin dephasing T_2^*. The quadrupolar coupling of nuclear spins induces oscillatory fluctuations of the Overhauser field at T_Q further reducing $\langle S_z \rangle$. The red line compares experimental results with analytical predictions based on the semi-classical model using Eq. (16.7). (b) Illustration of the electron spin evolution with initial spin S_0 in a frozen nuclear field $\mathbf{B}_n$

at zero external magnetic fields. The result of these experiments are presented in Figure 16.12a (black dots). Over the initial ~20 ns the average electron spin polarization exhibits a strong decay due to precession of the initial electron spin $\mathbf{S}_0$ around a frozen Overhauser field $\mathbf{B}_n$ (see Figure 16.12b). At these short timescales the Overhauser field experienced by the electron can be treated as being quasi-static within a single measurement cycle T_{cycle} but evolving between cycles during the few second integration time of our experiment. As discussed in Section 16.3 the magnitude and direction of $\mathbf{B}_n$ are described by a Gaussian distribution function $W(B_n) \propto \exp(B_n^2/2\sigma_n^2)$ with σ_n being the dispersion of the Overhauser field [66]. As a consequence of the field fluctuations between measurement cycles with a dispersion σ_n, the electron Larmor precession around the Overhauser field averaged over many precession frequencies lead to a characteristic dip in $\langle S_z \rangle$ reflecting the inhomogeneous dephasing time T_2^*. The initial degree of spin polarization after the electron has been initialized amounts to $S_0 \simeq 0.85$. This observation may be indicative of a weak magnetic interaction between the tunneling hole and the stored electron during the partial exciton ionization. This conclusion is supported by the fact that S_0 returns to ~1 when a weak static magnetic field ($|B_z| > 50$ mT) is applied.

In the second phase of spin relaxation observed in Figure 16.12a, which takes place from 20 ns to 1 μs, the degree of spin polarization is further reduced from $\langle S_z \rangle \sim 1/3$ to a small non-vanishing value $\langle S_z \rangle \sim 1/9$. We attribute this to time dependent changes of B_n due to the coupling of the nuclear spins I to a strain induced electric field gradient ∇E. Within a nucleus the charge distribution shows deviations from spherical symmetry, which can be described by a quadrupolar moment [96–98]. Due to the strain-driven formation of the QDs the crystal lattice is distorted away from cubic symmetry leading to electric field gradients, which couple to the quadrupolar moment of the nuclei. In the presence of such a quadrupolar mixing the Overhauser field acquires time-dependent components, which in turn modifies the central electron spin evolution [35]. Due to the much faster electron spin precession around the Overhauser field, the nuclei only see the long time average of the electron spin $\langle \mathbf{S} \rangle$ directed along $\mathbf{B}_n$. Since the electron spin follows the direction of the time dependent Overhauser field adiabatically and the time evolution of the Overhauser field itself is governed by quadrupolar couplings, a new electron spin depolarization channel becomes possible via the combined effect of both hyperfine and quadrupolar coupling. In analogy to the coherent character of the electron Larmor precession around a quasi-static nuclear field during the first dip, the presence of a second dip reflects the coherent character of nuclear spin evolution around a static quadrupolar field. After the second stage ($T_{\text{store}} > 1$ μs) only a small monotonic relaxation is observed up to 10 μs, as seen in Figure 16.12a (black dots).

To quantify the experimental data in Figure 16.12a we calculated $\langle S_z \rangle$ using the semi-classical model outlined in Section 16.3.1 in which the nuclear spins precess around the random static quadrupolar fields combined with a time-dependent hyperfine field of the central spin. In the presence of external magnetic fields $\mathbf{B}_{\text{ex}}$ the minimal Hamiltonian of the central spin interacting with a nuclear spin bath with quadrupolar couplings is given by

$$\hat{H} = g_e \mu_B \mathbf{B}_{\text{ex}} \cdot \hat{\mathbf{S}} + \sum_{i=1}^{N} \left(\gamma_H^i \hat{\mathbf{I}}^i \cdot \hat{\mathbf{S}} + g_n \mu_n \hat{\mathbf{B}}_{\text{ex}} \cdot \hat{\mathbf{I}}^i + \frac{\gamma_q^i}{2} \left(\hat{\mathbf{I}}^i \cdot \mathbf{n}^i \right) \right) \tag{16.13}$$

where $\hat{\mathbf{S}}$ and $\hat{\mathbf{I}}$ are operators for the central spin and the ith nuclear spin, γ_H^i and γ_q^i are the strengths of the hyperfine and the quadrupolar couplings of the ith nuclear spin, $I_i > 1/2$ are the sizes of the nuclear spins, e.g. I = 3/2 for Ga and I = 9/2 for most abundant In isotopes. The quadrupolar coupling of an ith nuclear spin is characterized by the direction of the coupling axis n_i and by the size of the energy level splitting γ_q^i along this quantization axis. In a self-assembled QD electric field gradients have a broad distribution of their direction and magnitude [98, 100]. At short timescales $t \ll 1$ μs one can disregard the temporal evolution of the nuclear field such that the effects of the electron's Knight field and the quadrupolar fields are neglected. Then, at zero external magnetic fields we obtain the familiar formula $\hat{H} = \sum_i \gamma_H^i \hat{\mathbf{I}}^i \cdot \hat{\mathbf{S}}$ described in Eq. (16.2) where we used $\gamma_H^i = v_0 A^i |\psi(\mathbf{r}_i)\rangle^2$. Using the Gaussian statistics of the Overhauser field and averaging over the electron spin motion we arrive at the equation for $\langle S_z(t) \rangle$, which is given by Eq. (16.7). The red line in Figure 16.12a along the first dip shows the fit of Eq. (16.7) from which we obtain $T_2^* = 2$ ns. The spin polarization curve $\langle S_z(t) \rangle$ has a minimum which develops at $T_{\min} = \frac{\sqrt{3}\hbar}{g_e \cdot \mu_B \cdot \sigma_n}$. A comparison with the experiment gives $T_{\min} = 3.4$ ns corresponding to an Overhauser field dispersion of $\sigma_n = 10.3$ mT. The effects of the quadrupolar coupling become apparent at relatively long times $t \sim 10^2$–10^3 ns. They can be modeled by the Hamiltonian in Eq. (16.13). In general, the unit vector along the direction of the quadrupolar coupling anisotropy $\mathbf{n}_i$ has a broad distribution inside a self-assembled QD. We will make an assumption that this distribution is uniform and the characteristic level splittings have Gaussian distribution throughout the spin bath. For simplicity, we also assume a bath of nuclear spins consisting of the same isotope and, hence, identical nuclear g-factors g_n. A similar Hamiltonian has been considered previously in [35] for application to spin relaxation in hole-doped QDs, as discussed later in this chapter. The major difference of electronic QDs is the nearly isotropic hyperfine coupling in Eq. (16.13), which follows from the contact exchange interaction and a relatively strong magnitude of this coupling comparable to typical quadrupolar coupling strengths in electron-doped QDs. This leads to drastically different relaxation curves for electron spins from the merely exponential relaxation of hole spins discussed in [35]. The analytical and even numerical treatment of evolution with Eq. (16.13) would be too complex to achieve for a realistic number of nuclear spins $N \sim 10^5$. In order to obtain analytical estimates for the central spin dynamics we use an observation made in [35] that essential effects of the quadrupolar coupling in Eq. (16.13) are captured by a much simpler model of a spin bath with spins-1/2 only:

$$\hat{H} = \mathbf{B} \cdot \hat{\mathbf{S}} + \sum_{i=1}^{N} \left(\gamma_H^i \hat{\mathbf{S}} \cdot \hat{\mathbf{s}^i} + \gamma_Q^i \hat{\mathbf{n}^i} \cdot \hat{\mathbf{s}^i} + \mathbf{b} \cdot \hat{\mathbf{s}}^i \right) \tag{16.14}$$

with $\mathbf{B} = g_e \mu_e \mathbf{B}_{\text{ex}}$ and $\mathbf{b} = g_N \mu_N \mathbf{B}_{\text{ex}}$ the effective Zeeman fields acting on, respectively, electron and nuclear spins. The spin-1/2 operators $\hat{\mathbf{S}}$ and $\hat{\mathbf{s}^i}$ represent the central spin and the ith nuclear spin, respectively. The quadrupolar coupling is mimicked here by introducing random static magnetic fields acting on nuclear spins with the same distribution of $\mathbf{n}^i$ as in Eq. (16.13), where the vector $\mathbf{n}^i$ points in a random direction and is different for each nuclear spin. The parameters γ_Q^i are connected to γ_q^i as $\gamma_Q^i \sim \gamma_q^i I/2$ and characterize the nearest energy level splitting of nuclear spins due to the quadrupolar coupling.

We are now in a position to show that in the presence of quadrupolar couplings $\langle S_z(t) \rangle$ will develop a second dip with a minimum at a fraction microsecond timescales. It is easiest to see this if we consider the case of a zero or weak external magnetic field and a strong quadrupolar coupling, $\gamma_Q \gg \gamma_H$. Each nuclear spin simply rotates then around the corresponding quadrupolar field axis. Within the model in Eq. (16.13) keeping only effects of quadrupolar coupling, the nuclear spin dynamics are given by

$$\mathbf{s}^i(t) = (\mathbf{s}_0^i \cdot \mathbf{n}^i)\mathbf{n}^i + (\mathbf{s}_0 - (\mathbf{s}_0^i \cdot \mathbf{n}^i)\mathbf{n}^i)\cos(\omega^i t) + (\mathbf{s}_0^i \times \mathbf{n}^i)\sin(\omega^i t) \tag{16.15}$$

where $\omega_i = |\gamma_Q^i\rangle$ is the coupling strength in units of angular frequency. Here, we introduced the vector $\gamma_Q^i \to \gamma_Q^i \mathbf{n}^i$ whose components were chosen from a Gaussian distribution $W(\gamma_Q) \propto \exp(-\frac{\gamma_Q^2}{2\sigma_Q^2})$, with σ_Q being the dispersion of quadrupolar fields. In the case of no external magnetic fields, we obtain the electron spin evolution for the second spin relaxation stage to be

$$\langle S_z(t) \rangle = S_0 \left(\frac{2\sqrt{1 - f(t)^2}}{3\pi f(t)} - \frac{2(1 - 2f(t)^2)}{3\pi f(t)^2} \right) \sin^{-1}(f(t)) \tag{16.16}$$

with $f(t) = \frac{1}{3}(1 + 2(1 - (\sigma_Q t)^2).\exp\left(\sigma_Q^2 t^2/2\right)$. For small times Eq. (16.16) saturates at $\langle S_z(t) \rangle = 1/3$ which connects to the first stage of relaxation, in our case ~20 ns.

On the other hand, $\langle S_z(t) \rangle$ saturates at the value $\sim 0.095S_0$ at long timescales and, prior to this, develops a second dip in the spin relaxation at a time corresponding to $T_Q = \sqrt{3}/\sigma_Q$. This behavior is confirmed experimentally in Figure 16.12a. A comparison to the experimental data gives $T_Q = 0.75$ μs, which corresponds to $\sigma_Q = 2.3$ μs^{-1} and is in good agreement with the quadrupolar splitting of ~1.4 μs^{-1} obtained in [98, 100] using optically detected nuclear magnetic resonance (ODNMR) spectroscopy at large magnetic fields. With the relation $\hbar\sigma_Q = g_n\mu_n B_Q$ one can transform this expression into an effective magnetic field B_Q210 mT with $g_n\mu_n \equiv 7.2 \times 10^9$ eV/T for As-isotopes.

However, the red line in Figure 16.12 shows the prediction of this model obtained in the limit that disregards the impact of the central spin on the nuclear spin dynamics. Even in this limit the model correctly captures the appearance of both relaxation dips. To capture the many-body co-flip effects beyond perturbative limits within the nuclear spin bath, we performed numerical simulations of our semi-classical model including up to $N = 50\,000$ spins. The result of these simulations at $\sigma_Q = 2\langle\gamma_H\rangle$ is presented in Figure 16.13a. These data demonstrate that complex many-body interactions, such as spin co-flips, do not remove either of the relaxation dips provided the quadrupolar coupling strength exceeds the hyperfine coupling $\sigma_Q > \langle\gamma_H\rangle$. Moreover, at even stronger values of hyperfine couplings the second dip disappears completely. Figure 16.13b presents the result of our numerical simulation taking $N = 900$ nuclear spins into account with a fixed normalized value for the hyperfine coupling $\langle\gamma_H^i\rangle = 1.0$ and for different values of the quadrupolar coupling σ_Q. Clearly, the presence of quadrupolar couplings significantly enhances the decay and leads to the appearance of the second dip in the spin relaxation. In contrast, without quadrupolar coupling ($\sigma_Q = 0$) the electron spin polarization is only slightly reduced owing to inhomogeneous nuclear spin

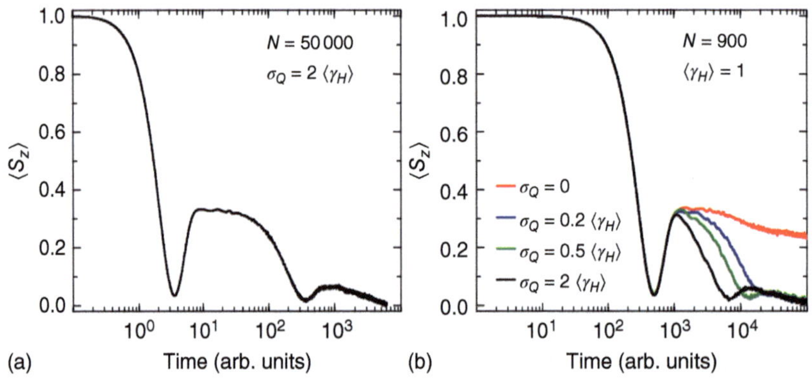

Figure 16.13 Numerical calculation of central spin relaxation at different values of quadrupolar couplings Q for (a) $N = 50\,000$ and (b) $N = 900$ nuclear spins. The energy scale is set by the average value of hyperfine coupling $\langle\gamma_H^i\rangle = 1.0$. For individual nuclear spins, the hyperfine coupling was chosen randomly from the interval $[0; 2\langle\gamma_H^i\rangle]$. The long relaxation tail for $\sigma_Q \neq 0$ at longer times is due to combined effects of hyperfine and quadrupolar coupling.

precession in the Knight field of the electron [35, 81]. The larger separation of both dips when taking $N = 50\,000$ into account (Figure 16.13a) as compared to the case of $N = 900$ (Figure 16.13b) arises from the fact that the precession of the electron spin in the Overhauser field is $\sqrt{N}$ times faster than the precession of a nucleus in the Knight field.

16.5 Hole Spin Relaxation in Single Quantum Dots

We continue to study the mechanisms responsible for the depolarization of a single hole spin at finite and zero magnetic field. Thereby time-resolved and magnetic field resolved pump-probe measurements are applied to probe hole spin dynamics over timescales ranging from 1 ns to 40 μs and magnetic fields of ±50 mT. The theoretical framework that describes the coupling of holes to their surrounding nuclear environment is similar to the case of electrons as discussed in Section 1.4, with a major caveat. As for electrons, it can be expressed classically as an interaction with the effective Overhauser field [66, 67, 85] arising from the nuclei. Unlike electron spins, the hyperfine coupling is strongly impacted by the anisotropy of the hyperfine coupling [35–38].

Hereby, we provide two phenomenological models which aim at developing a more physical description of the quadrupolar broadened nuclear spin spectrum than the simple isotropic Gaussian distribution that is often used to model the nuclear environment of electron qubits. These models reach a good agreement and allow the estimation of the average quadrupolar coupling strength and the hh–lh admixture in the valence band orbital.

16.5.1 Theory of the Anisotropic Hyperfine Coupling of Holes

The main difference between electron and hole spins is the p-type symmetry of the hole Bloch-wavefunction compared to an s-type symmetry of the electron. Due to this p-wave symmetry of the valence band hole, the three types of interaction, being the Fermi contact interaction, a dipole–dipole-like interaction and a coupling of the electron orbital angular momentum to the nucleus, have different relevance for electrons and hole, respectively [101]. While spin phenomena of electrons in a QDs nuclear ensemble are often sufficiently described using only the Fermi-contact interaction [36, 66], Fischer et al. [102] concluded that for a heavy hole, the Fermi-contact interaction is negligible while the two latter interactions dominate the coupling. The coupling is estimated to be an order of magnitude weaker [14, 102] than for the electron and the Hamiltonian takes on a simple Ising form

$$\hat{H} = \sum_j A_j^h \hat{S}_z \hat{I}_j z \tag{16.17}$$

with the sum running over all nuclei within the orbital QD wavefunction and the A_j^h denote the coupling strength of the hole to the jth nucleus. It should be noted that this strict classification only holds in theory, since in a realistic quantum dot valence band mixing of heavy and light hole states occurs [103–105]. Typically valence band mixing arises from quantum dot anisotropy like a reduction of symmetry due to the confinement geometry, the absence of an inversion symmetry in the growth direction due to its shape, and a non-homogeneous chemical profile [86, 102, 106]. The hole mixing is at the origin of a transverse term in the hole hf interaction, turning it from Ising-like to more or less Heisenberg-like, which allows spin flips between the mixed-hole states. It has also been recently pointed out that the non-zero contribution of the d-symmetry orbitals to the hf interaction leads to a transverse term of the HH hf interaction [100]. To introduce this mixing as a variable parameter in the theory, Testelin et al. expressed the dominant dipole–dipole Hamiltonian in the basis of light hole heavy hole mixed states $\tilde{\varphi}_{\pm 3/2}$, which are retrieved from the Bir–Pikus Hamiltonian of the valence band of strained GaAs [86, 103].

$$\tilde{\varphi}_{+3/2} = \frac{1}{\sqrt{1+\beta^2}} \left(\varphi_{+3/2} + \beta\ \varphi_{-1/2}\right) \tag{16.18}$$

$$\tilde{\varphi}_{-3/2} = \frac{1}{\sqrt{1+\beta^2}} \left(\varphi_{-3/2} + \beta^*\ \varphi_{+1/2}\right) \tag{16.19}$$

Here, $\varphi_{\pm 3/2} = \langle J = 3/2, J_z = \pm 3/2\rangle$ and $\varphi_{\pm 1/2} = \langle J = 3/2, J_z = \pm 1/2\rangle$ denote the pure heavy hole and light hole basis states, as discussed in Section 16.2.1. The degree of valence band mixing is given by β, defined as

$$\beta = \beta e^{i\phi} = i\,\frac{d_v\,\epsilon_{xy}}{\Delta_{\mathrm{lh}}} - \frac{\sqrt{3}}{2}\,\frac{b_v(\epsilon_{xx} - \epsilon_{yy})}{\Delta_{\mathrm{lh}}} \tag{16.20}$$

where d_v and b_v are the deformation potentials for the valence band, ϵ_{ij} denote the (ij) components of the strain tensor, and Δ_{lh} is the energy splitting between light and heavy hole bands due to the strong motional confinement and biaxial strain [86].

Under the assumption that the hyperfine coupling is much weaker than the valence band splitting ($\delta_{\mathrm{hf}} \ll \Delta_{\mathrm{lh}}$) the 4×4 dipole–dipole Hamiltonian describing the valence band is separable into 2×2 components of the light and heavy hole band. This then yields an individual Hamiltonian for each of the two pure hole bands, both exhibiting a strong and characteristic anisotropy. By summation over the interaction of all nuclei j coupled to the central hole with the wavefunction $\Psi(\mathbf{R}_j)$ it is possible to show that the Hamiltonian for a mixed hole state $\tilde{\varphi}_{\pm 3/2}$ takes the following form [35, 86]

$$\hat{H}_{\mathrm{hf}} = \Omega \sum_{j=1}^{N} M_j \Psi(\mathbf{R_j})^2 \left[\alpha(\hat{I}_{jx}\hat{S}_x + \hat{I}_{jy}\hat{S}_y) + \hat{I}_{jz}\hat{S}_z\right] \tag{16.21}$$

where the valence band mixing is captured by the anisotropy parameter $\alpha = 2\beta/\sqrt{3}$, Ω denotes the unit cell volume and M_j are the individual hyperfine constants for different nuclear species. For $\alpha = 1$, $\hat{H}_{\mathrm{hf}}$ becomes isotropic, thus describing the situation for the electron discussed earlier in this chapter. A heavy hole corresponds to $\alpha = 0$ where the Hamiltonian reduces to the Ising form of Eq. (16.17) and for $\alpha = 2$, the Hamiltonian describes the situation for a light hole. Small non-zero values of α are associated with mixed heavy hole light hole states. In the mean field approach, we average the individual contributions over a large number N of nuclei, thereby we can write an effective nuclear field that acts on the hole spin as

$$\mathbf{B}_N = \frac{\Omega}{g\ \mu_B} \sum_j M_j \Psi(\mathbf{R}_j)^2 \left[\alpha\ (I_{jx}\hat{e}_x + I_{jy}\hat{e}_y) + I_{jz}\hat{e}_z\right] \tag{16.22}$$

where $\hat{e}_i$ denotes the unit vector, g is the Landé g-factor of that state and μ_B is the Bohr magneton. As a consequence of our observations of nuclear field dynamics presented later in this chapter, we find that the components of the nuclear field follow a Gaussian distribution.

$$W(\mathbf{B}_N) = \frac{1}{\pi^{3/2}, \alpha^2\ \Delta_0^3} \exp\left[-\frac{B_{Nx}^2}{\alpha^2\Delta_0^2} - \frac{B_{Ny}^2}{\alpha^2\Delta_0^2} - \frac{B_{Nz}^2}{\Delta_0^2}\right] \tag{16.23}$$

Here, Δ_0, which is often written as $\sqrt{2}\sigma_n$ in the literature [36], characterizes the width of Overhauser field distribution and thus is the most relevant parameter that determines the central spin coherence decay on short time scales [38, 66, 107]. Note that contrary to the isotropic case $\alpha = 1$ of the electron spin, in the case of a hole spin this distribution reflects the anisotropy of the hyperfine coupling. Using the theoretical framework established so far, one can describe the hole spin relaxation dynamics during the first 100 ns. During this phase, as for the electron the Overhauser field itself is considered to be frozen due to the internal dynamics of the nuclear ensemble being much slower than the precession of the hole spin around $\mathbf{B}_N$ given by the magnitude of the hyperfine constants M_j [36, 51, 66]. In this case, since the correlation time of $\mathbf{B}_N$ is longer than the precession timescale of a single hole, the interaction remains coherent. However, the spin polarization, which is obtained by an ensemble average of a macroscopic number of sampling measurements, decays due to the averaging over all possible precession frequencies given by σ_n [36, 38, 66], as described in Section 16.3. For a spin coupled to an anisotropic hyperfine field and a macroscopic magnetic field along the growth direction $\mathbf{B} = B_z\ \hat{e}_z$ the equation of motion is given by [106]:

$$\langle S(t)\rangle = \left[S_{0x}R_x(t) - S_{0y}R_y(t)\right]\hat{e}_x + \left[S_{0x}R_y(t) - S_{0y}R_x(t)\right]\hat{e}_y + S_{0z}R_z(t)\hat{e}_z \tag{16.24}$$

$$\mathbf{S}_0 = S_{0x}\hat{e}_x + S_{0y}\hat{e}_y + S_{0z}\hat{e}_z \tag{16.25}$$

Here, $\mathbf{S}_0$ defines the spin state at $t = 0$ and $R_i(t)$ denotes the time dependent normalized spin component $S_i(t)/S_{0i}$ for $i = x, y, z$. In the following, we present experiments in which we detect hyperfine induced dephasing of a spin which is prepared and read out along the z-axis. It is, therefore, most relevant to study the $R_z(t)$ component under the assumption $\mathbf{S}_0 = S_{0z}\hat{e}_z$. Here, the equation of motion reduces to

$$\langle S_z(t)\rangle = S_{0z}\ R_z(t) \tag{16.26}$$

Moreover, the analytic form of $R_z(\tau, \delta, \alpha)$ is given by:

$$R_z(\tau,\ \delta,\ \alpha) = \int_{-\infty}^{\infty} \mathrm{d}x \int_{-\infty}^{\infty} \mathrm{d}y \int_{-\infty}^{\infty} \mathrm{d}z\ G(\delta + z,\ \alpha\ x,\ \alpha\ y,\ \tau)\ \mathrm{e}^{-(x^2+y^2+z^2)} \tag{16.27}$$

$$G(a,\ b,\ c,\ d) = \frac{1}{\pi^{3/2}} \frac{a^2 + (b^2 + c^2)\ \cos(d\ \sqrt{a^2 + b^2 + c^2})}{a^2 + b^2 + c^2} \tag{16.28}$$

The full derivation of this equation can be found in [86].

For further analysis of the results, it is convenient to use the normalized time $\tau = t/T_\Delta$ for which T_Δ is the characteristic spin decay time of a macroscopic ensemble of spins and is thus a function of σ_n

$$T_\Delta = \frac{\hbar}{g^h\ \mu_B\ \sigma_n} \tag{16.29}$$

with g^h denoting the hole g-factor [36, 86, 108]. T_Δ may be different from the inhomogeneous dephasing time T_2^*, in that it gives a time scale to the ensemble dephasing which arises due to this particular hyperfine Hamiltonian, whereas T_2^* denotes the time at which the projection of the spin ensemble has decayed to e^{-1} due to inhomogeneous broadening of the qubit splitting [109].

Figure 1.13a depicts a map of the spin component R_z at zero magnetic field as a function of normalized time τ and the anisotropy factor $\alpha \in [0,\ 2]$ resembling heavy hole, electron, and light hole spins as well as continuously mixed states. As illustrated by the red and white shaded areas in the figure, $\langle S_z \rangle$ does not follow a monotonic decay for most values of α but instead the relaxation experiences a dip and recovers to an asymptotic value dependent on the valence band mixing of the spin, according to Eq. (16.18). The isotropic Fermi-contact hyperfine coupling experienced by an electron leads to a fast decay to zero and a recovery of the polarization to 1/3 of its initial value. This behavior was predicted using a variety of calculation approaches [37, 66, 68, 110, 111], and was successfully observed in two independent experiments [36, 38], as presented earlier in this chapter. The emergence of the relaxation dip can be explained by the nuclear field fluctuations between the measurement cycles within the dispersion σ_n, leading to an averaging over the coherent Larmor precessions at many different frequencies. In the isotropic case, where the x-, y-, and z-components of the Overhauser field couple equally strongly to the central spin, the relaxation saturates at $\langle S_z(\tau \gg 1) \rangle = 1/3$, since only the two lateral components induce a precession away from the z-axis. The z-component B_{Nz} has no effect on the spin state initialized along the z-direction. In the case of a light hole, the polarization decays faster and can even reach net negative values due to the lateral components of $\mathbf{B}_N$ that dominate the spin stabilizing component along the z-direction, as can be deduced from the hyperfine Hamiltonian (Eq. (16.21)). However, light holes are widely disregarded in experimental literature since they can only be observed by applying a non-trivial strain tensor to the sample in order to overcome the ∼10 meV energy gap arising from the strong confinement and biaxial strain [54, 112].

Experimentally the most relevant case for holes in InGaAs/GaAs quantum dots formed by strain is a hole spin wavefunction exhibiting small valence band mixing $0 < \beta < 0.3$, which entails a *mainly* heavy hole character [97, 106, 113, 114]. The reduced depth of the relaxation dip in Figure 16.14b, for $\alpha = 0.3$ reflects the dominance of the out of plane component B_{Nz} in the anisotropic hyperfine Hamiltonian Eq. (16.21) over the lateral components B_{Nx} and B_{Ny} [35, 38, 86]. At the same time,

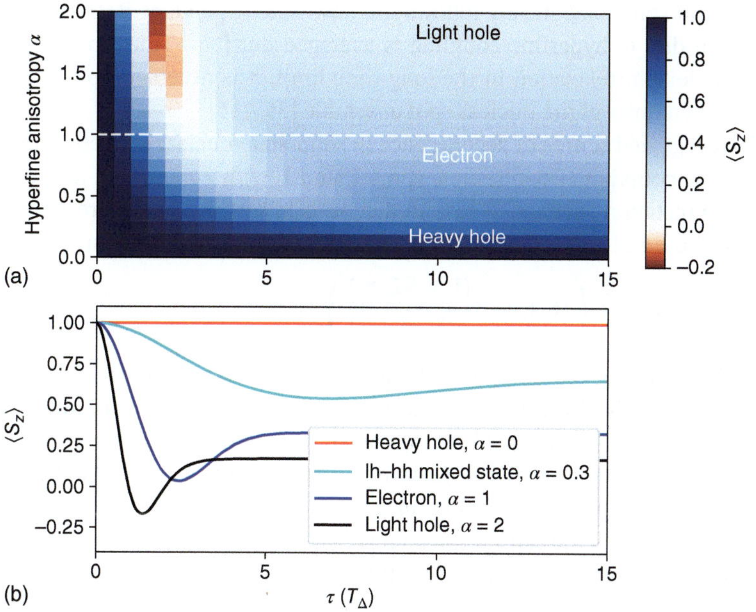

Figure 16.14 False color image of the ensemble spin polarization along the z-axis ($\langle S_z \rangle$) plotted as a function of time, normalized by characteristic decay time T_Δ. On the y-axis, α is from $\alpha = 0$ in the heavy hole limit through the isotropic coupling of an electron $\alpha = 1$ to the light hole limit $\alpha = 2$. The red and white shaded areas mark a dip in the decay of $\langle S_z \rangle$, which occurs at later time with increasing heavy hole character. Line cuts from (a), taken at different values of α to illustrate the behavior of the dip in the decay curve and the asymptotic value for large τ. For the light hole and electron, the decay exhibits a pronounced dip at $\tau \sim 2/\alpha$. Toward pure heavy holes, the spin relaxes more slowly and the dip signature becomes shallower as the lateral Overhauser components contribute less to the dynamics in the limit of Ising type interactions.

the position of the dip shifts to longer time $\tau \approx 2/\alpha$, thus the heavy hole-like state experiences slower decoherence, making its quantum information storage capabilities superior to the widely studied electron spin discussed earlier in the chapter [108, 114, 115]. In the Ising limit $\alpha = 0$ illustrated by the red line in Figure 16.14b, the ensemble spin projection is unaffected by nuclear noise due to the colinearity of the spin and nuclear operators in the Hamiltonian (16.17) [102].

16.5.2 Nuclear Spin Dynamics

In the Section 1.5.1, we treated $\mathbf{B}_N$ as quasi-static, which means that for every individual spin measurement the Overhauser field strength and orientation assumes a random value, but during the spin evolution its dynamics are frozen. This assumption is justified for short times, provided that the hyperfine coupling strength of the central spin to the nuclear ensemble is much larger than spin interactions, which cause dynamics of the nuclear spins, such as quadrupolar coupling or nuclear dipole–dipole coupling [66]. From this intermediate timescale onward, the central spin ensemble evolution is given by the time correlator of the Overhauser field, since the central spin adiabatically follows the now time-dependent $\mathbf{B}_N(t)$ and the fast precession due to hyperfine coupling is averaged out [36]. Thus, in order to understand hole spin relaxation in the long-time limit, it is necessary to describe the collective dynamics of the nuclear spin ensemble [35, 116].

A widely recognized source of decoherence in solid-state nuclear spin ensembles is dipole–dipole coupling of two remote spins $\mathbf{I}_i$ and $\mathbf{I}_j$ with $i \neq j$, which are separated by the translation vector $\mathbf{r}_{ij}$. This interaction is insensitive to the presence of an electron or hole and is described by the following coupling Hamiltonian [79, 117]

$$H_{dd} = \frac{\mu_N^2}{2} \sum_{i \neq j} \frac{g_i\, g_j}{\mathbf{r}_{ij}^3} \left((\mathbf{I}_i\ \mathbf{I}_j) - 3 \frac{(\mathbf{I}_i\ \mathbf{r}_{ij})(\mathbf{I}_j\ \mathbf{r}_{ij})}{\mathbf{r}_{ij}^2} \right) \tag{16.30}$$

with μ_N and g_i denoting the nuclear magneton and the g-factor of the ith nucleus, respectively. Effectively, this interaction captures the precession of each individual nucleus due to a local effective field, caused by all other nuclei of the ensemble. It enables nuclear spin exchange flip-flops, which accelerate nuclear spin bath decoherence. These flip-flops induce a random dynamic on the Overhauser field and thus causes dephasing of the central spin, a mechanism also termed "spectral diffusion" [63, 98].) However, it was shown that the presence of strain in self-assembled QDs in combination with the large spin number of atomic isotopic constituents, the emergence of quadrupolar coupling suppresses the effect of nuclear dipole–dipole coupling [98]. Quadrupolar coupling leads to an inhomogeneous broadening of the nuclear spin transitions, thus lifting the degeneracy among dipole–dipole coupled nuclear spin pairs, inhibiting co-flips (Eq. (16.30)). This mechanism renders dipole–dipole coupling inefficient for nuclear ensemble dephasing as nuclear bath coherence times in the millisecond regime have been reported, suggesting only a marginal impact on central spin coherence throughout the observable time frame [98, 117, 118].

Notably, quadrupolar coupling was identified to dominate the dynamics of the nuclear spin ensemble in recent investigations [36, 84, 98]. It emerges here, because the nuclear isotopes ^{115}In, ^{69}Ga, ^{71}Ga and ^{75}As each have nuclear spin quantum numbers larger than 1/2, with 9/2 for Indium and 3/2 for all the other isotopes [119]. For these high spin quantum numbers one finds a non-spherical nuclear charge distribution that can be represented by an electric *quadrupolar* moment. This in turn couples to local electric field gradients in the lattice [51]. Quantum dots produced in Stranski–Krastanov growth mode inherently exhibit a non-uniform strain profile which results in inhomogeneous local field gradients [96, 120]. In addition, the random alloying of In, Ga, and As enhances the formation of random field gradients. As a result, these static quadrupolar couplings, which are defined locally by the strength and orientation of the electric field gradient, lead to an individual dynamic for every nucleus, thus contributing to the time dependence of $\mathbf{B}_N$ [35, 121]. The Hamiltonian describing quadrupolar coupling of a single nucleus reads

$$\hat{H}_Q = \frac{h\, \nu_Q}{6} \left(3\mathbf{I}_{z'}^2 - \mathbf{I}^2 + \eta \left(\mathbf{I}_{x'}^2 - \mathbf{I}_{y'}^2 \right) \right) \tag{16.31}$$

where ν_Q denotes the characteristic quadrupolar energy splitting and η is the biaxiality parameter which is used to reflect the broken lateral symmetry [96]. Note, that the nuclear spin operators $\mathbf{I}_\alpha$ are transformed into the dashed coordinate system (x', y', z'), which is defined by the local quadrupolar field axis $\boldsymbol{n}_i$. To model the quadrupolar interaction for a nuclear ensemble of the size $\mathcal{O}(10^5)$, the coupling energy is taken as a Gaussian distribution which is unique for each atomic species [121, 122]. Furthermore, atomistic models of the quantum dot lattice under strain yield a non-trivial quadrupolar vector field that is dependent on the alloying profile and dot geometry [96].

16.5.3 Hole Spin Storage Devices

We continue to briefly outline how the concepts introduced earlier in the chapter are applied in order to determine $\langle S_z(t)\rangle$ via a sweep of the applied magnetic field. The sample is similar to the one used to study electron spin relaxation and discussed in Section 16.2, but with the difference that the AlGaAs barrier is above the QD layer, rather than below. This results in the trapping of optically generated *holes*, while electrons can tunnel out of the dot in the *n*-contact. The devices are processed for single dot measurements and mounted in Faraday geometry, meaning the growth direction or z-direction is oriented along the magnetic field axis of the superconducting solenoid ($\mathbf{B}_{\text{ext}} = B \cdot \hat{e}_z$). The time-dependent electric field function applied to the sample and the *pump* and *probe* pulse separation is fixed for the measurement of $\langle S_z(t)\rangle$ at a particular hole spin storage time. We probe hole-spin relaxation experimentally using similar samples to those used to study electron spin relaxation earlier (Section 16.2.2) and similar pulsed opto-electronic experimental approaches Section 16.2.3). We measure the spin polarization value by cycling the *reset-pump-probe-read* sequence for a few seconds, thus accumulating the signal of an ensemble of $\sim 10^6$ individual realizations of the experiment in the same quantum dot [36]. For a particular storage time, we measure $\langle S_z\rangle$ as a function of the external magnetic field in the range between $B = -50$ mT and $B = +50$ mT. This procedure allows us not only to normalize $\langle S_z\rangle_{B=0T}$ to the value of $\langle S_z\rangle_{B\gg 0T}$ at which the spin projection is conserved over long time, but it also allows us to obtain the Overhauser field dispersion σ_n.

We continue to present and discuss the measurement results and attempt to explain the different stages of dephasing using the theoretical concepts introduced in Section 1.5.1. A typical dataset of $\langle S_z\rangle$ at different storage times, recorded as a function of the external magnetic field is shown in Figure 16.15. The first panel is used to illustrate the fitting procedure used to extract the spin relaxation parameters from the measured data. Here, we numerically calculate the analytical formula for $R_z(\tau, \delta, \alpha)$ at a fixed time $\tau = 2/\alpha$ and for anisotropy $\alpha = 0.2$ [86]. The characteristic dip of $\langle S_z\rangle$ at zero magnetic field can be explained as follows: The spin precesses around the total magnetic field made up by the external magnetic field along the z-direction and the randomly oriented Overhauser field $\mathbf{B}_{\text{tot}} = \mathbf{B}_{\text{ext}} + \mathbf{B}_N$, as illustrated by the schematic in Figure 16.15. For large external magnetic fields ($B_z \gg B_N$), the external field effectively defines the central spin quantization axis and, therefore, the relaxation along the z-axis is suppressed. Consequently, $\langle S_z\rangle \approx 1$ for $B_z \geq 10$ mT as displayed by the measurement results in Figure 16.15 at various storage times. At zero magnetic field, the precession axis is solely defined by the

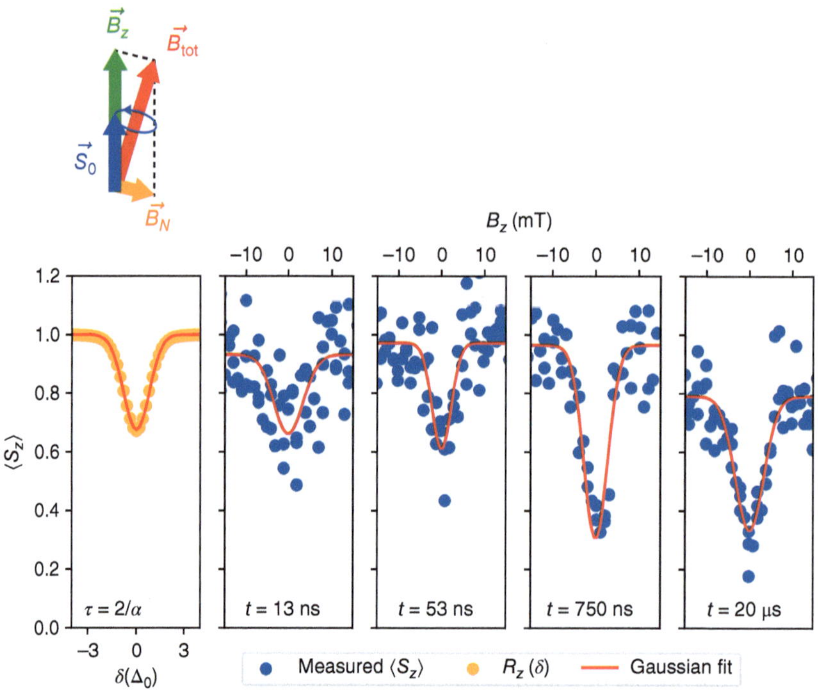

Figure 16.15 Hole spin polarization vs. magnetic field for specific spin storage times: The leftmost panel shows the spin polarization calculated according to Eq. (16.28) as a function of the normalized out of plane magnetic field δ, with the time fixed to $\tau = 2/\alpha$ and $\alpha = 0.2$. The other panels show the measurement results of $\langle S_z \rangle$ as a function of the external field B_z recorded at storage times of 13 ns, 53 ns, 750 ns, and 20 μs, respectively.

randomly oriented Overhauser field $\mathbf{B}_N$, causing the spin ensemble to lose its initial polarization over time. The experimental results displayed in Figure 16.15 clearly show a strong loss of polarization toward longer storage time and a reduction of polarization even at high magnetic fields for ultra long storage times in the tens of micro-seconds regime. The latter effect is due to T_1 processes like random spin flips starting to play a role. A theoretical analysis has shown that for valence band holes in InGaAs quantum dots, T_1 diverges when approaching zero magnetic field due to the suppression of SO mediated single phonon scatterings [123–125]. Experimentally, however, a saturation of T_1 at about 200 μs was reported at low fields, which largely agrees with our observation here [124].

The spread of the nuclear Overhauser field strength σ_n is expressed by the full width half maximum (FWHM) of the relaxation dip. However, this dependence is not constant but depends on the degree of anisotropy, as an examination of $R_z(\delta)$ at different anisotropy values α shows. Figure 16.16 shows the ratio of the FWHM and the actual σ_n as a function of the anisotropy, extracted from an analytical calculation of $R_z(\tau, \delta, \alpha)$ in the limit $\tau \gg 1$. Using this dependency, we can normalize the FWHM retrieved from the fits to the data in order to derive the actual magnitude of the Overhauser field dispersion. With the value for α, which is found by analyzing the spin

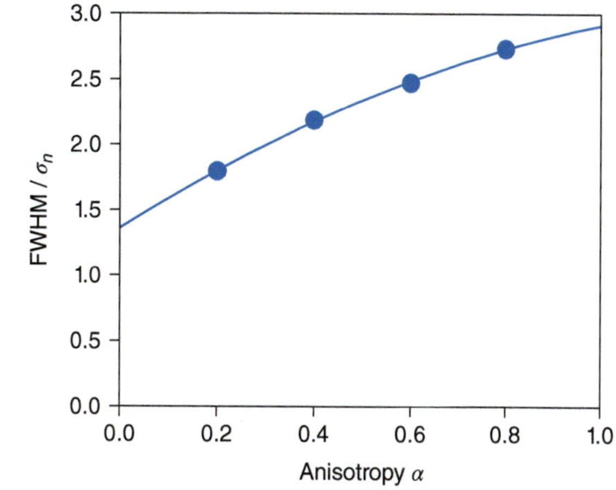

Figure 16.16 Overhauser distribution with anisotropy: The ratio between the FWHM of the relaxation dip and the dispersion of the Overhauser field σ_n calculated for some values of the hyperfine anisotropy factor α. The numerically calculated values are fitted using a parabola. This relation allows the determination of the true σ_n from the measurement, provided the degree of anisotropy is known.

decay dynamics below, we get the average dispersion $\langle\sigma_n\rangle = 3.7 \pm 0.5$ mT. This value is ~40% of the field dispersion experienced by an electron spin. As it was outlined in Section 16.5.1 α is yet to be determined from the spin relaxation dynamics. Thus, we briefly summarize recent theoretical findings about α from the literature: The degree of hyperfine coupling anisotropy for valence band holes remains a disputed quantity, not only in reference to the data set at hand, but generally in hole spin literature throughout the years [38, 86, 97, 102, 108, 113, 114, 126]. First theoretical calculations suggested $\alpha = 0$ in absence of heavy-hole light-hole mixing [102], meaning that heavy holes are free of transverse hyperfine couplings. Shortly after, the concept of valence band mixing in self-assembled quantum dots was added to the theory and experimental confirmation was given [86, 113]. Conclusions from later experiments hinted at the presence of transverse hyperfine coupling enhanced and even dominated by the admixture of d-orbital states to the valence band [97]. Based on this model, further experiments and theoretical works have found a wide range of coupling strengths from weakly anisotropic coupling of around $\alpha = 0.3$ [38, 113, 126] to extremely strong anisotropy with $\alpha = 0.01$ [114].

16.5.4 Fast Dynamics of the Central Spin

Figure 16.17 shows the spin polarization at zero magnetic field, measured from 1 ns to 40 μs. The data points are given by the value of $\langle S_z(B)\rangle$ at the center of the dip in a magnetic field scan, as presented in Figure 16.15. The error bars are calculated from the variance of $\langle S_z\rangle_{B=0}$ obtained typically across three or more individual field scans at the same storage time. At time zero, the central spin is initialized to the spin-up state, therefore the projection along the z-direction starts out close to 1 [66].

For the most part, spin polarization declines monotonically interrupted by two characteristic plateaus whose particular shape and position allow us to measure the parameters of hole spin hyperfine coupling and nuclear quadrupolar coupling [36]. The initial decay, up to ~50 ns, is attributed to the anisotropic hyperfine coupling detailed in Section 16.5.1. The green line represents a fit according to Eq. (16.26)

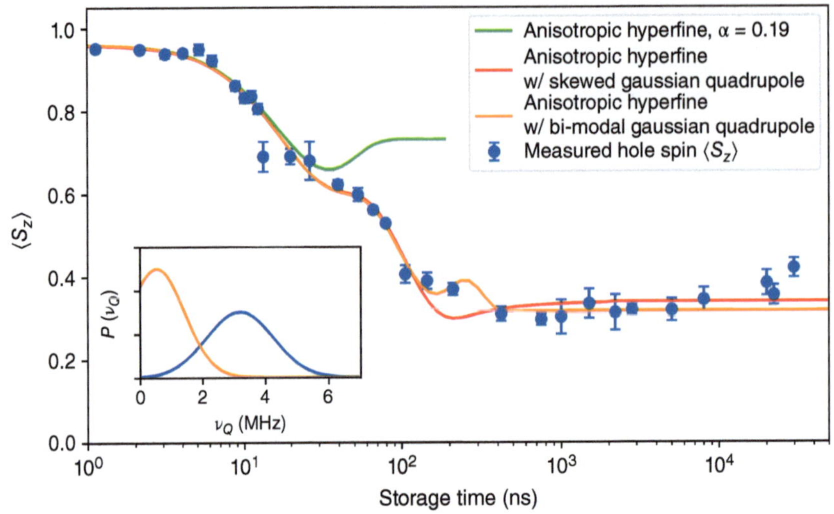

Figure 16.17 Hole spin relaxation at zero field: The spin polarization, measured at $B_z = 0$ T, is plotted as a function of the total storage time. The fast part of the relaxation is fitted using the anisotropic hyperfine coupling model shown in green (see Section 16.5.1). The slow part of the relaxation at $t > 80$ ns is fitted with one of two models, which are based on either a skewed Gaussian (red line) or bi-modal (orange line) distribution of quadrupolar frequencies ν_Q, respectively [35, 84, 122]. For the latter model, the inset shows the bi-modal quadrupolar frequency spectrum, which is found to produce the curve, which best fits the data set.

which yields an anisotropy factor $\alpha = 0.19$. This value lies well within the range suggested by previous experimental and theoretical works listed above. The local minimum of this partial relaxation appears at $T_{\text{min}} = 36.8$ ns. Since the spin is initialized in the z-direction, depolarization is a result of the in-plane components of hyperfine interaction. Therefore, the relaxation time gives an estimate of the in-plane coupling strength $\gamma_\perp$ via $T_{\text{min}} = \hbar/\gamma_\perp$ [38]. Consequently, we can derive the strength of the out-of-plane coupling using the relation [35, 86]

$$\alpha = \frac{\gamma_\perp}{\gamma_\parallel} \tag{16.32}$$

Applying Eq. (16.32) to our data we find $\gamma_\perp = 0.018$ μeV and $\gamma_\parallel = 0.095$ μeV.

Notably we find that the in-plane hole hyperfine coupling strength is about 5% of γ_e. This reduced coupling accounts for the more than ten-fold prolongation of the inhomogeneous broadening induced polarization decay [35, 36]. Reference to Figure 16.14b $\alpha < 1$ shows that the explicit shape of the initial decay of the hole spin polarization largely differs from the output of the theoretical model [86]. Unlike in this model, we observe no pronounced intermediate recovery of the polarization i.e. a shallow and elongated dip feature, since the slow relaxation due to quadrupolar coupling overlaps with the faster hyperfine induced decay. This produces a shoulder in the decay at $t \approx 30$ ns. A counter-example of this behavior was observed by Cogan et al. who present an experiment in which the emergence of the dip was aided by the reduced anisotropy $\alpha = 0.28$, despite a similar strength of quadrupolar coupling γ_Q [38]. The work of [37] is in better agreement with our observations. Their calculation employs a fully quantum mechanical treatment which is capable of describing the coupling to a nuclear spin ensemble of $N = \mathcal{O}(10 - 100)$ using the Chebyshev polynomial technique. There, quadrupolar coupling also overlaps with the initial Overhauser induced decay, obscuring the partial recovery of the polarization.

16.5.5 Slow Part of the Relaxation

Beyond the short timescale, where it is justified to treat the Overhauser field as frozen (Section 16.5.1), its dynamics start to play a role for central spin relaxation. The central spin rotates around the Overhauser field, and thus on the timescales larger than this rotation frequency, the central spin polarization follows the correlator of the Overhauser field component B_z:

$$S_z(t) \sim \left\langle \frac{B_z(t)B_z(0)}{B(t)B(0)} \right\rangle \tag{16.33}$$

This works roughly after 30 ns for hole doped QDs, i.e. after observing the first plateau in the relaxation curve. Analytical calculations of the correlator in Eq. (16.33) are impossible even in simplest models, due to the enlarged Hilbert space of 2^{2N+1} for a nuclear spin ensemble containing only spin $I = 3/2$. Hence, our goal here is to develop a more phenomenological but analytical formula that can fit experimental data with a few physically meaningful parameters. The theoretical approach was developed by N. Sinitsyn of Los Alamos National Labs. Different nuclear spins rotate around quadrupolar fields defined by n_i, independently of each other. The ith nuclear spin polarization is a constant plus an oscillation term:

$$\langle b_i(t)b_i(0)\rangle = A + B\cos(\omega_i\, t). \tag{16.34}$$

It is known from single quantum dot nuclear magnetic resonance (NMR) spectroscopy and accompanying theoretical works that each nuclear species contributes individually to form a complex nuclear frequency spectrum, which defines the distribution of ω_i [96, 121, 122, 127]. In addition the strain topology and chemical composition profile of the quantum dot strongly influence the quadrupolar mode composition [84, 96]. Thus, the quadrupolar broadened nuclear spin spectrum can exhibit arbitrary complexity and will generally be different for each quantum dot. We will therefore provide two simple approximations that provide different levels of accuracy in representing experimental data.

In Eq. (16.34), coefficients A and B are generally specific to the nucleus i, but we assume their *typical* values for simplicity. This is justified because A_i and B_i would be statistically independent of ω_i, so in our calculations they will get replaced by their average values when averaging over the frequency distribution. We can also safely disregard the much weaker Knight field acting back on the nuclear ensemble (see Section 16.5.2). Our first approximation is that different nuclear spins produce independent contributions to the correlator in Eq. (16.33). This is definitely not a reliable approximation but corrections are likely to be small since they emerge from the denominator that changes slowly with time. The correlator Eq. (16.33) is then, up to a constant, the sum of the correlators Eq. (16.34), i.e. the assumption states that each nuclear spin contributes with some constant plus an oscillating term.

For the first of the two ways of approximating the nuclear spin spectrum, we assume that quadrupolar fields $\boldsymbol{n}_i$ have a uniform direction distribution with a Gaussian magnitude distribution along each axis [35, 36]. A similar approach was successfully used to describe the long time relaxation of electron spins in a strained quantum dot [36]. This means that the probability distribution of $\omega_i = \sqrt{\omega_{ix}^2 + \omega_{iy}^2 + \omega_{iz}^2}$ is

$$P_{\text{uni}}(\omega_i) \sim \omega_i^2 \, \mathrm{e}^{-\omega_i^2/(2\omega_0^2)} \tag{16.35}$$

Here we show it is insufficient to capture the specific decay of a hole spin observed in the experiment due to the different ratio of hyperfine and quadrupolar coupling strengths compared to the electron. We simulate a less generic, qualitatively more realistic description of the nuclear spectrum to obtain an idea of the impact of different hyperfine noise spectra. It is reasonable to assume that many nuclear spins feel relatively weak anisotropy forces. Hence, the probability for averaging can have a smaller power near the Gaussian exponent, for example

$$P(\omega_i) \sim \omega_i^{2-r} \, \mathrm{e}^{-\omega_i^2/(2\omega_0^2)} \tag{16.36}$$

where r is a new parameter, whose positive values favor the bigger density of the weaker coupled nuclear spins. Figure 16.18a shows normalized examples of the modified distribution for different values of r. It becomes apparent that the distribution becomes skewed and the spectral weight shifts toward zero frequency for $r > 0$ which is equivalent with an increasing number of nuclear spins oscillating with low quadrupolar precession frequency. Averaging the correlator (16.34) over the frequency distribution Eq. (16.36), i.e. evaluating the integral

$$\int_0^\infty P(\omega_i)(A + B\cos(\omega_i\, t))\mathrm{d}\omega_i$$

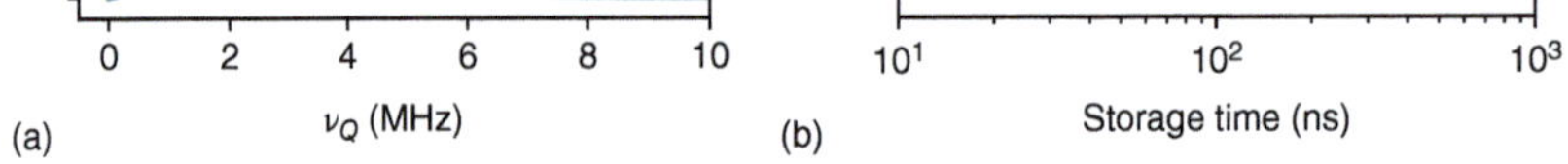

Figure 16.18 Modified Gaussian distribution of quadrupolar frequencies: Normalized distributions as defined in Eq. (16.36), for different levels of skewness quantified by r. The spectral weight is shifted toward zero frequency for $r > 0$. In the time domain, this produces a relaxation dip, which is less pronounced and spans across a wider part of the spin storage duration. Shifting the mean value of the frequency spectrum to lower values results in a longer relaxation time.

we find

$$\langle S_z(t)\rangle \equiv U_1(t) = a + b\cdot{}_1F_1\left[\frac{3-r}{2}, \frac{1}{2}, -\frac{(\omega_0 t)^2}{2}\right] \tag{16.37}$$

where ${}_1F_1[u, v, x]$ is the confluent hypergeometric function of the first kind [128, 129]. The values a, b are scaling parameters with which we can set $U_1(0) = 1$ by defining $b + a = 1$ and the saturation toward infinite time $U_1(t \to \infty) = a$. In Figure 16.18b, the behavior of the Function (16.37) is visualized at $r = 0$, $r = 1.0$ and $r = 1.7$, next to the corresponding frequency spectra. The blue curve, where $r = 0$ is the decay for the unmodified distribution (16.35) which yields a complete relaxation of and partial recovery of $\langle S_z\rangle$ as it was observed for the conduction band electron [36, 38]. With increasing r the relaxation dip becomes more shallow. Note that the time axis is logarithmic so the tail for $r = 1.7$ relaxes about an order of magnitude longer than for $r = 0$.

We can now proceed to fit the experimental data. Due to the separation of time scales of the fast and slow decay, we can fit the experimental curve by the product of the functions $R_z(t)$ and $U_1(t)$ that were defined in (16.28) and (16.37), respectively.

$$S_z(t) = \eta\, R_z(t)\, U_1(t) \tag{16.38}$$

where η is a rescaling parameter that makes $S_z(t)$ start at a non-unit value, if needed. Fitting this to the experimental values, we find a good qualitative agreement as illustrated by the red line in Figure 16.17. Except for data points at ultralong storage time $t > 10$ s, we are able to reproduce the experiment with this model and we find a saturation value $a = 0.35$, meaning that within the accessible temporal range we do not observe a complete relaxation of spin polarization. Neither do we observe the onset of the complete polarization decay induced by nuclear dipole–dipole coupling, which is observable for conduction band electrons beyond $t > 3$ [36, 66]. The minimum of the quadrupolar relaxation feature is located at $T_{\min} = 210$ ns, which roughly translates to a quadrupolar coupling strength of $\gamma_Q = 3.31$ nev. This relatively large value, when compared to a similar quantum dot, may emerge for two different reasons: The randomness of strain and composition profile can lead to large dot-to-dot variations of quadrupolar coupling γ_Q [37, 100] and secondly, the unspecific shape of the long-term decay between 100 ns and 10 μs prevents us from confidently pinpointing the characteristic quadrupolar relaxation time which results in the overestimation of γ_Q.

This theory also provides physical meaning of the time-dependent relaxation curve. Let us disregard a possible dependence of parameters A and B on i, i.e. the index of the nuclear spin. Then the time-dependent relaxation is up to a constant factor the Fourier transform of the probability distribution:

$$P(\omega) = \sum_i P(\omega_i)\delta(\omega - \omega_i) \tag{16.39}$$

which means that if we take the Fourier transform of the relaxation curve, we should obtain the power spectrum for nuclear spin bath quadrupolar fields. Thus, by taking the reverse path we are able to produce a fitting model for the time domain relaxation by iteratively guessing a suitable frequency spectrum and transforming it from the frequency domain into the time domain. In this sense we introduce the second model for the nuclear spin spectrum, which lends its structure from the experimentally reconstructed spectrum found by Munsch et al. by using an isotope-sensitive NMR method on a single quantum dot [122]. They found that the spectrum comprised of the isotopes ^{69}Ga, ^{71}Ga, ^{115}In, and ^{75}As approximately is a sum of Gaussian peaks of isotope-specific widths and center frequencies $\nu_Q > 0$. Due to the coarse temporal resolution of our dataset, we limit the number of individual peaks in our model spectrum to two. This approach is justified because in the spectrum reconstructed by Munsch et al. [122], the contributions of Gallium and Arsenic largely overlap and only Indium produces a distinct narrow-linewidth peak in the spectrum at low frequency. The reason for the dominant role of Indium in the spectrum is its large nuclear spin $I = 9/2$ combined with the large electric field gradients produced by this atomic species, which is further enhanced in quantum dots with *random* InGaAs alloy compositions [121]. In this case, the analytic expression of the normalized frequency spectrum is the sum of two Gaussian distributions and reads:

$$f(\omega) = \frac{1}{2}\sum_{i}^{2} A_i^{-1} \exp\left[-\frac{(\omega - \omega_i)^2}{2\sigma_i^2}\right] \tag{16.40}$$

with A_i denoting the integrated area of an individual truncated peak in the interval $\omega_i \in [0, \infty[$

$$A_i = \sqrt{\frac{\pi}{2}}, \sigma_i \left(1 + \text{Erf}\left(\frac{\omega_i}{\sqrt{2}\sigma_i}\right)\right) \tag{16.41}$$

ω_i and σ_i denote the center frequency and standard deviation of the frequency distribution, respectively. By taking the real part of the Fourier transform of $f(\omega)$ we obtain the fitting function. Again, we use the variables a and b for scaling to the values of $\langle S_z \rangle$ between 0 and 1.

$$U_2(t) \equiv a + b \cdot \text{Re}\left[\mathcal{F}\left(f(\omega)\right)\right] = a + b \cdot \text{Re}\left[\sum_{i}^{2} A_i^{-1}\sigma_i \exp\left(it\,\omega_i - \frac{t^2\sigma_i^2}{2}\right)\right] \tag{16.42}$$

We now use the function $S_z(t) = \eta\, R_z(t)\, U_2(t)$ for fitting and the result is shown by the orange line in Figure 16.17. The best fit is obtained when using the bi-modal frequency spectrum displayed in the inset of Figure 16.17, for which we find the distributions centered at the quadrupolar frequencies $\langle \nu_{Q1} \rangle = 0.5$ MHz and $\langle \nu_{Q2} \rangle = 3.2$ MHz, where the average is taken over all nuclei within the hole wavefunction. Respectively the dispersions are $\sigma_1 = 0.9$ MHz and $\sigma_2 = 1.1$ MHz. These numbers compare well to $\langle \nu_{\text{In}} = 1.5$ MHz and $\langle \nu_{\text{Ga+As}} \rangle = 2.18$ MHz with dispersions $\sigma_{\text{In}} = 0.4$ MHz and $\sigma_{\text{Ga+As}} \approx 1.95$ MHz reported by Munsch et al. [122]. Upon closer inspection of the fit in the time interval between 100 and 500, we observe a small oscillatory feature resulting from transformation the bi-modal frequency distribution. Even though due to this feature the fit is tracing a few data points more closely than the simple fit described before (see red line in Figure 16.17), the

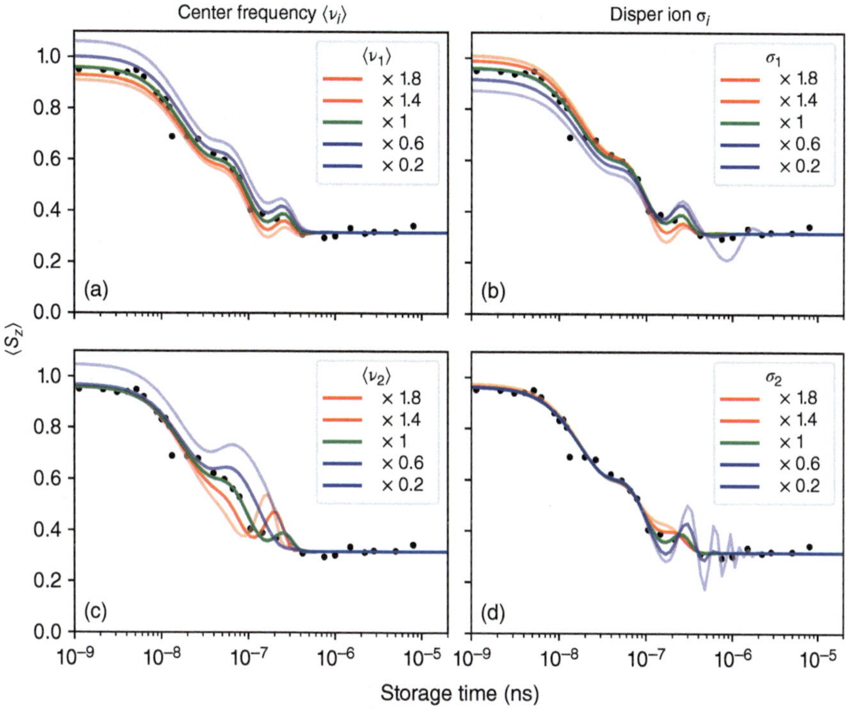

Figure 16.19 Quadrupolar parameter variation: the model based on a bi-modal quadrupolar frequency distribution is plotted vs. spin storage time. In each panel, one parameter of the model is varied by up to ±80%. (a, c) The effect of shifting the center frequencies $\langle \nu_1 \rangle$ and $\langle \nu_2 \rangle$, respectively. (b, d) The effect of varying the frequency dispersions σ_1 and σ_2. For reference, the best fit to the dataset (black dots) is plotted in green.

oscillatory feature is not present in the experimental data set. In order to determine how the parameters of the nuclear frequency spectrum affect the shape of the fitting curve in the time-domain, we vary $\langle \nu_i \rangle$ and σ_i of the two peaks in the spectrum individually. Figure 16.19a,c illustrate the influence of the average quadrupolar frequency $\langle \nu_1 \rangle$ and $\langle \nu_2 \rangle$, respectively. The red traces result from shifting to higher frequency by up to 80, while the blue traces correspond to a decrease in frequency. Note that the scaling coefficients a and b from Eq. (16.42) are kept constant for all traces in this figure and they are equal to the ones determined for the best fitting model displayed in Figure 16.17 (orange line). Shifting the lower frequency peak $\langle \nu_1 \rangle$ back and forth does not result in a visible change of the oscillatory behavior, since low-frequency components mostly affect the behavior on long timescales. The vertical shift of the time-domain decay curve can be attributed to the change of the zero-frequency component in the spectrum when the peak center approaches zero. From panel (c) we conclude that the position of the high frequency peak strongly influences the spin decay between ~20 ns and ~300 ns, since larger quadrupolar precession frequencies accelerate the spin decay via the dynamics of the Overhauser field. The oscillatory behavior in this time range is enhanced for larger quadrupolar frequencies (red curves), since in this case the peak represents a more *discrete* frequency component. Contrary to this, the total spectrum effectively takes the shape of a single low frequency wideband component when the peak is shifted to lower frequencies. Figure 16.19b,d shows that a reduced width of either peak in the spectrum generally leads to an enhancement of the oscillations between ~100 ns and ~2 μs, as we expect from a narrow band component in the spectrum. This confirms the absence of *discrete* spectral features in the actual quadrupolar spectrum at zero magnetic field, contrary to what is often observed at high magnetic fields where nuclear Zeeman transitions lead to narrow lines in the spectrum [84, 127].

Thc fact that the shape of the time-domain decay curve is a sensitive probe for at least four parameters that define the nuclear quadrupolar spectrum allows an important conclusion about this spin relaxation measurement on hole spins: The about tenfold decrease of the hyperfine coupling strength of holes compared to electrons not only prolongs the hole spin coherence but also strongly alters the signature of quadrupolar coupling in the central spin relaxation curve. Thus, the hole spin is a suitable probe of the underlying nuclear spin dynamics, i.e. the structure of quadrupolar broadened frequency spectrum via this time domain measurement principle. From single quantum dot NMR spectroscopy [127], transmission electron microscopy [120], and atomistic modeling [121], it is known that the spherically symmetric model for quadrupolar field distribution, which is often used to interpret electron spin relaxation, is oversimplified [36, 38]. Yet, the prominent quadrupolar relaxation dip found in these measurements does not permit a more granulated reconstruction of the topology of quadrupolar coupling, strain, or chemical composition. This is due to the strong isotropic hyperfine coupling of electrons lead to this relaxation mechanism to be completely dominated by strongly coupled nuclei with large spin. The reduced hyperfine coupling of hole spins to the dynamic nuclear ensemble in combination with increased sensitivity and temporal range of the experiment allow us to observe the actual implications of the rich nuclear spectrum, which is often found by other means, for example frequency domain measurements [122, 127, 130]. In addition, the nuclear spin ensemble is even more weakly perturbed by the Knight field produced by the central hole spin due to its reduced hyperfine coupling.

16.6 Summary

In this chapter, we performed pump-probe type of measurements on individual electron and hole spin qubits. For electron spin qubits, we demonstrated that the spin evolution is dominated by an additional intermediate relaxation mechanism, in contrast to earlier numerous theoretical predictions. An important and unexpected finding of our work is that at low magnetic fields the intrinsic coherence time of the electron is drastically reduced by combined effects of hyperfine and quadrupolar interactions, reducing T_2 to a value of $\sqrt{T_2^* T_Q}$, corresponding to only ~4 ns in our case. Only the application of a magnetic field stronger than the strength of the quadrupolar coupling could remove these effects. Our findings have major implications for extending the coherence time evolution of electron spin qubits at low magnetic fields. While the inhomogeneous dephasing time T_2^* can be reversed by the application of spin-echo methods, the strong back-action of quadrupolar couplings in the spin bath demonstrates the necessity of the development of strain engineered QD structures for extended coherence times T_2, which may also help pave the way toward creating useful semiconductor-based spin qubits that can be incorporated into quantum information devices.

For single hole spins preformed analogous experiments and probe the interaction with the surrounding nuclear spin ensemble. Hereby, we obtained insight on the underlying nuclear spin dynamics and probed how the hole spin decay evolves at long time scales. We confirm an $\sim$100× times longer spin dephasing time $T_2^* \approx 210$ ns for the hole spin over the electron spin. A promising number for applications that utilize the hole spin as a stationary storage qubit in long-distance quantum information protocols [102, 131]. The signature of anisotropic hyperfine coupling was observed in spin polarization decay, allowing us to quantify the degree of anisotropy $\alpha = 0.19$ which is fundamental to the character of the confined hole spin wave function. By modeling this behavior, we derived the degree of light-hole heavy-hole mixing, which is an essential mechanism for enabling hole spin dephasing and thus refining the description of hole hyperfine coupling beyond the initially suggested pure Ising form [86, 102].

Ackowledgments

J. J. F. would like to acknowledge many people for their tremendous contributions toward research into spin qubits and spin–photon interfaces using III–V nanostructures over the past decade in Munich. Beyond the authors of this chapter, on whose work the presented material is based, I would like to very gratefully acknowledge all current and past PhD, MSc, and BSc students in the SNQS chair at TUM (J. J. Finley), as well as the current and past members of the QEC group led by K. Mueller. I am also extremely grateful to our worldwide network of scientific collaborators, besides Nikolai A. Sinitsyn, J.J.F. is very grateful to Łukasz CYWIŃSKI (Polish Academy of Sciences, Warsaw) for the many enlightening and stimulating conversations on the theory of spins in condensed matter systems. Both you and Nik have undoubtedly forgotten more about spin qubits than I will ever know. I would also like to thank the highly talented postdoctoral researchers supported by the Alexander von Humboldt foundation who have come to Munich and made such vital and inspiring contributions. Last, but not least I am eternally grateful to the talented technicians and administrative staff at the Walter Schottky Institute. All of you have made important contributions to this research – thank you and please forgive me for not naming you all! Without funding there is no science. Here, JJF very gratefully acknowledges the German Science Foundation (Deutsche Forschungsgemeinschaft) for funding via the excellence clusters **MCQST** (EXC211 www.mcqst.de), **e-conversion** (EXC2089, www.e-conversion.de), **DIP** (FI 947-6), **SQAM** (FI947-5), and the BMBF via **QR.X**,

PhotonQ, **Q.Link** and **SPINNING**. Finally, JJF thanks the Bavarian Ministry for Science, Education and the Arts for funding via **Munich Quantum Valley** and the European Union for funding via **QLUSTER**.

References

1 Awschalom, D.D., Hanson, R., Wrachtrup, J., and Zhou, B.B. (2018). Quantum technologies with optically interfaced solid-state spins. *Nature Photonics* 12 (9): 516–527.

2 Kane, B.E. (1998). A silicon-based nuclear spin quantum computer. *Nature* 393 (6681): 133–137.

3 Lodahl, P., Mahmoodian, S., and Stobbe, S. (2015). Interfacing single photons and single quantum dots with photonic nanostructures. *Reviews of Modern Physics* 87 (2): 347–400.

4 Degen, C.L., Reinhard, F., and Cappellaro, P. (2017). Quantum sensing. *Reviews of Modern Physics* 89 (3): 035002.

5 Weiss, L., Gritsch, A., Merkel, B., and Reiserer, A. (2021). Erbium dopants in nanophotonic silicon waveguides. *Optica* 8 (1): 40.

6 Ferrando-Soria, J., Pineda, E.M., Chiesa, A. et al. (2016). A modular design of molecular qubits to implement universal quantum gates. *Nature Communications* 7 (1): 11377.

7 Doherty, M.W., Manson, N.B., Delaney, P. et al. (2013). The nitrogen-vacancy colour centre in diamond. *Physics Reports* 528 (1): 1–45.

8 Gottscholl, A., Diez, M., Soltamov, V. et al. (2021). Room temperature coherent control of spin defects in hexagonal boron nitride. *Science Advances* 7 (14): eabf3630.

9 Loss, D. and DiVincenzo, D.P. (1998). Quantum computation with quantum dots. *Physical Review A* 57 (1): 120–126.

10 Kroutvar, M., Ducommun, Y., Heiss, D. et al. (2004). Optically programmable electron spin memory using semiconductor quantum dots.*Nature* 432 (7013): 81–84.

11 Balasubramanian, G., Neumann, P., Twitchen, D. et al. (2009). Ultralong spin coherence time in isotopically engineered diamond. *Nature Materials* 8 (5): 383–387.

12 Mitchell, M., Lake, D.P., and Barclay, P.E. (2019). Realizing Q > 300 000 in diamond microdisks for optomechanics via etch optimization. *APL Photonics* 4 (1): 016101.

13 Liu, R.-B., Yao, W., and Sham, L.J. (2010). Quantum computing by optical control of electron spins. *Advances in Physics* 59 (5): 703–802.

14 Warburton, R.J. (2013). Single spins in self-assembled quantum dots. *Nature Materials* 12 (6): 483–493.

15 Gangloff, D.A., Zaporski, L., Bodey, J.H. et al. (2021). Witnessing quantum correlations in a nuclear ensemble via an electron spin qubit. *Nature Physics* 17 (11): 1247–1253.

16 Azuma, K., Tamaki, K., and Lo, H.-K. (2015). All-photonic quantum repeaters. *Nature Communications* 6 (1).

17 Briegel, H.J., Browne, D.E., Dür, W. et al. (2009). Measurement-based quantum computation. *Nature Physics* 5 (1): 19–26.

18 Raussendorf, R. and Briegel, H.J. (2001). A one-way quantum computer. *Physical Review Letters* 86 (22): 5188–5191.

19 Raussendorf, R., Browne, D.E., and Briegel, H.J. (2003). Measurement-based quantum computation on cluster states. *Physical Review A* 68 (2).

20 Nielsen, M.A. (2004). Optical quantum computation using cluster states. *Physical Review Letters* 93 (4).

21 Rudolph, T. (2017). Why I am optimistic about the silicon-photonic route to quantum computing. *APL Photonics* 2 (3): 030901.

22 Wang, J., Sciarrino, F., Laing, A., and Thompson, M.G. (2019). Integrated photonic quantum technologies. *Nature Photonics* 14 (5): 273–284.

23 Varnava, M., Browne, D.E., and Rudolph, T. (2006). Loss tolerance in one-way quantum computation via counterfactual error correction. *Physical Review Letters* 97 (12).

24 Knill, E., Laflamme, R., and Milburn, G.J. (2001). A scheme for efficient quantum computation with linear optics. *Nature* 409 (6816): 46–52.

25 Pichler, H., Choi, S., Zoller, P., and Lukin, M.D. (2017). Universal photonic quantum computation via time-delayed feedback. *Proceedings of the National Academy of Sciences of the United States of America* 114 (43): 11362–11367.

26 Li, F., Crooker, S.A., and Sinitsyn, N.A. (2016). Higher-order spin-noise spectroscopy of atomic spins in fluctuating external fields. *Physical Review A* 93 (3): 033814.

27 Unsleber, S., He, Y.-M., Gerhardt, S. et al. (2016). Highly indistinguishable on-demand resonance fluorescence photons from a deterministic quantum dot micropillar device with 74% extraction efficiency. *Optics Express* 24 (8): 8539.

28 Hanschke, L., Fischer, K.A., Appel, S. et al. (2018). Quantum dot single photon sources with ultra-low multi-photon error rate. *Conference on Lasers and Electro-Optics*. OSA.

29 Fischer, K.A., Hanschke, L., Wierzbowski, J. et al. (2017). Signatures of two-photon pulses from a quantum two-level system. *Nature Physics* 13 (7): 649–654.

30 De Greve, K., Yu, L., McMahon, P.L. et al. (2012). Quantum-dot spin–photon entanglement via frequency downconversion to telecom wavelength. *Nature* 491 (7424): 421–425.

31 Ardelt, P.-L., Gawarecki, K., Müller, K. et al. (2016). Coulomb mediated hybridization of excitons in coupled quantum dots. *Physical Review Letters* 116 (7): 077401.

32 Müller, K., Bechtold, A., Ruppert, C. et al. (2012). Electrical control of interdot electron tunneling in a double InGaAs quantum-dot nanostructure. *Physical Review Letters* 108 (19).

33 Loredo, J.C., Broome, M.A., Hilaire, P. et al. (2017). Boson Sampling with Single-Photon Fock States from a Bright Solid-State Source. *Physical Review Letters* 118 (13): 130503.

34 Wang, H., Duan, Z.-C., Li, Y.-H. et al. (2016). Near-transform-limited single photons from an efficient solid-state quantum emitter. *Physical Review Letters* 116 (21).

35 Sinitsyn, N.A., Li, Y., Crooker, S.A. et al. (2012). Role of nuclear quadrupole coupling on decoherence and relaxation of central spins in quantum dots. *Physical Review Letters* 109 (16): 166605.

36 Bechtold, A., Rauch, D., Li, F. et al. (2015). Three-stage decoherence dynamics of an electron spin qubit in an optically active quantum dot. *Nature Physics* 11 (12): 1005–1008.

37 Hackmann, J., Glasenapp, P., Greilich, A. et al. (2015). Influence of the nuclear electric quadrupolar interaction on the coherence time of hole and electron spins confined in semiconductor quantum dots. *Physical Review Letters* 115 (20): 207401.

38 Cogan, D., Kenneth, O., Lindner, N.H. et al. (2018). Depolarization of electronic spin qubits confined in semiconductor quantum dots. *Physical Review X* 8 (4): 041050.

39 Dyakonov, M.I. (2006). Introduction to spin physics in semiconductors. *Physica E: Low-dimensional Systems and Nanostructures* 35 (2): 246–250.

40 Pimpinelli, A. and Villain, J. (1998). *Physics of Crystal Growth*. Cambridge University Press.

41 Solomon, G.S., Trezza, J.A., Marshall, A.F., and Harris, J.S. Jr. (1996). Vertically aligned and electronically coupled growth induced InAs islands in GaAs. *Physical Review Letters* 76 (6): 952–955.

42 Stangl, J., Holý, V., and Bauer, G. (2004). Structural properties of self-organized semiconductor nanostructures. *Reviews of Modern Physics* 76 (3): 725–783.

43 Bimberg, D., Grundmann, M., and Ledentsov, N.N. (1998). *Quantum Dot Heterostructures*. Chichester, England: John Wiley & Sons.

44 Eto, M. (1997). Electronic structures of few electrons in a quantum dot under magnetic fields. *Japanese Journal of Applied Physics* 36 (Part 1, No. 6B): 3924–3927.

45 Warburton, R.J., Miller, B.T., Dürr, C.S. et al. (1998). Coulomb interactions in small charge-tunable quantum dots: a simple model. *Physical Review B* 58 (24): 16221–16231.

46 Bayer, M., Walck, S.N., Reinecke, T.L., and Forchel, A. (1998). Exciton binding energies and diamagnetic shifts in semiconductor quantum wires and quantum dots. *Physical Review B* 57 (11): 6584–6591.

47 Hawrylak, P. (1999). Excitonic artificial atoms: engineering optical properties of quantum dots. *Physical Review B* 60 (8): 5597–5608.

48 Michler, P. (ed.) (2003). *Single Quantum Dots, Topics in Applied Physics*, 2003 edition. Berlin, Germany: Springer-Verlag.

49 Heiss, D., Jovanov, V., Caesar, M. et al. (2009). Selective optical charge generation, storage, and readout in a single self-assembled quantum dot. *Applied Physics Letters* 94 (7).

50 Heiss, D., Jovanov, V., Klotz, F. et al. (2010). Optically monitoring electron spin relaxation in a single quantum dot using a spin memory device. *Physical Review B* 82 (24): 245316.

51 Urbaszek, B., Marie, X., Amand, T. et al. (2013). Nuclear spin physics in quantum dots: an optical investigation. *Reviews of Modern Physics* 85 (1): 79–133.

52 Jovanov, V. (2012). Optically probing spin-spin interactions in individual quantum dot nanostructures. PhD thesis. Technische Universität München.

53 Jovanov, V., Kapfinger, S., Bichler, M. et al. (2011). Direct observation of metastable hot trions in an individual quantum dot. *Physical Review B - Condensed Matter and Materials Physics* 84 (23): 1–5.

54 Bayer, M., Ortner, G., Stern, O. et al. (2002). Fine structure of neutral and charged excitons in self-assembled In(Ga)As/(Al)GaAs quantum dots. *Physical Review B* 65 (19): 195315.

55 Ardelt, P.-L., Koller, M., Simmet, T. et al. (2016). Optical control of nonlinearly dressed states in an individual quantum dot. *Physical Review B* 93 (16): 165305.

56 Greilich, A., Economou, S.E., Spatzek, S. et al. (2009). Ultrafast optical rotations of electron spins in quantum dots. *Nature Physics* 5 (4): 262–266.

57 Economou, S.E., Sham, L.J., Wu, Y., and Steel, D.G. (2006). Proposal for optical U(1) rotations of electron spin trapped in a quantum dot. *Physical Review B* 74 (20): 205415.

58 Economou, S.E. and Reinecke, T. (2007). Theory of fast optical spin rotation in a quantum dot based on geometric phases and trapped states. *Physical Review Letters* 99 (21): 217401.

59 Gao, W.B., Imamoglu, A., Bernien, H., and Hanson, R. (2015). Coherent manipulation, measurement and entanglement of individual solid-state spins using optical fields. *Nature Photonics* 9 (6): 363–373.

60 Emary, C., Xu, X., Steel, D.G. et al. (2007). Fast initialization of the spin state of an electron in a quantum dot in the voigt configuration. *Physical Review Letters* 98 (4): 047401.

61 Carter, S.G., Economou, S.E., Greilich, A. et al. (2014). Strong hyperfine-induced modulation of an optically driven hole spin in an InAs quantum dot. *Physical Review B* 89 (7): 075316.

62 Cywiński, L., Witzel, W.M., and Das Sarma, S. (2009). Electron spin dephasing due to hyperfine interactions with a nuclear spin bath. *Physical Review Letters* 102 (5): 1–4.

63 Witzel, W.M. and Das Sarma, S. (2006). Quantum theory for electron spin decoherence induced by nuclear spin dynamics in semiconductor quantum computer architectures: spectral diffusion of localized electron spins in the nuclear solid-state environment. *Physical Review B - Condensed Matter and Materials Physics* 74 (3): 1–24.

64 Bluhm, H., Foletti, S., Neder, I. et al. (2010). Dephasing time of GaAs electron-spin qubits coupled to a nuclear bath exceeding 200 μs. *Nature Physics* 7 (2): 109–113.

65 Zhao, N., Ho, S.-W., and Liu, R.-B. (2012). Decoherence and dynamical decoupling control of nitrogen vacancy center electron spins in nuclear spin baths. *Physical Review B* 85 (11).

66 Merkulov, I.A., Efros, A.L., and Rosen, M. (2002). Electron spin relaxation by nuclei in semiconductor quantum dots. *Physical Review B* 65 (20): 205309.

67 Khaetskii, A.V., Loss, D., and Glazman, L. (2002). Electron spin decoherence in quantum dots due to interaction with nuclei. *Physical Review Letters* 88 (18): 186802.

68 Al-Hassanieh, K.A., Dobrovitski, V.V., Dagotto, E., and Harmon, B.N. (2006). Numerical modeling of the central spin problem using the spin-coherent-state P representation. *Physical Review Letters* 97 (3): 1–4.

69 Faribault, A. and Schuricht, D. (2013). Integrability-based analysis of the hyperfine-interaction-induced decoherence in quantum dots. *Physical Review Letters* 110 (4): 1–5.

70 Braun, P.-F., Marie, X., Lombez, L. et al. (2005). Direct observation of the electron spin relaxation induced by nuclei in quantum dots. *Physical Review Letters* 94 (11).

71 Dou, X.M., Sun, B.Q., Jiang, D.S. et al. (2011). Electron spin relaxation in a single InAs quantum dot measured by tunable nuclear spins. *Physical Review B* 84 (3).

72 Johnson, A.C., Petta, J.R., Taylor, J.M. et al. (2005). Triplet-singlet spin relaxation via nuclei in a double quantum dot. *Nature* 435 (7044): 925–928.

73 Sallen, G., Kunz, S., Amand, T. et al. (2014). Nuclear magnetization in gallium arsenide quantum dots at zero magnetic field. *Nature Communications* 5: 3268.

74 Maletinsky, P., Badolato, A., and Imamoglu, A. (2007). Dynamics of quantum dot nuclear spin polarization controlled by a single electron. *Physical Review Letters* 99 (5).

75 Auer, T., Oulton, R., Bauschulte, A. et al. (2009). Measurement of the Knight field and local nuclear dipole-dipole field in an InGaAs/GaAs quantum dot ensemble. *Physical Review B - Condensed Matter and Materials Physics* 80 (20): 1–15.

76 Lai, C.W., Maletinsky, P., Badolato, A., and Imamoglu, A. (2006). Knight-field-enabled nuclear spin polarization in single quantum dots. *Physical Review Letters* 96 (16): 1–4.

77 Erlingsson, S.I. and Nazarov, Y.V. (2004). Evolution of localized electron spin in a nuclear spin environment. *Physical Review B - Condensed Matter and Materials Physics* 70 (20): 1–9.

78 Merkulov, I.A., Alvarez, G., Yakovlev, D.R., and Schulthess, T.C. (2010). Long-term dynamics of the electron-nuclear spin system of a semiconductor quantum dot. *Physical Review B - Condensed Matter and Materials Physics* 81 (11): 1–13.

79 Dyakonov, M.I. (2008). *Spin Physics in Semiconductors*. ISBN 9783540788195. http://link.springer.com/content/pdf/10.1007/978-3-540-78820-1.pdf.

80 Khaetskii, A.V. and Nazarov, Y.V. (2001). Spin-flip transitions between Zeeman sublevels in semiconductor quantum dots. *Physical Review B* 64 (12): 125316.

81 Hackmann, J. and Anders, F.B. (2014). Spin noise in the anisotropic central spin model. *Physical Review B* 89 (4): 045317.

82 Hackmann, J., Smirnov, D.S., Glazov, M.M., and Anders, F.B. (2014). Spin noise in a quantum dot ensemble: from a quantum mechanical to a semi-classical description. *Physica Status Solidi (B) Basic Research* 251 (6): 1270–1275.

83 Hackmann, J., Glasenapp, P., Greilich, A. et al. (2015). Influence of the nuclear electric quadrupolar interaction on the coherence time of hole and electron spins confined in semiconductor quantum dots. *Physical Review Letters* 115 (20): 30–32.

84 Stockill, R.H.J., Le Gall, C., Matthiesen, C. et al. (2016). Quantum dot spin coherence governed by a strained nuclear environment. *Nature Communications* 7: 12745.

85 Abragam, A. (1961). *The Principles of Nuclear Magnetism*. Oxford University Press Inc. ISBN 9780198520146.

86 Testelin, C., Bernardot, F., Eble, B., and Chamarro, M. (2009). Hole–spin dephasing time associated with hyperfine interaction in quantum dots. *Physical Review B* 79 (19): 195440.

87 Koppens, F.H.L., Nowack, K.C., and Vandersypen, L.M.K. (2008). Spin echo of a single electron spin in a quantum dot. *Physical Review Letters* 100 (23): 1–5.

88 Petta, J.R., Johnson, A.C., Taylor, J.M. et al. (2005). Coherent manipulation of coupled electron spins in semiconductor quantum dots. *Science* 309 (5744): 2180–2184.

89 Press, D., De Greve, K., McMahon, P.L. et al. (2010). Ultrafast optical spin echo in a single quantum dot. *Nature Photonics* 4 (6): 367–370.

90 Zhang, W., Dobrovitski, V., Al-Hassanieh, K. et al. (2006). Hyperfine interaction induced decoherence of electron spins in quantum dots. *Physical Review B* 74 (20): 205313.

91 Gangloff, D.A., Éthier-Majcher, G., Lang, C. et al. (2019). Quantum interface of an electron and a nuclear ensemble. *Science* 364 (6435): 62–66.

92 Gangloff, D.A., Zaporski, L., Bodey, J.H. et al. (2021). Witnessing quantum correlations in a nuclear ensemble via an electron spin qubit. *Nature Physics* 17 (11): 1247.

93 Chen, G., Bergman, D.L., and Balents, L. (2007). Semiclassical dynamics and long-time asymptotics of the central-spin problem in a quantum dot. *Physical Review B* 76 (4).

94 Meier, F. and Zakharchenya, B.P. (eds) (1984). *Optical Orientation* Modern Problems in Condensed Matter Science. London, England: S. Elsevier Science.

95 Ramanathan, C. (2008). Dynamic nuclear polarization and spin diffusion in nonconducting solids. *Applied Magnetic Resonance* 34 (3–4).

96 Bulutay, C. (2012). Quadrupolar spectra of nuclear spins in strained InGaAs quantum dots. *Physical Review B* 85 (11): 115313.

97 Chekhovich, E.A., Glazov, M.M., Krysa, A.B. et al. (2013). Element-sensitive measurement of the hole–nuclear spin interaction in quantum dots. *Nature Physics* 9 (2): 74–78.

98 Chekhovich, E.A., Hopkinson, M., Skolnick, M.S., and Tartakovskii, A.I. (2015). Suppression of nuclear spin bath fluctuations in self-assembled quantum dots induced by inhomogeneous strain. *Nature Communications* 6: 6348.

99 Jovanov, V., Eissfeller, T., Kapfinger, S. et al. (2011). Observation and explanation of strong electrically tunable exciton g factors in composition engineered In(Ga)As quantum dots. *Physical Review B* 83 (16): 161303.

100 Chekhovich, E.A., da Silva, S.F.C., and Rastelli, A. (2020). Nuclear spin quantum register in an optically active semiconductor quantum dot. *Nature Nanotechnology*.

101 Stoneman, A.M. (2001). *Theory of Defects in Solids*. Google Books.

102 Fischer, J., Coish, W., Bulaev, D.V., and Loss, D. (2008). Spin decoherence of a heavy hole coupled to nuclear spins in a quantum dot. *Physical Review B* 78 (15): 155329.

103 Pikus, G.E. and Pikus, F.G. (1994). The mechanism of heavy and light hole mixing in GaAs/AlAs superlattices. *Solid State Communications* 89 (4): 319–322.

104 Krizhanovskii, D.N., Ebbens, A., Tartakovskii, A.I. et al. (2005). Individual neutral and charged $In_xGa_{1-x}As - GaAs$ quantum dots with strong in-plane optical anisotropy. *Physical Review B* 72 (16): 161312.

105 Vidal, M., Durnev, M.V., Bouet, L. et al. (2016). Hyperfine coupling of hole and nuclear spins in symmetric (111)-grown GaAs quantum dots. *Physical Review B* 94 (12): 121302(R).

106 Kaji, R., Ohno, S., Hozumi, T., and Adachi, S. (2013). Effects of valence band mixing on hole spin coherence via hole-nuclei hyperfine interaction in InAlAs quantum dots. *Journal of Applied Physics* 113 (20): 1–7.

107 Bechtold, A. (2016). Quantum dynamics of spin qubits in optically active quantum dots. Phd thesis. Technische Universität München.

108 Huthmacher, L., Stockill, R.H.J., Clarke, E. et al. (2018). Coherence of a dynamically decoupled quantum-dot hole spin. *Physical Review B* 97 (24): 241413.

109 Yang, W., Ma, W.-L., and Liu, R.-b. (2017). Quantum many-body theory for electron spin decoherence in nanoscale nuclear spin baths. *Reports on Progress in Physics* 80 (1): 016001.

110 Faribault, A. and Schuricht, D. (2013). Spin decoherence due to a randomly fluctuating spin bath. *Physical Review B - Condensed Matter and Materials Physics* 88 (8): 1–12.

111 Fröhling, N. and Anders, F.B. (2017). Long-time coherence in fourth-order spin correlation functions. *Physical Review B* 96 (4): 045441.

112 Huo, Y.H., Witek, B.J., Kumar, S. et al. (2013). A light-hole exciton in a quantum dot. *Nature Physics* 10: 46–51.

113 Eble, B., Testelin, C., Desfonds, P. et al. (2009). Hole–nuclear spin interaction in quantum dots. *Physical Review Letters* 102 (14): 146601.

114 Prechtel, J.H., Kuhlmann, A.V., Houel, J. et al. (2016). Decoupling a hole spin qubit from the nuclear spins. *Nature Materials* 15 (9): 981–986.

115 Brunner, D., Gerardot, B.D., Dalgarno, P. et al. (2009). A coherent single-hole spin in a semiconductor. *Science (New York, NY)* 325 (5936): 70–72.

116 Chekhovich, E.A., Makhonin, M.N., Tartakovskii, A.I. et al. (2013). Nuclear spin effects in semiconductor quantum dots.*Nature Materials* 12 (6): 494–504.

117 Wüst, G., Munsch, M., Maier, F. et al. (2016). Role of the electron spin in determining the coherence of the nuclear spins in a quantum dot. *Nature Nanotechnology* 1–6.

118 Dzhioev, R.I. and Korenev, V.L. (2007). Stabilization of the electron-nuclear spin orientation in quantum dots by the nuclear quadrupole interaction. *Physical Review Letters* 99 (3): 1–4.

119 Fuller, G.H. (1976). Nuclear spins and moments. *Journal of Physical and Chemical Reference Data* 5 (4): 835–1092.

120 Sokolov, P.S., Petrov, M.Y., Mehrtens, T. et al. (2016). Reconstruction of nuclear quadrupole interaction in (In,Ga)As/GaAs quantum dots observed by transmission electron microscopy. *Phys. Rev. B* 93 (4): 045301.

121 Bulutay, C., Chekhovich, E.A., and Tartakovskii, A.I. (2014). Nuclear magnetic resonance inverse spectra of InGaAs quantum dots: atomistic level structural information. *Physical Review B - Condensed Matter and Materials Physics* 90 (20): 1–16.

122 Munsch, M., Wüst, G., Kuhlmann, A.V. et al. (2014). Manipulation of the nuclear spin ensemble in a quantum dot with chirped magnetic resonance pulses. *Nature Nanotechnology* 9 (9): 671–675.

123 Bulaev, D.V. and Loss, D. (2005). Spin relaxation and decoherence of holes in quantum dots. *Physical Review Letters* 95 (7): 1–4.

124 Heiss, D., Schaeck, S., Huebl, H. et al. (2007). Observation of extremely slow hole spin relaxation in self-assembled quantum dots. *Physical Review B* 76 (24): 241306.

125 Krzykowski, M., Gawarecki, K., and Machnikowski, P. (2020). Hole spin-flip transitions in a self-assembled quantum dot. *Physical Review B* 102 (20): 205301.

126 Machnikowski, P., Gawarecki, K., and Cywiński, L. (2019). Hyperfine interaction for holes in quantum dots: $\boldsymbol{k} \cdot \boldsymbol{p}$ model. *Physical Review B* 100 (8): 085305.

127 Chekhovich, E.A., Kavokin, K.V., Puebla, J. et al. (2012). Structural analysis of strained quantum dots using nuclear magnetic resonance. *Nature Nanotechnology* 7 (10): 646–650.

128 Abramowitz, M., Stegun, I.A., and Romer, R.H. (1988). Handbook of mathematical functions with formulas, graphs, and mathematical tables. *American Journal of Physics* 56 (10): 958.

129 Sun, C. and Sinitsyn, N.A. (2015). Exact transition probabilities for a linear sweep through a Kramers-Kronig resonance. *Journal of Physics A: Mathematical and Theoretical* 48 (50): 505202.

130 Gangloff, D.A., Éthier-Majcher, G., Lang, C. et al. (2018). Quantum interface of an electron and a nuclear ensemble. *Science* 364 (6435): 62–66.

131 McMahon, P.L. and De Greve, K. (2015). Towards quantum repeaters with solid-state qubits: spin-photon entanglement generation using self-assembled quantum dots. In *Engineering the Atom-Photon Interaction*, volume 2 (ed. A. Predojević and M. Mitchell), 365–402. Cham: Springer. ISBN 978-3-319-19230-7.

17

Ensemble-Based Quantum Memory: Principle, Advance, and Application

Bo Jing[1] *and Xiao-Hui Bao*[2]

[1] *School of Information Science and Technology, Southwest Jiaotong University, Xi-An Road No. 999, Chengdu, 611756, China*
[2] *Hefei National Laboratory for Physical Sciences at Microscale, University of Science and Technology of China, Jinzhai Road No. 96, Hefei 230026, China*

17.1 Introduction

Quantum memory [1–3], referring to a memory for photons in this chapter, is a device capable of mapping photonic quantum states into a memory medium or directly generating stored quantum states via emitting photons and capable of retrieving the stored states at a later time. Serving as one of the fundamental building blocks of quantum repeater [4, 5] and quantum network [6–9], it plays a significant role in a range of applications of quantum information technology, such as for long-distance quantum communication [10], linear-optical quantum computing [11], etc. In order to harness the full potential of these applications, it requires a high-performance quantum memory with high efficiency, long lifetime, high fidelity, large bandwidth, and multimode capacity. Over the past decades, a number of theoretical and experimental research efforts have been made for the development of such a quantum memory.

When it comes to build a quantum memory, there are mainly two considerations: one is the memory scheme and the other is the memory medium. The former one determines what working principle the quantum memory relies on, and here we note that it is a memory protocol independent to what the physical platforms we use, while the latter one refers to the physical platform we use to realize the transfer of photons into and out of the memories. For the memory schemes, there are optical delay line (ODL) [12, 13], electromagnetically induced transparency (EIT) [14, 15], atomic frequency comb (AFC) [16, 17], Duan–Lukin–Cirac–Zoller (DLCZ) [18–20], controlled reversible inhomogeneous broadening (CRIB) [21], far off-resonant Raman [22], Autler–Townes splitting (ATS) [23], etc. These schemes have been widely explored for storing single photons with discrete variables such as polarization, time-bin, and orbital angular momentum; actually, the schemes can be also applied for photons with continuous variables (CV) such as position and momentum, phase and amplitude, time and energy. For instance, storage of

Photonic Quantum Technologies: Science and Applications, First Edition. Edited by Mohamed Benyoucef.

squeezed light has been demonstrated with EIT scheme [24, 25]. In addition, a Faraday rotation-based quantum memory has ever been proposed for continuous variables and experimentally reported in 2004 [26]. The demonstration of quantum memory for photons with continuous variables is also attractive due to its valuable applications [27]. Nevertheless, due to limited space, these CV-type memories are beyond the scope of this chapter.

For the memory media, one of the main choices is the atomic ensembles like gas atomic ensembles [14, 15] and solid-state atomic ensembles [28]. A clear advantage of atomic ensembles is that the strong and controllable coupling between atoms and photons can be acquired, for instance, via the collective enhancement. An alternative choice for the memory medium is the single particles like single neutral atoms [29, 30], single trapped ions [31] and single color centers [32, 33], which interact weakly with single photons without a cavity, in comparison with atomic ensembles. Due to limited space, these single-particle memories are also beyond the scope of this chapter.

In the following, we will firstly introduce the widely used memory schemes. Next we will give criteria on judging the memory performances. Afterward, we will review the physical realizations, as well as their recent advances. Finally, we will discuss the applications.

17.2 Memory Schemes

17.2.1 Optical Delay Line

As the basic functionality of a quantum memory is to get photons delayed, the simplest way is using an ODL. For instance, optical fibers can be used to delay the photon and release it at a chosen time. However, the storage time after which the intensity of photons decreases to $1/e$ is limited to around 100 μs considering a fiber transmission loss of 0.2 dB/km. It is also difficult to be used as an on-demand quantum memory. A better way for delay line is to utilize a cavity and let the photons travel around it [12, 13], which is the so-called all optical loop quantum memory. Figure 17.1 shows a typical scheme for such kind of quantum memory. Once the photon enters the ring cavity via a polarization beam splitter (PBS), its polarization is flipped through an

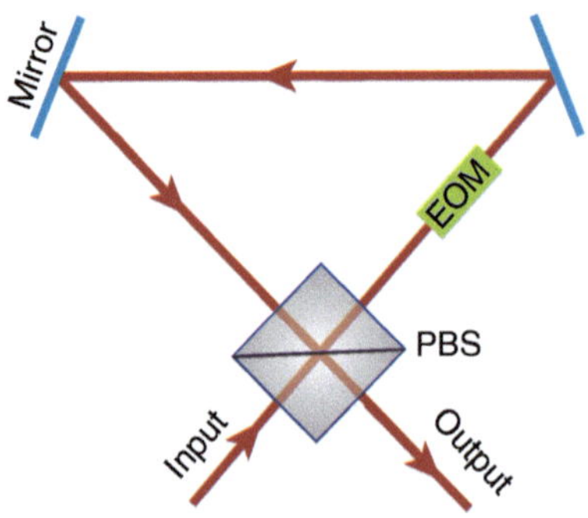

Figure 17.1 A typical scheme for cavity-assisted optical loop quantum memory.

electronic optical modulator (EOM), then the photon will fly in the cavity back and forth. After a chosen time, the polarization is flipped again, and thus the photon can be released from the cavity. The main requirement for the all optical loop quantum memory is ultralow transmission in each round trip, i.e. an ultrahigh-finesse cavity is needed, which is the main challenge especially there exist EOM, lenses, and other optical elements in the cavity. The storage time is determined by integer multiples of the flying time for each round trip of the photons. To acquire long storage time, the cyclic flying time is expected to be longer with a larger cavity length, which, however, reduces the flexibility to retrieve the photon on demand. A trade-off between the storage time and cyclic flying time should be considered.

As ODL has its limitations on the storage time and manipulations, more advantageous quantum memories directly store qubits in a medium, for instance, stationary atomic states, manipulating the atomic states and releasing the stored qubits via controllable optical pulses. This kind of quantum memories is what we mainly focus on and can be roughly divided into two classes, namely absorption type and embedded type. The former type, as the name suggests, relies on absorbing an extra photon for storage of the photonic quantum state. The most-often used schemes for this type of quantum memory include EIT and AFC. The latter type, also called built-in-type memory or emission-type memory, refers to a memory that directly generates the stored quantum state inside the memory. In contrast to the absorption-type quantum memory where the stored quantum state comes from input photons, it relies on emitting of a photon and thus preparing an atomic excitation as the stored quantum state, which can be easily encoded to a stored atomic qubit or used for photon–memory entanglement interface. A typical scheme for this type of quantum memory is the DLCZ scheme. Below we will mainly introduce give an introduction to the three most-often used schemes and a brief description on other less-often used schemes.

17.2.2 Electromagnetically Induced Transparency

EIT is a coherent optical nonlinear process where an opaque medium becomes transparent to light when extra control light is applied to the medium [34, 35]. It was experimentally observed by K.-J. Boller et al. with strontium in 1991 [36]. Although the medium becomes transparent, the controllable complex susceptibility χ will lead to a controllable group velocity of the transmitted light. For instance, the first experiment of group velocity slowed down to $c/165$ was reported in 1995 [37], and light was stopped later by several groups in 2001 [14, 15, 28]. In additional, group velocity exceeding speed of light in vacuum was also observed [38].

As shown in Figure 17.2a, we use a Λ-structure energy level to illustrate the EIT process. All atoms are initially prepared in a ground state $|g\rangle$. A weak signal light (frequency of ω_s, Rabi frequency of $\Omega_s = \boldsymbol{d_{eg}} \cdot \boldsymbol{E_s}$[1]) coupling the ground state $|g\rangle$ and excited state $|e\rangle$ with a detuning of Δ_s from $|e\rangle$ is sent to the atomic ensembles. Another strong control light (frequency of ω_c, Rabi frequency of $\Omega_c = \boldsymbol{d_{es}} \cdot \boldsymbol{E_c}$) coupling another ground state $|s\rangle$ and the excited state $|e\rangle$ with a detuning of Δ_c is

1 Let $\hbar = 1$ with natural units.

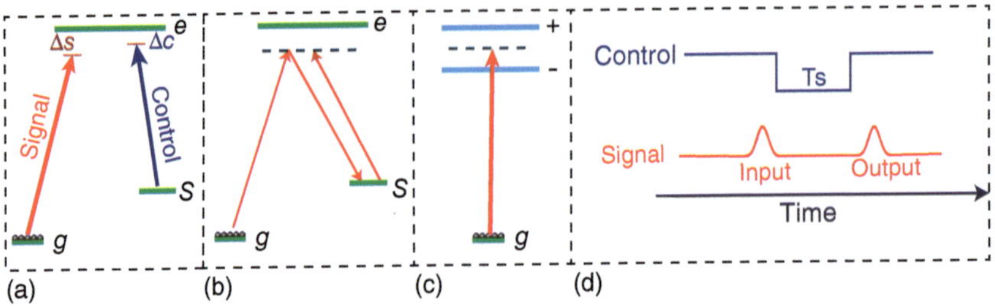

Figure 17.2 Scheme for EIT quantum memory. (a) A simple Λ-structure energy level for the EIT process. The transparency of the weak signal light can be controlled by the strong control light. (b) Illustration for different absorption pathways for the signal light absorbed to the bare excited state with the existence of strong control light. (c) Illustration for different absorption pathways for the signal light absorbed to the dressed states. (d) A time sequence for EIT quantum memory. The signal photon entering the memory medium will be stored as a spinwave when the control light is adiabatically decreased to zero. After a certain storage time, the control light is turn on again and the stored spin wave will be retrieved as a photon.

applied to control the EIT process. A simple understanding of EIT is the destructive interference between different transition pathways for the signal photons, i.e. $|g\rangle \rightarrow |e\rangle$ and $|g\rangle \rightarrow |e\rangle \rightarrow |s\rangle \rightarrow |e\rangle$, which inhibits the signal photon to be absorbed and thus leads to transparency [35], as shown in Figure 17.2b. Another explanation uses dressed state picture as follows. For simplicity, assuming the two lights propagate collinearly, for instance, in the z direction. Under the bases of $|g\rangle$, $|s\rangle$, and $|e\rangle$, the Hamiltonian can be written as:

$$H = -\frac{1}{2}\begin{pmatrix} 0 & 0 & \Omega_s \\ 0 & 2\delta & \Omega_c \\ \Omega_s & \Omega_c & 2\Delta_s \end{pmatrix} \tag{17.1}$$

where $\delta = \Delta_s - \Delta_c$ represents two-photon detuning. Considering a two-photon resonant case, i.e. $\Delta_s = \Delta_c = \Delta$, $\delta = 0$, the eigenstates can be rewritten in the term of the bare atomic eigenstates $|g\rangle$, $|s\rangle$, $|e\rangle$:

$$\begin{aligned} |+\rangle &= \sin\theta\sin\phi|g\rangle + \cos\phi|e\rangle + \cos\theta\sin\phi|s\rangle \\ |0\rangle &= \cos\theta|g\rangle - \sin\theta|s\rangle \\ |-\rangle &= \sin\theta\cos\phi|g\rangle - \sin\phi|e\rangle + \cos\theta\cos\phi|s\rangle \end{aligned} \tag{17.2}$$

with corresponding eigenenergy $-\Delta + \sqrt{\Delta^2 + \Omega_s^2 + \Omega_c^2}$, 0, $-\Delta - \sqrt{\Delta^2 + \Omega_s^2 + \Omega_c^2}$. Here θ, ϕ are the "mixing angles" and can be given by

$$\begin{aligned} \tan\theta &= \frac{\Omega_s}{\Omega_c} \\ \tan 2\phi &= -\frac{\sqrt{\Omega_s^2+\Omega_c^2}}{\Delta} \end{aligned} \tag{17.3}$$

For a common situation of weak signal light and continuous strong control light, which could induce a transparency of the signal photons, $\Omega_c \gg \Omega_s$, $\theta \rightarrow 0$. The eigenstates can be simplified as:

$$\begin{aligned} |+\rangle &= \cos\phi|e\rangle + \sin\phi|s\rangle \\ |0\rangle &= |g\rangle \\ |-\rangle &= -\sin\phi|e\rangle + \cos\phi|s\rangle \end{aligned} \tag{17.4}$$

It is now clear that the new eigenstates for the atoms dressed by the strong control light include a ground state $|g\rangle$ and two excited states $|+\rangle$ and $|-\rangle$, atoms in the ground state $|g\rangle$ can be absorbed via either of the transitions, i.e. $|g\rangle \to |+\rangle$ or $|g\rangle \to |-\rangle$, where the frequency of signal light with the two transitions has the same detunings but opposite signs from new eigenstates $|+\rangle$ and $|-\rangle$, as shown in Figure 17.2c. Interference of the two absorption pathways via either $|+\rangle$ or $|-\rangle$ is destructive, leading to transparency of the signal light.

For detailed analysis of how states of signal light and atoms in the medium evolve during the transmission, the dynamic process can be solved by Maxwell–Bloch equation and Lindblad equation [39, 40] as following:

$$\begin{aligned}\frac{\partial \Omega_s}{\partial z} + \frac{1}{c}\frac{\partial \Omega_s}{\partial t} &= i\frac{\text{od}\cdot\Gamma_e}{2L}\rho_{eg} \\ \frac{d\rho}{dt} &= \frac{1}{i}[H,\rho] + \frac{\Gamma_{eg}}{2}(2\sigma_{ge}\rho\sigma_{eg} - \sigma_{ee}\rho - \rho\sigma_{ee}) \\ &\quad + \frac{\Gamma_{es}}{2}(2\sigma_{se}\rho\sigma_{es} - \sigma_{ee}\rho - \rho\sigma_{ee}) + \frac{\Gamma_{sg}}{2}(2\sigma_{gs}\rho\sigma_{sg} - \sigma_{ss}\rho - \rho\sigma_{ss})\end{aligned} \tag{17.5}$$

where $\sigma_{ij} = |i\rangle\langle j|$ $(i,j: g,s,e)$, Γ_{ij} represents coherence relaxing rate between state $|i\rangle$ and $|j\rangle$, L is the length of atomic medium, od is the optical depth of atoms, and ρ is the density matrix of atomic states. By solving the equations, the responsive linear susceptibility of the atoms to the EIT process is given by:

$$\chi = -\frac{|d_{ge}|^2 n_0}{\epsilon_0}\frac{\delta + i\Gamma_{sg}/2}{(i\Delta_p - \Gamma_e/2)(i\delta - \Gamma_{sg}/2) + \Omega_c^2/4} = \chi' + i\chi'' \tag{17.6}$$

where n_0 is the atomic density, d_{ge} is the transition dipole moment. The imaginary part χ'' represents the absorption, and the real part χ' represents the dispersion corresponding to a refractive index $n = \sqrt{1+\chi'}$. The group velocity for the signal light is given by [35, 41]:

$$v_{gr} = \left|\frac{d\omega_s}{dk_s}\right|_{\delta=0} = \frac{c}{n + \omega_s(dn/d\omega_s)} \simeq \frac{c}{1 + n_0\sigma c\Gamma_{eg}/\Omega_c^2} \tag{17.7}$$

When the group velocity is controlled by the control light, the coherent transfer between photonic states and atomic states also happens. At this time, if the intensity of control light can be adiabatically decreased to zero, i.e. $\Omega_c \to 0$, the group velocity will also go to zero, the atomic state $|0\rangle$ in the dressed state picture will also adiabatically transfer from $|g\rangle$ to $|s\rangle$. It presents the possibility to freeze the signal light in the atoms, i.e. storing the signal light in the atomic medium. The basic idea of quantum memory with EIT process takes this as the starting point. The detailed process for the quantum memory can be explained with dark-state polaritons [42], where the dark-state polariton field is defined as:

$$\Psi(z,t) = \cos\phi\mathcal{E}_s(z,t) - \sin\phi\sqrt{n_0}\rho_{sg}(z,t)e^{i\Delta kz} \tag{17.8}$$

where $\tan\phi = \sqrt{n_0\sigma c\Gamma_{eg}}/\Omega_c$, $\Delta_k = k_c - k_s$ and $\mathcal{E}_s$ is the normalized, slowly varying signal field ($\boldsymbol{E_s}$) strength. Under assumptions of adiabatic approximations, Ψ obeys:

$$\left(\frac{1}{c}\frac{\partial}{\partial t} + \cos^2\phi\frac{\partial}{\partial z}\right)\Psi(z,t) = 0 \tag{17.9}$$

It describes propagation of polaritons with group velocity of $c \cdot \cos^2\phi$. Note that dark-state polaritons include signal photonic state, atomic state, and their coherent transfer, where the part of atomic states is the so-called *spinwave*. If decreasing Ω_c to zero, $\phi \to \pi/2$, only the part of atoms remains in Ψ, i.e. the photonic state is completely transferred to a spinwave. Thus, the main manipulation for EIT-based quantum memory is to adiabatically decrease the intensity of control light and transfer the signal photonic state into spinwave for storage. After a chosen storage time, the control light is turned on again, the spinwave will be converted back to the photonic state and finally leave the atomic medium, which is referred as the retrieval process. A typical time sequence for EIT-based quantum memory is shown in Figure 17.2d.

17.2.3 Atomic Frequency Comb

Quantum memories based on EIT have been widely utilized in atomic gases, while for rare earth ions–doped solids (REIDS), the AFC scheme [17] is more widely employed. In contrast to other schemes such as EIT, which allows od-dependent multimode quantum memory [43], the AFC scheme supports multimode storage of photons with stored mode numbers independent of the optical depth, which makes multimode quantum memories more realistic and efficient. Benefit from the wide inhomogeneous broadening in REIDS, storage of ultra-short wavepackets is also possible.

The AFC-based quantum memory works by absorbing photons and transferring the photonic state to a collective atomic excitation. As shown in Figure 17.3a, atomic ensembles are initially prepared to an periodic comb structure through frequency-selective optical pumping, i.e. forming an AFC. The AFC has a free spectral range (FSR) of Δ and linewidth of γ. Photons with a detuning of δ from the excited state will be absorbed by the AFC and a collective atomic state will thus be generated. The initial collective state is given by

$$|\psi\rangle_{\text{ini}} = \sum_j c_j \mathrm{e}^{\mathrm{i}2\pi\delta_j t} \mathrm{e}^{\mathrm{i}kz_j} |g, \dots e_j, \dots g\rangle \tag{17.10}$$

where N is the total number of atoms in the AFC, $|g_j\rangle$ and $e_j\rangle$ represent the ground state and the excited state of the jth atom respectively, k is the wave number of input

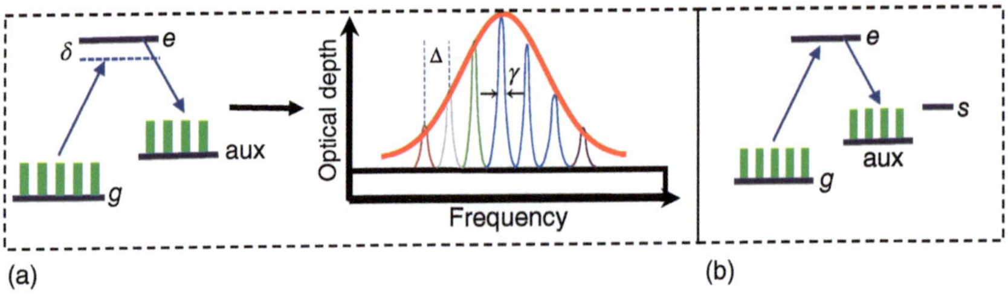

Figure 17.3 Scheme for AFC quantum memory. (a) Energy level for the AFC quantum memory (left figure). Atoms in ground state are frequency-selective pumped to an auxiliary state, forming an atomic frequency comb with periodic absorption line of photons. An absorption line for photons with different frequency is shown in the right figure. (b) Energy level for the spin AFC quantum memory assisted by an extra state $|s\rangle$.

light, δ_j is the detuning of input light from excited state, z_j is the position of the jth atom, and the complex amplitude c_j is determined by the frequency and position of the jth atom. After the photon absorbed via AFC, atoms with different detunings begin to evolve and accumulate different phases. Fortunately, due to the periodic structure of AFC, the atomic ensemble has a periodic density spectral distribution of $n(\delta)$ with a period of Δ, and δ_j can be replaced by $m_j\Delta$ with m_j an integer. Thus, all atoms will evolve back to the original phases (integer multiples of 2π) at time $t = 1/\Delta$, i.e. the process of rephasing, leading to re-emission of the input photon in the original state. $1/\Delta$ is also defined as the pre-set storage time.

Unlike EIT-based quantum memory, where storage process of signal light is under control of another strong light in a Λ-structure three-level atomic system, in AFC-based quantum memory, the input light couples a two-level atomic system (ground state $|g\rangle$ and excited state $|e\rangle$) and is directly absorbed by the atoms. The dynamic behavior of absorption and re-emission process can be explained by the following equations [17, 44]. For simplicity, the light field with a frequency of ω_0 is considered propagating in z direction and decomposed in forward and backward modes,

$$\mathcal{E}(z,t) = \mathcal{E}_f(z,t)\mathrm{e}^{\mathrm{i}\omega_0 z/c} + \mathcal{E}_b(z,t)\mathrm{e}^{-\mathrm{i}\omega_0 z/c} \tag{17.11}$$

The absorption process can be simply understood by Lambert–Beer law, where an effective optical depth $\tilde{\mathrm{od}} \simeq \mathrm{od}/\mathcal{F}$ is used for Gaussian shape peak. Here od is the peak optical depth of each frequency comb, and $\mathcal{F} = \Delta/\gamma$ is the finesse of AFC. During the transmission, $\mathcal{E}_f(z)$ is obtained as:

$$\mathcal{E}_f(z) = \mathcal{E}_f(0)\mathrm{e}^{-\tilde{\mathrm{od}}z/2} \tag{17.12}$$

For the dynamic evolution of input light fields and atomic states, atomic coherence σ_{ge} is used and defined as,

$$\begin{aligned}\sigma_{ge}(z,t,\delta) &= \frac{1}{N(\delta,z)}\sum_{j=1}^{N(\delta,z)} |g\rangle_{jj}\langle e| \\ &= \sigma_f(z,t;\delta)\mathrm{e}^{\mathrm{i}\omega_0 z/c} + \sigma_b(z,t;\delta)\mathrm{e}^{-\omega_0 z/c}\end{aligned} \tag{17.13}$$

where $N(\delta, z)$ is the total number of atoms coupled by the input light. It is related to a detuning of δ because of the nonuniform optical depth for photons with different frequency and also related to the position z because of the finite size of the medium or nonuniform atomic density distribution. Under the approximation that most atoms stay in the ground state, i.e. the number of excited atoms can be ignored compared to the total atom number, the dynamic evolution of the atomic coherence and input light are given by Maxwell–Bloch and Heisenberg–Langevin equations:

$$\begin{aligned}&\frac{\partial}{\partial t}\sigma_f(z,t;\delta) = -i\delta\sigma_f(z,t;\delta) + id\mathcal{E}_f(z,t)\\ &\left(\frac{\partial}{\partial t} + c\frac{\partial}{\partial z}\right)\mathcal{E}_f(z,t) = i\tilde{d}\int_{-\infty}^{\infty} d\delta n(\delta)\sigma_f(z,t;\delta)\\ &\frac{\partial}{\partial t}\sigma_b(z,t;\delta) = -i\delta\sigma_b(z,t;\delta) + id\mathcal{E}_b(z,t)\\ &\left(\frac{\partial}{\partial t} - c\frac{\partial}{\partial z}\right)\mathcal{E}_b(z,t) = i\tilde{d}\int_{-\infty}^{\infty} d\delta n(\delta)\sigma_b(z,t;\delta)\end{aligned} \tag{17.14}$$

Here d is the dipole moment of atomic transition, $\tilde{d} = g_0^2 d$ with $g_0 = \sqrt{\omega_0/(2\epsilon_0 V)}$, where V is the quantization volume.

As the storage time of AFC-based quantum memory is predetermined by the separation of adjacent combs (Δ), on-demand retrieval of stored photons is not directly allowed. To acquire an on-demand quantum memory with AFC, extra energy levels of ground states are needed. As shown in Figure 17.3b. the main idea is to apply a π-pulse of $|e\rangle \leftrightarrow |s\rangle$ transition to transfer the coherence between $|g\rangle$ and $|e\rangle$ to $|g\rangle$ and $|s\rangle$ at time t_1, and thus the phase evolution of atoms will be frozen and lasts to time t_2 until another π-pulse is applied to transfer the coherence back. A total storage time includes the rephasing time $1/\Delta$ and the time separation $t_2 - t_1$ between the two π-pulses. Thus the retrieval time can be controlled and on-demand quantum memory is allowed. For the main physical medium (REIDS) using AFC, Zeeman splitting of ground states is usually utilized. However, the split states are limited to two for Kramers ions such as neodymium, erbium, etc. The split states here have been used as $|g\rangle$ and $|aux\rangle$. The initial $|aux\rangle$ state is used to receive the atoms from $|g\rangle$ during frequency-selective pumping process. Once an input photon is absorbed by the AFC, it will result in an atom transferring from $|g\rangle$ to $|e\rangle$. In order to allow spin AFC memory, the atom in $|e\rangle$ needs to be transferred to another empty ground state, since $|g\rangle$ and $|aux\rangle$ have been used, a third ground spin state, i.e. $|s\rangle$ is needed to perform the transfer process. A possible way for this target may be to utilize the nuclear splitting states to form a double-Λ structure [45]. In most experiments, on-demand storage usually happens in non-Kramers ions such as praseodymium and europium, etc., where extra ground-state splitting exists.

17.2.4 DLCZ Scheme

In addition to EIT and AFC, the DLCZ scheme [18] is also widely used. The initial purpose of this scheme is not for storage of arbitrary photonic states but for the generation of remote atom-atom entanglement via photon interference for quantum repeater.

Different from the absorption-type schemes, in DLCZ, the quantum memories are built-in, which rely on directly generating a stored atomic excitation conditioned on emitting a Raman-scattered single photon, named as a write-out photon. As shown in Figure 17.4, a Λ-type level scheme is utilized, similar as EIT. All atoms are initially prepared in a ground state $|g\rangle$, an off-resonant write pulse with a detuning of Δ_w is applied to excite the atoms. Through Raman scattering, a write-out photon

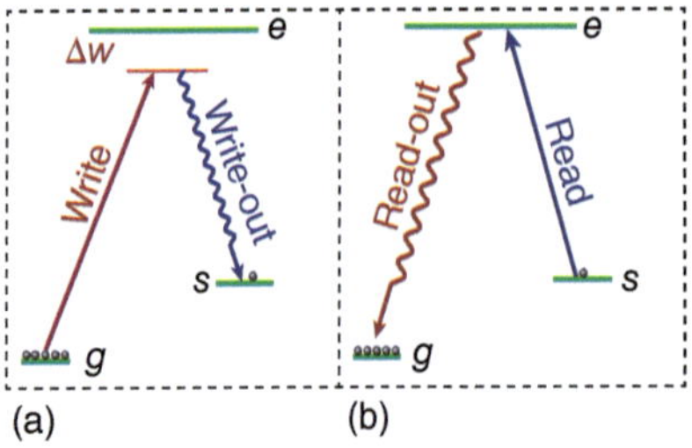

Figure 17.4 The DLCZ scheme. (a) The write process. An atom is transferred from the ground state to another state and stored as a spinwave via Raman-scattering a write-out photon. (b) The read process. A strong read light will convert the stored spinwave back to a read-out photon.

will be produced with a small probability χ_w in one mode, and simultaneously one atom gets transferred from $|g\rangle$ to $|s\rangle$, creating an atomic collective excitation. The non-normalized joint state between the atoms and the write-out field can be given as:

$$|\Psi\rangle = \left(1 + \sqrt{\chi_w}\sum_i \mathrm{e}^{\mathrm{i}\Delta k\cdot \boldsymbol{r}_i}\sigma_{sg}^i a_{\mathrm{wo}}^\dagger\right)|\mathrm{vac}\rangle \tag{17.15}$$

where multi-excitation events has been ignored in consideration of small χ_w, $\boldsymbol{\Delta k}=\boldsymbol{k_w}-\boldsymbol{k_{wo}}$ with the wavevectors of write ($\boldsymbol{k_w}$) and write-out ($\boldsymbol{k_{wo}}$) photons, $\boldsymbol{r_i}$ is the position of ith atom, $\sigma_{sg}^i=|s\rangle_{ii}\langle g|$, $a_{\mathrm{wo}}^\dagger$ is the creation operator for write-out photons, $|\mathrm{vac}\rangle$ is the initial state $|0\rangle_p \otimes |g,\ldots,g,\ldots,g\rangle_a$ with p stands for photons and a stands for atoms. Define an excitation operator for atoms as:

$$S^\dagger = \frac{1}{\sqrt{N}}\sum_i \mathrm{e}^{\boldsymbol{\Delta k}\cdot \boldsymbol{r}_i}\sigma_{sg}^i \tag{17.16}$$

Now $|\Psi\rangle$ can be represented as:

$$|\Psi\rangle = (1 + \sqrt{p_w}S^\dagger a_{\mathrm{wo}}^\dagger)|\mathrm{vac}\rangle \tag{17.17}$$

Here $p_w = N\chi_w$ is the probability of exciting one atom over all atoms and emitting one write-out photon. The state is a typical example of entangled state under Fock basis, where whether the atom is excited is entangled with the emission of write-out photons. To avoid multiple excitation events, which may lead to emission of multiple write-out photons, the write light is set weak enough to let only one atom be excited.

Once the write-out photon is emitted, a collective state of atoms is generated and stored. Conditioned on the detection of write-out photon, one can herald the existence of collective state, i.e. spinwave, given by:

$$|\Psi\rangle_a = S^\dagger|\mathrm{vac}\rangle_a \tag{17.18}$$

With a certain storage time after the write process, atoms move to a new position $\boldsymbol{r}_i'$, another strong read light can be applied to convert the spinwave to an anti-Stokes photon named as read-out photon, as shown in Figure 17.4b. Similar with the write process, we define an operator as:

$$S = \sum_i \mathrm{e}^{\mathrm{i}(\boldsymbol{k}_r-\boldsymbol{k}_{\mathrm{ro}})\cdot \boldsymbol{r}_i'}\sigma_{gs}^i \tag{17.19}$$

Now the joint state between atomic state and read-out photons can be given by:

$$\begin{aligned}|\Psi\rangle_f &= Sa_{\mathrm{ro}}^\dagger|\Psi\rangle_a|0\rangle_{\mathrm{ro}}\\ &= |\mathrm{vac}\rangle_a \otimes \frac{1}{\sqrt{N}}\sum_{i=1}^N \mathrm{e}^{\mathrm{i}[(\boldsymbol{k}_w-\boldsymbol{k}_{\mathrm{wo}})\cdot\boldsymbol{r}_i+(\boldsymbol{k}_r-\boldsymbol{k}_{\mathrm{ro}})\cdot\boldsymbol{r}_i']}a_{\mathrm{ro}}^\dagger|0\rangle_{\mathrm{ro}}\end{aligned} \tag{17.20}$$

which gives a description for the read-out photon as:

$$|\Psi\rangle_{\mathrm{ro}} = \frac{1}{\sqrt{N}}\sum_{i=1}^N \mathrm{e}^{\mathrm{i}[(\boldsymbol{k}_w-\boldsymbol{k}_{\mathrm{wo}})\cdot\boldsymbol{r}_i+(\boldsymbol{k}_r-\boldsymbol{k}_{\mathrm{ro}})\cdot\boldsymbol{r}_i']}a_{r0}^\dagger|0\rangle_{\mathrm{ro}} \tag{17.21}$$

with a probability P_{ro} of emitting the read-out photon in the direction of $\boldsymbol{k_{ro}}$:

$$P_{\mathrm{ro}} = \frac{1}{N}\left|\sum_{i=1}^N \mathrm{e}^{\mathrm{i}[(\boldsymbol{k}_w-\boldsymbol{k}_{\mathrm{wo}})\cdot\boldsymbol{r}_i+(\boldsymbol{k}_r-\boldsymbol{k}_{\mathrm{ro}})\cdot\boldsymbol{r}_i']}\right|^2 p_{\mathrm{ro}} \tag{17.22}$$

Here p_{ro} represents the probability of emitting the read-out photon in the direction of $\boldsymbol{k}_{ro}$ in the single atom case. If Ignoring the atom movements and the phase-matched condition is satisfied, i.e. $\boldsymbol{k}_w - \boldsymbol{k}_{wo} + \boldsymbol{k}_r - \boldsymbol{k}_{ro} = 0$, the probability of emitting the read-out photons along the direction of $\boldsymbol{k}_{ro}$ can be enhanced by N times compared to the single atom case, which is known as the collective enhancement and can significantly help to achieve high retrieval efficiency. Note that the read process in the DLCZ scheme is the same as EIT and can be solved by the methods mentioned in the previous Section 17.2.2.

17.2.5 Other Schemes

Quantum memory has attracted massive attention because of its significance for quantum communication and quantum computing, etc. Various schemes have been proposed for reliable quantum memories. Besides the schemes introduced in Sections 17.2.1, 17.2.2, 17.2.3, 17.2.4, there are several other schemes demonstrated successfully in experiments as well.

CRIB is based on the photon-echo technology [46]. Unlike AFC, it does not require the preparation of AFCs but makes use of a continuous atomic spectral distribution instead. When a photon is absorbed and stored by the atoms, the atomic phase evolution of $\int_0^t \delta_j \, dt$ for all atoms can evolve to original phase if the detuning δ_j can be adjusted to be flipped to $-\delta_j$ after sometime t_1. The phase evolution will be $\int_0^{t_1} \delta_j \, dt + \int_{t_1}^{t} (-\delta_j) \, dt$ for the jth atom and goes to original phase at $t = 2t_1$, leading to re-emission of the stored photons. Depending on the storage medium, methods to induce the atomic spectral distribution and flip the phase evolution vary. A specific example is the gradient echo memory (GEM) [47, 48] where the memory control is purely electro-optic, without requiring π pulses.

The far off-resonant Raman scheme was proposed mainly for high-speed and broadband quantum memory, it was experimentally realized in 2010 by K.F. Reim et al. [22]. The storage process of Raman quantum memory is similar as EIT, where the main difference is the detuning of the stored photons from the excited states. In EIT, the single photon detuning Δ_s is usually set to the level of MHz and the storage bandwidth is limited by the linewidth of energy level of excited states, while in the Raman scheme, the detuning is set to several GHz to inhibit the influence of excited states on storage bandwidth and thus allow broadband quantum memory. The large detuning also destroys the interference between different absorption pathways like EIT and results in no influence on the group velocity.

Recently, another scheme toward high-speed quantum memory was proposed – the ATS scheme, which mainly relies on absorbing and storing photons via the "Autler–Townes effect." It has been experimentally reported for storage of nanoseconds-long single-photon-level laser pulses [23, 49, 50].

17.3 Performance Criteria

To assess the performance of a quantum memory, there are several criteria being used such as working wavelength, efficiency, storage time, fidelity, bandwidth, multimode capacity, and integratability. Below we will give a description on these criteria.

17.3.1 Working Wavelength

The working wavelength of a quantum memory means the wavelength of photons that quantum memory can store, it varies with different quantum memory platforms. One of the most appealing wavelengths is telecom wavelength, which can support the lowest transmission loss in fibers and is significant for entangling distant quantum nodes. However, there are few mediums whose transition between atomic energy levels is at telecom band. The only element so far is erbium [51] whose separation of ground state and first excited state is at telecom wavelength. Although other elements such as rubidium [52], praseodymium [53] also have transition energy levels compatible with telecom band, the transition occurs among higher excited states. It remains challenge for quantum storage of telecom band photons because of the short lifetime, resulting in difficult initialization of atomic excited states, large decoherence rate, and so on.

Up to date, quantum storage of telecom band photons has been reported only in Er^{3+}-doped solid atomic ensembles with pre-set storage time [54, 55] or via a nanofabricated mechanical resonator [56]. Other physical media used for quantum memories, especially on-demand quantum memory, mainly have working wavelengths at visible, infrared, or near-infrared band such as ytterbium ions (369.5 nm), Eu^{3+}:YSO (580 nm), Pr $^{3+}$:YSO (606 nm), Nd^{3+}:YVO_4 (878 nm), NV centers (637 nm), rubidium atoms (780/795 nm), caesium atoms (852/895 nm), InGaAs quantum dots (966 nm), etc.

17.3.2 Efficiency

The efficiency of a quantum memory mainly includes storage efficiency and retrieval efficiency, it is a significant parameter for building quantum networks. Storage efficiency is defined as the number ratio of photons stored into the memory over input photons, which indicates how many the photons can be successfully stored into the memory. Retrieval efficiency is the number ratio of retrieved photons over the stored photons, which indicates how many the photons can be successfully retrieved from the memory.

The initial efficiency of a quantum memory with different schemes is determined by various mechanisms. For EIT, the storage efficiency can be close to one, while the retrieval efficiency is mainly determined by optical depth along with some other factors such as pulse shape, laser power,etc. [57–59]. With current state of the art, EIT memory with an intrinsic efficiency of 0.92 for coherent light [40] and 0.86 for single-photon qubit [60] is recently reported under large optical depth. For AFC, the coherent storage process relies on the high optical depth to absorb the photons completely, the retrieval process relies on the proportion of background absorption and rephasing of absorbed photonic state. By using an impedance-matched cavity [61], efficiency with 0.56 has been achieved [62]. For DLCZ, the storage is conditioned on emitting and detecting a single write-out photon, hence the conditioned storage efficiency is considered as unity. The retrieval efficiency is mainly determined by the optical depth as the same with EIT quantum memory. Intrinsic retrieval efficiency of 0.88 and overall detected efficiency exceeding 0.4 have been reported in cold atomic ensembles with od of 2 by utilizing a self-filtering ring-cavity [63].

17.3.3 Storage Time

The storage time of a quantum memory is the time interval between creating and releasing the stored atomic states. As the efficiency also depends on the storage time, an extra criterion for assessing quantum memories is storage lifetime, which is usually defined as the storage time when the efficiency decreases to $1/e$ or 50% of the initial efficiency. There are various mechanisms influencing the storage lifetime, which will simultaneously affect the efficiency while increasing storage time. For atomic-gas-based quantum memories, atomic motion is inevitable. EIT and DLCZ quantum memories are also sensitive with motion of spinwave and fluctuation of magnetic field, etc. For solid-state quantum memories, the main influence comes from the interaction of atoms and crystal field, including spin–spin flip, spin–lattice relaxation, spectral diffusion, etc. Besides, phonon relaxation process is also unavoidable, where stored atomic states will decay via phonons.

Long storage time is important for quantum memories especially when they are used to construct worldwide quantum repeaters [4], where simultaneously preparing entanglement between adjacent nodes needs the successfully prepared entanglement waiting until all entanglements are generated for further entanglement swapping. During the waiting process, a quantum memory with long storage time is necessary, which requires at least the time level as average time cost of successfully entangling quantum nodes [64]. With an ideal quantum memory with infinite storage time, the entanglement generation rate in quantum repeaters scales polynomially with distance L, while for quantum memories with finite storage time τ, the rate-over-cost scales, at best, as $e^{-\sqrt{(L/c)/\tau}}$ [65]. Currently, quantum memory with storage time up to hundreds of milliseconds has been achieved in cold atomic ensembles with the DLCZ scheme [66] by confining the atoms in a 3D optical lattice. Storage time over 1 hour for coherent light has also been reported in Eu^{3+}:YSO with AFC [67] by utilizing spinwave storage and dynamic decoupling, which can in principle support long time quantum storage of single photons.

17.3.4 Fidelity

Fidelity is used to evaluate the overlap of quantum states before and after storage. For a pure state $|\Psi_i\rangle$ to be stored, the retrieval state after storage is $|\Psi_o\rangle$, then the fidelity is defined as:

$$F = |\langle\Psi_o|\Psi_i\rangle|^2 = \mathrm{Tr}(\rho_i\rho_o) \tag{17.23}$$

where $\rho_i = |\Psi_i\rangle\langle\Psi_i|$ and $\rho_o = |\Psi_o\rangle\langle\Psi_o|$ are the density matrices. For an arbitrary quantum state with density matrix ρ_i, after the quantum memory, the output state becomes ρ_o, here a universal fidelity is usually calculated as [68]:

$$F = \left(\mathrm{Tr}\sqrt{\sqrt{\rho_o}\rho_i\sqrt{\rho_o}}\right)^2 \tag{17.24}$$

For a completely unknown quantum state, it needs to perform quantum tomography [69] to acquire the full density matrix; however, the tomography process seems cumbersome especially for multiple qubits in a large Hilbert space. Thus, in real experiments, if one just want to assess the overlap between a quantum state and a known target state (pure state like GHz state), Eq. (17.23) is used more often, where only a few local measurements are needed [70].

Fidelity is also used to confirm whether the storage process is quantum or classical. It is known that one can acquire what a single qubit is with a max probability of 2/3 through classical measurements [71], i.e. to confirm a quantum memory, the measured fidelity needs to be larger than 2/3, the lower bound for a quantum storage process.

As the quantum memories will not destroy the stored photonic states, the fidelity can reach as high as close to unity once the noise is suppressed enough. Up to date, fidelity exceeding 0.95 for quantum memories has been widely reported, among which a fidelity of 0.999 has been demonstrated for storage of photonic polarization qubit at single-photon level [72].

17.3.5 Bandwidth

The bandwidth of a quantum memory means the max spectral width of a single photonic wavepacket that can be totally stored, it determines how fast the storage process can occur and how many frequency multiplexed modes can be allowed for multimode quantum memory. The larger the bandwidth is, the shorter the photonic wavepackets can be used, the faster the stored pulse train can be.

The bandwidth is constrained by linewidth of excited energy levels in EIT and DLCZ quantum memories, thereby a promising way is to use far-off resonant Raman quantum memory. So far, quantum storage of GHz-bandwidth heralded single photons and time-bandwidth product exceeding 1000 has been reported [73], which proves the ability of the Raman scheme for broadband quantum memories. A THz-bandwidth quantum memory is also possible via larger detuning with hundreds of THz [74]. However, the capacity for storage of multiple photons based on Raman quantum memory is still determined by optical depth like EIT, where the storage number for multimodes scales as $\sqrt{\mathrm{od}}$.

Another promising way for broadband quantum memory is to harness the wide inhomogeneous broadening in rare-earth ions doped solids through the AFC scheme. Because of various species of doped ions and solid hosts, there are many choices for the memory medium. For instance, inhomogeneous broadening over thousands of GHz has been found in Er^{3+}-doped fibers, which can, in principle, support quantum storage of photonic wavepacket with spectral width of THz and temporal width of picoseconds.

Quantum memories with large bandwidth can not only allow high-speed quantum information processing but also allow the storage of frequency multiplexed modes by partition the broadening into several frequency sections. Currently, a total bandwidth of 16 GHz has been used to store heralded single photons and partitioned into six frequency sections for storing frequency multiplexed modes [75].

17.3.6 Multimode Capacity

Multimode capacity of a quantum memory means how many photonic modes can be stored simultaneously. The modes mainly include temporal modes, spectral modes, orbital angular momentum, spatial modes, other degrees of freedom, etc. A multiplexed quantum memory is important to develop on-chip quantum technology and multiplexed quantum information processing, which need to perform quantum operations with the least quantum devices.

Currently, the best scheme for multimode quantum memory is the AFC scheme, where the number of stored modes (N_s) is independent of optical depth in contrast with EIT (N_s scales as $\sqrt{\text{od}}$), CRIB (N_s scales as od) and other schemes [43]. For temporal multiplexing quantum storage, the number of modes that can be stored is often characterized by the time–bandwidth product, which is defined as the product of storage time and bandwidth. Time–bandwidth product determines the upper bound of how many temporal modes can be stored into quantum memories simultaneously. For spectral multiplexing quantum storage, multiple frequency sections need to be prepared by making using of Doppler broadening, inhomogeneous broadening, etc. By utilizing temporal and spectral multiplexed single photons, quantum storage of 6 spectral modes, 100 temporal modes, and 130 alternative modes have been realized with AFC in Er^{3+}-doped fibers [75], Nd^{3+}:YVO_4 [76] and Pr $^{3+}$:YSO [77], respectively. In addition, photonic modes with other degrees of freedom like spatial modes and orbital angular momentum modes can be also utilized to improve the multimode capacity [78, 79].

17.3.7 Integratability

Just like classical internet, whether the quantum memory device can be integrated is also an important question in future quantum networks. In the Moore era for information science, nanoscale device is necessary. Currently, the size of quantum memory devices such as atomic ensemble, bulk crystal is too large to support integration. It needs the realization of micro- or nano-photonic light-matter interface for further applications. To achieve the goal, on-chip waveguide with rare earth ions is one of the main quantum memory devices that can be integrated, which has been studied and reported for storage of single photons recently [77, 80, 81]. An alternative way by making using of the mature silicon fabrication technologies is to couple a Si-waveguide to thin film or crystal doped with rare earth ions via the evanescent field [82]. In addition, photonic crystal nanobeam resonator with rare earth ions is also under research for developing integrated quantum memories [83, 84].

17.4 Physical Realization

To build quantum memories, several physical platforms are available, including single particles like single atom and single color center, gas atomic ensembles like laser-cooled atoms and warm atomic vapors, solid state atomic ensembles like REIDS and some other physical mediums. Here we focus on the atomic ensembles used as the memory platforms.

17.4.1 Gas Atomic Ensemble

Gas atomic ensembles are currently one of the most popular mediums for quantum memories. In contrast to quantum memory with single particle where large N.A. lens or high finesse cavity is often used to collect the emitted photons for high efficiency, collective enhancement can happen in atomic ensembles and efficient collection of emitted photons is possible even without large N.A. lens or high finesse cavity. The most often used schemes are EIT or DLCZ. In 2001, experimental light storage with EIT was reported for the first time in cold atomic ensemble [15] and warm atomic ensemble [14], respectively. The experiments efficiently prove the feasibility of EIT-based quantum memory with gas atomic ensembles. In 2003, DLCZ quantum memory was experimentally realized in gas atomic ensembles [19, 20]. By utilizing the read-out photon in DLCZ quantum memory as single photon source, EIT-based quantum storage of single photons in gas atomic ensembles was experimentally demonstrated in 2005 [89, 90].

Alkali atoms such as rubidium and caesium are the main mediums used for quantum memories based on atomic gases, they work at a wavelength of near-infrared band and are often correlated with telecom band photons via quantum frequency conversion [52, 91]. The memory efficiency of EIT/DLCZ-based memory is mainly determined by the optical depth and can reach ~0.90 [40, 60] with large optical depth. For low optical depth, cavity can be utilized to improve the photon–atom interaction (or the effective optical depth) and thus high efficiency is still possible [92, 93]. By utilizing other schemes, high efficiency up to 0.87 [94] for GEM memory and 0.82 [95] for Raman memory have been also reported. The memory fidelity is influenced by the signal-to-noise ratio (SNR), once the efficient quantum memory allows large SNR, the fidelity is usually high, for instance, exceeding 0.99 [96]. For atomic gases, atomic motion has to be considered because it will prevent the achievement of long storage time. By cooling the atoms and trapping them in optical lattice to inhibit their movements, the storage time can reach hundreds of milliseconds [66]. Alternatively, by coating the vapor cells with paraffin, the atomic coherence properties can still preserve for a rather long time since the special coating can help atoms withstand collisions with the cell walls for tens of thousands of times, thus storage time up to several milliseconds in warm vapor cells is possible [26]. EIT or DLCZ memory bandwidth is importantly influenced by the natural linewidth of excited states, the quantum memories are limited to megahertz bandwidth. To improve the bandwidth, one way is to use the Raman scheme [22] in atomic gases, where gigahertz bandwidth has been reported [73]. Another way is to utilize AFC, which has great advantages over broadband quantum memory, realization of AFC quantum memory in atomic gases is also appealing [97]. The multimode capacity of quantum memories mainly depends on the memory scheme for enough atoms (Section 17.3.6), in quantum memories with atomic gases, most research is still based on storage for single photonic mode, only a few multiplexed quantum memories are reported through dividing a big atomic ensemble into several small ensemble cells [78, 98], harnessing multiple spatial modes of generated spinwaves [99–101] or manipulating temporally selective read-out process [102]. Considering the integratability of gas atomic ensembles-based quantum memories, warm atomic vapor cell is more likely to have smaller size and fabricated on a chip than cold atomic ensemble [103], cause the preparation of cold atom is sophisticated, and will result in a large size system. An research direction for micro-system cold atoms is the cold atomic chip, where the atoms are cooled and trapped on a small-scale chip. It has been widely researched for atoms-based precise measurement [104].

17.4.2 Solid-State Atomic Ensemble

Solid-state atomic ensembles, especially REIDS are also a favorable candidate for quantum memories because of their various working wavelength with different doped ions (for telecom band photons), long optical and spin coherence time [105] (for long storage time), large inhomogeneous broadening [54] (for large storage bandwidth). Different from atomic gases where embedded-type quantum memories have been widely used based on the DLCZ scheme, REIDS is often used as absorptive-type quantum memories based on photon echo technology such as CRIB and AFC. Currently, AFC has been the most often used scheme in REIDS quantum memories. The AFC quantum memory was experimentally observed with Nd^{3+}:YVO_4 in 2008 [16] and theoretically analyzed in 2009 [17]. Since then, more and more research studies are focusing on acquiring high-performance AFC quantum memories in REIDS. By improving the performance, entanglement between two multimode quantum memories has been demonstrated recently [106, 107].

As there are various species of rare ions that can be doped into solid-state host materials, the working wavelength for different ions varies. For instance, erbium ions can work at 1.5 μm for storing long distance transmitted photons while thulium ions can work at 795 nm approximate to where rubidium atomic gas works. By varying the host material such as $LiNbO_3$, Y_2SiO_5, $GdVO_4$, YVO_4, $Y_3Al_5O_{12}$, SiO_2, and even organic molecular materials, etc., the inhomogeneous broadening can vary from several MHz to THz, making broadband and frequency multiplexed quantum memories possible. With the current state of the art, the storage efficiency is a challenge to most AFC quantum memories with REIDS due to finite optical depth and non-zero background absorption. Despite the storage time being preset by the comb spacing of burned AFC, on-demand retrieval and long storage time can be acquired by transferring the optical coherence to spin coherence. As the 4f–4f transition of rare earth ions is shielded from the environment by the outer filled electron orbitals, coherence can preserve for rather long time and support a storage time of several seconds to hours. For multimode capacity, AFC has its unique advantages, where the number of stored modes is independent with the optical depth [43]. The integratability of solid-state materials seems much easier since these solid hosts can be easily fabricated to a micro- or nanoscale waveguide, thus on-chip waveguide based on REIDS is practical for chip-based quantum memories.

In contrast to the first proof of principle demonstration of AFC quantum memory for weak coherent light with wavelength of 880 nm, storage time of 250 ns, storage efficiency of 0.005, AFC quantum memories with REIDS have been widely studied for improving their performance. To let REIDS directly store photons at telecom band, erbium ions doped into different solid host materials are used as the memory medium [54, 81, 108]. To improve the memory efficiency, an impedance-matched cavity is suggested [61] and experimental realization of efficiency of 0.56 have been reported [62]. To improve the storage time, the optical excitation can be converted into a spinwave excitation for spinwave storage [109] and the stored photons are then retrieved by converting the spinwave excitation back to the optical excitation. By doing this, the AFC quantum memory can not only get longer storage time but also support on-demand retrieval process. Recently, storage time up to 1 hour has been reported for storage of coherent light [67]. For high-fidelity qubit memory, fidelity up to 0.999 for polarization-encoded qubit has been realized by designing the memory medium composing of two pieces of crystals sandwiching a half-wave plate [72]. To realize broadband quantum memory, medium with large inhomogeneous broadening is necessary since the storage bandwidth of AFC quantum memory mainly depends on the frequency width of burned AFC, which is directly influenced by the inhomogeneous broadening of the memory medium. By doping the rare earth ions into different solid hosts, which can provide various local electric field for the doped ions and lead to wide energy shift for doped ions [110], media with large inhomogeneous broadening can thus be found and used for broadband quantum memory. For instance, erbium-ions-doped fiber has inhomogeneous broadening up to THz and can be easily used as frequency multiplexed quantum memory with GHz storage bandwidth [75]. For temporal multimode quantum memory, AFC also suits well. The upper bound of stored modes is determined by the storage time–bandwidth product. With current state of the art, storage of 100 temporal modes [76], six spectral modes [75] of single photons have been demonstrated. By simultaneously multiplexing temporal and spectral modes, temporal and spectral quantum storage of 130 single photon modes has been reported [77]. To integrate the REIDS for quantum memory, several groups are right now developing on-chip quantum memories by fabricating the REIDS to a photonic crystal nanobeam resonator or a micro (nano)-scale waveguide.

Even though REIDS are mostly used for AFC quantum memories, REIDS are also available for storing photons with other schemes such as CRIB [46, 111], EIT [28, 112], DLCZ [113, 114].

Quantum memories with gas atomic ensemble and solid-state atomic ensemble have been widely studied and lots of application progresses have also been achieved. However, there isn't a high-performance quantum memory that behaves very well in all aspects, some research groups are still trying to develop other physical system for quantum memories such as single ions [85], single neutral atoms [86], NV centers [88], quantum dots [115–117], oscillators [56, 118] and molecules [119–121].

17.5 Applications

Quantum memory is a basic requirement for several applications of quantum information technologies like quantum computing and quantum communication. During the last decade, great efforts have been made toward the aim of satisfying the requirements for linear-optical quantum computation and quantum repeater. In addition, other prospective applications such as memory-assisted quantum key distribution and single-photon detection also emerged. Below we will review recent experimental achievements toward these applications. Assisted by the quantum memories, it has paved a big step to the development of quantum information processing.

17.5.1 Linear-Optical Quantum Computing

Quantum computation is believed to have the ability to solve questions that can't be solved efficiently via classical computers and is under research in various systems including ion-trap system [122], superconducting system [123], linear optical system [11], etc. Linear optical system is appealing for quantum computing because of its simple composition, convenient integration and good photon coherence.

Linear optical quantum computation [11, 124] is realized relying on multiple single photon states and the probabilistic quantum gates among the photons in a linear optical system. The optical system mainly includes linear optical elements such as waveplates, beam splitters, phase shifters, which can be easily integrated on a chip. To perform linear quantum computation, one of the most important steps is to generate multiple single photons simultaneously. By utilizing quantum memories, the generation rate for multiple photons can be effectively enhanced [125, 126], even though it still can't make the probabilistic photon source deterministic. Figure 17.5a displays a scheme to simultaneously generate multiple single photons via quantum memories. By storing probabilistic heralded single photons into quantum memories and retrieving them on demand, the generation rate can be enhanced by $1/p^{M-1}$ with an ideal quantum memory, where M is the number of photons and p is the photon generation probability for each individual single photons.

As seen from the aforementioned multi-photon generation scheme, a more basic application of quantum memories is also presented, i.e. single photon source realized with the DLCZ quantum memory. Conditioned on the detection of write-out photons, read-out photons can be retrieved as single photons. By doing this, multiple single photons with much narrower linewidth can be simultaneously generated via the DLCZ quantum memories(shown in Figure 17.5b) for interference [127]. Specially, by making use of the Rydberg blockade in atomic gases, deterministic single photon source is allowed with Rydberg atomic quantum memory [128, 129]. Making

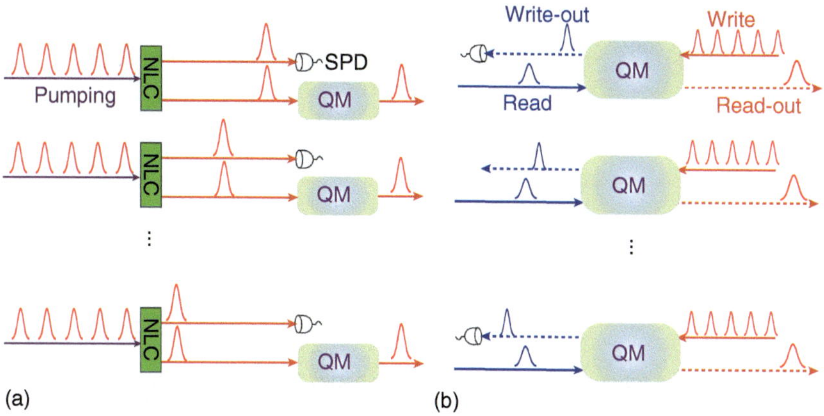

Figure 17.5 Scheme to generate multiple single photons synchronized by absorption-type (a) or embedded-type (b) quantum memories. (a) Multiple quantum correlated photon pairs are generated probabilistically via pumping non-linear crystals, one photon from the photon pairs is stored into the absorption-type quantum memories while the detection of the other photon acts as heralding signal to herald the existence of the stored photon. Once all pairs give heralding signals, the stored photons are all retrieved simultaneously to prepare multiple photons. (b) The stored atomic state in the embedded-type quantum memory is heralded by the detection of the write-out photon. Once all write-out photons from all quantum memories are successfully detected, multiple photons are prepared by applying the read light to retrieve the readout photons simultaneously. NLC: non-linear crystal; SPD: single-photon detector.

use of such a waiting mechanism, real-time ghost imaging was demonstrated with a quantum memory [130].

During the quantum computation, clock synchronization is also an important character to ensure effective information processing. The use of quantum memories can help store results and ensure all the parallel processing steps synchronized.

17.5.2 Quantum Repeater

Quantum communication is considered as an absolutely secure way of communication thanks to the no-cloning of quantum information. During photon transmission over noisy channel such as fiber, the probability of both photon loss and qubit decoherence grows exponentially with the length L of the fiber, which will influence the communication rate and fidelity of transmitted quantum information. An available solution is to generate quantum entanglement among distant quantum nodes and transmit quantum information among these distant nodes via quantum teleportation.

The key element is to generate long-distance quantum entanglement, however, direct entanglement distribution is also limited by the photon transmission loss. To still generate distant entanglement, quantum repeater scheme is proposed [4]. The basic idea is shown in Figure 17.6, to finally obtain entanglement between remote nodes A and Z, several sections are divided, and entanglement between adjacent nodes with shorter distance is generated, after entanglement in each section is

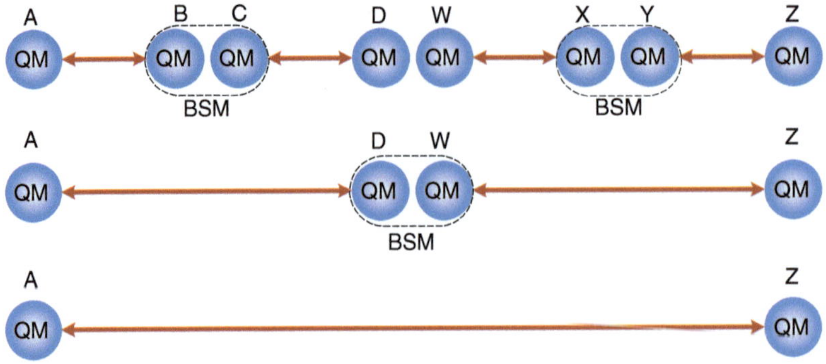

Figure 17.6 Scheme for quantum repeater. Distant nodes A and Z are entangled by dividing the link to several sections with short elementary links like A and B, ..., Y and Z. Entanglement is established between quantum memories A and B, ..., Y and Z. Then Bell-state measurements (BSMs) between B and C, X and Y are performed to swap the entanglement for preparing entanglement between A and D, W and Z. Finally, the same operation is applied to D and W for generating entanglement between distant nodes A and Z.

successfully prepared, Bell-state-measurement (BSM) is performed for entanglement swapping to obtain entanglement with longer distance. As a result, distant entanglement can be generated via this way. To guarantee the swapping, quantum memory is needed for storage of successful entanglement until all entanglements are prepared, i.e. photon–memory/memory–memory entanglement is the fundamental building block for quantum repeater. Under the utilization of quantum memories, the decrease of long-distance entanglement generation rate can be reduced from scaling as exponential to polynomial vs. the distance L, and the quantum communication over thousands of kilometers is possible with optical fiber network.

On the way to building photon-memory entanglement, quantum memories can not only directly store the entangled photons but also be used to emit single photons for directly generating photon–memory entanglement. For absorbed quantum memories with schemes such as EIT, AFC, etc., direct storage of entangled photons is utilized to generate entanglement between adjacent nodes. For embedded-type quantum memories with schemes such DLCZ, Raman scheme, two pairs of photon–memory entanglement can be directly generated, then memory–memory entanglement can be acquired by Bell state measurement performed on the two transmitted photons.

17.5.3 Quantum Key Distribution

Although quantum repeater can solve the problem of long-distance quantum communication, it strongly relies on the performance of quantum memories. For instance, a practical quantum repeater requires the memory storage time far exceeding the average time to successfully generate entanglement between quantum repeater nodes, which is actually difficult with current state of the art. To lower the harsh requirements on the high performances of quantum memories

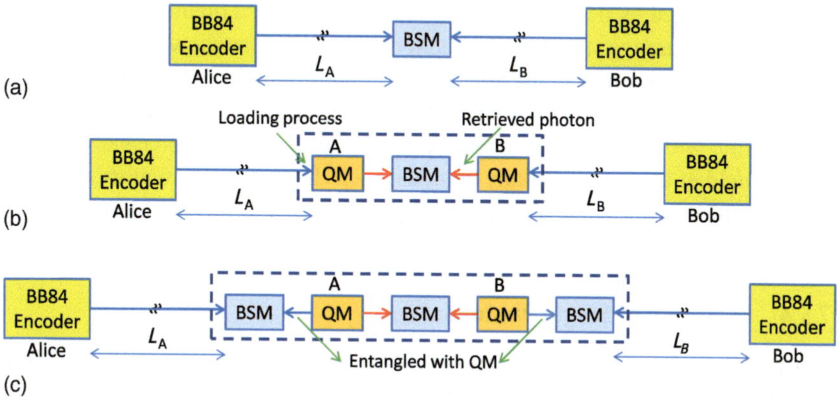

Figure 17.7 Scheme for measurement-device-independent quantum key distribution (MDI-QKD) with quantum memories. (a) The original MDI-QKD protocol. Alice and Bob send photons to the middle site to perform BSM. (b) MDI-QKD with heralding quantum memories. Alice and Bob send photons to their respective quantum memories (QMs) in the middle site. At each round, each QM attempts to store the incoming photons. Once they are both loaded via the heralding signal from the memories, we retrieve the QMs' states and perform a BSM on the retrieved photons. (c) MDI-QKD with non-heralding quantum memories. At each round, photons entangled with the QMs A and B are firstly generated. These photons interfere at the BSM modules next to the QMs with incoming photons from Alice and Bob. As soon as one of these BSMs succeeds, we stop the entanglement generation processes on the corresponding QM, and wait until another BSM related to the other QM also succeeds. In this case, QMs are not required to have the ability to herald the success of the storage process; a trigger event is declared by the success of the BSM located between the QM and the respective encoder. Source: Panayi et al. [131]/IOP Publishing/CC BY-3.0.

used for quantum communication, scheme for memory-assisted quantum key distribution (MA-QKD) is proposed [131], which is believed to have the potential to beat the highest quantum communication distance records with no-memory QKD.

Figure 17.7a shows the conventional measurement device independent quantum key distribution (MDI-QKD), which relies on sending encoded photons by Alice and Bob to a middle site for BSM and is more secure compared to the original BB84 protocol. The key generation rate depends on the successful interference probability of the transmitted photons, which is constrained by the low efficiency that photons can occur at the middle site. An undesired case is that photons from Alice successfully arrive at the middle site while photons from Bob fail or photons from Alice fail while photons from Bob success. Inspired by the concept of quantum repeater, if photons from Alice or Bob can be firstly stored into heralding quantum memories, which can herald whether photons are stored, and be retrieved for BSM until both photons success to arrive at the middle site, thus the unexpected cases is avoided, where the successfully arrived photons can be stored and wait. The scheme is shown in Figure 17.7b. By utilizing such a memory assisted MDI-QKD, the average time needed to get a successful event of BSM can thus be efficiently decreased, i.e. higher key rate is possible. Compared to the quantum repeater where classical communication time among repeater nodes is an important factor and limits the distribution period, in memory-assisted MDI-QKD, the repetition rate of the sending process is mainly determined by how fast the memories can be. For a quantum memory that can't herald the storage, an additional BSM can be placed, as shown in Figure 17.7c. Entanglement between photons and quantum memories is firstly generated, then the photon interferes with the photon from Alice (Bob), the residual procedures are the same with the scheme in Figure 17.7b. Recently, M.K. Bhaskar et al. and S. Langenfeld et al. experimentally demonstrated the quantum memory enhanced communication with single silicon-vacancy (SiV) color center and single neutral atom in a cavity [132, 133].

17.5.4 Detection of Single Photons

Highly efficient detection of single photons is an essential requirement for quantum information processing. By utilizing a quantum memory, one can firstly store single photons into the quantum memory with a storage efficiency reaching unity and then detect atomic state with an efficiency close to unity, thus the photon can be detected nearly 100% efficiency. For instance, photons at single photon level or single photons can be stored into a quantum memory, where storage efficiency around 90% [40, 60] has been demonstrated, then the stored photons can be mapped to an atomic state of cycle transition and measured via fluorescence detection with the cycle transition [134] (shown in Figure 17.8a), which could have a detection efficiency of unity. Furthermore, as the intensity of fluorescence is proportional to the number of atoms that participate in the cycle transition, if there exists multiple photons to be detected, by distinguishing the fluorescence intensity, the photon number can be also measured. Compared to the conventional single photon detectors, photon detection with quantum memories can in principle distinguish how many photons there are, i.e. it can act as a photon-number resolved detector. Another detection method is to store the photon into the atomic quantum memory and then ionize the atom along with using electron detectors to detect the charged particle [87] (shown in Figure 17.8b), the detection efficiency is also high.

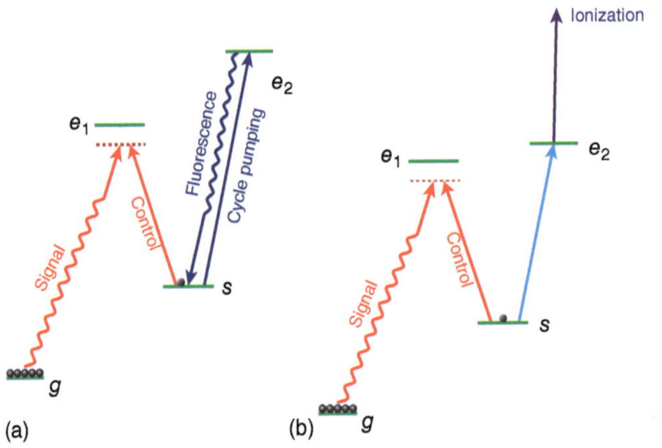

Figure 17.8 Example for single photon detection via quantum memories. (a) Photons are initially stored into quantum memories via EIT, then the atomic states are measured via fluorescence of cycle transition. (b) Atomic states are measured via ionization detection.

17.6 Summary and Outlook

In summary, we have given an overview of quantum memory for photons. Focusing on single-photon photon storage, we have reviewed the memory schemes, key figures of merit, physical realization, and the applications. Significant progress has been achieved in improving the memory performance in recent years, by demonstrating the capabilities of storing photons with high efficiency, long lifetime, high fidelity, large bandwidth, and high multimode capacity. However, these demanding capabilities are realized with different memory schemes or in different memory media. Exploring and developing a quantum memory with an overall high performance in a single platform is still a outstanding goal for its potential applications, such as in linear-optical quantum computing, quantum repeater, quantum key distribution, and detection of single photons. Along with further experimental investigations on the established systems, it is appealing as well to design new memory schemes and develop new memory mediums.

References

1 Lvovsky, A.I., Sanders, B.C., and Tittel, W. (2009). Optical quantum memory. *Nature Photonics* 3: 706–714.

2 Simon, C., Afzelius, M., Appel, J. et al. (2010). Quantum memories. *The European Physical Journal D* 58: 1–22.

3 Heshami, K., England, D.G., Humphreys, P.C. et al. (2016). Quantum memories: emerging applications and recent advances. *Journal of Modern Optics* 63: 2005–2028.

4 Briegel, H.-J., Dür, W., Cirac, J.I., and Zoller, P. (1998). Quantum repeaters: the role of imperfect local operations in quantum communication. *Physical Review Letters* 81: 5932.

5 Simon, C., De Riedmatten, H., Afzelius, M. et al. (2007). Quantum repeaters with photon pair sources and multimode memories. *Physical Review Letters* 98: 190503.

6 Kimble, H.J. (2008). The quantum internet. *Nature* 453: 1023–1030.

7 Simon, C. (2017). Towards a global quantum network. *Nature Photonics* 11: 678–680.

8 Wehner, S., Elkouss, D., and Hanson, R. (2018). Quantum internet: a vision for the road ahead. *Science* 362: eaam9288.

9 Wei, S.-H., Jing, B., Zhang, X.Y. et al. (2022). Towards real-world quantum networks: a review. *Laser & Photonics Reviews* 16 (3): 2100219.

10 Gisin, N. and Thew, R. (2007). Quantum communication. *Nature Photonics* 1: 165–171.

11 Kok, P., Munro, W.J., Nemoto, K. et al. (2007). Linear optical quantum computing with photonic qubits. *Reviews of Modern Physics* 79: 135.

12 Pittman, T.B. and Franson, J.D. (2002). Cyclical quantum memory for photonic qubits. *Physical Review A* 66: 062302.

13 Leung, P.M. and Ralph, T.C. (2006). Quantum memory scheme based on optical fibers and cavities. *Physical Review A* 74: 022311.
14 Phillips, D.F., Fleischhauer, A., Mair, A. et al. (2001). Storage of light in atomic vapor. *Physical Review Letters* 86: 783.
15 Liu, C., Dutton, Z., Behroozi, C.H., and Hau, L.V. (2001). Observation of coherent optical information storage in an atomic medium using halted light pulses. *Nature* 409: 490–493.
16 De Riedmatten, H., Afzelius, M., Staudt, M.U. et al. (2008). A solid-state light–matter interface at the single-photon level. *Nature* 456: 773–777.
17 Afzelius, M., Simon, C., De Riedmatten, H., and Gisin, N. (2009). Multimode quantum memory based on atomic frequency combs. *Physical Review A* 79: 052329.
18 Duan, L.M., Lukin, M.D., Cirac, J.I., and Zoller, P. (2001). Long-distance quantum communication with atomic ensembles and linear optics. *Nature* 414: 413–418.
19 Kuzmich, A., Bowen, W.P., Boozer, A.D. et al. (2003). Generation of nonclassical photon pairs for scalable quantum communication with atomic ensembles. *Nature* 423: 731–734.
20 van der Wal, C.H., Eisaman, M.D., André, A. et al. (2003). Atomic memory for correlated photon states. *Science* 301: 196–200.
21 Moiseev, S. and Kröll, S. (2001). Complete reconstruction of the quantum state of a single-photon wave packet absorbed by a Doppler-broadened transition. *Physical Review Letters* 87: 173601.
22 Reim, K.F., Nunn, J., Lorenz, V.O. et al. (2010). Towards high-speed optical quantum memories. *Nature Photonics* 4: 218–221.
23 Saglamyurek, E., Hrushevskyi, T., Rastogi, A. et al. (2018). Coherent storage and manipulation of broadband photons via dynamically controlled Autler–Townes splitting. *Nature Photonics* 12: 774–782.
24 Appel, J., Figueroa, E., Korystov, D. et al. (2008). Quantum memory for squeezed light. *Physical Review Letters* 100: 093602.
25 Honda, K., Akamatsu, D., Arikawa, M. et al. (2008). Storage and retrieval of a squeezed vacuum. *Physical Review Letters* 100: 093601.
26 Julsgaard, B., Sherson, J., Cirac, J.I. et al. (2004). Experimental demonstration of quantum memory for light. *Nature* 432: 482–486.
27 Seshadreesan, K.P., Krovi, H., and Guha, S. (2020). Continuous-variable quantum repeater based on quantum scissors and mode multiplexing. *Physical Review Research* 2: 013310.
28 Turukhin, A.V., Sudarshanam, V.S., Shahriar, M.S. et al. (2001). Observation of ultraslow and stored light pulses in a solid. *Physical Review Letters* 88: 023602.
29 Volz, J., Weber, M., Schlenk, D. et al. (2006). Observation of entanglement of a single photon with a trapped atom. *Physical Review Letters* 96: 030404.
30 Wilk, T., Webster, S.C., Kuhn, A., and Rempe, G. (2007). Single-atom single-photon quantum interface. *Science* 317: 488–490.

31 Blinov, B.B., Moehring, D.L., Duan, L.-M., and Monroe, C. (2004). Observation of entanglement between a single trapped atom and a single photon. *Nature* 428: 153–157.

32 Dutt, M.V.G., Childress, L., Jiang, L. et al. (2007). Quantum register based on individual electronic and nuclear spin qubits in diamond. *Science* 316: 1312–1316.

33 Nguyen, C., Sukachev, D.D., Bhaskar, M.K. et al. (2019). An integrated nanophotonic quantum register based on silicon-vacancy spins in diamond. *Physical Review B* 100: 165428.

34 Harris, S.E., Field, J.E., and Imamoğlu, A. (1990). Nonlinear optical processes using electromagnetically induced transparency. *Physical Review Letters* 64: 1107.

35 Fleischhauer, M., Imamoglu, A., and Marangos, J.P. (2005). Electromagnetically induced transparency: optics in coherent media. *Reviews of Modern Physics* 77: 633.

36 Boller, K.-J., Imamoğlu, A., and Harris, S.E. (1991). Observation of electromagnetically induced transparency. *Physical Review Letters* 66: 2593.

37 Kasapi, A., Jain, M., Yin, G.Y., and Harris, S.E. (1995). Electromagnetically induced transparency: propagation dynamics. *Physical Review Letters* 74: 2447.

38 Wang, L.J., Kuzmich, A., and Dogariu, A. (2000). Gain-assisted superluminal light propagation. *Nature* 406: 277–279.

39 Fleischhauer, M. and Lukin, M.D. (2000). Dark-state polaritons in electromagnetically induced transparency. *Physical Review Letters* 84: 5094.

40 Hsiao, Y.-F., Tsai, P.J., Chen, H.S. et al. (2018). Highly efficient coherent optical memory based on electromagnetically induced transparency. *Physical Review Letters* 120: 183602.

41 Harris, S.E., Field, J.E., and Kasapi, A. (1992). Dispersive properties of electromagnetically induced transparency. *Physical Review A* 46: R29.

42 Fleischhauer, M. and Lukin, M.D. (2002). Quantum memory for photons: Dark-state polaritons. *Physical Review A* 65: 022314.

43 Nunn, J., Reim, K., Lee, K.C. et al. (2008). Multimode memories in atomic ensembles. *Physical Review Letters* 101: 260502.

44 Sangouard, N., Simon, C., Afzelius, M., and Gisin, N. (2007). Analysis of a quantum memory for photons based on controlled reversible inhomogeneous broadening. *Physical Review A* 75: 032327.

45 Baldit, E., Briaudeau, S., Monnier, P. et al. (2010). Identification of Λ-like systems in Er^{3+}: Y_2SiO_5 and observation of electromagnetically induced transparency. *Physical Review B* 81: 144303.

46 Tittel, W., Afzelius, M., Chaneliere, T. et al. (2010). Photon-echo quantum memory in solid state systems. *Laser & Photonics Reviews* 4: 244–267.

47 Alexander, A.L., Longdell, J.J., Sellars, M.J., and Manson, N.B. (2006). Photon echoes produced by switching electric fields. *Physical Review Letters* 96: 043602.

48 Hétet, G., Longdell, J., Sellars, M. et al. (2008). Multimodal properties and dynamics of gradient echo quantum memory. *Physical Review Letters* 101: 203601.

49 Saglamyurek, E., Hrushevskyi, T., Cooke, L. et al. (2019). Single-photon-level light storage in cold atoms using the Autler-Townes splitting protocol. *Physical Review Research* 1: 022004.

50 Saglamyurek, E., Hrushevskyi, T., Rastogi, A. et al. (2021). Storing short single-photon-level optical pulses in Bose–Einstein condensates for high-performance quantum memory. *New Journal of Physics* 23: 043028.

51 Donlan, V.L. and Santiago, A.A. Jr. (1972). Optical spectra and energy levels of erbium-doped Yttrium orthoaluminate. *The Journal of Chemical Physics* 57: 4717–4723.

52 Radnaev, A.G., Dudin, Y.O., Zhao, R. et al. (2010). A quantum memory with telecom-wavelength conversion. *Nature Physics* 6: 894–899.

53 Zhou, B., Tao, L., Tsang, Y.H. et al. (2012). Superbroadband near-IR photoluminescence from Pr^{3+}-doped fluorotellurite glasses. *Optics Express* 20: 3803–3813.

54 Saglamyurek, E., Jin, J., Verma, V.B. et al. (2015). Quantum storage of entangled telecom-wavelength photons in an erbium-doped optical fibre. *Nature Photonics* 9: 83–87.

55 Jin, J., Saglamyurek, E., Verma, V. et al. (2015). Telecom-wavelength atomic quantum memory in optical fiber for heralded polarization qubits. *Physical Review Letters* 115: 140501.

56 Wallucks, A., Marinković, I., Hensen, B. et al. (2020). A quantum memory at telecom wavelengths. *Nature Physics* 16: 772–777.

57 Gorshkov, A.V., André, A., Fleischhauer, M. et al. (2007). Universal approach to optimal photon storage in atomic media. *Physical Review Letters* 98: 123601.

58 Gorshkov, A.V., André, A., Lukin, M.D., and Sørensen, A.S. (2007). Photon storage in Λ-type optically dense atomic media. I. Cavity model. *Physical Review A* 76: 033804.

59 Gorshkov, A.V., André, A., Lukin, M.D., and Sørensen, A.S. (2007). Photon storage in Λ-type optically dense atomic media. II. Free-space model. *Physical Review A* 76: 033805.

60 Wang, Y., Li, J., Zhang, S. et al. (2019). Efficient quantum memory for single-photon polarization qubits. *Nature Photonics* 13: 346–351.

61 Afzelius, M. and Simon, C. (2010). Impedance-matched cavity quantum memory. *Physical Review A* 82: 022310.

62 Sabooni, M., Li, Q., Kröll, S., and Rippe, L., (2013). Efficient quantum memory using a weakly absorbing sample. *Physical Review Letters* 110: 133604.

63 Jing, B., Wang, X.J., Yu, Y. et al. (2019). Entanglement of three quantum memories via interference of three single photons. *Nature Photonics* 13: 210–213.

64 Sangouard, N., Simon, C., De Riedmatten, H., and Gisin, N. (2011). Quantum repeaters based on atomic ensembles and linear optics. *Reviews of Modern Physics* 83: 33.

65 Razavi, M., Piani, M., and Lütkenhaus, N. (2009). Quantum repeaters with imperfect memories: Cost and scalability. *Physical Review A* 80: 032301.

66 Yang, S.-J., Wang, X.-J., Bao, X.-H., and Pan, J.-W. (2016). An efficient quantum light–matter interface with sub-second lifetime. *Nature Photonics* 10: 381–384.

67 Ma, Y., Ma, Y.-Z., Zhou, Z.-Q. et al. (2021). One-hour coherent optical storage in an atomic frequency comb memory. *Nature Communications* 12: 1–6.
68 Jozsa, R. (1994). Fidelity for mixed quantum states. *Journal of Modern Optics* 41: 2315–2323.
69 Altepeter, J.B., Jeffrey, E.R., and Kwiat, P.G. (2005). Photonic state tomography. *Advances in Atomic, Molecular, and Optical Physics* 52: 105–159.
70 Gühne, O. and Tóth, G. (2009). Entanglement detection. *Physics Reports* 474: 1–75.
71 Massar, S. and Popescu, S. (1995). Optimal extraction of information from finite quantum ensembles. *Physical Review Letters* 74: 1259–1263.
72 Zhou, Z.-Q., Lin, W.-B., Yang, M. et al. (2012). Realization of reliable solid-state quantum memory for photonic polarization qubit. *Physical Review Letters* 108: 190505.
73 Michelberger, P.S., Champion, T.F.M., Sprague, M.R. et al. (2015). Interfacing ghz-bandwidth heralded single photons with a warm vapour Raman memory. *New Journal of Physics* 17: 043006.
74 England, D.G., Fisher, K.A.G., MacLean, J.P.W. et al. (2015). Storage and retrieval of THz-bandwidth single photons using a room-temperature diamond quantum memory. *Physical Review Letters* 114: 053602.
75 Saglamyurek, E., Grimau Puigibert, M., Zhou, Q. et al. (2016). A multiplexed light-matter interface for fibre-based quantum networks. *Nature Communications* 7: 1–7.
76 Tang, J.-S., Zhou, Z.Q., Wang, Y.T. et al. (2015). Storage of multiple single-photon pulses emitted from a quantum dot in a solid-state quantum memory. *Nature Communications* 6: 1–7.
77 Seri, A., Lago-Rivera, D., Lenhard, A. et al. (2019). Quantum storage of frequency-multiplexed heralded single photons. *Physical Review Letters* 123: 080502.
78 Lan, S.-Y., Radnaev, A.G., Collins, O.A. et al. (2009). A multiplexed quantum memory. *Optics Express* 17: 13639–13645.
79 Yang, T.-S., Zhou, Z.Q., Hua, Y.L. et al. (2018). Multiplexed storage and real-time manipulation based on a multiple degree-of-freedom quantum memory. *Nature Communications* 9: 1–8.
80 Seri, A., Corrielli, G., Lago-Rivera, D. et al. (2018). Laser-written integrated platform for quantum storage of heralded single photons. *Optica* 5: 934–941.
81 Askarani, M.F., Lutz, T., Verma, V.B. et al. (2019). Storage and reemission of heralded telecommunication-wavelength photons using a crystal waveguide. *Physical Review Applied* 11: 054056.
82 Craiciu, I., Lei, M., Rochman, J. et al. (2021). Multifunctional on-chip storage at telecommunication wavelength for quantum networks. *Optica* 8: 114–121.
83 Zhong, T., Kindem, J.M., Bartholomew, J.G. et al. (2017). Nanophotonic rare-earth quantum memory with optically controlled retrieval. *Science* 357: 1392–1395.
84 Craiciu, I., Lei, M., Rochman, J. et al. (2019). Nanophotonic quantum storage at telecommunication wavelength. *Physical Review Applied* 12: 024062.

85 Duan, L.-M. and Monroe, C. (2010). Colloquium: quantum networks with trapped ions. *Reviews of Modern Physics* 82: 1209.

86 Reiserer, A. and Rempe, G. (2015). Cavity-based quantum networks with single atoms and optical photons. *Reviews of Modern Physics* 87: 1379.

87 Rosenfeld, W., Burchardt, D., Garthoff, R. et al. (2017). Event-ready bell test using entangled atoms simultaneously closing detection and locality loopholes. *Physical Review Letters* 119: 010402.

88 Ruf, M., Wan, N.H., Choi, H. et al. (2021). Quantum networks based on color centers in diamond. *Journal of Applied Physics* 130: 070901.

89 Chanelière, T., Matsukevich, D.N., Jenkins, S.D. et al. (2005). Storage and retrieval of single photons transmitted between remote quantum memories. *Nature* 438: 833–836.

90 Eisaman, M.D., André, A., Massou, F. et al. (2005). Electromagnetically induced transparency with tunable single-photon pulses. *Nature* 438: 837–841.

91 Albrecht, B., Farrera, P., Fernandez-Gonzalvo, X. et al. (2014). A waveguide frequency converter connecting rubidium-based quantum memories to the telecom c-band. *Nature Communications* 5: 1–6.

92 Simon, J., Tanji, H., Thompson, J.K., and Vuletić, V. (2007). Interfacing collective atomic excitations and single photons. *Physical Review Letters* 98: 183601.

93 Bao, X.-H., Reingruber, A., Dietrich, P. et al. (2012). Efficient and long-lived quantum memory with cold atoms inside a ring cavity. *Nature Physics* 8: 517–521.

94 Cho, Y.-W., Campbell, G.T., Everett, J.L. et al. (2016). Highly efficient optical quantum memory with long coherence time in cold atoms. *Optica* 3: 100–107.

95 Guo, J., Feng, X., Yang, P. et al. (2019). High-performance raman quantum memory with optimal control in room temperature atoms. *Nature Communications* 10: 1–6.

96 Vernaz-Gris, P., Huang, K., Cao, M. et al. (2018). Highly-efficient quantum memory for polarization qubits in a spatially-multiplexed cold atomic ensemble. *Nature Communications* 9: 1–6.

97 Main, D., Hird, T., Gao, S. et al. (2021). Room temperature atomic frequency comb storage for light. *Optics Letters* 46: 2960–2963.

98 Pu, Y., Jiang, N., Chang, W. et al. (2017). Experimental realization of a multiplexed quantum memory with 225 individually accessible memory cells. *Nature Communications* 8: 1–6.

99 Dai, H.-N., Zhang, H., Yang, S.J. et al. (2012). Holographic storage of biphoton entanglement. *Physical Review Letters* 108: 210501.

100 Tian, L., Xu, Z., Chen, L. et al. (2017). Spatial multiplexing of atom-photon entanglement sources using feedforward control and switching networks. *Physical Review Letters* 119: 130505.

101 Parniak, M., Dąbrowski, M., Mazelanik, M. et al. (2017). Wavevector multiplexed atomic quantum memory via spatially-resolved single-photon detection. *Nature Communications* 8: 1–9.

102 Albrecht, B., Farrera, P., Heinze, G. et al. (2015). Controlled rephasing of single collective spin excitations in a cold atomic quantum memory. *Physical Review Letters* 115: 160501.

103 Kitching, J. (2018). Chip-scale atomic devices. *Applied Physics Reviews* 5: 031302.

104 Keil, M., Amit, O., Zhou, S. et al. (2016). Fifteen years of cold matter on the atom chip: promise, realizations, and prospects. *Journal of Modern Optics* 63: 1840–1885.

105 Zhong, M., Hedges, M.P., Ahlefeldt, R.L. et al. (2015). Optically addressable nuclear spins in a solid with a six-hour coherence time. *Nature* 517: 177–180.

106 Liu, X., Hu, J., Li, Z.F. et al. (2021). Heralded entanglement distribution between two absorptive quantum memories. *Nature* 594: 41–45.

107 Lago-Rivera, D., Grandi, S., Rakonjac, J.V. et al. (2021). Telecom-heralded entanglement between multimode solid-state quantum memories. *Nature* 594: 37–40.

108 Lauritzen, B., Minář, J., De Riedmatten, H. et al. (2011). Approaches for a quantum memory at telecommunication wavelengths. *Physical Review A* 83: 012318.

109 Afzelius, M., Usmani, I., Amari, A. et al. (2010). Demonstration of atomic frequency comb memory for light with spin-wave storage. *Physical Review Letters* 104: 040503.

110 Thiel, C., Böttger, T., and Cone, R. (2011). Rare-earth-doped materials for applications in quantum information storage and signal processing. *Journal of Luminescence* 131: 353–361.

111 Nilsson, M. and Kröll, S. (2005). Solid state quantum memory using complete absorption and re-emission of photons by tailored and externally controlled inhomogeneous absorption profiles. *Optics Communications* 247: 393–403.

112 Longdell, J.J., Fraval, E., Sellars, M.J., and Manson, N.B. (2005). Stopped light with storage times greater than one second using electromagnetically induced transparency in a solid. *Physical Review Letters* 95: 063601.

113 Kutluer, K., Mazzera, M., and de Riedmatten, H. (2017). Solid-state source of nonclassical photon pairs with embedded multimode quantum memory. *Physical Review Letters* 118: 210502.

114 Laplane, C., Jobez, P., Etesse, J. et al. (2017). Multimode and long-lived quantum correlations between photons and spins in a crystal. *Physical Review Letters* 118: 210501.

115 Gao, W., Fallahi, P., Togan, E. et al. (2012). Observation of entanglement between a quantum dot spin and a single photon. *Nature* 491: 426–430.

116 Shou, C., Zhang, Q., Luo, W., and Huang, G. (2021). Photon storage and routing in quantum dots with spin-orbit coupling. *Optics Express* 29: 9772–9785.

117 Neuwirth, J., Basset, F.B., Rota, M.B. et al. (2021). Quantum dot technology for quantum repeaters: from entangled photon generation towards the integration with quantum memories. *Material Quantum Technology* 1: 043001.

118 Riedinger, R., Hong, S., Norte, R.A. et al. (2016). Non-classical correlations between single photons and phonons from a mechanical oscillator. *Nature* 530: 313–316.

119 Bustard, P.J., Lausten, R., England, D.G., and Sussman, B.J. (2013). Toward quantum processing in molecules: a THz-bandwidth coherent memory for light. *Physical Review Letters* 111: 083901.

120 Gerasimov, K., Moiseev,S., Morosov, V., and Zaripov, R. (2014). Room-temperature storage of electromagnetic pulses on a high-finesse natural spin-frequency comb. *Physical Review A* 90: 042306.

121 Bustard, P.J., Erskine, J., England, D.G. et al. (2015). Nonclassical correlations between terahertz-bandwidth photons mediated by rotational quanta in hydrogen molecules. *Optics Letters* 40: 922–925.

122 Häffner, H., Roos, C.F., and Blatt, R. (2008). Quantum computing with trapped ions. *Physics Reports* 469: 155–203.

123 Huang, H.-L., Wu, D., Fan, D., and Zhu, X. (2020). Superconducting quantum computing: a review. *Science China Information Sciences* 63: 1–32.

124 Knill, E., Laflamme, R., and Milburn, G.J. (2001). A scheme for efficient quantum computation with linear optics. *Nature* 409: 46–52.

125 Nunn, J., Langford, N.K., Kolthammer, W.S. et al. (2013). Enhancing multiphoton rates with quantum memories. *Physical Review Letters* 110: 133601.

126 Kaneda, F., Xu, F., Chapman, J., and Kwiat, P.G. (2017). Quantum-memory-assisted multi-photon generation for efficient quantum information processing. *Optica* 4: 1034–1037.

127 Wang, X.-J., Jing, B., Sun, P.F. et al. (2018). Experimental time-resolved interference with multiple photons of different colors. *Physical Review Letters* 121: 080501.

128 Dudin, Y. and Kuzmich, A. (2012). Strongly interacting rydberg excitations of a cold atomic gas. *Science* 336: 887–889.

129 Ripka, F., Kübler, H., Löw, R., and Pfau, T. (2018). A room-temperature single-photon source based on strongly interacting rydberg atoms. *Science* 362: 446–449.

130 Mazelanik, M., Leszczyński, A., Lipka, M. et al. (2021). Real-time ghost imaging of Bell-nonlocal entanglement between a photon and a quantum memory. *Quantum* 5: 493.

131 Panayi, C., Razavi, M., Ma, X., and Lütkenhaus, N., (2014). Memory-assisted measurement-device-independent quantum key distribution. *New Journal of Physics* 16: 043005.

132 Bhaskar, M.K., Riedinger, R., Machielse, B. et al. (2020). Experimental demonstration of memory-enhanced quantum communication. *Nature* 580: 60–64.

133 Langenfeld, S., Thomas, P., Morin, O., and Rempe, G. (2021). Quantum repeater node demonstrating unconditionally secure key distribution. *Physical Review Letters* 126: 230506.

134 Imamoḡlu, A. (2002). High efficiency photon counting using stored light. *Physical Review Letters* 89: 163602.

Part IV

Solid-State and van der Waals Material Platforms: From Single Quantum Emitters to Hybrid Integration

18

Telecom Wavelengths InP-Based Quantum Dots for Quantum Communication

Mohamed Benyoucef[1] and Anna Musiał[2]

[1] *Institute of Physics, Center for Interdisciplinary Nanostructure Science and Technology (CINSaT), University of Kassel, Heinrich-Plett-Str. 40, 34132 Kassel, Germany*
[2] *Laboratory for Optical Spectroscopy of Nanostructures, Department of Experimental Physics, Faculty of Fundamental Problems of Technology, Wrocław University of Science and Technology, Wybrzeże Wyspiańskiego 27, 50-370 Wrocław, Poland*

18.1 Introduction

Worldwide, researchers are fascinated by the rich physics of semiconductor quantum dots (QDs) and their high potential for applications in photonics and quantum information technologies. QDs are small objects with a size in the range of a few tens of nanometers that are made up of thousands to a million atoms. They are often called "artificial atoms" because they provide local spatial confinement of single charges, which results in a fully discrete energy spectrum analogous to the energy levels of natural atoms. In contrast to atoms, QDs offer broad engineering opportunities in terms of electronic structure and therefore, optical properties. These can be post-growth tailored by external factors providing a basis for the realization of tunable quantum devices. However, it is challenging to obtain identical QDs, which is an issue for mass production and scalability.

Progress in realization of quantum devices and circuits enables breakthroughs in experimental tests of quantum physics and in practical applications of quantum technologies in imaging, secure communication, ultrasensitive metrology, and quantum computing [1]. Bringing quantum communication to a practical long-haul level beyond proof-of-principle demonstrations in a laboratory environment requires the development of quantum light sources operating in the telecommunication wavelengths, compatible quantum memories, and quantum repeaters corresponding to signal amplification in a classical network. Hereby, we focus on quantum light sources. In the quantum network, each node needs a transmitter and a receiver – an efficient source of single photons acting as flying qubits and an efficient detection system with sensitivity on the level of single photons to encode the message. To ensure the security of the transmission, quantum key distribution employing either a single (e.g. BB84 [2]) or pair of entangled photons (e.g. Ekert91 [3]) can be used.

Photonic Quantum Technologies: Science and Applications, First Edition. Edited by Mohamed Benyoucef.

Fiber optics is the technology used to transmit information over long distances through fiber made of glass or plastic. The O-band (1260–1360 nm) was historically the first wavelength band used for optical communication because available optical fibers at that time showed their lowest loss within this band. Nowadays, optical fibers show their lowest loss in the C-band (1530–1565 nm). To make quantum networks truly practical, quantum light sources need to fulfill certain requirements. The ideal source would operate at 1550 nm at room temperature, be electrically triggered with on-demand emission, provide a high generation rate of Fourier-transform-limited photons into a single Gaussian mode with high directionality (into numerical aperture of a standard telecom single-mode fiber), be deterministically fabricated with high yield, and feature high single-photon purity (no multiphoton emission events) and indistinguishability (all emitted photons are identical) or high entanglement fidelity (maximally entangled states are generated). Preferred are sources realized within compact portable designs, integrated on a photonic chip, and fiber-coupled for easy integration with existing fiber networks. So far, there is no physical system meeting all the abovementioned conditions. Therefore, different approaches are developed in parallel and optimized for certain applications for which some of the requirements are loosened.

One of the possible resources for quantum light emitters is epitaxial QDs. These have proven excellent performance at shorter wavelengths (below 1000 nm), which allowed for the realization of nearly ideal single-photon sources [4–14] with the lowest probability of multiphoton emission events among all physical systems both for telecom – $(4.4 \pm 0.2) \times 10^{-4}$ [15] and shorter wavelengths – $(7.5 \pm 1.6) \times 10^{-5}$ [16, 17]. They provide truly single-photon emission from a two-level system as opposed to attenuated laser beams with Poissonian emission statistics with a very low mean number of photons achieved only for a low single-photon rate. Additionally, single-photon statistics have proven higher transmission distance in the case of lossy optical channels [18]. In the case of attenuated laser beams, error correction protocols need to be implemented to account for the inherent errors in the source.

Epitaxial QDs offer triggered emission (under both optical and electrical excitation) as opposed to the probabilistic nature of nonlinear processes used to generate single and/or entangled photon states, i.e. spontaneous parametric downconversion and four-wave mixing. In that case, heralding or storing emitted photons for a certain time can make this approach deterministic, but these are more susceptible to losses and decoherence, respectively. The advantages of the alternative approaches mentioned so far are room temperature operation at 1550 nm at the expense of low emission rates for which reasonably high single-photon purity is achieved. Room-temperature operation is not achievable in the case of the III-V QDs emitting in the telecom spectral range due to the details of confining potential, which is not deep enough in these material systems. Room-temperature emission is possible in II–VI or GaN-based QDs emitting in the visible and ultraviolet range, respectively. Regarding the telecommunication spectral range, the strain-engineered GaAs-based nanostructures offer single-photon emission at 1.3 μm only to temperatures below 120 K [19] and InP-based QDs in addition to emission of entangled photon pairs at 1550 nm up till 93 K [20], in both cases not suitable for thermo-electric cooling. However, the best performance in terms of the quality of quantum light (single-photon purity, indistinguishability, and entanglement fidelity) is achieved at low temperatures. Despite that epitaxial QDs are prominent candidates to fulfill the requirements of quantum communication applications [21, 22] due to their compatibility with well-developed semiconductor technology, easy integration into photonic chips and circuits [23], engineering flexibility during growth and post-growth tuning possibilities [24–27] as well as electrical driving (see e.g. Refs. [20, 28] and references therein). Plug&play single-photon sources operating at low temperatures are already available [29–31]. Offered advantages make QD-based quantum light sources worth developing as the required low temperature can be provided with more and more compact, easy, and cheap closed-cycle cryocoolers.

Semiconductor material systems suitable for epitaxial growth of nanostructures providing the most direct and technologically easiest solutions to obtain emission at 1550 nm are InAs QDs grown on either GaAs or InP substrate. The GaAs-based nanostructures emitting at shorter (<1 μm) wavelengths feature close to ideal structural quality and high optical quality due to very mature well-developed technology. The deterministic fabrication techniques are well established [8, 32, 33], QD morphology can be well controlled, and it is easy to find lattice-matched materials providing high refractive index contrast for distributed Bragg reflectors (DBRs) and further – microcavities. However, the lattice mismatch between InAs and GaAs is high (~7.2%), and therefore, the strain in the QD material is high. It prevents realization of confining potential landscape and growth of large enough nanostructures to obtain emission at 1550 nm. To reach telecom wavelengths, strain engineering needs to be employed, and the layer design of the actual structure has to be properly designed. The approaches used for that are: capping QDs with additional InGaAs strain-reducing layer with In content lower in comparison to QD material [31, 34–41], growing InAs QDs on a virtual substrate provided by a metamorphic buffer layer of InGaAs with gradient In content grown on GaAs substrate [26, 42–45] or double layer of InGaAs QDs where the first QD layer acts as both a seeding layer for the upper QD layer and relaxes strain for its growth. From these approaches, metamorphic buffer layer has emerged to be the one providing tunable single-photon and entangled photon-pair sources at 1550 m [26, 27, 44, 45].

Alternative to GaAs-based QDs are InP-based nanostructures. As can be seen in the inset of Figure 18.1, InP-based QDs cover the entire III telecom spectral window. This is possible due to low lattice mismatch between InP and InAs of 3.2% – more than two times smaller compared to GaAs-based InAs QDs, and deep confining potential. Therefore, the emission wavelength of 1550 nm is easily achievable in this material system making InAs QDs grown on InP substrates suitable active region of light sources for long-distance fiber-based optical communication [15, 20, 46–50].

In this chapter, we report our recent progress in C- and L-telecom bands InP-based QDs and photonic structures for quantum communication applications. In Section 18.2, we review the basic concepts of semiconductor QDs. Section 18.3 and related subsections discuss low-density nanostructures with quaternary (InGaAlAs) and binary (InP) barriers. The growth of low-density QDs on DBR structures is discussed

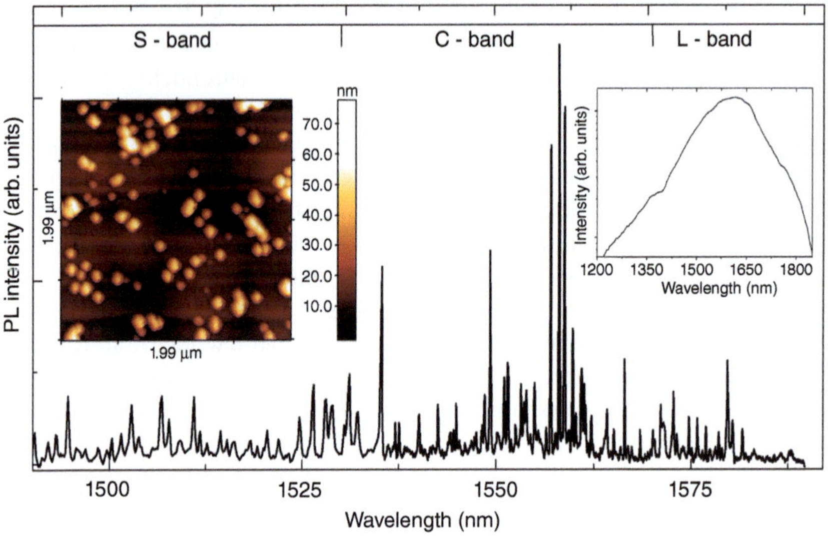

Figure 18.1 Micro-photoluminescence (μPL) spectrum from a single InAs/InP QD (T = 4.2 K, CW non-resonant excitation) in a broad spectral range. Insets are an atomic force microscopy (AFM) image of the QD sample (left) and the low-temperature macro-PL spectrum of low density QDs, the intensity is plotted on a logarithmic scale (right).

in Section 18.3.1. The next two subsections discuss QD growth and structural properties in quaternary systems. The spin dynamics and g-factors of electrons and holes are discussed in Section 18.2.3. In Section 18.3, a special growth technique for obtaining symmetric QDs with low density is described. For these structures, we also discuss excitonic complexes, magneto-optical properties, radiative lifetimes, thermal stability, and spin memory effects. In order to enhance photon extraction efficiency, the QDs are either embedded in photonic crystal microcavities (Section 18.3.3) or grown on top of DBRs. To increase the directionality of the emission by minimizing the in-plane emission losses, micrometer mesa photonic structures are fabricated in DBR structures (Section 18.3.4). The realization of QDs with vanishing fine-structure splitting, which are suitable for generating polarization-entangled photon pairs, and the demonstration of triggered high-purity single-photon emission in both C- and L- telecom bands are described in Section 18.4. Section 18.5 concludes the chapter with a brief overview of challenges and future directions.

18.2 Basic Concepts

18.2.1 Artificial and Real Atoms

QDs are often referred to as "artificial atoms," as their electron motion is confined in all three spatial directions, resulting in a fully discrete energy-level spectrum first observed in atoms. They constitute a quantum system, which is integrated into robust, monolithic semiconductor devices during growth and can be engineered to have desired properties from a broad range. However, it is subjected to interaction with a solid-state matrix, and it is challenging to grow identical QDs due to the influence of the local environment on their formation and their electronic and optical properties afterward. In contrast, all atoms of a given chemical element are identical, and their properties can neither be engineered at will nor easily altered, but the interactions within atomic vapors can be neglected (it can be treated as an ensemble of isolated atoms). Due to the small size of real atoms (~0.1 nm), the shell structure of their energy eigenstates reveals high-energy splittings in the 1–10 eV range, whereas these splittings in QDs are significantly lower (<100 meV). As a direct consequence, the emission properties of QDs are sensitive to the lattice temperature in the sense that bound carriers might be thermally activated into unbound states of the barrier material. Best performance of these structures is therefore obtained at cryogenic temperatures. However, operation at room temperature is desired for practical applications. Recent progress in elevated-temperature single-photon generation with QDs allowed to use relatively cheap, compact, and portable closed-cycle cryocooling solutions, e.g. Stirling cryocooler [29, 31].

Semiconductor-based QDs made of a material *A* provide a smaller bandgap energy than the surrounding semiconductor matrix of material *B*. Once the QDs are fabricated, their positions in the host matrix are permanently defined. In the case of most common self-assembled growth mode offering the highest optical quality of the nanostructures, these positions are random. However, the matrix containing QDs can be further processed and functionalized using well-developed semiconductor technology, which will allow deterministic incorporation of QDs in semiconductor devices. Epitaxial QDs have several advantages with respect to real atoms for potential applications, which can be summarized as follows: (i) the emission properties of a QD can be adjusted over a broad spectral range from the near-infrared to UV by influencing its structural properties (shape, size, composition, strain, and material system) during growth, (ii) post-growth tuning possibilities available (e.g. strain-tuning), (iii) excellent long- and short-term stability (no emission bleaching or blinking), (iv) high oscillator strengths of optical transitions, which lead to high emission rates, (v) possibility of electrical injection of carriers (electrons and holes), (vi) easy integration within a device structure, (vii) they exhibit radiative lifetime limited emission lines at low temperature. Hence, semiconductor QDs have emerged as promising candidates for studying quantum optical phenomena. The incredible progress in growth and fabrication (structural quality) entailed the advances in optical investigations for fundamental study and applications in physics, engineering, biology, and chemistry.

Employing quantum light states in photonic applications allows functionalities that are not possible using classical light. For example, single photons on demand are an important resource in various areas of emerging quantum technologies such as quantum key distribution (QKD) and all-optical quantum information processing [22]. They are the basic prerequisite for unconditional security in quantum key distribution protocols [51–53] and a key ingredient for fault-tolerant quantum computing schemes. Although secure QKD systems based on weak laser pulses have already been realized for simple point-to-point links, true single-photon sources would improve their performance [18] as they are beneficial in the case of lossy optical channels and in principle could be free from multiphoton emission events. Furthermore, quantum light sources are important for future quantum communication protocols, such as quantum teleportation [54].

Key performance measures for such sources are the efficiency (also referred to as brightness), defined as the fraction of photons collected in the experiment (first lens) per excitation pulse, translating into photon emission rate, and the second-order correlation function $g^{(2)}(t)$ at zero time delay [53]. The second-order correlation function at zero time delay is essentially a measure of the multiphoton emission probability. Photon anti-bunching, the tendency of a quantum source to emit individual photons separated in time, was first demonstrated in the resonance fluorescence of a low-density vapor of sodium atoms [55] and subsequently for a single ion [56].

Remarkable progress toward realizing quantum communication has been achieved using color centers, molecules, natural atoms, ions (details can be found in the corresponding chapters of this book), and quantum dots (this chapter). On the one hand, natural atoms (such as neutral atoms and ions) have long coherence times ideal for quantum memories. On the other hand, artificial atoms (such as QDs) have the advantage of custom-designed properties. Natural and artificial atoms can be coupled with each other and can also be interfaced with photons for long-distance communications, e.g. when embedded into a microcavity. The artificial atoms are beneficial solution in terms of on-chip integration. Hybrid devices made of natural/artificial atoms and photons may provide the next-generation design for quantum repeaters for quantum communication.

18.2.2 Formation of Quantum Dots

The most common way to create optically active QDs is via a bottom-up approach. This is based on a self-assembled QD formation. It can be realized by epitaxial growth technique, i.e. by deposition of QD material with lower bandgap energy layer by layer onto a crystalline substrate with higher bandgap energy. QDs are made of the fundamental building blocks of the crystal, i.e. atoms or molecules. In the last decades, various sophisticated semiconductor growth methods have been developed and refined such as metalorganic vapor-phase epitaxy (MOVPE) or molecular beam epitaxy (MBE), in order to gain a high degree of control on the exact size and composition of the as-grown nanoislands.

One well-established method for self-assembled growth is the Stranski–Krastanov (SK) growth mode in which the formation of islands is achieved via surface energy minimization of a low-bandgap material *B* on top of material *A*. Then the islands are overgrown by material *A* generating a fully embedded QDs in the host material *B* as shown in Figure 18.2. The bandgap difference between the two materials provides the confinement potential for the charge carriers. The advantage of using SK growth is to produce dislocation-free semiconductor QDs with tunable optoelectronic properties. In this mode, the growth relies on a slight lattice mismatch between the epilayer and substrate for InAs on GaAs. The growth initially proceeds

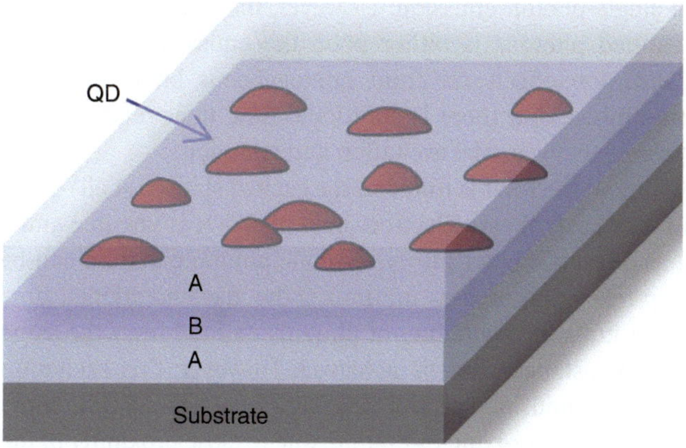

Figure 18.2 Schematic of epitaxially self-assembled growth of embedded QDs. Note the variation in QD size and their random placement on the surface.

layer by layer forming a planar wetting layer. When the thickness of the deposited film exceeds a certain critical value (e.g. 1.7 monolayer for InAs grown on GaAs), islands start forming. This occurs because the energy for island formation is lower than the strain energy to keep a planar wetting layer. Most QD systems are grown in this mode using either MBE or metal-organic chemical vapor deposition (MOCVD).

18.2.3 Excitons in Quantum Dots

Carriers in a QD are usually created either through electrical injection or photo-excitation. Excitons are formed in the QD either by the capture of a free electron and a free hole from the surrounding matrix or by direct optical excitation of the QD states (resonant excitation). In the case of doped structures, a charge carrier can be trapped in the QD without optical excitation. In the presence of such residual carrier, a single carrier of an opposite sign originating from optical excitation is enough to create an exciton. As far as the excitation energy is considered, the QDs can be optically excited in three ways, above the host material bandgap, quasi-resonant, and on resonance with target transition within the QD.

In the above-band excitation, the laser is tuned above the matrix bandgap. In such conditions, electrons and holes are thus generated in the matrix surrounding the QDs. A fraction of these carriers is captured first by the wetting layer, which is essentially a thin quantum well, but with the band edge and confined levels differing from those in QDs due to different strain and only one-dimensional confinement of carriers. Then carriers relax down to the lowest energy levels of the QD through non-radiative processes occurring on a timescale of around 10–100 ps and dominated by longitudinal-optical phonon scattering [57]. With this excitation technique, even when a single QD is isolated, several spectral lines originating from different excitonic complexes are typically seen in the photoluminescence (PL) spectrum. These lines have different frequencies and can be spectrally filtered to isolate emission from a single quantum transition (resembling two level system) to obtain single-photon emission. However, the indistinguishability of the consecutively emitted photons is rather poor, the emission linewidth is governed by spectral diffusion effects far from homogeneous broadening and being Fourier-transform-limited, and there is additional time jitter related to the multistage relaxation process, between the excitation and emission.

In quasi-resonant excitation, the laser is tuned to the excited state (p-shell) of the QD. In this scheme, a better control over electron–hole injection is possible and a cleaner spectrum and, as a result, higher purity of single-photon emission can be obtained. The single-step relaxation process minimizes the abovementioned time jitter and higher degree of indistinguishability of photons emitted from the same transition at different time instances can be achieved. However, this excitation scheme requires tunable laser, and the spectral filtering of the excitation is more challenging when compared to above-band excitation, due to close proximity of the excitation and emission energies. Additionally, the absorption of discrete energy levels is limited.

The most desirable scheme of excitation is the resonant optical excitation technique since no additional relaxation process from a higher excited state is necessary before the photon is emitted. Therefore, a single electron–hole pair is created directly inside the QD and in the state that will further emit. Although this scheme relies on very low absorption of a single QD, it prevents the creation of multiple electron–hole pairs inside or in the vicinity of the QD, giving a cleaner spectrum, minimizing decoherence processes and avoiding re-pumping. This excitation scheme presents the same advantages as quasi-resonant excitation. Moreover, no relaxation mechanism between the excited and emitting state is needed in the case of resonant excitation, which reduces the uncertainty in the emission time of the photon. The disadvantage of using this scheme is that the excitation laser needs to be filtered out using spatial, polarization, or time-domain filtering since the excitation laser and signal have the same wavelength. The most commonly used polarization filtering results in reduced efficiency of the single-photon source by about 50% and poses an experimental challenge. Nevertheless, novel methods for resonant excitation have been recently developed in order to eliminate the filtering constraints, such as coherent two-photon excitation to drive the cascaded biexciton–exciton transition in a QD [58].

As previously discussed, the QD consists of a lower-bandgap semiconductor (B) embedded in a higher-bandgap semiconductor (A). In the case of type I band alignment, this leads to three-dimensional electronic confinement for holes and electrons in the same spatial region. Initially, electrons are present in the valence band, and the optical excitation can cause an electron to be excited to the conduction band, leaving a hole in the valence band. These electron–hole pairs can be trapped by the QD and quickly decay into the first excited state of the QD, forming an exciton (X) state. Radiative decay of this X leads to the emission of a single photon. Different carrier configurations (number and sign) can be confined within the QD upon electron–hole pair recombination and are referred to as different excitonic complexes. Higher-order multiexciton and charged states can also be observed in the QD spectrum when the QD is pumped incoherently. Charged excitons can be formed either due to residual carriers present in the QD before optical excitation or

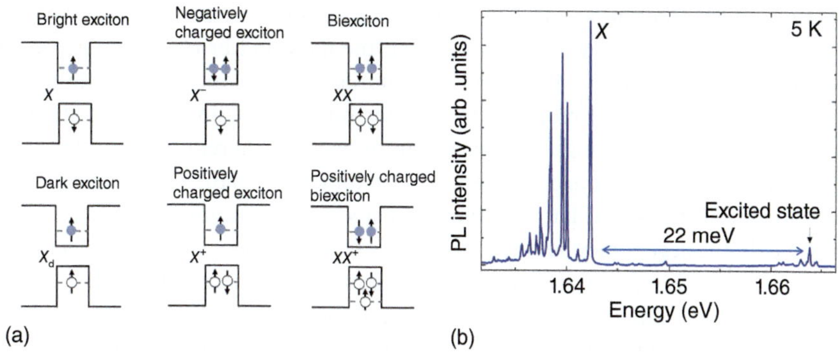

Figure 18.3 (a) Schematic representation of exemplary lowest energy excitonic complexes (carrier configurations) confined within a QD. Arrows indicate the directions of the electron's spin and the hole's angular momentum projections on the growth axis. (b) μPL spectrum of a single GaAs/AlGaAs QD at 5 K taken at saturation power of the exciton line (X).

their formation can be driven by different relaxation/capture rates of electrons and holes leading to imbalanced occupation of QDs with carriers of a different sign.

Different states of a QD can be classified depending on the number of carriers and their spin orientation (for examples, see Figure 18.3). The projection of the electron spin on the z-axis (growth axis) of a QD is either 1/2 or – 1/2, whereas for the heavy hole, the projection of the total angular momentum is either 3/2 or – 3/2. As far as compressively strained self-assembled QDs are concerned, the lowest energy state in the valence band is of a heavy-hole character and the splitting between the heavy-hole and light-hole states allows us to neglect the latter. For simplicity, we will first also neglect the mixing between the different subbands of the valence and conduction bands. This results in four distinct spin configurations of one electron–hole pair and therefore, four values of total angular momentum projections on the growth axis for exciton in the ground state. The two so-called bright exciton states $|-1/2, 3/2\rangle$ and $|1/2, -3/2\rangle$ (the notation with z-axis spin projection of an electron at the first position and the projection of a hole total angular momentum at the second position) have total angular momentum projection of ± 1 and are coupled to the light field due to possibility of fulfilling the angular momentum conservation principle by a single photon, whereas the other two states $|1/2, 3/2\rangle$ and $|-1/2, -3/2\rangle$ have a total angular momentum projection in the z direction of ± 2 and are therefore optically decoupled, i.e. dark, due to the selection rules for dipole transitions. Dark excitons could decay via a pair of photons, but as there are no intermediate energy levels, this process is highly suppressed.

In the case of QDs with confining potential exhibiting full rotational symmetry with respect to the growth axis, the bright excitons are energetically degenerate and separated from the doublet of dark excitons. The splitting between the excitonic states is driven by the electron–hole exchange interaction, and the splitting between bright and dark states is typically an order of magnitude larger than the energy difference between the two dark states. If no exchange interaction was present, all four excitonic states would be degenerate and eigenstates of the angular momentum operators with different angular momentum projections. In the case of symmetric confining potential, the degeneracy between the dark and bright states as well as between the dark states is lifted, and the latter are mixed, and they are no longer eigenstates of the angular momentum operator, but they are linear combinations of the two projections for angular momentum of two and therefore, keep their dark character. In the real physical systems, the symmetry is hardly ever maintained. Asymmetry of the crystal lattice, strain, or shape without asymmetry is enough to mix and lift the degeneracy between the bright excitonic states. The splitting between the bright states is referred to as a fine structure splitting (FSS), and it originates from the anisotropic exchange interaction. Mixing of the two bright excitonic states has important consequences for the polarization of the emitted photons. In the symmetric case, they are circularly polarized, and if the confining potential is asymmetric, two (in the first approximation orthogonal) linearly polarized photons are emitted. The lack of fine structure splitting allows for realization of a source of pairs of polarization entangled photons from the biexciton–exciton cascade, and therefore, growth of symmetric nanostructures is of high interest for quantum technology applications.

The biexciton (XX) consists of two electrons and two holes with anti-parallel spins that occupy the first quantized state of the conduction and the valence band in the QD, respectively. Due to Fermi exclusion principle, there is only one possible spin configuration for the XX in the ground state. However, the final state for its recombination is an exciton or more precisely one of the two bright excitonic states split by fine structure splitting. This is why in the optical spectrum of XX, a doublet of linearly polarized emission lines with orthogonal polarizations appears. The energetic order is opposite to the one in the bright exciton doublet, and the splitting is equal to the FSS. Emission energy of electron–hole pair in the presence of another electron–hole pair (XX) differs from emission energy of the exciton by biexciton binding energy, which is typically in the range of few meV.

Other excitonic complexes consist of X or XX states with additional carriers (either electrons or holes) known as charged excitons ($X^{\pm}$) and charged biexcitons ($XX^{\pm}$), respectively. The simplest $X^{\pm}$ configuration is given by one X plus a single carrier (electron or hole). As an example, for the negatively charged exciton X^{-}, the two electrons must have opposite spins due to the Pauli exclusion principle, whereas both spin orientations are allowed for the hole. The two possible trion configurations in the ground state are degenerate in energy, and emitted photons are orthogonally circularly polarized. Admixture of a light-hole state into the heavy-hole state results in ellipticity of the emission polarization. Trion states are good candidates for sources of single photons due to lack of dark states in the ground state resulting in higher emission rates of a final device when compared to the neutral exciton. An example of μPL spectrum is shown in Figure 18.3b for a GaAs/AlGaAs QD in AlGaAs layer on GaAs substrate [59]. Two group of lines resembling the shell structure of natural atoms can be clearly identified in the spectrum. The splitting between the excited state (p-shell) and the ground state (s-shell) equals to 22 meV and can be related to the size of the QD. The s-shell emission consists of many sharp single emission lines originating from recombination of different excitonic complexes (carrier configurations) in a QD. In the presented case, all excitonic complexes are of binding character – their emission energy is lower than emission energy of the neutral exciton (marked as X in the Figure). The relative intensity between the s- and p-shell emission depends on the excitation power. The s-shell emission saturates, and the p-shell emission takes over at higher excitation conditions if the recombination from the excited states is faster than relaxation to the ground state, which is blocked if the ground state is occupied.

18.3 Low-Density InP-Based Quantum Dots

This section deals with the growth of low-density InP-based QDs and their structural and optical characteristics. InAs QDs are either grown on quaternary surface lattice-matched to InP and/or on binary materials. III-V compound semiconductors are formed by combing elements from group III of the periodic table (e.g. Indium, Aluminum, Gallium) with group V elements (e.g. Phosphorus and Arsenic). So, binary compounds are formed by combining one element from each group (e.g. InP). Whereas quaternary materials are formed from four elements (e.g. GaInAlAs). For InP-based growth, the lattice constant of the quaternary layer should be matched with the InP substrate. Important issue in the InP- based systems was the lack of binary compounds lattice-matched to InP, which could provide high enough refractive index contrast to realize high-reflectivity DBRs. This issue was addressed by using quaternary InGaAlAs materials. Even though they require better control of the epitaxial growth process, the DBR structures have been successfully realized in both MBE [47, 60, 61] and MOVPE [20] growth techniques with reflectivity reaching 99% and 95%, respectively. These have been used to increase the extraction efficiency of single QD emission by even an order of magnitude [61].

Over the past years, different techniques and growth conditions are investigated for this material system providing emission covering the telecommunication bands. In contrast to GaAs-based material system, it does not require strain engineering to reach this spectral range. This is related to smaller lattice mismatch between InAs and InP as compared to GaAs, therefore, lower strain at the interface. This changes the confining potential landscape and enables growth of larger nanoislands. Additionally, InAs/InP material system offers deeper confining potential for both types of carriers. At the beginning, the control of the growth and QD morphology was challenging, and it took some time to realize low-density dot ensembles. Without any growth modification, InAs on InP or lattice-matched alloys forms quantum dashes (QDashes) strongly elongated in [1–10] crystallographic direction. This is due to the strong anisotropy of the diffusion coefficients of adatoms on the substrate surface. Term "QDashes" has been used for the first time at the end of the twentieth century for elongated (in-plane aspect ratio of 2.5) InSb/InP structures [62] and was further adopted for InAs/InGaAlAs/InP structures.

The quaternary barrier serves multiple purposes: increasing the barrier height for the QDs and providing waveguiding effect important for edge-emitting laser structures. The useful feature of QDashes is relatively ease to control emission wavelength by varying the amount of deposited InAs material. Linear scaling between the cross-sectional sizes and InAs amount with preserved shape has been shown [63] and provides an easy way to engineer the average properties of the QDash ensemble during growth, e.g. shifting emission wavelength over hundreds of nanometers. High spatial QDash densities on the level of $10^{11}/cm^2$ allowed studying lateral coupling of nanostructures within the ensemble [64, 65]. Despite that, high inhomogeneity of QDashes and the existence of various QDash families within one structure allowed to isolate single QD emission and therefore, to study properties of different excitonic complexes confined in a single QDash [66–68] and to observe single-photon emission from both neutral and charged excitons with a probability of multiphoton events lower than 0.01 up to 80 K [69, 70]. Due to strong elongation, QDashes feature a very large fine structure splitting on the order of hundred μeV. This makes the two bright excitons well separated spectrally and therefore allows to realize single-photon sources with well-defined linear polarization. If instead of typically flat geometry, columnar QDashes are grown, the degree of linear polarization exceeding 0.9 can be achieved [71]. Polarization properties of emission can be used as a probe of the symmetry of confining potential, thus verifying whether the wave function is confined within the whole QDash volume or trapped within a potential fluctuation [72, 73]. On the other hand, if no degree of linear polarization is desired, the polarization properties of emission from QDashes can be tailored via a photonic environment (e.g. asymmetric mesa structures) [74]. Also, the fine structure splitting can be brought to zero by the in-plane magnetic field due to reversed ordering of the excitonic states in comparison to common QDs – bright exciton polarized along [110] crystallographic direction has lower energy than bright exciton polarized along [1–10] direction [75, 76]. Therefore, QDashes exhibit vast engineering capabilities of structural and, as a consequence, optical properties. Below, we will discuss our recently developed growth technique to obtain low-density round-shaped QDs with MBE using valved cracker cells for group-V elements.

18.3.1 Low-Density InAs QDs on Distributed Bragg Reflectors

To enhance the light emission from single QDs, a DBR consisting of 20 pairs of AlGaInAs/InP layers is introduced below the QDs. Each quarter-wave DBR pair consists of a 123 nm thick InP and 110 nm thick AlGaInAs lattice-matched to the InP substrate. The measured reflectivity is more than 95% at 1550 nm, which proves high interface quality between the DBR layers, good DBR design, and high level of growth control enabling matching the designed thicknesses of DBR layers and maintaining them during growth of the whole DBR structure, as well as correct choice and realization of the composition of the quaternary material to provide high refractive index contrast. It should be noted that for the PL intensity of QDs grown on DBR, an enhancement factor of more than 10 was found in comparison to the PL intensity of the as-grown QD sample [60, 61] and further confirmed in the extraction efficiency measurements [50, 77]. Based on comparison between the results of the reflectivity measurements of the DBR structure and reflectivity spectrum modeled within transfer matrix approach, it is possible to determine the temperature dependence of the refractive index of the quaternary barrier [78, 79]. This is a very important parameter in design and optimization of photonic structures, which operate at low temperature, and it has been so far only approximated based on refractive indices of the binary compounds and energy bandgap dependence on temperature, because most of the standard techniques to measure refractive index contrast are not applicable at low temperatures. Moreover, a specific composition of the quaternary barrier is required to provide high enough refractive index contrast for the DBR. The refractive index value decreases with temperature and saturates at low temperatures. The temperature dependence can be approximated by a third- or fourth-order polynomial. In the case of $In_{0.53}Al_{0.1}Ga_{0.37}As$, the low temperature value equals 3.462 ± 0.005 compared to 3.535 at room temperature. The differences in the temperature dependence of the refractive index between the two materials of the DBR cause variation in the refractive index contrast and therefore the DBR performance with temperature. In the case of InP/InGaAlAs DBR, it results in decrease of the refractive index contrast from 0.36 at 300 K to 0.33 at 10 K [79]. At the same time, the center of the stopband blue-shifts and narrows.

The QDs on the DBR structure were characterized by micro-photoluminescence (μPL) and correlation measurements. Due to the low dot density, no mesa structures were required for the measurements. Low-temperature μPL spectra exhibit sharp excitonic emission lines from single QDs. The combination of excitation power-dependent and polarization-resolved PL measurements reveals a characteristic exciton–biexciton behavior with biexciton binding energies in the range of 3.5–4 meV. The values are comparable with InGaAs/GaAs QDs [80]. Relatively low FSS values down to 20 μeV were obtained, higher than the ones predicted theoretically for this material system [81]. The FSS could be further decreased by external tuning knobs or by adjusting the growth parameters, e.g. optimization of the underlying buffer layer as can be shown in the next sections. Furthermore, single-photon emission at telecom C-band wavelength from single InAs QDs embedded in an AlGaInAs matrix grown on a DBR structure was demonstrated [47].

18.3.2 Quantum Dots in Quaternary Barriers: Round-Shaped Dots

18.3.2.1 Epitaxial Quantum Dot Growth

The growth of self-assembled InAs QDs was performed on (100) InP substrate using solid source MBE. The system has two-valved solid source cracker cells to supply the group-V materials – arsenic and phosphorus. This allows precise control and rapid change in the beam fluxes. Two sets of samples were grown, the first one without capping layer, which was used for the morphological investigations by atomic force microscopy (AFM), and the other set with InP cap layer that was used for PL investigations. The grown samples consist of an InP buffer layer followed by a lattice-matched InAlGaAs layer, then the QD layer is grown and capped with an InAlGaAs layer. The mismatch is quantified by ex situ analysis of the X-ray diffraction (XRD) diffraction pattern.

First, the effect of growth parameters on the density and shape of InAs QDs grown on lattice-matched AlGaInAs/InP material was investigated. It was found that the

QD density increases with increasing the nominal thickness of InAs material, and the QD emission wavelength is red-shifted. In order to reduce the dot density, a modification in the dot growth is introduced. Hence, the low QD density is obtained by careful control of the QD growth using a temperature cooling process. It permits access to single QDs without the need for multiprocessing steps. Further details are reported in [47, 60]. To investigate its influence on the size and shape of the QDs, a series of samples were grown without and with a capping layer for morphological and optical investigations using AFM and PL, respectively. In the case of the samples for AFM measurements, the QDs were grown on top of the AlGaInAs capped layer. In the samples for the PL measurements, the QDs were additionally capped with AlGaInAs. By cooling down, dots are formed, and a ripening process is started leading to the formation of low density larger QDs. The density of the large QDs decreases to around $5 \times 10^8\ \text{cm}^{-2}$ for a cooling temperature of 413 °C. The formation of low density larger QDs emitting around 1.55 μm was also confirmed by low-temperature macro-PL measurements on the QD ensemble. Additionally, it was observed that the integrated PL intensity decreases with the cooling temperature due to the decrease in the QD density [47, 60].

18.3.2.2 Structural Properties of Quantum Dots

To attain a deeper understanding of the optical properties, a comprehensive high-resolution scanning transmission electron microscopy (HRSTEM) and theoretical investigations of the grown QD structures were carried out, more details of which can be found in [82]. The QDs exhibit a pyramidal shape with (1–13) side facets. A higher angle annular dark field intensity on top of their side facets is observed. That suggests an out-diffusion of *In* during capping of the QDs. To quantify the composition within the dots, quantitative high-resolution scanning transmission electron microscopy as outlined in [83] was employed. Strain and intensity were measured from HRSTEM images and then compared with the reference dataset to measure both compositions. The *In*- and *Al*-concentration distributions within a QD as well as composition profiles through the center of the dot were measured. Both composition distributions follow the pyramidal shape of the QDs. The *In*-concentration profile increases abruptly to a constant value of about 86%, whereas the *Al*-concentration drops toward about 7% suggesting equal *Al*- and *Ga* concentrations. Since the pyramidal QD is buried in the InAlGaAs matrix, the actual concentration within the dot should increase linearly from bottom to top.

Measured composition and morphology were then used in atomistic tight-binding simulations for modeling the inhomogeneous broadening of the PL of the dot ensemble as shown in Figure 18.4a and to investigate the influence of size and alloy concentration on the emission wavelength. Simulations were carried out as a function of the size of the pyramid base and variation of *In* concentration with assumption of an equal distribution of the remaining lattice sites by *Al* and *Ga* atoms. A comparison with the experimentally determined PL spectrum shows that the inhomogeneous broadening of the QD ensemble is caused by a strong contribution of *In*-concentration fluctuations in addition to size variations. For the

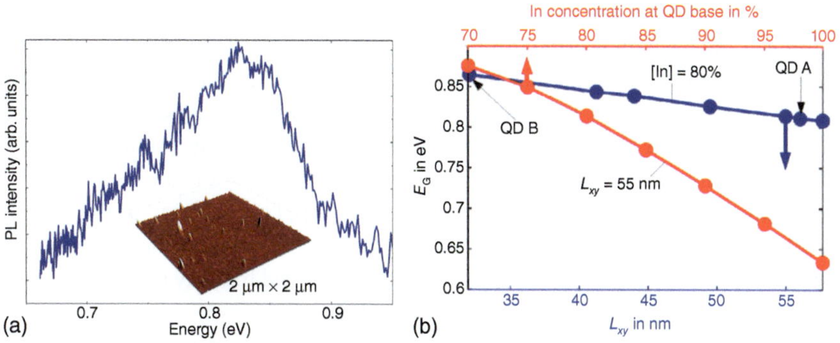

Figure 18.4 (a) Low-temperature (10 K) macro-PL spectrum taken at relatively low excitation power of 52 W/cm^2. The inset: 3D AFM image of an uncapped QD structure showing a minimum (6.47 nm) and a maximum (16.8 nm) height of the QDs. (b) Variation of quasiparticle band-gap E_G for different QD side lengths of the rectangular basis with a constant diameter-to-height ratio (bottom axis) and for different In concentrations (top axis) at the base of the pyramid. Large and small QDs are indicated by A and B, respectively. Source: Carmesin et al. [82]/with permission of American Physical Society.

analysis of size variations, one considers a constant diameter-to-height ratio for all QDs. According to Figure 18.4b, fluctuations of the QD size within a range of 25% lead to an energy shift of 30 meV. However, a shift of about 200 meV by varying In concentration was found. Therefore, the measured width of 150 meV from the ensemble of QDs can be attributed mainly to the variation of the *In* concentration. Due to the rather large QDs, size fluctuations have a slightly reduced influence. More details are reported in [82].

18.3.2.3 Charged Quantum Dots

Spin physics has attracted great attention in recent years, inspired by the possibility of using electron or hole spins for storing and encoding quantum information [84]. For example, charged QDs are interesting candidates for building quantum repeaters because they provide both a stationary qubit (as a memory) and a fast optical interface. For QDs in an external magnetic field, a trapped electron's spin states can be used to encode a qubit due to lifting their degeneracy via Zeeman effect. The spatial confinement of electrons in QDs suppresses the most efficient spin relaxation mechanisms [85] and results in long spin coherence times [86]. One of the most important parameters for spin control is the g-factor, which characterizes the susceptibility of a spin to a magnetic field. Few optical methods have been used to determine the g-factor, such as spin noise spectroscopy and spin-flip Raman scattering. Particularly, high precision in the measurement can be achieved with optical pump-probe spectroscopy, where the g-factor is determined from the frequency of spin polarization oscillations in a perpendicular magnetic field [87–89]. An electron spin in a QD is considered as a qubit, which can be efficiently manipulated by light pulses that give the possibility of easy integration into existing optical telecommunication networks. Although the spin properties for III–V QDs emitting below 1 µm have been intensively studied in the recent years, information on the spin dynamics of QDs emitting at the telecom wavelength range is limited.

We investigated spin dynamics in InAs/InAlGaAs/InP self-assembled QDs with emission wavelengths of 1.5–1.6 μm. The QD structure was placed in a helium bath cryostat with a split-coil superconducting magnet with the possibility to vary the temperature from 5 to 260 K. Magnetic fields up to $B = 2\,\text{T}$ are applied either in Faraday (parallel to the light propagation direction, coinciding with the sample growth axis) or in Voigt configuration (perpendicular to the light propagation direction) geometry. Pump-probe Faraday rotation is employed to study the carriers' spin dynamics in the QDs; details are reported in [89].

Pronounced oscillations on two different frequencies, corresponding to the QD electron and hole spin precession around the field direction, are observed [87]. The obtained electron ($|g_e| \approx 1.9$) and hole ($|g_h| \approx 0.6$) g-factors have the largest absolute values measured so far for III–V QDs by optical methods. All components of the g-factor tensors are measured, including their spread in the QD ensemble [88]. Surprisingly, the electron g-factor shows a large out-of-plane anisotropy ($|g_{||} - g_{\perp}| \sim 1$). The hole g-factor shows even larger anisotropy, which strongly increases for lower QD emission energies. On the other hand, the in-plane anisotropies of electron and hole *g*-factors are small. The pronounced out-of-plane anisotropy is also observed for the spread of the g-factors, determined from the spin dephasing time. The measured hole longitudinal g-factors allow estimation of the QD parameters using a theoretical model. It was found that the QD height-to-diameter ratio increases while the indium composition decreases with increasing QD emission energy. A comprehensive investigation of the g-factor anisotropy of electrons and holes in this type of QDs is reported in [88].

Next, we concentrate on evaluating the longitudinal spin relaxation time T_1 and studying its temperature dependence. Magnetic fields up to B = 2 T are applied in Faraday configuration. Figure 18.5 compares the temperature dependence of longitudinal electron-spin relaxation time (T_1) to that of inhomogeneous transverse spin dephasing times (T_2^*) for electrons and holes (solid squares and circles, respectively) [89]. It was found that at low temperatures, T_2^* is determined by the nuclear field and the spread of g-factors, whereas, at higher temperatures, the homogeneous dephasing mechanisms become important, and T_2^* decreases with temperature. At very low temperatures, it was found that T_1 is determined by the hyperfine interaction with nuclear spins (horizontal solid line in Figure 18.5), while at relatively high temperatures, the spin relaxation is governed by the interaction with longitudinal optical phonons. As a summary, at weak magnetic fields, the major fraction of the total spin polarization decays on the nanosecond timescale due to the precession of the individual spins in random nuclear fields. At the increased longitudinal magnetic field, at low temperatures, the spin polarization decays during the submicrosecond time T_1, which decreases by three orders of magnitude when approaching room temperature due to optical phonon scattering through spin–orbit interaction.

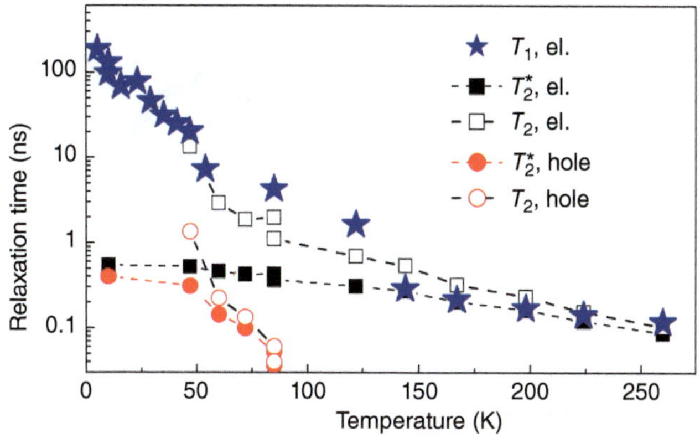

Figure 18.5 (a) Temperature dependencies of longitudinal electron-spin relaxation time T_1 (stars), inhomogeneous transverse spin relaxation times T_2^* for electrons (el.) and holes (solid squares and circles, respectively), and homogeneous transverse spin relaxation times T_2 for electrons and holes (open squares and circles, respectively). Dashed lines are guide to the eye. Source: Reproduced from Mikhailov et al. [89]/with permission of American Physical Society.

18.3.3 Quantum Dots in Binary Barriers: Symmetric Dots

In this section, we highlight the properties of InAs QDs embedded in binary systems grown directly on InP substrates. The growth idea is like that for the quaternary system. During InP buffer layer growth, the MBE chamber is pumped from residual gases after oxide desorption. Thus, to achieve a low-density round-shaped QDs emitting at telecom wavelength, a ripening technique is used. The ripening effect is confirmed by AFM analysis and low-temperature PL measurements. Furthermore, in situ confirmation of the ripening process is also obtained via reflection high-energy electron diffraction (RHEED) measurements. The evolution of the RHEED diffraction pattern during the ripening process shows a streaky pattern during the InP layer growth, indicating an atomically flat surface. Directly after deposition of 2 MLs of InAs, a strained smooth layer is formed. During ripening time, the substrate temperature is decreasing under the arsenic overpressure, and ripening is visible by gradual RHEED pattern transition from streaky to a spotty one, indicating QDs formation. Results showed that during the ripening process, QD density is decreasing, and the emission is shifted toward telecom C-band wavelength. Using the growth procedure described above, a relatively low dot density of about 10^9 cm^{-2} with emission red-shifted to the telecom C-band wavelength was obtained.

After the growth optimization, the QDs are grown on top of InGaAlAs/InP DBR structure to enhance photon extraction efficiency due to decreased photon loss into the substrate and increase emission rate due to Purcell effect related to weak cavity formed by the DBR and semiconductor/air interface. The DBR structure consists of 25 pairs of 123 nm thick InP and 110 nm thick $In_{0.53}Al_{0.1}Ga_{0.37}As$. For enhanced

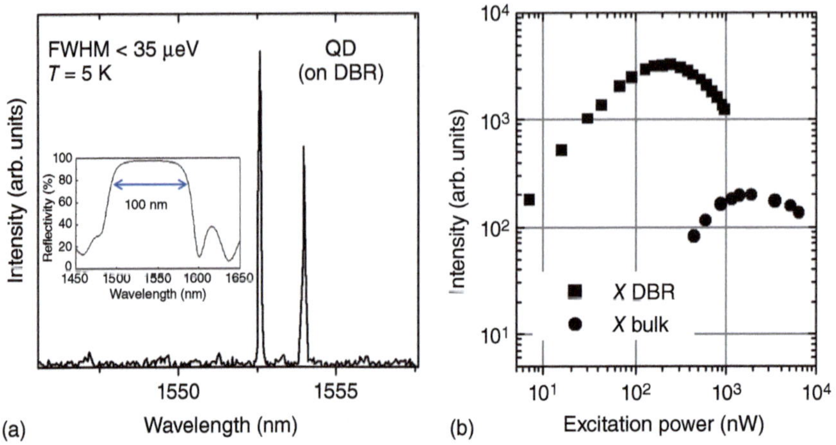

Figure 18.6 (a) Low-temperature (5 K) μPL spectrum of single QD on DBR from a planar sample, the inset is room-temperature reflectivity spectrum. (b) Comparison of excitation power dependence of the excitonic line emission intensities from single QD structures without (circles) and with (squares) bottom DBR. Source: The figures are slightly modified and reproduced from [61], with the permission of AIP Publishing.

spontaneous emission, QDs are placed in the middle of an InP cavity where the electric field of the cavity mode is maximal. The good optical quality of the DBR structure is confirmed with reflectivity measurements. Reflectivity values above 97% with a broad stopband of 100 nm centered at 1.55 μm are demonstrated (in inset of Figure 18.6a). High-resolution PL spectra exhibit pronounced background-free emission excitonic lines at around 1.55 μm with a spectral linewidth below 35 μeV (Figure 18.6a). The QDs on the DBR structure exhibit an order of magnitude higher emission intensity compared to a reference sample (without DBR) as shown in Figure 18.6b.

18.3.3.1 Optical Properties

In the case of symmetric QDs, identification of excitonic complexes is not straightforward due to high in-plane symmetry resulting in ultralow fine structure splitting [61]. In that case, it is not possible to distinguish between neutral and charged excitonic complexes using standard methods, i.e. linear polarization-resolved and excitation power-dependent μPL. The emission intensity could vary sub-linearly with excitation power as a result of either the fact that data points for the lowest excitation powers are missing due to detection limit or due to similar probability of the same dot being charged and neutral. In the latter case, the sum of emission intensity of neutral and charged complexes in their ground states should increase linearly with excitation power. Either way, it is not possible to tell whether a given emission line originates from recombination of neutral or charged exciton, if its formation results from the presence of residual charges within the dot prior to optical excitation. Also, polarization-resolved measurements on this type of QDs cannot bring extra information. One cannot identify whether this line is a neutral exciton with ultralow fine structure splitting (below the resolution of the experimental setup) or charged exciton with two degenerate transitions due to two different possible spin configurations.

Therefore, another approach to identify emission lines in the spectrum is needed. In the case of biexciton and exciton, cascaded emission is evidenced in the results of cross-correlation measurements by large bunching accompanied by antibunching with much lower amplitude. However, this behavior is not specific for neutral biexciton–exciton cascade, but in the case of slow spin relaxation, the difference to charged biexciton–trion cascade is only quantitative, and therefore, both would have to be investigated to make a distinction [90]. Direct observation of biexciton–exciton cascade allowed to determine the biexciton binding energy to be on the level of 3.5 meV similarly to other InP-substrate-based nanostructures even with much different geometry, like QDashes [91]. This can be related to the fact that both nanostructure groups have a relatively large volume, and it would suggest that the size in the growth direction – QD(ash) height – is not determining the biexciton binding energy in this confinement regime [92] (intermediate one argued for QDashes [93]). Cross-correlation measurements can be also useful to identify emission lines originating from recombination of various carrier configurations confined within the same QD in the spectrum but are not specific enough for unambiguous identification of different excitonic complexes. For example, in the case of neutral and charged exciton cross-correlation, one expects asymmetric antibunching, but in certain situations also slight bunching can be observed[50, 94].

An extra means for the identification of excitonic complexes need to be applied. For diode structures with doped layers, controllable charging of the dot depending on the bias voltage can be used for this purpose with the advantage of distinguishing between charged complexes of a different sign. However, it requires well-controllable doping of different types while maintaining the optical quality of the QDs and is a level more difficult than growing a standard sample for purely optical investigation. The electrical injection has so far been only shown for InAs/InP QDs MOCVD-grown within double cap technique [28]. For optical samples, the specific behavior and polarization selection rules of trions in the external magnetic field in Voigt configuration can be used to distinguish them from neutral excitons [95–98] as will be shown in the next section.

18.3.3.2 Magneto-Optical Studies

As mentioned previously, the linear polarization-resolved spectra do not fully allow to distinguish charged excitons from neutral ones with ultralow fine structure splitting (upper panel of Figure 18.7a). The emission line in the polarization map without a magnetic field was further identified as charged exciton based on the linear polarization-resolved emission pattern in the in-plane magnetic field (Voigt configuration) – lower polarization map in Figure 18.7a and corresponding spectra in Figure 18.7b. The disadvantage of this method is that it is not sensitive to the sign of the trion, so positively and negatively charged excitons cannot be distinguished. In the case without magnetic field, the initial and final states are twofold degenerate, and as a result, there are two degenerate transitions with circular (in the case

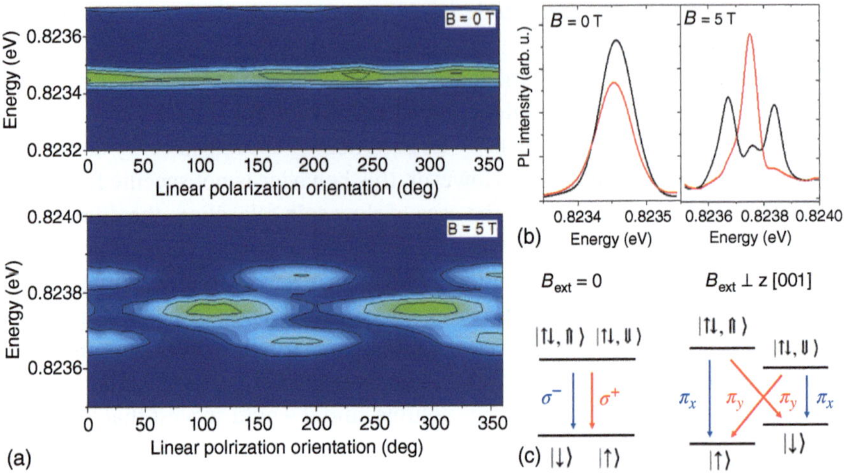

Figure 18.7 (a) Linear polarization-resolved μPL maps for exemplary single QD trion transition in a micrometer-sized mesa with normalized PL intensity color-coded, without external magnetic field (upper map) and with magnetic field of 5 T applied in Voigt configuration (lower map); (b) PL spectra for two orthogonal linear polarizations without magnetic field (left panel) and in magnetic field of 5 T applied in Voigt configuration (right panel); (c) Level scheme and spin configurations of carriers for negatively charged trion with allowed optical transitions and their respective linear polarizations marked as either *V* – vertical and *H* – horizontal without magnetic field (left) and in in-plane magnetic field (right).

of purely heavy-hole states) or elliptical (for mixed light-heavy hole states) polarization. If the external magnetic field is applied, both the initial and final states for the recombination process are Zeeman-split resulting in four optical transitions, which are linearly polarized as can be seen in the level scheme in Figure 18.7c. Two horizontally polarized transitions are energy-degenerate, and their transition energy lies in between the other two transitions, which are horizontally polarized. This results in the characteristic three-peak pattern presented in Figure 18.7b. This is true for both negatively and positively charged exciton. The differences are rather subtle as they are related to differences in the respective g-factors, and this would again require comparing quantitatively results for trions of different sign from the same QD to avoid misinterpretation due to dot-to-dot differences [95].

The QDs for which neutral exciton and trion from the same QD have been identified, the trion binding energy can be determined to be on the level of (0.6–1) meV. The distinction between the positive and negative trion can possibly be made utilizing the measurements of the Hanle curve with the careful characterization of the direction of magnetic field and helicity of the circular polarization under quasi-resonant circularly polarized excitation, but this requires much more experimental effort. Also, comparison with a calculated pattern of excitonic complexes might be useful if the binding energies of the charged excitons with opposite sign differ substantially.

The magneto-optical study provides other interesting insights into the physics of investigated nanostructures. The sensitivity of the emission energy to the external magnetic field brings information on the strength of the confining potential due to competition between the magnetic field and spatial confinement. Larger diamagnetic shift is a signature of larger wave function extension. However, the exact interpretation of derived values is not straightforward in the intermediate and weak confinement regime. It is clear for the limiting cases: for strong confinement – wave function extension and for 2D structures – the distance between the maximum of the probability density for electron and hole forming an exciton. For relatively large InAs/InP QDs considered here, the standard procedure, which is utilized for QDs [99, 100], gives wave function extension of (5–8) nm in the growth direction (QD height) and (12–13) nm in-plane. The wave function extension seems to correspond well with the QD height as determined from the structural data, whereas the in-plane extension is much smaller than expected from the QD size. For sure this is partially related to the fact that the wave function extension is not scaling linearly with the physical nanostructure size [73], and its extension should not be compared with the base diameter, but rather QD cross-section diameter at a height corresponding to maximum of the probability density function. Approaching the exciton Bohr radius of the bulk material with the nanostructure size changes the interpretation of the number determined from diamagnetic coefficient.

What is important to notice is that the in-plane wave function extension is similar for the investigated symmetric structures as well as for strongly anisotropic QDashes in the same material system (however, with different barrier matrix) with a smaller width and much larger length when compared to the base diameter of the investigated symmetric QDs. To determine the strength of the confining potential for these dots is a very interesting open question that requires comprehensive experimental study of various optical characteristics supported by modeling including excitonic effects and preferentially proper description of the exchange interactions. The latter is a very challenging task due to relatively large sizes of the nanostructures making atomistic approaches very time-consuming and even in the $\mathbf{k}{\cdot}\mathbf{p}$ approach followed by configuration interaction method, the basis of electron and hole states required to reach convergence of, e.g. binding energies of basic excitonic complexes, exceeds 50 states of each carrier type, which makes it very time- and resource-consuming. Therefore, more structural data would be very beneficial to limit the parameter space that needs to be considered in the calculations.

The intraband level spacing is in the range of a few meV for the heavy holes and over a dozen meV for electrons. The size and depth of the confining potential allow for a dense ladder of states to be confined within a single QD. To probe the excited states experimentally, the photoluminescence excitation (PLE) measurements for many single QDs have been performed [101]. The summary of the results – the energy difference between the excitation energy and the emission energy vs the latter – is presented in Figure 18.8b. In the PLE experiments, spectral filters used to reject the excitation limit the smallest energy distance to which one can approach the emitting state, and even more, this distance depends on the emission energy. The available experimental conditions are marked with the shaded areas in Figure 18.8b.

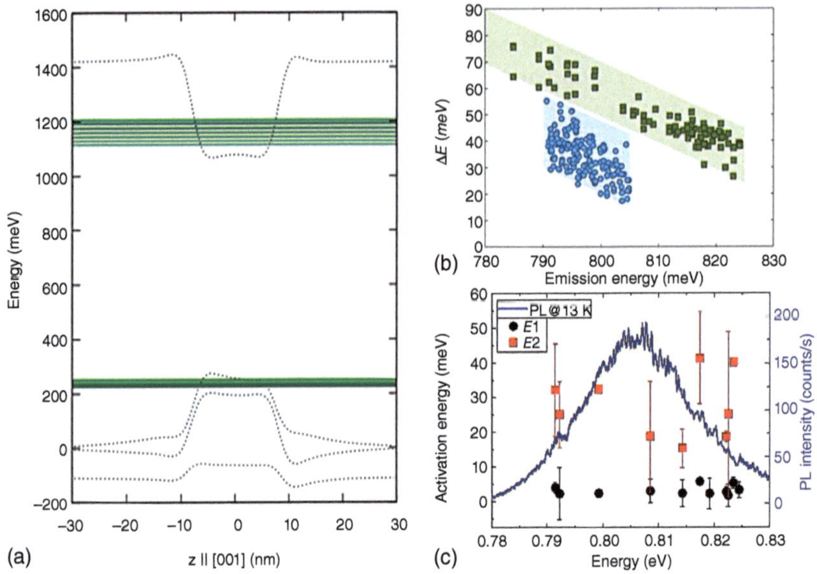

Figure 18.8 (a) Energy of electron and heavy hole levels shown together with the band edges in the QD region in the growth direction calculated within the 8-band kp approach following the structural parameters and with the restriction of the ground state QD transition to be equal to the emission energy of the maximum PL intensity of the QD ensemble (courtesy of Michał Gawełczyk, Department of Theoretical Physics, Wrocław University of Science and Technology); (b) Energy difference between the excitation and detection energy as a function of the emission energy determined based on the results of the μPL excitation experiments with shaded boxes marking the experimentally available range of detected and excitation energies (Source: Podemski et al. [101]/Optical Society of America/CC BY-4.0); (c) Two activation energies (E1 – black dots and E2 – red squares) related to two PL quenching processes (left hand side vertical axis) determined from thermal PL quenching of many single QD lines overlaid with low-temperature (13 K) low-excitation macro-PL spectrum (blue solid line – right hand side vertical axis). Source: Smołka et al. [77]/MDPI/CC BY-4.0.

Obtained results confirm the dense ladder of higher energy states confined within QDs, but calculated energy level spacing suggests that the first excited state might not be probed experimentally. To verify that, activation energies for carrier thermal escape leading to PL quenching have been determined from temperature-dependent measurements of emission from single QDs [77] as shown in Figure 18.8c. These energies correspond well with the level spacing in the valence band (E1) and in the conduction band (E2).

18.3.3.3 Photon Extraction Efficiency

An important figure of merit to assess for each new generation of nanostructures is the optical quality. A commonly used quantitative measure of optical quality is the internal quantum efficiency defined as the percentage of generated electron–hole pairs that recombine radiatively. In the case of quantum emitters embedded in the solid state, it is not straightforward to measure it as the number of detected photons depends critically on the extraction efficiency.

The extraction efficiency of emission remains a challenge for QDs operating at telecom wavelengths in general, no matter what the material system. The highest value reported so far is 36% for InAs/InP QDs in photonic crystal cavity, but for 1310 nm (the second telecommunication window) [102]. In this spectral range, values up to 10% are reported for non-resonant photonic structures with broadband extraction efficiency enhancement [103, 104], even in the case of deterministic fabrication of photonic mesa structures [105]. At 1550 nm, extraction efficiencies of up to 13% have been measured for QD in a mesa (non-deterministic fabrication) into the numerical aperture of 0.4 [50], 10.9% for a QD in a horn structure into the numerical aperture of 0.55 [106], and 10% for structures with a metallic mirror [107]. These photonic structures offer broadband enhancement of the extraction efficiency, but with maximally achievable extraction efficiency values lower in comparison to microcavity approaches. The maximal theoretically predicted extraction efficiency for QD in a mesa on DBR for InP-based QDs yields 25% (30%) into the numerical aperture of 0.4 (0.55) for a broad spectral range exceeding 60 nm [108]. Regarding microcavities for InAs/InP nanostructures, the design of choice has so far been photonic crystal cavities, but circular Bragg grating cavity designs for this spectral range have been recently proposed [109] and realized for telecom O-band [110]. This is a very promising approach providing photon extraction efficiency exceeding 80%.

The extraction efficiency is typically defined as the ratio between the number of photons that can be collected by the first optical element in the experimental setup and the number of all emitted photons. At the end, the number of detected photons is the product of the excitation efficiency (what percentage of excitation pulses leads to the occupation of the target QD), internal quantum efficiency (what percentage of electron–hole pairs is recombining radiatively), extraction efficiency (what percentage of emitted photons can be collected by the first optical element in the setup), and setup efficiency (what percentage of collected photons are detected). This makes the interpretation of the number of detected photons troublesome. The setup efficiency can be measured in an independent experiment, i.e. one can use a laser source with the wavelength tuned to the QD emission wavelength and use such a strongly attenuated beam to mimic QD emission. The attenuation has to be well characterized, and the spatial characteristics of the laser beam should be as close to the QD emission as possible. The latter is not an easy condition to fulfill especially if the far-field emission pattern of the QD is not known and in the case of complex optical setup with a long optical axis in which the divergence of the laser beam is becoming an issue.

To assess the setup efficiency, one measures the excitation power before the attenuation, recalculates it into number of photons, which is further divided by the known attenuation. The next step is to measure the number of photons on the single-photon detectors and calculate what percentage of photons are lost in the experimental setup. This allows the determination of the number of photons collected by the first optical element in the experimental setup once the number of photons on the single-photon detector is measured. In this way, we can characterize and therefore, eliminate one of the factors influencing the number of detected photons – the transmission of the experimental setup and make the result independent of the used equipment. Of course, it does not make it independent of the adjustment of the experimental setup, and the described procedure assumes that the adjustment does not change between the measurement with the attenuated laser beam and the actual experiment with QDs.

One can also directly measure the transmission of each element in the experimental setup and include the total quantum efficiency of the detectors to get the same information as a product of these values. In both methods, the most problematic part is evaluating the percentage of light coupled into the single-mode fiber, which is the necessary element of the optical setup as all single-photon detectors efficient at telecommunication wavelengths are fiber-coupled. The second user-dependent factor is provided excitation. To ensure the closest to ideal excitation conditions with the occupation of the target state within each excitation pulse, the saturation power and the π-pulse are used in the case of non-resonant and coherent driving, respectively. In such conditions, one expects one photon emitted from a QD state per pulse if the internal quantum efficiency is 1. Therefore, the repetition frequency of the excitation laser is commonly taken as a measure of the emission rate of the QD. This has important consequences for the determined extraction efficiency. It is only a lower limit of the actual extraction efficiency and can be strongly underestimated if the internal quantum efficiency is far from 1. Hence, it is desirable to measure the internal quantum efficiency independently, as this would make the measurements of the extraction efficiency more reliable. Unfortunately, in the case of embedded QDs, it is not straightforward to measure internal quantum efficiency due to the limited extraction efficiency of photons. Additionally, the measurements in most cases should be carried out at low temperatures, and these are two reasons why it is not possible to use integrating spheres.

The existing solution is based on the time-resolved PL measurements for a few different photonic environments varied in a controllable way [41, 111–113]. This allows to determine radiative and non-radiative recombination rates for investigated QDs and to remove the extraction efficiency from considerations. However, it requires specific patterning of the sample surface to realize different photonic environments (local density of states) for the QD emission, and the same set of measurements must be repeated on each sample part to obtain one data point on the decay rate vs distance to interface curve, which is further fitted with the results of a model. This assumes high homogeneity of QDs with respect to the internal quantum efficiency or requires processing steps of the same sample piece in between the measurement steps. Thus, we use a slightly easier approach, which, however, should be treated only as an estimation of the internal quantum efficiency.

The idea is to determine the internal quantum efficiency from the comparison between the extraction efficiency measured and calculated for QDs in a planar sample without any surface structurization and later use this estimation of internal quantum efficiency to know how much the extraction efficiency might be underestimated in the case of QDs in a patterned structure. In the case of a planar structure, the extraction efficiency can be reliably calculated within different approaches (finite difference time domain – FDTD [108] (Available from: https://www.comsol.com/. 01.02.2022; Available from: https://www.lumerical.com/. 01.02.2022), finite element method – FEM [114], modal method [108, 115] etc.). However, these do not consider the details of a QD as an emitter but assume some kind of ideal light source at the position of a QD (point, dipole source, etc.). Therefore, the internal quantum efficiency is again indirectly assumed to be 1, and this is the difference between the calculated and measured extraction efficiency, and therefore, comparing the two should give one an estimate of the internal quantum efficiency. Measuring a single QD would be an oversimplification, but measuring many QDs emitting within a spectral range of interest would justify derived conclusions.

The abovementioned comparison is presented in Figure 18.9, where the experimental data points are marked in black (squares) and the respective calculations in blue (dots) [77]. Both emission wavelength dependences are fitted with Gaussian distribution providing the best fit to experimental as well as to calculated results with the maximum of extraction efficiency on the level of 6% at 1535 nm emission wavelength. It is already clear on the qualitative level that investigated QDs exhibit very high optical quality, and the quantitative analysis gives an estimation of 0.92 for internal quantum efficiency.

Extraction efficiency values were obtained for a numerical aperture of 0.4 placed in the far field. It is instructive to examine the dependence of the maximal extraction efficiency (achieved for 1535 nm emission wavelength in our case) on the numerical aperture of the collection optics. This gives insight into the directionality of the emission and answers two important practical questions: (i) how high improvement in the optical signal one can achieve by simply exchanging the collection optics? (ii) how high extraction efficiency one could expect into a standard single-mode fiber (numerical aperture of 0.11–0.13)? In the case of the QDs in planar structure,

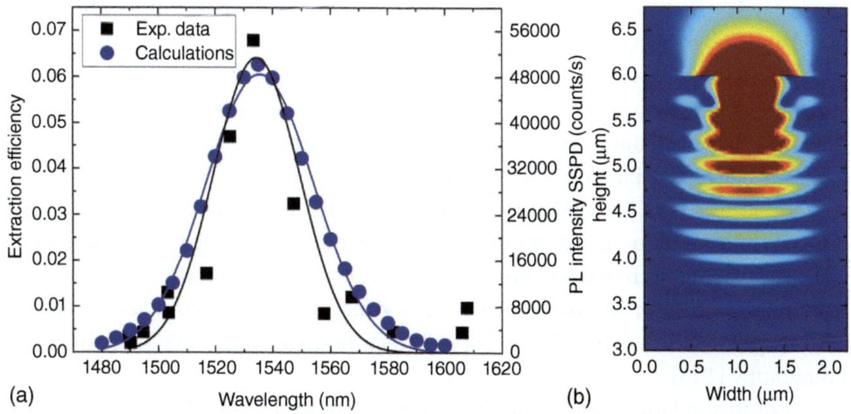

Figure 18.9 (a) Experimental (black squares) and calculated (blue dots) extraction efficiency for NA = 0.4 (left axis) with Gaussian fits (solid lines), right axis – PL intensity under non-resonant pulsed excitation with saturation power measured on single-photon detectors for the temperature of 5 K corresponding to experimentally obtained extraction efficiency. Source: Smołka et al. [77]/MDPI/CC BY-4.0. (b) Calculated distribution of electromagnetic field (the intensity of electric field component is color-coded and presented in a logarithmic scale) for 1535 nm wavelength, for which maximal extraction efficiency was obtained, and 0.4 numerical aperture. Source: Smołka et al. [78] / MDPI / CC BY 4.0.

the expected directionality of emission is low, and the extraction efficiency increases steadily starting from the numerical aperture of 0.2, and it saturates only above 0.8. This can be understood from the distribution of the electric field color-coded in Figure 18.9b. Emission to the substrate is efficiently blocked by the DBR, but there is still some leakage in-plane, and the far-field pattern has a rather hemispherical profile.

Many approaches to increase the extraction efficiency of embedded QDs have recently been developed. These can be categorized into spectrally broadband approaches (e.g. mesas [105], microlenses [8, 116]) in contrast to narrow-band cavity-based approaches (e.g. micropillars [6, 14, 117–119], circular Bragg grating cavities [120, 121]). The latter promises much higher extraction efficiencies at the expense of the requirement of spatial and spectral matching to the cavity mode. Another drawback of the cavities is the ability to enhance extraction from only a single transition unless the quality factor is kept at the moderate level to make, e.g. simultaneous coupling of excitonic and biexcitonic transition to the cavity mode possible. All the abovementioned approaches have been very successful in boosting the extraction efficiency of emission from single QDs in the visible and near-infrared (below 1000 nm) spectral range. However, it still seems to be challenging to achieve high extraction efficiency for QDs emitting in the telecommunication spectral range, even if the cavity approach is utilized – maximum of 36% for the extraction efficiency has been reported for a single InAs/InP QD emitting at 1310 nm embedded in a photonic crystal cavity [102].

Our idea was to achieve a broadband enhancement of the extraction efficiency with well-developed fabrication methods. This is possible with a simple cylindrical mesa design [108] within a 60 nm full width at half maximum (FWHM) spectral range. Such structures can be fabricated within a standard e-beam lithography combined with dry etching. These were fabricated and afterward inspected within atomic force microscopy to verify the shape, sharpness of the edges, and size of the mesas. Once the geometry of a particular mesa and emission wavelength of the QD inside are known, the calculations can be repeated with these experimental input parameters for better comparison with experimentally determined extraction efficiency. This amounts to $(13.3 \pm 2)\%$ into 0.4 numerical aperture, which translates into photon generation rate on the order of 5 MHz (under 80 MHz optical excitations) [50]. This value is higher than what has been achieved so far at the 1550 nm spectral range and could be further improved if the mesa structure was placed deterministically with respect to the target QD, e.g. via in situ cathodoluminescence [122] or optical lithography [123].

Deterministic fabrication technology allows to place QD exactly in the middle of the photonic structure (in-plane) and therefore, maximize the extraction efficiency. In the case of structures fabricated non-deterministically, this is the main reason for the discrepancy between the calculated and measured extraction efficiency values, and hence, many QDs have to be investigated to verify what is the maximal achievable extraction efficiency. This can be used only for fundamental study, but the yield is not satisfactory given quantum devices and their applications. The second effect contributing to lowering the extraction efficiency is the deterioration of the QD internal quantum efficiency due to patterning of the sample surface, which might introduce non-radiative recombination channels in the QD vicinity. From the comparison of the emission intensity between the planar sample with and without DBR, one can see how important it is to introduce the lower DBR section as one order of magnitude increase has been observed [61]. On the other hand, introducing mesa structures to increase the directionality in emission allows for more than twofold enhancement of the photon extraction efficiency [50]. This is still two times lower value than the expected one according to the calculations [108]. As discussed above, it can be traced back either to the non-centric position of the QD within the mesa (verification of this hypothesis requires high-resolution PL mapping, which is not straightforward to realize at 1550 nm) or influence of the mesa patterning on the internal quantum efficiency, which might overshadow enhancement of the extraction efficiency.

18.3.3.4 Quantum Dots in Photonic Crystal Cavities

In this section, we discuss the fabrication and the characterization of InP-based photonic crystal (PhC) cavities with embedded QDs emitting at the telecom C-band. As indicated above, the spontaneous emission of photons from a QD is distributed over the full solid angle. Due to the high refractive index contrast between the semiconductor material and air, only a fraction of the emitted light can leave the sample. PhC microcavity confines light in a small volume and affects the emission characteristics of embedded emitters in terms of spatial emission pattern and emission lifetime. Combining very small mode volume and a high-quality cavity enhances the spontaneous emission rate into that mode; this phenomenon is known as the Purcell effect [124–126].

Suitable cavities for operation at 1.55 μm are designed using FDTD numerical simulation method [127, 128]. It was found that the L3 PhC cavity designs with adjusted inner hole shifts result in cavity modes with room-temperature M1 ground mode emission at 1.55 μm and theoretical quality factor of about 10^5 [127]. The simulated far-field properties of the M1 mode were found to be unfavorable, with large-intensity portions not being collected by an objective of NA = 0.7. The far-field emission of the M1 mode was changed toward a Gaussian intensity distribution by applying modifications, which consisted of specific enlarged holes around the cavity. The simulated Q-factor for a cavity modified in such a way was decreased. The modal volume of the M1 ground mode for the investigated L3 cavity designs was found to be about $0.8\times(\lambda/n)^3$ [127]. Alternative InP-based L4/3 PhC cavities with embedded QDs were also investigated. These PhC cavities were designed for the M1 ground mode at telecom C-band wavelengths. The simulated M1 Q-factor values exceed 10^6, while the M1 mode volume is found to be $0.33\times(\lambda/n)^3$, which is less than half the value of the M1 mode volume of an L3 cavity [128].

High-quality InP-based PhC microcavities with InAs/InP QDs embedded have been successfully fabricated using electron beam lithography, inductively coupled plasma reactive ion etching, and wet etching techniques [127–129]. QDs are grown at the center of an InP membrane on top of a thick InGaAs sacrificial layer. QDs with a medium density and a broad emission spectrum, covering telecom C-band window of interest, are grown using the ripening technique allowing the control of QDs density and emission wavelength. μPL measurements of PhC samples reveal sharp cavity modes also at room temperature. M1 Q-factor values on the order of 10^4 with emission wavelengths around 1.55 μm were measured. At low temperature and with lower excitation powers, the PL spectrum presents sharp lines, which originate from a single QD transitions. An enhancement factor of the PL intensity of about 40 was determined for the QD emission at telecom C-band coupled to the cavity mode in comparison to the PL intensity of QDs in bulk semiconductor [129]. FSS values below 5 μeV were extracted from the fitting of the experimental data. This indicates that such QDs are an ideal candidate to generate polarization entangled photon pairs.

The influence of geometrical parameters on the quality factor, light enhancement, and mode profiles of PhC cavities was also investigated. Figure 18.10(a) shows low-temperature PL spectra of the fabricated L4/3 PhC cavities with varying hole radius $r_1 \sim 110\,\text{nm} > r_2 \sim 100\,\text{nm} > r_3 \sim 90\,\text{nm}$ taken at high excitation powers. By appropriate choice of the hole radius (in this case r_3) cavity ground mode M1

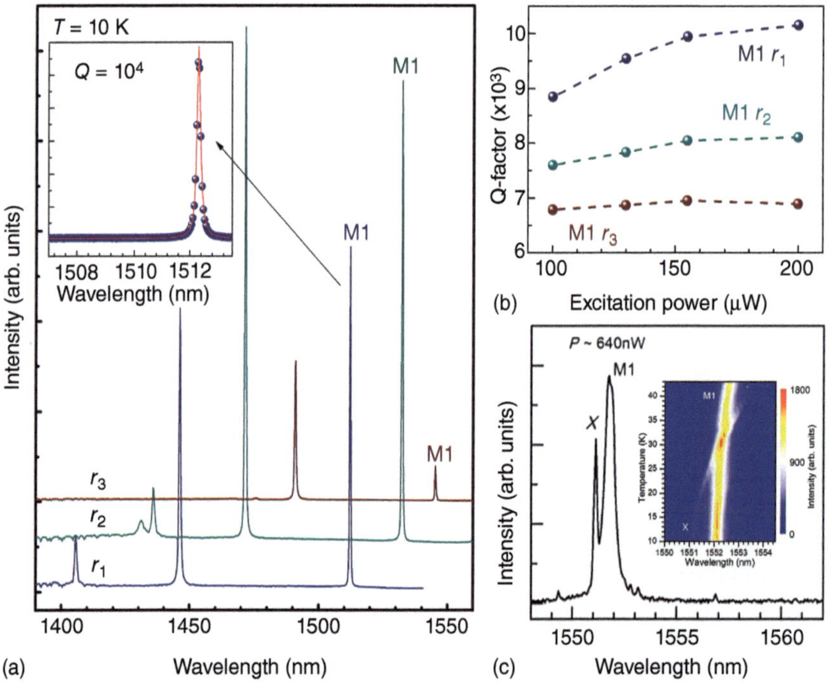

Figure 18.10 (a) Low-temperature PL spectra. Inset: High-resolution PL spectrum recorded at 10 K and 200 μW excitation power for the cavity with hole radius r_1. The M1 Q-factor exceeds 10^4. (b): M1Q factor of the PhC cavities with varying hole radius ($r_1 \sim 110\,\text{nm} > r_2 \sim 100\,\text{nm} > r_3 \sim 90\,\text{nm}$) versus excitation power. (c): High-resolution μPL spectra emitting around 1.55 μm from another L3 PhC cavity. Low power μPL spectrum showing the M1 cavity mode and the X-line. Inset: Temperature tuning of the X-line and the M1 mode showing a weak coupling regime. Source: Rickert et al. [128]/with permission of AIP Publishing LLC.

emission could be shifted to telecom C-band wavelength of 1.55 μm. The measured M1 Q-factors of the cavities with increasing excitation powers for the three different radii are displayed in Figure 18.10b. The Q-factor values increase with excitation powers; this is due to the saturation of QD transitions [130]. A high-resolution PL spectrum of the M1 mode from the cavity with radius r_1 is shown as an inset of Figure 18.10a, where a Q-factor above 10 000 is measured at excitation power of about 200 μW. Note that the variations in Q-factor for cavities with these radii could also be related to the fabrication processes.

For the L3 cavity, different far-field properties have been observed, where the vertical collection of the high-Q ground mode M1 is much lower compared to higher energy cavity modes that have a significantly lower Q-factor. Modification of the far-field properties of the L3 cavity ground mode is possible by using a grating effect due to modification of air holes surrounding the cavity. This will cause the mode intensity to be folded back into the center of the light cone, which results in a near-Gaussian far-field pattern for M1 that can be efficiently collected vertically. This will be on the expense of Q-factor reduction for the optimized mode [127]. We have applied the grating effect proposed in the form of the extractor (hole radius enlargement) structures for the first time for the InP-based PhC cavities to enhance the collection efficiency of M1 mode emitting around 1.55 μm. It was found that increasing hole radius of the extractor structures leads to changes in the M1 Gaussian shaped far-field pattern, which results in an increased emission intensity. Based on FDTD simulation, the Q-factor value is reduced by a factor of 13 for a hole radius enlargement by 20 nm around the cavity compared to unchanged hole radius around the cavity, while the field intensity is increased by a factor of 26, whereas the modal volume of about 0.8 $(\lambda/n)^3$ remained almost unchanged. For the investigated cavities, an increase in mean M1 mode intensity by a factor of 20 between unchanged and changed (20 nm hole enlargement) design is observed, which agrees with the simulated data within the standard deviation errors. The measured mean Q-factor is as expected decreased; however, only by 30% that differs from the factor of 13 found in the simulations. We attribute this to the overall much lower Q-factor in the experiment compared with the theoretical values [127].

Figure 18.10c displays a μPL spectrum taken at low excitation power with a sharp emission line near the M1 ground mode. This peak is assigned to an excitonic transition based on the excitation power-dependent measurements. The biexcitonic transition is not observed in this cavity, possibly due to initial spectral overlap with the cavity mode at low temperature. The inset shows temperature-dependent μPL spectra of M1 and the X line, showing the characteristics of a weak coupling of the QD to the cavity mode. The peak intensity of the X line at 30 K is enhanced by a factor of seven compared to the peak intensity at 10 K.

18.3.3.5 Radiative Lifetime

Besides photon extraction efficiency, another fundamental limit for photon generation rate is the radiative lifetime of electrons in the conduction band states. It can be tailored by engineering the spontaneous emission lifetime in microcavities via the

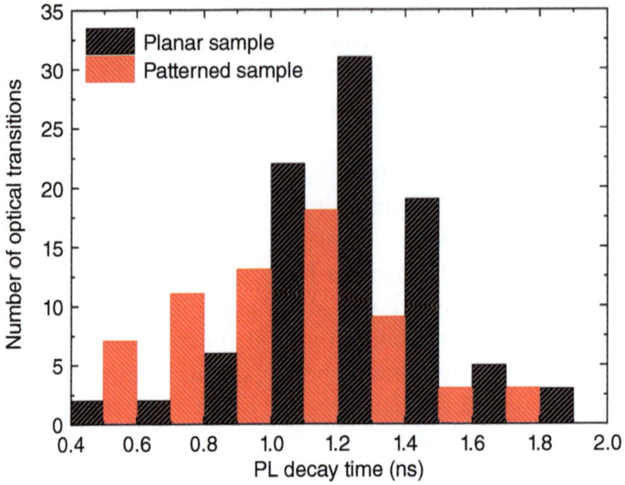

Figure 18.11 Low-temperature (5 K) PL decay time distribution for single QDs in a planar (black) and patterned (red) structure.

Purcell effect, but it cannot be overcome. So far, radiative lifetime was not influenced by external means in the investigated structures. Detailed investigation of many single QDs in the planar structure and another one with non-deterministically fabricated mesas allows to derive the following conclusions (Figure 18.11): (i) the oscillator strength of optical transitions does not change much within the QD ensemble; (ii) the average (typical) PL decay time is similar to other groups of InP-substrate-based nanostructures – in the range of (1–1.5) ns [67, 131–134]; (iii) sample patterning influences the distribution of observed lifetimes – it gets broader, with average shifted toward shorter values and increased percentage of the non-monoexponential decays.

In both cases, there is no energy dispersion visible within the experimental data [77]. The very broad distribution of the PL decay times both in general and for given emission energy can simply be related to the fact that QDs differ in size, geometry, As—P intermixing, and therefore, due to alloy randomness. Additionally, each QD can support various excitonic complexes (carrier configurations) and their respective ladder of higher energy states, which can differ in lifetimes even by a few times [135, 136].

18.4 Symmetric InP-Based Quantum Dots as Quantum Light Sources

A lot of effort has been devoted to the realization of symmetric QDs in the InAs/InP and InAs/InGaAlAs/InP material systems. One of the possible approaches is to modify the growth conditions or growth mode to change the kinetics of atomic species on the sample surface. Exchanging arsenic source in the effusion cell of the MBE reactor has enabled to decrease the lateral aspect ratio below 2 (in comparison to 3–5

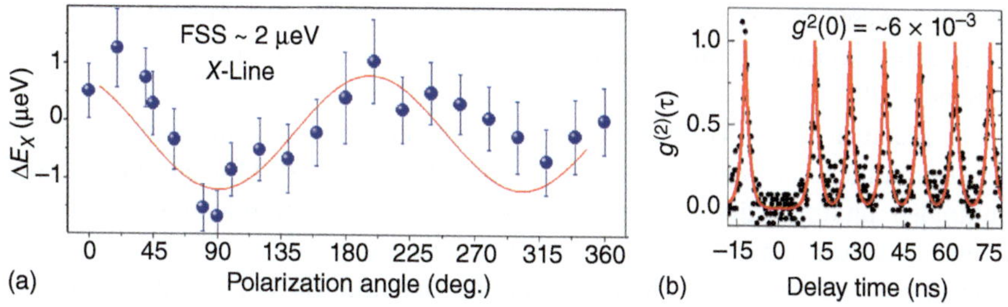

Figure 18.12 (a) Emission energies of exciton obtained by Lorentzian fits of the spectra as a function of polarization angles showing a very small excitonic FSS value of ~2 μeV. (b) Autocorrelation histogram measured under non-resonant pulsed excitation (black data points). Solid line is a fit to the experimental data.

in the case of QDashes) [137]. However, this was not enough to achieve low-density round-shaped QDs emitting at the telecom C-band range.

As previously discussed, a new growth method based on the ripening technique during the MBE growth needed to be developed [47, 60], which now permits the formation of nearly round-shaped QDs in InAs/InGaAlAs/InP material system [47, 60, 138]. For the case of InAs QDs embedded directly in the InP matrix, it has allowed for realization of symmetric QDs featuring negligible exciton FSS of about 2 μeV (see Figure 18.12) without any post-growth tuning [61] and spatial QD density decreased to the level of $5 \times 10^8 - 2 \times 10^9/\text{cm}^2$ [60, 61]. Obtaining low QD spatial density below $10^9/\text{cm}^2$ – has been also possible within the MOCVD growth [133]. The obtained FSS value is on the order of the lifetime-limited linewidth, which is a key parameter in realization of entangled photon pairs for quantum communication applications. Therefore, this new type of quantum emitters is an excellent candidate for an entangled photon sources without a need for external tuning.

Note that the FSS results from electron–hole exchange interaction, which depends on symmetry of the confining potential. In a QD, there are two possible decay channels for the biexciton. In ideal semiconductor QD with zero FSS, the two degenerate intermediate X states lead to two indistinguishable decay paths, which results in a pair of maximally entangled polarization states. In a real QD, due to an electron hole exchange interaction in the bright states, this leads to a degradation or even prohibiting the entanglement since the polarization information is encoded in the energy of the photons. Since in as-grown QDs, the symmetry of the confinement potential can be reduced by asymmetry in QD shape, strain, chemical composition, and piezoelectricity, the FSS is always nonzero, and as a result, emitted photons feature linear polarizations.

Symmetric QDs have also been obtained using the droplet-epitaxy growth method [139]. This has been proven successful in the case of MOCVD-grown InAs QDs in the InP matrix [49, 140, 141]. Beyond observation of single-photon emission [49, 141], droplet epitaxy grown InP-based QDs have been used as an active region of a quantum light-emitting diode [20]. In this case, QDs were embedded in a weak cavity with only three pairs of DBR on top and a doped structure to realize electrical injection of carriers into the active region. With this device, emission of pairs of entangled photons preserved up to the temperature of 93 K has been demonstrated. These QDs were also used to implement quantum teleportation protocol in the telecom C-band [142].

For practical use, it is desirable to realize symmetric dots that can operate at telecommunication wavelengths of 1.55 μm, which offers efficient on-demand entangled photon pairs for long-distance quantum communication via optical fibers. Quantum device widely applicable in quantum communication and cryptography is the source of single photons on demand. So far, it was possible to demonstrate high-purity triggered single-photon emission realized even under non-resonant excitation (Figure 18.12) and quasi-resonant excitation [48]. Such single-photon emitters pave the way toward practical implementation in quantum communication schemes in the long-haul fiber networks with the potential for high photon generation rates.

18.5 Challenges and Future Directions

Although tremendous technological advances have been made in the development of telecom quantum emitters, there are still many challenges to overcome before photonic quantum technology can be fully exploited for its intended applications. The success of future long-distance quantum communications and global quantum key distribution systems depends on the development of efficient quantum repeaters. Many efforts have been devoted to the development of solid-state telecom wavelength qubits, which represent promising candidates for quantum repeaters. Solid-state implementations based on optically controlled electron spins in semiconductor QDs are particularly appealing due to their inherent ultrafast gate operation times and their natural interface to optical fiber communication networks. However, for quantum repeater applications, spontaneously emitted single photons should be Fourier-transform-limited and should exhibit negligible timing jitter. These are ongoing research endeavors for QDs emitting at telecom wavelengths.

Currently, the most satisfactory results in terms of indistinguishability are obtained with single-photons downconverted to telecom wavelengths. The approach is based on nearly perfect quantum emitters – InAs/GaAs QDs – and also takes full advantage of the fact that differences between two remote emitters can be compensated during the frequency conversion. These processes are more complex, requiring nonlinear media and additional laser sources. One must also consider the conversion efficiency itself, which represents an additional loss. These could be overcome by QDs emitting at telecom wavelengths, but their structure and subsequently optical properties, including coherence, have to be further improved. This can be achieved by optimizing the layers and growth parameters of telecom QDs or developing novel growth concepts. Moreover, resonant excitation schemes such as resonance fluorescence and two-photon resonant excitation must be established and optimized to maximize photon indistinguishability.

The extraction efficiency of photons from quantum emitters is another ongoing research area. At telecom C-band, the experimentally demonstrated values do not exceed 15%. This could be due to the self-organized growth process, in which single

QDs have random positions and wavelengths, which leads to low yields. The photon extraction efficiency can be significantly increased for single InAs/InP QDs emitting at 1550 nm by utilizing deterministic fabrication techniques and microcavity designs, such as PhC or bullseye structures. The combination of microcavity-based fabrication with electrical carrier injection for electrically driven source, which has been recently proposed [109], is another feature desired in view of applications. Electrically driven single-photon sources based on InAs/InP QDs were first realized in 2008 [28]. However, this was rather a proof-of-principle demonstration with a relatively high probability of multiphoton events on the level of 0.36 and therefore not application-relevant. These QDs embedded into the optical horn structure have proven to be a very prominent resource for quantum communication exhibiting the lowest probability of multiphoton events $(4.4 \pm 0.2) \times 10^{-4}$ [15] and exceeding 10% photon extraction efficiency [106]. They were further used to demonstrate quantum key distribution over 120 km and achieved emission efficiency of 5.8% (corrected for non-ideal $g^{(2)}(0)$ value) and emission rate of single-photon pulses exceeding 3 MHz [143, 144]. This indicates that it is possible to realize electrical devices with InAs/InP QDs as an active region, and their superb single-photon properties as well as potential to emit pairs of entangled photons have been demonstrated for MOCVD-grown QDs. A preliminary study on InAs/InP QDs grown in MBE technology reveals that the entanglement fidelity of photons emitted from the biexciton–exciton cascade is above 0.5. Another interesting research field that should be also further developed is the heterogeneous or/and monolithic integration of optically active InP-based QDs with Si-based photonic platforms, which have been considered promising solutions to achieve fully integrated photonic circuits with reduced costs.

Finally, the development of modular bright single-photon sources operating in the telecom C-band, as well as efficient on-chip coupling to single-mode fibers, will be crucial for the implementation of fiber-based long-distance quantum communication.

Acknowledgments

The work described here was the result of collaborations with several researchers. We would like to thank the present and former members of the Nano Optics Group and Technological Physics Department at the University of Kassel (UKS), and Laboratory for Optical Spectroscopy of Nanostructures at Wroclaw University of Science and Technology for their contributions in carrying out the experiments described in this chapter. Thanks to J. P. Reithmaier for our successful collaboration over the last years. Also, I would like to thank all the physics colleagues and CINSaT members at the UKS. A special thanks should also be extended to the INA's technicians and administrative staff at the UKS. Thanks to all collaborators, in particular (in alphabetical order): M. Bayer, V. Belykh, M. Gawełczyk, F. Jahnke, P. Michler, S. Reitzenstein, A. Rosenauer, M. Schowalter, A. Zrenner, – please accept my apologies for not naming everyone –, and all (PhD, Master and Bachelor) students who contributed to the works described here. Their skills and support are greatly appreciated. This chapter has been based on a variety of studies that were funded by the German Research Foundation-DFG (DeliCom, Heisenberg grant BE 5778/4-1), the Federal Ministry of Education and Research (Germany) – BMBF (QuaHL-Rep, Q.Com, Q.Link.X, QR.X), as well as National Science Centre (Poland) within Project No. 2020/39/D/ST5/02952, and via the "Quantum dot-based indistinguishable and entangled photon sources at telecom wavelengths" project, carried out within the HOMING program of the Foundation for Polish Science co-financed by the EU under the ERDF.

References

1 Benyoucef, M., Bennett, A., Götzinger, S., and Lu, C.-Y. (2020). Photonic quantum technologies. *Advanced Quantum Technologies* 3 (2): 2000007.

2 Bennett, C.H. and Brassard, G. (1984). Quantum cryptography: quantum key distribution and coin tossing. In: *Proceedings of the IEEE International Conference on Computers, Systems and Signal Processing, Bangalore, India*, 175–179.

3 Ekert, A.K. (1991). Quantum cryptography based on Bell's theorem. *Physical Review Letters* 67 (6): 661–663.

4 Huber, D., Reindl, M., Covre da Silva, S.F. et al. (2018). Strain-tunable GaAs quantum dot: a nearly dephasing-free source of entangled photon pairs on demand. *Physical Review Letters* 121: 33902.

5 Somaschi, N., Giesz, V., de Santis, L. et al. (2016). Near-optimal single-photon sources in the solid state. *Nature Photonics* 10 (5): 340–345.

6 Unsleber, S., He, Y.M., Gerhardt, S. et al. (2016). Highly indistinguishable on-demand resonance fluorescence photons from a deterministic quantum dot micropillar device with 74% extraction efficiency. *Optics Express* 24 (8): 8539.

7 Schlehahn, A., Gaafar, M., Vaupel, M. et al. (2015). Single-photon emission at a rate of 143 MHz from a deterministic quantum-dot microlens triggered by a mode-locked vertical-external-cavity surface-emitting laser. *Applied Physics Letters* 107 (4): 41105.

8 Gschrey, M., Thoma, A., Schnauber, P. et al. (2015). Highly indistinguishable photons from deterministic quantum-dot microlenses utilizing three-dimensional in situ electron-beam lithography. *Nature Communications* 6 (1): 7662.

9 Gazzano, O. et al. (2013). Bright solid-state sources of indistinguishable single photons. *Nature Communications* 4 (1): 1425.

10 Claudon, J. et al. (2010). A highly efficient single-photon source based on a quantum dot in a photonic nanowire. *Nature Photonics* 4: 174.

11 Heindel, T., Schneider, C., Lermer, M. et al. (2010). Electrically driven quantum dot-micropillar single photon source with 34% overall efficiency. *Applied Physics Letters* 96 (1): 11107.

12 Benyoucef, M., Wang, L., Rastelli, A., and Schmidt, O.G. (2009). Toward quantum interference of photons from independent quantum dots. *Applied Physics Letters* 95 (26): 261908.

13 Benyoucef, M. et al. (2009). Wavelength tunable triggered single-photon source from a single CdTe quantum dot on silicon substrate. *Nano Letters* 9 (1): 304.

14 Ding, X., He, Y., Duan, Z.-C. et al. (2016). On-demand single photons with high extraction efficiency and near-unity indistinguishability from a resonantly driven quantum dot in a micropillar. *Physical Review Letters* 116 (2): 20401.

15 Miyazawa, T., Takemoto, K., Nambu, Y. et al. (2016). Single-photon emission at 1.5 μm from an InAs/InP quantum dot with highly suppressed multi-photon emission probabilities. *Applied Physics Letters* 109 (13): 132106.

16 Schweickert, L., Jöns, K.D., Zeuner, K.D. et al. (2018). On-demand generation of background-free single photons from a solid-state source. *Applied Physics Letters* 112: 93106.

17 Hanschke, L. et al. (2018). Quantum dot single-photon sources with ultra-low multi-photon probability. *npj Quantum Information* 4 (1): 43.

18 Waks, E., Inoue, K., Santori, C. et al. (2002). Secure communication: quantum cryptography with a photon turnstile. *Nature* 420 (6917): 762.

19 Xue, Y., Chen, Z., Ni, H. et al. (2017). 1.3 μm single-photon emission from strain-coupled bilayer of InAs/GaAs quantum dots at the temperature up to 120 K. *Applied Physics Letters* 111 (18): 182102.

20 Müller, T., Skiba-Szymanska, J., Krysa, A.B. et al. (2018). A quantum light-emitting diode for the standard telecom window around 1,550 nm. *Nature Communications* 9 (1): 862.

21 Arakawa, Y. and Holmes, M.J. (2020). Progress in quantum-dot single photon sources for quantum information technologies: a broad spectrum overview. *Applied Physics Reviews* 7 (2): 21309.

22 Michler, P. (2017). *Quantum Dots for Quantum Information Technologies*. Springer International Publishing.

23 Rodt, S. and Reitzenstein, S. (2021). Integrated nanophotonics for the development of fully functional quantum circuits based on on-demand single-photon emitters. *APL Photonics* 6 (1): 10901.

24 Trotta, R. and Rastelli, A. (2015). Engineering of quantum dot photon sources via electro-elastic fields. In: *Engineering the Atom-Photon Interaction* (ed. A. Predojević and M.W. Mitchell), 277–302. Cham: Springer International Publishing (Nano-Optics and Nanophotonics).

25 Schmidt, M., Helversen, M.V., Fischbach, S. et al. (2020). Deterministically fabricated spectrally-tunable quantum dot based single-photon source. *Optical Materials Express* 10 (1): 76.

26 Zeuner, K.D., Paul, M., Lettner, T. et al. (2018). A stable wavelength-tunable triggered source of single photons and cascaded photon pairs at the telecom C-band. *Applied Physics Letters* 112 (17): 173102.

27 Lettner, T., Gyger, S., Zeuner, K.D. et al. (2021). Strain-controlled quantum dot fine structure for entangled photon generation at 1550 nm. *Nano Letters* 21 (24): 10501–10506.

28 Miyazawa, T., Okumura, S., Hirose, S. et al. (2008). First demonstration of electrically driven 1.55 μm single-photon generator. *Japanese Journal of Applied Physics* 47 (4): 2880–2883.

29 Schlehahn, A., Krüger, L., Gschrey, M. et al. (2015). Operating single quantum emitters with a compact Stirling cryocooler. *Review of Scientific Instruments* 86 (1): 13113.

30 Schlehahn, A., Fischbach, S., Schmidt, R. et al. (2018). A stand-alone fiber-coupled single-photon source. *Scientific Reports* 8 (1): 1340.

31 Musiał, A., Żołnacz, K., Srocka, N. et al. (2020). Plug&Play fiber-coupled 73 kHz single-photon source operating in the telecom O-band. *Advanced Quantum Technologies* 3 (6): 2000018.

32 Rodt, S., Reitzenstein, S., and Heindel, T. (2020). Deterministically fabricated solid-state quantum-light sources. *Journal of Physics: Condensed Matter* 32 (15): 153003.

33 Senellart, P. (2012). Deterministic light–matter coupling with single quantum dots. In: *Quantum Dots* (ed. A. Tartakovskii), 137–152. Cambridge: Cambridge University Press.

34 Guffarth, F., Heitz, R., Schliwa, A. et al. (2001). Strain engineering of self-organized InAs quantum dots. *Physical Review B* 64 (8): 085305.

35 Alloing, B., Zinoni, C., Zwiller, V. et al. (2005). Growth and characterization of single quantum dots emitting at 1300 nm. *Applied Physics Letters* 86 (10): 101908.

36 Olbrich, F., Kettler, J., Bayerbach, M. et al. (2017). Temperature-dependent properties of single long-wavelength InGaAs quantum dots embedded in a strain reducing layer. *Journal of Applied Physics* 121: 184302.

37 Zhukov, A.E., Kovsh, A.R., Maleev, N.A. et al. (1999). Long-wavelength lasing from multiply stacked InAs/InGaAs quantum dots on GaAs substrates. *Applied Physics Letters* 75 (13): 1926–1928.

38 Seravalli, L., Minelli, M., Frigeri, P. et al. (2003). The effect of strain on tuning of light emission energy of InAs/InGaAs quantum-dot nanostructures. *Applied Physics Letters* 82 (14): 2341–2343.

39 Dusanowski, Ł. (2017). Triggered high-purity telecom-wavelength single-photon generation from p-shell-driven InGaAs/GaAs quantum dot. *Optics Express* 25 (25): 31122.

40 Holewa, P., Burakowski, M., Musiał, A. et al. (2020). Thermal stability of emission from single InGaAs/GaAs quantum dots at the telecom O-band. *Scientific Reports* 10 (1): 21816.

41 Große, J., Mrowiński, P., Srocka, N., and Reitzenstein, S. (2021). Quantum efficiency and oscillator strength of InGaAs quantum dots for single-photon sources emitting in the telecommunication O-band. *Applied Physics Letters* 119 (6): 61103.

42 Semenova, E.S., Hostein, R., Patriarche, G. et al. (2008). Metamorphic approach to single quantum dot emission at 1.55 μm on GaAs substrate. *Journal of Applied Physics* 103 (10): 103533.

43 Seravalli, L., Frigeri, P., Nasi, L. et al. (2010). Metamorphic quantum dots: quite different nanostructures. *Journal of Applied Physics* 108 (6): 64324.

44 Nawrath, C., Vural, H., Fischer, J. et al. (2021). Resonance fluorescence of single In(Ga)As quantum dots emitting in the telecom C-band. *Applied Physics Letters* 118 (24): 244002.

45 Wroński, P.A., Wyborski, P., Musiał, A. et al. (2021). Metamorphic buffer layer platform for 1550 nm single-photon sources grown by MBE on (100) GaAs substrate. *Materials (Basel)* 14 (18): 5221.

46 Usuki, T., Sakuma, Y., Hirose, S. et al. (2006). Single-photon generator for optical telecommunication wavelength. *Journal of Physics: Conference Series* 38: 140–143.

47 Benyoucef, M., Yacob, M., Reithmaier, J.P. et al. (2013). Telecom-wavelength (1.5 μm) single-photon emission from InP-based quantum dots. *Applied Physics Letters* 103 (16): 162101.

48 Musiał, A., Holewa, P., Wyborski, P. et al. (2020). High-purity triggered single-photon emission from symmetric single InAs/InP quantum dots around the telecom C-band window. *Advanced Quantum Technologies* 3 (2): 1900082.

49 Holewa, P., Kadkhodazadeh, S., Gawełczyk, M. et al. (2022). Droplet epitaxy symmetric InAs/InP quantum dots for quantum emission in the third telecom window: morphology, optical and electronic properties. *Nanophotonics* 11 (8): 1515–1526.

50 Musiał, A., Mikulicz, M., Mrowiński, P. et al. (2021). InP-based single-photon sources operating at telecom C-band with increased extraction efficiency. *Applied Physics Letters* 118 (22): 221101.

51 Gisin, N., Ribordy, G., Tittel, W., and Zbinden, H. (2002). Quantum cryptography. *Reviews of Modern Physics* 74 (1): 145–195.

52 Dušek, M., Lütkenhaus, N., and Hendrych, M. (2006). Chapter 5 Quantum cryptography. *Progress in Optics* 49: 381–454.

53 Shields, A.J. (2007). Semiconductor quantum light sources. *Nature Photon* 1 (4): 215–223.

54 Bouwmeester, D., Pan, J.-W., Mattle, K. et al. (1997). Experimental quantum teleportation. *Nature* 390 (6660): 575–579.

55 Kimble, H.J., Dagenais, M., and Mandel, L. (1977). Photon antibunching in resonance fluorescence. *Applied Physics Letters* 39: 691–695.

56 Diedrich, F. and Walther, H. (1987). Nonclassical radiation of a single stored ion. *Physical Review Letters* 58 (3): 203–206.

57 Heitz, R., Grundmann, M., Ledentsov, N.N. et al. (1996). Multiphonon-relaxation processes in self-organized InAs/GaAs quantum dots. *Applied Physics Letters* 68 (3): 361.

58 He, Y.-M., Wang, H., Wang, C. et al. (2019). Coherently driving a single quantum two-level system with dichromatic laser pulses. *Nature Physics* 15 (9): 941–946.

59 Benyoucef, M., Rastelli, A., Schmidt, O.G. et al. (2006). Temperature dependent optical properties of single, hierarchically self-assembled GaAs/AlGaAs quantum dots. *Nanoscale Research Letters* 1 (2): 172–176.

60 Yacob, M., Reithmaier, J.P., and Benyoucef, M. (2014). Low-density InP-based quantum dots emitting around the 1.5 μm telecom wavelength range. *Applied Physics Letters* 104 (2): 22113.

61 Kors, A., Reithmaier, J.P., and Benyoucef, M. (2018). Telecom wavelength single quantum dots with very small excitonic fine-structure splitting. *Applied Physics Letters* 112 (17): 172102.

62 Utzmeier, T., Postigo, P.A., Tamayo, J. et al. (1996). Transition from self-organized InSb quantum-dots to quantum dashes. *Applied Physics Letters* 69 (18): 2674–2676.

63 Sauerwald, A., Kümmell, T., Bacher, G. et al. (2005). Size control of InAs quantum dashes. *Applied Physics Letters* 86 (25): 253112.

64 Maryński, A., Syperek, M., Pieczarka, M. et al. (2017). Lateral interdot coupling among dense ensemble of InAs quantum dots grown on InP substrate observed at cryogenic temperatures. *Journal of Physics: Conference Series* 906: 12008.

65 Ryczko, K., Sęk, G., and Misiewicz, J. (2013). Lateral coupling within the ensemble of InAs/InGaAlAs/InP quantum dashes. *Acta Physica Polonica A* 124 (5): 805–808.

66 Sęk, G., Podemski, P., Musiał, A. et al. (2009). Exciton and biexciton emission from a single InAs/InP quantum dash. *Journal of Applied Physics* 105 (8): 86104.

67 Dusanowski, Ł., Syperek, M., Rudno-Rudziński, W. et al. (2013). Exciton and biexciton dynamics in single self-assembled InAs/InGaAlAs/InP quantum dash emitting near 1.55 μm. *Applied Physics Letters* 103 (25): 253113.

68 Dusanowski, Ł., Gawełczyk, M., Misiewicz, J. et al. (2018). Strongly temperature-dependent recombination kinetics of a negatively charged exciton in asymmetric quantum dots at 1.55 μm. *Applied Physics Letters* 113 (4): 43103.

69 Dusanowski, Ł. (2016). Single-photon emission of InAs/InP quantum dashes at 1.55 μm and temperatures up to 80 K. *Applied Physics Letters* 108 (16): 163108.

70 Dusanowski, Ł., Syperek, M., Mrowiński, P. et al. (2014). Single photon emission at 1.55 μm from charged and neutral exciton confined in a single quantum dash. *Applied Physics Letters* 105 (2): 21909.

71 Musiał, A., Podemski, P., Sęk, G. et al. (2012). Height-driven linear polarization of the surface emission from quantum dashes. *Semiconductor Science and Technology* 27 (10): 105022.

72 Musiał, A., Kaczmarkiewicz, P., Sęk, G. et al. (2012). Carrier trapping and luminescence polarization in quantum dashes. *Physical Review B* 85 (3).

73 Dusanowski, Ł., Musiał, A., Maryński, A. et al. (2014). Phonon-assisted radiative recombination of excitons confined in strongly anisotropic nanostructures. *Physical Review B* 90 (12): 125424.

74 Mrowiński, P., Tarnowski, K., Olszewski, J. et al. (2016). Tailoring the photoluminescence polarization anisotropy of a single InAs quantum dash by a post-growth modification of its dielectric environment. *Journal of Applied Physics* 120 (7): 74303.

75 Young, R.J., Stevenson, R.M., Shields, A.J. et al. (2005). Inversion of exciton level splitting in quantum dots. *Physical Review B* 72 (11): 113305.

76 Stevenson, R.M., Young, R.J., See, P. et al. (2006). Magnetic-field-induced reduction of the exciton polarization splitting in InAs quantum dots. *Physical Review B* 73 (3): 033306.

77 Smołka, T., Posmyk, K., Wasiluk, M. et al. (2021). Optical quality of InAs/InP quantum dots on distributed Bragg reflector emitting at 3rd telecom window grown by molecular beam epitaxy. *Materials* 14 (21): 6270.

78 Talghader, J. and Smith, J.S. (1995). Thermal dependence of the refractive index of GaAs and AlAs measured using semiconductor multilayer optical cavities. *Applied Physics Letters* 66 (3): 335–337.

79 Zielińska, A., Musiał, A., Wyborski, P. et al. (2022). Temperature dependence of refractive indices of $Al_{0 \cdot 9}Ga_{0 \cdot 1}As$ and $In_{0 \cdot 53}Al_{0 \cdot 1}Ga_{0 \cdot 37}As$ in the telecommunication spectral range. *Optics Express* 30 (12): 20225–20240.

80 Rodt, S., Heitz, R., Schliwa, A. et al. (2003). Repulsive exciton-exciton interaction in quantum dots. *Physical Review B* 68 (3): 035331.

81 He, L., Gong, M., Li, C.-F. et al. (2008). Highly reduced fine-structure splitting in InAs/InP quantum dots offering an efficient on-demand entangled 1.55-microm photon emitter. *Physical Review Letters* 101 (15): 157405.

82 Carmesin, C., Schowalter, M., Lorke, M. et al. (2017). Interplay of morphology, composition, and optical properties of InP-based quantum dots emitting at the 1.55 μm telecom wavelength. *Physical Review B* 96 (23): 235309.

83 Rosenauer, A., Gries, K., Müller, K. et al. (2009). Measurement of specimen thickness and composition in Al(*x*)Ga(1-*x*)N/GaN using high-angle annular dark field images. *Ultramicroscopy* 109 (9): 1171–1182.

84 Dyakonov, M.I. (ed.) (2008). *Spin Physics in Semiconductors*, Springer Series in Solid-State Sciences, vol. 157. Berlin, Heidelberg: Springer http://gbv.eblib.com/patron/FullRecord.aspx?p=367527.

85 Khaetskii, A.V. and Nazarov, Y.V. (2000). Spin relaxation in semiconductor quantum dots. *Physical Review B* 61 (19): 12639–12642.

86 Greilich, A., Yakovlev, D.R., Shabaev, A. et al. (2006). Mode locking of electron spin coherences in singly charged quantum dots. *Science* 313 (5785): 341–345.

87 Belykh, V.V., Greilich, A., Yakovlev, D.R. et al. (2015). Electron and hole g factors in InAs/InAlGaAs self-assembled quantum dots emitting at telecom wavelengths. *Physical Review B* 92 (16): 165307.

88 Belykh, V.V., Yakovlev, D.R., Schindler, J.J. et al. (2016). Large anisotropy of electron and hole g factors in infrared-emitting InAs/InAlGaAs self-assembled quantum dots. *Physical Review B* 93 (12): 125302.

89 Mikhailov, A.V., Belykh, V.V., Yakovlev, D.R. et al. (2018). Electron and hole spin relaxation in InP-based self-assembled quantum dots emitting at telecom wavelengths. *Physical Review B* 98 (20): 205306.

90 Kettler, J., Paul, M., Olbrich, F. et al. (2016). Neutral and charged biexciton-exciton cascade in near-telecom-wavelength quantum dots. *Physical Review B* 94 (4): 045303.

91 Mrowiński, P., Zieliński, M., Świderski, M. et al. (2016). Excitonic fine structure and binding energies of excitonic complexes in single InAs quantum dashes. *Physical Review B* 94 (11): 115434.

92 Narvaez, G.A., Bester, G., and Zunger, A. (2005). Excitons, biexcitons, and trions in self-assembled (In,Ga)As/GaAs quantum dots: recombination energies, polarization, and radiative lifetimes versus dot height. *Physical Review B* 72 (24): 245318.

93 Dusanowski, Ł., Mrowiński, P., Syperek, M. et al. (2017). Confinement regime in self-assembled InAs/InAlGaAs/InP quantum dashes determined from exciton and biexciton recombination kinetics. *Applied Physics Letters* 111 (25): 253106.

94 Reimer, M.E., Bulgarini, G., Fognini, A. et al. (2016). Overcoming power broadening of the quantum dot emission in a pure wurtzite nanowire. *Physical Review B* 93 (19): 195316.

95 Rudno-Rudziński, W., Burakowski, M., Reithmaier, J.P. et al. (2021). Magneto-optical characterization of trions in symmetric InP-based quantum dots for quantum communication applications. *Materials (Basel)* 14 (4): 942.

96 Bennett, A.J., Pooley, M.A., Cao, Y. et al. (2013). Voltage tunability of single-spin states in a quantum dot. *Nature Communications* 4: 1522.

97 Koudinov, A.V., Akimov, I.A., Kusrayev, Y.G., and Henneberger, F. (2004). Optical and magnetic anisotropies of the hole states in Stranski-Krastanov quantum dots. *Physical Review B* 70 (24): 241305(R).

98 Yamamoto, Y., Ladd, T.D., Press, D. et al. (2009). Optically controlled semiconductor spin qubits for quantum information processing. *Physica Scripta* T137: 14010.

99 Walck, S.N. and Reinecke, T.L. (1998). Exciton diamagnetic shift in semiconductor nanostructures. *Physical Review B* 57 (15): 9088–9096.

100 Reitzenstein, S., Münch, S., Franeck, P. et al. (2009). Control of the strong light-matter interaction between an elongated In_{0.3}Ga_{0.7}As quantum dot and a micropillar cavity using external magnetic fields. *Physical Review Letters* 103 (12): 127401.

101 Podemski, P., Gawełczyk, M., Wyborski, P. et al. (2021). Spin memory effect in charged single telecom quantum dots. *Optics Express* 29 (21): 34024–34034.

102 Kim, J.-H., Cai, T., Richardson, C.J.K. et al. (2016). Two-photon interference from a bright single-photon source at telecom wavelengths. *Optica* 3 (6): 577.

103 Zinoni, C., Alloing, B., Monat, C. et al. (2006). Time-resolved and antibunching experiments on single quantum dots at 1300 nm. *Applied Physics Letters* 88 (13): 131102.

104 Chen, Z.-S., Ma, B., Shang, X.-J. et al. (2017). Bright single-photon source at 1.3 μm based on InAs bilayer quantum dot in micropillar. *Nanoscale Research Letters* 12 (1): 378.

105 Srocka, N., Musiał, A., Schneider, P.-I. et al. (2018). Enhanced photon-extraction efficiency from InGaAs/GaAs quantum dots in deterministic photonic structures at 1.3 μm fabricated by in-situ electron-beam lithography. *AIP Advances* 8 (8): 85205.

106 Takemoto, K., Takatsu, M., Hirose, S. et al. (2007). An optical horn structure for single-photon source using quantum dots at telecommunication wavelength. *Journal of Applied Physics* 101 (8): 81720.

107 Holewa, P., Sakanas, A., Gür, U.M. et al. (2022). Bright quantum dot single-photon emitters at telecom bands heterogeneously integrated on Si. *ACS Photonics* 9 (7): 2273–2279.

108 Mrowiński, P. and Sęk, G. (2019). Modelling the enhancement of spectrally broadband extraction efficiency of emission from single InAs/InP quantum dots at telecommunication wavelengths. *Physica B: Condensed Matter* 562: 141–147.

109 Barbiero, A., Huwer, J., Skiba-Szymanska, J. et al. (2022). Design study for an efficient semiconductor quantum light source operating in the telecom C-band based on an electrically-driven circular Bragg grating. *Optics Express* 30 (7): 10919–10928.

110 Kolatschek, S., Nawrath, C., Bauer, S. et al. (2021). Bright purcell enhanced single-photon source in the telecom O-band based on a quantum dot in a circular Bragg grating. *Nano Letters* 21 (18): 7740–7745.

111 Johansen, J., Stobbe, S., Nikolaev, I.S. et al. (2007). Quantum efficiency of self-assembled quantum dots determined by a modified optical local density of states. In: *Conference on Lasers and Electro-Optics/Quantum Electronics and Laser Science Conference and Photonic Applications Systems Technologies*. Optica Publishing Group QMK5.

112 Albert, F., Stobbe, S., Schneider, C. et al. (2010). Quantum efficiency and oscillator strength of site-controlled InAs quantum dots. *Applied Physics Letters* 96 (15): 151102.

113 Leistikow, M.D., Johansen, J., Kettelarij, A.J. et al. (2009). Size-dependent oscillator strength and quantum efficiency of CdSe quantum dots controlled via the local density of states. *Physical Review B* 79 (4): 045301.

114 Schneider, P.-I., Srocka, N., Rodt, S. et al. (2018). Numerical optimization of the extraction efficiency of a quantum-dot based single-photon emitter into a single-mode fiber. *Optics Express* 26 (7): 8479–8492.

115 Gür, U.M., Arslanagić, S., Mattes, M., and Gregersen, N. (2021). Open-geometry modal method based on transverse electric and transverse magnetic mode expansion for orthogonal curvilinear coordinates. *Physical Review E* 103 (3–1): 33301.

116 Sartison, M., Engel, L., Kolatschek, S. et al. (2018). Deterministic integration and optical characterization of telecom O-band quantum dots embedded into wet-chemically etched Gaussian-shaped microlenses. *Applied Physics Letters* 113: 32103.

117 Liu, S., Wei, Y., Su, R. et al. (2017). A deterministic quantum dot micropillar single photon source with 65% extraction efficiency based on fluorescence imaging method. *Scientific Reports* 7 (1): 13986.

118 Ginés, L., Moczała-Dusanowska, M., Dlaka, D. et al. (2022). High extraction efficiency source of photon pairs based on a quantum dot embedded in a broadband micropillar cavity. *Physics Review Letter* 129 (3): 33601.

119 Schlehahn, A., Thoma, A., Munnelly, P. et al. (2016). An electrically driven cavity-enhanced source of indistinguishable photons with 61% overall efficiency. *APL Photonics* 1 (1): 11301.

120 Wang, H., He, Y.-M., Chung, T.-H. et al. (2019). Towards optimal single-photon sources from polarized microcavities. *Nature Photonics* 13 (11): 770–775.

121 Rickert, L., Kupko, T., Rodt, S. et al. (2019). Optimized designs for telecom-wavelength quantum light sources based on hybrid circular Bragg gratings. *Optics Express* 27 (25): 36824–36837.

122 Gschrey, M., Gericke, F., Schüßler, A. et al. (2013). In situ electron-beam lithography of deterministic single-quantum-dot mesa-structures using low-temperature cathodoluminescence spectroscopy. *Applied Physics Letters* 102 (25): 251113.

123 Dousse, A., Lanco, L., Suffczyński, J. et al. (2008). Controlled light-matter coupling for a single quantum dot embedded in a pillar microcavity using far-field optical lithography. *Physical Review Letters* 101 (26): 267404.

124 Purcell, E. (1946). Spontaneous emission probabilities at radio frequencies. *Physics Review* 69 (11–12): 674.

125 Vahala, K.J. (2003). Optical microcavities. *Nature* 424 (6950): 839–846.

126 Lodahl, P., van Floris, D.A., Nikolaev, I.S. et al. (2004). Controlling the dynamics of spontaneous emission from quantum dots by photonic crystals. *Nature* 430 (7000): 654–657.

127 Rickert, L., Reithmaier, J.P., and Benyoucef, M. (2020). Telecom wavelength InP-based L3 photonic crystal cavities: properties of the cavity ground mode. In: *Fifth International Conference on Quantum Technologies (ICQT-2019)*, 20004. AIP Publishing [AIP Conference Proceedings].

128 Rickert, L., Fritsch, B., Kors, A. et al. (2020). Mode properties of telecom wavelength InP-based high-(Q/V) L4/3 photonic crystal cavities. *Nanotechnology* 31 (31): 315703.

129 Kors, A., Fuchs, K., Yacob, M. et al. (2017). Telecom wavelength emitting single quantum dots coupled to InP-based photonic crystal microcavities. *Applied Physics Letters* 110 (3): 31101.

130 Gayral, B. and Gérard, J.M. (2008). Photoluminescence experiment on quantum dots embedded in a large Purcell-factor microcavity. *Physical Review B* 78 (23): 681.

131 Syperek, M., Dusanowski, Ł., Andrzejewski, J. et al. (2013). Carrier relaxation dynamics in InAs/GaInAsP/InP(001) quantum dashes emitting near 1.55 μm. *Applied Physics Letters* 103 (8): 83104.

132 Gawełczyk, M., Syperek, M., Maryński, A. et al. (2017). Exciton lifetime and emission polarization dispersion in strongly in-plane asymmetric nanostructures. *Physical Review B* 96 (24): 245425.

133 Holewa, P., Gawełczyk, M., Ciostek, C. et al. (2020). Optical and electronic properties of low-density InAs/InP quantum-dot-like structures designed for single-photon emitters at telecom wavelengths. *Physical Review B* 101 (19): 195304.

134 Labbé, C., Cornet, C., Folliot, H. et al. (2006). Carrier relaxation dynamics 1.55 μm InAs/InP quantum dots under high resonant excitation. In: *AIP Conference Proceedings*, 991–992. AIP.

135 Wimmer, M., Nair, S.V., and Shumway, J. (2006). Biexciton recombination rates in self-assembled quantum dots. *Physical Review B* 73 (16): 165305.

136 Narvaez, G.A., Bester, G., Franceschetti, A., and Zunger, A. (2006). Excitonic exchange effects on the radiative decay time of monoexcitons and biexcitons in quantum dots. *Physical Review B* 74 (20): 205422.

137 Maryński, A., Sęk, G., Musiał, A. et al. (2013). Electronic structure, morphology and emission polarization of enhanced symmetry InAs quantum-dot-like structures grown on InP substrates by molecular beam epitaxy. *Journal of Applied Physics* 114 (9): 94306.

138 Holewa, P., Gawełczyk, M., Maryński, A. et al. (2020). Optical and electronic properties of symmetric InAs/(In, Al, Ga)As/InP quantum dots formed by ripening in molecular beam epitaxy: a potential system for broad-range single-photon telecom emitters. *Physical Review Applied* 14 (6): 064054.

139 Liu, X., Ha, N., Nakajima, H. et al. (2014). Vanishing fine-structure splittings in telecommunication-wavelength quantum dots grown on (111)A surfaces by droplet epitaxy. *Physical Review B* 90 (8): 081301(R).

140 Skiba-Szymanska, J., Stevenson, R.M., Varnava, C. et al. (2017). Universal growth scheme for quantum dots with low fine-structure splitting at various emission wavelengths. *Physical Review Applied* 8 (1): 014013.

141 Ha, N., Mano, T., Dubos, S. et al. (2020). Single photon emission from droplet epitaxial quantum dots in the standard telecom window around a wavelength of 1.55 μm. *Applied Physics Express* 13 (2): 25002.

142 Anderson, M., Müller, T., Huwer, J. et al. (2020). Quantum teleportation using highly coherent emission from telecom C-band quantum dots. *Quantum Information* 6 (1): 14.

143 Takemoto, K., Nambu, Y., Miyazawa, T. et al. (2010). Transmission experiment of quantum keys over 50 km using high-performance quantum-dot single-photon source at 1.5 μm wavelength. *Applied Physics Express* 3 (9): 92802.

144 Takemoto K, Nambu Y, Miyazawa T, Sakuma Y, Yamamoto T, Yorozu S et al. Quantum key distribution over 120 km using ultrahigh purity single-photon source and superconducting single-photon detectors. *Scientific Reports* 2015; 5(1):14383. doi:https://doi.org/10.1038/srep14383.

19

Quantum Optics with Solid-State Color Centers

Joseph H.D. Munns, Laura Orphal-Kobin, Gregor Pieplow, and Tim Schröder

Department of Physics, Humboldt-Universität zu Berlin, Newtonstr. 15, Berlin, 12489, Germany

19.1 Introduction

Quantum optics has developed from the pure investigation of quantum mechanics in the optical domain to proof-of-principle implementations of quantum information processing based on quantum states of light. From providing a platform for testing the fundamentals of quantum mechanics [1–4], a multitude of promising technological applications emerged – in sensing [5, 6]; communication [7, 8]; simulation [9]; and computation [10–12] – that all may be enhanced with respect to classical limits through exploiting the laws of quantum mechanics.

Because photons barely interact at optical frequencies (100s THz) and since there are few contributing noise sources in this energy range, information encoded in the quantum state of light can be long-lived and in principle traverse cosmic distances [13]. Additionally, photonics is appealing due to the relative technical simplicity of performing linear operations such as polarization rotations or temporal delays of single photons [10].

There exists, however, a primary limitation when interfacing photons with other physical systems, for example, when coupling an emitter to a single mode, or when light is propagated through fibers or a device: losses. For instance in fibers, transmission of single photons is limited to a length scale of a few tens of km in the low-loss telecom band [14]. Many color centers, for example defect centers in diamond, emit in the visible spectrum and are therefore subject to much higher losses when emission is not converted to telecom wavelengths (A description of light sources operating directly at telecommunication wavelengths is given in Chapter 18). A further challenge with photons is to induce interactions between them: coupling is limited due to the typically low energy densities and weak interaction with the higher-order (nonlinear) dielectric response of materials. One approach to overcome weak interactions is to implement a *single photon nonlinearity* mediated by an atom-like system. In this case, the presence of a single photon can induce a nonlinear response of the system, such as in the recently demonstrated spin dependent cavity reflection example [15], discussed in section 19.3.3. Alternatively,

Photonic Quantum Technologies: Science and Applications, First Edition. Edited by Mohamed Benyoucef.

such nonlinearities can be implemented *probabilistically* through measurement, where obtaining a particular outcome projects the system into a desired target state [16].

Thematically, we focus on the use of solid-state color centers in networked quantum protocols, where building up a quantum network [17] is an appealing vision, not only from the perspective of secure communication, but also as a means for enabling distributed quantum computation. This architecture is also technologically interesting from a perspective of modularity, whereby particular tasks can be distributed to particular nodes, which may utilize different physical platforms that are best suited to perform the task, while the overall protocol is agnostic to the physical implementation. This approach is also robust in the sense that nodes can be added or removed without impacting the rest of the network. For quantum technologies, desirable properties of such a node is a localized, controllable, and optically addressable quantum bit – a qubit – that can interact (optically) with distant systems, as well as the ability to couple with local ancilla qubits, and finally, provide access to a long-lived memory qubit.

A significant milestone toward the implementation of a quantum network is the generation of entanglement shared between remote nodes. A precursor to entanglement networks is quantum communication protocols, which only necessitate the transmission of a quantum state (see section 19.3.1), such as BB84 [18]. Indeed, there are numerous demonstrations of quantum key distribution (QKD); however, despite commercial availability, it is still yet to be deployed in large scale real-world applications.

As a requirement for photonic quantum-technology applications, the first ingredient is an optically addressable qubit system. For networked quantum technologies, wavelength conversion to the telecom band is also desirable to enable compatibility with existing fiber networks and facilitate large-scale node-to-node connections. From a technological standpoint, scalability is a key consideration, meaning that solid-state emitters are an appealing platform. In principle, color centers can be integrated with a high density into devices, which offer the prospect of chip-scale technologies; however, they must be kept at cryogenic temperatures.

For combining and interfacing different devices, repeatability is also required for scalability. In contrast to isolated atomic systems, solid-state systems are subject to both static and time-varying changes in their local environment, which can, for example, have an effect on the emission frequency of photons. Actively compensating for these interactions is an additional technical challenge.

In practice, the optical requirements for an effective qubit/solid-state emitter for use in quantum information processing applications, such as large-scale quantum networks, can be summarized by three points:

i) transform-limited emission: photons from the same emitter must be indistinguishable – the presence of inhomogeneous broadening results in spectral distinguishability between photons, which limits the fidelity of two-photon interference, the fundamental element of a projective Bell measurement (see section 19.3.3.3). Related to this,
ii) uniformity between different emitters such that multiple devices can be interfaced with one another; and

iii) efficiency: the probability of obtaining a coherent photon in a specific single mode, following excitation.

For the matter qubit itself, the coherence time is of importance. Quantum dots (QDs) are among the brightest and purest single photon sources, and are to date the only solid-state *spin–photon interface* to generate photonic cluster states, which are multi-qubit entangled states that are an essential resource for several quantum technology applications [19, 20]. Photonic cluster state schemes that do not use a spin–photon interface have been demonstrated, for example, by Wang et al. [21] and Asavant et al. [22]. QDs, however, face significant challenges in coherence times: typically freely evolving spin dephasing times (T_2^*) are limited to few ns for electrons and hundreds of ns for hole spins [23], while spin coherence times (T_2) can reach a few μs [24], which is still orders of magnitude smaller than coherence times achievable in other material platforms such as nitrogen-vacancy (NV) defect centers in diamond. Many color centers in wide-bandgap materials exhibit significantly longer dephasing and coherence times while also improving in the generation of indistinguishable single photons as will be discussed in Section 19.3.2.

Quantum memory protocols (section 19.3.2), as well as large-scale quantum computation requires access to multiple qubits. In particular, the ability to locally couple an optically active qubit with ancilla qubits that have intrinsically long coherence times is very advantageous. Ancilla qubits can either be used to enlarge the memory of the local node, independently of the coherence time of the emitter itself, or as additional resource qubits for certain quantum algorithms (section 19.3.3.2).

In the current field of quantum technologies research, there are a number of promising physical systems. Among these, color centers in wide-bandgap materials, QDs, trapped ions, and superconducting qubits are prominent candidates. In this chapter, the focus lies on applications in quantum optics and optical quantum technologies using color centers. The term color center can broadly be defined as intra-bandgap defect within an optically transparent crystal lattice, leading to absorption and emission of light via optically active electronic transitions.

After introducing material considerations and a selection of color centers that are the subject of present research in section 19.2, the chapter will put them into context of particular technological applications. First, the use of color centers as a single photon source (section 19.3.1) and their capability to store a quantum state (section 19.3.2) is introduced. Together, these properties provide the foundation for implementing quantum gates between two qubits (section 19.3.3). These two-qubit gates include the generation of spin–photon entanglement (section 19.3.3.1) and the coupling to additional solid-state ancilla qubits, which is the basis for applications that utilize local spin–spin interactions (section 19.3.3.2). Finally, these concepts are combined for the generation of remote spin–spin entanglement (section 19.3.3.3), which constitutes the basis for a distributed quantum network.

19.2 Color Centers

Color centers in solid-state materials, particularly in diamond, have been investigated for more than 100 years [25]. They have fascinated scientist for their ability to absorb and emit light in crystals. An overview of the spectroscopic activities focused on diamond can be found in Ref. [26], in which the discovery of more than 500 electronic optical centers in diamond is reported. In 1997, the optical detection of single color centers in diamond, NV centers, via optically detected magnetic resonance spectroscopy initialized a huge interest in color centers with paramagnetic ground states [27]. Soon after, second-order auto-correlation measurements provided a direct proof of single photon emission [28, 29]. Research in the field of quantum optics with color centers in solid-state materials has intensified ever since. While the NV is still the most explored and developed color center, today more than 10 color centers in diamond have been identified as single photon emitters, along with many more in other wide-bandgap materials. From a performance and application perspective, the NV is presently challenged by the silicon-vacancy (SiV) center in diamond. Progress in SiC-based color centers has also accelerated. In this section, the most important color centers and their properties relevant for quantum optical experiments and applications are introduced.

19.2.1 Diamond

19.2.1.1 Material Considerations

Physical Properties Diamond is an appealing platform for quantum optics, as it is an optically transparent medium with a particularly wide bandgap ($\sim$ 5.5 eV), which means that a large variety of different optically active emitters – color centers – can be hosted within the gap. Such defect centers result from a combination of lattice vacancies, displacements, and substitutional atoms and can create a localized atom-like system that may involve a spin-qubit state, which can be optically addressed.

The high Debye temperature of diamond (low phonon density at a given temperature), and the low spin density (due to the natural abundance of spin-0 ^{12}C of around 99%) provides a relatively clean environment, which is favorable for the coherence properties of the spin state of an embedded color center. Thus, compared to other host materials, decoherence through interactions with phonons, and magnetic noise from fluctuating spins is suppressed.

Some challenges with diamond, such as low photon collection efficiencies, are introduced due to the relatively high refractive index of 2.41, which results in a critical angle of 24.5° for total internal reflection at the diamond-air interface. Low collection efficiencies are particularly problematic for the NV which additionally has a low Debye–Waller factor (DWF), i.e. only around 3% of the photons are emitted into the spectrally narrow zero-phonon line (ZPL). The bulk of photons is emitted into the broad phonon-sideband (PSB). For centrosymmetric emitters such as group IV vacancies (G4V) the DWF is typically higher: 70–80% for the SiV [30], and 60% for the SnV [31]. One approach to enhance photon collection efficiencies and to control the emission of photons into specific optical modes is the fabrication of nanostructures [32] into the diamond, which also enhance the light–matter interaction so that the spin state of the color center can be more efficiently accessed (see section 19.2.4).

Color Center Formation Artificial diamond substrates are fabricated by chemical vapor deposition (CVD) or by applying the high-pressure high-temperature (HPHT) technique. Variable initial defect concentrations are provided by adding ions, for example N or Si, in the chamber during growth. The substrates are classified by the type and level of their impurities. A more sophisticated method of CVD is delta-doping, which enables nanometer-precision depth control of defect centers as demonstrated for NVs L. Hughes et al., APL Materials 11, 021101 (2023); https://doi.org/10.1063/5.0133501 [33, 34]. To increase the formation yield of defect centers, the sample is irradiated with electrons, which create additional vacancies. Those are likely to assist in the formation of an atom-vacancy defect that often become active color centers [35]. Despite native defects, ion implantation enables targeted creation of NVs as well as formation of group IV and other defects in pure diamond substrates. Thermal annealing after electron irradiation or ion implantation induces vacancy diffusion, subsequent defect center formation as well as healing of lattice defects. However, at high temperatures above 1200°C NV dissociation emerges before the crystal lattice is completely recovered. This effect can be countered by high-pressure high-temperature annealing, which not only heals lattice defects but also avoids graphitization.

Optimizing defect formation is still the subject of active research. Recently, Coulomb-driven defect engineering [36] was demonstrated, a novel approach that uses donor and acceptor doping of diamond. Here, co-implantation of donors led to a high creation yield of NV, SnV, and MgV defects as well as favored negative charge state formation and improved spin coherence for the NV.

Compared to defects created by ion implantation, native defect centers usually have superior optical properties due to reduced damage in the local lattice environment (van Dam et al. [37], see also section 19.3.1.2), and they are less exposed to strain.

19.2.1.2 Nitrogen-Vacancy Defect Center

Since the first demonstrations of optically detected magnetic resonance (ODMR) with NVs [27, 38], the NV^-, the negative charge state of the NV, is the most well studied color center in diamond and has proved itself as promising candidate for quantum information applications. The NV^- has also become a well-established platform for sensors for magnetic fields or temperature [35, 39–42].

Structure The NV is a point defect consisting of adjacent nitrogen and vacancy in the diamond lattice, resulting in a C_{3v} symmetry along the $\langle 111 \rangle$ direction. In the neutral charge state (NV^0), one electron is not bound resulting in a quantum system with spin quantum number $S = 1/2$. In the NV^-, two electrons are not bound, leading to a $S = 1$ system.

The ground state of the NV^- is represented by the ${}^3A_2$ spin triplet. The spin-spin interaction splits the ground state spin sublevels by ~ 2.87 GHz. The ${}^3E$ excited state is an orbital doublet, its degeneracy is lifted by non-axial strain [43, 44]. Each orbital branch comprises a spin triplet with sublevels that are optically coupled to the ground state manifold through spin preserving selection rules leading to six zero phonon line transitions at ~ 637 nm (absorption and emission). In addition, an optical dipole transition at ~ 1042 nm connects two singlet states ${}^1A_1 \rightarrow {}^1E$ energetically located between the ground and excited state triplet [45]. The ${}^1A_1$

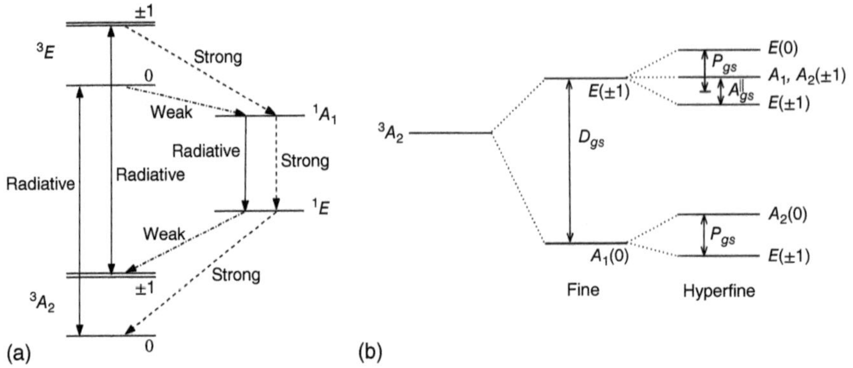

Figure 19.1 (a) Schematic electronic structure levels of the NV^- indicating the intersystem crossing. (b) Hyperfine structure of the $^{14}NV^-$ $^3A_2$ ground states. Source: Reproduced from Doherty et al. [46]/with permission from Elsevier.

state is relatively long-lived. An electronic decay via this metastable path is spin non-conserving, which allows spin initialization through simple non-resonant optical pumping. The energy degeneracy of the $m_s = \pm 1$ spin sublevels of the $^3A_2$ spin triplet can be lifted through an external magnetic field via the Zeeman effect. The resulting splitting directly alters the $\sim$ 2.87 GHz zero-field transition frequency, making the NV^- magnetic field sensitive, and thereby forming the basis of many sensing protocols. The schematic structure of the NV^- is illustrated in Figure 19.1(a). For a more detailed model of the NV energetic structures, see, e.g. [44, 46–49]. In natural abundance, NVs are usually formed with ^{14}N (with $I = 1$) resulting in a hyperfine splitting on the order of 2 – 5 MHz [46], shown in Figure 19.1(b).

Charge State Stability During resonant excitation, the NV^- can be ionised to NV^0 through a two-photon process [50]. The charge state can be reinitialized through application of shorter wavelength light, typically 532 nm, which is able to excite electrons from the valence band to recharge the NV. However, the excitation with high energy laser light also ionizes other defect states in the vicinity of the NV leading to fluctuations in the charge environment of the NV resulting in spectral diffusion – a varying ZPL position over time [46]. The possibility of the charge state switching means that typically a "check" stage is included in control protocols to verify that the NV is in the correct charge state. For the "check" stage, an optical pulse at 532 nm is used to re-initialize the charge state, followed by a verification pulse, which is, for example, resonant with the E_x or A_1 transitions. Measuring a fluorescence signal above some threshold indicates the correct charge state (this is followed by a spin-initialization pulse resonant with A_1, see section 19.3.2.2). Although commonly 532 nm is used, by exciting at 575 nm resonant to the NV^0 ZPL transition, improved spectral stability of the NV is achieved, since excitation and recharging of the NV^0 require lower photon energies and lower overall power [44, 50].

19.2.1.3 Group IV Defect Centers

Structure G4Vs form an inversion symmetric center with D_{3d} symmetry, with the dopant atom at an interstitial site between two vacancies. The increased symmetry of G4Vs compared to the NV results in a reduced sensitivity to electric field fluctuations

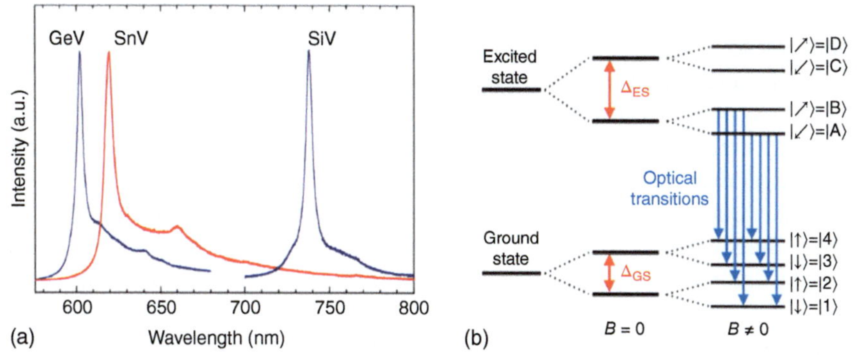

Figure 19.2 Group IV vacancies in diamond: (a) Photoluminescence spectra of different group IV vacancy defect centers at room temperature. (b) Schematic sketch of the electronic level structure of the spin-1/2 group IV vacancies. Source: (a) Iwasaki et al. [54]/With permission of American Physical Society; (b) Bradac et al. [55]/Springer Nature/CC BY-4.0.

[51]. Therefore, G4Vs are considered promising for integration with nanostructures where proximity to surface charges could otherwise be detrimental. This structure also gives rise to strong emission into ZPLs in comparison to the NV (> 50% vs. 3%, respectively); however, this effect is less pronounced for larger dopant atoms due to distortion of the lattice [52].

$G4V^-$ centers have a pair of ground and excited orbitals (^2E_g and ^2E_u, respectively). The orbital degeneracy is lifted by spin-orbit and Jahn-Teller interactions. Each of these four orbitals is spin degenerate (spin 1/2), which can be lifted by a magnetic field [53]. Figure 19.2 shows PL spectra for G4V ensembles, and a general schematic of the electronic level structure.

Conversely to the NV, in addition to magnetic field fluctuations from the spin bath contributing to decoherence, systems with $S = 1/2$ are subject to resonant dipolar coupling with bath spins, e.g. *P*1 centers. Such coupling is Markovian, and thus cannot be rephased through dynamical decoupling [56].

Silicon-Vacancy The SiV^- is the most well studied G4V and was first observed in the 80s, and unambiguously identified in the 90s. It was then revisited as a single-photon quantum emitter in 2006 [57], and subsequently started to attract more attention in 2011 [30] after becoming the brightest diamond based single-photon emitter at the time, with 4.8 Mcps at saturation at room temperature.

Early studies [58, 59] revealed that the fine structure consists of four zero phonon lines visible at liquid He temperatures with wavelengths around 737 nm, corresponding to transitions between the two ^2E_u levels and two ^2E_g levels. The spin-orbit splitting of the ^2E_g branches is around 50 GHz. A more detailed model for a widely accepted picture of the electronic structure was then presented [53].

Germanium-Vacancy After promising results from the SiV, other group IV defects have attracted interest because they share the same symmetry and therefore offer the same resilience to electric field variations. The increase of the size of the dopant atom is accompanied by an increase of the corresponding spin-orbit splitting. A larger splitting leads to a reduced thermal phonon density of states at the splitting energy at a given temperature and in turn reduces the dephasing of the spin states.

The GeV^- was first investigated as a quantum emitter in 2015 [60, 61], showing ~60% ZPL emission at ~ 602 nm, and a ground state splitting ~ 150 GHz [62] three times larger than for the SiV.

A higher quantum efficiency compared to the SiV is found for the GeV [63]. Even though the SiV has a higher DWF of 80%, relaxation from the excited state is dominated by non-radiative processes, which is less pronounced for the GeV [58].

Tin-Vacancy The SnV^- has also attracted attention, first studied in 2017 [54, 64]. The SnV has shown similar properties to the SiV and GeV. The larger dopant ion results in a larger spin-orbit coupling, and thus, phonon-mediated dephasing can be avoided at much more readily accessible temperatures. A reduction of dephasing comes at the expense of introducing more strain in the lattice due to the larger size of the dopant. The distortion of the lattice reduces the Debye–Waller factor compared to the SiV to ~60% [31]. The larger impurity atom may also result in increased lattice damage during implantation, which creates a more noisy environment. The additional damage may be reduced with HPHT treatment in combination with sufficiently low implantation densities.

Even though the SiV has been extensively studied, and the SnV shares many of its features [31], there still remain open questions regarding the electronic structure and dynamics of the SnV. For shallow (~ 20 nm) SnVs, in addition to the predicted ZPL at around 619 nm, other transitions (for the same emitter, as verified through $g^{(2)}$ measurements) at ~ 630 nm and ~ 647 nm have been observed. One working hypothesis, however, is that not the number of electronic levels is altered but that the discrete additional levels are induced by varying charge environment. Oxygen termination of the diamond surface [65] suppressed these transitions.

For a review of group IV vacancies in quantum nanophotonic technologies, see Bradac et al. [55].

19.2.1.4 Other Color Centers in Diamond

Numerous other defects in diamond have been observed, although not all of them have been unambiguously identified. These include chromium [66], xenon [67], and the nickel-nitrogen complex, NE8 [68]. Near-infrared optical activity [69, 70] was attributed to the nickel-vacancy center. A yet unidentified defect with a ZPL transition at 550 nm was observed in [71] and named ST1 center. It was used to manipulate a ^{13}C spin. Given the more than 500 identified optically active impurities in diamond, it is expected that many more defect centers will be identified exhibiting single photon emission and optically accessible spin ground states.

19.2.2 Silicon Carbide

19.2.2.1 Material Properties

Silicon carbide is an optically transparent material with a refractive index of about 2.6 at around 1550 nm, and exhibits a relatively large bandgap of 2.4–3.2 eV depending on its polytype [72]. More importantly for quantum optics, SiC can host a wide range of color centers and exhibits a low two-photon absorption probability.

SiC is compatible with complementary metal-oxide-semiconductor (CMOS) materials [73]. Due to significant commercial investment over the past 20 years for applications based on semiconductors, the manufacturing of high-quality single crystal SiC was developed on the wafer scale [72] for applications in high-power electronics and MOSFET technologies. The commercial availability of high-purity single crystal SiC also makes it a promising platform for electronically controlled quantum devices. Recently, the efforts in growth were extended to SiC wafers suitable for photonic applications [74, 75]. Besides hosting color centers, SiC is also thermally conductive and exhibits a second-order nonlinearity, which makes it suitable for nonlinear photonics. Until recently, however, the realization of high-performance optical devices has been challenging, but is now being addressed in several worldwide efforts toward photonic quantum technologies.

SiC exists in a number of polytypes, most commonly as cubic lattice (3C-SiC) and hexagonal lattice (4H- and 6H-SiC), among others. These structures result in non-equivalent lattice sites, i.e. cubic-like (k), or hexagonal-like (h), which can be occupied by vacancies or dopant atoms. Within the lattice, the presence of a certain defect may result in a family of similar but distinct color centers, depending on the particular site position and defect orientation. A few studied examples include single defects or complexes consisting of multiple defects, such as V_{Si}^- and the divacancy $V_{Si}V_C^-$. Many other optically active impurities based on transition metals have also been observed.

As with diamond, defects are present natively, but can be also induced by doping during growth, or post-growth implantation followed by annealing to heal out the lattice. Alternatively, color centers can also be created with direct femtosecond laser writing, which was demonstrated for the silicon vacancy center V_{Si} [76, 77]. Implantation damage – which can impair the coherence properties – can be limited by using lighter projectiles (electrons or protons), implanted into the pre-doped SiC substrate, which subsequently combine during annealing. SiC is also more robust than diamond against effects such as graphitization *during irradiation* [72].

ZPL emission of SiC color centers have been identified between 400-1600 nm depending on the particular defect (see Figure 19.3); however, as discussed previously, for a number of networked applications, access to telecom band for low absorption in optical fibers is advantageous. It is worth noting that quantum frequency conversion via difference-frequency generation (DFG) from visible to telecom (1.5 μm) can suffer from parasitic spontaneous parametric down-conversion (SPDC), which becomes more problematic the shorter the signal wavelengths: the longer wavelengths available from many SiC defects enables higher-fidelity quantum frequency conversion [78].

For recent review articles discussing the applications of SiC color centers in quantum optics, see [72, 79–82].

19.2.2.2 Prominent Color Centers

The Neutral Divacancy ($V_{Si}V_C{}^0$) has been observed in six configurations (PL1-PL6) with emission frequencies in the spectral region around 1100 nm, depending on

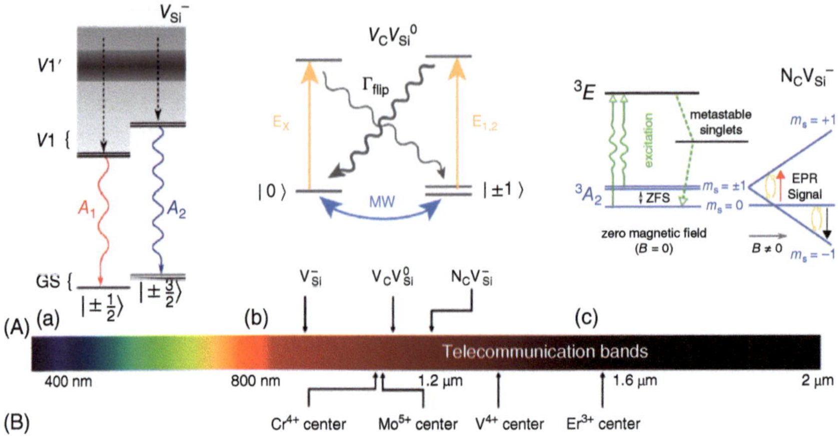

Figure 19.3 (A) (a) negatively charged silicon vacancy in SiC; (b) neutral divacancy in SiC; (c) negative nitrogen vacancy; (B) ZPL positions of selected color centers in SiC. Source: (top-l) Morioka et al. [89]/Springer Nature/CC BY-4.0; (top-c) Anderson et al. [84]/American Association for the Advancement of Science/CC BY-4.0; (top-r) von Bardeleben et al. [93]/With permission of American Chemical Society; (bottom) Majety et al. [81]/With permission of AIP Publishing.

the symmetry of the defect. Electronically, it is similar to the NV in diamond and constitutes a spin-1 system, with ~ 7% ZPL emission, high-fidelity initialization, and spin-state readout [83], and has demonstrated spin-coherence times up to 5 *s* and spin-to-charge conversion [84]. In a device based on a p-i-n diode created through commercial growth of doped SiC, optical transitions with near lifetime-limited linewidth resulting from charge depletion and a broad 850 GHz tuning range via Stark shift were shown [85]. Control of the spin state was demonstrated [86, 87] as well as coupling to neighboring nuclear spins [88].

The Silicon-Vacancy V_{Si}^- has been identified in multiple configurations with ZPLs around 900 nm. It forms a spin-3/2 system. An intersystem crossing in the excited manifold allows for high-fidelity spin initialization. Near lifetime limited lines ~ 50 MHz have been observed [89, 90], even for V_{Si}^- in nanostructures, which were created via ion implantation [91, 92].

The Nitrogen-Vacancy Emits in the region around 1200 nm. Its ZPL emission is, therefore, close to a telecom-compatible near-infrared band. The nitrogen vacancy remains under study due to its low brightness compared to the NV in the diamond. The drop in brightness is likely due to a nonradiative intersystem crossing or a metastable state that is long-lived [80].

19.2.3 Other Host Materials

Besides diamond and silicon carbide, a large variety of other wide-bandgap semiconductor materials have been identified as hosts for color centers. This is not surprising as impurities alter the crystalline structure of a host material, breaking its periodicity, and thereby modify the material's electrical-, optical-, and mechanical properties. This can, for example, introduce allowed electronic energy levels in the bandgap, which can feature optically active dipole transitions. In principle, any wide-bandgap material should feature color centers, and indeed, a large number was reported on in group II-VI and in group III-V semiconductor materials, including ZnO [94–96] and GaN [97–99].

Recently, impurities in quasi-two-dimensional van der Waals materials were demonstrated to provide single photon generation. This is a very active field of investigation and includes transition metal dichalcogenides, for example, WSe_2 [100–104] and $MoSe_2$ [105], and hexagonal boron nitride (hBN) [106–108]. A full theoretical description of the origin and electronic structures of these "color centers" has not been developed for most of these quantum systems but is under investigation. For more details and overview, please refer to recent review articles, for example [109].

Another exciting "class" of optically active quantum systems in solid-state materials is the large assortment of single rare-earth-ion material systems. Single rare-earth-ion systems often behave very similar to real atoms, giving them promising properties despite being embedded in a solid-state lattice. A significant effort has been directed to fast and efficient ensemble-based quantum memories [110], but also quantum sensing applications [111]. More recently, single rare earth ions have been used for the generation of single photons [112, 113] and addressed as single ion quantum memory [114, 115], enabled by embedding the ion in a photonic crystal cavity to reduce the long ion spontaneous emission decay times of typically milliseconds by orders of magnitude making them accessible by common quantum spectroscopy methods. For a better overview, several review articles can be considered, for example, [116, 117].

19.2.4 Optical Micro- and Nanostructures for Controlled Light-Matter Interaction

In this section, a brief overview of the fabrication and functionality of optical micro- and nanostructures is given. Structuring of materials is used for enhanced light–matter interaction, for example, interaction of a spin-qubit with a photon; optimized photon collection efficiencies; and photonic integration. The article "Quantum nanophotonics in diamond" by Schröder et al. [32] gives a comprehensive introduction into the large variety of structures and devices that have been designed and fabricated in diamond. The article also provides details of methods for the (targeted) creation of color centers, the fabrication of micro- and nanodevices, and of device performance.

In bulk solid-state materials, the optically active electronic transitions of color centers are mainly of dipolar type leading to the well-known dipole emission and absorption mode pattern that is not optimal for efficient light–matter interaction. Emission and absorption characteristics can be controlled and altered by means of photonic structures; or, phrased differently, photonic structures allow for the individual control of key figures of merit for enhanced light–matter interaction.

The most important figures of merit are based upon the modification of the local density of states (LDOS) of optical modes [118]:

- The **beta factor** β quantifies the fraction of photons emitted into a desired mode. It is an important figure of merit to determine the single mode system efficiency; however, it does not influence the transition rates or selection rules of the color center.
- The **Purcell factor F** quantifies the enhancement of emission rate with respect to free space [119]. It is an important figure of merit for the direct influence of a photonic structure on the color center's internal dipole transition dynamics. Enhancing optical transition rates increases the effective quantum efficiency of the emitter, i.e. the ratio between radiative and non-radiative transition rates. The process of selective rate enhancement also allows for control of the transition selection rules and thereby an increase of the DWF, corresponding to the ratio of emission channeled into the zero-phonon line. Finally, Purcell enhancement can result in overall lifetime reduction and thereby natural linewidth broadening, which, in addition, reduces susceptibility to spectral diffusion because shifts of the transition frequency are relatively smaller for natural transition lines with a wider profile.
- The **cooperativity parameter C** quantifies the ratio of atom-cavity coupling to other dissipative losses [120]. It encapsulates the effective strength of the atom–cavity interaction and plays an important role for two-qubit gate operations between the color center's spin and an external photon.

The greatest level of control can be achieved with resonant cavity structures and devices. Resonant structures enable the suppression or enhancement of optical transitions by modifying the density of photonic modes [121, 122]. The modification allows, for example, for the Purcell induced enhancement of optical transition rates and thereby the redistribution of the optical emission between zero-phonon and phonon-sideband transitions ([123], see Figure 19.4). Such a redistribution is particularly important for the NV color center in diamond, which has a low DWF of 3%, where, for example, a redistribution from 3% to 54% was demonstrated with a nanobeam waveguide photonic crystal cavity [123, 124].

Optical cavities require three-dimensional photonic confinement down to mode volumes on the order of λ^3. Such confinement can be achieved with one- or two-dimensional photonic crystal cavities in combination with total internal reflection in the two or one remaining spatial dimension(s), respectively [122]. Depending on the solid-state material, different concepts for nanofabrication of optical cavities need to be applied as detailed below.

Typically, two cases are considered when designing a photonic structure for the control and guiding of emitted light: either (i) the coupling from the micro- or nanostructured crystal to a free space mode, and subsequently to a guided – usable – mode; or (ii) the direct coupling from an emitter to a guided mode.

Of these, (i) is technically more direct, but is limited in its efficiency compared to (ii). In the absence of any photonic structure, the coupling from a bulk crystal to air modes is limited to a few percent [125]. Macro- and microstructured solid immersion lenses can enable roughly 10-fold efficiency enhancement [126], but do not allow for control of the emission pattern that can, for example, maximize the mode overlap to fiber modes. To overcome this limitation, a large variety of photonic micro- and nanostructures have been developed and implemented, ranging from nanopillars [127] to parabolic reflectors [128] to circular "bullseye" gratings [129] with functionalities ranging from increased bulk crystal to air mode coupling to the redistribution of the emission pattern that matches fiber modes [125].

With (ii), color centers couple directly to the mode of a waveguide skipping free-space propagation: considering a refractive index of $n = 2.4$ for diamond, up to $\beta = 86\%$ of emission around 637 nm can be collected and directly guided to a photonic integrated circuit (PIC) for further manipulation of the emitted photon [130]. In solid-state materials with larger refractive indices $n > 3$ coupling efficiencies can increase beyond 90%. On-chip waveguide modes can then also be directly coupled to single mode fibers for delivery to detectors or long-distance transfer. Recently, the coupling of diamond cavity photons to a fiber with an overall system detection efficiency of 85% was demonstrated [15].

19.2.4.1 Fabrication and Nanostructures in Diamond

In contrast to most dielectric photonic material platforms, mono-crystalline diamond cannot be grown in thin films on non-diamond substrates, since more than a micrometer growth is required after nucleation to form a continuous crystal, and several ten micrometers to reduce the lattice defect density until it is suitable for quantum optical applications [131]. Instead, diamond thin films are fabricated by ion slicing and subsequent overgrowth [132, 133], or by etching bulk-like diamond membranes of about 20 μm down to several hundred nanometers [134]. Alternatively, for total internal reflection confinement, angled etching [135–137], or quasi-isotropic underetching methods [138, 139] have been developed. The latter two enable the carving of free-standing structures into bulk diamond and make nanofabrication independent of thin film fabrication.

For the fabrication of microstructures, the main method used is focused ion beam (FIB) milling. For example, a gallium ion beam is used to physically remove diamond in a predefined pattern [126]. In the early stages of the development of diamond nanophotonics, FIB milling was also applied for the fabrication of photonic crystal cavities [140]; however, this was not widely adapted. Alternatively, and with higher quality, fabricating nanostructures into diamond can be done by transferring an electron beam lithography defined mask pattern into diamond. The pattern transfer is then achieved by dry etching methods such as reactive ion etching (RIE), inductively coupled plasma (ICP) etching, or ion beam etching (IBE). An example of diamond photonic crystal nanocavities, fabricated with RIE and incorporating single NVs, is shown in Figure 19.4.

Over the past decade, there has been a tremendous improvement in applying these methods to diamond nanofabrication. However, to reach silicon-like scalability and reproducibility still is a long way to go. Besides these very technical challenges, not all fundamental questions have been solved. While group

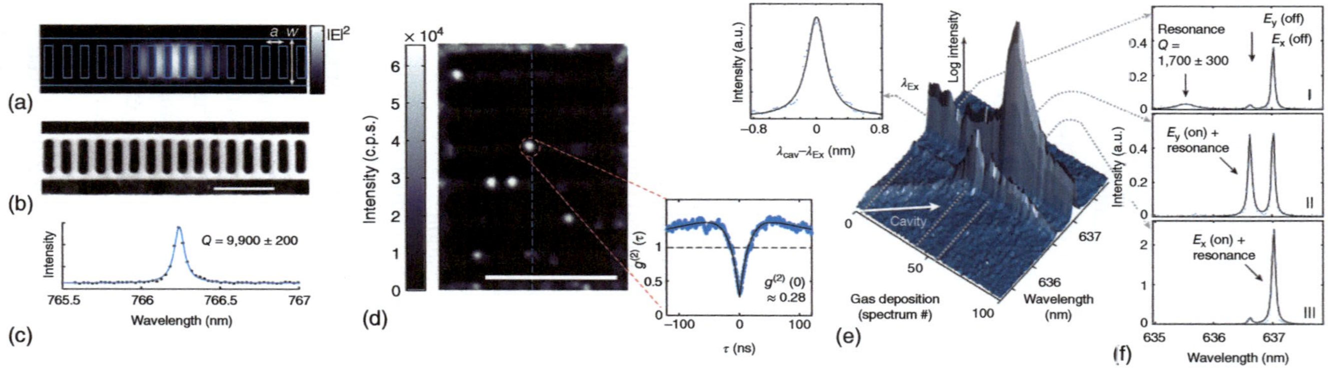

Figure 19.4 (a) For a diamond photonic crystal cavity (PhC), the electric field intensity is simulated for the optimized fundamental mode. The lattice constant *a* varies from 198 nm at the center to 220 nm over five periods. (b) Scanning electron microscope image of a diamond cavity structure fabricated by reactive ion etching. The scale bar corresponds to a dimension of 1 mm. (c) Measured cavity resonance (dots) described by a Lorentzian function (line) with a quality factor of Q∼9900 ± 200. (d) Photoluminescence confocal image of diamond PhC structures incorporating single NVs (white spots). The scale bar represents 5 mm. Inset: second-order autocorrelation measurement showing antibunching of a single NV close to the cavity center. (e) The cavity resonance can be tuned by gas deposition. The logarithmic plot shows the cavity resonance, which is red-shifted by gas condensation, sequentially enhancing the two strain-split zero-phonon line branches E_x and E_y from a single NV. Inset: Intensity of the E_x zero-phonon line transition as a function of cavity detuning. The Lorentzian fit shows the expected dependence of the Purcell enhancement and a constant cavity Q factor throughout the tuning process. (f) Spectra of the considered system in the uncoupled (I) and coupled cases with λ_{cav} tuned into resonance with E_y (II) and E_x (III). Lorentzian fits to the data yield a cavity quality factor of Q = 1700 ± 300. Source: Li et al. [123]/with permission of Springer Nature.

IV color center nano-integration is possible without constraining optical properties drastically, NV color centers in nanostructures suffer from significant spectral diffusion. Only recently it was demonstrated that by applying suitable sample properties, fabrication and control methods spectral diffusion can be mitigated for individual NV-nano-devices [141]. Despite the aforementioned limitations, fabrication methods are now sufficiently mature to approach large-scale fabrication of diamond photonic integrated circuits. Recently, the integration of more than 100 color centers into an on-chip photonic circuit was demonstrated [142]. Still, the pre-selection and subsequent deterministic incorporation of color centers into diamond nanostructures are still an open challenge. In contrast to InGaAs QD devices, for which such incorporation was demonstrated [143, 144], this method is still being developed. While the fabrication of a large number of optical cavities remains challenging, the quality of post-selected optical nanocavities has improved to such a degree that device cooperativities of C= 105 have been demonstrated for the SiV and the strong coupling regime is now in reach [15]. These improvements have made a large range of quantum information protocols feasible as discussed in detail later in this chapter.

19.2.4.2 Fabrication and Nanostructures in SiC

SiC photonics can rely on mature material growth methods providing wafer-sized substrates. While more than 200 polytypes with different crystalline structure exist, 3C-, 4H-, and 6H-SiC are the most common polytypes. Of those, 3C-SiC is the most commonly used polytype as it can be epitaxially grown in films on silicon substrates, while 4H- and 6H-SiC are only available as crystalline bulk material. Therefore, the smart cut process has been applied to the latter polytypes [145]. During the smart cut process, the SiC quality can suffer from damage introduced by high density of ions implanted during the process. In contrast to Si, which can soften and heal out at modest annealing temperatures, SiC is highly refractory and damage is difficult to be effectively healed out. Despite these challenges and using advanced material processing, high-quality, wafer-sized substrates for photonic applications were recently demonstrated for 4H-SiC [75].

SiC is inert and hard to wet-etch, yet dry etching techniques have been used to fabricate optical nanocavities. Here, residual damage, such as surface roughness in dry etching and lattice mismatch in hetero-epitaxial growth on silicon substrates, limits the achievable coherence and gives rise to optical losses.

For many fabrication techniques, the intrinsic optical loss is a limiting factor for high-quality optical devices [146, 147]. Although in the case of smart cut method, some improvements may be possible through better healing of damage during implantation. However, the intrinsic absorption of 4H-SiC in a sublimation growth sample was found to be 0.02 dB/cm [148]. If this can be combined with nanostructure fabrication, high Q devices will be possible. Low-strain crystal growth enabled the relatively low inhomogeneous distribution (±10 GHz) of V_{Si}^- center resonances [149].

Recent work has demonstrated the high-quality fabrication of SiC waveguides with integrated V_{Si}^- centers, enabling high waveguide coupling efficiencies while maintaining nearly lifetime-limited photon emission and high spin-coherence times [91]. Fabrication of microscale ring resonators with cavity quality factors of

$1.3 \cdot 10^5$ can also maintain close to lifetime-limited linewidths and has enabled the demonstration of optical superradiance of a pair of V_{Si}^- centers [92].

A recent review summarizes the main progress that has been made toward realizing high-quality photonic components. The review also details strategies to overcome the current challenges in SiC quantum photonics [82].

19.3 Applications

19.3.1 Color Centers as Quantum Emitters

A variety of non-classical states of light, such as squeezed states [150] or single photon states [151], can be used for quantum applications. Naturally, the use of quantum photonics requires a robust source of such states. Focusing on single photons, the simplest scheme for a single photon emitter consists of atom-like energy levels, which may be coupled to an electromagnetic field that drives population into an excited energy state. From the excited state the population decays, which is accompanied by populating photonic modes, leading to the emission of a single photon. In the ideal case, the single photon emitter should operate in a deterministic fashion – one excitation results in a single photon in a desired single mode. The efficiency of this process is captured by the *overall single mode collection efficiency*. It includes the *quantum efficiency* of the emitter, which is the probability of emitting a photon into a radiative mode after excitation. In general, this emission is required to repeatedly produce a photon in the same mode. In the solid state, interactions with phonons give rise to the possibility of decaying radiatively into a broad red-shifted PSB rather than producing a photon at the ZPL-frequency. The interplay between direct and indirect radiative processes can be quantified in terms of either the Huang-Rhys (HR) [152] or Debye-Waller factor (DWF) [153]. Furthermore, the photon must be coupled into a mode that allows for further operations on the photon, rather than into an arbitrary mode (as for a free space emitter with a dipole radiation pattern) – which includes "downstream modes" in the same or other devices, and eventually the detection mode – resulting in an overall system efficiency. Tremendous efforts are undertaken to address the challenge of redistributing the dipole radiation into a single (guided) mode. Then finally, the efficient detection of single photons is, in itself, challenging; here device figures of merit such as the quantum efficiency, temporal jitter, dark counts, and possibly photon number resolution must be considered. Today, near-unit efficiency is achievable with superconducting single photon detectors, with several commercial vendors offering off-the-shelf solutions.

The presence of a single photon is typically verified through measurement of the second-order autocorrelation function $g^{(2)}(\tau) = \langle a^\dagger(t)a^\dagger(t-\tau)a(t-\tau)a(t)\rangle / \langle a^\dagger(t)a(t)\rangle^2$ [154] either with a fast single photon detector, or alternatively at two output ports of a beam-splitter in a Hanbury–Brown–Twiss setup [155]. In the latter, the mode is incident on a beam-splitter with single-photon detectors at the two output ports. A single photon will exit at only one of the ports, and only one of the detectors will "click" at any given time yielding $g^{(2)}(0) = 0$. The temporal dependence of the $g^{(2)}$ curve shows additional information about the dynamics of the photon emission (as well as the detector response) – the shape of the $g^{(2)}(\tau)$ near its minimum is determined by the emission decay timescales, and it reveals if other decay pathways with different timescales are present [156].

For the practical use of single photons, the emission wavelength is also an important consideration. As well as (indeed, as a consequence of) telecom wavelengths being advantageous for transmission in fiber, advanced optical devices have been developed for the telecommunications industry. Nevertheless, even in the low loss transmission region for optical fibers at around 1550 nm, the best attenuation record to date of $\sim$ 0.14 dB/km [14] still results in 50% loss after about 20 km. One route to address losses of single photons carrying quantum information is by employing quantum repeater stations (section 19.4). Many quantum emitters operate in the visible or near-infrared range, and thus a means of wavelength conversion is necessary in order to access the low loss telecom band.

19.3.1.1 Quantum Emitters at Room Temperature

In ambient conditions, the interaction of an emitter with its environment results in complicated dynamics. Main contributors to these dynamics are phonon-mediated processes, which result in a broadened spectrum. Nevertheless, quantum emitters have a number of useful properties for applications under these conditions.

One example are QKD protocols that only necessitate the transmission of a qubit. Here, indistinguishability between successive photons is not fundamentally required, provided that their differences do not interfere with the state preparation, transmission, and measurement of the exchanged qubits. As stated above, QKD is a near term deployable quantum technology. Indeed, there have been numerous demonstrations of QKD, including with color centers [157–159]. However, scaling QKD in a secure manner to a metropolitan or global level necessitates overcoming challenges of photon loss so that quantum states can be successfully and faithfully transmitted. One route to addressing loss is the implementation of a quantum repeater network (discussed further in section 19.4).

19.3.1.2 Quantum Emitters at Cryogenic Temperatures

At cryogenic temperatures, with a suppressed thermal phonon population, solid-state defects can become more similar to an isolated atom-like system. This means that the likelihood of a direct optical transition between the energy levels is increased, and therefore, photons are emitted into a narrower spectrum defined by the lifetime limit of the excited state.

Photons emitted into a single spectral mode enable further applications that make use of the quantum statistics of indistinguishable particles. These statistics can be observed in the quantum interference of single photons. Applications include quantum simulation, such as boson sampling, as well as communication and computation protocols that make use of entanglement.

Indistinguishability Quantum interference requires that the particles cannot be distinguished from one another. Indistinguishability for interfering photons requires them to be identical in all degrees of freedom, for instance in arrival time, frequency, polarization, and spatial mode distribution. Hence quantum emitters that generate such indistinguishable single photons can be used as resource for implementing a two-photon Hong–Ou–Mandel type interference experiment, which in turn can also be used as proof of indistinguishability of the photons [160]. In such an experiment, the two photons are incident on the two input ports of a beam splitter and detection events at the two output ports are monitored. A beam splitter implements a unitary transformation of the two input modes, where, bosonic commutation relations of photons dictate that the two photons populate the same detection mode. A simple calculation based on a 50:50 beam-splitter transformation (with zero phase difference) yields $a_{\mathrm{in}}^{\dagger} b_{\mathrm{in}}^{\dagger} \rightarrow \left(a_{\mathrm{out}}^{\dagger} - b_{\mathrm{out}}^{\dagger}\right)\left(a_{\mathrm{out}}^{\dagger} + b_{\mathrm{out}}^{\dagger}\right)$ where a and b denote the spatially orthogonal modes at the in- or output ports of the beam-splitter ($\left[a, b\right] = 0$). The output modes then are $a_{\mathrm{out}}^{\dagger} a_{\mathrm{out}}^{\dagger} - b_{\mathrm{out}}^{\dagger} b_{\mathrm{out}}^{\dagger}$, indicating that for perfect photon indistinguishability only one of the two detectors produces detection events. If the photons are partly distinguishable, both detectors can count simultaneous events and reduce the visibility of the interference effect.

Spectral Diffusion For solid-state quantum emitters the generation of coherent single photons with lifetime-limited linewidth is, after many years of research, a persisting challenge. While homogeneous broadening effects are mainly suppressed at cryogenic temperatures, the effect of spectral diffusion leads to an inhomogeneously broadened transition linewidth. On short timescales, quantum emitters can exhibit near lifetime-limited emission linewidths, but on longer time scales, the frequency of emitted photons fluctuates, referred to as spectral diffusion. Spectral diffusion can be caused by a fluctuating electric field (DC Stark effect) due to charge noise in the environment of the quantum emitter. In optical control protocols, quantum emitters are excited for system initialization or state read-out by laser light with energies inducing photoionization of nearby bulk or surface crystal lattice defects.

In particular, in nanostructures the quantum emitter is located close to rough-etched surfaces with large densities of surface defects as illustrated in Figure 19.5(a). In addition to the charge environment, intrinsic properties of the quantum emitter determine the impact of charge noise on the emission linewidth. For example, the NV in diamond, exhibiting no crystallographic inversion symmetry and therefore a permanent electric dipole moment, is very susceptible to spectral diffusion [141, 162]. In contrast, inversion symmetric G4V defects in diamond [51] are not as susceptible as they exhibit almost no permanent, and therefore mostly induced dipole moments. As discussed in section 19.2.1.2, the charge state of the emitter itself may also vary under resonant excitation. The wavelength selected for re-initialization should be chosen so that the efficiency of the recharging process is maximized, while keeping the overall power requirements low [50, 163]. In some cases, the ionization of defects in the emitter's environment can be minimized by choosing an appropriate excitation wavelength [141].

A passive approach to mitigate spectral diffusion is to improve fabrication methods, involving defect formation techniques (native defects, formation during growth, ion implantation) [37], micro- and nanostructuring methods [164, 165] as well as surface termination [166]. In addition to passive mitigation, active control schemes have been developed for suppressing spectral diffusion. In an active feedback scheme, the emission frequency of photons is adjusted by applying an external electric field, which induces DC Stark tuning [161]. Furthermore, a pulsed coherent control protocol was proposed by Fotso et al. [167] that cancels spectral detuning and stabilizes the emission frequency of a two-level solid-state emitter at a certain target frequency – the carrier frequency of the pulses.

Active Tuning For single photon sources used for certain quantum applications not only the spectral stability of individual emitters need to be ensured, but moreover, photons emitted from different devices need to be spectrally aligned. Here, similar to the correction of spectral diffusion, DC Stark tuning of electronic transitions is achieved by applying external electric fields [161] as shown in Figure 19.5(b)-(d). Another method is the application of strain, which changes the crystal field surrounding the emitter [168, 169]. The particular choice of the tuning scheme depends on the achievable tuning range with regard to the susceptibility of the emitter, the tuning rate (how fast the applied field can be varied), as well as achievable tuning stability and resolution.

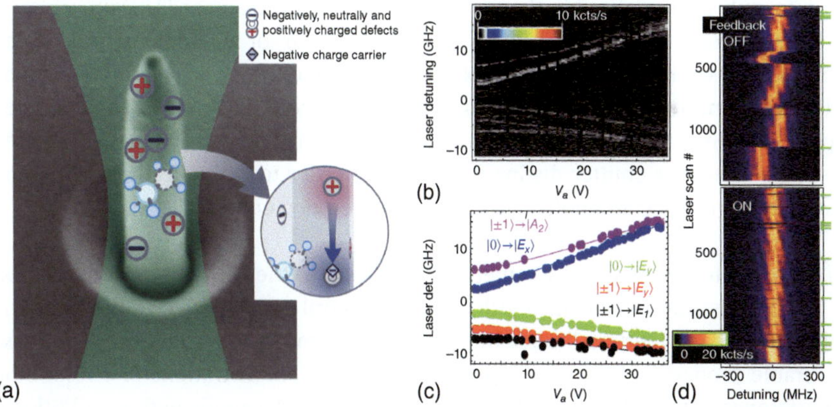

Figure 19.5 (a) Schematics illustrating the process of spectral diffusion of an NV in a diamond nanopillar with a radius of 125 nm and a height of 1.6 μm (scanning electron microscopy image). At rough etched surfaces, the density of unwanted defects that act as charge traps and electron donors is very large. Green laser irradiation at about 532 nm causes photoionization of defects and diffusing charges, in turn leading to fluctuations of the electrostatic environment of the NV and spectral diffusion of the ZPL resonances (DC Stark shift). (b) Active tuning of the NV ZPL transition frequencies can be achieved by applying an external voltage. Here, photoluminescence excitation (PLE) spectra as a function of applied voltage are shown. (c) The transition frequencies can be modeled with good agreement. (d) PLE spectra of the ZPL E_x transition of a single NV with and without applied voltage feedback. Green arrows indicate when green initialization pulses were applied. Source: Acosta et al. [161]/American Physical Society / CC BY 3.0.

19.3.1.3 State-of-the-art

NV at Room Temperature The NV enabled the first stable and long-term generation of single photon states at room temperature in 2000 [28, 29]. At that time, NVs in bulk diamond without optimized outcoupling structures were investigated, leading to single photon detection rates of only tens of thousands detected photons per second. Since then, the rate of the detected photons has been increased by fabricating micro- and nanodevices into diamond. Through pulsed excitation the overall detection efficiency per pulse can be determined, which is an important figure of merit for the applicability of the emitter as single photon source. Recently, a raw experimental detection efficiency of $(12 \pm 1)\%$ with fluorescence detection rates as high as 4×10^6 counts per second was demonstrated [128].

NV at Cryogenic Temperature For quantum information applications, the spin state of the NV qubit can be selectively interfaced with single photons. The selectivity of the interaction allows for applications that go beyond single photon generation such as spin–spin entanglement of distant quantum nodes heralded by two-photon interference (section 19.3.3.3).

Two-photon interference between photons from two different NVs was demonstrated by two groups at a similar time in 2012 [170, 171]; both of which were realized in solid-immersion-lens samples, using DC stark tuning in order to match the resonances. However, extending this to the generation of spin-spin entanglement has, to date, only been realized in the group led by Ronald Hanson at TU Delft.

Recent work has demonstrated the preservation of entanglement between converted telecom photons and an NV spin [172] addressing the challenge of long distance communication via telecommunication frequency bands in the infrared spectral regime. Subsequently, the quantum frequency conversion stage was used to compensate for emitter distinguishability in order to facilitate the two-photon interference of telecom photons from separate detuned emitters [173].

SiV Owing to its enhanced symmetry with respect to the NV, SiVs typically exhibit a narrow distribution of linewidths. Large spectral overlap of different emitters exhibiting nearly lifetime-limited linewidths was demonstrated in bulk [174]. These favorable properties lead very rapidly to the first demonstrations of two-photon interference of two SiVs in 2014 from emitters in a bulk sample with no additional tuning [175]. Moreover, even in nanodiamonds as well as in nanostructures almost lifetime-limited linewidths were observed [176, 177]. The structures were either post-fabricated around SiVs [178] or pre-fabricated with subsequent SiV implantation via masking [179].

Tuning of emission from SiVs has been demonstrated via selective cavity enhancement [180], as well as strain [169, 181]. Very recently, two-photon interference was successfully shown between SiVs in optical nanocavities [181].

GeV In contrast to the SiV and SnV, the GeV has been explored to a lesser extent. The GeV shows a sharp and strong luminescence band with a ZPL at around 602 nm and has single photon emission capability at room temperature [61]. Its DWF was determined to be about 60% [62]. The GeV is of the same symmetry as the SiV and SnV, and should therefore also be less sensitive to electric noise compared to the NV. The GeV possesses a ground state splitting of about 170 GHz [61, 182], which is advantageous compared to the SiV, due to the reduced thermal photonic relaxation processes. However, it does not outperform the SnV, which has a ground state splitting of about 850 GHz [183]. Optical and microwave control of the GeV spin ground states was demonstrated by Siyushev et al. [62].

SnV In the first work on SnVs, single emitters in nanopillars with narrow emission linewidth < 250 MHz were demonstrated [184], and in bulk, lifetime-limited linewidths of about 35 MHz together with spin coherence times of $T_2^* = 540$ ns at 2 K were reported [183]. However, due to the large size of the Sn atom, damage introduced during implantation can make it challenging to find high-quality emitters. For SnVs created by Sn implantation and successive vacuum annealing, the yield of close to lifetime limited emission linewidth is on the order of 0.1% to 1%.

Novel fabrication techniques such as shallow implantation and subsequent diamond overgrowth [185] have been introduced in order to enable nanostructure fabrication with narrow near lifetime-limited lines [186]. This has been extended recently to demonstrations of photonic crystal cavity integration with Purcell factors of 25 for the coupled system, resulting in a coupling of 90% of the emitted photons into the cavity mode [187]. Also in other works coupling of single SnVs to photonic crystal cavities with measured quality factors as large as 11000 and Purcell factors for the SnV ZPL of 37 together with a high cavity coupling efficiency β of 95% have been achieved [188]. Even though G4V defects are less sensitive to electric fields than NVs, electrical tuning via Stark effect of ZPL resonances of nanopillar-coupled SnVs has been demonstrated, covering a range exceeding 1.7 GHz [189].

Similarly to the NV^- and many other color centers, the SnV is also subject to "blinking" under continuous resonant excitation attributed to a change of charge state [183]. Due to its position within the bandgap, this likely corresponds to the charging of the $SnV^- \rightarrow SnV^{2-}$. Initialization as well as constant charge stabilization can be maintained by adding a laser of shorter wavelengths, e.g. of about 445 nm, that induces recharging of the neutral divacancy which is involved in the charge cycle while also avoiding direct excitation of the SnV [163].

SiC at Room Temperature Since the first demonstration of single photon generation with a color center in SiC in 2014 [190], for which 4H-SiC was used, a large variety of different types of single color centers were discovered both in hexagonal as well as cubic SiC. Besides this color center attributed to the carbon antisite-vacancy pair defect, silicon vacancy centers (V1) [191, 192], divacancies [193], and not yet identified bright and photostable emitters [194] among others were demonstrated. The latter exhibit count rates up to 2 Mcps at saturation at room temperature from unpatterned bulk material. Similar to diamond, optical and spin properties are superior at cryogenic temperatures, and most recent work has focused on optically accessing and controlling spin states at 4 K.

SiC at Cryogenic Temperature The most investigated color center in SiC is the negatively charged silicon vacancy center at hexagonal lattice site (h-V_{Si}) in the 4H-SiC polytype. It can be formed through electron irradiation, for example, in high purity SiC. It is a defect spin-3/2 system with strongly spin preserving optical transitions. It features a non-radiative intersystem crossing that can be used for spin initialization and has a DWF$\sim$ 80%. It was shown to have narrow optical transition lines which are $\sim 2\times$ of the lifetime-limited linewidth below 6.9 K, and was used to demonstrate a Hong-Ou-Mandel (HOM) visibility of 70% between successively emitted photons ($\tau \approx 50$ ns) [89]. Their temperature-dependent properties have also been analyzed theoretically [195].

For waveguide-coupled V_{Si}, nearly lifetime-limited photon emission was shown [91]. A pair of V_{Si} integrated into a ring resonator demonstrated optical superradiance [92]. Also Stark tuning over $\pm$100 GHz can be applied to the V_{Si}, both for DC [196] and GHz frequencies [197]. Besides promising optical properties, single h-V_{Si} centers feature electron spin state with excellent properties [198] for which optical coherent control was implemented to quantify its internal spin-optical dynamics [199].

For further recent progress on optical aspects of SiC color centers, reference [72] gives a nice overview.

19.3.2 Color Centers as Quantum Memories

The local storage of a quantum state requires the ability to control the designated memory states that are provided by the color center. This immediately leads to the question of what degrees of freedom can be used for storing the quantum information. In other words: Which physical states of the system can be controlled, are long-lived, and allow the retrieval of the quantum state. For example, the memory qubit could be a pair of ground and excited states, two orbital ground states, or two spin states.

Control generally entails being able to initialize the state of the qubit into a desired quantum state (section 19.3.2.). Likewise, the inverse of this process – the ability to read out the qubit state – must be possible (section 19.3.2.). A key figure of merit of a quantum memory is the memory storage time, which is determined by its coherence time (section 19.3.2.). Beyond these essential features, it is important to map a qubit state into the memory, which requires a joint operation between the target qubit and memory qubit. Joint gates between multiple spins, as well as the interaction of a photonic qubit with a quantum memory are considered in section 19.3.3.

19.3.2.1 Considerations and Figures of Merit

Commonly, the spin degree of freedom provides long coherence times compared to the other physical states considered above and therefore is selected as the memory system that is the focus of this section. In the following, state initialization and coherent control, state read-out, and coherence times are discussed as the key elements in evaluating a system for a quantum memory.

State Initialization and Coherent Control For the NV, even at room temperature, spin state initialization is possible due to a spin non-conserving transition [27]. For many other defects, resonant spin pumping must be used [200].

The coherent manipulation of the spin state of an atom-like system is often achieved through microwave control, which requires that the spin states are non-degenerate such that a resonant microwave can drive Rabi oscillations. Microwaves can be produced by an antenna near the emitter. The resonance frequency can be identified through ODMR, where the microwave frequency is swept while the fluorescence from one of the levels is monitored. A change in the steady state fluorescence is observed when the microwave field is in resonance and drives oscillations between the spin states. This protocol is also used for some magnetic sensing techniques because the Zeeman splitting of the spin states is determined by external magnetic fields [42]. Extending microwave control to full qubit control requires the ability to perform rotations about multiple axes on the Bloch sphere. These rotations can be achieved through phase modulation of the applied driving fields. Moreover, the control can be extended to perform rotations for dynamical decoupling, see section 19.3.2.

Another approach of fast *optical* coherent control was demonstrated, for example, for the NV [201], where the authors use ultrafast pulses to control the ground-state spin.

State Read-Out Typically, state read-out is done optically, where irradiation with a transition-resonant laser field will only excite the color center if it is in a specific state. In this measurement scheme, the outcomes are "bright" or "dark" depending on whether an excitation occurred *and* was detected [202]. If the excited color center has a high probability of relaxing to the initial state (a *cycling transition*), then the excitation can be repeated many times before spin mixing occurs. This increases the contrast between the "bright" and "dark" states and thereby can increase the fidelity of the measurement up to the point at which mixing of the states leads to spurious read out. Access to a cycling transition enables *single-shot* read out (in the sense that many consecutive excitation events can be used to gain sufficient photon statistics without having to repeat the entire protocol). Another critical factor in read-out fidelity is therefore the efficiency of photon collection and detection. Besides optical read-out, also state-to-charge conversion was used in [203] with high fidelities.

Of course, during the emission process, the phase is not preserved between the two qubit states, and so repeated measurement sequences are required. In order to read out the full qubit state, repeated measurements in different bases are needed, i.e. a rotation is performed prior to the read-out in the accessible basis.

Coherence Time To quantify coherence times of a qubit system, the longitudinal and transverse relaxation times (T_1 and T_2, respectively) are generally employed. In practice, the coherence time of a qubit system can be influenced by additional control processes, which is captured by the distinction between T_2^* and T_2, where T_2^* gives the bare coherence time with no external control processes, versus the T_2, which quantifies the coherence time including active control.

The T_1 captures the population decay between the two qubit states: if a qubit is initialized into one state, the remaining population is recorded after some waiting time. In general, this is not the same for the two non-degenerate levels used for the memory qubit since thermalisation imposes a directionality. This is most apparent when considering optical transitions, where thermal excitation is negligible at the cryogenic temperatures while the spontaneous decay time is on the order of nanoseconds. The overall coherence between the two states is commonly measured through a Ramsey sequence [204] and yields T_2^*. In order to protect the coherence of a spin state, dynamical decoupling (DD) such as a Hahn echo sequence may be employed [130]. The extended lifetime that is achievable with additional decoupling yields the T_2 of the qubit. T_2^* and T_2 have an effect on the inhomogeneous and homogeneous linewidth of the emitter respectively. The principle of the Hahn echo sequence can be extended to a more generalized DD sequence. For example Carr–Purcell Meiboom–Gill (CPMG) sequences are a standard technique [129] consisting of some number N of π pulses during the waiting time τ [205]. For this sequence, the timescale of fluctuations must be slower than the time it takes to apply two π pulses ($2\tau/N$). Furthermore, using rotations about multiple axes (e.g. XY sequences) improves robustness against pulse errors, as well as increasing the range of states, which are preserved during the decoupling process [206].

Spin Environment in Diamond Although the spin density of diamond is usually low, a residual paramagnetic bath is present, which can be comprised of ^{13}C spins, $P1$ centers (nitrogen impurities) and others. The presence of spins in the environment constitutes a random, fluctuating, magnetic environment, which has the effect of perturbing the splitting of the NV levels, thus resulting in dephasing between the qubit states. Targeted compensation for the dephasing requires tailored pulse sequences, which extend the T_2 of the NV spin via dynamical decoupling, see, e.g. [207] and Figure 19.6.

19.3.2.2 Physical Considerations on Cycling Transitions: Spin Initialization and State Read-out of Qubits in Diamond

In the NV^- in diamond, transitions between the $^3A_2$ and ^3E manifolds are spin selective. Of these, the E_x transition between the $m_s = 0$ states is primarily cycling. Driving the A_1 transition between $m_s = \pm 1$, however, is accompanied by an increased probability of decaying via the meta-stable singlet state, which likely causes spin-mixing. This process eventually leads to the decay of the system into any of the spin ground states [208], as shown in Figure 19.1(a). The qubit is typically encoded between the $m_s = 0$ and one of the $m_s = \pm 1$ levels for quantum information applications. The actual dynamics of the intersystem crossing in a given sample can also be influenced by phonon interactions and strain [209, 210].

Non-spin conserving transitions can be used for a high fidelity spin initialization via optical pumping (e.g. $> 99\%$ is estimated in [211] by resonant A_1 excitation. Off-resonant excitation produces typically lower fidelities around $70--90\%$ [212]). Single-shot read-out can similarly be achieved by exciting with higher energies than the ZPL transition and thresholding on the number of counts: shelving in

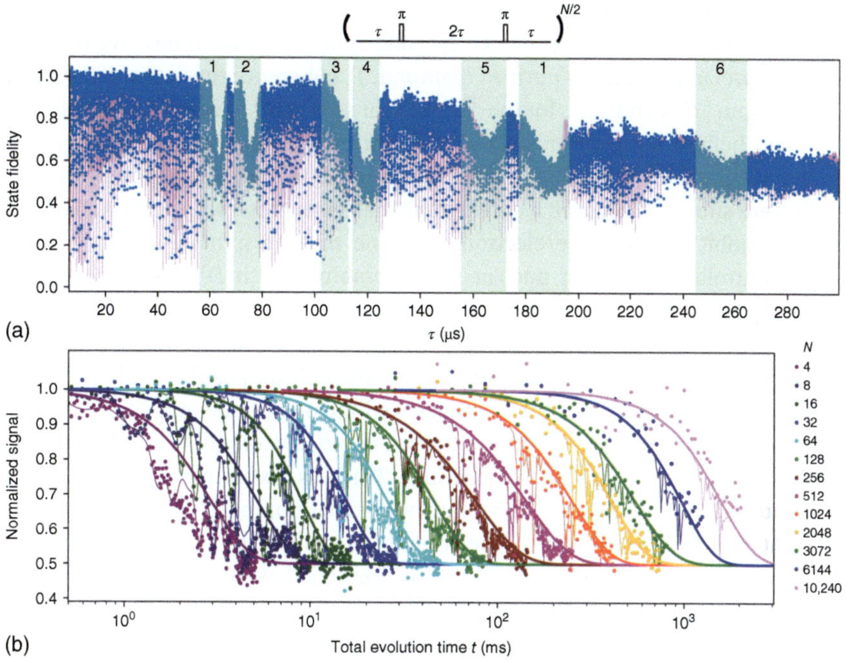

Figure 19.6 (a) Dynamical decoupling spectroscopy reveals structure in the electron coherence from resonances with individual ^{13}C nuclei. Dips indicate loss of coherence from interaction with nuclear spins: sharp dips from coupling with individual nuclei and broader features from interaction with strongly coupled spin pairs. (b) Dynamical decoupling sequences tailored to a specific ^{13}C environment enabling over 1 s coherence time of the NV electron spin. Source: Abobeih et al. [207]/Springer Nature/CC BY-4.0.

the meta-stable state, e.g. over ~ 200 ns [213] until the $m_s = 0$ ground state gradually repopulates, provides initial contrast in the PL signal. Resonant excitation on E_x, however, allows a longer readout probe time to be used, e.g. 40 μs, during which repeated cycling can produce high fidelities of, e.g. 93.2% in [211]. Repeated cycling will eventually be destructive due to optical pumping. Conversely, using shorter read-out periods can be nondestructive, but limits the achievable read-out fidelity.

A review on spin read-out techniques of the NV can be found in [213]. In contrast to “traditional” optical off-resonant spin-state read-out, spin-to-charge conversion can be used to increase read-out robustness of the spin-state [214].

For the G4Vs, with no shelving state and no intersystem crossing, spin-state initialization is more challenging. In the absence of strain or applied magnetic fields, there is no mechanism for spin mixing [53] and the optical transitions are spin preserving. An on axis magnetic field can therefore be used for spectrally addressing individual spin states, but does not allow for optical spin pumping. Typically a slight off-axis field is used, providing access to both mechanisms. Bigger dopant atoms such as the Sn lead to larger spin orbit coupling term and thus transitions are more strongly spin preserving [163] even in an off-axis field; however, as before, the presence of residual strain (which can be partially remedied through HPHT treatment) can reduce the read-out contrast.

19.3.2.3 State-of-the-Art

NV in Diamond The NV is renowned for long spin coherence times, even at room temperature. In isotopically purified diamond, T_2 coherence times of > 1 ms for the electron spin [215] and > 1 s for the ^{13}C nuclear spin [216] have been demonstrated. At liquid helium temperatures of 3.7 K, electron spin relaxation times T_1 of one hour and coherence times $T_2 > 1$ s using tailored dynamical decoupling sequences [207] were shown. Moreover, a single-qubit state was preserved for more than 75 s in a ten-qubit register (one electron and nine nuclear spins) applying selective phase-controlled driving of nuclear spins combined with DD sequences on the electron spin [217].

SiV in Diamond Within the SiV manifold there are multiple possible choices for qubit encoding. Those that have been demonstrated include orbital and spin degrees of freedom. Spin control of the SiV has attracted significant attention, due to the ability to isolate spin preserving optical transitions. This was first demonstrated in 2014 by two groups [218, 219], showing coherent population trapping (CPT) between the two spin states of the upper branch of the ground state, with a $T_2^* \sim 35 - 45$ ns. In order to realize a closed Λ system, a slight off-axis magnetic field must be applied in order to produce mixing of the spin states, since the excited- and ground-state manifolds have different spin-orbit coupling strengths [53]. The T_2^* obtained in these measurements is limited by electron–phonon processes, although they are spin-preserving [220], as illustrated in Figure 19.7(a). While orbital relaxation can be reduced to timescales of $T_1 \sim$ ms [220], the orbital-$T_1^{\uparrow} \sim$ spin-$T_2^* \sim \mathcal{O}(10^1)$ns is still rather short lived (see Figure 19.7(b)–(c)). To address the short spin-coherence times, mK temperatures are required – achievable with dilution refrigerators – in order to suppress the thermal occupation of phonons that can interact with the qubit system.

The use of microwave control instead of an optical Λ-scheme allows flexibility in the magnetic field direction, which can be chosen such that spin cycling transitions become available resulting in high fidelity read-out. Microwave control was first demonstrated in ODMR [221]. The magnetic field orientation comes at a trade-off between initialization and read-out state efficiencies: initialization in an aligned magnetic field via spin-pumping needs 150 ms, but enables a read-out window as long as $\sim$ 10 ms [200].

Note that there is an optimal regime for microwave control – a near aligned magnetic field favors spin-conserving transitions, but with the addition of slight strain to allow individual optical addressability (frequency resolution) of spin conserving channels [222]. Furthermore, strain is needed for dipole allowed magnetic transitions, which can be addressed using microwaves.

Combining mK temperatures and microwave control with CPMG32 sequence was used to demonstrate coherence times of $T_2 \sim 10$ ms in a high purity (0.001% ^{13}C) sample [200], see Figure 19.7(d)–(g). In this work a bimodal noise source (one fast, one slow) was observed and a linear scaling with pulse number was found.

However, microwave control may introduce the risk of heating [222, 223], depending for instance, on the efficiency of the microwave delivery system or the amount of strain in the sample – and thus on the strength of the magnetic dipole transition.

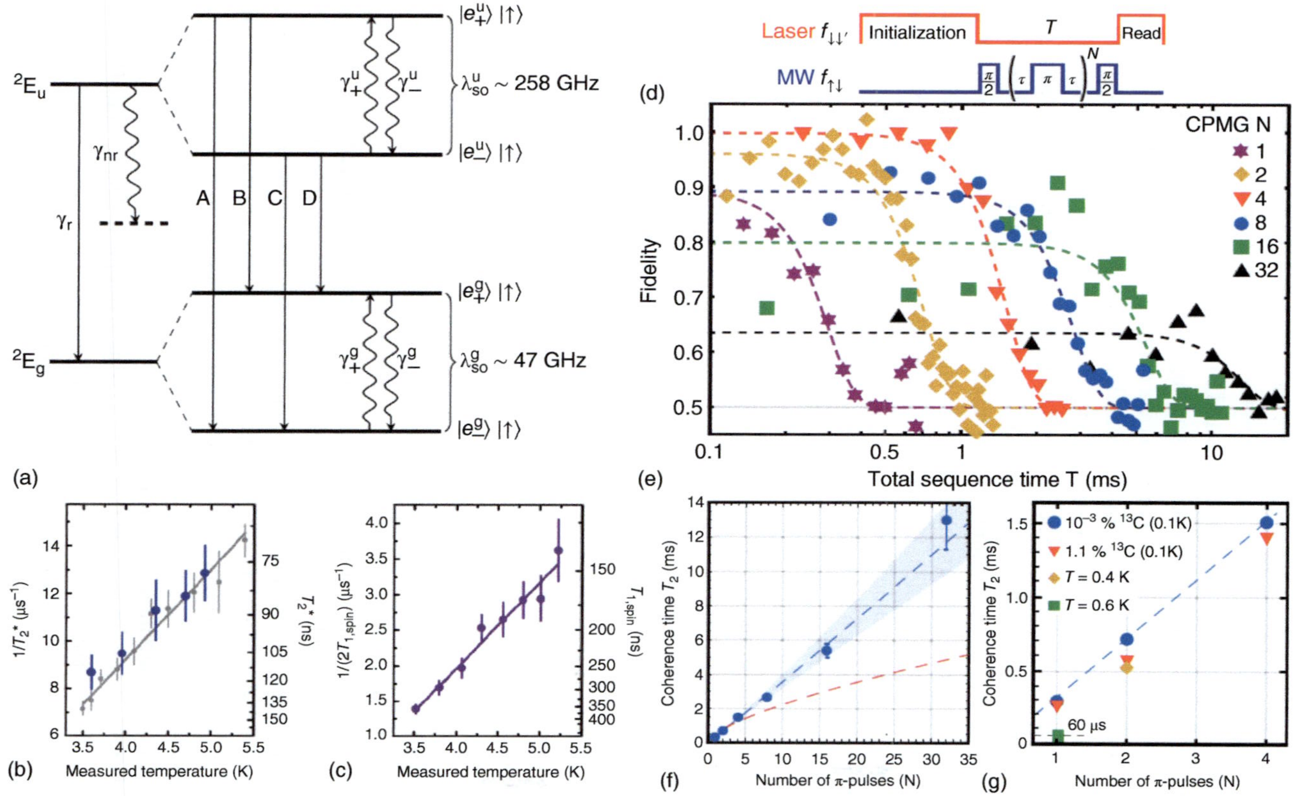

Figure 19.7 (a) Structure of the SiV with labelled optical and phonon mediated transitions. (b) Scaling of the SiV spin dephasing rate (T_2^*) with temperature, grey shows the orbital decay rate ($2T_{1,orbital}$); (c) Scaling of the spin decay rate. The linear scaling of the orbital decay rate (b) is indicative of single phonon processes. (d) CPMG-*N* pulse sequence applied to the SiV spin: (e-g) Spin coherence measurements with varying number of rephrasing pulses maintaining coherence up to 13ms. Source: (a) Jahnke et al. [220] licensed under CC BY 3.0; (b, c) Pingault et al. [221]/Springer Nature/CC BY-4.0; (d-g) Sukachev et al. [200]/With permission of American Physical Society.

As an alternative approach to accommodate short lifetimes, one may consider fast broadband optical control via Raman rotations between orbital states [224], or widely separated (5 GHz) spin states [56]. In the latter work, a magnetic field aligned at $\sim 70°$ to the symmetry axis was used to enable cross coupling (initialization fidelity 99.93%) at the expense of single-shot read-out. In this regime phonon processes freeze out at mK temperatures and coherence times become limited by the spin bath. The longest T_2 achieved with a single spin echo was ~ 140 ns at 40 mK: This experiment was conducted in HPHT IIa diamond with a high concentration of substitutional nitrogen, where dipolar coupling cannot be rephrased through DD.

Manipulating the strain environment of the SiV not only can be used for resonance tuning [169, 181], but can also be exploited to enhance coherence: by increasing the ground state splitting, which reduces the effect of thermal phonons that drive transitions in the ground state manifold. In this way, a spin $T_2^* = 0.25$ μs was demonstrated at 4 K [225], which is comparable to low strain SiV at 100 mK, or the NV without dynamical decoupling [215]. However, increased strain may affect the spin mixing and therefore the selection rules, as well as potentially having implications for other optical properties such as the quantum efficiency.

Note also, that incorporation into nanostructures additionally influences the scaling of the phononic relaxation rates, which typically follow a power law scaling Δ_{GS}^n wiht $n = 3$ in bulk, and $n = 2$ close to a surface, where Δ_{GS} is the ground state splitting. The local geometry can influence the scaling behavior as well. The authors of [169] showed a $n = 1.9$ scaling, whereas in [180] a scaling of $n = 0.3$ was shown.

For a review of optical control of the SiV, see the work of Becker and Becher [226].

SnV in Diamond Due to the increased ground state splitting compared to the SiV, the SnV is expected to offer enhanced spin coherence properties at 4 K [54], and 2 K, which are considered technically reasonable working points in temperature.

A demonstration of coherent population trapping in the SnV [227] found a branching ratio between spin flipping and spin conserving transitions of $\eta \sim 80$ in a moderate strain nanopillar sample with an applied off-axis magnetic field of 200 mT at a 54.7° tilt to the symmetry axis.

In comparison, a low-strain HPHT sample (without nanostructuring) with an off-axis magnetic field of 200 mT at a 54.7° tilt to the symmetry axis exhibited a branching ratio $\eta \sim 650$ and was able to demonstrate single-shot readout with an estimated fidelity of 74% [163]. The fidelity can be further increased by enhancing the collection efficiency (via nanostructuring) and by improving detection efficiencies.

The CPT measurements yielded comparable T_2^* times of ~ 1 μs [227] and 5 μs [163] both at 1.7 K and 200 mT at 54.7°. Using dynamical decoupling, Debroux et al. [227] obtained $T_2 = 0.33$ ms with (only) 4-pulse DD, which could have been increased with an aligned field. The observed lifetimes in this case were consistent with the natural ^{13}C abundance.

Silicon Carbide Compared to diamond, the structure of SiC is favorable in terms of coupling to a residual spin bath due to the longer bond lengths (i.e. increased separation) and the binary lattice (i.e. reduced likelihood of strongly coupled homo-nuclear spin pairs) as investigated in [228]. Furthermore, partially due to its mirror symmetry planes, the basal *kh* divacancy exhibits first-order insensitivity to magnetic fields near $\mathbf{B} = 0$ (a clock transition).

This property was exploited to demonstrate a decoherence protected qubit by additionally applying microwave dressing to the spin-1 system [229] and showing $T_2 = 64$ ms with a single rephrasing pulse. The polar structure does, however, cause increased susceptibility to charge noise.

Recently, a coherence time of $T_2 = 5\ s$ was measured for a basal divacancy in isotopically purified SiC using $\sim 10^4$ DD pulses [84], without showing saturation at an increased pulse number. This work additionally demonstrated improved single-shot state read-out fidelities by using state-to-charge conversion.

In a different configuration the *c*-oriented *kk* divacancy is of the same C_3v symmetry group as the NV. This configuration has no clock transition; however, it has a reduced susceptibility to charge noise [88]. In general, isotopic purification of the host material increases coherence times, which together with increased decoupling pulses can also produce second-timescale coherence times [88].

19.3.3 Color Centers for 2-Qubit Quantum Gates

In this section, color centers are considered as interface nodes for facilitating two-qubit quantum gates. In such an interface node a color center is used to mediate a joint operation on two qubits. The target qubits may be photonic qubits (emitted or coming from another emitter), or encoded in local ancillary solid-state spins.

For example, in the series of works [230, 231], a cavity coupled SiV was used to implement a switch at the single-photon level. First, in [230] orbital states of the SiV were used, and later work [231] employed the spin degree of freedom. Additionally, in both cases photon-mediated interactions between two SiVs in a single nanocavity were shown. It was highlighted in this work, that in spite of a high DWF, the quantum efficiency of SiV was estimated at around 10% [230]. A higher quantum efficiency of $\sim 80\%$ was estimated for the SnV [54], and in combination with improved fabrication techniques, the improvements can be readily applied to the SiV.

19.3.3.1 Color Centers for Spin–Photon Entanglement

The first step for distributing entanglement is the generation of a photon whose quantum state is correlated with the state of the emitter. For example, $|\psi\rangle_{\text{em.}} = \alpha|0\rangle_{\text{em.}} + \beta|1\rangle_{\text{em.}} \overset{\text{gen.}}{\rightarrow} \alpha|0\rangle_{\text{em.}}|0\rangle_{\text{ph.}} + \beta|1\rangle_{\text{em.}}|1\rangle_{\text{ph.}}$ where the photon generation process (gen.) depends on the details of the emitter and the photon encoding. An illustrative example is shown in Figure 19.8(a), where a time-bin spin–photon entanglement protocol is outlined. A spin/time-bin encoding necessitates that only one of the qubit states of the emitter is coupled to an excited state and emits a photon. For this scheme to work efficiently, the spontaneous emission must be spin preserving. The population in the two qubit basis states is then inverted by a π pulse, and the excitation repeated. This results in a photon in the early or late time bin being correlated with population in one or the other qubit state [11].

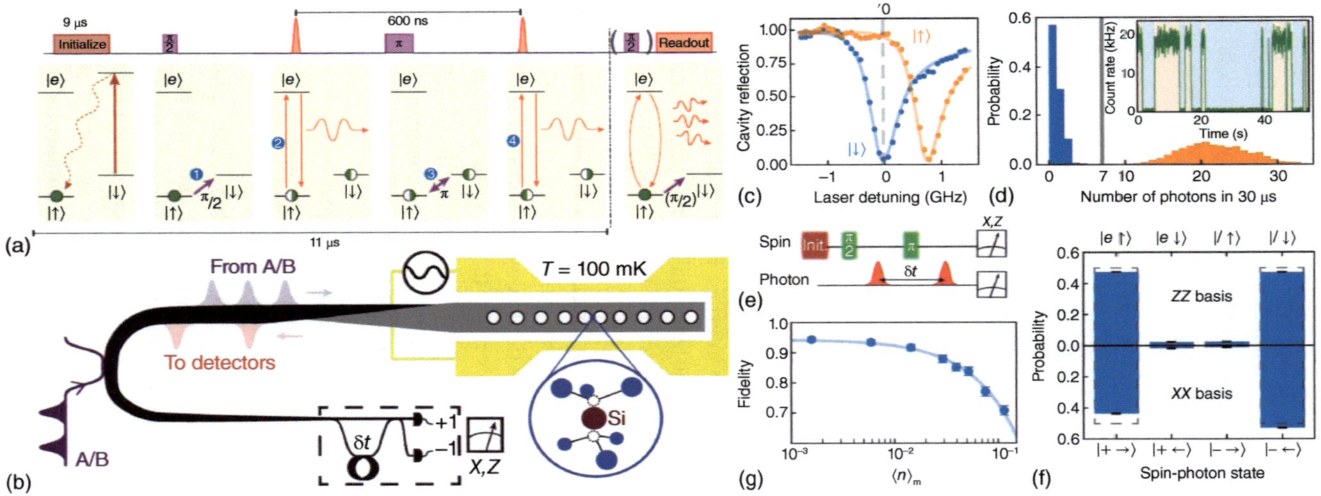

Figure 19.8 (a) Spin–photon entanglement generation scheme using the NV; (b) Schematic of spin-dependent cavity reflection scheme for writing a photonic time-bin qubit into an SiV spin qubit; (c) Spin-dependent cavity reflection measurements; (d) Single-shot thresholds for reading out the spin state of the SiV. (e) Control sequence to characterize spin photon entanglement. (f) Correlations between spin and time-bin photonic qubits. (g) Spin–photon entanglement fidelity measured using (e). Source: (a) Bernien et al. [232]/With permission of Springer Nature; (b-g) Bhaskar et al. [15]/With permission of Springer Nature.

Verifying the entanglement between the emitter and photon requires performing a tomographic reconstruction of the joint state, via individual measurements of the two qubits in different combinations of bases.

NV in Diamond Since the first demonstration of spin–photon entanglement from an NV, in 2010 [233], the last decade has shown a rapid advance in techniques enabling entanglement of multiple remote NVs and local register qubits [232, 234, 235], discussed further in Sections 19.3.3.2 and 19.3.3.3.

SiV in Diamond As an alternative strategy, Nguyen et al. [222, 236] used the spin-dependent reflection of an SiV critically coupled to a nanocavity to write a time-bin photonic qubit into the SiV. Interfacing with the SiV was achieved using adiabatic transfer between the waveguide and a tapered fiber. Targeted SiV incorporation with nanostructures can be achieved via masked implantation, together with gas tuning of the cavity, a cooperativity of $C \sim 38$ was achieved. The protocol, illustrated in Figure 19.8(b), follows the same principle as generating a spin/time-bin entangled state: the SiV is initialized in an equal superposition state. The time-bin encoded photonic qubit is then reflected off the cavity. A π-rotation is applied to the SiV in between the time-bins, which results in a spin-photon entangled state. After measuring the photonic qubit in an unbiased basis (removing time-bin information), using an imbalanced Mach–Zehnder interferometer, its state can be mapped to the SiV. Here, microwave control of the spin qubit was employed and showed $T_2 > 1.5$ ms using an XY DD sequence ($N = 64$). The observed coherence timescaling was found to be similar to NVs near a surface [237] differing from the scaling from an SiV in bulk [200], indicating that the proximity of the surface still influences the SiV properties. Subsequent improvements [15] enabled non-destructive spin-state readout with $> 99.9\%$ fidelity, in a cavity with a cooperativity of $C = 105$. A spin–photon entanglement fidelity after the first reflection of $\sim 94\%$ was reported. In the same series of works, coupling to an ancillary ^{13}C spin register was also explored, which allows access to longer-lived memory qubits, as well as potential for more complex multi-qubit protocols, which are discussed in Sections 19.3.3.2, 19.3.3.3, and 19.4.

19.3.3.2 Color Centers in Solid-State Multi-Qubit Registers

Even though nearby spins cause a potentially detrimental fluctuating magnetic environment, if controlled, they can also be used as additional long-lived qubits: Due to the difference in the nuclear- and Bohr magneton, compared to the electron spin, nuclear spins are significantly less affected by the same magnetic field and thus exhibit much longer coherence times. Using this property, a system of multiple coupled qubits may be used as a multi-qubit register for the implementation of quantum algorithms. In this scenario the color center acts as an optically active interface qubit, which can then be coupled to nearby spins that constitute weakly interacting ancilla qubits. Through performing DD sequences with varying waiting times, resonance with nuclear precession times reveals periodic revivals that indicate coherent coupling between spins. The latter method can resolve individual nuclei [238–240].

Coupling can be mediated through microwave pulse sequences tuned to the respective hyperfine coupling strengths.

An additional application of coupling with ancilla qubits has been demonstrated to enable improved read-out robustness through coherent feedback and repetitive measurement [241], conducted in room temperature nanodiamond. The scheme resulted in higher fidelity read-out in a regime where usually only 0.02 photons could be detected before the spin is depolarized. Controllable coupling to ancilla spin systems also offers enhancements in sensing protocols [242, 243].

NV in Diamond The NV electronic spin state determines the nuclear spin precession axis as demonstrated in [244], and thus can be used to conditionally manipulate the evolution of the nuclear spin. Alternatively, controllable interactions can be realized via state transfer by coupling the spin sublevels with a Landau-Zener crossing when sweeping the magnetic field [245] (achievable state transfer in ca. 120 ns). In either case, the basic idea is to perform an operation on the NV electron spin, then a swap operation between NV and register qubit, and reset the NV. When the electron is in the $m_s = 0$ state, the nuclear spin precesses at the Larmor frequency, whereas when in the $m_s = -1$ state, the hyperfine splitting prevents any precession.

Such techniques can be exploited to encode a logical qubit consisting of multiple physical qubits, for example in a decoherence free subspace composed of multiple physical qubits [246] with a T_2^* on the order of minutes.

In diamond, there are a number of possible ancilla qubits that may be used. The nearest ancilla is the intrinsic ^{14}N nuclear spin. More commonly, however, ^{13}C nuclei are employed, first proposed in 2001 [247], and demonstrated in 2007 [244]. Diamond has a natural abundance of 1.1% ^{13}C with $I = 1/2$, compared to ^{12}C with $I = 0$.

Coherent coupling to nearby spins is achieved through a combination of microwave pulse sequences with time-based gates [248]. Additionally, the control sequence needs to be interleaved with decoupling sequences, in order to decouple the target spins from the remaining spin bath. Since control of a ^{13}C nucleus mediated by the NV electron spin was first demonstrated in 2007 [244], subsequent improvements in fidelities, achieved through (i) state preparation using resonant pumping; and (ii) read-out with higher collection efficiencies using deterministic fabrication of a SIL [211], as well as control techniques such as dynamical decoupling have helped to enable scaling up the number of addressable ancilla qubits resulting in a ten-qubit spin register [217].

As well as ^{13}C nuclei, nearby NVs can also be used for entanglement generation, via performing a parity measurement. The latter has been demonstrated between two NV nuclear spins, mediated via optical control of the electron spin of one of the NVs [249].

More involved protocols have also been demonstrated, including error correction schemes [250, 251]; encoding of a logical qubit in a decoherence free subspace consisting of a pseudo-spin of a pair of ^{13}C nuclei $|\Uparrow\rangle = |\uparrow\downarrow\rangle$, $|\Downarrow\rangle = |\downarrow\uparrow\rangle$ obtaining $T_2^* = 1.9(3)$ min [246]; and recently, demonstrating a flag-fault-tolerant encoding with a logical qubit comprised of five ^{13}C nuclei and a flag qubit in the native ^{14}N spin, which were interfaced by the ancilla NV electron spin [252].

Additional candidates for ancilla qubits are *P*1 center electron spins, e.g. in [253] entanglement of two *P*1 centers' electron spins mediated by a NV center was demonstrated. Control of both *P*1 electron spin, and the ^{14}N nuclear spin was shown, with $T_2 = 1$ and 4 ms, respectively.

Adding a second optically addressable spin qubit to the register could enable advanced write and read-out schemes, and possibly allow for higher-fidelity operations. Toward this goal, NV-NV entanglement was demonstrated in [254] based on magnetic dipolar coupling of the NV's electron spins. Remarkably, the two-qubit entanglement generation was shown at room temperature. Entanglement fidelities were subsequently increased using optimal control [255].

In summary, to realize these multi-qubit registers, one requires high-fidelity read-out, decoupling from the residual spin bath, and the ability to perform high-fidelity gate operations between the interface and ancilla qubits. With these, color centers – such as the NV – can be used as an optically active interface node, with multiple memory qubits, which is, for instance, a fundamental ingredient for several proposals for generation of high-dimensional cluster states [256–258]. Furthermore, in addition to coherent control of individual qubits, multi-qubit gates are necessary for the realization of multi-qubit quantum algorithms. The ability to couple to a local, long-lived register of qubits is vital for the implementation of error correcting logical qubits [246, 252].

G4V in Diamond Coupling G4Vs to nuclear spins requires an alternative approach compared to using the NV. For weakly coupled spin-1/2 systems, the resonance conditions is, to first order, insensitive to the particular coupling terms between two spins (it is linear for the spin-1 NV system). However, once identified, more strongly coupled nuclei can be addressed. Initialization and read-out of a ^{13}C nuclear spin were first demonstrated using the SiV by Metsch et al. [259]. Here, for the choice of a magnetic field the orientation with respect to the SiV need to be considered (for spin initialization via pumping vs. spin state read-out as discussed in section 19.3.2), as well as the orientation with respect to the position of the nearby nuclear spins [236].

SiC The neutral divacancy in SiC is electronically similar to the NV^- in diamond, and analogous techniques can be used for coupling to ^{13}C and ^{29}Si nuclei. In [88], coupling to nuclear spins was demonstrated via electron spin polarization and a swap gate (see, e.g. [244]) with an estimated initialization fidelity of 93%. Similarly, but waveguide integrated, coherent coupling to nuclear spins as well as nuclear spin quantum gates with 95% fidelity were demonstrated [91].

19.3.3.3 Color Centers for Remote Spin–Spin Entanglement

Remote Spin–Spin Entanglement Having generated a photonic qubit entangled with the spin state of a quantum emitter, the emitted photons can then be transmitted across a communication channel (e.g. fiber link or free-space); and then, through a projective measurement on the photons, a fusion operation can be implemented that results in the entanglement of the two emitters. While linear operations on photons are "cheap," implementing photon interactions – necessary for logical operations – are challenging. With linear optics and photon detection this can only be achieved probabilistically [16], with successful events heralded by a particular combination of detection events. Although, the success probability can be boosted through the use of additional resources [260, 261].

A simple variant of this scheme implements a parity measurement between two input photon modes on a beam-splitter and can discriminate the antisymmetric singlet state $|\Psi^-\rangle = \frac{1}{\sqrt{2}}(|01\rangle - |10\rangle)$ from the symmetric triplet states, although this can readily be extended to also discern the state $|\Psi^-\rangle = \frac{1}{\sqrt{2}}(|01\rangle - |10\rangle)$, but is unable to distinguish the $|\Phi^{\pm}\rangle = \frac{1}{\sqrt{2}}(|00\rangle \pm |11\rangle)$ states. In all these cases, photon interference requires indistinguishable photons for both input modes, where the most obvious indistinguishability criterion is that they arrive at the same time (however, they have to be indistinguishable in all degrees of freedom).

The time-bin protocol based on Barrett and Kok [11] succeeds when both photons – one from each emitter – are detected, with one in each time-bin. A reduced scheme using only a single time-bin that, conditional on detecting *only one* photon, erases the which-path information provided that a second photon was not emitted and then lost [234]. The latter protocol is more efficient, even though it is more susceptible to loss; and additionally, requires phase stability of the paths between each emitter and detection to yield a well-defined state.

Current Records Although the first spin–photon entanglement was shown with polarization encoding [233], alternative encodings have since become more dominant. Spin–spin entanglement schemes with the NV in diamond typically now make use of either a time-bin encoding [232, 262] or photon/vacuum encoding [234, 235, 263].

The first demonstration of spin–spin entanglement with the NV using a two-photon scheme [232] was followed up with a scaled-up version in order to perform a loophole-free Bell test over a distance of 1.3 km [262] with an entanglement rate of one event per hour due to the loss overheads. Since then, later work moved to photon/vacuum encoding, in which only one photonic qubit is transmitted down the channel in a successful entanglement event. Furthermore, the choice of the initial spin qubit state can be selected in order to help balance loss and was used to demonstrate delivery of entanglement at a rate faster than decoherence [234]. This scheme was designed to deliver an entangled spin–spin pair after a fixed number of attempt clock cycles, where after heralding a successful entanglement event, the state was preserved via DD for the duration of the delivery clock cycle.

With the additional ability to couple to ancilla qubits in nuclear spins, these techniques were applied to demonstrations of entanglement distillation [263] and three-node entanglement [235].

In an alternative scheme that builds on the results of the spin-dependent reflection quantum memory scheme (illustrated in Figure 19.8(b)), the concept was then developed to implement an asynchronous Bell state measurement [15].

19.4 Proposals and Perspectives

19.4.1 Applications

Having a reliable single photon quantum emitter opens the door to wider quantum technology applications. Secure quantum communication is considered a near-term deliverable quantum technology. Some QKD protocols such as the BB84 protocol [18] only require the transmission of qubits between two parties (Alice and Bob). As a resource, the ability to generate a high-purity photon in a specific mode is required, and then, through preparation and read-out of the photonic qubit in two complementary bases a secure key can be shared when the preparation and measurement bases correspond to each other. However, transmitting a string of qubits through some channel will be subject to loss. In classical communication networks, this can be overcome through periodic amplifier stations. For a quantum state, however, the no-cloning theorem applies, which forbids the amplification of the state. Similarly, a "measure and prepare" strategy (measuring the qubit in some basis and preparing a new copy) would result in a fidelity limited to 50%.

To overcome the fundamental challenge of photon losses, a quantum repeater approach was proposed in [264], where a transmission channel was divided into several shorter segments that share entanglement, as illustrated in Figure 19.9. Since both the transmission and the entanglement operations are probabilistic, it is necessary that the stationary node qubits exhibit sufficiently long lifetimes, such that once entanglement is successfully established across one link, the state can be preserved for long enough so that neighboring links can also, with a high probability, be established within the memory's lifetime. Then, through entanglement swap operations, the entanglement can be distributed between distant nodes.

Once shared entanglement exists between two parties, a qubit can be transmitted via a teleportation operation, where the sender performs a joint measurement on the qubit to be sent and on half of the entangled pair. After communicating the two measurement outcomes (two classical bits), the quantum state can be faithfully retrieved at the other end of the channel. Recent theoretical work quantitatively analyzes achievable rates in such and similar quantum repeater schemes [266–268].

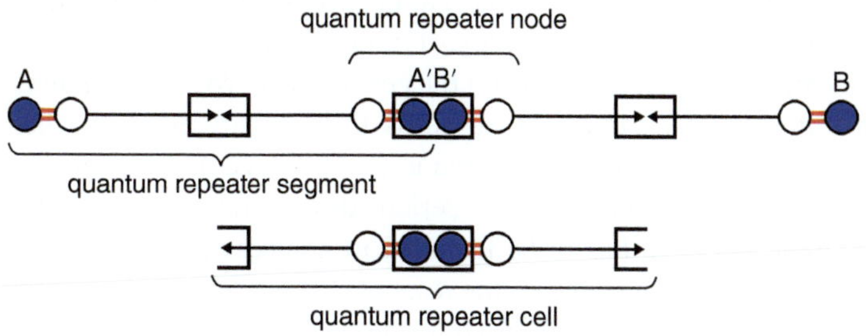

Figure 19.9 Quantum repeaters overcome photon losses by establishing pairwise entanglement links between repeater stations, which divide the communication path into shorter segments. Once all stations are linked, a quantum state can be teleported from Alice (A) to Bob (B) through a swap operation. Source: Loock et al. [265]/John Wiley & Sons/CC BY-4.0.

There also exist extensions to the traditional quantum repeater approach [269–271], which are based on the generation of highly entangled resource states, for example photonic cluster states, which are more tolerant to losses; however, the principle of breaking the channel into shorter links still applies. The generation of such large entangled resource states is not only relevant for quantum communication, but also finds application in quantum computation schemes, such as measurement-based quantum computation (MBQC) [272], and error correction [273]. Indeed, present research explores the application of such resource states for the development of quantum processors and quantum computers.

19.4.2 Directions and Remaining Challenges

Even though a lot of progress has been made to harness color centers for quantum applications, there are still challenges that need to be overcome to stand out from competing quantum architectures based on, for example, ultra-cold atoms [274], QD [275], ions [276], or transmons [277]. A major part of this challenge is scalability. Color centers are ideal candidates for a scalable quantum platform because they are isolated, optically addressable atom-like quantum systems in the solid state. As mentioned before, there are two prominent research strands under the umbrella of quantum information science, quantum computing, and quantum communication, which would immensely benefit from such a scalable resource. In both fields color centers act as a solid-state system that interacts with photonic degrees of freedom either for producing large photonic resource states [278], for measurement-based quantum computation, error correction [273], for one-way quantum repeaters [271] or by acting as a quantum memory for photonic qubits [15, 279]. The two most promising approaches (outlined in section 19.3.3) for facilitating such an interaction rely on the scattering or emission of photons from a color center coupled to a resonator, which is a necessary requirement to enhance the otherwise weak interaction of the color center with single photons. To achieve this in a scalable manner, color centers need to be integrated into photonic nanostructures. One of the biggest challenges for color centers with a lack of inversion symmetry in a nanostructure is spectral diffusion (see section 19.3.1), which is caused by charge fluctuations of its environment. This noise can be enhanced if the color center is close to surfaces [141] and will therefore require more sophisticated strategies to suppress its effect on the generation of coherent photons. Beyond spectral diffusion, the nanostructure needs to be highly efficient when coupling photons to the color center. Any photon loss will reduce the performance of the device. Minimizing photon loss poses a serious design challenge and every aspect of the photonic interface has to be included in a successful design. This includes the coupling of the color center to the resonator, the resonator waveguide coupling, and the coupling of a waveguide to a transmission fiber [173] or free space propagation [280].

In addition to the design of the photonic interfaces, the local control of the color centers can be advanced, making it faster, less prone to cause unwanted heating or require filtering. This includes novel resonant excitation techniques [281] as well as optical control of the spin [223]. Another promising path of investigation is the use of color centers as a quantum bus between even longer-lived memory spins [246] in the center's environment, which among other things enables fault tolerant operations on a qubit [252]. A recent experiment demonstrated multinode entanglement [235], already showing that color enters can act as nodes in a quantum network.

Finally, maintaining the coherence of a quantum memory or quantum register for arbitrarily large times would enable the permanent storage of quantum states, significantly expanding the use of quantum memories in quantum technologies: for example, for storing and retrieving a quantum token for secure identification [282], for storing unforgeable quantum money [283], for storing outcomes of quantum computation, or for enabling entanglement purification [284]. Recent 1-minute storage times [217, 252] demonstrate that by building even more complex quantum registers the permanent storage of quantum states will be feasible in the future.

In summary, even though many challenges remain, color centers have bright prospects for becoming a viable platform for future quantum applications, which require local quantum operations as well as remote quantum connections based both on spin and photonic quantum states, respectively. With an impressive rate of progress, large-scale quantum technological applications such as quantum networks and quantum information processing with color centers seem realistic future goals.

References

1 Ou, Z.Y., Pereira, S.F., Kimble, H.J., and Peng, K.C. (1992). Realization of the Einstein-Podolsky-Rosen paradox for continuous variables. *Physical Review Letters* 68 (25): 3663–3666.

2 Giustina, M., Mech, A., Ramelow, S. et al. (2013). Bell violation using entangled photons without the fair-sampling assumption. *Nature* 497 (7448): 227–230.

3 Christensen, B.G., McCusker, K.T., Altepeter, J.B. et al. (2013). Detection-loophole-free test of quantum nonlocality, and applications. *Physical Review Letters* 111 (13): 1–5.

4 Shalm, L.K., Meyer-Scott, E., Christensen, B.G. et al. (2015). Strong loophole-free test of local realism. *Physical Review Letters* 115 (25): 1–10.

5 Caves, C.M. (1981). Quantum-mechanical noise in an interferometer. *Physical Review D* 23 (8): 1693–1708.

6 Tse, M., Yu, H., Kijbunchoo, N. et al. (2019). Quantum-enhanced advanced LIGO detectors in the era of gravitational-wave astronomy. *Physical Review Letters* 123: 231107.

7 Liao, S.-K., Cai, W.-Q., Liu, W.-Y. et al. (2017). Satellite-to-ground quantum key distribution. *Nature* 549 (7670): 43–47.

8 Yin, J., Cao, Y., Li, Y.-H. et al. (2017). Satellite-to-ground entanglement-based quantum key distribution. *Physical Review Letters* 119 (20): 200501.

9 Gard, B.T., Motes, K.R., Olson, J.P. et al. (2014). An introduction to boson-sampling. In: *From Atomic to Mesoscale*, 167–192. World Scientific.

10 Knill, E., Laflamme, R., and Milburn, G.J. (2001). A scheme for effcient quantum computation with linear optics. *Nature* 409: 46–52.

11 Barrett, S. and Kok, P. (2005). Efficient high-fidelity quantum computation using matter qubits and linear optics. *Physical Review A* 71 (6): 060310.

12 Kok, P., Nemoto, K., Ralph, T.C. et al. (2007). Linear optical quantum computing with photonic qubits. *Reviews of Modern Physics* 79 (1): 135–174.

13 Martin, J. and Vennin, V. (2021). Real-space entanglement in the Cosmic Microwave Background. *Journal of Cosmology and Astroparticle Physics* 2021 (10): 036.

14 Li, M.-J. and Hayashi, T. (2020). Advances in low-loss, large-area, and multicore fibers. In: *Optical Fiber Telecommunications VII* (ed. A.E. Willner), 3–50. Elsevier.

15 Bhaskar, M.K., Riedinger, R., Machielse, B. et al. (2020). Experimental demonstration of memory-enhanced quantum communication. *Nature* 580 (7801): 60–64.

16 Lütkenhaus, N., Calsamiglia, J., and Suominen, K.-A. (1999). Bell measurements for teleportation. *Physical Review A* 59 (5): 3295–3300.

17 Kimble, H.J. (2008). The quantum internet. *Nature* 453 (7198): 1023–1030.

18 Bennett, C.H. and Brassard, G. (2014). Quantum cryptography: public key distribution and coin tossing. *Theoretical Computer Science* 560: 7–11.

19 Schwartz, I., Cogan, D., Schmidgall, E.R. et al. (2016). Deterministic generation of a cluster state of entangled photons. *Science* 354 (6311): 434–437.

20 Istrati, D., Pilnyak, Y., Loredo, J.C. et al. (2020). Sequential generation of linear cluster states from a single photon emitter. *Nature Communications* 11 (1): 5501.

21 Wang, X.L., Chen, L.K., Li, W. et al. (2016). Experimental ten-photon entanglement. *Physical Review Letters* 117 (21): 1–6.

22 Asavanant, W., Shiozawa, Y., Yokoyama, S. et al. (2019). Generation of time-domain-multiplexed two-dimensional cluster state. *Science* 366 (6463): 373–376.

23 Huthmacher, L., Stockill, R., Clarke, E. et al. (2018). Coherence of a dynamically decoupled quantum-dot hole spin. *Physical Review B* 97 (24): 241413.

24 Stockill, R., Le Gall, C., Matthiesen, C. et al. (2016). Quantum dot spin coherence governed by a strained nuclear environment. *Nature Communications* 7 (1): 12745.

25 Tilley, R.J.D. (2008). *Defects in Solids*. John Wiley & Sons.

26 Zaitsev, A.M. (2001). *Optical Properties of Diamond*. Berlin, Heidelberg: Springer-Verlag.

27 Gruber, A., Dräbenstedt, A., Tietz, C. et al. (1997). Scanning confocal optical microscopy and magnetic resonance on single defect centers. *Science* 276 (5321): 2012–2014.

28 Brouri, R., Beveratos, A., Poizat, J.-P., and Grangier, P. (2000). Photon antibunching in the fluorescence of individual color centers in diamond. *Optics Letters* 25 (17): 1294.

29 Kurtsiefer, C., Mayer, S., Zarda, P., and Weinfurter, H. (2000). Stable solid-state source of single photons. *Physical Review Letters* 85 (2): 290–293.

30 Neu, E., Steinmetz, D., Riedrich-Möller, J. et al. (2011). Single photon emission from silicon-vacancy colour centres in chemical vapour deposition nano-diamonds on iridium. *New Journal of Physics* 13 (2): 025012.

31 Görlitz, J., Herrmann, D., Thiering, G. et al. (2020). Spectroscopic investigations of negatively charged tin-vacancy centres in diamond. *New Journal of Physics* 22 (1): 013048.

32 Schröder, T., Mouradian, S.L., Zheng, J. et al. (2016). Quantum nanophotonics in diamond [Invited]. *Journal of the Optical Society of America B* 33 (4): B65.

33 Hughes, L. et al. (2023). *APL Materials* 11: 021101.

34 Ohno, K., Heremans, F.J., Bassett, L.C. et al. (2012). Engineering shallow spins in diamond with nitrogen delta-doping. *Applied Physics Letters* 101 (8): 082413.

35 Acosta, V.M., Bauch, E., Ledbetter, M.P. et al. (2009). Diamonds with a high density of nitrogen-vacancy centers for magnetometry applications. *Physical Review B - Condensed Matter and Materials Physics* 80 (11): 1–15.

36 Lühmann, T., John, R., Wunderlich, R. et al. (2019). Coulomb-driven single defect engineering for scalable qubits and spin sensors in diamond. *Nature Communications* 10 (1): 4956.

37 van Dam, S.B., Walsh, M., Degen, M.J. et al. (2019). Optical coherence of diamond nitrogen-vacancy centers formed by ion implantation and annealing. *Physical Review B* 99 (16): 161203.

38 van Oort, E., Manson, N.B., and Glasbeek, M. (1988). Optically detected spin coherence of the diamond N-V centre in its triplet ground state. *Journal of Physics C: Solid State Physics* 21 (23): 4385–4391.

39 Grinolds, M.S., Hong, S., Maletinsky, P. et al. (2013). Nanoscale magnetic imaging of a single electron spin under ambient conditions. *Nature Physics* 9 (4): 215–219.

40 Schirhagl, R., Chang, K., Loretz, M., and Degen, C.L. (2014). Nitrogen-vacancy centers in diamond: nanoscale sensors for physics and biology. *Annual Review of Physical Chemistry* 65: 83–105.

41 Jeske, J., Cole, J.H., and Greentree, A.D. (2016). Laser threshold magnetometry. *New Journal of Physics* 18 (1): 013015.

42 Clevenson, H., Trusheim, M.E., Teale, C. et al. (2015). Broadband magnetometry and temperature sensing with a light-trapping diamond waveguide. *Nature Physics* 11 (5): 393–397.

43 Batalov, A., Jacques, V., Kaiser, F. et al. (2009). Low temperature studies of the excited-state structure of negatively charged nitrogen-vacancy color centers in diamond. *Physical Review Letters* 102 (19): 195506.

44 Siyushev, P., Pinto, H., Vörös, M. et al. (2013). Optically controlled switching of the charge state of a single nitrogen-vacancy center in diamond at cryogenic temperatures. *Physical Review Letters* 110: 167402.

45 Rogers, L.J., Armstrong, S., Sellars, M.J., and Manson, N.B. (2008). Infrared emission of the NV centre in diamond: Zeeman and uniaxial stress studies. *New Journal of Physics* 10 (10): 103024.

46 Doherty, M.W., Manson, N.B., Delaney, P. et al. (2013). The nitrogen-vacancy colour centre in diamond. *Physics Reports* 528 (1): 1–45.

47 Doherty, M.W., Manson, N.B., Delaney, P., and Hollenberg, L.C.L. (2011). The negatively charged nitrogen-vacancy centre in diamond: the electronic solution. *New Journal of Physics* 13 (2): 025019.

48 Maze, J.R., Gali, A., Togan, E. et al. (2011). Properties of nitrogen-vacancy centers in diamond: the group theoretic approach. *New Journal of Physics* 13 (2): 025025.

49 Manson, N., Harrison, J., and Sellars, M. (2006). Nitrogen-vacancy center in diamond: model of the electronic structure and associated dynamics. *Physical Review B* 74 (10): 104303.

50 Beha, K., Batalov, A., Manson, N.B. et al. (2012). Optimum photoluminescence excitation and recharging cycle of single nitrogen-vacancy centers in ultrapure diamond. *Physical Review Letters* 109 (9): 097404.

51 De Santis, L., Trusheim, M.E., Chen, K.C., and Englund, D.R. (2021). Investigation of the stark effect on a centrosymmetric quantum emitter in diamond. *Physical Review Letters* 127 (14): 147402.

52 Thiering, G. and Gali, A. (2018). Ab initio magneto-optical spectrum of group-IV vacancy color centers in diamond. *Physical Review X* 8 (2): 021063.

53 Hepp, C., Müller, T., Waselowski, V. et al. (2014). Electronic structure of the silicon vacancy color center in diamond. *Physical Review Letters* 112 (3): 036405.

54 Iwasaki, T., Miyamoto, Y., Taniguchi, T. et al. (2017). Tin-vacancy quantum emitters in diamond. *Physical Review Letters* 119 (25): 1–6.

55 Bradac, C., Gao, W., Forneris, J. et al. (2019). Quantum nanophotonics with group IV defects in diamond. *Nature Communications* 10 (1): 1–13.

56 Becker, J.N., Pingault, B., Groß, D. et al. (2018). All-optical control of the silicon-vacancy spin in diamond at millikelvin temperatures. *Physical Review Letters* 120 (5): 053603.

57 Wang, C., Kurtsiefer, C., Weinfurter, H., and Burchard, B. (2006). Single photon emission from SiV centres in diamond produced by ion implantation. *Journal of Physics B: Atomic, Molecular and Optical Physics* 39 (1): 37–41.

58 Neu, E., Hepp, C., Hauschild, M. et al. (2013). Low-temperature investigations of single silicon vacancy colour centres in diamond. *New Journal of Physics* 15 (4): 043005.

59 Müller, T., Hepp, C., Pingault, B. et al. (2014). Optical signatures of silicon-vacancy spins in diamond. *Nature Communications* 5 (1): 3328.

60 Palyanov, Y.N., Kupriyanov, I.N., Borzdov, Y.M., and Surovtsev, N.V. (2015). Germanium: a new catalyst for diamond synthesis and a new optically active impurity in diamond. *Scientific Reports* 5: 1–8.

61 Iwasaki, T., Ishibashi, F., Miyamoto, Y. et al. (2015). Germanium-vacancy single color centers in diamond. *Scientific Reports* 5 (1): 12882.

62 Siyushev, P., Metsch, M.H., Ijaz, A. et al. (2017). Optical and microwave control of germanium-vacancy center spins in diamond. *Physical Review B* 96 (8): 1–5.

63 Bhaskar, M.K., Sukachev, D.D., Sipahigil, A. et al. (2017). Quantum nonlinear optics with a germanium-vacancy color center in a nanoscale diamond waveguide. *Physical Review Letters* 118 (22): 223603.

64 Tchernij, S.D., Herzig, T., Forneris, J. et al. (2017). Single-photon-emitting optical centers in diamond fabricated upon Sn implantation. *ACS Photonics* 4 (10): 2580–2586.

65 Corte, E., Sachero, S., Tchernij, S.D. et al. (2021). Spectral emission dependence of tin-vacancy centers in diamond from thermal processing and chemical functionalization. *Advanced Photonics Research* 3 (1): 2100148.

66 Aharonovich, I., Castelletto, S., Simpson, D.A. et al. (2010). Photophysics of chromium-related diamond single-photon emitters. *Physical Review A* 81 (4): 043813.

67 Deshko, Y. and Gorokhovsky, A.A. (2010). Spectroscopy and micro-luminescence mapping of Xe-implanted defects in diamond. *Low Temperature Physics* 36 (5): 465–471.

68 Gaebel, T., Popa, I., Gruber, A. et al. (2004). Stable single-photon source in the near infrared. *New Journal of Physics* 6: 98–98.

69 Siyushev, P., Jacques, V., Aharonovich, I. et al. (2009). Low-temperature optical characterization of a near-infrared single-photon emitter in nanodiamonds. *New Journal of Physics* 11 (11): 113029.

70 Wu, E., Jacques, V., Zeng, H. et al. (2006). Narrow-band single-photon emission in the near infrared for quantum key distribution. *Optics Express* 14 (3): 1296.

71 Lee, S.-Y., Widmann, M., Rendler, T. et al. (2013). Readout and control of a single nuclear spin with a metastable electron spin ancilla. *Nature Nanotechnology* 8 (7): 487–492.

72 Castelletto, S. and Boretti, A. (2020). Silicon carbide color centers for quantum applications. *Journal of Physics: Photonics* 2 (2): 022001.

73 Liu, G., Tuttle, B.R., and Dhar, S. (2015). Silicon carbide: a unique platform for metal-oxide-semiconductor physics. *Applied Physics Reviews* 2 (2): 021307.

74 Xing, P., Ma, D., Ooi, K.J.A. et al. (2019). CMOS-compatible PECVD silicon carbide platform for linear and nonlinear optics. *ACS Photonics* 6 (5): 1162–1167.

75 Yi, A., Zheng, Y., Huang, H. et al. (2020). Wafer-scale 4H-silicon carbide-on-insulator (4H–SiCOI) platform for nonlinear integrated optical devices. *Optical Materials* 107: 109990.

76 Castelletto, S., Almutairi, A.F.M., Kumagai, K. et al. (2018). Photoluminescence in hexagonal silicon carbide by direct femtosecond laser writing. *Optics Letters* 43 (24): 6077–6080.

77 Chen, Y.-C., Salter, P.S., Niethammer, M. et al. (2019). Laser writing of scalable single color centers in silicon carbide. *Nano Letters* 19 (4): 2377–2383.

78 Pelc, J.S., Langrock, C., Zhang, Q., and Fejer, M.M. (2010). Influence of domain disorder on parametric noise in quasi-phase-matched quantum frequency converters. *Optics Letters* 35 (16): 2804.

79 Son, N.T., Anderson, C.P., Bourassa, A. et al. (2020). Developing silicon carbide for quantum spintronics. *Applied Physics Letters* 116 (19): 190501.

80 Lukin, D.M., Guidry, M.A., and Vučković, J. (2020). Integrated quantum photonics with silicon carbide: challenges and prospects. *PRX Quantum* 1 (2): 1–19.

81 Majety, S., Saha, P., Norman, V.A., and Radulaski, M. (2022). Quantum information processing with integrated silicon carbide photonics. *Journal of Applied Physics* 131 (13): 130901.

82 Castelletto, S., Peruzzo, A., Bonato, C. et al. (2022). Silicon carbide photonics bridging quantum technology. *ACS Photonics* 9 (5): 1434–1457.

83 Christle, D.J., Klimov, P.V., de las Casas, C.F. et al. (2017). Isolated spin qubits in SiC with a high-fidelity infrared spin-to-photon interface. *Physical Review X* 7 (2): 021046.

84 Anderson, C.P., Glen, E.O., Zeledon, C. et al. (2022). Five-second coherence of a single spin with single-shot readout in silicon carbide. *Science Advances* 8 (5): eabm5912.

85 Anderson, C.P., Bourassa, A., Miao, K.C. et al. (2019). Electrical and optical control of single spins integrated in scalable semiconductor devices. *Science* 366 (6470): 1225–1230.

86 Whiteley, S.J., Wolfowicz, G., Anderson, C.P. et al. (2019). Spin-phonon interactions in silicon carbide addressed by Gaussian acoustics. *Nature Physics* 15 (5): 490–495.

87 Falk, A.L., Klimov, P.V., Buckley, B.B. et al. (2014). Electrically and mechanically tunable electron spins in silicon carbide color centers. *Physical Review Letters* 112 (18): 187601.

88 Bourassa, A., Anderson, C.P., Miao, K.C. et al. (2020). Entanglement and control of single nuclear spins in isotopically engineered silicon carbide. *Nature Materials* 19 (12): 1319–1325.

89 Morioka, N., Babin, C., Nagy, R. et al. (2020). Spin-controlled generation of indistinguishable and distinguishable photons from silicon vacancy centres in silicon carbide. *Nature Communications* 11 (1): 2516.

90 Banks, H.B., Soykal, O.O., Myers-Ward, R.L. et al. (2019). Resonant optical spin initialization and readout of single silicon vacancies in 4*H*-SiC. *Physical Review Applied* 11 (2): 024013.

91 Babin, C., Stöhr, R., Morioka, N. et al. (2022). Fabrication and nanophotonic waveguide integration of silicon carbide colour centres with preserved spin-optical coherence. *Nature Materials* 21 (1): 67–73.

92 Lukin, D.M., Guidry, M.A., Yang, J. et al. (2022). Optical superradiance of a pair of color centers in an integrated silicon-carbide-on-insulator microresonator. arXiv:2202.04845 [physics, physics:quant-ph].

93 von Bardeleben, H.J., Cantin, J.-L., Gerstmann, U. et al. (2021). Spin polarization, electron–phonon coupling, and zero-phonon line of the NV center in 3*C*-SiC. *Nano Letters* 21 (19): 8119–8125.

94 Morfa, A.J., Gibson, B.C., Karg, M. et al. (2012). Single-photon emission and quantum characterization of zinc oxide defects. *Nano Letters* 12 (2): 949–954.

95 Neitzke, O., Morfa, A., Wolters, J. et al. (2015). Investigation of line width narrowing and spectral jumps of single stable defect centers in ZnO at cryogenic temperature. *Nano Letters* 15 (5): 3024–3029.

96 Linpeng, X., Viitaniemi, M.L.K., Vishnuradhan, A. et al. (2018). Coherence properties of shallow donor qubits in ZnO. *Physical Review Applied* 10 (6): 064061.

97 Berhane, A.M., Jeong, K.-Y., Bodrog, Z. et al. (2017). Bright room-temperature single-photon emission from defects in gallium nitride. *Advanced Materials* 29 (12): 1605092.

98 Lyons, J.L. and Van de Walle, C.G. (2017). Computationally predicted energies and properties of defects in GaN. *npj Computational Materials* 3 (1): 12.

99 Zhou, Y., Wang, Z., Rasmita, A. et al. (2018). Room temperature solid-state quantum emitters in the telecom range. *Science Advances* 4 (3): aar3580.

100 Chakraborty, C., Kinnischtzke, L., Goodfellow, K.M. et al. (2015). Voltage-controlled quantum light from an atomically thin semiconductor. *Nature Nanotechnology* 10 (6): 507–511.

101 He, Y.-M., Clark, G., Schaibley, J.R. et al. (2015). Single quantum emitters in monolayer semiconductors. *Nature Nanotechnology* 10 (6): 497–502.

102 Koperski, M., Nogajewski, K., Arora, A. et al. (2015). Single photon emitters in exfoliated WSe_2 structures. *Nature Nanotechnology* 10 (6): 503–506.

103 Srivastava, A., Sidler, M., Allain, A.V. et al. (2015). Optically active quantum dots in monolayer WSe_2. *Nature Nanotechnology* 10 (6): 491–496.

104 Tonndorf, P., Schmidt, R., Schneider, R. et al. (2015). Single-photon emission from localized excitons in an atomically thin semiconductor. *Optica* 2 (4): 347.

105 Chakraborty, C., Goodfellow, K.M., and Vamivakas, A.N. (2016). Localized emission from defects in $MoSe_2$ layers. *Optical Materials Express* 6 (6): 2081.

106 Tran, T.T., Bray, K., Ford, M.J. et al. (2016). Quantum emission from hexagonal boron nitride monolayers. *Nature Nanotechnology* 11 (1): 37–41.

107 Jungwirth, N.R., Calderon, B., Ji, Y. et al. (2016). Temperature dependence of wavelength selectable zero-phonon emission from single defects in hexagonal boron nitride. *Nano Letters* 16 (10): 6052–6057.

108 Chejanovsky, N., Rezai, M., Paolucci, F. et al. (2016). Structural attributes and photodynamics of visible spectrum quantum emitters in hexagonal boron nitride. *Nano Letters* 16 (11): 7037–7045.

109 Abdi, M., Chou, J.-P., Gali, A., and Plenio, M.B. (2018). *Color centers in hexagonal boron nitride monolayers: a group theory and ab initio analysis.* 5 (5): 1967–1976.

110 Tittel, W., Afzelius, M., Chaneliére, T. et al. (2010). Photon-echo quantum memory in solid state systems. *Laser & Photonics Reviews* 4 (2): 244–267.

111 Degen, C.L., Reinhard, F., and Cappellaro, P. (2017). Quantum sensing. *Reviews of Modern Physics* 89 (3): 035002.

112 Dibos, A.M., Raha, M., Phenicie, C.M., and Thompson, J.D. (2018). Atomic source of single photons in the telecom band. *Physical Review Letters* 120 (24): 243601.

113 Zhong, T., Kindem, J.M., Bartholomew, J.G. et al. (2018). Optically addressing single rare-earth ions in a nanophotonic cavity. *Physical Review Letters* 121 (18): 183603.

114 Kindem, J.M., Ruskuc, A., Bartholomew, J.G. et al. (2020). Control and single-shot readout of an ion embedded in a nanophotonic cavity. *Nature* 580 (7802): 201–204.

115 Chen, S., Raha, M., Phenicie, C.M. et al. (2020). Parallel single-shot measurement and coherent control of solid-state spins below the diffraction limit. *Science* 370 (6516): 592–595.

116 Kunkel, N. and Goldner, P. (2018). Recent advances in rare earth doped inorganic crystalline materials for quantum information processing. *Zeitschrift für anorganische und allgemeine Chemie* 644 (2): 66–76.

117 Zhong, T. and Goldner, P. (2019). Emerging rare-earth doped material platforms for quantum nanophotonics. *Nanophotonics* 8 (11): 2003–2015.

118 Fox, M. (2006). *Quantum Optics: An Introduction*. Oxford, UK: Oxford University Press, Inc.

119 Purcell, E.M. (1946). Spontaneous emission probabilities at radio frequencies. *Physical Review* 69 (11–12): 681.

120 Luo, L., Hayes, D., Manning, Ta. et al. (2009). Protocols and techniques for a scalable atom-photon quantum network. *Fortschritte der Physik* 57 (11–12): 1133–1152.

121 J. Bopp et al. https://arxiv.org/abs/2210.04702

122 Joannopoulos, J.D., Johnson, S.G., Winn, J.N., and Meade, R.D. (2011). *Photonic Crystals*, 2e. Princeton University Press.

123 Li, L., Schröder, T., Chen, E.H. et al. (2015). Coherent spin control of a nanocavity-enhanced qubit in diamond. *Nature Communications* 6 (1): 6173.

124 Faraon, A., Santori, C., Huang, Z. et al. (2012). Coupling of nitrogen-vacancy centers to photonic crystal cavities in monocrystalline diamond. *Physical Review Letters* 109: 033604.

125 Torun, C.G., Schneider, P.-I., Hammerschmidt, M. et al. (2021). Optimized diamond inverted nanocones for enhanced color center to fiber coupling. *Applied Physics Letters* 118 (23): 234002.

126 Hadden, J.P., Harrison, J.P., Stanley-Clarke, A.C. et al. (2010). Strongly enhanced photon collection from diamond defect centers under microfabricated integrated solid immersion lenses. *Applied Physics Letters* 97 (24): 241901.

127 Babinec, T.M., Hausmann, B.J.M., Khan, M. et al. (2010). A diamond nanowire single-photon source. *Nature Nanotechnology* 5 (3): 195–199.

128 Wan, N.H., Shields, B.J., Kim, D. et al. (2018). Efficient extraction of light from a nitrogen-vacancy center in a diamond parabolic reflector. *Nano Letters* 18 (5): 2787–2793.

129 Li, L., Chen, E.H., Zheng, J. et al. (2015). Efficient photon collection from a nitrogen vacancy center in a circular bullseye grating. *Nano Letters* 15 (3): 1493–1497.

130 Mouradian, S.L., Schröder, T., Poitras, C.B. et al. (2015). Scalable integration of long-lived quantum memories into a photonic circuit. *Physical Review X* 5: 031009.

131 Schreck, M., Gsell, S., Brescia, R., and Fischer, M. (2017). Ion bombardment induced buried lateral growth: the key mechanism for the synthesis of single crystal diamond wafers. *Scientific Reports* 7 (1): 44462.

132 Gaathon, O., Hodges, J.S., Chen, E.H. et al. (2013). Planar fabrication of arrays of ion-exfoliated single-crystal-diamond membranes with nitrogen-vacancy color centers. *Optical Materials* 35 (3): 361–365.

133 Lee, J.C., Magyar, A.P., Bracher, D.O. et al. (2013). Fabrication of thin diamond membranes for photonic applications. *Diamond and Related Materials* 33: 45–48.

134 Li, L., Bayn, I., Lu, M. et al. (2015). Nanofabrication on unconventional substrates using transferred hard masks. *Scientific Reports* 5: 7802.

135 Burek, M.J., De Leon, N.P., Shields, B.J. et al. (2012). Free-standing mechanical and photonic nanostructures in single-crystal diamond. *Nano Letters* 12 (12): 6084–6089.

136 Bayn, I., Meyler, B., Salzman, J., and Kalish, R. (2011). Triangular nanobeam photonic cavities in single-crystal diamond. *New Journal of Physics* 13 (2): 025018.

137 Bayn, I., Mouradian, S., Li, L. et al. (2014). Fabrication of triangular nanobeam waveguide networks in bulk diamond using single-crystal silicon hard masks. *Applied Physics Letters* 105 (21): 211101.

138 Khanaliloo, B., Mitchell, M., Hryciw, A.C., and Barclay, P.E. (2015). High-*Q/V* monolithic diamond microdisks fabricated with quasi-isotropic etching. *Nano Letters* 15 (8): 5131–5136.

139 Mouradian, S., Wan, N.H., Schröder, T., and Englund, D. (2017). Rectangular photonic crystal nanobeam cavities in bulk diamond. *Applied Physics Letters* 111 (2): 021103.

140 Riedrich-Möller, J., Kipfstuhl, L., Hepp, C. et al. (2012). One-and two-dimensional photonic crystal microcavities in single crystal diamond. *Nature Nanotechnology* 7 (1): 69–74.

141 Orphal-Kobin, L., Unterguggenberger, K., Pregnolato, T. et al. (2022). Optically coherent nitrogen-vacancy defect centers in diamond nanostructures. arXiv:2203.05605. arXiv:2203.05605v1. Accepted in Physical Review X.

142 Wan, N.H., Lu, T.-J., Chen, K.C. et al. (2020). Large-scale integration of artificial atoms in hybrid photonic circuits. *Nature* 583 (7815): 226–231.

143 Liu, J., Davanço, M.I., Sapienza, L. et al. (2017). Cryogenic photoluminescence imaging system for nanoscale positioning of single quantum emitters. *Review of Scientific Instruments* 88 (2): 023116.

144 Pregnolato, T., Chu, X.-L., Schröder, T. et al. (2020). Deterministic positioning of nanophotonic waveguides around single self-assembled quantum dots. *APL Photonics* 5 (8): 086101.

145 Di Cioccio, L., Letertre, F., Le Tiec, Y. et al. (1997). Silicon carbide on insulator formation by the Smart-Cut® process. *Materials Science and Engineering: B* 46 (1): 349–356.

146 Zheng, Y., Pu, M., Yi, A. et al. (2019). High-quality factor, high-confinement microring resonators in 4H-silicon carbide-on-insulator. *Optics Express* 27 (9): 13053.

147 Fan, T., Wu, X., Eftekhar, A.A. et al. (2020). High-quality integrated microdisk resonators in the visible-to-near-infrared wavelength range on a 3C-silicon carbide-on-insulator platform. *Optics Letters* 45 (1): 153.

148 Guidry, M.A., Yang, K.Y., Lukin, D.M. et al. (2020). Optical parametric oscillation in silicon carbide nanophotonics. *Optica* 7 (9): 1139.

149 Nagy, R., Dasari, D.B.R., Babin, C. et al. (2021). Narrow inhomogeneous distribution of spin-active emitters in silicon carbide. *Applied Physics Letters* 118 (14): 144003.

150 Andersen, U.L., Gehring, T., Marquardt, C., and Leuchs, G. (2016). 30 years of squeezed light generation. *Physica Scripta* 91 (5): 053001.

151 Takeuchi, S. (2014). Recent progress in single-photon and entangled-photon generation and applications. 53 (3): 030101.

152 Huang, K. and Rhys, A. (2000). Theory of light absorption and non-radiative transitions in F-centres. In: *Selected Papers of Kun Huang, World Scientific Series in 20th Century Physics*, vol. 23, pp. 74–92. World Scientific.

153 Alkauskas, A., Buckley, B.B., Awschalom, D.D., and Van de Walle, C.G. (2014). First-principles theory of the luminescence lineshape for the triplet transition in diamond NV centres. *New Journal of Physics* 16 (7): 073026.

154 Loudon, R. (2004). *The Quantum Theory of Light*, 3e. Oxford: Oxford Science Publications. Oxford University Press.

155 Brown, R.H. and Twiss, R.Q. (1994). Correlation between photons in two coherent beams of light. *Journal of Astrophysics and Astronomy* 15 (1): 13–19.

156 Berthel, M., Mollet, O., Dantelle, G. et al. (2015). Photophysics of single nitrogen-vacancy centers in diamond nanocrystals. *Physical Review B* 91 (3): 035308.

157 Beveratos, A., Brouri, R., Gacoin, T. et al. (2002). Single photon quantum cryptography. *Physical Review Letters* 89 (18): 187901.

158 Alléaume, R., Treussart, F., Messin, G. et al. (2004). Experimental open-air quantum key distribution with a single-photon source. *New Journal of Physics* 6: 92.

159 Leifgen, M., Schröder, T., Gädeke, F. et al. (2014). Evaluation of nitrogen- and silicon-vacancy defect centres as single photon sources in quantum key distribution. *New Journal of Physics* 16 (2): 023021.

160 Hong, C.K., Ou, Z.Y., and Mandel, L. (1987). Measurement of subpicosecond time intervals between two photons by interference. *Physical Review Letters* 59 (18): 2044–2046.

161 Acosta, V.M., Santori, C., Faraon, A. et al. (2012). Dynamic stabilization of the optical resonances of single nitrogen-vacancy centers in diamond. *Physical Review Letters* 108 (20): 206401.

162 Wolters, J., Sadzak, N., Schell, A.W. et al. (2013). Measurement of the ultrafast spectral diffusion of the optical transition of nitrogen vacancy centers in nano-size diamond using correlation interferometry. *Physical Review Letters* 110 (2): 027401.

163 Görlitz, J., Herrmann, D., Fuchs, P. et al. (2022). Coherence of a charge stabilised tin-vacancy spin in diamond. *npj Quantum Information* 8 (1): 1–9.

164 Lekavicius, I., Oo, T., and Wang, H. (2019). Diamond Lamb wave spin-mechanical resonators with optically coherent nitrogen vacancy centers. *Journal of Applied Physics* 126 (21): 214301.

165 Ruf, M., IJspeert, M., van Dam, S. et al. (2019). Optically coherent nitrogen-vacancy centers in micrometer-thin etched diamond membranes. *Nano Letters* 19 (6): 3987–3992.

166 Sangtawesin, S., Dwyer, B.L., Srinivasan, S. et al. (2019). Origins of diamond surface noise probed by correlating single-spin measurements with surface spectroscopy. *Physical Review X* 9: 031052.

167 Fotso, H.F., Feiguin, A.E., Awschalom, D.D., and Dobrovitski, V.V. (2016). Suppressing spectral diffusion of emitted photons with optical pulses. *Physical Review Letters* 116 (3): 033603.

168 Martín-Sánchez, J., Trotta, R., Mariscal, A. et al. (2018). Strain-tuning of the optical properties of semiconductor nanomaterials by integration onto piezoelectric actuators. *Semiconductor Science and Technology* 33 (1): 013001.

169 Meesala, S., Sohn, Y.-I., Pingault, B. et al. (2018). Strain engineering of the silicon-vacancy center in diamond. *Physical Review B* 97 (20): 205444.

170 Sipahigil, A., Goldman, M.L., Togan, E. et al. (2012). Quantum interference of single photons from remote nitrogen-vacancy centers in diamond. *Physical Review Letters* 108 (14): 143601.

171 Bernien, H., Childress, L., Robledo, L. et al. (2012). Two-photon quantum interference from separate nitrogen vacancy centers in diamond. *Physical Review Letters* 108 (4): 043604.

172 Tchebotareva, A., Hermans, S.L.N., Humphreys, P.C. et al. (2019). Entanglement between a diamond spin qubit and a photonic time-bin qubit at telecom wavelength. *Physical Review Letters* 123 (6): 063601.

173 Stolk, A., van der Enden, K.L., Roehsner, M.-C. et al. (2022). Telecom-band quantum interference of frequency-converted photons from remote detuned NV centers. *Physical Review X Quantum* 3: 020359.

174 Rogers, L.J., Jahnke, K.D., Teraji, T. et al. (2014). Multiple intrinsically identical single-photon emitters in the solid state. *Nature Communications* 5 (1): 4739.

175 Sipahigil, A., Jahnke, K.D., Rogers, L.J. et al. (2014). Indistinguishable photons from separated silicon-vacancy centers in diamond. *Physical Review Letters* 113 (11): 113602.

176 Li, K., Zhou, Y., Rasmita, A. et al. (2016). Nonblinking emitters with nearly lifetime-limited linewidths in CVD nanodiamonds. *Physical Review Applied* 6 (2): 024010.

177 Jantzen, U., Kurz, A.B., Rudnicki, D.S. et al. (2016). Nanodiamonds carrying silicon-vacancy quantum emitters with almost lifetime-limited linewidths. *New Journal of Physics* 18 (7): 073036.
178 Evans, R.E., Sipahigil, A., Sukachev, D.D. et al. (2016). Narrow-linewidth homogeneous optical emitters in diamond nanostructures via silicon ion implantation. *Physical Review Applied* 5 (4): 1–8.
179 Schröder, T., Trusheim, M.E., Walsh, M. et al. (2017). Scalable focused ion beam creation of nearly lifetime-limited single quantum emitters in diamond nanostructures. *Nature Communications* 8: 15376.
180 Sun, S., Zhang, J.L., Fischer, K.A. et al. (2018). Cavity-enhanced raman emission from a single color center in a solid. *Physical Review Letters* 121 (8): 083601.
181 Machielse, B., Bogdanovic, S., Meesala, S. et al. (2019). Quantum interference of electromechanically stabilized emitters in nanophotonic devices. *Physical Review X* 9 (3): 031022.
182 Fan, J.-W., Cojocaru, I., Becker, J. et al. (2018). Germanium-vacancy color center in diamond as a temperature sensor. *ACS Photonics* 5 (3): 765–770.
183 Trusheim, M.E., Pingault, B., Wan, N.H. et al. (2020). Transform-limited photons from a coherent tin-vacancy spin in diamond. *Physical Review Letters* 124 (2): 023602.
184 Rugar, A.E., Dory, C., Sun, S., and Vučković, J. (2019). Characterization of optical and spin properties of single tin-vacancy centers in diamond nanopillars. *Physical Review B* 99 (20): 205417.
185 Rugar, A.E., Lu, H., Dory, C. et al. (2020). Generation of tin-vacancy centers in diamond via shallow ion implantation and subsequent diamond overgrowth. *Nano Letters* 20 (3): 1614–1619.
186 Rugar, A.E., Dory, C., Aghaeimeibodi, S. et al. (2020). Narrow-linewidth tin-vacancy centers in a diamond waveguide. *ACS Photonics* 7 (9): 2356–2361.
187 Rugar, A.E., Aghaeimeibodi, S., Riedel, D. et al. (2021). Quantum photonic interface for tin-vacancy centers in diamond. *Physical Review X* 11 (3): 1–7.
188 Kuruma, K., Pingault, B., Chia, C. et al. (2021). Coupling of a single tin-vacancy center to a photonic crystal cavity in diamond. *Applied Physics Letters* 118 (23): 1–12.
189 Aghaeimeibodi, S., Riedel, D., Rugar, A.E. et al. (2021). Electrical tuning of tin-vacancy centers in diamond. *Physical Review Applied* 15 (6): 064010.
190 Castelletto, S., Johnson, B.C., Ivády, V. et al. (2014). A silicon carbide room-temperature single-photon source. *Nature Materials* 13 (2): 151–156.
191 Widmann, M., Lee, S.-Y., Rendler, T. et al. (2015). Coherent control of single spins in silicon carbide at room temperature. *Nature Materials* 14 (2): 164–168.
192 Fuchs, F., Stender, B., Trupke, M. et al. (2015). Engineering near-infrared single-photon emitters with optically active spins in ultrapure silicon carbide. *Nature Communications* 6 (1): 7578.
193 Christle, D.J., Falk, A.L., Andrich, P. et al. (2015). Isolated electron spins in silicon carbide with millisecond coherence times. *Nature Materials* 14 (2): 160–163.
194 Lienhard, B., Schröder, T., Mouradian, S. et al. (2016). Bright and photostable single-photon emitter in silicon carbide. *Optica* 3 (7): 768.

195 Udvarhelyi, P., Thiering, G.H., Morioka, N. et al. (2020). Vibronic states and their effect on the temperature and strain dependence of silicon-vacancy qubits in 4*H*-SiC. *Physical Review Applied* 13 (5): 054017.

196 Rühl, M., Bergmann, L., Krieger, M., and Weber, H.B. (2020). Stark tuning of the silicon vacancy in silicon carbide. *Nano Letters* 20 (1): 658–663.

197 Lukin, D.M., White, A.D., Trivedi, R. et al. (2020). Spectrally reconfigurable quantum emitters enabled by optimized fast modulation. *npj Quantum Information* 6 (1): 1–9.

198 Nagy, R., Niethammer, M., Widmann, M. et al. (2019). High-fidelity spin and optical control of single silicon-vacancy centres in silicon carbide. *Nature Communications* 10 (1): 1954.

199 Morioka, N., Liu, D., Soykal, O.O. et al. (2022). Spin-optical dynamics and quantum efficiency of single V1 center in silicon carbide. *Physical Review Applied* 17: 054005.

200 Sukachev, D.D., Sipahigil, A., Nguyen, C.T. et al. (2017). Silicon-vacancy spin qubit in diamond: a quantum memory exceeding 10 ms with single-shot state readout. *Physical Review Letters* 119 (22): 223602.

201 Bassett, L.C., Heremans, F.J., Christle, D.J. et al. (2014). Ultrafast optical control of orbital and spin dynamics in a solid-state defect. *Science* 345 (6202): 1333–1337.

202 Gao, W.B., Imamoglu, A., Bernien, H., and Hanson, R. (2015). Coherent manipulation, measurement and entanglement of individual solid-state spins using optical fields. *Nature Photonics* 9 (6): 363–373.

203 Zhang, Q., Guo, Y., Ji, W. et al. (2021). High-fidelity single-shot readout of single electron spin in diamond with spin-to-charge conversion. *Nature Communications* 12 (1): 1529.

204 Trusheim, M.E., Li, L., Laraoui, A. et al. (2014). Scalable fabrication of high purity diamond nanocrystals with long-spin-coherence nitrogen vacancy centers. *Nano Letters* 14 (1): 32–36.

205 Carr, H.Y. and Purcell, E.M. (1954). Effects of diffusion on free precession in nuclear magnetic resonance experiments. *Physical Review* 94 (3): 630–638.

206 de Lange, G., Wang, Z.H., Ristè, D. et al. (2010). Universal dynamical decoupling of a single solid-state spin from a spin bath. *Science* 330 (6000): 60–63.

207 Abobeih, M.H., Cramer, J., Bakker, M.A. et al. (2018). One-second coherence for a single electron spin coupled to a multi-qubit nuclear-spin environment. *Nature Communications* 9 (1): 1–8.

208 Tamarat, Ph., Manson, N.B., Harrison, J.P. et al. (2008). Spin-flip and spin-conserving optical transitions of the nitrogen-vacancy centre in diamond. *New Journal of Physics* 10 (4): 045004.

209 Rogers, L.J., Doherty, M.W., Barson, M.S.J. et al. (2015). Singlet levels of the NV- centre in diamond. *New Journal of Physics* 17: 013048.

210 Goldman, M.L., Sipahigil, A., Doherty, M.W. et al. (2015). Phonon-induced population dynamics and intersystem crossing innitrogen-vacancy centers. *Physical Review Letters* 114 (14): 145502.

211 Robledo, L., Childress, L., Bernien, H. et al. (2011). High-fidelity projective read-out of a solid-state spin quantumregister. *Nature* 477 (7366): 574–578.

212 Gupta, A., Hacquebard, L., and Childress, L. (2016). Efficient signal processing for time-resolved fluorescence detectionof nitrogen-vacancy spins in diamond. *Journal of the Optical Society of America B* 33 (3): B28.

213 Hopper, D.A., Shulevitz, H.J., and Bassett, L.C. (2018). Spin readout techniques of the nitrogen-vacancy center in diamond. *Micromachines* 9 (9): 1–30.

214 Irber, D.M., Poggiali, F., Kong, F. et al. (2021). Robust all-optical single-shot readout of nitrogen-vacancy centers in diamond. *Nature Communications* 12 (1): 532.

215 Balasubramanian, G., Neumann, P., Twitchen, D. et al. (2009). Ultralong spin coherence time in isotopically engineered diamond. *Nature Materials* 8 (5): 383–387.

216 Maurer, P.C., Kucsko, G., Latta, C. et al. (2012). Room-temperature quantum bit memory exceeding one second. *Science (New York, N.Y.)* 336 (6086): 1283–1286.

217 Bradley, C.E., Randall, J., Abobeih, M.H. et al. (2019). A ten-qubit solid-state spin register with quantum memory up to one minute. *Physical Review X* 9 (3): 31045.

218 Pingault, B., Becker, J.N., Schulte, C.H.H. et al. (2014). All-optical formation of coherent dark states of silicon-vacancy spins in diamond. *Physical Review Letters* 113 (26): 263601.

219 Rogers, L.J., Jahnke, K.D., Metsch, M.H. et al. (2014). All-optical initialization, readout, and coherent preparation of single silicon-vacancy spins in diamond. *Physical Review Letters* 113 (26): 263602.

220 Jahnke, K.D., Sipahigil, A., Binder, J.M. et al. (2015). Electron-phonon processes of the silicon-vacancy centre in diamond. *New Journal of Physics* 17 (4): 043011.

221 Pingault, B., Jarausch, D.-D., Hepp, C. et al. (2017). Coherent control of the silicon-vacancy spin in diamond. *Nature Communications* 8 (1): 15579.

222 Nguyen, C.T., Sukachev, D.D., Bhaskar, M.K. et al. (2019). An integrated nanophotonic quantum register based on silicon-vacancy spins in diamond. *Physical Review B* 100 (16): 165428.

223 Takou, E. and Economou, S.E. (2021). Optical control protocols for high-fidelity spin rotations of single SiV^- and SnV^- centers in diamond. *Physical Review B* 104 (11): 115302.

224 Becker, J.N., Görlitz, J., Arend, C. et al. (2016). Ultrafast all-optical coherent control of single silicon vacancy colour centres in diamond. *Nature Communications* 7: 1–6.

225 Sohn, Y.I., Meesala, S., Pingault, B. et al. (2018). Controlling the coherence of a diamond spin qubit through its strain environment. *Nature Communications* 9 (1): 17–22.

226 Becker, J.N. and Becher, C. (2017). Coherence properties and quantum control of silicon vacancy color centers in diamond. *Physica Status Solidi A* 214 (11): 1700586.

227 Debroux, R., Michaels, C.P., Purser, C.M. et al. (2021). Quantum control of the tin-vacancy spin qubit in diamond. *Physical Review X* 11 (4): 041041.

228 Seo, H., Falk, A.L., Klimov, P.V. et al. (2016). Quantum decoherence dynamics of divacancy spins in silicon carbide. *Nature Communications* 7 (1): 12935.

229 Miao, K.C., Blanton, J.P., Anderson, C.P. et al. (2020). Universal coherence protection in a solid-state spin qubit. *Science* 369 (6509): 1493–1497.

230 Sipahigil, A., Evans, R.E., Sukachev, D.D. et al. (2016). An integrated diamond nanophotonics platform for quantum-optical networks. *Science* 354 (6314): 847–850.

231 Evans, R.E., Bhaskar, M.K., Sukachev, D.D. et al. (2018). Photon-mediated interactions between quantum emitters in a diamond nanocavity. *Science* 362 (6415): 662–665.

232 Bernien, H., Hensen, B., Pfaff, W. et al. (2013). Heralded entanglement between solid-state qubits separated by three metres. *Nature* 497 (7447): 86–90.

233 Togan, E., Chu, Y., Trifonov, A.S. et al. (2010). Quantum entanglement between an optical photon and a solid-state spin qubit. *Nature* 466 (7307): 730–734.

234 Humphreys, P.C., Kalb, N., Morits, J.P.J. et al. (2018). Deterministic delivery of remote entanglement on a quantum network. *Nature* 558 (7709): 268–273.

235 Pompili, M., Hermans, S.L.N.N., Baier, S. et al. (2021). Realization of a multinode quantum network of remote solid-state qubits. *Science* 372 (6539): 259–264.

236 Nguyen, C.T., Sukachev, D.D., Bhaskar, M.K. et al. (2019). Quantum network nodes based on diamond qubits with an efficient nanophotonic interface. *Physical Review Letters* 123 (18): 1–6.

237 Myers, B.A., Das, A., Dartiailh, M.C. et al. (2014). Probing surface noise with depth-calibrated spins in diamond. *Physical Review Letters* 113 (2): 1–6.

238 Kolkowitz, S., Unterreithmeier, Q.P., Bennett, S.D., and Lukin, M.D. (2012). Sensing distant nuclear spins with a single electron spin. *Physical Review Letters* 109 (13): 1–5.

239 Taminiau, T.H., Wagenaar, J.J.T., van der Sar, T. et al. (2012). Detection and control of individual nuclear spins using a weakly coupled electron spin. *Physical Review Letters* 109 (13): 137602.

240 Zhao, N., Honert, J., Schmid, B. et al. (2012). Sensing single remote nuclear spins. *Nature Nanotechnology* 7 (10): 657–662.

241 Holzgrafe, J., Beitner, J., Kara, D. et al. (2019). Error corrected spin-state readout in a nanodiamond. *npj Quantum Information* 5 (1): 13.

242 Lovchinsky, I., Sushkov, A.O., Urbach, E. et al. (2016). Nuclear magnetic resonance detection and spectroscopy of single proteins using quantum logic. *Science* 351 (6275): 836–841.

243 Rosskopf, T., Zopes, J., Boss, J.M., and Degen, C.L. (2017). A quantum spectrum analyzer enhanced by a nuclear spin memory. *npj Quantum Information* 3 (1): 33.

244 Gurudev Dutt, M.V., Childress, L., Jiang, L. et al. (2007). Quantum register based on individual electronic and nuclear spin qubits in diamond. *Science* 316 (5829): 1312–1316.

245 Fuchs, G.D., Burkard, G., Klimov, P.V., and Awschalom, D.D. (2011). A quantum memory intrinsic to single nitrogen-vacancy centres in diamond. *Nature Physics* 7 (10): 789–793.

246 Bartling, H.P., Abobeih, M.H., Pingault, B. et al. (2021). Coherence and entanglement of inherently long-lived spin pairs in diamond. arXiv:2103.07961. arXiv:2103.07961v1.

247 Wrachtrup, J., Kilin, S.Y., and Nizovtsev, A.P. (2001). Quantum computation using the 13C nuclear spins near the single NV defect center in diamond. *Optics and Spectroscopy* 91 (3): 429–437.

248 Childress, L., Dutt, M.V.G., Taylor, J.M. et al. (2006). Coherent dynamics of coupled electron and nuclear spin qubits in diamond. *Science* 314 (5797): 281–285.

249 Pfaff, W., Taminiau, T.H., Robledo, L. et al. (2013). Demonstration of entanglement-by-measurement of solid-state qubits. *Nature Physics* 9 (1): 29–33.

250 Taminiau, T.H., Cramer, J., Van Der Sar, T. et al. (2014). Universal control and error correction in multi-qubit spin registers in diamond. *Nature Nanotechnology* 9 (3): 171–176.

251 Waldherr, G., Wang, Y., Zaiser, S. et al. (2014). Quantum error correction in a solid-state hybrid spin register. *Nature* 506 (7487): 204–207.

252 Abobeih, M.H., Wang, Y., Randall, J. et al.. Fault-tolerant operation of a logical qubit in a diamond quantum processor. *Nature* 606: 884–889.

253 Degen, M.J., Loenen, S.J.H., Bartling, H.P. et al. (2021). Entanglement of dark electron-nuclear spin defects in diamond. *Nature Communications* 12 (1): 3470.

254 Dolde, F., Jakobi, I., Naydenov, B. et al. (2013). Room-temperature entanglement between single defect spins in diamond. *Nature Physics* 9 (3): 139–143.

255 Dolde, F., Bergholm, V., Wang, Y. et al. (2014). High-fidelity spin entanglement using optimal control. *Nature Communications* 5: 3371.

256 Buterakos, D., Barnes, E., and Economou, S.E. (2017). Deterministic generation of all-photonic quantum repeaters from solid-state emitters. *Physical Review X* 7 (4): 1–10.

257 Russo, A., Barnes, E., and Economou, S.E. (2018). Photonic graph state generation from quantum dots and color centers for quantum communications. *Physical Review B* 98 (8): 085303.

258 Michaels, C.P., Martínez, J.A., Debroux, R. et al. (2021). Multidimensional cluster states using a single spin-photon interface coupled strongly to an intrinsic nuclear register. *Quantum* 5: 565.

259 Metsch, M.H., Senkalla, K., Tratzmiller, B. et al. (2019). Initialization and readout of nuclear spins via a negatively charged silicon-vacancy center in diamond. *Physical Review Letters* 122 (19): 190503.

260 Grice, W.P. (2011). Arbitrarily complete Bell-state measurement using only linear optical elements. *Physical Review A* 84 (4): 042331.

261 Ewert, F. and van Loock, P. (2014). 3/4-efficient Bell measurement with passive linear optics and unentangled ancillae. *Physical Review Letters* 113 (14): 140403.

262 Hensen, B., Bernien, H., Drea, A.E. et al. (2015). Loophole-free Bell inequality violation using electron spins separated by 1.3 kilometres. *Nature* 526 (7575): 682–686.

263 Kalb, N., Reiserer, A.A., Humphreys, P.C. et al. (2017). Entanglement distillation between solid-state quantum network nodes. *Science* 356 (6341): 928–932.

264 Briegel, H.J., Dür, W., Cirac, J.I., and Zoller, P. (1998). Quantum repeaters: the role of imperfect local operations in quantum communication. *Physical Review Letters* 81 (26): 5932–5935.

265 Loock, P., Alt, W., Becher, C. et al. (2020). Extending quantum links: modules for fiber- and memory-based quantum repeaters. *Advanced Quantum Technologies* 3 (11): 1900141.

266 Muralidharan, S., Li, L., Kim, J. et al. (2016). Optimal architectures for long distance quantum communication. *Scientific Reports* 6: 20463.

267 van Loock, P., Alt, W., Becher, C. et al. (2020). Extending quantum links: modules for fiber- and memory-based quantum repeaters. *Advanced Quantum Technologies* 3 (11): 1900141.

268 Kamin, L., Shchukin, E., Schmidt, F., and van Loock, Peter. (2022). Exact rate analysis for quantum repeaters with imperfect memories and entanglement swapping as soon as possible. arXiv:2203.10318 arXiv:2203.10318 [quant-ph]. Accepted in Physical Review Research.

269 Azuma, K., Tamaki, K., and Lo, H.K. (2015). All-photonic quantum repeaters. *Nature Communications* 6: 1–7.

270 Zwerger, M., Briegel, H.J., and Dür, W. (2016). Measurement-based quantum communication. *Applied Physics B* 122 (3): 50.

271 Borregaard, J., Pichler, H., Schröder, T. et al. (2020). One-way quantum repeater based on near-deterministic photon-emitter interfaces. *Physical Review X* 10 (2): 021071.

272 Raussendorf, R. and Briegel, H.J. (2001). A one-way quantum computer. *Physical Review Letters* 86 (22): 5188–5191.

273 Varnava, M., Browne, D.E., and Rudolph, T. (2006). Loss tolerance in one-way quantum computation via counterfactual error correction. *Physical Review Letters* 97: 120501.

274 Gross, C. and Bloch, I. (2017). Quantum simulations with ultracold atoms in optical lattices. *Science* 357 (6355): 995–1001.

275 Loss, D. and DiVincenzo, D.P. (1998). Quantum computation with quantum dots. *Physical Review A* 57 (1): 120–126.

276 Bruzewicz, C.D., Chiaverini, J., McConnell, R., and Sage, J.M. (2019). Trapped-ion quantum computing: progress and challenges. *Applied Physics Reviews* 6 (2): 021314.

277 Wang, C., Li, X., Xu, H. et al. (2022). Towards practical quantum computers: transmon qubit with a lifetime approaching 0.5 milliseconds. *npj Quantum Information* 8 (1): 3.

278 Shi, Y. and Waks, E. (2021). Deterministic generation of multidimensional photonic cluster states using time-delay feedback. *Physical Review A* 104: 013703.

279 Chen, K.C., Dai, W., Errando-Herranz, C. et al. (2021). Scalable and high-fidelity quantum random access memory in spin-photon networks. *PRX Quantum* 2: 030319.

280 Knall, E.N., Knaut, C.M., Bekenstein, R. et al. (2022). Efficient source of shaped single photons based on an integrated diamond nanophotonic system. *Physical Review Letters* 129, 053603.

281 Bracht, T.K., Cosacchi, M., Seidelmann, T. et al. (2021). Swing-up of quantum emitter population using detuned pulses. *Physical Review X Quantum* 2(4): 040354.

282 Ben-David, S. and Sattath, O. (2023). Quantum tokens for digital signatures. Quantum 7: 901.

283 Wiesner, S. (1983). Conjugate coding. *ACM SIGACT News* 15 (1): 78–88.

284 Razavi, M., Piani, M., and Lütkenhaus, N. (2009). Quantum repeaters with imperfect memories: cost and scalability. *Physical Review A* 80(3): 032301.

20

Quantum Photonics with 2D Semiconductors

Mauro Brotons-Gisbert and Brian D. Gerardot

Institute of Photonics and Quantum Sciences, SUPA, Heriot-Watt University, EH14 4AS Edinburgh, UK

20.1 Introduction

The simple crystal lattice of 2D materials offers unprecedented opportunities for quantum photonics: the atomically thin nature offers potential for highly efficient light–matter interaction, novel ways to design quantum heterostructures, and facile coupling to hybrid integrated photonic chips. Zero-dimensional quantum confinement in bulk semiconductors is typically achieved by a point defect in the crystal lattice, such as electrons bound to donor impurities in diamond [1], or a local change in electronic bandgap such that an electron (or hole, or Coulomb bound electron–hole complex called an exciton) can be spatially confined, such as in a self-assembled quantum dot [2]. Both regimes for quantum confinement are possible in 2D materials, including some novel means not possible in conventional materials. In any case, highly efficient (optically bright) single-photon emission can be obtained with 2D QEs. The spin and optical properties of a QE are determined by its intrinsic selection rules as well its interaction with the surrounding environment. For instance, the photon polarization and any fine-structure splitting in the optical spectra are governed by the intrinsic qualities, while the electron–phonon and spin coherence properties are governed by the crystalline environment [3]. With the importance of the crystal environment in mind, we introduce the most prominent types of crystals for 2D quantum photonics.

20.1.1 Hexagonal Boron Nitride

Hexagonal boron nitride (hBN) is a layered van der Waals material in which the boron and nitrogen atoms are arranged in a hexagonal (honeycomb) structure in the layer plane (Figure 20.1(a)). Within each layer, each boron atom is covalently bound to three nitrogen atoms, and vice versa, while the different layers are held together by the much weaker van der Waals forces [4]. The presence of two different atomic elements in the unit cell of hBN breaks the sublattice symmetry, which

Photonic Quantum Technologies: Science and Applications, First Edition. Edited by Mohamed Benyoucef.

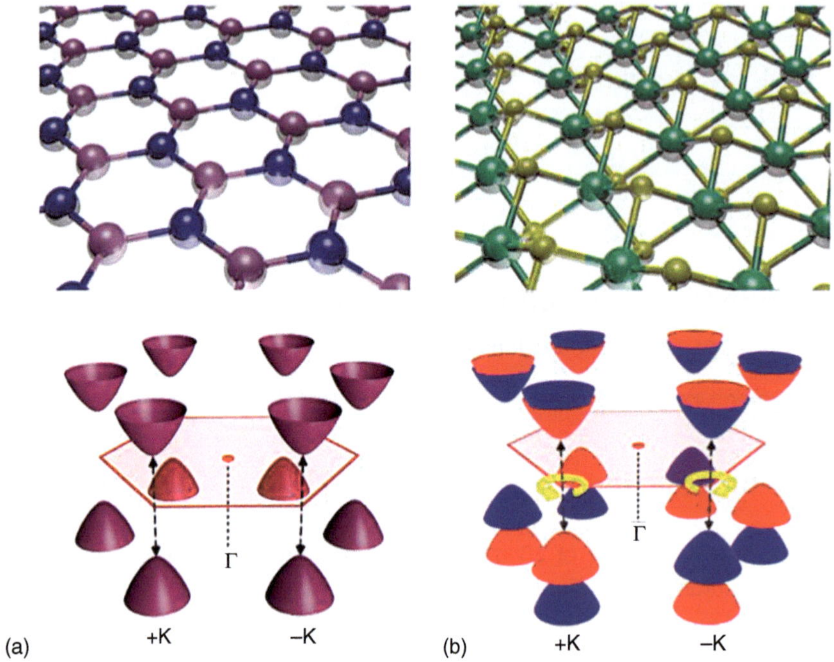

Figure 20.1 Crystal structures (top panel) and band structures (bottom panel) at the corners of the first Brillouin zone of monolayer (a) hexagonal boron nitride and (b) the transition metal dichalcogenides (TMDs). The presence of two different atomic elements in the unit cell of hBN breaks the sublattice symmetry, resulting in a large bandgp of 5.5 eV. For TMDs, due to the combination of broken inversion symmetry and strong spin–orbit coupling, two spin-polarized bands (red and blue) arise, which couple to optical polarization via the optical selection rules. For TMDs, the sign of the spin splitting in the conduction band depends on the metal atom (Mo or W), creating optically bright (dark) ground states. The band structure in (b) shows transition metal the case for MoX_2 monolayers, which have optically bright ground states.

results in a large bandgap of ~5.5 eV [4, 5]. These properties have positioned hBN as one of the main ingredients of the van der Waals material platform: an atomically flat and transparent layered insulator with good dielectric and thermal conductivity properties, which can be used as a substrate or to encapsulate other active material for both protection and device functionality. Encapsulation of graphene or TMDs with hBN has shown to significantly reduce environmental disorder [6] to yield significantly enhanced transport [7, 8] and optical [9, 10] properties. Additionally, hBN has proved to be an excellent dielectric spacer for engineering the optical properties of other 2D materials [11–14].

Besides its excellent properties as a passive layer, hBN has also attracted a lot attention as an active layer for quantum photonic applications. The large band gap of hBN gives rise to deep-level point defects with atomic-like wavefunctions that can form QEs [15]. Although these QEs are beyond the topic of the current chapter, here we briefly introduce the platform as active, rather than passive, component's in a quantum photonic device. Room-temperature, single-photon emission has been observed in bulk crystalline and monolayer hBN [15, 16], with the emission covering quite a broad spectral range (1.6–2.5 eV). Moreover, these QEs can have high quantum efficiencies [17] and count rates exceeding a million counts per second. Although the precise identification of the point defect responsible for the different quantum emitters (QEs) in hBN is not yet conclusive, several proposals have been investigated by ab initio methods, including the antisite nitrogen vacancy N_BV_N [15], the carbon substitutional impurity at nitrogen site (C_N) [18], and the carbon antisite (C_BV_N) [19] defects and dangling bonds. Moreover, an important step toward coherent optical control of QEs in 2D materials has been shown in the hBN platform. Rabi oscillations [20] and the coherent control of spin defects [21] are possible in hBN QEs.

20.1.2 Transition Metal Dichalcogenide Semiconductors

Group-VI transition metal dichalcogenides semiconductors (TMDs) are van der Waals layered materials of the form MX_2, where M is a transition metal element, and X is a chalcogen element of the group VI [22, 23]. Each TMD layer shows an X-M-X structure where a plane of metal atoms (M) is sandwiched by two planes of chalcogen atoms (X) with hexagonal symmetry (Figure 20.1b) [22, 23]. Each metal atom is covalently bonded to six chalcogen atoms in the neighboring atomic layers, while adjacent TMD layers are held together by van der Waals forces. Bulk TMDs exhibit a wide variety of polytypes and stacking polymorphs because of the two possible local coordination of the metal species in an individual MX_2 layer. The most commonly encountered polytypes are 1T, 2H, and 3R, where the letters stand for trigonal, hexagonal, and rhombohedral, respectively, and the digits indicate the number of X-M-X units in the unit cell (that is, the number of layers in the stacking sequence) [24]. The coordination of metal atoms is either trigonal prismatic or octahedral [23].

Among the different TMDs, the combinations of X = Mo or W and X = S, Se, or Te have attracted a lot of attention due to their appealing properties for a large palette of optoelectronics, spintronics, and photonics applications. Despite their layer- and strain-dependent indirect band gaps in their few-layer and bulk forms [25, 26], monolayer TMDs exhibit momentum-direct optical band gaps with energies in the visible to near-infrared spectral range (~1 –2.5 eV [27]), with the band edges located at the degenerate but inequivalent corners of the hexagonal Brillouin zone, typically referred to as ±K valleys (Figure 20.1b). Carriers occupying the conduction and valence band edges at ±K form hydrogen-like states known as excitons, in which the electron and hole are strongly bound together (see Figure 20.3(a)) with a typical binding energy on the order of 0.5 eV [28, 29]. Due to their large binding energy, excitons dominate the optical response of TMDs at both cryogenic and room temperatures. Excitons can be localized, trapped, for example by disorder or defects, to create single-photon sources [30–35]. Moreover, the strong spin–orbit coupling induced by the heavy transition-metal atoms and lack of inversion symmetry in the TMDs lattice unit cell result in an effective coupling between the valley index and spin of the carriers at the ±K corners, typically referred to as spin–valley locking [36]. Such effective spin–valley coupling results in valley-dependent optical selection rules [37]: excitonic transitions at ±K involve $\sigma^{\pm}$-polarized photons, respectively, enabling optical control of excitons [38, 39].

Moreover, a number of 2D TMDs have now been shown to host QEs. Quantum light emission in TMDs was first demonstrated for monolayer and bilayer WSe_2 [30–35] and monolayer WS_2 soon after [40]. Subsequently, QEs have been conclusively demonstrated in $MoSe_2$ [41], $MoTe_2$ [42], MoS_2 [43], and $MoSe_2/WSe_2$ moire heterostructures [44]. In general, these QEs exhibit inhomogeneous spectral features with emission linewidths as narrow as a few tens of μeV (full width at half maximum) (Figure 20.2(a)), with varying spectral and magneto-optical properties, as described below.

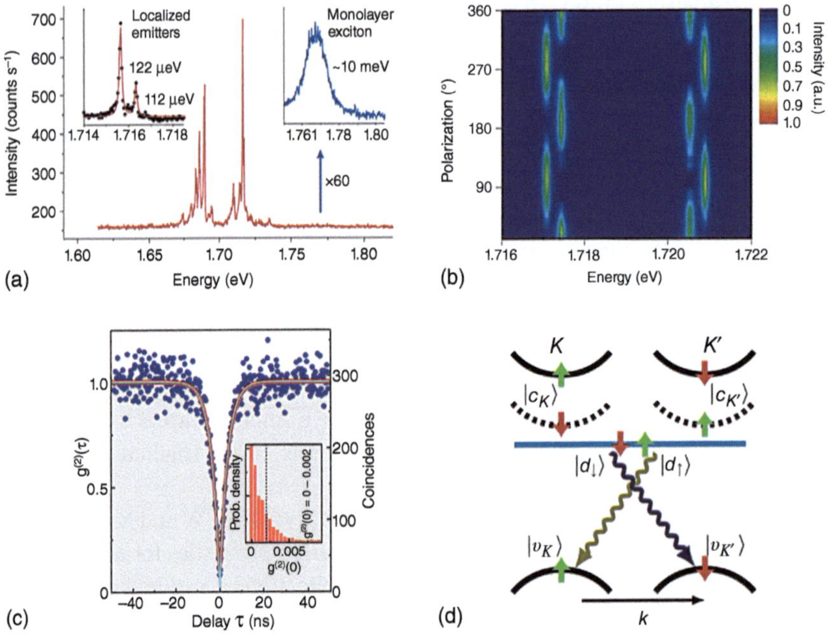

Figure 20.2 (a) Typical photoluminescence (PL) spectrum of semiconductor quantum emitters (QEs) in monolayer WSe_2. The left inset shows a high-resolution spectrum of the brightest peak, which highlights the presence fine-structure splitting emission doublet. The right inset shows the PL peak corresponding to the emission from free excitons. Reproduced with permission from Ref. [31]. (b) Linear-polarization-resolved PL spectra from a biexciton and a neutral exciton of a single QE in WSe_2, where the fine-structure splitting and the out-of-phase orthogonal linear polarizations of each both exciton states can be observed. Reproduced with permission from Ref. [45]. (c) $g^{(2)}(\tau)$ of a QE in WSe_2 under CW laser excitation. Reproduced with permission from Ref. [46]. (d) Reciprocal space sketch of the origin of QE formation in WSe_2 proposed in Ref. [47]. The $|c_{K(K')}\rangle$ and $|\upsilon_{K(K')}\rangle$ labels represent strain-localized electronic states originating from the conduction and valence band states at the corners of the Brillouin zone, respectively. The green and red arrows indicate the electron spin of each electronic state. Hybridization of the strain-localized dark exciton with a defect state $|d_{\uparrow,\downarrow}\rangle$ leads to efficient single photon emission (dark blue and yellow wavy arrows). Reproduced with permission from Ref. [47].

20.1.3 Van der Waals Heterostructures

The van der Waals material platform has opened unprecedented possibilities for band-structure and quantum material engineering due to its ability to stack individual layers of different 2D crystals to create heterostructures, which are not limited by lattice mismatch between adjacent layers, as in conventional semiconductor devices fabricated via epitaxial techniques [48, 49]. Such remarkable flexibility in material combinations and tunability of their physical parameters positions van der Waals heterostructures as a promising platform to realize atomically thin devices in which novel quantum states can be accessed and engineered. Some recent examples include, for instance, the Hofstadter butterfly effect in graphene-hBN heterostructures [50], the realization of superconductivity in magic-angle twisted bilayer graphene [51], and the observation of strongly correlated states [52–54] and moiré trapped excitons [44, 55, 56] in TMD heterobilayers. Moreover, optoelectronic and electronic devices for quantum photonics can be realized, such as single-photon LEDs [40], tuning quantum emitters with electric fields [57, 58], or Coulomb blockade of single electrons or holes in quantum emitters [59].

20.2 Semiconductor Quantum Emitters

Historically, QEs were first discovered in monolayer W-based TMDs. Typically, the localized neutral excitons in these QEs present a fine-structure split emission doublet, with orthogonal linear polarization and a splitting of 0.4–0.8 μeV (20.1.2(a)). Additionally, the localized excitons in these TMDs exhibit Landé g-factors of ~8–10 under out-of-plane magnetic fields (Faraday geometry) [30–35]. At low temperature ($T < 10K$), time-resolved PL measurements reveal lifetimes of a few ns to a few tens of ns [30–35], while second-order autocorrelation measurements have shown high-purity single-photon emission ($g^2(0) < 1\%$). Although the precise origin of TMD-based QEs has been subject to debate [3], a theoretical proposal in which the WSe_2 QEs are assumed to be intervalley defect excitons arising from hybridization of the dark excitons and defect states associated with intrinsic Se vacancies [47] (Figure 20.2(d)) has successfully accounted for several experimental features (such as the g-factors, FSS, radiative lifetimes, and polarization dependence).

20.3 Engineering 2D Semiconductor Quantum Emitters

Figure 20.3 provides the primary considerations in the design and engineering of 2D semiconductor QEs and a few specific examples. In Figure 20.3, the boxes labeled at the top of the figure correspond to the regions of the sample schematic below. In addition, the surface topography, electrostatic potential, and dielectric environment are highlighted below each region of the sample schematic. The legend provides the relevant information to explain the schematic.

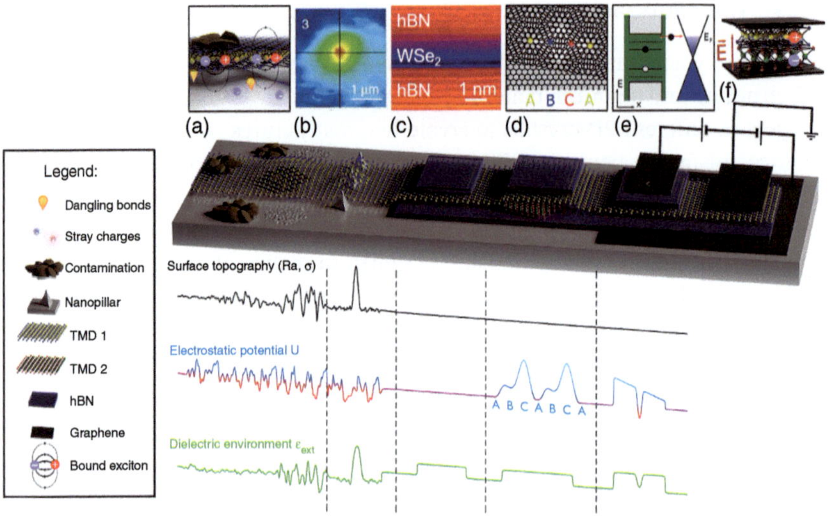

Figure 20.3 Main considerations in the design and engineering of 2D semiconductor QEs. The top panels show some specific examples of the more general sample schematics shown in the middle row, while the bottom panel provides a qualitative description of the surface topography, electrostatic potential, and dielectric environment corresponding to each region of the sample schematic. (a) The electric field lines joining the bound electron–hole pairs in 2D excitons spread outside the 2D layer, which makes them very sensitive to their surrounding dielectric environment. (b) Deterministic creation and positioning of QEs in TMDs transferred on top of substrates with defined stressors that create point-like strain perturbations. From Branny et al. [60]/Springer Nature/CC BY 4.0. (c) Cross-sectional scanning transmission electron microscope image of a WSe_2 layer fully encapsulated by hBN exhibiting pristine interfaces. Adapted Rooney et al. [61]/with permission of American Chemical Society. (d) Top- and 3D-view of a TMD moiré heterobilayer encapsulated by hBN (top and middle panels, respectively). The yellow, blue, and red dots in the top panel show the three moiré trapping sites (A, B and C) with different atomic registries. (e) Van der Waals device showing gate-tunable tunnel coupling between a 2D QE and a tunable Fermi reservoir in graphene (top panel). Adapted with permission from Ref. [59]. The sample schematics shows a top graphene layer with an etched hole, which can be used to achieve quantum confinement of carriers [62]. (f) Sketch of a van der Waals heterostructure that enables DC Stark tuning of interlayer excitons via vertical applied electric fields [44].

20.3.1 Dielectric and Electrostatic Engineering

The environment of a 2D QE is a crucial consideration. First, the substrate which the 2D material rests on can contain dirt, polymer residue, dangling bonds, and roughness. All of these features likely present on the surface of a SiO_2 substrate as shown in 20.3(a), can affect the excitons, which are highly sensitive to their environment [14]. Hence, encapsulation of the active QE layer by an atomically smooth van der Waals material such as hBN [9] (see Figure 20.3(c)) or MoO_3 [63] can significantly reduce the inhomogeneous broadening in the optical emission. Furthermore, the dielectric environment can be engineered to tune the emission energy of the emission. Recent evidence also suggests novel ways to create quantum confinement of carriers via local changes in the dielectric environment [62], as highlighted by the etched hole in the graphene layer in Figure 20.3(e). Changes in the dielectric environment are inevitably accompanied by a change in the local electrostatics, which can more actively be controlled. A longstanding dream in quantum photonics is to achieve efficient electrostatic control of optically active QEs, similar to what has been achieved in transport measurements [64]. In this light, there are renewed opportunities in 2D materials due to the reduced dielectric screening. Recent progress has shown the ability to electrostatically direct the flow of excitons [65, 66] and even achieve confinement in the quantum regime [67], although the ultimate proof of quantum emission with this technique remains an outstanding challenge.

20.3.2 Strain-Tuning

Strain engineering of 2D materials has proven to be an effective approach to induce and tune the properties of 2D QEs. In TMDs, delocalized 2D excitons created optically in the vicinity of a localized stressor can diffuse to the potential minimum, resulting in an efficient exciton funneling effect that can result in the creation of a QE [68], as depicted in Figure 20.3(b). The spatial correlation between localized strain pockets and the formation of QEs in 2D TMDs was observed early on [34, 35], which led to different efforts to exploit local stressors to deterministically create QEs in this platform. For example, monolayer WSe_2 [60, 69] and WS_2 [69] have shown to be able to host 2D arrays of QEs when transferred on top of lithographically defined nanopillars that create point-like strain perturbations, which locally modify the electronic band gap of the host material. As in the case of W-based TMDs, it is also possible to activate QEs in hBN by the deformation potential wells created by inducing a local curvature with a nanopillar [70]. More recently, it has also been demonstrated that substrates patterned with nanoscale depressions can be exploited to create 2D arrays of QEs in other TMDs, such as monolayer $MoSe_2$ [41].

Besides the deterministic creation and positioning of 2D QEs, strain tuning has also been successfully employed to tune some of the intrinsic properties of the 2D quantum emitters. For example, the strain originating along the edges of prepatterned square-shaped substrate protrusions has been exploited to induce quasi-one-dimensional QEs in WSe_2 monolayers with highly linearly polarized quantum emission [71]. Other examples of strain engineering include emission energy [72–74], fine-structure splitting [73], and emission axis tuning [74] of quantum emitters in 2D WSe_2. Finally, strain tuning can potentially be used to modify the moiré superlattice geometry [75], which can have an important impact on the particle interactions within the lattice and the type of excitonic confinement [76].

20.3.3 Site-Controlled Implantation

Beyond strain-induced QE localization techniques, alternative site-selective approaches have been shown to result in the deterministic creation and spatial positioning of QEs in several 2D materials. For example, deterministic spatial positioning of QEs in TMDs such as monolayer MoS_2 has recently been shown with a lateral precision of ~9 nm by using He ion beam lithography [43]. In hBN, methods such as focused ion beam [77], irradiation with electron beam [78], and ultrafast laser pulses [79] have also been demonstrated. Among all these techniques, electron-beam irradiation has shown to result into the activation of hBN QEs with an improved control of their spatial position and emission wavelength [80].

20.3.4 Moiré Heterostructures

TMD heterostructures represent a promising nanoscale system to engineer 2D QEs. Vertically stacked TMD heterobilayers present type II band alignment with atomically sharp interfaces [81], favoring the formation of interlayer excitons, in which the Coulomb-bound electron and hole reside in different layers. The spatial separation of the exciton carriers results in a large, permanent out-of-plane electric dipole moment that enables a large tunability of the exciton transition energy by externally applied electric fields [82]. Interestingly, despite their out-of-plane permanent electric dipole, interlayer excitons in TMD heterobilayers present optical transitions with in-plane optical dipoles [83], similar to the bright exciton transitions in the bare monolayer TMDs [84, 85]. Moreover, local stressors induced by prepatterned substrates can also be used to achieve exciton trapping, where the mean number of trapped interlayer excitons can be controlled through the optical excitation level [86, 87].

Additionally, heterobilayers with a relative angle twist and/or lattice mismatch between the constituent monolayers feature a moiré superlattice with a spatial periodicity that can be tuned by the relative twist angle of the individual layers [88]. The moiré superlattice gives rise to a periodic potential landscape for interlayer excitons (dependent on the atomic registry, as shown in Figure 20.3(d)) in the heterobilayer [89] and has led to experimental demonstrations of interlayer exciton trapping in $MoSe_2/WSe_2$ heterobilayers with twist angles of around 0°, 21.8° and 60° at cryogenic temperatures [44, 55, 56]. These moiré-trapped interlayer excitons present spectral linewidths below 100 μeV [44, 55], single-photon emission with strong circular polarization [44], and spin properties that can be tuned by both twist and layer engineering [56], raising the prospect of creating 2D arrays of highly tunable QEs. Moreover, controlled electron and hole doping of the $MoSe_2/WSe_2$ heterobilayers results in the creation of charged interlayer excitons that can also be trapped in the moiré potential [90–92], and therefore enable charge control of these novel 2D QEs.

20.3.5 Heterostructure Device Functionalization

In addition to the superior opportunities for integration of 2D QEs into photonic devices, the van der Waals platform offers an unprecedented potential to engineer devices with tailored tuning of the quantum emitter properties. The 2D host material and its quantum emitters can be easily combined with a wide range of atomic layers with different intrinsic properties to create atomically thin heterostructure devices with engineered functionalities [48]. Due to the absence of broken covalent bonds, the interfaces of van der Waals heterostructures normally present high-quality optoelectronic properties. Graphene layers can be used as sub-nanometre-thick semi-metallic electrodes for electrically tunable quantum photonics devices [48]. Ultra-flat, large band-gap insulators such as hBN can be employed both to protect the active region of the heterostructure from the environment and as a tunnel barrier for electrically injected carriers from the electrodes into the 2D QEs, with a tunneling probability that can be precisely controlled by the number of atomic layers [93]. Moreover, the van der Waals material platform enables the realization of Fermi reservoirs beyond the conventional flat density of states. A single graphene layer can provide a reservoir of carriers that can be electrically tuned from n-type (electron sea) to p-type (hole sea). Fermi reservoirs of spin-polarized carriers can also be achieved by employing ferromagnetic van der Waals materials, such as semiconducting CrI_3 [94] or metallic $Cr_2Ge_2Te_6$ [95].

Owing to the prospects that the van der Waals platform offers to engineer the interactions between a localized quantum state and its environment, achieving gate-control of van der Waals nanostructures at the single particle level has been a prime goal of the 2D material community. For instance, gate-tunable tunnel coupling between a 2D QE and a tunable Fermi reservoir was reported in a van der Waals heterostructure consisting of a WSe_2 QE coupled to a graphene contact through an atomically thin hBN tunnel barrier [59], as depicted in Figure 20.3(e). This device enabled Coulomb blockade, the mechanism whereby electrons or holes can be loaded one-by-one into a quantum dot. Thanks to the tunable Fermi reservoir in the few-layer graphene contact, the device allowed the deterministic loading of either a single electron or a single hole into the WSe_2 quantum emitter, enabling the observation and magneto-optical study of both negatively- and positively charged excitons in these quantum emitters. Moreover, the atomically thin hBN tunnel barrier leads to strong and spin-conserving tunnel coupling between the quantum emitter and the Fermi reservoir, resulting into the formation of gate-tunable hybrid excitons composed of localized quantum dot states and delocalized continuum states. These results confirm the potential of van der Waals heterostructures as a platform to engineer devices with tailored interactions between a 2D QE and its environment.

Besides engineering the interactions between a quantum emitter and its environment, van der Waals heterostructure devices also enable a tailored tuning of the intrinsic properties of 2D QEs. For example, by incorporating moiré-trapped interlayer excitons into a van der Waals heterostructure device, which enables the application of vertical electric fields, it is possible to exploit the large permanent dipole of these quantum emitters to achieve a large DC Stark tuning of their emission energy [44], as depicted in Figure 20.3(f). These results open opportunities to precisely control the emission energy of the quantum emitters in a broad spectral range, raising the prospect to energetically tune QEs in 2D materials into resonance with cavity modes or other emitters.

20.3.6 Integrated Photonics with 2D Quantum Emitters

The broad choice of 2D semiconductor QEs (e.g. WSe_2, WS_2, $MoSe_2$, MoS_2, and $MoTe_2$), together with the atomically thin nature of their host crystals, has allowed the integration of such QEs with a large variety of optical cavity architectures with different intrinsic properties. Figure 20.4 shows a schematic representation of the main cavity architectures used for 2D QEs: engineered substrate, distributed-Bragg-reflector cavity, open dielectric cavity, plasmonic cavity, ridge waveguide, and a photonic crystal nanobeam. The color bars show a qualitative comparison of the main properties of each cavity architecture: mode volume (V), Purcell enhancement factor (F_P), and spontaneous emission coupling factor (β). Among the different cavity architectures, engineered substrates have shown to be the easiest approach to shape the far-field radiation pattern of both 2D QEs [12, 44] and delocalized 2D excitons [13], while achieving moderate enhancements of their radiative lifetimes ($F_P < 1.5$). Tunable open cavities have also been used to realize 2D QEs with Purcell enhancement in WSe_2 [96], and coupling efficiencies of 10% [97], and $F_P \sim 4$ [98] in hBN QEs. Further, dielectric nanoantennas have been established with enhanced quantum efficiencies [99]. In addition to dielectric cavities, plasmonic metal nanocavities have also been exploited to increase the brightness of WSe_2 QEs without affecting their single-photon purity [100]

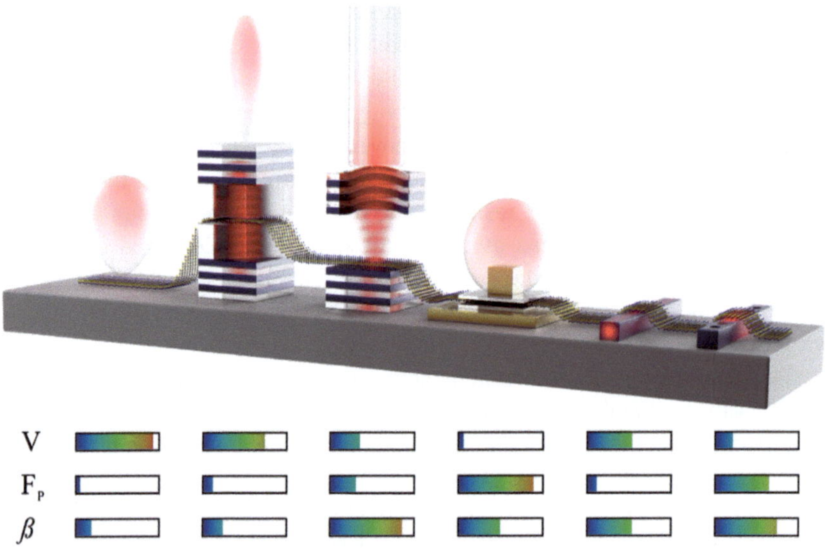

Figure 20.4 Sketch summarizing the main cavity architectures used for 2D semiconductor QEs: engineered substrate, distributed-Bragg-reflector cavity, open dielectric cavity, plasmonic cavity, ridge waveguide, and a photonic crystal nanobeam. The color bars show a qualitative comparison of the main properties of each cavity architecture: mode volume (V), Purcell enhancement factor (F_P), and spontaneous emission coupling factor (β). The red-shaded lobes represent the typical far-field directional radiation patterns associated to each cavity. Reproduced with permission from Ref. [3].

Besides enhancing the QEs emission, several efforts have also been made to couple the emitted quantum light into photonic circuitry. Some examples include the transfer of a WSe_2 flake on the facet of a titanium-diffused lithium niobate waveguide and directional coupler [101] and the coupling of the PL from a WSe_2 monolayer into the waveguide mode of a SiN waveguide [102]. Recently, the quantum nature of the photons coupled to a SiN waveguide has been demonstrated [103]. Nevertheless, in all these first demonstrations of QE coupling, the coupling efficiency into the waveguide mode was below 10%.

20.4 Outlook

2D materials can host bright QEs with novel prospects for integrated electronic and photonic devices. To further realize their promise, several obstacles must first be overcome. Perhaps the most important challenge is a robust understanding of the various dephasing mechanisms, both of the optical emission and the spin coherence. To date, successful generation of highly indistinguishable single photons has not been demonstrated 2D QEs. Success in this challenge, along with the potential of a coherent spin–photon interface in some 2D QE platforms, will raise new opportunities in quantum networking applications [3]. Beyond such mainstream quantum applications, 2D QEs are likely best positioned for applications that can take advantage of their unique atomically thin nature and the ease with which they can be combined into novel functional heterostructure devices (see Figure 20.3) or photonic chips (see Figure 20.4).

Acknowledgments

We thank Artur Branny and Hyeonjun Baek for the fruitful discussions and their assistance in developing Figures 20.3 and 20.4. M.B.-G. acknowledges the Royal Society for support via a University Research Fellowship. B.D.G. acknowledges the Royal Society for a Wolfson Merit Award and the Royal Academy of Engineering for a Chair in Emerging Technology.

References

1 Aharonovich, I., Castelletto, S., Simpson, D.A. et al. (2011). Diamond-based single-photon emitters. *Reports on Progress in Physics* 74 (7): 076501.
2 Petroff, P.M., Lorke, A., and Imamoglu, A. (2001). Epitaxially self-assembled quantum dots. *Physics Today* 54 (5): 46–52.
3 Turunen, M., Brotons-Gisbert, M., Dai, Y. et al. (2022). Quantum photonics with layered 2D materials. *Nature Reviews Physics* 4 (4): 219–236.
4 Song, L., Ci, L., Lu, H. et al. (2010). Large scale growth and characterization of atomic hexagonal boron nitride layers. *Nano Letters* 10 (8): 3209–3215.

5 Watanabe, K., Taniguchi, T., and Kanda, H. (2004). Direct-bandgap properties and evidence for ultraviolet lasing of hexagonal boron nitride single crystal. *Nature Materials* 3 (6): 404–409.

6 Rhodes, D., Chae, S.H., Ribeiro-Palau, R., and Hone, J. (2019). Disorder in van der Waals heterostructures of 2D materials. *Nature Materials* 18 (6): 541–549.

7 Dean, C.R., Young, A.F., Meric, I. et al. (2010). Boron nitride substrates for high-quality graphene electronics. *Nature Nanotechnology* 5 (10): 722–726.

8 Bandurin, D.A., Tyurnina, A.V., Yu, G.L. et al. (2017). High electron mobility, quantum Hall effect and anomalous optical response in atomically thin InSe. *Nature Nanotechnology* 12 (3): 223–227.

9 Cadiz, F., Courtade, E., Robert, C. et al. (2017). Excitonic linewidth approaching the homogeneous limit in MoS_2-based van der Waals heterostructures. *Physical Review X* 7 (2): 021026.

10 Ajayi, O.A., Ardelean, J.V., Shepard, G.D. et al. (2017). Approaching the intrinsic photoluminescence linewidth in transition metal dichalcogenide monolayers. *2D Materials* 4 (3): 031011.

11 Gerber, I.C. and Marie, X. (2018). Dependence of band structure and exciton properties of encapsulated WSe_2 monolayers on the hBN-layer thickness. *Physical Review B* 98 (24): 245126.

12 Brotons-Gisbert, M., Martínez-Pastor, J.P., Ballesteros, G.C. et al. (2018). Engineering light emission of two-dimensional materials in both the weak and strong coupling regimes. *Nanophotonics* 7 (1): 253–267.

13 Fang, H.H., Han, B., Robert, C. et al. (2019). Control of the exciton radiative lifetime in van der waals heterostructures. *Physical Review Letters* 123 (6): 067401.

14 Raja, A., Chaves, A., Yu, J. et al. (2017). Coulomb engineering of the bandgap and excitons in two-dimensional materials. *Nature Communications* 8 (1): 1–7.

15 Tran, T.T., Bray, K., Ford, M.J. et al. (2016). Quantum emission from hexagonal boron nitride monolayers. *Nature Nanotechnology* 11 (1): 37–41.

16 Tran, T.T., Zachreson, C., Berhane, A.M. et al. (2016). Quantum emission from defects in single-crystalline hexagonal boron nitride. *Physical Review Applied* 5 (3): 034005.

17 Nikolay, N., Mendelson, N., Özelci, E. et al. (2019). Direct measurement of quantum efficiency of single-photon emitters in hexagonal boron nitride. *Optica* 6 (8): 1084–1088.

18 Bourrellier, R., Meuret, S., Tararan, A. et al. (2016). Bright UV single photon emission at point defects in h-BN. *Nano Letters* 16 (7): 4317–4321.

19 Tawfik, S.A., Ali, S., Fronzi, M. et al. (2017). First-principles investigation of quantum emission from hBN defects. *Nanoscale* 9 (36): 13575–13582.

20 Konthasinghe, K., Chakraborty, C., Mathur, N. et al. (2019). Rabi oscillations and resonance fluorescence from a single hexagonal boron nitride quantum emitter. *Optica* 6 (5): 542–548.

21 Gottscholl, A., Kianinia, M., Soltamov, V. et al. (2020). Initialization and read-out of intrinsic spin defects in a van der Waals crystal at room temperature. *Nature Materials* 19 (5): 540–545.

22 Wang, Q.H., Kalantar-Zadeh, K., Kis, A. et al. (2012). Electronics and optoelectronics of two-dimensional transition metal dichalcogenides. *Nature Nanotechnology* 7 (11): 699–712.

23 Chhowalla, M., Shin, H.S., Eda, G. et al. (2013). The chemistry of two-dimensional layered transition metal dichalcogenide nanosheets. *Nature Chemistry* 5 (4): 263–275.

24 Wang, Z.M., Ren, C., Tian, H. et al. (2013). *MoS_2: Materials, Physics, and Devices*. Springer.

25 Mak, K.F., Lee, C., Hone, J. et al. (2010). Atomically thin MoS_2: a new direct-gap semiconductor. *Physical Review Letters* 105 (13): 136805.

26 Brotons-Gisbert, M., Segura, A., Robles, R. et al. (2018). Optical and electronic properties of 2H-MoS_2 under pressure: revealing the spin-polarized nature of bulk electronic bands. *Physical Review Materials* 2 (5): 054602.

27 Xia, F., Wang, H., Xiao, D. et al. (2014). Two-dimensional material nanophotonics. *Nature Photonics* 8 (12): 899–907.

28 Chernikov, A., Berkelbach, T.C., Hill, H.M. et al. (2014). Exciton binding energy and nonhydrogenic Rydberg series in monolayer WS_2. *Physical Review Letters* 113 (7): 076802.

29 He, K., Kumar, N., Zhao, L. et al. (2014). Tightly bound excitons in monolayer WSe_2. *Physical Review Letters* 113 (2): 026803.

30 Chakraborty, C., Kinnischtzke, L., Goodfellow, K.M. et al. (2015). Voltage-controlled quantum light from an atomically thin semiconductor. *Nature Nanotechnology* 10 (6): 507–511.

31 He, Y.-M., Clark, G., Schaibley, J.R. et al. (2015). Single quantum emitters in monolayer semiconductors. *Nature Nanotechnology* 10 (6): 497–502.

32 Koperski, M., Nogajewski, K., Arora, A. et al. (2015). Single photon emitters in exfoliated WSe_2 structures. *Nature Nanotechnology* 10 (6): 503–506.

33 Srivastava, A., Sidler, M., Allain, A.V. et al. (2015). Optically active quantum dots in monolayer WSe_2. *Nature Nanotechnology* 10 (6): 491–496.

34 Tonndorf, P., Schmidt, R., Schneider, R. et al. (2015). Single-photon emission from localized excitons in an atomically thin semiconductor. *Optica* 2 (4): 347–352.

35 Kumar, S., Kaczmarczyk, A., and Gerardot, B.D. (2015). Strain-induced spatial and spectral isolation of quantum emitters in mono-and bilayer WSe_2. *Nano Letters* 15 (11): 7567–7573.

36 Xiao, D., Liu, G.-B., Feng, W. et al. (2012). Coupled spin and valley physics in monolayers of MoS_2 and other group-VI dichalcogenides. *Physical Review Letters* 108 (19): 196802.

37 Xu, X., Yao, W., Xiao, D., and Heinz, T.F. (2014). Spin and pseudospins in layered transition metal dichalcogenides. *Nature Physics* 10 (5): 343–350.

38 Mak, K.F., He, K., Shan, J., and Heinz, T.F. (2012). Control of valley polarization in monolayer MoS_2 by optical helicity. *Nature Nanotechnology* 7 (8): 494–498.

39 Zeng, H., Dai, J., Yao, W. et al. (2012). Valley polarization in MoS_2 monolayers by optical pumping. *Nature Nanotechnology* 7 (8): 490–493.

40 Palacios-Berraquero, C. (2018). Atomically-thin quantum light emitting diodes. *Quantum Confined Excitons in 2-Dimensional Materials*, pp. 71–89. Springer.
41 Yu, L., Deng, M., Zhang, J.L. et al. (2021). Site-controlled quantum emitters in monolayer $MoSe_2$. *Nano Letters* 21 (6): 2376–2381.
42 Zhao, H., Pettes, M.T., Zheng, Y., and Htoon, H. (2021). Site-controlled telecom-wavelength single-photon emitters in atomically-thin $MoTe_2$. *Nature Communications* 12 (1): 1–7.
43 Klein, J., Lorke, M., Florian, M. et al. (2019). Site-selectively generated photon emitters in monolayer MoS_2 via local helium ion irradiation. *Nature Communications* 10 (1): 1–8.
44 Baek, H., Brotons-Gisbert, M., Koong, Z.X. et al. (2020). Highly energy-tunable quantum light from moiré-trapped excitons. *Science Advances* 6 (37): eaba8526.
45 He, Y.-M., Iff, O., Lundt, N. et al. (2016). Cascaded emission of single photons from the biexciton in monolayered WSe_2. *Nature Communications* 7 (1): 1–6.
46 Kumar, S., Brotóns-Gisbert, M., Al-Khuzheyri, R. et al. (2016). Resonant laser spectroscopy of localized excitons in monolayer WSe_2. *Optica* 3 (8): 882–886.
47 Linhart, L., Paur, M., Smejkal, V. et al. (2019). Localized intervalley defect excitons as single-photon emitters in WSe_2. *Physical Review Letters* 123 (14): 146401.
48 Geim, A.K. and Grigorieva, I.V. (2013). Van der Waals heterostructures. *Nature* 499 (7459): 419–425.
49 Novoselov, K.S., Mishchenko, A., Carvalho, A., and Neto, A.H.C. (2016). 2D materials and van der waals heterostructures. *Science* 353 (6298): aac9439.
50 Hunt, B., Sanchez-Yamagishi, J.D., Young, A.F. et al. (2013). Massive dirac fermions and hofstadter butterfly in a van der Waals heterostructure. *Science* 340 (6139): 1427–1430.
51 Cao, Y., Fatemi, V., Fang, S. et al. (2018). Unconventional superconductivity in magic-angle graphene superlattices. *Nature* 556 (7699): 43–50.
52 Tang, Y., Li, L., Li, T. et al. (2020). Simulation of Hubbard model physics in WSe_2/WS_2 moiré superlattices. *Nature* 579 (7799): 353–358.
53 Regan, E.C., Wang, D., Jin, C. et al. (2020). Mott and generalized Wigner crystal states in WSe_2/WS_2 moiré superlattices. *Nature* 579 (7799): 359–363.
54 Shimazaki, Y., Schwartz, I., Watanabe, K. et al. (2020). Strongly correlated electrons and hybrid excitons in a moiré heterostructure. *Nature* 580 (7804): 472–477.
55 Seyler, K.L., Rivera, P., Yu, H. et al. (2019). Signatures of moiré-trapped valley excitons in $MoSe_2/WSe_2$ heterobilayers. *Nature* 567: 66–70.
56 Brotons-Gisbert, M., Baek, H., Molina-Sánchez, A. et al. (2020). Spin-layer locking of interlayer excitons trapped in moiré potentials. *Nature Materials* 19: 630.
57 Chakraborty, C., Goodfellow, K.M., Dhara, S. et al. (2017). Quantum-confined Stark effect of individual defects in a van der Waals heterostructure. *Nano Letters* 17 (4): 2253–2258.
58 Xia, Y., Li, Q., Kim, J. et al. (2019). Room-temperature giant Stark effect of single photon emitter in van der Waals material. *Nano Letters* 19 (10): 7100–7105.

59 Brotons-Gisbert, M., Branny, A., Kumar, S. et al. (2019). Coulomb blockade in an atomically thin quantum dot coupled to a tunable Fermi reservoir. *Nature Nanotechnology* 14 (5): 442–446.

60 Branny, A., Kumar, S., Proux, R., and Gerardot, B.D. (2017). Deterministic strain-induced arrays of quantum emitters in a two-dimensional semiconductor. *Nature Communications* 8 (1): 1–7.

61 Rooney, A.P., Kozikov, A., Rudenko, A.N. et al. (2017). Observing imperfection in atomic interfaces for van der Waals heterostructures. *Nano Letters* 17 (9): 5222–5228.

62 Shanks, D.N., Mahdikhanysarvejahany, F., Muccianti, C. et al. (2021). Nanoscale trapping of interlayer excitons in a 2D semiconductor heterostructure. *Nano Letters* 21 (13): 5641–5647.

63 Andres-Penares, D., Brotons-Gisbert, M., Bonato, C. et al. (2021). Optical and dielectric properties of MoO_3 nanosheets for van der waals heterostructures. *Applied Physics Letters* 119 (22): 223104.

64 Hanson, R., Kouwenhoven, L.P., Petta, J.R. et al. (2007). Spins in few-electron quantum dots. *Reviews of Modern Physics* 79 (4): 1217.

65 Unuchek, D., Ciarrocchi, A., Avsar, A. et al. (2018). Room-temperature electrical control of exciton flux in a van der Waals heterostructure. *Nature* 560 (7718): 340–344.

66 Liu, Y., Dini, K., Tan, Q. et al. (2020). Electrically controllable router of interlayer excitons. *Science Advances* 6 (41): eaba1830.

67 Thureja, D., Imamoglu, A., Smoleński, T. et al. (2022). Electrically tunable quantum confinement of neutral excitons. *Nature* 606 (7913): 298–304.

68 Feng, J., Qian, X., Huang, C.-W., and Li, J. (2012). Strain-engineered artificial atom as a broad-spectrum solar energy funnel. *Nature Photonics* 6 (12): 866–872.

69 Palacios-Berraquero, C., Kara, D.M., Montblanch, A.R.-P. et al. (2017). Large-scale quantum-emitter arrays in atomically thin semiconductors. *Nature Communications* 8 (1): 1–6.

70 Proscia, N.V., Shotan, Z., Jayakumar, H. et al. (2018). Near-deterministic activation of room-temperature quantum emitters in hexagonal boron nitride. *Optica* 5 (9): 1128–1134.

71 Wang, Q., Maisch, J., Tang, F. et al. (2021). Highly polarized single photons from strain-induced quasi-1D localized excitons in WSe_2. *Nano Letters* 21 (17): 7175–7182.

72 Iff, O., Tedeschi, D., Martín-Sánchez, J. et al. (2019). Strain-tunable single photon sources in WSe_2 monolayers. *Nano Letters* 19 (10): 6931–6936.

73 Kim, H., Moon, J.S., Noh, G. et al. (2019). Position and frequency control of strain-induced quantum emitters in WSe_2 monolayers. *Nano Letters* 19 (10): 7534–7539.

74 Chakraborty, C., Mukherjee, A., Moon, H. et al. (2020). Strain tuning of the emission axis of quantum emitters in an atomically thin semiconductor. *Optica* 7 (6): 580–585.

75 Kögl, M., Soubelet, P., Brotons-Gisbert, M. et al. (2022). Moiré straintronics: a universal platform for reconfigurable quantum materials. *arXiv preprint arXiv:2207.12115.*

76 Bai, Y., Zhou, L., Wang, J. et al. (2020). Excitons in strain-induced one-dimensional moiré potentials at transition metal dichalcogenide heterojunctions. *Nature Materials* 19 (10): 1068–1073.

77 Ziegler, J., Klaiss, R., Blaikie, A. et al. (2019). Deterministic quantum emitter formation in hexagonal boron nitride via controlled edge creation. *Nano Letters* 19 (3): 2121–2127.

78 Choi, S., Tran, T.T., Elbadawi, C. et al. (2016). Engineering and localization of quantum emitters in large hexagonal boron nitride layers. *ACS Applied Materials & Interfaces* 8 (43): 29642–29648.

79 Hou, S., Birowosuto, M.D., Umar, S. et al. (2017). Localized emission from laser-irradiated defects in 2D hexagonal boron nitride. *2D Materials* 5 (1): 015010.

80 Fournier, C., Plaud, A., Roux, S. et al. (2021). Position-controlled quantum emitters with reproducible emission wavelength in hexagonal boron nitride. *Nature Communications* 12 (1): 1–6.

81 Wilson, N.R., Nguyen, P.V., Seyler, K. et al. (2017). Determination of band offsets, hybridization, and exciton binding in 2D semiconductor heterostructures. *Science Advances* 3 (2): e1601832.

82 Ciarrocchi, A., Unuchek, D., Avsar, A. et al. (2019). Polarization switching and electrical control of interlayer excitons in two-dimensional van der Waals heterostructures. *Nature Photonics* 13: 131–136.

83 Sigl, L., Troue, M., Katzer, M. et al. (2022). Optical dipole orientation of interlayer excitons in $MoSe_2$-WSe_2 heterostacks. *Physical Review B* 105 (3): 035417.

84 Schuller, J.A., Karaveli, S., Schiros, T. et al. (2013). Orientation of luminescent excitons in layered nanomaterials. *Nature Nanotechnology* 8 (4): 271–276.

85 Brotons-Gisbert, M., Proux, R., Picard, R. et al. (2019). Out-of-plane orientation of luminescent excitons in two-dimensional indium selenide. *Nature Communications* 10 (1): 1–10.

86 Li, W., Lu, X., Dubey, S. et al. (2020). Dipolar interactions between localized interlayer excitons in van der Waals heterostructures. *Nature Materials* 19: 624.

87 Kremser, M., Brotons-Gisbert, M., Knörzer, J. et al. (2020). Discrete interactions between a few interlayer excitons trapped at a $MoSe_2$-WSe_2 heterointerface. *npj 2D Materials and Applications* 4 (1): 1–6.

88 Kang, J., Li, J., Li, S.-S. et al. (2013). Electronic structural Moiré pattern effects on $MoS_2/MoSe_2$ 2D heterostructures. *Nano Letters* 13 (11): 5485–5490.

89 Yu, H., Liu, G.-B., Tang, J. et al. (2017). Moiré excitons: from programmable quantum emitter arrays to spin-orbit–coupled artificial lattices. *Science Advances* 3 (11): e1701696.

90 Liu, E., Barré, E., van Baren, J. et al. (2021). Signatures of moiré trions in $WSe_2/MoSe_2$ heterobilayers. *Nature* 594 (7861): 46–50.

91 Baek, H., Brotons-Gisbert, M., Campbell, A. et al. (2021). Optical read-out of Coulomb staircases in a moiré superlattice via trapped interlayer trions. *Nature Nanotechnology* 16: 1237–1243.

92 Brotons-Gisbert, M., Baek, H., Campbell, A. et al. (2021). Moiré-trapped interlayer trions in a charge-tunable $WSe_2/MoSe_2$ heterobilayer. *Physical Review X* 11 (031033): 1–12.

93 Britnell, L., Gorbachev, R.V., Jalil, R. et al. (2012). Electron tunneling through ultrathin boron nitride crystalline barriers. *Nano Letters* 12 (3): 1707–1710.

94 Mukherjee, A., Shayan, K., Li, L. et al. (2020). Observation of site-controlled localized charged excitons in CrI_3/WSe_2 heterostructures. *Nature Communications* 11 (1): 1–8.

95 Shayan, K., Liu, N., Cupo, A. et al. (2019). Magnetic proximity coupling of quantum emitters in WSe_2 to van der Waals ferromagnets. *Nano Letters* 19 (10): 7301–7308.

96 Flatten, L.C., Weng, L., Branny, A. et al. (2018). Microcavity enhanced single photon emission from two-dimensional WSe_2. *Applied Physics Letters* 112 (19): 191105.

97 Schell, A.W., Takashima, H., Tran, T.T. et al. (2017). Coupling quantum emitters in 2D materials with tapered fibers. *ACS Photonics* 4 (4): 761–767.

98 Vogl, T., Lecamwasam, R., Buchler, B.C. et al. (2019). Compact cavity-enhanced single-photon generation with hexagonal boron nitride. *ACS Photonics* 6 (8): 1955–1962.

99 Sortino, L., Brooks, M., Zotev, P.G. et al. (2020). Dielectric nanoantennas for strain engineering in atomically thin two-dimensional semiconductors. *ACS Photonics* 7 (9): 2413–2422.

100 Luo, Y., Shepard, G.D., Ardelean, J.V. et al. (2018). Deterministic coupling of site-controlled quantum emitters in monolayer WSe_2 to plasmonic nanocavities. *Nature Nanotechnology* 13 (12): 1137–1142.

101 White, D., Branny, A., Chapman, R.J. et al. (2019). Atomically-thin quantum dots integrated with lithium niobate photonic chips. *Optical Materials Express* 9 (2): 441–448.

102 Peyskens, F., Chakraborty, C., Muneeb, M. et al. (2019). Integration of single photon emitters in 2D layered materials with a silicon nitride photonic chip. *Nature Communications* 10 (1): 1–7.

103 Errando-Herranz, C., Schll, E., Picard, R. et al. (2021). Resonance fluorescence from waveguide-coupled, strain-localized, two-dimensional quantum emitters. *ACS Photonics* 8 (4): 1069–1076.

21

Nano-Opto-Electro-Mechanical Systems for Integrated Quantum Photonics

Leonardo Midolo and Celeste Qvotrup

Niels Bohr Institute, University of Copenhagen, Blegdamsvej 17, 2100-DK, Copenhagen, Denmark

21.1 Introduction and Overview

The widespread adoption of integrated circuits to scale up quantum technologies has produced a tremendous impact in the quantum photonics community over the past decade. The combination of improved epitaxial growth techniques and advanced fabrication methods, as well as deeper insight into the field of material science, enables the development of ever more advanced nano-scale devices, opening new avenues in photonic integrated technologies [1]. The most important technological development in this regard is the possibility to miniaturize and integrate a multitude of photonic devices in a single small-scale chip, thereby reducing costs and number of production cycles, as well as closing the gap between laboratory-based demonstrations of quantum devices and their applications in real-life scenarios. In fact, quantum photonic integrated circuits (QPIC) enable routing and controlling light with unprecedented flexibility, without resorting to expensive and bulky free-space optical components. For example, the well-known boson sampling experiments [2], which constitute an important benchmark of our ability to control an ever-increasing number of photons and optical paths, are ultimately limited by the number of optical elements that we can possibly manipulate, by the footprint of the experiments, the costs, and potentially the power dissipation. Recent demonstrations that include over 100 distinct optical modes [3] are already prohibitively expensive for most laboratories, since photon sources, detectors, nonlinear devices, and reconfigurable optical routers are not yet integrated. It is indeed the integration of light sources with large multi-path interferometers and photon detectors into a common, unified, platform that could drastically improve the scaling of experiments such as boson sampling and enable a more accessible platform for photonic quantum technologies. Recently, a large effort has been devoted to the integration of light switches and routers in QPICs, especially in the mature silicon-on-insulator (SOI) waveguide platform [4–6]. Such efforts stem from the development of classical photonic circuits, where fast and low-power switching is a key asset for electro-optical communications (e.g. optical packet routing) and where several (including commercial) well-established solutions exist.

Photonic Quantum Technologies: Science and Applications, First Edition. Edited by Mohamed Benyoucef.

Integrated optical routers require some means of changing the refractive index of the medium where light propagates. The most common and simple method for routing builds on the thermo-optic effect, i.e. the change of the refractive index occurring in dielectric materials upon the change of temperature [7]. Being a direct consequence of thermal expansion in solids, this method has a known drawback: it prevents its direct integration with emitters and detectors, which operate at cryogenic temperatures, at which thermal expansion is negligible. On top of this, thermal cross talk between different switches mediated by the chip substrate is unavoidable, which makes the independent control of each router in the circuit more challenging. Yet, to date, the vast majority of experimental demonstrations featuring reconfigurable QPICs use thermo-optic switches, as they are small and simple to manufacture (a thin metallic strip line in proximity of a waveguide with a current flowing in can act as a local heater). While several alternatives exist, such as electro-optic switches using the Pockels effect [8], free-carrier injection [9], or electro-absorption [10], they often rely on intrinsic material properties that only occurs in some materials (for example the Pockels effect occurs in materials lacking inversion symmetry). For this reason, the most common switching mechanism in the SOI or silicon nitride (SiN) platforms (which are inversion symmetric) is the thermo-optic effect.

At the time of writing, a wide range of materials and platforms are being investigated, yet none of them seem to offer the required adaptability to implement simultaneously the functionalities for generating, detecting, routing, and manipulating photons at the quantum level. A very promising solution is the hybrid integration of different materials into a common substrate, which, however, requires novel and more advanced manufacturing techniques such as transfer printing [11], wafer bonding [12], or flip-chip assembly [13]. Alternatively, by focusing on materials capable of containing single-photon sources, such as direct bandgap semiconductors or other materials with quantum emitters, the attention could be turned toward identifying new methods for switching and routing light in these materials. By doing so, existing platforms with mature single-photon sources, such as gallium arsenide (GaAs), diamond, or indium phosphide (InP), could be enriched to implement the routing and switching functionalities required for quantum photonics and potentially further enhanced by the aforementioned hybrid integration techniques.

The topic of this chapter concerns the emerging technology of nano-opto-electro-mechanical systems, or NOEMS for short [14, 15]. In NOEMS, devices and circuits are designed such that the mechanical, optical, and electrical degrees of freedom are coupled, as shown schematically in Fig. 21.1(a). For example, a device could contain movable parts controlled by electrical forces, which in turn influence the optical properties of the device. The mechanical degree of freedom thus acts as a mediator between the electrical and optical degrees of freedom, creating an *effective* electro-optic interaction. Figure 21.1(b) shows an example of NOEMS with an artistic illustration of a waveguide-integrated device for controlling the wavelength tuning of a photonic crystal nano-cavity. In this case, an electrostatic force is produced by a capacitive actuator that changes the distance between two one-dimensional coupled photonic crystal resonators. The cavity wavelength shifts as a result of the change of coupling strength between the two resonators, as in standard cavity opto-mechanics experiments [16].

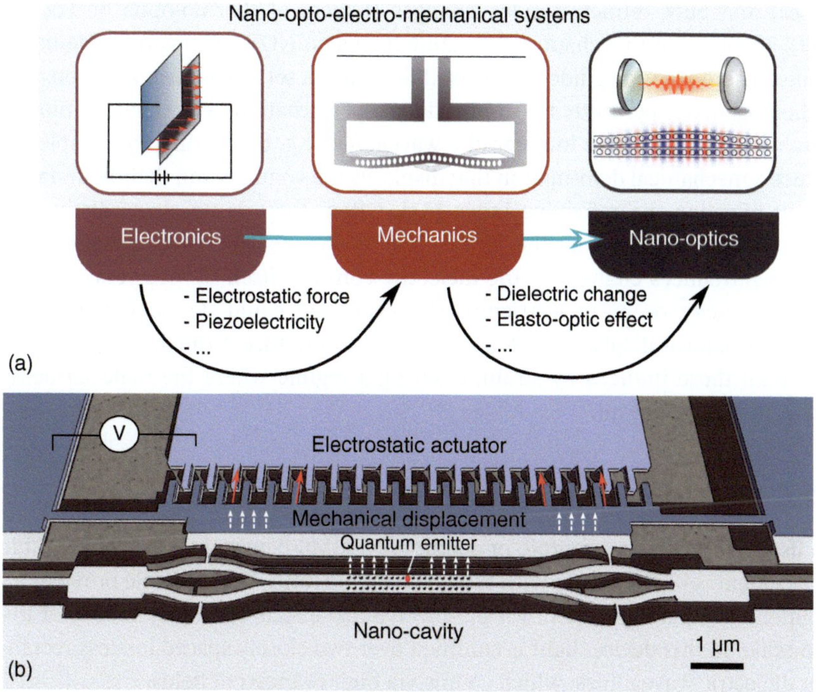

Figure 21.1 (a) Physical mechanisms involved in nano-opto-electro-mechanical systems that enable an effective electro-optic interaction mediated by mechanical degrees of freedom. Source: Leonardo Midolo. (b) Concept illustration of an NOEMS-based tunable photonic crystal cavity with embedded quantum emitter in a waveguide circuit, which could be used for, e.g. cavity quantum electrodynamics or as a tunable filter. An electrostatic comb drive with an applied bias V produces a force (red arrows) that results in a nano-mechanical displacement (white dashed arrows). The motion changes the resonant wavelength of a photonic crystal nano-cavity.

We focus on the development of chip-scale and integrated reconfigurable devices based on NOEMS, including photon routers, tunable cavities, mirrors, and filters and discuss their role in integrated quantum photonics as well as some of the challenges in their manufacturing. While NOEMS are still an emerging technology, with few demonstrations in the literature, the interest in their development is rapidly growing, and its adoption in the quantum domain is likely to become a reality in the next decade.

21.2 Device Principles

Light propagation in dielectric media is prescribed and controlled by the spatial distribution of the complex dielectric permittivity $\varepsilon(r)$. To dynamically change how light propagates, ε must vary to some extent, to introduce phase shifts, scattering, or reflections by means of an external control. Here, a distinction is required between so-called spatial changes of the dielectric boundaries (typical of mechanical devices) and bulk refractive-index changes (typical of thermo-optic or Pockels effect). Throughout this chapter, we mainly focus on NOEMS in waveguide-based circuits, where the waveguide and its cladding form a set of volumes with constant and homogeneous permittivity. To introduce a perturbation, we can either modify the value of the refractive index in the waveguide (e.g. by heating the sample) or produce a mechanical deformation that displaces the volumes and their boundaries without affecting the refractive index [17]. Often, there is no sharp distinction between these methods, as the strain due to the mechanical displacement of volumes introduces changes to the dielectric constant itself or vice versa (due to thermo-elasticity or piezoelectricity). However, one could devise structures in which mechanical displacements are designed to produce a much more sizeable effect than those induced by strain, entering a regime where the material being used is, in fact, irrelevant.

21.2.1 Tunable Beam Splitters

It is useful to make an example of a simple NOEMS device, based on waveguide displacement: a beam splitter with tunable splitting ratio. This example provides an enlightening explanation of how opto-mechanical effects can be enhanced at the nano-scale. In this device, light is confined over two closely spaced lossless rectangular dielectric waveguides, which couple via their evanescent fields.

We denote the scalar complex amplitude of the field, propagating along the z direction, in each waveguide with $a_1(z)$ and $a_2(z)$ and their respective propagation constants with β_1 and β_2. For waveguides with infinite separation, a_1 and a_2 do not vary along z. However, if the waveguides are close enough that their evanescent fields have a significant overlap, the light confined in one of the waveguides will tunnel to the other waveguide and back again, with a final distribution between the waveguides depending on the length of the coupling section as illustrated in Fig. 21.2(a).

According to coupled-mode theory [18], which is a perturbation theory that holds in the limit of lossless waveguides and small coupling, the interaction between

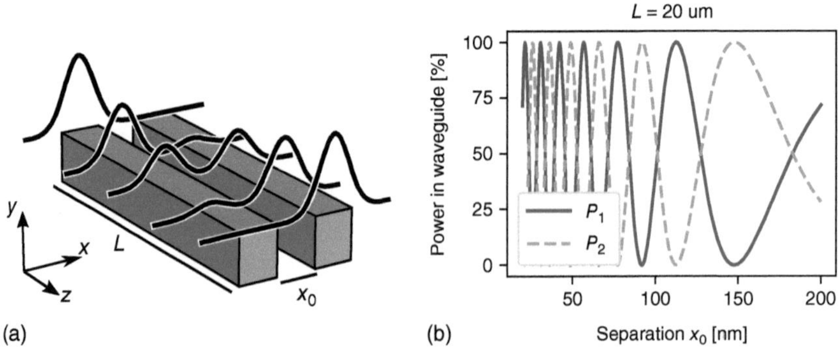

Figure 21.2 (a): Sketch of tunable beam splitter with light oscillating between the waveguides. (b): Light routing behavior in GaAs-in-air waveguides with waveguide width $w = 200$ nm and fixed coupling length $L = 20$ micron. Source: Celeste Qvotrup.

waveguides is linear and characterized by a real-valued coupling strength g. For evanescently coupled waveguides, a good approximation of g is given by

$$g = g_0 e^{-\gamma x}, \tag{21.1}$$

where g_0 is a constant dependent on the material properties and dimensions of the waveguides, γ is the photonic tunneling constant (also given by the waveguide shape), and x is the distance between the guides. The field evolution in the two waveguides is given, in matrix form, by the first-order differential equation:

$$\frac{d}{dt}\begin{bmatrix} a_1(z) \\ a_2(z) \end{bmatrix} = i\begin{pmatrix} \beta_1 & g \\ g & \beta_2 \end{pmatrix}\begin{bmatrix} a_1(z) \\ a_2(z) \end{bmatrix}. \tag{21.2}$$

Integrating equation 21.2 over the length of the device L allows calculating the transmission across the device. The coupling matrix $\mathbf{A}$ can be diagonalized and written in the form $\mathbf{A} = \mathbf{PDP}^{-1}$ where $\mathbf{D}$ is the eigenvalue matrix and $\mathbf{P}$ is the eigenvector matrix. The solution is

$$\begin{bmatrix} a_1(L) \\ a_2(L) \end{bmatrix} = \mathbf{P}e^{i\mathbf{D}L}\mathbf{P}^{-1}\begin{bmatrix} a_1(0) \\ a_2(0) \end{bmatrix}, \tag{21.3}$$

which holds even for asymmetric waveguides. If the waveguides are perfectly symmetric ($\beta_1 = \beta_2 = \beta$), full power transfer between the two waveguides is possible and the solution simplifies greatly. The two eigenvalues are the propagation constants of the symmetric (or bonding) super-mode $\beta_S = \beta + g$ and anti-symmetric (or anti-bonding) super-mode $\beta_{AS} = \beta - g$. The eigenvectors for the symmetric [1, 1] and anti-symmetric [−1, 1] modes indicate that at the entrance and at the output of the coupling section, the power is equally split between the two super-modes of the coupled system. Assuming light is initially entering from waveguide 1, the resulting field at the output is simply

$$a_1(L) = e^{i\beta L}\cos(gL)a_1(0), \quad a_2(L) = ie^{i\beta L}\sin(gL)a_1(0). \tag{21.4}$$

The output power on waveguide 1 is therefore given by $\cos^2(gL)$. Figure 21.2(b) shows a plot of the expected power output as a function of the distance between two 200-nm-wide gallium arsenide waveguides (index $n = 3.5$) for a fixed coupling length of 20 μm. As expected, the smaller the distances involved, the stronger the opto-mechanical interaction becomes.

We observe that the product $g \cdot L$ represents the equivalent of the phase shift between the arms of a Mach–Zehnder interferometer. Such equivalence allows us to draw a direct comparison between a gap-variable beam splitter and thermo-optic or electro-optic phase shifters, where the phase shift per unit length induced by a variation of refractive index Δn is given by $\Delta\phi/L = 2\pi\Delta n/\lambda$ (here λ is the free-space wavelength). By linearizing the coupling factor g around a working point x_0, i.e. the initial separation between two waveguides, we note that the phase shift per unit length as a function of the displacement Δx is

$$\Delta\phi/L = g_0 e^{-x_0\gamma}\gamma\Delta x \tag{21.5}$$

which can be greatly enhanced at small values of x_0. For example, with the geometry of Fig. 21.2,(b) a separation of 100 nm results in phase shifts per unit length in the order of 0.2–0.5 rad/μm with displacements Δx in the order of 50 nm, which means a π-shift can be achieved with a coupling length as short as 10 μm. This corresponds to a bulk index change in the order of $\Delta n \sim 0.02$ or larger, which is very challenging to achieve conventionally. For example it would require over 100 K temperature change in Silicon to obtain a comparable phase shift per unit length at room temperature, where the thermo-optic coefficient $dn/dT = 1.8 \cdot 10^{-4}\ \mathrm{K}^{-1}$ [7].

21.2.2 Device Speed

Another important aspect of NOEMS is its dynamical response. Neglecting the response time of the electrical circuit used for controlling a switch, its response time is ultimately limited by the type of actuation involved, and whether it is purely a bulk effect or if it involves mechanical displacements. Figure 21.3 shows how the various effects discussed so far are typically designed to perform in terms of speed. Bulk electro-optic effects are usually the fastest (Pockels effect can reach picoseconds response time) enabling GHz switching speed, while piezoelectric devices can be operated up to tens of MHz and electrostatic device in the few MHz range. In electro-mechanical actuation, the speed is limited by the fundamental mechanical frequency of the device. Such frequency scales as $\omega = \sqrt{k/m_{\mathrm{eff}}}$, where k is the lumped stiffness of the mechanical mode, and m_{eff} is its effective mass (assuming the mechanical device can be reduced to a simple mass-spring system) [14]. At the nano-scale, both the stiffness and the mass are reduced, while their ratio scales proportionally with the square of the characteristic device length L, yielding a linear increase in frequency $\omega \propto L$. Thus, smaller devices enable faster actuation. A characteristic of NOEMS that distinguishes it from standard micro-electro-mechanical systems (MEMS) is the ability to push the actuation rate below the microsecond timescales. We note that this timescale is particularly significant in quantum

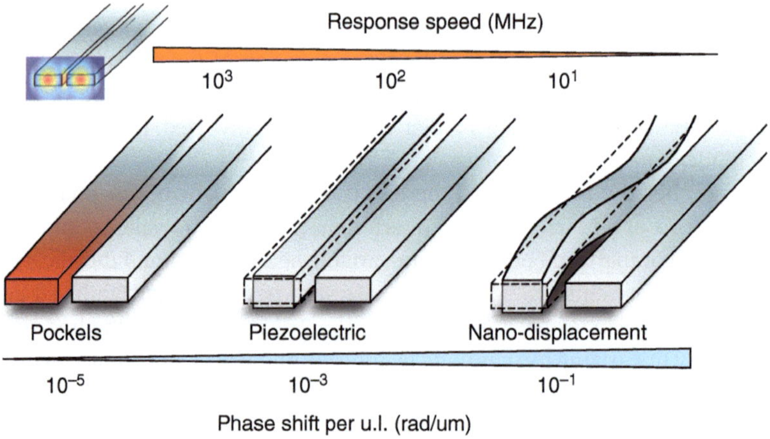

Figure 21.3 Comparison of various switching methods using two parallel waveguides. The optical mode is shown in the upper left inset. An indicative device response speed is provided together with the magnitude of the expected optical response, expressed in phase shift per unit length. Source: Leonardo Midolo.

photonics, as it corresponds to the indistinguishability window of most quantum emitters. For example, trains of highly indististinguishable (> 97%) single photons emitted from InAs quantum dots over >1 μs have been recently demonstrated in photonic crystal waveguides [19]. A photon router capable of switching beyond 1 MHz rate would then prove very useful for tasks such as demultiplexing these photons and building parallel streams of indistinguishable photons [20–22].

21.2.3 Electro-mechanical Actuation

Having established how mechanical displacements affect light propagation in NOEMS, we shift the discussion to the electro-mechanical control that enables connecting electrical and optical degrees of freedom. In pure opto-mechanical systems, optical forces are exploited to induce mechanical displacements. These forces arise from the Lorenz force and are usually computed via the Maxwell stress tensor. However, radiation pressure itself is very weak, for example the linear force density between two parallel evanescently coupled waveguides carrying approximately 1 mW of infrared light is in the order of few pN/μum, which is usually insufficient to achieve any sizeable effect [17]. Using static or quasi-static electric fields to achieve mechanical deformations enables producing much stronger forces, in the order of tens to hundreds of nN, and displaces objects irrespective of their material composition. Moreover, at the nano-scale, electro-mechanical effects are also greatly amplified. For example, the force per unit area F_A between two parallel plates in vacuum with an applied voltage V reads

$$F_A = \frac{\varepsilon_0 V^2}{(x_0 - \Delta x)^2}, \tag{21.6}$$

where ε_0 is the permittivity of vacuum, and Δx (x_0) is the displacement (initial distance) between plates. It is evident that the force is inversely proportional to the square of the gap distance between the plates, thus smaller devices enable larger deformations at lower applied bias. For most NOEMS designs such as the example of Fig. 21.1(b), a full three-dimensional numerical simulation is required to account for fringing effects; however, this does not constitute a limitation for the miniaturization of electrostatic drives [23]. In the limit of linear elasticity, the mechanical displacement is given by Hooke's law $F = -k\Delta x$, where we assume that the entire device can be lumped into a single spring with constant stiffness k. For example, the typical stiffness of 20 μm-long dielectric waveguides clamped at both ends is in the order of 1 N/m, thus a force of roughly 50 nN is required to achieve 50 nm displacement. Such force can be easily obtained with comb-drive geometries or simple parallel metallic wires with moderate voltages in the 5 – 20 V range and gaps x_0 around 100 – 300 nm.

Several other electromechanical effects are available. While their adoption is somewhat less frequent in the literature, it is worth mentioning them, as they could be employed in future generations of NOEMS.

1. Piezoelectricity is a material-dependent effect which, as the Pockels effect, requires a lack of inversion symmetry in the crystal [24]. The piezoelectric effect

creates an electric charge as a reaction to mechanical stress or conversely induces a strain as a response to an applied electric field. The components of the strain that induces an electric response - and the direction of such response - depends highly on the structure and degree of symmetry in the crystal. With proper device design, e.g. using bended devices, the piezoelectric effect can provide enough force to achieve large optical response in NOEMS at lower voltages than electrostatic devices [25] and boost the opto-electro-mechanical response speed to the GHz [26].

2. Electrostriction is similar to piezoelectricity, in the sense that it enables mechanical motion by applying an electric field. The main difference with piezoelectricity is that it does not require a lack of inversion symmetry, and it is thus available in most materials. However, the effect is rather weak and scales with the square of the applied field, thus it is often compared to radiation pressure as it can be induced by a strong propagating optical beam [27].
3. Flexoelectricity [28] is an effect in which a strain gradient induces a linear electric field. In comparison to piezoelectricity, which requires a lack of inversion symmetry, an object that exhibits a strain gradient across a certain direction provides an intrinsic symmetry breaking and results in an electric field, irrespective of the materials. While the flexoelectric effect tends to be small compared to the piezoelectric effect, recent years have seen a renewed interest in the topic. As the strain gradient for a bent object increases as it is uniformly scaled down, the flexoelectric contribution is becoming more and more sizeable [29]. Similarly to piezoelectricity, flexoelectricity, the components of strain gradient that induces a field, and which directions the field is induced in, are heavily material-dependent. The inverse flexoelectric effect (whereby an external field induces a mechanical deformation) is however challenging to implement, and further research is needed to exploit such mechanism in practical devices.

21.3 NOEMS Fabrication

The manufacturing of NOEMS requires a combination of bulk and surface micromachining and the growth or deposition of layers of sacrificial materials, which will be etched away at the end of the process, releasing the upper layers and allowing them to freely move in space. Figure 21.4 highlights the main processing steps required for the fabrication of NOEMS in an epitaxially grown semiconductor stack of GaAs. Without loss of generality, the steps can be summarized as follows: 1) Electrodes and pattern definition via electron-beam lithography. 2) Dry etching. 3) Sacrificial undercut and drying.

Owing to the nanometer size of the features involved, NOEMS require pattern definition via electron beam lithography, instead of more conventional optical lithography. Sub-10 nanometer resolution is often required in waveguide-based routers and switches or for the realization of periodic elements such as circles making up photonic crystal mirrors or cavities. The resulting dimensions need to be accurate within few nanometers, to avoid drifting from the design parameters and achieve

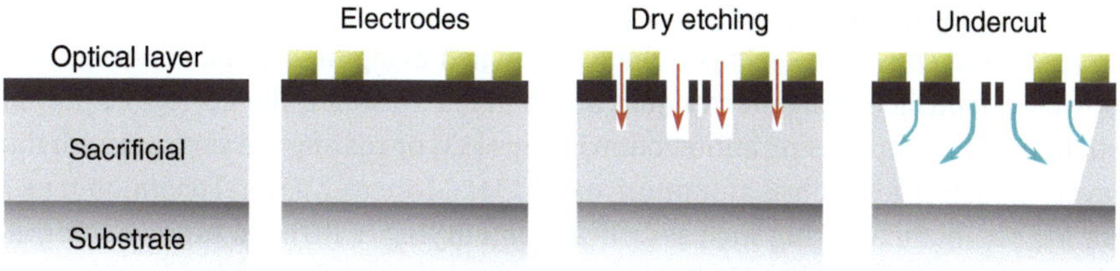

Figure 21.4 Typical NOEMS fabrication procedure. From left to right: Starting with a wafer consisting of an optical waveguide layer on top of a sacrificial layer grown on top of a substrate. Electrodes are defined with electron beam lithography and deposited via physical vapor deposition. Waveguides and other features on the surface wafer are defined using deep reactive ion etching. Finally, the sacrificial layer is removed using a wet etching procedure followed by careful drying to avoid sticking parts. Source: Leonardo Midolo.

the desired opto-mechanical or electro-mechanical coupling. Typical electron-beam lithography tools required for NOEMS fabrication provide a large acceleration voltage (> 50 keV) and state-of-the-art proximity effect correction software [30] that compensates for both short-range and long-range electron spread in the substrate. NOEMS constitute in fact an excellent test bed and benchmark for ultra-precise nanolithography.

Transferring the designed features on the optical layer requires methods such as reactive ion etching, usually in combination with an inductively coupled plasma to increase the verticality and anisotropy of the structures. To achieve a good fidelity between the lithographically defined pattern and the resulting etched pattern, materials with good resistance to the etching chemicals should be used (also known as hard masks) [31]. Silicon nitride or silicon dioxide thin films or Cr layers offer the best protection to plasma damage and suitable hard masks for most etching processes. Alternatively, electron beam resists like ZEP520 can be used as protective films at the expense of a more complex cleaning or stripping procedure [32].

Following the dry etching, a wet etching process is usually carried out to perform a so-called sacrificial layer undercut, i.e. the selective removal of one or more layers to release suspended devices such as waveguides or membranes. The choice of a wet process as opposed to dry etching process is usually due to its higher selectivity between materials, which enables undercutting electrodes, waveguides, and other suspended features. Well-known examples of selective wet etching are the removal of silicon dioxide from silicon, or AlGaAs from GaAs using hydrofluoric acid, or InP from InGaAsP in hydrochloric acid, or silicon from silicon nitride in potassium hydroxide. Wet etching and the subsequent drying are the most delicate steps in NOEMS manufacturing as mechanical failures are very likely to occur. Stiction, i.e. the permanent collapse of suspended devices into each other, is the most common failure observed at this stage. There are a wide range of physical mechanisms that cause stiction, from Casimir or short-range Van der Waals forces to electrostatic forces or from turbulence in the etching fluids to capillary forces during drying. We refer the reader to Ref. [33] for a review on some of these mechanisms and their physical origin.

Controlling the wet etching and the drying steps is crucial in the realization of NOEMS and constitutes a major challenge in achieving high yield. However, a set of well-established design techniques are available to stiffen the devices or prevent stiction, for example using anti-stiction bumps [33] or coating the surfaces with thin layers of dielectrics such as aluminum oxide [34]. Moreover, several techniques used for MEMS manufacturing such as critical point drying (CPD) and vapor etching are also commonly applied to NOEMS. CPD is the process of removing liquid from a chip without forming any liquid–air interface, which is usually responsible for stiction due to capillary forces. It is a technique widely used for drying delicate biological specimens for scanning electron microscope imaging. In a CPD chamber, highly pressurized liquid CO_2 is gradually introduced to replace the previous liquid (usually acetone or ethanol) and subsequently heated to transition into the supercritical state. By gradually reducing the pressure, CO_2 undergoes a continuous rather than a discrete phase transition from liquid to gas, eliminating the surface tension associated with crossing a phase boundary. Alternatively, vapor etching can be used instead of wet etching by exposing the sample to a controlled flow of acid vapors. With the proper etching rate, no liquid is introduced, and the release of the mechanical parts can be performed without stiction. Finally, hybrid techniques exist, where additional sacrificial layers (e.g. a polymer or a thin dielectric film) are deposited prior to wet etching to stiffen the structure and avoid its collapse [35].

21.4 Application of NOEMS to Quantum Photonics

The possibility to reconfigure an optical element dynamically and controllably in a photonic integrated circuit is a key asset in quantum applications. Tunable elements not only enable compensating for unavoidable fabrication errors and inconsistencies, but also provide an additional knob for developing new functionalities. We will now review some of the important applications of NOEMS in QPIC demonstrated in the recent literature, which are schematically illustrated in Fig. 21.5(a).

21.4.1 Routing and Switching Single Photons

The tunable directional coupler discussed in this chapter is a paradigmatic example of an individual element, which, if scaled, could produce a large, programmable multi-mode interferometer with tens or hundreds of ports. As pointed out earlier, a large network of beam splitters, as found in experiments such as boson sampling, quantum simulators, or neuromorphic computers, requires potentially a large footprint, which cannot be ignored when approaching the hundreds of optical modes. Replacing each tunable element with NOEMS could result in up to a factor of 1000 reduction in length and a significant reduction of power consumption.

Several implementations of NOEMS-based routers have been demonstrated. In GaAs, a mechanical beam splitter (cf. Fig. 21.5(e)) for routing photons emitted by single quantum dots has been realized [38]. Single photons at different wavelengths have been dynamically routed between two waveguides using a bias

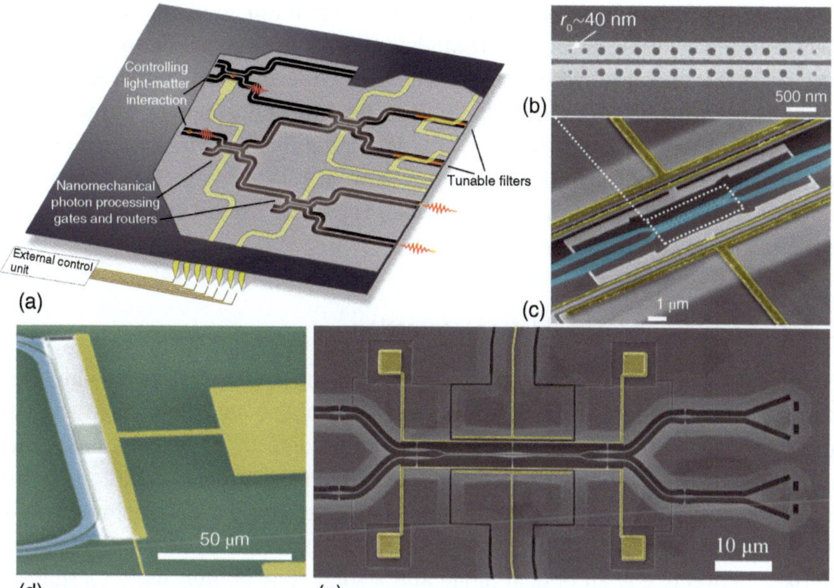

Figure 21.5 Applications of NOEMS in quantum photonic integrated circuits and microscope images of devices. (a) Illustration of a QPIC with tunable functionalities implemented with NOEMS. Source: Leonardo Midolo. (b) Scanning electron microscope (SEM) micrograph of coupled photonic crystal cavity (detail) and (c) its integration in a wavelength-tunable filter in GaAs. Source: Adapted with permission from Zhou et al. [36]/ John Wiley & Sons, Inc. (d) False-color SEM micrograph of a phase shifter realized in silicon nitride waveguides. Adapted with permission Poot et al. [37]/AIP Publishing. (e) A nanomechanical single-photon router in GaAs (source: Leonardo Midolo). Metal electrodes are highlighted in yellow in all the above examples.

voltage up to 10 V. Such device implements a tunable 1-axis rotation gate with switching times potentially up to 370 ns when the device is operated resonantly. An alternative approach uses out-of-plane motion of waveguides to switch on and off the coupling [39, 40]. By using a vertical actuation mechanism between the waveguide and the substrate, an arm of a directional coupler can be entirely decoupled. This effect has been demonstrated both in SOI and in GaAs with single quantum emitters. More advanced designs for signal routing have been suggested and realized using double membranes, which use the out-of-plane coupling between two thin waveguide slab to obtain tunable devices [41]. NOEMS switches have also been developed using plasmonic disk resonators [42]. Unlike broadband devices, resonators coupled to two waveguides can be used as switching elements (e.g. in add-drop configurations) for a single frequency of light by tuning the resonator via NOEMS. In their work, the authors have also demonstrated a full matrix switches capable of routing signals between different ports in a complementary metal-oxide semiconductor (CMOS) compatible process. In this case, the device employs electrostatic forces to pull a metallic disk located 35 nm above a Silicon disk resonator. Plasmonic effects in the tiny gap cause a sizable wavelength shift that detunes the resonance by over 6 nm, enabling the switching effect between multiple ports. A NOEMS-based tunable phase shifter has also been realized in the silicon nitride platform [37]. In such implementation, a waveguide is slightly perturbed from the side, to avoid dissipating optical power, by means of a planar electrostatic actuator (see Fig. 21.5(d)). A π shift was demonstrated using a total device length of 100 μm and a voltage drive of 10 V. Even better performance has been recently reported in silicon nitride [43] where up to 13π phase shift was demonstrated, albeit in longer devices.

21.4.2 Controlling Light–Matter Interaction

Dynamical device reconfiguration becomes even more attractive in combination with quantum emitters, as it offers a versatile tool for controlling light–matter interaction. Quantum emitters such as quantum dots or color centers are not only excellent single-photon sources, but also enable nonlinear operations at the single-photon level or long-lived quantum memories (for example using the spin degree of freedom) [44]. Building a controlled light–matter or photon–emitter interface between, e.g. an optical mode and a solid state emitter requires shaping the dielectric environment surrounding the emitter to control the local density of optical states (LDOS) [45]. As LDOS is a combined property of the emitter and its environment, it is evident that NOEMS could play an important role in shaping light–matter interaction without affecting the emitter. Moreover, they would allow a more deterministic fabrication of single-photon sources, with higher yield and scalability. For example, self-assembled quantum dots nucleate at random positions in the wafer and emit at random energies over several tens of meV. Various techniques have been employed to adjust the emission wavelength, e.g. via the quantum-confined Stark effect [46] or via strain tuning [47, 48], but the amount of control is typically very limited. Pre-characterization and pre-localization of the emitters before fabrication are also a widely used strategy. Yet, fabrication errors always occur, and small energy drifts over time are always to be expected.

Combining the emitters with NOEMS greatly simplifies the process of controlling light–matter interaction and could be performed in a local and scalable fashion. An interesting example is given by the tunable cavities realized with photonic crystals. The first demonstrations of mechanically programmable or tunable cavities were performed with parallel one-dimensional photonic crystals [49, 50], slotted photonic crystals [51], and in 2D cavities in GaAs with single quantum emitters [52]. Here, a double-membrane photonic crystal cavity was tuned in and out of resonance with a quantum dot resulting in a tunable and controllable emitter lifetime, with up to a fourfold Purcell enhancement. The development of mechanically tunable cavities is likely to have a large impact in quantum photonics, especially if more conventional structures such as ring or disk resonators could be realized as well, for example using double-membrane devices.

Tunable cavities could also play a major role in suppressing phonon side-band noise in solid-state quantum emitters. The interaction with phonons at a finite temperature typically produces emission side bands at different energies, which make the photons distinguishable and thus reduce the performance of single-photon sources. To avoid such undesired effect, the source can either be Purcell-enhanced with a nano-cavity or filtered to suppress the unwanted side bands. Filtering is usually performed with external high-finesse etalons or gratings. However, scaling up the technology further will require the implementation of on-chip tunable filters [53]. Recently, an NOEMS photonic crystal cavity whose wavelength can be continuously tuned electrostatically over 10 nm has been demonstrated [36] (Figs. 21.5(b) and 21.5(c)). While the insertion loss and quality factor are still relatively modest for filtering phonon sidebands, the device has been used to filter the above band excitation laser with comparable extinction performance as standard free-space grating filters.

21.5 Challenges and Perspectives

NOEMS have a unique standpoint in quantum photonics: they enable controlling light without resorting to a specific choice of material, and they are compatible with cryogenic operation, which is essential for the integration with emitters and detectors. To conclude this chapter, we highlight some of the challenges that NOEMS technology is likely to face in the area of quantum photonics, as well as some exciting perspectives.

The combination of phase shifters with a tunable beam splitter will enable implementing arbitrary rotations of path-encoded photonic qubits based exclusively on NOEMS. Here, a key aspect that requires deeper investigation is the stochastic gate noise associated with nano-mechanical motion. While it has been pointed out that any systematic error in programming a quantum gate can be ideally corrected in post-processing, the stochastic noise usually cannot be accounted for and constitutes an important parameter to evaluate the quality of a switch [54]. Stochastic noise in NOEMS originates from various sources: Brownian motion of the mechanical system due to the fluctuation-dissipation theorem, optomechanical back-action, voltage noise, etc. The noise that stems from intense optical beams such as opto-mechanical back-action, photoelastic effects or thermo-refractive effects are usually of little concern in this case, as the devices are operated at the single-photon level. At cryogenic temperatures (T =1–10 K), the average thermo-mechanical noise $\langle \delta x \rangle$ associated with small nano-mechanical waveguides is expected to be in the order of 10^{-3}–10^{-1} nm, while it could vary significantly depending on the specific implementation. With the definition of phase provided in equation 21.5, we can estimate a phase sensitivity as $L\frac{dg}{dx}$, which for the parameters used in Fig. 21.2 amounts to $\simeq$ 20 mrad/nm [38]. Therefore, the amount of root mean square stochastic noise $\langle \delta \phi \rangle$ due to thermo-mechanical noise is expected to fall well below $\langle \delta \phi \rangle < 10^{-5}$ rad, although a full characterization is needed to confirm this. Such values suggest that stochastic phase noise of quantum gates implemented with NOEMS could be on par or lower than what is achieved with conventional thermo-optic devices. Additionally, an electro-mechanical noise can arise from fluctuations in the voltage source used to drive the electrostatic actuator.

Depending on the device construction and geometry, a sub-mV rms voltage noise is required to suppress detectable noise in current implementations.

Another important challenge associated with NOEMS devices lies in their fabrication: in a photonic integrated circuit, the integration of movable parts is likely to cause additional issues in manufacturing and packaging. A direct consequence to this is that building a back-end of layer with electrical interconnects [55] that encapsulates the whole wafer surface is not trivial. Techniques such as chemical mechanical polishing (used for planarization or multi-layer interconnect fabrication) are incompatible with NOEMS fabrication. Packaging a chip with many integrated nano-mechanical devices will require the development of new methods. Flip-chip techniques seem to be the most promising to avoid developing electrical interconnects with wire-to-wire and wire-to-waveguide crossings, but only few specialized laboratories worldwide offer such capabilities. Fortunately, the challenges are not much dissimilar than those encountered for traditional MEMS devices, but additional steps such as fiber pigtailing and cryogenic-compatible packaging might prove hard to solve. In this direction, heterogeneous integration of NOEMS on different substrates, where traditional photonic packaging is more established, could offer a promising solution.

Finally, an interesting aspect related to NOEMS is the implementation of low-noise, low-power, and low-voltage electronic drivers. Several works in the literature have stressed the importance of reducing voltage to CMOS-compatible levels (i.e. < 3 V) [25, 41, 42]. However, we should also stress that pushing the opto-mechanical sensitivity further to reduce the voltage comes at the expense of larger phase noise. Moreover, CMOS-controlled analog driving electronic with proper voltage amplification (and potentially readout feedback) could be implemented as well and directly integrated in the proximity of the electrostatic device. In this sense, GaAs enables developing high-voltage metal-semiconductor transistors embedded in the epitaxial structure, and it is therefore quite likely that future implementations of NOEMS will include the driving electronics on-chip.

References

1 Wang, J., Sciarrino, F., Laing, A., and Thompson, M.G. (2020). Integrated photonic quantum technologies. *Nature Photonics* 14: 273–284.

2 Spring, J.B., Metcalf, B.J., Humphreys, P.C. et al. (2012). Boson sampling on a photonic chip. *Science* 339 (6121): 798–801.

3 Zhong, H.-S., Wang, H., Deng, Y.-H. et al. (2020). Quantum computational advantage using photons. *Science* 370 (6523): 1460–1463.

4 Harris, N.C., Steinbrecher, G.R., Prabhu, M. et al. (2017). Quantum transport simulations in a programmable nanophotonic processor. *Nature Photonics* 11: 447–452.

5 Qiang, X., Zhou, X., Wang, J. et al. (2018). Large-scale silicon quantum photonics implementing arbitrary two-qubit processing. *Nature Photonics* 12: 534–539.

6 Paesani, S., Ding, Y., Santagati, R. et al. (2019). Generation and sampling of quantum states of light in a silicon chip. *Nature Physics* 15: 925–929.

7 Komma, J., Schwarz, C., Hofmann, G. et al. (2012). Thermo-optic coefficient of silicon at 1550 nm and cryogenic temperatures. *Applied Physics Letters* 101: 041905.

8 Wang, C., Zhang, M., Stern, B. et al. (2018). Nanophotonic lithium niobate electro-optic modulators. *Optics Express* 26 (2): 1547–1555.

9 Dupuis, N., Lee, B.G., Rylyakov, A.V. et al. (2015). Design and fabrication of low-insertion-loss and low-crosstalk broadband 2×2 Mach–Zehnder silicon photonic switches. *Journal of Lightwave Technology* 33 (17): 3597–3606.

10 Kuo, Y.-H., Chen, H.-W., and Bowers, J.E. (2008). High speed hybrid silicon evanescent electroabsorption modulator. *Optics Express* 16 (13): 9936–9941.

11 Goyvaerts, J., Kumari, S., Uvin, S. et al. (2020). Transfer-print integration of GaAs p-i-n photodiodes onto silicon nitride waveguides for near-infrared applications. *Optics Express* 28 (14): 21275–21285.

12 Bauters, J.F., Heck, M.J.R., John, D.D. et al. (2011). Planar waveguides with less than 0.1 dB/m propagation loss fabricated with wafer bonding. *Optics Express* 19 (24): 24090–24101.

13 Zimmermann, L., Preve, G.B., Tekin, T. et al. (2011). Packaging and assembly for integrated photonics - a review of the ePIXpack photonics packaging platform. *IEEE Journal of Selected Topics in Quantum Electronics* 17 (3): 645–651.

14 Midolo, L., Schliesser, A., and Fiore, A. (2018). Nano-opto-electro-mechanical systesms. *Nature Nanotechnology* 13: 11–18.

15 Xu, N., Cheng, Z.D., Tang, J.D. et al. (2021). Recent advances in nano-opto-electro-mechanical systems. *Nanophotonics* 10 (9).

16 Aspelmeyer, M., Kippenberg, T.J., and Marquardt, F. (2014). Cavity optomechanics. *Reviews of Modern Physics* 86 (4): 1391.

17 Rakich, P.T., Popović, M.A., and Wang, Z. (2009). General treatment of optical forces and potentials in mechanically variable photonic systems. *Optics Express* 17 (20): 18116–18135.

18 Haus, H., Huang, W., Kawakami, S. et al. (1987). Coupled-mode theory of optical waveguides. *Journal of Lightwave Technology* 5 (1): 16–23.

19 Uppu, R., Pedersen, F.T., Wang, Y. et al. (2020). Scalable integrated single-photon source. *Science Advances* 6 (50).

20 Lenzini, F., Haylock, B., Loredo J.C. et al. (2017). Active demultiplexing of single photons from a solid-state source. *Laser & Photonics Reviews* 11 (3): 1600297.

21 Wang, H., He, Y., Li Y.-H. et al. (2017). High-efficiency multiphoton boson sampling.*Nature Photonics* 11: 361–365.

22 Hummel, T., Ouellet-Plamondon, C., Ugur, E. et al. (2019). Efficient demultiplexed single-photon source with a quantum dot coupled to a nanophotonic waveguide. *Applied Physics Letters* 115: 021102.

23 Tsoukalas, K., Lahijani, B.V., and Stobbe, S. (2020). Impact of transduction scaling laws on nanoelectromechanical systems. *Physical Review Letters* 124 (22): 223902.

24 Sun, Y., Thompson, S.E., and Nishida, T. (2009). *Strain Effect in Semiconductors: Theory and Device Applications*. Springer Science & Business Media.

25 Jiang, W., Mayor, F.M., Patel, R.N. et al. (2020). Nanobenders as efficient piezoelectric actuators for widely tunable nanophotonics at CMOS-level voltages. *Communications Physics* 3 (1): 1–9.

26 Sun, X., Zhang, X., Poot, M. et al. (2012). A superhigh-frequency optoelectromechanical system based on a slotted photonic crystal cavity. *Applied Physics Letters* 101 (22): 221116.

27 Rakich, P.T., Davids, P., and Wang, Z. (2010). Tailoring optical forces in waveguides through radiation pressure and electrostrictive forces. *Optics Express* 18 (14): 14439–14453.

28 Yudin, P.V. and Tagantsev, A.K. (2013). Fundamentals of flexoelectricity in solids. *Nanotechnology* 24 (43): 432001.

29 Nguyen, T.D., Mao, S., Yeh, Y.-W. et al. (2013). Nanoscale flexoelectricity. *Advanced Materials* 25 (7): 946–974.

30 Owen, G. (1990). Methods for proximity effect correction in electron lithography. *Journal of Vacuum Science and Technology B: Microelectronics Processing and Phenomena* 8 (6): 1889–1892.

31 Lončar, M., Doll, T., Vučković, J. et al. (2000). Design and fabrication of silicon photonic crystal optical waveguides. *Journal of Lightwave Technology* 18 (10): 1402.

32 Midolo, L., Pregnolato, T., Kiršanskė, G. et al. (2015). Soft-mask fabrication of gallium arsenide nanomembranes for integrated quantum photonics. *Nanotechnology* 26 (48): 484002.

33 Tas, N., Sonnenberg, T., Jansen, H. et al. (1996). Stiction in surface micromachining. *Journal of Micromechanics and Microengineering* 6 (4): 385.

34 Petruzzella, M., Zobenica, Ž., Cotrufo, M. et al. (2018). Anti-stiction coating for mechanically tunable photonic crystal devices. *Optics Express* 26 (4): 3882–3891.

35 Kobayashi, D., Kim, C.J., and Fujita, H. (1993). Photoresist-assisted release of movable microstructures. *Japanese Journal of Applied Physics* 32 (11A): L1642.

36 Zhou, X., Uppu, R., Liu, Z. et al. (2020). On - chip nanomechanical filtering of quantum - dot single - photon sources. *Laser & Photonics Reviews* 14 (7): 1900404.

37 Poot, M. and Tang, H.X. (2014). Broadband nanoelectromechanical phase shifting of light on a chip. *Applied Physics Letters* 104 (6): 061101.

38 Papon, C., Zhou, X., Thyrrestrup, H. et al. (2019). Nanomechanical single-photon routing. *Optica* 6 (4): 524–530.

39 Gyger, S., Zichi, J., Schweickert, L. et al. (2021). Reconfigurable photonics with on-chip single-photon detectors. *Nature Communications* 12 (1): 1–8.

40 Bishop, Z.K., Foster, A.P., Royall, B. et al. (2018). Electro-mechanical control of an on-chip optical beam splitter containing an embedded quantum emitter. *Optics Letters* 43 (9): 2142–2145.

41 Liu, T., Pagliano, F., van Veldhoven R. et al. (2019). Low-voltage MEMS optical phase modulators and switches on a indium phosphide membrane on silicon. *Applied Physics Letters* 115 (25): 251104.

42 Haffner, C., Joerg, A., Doderer, M. et al. (2019). Nano-opto-electro-mechanical switches operated at CMOS-level voltages. *Science* 366 (6467): 860–864.

43 Grottke, T., Hartmann, W., Schuck, C., and Pernice, W.H.P. (2021). Optoelectromechanical phase shifter with low insertion loss and a 13π tuning range. *Optics Express* 29 (4): 5525–5537.

44 Press, D., Ladd, T.D., Zhang, B., and Yamamoto, Y. (2008). Complete quantum control of a single quantum dot spin using ultrafast optical pulses. *Nature* 456 (7219): 218–221.

45 Lodahl, P., Mahmoodian, S., and Stobbe, S. (2015). Interfacing single photons and single quantum dots with photonic nanostructures. *Reviews of Modern Physics* 87 (2): 347.

46 Bennett, A.J., Patel, R.B., Skiba-Szymanska, J. et al. (2010). Giant Stark effect in the emission of single semiconductor quantum dots. *Applied Physics Letters* 97 (3): 031104.

47 Meesala, S., Sohn, Y.I., Pingault, B. et al. (2018). Strain engineering of the silicon-vacancy center in diamond. *Physical Review B* 97 (20): 205444.

48 Grim, J.Q., Bracker, A.S., Zalalutdinov, M. et al. (2019). Scalable in operando strain tuning in nanophotonic waveguides enabling three-quantum-dot superradiance. *Nature Materials* 18 (9): 963–969.

49 Frank, I.W., Deotare, P.B., McCutcheon, M.W., and Lončar, M. (2010). Programmable photonic crystal nanobeam cavities. *Optics Express* 18 (8): 8705–8712.

50 Perahia, R., Cohen, J.D., Meenehan, S. et al. (2010). Electrostatically tunable optomechanical "zipper" cavity laser. *Applied Physics Letters* 97 (19): 191112.

51 Winger, M., Blasius, T.D., Alegre, T.P.M. et al. (2011). A chip-scale integrated cavity-electro-optomechanics platform. *Optics Express* 19 (25): 24905–24921.

52 Midolo, L., Pagliano, F., Hoang, T.B. et al. (2012). Spontaneous emission control of single quantum dots by electromechanical tuning of a photonic crystal cavity. *Applied Physics Letters* 101 (9): 091106.

53 Deotare, P.B., Kogos, L.C., Bulu, I., and Lončar, M. (2012). Photonic crystal nanobeam cavities for tunable filter and router applications. *IEEE Journal of Selected Topics in Quantum Electronics* 19 (2): 3600210.

54 Rudolph, T. (2017). Why I am optimistic about the silicon-photonic route to quantum computing. *APL Photonics* 2 (3): 030901.

55 Atabaki, A.H., Moazeni, S., Pavanello, F. et al. (2018). Integrating photonics with silicon nanoelectronics for the next generation of systems on a chip. *Nature* 556 (7701): 349–354.

22

Silicon Quantum Photonics - Platform and Applications

Cagin Ekici, Elizaveta Semenova, Davide Bacco, and Yunhong Ding

Department of Electrical and Photonics Enginering, Technical University of Denmark, 2800, Kgs. Lyngby, Orsteds Plads, Denmark

After reading this chapter you should be able to:

- Have an overview of silicon quantum photonics
- Understand the basic building blocks of silicon quantum photonics platform
- Quantum state of light on silicon photonics platform
- Quantum information processing on silicon photonics platform
- Quantum communication on silicon photonics platform

22.1 Introduction

Quantum information technologies are expected to lead to a step change in human capabilities in a variety of applications. Quantum computing enables to solve computational-hard problems orders of magnitude faster than world's most powerful supercomputers, quantum communication enables ultimate secure communication with information protected by quantum mechanism, and quantum sensing enables to detect weak signal below shot-noise limit with extremely high sensitivity. Given such great potentials, billions of dollars in both academic and industrial community have been invested globally to develop quantum information technologies. There are a few candidates could carry quantum information, e.g. electrons, trapped ions, superconducting Josephson junctions, and photons. So far those platforms have been in a competition and there is no winner yet, and significant progress has been made in the past a few years, including loophole free Bell-state tests [1–3], over thousands kilometers quantum communication via satellites [4–6], and quantum advantage over the world's most powerful supercomputers (for one specific computational task) [7–9].

Photonic Quantum Technologies: Science and Applications, First Edition. Edited by Mohamed Benyoucef.

Photon is at the core of those quantum technologies, as it travels at the fastest possible speed and readily protected from decoherence. Scaling up quantum photonic systems, which increases computational space, or communication distance, or measurement sensitivity, is the key for quantum photonic platform. Though free-space optics has significantly pushed the research and development of quantum photonics technologies, it faces more and more challenges in terms of stability, controllability, and flexibility when the quantum system is scaling up, and integrated quantum photonics has been considered to be the only path toward further scaling. With the standard lithography based nano-fabrication process, thousands of quantum chips can be fabricated simultaneously. Quantum photonic experiments can be implemented in a compact quantum chip of cm^2 scale integrating millions of components. Despite the great potential, integrated quantum photonics still faces significant challenges, particularly limited efficiency and purity of photon sources and optical loss. Today, the largest quantum photonics systems are still demonstrated on bulk optics, for instance 20 simultaneous photons in 60 × 60 a interferometer [10], 12-photon entanglement [11], 18 entangled qubits [12], and large-scale Boson sampling [7–9].

Among different integrated quantum photonics platform, silicon quantum photonics is the leading candidate. The high refractive index 3.5 of silicon provides high light confinement and capability of ultradense integration. The complementary metal–oxide semiconductor (CMOS) compatible fabrication makes it possible for electric-photonic co-integration [13]. In the meantime, classical silicon photonics continues to lead the way in large-scale photonic-integrated circuits, including ultradense integration with tens of thousands of optical components [14], and electrical/photonics hybrid integration [15, 16]. All this technological breakthrough of classical silicon photonics would lead to further breakthroughs in quantum photonics applications.

Since the quantum advantage, i.e. quantum device outperformed the state-of-the-art supercomputers though for a specific computational task, has been announced in 2020, a new era of so-called noisy intermediate-scale quantum (NISQ) [17] technology has begun. In this era, it is extremely important to develop medium-scale applications, such as quantum simulators [18–20], machine learning [21], and cluster-state quantum computing [22, 23]. Silicon photonics will provide a powerful platform for implementation of those quantum technologies. However, to enter the fault-tolerant (i.e. a quantum system is able to continuously operate despite of errors), breakthroughs on the photon sources, optical loss, and quantum process fidelity (larger than 0.99) have to be achieved before silicon-integrated quantum photonics can really achieve a useful scale.

In the meantime, quantum photonic communications have achieved significant progress. Photonics is the only viable choice for low-loss propagation, which promises long-distance quantum communication with security beyond any classical means today. Silicon photonics has already steadily played a more and more important roles for quantum communication systems. Compact, reliable, and energy efficient silicon quantum photonics transmitters and receivers have already been widely investigated for quantum key distribution (QKD) [24–27]. Preparation and distribution of entangled states over fiber links have also been studied aiming to future large-scale quantum networks [28, 29].

In this chapter, we will present the overview of the state of the art of silicon quantum photonics. The chapter is structured as follows. In Section 22.2, we will discuss the fundamentals of silicon quantum photonics, including the fabrication process, and necessary passive and active components. In Section 22.3, We will further present an overview of different quantum light sources on silicon, including quantum dot (QD) sources on silicon and nonlinear sources. In Section 22.4, We will discuss applications in processing of quantum information on silicon and quantum communication. Finally, we will give an outlook of silicon photonics and discuss the challenges for further scaling.

22.1.1 Fabrication Process

Fabrication of silicon photonic-integrated circuits follows standard lithography-based nanofabrication. A typical fabrication process of thermal-tunable silicon photonics chips is presented in Figure 22.1. It includes four main fabrication sections. The first fabrication section is lithography. Taking electron beam lithography (EBL) for example, after a thin layer of electron beam (ebeam) resist is spin-coated on the silicon on insulator (SOI) sample, electron beam is focused on the sample and expose the defined structures (Figure 22.1(a)). The property of the exposed resist is changed by the high-energy electron beam. After the developing process (Figure 22.1(b)), the exposed resist will be dissolved by the developer for positive ebeam resist (for negative resist, the unexposed area will be dissolved by the developer), forming the etching mask (Figure 22.1(c)) for the etching process. The second fabrication section is the etching process that transfers the patterns on the resist layer to silicon layer (Figure 22.1(d) and (e)), typically inductively coupled plasma (ICP) etching process is used. The etching process can be optimized by tuning the coil power, platen power, gas flow, and temperature. In order to achieve vertical etching profile, Bosch process is typical used, which uses alternation between an etch cycle and a passivation cycle. Introducing a passivation step in an etch process allows to cover the sidewalls with a protective layer that suppresses the isotropic etching. After the etching process, a silicon dioxide (SiO2) layer is deposited on the wafer to form a cladding and protecting layer before heater fabrication (Figure 22.1(f)). The thickness of the cladding SiO2 layer should be large enough (typically more than 1 μm) to avoid the potential metallic loss by the metal heater [29, 30]. The third fabrication section is the heater fabrication process, the ultra-violet (UV) lithography based on mask aligner or direct-laser-writing-based maskless exposure can be used (Figure 22.1(g)). Typically, negative resist is preferred to form a undercut in the bottom of the opening window after developing process (Figure 22.1(h)). After the lithography, heater metal material (typically titanium (Ti) [29–31]) is deposited on the sample (Figure 22.1(i)). The metal and resist will be washed away after the liftoff process, forming the final heater structures on the wafer (Figure 22.1(j)). The final fabrication section is the metal wire/pad fabrication. The heaters are typically connected by electrical wires with low resistance (typically Au) to the pads

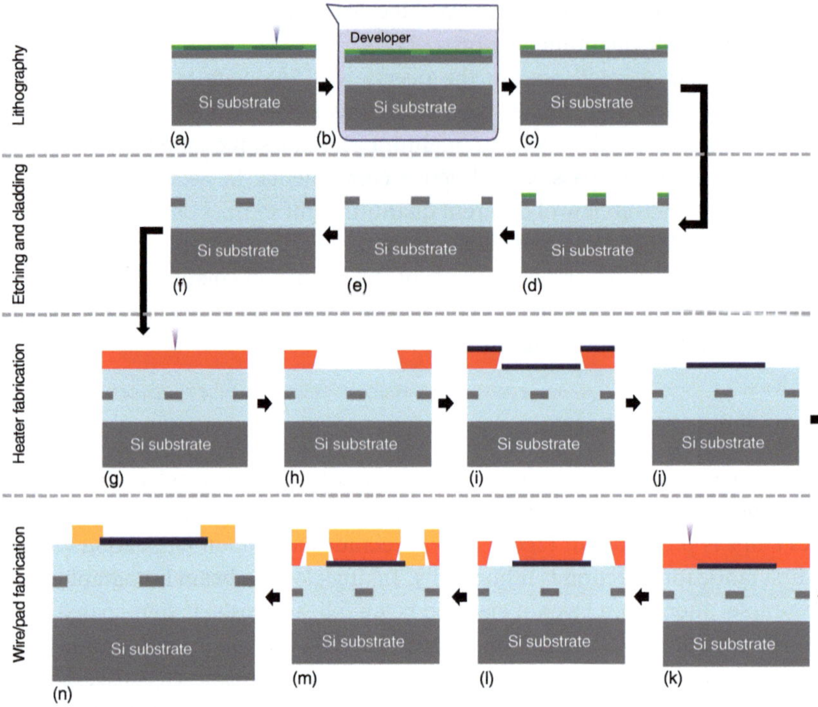

Figure 22.1 Typical fabrication process of silicon photonic devices with thermal tuning. The fabrication involves lithography section: (a) ebeam exposure, (b) development, and (c) forming final resist mask after developing; etching and cladding section: (d) ICP etching process with resist etching mask, (e) resist striping after etching, (f) SiO2 deposition for upper cladding; heater fabrication section: (g) UV maskless exposure, (h) heater pattern window opening after developing, (i) heater metal deposition, (j) final heater achieved after liftoff process; and wire/pad fabrication section: (n) UV maskless exposure, (l) wire/pad pattern window-opening after developing, (m) wire/pad metal deposition, (k) final wire/pad obtained after liftoff process.

on the edge of the chip, which would be further wire bonded to a printed circuit board (PCB) in the packaging process. Here, another UV lithography is applied to define the wires and pads structures (Figure 22.1(k)–(l))), followed by thick metal deposition Figure 22.1(m)) and liftoff process (Figure 22.1(n)), and finally results in the fabricated devices.

In silicon photonics, other active components such as high-speed depletion-based electro-optical (E/O) modulators, superconductive nanowire single-photon detectors (SNSPDs) would be also integrated. Typical fabrication process involves ion-implantation, annealing, lithography, and liftoff process. More details we refer to [32, 33].

22.1.2 Passive Components

Passive components are basic building blocks for integrated quantum photonic systems. These includes fiber-to-chip couplers, pump rejection filters, cross intersections, and delay lines. As data rate of a quantum information processing system is exponentially sensitive to the optical loss, it is critical for those passive components to work with utmost performances in terms of extremely low insertion loss and crosstalk.

Fiber-to-chip couplers play a critical role for today's integrated quantum photonic systems since the coupler loss directly limit the final system scale. Currently, there are mainly two kinds of fiber-to-chip couplers: inverse adiabatic taper-based edge couplers (Figure 22.2(a) and (c)) [34, 36, 43, 44] and scattering grating-based vertical couplers [35, 45] (Figure 22.2(b)). The edge couplers can be realized by single [34, 43] or dual inverse tapers [44], it features large coupling bandwidth and potentially extremely high coupling efficiency [36, 43]. Currently, the record high coupling efficiency of more than 95% has been reported by adiabatic tapering both the nanowavegudie and fiber [36]. The edge couplers, however, require high-quality cleaving of the chip that would potentially restrict the device layout. Moreover, the edge couplers are typically coupled with lensed fiber, making the coupling sensitive to the misalignment. On the other hand, scattering grating-based vertical couplers do not require chip cleaving, making it very convenient for wafer-scale test. The coupling efficiency is typically around 3dB [30, 46], which is restricted to the leakage loss to the substrate. The coupling efficiency can be boosted by introducing an Al mirror [47] or Bragg grating below the grating coupler [48, 49]. Record coupling efficiency of 0.58 dB of fully etched grating coupler [47] and 0.62 dB for shallowly etched grating coupler [50] have been reported. Standard one-dimensional grating couplers are typically polarization sensitive, which restricts their application in polarization encoded quantum systems. In order to mitigate the restriction, two-dimensional grating couplers can be used to efficiently couple the two orthogonal polarization to two separate waveguides on the same polarization [28, 29]. The same strategy of coupling efficiency boosting in standard one-dimensional grating couplers can be also applied [51].

For nonlinear source, pump light is used for single-photon pair generation, and it must be removed from the generated single photon pairs before they are detected. In order to achieve sufficient rejection of the pump, the extinction ratio (ER) between the pump light the generated photons has to be at least more than 100 dB. At the same time, the insertion loss at the single-photon wavelength bands has to be kept minimal to maintain the high transmission of the generated single photons. Coupled resonator optical waveguides (CROWs) [52], and cascaded asymmetric MZIs [53] have been used to achieve more than 50 dB extinction on a single chip. The extinction ratio was limited by pump light scattered in to the chip substrate, increasing the grating coupler efficiency is expected to improve the extinction ratio since the scattered pump light is reduced. The first demonstration of simultaneous pump rejection and single-photon generation was reported in 2018 [37]. Here four second-order cascaded micro-ring resonators are applied to isolate the generated signal-idler pair from the pump light. Cascading two chips is able to produce more than 100dB extinction ratio [52–54]. Cascaded resonators typically require independent resonance control of the each resonator to achieve high extinction ratio, due to the uncontrollable fabrication error introduced to each resonator.

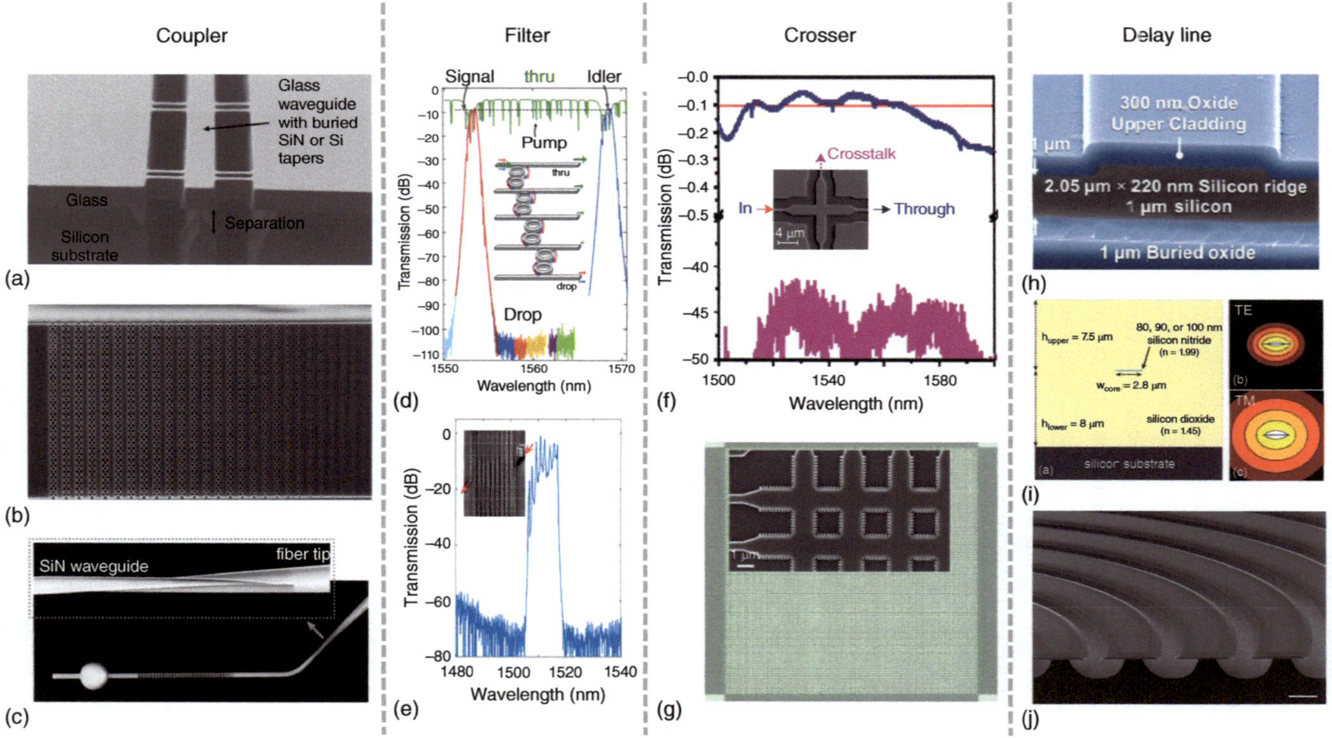

Figure 22.2 Typical passive components used in quantum photonic circuits, including couplers for fiber to chip interface: (a) spot size converter lateral coupler. Source: Chen et al. [34]/with permission of IEEE. (b) photonic crystal based grating coupler. Ding et al. [35]/with permission of Optical Society of America. and (c) extremely efficient coupler with tapered fiber on a tapered SiN.Source: Tiecke et al. [36]/with permission of Cornell University; filters for pump light rejection: (d) four stage of second-order cascaded micro-ring resonators achieve simultaneous pump rejection and single-photon generation. Source: Gentry et al. [37]/with permission of Optical Society of America. (e) low-loss and high extinction ratio CROW fano nanobeam resonators. Source: Cheng et al. [38]/with permission of Optical Society of America.; crossers: (f) MMI-based cross intersection with low crosstalk and insertion loss. Source: Wang et al. [30]/with permission of American Association for the Advancement of Science - AAAS. and (g) Bloch mode employed to further reduce the loss of cross intersection. Source: Zhang et al. [39]/with permission of Optical Society of America; delay lines to rout the photons with extremely low-loss: (h) shallowely etched ultralow loss silicon waveguide. Source: Biberman et al. [40]/with permission of Optical Society of America, (i) ultralow loss slab SiN waveguide. Source: Puckett et al. [41]/With permission of IEEE. and (j) slight angle wedged silica waveguide. Source: Lee et al. [42]/with permission of Springer Nature.

CROW fano nanobeam resonators feature extremely small footprint, making it possible for a uniform fabrication of the resonators, exhibiting pure passive filter with extinction ratio more than 70dB and insertion loss lower than 1 dB [38]. In addition, cascaded grating-assisted contradirectional couplers with more than 70dB ER [55] and a third-order ring resonator with more than 80 dB ER [56] have also been reported both in silicon nitride platform.

In quantum photonic-integrated circuits, it is often inevitable to have waveguides cross each other, particularly in the path-encoded quantum systems [30, 57]. It is of outmost importance for the cross intersections to have extremely low crosstalk to reduce qubit errors and at same time achieve ultralow insertion loss to maintain the data rate. Promising cross intersections have been reported based on MMI structures [30, 58], shaped waveguides [59–61], tilted waveguides [62, 63], and subwavelength gratings [64], with typical insertion loss of around 0.1 dB. The insertion loss can be further reduced by Bloch mode structures [39, 65], featuring record low 0.01 dB insertion loss and crosstalk below −40 dB.

Low-loss waveguide is essential for integrated silicon quantum photonics. Demultiplexing single-photon sources and spatial multiplexing nonlinear sources requires extremely low-loss photon routing. Temporal multiplexing nonlinear sources relies on or low-loss delay line for photon storage [66, 67]. Currently, the typical propagation loss of silicon single mode waveguide has propagation loss of around 2dB/cm, which is too high for scaling quantum systems. Further fabrication optimization can be applied to further reduce the propagation loss, such as oxidization [68, 69], etchless waveguide fabrication [70], hydrogen thermal annealing [71], and shallow waveguides [40, 72]. The record-low loss of 2.7 dB/m (Figure 22.2(h)) [40] has been reported. In order to further reduce the propagation loss on silicon, hybrid integration with the other material platforms, such as thin silicon nitride slab waveguide with propagation loss of 0.123 dB/m [41] (Figure 22.2(i)) and slight angle wedged silica waveguide with propagation loss of 0.034 dB/m [42] (Figure 22.2(j)) would be needed.

22.1.3 Active Components

Active components are critical devices that are able to dynamically manipulate or detect the quantum state on silicon chips. Those includes phase shifters, high-speed electro-optical (E/O) modulators, and single-photon detectors (SPDs). Similar to passive components, optical insertion loss remains significantly important. For SPDs, the loss can be considered as part of the detection efficiency.

Phase shifters are typically used to reconfigure the silicon quantum circuit for preparation and measurement of the quantum state on different basis. As these operations might not require high-speed operation, thermo-optic (TO) phase shifters are typically used. The TO phase shifter is typically realized by metallic heaters on the silicon waveguide [30] (Figure 22.3(a)). The current flowing through the metal generates heat that spreads to the waveguide underneath and further changes the refractive index. In order to isolate the metallic loss of the heater, a layer of cladding between the heater and silicon waveguide (normally SiO2 around 1 μm thick) is

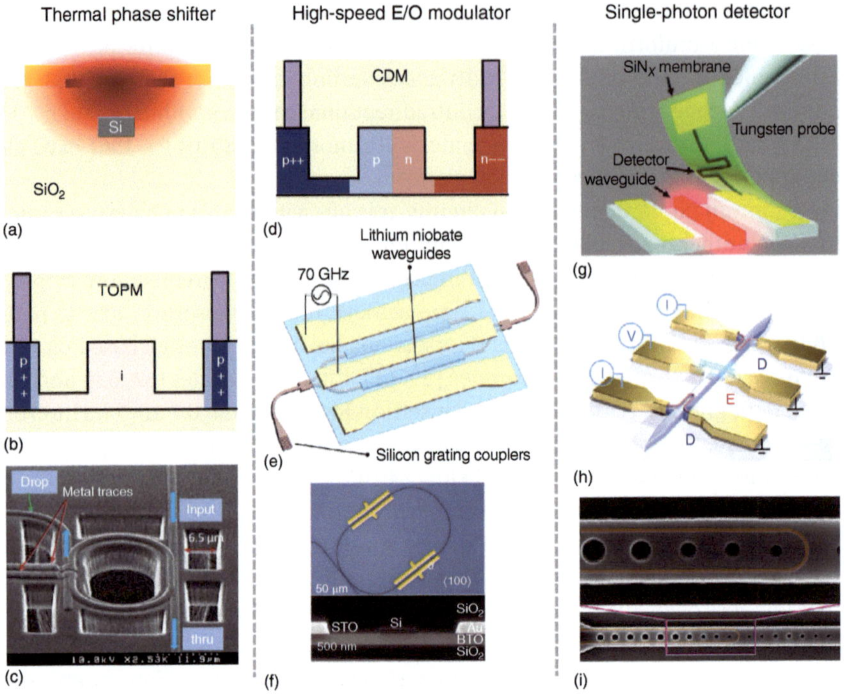

Figure 22.3 Typical active components used in quantum photonic systems, including thermal phase shifters: (a) heater-based TO phase shifter, (b) pip junction-based TO phase shifter. Source: Sibson et al. [24]/Optical Society of America/CC BY 4.0. (c) trench and undercut introduced on a TO tunable microring to increase the TO efficiency. Source: Dong et al. [73]/with permission of Optical Society of America; high-speed E/O modulators: (d) pn junction-based E/O depletion modulator. Source: Sibson et al. [24]/Optical Society of America/CC BY 4.0. (e) LN modulator on silicon. Source: He et al. [74]/with permission of Springer Nature. (f) BTO modulator on silicon. Source: Abel et al. [75]/with permission of Springer Nature; single-photon detectors: (g) transfer-print SNSPD on silicon nitride waveguide. Source: Najafi et al. [76] / Springer Nature / CC BY 4.0. (h) SNSPDs on silicon integrated with carbon nanotube single-photon emitters. Source: Khasminskaya et al. [77]/with permission of Springer Nature. (i) SNSPD on silicon photonic crystal cavity. Source: Vetter et al. [78]/with permission of Springer Nature.

needed. The detailed fabrication process can be found in Section 22.1. Typically, about tens of milliwatts (mW) is needed to achieve 2π phase shift [29–31]. This thermal heater has an advantage of low loss (metal loss can be negligible with sufficient thick upper cladding layer), and the tuning speed is typically in kHz scale due to slow heat dissipation. Alternatively, the TO phase shifter can be realized by a pip junction [24] (Figure 22.3(b)), where the current is flowing through the silicon waveguide directly and generating heat. In order to achieve the low optical loss, the waveguide is kept intrinsic, resulting in large resistance, and typically more than 20V is needed to achieve 2π phase shift. Another novel method to isolate the heater loss is to take advantage of the bending effect. In a multimode bending section, the heater can be made in the inner bending side with a proper doping level, and the optical mode tends to propagate on the outer side, which decouples from the heater- induced loss [79]. The advantage of the doping-based heater is the speed can be hundreds of kHz, which is much faster than standard metallic heaters. In order to increase the heater efficiency, trenches and undercuts can be made to concentrate the heat, featuring a high efficiency of 2.4 mW/2π [73](Figure 22.3(c)). Another method to increase the efficiency is to recycle the residual heat, and spiral waveguides [80] and multimode waveguide [81] have been utilized to increase efficiency to a few mW/2π. Novel heater material such as graphene has also been applied for efficient TO phase shifter, exhibiting both high efficiency and sub−μs tuning speed [82]. Another novel scheme is to utilize the silicon nanowire as waveguide and heater simultaneously, resulting in sub-mW/2π efficiency [83].

E/O modulators are able to prepare the quantum state in high speed, which is critical for optical quantum communications. In silicon photonics, the E/O modulation is typically based on plasma dispersion effect in a pn junction [24] (Figure 22.3(d)), which is attractive due to the CMOS compatible fabrication process and large bandwidth of up to tens of GHz [32]. However, the plasma effect is weak, leading to long device length and large energy consumption. Moreover, the free-carrier absorption (FCA) results in extra optical loss. Recently, electrical-field-induced second-order nonlinear optical effect of silicon waveguide has been studied [84]. It produces direct-current (d.c.) Kerr effect results in high-speed phase-only modulation. The modulator exhibits VπL of 2.8 V.cm [84]. Another phase-only modulation is Pockels effect, which is more attractive since it is able to achieve efficient and high-speed E/O modulation without introducing extra optical loss [85]. Materials such as lithium niobate (LN) [74, 86–89] (Figure 22.3(e)) and barrium titanate (BTO) [75, 90] (Figure 22.3(f)) have strong Pockels effect and have been hybrid-integrated on silicon platform. The state-of-the art TFLN has achieved extremely low propagation loss of around 2.7 dB/m [91, 92], and more than 100 GHz modulation bandwidth has been achieved on TFLN platofrm [93], indicating a great potential of Si-LN hybrid integration [74, 94, 95]. BTO features the highest Pockels effect coefficient [75, 90], having the potential of efficient E/O modulation with ultra-compact device size [96] compared to TFLN devices. Though BTO is still in the preliminary stage, tens of GHz modulation bandwidth has been demonstrated [90]. However, thin-film BTO material on insulator and low-loss BTO waveguide remain challenging.

SPDs are used to count the photons, and they are indispensable devices in quantum photonic systems. Silicon photonics is typically optimized to work around C-band (1530–1560 nm) to align with the low-loss window of single mode fiber, which limits the options of SPDs. InGaAs detectors offer a cost-effective solution [97, 98] and could potentially be integrated on silicon with bonding process. The count rate can be a few to tens of MHz with dark count of a few hundreds Hz. However, the detection efficiency is typically lower than 10%. Superconductive nanowire SPDS (SNSPDs) working at cryogenic temperature (typically lower than 4K) offers much better performance [99, 100], with system efficiency up to 95.5% [101], dard-count rate less than 100 Hz, and count rate of tens of MHz [102]. Moreover, SNSPDs can be directly hybrid integrated on silicon photonics platform, demonstrating great potential for future on-chip quantum photonics systems. For example, membrane-transferring SNSPDs on silicon nitride has been reported, with successful demonstration of on-chip $g^{(2)}$ experiment [76] (Figure 22.3(g)). SNSPDs on silicon integrating carbon nanotube single-photon emitters has also been reported [77] (Figure 22.3(h)). Typically, SPSPDS has a recovery time of tens of nanoseconds, limiting the counting rate to be tens of MHz. Silicon photonic crystal cavities have been discovered to enhance the absorption for shorter detector, resulting in a much faster recovery time of only 500ps [78] (Figure 22.3(i)), allowing GHz counting rate. Another important parameter of SPDs is timing jitter (temporal resolution), which limits the repetition rate of quantum communication. The jitter of commercial SNSPDs is in a scale of tens of picosecond. Recently, record-low 3 ps resolution was demonstrated employing a 5 μm detector and cryogenic electronics [103]. In some quantum systems, for example Gaussian boson sampling [31] and universal quantum photonic gates [104], photon number resolving detectors are needed. Space- or time-multiplexing SNSPDs [105–109] have been investigated to realize photon number resolution. Titanium-based superconducting transition-edge detectors (TESs) [110, 111] provide the best number resolution so far with excellent discrimination up to 15 photons in real time, though it suffers from limited repetition rate of kHz. However, so far, the integration of photon number resolving detectors on silicon photonic platform has not reported yet.

22.2 Quantum State of Light on Silicon Photonics Platform

Quantum states of light are essential for quantum photonic systems. Quantum states of light include single-photon states and squeezed states of light. Single-photon states are used in discrete variable (DV) quantum systems, while squeezed states are used in continuous variable (CV) quantum systems. In this section, the solid-state single-photon sources and nonlinear light sources on silicon will be discussed in detail.

22.2.1 Toward Deterministic and Efficient Quantum Light Sources on Silicon Platform

While enabling a low loss (about 2.7 dB/m [112]) and a phase-stable environment for optical signal propagation, silicon is a very inefficient material for light emission due to the indirect band gap. Highly efficient emission of quantum light is an essential requirement for optical quantum information processors as it determines the depth of the architecture and thus the complexity of the problems that can be address [113]. It is therefore required to integrate Si with other materials that are suitable for the efficient emission of quantum light.

Ideally, an on-demand quantum emitter should produce either one photon or a pair of entangled photons into a spatial-temporal mode at a time. All photons have to possess a Fourier-limited linewidth and be indistinguishable, meaning they should be identical in all degrees of freedom including frequency and polarization. An essential requirement for the integration of quantum light emitters into a Si photonic circuit is that the emission wavelength exceeds the absorption band of silicon at ~ 1.1 μm. Moreover, operation at the telecom wavelength bands would open up the potential integration with a global quantum network via fiber-based long-distance telecommunication. Until now, there is no universal solution for scalable integration of a quantum emitter onto the Si-based quantum photonic platform. The most promising approaches to fabricate CMOS compatible quantum emitters include epitaxially grown ODs [114–117], (SiC) [118–120] and in gallium nitride (GaN) [121] and single rare-earth atoms (e.g. erbium) [122, 123]. The feasibility for integration of some of those quantum light emitters onto the SOI platform was already demonstrated [124, 125].

The main advantage suggested by solid-state quantum emitters in SiC and GaN platforms is their operation at room temperature [118–120, 126]. Though those quantum sources lack control of the emission wavelength, and high photon indistinguishability was not achieved. Nevertheless, epitaxially grown semiconductor QDs prove their high potential for leading the competition for single-photon emitters and sources of entangled photon pairs. Epitaxial QDs possess all the essential properties required for this application, in particular, Fourier limited emission lines, near deterministic operation regime, photon extraction efficiency, high brightness, close to unity photon indistinguishability, and fast radiative relaxation [127–135]. Moreover, QDs with an extremely low fine structure splitting were demonstrated, which is a requirement for generating pairs of entangled photons [136–138]. Most III-V compound semiconductors have a direct bandgap structure and are very suitable bandgap and thus emission wavelength engineering. For these reasons, III-V semiconductor QDs are an excellent choice for optically active devices integrated with Si photonics.

Usually, epitaxial QDs are randomly distributed across the wafer as there are spontaneously formed in a self-assembled nucleation process. This applies to both the Stranski–Krastanov growth mode and QD droplet epitaxy. Furthermore, to increase the efficiency [127] and indistinguishability [139] of single-photons emitted by a QD, it should be coupled to a carefully designed cavity or waveguide [140]. As the QDs are randomly distributed, several high precision methods for the deterministic fabrication of a cavity or a waveguide were developed, enabling the efficient coupling of a QD to a desired optical mode. Those methods include: scanning electron microscopy to localize QD position [141], in situ far-field optical lithography [142], *in situ* electron-beam lithography [143], and photoluminescence imaging [144]. For on-chip integration, several issues have to be addressed: precise matching in spatial and frequency domains of a QD emitter with a hosting cavity or waveguide, frequency, phase, etc., matching between two different QD emitters, and integration of a QD emitters with desired properties onto a Si-photonic circuit.

Tuning of the QD emission wavelength can be realized by adjusting the local temperature [145], application of local strain [146], or magnetic [147] and electric [148, 149] fields. The application of an electric field, e.g. by placing the QD inside a p-i-n structure can further facilitate the suppression of dephasing caused by the local charge environment of the QD [150] and used to control the exciton–biexciton splitting [148]. Furthermore, the combination of mechanical and electric field tuning allows independent control of exciton and biexciton properties [151]. All approaches mentioned are highly efficient. However, the feasibility of temperature and strain tuning of individual emitters integrated on-chip is questionable. Thus, the most prospective approach to enable multiphoton interference experiments on-chip is to use the field-induced Stark effect [151, 152] to tune the emission wavelength of individual QDs with high precision.

Most of the research has been conducted for QDs in GaAs-based material system operating in the wavelength range below or near 1 μm that is incompatible with the SOI platform. An alternative path for this wavelength range is on-chip integration with higher bandgap energy materials, such as III-V platform [153, 154] and Si_3N_4 photonics [144, 155–157]. Optimized Si_3N_4 waveguides can possess ultralow losses down to 0.1 dB/m [158] with a transparency band above 400 nm. This material is recently included in the SOI foundry process, allowing for 3D integration [159]. The price for using Si_3N_4 is a large device footprint since the bend radius is larger than for Si devices, which impacts on-chip delay lines [160]. Using III-V photonic platform [154] will cost, in addition to other integration disadvantages and fabrication challenges, a higher propagation loss of about 2 dB/cm [161]. All above-mentioned issues motivate the community to find a solution for integrating III-V semiconductor QDs onto Si photonic chips as this will allow for improving the efficiency, scalability, stability, and functionality of a quantum photonic system [125].

To enable integration to the Si-photonic platform, the emission wavelength of QDs should be above the absorption band of Si and preferably within the telecom wavelength band. This can be done by applying the metamorphic growth approach [162, 163]. The metamorphic growth approach extends the operation of the GaAs-based material system to the wavelength spectral range close to 1.55 μm or even allows for the grow of GaAs directly on Si [164]. The consequences of this method are high dislocation density in the III-V device layers caused by strain relaxation. An alternative path is to employ the InP-based material system.

InAs QDs in InP are not developed to the same extend as InAs/GaAs QDs. However, the reasons for this are historical rather than fundamental. Several examples of single-photon emission from InAs/InP QDs emitting in the telecom C-band already exist [115–117, 165, 166], demonstrating high-purity photons with an autocorrelation function $g^2(0) = 4.4 \cdot 10^{-4}$ at 4K temperature [167] and at elevated temperatures of 50K with $g^2(0) < 0.25$ [117]. Moreover, single-photon emission with fidelity of about 87% (state purity) by an electrically driven light-emitting diode at 93K was demonstrated [168] in this material system.

Now, the question of practical integration of QD quantum emitters into Si photonic circuits arises. Ultimate scalability would be suggested by monolithic integration, meaning direct epitaxial growth of defect-free III-V semiconductor QDs devices on Si. In practice, the difference in lattice parameters and thermal expansion coefficients between those two classes of materials and the challenges related to the epitaxy of polar on nonpolar material usually result in the formation of various types of crystal defects, yielding properties that are incompatible with applications in quantum photonoics [169]. Various solutions for defect filtering exist [169, 170] such the addition of a micrometer-scale buffer layer [164]. Nonetheless, most of these solutions accept the presence of crystal defects and aim mainly to localize them in a buffer layer, thus increasing the distance and reducing the coupling efficiency with the Si photonic circuits underneath. The monolithic integration of III-V material on Si for quantum photonic applications would be possible only via the reliable epitaxy of defect-free III-V on Si. The most promising approaches are based on selective epitaxy with a small footprint of the III-V/Si heterointerface [171–173], but yet these approaches are not mature.

Until now, heterogeneous integration of III-V onto Si plays a key role [125] in the fabrication of hybrid photonic circuits. A III-V semiconductor device is wholly or partially grown on a III-V substrate and then transferred to the hosting SOI. The three main approaches are transfer printing [174, 175], the pick-and-place technique [124, 157], and wafer bonding [143].

The pick-and-place technique allows for broad flexibility and functionality of fabricated systems. A III-V quantum device is characterized in advance, preselected and precisely transferred onto a specific position of the hosting photonic circuit by a microprobe. Using this approach, a recent example demonstrated the deterministic integration of InAs/InP QDs onto Si photonic chip with nanoscale precision. The integration was performed in a scanning electron microscope system combined with a focused ion beam and a microprobe manipulator. The integration process is illustrated in Figure 22.4. The on-chip Hanbury–Brown and Twiss experiment reveal a second-order autocorrelation function at zero delay time about $g^2(0) = 0.33$ at continuous-wave regime [124]. Later, the quantum-confined Stark effect allowed for tuning of the emission wavelength of integrated InAs/InP QDs up to 5.1 meV [176].

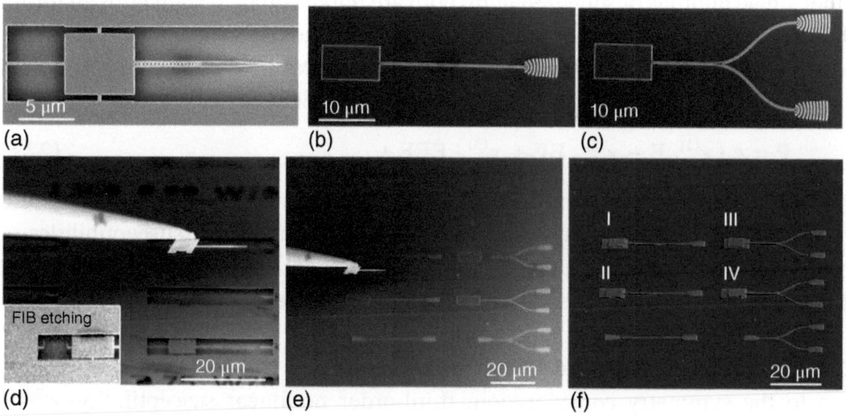

Figure 22.4 Scanning electron microscope images illustrating heterogeneous integration via pick-and-place techniques of III-V QD based single single-photon emitter on to Si waveguide. (a) Fabricated III-V device, (b,c) fabricated Si devices, (d,f) transfer of the III-V device onto Si using microprobe. To release III-V device from III-V wafer the tether were cut by a focused ion beam as shown on insert. (f) Integrated QD emitters and silicon waveguide devices. False-colors red, yellow, and navy indicate the III-V, silicon, and SiO2 top surface, respectively. Source: Kim et al. [124]/with permission of American Chemical Society.

Transfer printing could be considered as a variant of the pick-and-place technique. The III-V device is transferred by a transparent adhesive stamp. The user monitors the alignment in real time through the stamp in the optical microscope. Thus, the accuracy of this method is limited by the diffraction limit and about a few hundred nanometers. To improve the accuracy, one could introduce additional alignment marcs [175, 177]. A QD single-photon source integrated on a glass-clad silicon photonic waveguide was thus fabricated by a CMOS foundry. The single-photon emission purity was measured by the second-order autocorrelation function at zero delay time about $g^2(0) = 0.3$. The total single-photon coupling efficiency to the Si waveguide was estimated to be about 70 %. [175]

The overall drawback of the pick-and-place method is the lack of scalability since every QD emitter has to be transferred one by one. The advantage of the wafer bonding technique is the resulting wafer-scale areas of very high-quality III-V material on Si [178]. The epilayers can be grown either prior to or after bonding with the limitation of III-V thickness in the range of a few hundred nanometers [179, 180]. A sufficient difference in the thermal expansion coefficients of III-V material and Si could cause defect formation in III-V or even peel off the III-V layer at the growth temperatures. The alignment precision of this method is a few tens of nm [181]. The challenge is extreme requirements to the surface smoothness, which should be at the monoatomic level. So far, there has been no demonstration of a functional QD single-photon emitter that was heterogeneously integrated into the Si photonic circuit via direct wafer bonding.

22.2.2 Nonlinear Sources on Silicon

Nonlinear sources are another type of quantum light sources that is able to provide either single photons or squeezed light, both are important light sources in quantum photonic systems. Nonlinear optical behavior of the medium arises as a result of the response of an electron to the optical field. An optical field with an electric field component **E** interacting with a medium induces a polarization **P** [182]

$$\mathbf{P} = \epsilon_0(\chi^{(1)}. \mathbf{E} + \chi^{(2)} : \mathbf{EE} + \chi^{(3)} : \mathbf{EEE} + ...), \tag{22.1}$$

where ϵ_0 is the vacuum permittivity, $\chi^{(1)}$ is the linear susceptibility, $\chi^{(2)}$, $\chi^{(3)}$, and so on are called second order, third order, and so on nonlinear susceptibilities of the medium, respectively. This expansion illustrates that electromagnetic fields with different frequencies can interact and may give rise to new field(s).

22.2.2.1 Fundamentals of Nonlinear Photon Sources

Due to the symmetry consideration, third-order nonlinear susceptibility $\chi^{(3)}$ is responsible for nonlinear processes to take place in silicon nonlinear sources. At the quantum level, photons can be generated in crystalline silicon via spontaneous four wave-mixing (SFWM) in which two photons from single pump pulse or one from each distinct pump pulses are absorbed virtually to create photon pairs, known as signal(s) and idler(i) for historical reasons, with some probability. Here a catalytic role is played by the medium and in order to have appreciable level of photon generation probability, energy conservation ($\omega_{p_1} + \omega_{p_2} = \omega_s + \omega_i$) and phase-matching condition ($k_{p_1}(\omega_{p_1}) + k_{p_2}(\omega_{p_1}) = k_s(\omega_s) + k_i(\omega_i)$) should be fulfilled. The quantum state emerging from a single-mode silicon photon sources is well described by squeezed vacuum state [183],

$$|\Psi\rangle = \sqrt{1 - |\zeta|^2} \sum_{n=0}^{\infty} (-\zeta)^n |n, n\rangle_{s,i} \tag{22.2}$$

where $\zeta = ie^{i \arg \xi} \tanh |\xi|$ for squeezing parameter $|\xi|$ that depends mainly on pump power and n denotes number of signal and idler photons. Note that the general multimode state is expressed as a tensor product of individual single-mode squeezers. The details of this modification are beyond the scope of the simple treatment outlined in this chapter; however, the discussion can be found in [184].

Because of the stochastic nature of spontaneous processes, the state consists of number of terms with increasing photon number in the absence of loss. This is the drawback that all nonlinear sources suffer from as opposed to the true single-photon emitters. However, number of photons in the signal and idler modes are perfectly correlated, and therefore detection of the heralding (idler) photon(s) removes vacuum contribution and announces existence of the heralded (signal) photon(s). Because of the aforementioned characteristics, nonlinear photon sources are also known as "heralded (probabilistic) photon sources" [185], see Figure 22.5a).

In order to suppress the multiple-pair contamination (photon number impurity), which deteriorates many quantum information implementations, squeezing parameter should be kept low; that is, the state is dominated by vacuum emission $|0{,}0\rangle_{s,i}$ with an $\mathcal{O}(\zeta)$ component corresponding to a single-photon pair-state $|1{,}1\rangle_{s,i}$. Contribution from the higher order terms $\leq \mathcal{O}(\zeta^2)$ can be made arbitrarily small by simply reducing the input pump power. This, however, limits the brightness of the source, which cannot always be tolerable. From experimental perspective, the photon-number purity can be measured via heralded, i.e. conditioned on the detection of heralding (say idler) photon, second-order correlation function $g_c^{(2)}(t)$, typically in a Hanbury–Brown and Twiss (HBT) setup, see Figure 22.5(b)).

Photon pair generation in silicon waveguides was modeled theoretically by Lin et. al. [186] and was first demonstrated experimentally by Sharping et al in 2006 [187]. Since then, a number of experimental works have been reported, resulting in rapid advancement in silicon photonics [188–190]. Various types of resonator

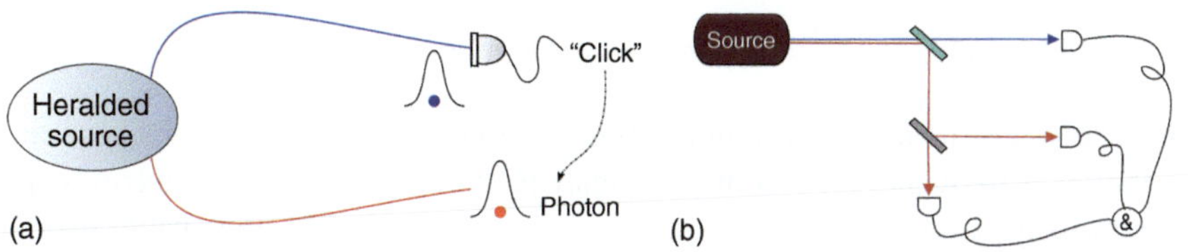

Figure 22.5 (a) Heralded Photon Pair Source: Detection of one of the member of a pair heralds the existence of the other in the absence of losses. (b) Basic HBT setup based on the detection of heralding photon, namely conditional $g_c^{(2)}(t)$. The pairs are deterministically separated into two separate channels. Upon detection of the photons (say idler), the statistics can be obtained following $g^{(2)}(t)$ measurement of sibling (signal) photons.

have been examined: single-ring resonator demonstrating an improvement in the pair generation rate due to reverse biasing [191], including the studies of reducing pump power requirements to submilliwatt levels by decreasing mode volume; high-Q microdisks [192]; and photonic crystal resonators [193].

Effective two-photon interaction has been a serious problem to perform computational operations in any photonics quantum processing scheme owing to limited interaction of photons until the protocol proposed by Knill, Laflamme, and Milburn (KLM) [194], which relies on Hong-Ou-Mandel (HOM) interference [195]. The underlying notion of standard HOM interference is that once two pure and indistinguishable photons interfere on a beamsplitter, they will always emerge in the same output port. Therefore, the HOM setup quantifies the level of indistinguishability in all degrees of freedom such as spatial mode, polarization, spectrum, and arrival time. The quality of the overlap, from experimental point of view, is defined by the visibility $0 < V < 1$

$$V_{HOM} = \frac{x - y}{x} \tag{22.3}$$

where x is the coincidence level when input photons are asynchronous in time, whereas y is the coincidence minimum corresponds to fine-tuned time synchronized input photons. The above idea can be projected into the integrated photonics application. However, due to naturally matched path lengths in on-chip applications HOM interference can be measured in a Mach–Zehnder interferometer (MZI), which acts as a completely tunable beamsplitter [196]. In this experiment, photons are fed into an MZI, and coincidence fringe as function of MZI phase can be used to establish standard HOM equivalent visibility,

$$V_{HOM} = \frac{a - 2b}{a} \tag{22.4}$$

where $a = 2x$ is the maxima of the fringe, while $b = y$ is the fringe minimum. In literature, there are some different expressions to quantify HOM interference in an MZI, such as $C = \frac{a-b}{a+b}$ and $V' = \frac{a-b}{a}$. Although they are quantifiers of quantum interference, they do not correspond to V_{HOM}. For example, for completely distinguishable photons, $V' = \frac{1}{2}$ and $C = \frac{1}{3}$, whereas $V_{HOM} = 0$.

Provided that the low-power pump pulse having finite spectral bandwidth is used to seed the silicon source, the expansion in Eq. 22.2 can be restricted up to first order. Hence, considering only spectral degree of freedom, the state can be written as

$$|\Psi\rangle \approx |0,0\rangle_{s,i} + C \iint d\omega_s d\omega_i f(\omega_s, \omega_i) \hat{a}_s^\dagger(\omega_s) \hat{a}_i^\dagger(\omega_s) |0,0\rangle_{s,i} \tag{22.5}$$

where $\hat{a}^\dagger(\omega)$ is the usual continuous-mode photon creation operator that generates a photon at frequency ω, C is the efficiency parameter related to the intrinsic $\chi^{(3)}$ nonlinearity strength of the silicon source, pump amplitude, and spatial overlap integral of the involved modes in the case of waveguide sources. The function $f(\omega_s, \omega_i)$ is the joint spectral amplitude (JSA), which has a simple physical meaning: the absolute square of the function, namely joint spectral intensity (JSI) $|f(\omega_s, \omega_i)|^2$, gives a joint spectral probability density of possible output pairs. In a straight SFWM waveguide sources, the shape of JSA is defined by the pump envelope function and the phase-matching function that depends on the interaction length and dispersive properties of the source. The pump envelope function is usually assumed to be Gaussian. Meanwhile, in a negligible temporal walk-off regime, i.e. two-pump pulses propagate with almost identical group velocities (or single-pump pulse input) phase-matching function ∝ sine cardinal (*sinc*) function exhibits oscillatory behavior, whose arguments are phase mismatch and effective interaction length. In the opposite regime (so-called complete temporal walk-off regime) based on nondegenerate pump pulses, phase-matching function simplified to a Gaussian, leading to a joint spectrum with suppressed residual correlations [197]. The shape of the phase-matching function can intuitively be understood in the time domain. In a negligible temporal walk-off regime, interaction of the medium and pump field(s) is suddenly turned on at the beginning and turned off instantly when pump field(s) exits the medium, which can be described by a step function whose Fourier transform is clearly a *sinc* function. In contrast, nondegenerate scheme offers a Gaussian-shaped interaction whose Fourier transform is still a Gaussian profile. These type of interactions can be achieved as a result of injection of two distinct pump pulses into the medium with an inserted time delay as slower one is sent in front of the fast one. Hence, the interaction gradually increases until fast one catches up the slower one, then it starts decreasing and completely vanishes once the pump pulses separate. The disadvantage of this scheme is the reduced interaction length and therefore the lower efficiency. Despite this downside, Gaussian-shaped phase-matching function potentially provides higher spectral purity [198]. In the context of resonator-based sources, on the other hand, JSA is a function of Lorentzian resonance lines along with the pump spectrum and the phase-matching function [199].

The magnitude of the JSA can be measured simply by using two physically separated scanning monochromator [200], see Figure 22.6(a)) for basic underlying principle, by fiber spectroscopy [201], by Fourier spectroscopy [202], or by stimulated emission tomography [203]. A fully phase-sensitive characterization of a JSA has also been demonstrated [204, 205]. Finally, the overlap between the

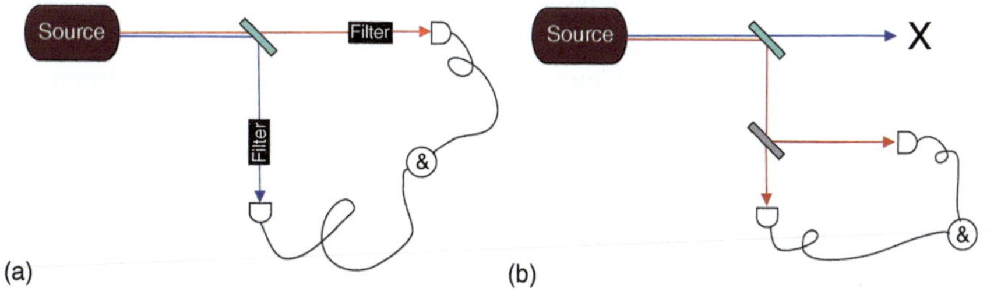

Figure 22.6 Experimental setups for spectral purity estimation: (a) Typical JSI measurement setup. Frequency scan is performed by the two tunable filters corresponding to a two-dimensional sweep of the signal and idler frequencies. Hence, JSI can be reconstructed by recording the coincidence rate. (b) Relatively easier marginal second-order correlation measurement $g_m^{(2)}(0)$ does not rely on heralding.

JSAs from two different photon pair sources has been examined via reverse HOM interference [206].

By virtue of the energy- and phase-matching requirements, the generated signal and idler photons usually exhibit strong correlations in their spectra. This correlation reveals itself in the process of "heralding" with nonfrequency resolving standard single-photon detectors, which projects the heralded single photon into an incoherent mixture of spectral modes called a mixed state. The correlation can be quantified by decomposing JSA into spectral Schmidt modes [207].

$$f(\omega_s, \omega_i) = \sum_j c_j \, \varphi_j(\omega_s) \, \phi_j(\omega_i) \tag{22.6}$$

where c_j's are the real amplitude coefficients, and $\{\varphi_j(\omega_s)\}_j$ and $\{\phi_j(\omega_i)\}_j$ are a set of orthogonal spectral modes for signal and idler photons. The spectral purity $\mathcal{P}$ is quantified by the following equation

$$\mathcal{P} = \sum_j c_j^4, \tag{22.7}$$

which identifies how many spectral Schmidt modes are active in a photon-pair state, hence it is a measure of entanglement. The state exhibits spectral entanglement if there is more than one eigenvalues in the decomposition, otherwise the state can be factorized into two functions. Relatively easier method to characterize spectral correlation is the marginal, i.e. only one (say signal) arm of the source is measured, second-order correlation measurement $g_m^{(2)}(0)$, see Figure 22.6(b)). Relation between $\mathcal{P}$ and $g_m^{(2)}(0)$ is governed by the following simple equation in a low squeezing regime [208]

$$g_m^{(2)}(0) = 1 + \mathcal{P} \tag{22.8}$$

This can be understood intuitively by the fact that spectrally pure photons mimic thermal statistics, leading to $g_m^{(2)}(0) \to 2$, whereas spectrally very impure photons exhibit Poissonian statistics, resulting in $g_m^{(2)}(0) \to 1$.

The most common way to suppress this correlation is to apply spectral filter, but it comes at the expense of reduced count rates [209]. The more suitable approach relies on generating photons in one joint spectral mode by orchestrating pump field(s) properties (envelope shape, chirp, etc..) and/or medium features [210]. In silicon photonics, the micro-resonator structure can be utilized to generate spectrally pure photons, see Figure 22.7(b)). In such a resonance structures, generated photon pairs are restricted in a spectral band, resulting in an appreciable enhancement in a corresponding SFWM process [189]. Furthermore, these resonances can be engineered to obtain more symmetric JSA. Although in conventional micro-resonator systems, spectral purity is bounded up to 0.93 [199], two different theoretical proposals showed that high spectral purity > 0.99 can be achieved via interferometrically coupled dual channel configuration [211] and via manipulating the pump pulse [212]. An experimental study has reported in [213] has verified a spectral purity $\mathcal{P} = 0.95$ exploiting the technique proposed in [211]. Relatively recently, spectral purity $\mathcal{P} = 0.99$ and spectral overlap 0.98 between separate sources have reached in a silicon waveguide supporting multiple transverse mode at the

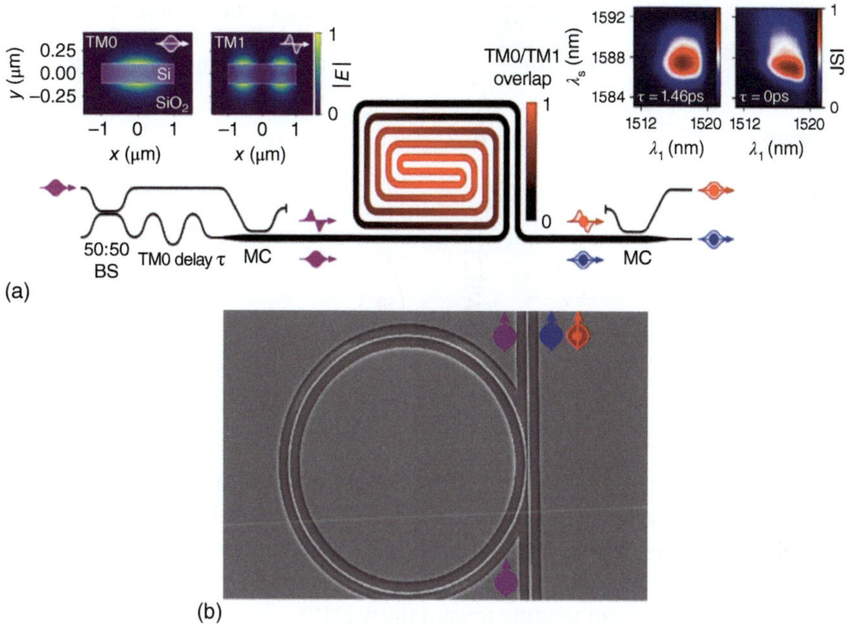

Figure 22.7 Enhancment Spectral Purity: (a) Exploiting nondegenerate scheme enables Gaussian phasematching profile and measured JSIs with and without inserted delay. Source: Stefano Paesani et al. [206]/Springer Nature/CC BY 4.0. (b) Top-view picture of microring resonator.

wavelength of interest [206], see Figure 22.7(a)). This promising approach is based on complete temporal walk-off scheme as mentioned above. Two lowest order transverse magnetic (TM) modes, namely TM_0 and TM_1, illuminate the waveguide sequentially depending on the group velocities ($v_{g_0} > v_{g_1}$) and generate photon pairs via intermodal phase matching [214].

It is worth stressing, to avoid confusion, that purity can have different meanings: i) Purity in single-photon source literature describes deviation of the state from single-photon output, which we have called photon-number purity. ii) Purity in photon pair source literature is used to quantify the spectral (and/or spatial) mode correlation between individual photons forming the photon-pair state, which we have labeled as spectral (and/or spatial) purity.

22.2.2.2 Multiplexing for Deterministic Photon Generation

In order to suppress multiphoton generation for silicon-based SFWM, the single-photon output probability has to be kept low, typically $0.01 < p < 0.05$. These heralded photon pair sources scale down by p^n as quantum information and communication applications scale up, which makes them unsuitable for more complex architectures. A promising avenue to overcome this issue is to multiplex number of sources pumped at low powers into a single output to have deterministic photon pair source, since scalable quantum information applications require on-demand (or at least with high predictability) generation of single photons. The basic idea

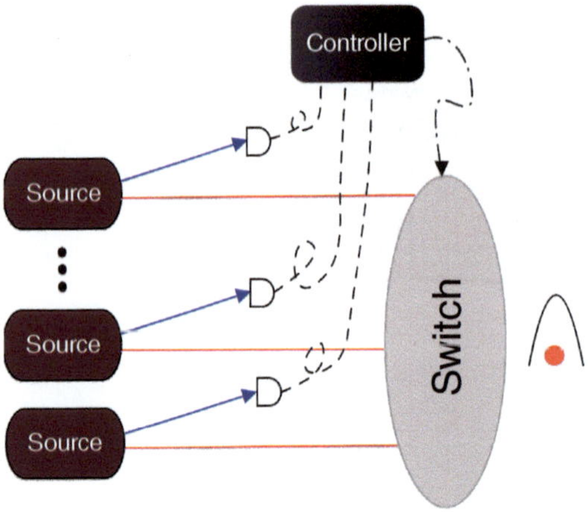

Figure 22.8 Spatial-multiplexing setup: If any detector fires, controller triggers the switch.

is to use a number of photon pair sources and exploit heralding for routing the heralded photon to a single output [215, 216], see Figure 22.8. In the absence of loss, deterministic single-photon output can be achieved as the number of pair sources increases. However, the loss is the limiting factor of number of sources and hinders the possible enhancement tending toward unity [217].

Any orthogonal degree of freedom of photon can be exploited for multiplexing. Spatial multiplexing requires multiple pair generation events to occur in different spatial modes generally in physically separated sources [218, 219], whereas temporal multiplexing schemes provide an advantage in terms of scaling the physical resources. In this scheme, a photon pair source is pumped by fast, periodic pump and upon heralding single photons multiplexed into the single output [66, 67, 220, 221]. Furthermore, combination of spatial and temporal, namely spatio-temporal, multiplexing has been realized to enable an additional enhancement to the single-photon emission probability [222]. Meanwhile, spectral multiplexing schemes use the spectrally impure (entangled) photon pairs and detection of heralding (idler) photons with a frequency resolved detection technique [223, 224].

Hybrid integration of pair sources in silicon photonics and thin-film LN (or barium titanate) can establish a fruitful association for multiplexing schemes [225], since these modulators offer modulation speed order of GHz and limited propagation loss as low as 2.7 dB/m. Another important limiting factor is due to the detect-switch-release latency, which can be compensated via routing photons off the low-loss guiding medium in spite of the fact that grating couplers add extra losses to system. However, low-loss and low-latency achievements can be gained to increase single-photon emission probability by tracking advancements in classical silicon photonics [13].

22.2.2.3 Parasitic Effects

Although nonlinear photon sources workhorse for most of the experiments, additional nonlinear effects come into play and interfere with these sources. Silicon photon sources at optical communication band (~1550 nm) suffer from intrinsic two-photon absorption (TPA) exciting electron into conduction band, leading to a reduction in heralding efficiency [226]. One solution is to operate in which total energy of the photons does not match with the bandgap of silicon, namely in the mid infrared, to avoid TPA effects. Recently, potential TPA roll-off has been shown by translating silicon quantum photonics to the mid-infrared regime [227]. Apart from TPA, the fields experience other well-known nonlinear effects, such as self-phase modulation and cross-phase modulation, as well as absorption and dispersion. Although these effects are well known to nonlinear optics community, their modelings are nontrivial at least in full generality [228]. Meanwhile, the SFWM simulation based on split-step Fourier method has been detailed, with a prescription for including silicon-specific effects such as TPA [229]. Moreover, due to the crystallinity of silicon, Raman contamination is restricted to a narrow bandwidth far from the pump (15.6 THz), making it easier to manage in comparison with silica fiber and chalcogenide glasses [230].

22.3 Applications

22.3.1 Quantum Information Processing

The concept of a single photon may appear to be easily understandable at first glance; however, a photon is itself a tricky entity. Single-photon features broad range of different degrees of freedom such as path, polarization, frequency, and spatial modes that can be the basic logical unit of quantum information, qubit $\alpha\,|0\rangle + \beta\,|1\rangle$, where α and β are the complex amplitudes. Moreover, several quantum particles may share (quantum) correlation called entanglement, which is one of the most prominent findings of quantum mechanics [231], lies at the heart of quantum information processing.

The primary encoding scheme in silicon photonics is path encoding due to the reconfigurability, universality, and ease in manipulation by integrated linear optical components with high fidelity. Multidimensional (up to 15×15) path-encoded entanglement on silicon chip was realized with 16 identical photon pair sources [30], see Figure 22.9(b)). Although other types of encoding schemes on chip, such as polarization [233], discrete frequency bins (up to 10 dimensions)[232], see Figure 22.9(a)), and transverse-spatial-mode [234], were proposed, they are less common compared to path encoding schemes.

A multiphoton Greenberger-Horne–Zeilinger (GHZ) entangled state was proposed [235], leading to extended tests of nonlocality [236]. Although 12-photon GHZ entangled state was demonstrated [11] in bulk optics scheme, on-chip genuine multipartite GHZ entangled states is limited to 4 photons. First on-chip generation of 4-photon GHZ entanglement was reported by Adcock et al. [196] in 2019, see Figure 22.10(a), and by Llewellyn et al. [29] in 2020, see Figure 22.10(b).

What makes photons ideal carries of quantum information in applications such as communication is their limited interaction with medium and with one another. For computation, however, although the limited interactions seem to be a deal breaker, various ingenious schemes were proposed [237]. Furthermore, due to high precision

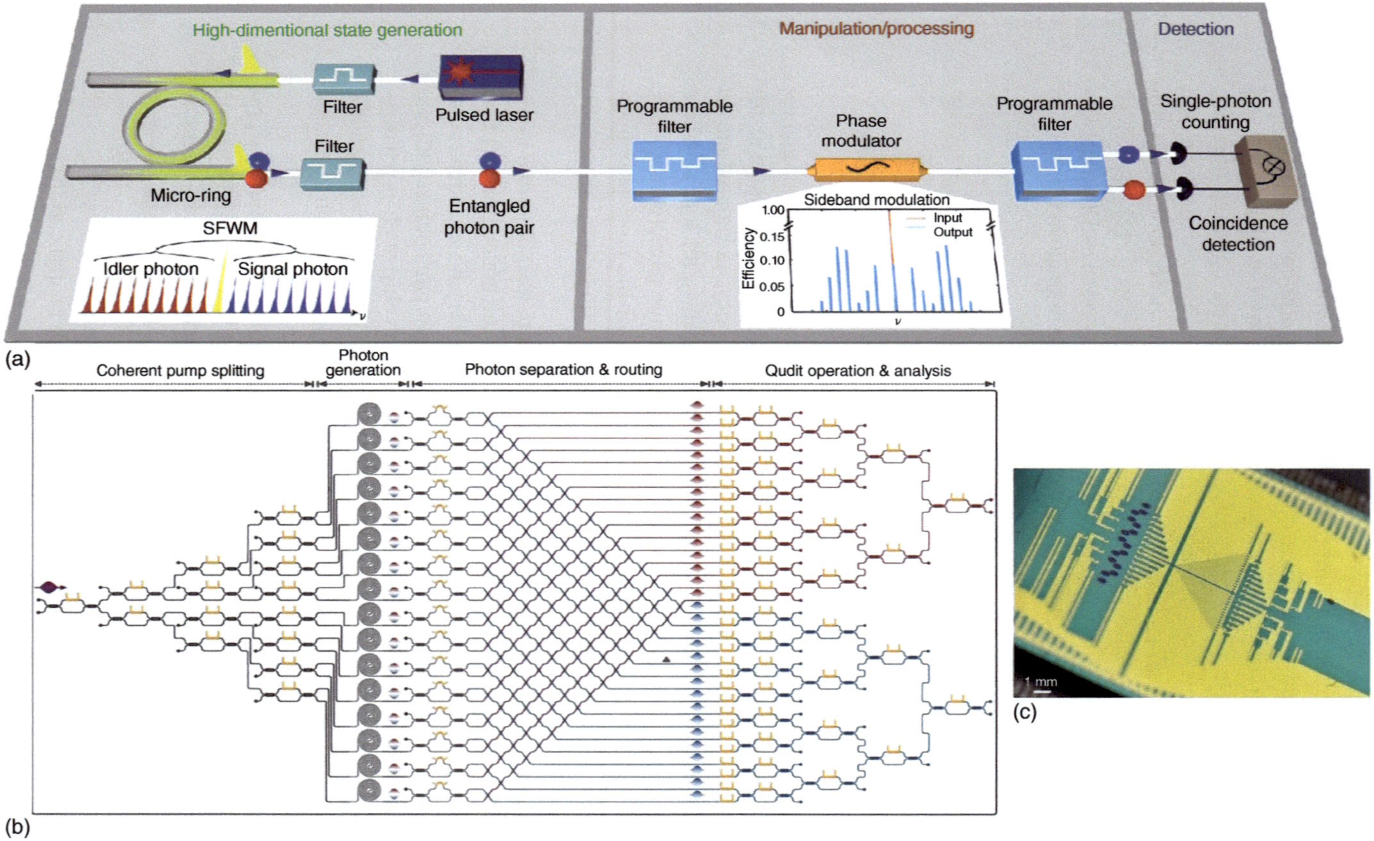

Figure 22.9 Multidimensional entanglement generation on chip. (a) Frequency bin encoded multidimensional entangled state. Source: Kues et al. [232]/With permission of Springer Nature. (b) Path-encoded entanglement generation on a programmable photonics chips having more than 550 photonics components. (c) Photograph of the chip whose diagram shown in (b). Source: Wang et al. [30]/with permission of American Association for the Advancement of Science - AAAS.

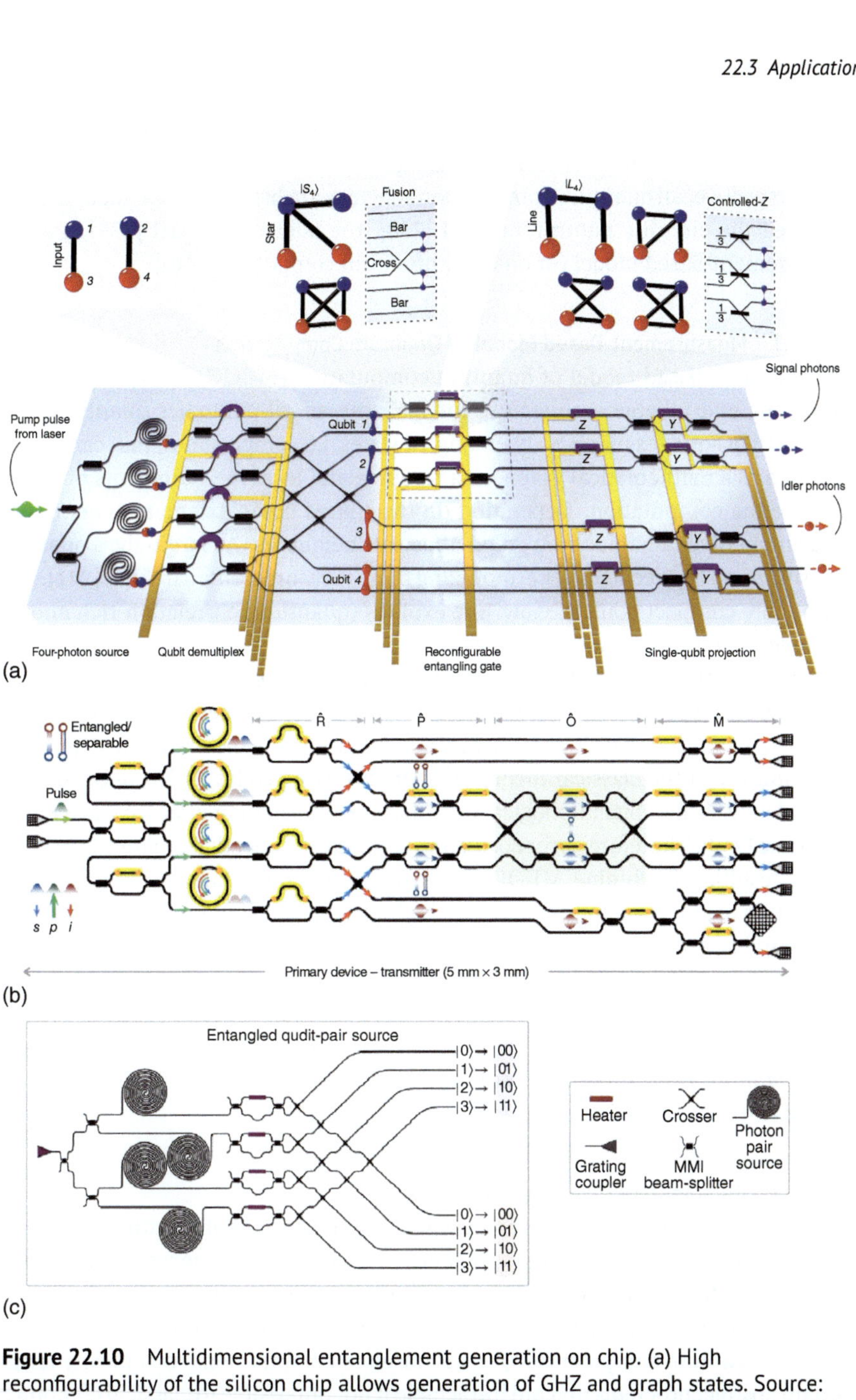

Figure 22.10 Multidimensional entanglement generation on chip. (a) High reconfigurability of the silicon chip allows generation of GHZ and graph states. Source: Adcock et al. [196]/Springer Nature/CC BY 4.0. (b) Generation of four-photon four-qubit genuine GHZ entangled states via microresonator. Source: Llewellyn et al. [29]/With permission of Springer Nature. (c) Graph state generation on a silicon chip. Reconfigurable graph states are realized with and without error correction to compare quantum information processing tasks. The success rate would improve from 62.5% to 95.8% owing to error correction. Source: Vigliar et al. [57]/With permission of Springer Nature.

manipulation of photons and scalability (in principle) of the system makes optical quantum system prominent candidate for quantum computation. From more general perspective, all quantum computation schemes are based on three main models, so called the circuit ("natural") model [238], the adiabatic model [239], and the measurement-based model (or one-way quantum computing) [240].

22.3.1.1 Measurement-Based Model of Quantum Computation

Measurement-based model of quantum computation (MBQC) is considered to be promising and reliable architecture for realizing an all-photonics quantum computer. MBQC utilizes many-body entangled states, i.e., graph states, that can be represented as a mathematical graph, as a resource and local measurements on qubits to drive the computation. Depending on the shared complex entanglement, local measurement on a qubit of the graph affects its neighbors. Accordingly, a quantum algorithm can be executed by harmonizing these measurements carefully [241]. Taking a more compact point of view, one exploits (quantum) correlation rich enough to implement universal logical processing.

The most important obstacles to unleash full power of quantum computers are errors, which will start to dominate the computation as system size grows. Quantum error correction (QEC) codes come into play following the marriage of quantum mechanics with the classical theory of error correcting codes. QEC codes combine multiple physical qubits into logical ones and the information is stored in an entangled state [242]. Therefore, error correction codes can be encoded inherently in graph states, resulting in shining MBQC out in quantum computation world. Recently, generation of graph entangled states via high-dimensional entangling gates on a silicon chip has been reported [57], see Figure 22.10(c).

Although QEC schemes boost the future prospects of quantum computation, they are not sufficient alone that a noisy quantum computer works reliably [243]. Remarkably, if individual error rates can be kept under a threshold value *fault-tolerant*, quantum computation is possible in conjunction with the implementation of QEC codes [244, 245].

22.3.1.2 Rudimentary Quantum Computers

Building universal quantum computers with photons is possible with large cluster states – particular instance of graph states – [246], but is immensely challenging. On the other hand, it is strongly believed that raw computational power of quantum computers can be demonstrated with modest experimental requirements via rudimentary quantum computers (RQCs). Phrased in computational complexity terms, certain abstract problems that are intractable if not impossible to classical computers can be solved via RQCs.

Photonics has its own strong candidate to demonstrate "quantum computational supremacy": Boson Sampling (BS), which was formulated by Aaronson and Arkhipov (AA) in 2011 [247]. Full complexity proof of computational hardness is complicated, it is based essentially on calculating permanents of complex-valued matrices, which is hard to simulate classically. In this task, photons pass through the passive linear

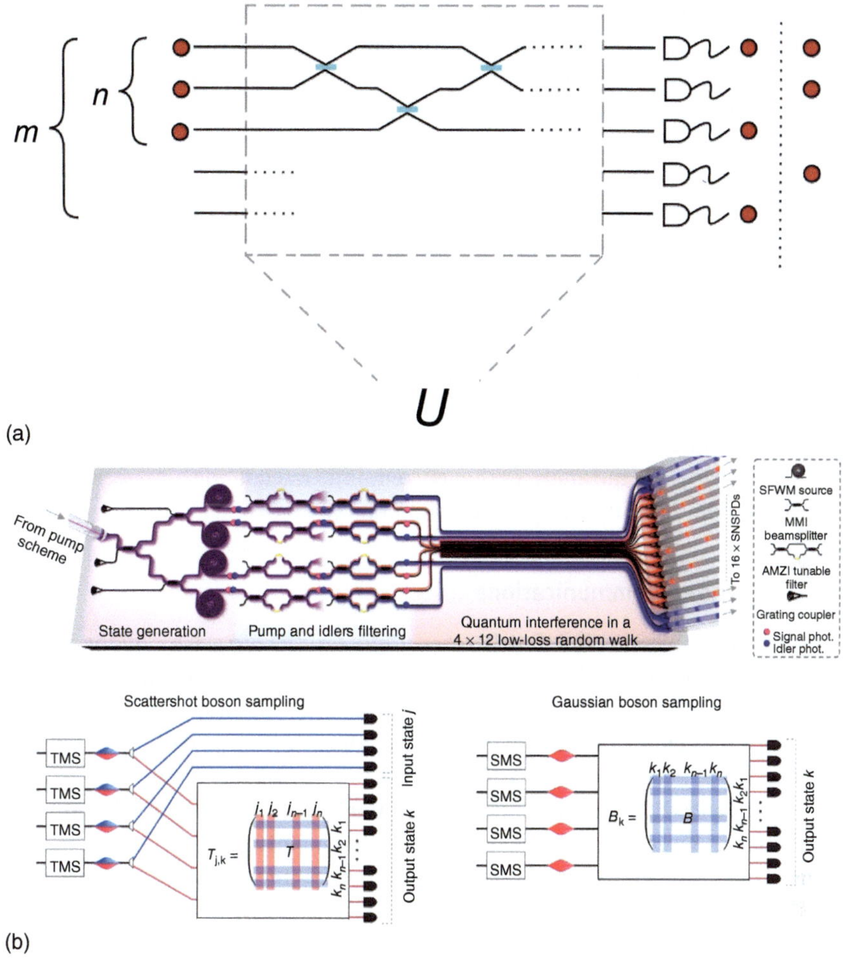

Figure 22.11 (a) Conceptual scheme of BS task: m-mode linear interferometer described by the linear transformation U is fed with photons, after number of trials output statistics is built up. These transitions correspond to permanent of n×n submatrices of U, obtained by selecting rows and columns of the matrix U. (b) On-chip realization of both SBS and GBS up to eight degenerate photons with a fixed unitary. Source: Paesani et al. [31]/With permission of Springer Nature.

optical interferometer described by a unitary matrix, then are measured via coincidence postdetection, leading to reproducing the approximate distribution governing the experiment, see Figure 22.11(a) for a conceptual scheme. Note that beating classical computer highly depends on number of occupied input modes and total number of interferometer mode; however, increasing either of these is challenging in practice [248]. Nevertheless, recently, breakthrough bulk-optical demonstration of BS has been implemented with 20 input photons (of which 14 are detected) and a 60-mode interferometer [10]. Establishing quantum supremacy, however, needs increased problem size [249], which is challenging for bulk-optical implementation.

On-chip silicon photonics, on the other hand, offers a scalable technology and may pave the way for demonstration of quantum computational advantage.

Initial proposal of AA, namely standard BS, has led to new proposals, such as Scattershot BS (SBS), which is suitable for implementation using heralded single-photon sources, relies on two-mode squeezing (TMS) vacuum state [250] and Gaussian BS (GBS) exploits single-mode squeezed (SMS) states, and result in disclosing a connection between the output probabilities and the hafnian of submatrices [251].

To perform a BS experiment photon sources, low-loss linear interferometer and high-efficiency photon detectors are needed. The first demonstrations share these three main building blocks [252, 253]. Furthermore, experimental studies have been reported in femtosecond laser writing [254] or silica-on-silicon waveguide circuits [255]. Recently, SBS and GBS on a silicon chip have been realized with states up to eight photons [31], see Figure 22.11(b). In this scheme, single color or two-color laser is used to generate input fields for implementing SBS and GBS. Most importantly, it was shown based on the analyses that current integrated fabrication technologies would allow demonstration of quantum supremacy.

22.3.2 Quantum Communications

Quantum communication is defined as *the faithful transfer of a generic quantum state between remote locations* [256] independently from the quantum channel, e.g. fiber, free-space, or underwater channels, see Figure 22.12(a). Quantum communications exploit multiple degrees of freedom for encoding the quantum states, e.g. time-bin or phase encoding, polarization or spatial modes, orbital angular momentum (OAM) or frequency encoding can be adopted depending on the different application and environment. As a concrete example in fiber-based quantum communication, time bin enconding is one of the most preferred degrees of freedom since intrinsic fiber effects like birifrangence and polarization dispersion are not destroying the quantum states. On the contrary, in free-space link, polarization is the most used one since this degree of freedom is well preserved in a satellite or intra-city link.

In the past ten years, different research groups have achieved incredible contributions in this area, e.g. the distribution of entangled states over 1200 km [257], ultra long distance QDK protocol [258], proof-of-concept experiment of underwater quantum communication [259, 260] and large-scale multiplexing of quantum and classical signal [261, 262]. It is important to be highlighted that to accomplish a quantum communication protocol, it is necessary to have an additional classical telecommunication channel that can be used for example as information

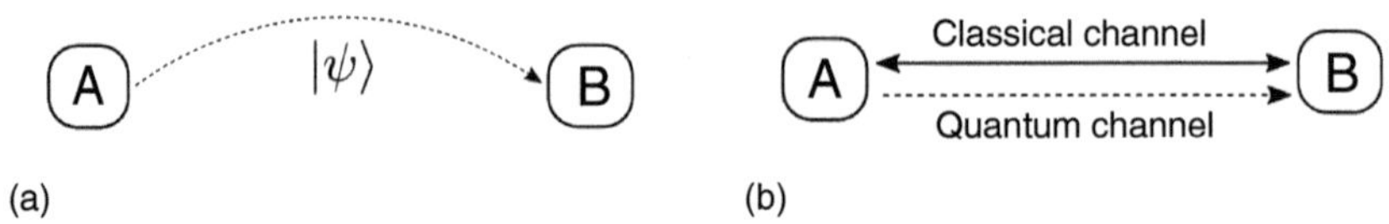

Figure 22.12 Generic scheme of quantum communications. (a) Definition of quantum communication between locations A and B. (b) Example of quantum communication channel including the quantum channel and the classical channel.

reconciliation protocol in QKD systems or for information exchange in more advance quantum communication primitives, e.g. entanglement distribution. This classical channel as reported in Figure 22.12(b) can be implemented in a different way compared to the quantum one, thus there are no restrictions in terms of types of communication (Internet, optical, radio, etc.) but it is usually a full-duplex communication.

In this overall context in which quantum communications are vastly expanding, we want to focus our attention into the practical realization of these systems. Most of them used bulk-optical components, which limit the range of applicability, the stability of the subsystems and most important the possibility of a vast implementation of the quantum technologies. On the contrary, integrated photonics offers a route toward cost-effective quantum communication systems, deployed at large scale. In the following Sections 22.3.2.1 and 22.3.2.2, we will carefully describe the state-of-the-art of quantum communication exploiting silicon photonic highlighting the benefits and the drawbacks of these devices.

22.3.2.1 Quantum Key Distribution

QKD, the first commercial application of the quantum technologies, allows to distribute symmetric keys between two or more users. QKD based its security on the fundamental laws of quantum physics and in particular it exploits measurement-induced collapse of superposition states to detect eavesdroppers on a quantum channel. In this way, it is possible to establish secure keys in an information theoretic secure way.

In order to distribute a symmetric key, it is necessary to follow a well-known procedure. In particular, in Figure 22.13, it is reported the full stack of the QKD protocol. In the following, we are describing the different actions in case of a single-photon source.

1. Alice randomly prepares N quantum states, in two different mutually unbiased bases, using a quantum random number generator (S1, …, SN).
2. Alice sends the N states to Bob through the quantum channel, i.e. free space, underwater, satellite, or fiber. It is important to stress the hypothesis that the quantum channel is under Eve's total control.
3. Bob measures the quantum states one by one choosing independently which is the basis to use for the measurement. The ratio between the two bases can be optimized as a function of the overall channel loss. Due to channel and receiver losses, only a few states reach Bob's apparatus producing a valid detection. Furthermore, many states contain zero photons and are therefore lost regardless. All these events will be discarded in the next step of the protocol (O1, …, ON).
4. Once all the states reached on Bob have been measured (and therefore destroyed), Alice and Bob begin to communicate on the service channel (classic channel). By hypothesis, the classic channel is authenticated, but the information exchanged here is public and not encrypted.
5. Sifting process: Bob communicates the subset of indices (1, 2, …, N) that produced a non"empty" result, also specifying the basis of the measurement

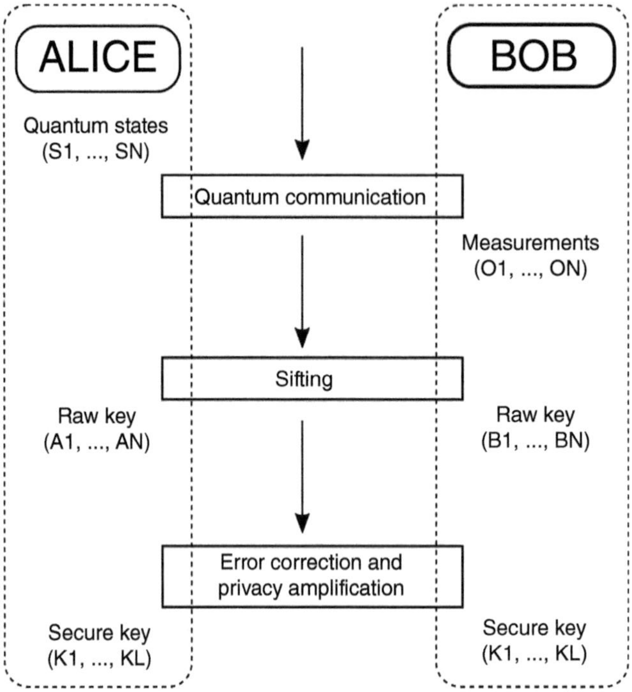

Figure 22.13 Quantum key distribution protocol. Full stack of the QKD protocol with the different actions required to extract a secret key rate.

(raw key). The raw keys of Alice and Bob have, by construction, the same length R. The percentage of bits that differ between two raw keys is called QBER (quantum bit error rate).

6. Error correction. Alice and Bob perform the error correction algorithm, exchanging communications on the service channel. Different types of error correction code can be adopted and implemented, e.g. cascade, winnow, and LPDC codes.
7. Error verification. To verify that the error correction procedure was successful, Alice and Bob each perform a hash of their corrected key.
8. Privacy amplification. Alice and Bob hash their corrected key, condensing it into L bits. These bits constitute the final key.

In the worldwide context, the application of QKD has to be compatible with current telecommunication network. First of all, the operation, i.e. the quantum channel should be at 1550 nm (C-band) or 1310 nm (O-band). Indeed in this range, the losses are quite low (∘ 0.2 or 0.3 dB/km) and allows the communication in the long range. However, today is not possible to extend the quantum distance over 500 km in the lab environment and about 100 km in deployed fiber links [263]. In this context, recently there have been multiple demonstrations using integrated silicon optics with weak coherent states (attenuated laser) [24, 26, 264]. More in particular, focusing on the DVs approach, Ref. [24] achieves chip-to-chip QKD using silicon photonics at 1550 nm. In this work, various QKD protocols were demonstrated (COW, BB84-polarization encoding and BB84-time-bin encoding) by exploiting 10 GHz bandwidth carrier-depletion modulation. These were combined with DC-operated TO modulators to minimize phase-dependent loss errors. In the same direction, two field-trial experiments, one over free space in daylight conditions [26] and one in fiber [265], have recently been realized using integrated photonics employing weak coherent states at telecommunications wavelengths. Here, the use of silicon photonics enabled the experiments high secret key rates and increased the stability of the system, particularly by correcting for polarization drift. One recent innovation comes from a polarization-based QKD transceiver device, where the same circuit traversed in different directions gives provides transmitter or receiver functionality [266].

Increasing secret key rates is crucial for developing quantum communications networks and for enlarging the range of applicability of QKD and quantum communication more in general.

One way to increase the secret key rate is to use high-dimensional quantum states,[268] exploiting the higher photon information efficiency of these special quantum states. In particular, in (Figure 22.14(a)), it is reported a first demonstration of high-dimensional QKD exploiting path encoded and multicore fiber. In [27], two silicon photonic chips are used for preparing high-dimensional quantum states in the path domain and then a multicore fiber was used for transporting these quantum states. In the experiment, a four-dimensional quantum state has been compared to a two-dimensional state and proved that higher dimensionality was able to guarantee an higher secret key rate. More recently, this transmission distance was extended to kilometer length scales using phase-locked loops [269, 270].

In addition, by using the same technology and by taking inspiration from the classical telecommunication technology, secret key rate can be increased by multiplexing multiple keys. For example, space division multiplexing was recently demonstrated by use of a multicore fiber (Figure 22.14(b)) [271]. In the experiment, keys were generated by a parallel silicon photonics transmitters and coupled to a multicore fiber by a circular grating coupler array, with a similar silicon chip featuring parallel receiver used to decode the stream of qubits. The same concept was scaled up to 37 cores of a multicore fiber and used to demonstrate 105 Mbit/s of secret key rate in 8 km long fiber link [272].

So far we have discussed DV QKD; however, other qubit formats can also be used, for example CV QKD has recently gained attention [273, 274]. This format is preferred for short links (< 100 km), due to limited signal-to-noise ratio at long distance. Silicon quantum photonics has application here, too, with scalable on-chip homodyne detection, based on high-quantum efficiency photodiodes [275]. Recently, a silicon chip has been employed to generate a Gaussian modulation scheme reaching an secret key rate of 0.14 kbps over 100 km of fiber (Figure 22.12(d)) [25]. A local oscillator and the quantum signal – which have orthogonal polarization – were coupled together into an optical fiber before demultiplexing and measurement through homodyne detection on the device. Recent theoretical advances in CV QKD have proved a the same level of security as in the DV approach [273, 276], setting the stage for future developments on silicon and otherwise.

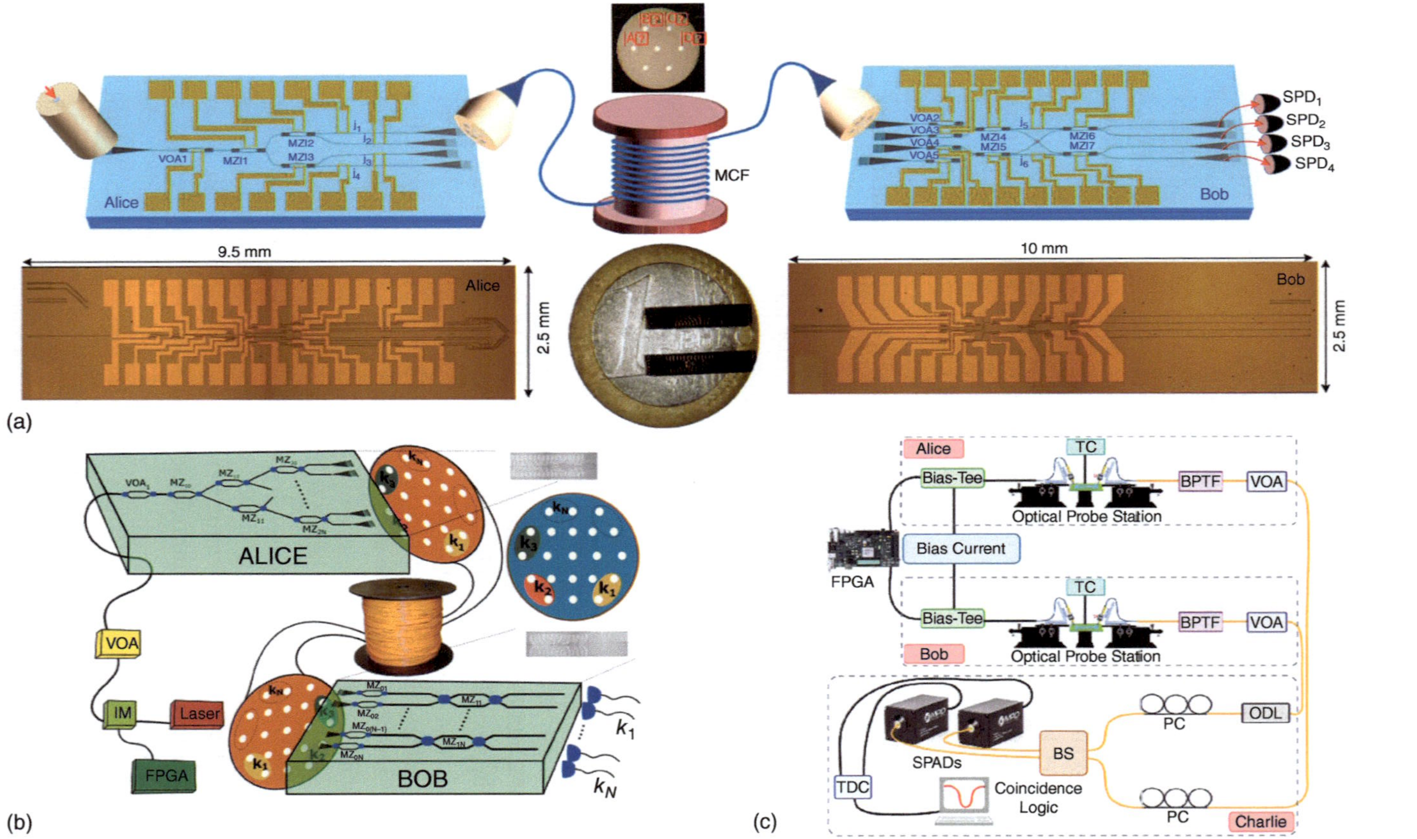

Figure 22.14 Examples of silicon PIC for QKD application. (a) Schematic of the HD-QKD based on multicore fiber using silicon PIC for Alice and Bob. The inset shows the cross section of the multicore fiber, where four cores are used. We also report the fabricated silicon PIC for Alice and Bob, respectively and the comparison with a 1 euro coin, indicating the compact size of the silicon PICs. (b) Experimental scheme of space division multiplexing in multicore fiber. The rectangle insets show Alice's (9.5 2.5 mm) and Bob's (10 2.5 mm) silicon chips. (c) Experimental setup used to perform the study of the HOM interference between weak coherent pulses generated by gain-switched III-V on silicon waveguide-integrated lasers. More details are reported in [267]. Source: Ding et al. [271]/Springer Nature/CC BY 4.0; Agnesi et al. [267]/with permission of Optical Society of America.

Measurement-device-independent (MDI) protocols allow the creation of a quantum key without relying on a trusted receiver, improving the security. Toward this goal, 0.46 visibility HOM interference – crucial for MDI protocols and limited to 0.5 for coherent states [277, 278] – was recently demonstrated (Figure 22.12(b)) [267]) between two integrated distributed feedback lasers on discrete silicon chips – simultaneously with a result using indium phosphide devices [279]. It is worth to be mentioned that one of the most critical devices in a QKD system (or more in general) in DV quantum communications is the single-photon detector. Superconductive nanowire single-photon detectors (SNSPDs) are among the most costly and least portable resource in quantum photonics – their availability is often a bottleneck in quantum photonics research. A solution can be offered by time multiplexing detectors between multiple users and was recently demonstrated in a QKD setting with a silicon chip [280]. In addition, very recently a silicon chip integrating SNSPDS directly patterned on to a lithographically defined silicon photonic chip was demonstrated, with application to untrusted MDI QKD nodes [281].

Future application of silicon photonic QKD links will depend on the availability of efficient light sources for the generation of weak coherent pulses on the platform. Though progress is being made [282], today, solutions based on hybrid techniques are prevalent, for example by flip-chip bonded III-IV devices [267, 283] also reported in Figure 22.14c). For example, a silicon photon pair source pumped by a silicon photonic with heterogeneously integrated laser [284] has been reported.

Another important degrees of freedom to be mentioned is OAM. OAM degree of freedom of light constitutes an important resource for both classical and quantum information technologies. OAM mode multiplexing significantly increases data transmission capacity in classical and quantum communications, while OAM unbounded nature allows for generating high-dimensional quantum states that are useful to increase the secret key rate or resilience against errors in QKD [268]. In Figure 22.15(a), it is reported a silicon-based photonic-integrated chip, able to excite

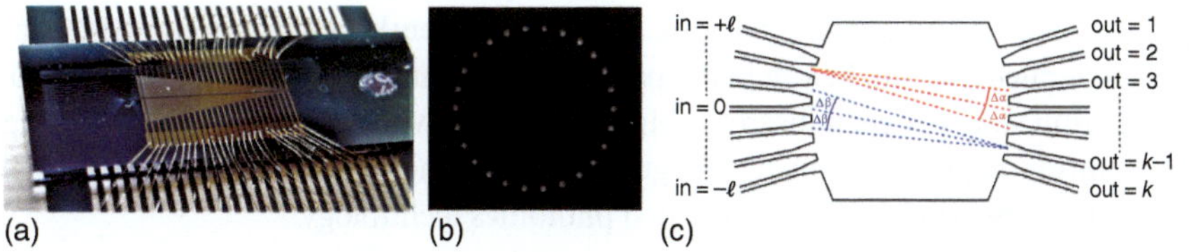

Figure 22.15 Examples of silicon PIC exploiting orbital angular momentum modes. (a) Picture shows the photonic-integrated device used in the experiment for the generation of multiplexed quantum OAM modes. (b) Infrared image of the grating couplers output. The 26 output are arranged on a ring of 325.5 μm. Source: Zahidy et al. [285]/with permission of Cornell University. (c) Schematic of the star-coupler structure. More detailed are reported in [285]. Source: Zahidy et al. [285]/De Gruyter/CC BY 4.0. (d) Schematic idea for the generation of quantum random numbers exploiting the superposition of OAM modes. To generate the superposition the light enters into the ring core fiber (RCF) with a tilted input. (A: standard input mode which will not experience distortion as its wavefront is aligned to the RCF; however, input mode B is transformed to a superposition of different modes due to the coupling angle. Various modes are separated along the fiber due to their different group velocities.

OAM modes in a ring-core fiber. The chip is proved to perform parallel QKD using 2 and 3 different OAM modes simultaneously. A ring-core fiber of 800 m length was used as the quantum channel. An overall crosstalk in the range of −10 to −18 dB between modes that allowed us to perform QKD with a quantum bit error rate of 1–4%. In Figure 22.15(b) and (c), it is reported the infrared image of the grating couplers output and the schematic of the star-coupler structure. More details are reported in [285]. In addition, by using the same device, in Figure 22.15(d), a new method to generate genuine random numbers exploiting an undesirable drawback of the channel, thus OAM mode crosstalk. Random numbers are an invaluable resource with many applications in science, cryptography, etc. In particular, QKD requires that generation of states is completely random, hence, a true random number generator is necessary to guarantee successful transmission of a secure key. Measurement result in quantum mechanics is intrinsically random. By introducing a partially controlled crosstalk, a superposition of OAM modes is generated upon coupling light in the ring-core fiber. The modes undergo a mode-to-time bin transformation that helps us to separately detect them. Performing time-of-arrival measurement in single-photon regime allows to generate random numbers with a rate of higher than 10 Mbit/s after privacy amplification.

22.3.2.2 Future Quantum Networks

In the last paragraph, we have seen how QKD enables the sharing of a secret symmetric key between two parties. More in general, to distribute a quantum key with more users and to expand the type of protocol we are using, nonclassical states of light are required [286, 287]. For example, the distribution of entangled states enables two quantum states, to be "teleported" between the remote locations. In this way, it is possible to establish a connection with multiple and different users around the world, who can utilize nonclassical correlation for various applications, from digital transaction and secure communications to coordination agreements protocols [288].

The first chip-to-chip transportation of entangled quantum states was achieved using two silicon photonic devices via coherent path-to-polarization state conversion grating coupler and optical fiber link [28]. More recently, a four-photon GHZ state was generated on-chip before being partially distributed and being used to teleport a quantum state between discrete chips [29]. These works extend the overall range of quantum applications using integrated silicon photonics and pave the way for quantum networks based on integrated photonics technology.

22.4 Outlook

Today, prolific field of on-chip silicon quantum photonics is being subjected to wide researches both theoretically and experimentally all around the world. At its core, silicon photonics technology allows myriad of optical components to be integrated on a single chip. Due to this natural privilege, in the future, it is expected to enable schemes for building all-photonics quantum computers, for actively multiplexed on-demand single photon sources with fast, low-loss switches and low-latency electronic control, and for demonstrating quantum computational advantage by e.g. on-chip reconfigurable BS schemes with full control over the unitary transformation.

It is quite clear that the 'holy grail' of limitless quantum communications will be enabled by quantum repeaters. Although there are different ways of building a quantum repeater, some of the technology will be the same shared for scaling quantum photonic information processors. In fact, the mechanics of a quantum repeater – teleporting a qubit from one location to another – are a core feature of measurement-based quantum computation. Eventually, the transmission of quantum data between quantum computers will require quantum interconnects between computer and network qubits. In this context, all-photonic repeater schemes using graph states [287] has already been proved in bulk optics [289]. Silicon photonics has recently demonstrated to be a powerful platform for graph state generation [196]. Future all-photonic repeaters will likely require integrated photonics and will form the basis of an entanglement-based quantum internet [288, 290].

The success of silicon quantum photonics in future relies on solving the challenges of all the components discussed in this chapter. As the limitations of silicon alone is obvious, hybrid integration with the other materials, e.g. III-V materials, LN, BTO, silicon nitride, etc., would be the goal, with every advance on the same platform, enabling new ground-breaking applications. No doubt, silicon quantum photonics will continue to be an extremely active field, tremendous effort will be put to address the challenges in light sources, components, and investigating new applications, new communication protocols, and new algorithms. Unprecedented advances in silicon quantum photonics are expected in the coming years.

References

1 Giustina, M., Versteegh, M.A.M., Wengerowsky, S. et al. (2015). Significant-loophole-free test of Bell's theorem with entangled photons. *Physical Review Letters* 115 (25): 250401.

2 Shalm, L.K., Meyer-Scott, E., Christensen, B.G. et al. (2015). Strong loophole-free test of local realism. *Physical Review Letters* 115 (25): 250402.

3 Hensen, B., Bernien, H., Dreau, A.E. et al. (2015). Loophole-free Bell inequality violation using electron spins separated by 1.3 kilometres. *Nature* 526 (7575): 682.

4 Ren, J.-G., Xu, P., Yong, H.-L. et al. (2017). Ground-to-satellite quantum teleportation. *Nature* 549 (7670): 70.

5 Liao, S.-K., Cai, W.-Q., Liu, W.-Y. et al. (2017). Satellite-to-ground quantum key distribution. *Nature* 549 (7670): 43–47.

6 Liao, S.-K., Cai, W.-Q., Handsteiner, J. et al. (2018). Satellite-relayed intercontinental quantum network. *Physical Review Letters* 120 (3): 030501.

7 Arute, F., Arya, K., Babbush, R. et al. (2019). Quantum supremacy using a programmable superconducting processor. *Nature* 574 (7779): 505–510.

8 Zhong, H.-S., Wang, H., Deng, Y.-H. et al. (2020). Quantum computational advantage using photons. *Science* 370 (6523): 1460–1463.

9 Madsen, L.S., Laudenbach, F., Askarani, M.F. et al. (2022). The theory of variational hybrid quantum-classical algorithms. *Nature* 606: 75–81.

10 Wang, H., Qin, J., Ding, X. et al. (2019). Boson sampling with 20 input photons and a 60-mode interferometer in a 10^{14}-dimensional hilbert space. *Physical Review Letters* 123 (25): 250503.

11 Zhong, H.-S., Li, Y., Li, W. et al. (2018). 12-photon entanglement and scalable scattershot boson sampling with optimal entangled-photon pairs from parametric down-conversion. *Physical Review Letter* 121: 250505.

12 Wang, X.-L., Luo, Y.-H., Huang, H.-L. et al. (2018). 18-qubit entanglement with six photons' three degrees of freedom. *Physical Review Letters* 120 (26): 260502.

13 Atabaki, A.H., Moazeni, S., Pavanello, F. et al. (2018). Integrating photonics with silicon nanoelectronics for the next generation of systems on a chip. *Nature* 556 (7701): 349–354.

14 Seok, T.J., Kwon, K., Henriksson, J. et al. (2019). 240 × 240 wafer-scale silicon photonic switches. *Optical Fiber Communication Conference*. pp. Th1E–5. Optical Society of America.

15 Chung, S.W., Abediasl, H., and Hashemi, H. (2018). A monolithically integrated large-scale optical phased array in silicon-on-insulator CMOS. *IEEE Journal of Solid-State Circuits* 53 (1): 275–296.

16 Sun, C., Wade, M.T., Lee, Y. et al. (2015). Single-chip microprocessor that communicates directly using light. *Nature* 528 (7583): 534.

17 Preskill, J. (2018). Quantum computing in the NISQ era and beyond. *Quantum* 2: 79.

18 Aspuru-Guzik, A. and Walther, P. (2012). Photonic quantum simulators. *Nature Physics* 8 (4): 285.

19 McArdle, S., Endo, S., Aspuru-Guzik, A. et al. (2020). Quantum computational chemistry. *Reviews of Modern Physics* 92 (1): 015003.

20 Sparrow, C., Martin-Lopez, E., Maraviglia, N. et al. (2018). Simulating the vibrational quantum dynamics of molecules using photonics. *Nature* 557 (7707): 660.

21 Steinbrecher, G.R., Olson, J.P., Englund, D., and Carolan, J. (2019). Quantum optical neural networks. *npj Quantum Information* 5: 60.

22 Bromley, T.R., Arrazola, J.M., Jahangiri, S. et al. (2020). Applications of near-term photonic quantum computers: software and algorithms. *Quantum Science and Technology* 5: 034010.

23 Yamamoto, Y., Aihara, K., Leleu, T. et al. (2017). Coherent Ising machines-optical neural networks operating at the quantum limit. *npj Quantum Information* 3 (1): 1–15.

24 Sibson, P., Kennard, J.E., Stanisic, S. et al. (2017). Integrated silicon photonics for high-speed quantum key distribution. *Optica* 4 (2): 172–177.

25 Zhang, G., Haw, J.Y., Cai, H. et al. (2019). An integrated silicon photonic chip platform for continuous-variable quantum key distribution. *Nature Photonics* 13 (12): 839–842.

26 Avesani, M., Calderaro, L., Schiavon, M. et al. (2021). Full daylight quantum-key-distribution at 1550 nm enabled by integrated silicon photonics. *npj Quantum Information* 7: 93.

27 Ding, Y., Bacco, D., Dalgaard, K. et al. (2017). High-dimensional quantum key distribution based on multicore fiber using silicon photonic integrated circuits. *npj Quantum Information* 3 (1): 1–7.

28 Wang, J., Bonneau, D., Villa, M. et al. (2016). Chip-to-chip quantum photonic interconnect by path-polarization interconversion. *Optica* 3 (4): 407–413.

29 Llewellyn, D., Ding, Y., Faruque, I.I. et al. (2020). Chip-to-chip quantum teleportation and multi-photon entanglement in silicon. *Nature Physics* 16 (2): 148–153.

30 Wang, J., Paesani, S., Ding, Y. et al. (2018). Multidimensional quantum entanglement with large-scale integrated optics. *Science* 360: 285–291.

31 Paesani, S., Ding, Y., Santagati, R. et al. (2019). Generation and sampling of quantum states of light in a silicon chip. *Nature Physics* 15: 925–929.

32 Reed, G.T., Mashanovich, G., Gardes, F.Y., and Thomson, D.J. (2010). Silicon optical modulators. *Nature Photonics* 4: 518–526.

33 Ferrari, S., Schuck, C., and Pernice, W. (2018). Waveguide-integrated superconducting nanowire single-photon detectors. *Nanophotonics* 7 (11): 1725–1758.

34 Chen, L., Doerr, C.R., Chen, Y., and Liow, T. (2010). Low-loss and broadband cantilever couplers between standard cleaved fibers and high-index-contrast Si_3N_4 or Si waveguides. *IEEE Photonics Technology Letters* 22 (23): 1744–1746.

35 Ding, Y., Ou, H., and Peucheret, C. (2013). Ultrahigh-efficiency apodized grating coupler using fully etched photonic crystals. *Optics Letters* 38 (15): 2732–2734. https://doi.org/10.1364/OL.38.002732.

36 Tiecke, T.G., Nayak, K.P., Thompson, J.D. et al. (2015). Efficient fiber-optical interface for nanophotonic devices. *Optica* 2 (2): 70–75. https://doi.org/10.1364/OPTICA.2.000070.

37 Gentry, C.M., Magana-Loaiza, O.S., Wade, M.T. et al. (2018). Monolithic source of entangled photons with integrated pump rejection. *CLEO: QELS_Fundamental Science*, p. JTh4C–3. Optical Society of America.

38 Cheng, Z., Zhang, J., Dong, J., and Ding, Y. (2021). Compact high-contrast silicon optical filter using all-passive and CROW Fano nanobeam resonators. *Optics Letters* 46 (16): 3873–3876. https://doi.org/10.1364/OL.433850.

39 Zhang, Y., Hosseini, A., Xu, X. et al. (2013). Ultralow-loss silicon waveguide crossing using Bloch modes in index-engineered cascaded multimode-interference couplers. *Optics Letters* 38 (18): 3608–3611. https://doi.org/10.1364/OL.38.003608.

40 Biberman, A., Shaw, M.J., Timurdogan, E. et al. (2012). Ultralow-loss silicon ring resonators. *Optics Letters* 37 (20): 4236–4238. https://doi.org/10.1364/OL.37.004236.

41 Puckett, M.W., Wang, J., Bose, D. et al. (2019). Silicon nitride ring resonators with 0.123 dB/m loss and q-factors of 216 million for nonlinear optical applications. *2019 Conference on Lasers and Electro-Optics Europe and European Quantum Electronics Conference*, p. ce-11_3. Optical Society of America.

42 Lee, H., Chen, T., Li, J. et al. (2012). Ultra-low-loss optical delay line on a silicon chip. *Nature Communications* 3: 867.

43 Pu, M., Liu, L., Ou, H. et al. (2010). Ultra-low-loss inverted taper coupler for silicon-on-insulator ridge waveguide. *Optics Communications* 283 (19): 3678–3682. https://doi.org/10.1016/j.optcom.2010.05.034.

44 Wang, J., Xuan, Y., Lee, C. et al. (2016). Low-loss and misalignment-tolerant fiber-to-chip edge coupler based on double-tip inverse tapers. *2016 Optical Fiber Communications Conference and Exhibition (OFC)*, pp. 1–3. IEEE.

45 Taillaert, D., Bienstman, P., and Baets, R. (2004). Compact efficient broadband grating coupler for silicon-on-insulator waveguides. *Optics Letters* 29 (23): 2749–2751. https://doi.org/10.1364/OL.29.002749.

46 Van Laere, F., Claes, T., Schrauwen, J. et al. (2007). Compact focusing grating couplers for silicon-on-insulator integrated circuits. *IEEE Photonics Technology Letters* 19 (23): 1919–1921.

47 Ding, Y., Peucheret, C., Ou, H., and Yvind, K. (2014). Fully etched apodized grating coupler on the SOI platform with −0.58 dB coupling efficiency. *Optics Letters* 39 (18): 5348–5350.

48 Zhang, H., Li, C., Tu, X. et al. (2014). Efficient silicon nitride grating coupler with distributed bragg reflectors. *Optics Express* 22 (18): 21800–21805. https://doi.org/10.1364/OE.22.021800.

49 Hong, J., Spring, A.M., and Alán Qiu, F. (2019). Silicon optical modulators. *Scientific Reports* 9: 12988.

50 Zaoui, W.S., Kunze, A., Vogel, W. et al. (2014). Bridging the gap between optical fibers and silicon photonic integrated circuits. *Optics Express* 22 (2): 1277–1286. https://doi.org/10.1364/OE.22.001277.

51 Luo, Y., Nong, Z., Gao, S. et al. (2018). Low-loss two-dimensional silicon photonic grating coupler with a backside metal mirror. *Optics Letters* 43 (3): 474–477. https://doi.org/10.1364/OL.43.000474.

52 Ong, J.R., Kumar, R., and Mookherjea, S. (2013). Ultra-high-contrast and tunable-bandwidth filter using cascaded high-order silicon microring filters. *IEEE Photonics Technology Letters* 25 (16): 1543–1546.

53 Piekarek, M., Bonneau, D., Miki, S. et al. (2017). High-extinction ratio integrated photonic filters for silicon quantum photonics. *Optics Letters* 42 (4): 815–818. https://doi.org/10.1364/OL.42.000815.

54 Kumar, R.R., Wu, X., and Tsang, H.K. (2020). Compact high-extinction tunable crow filters for integrated quantum photonic circuits. *Optics Letters* 45 (6): 1289–1292. https://doi.org/10.1364/OL.384187.

55 Nie, X., Turk, N., Li, Y. et al. (2019). High extinction ratio on-chip pump-rejection filter based on cascaded grating-assisted contra-directional couplers in silicon nitride rib waveguides. *Optics Letters* 44 (9): 2310–2313. https://doi.org/10.1364/OL.44.002310.

56 Huffman, T., Baney, D., and Blumenthal, D.J. (2017). High extinction ratio widely tunable low-loss integrated Si_3N_4 third-order filter. *Arxiv preprint arXiv:1708.06344.*

57 Vigliar, C., Paesani, S., Ding, Y. et al. (2021). Error-protected qubits in a silicon photonic chip. *Nature Physics* 17 (10): 1137–1143.

58 Chen, H. and Poon, A.W. (2006). Low-loss multimode-interference-based crossings for silicon wire waveguides. *IEEE Photonics Technology Letters* 18 (21): 2260–2262.

59 Fukazawa, T., Hirano, T., Ohno, F., and Baba, T. (2004). Low loss intersection of Si photonic wire waveguides. *Japanese Journal of Applied Physics* 43 (2): 646–647. https://doi.org/10.1143/jjap.43.646.

60 Sanchis, P., Villalba, P., Cuesta, F. et al. (2009). Highly efficient crossing structure for silicon-on-insulator waveguides. *Optics Letters* 34 (18): 2760–2762. https://doi.org/10.1364/OL.34.002760.

61 Ma, Y., Zhang, Y., Yang, S. et al. (2013). Ultralow loss single layer submicron silicon waveguide crossing for SOI optical interconnect. *Optics Express* 21 (24): 29374–29382. https://doi.org/10.1364/OE.21.029374.

62 Xie, Y., Xu, J., and Zhang, J. (2011). Elimination of cross-talk in silicon-on-insulator waveguide crossings with optimized angle. *Optical Engineering* 50 (6): 1–5. https://doi.org/10.1117/1.3593153.

63 Kim, S.-H., Cong, G., Kawashima, H. et al. (2014). Tilted MMI crossings based on silicon wire waveguide. *Optics Express* 22 (3): 2545–2552. https://doi.org/10.1364/OE.22.002545.

64 Bock, P.J., Cheben, P., Schmid, J.H. et al. (2010). Subwavelength grating crossings for silicon wire waveguides. *Optics Express* 18 (15): 16146–16155. https://doi.org/10.1364/OE.18.016146.

65 Liu, Y., Shainline, J.M., Zeng, X., and Popović, M.A. (2014). Ultra-low-loss CMOS-compatible waveguide crossing arrays based on multimode bloch waves and imaginary coupling. *Optics Letters* 39 (2): 335–338. https://doi.org/10.1364/OL.39.000335.

66 Adcock, J.C., Bacco, D., and Ding, Y. (2022). Enhancement of a silicon waveguide single photon source by temporal multiplexing. *Quantum Science and Technology* 7 (2): 025025. https://doi.org/10.1088/2058-9565/ac57f2.

67 Kaneda, F. and Kwiat, P.G. (2019). High-efficiency single-photon generation via large-scale active time multiplexing. *Science Advances* 5 (10): eaaw8586.

68 Lee, K.K., Lim, D.R., Kimerling, L.C. et al. (2001). Fabrication of ultralow-loss Si/SiO_2 waveguides by roughness reduction. *Optics Letters* 26 (23): 1888–1890. https://doi.org/10.1364/OL.26.001888.

69 Sparacin, D.K., Spector, S.J., and Kimerling, L.C. (2005). Silicon waveguide sidewall smoothing by wet chemical oxidation. *Journal of Lightwave Technology* 23 (8): 2455.

70 Cardenas, J., Poitras, C.B., Robinson, J.T. et al. (2009). Low loss etchless silicon photonic waveguides. *Optics Express* 17 (6): 4752–4757. https://doi.org/10.1364/OE.17.004752.

71 Bellegarde, C., Pargon, E., Sciancalepore, C. et al. (2018). Improvement of sidewall roughness of submicron SOI waveguides by hydrogen plasma and annealing. *IEEE Photonics Technology Letters* 30 (7): 591–594.

72 Dong, P., Qian, W., Liao, S. et al. (2010). Low loss shallow-ridge silicon waveguides. *Optics Express* 18 (14): 14474–14479. https://doi.org/10.1364/OE.18.014474.

73 Dong, P., Qian, W., Liang, H. et al. (2010). Thermally tunable silicon racetrack resonators with ultralow tuning power. *Optics Express* 18 (19): 20298–20304. https://doi.org/10.1364/OE.18.020298.

74 He, M., Xu, M., Ren, Y. et al. (2019). High-performance hybrid silicon and lithium niobate Mach–Zehnder modulators for 100 Gbit s^{-1} and beyond. *Nature Photonics* 13: 359–364.

75 Abel, S., Eltes, F., Ortmann, J.E. et al. (2019). Large pockels effect in micro- and nanostructured barium titanate integrated on silicon. *Nature Materials* 18: 42–47.

76 Najafi, F., Mower, J., Harris, N.C. et al. (2015). On-chip detection of non-classical light by scalable integration of single-photon detectors. *Nature Communications* 6: 5873.

77 Khasminskaya, S., Pyatkov, F., Słowik, K. et al. (2016). Fully integrated quantum photonic circuit with an electrically driven light source. *Nature Photonics* 10 (11): 727.

78 Vetter, A., Ferrari, S., Rath, P. et al. (2016). Cavity-enhanced and ultrafast superconducting single-photon detectors. *Nano Letters* 16 (11): 7085–7092.

79 Watts, M.R., Sun, J., DeRose, C. et al. (2013). Adiabatic thermo-optic Mach-Zehnder switch. *Optics Letters* 38 (5): 733–735.

80 Qiu, H., Liu, Y., Luan, C. et al. (2020). Energy-efficient thermo-optic silicon phase shifter with well-balanced overall performance. *Optics Letters* 45 (17): 4806–4809. https://doi.org/10.1364/OL.400230.

81 Miller, S.A., Chang, Y.-C., Phare, C.T. et al. (2020). Large-scale optical phased array using a low-power multi-pass silicon photonic platform. *Optica* 7 (1): 3–6. https://doi.org/10.1364/OPTICA.7.000003.

82 Yan, S., Zhu, X., Frandsen, L. et al. (2017). Slow-light-enhanced energy efficiency for graphene microheaters on silicon photonic crystal waveguides. *Nature Communications* 8: 14411.

83 Muthali, A.L., Adcock, J.C., Yvind, K., and Ding, Y. (2022). A Sub-mW silicon waveguide microheater. *CLEO: QELS_Fundamental Science*, p. JW3B.186. Optical Society of America.

84 Timurdogan, E., Poulton, C., Byrd, M. et al. (2017). Electric field-induced second-order nonlinear optical effects in silicon waveguides. *Nature Photonics* 11: 200–206.

85 Li, M. and Tang, H.X. (2019). Strong pockels materials. *Nature Materials* 18: 9–11.

86 Wang, C., Zhang, M., Stern, B. et al. (2018). Nanophotonic lithium niobate electro-optic modulators. *Optics Express* 26 (2): 1547–1555. https://doi.org/10.1364/OE.26.001547.

87 Zhang, M., Wang, C., Chen, X. et al. (2018). Ultra-high bandwidth integrated lithium niobate modulators with record-low v_π. *Optical Fiber Communication Conference Postdeadline Papers*, p. Th4A.5. Optical Society of America. https://doi.org/10.1364/OFC.2018.Th4A.5.

88 Wang, C., Zhang, M., Chen, X. et al. (2018). Integrated lithium niobate electro-optic modulators operating at CMOS-compatible voltages. *Nature* 562 (7725): 101–104.

89 Rao, A. and Fathpour, S. (2018). Compact lithium niobate electrooptic modulators. *IEEE Journal of Selected Topics in Quantum Electronics* 24 (4): 1–14.

90 Eltes, F., Mai, C., Caimi, D. et al. (2019). A $BaTiO_3$-based electro-optic pockels modulator monolithically integrated on an advanced silicon photonics platform. *Journal of Lightwave Technology* 37 (5): 1456–1462.

91 Desiatov, B., Shams-Ansari, A., Zhang, M. et al. (2019). Ultra-low-loss integrated visible photonics using thin-film lithium niobate. *Optica* 6 (3): 380–384. https://doi.org/10.1364/OPTICA.6.000380.

92 Zhang, M., Wang, C., Cheng, R. et al. (2017). Monolithic ultra-high-Q lithium niobate microring resonator. *Optica* 4 (12): 1536–1537. https://doi.org/10.1364/OPTICA.4.001536.

93 Weigel, P.O., Zhao, J., Fang, K. et al. (2018). Bonded thin film lithium niobate modulator on a silicon photonics platform exceeding 100 GHz 3-dB electrical modulation bandwidth. *Optics Express* 26 (18): 23728–23739. https://doi.org/10.1364/OE.26.023728.

94 Sun, S., He, M., Xu, M. et al. (2020). 120 Gbs^{-1} hybrid silicon and lithium niobate modulators with on-chip termination resistor. *Optical Fiber Communication Conference (OFC) 2020*, p. M2B.7. Optical Society of America. https://doi.org/10.1364/OFC.2020.M2B.7.

95 Gao, S., Xu, M., He, M. et al. (2019). Fast polarization-insensitive optical switch based on hybrid silicon and lithium niobate platform. *IEEE Photonics Technology Letters* 31 (22): 1838–1841.

96 Geler-Kremer, J., Eltes, F., Stark, P. et al. (2022). A ferroelectric multilevel non-volatile photonic phase shifter. *Nature Photonics*. https://doi.org/10.1038/s41566-022-01003-0.

97 Namekata, N., Adachi, S., and Inoue, S. (2009). 1.5 GHz single-photon detection at telecommunication wavelengths using sinusoidally gated InGaAs/InP avalanche photodiode. *Optics Express* 17 (8): 6275–6282. https://doi.org/10.1364/OE.17.006275.

98 Zhang, J., Itzler, M.A., Zbinden, H., and Pan, J.-W. (2015). Advances in InGaAs/InP single-photon detector systems for quantum communication. *Light: Science & Applications* 4: e286.

99 Natarajan, C.M., Tanner, M.G., and Hadfield, R.H. (2012). Superconducting nanowire single-photon detectors: physics and applications. *Superconductor Science and Technology* 25 (6): 063001. https://doi.org/10.1088/0953-2048/25/6/063001.

100 Rosenberg, D., Kerman, A.J., Molnar, R.J., and Dauler, E.A. (2013). High-speed and high-efficiency superconducting nanowire single photon detector array. *Optics Express* 21 (2): 1440–1447. https://doi.org/10.1364/OE.21.001440.

101 Photon Spot, Inc. (2018). https://www.photonspot.com/ (accessed 8 December 2022).

102 Zhang, W.J., You, L.X., Li, H. et al. (2017). NbN superconducting nanowire single photon detector with efficiency over 90% at 1550 nm wavelength operational at compact cryocooler temperature. *Science China Physics, Mechanics & Astronomy* 60: 120314. https://doi.org/10.1007/s11433-017-9113-4.

103 Korzh, B., Zhao, Q.-Y., Allmaras, J.P. et al. (2020). Demonstration of sub-3 ps temporal resolution with a superconducting nanowire single-photon detector. *Nature Photonics* 14 (4): 250–255.

104 Bartlett, S.D. and Sanders, B.C. (2002). Universal continuous-variable quantum computation: requirement of optical nonlinearity for photon counting. *Physical Review A* 65: 042304. https://doi.org/10.1103/PhysRevA.65.042304.

105 Divochiy, A., Marsili, F., Bitauld, D. et al. (2008). Superconducting nanowire photon-number-resolving detector at telecommunication wavelengths. *Nature Photonics* 2 (5): 302.

106 Allman, M.S., Verma, V.B., Stevens, M. et al. (2015). A near-infrared 64-pixel superconducting nanowire single photon detector array with integrated multiplexed readout. *Applied Physics Letters* 106 (19): 192601.

107 Sahin, D., Gaggero, A., Zhou, Z. et al. (2013). Waveguide photon-number-resolving detectors for quantum photonic integrated circuits. *Applied Physics Letters* 103 (11): 111116.

108 Fitch, M.J., Jacobs, B.C., Pittman, T.B., and Franson, J.D. (2003). Photon-number resolution using time-multiplexed single-photon detectors. *Physical Review A* 68: 043814. https://doi.org/10.1103/PhysRevA.68.043814.

109 Mattioli, F., Zhou, Z., Gaggero, A. et al. (2016). Photon-counting and analog operation of a 24-pixel photon number resolving detector based on superconducting nanowires. *Optics Express* 24 (8): 9067–9076.

110 Fukuda, D., Fujii, G., Numata, T. et al. (2011). Titanium-based transition-edge photon number resolving detector with 98% detection efficiency with index-matched small-gap fiber coupling. *Optics Express* 19 (2): 870–875. https://doi.org/10.1364/OE.19.000870.

111 Calkins, B., Mennea, P.L., Lita, A.E. et al. (2013). High quantum-efficiency photon-number-resolving detector for photonic on-chip information processing. *Optics Express* 21 (19): 22657–22670.

112 Biberman, A., Shaw, M.J., Timurdogan, E. et al. (2012). Ultralow-loss silicon ring resonators. *The 9th International Conference on Group IV Photonics (GFP)*, Volume 37, pp. 39–41. IEEE. ISBN 978-1-4577-0825-1. https://doi.org/10.1109/GROUP4.2012.6324079.

113 Chung, S.W., Abediasl, H., and Hashemi, H. (2017). A 1024-element scalable optical phased array in 0.18 μm SOI CMOS. *2017 IEEE International Solid-State Circuits Conference (ISSCC)*, pp. 262–263. IEEE, February 2017. ISBN 978-1-5090-3758-2. https://doi.org/10.1109/ISSCC.2017.7870361.

114 Liu, X., Akahane, K., Jahan, N.A. et al. (2013). Single-photon emission in telecommunication band from an InAs quantum dot grown on InP with molecular-beam epitaxy. *Applied Physics Letters* 103 (6): 061114. https://doi.org/10.1063/1.4817940.

115 Benyoucef, M., Yacob, M., Reithmaier, J.P. et al. (2013). Telecom-wavelength (1.5 μm) single-photon emission from InP-based quantum dots. *Applied Physics Letters* 103 (16): 162101. https://doi.org/10.1063/1.4825106.

116 Anderson, M., Müller, T., Skiba-Szymanska, J. et al. (2021). Coherence in single photon emission from droplet epitaxy and Stranski–Krastanov quantum dots in the telecom C-band. *Applied Physics Letters* 118 (1): 014003. https://doi.org/10.1063/5.0032128.

117 Holewa, P., Sakanas, A., Gür, U.M. et al. (2022). Bright quantum dot single-photon emitters at telecom bands heterogeneously integrated on Si. *ACS Photonics* https://doi.org/10.1021/acsphotonics.2c00027.

118 Wang, J., Zhou, Y., Wang, Z. et al. (2018). Bright room temperature single photon source at telecom range in cubic silicon carbide. *Nature Communications* 9 (1): 4106. https://doi.org/10.1038/s41467-018-06605-3.

119 Wang, J.-F., Yan, F.-F., Li, Q. et al. (2020). Coherent control of nitrogen-vacancy center spins in silicon carbide at room temperature. *Physical Review Letters* 124 (22): 223601. https://doi.org/10.1103/PhysRevLett.124.223601.

120 Wolfowicz, G., Anderson, C.P., Diler, B. et al. (2020). Vanadium spin qubits as telecom quantum emitters in silicon carbide. *Science Advances* 6 (18): 2–10. https://doi.org/10.1126/sciadv.aaz1192.

121 Zhou, Y., Wang, Z., Rasmita, A. et al. (2018). Room temperature solid-state quantum emitters in the telecom range. *Science Advances* 4 (3): 1–7. https://doi.org/10.1126/sciadv.aar3580.

122 Dibos, A.M., Raha, M., Phenicie, C.M., and Thompson, J.D. (2018). Atomic source of single photons in the telecom band. *Physical Review Letters* 120 (24): 243601. https://doi.org/10.1103/PhysRevLett.120.243601.

123 Alizadehkhaledi, A., Frencken, A.L., van Veggel, F.C.J.M., and Gordon, R. (2020). Isolating nanocrystals with an individual erbium emitter: a route to a stable single-photon source at 1550 nm wavelength. *Nano Letters* 20 (2): 1018–1022. https://doi.org/10.1021/acs.nanolett.9b04165.

124 Kim, J.-H., Aghaeimeibodi, S., Richardson, C.J.K. et al. (2017). Hybrid integration of solid-state quantum emitters on a silicon photonic chip. *Nano Letters* 17 (12): 7394–7400.

125 Kim, J.-H., Aghaeimeibodi, S., Carolan, J. et al. (2020). Hybrid integration methods for on-chip quantum photonics. *Optica* 7 (4): 291. https://doi.org/10.1364/OPTICA.384118.

126 Aharonovich, I., Englund, D., and Toth, M. (2016). Solid-state single-photon emitters. *Nature Photonics* 10 (10): 631–641. https://doi.org/10.1038/nphoton.2016.186.

127 Gregersen, N., Kaer, P., and Mork, J. (2013). Modeling and design of high-efficiency single-photon sources. *IEEE Journal of Selected Topics in Quantum Electronics* 19 (5): 1–16. https://doi.org/10.1109/JSTQE.2013.2255265.

128 Senellart, P., Solomon, G., and White, A. (2017). High-performance semiconductor quantum-dot single-photon sources. *Nature Nanotechnology* 12 (11): 1026–1039. https://doi.org/10.1038/nnano.2017.218.

129 Wang, H., Hu, H., Chung, T.-H. et al. (2019). On-demand semiconductor source of entangled photons which simultaneously has high fidelity, efficiency, and indistinguishability. *Physical Review Letters* 122 (11): 113602. https://doi.org/10.1103/PhysRevLett.122.113602.

130 Rodt, S. and Reitzenstein, S. (2021). Integrated nanophotonics for the development of fully functional quantum circuits based on on-demand single-photon emitters. *APL Photonics* 6 (1): 010901. https://doi.org/10.1063/5.0031628.

131 Vural, H., Portalupi, S.L., and Michler, P. (2020). Perspective of self-assembled InGaAs quantum-dots for multi-source quantum implementations. *Applied Physics Letters* 117 (3): 030501. https://doi.org/10.1063/5.0010782.

132 Pan, J.-W., Chen, Z.-B., Lu, C.-Y. et al. (2012). Multiphoton entanglement and interferometry. *Reviews of Modern Physics* 84 (2): 777–838. https://doi.org/10.1103/RevModPhys.84.777.

133 Uppu, R., Eriksen, H.T., Thyrrestrup, H. et al. (2020). On-chip deterministic operation of quantum dots in dual-mode waveguides for a plug-and-play single-photon source. *Nature Communications* 11 (1): 3782. https://doi.org/10.1038/s41467-020-17603-9.

134 Shields, A.J. (2007). Semiconductor quantum light sources. *Nature Photonics* 1 (4): 215–223. https://doi.org/10.1038/nphoton.2007.46.

135 Uppu, R., Pedersen, F.T., Wang, Y. et al. (2020). Scalable integrated single-photon source. *Science Advances* 6 (50). https://doi.org/10.1126/sciadv.abc8268.

136 Liu, J., Su, R., Wei, Y. et al. (2019). A solid-state source of strongly entangled photon pairs with high brightness and indistinguishability. *Nature Nanotechnology* 14 (6): 586–593. https://doi.org/10.1038/s41565-019-0435-9.

137 Schimpf, C., Reindl, M., Basset, F.B. et al. (2021). Quantum dots as potential sources of strongly entangled photons: perspectives and challenges for applications in quantum networks. *Applied Physics Letters* 118 (10): 100502. https://doi.org/10.1063/5.0038729.

138 Basset, F.B., Valeri, M., Roccia, E. et al. (2021). Quantum key distribution with entangled photons generated on demand by a quantum dot. *Science Advances* 7 (12): 1–8. https://doi.org/10.1126/sciadv.abe6379.

139 Wang, B.-Y., Denning, E.V., Gür, U.M. et al. (2020). Micropillar single-photon source design for simultaneous near-unity efficiency and indistinguishability. *Physical Review B* 102 (12): 125301. https://doi.org/10.1103/PhysRevB.102.125301.

140 Grange, T., Somaschi, N., Antón, C. et al. (2017). Reducing phonon-induced decoherence in solid-state single-photon sources with cavity quantum electrodynamics. *Physical Review Letters* 118 (25): 253602. https://doi.org/10.1103/PhysRevLett.118.253602.

141 Badolato, A., Hennessy, K., Atatüre, M. et al. (2005). Deterministic coupling of single quantum dots to single nanocavity modes. *Science* 308 (5725): 1158–1161. https://doi.org/10.1126/science.1109815.

142 Dousse, A., Lanco, L., Suffczyński, J. et al. (2008). Controlled light-matter coupling for a single quantum dot embedded in a pillar microcavity using far-field optical lithography. *Physical Review Letters* 101 (26): 267404. https://doi.org/10.1103/PhysRevLett.101.267404.

143 Gschrey, M., Thoma, A., Schnauber, P. et al. (2015). Highly indistinguishable photons from deterministic quantum-dot microlenses utilizing three-dimensional in situ electron-beam lithography. *Nature Communications* 6 (1): 7662. https://doi.org/10.1038/ncomms8662.

144 Liu, J., Davanço, M.I., Sapienza, L. et al. (2017). Cryogenic photoluminescence imaging system for nanoscale positioning of single quantum emitters. *Review of Scientific Instruments* 88 (2): 023116. https://doi.org/10.1063/1.4976578.

145 Katsumi, R., Ota, Y., Osada, A. et al. (2020). In situ wavelength tuning of quantum-dot single-photon sources integrated on a CMOS-processed silicon waveguide. *Applied Physics Letters* 116 (4): 041103. https://doi.org/10.1063/1.5129325.

146 Grim, J.Q., Bracker, A.S., Zalalutdinov, M. et al. (2019). Scalable in operando strain tuning in nanophotonic waveguides enabling three-quantum-dot superradiance. *Nature Materials* 18 (9): 963–969. https://doi.org/10.1038/s41563-019-0418-0.

147 Stevenson, R.M., Young, R.J., See, P. et al. (2006). Magnetic-field-induced reduction of the exciton polarization splitting in InAs quantum dots. *Physical Review B* 73 (3): 033306. https://doi.org/10.1103/PhysRevB.73.033306.

148 Kaniber, M., Huck, M.F., Müller, K. et al. (2011). Electrical control of the exciton-biexciton splitting in self-assembled InGaAs quantum dots. *Nanotechnology* 22 (32): 325202. https://doi.org/10.1088/0957-4484/22/32/325202.

149 Reimer, M.E., van Kouwen, M.P., Hidma, A.W. et al. (2011). Electric field induced removal of the biexciton binding energy in a single quantum dot. *Nano Letters* 11 (2): 645–650. https://doi.org/10.1021/nl1037424.

150 Somaschi, N., Giesz, V., De Santis, L. et al. (2016). Near-optimal single-photon sources in the solid state. *Nature Photonics* 10 (5): 340–345. https://doi.org/10.1038/nphoton.2016.23.

151 Trotta, R., Zallo, E., Magerl, E. et al. (2013). Independent control of exciton and biexciton energies in single quantum dots via electroelastic fields. *Physical Review B* 88 (15): 155312. https://doi.org/10.1103/PhysRevB.88.155312.

152 Nowak, A.K., Portalupi, S.L., Giesz, V. et al. (2014). Deterministic and electrically tunable bright single-photon source. *Nature Communications* 5 (1): 3240. https://doi.org/10.1038/ncomms4240.

153 Javadi, A., Söllner, I., Arcari, M. et al. (2015). Single-photon non-linear optics with a quantum dot in a waveguide. *Nature Communications* 6 (1): 8655. https://doi.org/10.1038/ncomms9655.

154 Liu, F., Brash, A.J., O'Hara, J. et al. (2018). High Purcell factor generation of indistinguishable on-chip single photons. *Nature Nanotechnology* 13 (9): 835–840. https://doi.org/10.1038/s41565-018-0188-x.

155 Sapienza, L., Davanço, M., Badolato, A., and Srinivasan, K. (2015). Nanoscale optical positioning of single quantum dots for bright and pure single-photon emission. *Nature Communications* 6 (1): 7833. https://doi.org/10.1038/ncomms8833.

156 Schnauber, P., Singh, A., Schall, J. et al. (2019). Indistinguishable photons from deterministically integrated single quantum dots in heterogeneous GaAs/Si_3N_4 quantum photonic circuits. *Nano Letters* 19 (10): 7164–7172. https://doi.org/10.1021/acs.nanolett.9b02758.

157 Zadeh, I.E., Elshaari, A.W., Jöns, K.D. et al. (2016). Deterministic integration of single photon sources in silicon based photonic circuits. *Nano Letters* 16 (4): 2289–2294. https://doi.org/10.1021/acs.nanolett.5b04709.

158 Bauters, J.F., Heck, M.J.R., John, D. et al. (2011). Ultra-low-loss high-aspect-ratio Si_3N_4 waveguides. *Optics Express* 19 (4): 3163. https://doi.org/10.1364/OE.19.003163.

159 Sacher, W.D., Mikkelsen, J.C., Dumais, P. et al. (2017). Tri-layer silicon nitride-on-silicon photonic platform for ultra-low-loss crossings and interlayer transitions. *Optics Express* 25 (25): 30862. https://doi.org/10.1364/OE.25.030862.

160 Huffman, T., Davenport, M., Belt, M. et al. (2017). Ultra-low loss large area waveguide coils for integrated optical gyroscopes. *IEEE Photonics Technology Letters* 29 (2): 185–188. https://doi.org/10.1109/LPT.2016.2620433.

161 Smit, M., Leijtens, X., Ambrosius, H. et al. (2014). An introduction to InP-based generic integration technology. *Semiconductor Science and Technology* 29 (8): 083001. https://doi.org/10.1088/0268-1242/29/8/083001.

162 Semenova, E.S., Hostein, R., Patriarche, G. et al. (2008). Metamorphic approach to single quantum dot emission at 1.55 μm on GaAs substrate. *Journal of Applied Physics* 103 (10): 103533. https://doi.org/10.1063/1.2927496.

163 Portalupi, S.L., Jetter, M., and Michler, P. (2019). InAs quantum dots grown on metamorphic buffers as non-classical light sources at telecom C-band: a review. *Semiconductor Science and Technology* 34 (5): 053001. https://doi.org/10.1088/1361-6641/ab08b4.

164 Luxmoore, I.J., Toro, R., Del Pozo-Zamudio, O. et al. (2013). III-V quantum light source and cavity-QED on Silicon. *Scientific Reports* 3 (1): 1239. https://doi.org/10.1038/srep01239.

165 Holewa, P., Gawełczyk, M., Ciostek, C. et al. (2020). Optical and electronic properties of low-density InAs/InP quantum-dot-like structures designed for single-photon emitters at telecom wavelengths. *Physical Review B* 101 (19): 195304. https://doi.org/10.1103/PhysRevB.101.195304.

166 Sala, E.M., Na, Y.I., Godsland, M. et al. (2020). InAs/InP quantum dots in etched pits by droplet epitaxy in metalorganic vapor phase epitaxy. *physica Status Solidi (RRL) - Rapid Research Letters* 14 (8): 2000173. https://doi.org/10.1002/pssr.202000173.

167 Miyazawa, T., Takemoto, K., Nambu, Y. et al. (2016). Single-photon emission at 1.5 μm from an InAs/InP quantum dot with highly suppressed multi-photon emission probabilities. *Applied Physics Letters* 109 (13): 132106. https://doi.org/10.1063/1.4961888.

168 Müller, T., Skiba-Szymanska, J., Krysa, A.B. et al. (2018). A quantum light-emitting diode for the standard telecom window around 1,550 nm. *Nature Communications* 9 (1): 862. https://doi.org/10.1038/s41467-018-03251-7.

169 Kunert, B., Mols, Y., Baryshniskova, M. et al. (2018). How to control defect formation in monolithic III/V hetero-epitaxy on (100) Si? A critical review on current approaches. *Semiconductor Science and Technology* 33 (9): 093002. https://doi.org/10.1088/1361-6641/aad655.

170 Supplie, O., Romanyuk, O., Koppka, C. et al. (2018). Metalorganic vapor phase epitaxy of III-V-on-silicon: experiment and theory. *Progress in Crystal Growth and Characterization of Materials* 64 (4): 103–132. https://doi.org/10.1016/j.pcrysgrow.2018.07.002.

171 Schmid, H., Borg, M., Moselund, K. et al. (2015). Template-assisted selective epitaxy of III-V nanoscale devices for co-planar heterogeneous integration with Si. *Applied Physics Letters* 106 (23): 233101. https://doi.org/10.1063/1.4921962.

172 Wen, P., Tiwari, P., Mauthe, S. et al. (2022). Waveguide coupled III-V photodiodes monolithically integrated on Si. *Nature Communications* 13 (1): 909. https://doi.org/10.1038/s41467-022-28502-6.

173 Viazmitinov, D.V., Berdnikov, Y., Kadkhodazadeh, S. et al. (2020). Monolithic integration of InP on Si by molten alloy driven selective area epitaxial growth. *Nanoscale* 12 (46): 23780–23788. https://doi.org/10.1039/D0NR05779G.

174 Kou, R., Hiratani, T., Yagi, H. et al. (2020). Inter-layer light transition in hybrid III-V/Si waveguides integrated by μ-transfer printing. *Optics Express* 28 (13): 19772. https://doi.org/10.1364/OE.394492.

175 Katsumi, R., Ota, Y., Osada, A. et al. (2019). Quantum-dot single-photon source on a CMOS silicon photonic chip integrated using transfer printing. *APL Photonics* 4 (3): 036105. https://doi.org/10.1063/1.5087263.

176 Aghaeimeibodi, S., Lee, C.-M., Buyukkaya, M.A. et al. (2019). Large stark tuning of InAs/InP quantum dots. *Applied Physics Letters* 114 (7): 071105. https://doi.org/10.1063/1.5082560.

177 Katsumi, R., Ota, Y., Kakuda, M. et al. (2018). Transfer-printed single-photon sources coupled to wire waveguides. *Optica* 5 (6): 691. https://doi.org/10.1364/OPTICA.5.000691.

178 Sahoo, H.K., Ottaviano, L., Zheng, Y. et al. (2018). Low temperature bonding of heterogeneous materials using Al_2O_3 as an intermediate layer. *Journal of Vacuum Science & Technology B, Nanotechnology and Microelectronics: Materials, Processing, Measurement, and Phenomena* 36 (1): 011202. https://doi.org/10.1116/1.5005591.

179 Zhang, C., Liang, D., and Bowers, J.E. (2013). MOCVD regrowth of InP on hybrid silicon substrate. *ECS Solid State Letters* 2 (11): Q82–Q86. https://doi.org/10.1149/2.008311ssl.

180 Sakanas, A., Yu, Y., Semenova, E. et al. (2017). Photonic crystal with buried heterostructure platform for laser devices directly bonded to Si. *Frontiers in Optics 2017*, p. FW5A.2. Washington, DC: OSA. ISBN 978-1-943580-33-0. https://doi.org/10.1364/FIO.2017.FW5A.2.

181 Sakanas, A., Semenova, E., Ottaviano, L. et al. (2019). Comparison of processing-induced deformations of InP bonded to Si determined by e-beam metrology: direct vs. adhesive bonding. *Microelectronic Engineering* 214: 93–99. https://doi.org/10.1016/j.mee.2019.05.001.

182 Butcher, P.N. and Cotter, D. (1990). *The Elements of Nonlinear Optics*, Number 9. Cambridge University Press.

183 Adcock, J.C., Bao, J., Chi, Y. et al. (2021). Advances in silicon quantum photonics. *Journal of Selected Topics in Quantum Electronics* 27 (2): 1–24.

184 Mauerer, W., Avenhaus, M., Helwig, W., and Silberhorn, C. (2009). How colors influence numbers: photon statistics of parametric down-conversion. *Physical Review A* 80 (5): 053815.

185 Eisaman, M.D., Fan, J., Migdall, A., and Polyakov, S.V. (2011). Single-photon sources and detectors. *Review of Scientific Instruments* 82 (7): 071101.

186 Lin, Q. and Agrawal, G.P. (2006). Silicon waveguides for creating quantum-correlated photon pairs. *Optics Letters* 31 (21): 3140–3142.

187 Sharping, J.E., Lee, K.F., Foster, M.A. et al. (2006). Silicon waveguides for creating quantum-correlated photon pairs. *Optics Express* 14 (25): 12388–12393.

188 Takesue, H., Fukuda, H., Tsuchizawa, T. et al. (2008). Generation of polarization entangled photon pairs using silicon wire waveguide. *Optics Express* 16 (8): 5721–5727.

189 Clemmen, S., Huy, K.P., Bogaerts, W. et al. (2009). Continuous wave photon pair generation in silicon-on-insulator waveguides and ring resonators. *Optics Express* 17 (19): 16558–16570.

190 Ma, C., Wang, X., Anant, V. et al. (2017). Silicon photonic entangled photon-pair and heralded single photon generation with car > 12,000 and $g^{(2)}(0) < 0.006$. *Optics Express* 25 (26): 32995–33006.

191 Engin, E., Bonneau, D., Natarajan, C.M. et al. (2013). Photon pair generation in a silicon micro-ring resonator with reverse bias enhancement. *Optics Express* 21 (23): 27826–27834.

192 Lu, X., Lee, J.Y., Feng, P.X.-L., and Lin, Q. (2014). High Q silicon carbide microdisk resonator. *Applied Physics Letters* 104 (18): 181103.

193 Azzini, S., Grassani, D., Galli, M. et al. (2013). Stimulated and spontaneous four-wave mixing in silicon-on-insulator coupled photonic wire nano-cavities. *Applied Physics Letters* 103 (3): 031117.

194 Knill, E., Laflamme, R., and Milburn, G.J. (2001). A scheme for efficient quantum computation with linear optics. *Nature* 409 (6816): 46–52.

195 Hong, C.K., Ou, Z.-Y., and Mandel, L. (1987). Measurement of subpicosecond time intervals between two photons by interference. *Physical Review Letters* 59 (18): 2044.

196 Adcock, J.C., Vigliar, C., Santagati, R. et al. (2019). Programmable four-photon graph states on a silicon chip. *Nature Communications* 10 (1): 1–6.

197 Fang, B., Cohen, O., Moreno, J.B., and Lorenz, V.O. (2013). State engineering of photon pairs produced through dual-pump spontaneous four-wave mixing. *Optics Express* 21 (3): 2707–2717.

198 Ekici, c.C. and Dinleyici, M.S. (2020). Graded-index optical fiber transverse-spatial-mode entanglement. *Physical Review A* 102 (1): 013702.

199 Helt, L., Yang, Z., Liscidini, M., and Sipe, J.E. (2010). Spontaneous four-wave mixing in microring resonators. *Optics Letters* 35 (18): 3006–3008.

200 Garay-Palmett, K., Cruz-Delgado, D., Dominguez-Serna, F. et al. (2016). Photon-pair generation by intermodal spontaneous four-wave mixing in birefringent, weakly guiding optical fibers. *Physical Review A* 93 (3): 033810.

201 Avenhaus, M., Eckstein, A., Mosley, P.J., and Silberhorn, C. (2009). Fiber-assisted single-photon spectrograph. *Optics Letters* 34 (18): 2873–2875.

202 Wasilewski, W., Wasylczyk, P., Kolenderski, P. et al. (2006). Joint spectrum of photon pairs measured by coincidence Fourier spectroscopy. *Optics Letters* 31 (8): 1130–1132.

203 Liscidini, M. and Sipe, J.E. (2013). Stimulated emission tomography. *Physical Review Letters* 111 (19): 193602.

204 Jizan, I., Bell, B., Helt, L.G. et al. (2016). Phase-sensitive tomography of the joint spectral amplitude of photon pair sources. *Optics Letters* 41 (20): 4803–4806.

205 Bell, B.A., Garces, G.T., and Walmsley, I.A. (2020). Diagnosing phase correlations in the joint spectrum of parametric downconversion using multi-photon emission. *Optics Express* 28: 34246–34254.

206 Paesani, S., Borghi, M., Signorini, S. et al. (2020). Near-ideal spontaneous photon sources in silicon quantum photonics. *Nature Communications* 11 (1): 1–6.

207 Law, C.K., Walmsley, I.A., and Eberly, J.H. (2000). Continuous frequency entanglement: effective finite hilbert space and entropy control. *Physical Review Letters* 84 (23): 5304.

208 Christ, A., Laiho, K., Eckstein, A. et al. (2011). Probing multimode squeezing with correlation functions. *New Journal of Physics* 13 (3): 033027.

209 Meyer-Scott, E., Montaut, N., Tiedau, J. et al. (2017). Limits on the heralding efficiencies and spectral purities of spectrally filtered single photons from photon-pair sources. *Physical Review A* 95 (6): 061803.

210 Ansari, V., Donohue, J.M., Brecht, B., and Silberhorn, C. (2018). Tailoring nonlinear processes for quantum optics with pulsed temporal-mode encodings. *Optica* 5 (5): 534–550.

211 Vernon, Z., Menotti, M., Tison, C.C. et al. (2017). Truly unentangled photon pairs without spectral filtering. *Optics Letters* 42 (18): 3638–3641.

212 Christensen, J.B., Koefoed, J.G., Rottwitt, K., and McKinstrie, C.J. (2018). Engineering spectrally unentangled photon pairs from nonlinear microring resonators by pump manipulation. *Optics Letters* 43 (4): 859–862.

213 Liu, Y., Wu, C., Gu, X. et al. (2020). High-spectral-purity photon generation from a dual-interferometer-coupled silicon microring. *Optics Letters* 45 (1): 73–76.

214 Stolen, R.H., Bjorkholm, J.E., and Ashkin, A. (1974). Phase-matched three-wave mixing in silica fiber optical waveguides. *Applied Physics Letters* 24 (7): 308–310.

215 Migdall, A.L., Branning, D., and Castelletto, S. (2002). Tailoring single-photon and multiphoton probabilities of a single-photon on-demand source. *Physical Review A* 66 (5): 053805.

216 Pittman, T.B., Jacobs, B.C., and Franson, J.D. (2002). Single photons on pseudodemand from stored parametric down-conversion. *Physical Review A* 66 (4): 042303.

217 Bonneau, D., Mendoza, G.J., O'Brien, J.L., and Thompson, M.G. (2015). Effect of loss on multiplexed single-photon sources. *New Journal of Physics* 17 (4): 043057.

218 Collins, M.J., Xiong, C., Rey, I.H. et al. (2013). Integrated spatial multiplexing of heralded single-photon sources. *Nature Communications* 4: 2582.

219 Ma, X.-s., Zotter, S., Kofler, J. et al. (2011). Experimental generation of single photons via active multiplexing. *Physical Review A* 83 (4): 043814.

220 Mower, J. and Englund, D. (2011). Efficient generation of single and entangled photons on a silicon photonic integrated chip. *Physical Review A* 84 (5): 052326.

221 Xiong, C., Zhang, X., Liu, Z. et al. (2016). Active temporal multiplexing of indistinguishable heralded single photons. *Nature Communications* 7: 10853.

222 Mendoza, G.J., Santagati, R., Munns, J. et al. (2016). Active temporal and spatial multiplexing of photons. *Optica* 3 (2): 127–132.

223 Joshi, C., Farsi, A., Clemmen, S. et al. (2018). Frequency multiplexing for quasi-deterministic heralded single-photon sources. *Nature Communications* 9 (1): 1–8.

224 Puigibert, M.G., Aguilar, G.H., Zhou, Q. et al. (2017). Heralded single photons based on spectral multiplexing and feed-forward control. *Physical Review Letters* 119 (8): 083601.

225 Adcock, J.C. and Ding, Y. (2022). Quantum prospects for hybrid thin-film lithium niobate on silicon photonics. *Frontiers of Optoelectronics* 15 (1): 1–7.

226 Husko, C.A., Clark, A.S., Collins, M.J. et al. (2013). Multi-photon absorption limits to heralded single photon sources. *Scientific Reports* 3: 3087.

227 Rosenfeld, L.M., Sulway, D.A., Sinclair, G.F. et al. (2020). Mid-infrared quantum optics in silicon. *Optics Express* 28 (25): 37092–37102.

228 Ma, C. and Mookherjea, S. (2020). Prospects for photon-pair generation using silicon microring resonators with two photon absorption and free carrier absorption. *OSA Continuum* 3 (5): 1138–1153.

229 Koefoed, J.G. and Rottwitt, K. (2019). Complete evolution equation for the joint amplitude in photon-pair generation through spontaneous four-wave mixing. *Physical Review A* 100 (6): 063813.

230 Silverstone, J.W., Bonneau, D., O'Brien, J.L., and Thompson, M.G. (2016). Silicon quantum photonics. *IEEE Journal of Selected Topics in Quantum Electronics* 22 (6): 390–402.

231 Einstein, A., Podolsky, B., and Rosen, N. (1935). Can quantum-mechanical description of physical reality be considered complete? *Physical Review* 47 (10): 777.

232 Kues, M., Reimer, C., Roztocki, P. et al. (2017). On-chip generation of high-dimensional entangled quantum states and their coherent control. *Nature* 546 (7660): 622.

233 Matsuda, N., Le Jeannic, H., Fukuda, H. et al. (2012). A monolithically integrated polarization entangled photon pair source on a silicon chip. *Scientific Reports* 2: 817.

234 Feng, L.-T., Zhang, M., Xiong, X. et al. (2019). On-chip transverse-mode entangled photon pair source. *npj Quantum Information* 5 (1): 1–7.

235 Greenberger, D.M., Horne, M.A., and Zeilinger, A. (1989). Going beyond bell's theorem. In: *Bell's Theorem, Quantum Theory and Conceptions of the Universe, Fundamental Theories of Physics*, vol. 37 (ed. M. Kafatos), 69–72. Dordrecht: Springer.

236 Mermin, N.D. (1990). Extreme quantum entanglement in a superposition of macroscopically distinct states. *Physical Review Letters* 65 (15): 1838.

237 Barz, S. (2015). Quantum computing with photons: introduction to the circuit model, the one-way quantum computer, and the fundamental principles of photonic experiments. *Journal of Physics B: Atomic, Molecular and Optical Physics* 48 (8): 083001.

238 Deutsch, D.E. (1989). Quantum computational networks. *Proceedings of the Royal Society of London. Series A* 425 (1868): 73–90.

239 Farhi, E., Goldstone, J., Gutmann, S., and Sipser, M. (2000). Quantum computation by adiabatic evolution. *arXiv preprint quant-ph/0001106*.

240 Raussendorf, R. and Briegel, H.J. (2001). A one-way quantum computer. *Physical Review Letters* 86 (22): 5188.

241 Briegel, H.J., Browne, D.E., Dür, W. et al. (2009). Measurement-based quantum computation. *Nature Physics* 5 (1): 19–26.

242 Knill, E. and Laflamme, R. (1997). Theory of quantum error-correcting codes. *Physical Review A* 55 (2): 900.

243 Preskill, J. (1998). Fault-tolerant quantum computation. *Introduction to Quantum Computation and Information*, pp. 213–269. World Scientific.

244 Shor, P.W. (1996). Fault-tolerant quantum computation. *37th Annual Symposium on Foundations of Computer Science, 1996. Proceedings*, pp. 56–65. IEEE.

245 Aharonov, D. and Ben-Or, M. (2008). Fault-tolerant quantum computation with constant error rate. *SIAM Journal on Computing* 38 (4): 1207–1282.

246 O'Brien, J.L. (2007). Optical quantum computing. *Science* 318 (5856): 1567–1570.

247 Aaronson, S. and Arkhipov, A. (2011). The computational complexity of linear optics. *Proceedings of the 43rd Annual ACM Symposium on Theory of Computing*, pp. 333–342. ACM.

248 Lund, A.P., Bremner, M.J., and Ralph, T.C. (2017). Quantum sampling problems, bosonsampling and quantum supremacy. *npj Quantum Information* 3 (1): 1–8.

249 Clifford, P. and Clifford, R. (2020). Faster classical boson sampling. *arXiv preprint arXiv:2005.04214*.

250 Lund, A.P., Laing, A., Rahimi-Keshari, S. et al. (2014). Boson sampling from a gaussian state. *Physical Review Letters* 113 (10): 100502.

251 Hamilton, C.S., Kruse, R., Sansoni, L. et al. (2017). Gaussian boson sampling. *Physical Review Letters* 119 (17): 170501.

252 Tillmann, M., Dakic, B., Heilmann, R. et al. (2013). Experimental boson sampling. *Nature Photonics* 7 (7): 540.

253 Spring, J.B., Metcalf, B.J., Humphreys, P.C. et al. (2013). Boson sampling on a photonic chip. *Science* 339 (6121): 798–801.

254 Crespi, A., Osellame, R., Ramponi, R. et al. (2013). Integrated multimode interferometers with arbitrary designs for photonic boson sampling. *Nature Photonics* 7 (7): 545–549.

255 Carolan, J., Harrold, C., Sparrow, C. et al. (2015). Universal linear optics. *Science* 349 (6249): 711–716.

256 Gisin, N. and Thew, R. (2007). Quantum communication. *Nature Photonics* 1 (3): 165.

257 Yin, J., Cao, Y., Li, Y.-H. et al. (2017). Satellite-based entanglement distribution over 1200 kilometers. *Science* 356 (6343): 1140–1144.

258 Chen, J.-P., Zhang, C., Liu, Y. et al. (2020). Sending-or-not-sending with independent lasers: secure twin-field quantum key distribution over 509 km. *Physical Review Letters* 124 (7): 070501.

259 Hufnagel, F., Sit, A., Bouchard, F. et al. (2020). Underwater quantum communication over a 30-meter flume tank. *arXiv preprint arXiv:2004.04821*.

260 Tarantino, S., Da Lio, B., Cozzolino, D., and Bacco, D. (2020). Feasibility study of quantum communications in aquatic scenarios. *Optik* 216: 164639. https://doi.org/10.1016/j.ijleo.2020.164639.

261 Bacco, D., Da Lio, B., Cozzolino, D. et al. (2019). Boosting the secret key rate in a shared quantum and classical fibre communication system. *Communications Physics* 2 (1): 1–8.

262 Eriksson, T.A., Hirano, T., Puttnam, B.J. et al. (2019). Wavelength division multiplexing of continuous variable quantum key distribution and 18.3 Tbit/s data channels. *Communications Physics* 2 (1): 1–8.

263 Ribezzo, D., Zahidy, M., Vagniluca, I. et al. (2022). Deploying an inter-european quantum network. *Advanced Quantum Technologies*: 2200061.

264 Ma, C., Sacher, W.D., Tang, Z. et al. (2016). Silicon photonic transmitter for polarization-encoded quantum key distribution. *Optica* 3 (11): 1274–1278.

265 Bunandar, D., Lentine, A., Lee, C. et al. (2018). Metropolitan quantum key distribution with silicon photonics. *Physical Review X* 8 (2): 021009.

266 Cai, H., Long, C.M., DeRose, C.T. et al. (2017). Silicon photonic transceiver circuit for high-speed polarization-based discrete variable quantum key distribution. *Optics Express* 25 (11): 12282–12294.

267 Agnesi, C., Da Lio, B., Cozzolino, D. et al. (2019). Hong–Ou–Mandel interference between independent III–V on silicon waveguide integrated lasers. *Optics Letters* 44 (2): 271–274.

268 Cozzolino, D., Da Lio, B., Bacco, D., and Oxenløwe, L.K. (2019). High-dimensional quantum communication: benefits, progress, and future challenges. *Advanced Quantum Technologies* 2 (12): 1900038.

269 Ca nas, G., Vera, N., Cari ne, J. et al. (2017). High-dimensional decoy-state quantum key distribution over multicore telecommunication fibers. *Physical Review A* 96 (2): 022317.

270 Da Lio, B., Bacco, D., Cozzolino, D. et al. (2019). Stable transmission of high-dimensional quantum states over a 2-km multicore fiber. *IEEE Journal of Selected Topics in Quantum Electronics* 26 (4): 1–8.

271 Bacco, D., Ding, Y., Dalgaard, K. et al. (2017). Space division multiplexing chip-to-chip quantum key distribution. *Scientific Reports* 7 (1): 1–7.

272 Da Lio, B., Bacco, D., Cozzolino, D. et al. (2018). Record-high secret key rate for joint classical and quantum transmission over a 37-core fiber. *2018 IEEE Photonics Conference (IPC)*, pp. 1–2. IEEE.

273 Pirandola, S., Andersen, U.L., Banchi, L. et al. (2019). Advances in quantum cryptography. *Advances in Optics and Photonics* 12: 1012–1236.

274 Xu, F., Ma, X., Zhang, Q. et al. (2019). Quantum cryptography with realistic devices. *arXiv preprint arXiv:1903.09051*.

275 Raffaelli, F., Ferranti, G., Mahler, D.H. et al. (2018). A homodyne detector integrated onto a photonic chip for measuring quantum states and generating random numbers. *Quantum Science and Technology* 3 (2): 025003.

276 Ziebell, M., Persechino, M., Harris, N. et al. (2015). Towards on-chip continuous-variable quantum key distribution. *2015 European Conference on Lasers and Electro-Optics - European Quantum Electronics Conference*, p. JSV_4_2. Optical Society of America. http://www.osapublishing.org/abstract.cfm?URI=EQEC-2015-JSV_4_2 (accessed 8 December 2022).

277 Wei, K., Li, W., Tan, H. et al. (2019). High-speed measurement-device-independent quantum key distribution with integrated silicon photonics. *Physical Review X* 10: 03103.

278 Semenenko, H., Sibson, P., Hart, A. et al. (2020). Chip-based measurement-device-independent quantum key distribution. *Optica* 7 (3): 238–242.

279 Semenenko, H., Sibson, P., Thompson, M.G., and Erven, C. (2019). Interference between independent photonic integrated devices for quantum key distribution. *Optics Letters* 44: 275–278.

280 Kong, L., Li, Z., Li, C. et al. (2020). Photonic integrated quantum key distribution receiver for multiple users. *Optics Express* 28 (12): 18449–18455.

281 Zhang, P., Zheng, X., Ge, R. et al. (2019). Integrated relay server for measurement-device-independent quantum key distribution. *Arxiv preprint arXiv:1912.09642*.

282 Zhou, Z., Yin, B., and Michel, J. (2015). On-chip light sources for silicon photonics. *Light: Science & Applications* 4 (11): e358.

283 Song, B., Stagarescu, C., Ristic, S. et al. (2016). 3D integrated hybrid silicon laser. *Optics Express* 24 (10): 10435–10444.

284 Wang, X., Ma, C., Kumar, R. et al. (2018). Photon pair generation using a silicon photonic hybrid laser. *APL Photonics* 3 (10): 106104.

285 Zahidy, M., Liu, Y., Cozzolino, D. et al. (2022). Photonic integrated chip enabling orbital angular momentum multiplexing for quantum communication. *Nanophotonics* 11 (4): 821–827.

286 Markham, D. and Sanders, B.C. (2008). Graph states for quantum secret sharing. *Physical Review A* 78 (4): 042309.

287 Azuma, K., Tamaki, K., and Lo, H.-K. (2015). All-photonic quantum repeaters. *Nature Communications* 6: 6787.

288 Kimble, H.J. (2008). The quantum internet. *Nature* 453 (7198): 1023–1030.

289 Li, Z.-D., Zhang, R., Yin, X.-F. et al. (2019). Experimental quantum repeater without quantum memory. *Nature Photonics* 13 (9): 644–648.

290 Lee, Y., Bersin, E., Dahlberg, A. et al. (2022). A quantum router architecture for high-fidelity entanglement flows in quantum networks. *npj Quantum Information* 8: 75.

Part V

Emerging Quantum Technologies: Challenges and Potential Applications

23

Photonic Realization of Qudit Quantum Computing

Yuchen Wang, Zixuan Hu, and Sabre Kais

Department of Chemistry, Department of Physics and Purdue Quantum Science and Engineering Institute, Purdue University, 560 Oval Dr, West Lafayette, IN 47907, USA

23.1 Introduction to Qudit Quantum Computing

Qudit is an alternative to qubit as the basic computational element for a quantum computer. A *qudit* is a quantum version of d-ary digits whose state can be described by a vector in the d dimensional Hilbert space $\mathscr{H}_d$ [1]. A set of orthonormal basis vectors $\{|0\rangle, |1\rangle, |2\rangle, \ldots, |d-1\rangle\}$ is used to span this space. A general form of the qudit state is

$$|\alpha\rangle = \alpha_0 |0\rangle + \alpha_1 |1\rangle + \alpha_2 |2\rangle + \cdots + \alpha_{d-1} |d-1\rangle = \begin{pmatrix} \alpha_0 \\ \alpha_1 \\ \alpha_2 \\ \vdots \\ \alpha_{d-1} \end{pmatrix} \in \mathbb{C}^d \tag{23.1}$$

where $|\alpha_0|^2 + |\alpha_1|^2 + |\alpha_2|^2 + \cdots + |\alpha_{d-1}|^2 = 1$. This section introduces the basic knowledge and concepts of the qudit quantum computing that includes qudit gates and universality in Section 23.1.1 and the important qudit algorithms such as quantum Fourier transform (QFT) and Phase Estimation Algorithm (PEA) in Section 23.1.2. The qudit system, with a state space larger than that of a qubit to store and process information, can reduce the circuit complexity, simplify the experimental setup, and enhance the algorithm efficiency [2].

23.1.1 Universality and Examples of Qudit Gates

Qudit gates are used to alter the state of a qudit and thus perform the computational operations. In quantum computation, the universal quantum gate set is defined as a set of matrices $U_k \in U(d^n)$ such that any arbitrary unitary transformation U of the Hilbert space $\mathscr{H}_d^{\otimes n}$ can be approximated via the product of its elements within the tolerance of error [3]. The idea of *universality* can be applied to the qudit logic [4–9] and there are several proposals and discussions of the standards for a universal qudit

Photonic Quantum Technologies: Science and Applications, First Edition. Edited by Mohamed Benyoucef.

gate set. In order to precisely simulate any unitary $U \in U(d^n)$ and thus set the criteria for the exact qudit universality, Vlasov shows that two noncommuting single qudit gates and a two-qudit gate are enough [3], while Brylinski and Brylinski show that a two-qudit gate with entangled qudits combined with some random single qudit gates are enough [1]. With these recipes any qudit gates can be reduced to sequences of elementary qudit gates of lower-dimension [10, 11]. Muthukrishnan and Stroud provide a physically workable procedure for building a universal qudit gate set, which implements the spectral decomposition of unitary transformations and involves a gate library of continuous parameter gates [12]. Luo and Wang propose a modified set comprising one-qudit general rotation gates and two-qudit controlled extensions of rotation gates that has gained a reduction in the gate requirement [13]. Brennen et al. provide a universal gate library with a fixed set of one qudit operations and a single free parameter "controlled phase" gate that utilize the QR decomposition of unitary transformations [14].

Many of the qubit gate either has a qudit variant or can be generalized to d dimension. For examples, the d dimension Pauli Z gate and Pauli X gate are in the matrix forms given by: [15–18]

$$X_d = \begin{pmatrix} 0 & 0 & \cdots & 0 & 1 \\ 1 & 0 & \cdots & 0 & 0 \\ 0 & 1 & \cdots & 0 & 0 \\ \vdots & \vdots & \ddots & \vdots & \vdots \\ 0 & 0 & \cdots & 1 & 0 \end{pmatrix}, \quad Z_d = \begin{pmatrix} 1 & 0 & 0 & \cdots & 0 \\ 0 & \omega & 0 & \cdots & 0 \\ 0 & 0 & \omega^2 & \cdots & 0 \\ \vdots & \vdots & \vdots & \ddots & 0 \\ 0 & 0 & 0 & \cdots & \omega^{d-1} \end{pmatrix} \tag{23.2}$$

in which ω is the dth root of unity. Other gates such as the qudit version of $\pi/8$ gate [19], SWAP gate [20–27], Toffoli gate [28, 29] have been proposed and reviewed in Ref. [2]. From the discussion of qudit universality, we know that the two-qudit gate, such as a qudit-controlled gate, is an indispensable component for universality. Since qudits have multiple states to be utilized, a qudit-controlled gate can accomplish operations of greater complexity than its qubit counterpart [30]. To fully utilize the d states on the control qudit, we discuss the *multi-value-controlled gate* (*MVCG*) for qudits. The *MVCG* can assign a unique operation to the target qudit for each corresponding state of the control qudit in one shot and within a single controlled-gate [31]. For a d-dimensional qudit system, the mathematical form of a two-qudit MVCG is a $d^2 \times d^2$ matrix given by:

$$\text{MVCG} = \begin{pmatrix} U_0 & 0 & 0 & \cdots & 0 \\ 0 & U_1 & 0 & \cdots & 0 \\ 0 & 0 & U_2 & \cdots & 0 \\ \vdots & \vdots & \vdots & \ddots & 0 \\ 0 & 0 & 0 & \cdots & U_{d-1} \end{pmatrix} \tag{23.3}$$

where each U_i $(i = 0, 1, \ldots, d-1)$ can be any given single-qudit unitary operation. The U_i operation is applied to the target qudit when the control qudit is in the $|i\rangle$ state. In Section 23.2.1, we show how to implement *MVCG* on a photonic system and how the *MVCG* can improve the efficiency of the circuit by reducing the gate requirement.

23.1.2 Examples of Qudit Quantum Algorithms

The quantum Fourier transform algorithm is the key component of many quantum algorithms [32]. It can be generalized for a qudit system [33, 34]. Given an N-dimensional system that is described with n d-dimensional qudits, the QFT, $F(d, N)$, where $N = d^n$, transforms the computational basis

$$\{|0\rangle, |1\rangle, \ldots, |N-1\rangle\} \tag{23.4}$$

into a new basis set [35]

$$F(d,N)\,|j\rangle = \frac{1}{\sqrt{N}}\sum_{k=0}^{N-1} e^{2\pi ijk/N}\,|k\rangle \tag{23.5}$$

As a convention, an integer j is rephrased in a base-d form. If $j > 1$ then

$$j = j_1 j_2 \ldots j_n = j_1 d^{n-1} + j_2^{n-2} + \cdots + j_n d^0 \tag{23.6}$$

and, if $j < 1$, then

$$j = 0.j_1 j_2 \ldots j_n = j_1 d^{-1} + j_2 d^{-2} + \cdots + j_n d^{-n} \tag{23.7}$$

The QFT applied to a state $|j\rangle$ can be represented in a product form as

$$\begin{aligned}
|j\rangle = |j_1 j_2 \ldots j_n\rangle &\mapsto \frac{1}{d^{n/2}}\sum_{k=0}^{d^n-1} e^{2\pi ijk/d^n}\,|k\rangle \\
&= \frac{1}{d^{n/2}}\sum_{k_1=0}^{d-1}\cdots\sum_{k_n=0}^{d-1} e^{2\pi ij(\sum_{l=1}^{n} k_l d^{-l})}\,|k_1 k_2 \ldots k_n\rangle \\
&= \frac{1}{d^{n/2}}\sum_{k_1=0}^{d-1}\cdots\sum_{k_n=0}^{d-1}\bigotimes_{l=1}^{n} e^{2\pi ijk_l d^{-l}}\,|k_l\rangle \\
&= \frac{1}{d^{n/2}}\bigotimes_{l=1}^{n}\left[\sum_{k_l=0}^{d-1} e^{2\pi ijk_l d^{-l}}\,|k_l\rangle\right]
\end{aligned}$$

The quantum circuit illustration of the process is shown in Figure 23.1, and the explicit expression of the product form is listed in Figure 23.1 on the right.

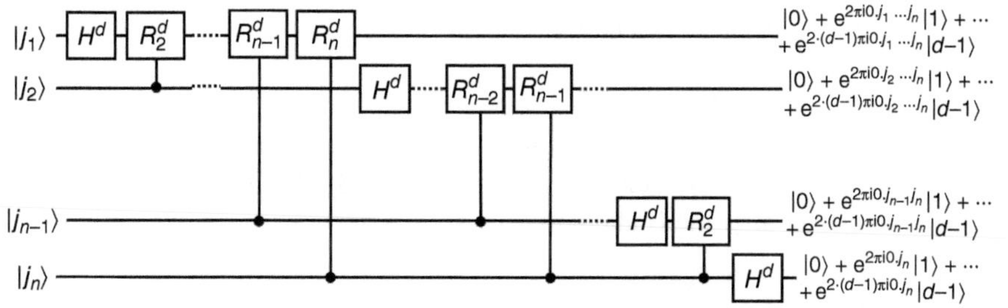

Figure 23.1 Circuit schematic of qudit quantum Fourier transform. The d-dimensional Hadamard gate is represented as H^d and the phase gate R^d is shown in Eq. (23.10). Explicit expressions of the qudit states are on the right. Source: Wang et al. [2], (figure 10) / Frontiers Media S.A / CC BY-4.0.

The generalized Hadamard gate H^d is equivalent to single qudit QFT as $H^d := F(d,d)$, which does the operation

$$H^d|j_n\rangle = |0\rangle + e^{2\pi i 0.j_n}|1\rangle + \cdots + e^{2(d-1)\pi i 0.j_n}|d-1\rangle \tag{23.8}$$

The matrix representation of H^d is

$$\begin{pmatrix} 1 & 1 & \cdots & 1 \\ 1 & e^{2\pi i 0.1} & \cdots & e^{2\pi i 0.(d-1)} \\ \vdots & \vdots & \ddots & \vdots \\ 1 & e^{2(d-1)\pi i 0.1} & \cdots & e^{2(d-1)\pi i 0.(d-1)} \end{pmatrix} \tag{23.9}$$

In the circuit the R_k^d gate is a phase gate that has the expression

$$R_k^d = \begin{pmatrix} 1 & 0 & \cdots & 0 \\ 0 & e^{2\pi i/d^k} & \cdots & 0 \\ \vdots & \vdots & \ddots & \vdots \\ 0 & 0 & \cdots & e^{2\pi i(d-1)/d^k} \end{pmatrix} \tag{23.10}$$

The black dots in the circuit are the MVCGs. The operation of which is applying R_k^d to the target qudit j times, when a control qudit is in state $|j\rangle$. In order to fix the sequence of $j_1 j_2 \dots j_n$, a series of qudit SWAP gates are applied at the end, which are not explicitly drawn in Figure 23.1.

The qudit QFT provides a crucial subroutine for many qudits algorithms. Qudit QFT offers superior approximations where the error bound decreases exponentially with d [34], which outperforms its binary counterpart [36].

Once the qudit QFT is available, quantum PEA can be generalized to the qudit version [35]. The goal of qudit PEA is to obtain the phase on the eigenvalue of $|u\rangle$ under U, i.e. the phase factor r in $U|u\rangle = e^{2\pi i r}|u\rangle$. The PEA in the qudit system can be decomposed to two registers of qudits. The first register contains t qudits. To successfully obtain the phase factor r accurate to the s dits with a success probability of at least $1-\epsilon$, we choose t to be $t = s + \log_d(\frac{1}{2\epsilon} + 2)$ [35]. The assumption of performing an arbitrary number of times of the unitary operation U and the ability to initialize eigenvector $|u\rangle$ with the second register's qudits are similar to those of the qubit PEA [37].

As a convention, we rewrite the rational number r in a base-d form as

$$r = R/d^t = \sum_{l=1}^{t} \overline{R_l}/d^l = 0.\overline{R_1}\,\overline{R_2}\dots\overline{R_t} \tag{23.11}$$

As shown in Figure 23.2a, we first apply the generalized Hadamard gates $H \equiv F(d,d)$ to each qudit in the first register. The following derivation of the explicit expressions of the PEA are summarized from Ref. [35]. The state of the lth qudit of the first register after the Hadamard gate is

$$F(d,d)|0_l\rangle = \frac{1}{\sqrt{d}}\sum_{k_l=0}^{d-1}|k_l\rangle \tag{23.12}$$

Then the lth qudit controls the operation $U^{d^{t-l}}$ applied on the target register qudits with the state $|u\rangle$. The controlled gate gives

$$CU^{d^{l-1}}|k\rangle \otimes |u\rangle = |k\rangle (U^{d^{t-l}})^k |u\rangle = e^{2\pi i k d^{t-l} r}|k\rangle \otimes |u\rangle \tag{23.13}$$

Notice that the controlled gate here is a MVCG such that the phase $e^{2\pi i k d^{t-l} r}$ is applied to $|u\rangle$ according to the state number k of the controlled qudit.

After all the controlled operations the resulting qudit system states become

$$\left(\prod_{l=1}^{t} \otimes \frac{1}{\sqrt{d}} \sum_{k_l=0}^{d-1} e^{2\pi i k_l d^{t-l} r} \left|k_l\right\rangle\right) \otimes |u\rangle \tag{23.14}$$

Then through a process called the "phase kick-back," the phase factor generated in the second register is passed on to the state of the first register such that

$$|\text{Register } 1\rangle = \frac{1}{d^{t/2}} \sum_{k=0}^{d^t-1} e^{2\pi i r k} |k\rangle \tag{23.15}$$

We can get the approximated eigenvalue r from the state number R of $|R\rangle$ by applying the inverse QFT to the qudits in the first register:

$$F^{-1}(d, d^t)\,|\text{Register } 1\rangle = |R\rangle \tag{23.16}$$

The full process of PEA is shown in Figure 23.2b.

Introducing qudits to the PEA not only reduces the number requirement of qudits and qudit gates but also decreases the error rate exponentially as the qudit dimension increases [38]. The qudit PEA enables the qudit variations of a long

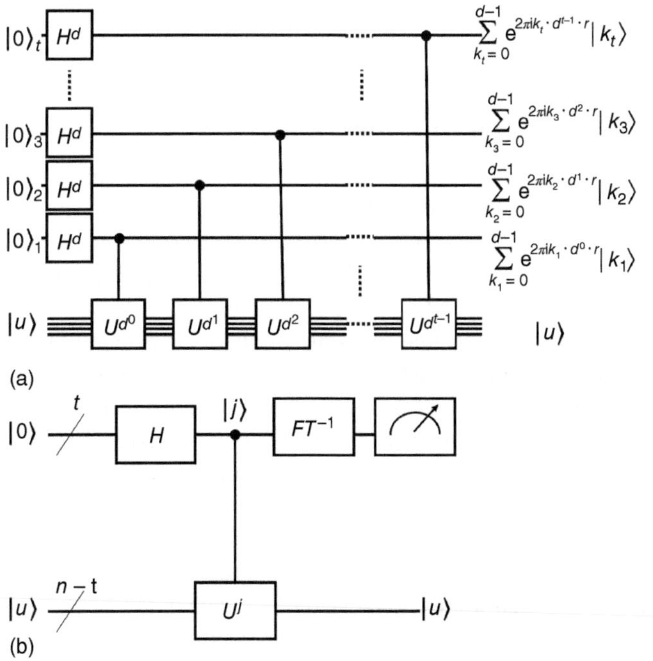

Figure 23.2 (a) First half of the PEA where the "phase kick-back" happens. The explicit states of the first register qudits after the U^j operations are giving on the right. (b) The simplified circuit of the PEA. After the Hadamard gate and controlled-U^j operations, the inverse Fourier Transform (FT^{-1}) is performed and the phase factor is obtained by measuring the first register qudits states. Source: Wang et al. [2], (figure 11) / Frontiers Media S.A / CC BY-4.0.

list of applications that includes Shor's factorization algorithm [39], simulation of quantum systems [40], quantum algorithm for solving linear equations [41, 42], and quantum counting [43]. Recently, a quantum algorithm solving a linear system using the qutrit version of the PEA has been proposed [44]. The qudit PEA has the potential to out-perform its qubit counterpart in those applications.

23.2 Qudit Implementation on Photonic Systems

Photonic system provides a good platform for quantum computing because photons have a comparatively long decoherence time. It is also a good candidate for qudit quantum computing since many of the physical properties of the photon, such as the orbital angular momentum [45, 46], frequency-bin [47–50] and time-bin [51, 52], have more than two states. There are examples of photonic qudit quantum computing such as the experimental realization of arbitrary multidimensional multiphotonic transformations [53] and experimental entanglement of high-dimensional qudits [47]. In Section 23.2.1, we review a photonic qudit computation scheme that makes use of the time and frequency degrees of freedom (DoF). To make connection between the reviewed photonic system to practical physical chemistry problems, we introduce an algorithm for open quantum dynamics in the last part of Section 23.2.1. In Section 23.2.2 we discuss the bosonic qudit in microwave cavity and its application in simulating the molecular vibration modes, an inherently multidimensional system.

23.2.1 Qudits in Time and Frequency Degrees of Freedom

In this section, we review a single photon system that integrated on-chip modulators, pulse shapers and other well-established fiber-optic components. This system can achieve the high-dimensional($d > 2$) high-fidelity Fourier-transform pulse shaping [48], acquire the two-qudit SUM gate via the time and frequency DoF [50] as well as demonstrate a proof-of-principle qutrit PEA [31].

First, we review a parallelizable scalable photonic frequency-bin manipulation platform [48]. It can realize qudit quantum computing based on electro-optic phase modulation and pulse shaping. The qudit in this scheme is represented by a single photon spread over d narrow-band modes with frequency centered at $\omega_n = \omega_0 + n\Delta\omega$ [54]. The operations applying on the qudit are implemented via a frequency multiport V connecting the input $\hat{a}_n^{(\text{in})}$ and output $\hat{a}_m^{(\text{out})}$ in a way $\hat{a}_m^{(\text{out})} = \sum_n V_{mn}\hat{a}_n^{(\text{in})}$ so the photon is manipulated in the desired fashion. The first crucial component to develop a fully functional qudit quantum circuit for the qudit PEA is the discrete QFT. QFT is equivalent to the high-dimensional Hadamard gate in the single qudit case. This step is realized with a balanced frequency tritter, the three-mode extension of the electro-optic-base frequency beam splitter. The single qutrit QFT achieved is in the form:

$$U = \frac{1}{\sqrt{3}} \begin{pmatrix} 1 & 1 & 1 \\ 1 & e^{2\pi i/3} & e^{4\pi i/3} \\ 1 & e^{4\pi i/3} & e^{2\pi i/3} \end{pmatrix} \tag{23.17}$$

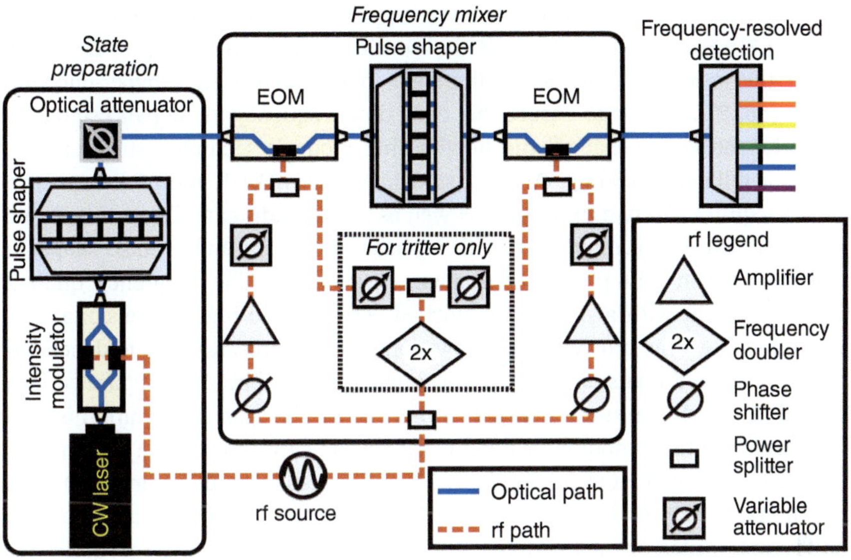

Figure 23.3 The experimental setup for the frequency beam splitter. Source: Reprinted from Lu et al. "Electro-optic frequency beam splitters and tritters for high-fidelity photonic quantum information processing." Physical review letters 120.3 (2018): 030502. Lu et al. [48] with the permission of American Physical Society under license RNP/22/FEB/050615.

The frequency beam splitter (whose schematic illustration is given in Figure 23.3) has three stages: State preparation, frequency mixer (build by a pulse shaper [PS] sandwiched between two electro-optic phase modulator [EOM]) and frequency-resolved detection. A radio-frequency (rf) oscillator provides a 25-GHz drive signal to each EOM. Combining with amplifiers and delay lines, it sets the amplitude and phase properly for each waveform. The central pulse shaper is used to implement the spectral phase pattern, which is optimized numerically, in order to perform the Hadamard gate. The frequency tritter is attained by incorporating an additional harmonic in the microwave drive signal with a rf frequency doubler as shown in the setup. The pulse shaper in the middle of two EOMs imparts the spectral phase that makes sure all the sidebands generated by the first EOM are reabsorbed to the computational space after the second one. This can circumvent the unavoidable scattering of the input photon causing by the EOM and assure the fully deterministic nature of the proposed frequency beam splitter[54]. This first ever demonstrated balanced frequency tritter acquires a fidelity as high as (0.9989 ± 0.0004) and provides the important building block toward scalable, high-fidelity qudit quantum computing based on frequency encoding [48].

The second crucial component of the qudit PEA is the two-qudit gate. With standard linear optics and photon counting, the interacting of the two photons is not deterministic and has probabilistic results [55], adding the difficulty to build a reliable controlled gate for the photonic qudits. The proposed photonic system bypasses this difficulty by encoding the qudits to two DoFs on a single photon – i.e. the time-bin and frequency-bin [50]. In this multidimensional time-bin and frequency-bin settings, the distance of the frequency spacing (Δf) and the time-bin spacing (Δt) are chosen to surpass the Fourier Transform limit (i.e. $\Delta f \Delta t > 1$) so that the two DoFs can be manipulated independently in a hyper-encoding fashion [56, 57]. This is to ensure that each time-frequency mode pair establishes a well-defined entity that all the neighboring mode is separated sufficiently to achieve a reliable encoding. As discussed in Section 23.1.1, the qudit universal gate set requires the two-qudit gate as well as single qudit gate like X and Z gate. The Z gate can be accomplished with a phase modulator and a pulse shaper in the time domain and frequency domain, respectively. The frequency qudit X gate can be built with a Z gate sandwiched between two high-dimensional QFT gates [48]. The time-bin X gate can be done by the process of state-dependent delay. In a three-dimensional case for example, the time-bin $|0\rangle_t$ is separated from $|1\rangle_t$ and $|2\rangle_t$, and its route is delayed by three bins. The spatial separation is done by a Mach–Zehnder modulator (MZM) switch, and the illustration is shown in Figure 23.4. The gate output is measured by a single-photon detector (SPD) and a time interval analyzer. For the two-qudit gates shown in Figure 23.5, the control qudit is represented by the frequency DoF and the time DoF is the target qudit. The true single photon (instead of weak coherent states) used for the two-qudit gate is obtained from a frequency-bin entangled pair generated via spontaneous four-wave

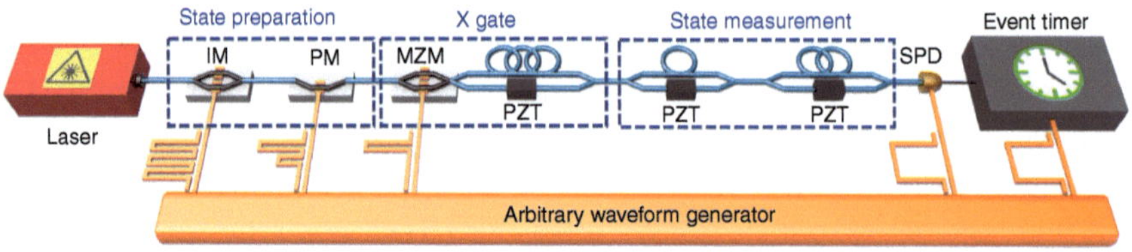

Figure 23.4 The full setup for the time-bin X gate. The components in the experiment are phase modulator (PM), intensity modulator (IM), piezo-electric-phase shifter (PZT), Mach–Zehnder modulator (MZM), and single-photon detector (SPD). The time-bin delay is indicated by the circle-shaped fibers. Source: Imany et al. [50], (figure 2a) / Springer Nature / CC BY-4.0.

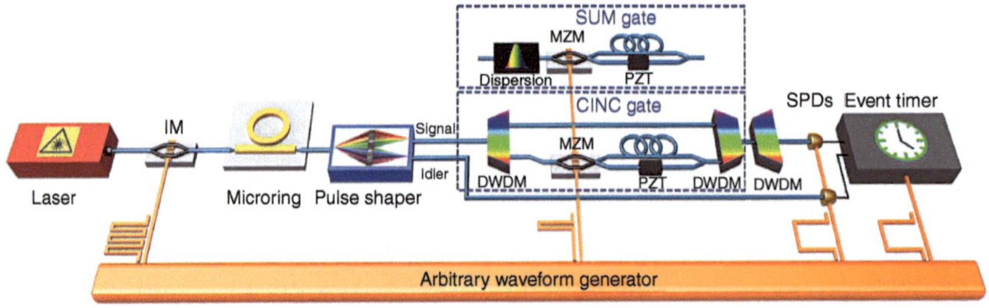

Figure 23.5 The illustration of the CINC and SUM gate experimental setup. In the CINC gate, the time bin $|2\rangle_t$ is separated from time bins $|0\rangle_t$ and $|1\rangle_t$ by the MZM. For the SUM gate, the outside time bins $|3\rangle_t$ and $|4\rangle_t$ are separated from the time bins $|0\rangle_t$, $|1\rangle_t$, and $|2\rangle_t$ inside the computational space by the MZM. The dense wavelength-division multiplexer is DWDM. Source: Imany et al. [50], (figure 3a) / Springer Nature / CC BY-4.0.

mixing in an on-chip silicon nitride microresonator. The time-bins are characterized by intensity modulation of the pump and couple into a microring resonator (with free spectral range $\Delta f = 380$ GHz and resonance line widths $\delta f \simeq 250$ MHz) to generate a biphoton frequency comb. The signal and idler photons from the first three comb-line pairs are selected out with a commercial pulse shaper. While the idler photons are sent to a SPD for heralding, the signal photons with the time and frequency qudits can be prepared to any product state $|m\rangle_t|n\rangle_f$ $(m, n = 0,1,2)$ and provide a full computational basis set.

The two types of two-qudit gates are the controlled increment (CINC) operation and the more complex SUM gate – a generalized controlled-NOT gate [58]. The CINC gate applies the X gate to time-bin qudit only when the frequency qudit is in the state $|2\rangle_f$. The CINC gate is implemented by separating $|2\rangle_f$ with a dense wavelength-division multiplexing (DWDM) filter, routing it to a time-bin X gate, and keeping other two frequency-bins unchanged. Another DWDM is used to bring back the frequency-bins with zero relative delay. The SUM gate adds the value of the control qudit to the value of the target qudit, modulo 3. It applies a cyclic shift of 1 slot delay to the time-bins associated with $|1\rangle_f$ and 2 slots delay to the time-bins corresponding to $|2\rangle_f$ and keeps those related to $|0\rangle_f$ unchanged. To employ the frequency-dependent delay to the time-bins, a chirped fiber Bragg grating (CFBG) is used to induce a dispersion of −2 ns/nm to the photon. The delay operations are linear and not cyclic and thus move some of the time-bins outside of the computational space. These time-bins can be returned to the computational space using principles identical to the time-bin X gate with a relative delay. In order to measure the transformation matrix of these gates in the computational basis, all nine combinations of single time-bins and frequency-bins are prepared with an intensity modulator (IM) and a pulse shaper as the input states. The signal counts in all possible output time-bin/frequency-bin pairs are recorded with the three different setups of DWDMs in the path of the signal photons to identify the different frequency-bins. The measured transformation matrix for the CINC gate has a so-called "classical" fidelity of $\mathscr{F}_C = 0.90 \pm 0.01$ and for the SUM gate the fidelity is $\mathscr{F}_C = 0.92 \pm 0.01$ [50, 59]. This fully deterministic photonic two-qudit gate setup can be integrated with on-chip components in a low cost, room temperature, and scalable manner. It lays down a foundation for building the qudit universal gate set and the high-dimensional quantum circuit in an integrated on-chip photonic platform.

Lastly, with the single qutrit QFT and two-qudit gate in hand, we review a demonstration of a proof-of-principle qutrit PEA [31]. Following the setup of the previous paragraph, the frequency DoF carries the control qutrit, and the time DoF carries the target qutrit. Some of the well-established techniques and fiber-optic components are used to assemble the quantum circuit including: phase modulator (PM), intensity modulator (IM), continuous-wave (CW) laser source, pulse shaper (PS), and chirped fiber Bragg grating (CFBG). The qutrit operations, as illustrated in Figure 23.6, are performed in three stages [31]: (i) a state preparation stage where the frequency-bin and time-bin of the input photon are initialized by a PM followed by a PS and an IM; (ii) a controlled-unitary operation stage where the MVCG gate (as described in

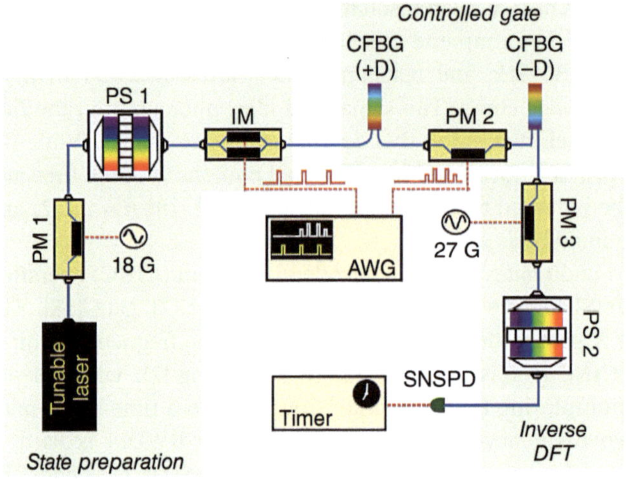

Figure 23.6 Experimental setup. Component (abbreviate) is, correspondingly, Fourier-transform pulse shaper (PS); Electro-optic phase/intensity modulator (PM/IM); Chirped fiber Bragg grating (CFBG); Arbitrary waveform generator (AWG); Superconducting nanowire single-photon detector (SNSPD). The AWG has a 10 MHz reference clock that is synchronized to both radio-frequency oscillators (18 and 27 GHz).) Source: Lu et al. [31]/John Wiley & Sons.

Section 23.1.1) is built with a PM sandwiched by two CFBGs; and (iii) an inverse Fourier Transformation and measuring stage where the phase information of the photon is extracted by a superconducting nanowire single-photon detector (SNSPD) connected to a timer. The inverse QFT comprises a PM and a PS. It is worth noting that the MVCG gate in the circuit applies different operations to the target qutrit based on the three unique states of the control qutrit. In the experiment, frequency and time-bin qutrits are sent through the control and target registers, respectively. In the measuring stage, the state of the control register qutrits is measured, and state distribution is recorded to obtain the phase information. The eigenphases obtained by this photonic PEA setup can be retrieved with 98% fidelity [31].

In Table 23.1 we provide results generated by the photonic PEA experiment. To demonstrate the ability of implementing qutrit PEA with the simplest unitary operation possible, the first experiment encoding the unitary operation as follow:

$$\hat{U}_1 = \mathrm{diag}(1, \omega, \omega^2) \tag{23.18}$$

with ω being the cubic root of unity. Then to utilize the repetition advantage of the photonic setup, i.e. the experiment can repeat the same setup at a relatively low cost, another more complex unitary is encoded in the circuit as:

$$\hat{U}_2 = \mathrm{diag}\left(1, \mathrm{e}^{\mathrm{i}0.351\pi}, \mathrm{e}^{\mathrm{i}1.045\pi}\right) \tag{23.19}$$

After a given number of experimental shots, we can obtain the statistical inference of the phase based on the numerical data. To do this, the statistical relation between the photon state distribution and eigenphase can be summarized as follows: given a eigenphase ϕ corresponding to one of the eigenstate, the probability of measuring

Table 23.1 Normalized photon counts and comparison of the true phase ϕ and the experimentally estimated phase ϕ' for each eigenstate of $\hat{U}_1$ (Eq. 23.18) and $\hat{U}_2$ (Eq. 23.19).

	$\hat{U}_1$		
Eigenstate	$\|0\rangle_t$	$\|1\rangle_t$	$\|2\rangle_t$
E_0	0.9948 ± 0.0004	0.0101 ± 0.0004	0.0122 ± 0.0005
E_1	0.0023 ± 0.0002	0.9805 ± 0.0009	0.0120 ± 0.0005
E_2	0.0029 ± 0.0002	0.0094 ± 0.0004	0.9758 ± 0.0010
True phase, ϕ	0	$2\pi/3$	$4\pi/3$
Est. phase, $\tilde{\phi}$	1.972π	0.612π	1.394π
Error, $\frac{\|\phi-\tilde{\phi}\|}{2\pi}$	1.4%	2.7%	3.0%
	$\hat{U}_2$		
Eigenstate	$\|0\rangle_t$	$\|1\rangle_t$	$\|2\rangle_t$
E_0	0.878 ± 0.002	0.316 ± 0.003	0.143 ± 0.002
E_1	0.032 ± 0.001	0.530 ± 0.003	0.318 ± 0.003
E_2	0.090 ± 0.002	0.154 ± 0.002	0.539 ± 0.003
True phase, ϕ	0	0.3511π	1.045π
Est. phase, $\tilde{\phi}$	1.859π	0.377π	1.045π
Error, $\frac{\|\phi-\tilde{\phi}\|}{2\pi}$	7.1%	1.3%	0.0%

Source: Ref. [31]/John Wiley & Sons.

the output state into $|n\rangle$, where $n \in \{0,1,2\}$, is

$$C(n, \phi) = \frac{1}{9}\left|1 + e^{i(\phi-\frac{n2\pi}{3})} + e^{i2(\phi-\frac{n2\pi}{3})}\right|^2 \tag{23.20}$$

Define E_0, E_1, and E_2 to be the probability of detecting the photons that fell into $|0\rangle_f$, $|1\rangle_f$, and $|2\rangle_f$. The estimated phase, denoted $\tilde{\phi}$, is the phase that has the smallest mean-square error between the measured and theoretical results:

$$\min_{\tilde{\phi}} \sum_{n=0}^{2} (E_n - C(n, \tilde{\phi}))^2 \tag{23.21}$$

The estimated phases for $\hat{U}_1$ (23.18) and $\hat{U}_2$ (23.19) are shown in Table 23.1 [31]. Although in the second experiment (with U_2) the eigenphases are chosen to be arbitrary values (not a fraction of π), they are derivable from the statistical distributions of the photon counts.

In many subjects of physical sciences such as physical chemistry, chemical physics, and materials science, the physical system of interest is often interacting with a larger environment whose scale and complexity prevent exact treatment by the time-dependent Schrodinger equation. To this end, the theory of open quantum dynamics uses a variety of approximations to average out the effects of environments and focus on the core system [60]. Recently, the open quantum dynamics formalism has been adapted to a quantum algorithm that can simulate excitonic dynamics on molecular systems [61, 62]. As the quantum algorithm for open quantum dynamics relies on projection measurements on the output states to extract physical information, it is a natural candidate for implementation with the electro-optical system just described. Suppose the density matrix of the initial physical system can be described by a sum of different pure quantum states multiplied by the corresponding probabilities:

$$\rho = \sum_i p_i |\phi_i\rangle\langle\phi_i| \tag{23.22}$$

where each p_i denotes the probability of detecting each $|\phi_i\rangle$ in the mixed state of ρ. The dynamics is represented via the *Kraus operators* M_k in an operator sum representation [61]:

$$\rho(t) = \sum_k M_k \rho M_k^\dagger \tag{23.23}$$

Then each $|\phi_i\rangle$ can be prepared as an input state vector v_i of a given basis and evolved as follows:

$$|\phi_{ik}(t)\rangle = M_k v_i \xrightarrow{\text{unitary dilation}} U_{M_k}(v_i^T, 0, \ldots, 0)^T \tag{23.24}$$

In Eq. (23.24), the M_k is undergoing the 1-dilation process to become $U_{M_k} = \begin{pmatrix} M_k & D_{M_k^\dagger} \\ D_{M_k} & -M_k^\dagger \end{pmatrix}$ [61, 63], where $D_{M_k} = \sqrt{I - M_k^\dagger M_k}$ is the defect operator of M_k.

After the time evolution of $|\phi_{ik}(t)\rangle$, the population of each basis state can be retrieved by calculating the diagonal vector:

$$\text{diag}(\rho(t)) = \sum_{ik} p_i \cdot \text{diag}(|\phi_{ik}\rangle\langle\phi_{ik}|) \tag{23.25}$$

The $\text{diag}(|\phi_{ik}(t)\rangle\langle\phi_{ik}(t)|)$ is efficiently obtainable via the projection measurements on the first half subspace of $U_{M_k}(v_i^T, 0, \ldots, 0)^T$. This algorithm also supports the calculation of the expectation value of an observable $\langle A\rangle = \text{Tr}(A\rho(t))$. To ensure the contraction and positive-semidefinite requirement, we consider the operator $\tilde{A} = \frac{A+I||A||}{2||A||}$ with the Cholesky decomposition $\tilde{A} = LL^\dagger$ [64]. Then the evolution of the observable is

$$L^\dagger|\phi_{ik}(t)\rangle = L^\dagger M_k v_i \xrightarrow{\text{unitary dilation}} U_{L^\dagger} U_{M_k}(v_i^T, \ldots, 0)^T \tag{23.26}$$

and $\langle\tilde{A}\rangle$ can be calculated by:

$$\langle\tilde{A}\rangle = \text{Tr}(\tilde{A}\rho(t)) = \sum_{i,k} \text{Tr}(p_i \cdot L^\dagger|\phi_{ik}(t)\rangle\langle\phi_{ik}(t)|L) \tag{23.27}$$

The trace of $L^\dagger|\phi_{ik}(t)\rangle\langle\phi_{ik}(t)|L$ is obtainable by applying the projection measurements into the first N-dimensional space of $U_{L^\dagger} U_{M_k}(v_i^T, 0, \ldots, 0)^T$ [61]. $\langle A\rangle$ can then be derived from $\langle\tilde{A}\rangle$ by $\langle A\rangle = 2||A||\langle\tilde{A}\rangle - ||A||$. In the algorithm, the projection operations facilitate the measurements of both the density matrix and the observable. Hu et al. [61] proposed and demonstrated this general quantum algorithm to evolve open quantum dynamics on quantum computing devices. Within the algorithm, Kraus operators are converted into unitary matrices with minimal dilation guaranteed by the Sz.-Nagy theorem [63]. The algorithm is demonstrated on an amplitude damping channel using the IBM Qiskit quantum simulator and the IBM Q 5 Tenerife quantum device [61]. This general algorithm does not require particular models of dynamics or decomposition of the quantum channel and thus can be easily generalized to other open quantum dynamical models. Since projection operations can be performed naturally and efficiently on the electro-optical system, the open quantum dynamics algorithm is compatible and viable for implementation on this platform.

23.2.2 Superconducting Bosonic Processor

In this section, we review a photonic system that consists of microwave cavities and transmon qubits to implement a two-mode superconducting bosonic processor that enables the scalable simulation of molecular vibronic spectra [65]. The microwave cavity with its underlying bosonic structure provides a natural bosonic platform to simulate the inherently high-dimensional molecular vibrational modes. Although the read-out apparatus consists of a transmon qubit, the microwave cavity as the central component is simulating a multidimensional system and can be treated as a qudit in general.

The simulation of molecular vibronic spectra is to study the electronic transition accompanied with the vibrational dynamics. Keeping the adiabatic Born–Oppenheimer approximation that separates the nuclear and electronic DoF, each of the distinct electronic states forms a potential-energy surface (PES) that carries a unique manifold of vibrational eigenstate. The normal modes of the electronic vibration can be acquired with a harmonic approximation: keeping only the quadratic terms of the displacement of the PES about the equilibrium. The corresponding transformation of the set of creation and annihilation operators $\hat{\mathbf{a}} = (\hat{a}_1, \dots, \hat{a}_N)$ for N vibrational modes can be described with the Duschinsky transformation [66]. The Duschinsky transformation can be disintegrated into Gaussian operations in terms of the Doktorov operator [67]

$$\hat{\mathbf{a}} \to \hat{U}_{\text{Dok}} \hat{\mathbf{a}} \hat{U}^{\dagger}_{\text{Dok}}, \quad \hat{U}_{\text{Dok}} = \hat{\mathbf{D}}(\boldsymbol{\alpha}) \hat{\mathbf{S}}^{\dagger}(\boldsymbol{\zeta}') \hat{\mathbf{R}}(U) \hat{\mathbf{S}}(\boldsymbol{\zeta}) \tag{23.28}$$

where

$$\hat{\mathbf{D}}(\boldsymbol{\alpha}) = \hat{D}(\alpha_1) \otimes \hat{D}(\alpha_1) \otimes \cdots \otimes \hat{D}(\alpha_N) \tag{23.29}$$

$$\hat{\mathbf{S}}^{\dagger}(\boldsymbol{\zeta}') = \hat{S}^{\dagger}(\zeta_1') \otimes \hat{S}^{\dagger}(\zeta_2') \otimes \cdots \otimes \hat{S}^{\dagger}(\zeta_N') \tag{23.30}$$

$\hat{\mathbf{D}}(\boldsymbol{\alpha})$ and $\hat{\mathbf{S}}(\boldsymbol{\zeta})$ describe a tensor product of single-mode displacement and squeezing operations across all N modes correspondingly. $\hat{\mathbf{R}}(U)$, the N-mode rotation operator, is represented by an $N \times N$ rotation matrix U which can be generated with a product of two-mode beam splitter operations [10]. The set of dimensionless Doktorov parameters $\boldsymbol{\alpha} = (\alpha_1 \dots \alpha_N)$, $\boldsymbol{\zeta} = (\zeta_1 \dots \zeta_N)$, $\boldsymbol{\zeta}' = (\zeta_1' \dots \zeta_N')$ can be derived from molecular structural information in the different electronic configurations. In the $N = 2$ case , U has the form of a two-dimensional rotation matrix with a single free parameter, i.e. the angle θ. The transitions between the vibrational states due to a

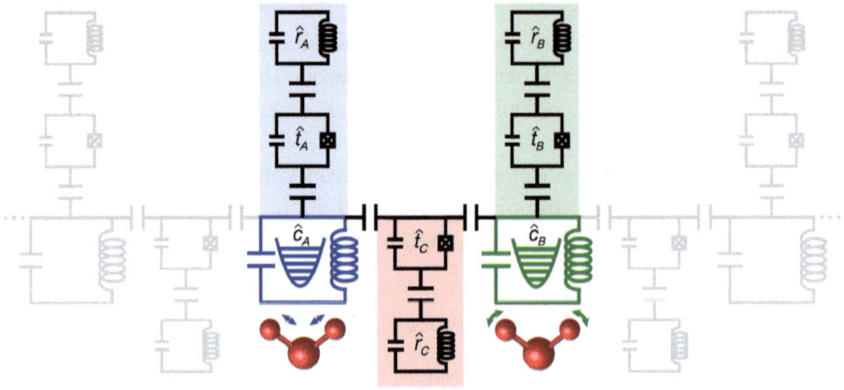

Figure 23.7 The superconducting bosonic processor circuit. Two microwave cavity modes are denoted in blue $\hat{c}_A$, and green $\hat{c}_B$. Ancilla measurement and control modules consist of a transmon qubits ($\hat{t}_A, \hat{t}_B$) and readout resonators ($\hat{r}_A, \hat{r}_B$) coupled to each cavity mode have blue and green shades correspondingly. The transmon qubit serving as the coupler module $\hat{t}_C$ and readout resonator $\hat{r}_C$ is shaded red. To illustrate the scalability of this configuration, a linear array of *N* cavity modes (with *N* ancillary modules) connected by coupler modules with its nearest-neighbor have light gray shades. Source: Wang et al. [65], (figure 1) / American Physical Society / CC BY-4.0.

sudden change in electronic PES is emulated as applying $\hat{U}_{\text{Dok}}$ to an initial state $|\psi_0\rangle$ of the bosonic processor. This projection produces the Franck–Condon factors (FCFs) which are the vibrational overlap integrals of an initial pretransition vibrational eigenstate, $|\hat{n}\rangle$ with a final post-transition vibrational eigenstate, $\left|\hat{n'}\right\rangle$:

$$\text{FCF}|_{\hat{n},\hat{n'}} = |\left\langle \hat{n'}\right| \hat{U}_{\text{Dok}} |\hat{n}\rangle|^2 \tag{23.31}$$

The FCFs have practical importance that can provide insights of the structure and dynamic nature of the excited electronic states.

The superconducting bosonic processor (as shown in Figure 23.7) is designed to employ the bosonic modes of two microwave cavities $\hat{c}_A$ and $\hat{c}_B$. The two cavities are dispersively coupled by a coupler transmon $\hat{t}_C$ which facilitates the beam splitter and squeezing operations through driven four-wave mixing processes [68]. A readout resonator $\hat{r}_C$ is dispersively connected to the coupler transmon. The cavity modes displacement is operated via resonant drives through local coupling ports. Two ancillary transmon-readout systems $\hat{t}_A, \hat{r}_A$ and $\hat{t}_B, \hat{r}_B$ are added to the device and coupled to each cavity to execute state preparation and tomography. The photonic state of the quantum processor is used to simulate the transformation of a molecular vibrational state caused by the electronic transition. In the state initialization process, both bosonic modes are prepared in vacuum state via feedback cooling protocols and initialized to the Fock state $|n, m\rangle$ using optimal control techniques [69]. This bosonic processor has the ability to synthesize arbitrary Fock states reliably allowing the simulation of FCFs starting from vibrationally excited states which outperforms most other bosonic simulators. The Doktorov transformation is then implemented by triggering a basis change to that of the post-transition vibrational Hamiltonian. In the $N = 2$ case, a single beam splitter is used to apply the rotation operation. The four-wave mixing capabilities of the coupler transmon help enforce the single-mode squeezing and beam splitter operations. Two pump tones (with frequencies $\omega_{1/2}$) couple to the coupler transmon through a port to matching the appropriate frequency condition such that $\omega_1 + \omega_2 = 2\omega_{A/B}$ for the squeezing operation and $\omega_2 - \omega_1 = \omega_B - \omega_A$ where $\omega_{A/B}$ are the cavity frequencies. This realizes the desired Hamiltonians [68]

$$\hat{H}_{\mathrm{BS}}/\hbar = g_{\mathrm{BS}}(t)(\mathrm{e}^{\mathrm{i}\varphi}\hat{c}_A\hat{c}_B^\dagger + \mathrm{e}^{-\mathrm{i}\varphi}\hat{c}_A^\dagger\hat{c}_B) \tag{23.32}$$

$$\hat{H}_{\mathrm{sq},i}/\hbar = g_{\mathrm{sq},i}(t)(\mathrm{e}^{\mathrm{i}\phi_i}\hat{c}_i^2 + \mathrm{e}^{-\mathrm{i}\phi_i}\hat{c}_i^{\dagger 2}) \tag{23.33}$$

where $\hat{c}$ denotes the creation operator of the microwave mode and $i \in \{A, B\}$, g_{BS} and $g_{\mathrm{sq},i}$ are the coupling coefficients for beam splitters and squeezing operations where the explicit form can be found in Ref. [68]. It is worth noticing that the phase factors of the Hamiltonians $\{\varphi, \phi_i\}$ are controlled by the phase of the pump tones that facilitate the performance of $\hat{U}_{\mathrm{Dok}}$ with the correct family of beam splitters and squeezing operations. A post-selecting process in terms of transmon measurement rejects the heating events of the transmons that can dephase the cavities and halt the pumped operations. This postselection also initializes the ancillas to their ground states for the subsequent measurement of the cavities. Finally, the desired FCFs are derived from the average photon numbers of the repeating cavities measurements.

There are two complementary measurement schemes for extracting FCFs from the final state. The shared basic idea of the two measurement schemes is to make use of the dispersive coupling of each microwave cavity to its ancillary transmon-readout system [65]:

$$\hat{H}_{\mathrm{int}}/\hbar = -\sum_{i\in\{A,B\}} \chi_i \hat{c}_i^\dagger \hat{c}_i \hat{t}_i^\dagger \hat{t}_i \tag{23.34}$$

where $\hat{H}_{\mathrm{int}}$ is the interaction Hamiltonian and $\hat{c}$ and $\hat{t}$ denote the creation operator of the microwave mode and transmon qubit respectively. The first scheme, single-bit extraction, uses the state-selective π pulses to build the mapping between a given joint cavity photon number population $\{n', m'\}$ and the joint state of the two transmons. The frequencies of the pulses are $\omega_{t_A} = \omega_{t_A}^0 - n'\chi_A + (n'^2 - n')(\chi_A'/2)$ and $\omega_{t_B} = \omega_{t_B}^0 - m'\chi_B + (m'^2 - m')(\chi_B'/2)$. In this case the small second-order dispersive shift is also taken into account:

$$\hat{H}_{\mathrm{int}}'/\hbar = \sum_{i\in\{A,B\}} (\chi_i'/2)\hat{c}_i^\dagger \hat{c}_i^\dagger \hat{c}_i \hat{c}_i \hat{t}_i^\dagger \hat{t}_i \tag{23.35}$$

Here in this experiment $\chi_A' = 2\pi \times 1.31$ kHz and $\chi_B' = 2\pi \times 1.35$ kHz [65]. The read-out of the individual transmons uses the standard dispersive techniques, and the results are obtained in a shot-by-shot basis where a single bit of information is extracted from each joint photon number state been probed. Notice that this single-bit extraction of FCFs is not scalable as the bosonic Hilbert space grows exponentially with the number of modes N. To overcome the scalability issue, the second measurement scheme implements single-shot photon number resolving detection. Based on the dispersive Hamiltonian between the cavity mode and transmon ancilla, a quantum nondemolition (QND) mapping of arbitrary binary valued operator of the cavity Hilbert space to the state of transmon can be achieved by driving the transmon with numerically optimized waveforms $\epsilon(t)$ [69, 70],

$$\hat{H}_{\text{drive}}/\hbar = \epsilon^*(t)\hat{t} + \epsilon(t)\hat{t}^\dagger \tag{23.36}$$

Here QND means the measurement process preserves the physical integrity of the measured system [71]. Therefore, the number of photons for a given cavity state can be measured in the binary form $|n\rangle = |\prod_{k=0}^{k_{\max}} b_k\rangle$, where $n = \prod_{k=0}^{k_{\max}} 2^{b_k}$ and $\{b_k\} \in \{0,1\}$ and $k_{\max}$ is the number of binary digits in n. Each bit b_k can be measured sequentially in a given run of the experiment. The QND nature of each measuring operations allows the sequential application of the mapping pulses to the cavity states followed by transmon measurements. These mappings project an initial cavity state $|\psi\rangle = \sum c_n|n\rangle$ with n up to 16 into a definite Fock state $|n\rangle$ with the probability $|c_n|^2$. The advantage of the photon number resolved sampling protocol compared to the single-bit extraction is a notable reduction of the simulator runs required within a desired statistical error on all output probabilities.

The FCFs of a given electronic transition can be used to derive the relative intensities of the progressions in photo-electron vibronic spectra. The simulated FCFs are under the harmonic approximation and may deviate from the measured actual molecular spectra due to the vibrational anharmonicity in the true PES. In order to show the simulator's capability of large photon numbers generation and measurement, we review the simulation of the FCFs of water molecule (as shown in Figure 23.8) as an example of a long vibronic progressions simulation [65]. The water molecule starting in the vacuum state ($|\psi_0\rangle = |n, m\rangle$ $n = 0, m = 0$) transforms under photoionization to the second excited state ($\tilde{B}^2B_2$) of doublet spin multiplicity, and there is B_2 symmetry to the attendant electronic wave function. The process can be described as

$$H_2O \xrightarrow{h\nu} H_2O + (\tilde{B}^2B_2) + e^- \tag{23.37}$$

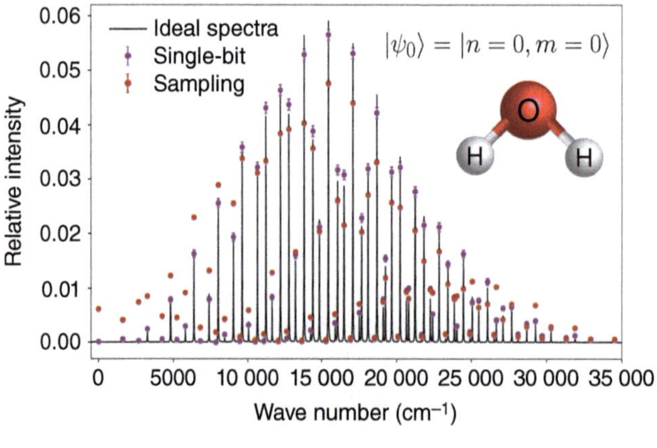

Figure 23.8 Franck–Condon factors for the photoionization of the water molecule. The theoretical FCFs, artificially broadened with Lorentzian profiles (10 cm^{-1} FWHM), is depicted by the solid lines. Purple and red circles represent experimental data using the single-bit extraction and sampling measurement scheme, respectively. Source: Wang et al. [65], (figure 3a) / American Physical Society / CC BY-4.0.

To quantify the quality of the quantum simulation, we introduce the distance $D = \frac{1}{2}\sum_{i=0}^{n_{\max}}\sum_{j=0}^{n_{\max}}|p_{ij}^{\text{meas}} - p_{ij}^{\text{ideal}}|$ between the measured probabilities p_{ij}^{meas} and the ideal distribution p_{ij}^{ideal}. The distance for the water molecule simulation is 0.049. Another relevant metric is run time. In the experiment with repetition rate of roughly 300 Hz, the data acquisition run time of 3×10^4 samples of the photoionization of water with single bit extraction and photon number resolved sampling is 7 hours vs. 100 seconds, respectively. Furthermore, because of the postselection of transmon heating events, all the metrics are accompanied with a success probability of 95%. Other detailed simulation data of the experiment could be found in the supplementary materials of Ref. [65].

The bosonic system has an inherent advantage in simulating the transformation of a bosonic Hamiltonian compared to the conventional qubit processor. It can encode the vibrational dynamics of the system naturally and decompose the Doktorov transformation into Gaussian operations efficiently. A recent proposal of obtaining FCFs on a spin-1/2 quantum computer requires $n_g = O[N^2 n_{\max}^2 \log^3(1/\varepsilon)]$ gates to implement $\hat{U}_{\text{Dok}}$ with a universal qubit gate set within an error of ε [72]. The choice of $n_{\max}$ depends on the initial state as well as the magnitude of the displacement and squeezing. As a comparison, this native N modes bosonic simulator only requires $2N$ squeezing operations, N displacement operations, and a maximum of $N(N-1)/2$ nearest-neighbor beam splitter operations [73]. This equivalent to a total of $O[N^2]$ operations and a corresponding circuit depth of $O[N]$ when non-overlapping beam splitters are applied simultaneously [65]. To sum up, the superconducting bosonic processor has the potential to integrate all of the necessary components of qudit quantum computing to implement high-fidelity, scalable, and resource-efficient simulation of molecular vibronic spectra and outperform its conventional qubit counterpart.

23.3 Summary and Future Outlooks

In quantum computing, qudit is an alternative computational unit that uses multiple states for information storage and processing. Photonic systems make natural platforms for qudit quantum computing given the multidimensional nature in many photon properties. This review chapter first introduces the basic ideas and methods of qudit quantum computing that include qudit gates, universality, and important qudit algorithms such as the QFT and the PEA. Next, it introduces two physical realizations of qudit computing with photonic systems. The first system is an electro-optic platform that integrates on-chip modulators, pulse shapers, and other well-established fiber-optic components to build all the crucial components for a proof-of-principle qutrit PEA. The second system is a two-mode superconducting bosonic processor consisting of microwave cavities and transmon qubits to simulate molecular vibronic spectra.

Compared to the conventional qubits, qudits can reduce circuit complexity, simplify experimental setup, and enhance quantum algorithm efficiency. For example, the number of gates required to build the universal qudit set under

Muthukrishnan and Stroud's proposal achieves a $(\log_2 d)^2$ scaling advantage compared to that of the qubits [12, 32]. This number is reduced by another factor of n (the number of qudits) in Luo and Wang's proposed scheme [13]. The MVCG gate (as shown in Section 23.1.1) can reduce the circuit depth by lowering the number of controlled gates. The electro-optic photonic platform is the first physical system that has been demonstrated to perform the MVCG gate, which compressed multiple control operations in a single shot. The two-qudits-out-of-one-photon scheme makes sure the controlled-gate is deterministic and increases the fidelity of the setup. All the fiber-optic components in this scheme can be integrated on-chip making it compact and scalable. Furthermore, this platform can process and measure thousands of photons simultaneously, which enables the fast generation of statistical patterns: this feature enables the electro-optic platform to realize a proposed statistical phase estimation that loses the prior knowledge requirement of the eigenstates and thus offers more flexibility in searching eigenphase-eigenstate pairs within some specified range of interest [74]. As for the bosonic processor discussed in Section 23.2.2, it is shown that simulating a multidimensional system with a qudit-like system such as the microwave cavity can reduce the operation requirement by a large amount. The single-shot photon number resolving detection scheme can resolve the photon numbers with a notable reduction of the simulator runs. This bosonic processor is expandable by coupling more cavities where each of them is combined with ancilla measurement and control modules. The scalability of the system grants it the potential to simulate more complicated many-boy complex systems [75].

Acknowledgments

We acknowledge the financial support of National Science Foundation under award number 2124511, CCI Phase I: NSF Center for Quantum Dynamics on Modular Quantum Devices (CQD-MQD), and Award No. 2124511-CHE, NSF 1839191-ECCS.

References

1 Brylinski, J.-L. and Brylinski, R. (2002). Universal quantum gates. In: *Mathematics of quantum computation*, (ed. R.K. Brylinski, G. Chen) 117–134. Chapman and Hall/CRC.

2 Wang, Y., Hu, Z., Sanders, B.C., and Kais, S. (2020). Qudits and high-dimensional quantum computing. *Frontiers in Physics* 8: 479.

3 Vlasov, A.Y. (2002). Noncommutative tori and universal sets of nonbinary quantum gates. *Journal of Mathematical Physics* 43 (6): 2959–2964.

4 Gottesman, D. (1999). Fault-tolerant Quantum computation with higher-dimensional systems. In: *Quantum Computing and Quantum Communications*, vol. 1509 (ed. C.P. Williams) *NASA International Conference on Quantum Computing and Quantum Communications*, 302–313. Berlin: Springer-Verlag.

5 Daboul, J., Wang, X., and Sanders, B.C. (2003). Quantum gates on hybrid qudits. *Journal of Physics A: Mathematical and General* 36: 2525–2536.
6 Zhou, D.L., Zeng, B., Xu, Z., and Sun, C.P. (2003). Quantum computation based on *d*-level cluster state. *Physical Review A* 68: 062303.
7 Bullock, S.S., O'Leary, D.P., and Brennen, G.K. (2005). Asymptotically optimal quantum circuits for *d*-level systems. *Physical Review Letters* 94: 230502.
8 Li, W.-D., Gu, Y.-J., Liu, K. et al. (2013). Efficient universal quantum computation with auxiliary Hilbert space. *Physical Review A* 88: 034303.
9 Mischuck, B. and Mølmer, K. (2013). Qudit quantum computation in the Jaynes-Cummings model. *Physical Review A* 87: 022341.
10 Reck, M., Zeilinger, A., Bernstein, H.J., and Bertani, P. (1994). Experimental realization of any discrete unitary operator. *Physical Review Letters* 73: 58–61.
11 Rowe, D.J., Sanders, B.C., and de Guise, H. (1999). Representations of the Weyl group and Wigner functions for SU(3). *Journal of Mathematical Physics* 40 (7): 3604–3615.
12 Muthukrishnan, A. and Stroud, C.R. (2000). Multivalued logic gates for quantum computation. *Physical Review A* 62: 052309.
13 Luo, M. and Wang, X. (2014). Universal quantum computation with qudits. *Science China Physics, Mechanics & Astronomy* 57: 1712–1717.
14 Brennen, G.K., O'Leary, D.P., and Bullock, S.S. (2005). Criteria for exact qudit universality. *Physical Review A* 71: 052318.
15 Patera, J. and Zassenhaus, H. (1988). The Pauli matrices in n dimensions and finest gradings of simple lie algebras of type A_{n-1}. *Journal of Mathematical Physics* 29 (3): 665–673.
16 Gottesman, D., Kitaev, A., and Preskill, J. (2001). Encoding a qubit in an oscillator. *Physical Review A* 64 (1): 012310.
17 Bartlett, S.D., de Guise, H., and Sanders, B.C. (2002). Quantum encodings in spin systems and harmonic oscillators. *Physical Review A* 65: 052316.
18 Nielsen, M.A., Bremner, M.J., Dodd, J.L. et al. (2002). Universal simulation of Hamiltonian dynamics for quantum systems with finite-dimensional state spaces. *Physical Review A* 66: 022317.
19 Howard, M. and Vala, J. (2012). Qudit versions of the qubit $\pi/8$ gate. *Physical Review A* 86: 022316.
20 Wilmott, C.M. (2011). On swapping the states of two qudits. *International Journal of Quantum Information* 09 (06): 1511–1517.
21 Wilmott, C.M. and Wild, P.R. (2012). On a generalized quantum swap gate. *International Journal of Quantum Information* 10 (03): 1250034.
22 Mermin, N.D. (2001). From classical state swapping to quantum teleportation. *Physical Review A* 65: 012320.
23 Fujii, K. (2003). Exchange gate on the qudit space and Fock space. *Journal of Optics B: Quantum and Semiclassical Optics* 5: S613–S618.
24 Paz-Silva, G.A., Rebić, S., Twamley, J., and Duty, T. (2009). Perfect mirror transport protocol with higher dimensional quantum chains. *Physical Review Letters* 102: 020503.

25 Wang, X. (2001). Continuous-variable and hybrid quantum gates. *Journal of Physics A: Mathematical and General* 34: 9577–9584.

26 Alber, G., Delgado, A., Gisin, N., and Jex, I. (2001). Efficient bipartite quantum state purification in arbitrary dimensional Hilbert spaces. *Journal of Physics A: Mathematical and General* 34: 8821–8833.

27 Garcia-Escartin, J.C. and Chamorro-Posada, P. (2013). A swap gate for qudits. *Quantum Information Processing* 12: 3625–3631.

28 Ralph, T.C., Resch, K.J., and Gilchrist, A. (2007). Efficient Toffoli gates using qudits. *Physical Review A* 75: 022313.

29 Kiktenko, E.O., Nikolaeva, A.S., Xu, P. et al. (2020). Scalable quantum computing with qudits on a graph. *Physical Review A* 101: 022304.

30 Di, Y.-M. and Wei, H.-R. (2013). Synthesis of multivalued quantum logic circuits by elementary gates. *Physical Review A* 87: 012325.

31 Lu, H.-H., Hu, Z., Alshaykh, M.S. et al. (2019). Quantum phase estimation with time-frequency qudits in a single photon. *Advanced Quantum Technologies* 0 (0): 1900074.

32 Nielsen, M.A. and Chuang, I.L. (2011). *Quantum Computation and Quantum Information: 10th Anniversary Edition*, 10e. New York: Cambridge University Press.

33 Stroud, A.M.C.R. (2002). Quantum fast Fourier transform using multilevel atoms. *Journal of Modern Optics* 49 (13): 2115–2127.

34 Zilic, Z. and Radecka, K. (2007). Scaling and better approximating quantum Fourier transform by higher radices. *IEEE Transactions on Computers* 56: 202–207.

35 Cao, Y., Peng, S.-G., Zheng, C., and Long, G.-L. (2011). Quantum Fourier transform and phase estimation in qudit system. *Communications in Theoretical Physics* 55: 790–794.

36 Coppersmith, D. (2002). An approximate Fourier transform useful in quantum factoring. *arXiv preprint quant-ph/0201067*.

37 Bocharov, A., Roetteler, M., and Svore, K.M. (2017). Factoring with qutrits: Shor's algorithm on ternary and metaplectic quantum architectures. *Physical Review A* 96: 012306.

38 Parasa, V. and Perkowski, M. (2011). Quantum phase estimation using multivalued logic. *2011 41st IEEE International Symposium on Multiple-Valued Logic*, pp. 224–229, May 2011.

39 Shor, P.W. (1994). Algorithms for quantum computation: discrete logarithms and factoring. *Proceedings 35th Annual Symposium on Foundations of Computer Science*, pp. 124–134, November 1994.

40 Abrams, D.S. and Lloyd, S. (1999). Quantum algorithm providing exponential speed increase for finding eigenvalues and eigenvectors. *Physical Review Letters* 83: 5162–5165.

41 Harrow, A.W., Hassidim, A., and Lloyd, S. (2009). Quantum algorithm for linear systems of equations. *Physical Review Letters* 103: 150502.

42 Pan, J., Cao, Y., Yao, X. et al. (2014). Experimental realization of quantum algorithm for solving linear systems of equations. *Physical Review A* 89 (2): 022313.

43 Tonchev, H.S. and Vitanov, N.V. (2016). Quantum phase estimation and quantum counting with qudits. *Physical Review A* 94: 042307.

44 Sawerwain, M. and Leoński, W. (2013). Quantum circuits based on qutrits as a tool for solving systems of linear equations. *arXiv:1309.0800*.

45 Babazadeh, A., Erhard, M., Wang, F. et al. (2017). High-dimensional single-photon quantum gates: concepts and experiments. *Physical Review Letters* 119: 180510.

46 Erhard, M., Fickler, R., Krenn, M., and Zeilinger, A. (2018). Twisted photons: new quantum perspectives in high dimensions. *Light: Science & Applications* 7: 17146.

47 Kues, M., Reimer, C., Roztocki, P. et al. (2017). On-chip generation of high-dimensional entangled quantum states and their coherent control. *Nature* 546 (6760): 622–626.

48 Lu, H.-H., Lukens, J.M., Peters, N.A. et al. (2018). Electro-optic frequency beam splitters and tritters for high-fidelity photonic quantum inf. process. *Physical Review Letters* 120 (3): 030502.

49 Imany, P., Jaramillo-Villegas, J.A., Odele, O.D. et al. (2018). 50-GHz-spaced comb of high-dimensional frequency-bin entangled photons from an on-chip silicon nitride microresonator. *Optics Express* 26 (2): 1825–1840.

50 Imany, P., Jaramillo-Villegas, J.A., Alshaykh, M.S. et al. (2019). High-dimensional optical quantum logic in large operational spaces. *npj Quantum Information* 5: 59.

51 Islam, N.T., Lim, C.C.W., Cahall, C. et al. (2017). Provably secure and high-rate quantum key distribution with time-bin qudits. *Science Advances* 3 (11): e1701491.

52 Humphreys, P.C., Metcalf, B.J., Spring, J.B. et al. (2013). Linear optical quantum computing in a single spatial mode. *Physical Review Letters* 111: 150501.

53 Gao, X., Erhard, M., Zeilinger, A., and Krenn, M. (2020). Computer-inspired concept for high-dimensional multipartite quantum gates. *Physical Review Letters* 125: 050501.

54 Lukens, J.M. and Lougovski, P. (2017). Frequency-encoded photonic qubits for scalable quantum information processing. *Optica* 4: 8–16.

55 Knill, E., Laflamme, R., and Milburn, G.J. (2001). A scheme for efficient quantum computation with linear optics. *Nature* 409 (6816): 46–52.

56 Fang, W.-T., Li, Y.-H., Zhou, Z.-Y. et al. (2018). On-chip generation of time-and wavelength-division multiplexed multiple time-bin entanglement. *Optics Express* 26 (10): 12912–12921.

57 Humphreys, P.C., Kolthammer, W.S., Nunn, J. et al. (2014). Continuous-variable quantum computing in optical time-frequency modes using quantum memories. *Physical Review Letters* 113 (13): 130502.

58 Wang, X., Sanders, B.C., and Berry, D.W. (2003). Entangling power and operator entanglement in qudit systems. *Physical Review A* 67: 042323.

59 De Greve, K., McMahon, P.L., Yu, L. et al. (2013). Complete tomography of a high-fidelity solid-state entangled spin–photon qubit pair. *Nature Communications* 4 (1): 1–7.

60 Breuer, H.-P. and Petruccione, F. (2002). *The Theory of Open Quantum Systems*. Oxford University Press on Demand.

61 Hu, Z., Xia, R., and Kais, S. (2020). A quantum algorithm for evolving open quantum dynamics on quantum computing devices. *Scientific Reports* 10 (1): 1–9.

62 Hu, Z., Head-Marsden, K., Mazziotti, D.A. et al. (2021). A general quantum algorithm for open quantum dynamics demonstrated with the Fenna-Matthews-Olson complex. *arXiv preprint arXiv:2101.05287*.

63 Levy, E. and Shalit, O.M. (2014). Dilation theory in finite dimensions: the possible, the impossible and the unknown. *Rocky Mountain Journal of Mathematics* 44 (1): 203–221.

64 Krishnamoorthy, A. and Menon, D. (2013). Matrix inversion using Cholesky decomposition. *2013 Signal Processing: Algorithms, Architectures, Arrangements, and Applications (SPA)*, pp. 70–72, IEEE.

65 Wang, C.S., Curtis, J.C., Lester, B.J. et al. (2020). Efficient multiphoton sampling of molecular vibronic spectra on a superconducting bosonic processor. *Physical Review X* 10: 021060.

66 Duschinsky, F. (1937). The importance of the electron spectrum in multi atomic molecules. Concerning the Franck-Condon principle. *Acta Physicochimica URSS* 7: 551–566.

67 Doktorov, E., Malkin, I., and Man'Ko, V. (1977). Dynamical symmetry of vibronic transitions in polyatomic molecules and the Franck-Condon principle. *Journal of Molecular Spectroscopy* 64 (2): 302–326.

68 Gao, Y.Y., Lester, B.J., Zhang, Y. et al. (2018). Programmable interference between two microwave quantum memories. *Physical Review X* 8: 021073.

69 Heeres, R.W., Reinhold, P., Ofek, N. et al. (2017). Implementing a universal gate set on a logical qubit encoded in an oscillator. *Nature Communications* 8 (1): 1–7.

70 Krastanov, S., Albert, V.V., Shen, C. et al. (2015). Universal control of an oscillator with dispersive coupling to a qubit. *Physical Review A* 92: 040303.

71 Johnson, B., Reed, M., Houck, A.A. et al. (2010). Quantum non-demolition detection of single microwave photons in a circuit. *Nature Physics* 6 (9): 663–667.

72 Sawaya, N.P. and Huh, J. (2019). Quantum algorithm for calculating molecular vibronic spectra. *The Journal of Physical Chemistry Letters* 10 (13): 3586–3591.

73 Low, G.H. and Chuang, I.L. (2017). Optimal Hamiltonian simulation by quantum signal processing. *Physical Review Letters* 118 (1): 010501.

74 Moore, A.J., Wang, Y., Hu, Z. et al. (2021). Statistical approach to quantum phase estimation. *New Journal of Physics* 23: 113027.

75 Kais, S., Whaley, K., Dinner, A., and Rice, S. (2014). Quantum information and computation for chemistry. In: *Advances in Chemical Physics*, ed. S.A. Rice, A.R. Dinner). Wiley.

24

Fiber-Based Quantum Repeaters

Mohsen Razavi

School of Electronic and Electrical Engineering, University of Leeds, Leeds LS2 9JT, UK

One of the key features of today's Internet is the ability to send classical data, to arbitrarily long distances, in a reliable and efficient way. Future quantum communications networks should also offer the same service but, notably, for *quantum data transfer*. Such a capability is crucial to the operation of the quantum Internet [1], in which quantum nodes, of computing, communications, and/or sensing nature, can efficiently exchange quantum states in a faithful way. The enabling platform for this transformative feature is known as *quantum repeaters* [2, 3]. Quantum repeaters not only enable trust-free end-to-end security for our communications networks [4], but can also be used for distributed quantum computing [5] and high-precision quantum sensing [6], among other applications [7]. This chapter takes a look at the road ahead for building such an infrastructure by going through different concepts that can be employed to implement quantum repeaters. While most of these concepts can also be used in space settings, involving satellites [8], in this chapter, we only focus on fiber-based networks for the description of different classes of quantum repeaters.

The problem of transferring a quantum state from one place to another has been of interest to the quantum community from early on. In fact, one of the key techniques for quantum data transfer (QDT) is based on the well-known concept of *teleportation* [9], whose notion has been featured in several science-fiction movies. While the televised version of teleportation may seem quite magical, its quantum information formulation can clarify the physical resources needed for this operation. At the core of teleportation, we use one of the most intriguing resources in quantum mechanics, i.e. entanglement, which plays the role of a one-time wire for our communication purposes. The problem of reliable QDT would then turn into how entangled states can be shared in an efficient and reliable way between remote nodes. Many classes of quantum repeaters are then designed to answer this question.

There are several practical challenges that we need to overcome in order to implement quantum repeaters. One key issue is the problem of channel loss. Almost all relevant entanglement distribution schemes rely on light transmission across an

Photonic Quantum Technologies: Science and Applications, First Edition. Edited by Mohamed Benyoucef.

optical channel [10], where their chance of success is often proportional to the channel transmissivity. In fiber-optic channels, this success probability goes down exponentially with distance, as a typical fiber-optic channel suffers 0.2 dB/km of loss. Entanglement distribution would then need to be first tried over shorter segments and, if successful, be extended to the remote nodes. This would, however, introduce new challenges. First, entanglement distribution needs to be repeated until success, which could slow down the process. Secondly, upon success, the entanglement needs to be stored in quantum memories to be used for further extension. Quantum repeaters that rely on quantum memories will then need to overcome the challenges of building and controlling efficient interfaces with light, while maintaining low noise and long coherence times for proper memory operation. The latter requirements would often necessitate entanglement distillation techniques [11–13] that can keep the quality of the final entangled state at the target level.

An alternative way to look at QDT, in lossy and noisy channels, is via a classical approach to data communications. In classical communications, whenever the channel is noisy, we can use error correction techniques and several relay nodes to transmit our data from one node to another until it reaches its final destination. Interestingly, the same concept can be replicated in quantum mechanics by using entanglement-based structures known as quantum error correction (QEC) codes. In quantum communications, in addition to errors and noise, we are often heavily impacted by loss in the channel too. It turns out that QEC can be used to address both issues, based on which, some new classes of quantum repeaters have emerged, as we will discuss later in this chapter.

Based on different ways in which quantum repeaters can be implemented and then used, we can think of different classes of such systems. Table 24.1 gives one possible way, which I will explain later in this chapter, for such a classification. In general, we can first distinguish between quantum repeater protocols that rely on repeat-until-success (RUS) entanglement generation, versus those that rely on QEC for combating loss. Within each group then there are different examples of quantum repeaters based on their approach to noise and entanglement distillation, and whether they rely on deterministic or probabilistic gates. Quantum repeaters can also be distinguished based on whether an additional teleportation operation is

Table 24.1 Different classes of quantum repeaters, and how they each handle loss and error, perform their measurements, and the method they use for QDT.

Repeater type	Loss handling	Error handling	Measurements	QDT
Probabilistic	RUS	Post-selection	Probabilistic	Teleportation
Semi-probabilistic	RUS	Probabilistic	Deterministic	Teleportation
Encoded	RUS	QEC	Deterministic	Teleportation
All-photonic	QEC	QEC	Deterministic	Teleportation
One-way	QEC	QEC	Deterministic	Direct

Note that the terminologies introduced here are not unique or exclusive; for their exact description, please see Sections 24.2 and 24.3.

needed for QDT, or whether quantum messages can be encoded and sent directly over the network. Please note that the terminologies we have used for the five classes of repeaters in Table 24.1 are not necessarily unique or exclusive. The terms "Encoded" and "All-photonic," for instance, have been borrowed from the key example papers that introduce such schemes, while the notion of using encoded and/or all-photonic states can also appear in one-way class of repeaters as well.

There is another classification known in the literature [14], which refer to the first two classes in Table 24.1 as first-generation quantum repeaters, the encoded ones as the second generation, and the last two as the third generation. While this has been a useful way to quickly convey what we are referring to in scientific discussions, the author of this chapter has reservations about using the term "generation" in this context. Perhaps the most well-known use of this term is in mobile communications networks, where terms such as 5G and 6G have been coined primarily by industry and standardization bodies to address the new services that are going to be offered in imminent future. Unfortunately, for quantum repeaters, there could take a while until any of the above ideas can be seen in a commercial platform, and, by that time, another form of numbering/labeling may be suitable. This could cause confusion later on if there is a discrepancy between the usage of the term generation in scientific literature and industry/standardization documents. In this chapter, I therefore use the terms in Table 24.1 to refer to different classes of repeaters discussed here.

The rest of this chapter is organized as follows. In Section 24.1, we first introduce a set of tools that will be required to build quantum repeaters. We then take a step-by-step approach to explain the key concepts behind the operation of different classes of quantum repeaters. We first go, in Section 24.2, over examples of quantum repeaters that rely on repeat-until-success entanglement generation and then, in Section 24.3, cover those that rely on QEC to combat loss. Along the way, we point out the resources that they each need versus the performance they can potentially offer. We conclude the chapter in Section 24.4 by summarizing the current developments and future directions in this field.

24.1 Quantum Repeater Toolbox

There are certain techniques and ingredients that need to be implemented for quantum repeaters to work. Here, we give a brief summary of them, on which basis, in Sections 24.2 and 24.3, the structure of full repeater systems will be explained.

To exemplify what components or functionalities may be needed for quantum repeaters, let us first focus on teleportation as an application. To teleport the state of an unknown qubit C at point A to point B, we first need to create a maximally entangled state between A and B; see Section 24.1.1 to see how this can be done. Ideally, this entangled state needs to be stored in an auxiliary pair of qubits A and B, so that qubit C can be teleported whenever needed; see Section 24.1.2 for some examples of quantum memories. We then need to perform a Bell-state measurement (BSM) on qubits A and C and classically transfer the results to point B; Section 24.1.3 explains how this can be implemented in practice. The generated entangled state over a short distance would then need to be extended to longer distances, and, along the way, may need to be distilled. We cover the basic ideas behind entanglement distillation in Sections 24.1.4–24.1.5.

One key application of teleportation in quantum repeaters is *entanglement swapping*. If qubit C in our teleportation setting is itself entangled with another qubit D, by teleporting the state of C to B, we effectively create an entangled state between qubits D and B. As we will show in Section 24.2, this idea can be used in a nested way to create entangled states between remote nodes.

24.1.1 Entanglement Distribution

A key task in many quantum repeaters is to create an entangled state between two remote quantum memories. Ideally, this state should be a maximally entangled state, e.g. a Bell state. There are various techniques that can be used for entanglement distribution. Here, we explain a popular technique that is based on entanglement swapping. The key idea is to locally generate entanglement between the quantum memory unit and an optical mode of light. If we do this process at both memories, and direct the corresponding optical signals to interfere at a BSM module in the middle of the link, we can then create entanglement between the two memories.

A few points are worth mentioning here. First, given that the optical signals suffer from channel loss, we should either accept some noise in our generated entangled state, and/or, most often, be prepared to repeat the whole entanglement distribution scheme, if the corresponding photons fail to reach the BSM module, or if the BSM operation is unsuccessful. The second point is that, in the latter two cases, we need a *heralding* mechanism for acknowledging the success or failure of our entanglement distribution attempt. In such cases, there would be a limitation on how fast we can repeat the entanglement distribution scheme, as the next trial should happen after we have learned about the result of the previous attempt. For a channel of distance L_0, this implies that shortest repetition period is L_0/c, where c is the speed of light in that channel.

24.1.2 Quantum Memories

While quantum storage is not a requirement for certain classes of quantum repeaters, a quantum memory will be needed whenever we use a heralded repeat-until-success approach to entanglement distribution as was mentioned above. There are several memory candidates that are being considered for implementing quantum repeaters. This includes ensembles of atoms [10, 15, 16], as well as trapped single atoms/ions [17–20], or quasi single-atom units, such as nitrogen vacancy (NV) centers in diamond [21–26], or silicon vacancies [27]. Each candidate offers certain advantages and disadvantages, and they may lend themselves to certain classes of quantum repeaters better than the others. But, typical characteristics that are expected from a quantum memory are to have an efficient interface with light, at the single-photon level, as well as long coherence times. Given that these two requirements are often in competition with each other, two-qubit memories, where one qubit acts as the optical interface, and the other is used for storage purposes, can be quite promising. For instance, the electron spin in NV centers can be used for interaction with light, whereas relevant nuclear spins can be used for long-time storage. Such a structure also makes the efficient implementation of certain joint operations, such as a BSM, possible within one memory unit [28–31].

24.1.3 Bell-State Measurement

A BSM is a joint operation on two qubits, and in that sense it requires some form of interaction between them. In theory, a BSM is modeled by a controlled-not (CNOT) gate on both qubit, followed by a Hadamard gate on the control qubit, and single-qubit measurements in the standard basis. In practice, whether we can directly perform a CNOT gate, or equivalent, on our two qubits is primarily a function of the physical system that contains our quantum information. As was mentioned, in certain memories, such as NV centers, or trapped ions, it is possible to implement this operation in a deterministic way, which is the preferred option for quantum repeater purposes. If the platform does not allow for such an option, then an alternative route, albeit imperfect, is to convert the state of quantum memories into photons, and then do a BSM on the optical modes. This is done by interfering the two optical modes at a beam splitter and then using single-photon detectors to post-select successful cases. Given that we may get no click on the detectors, this approach may result in inconclusive outcomes, in which case the BSM has failed. Even in the case of success, such optical implementations of the BSM may only distinguish some of the Bell states and not all of them [32]. In that sense, the optical BSM is often probabilistic, although there are certain schemes in which, by using auxiliary resources, one can increase the chance of success [33–37].

24.1.4 Entanglement Distillation

For a variety of practical reasons, the entanglement shared between two remote nodes may be far from a maximally entangled state. A well-known technique to remedy this problem, and generate higher-quality entangled states, is to use entanglement distillation techniques. The key objective is to convert N copies of an imperfect entangled state to $M < N$ entangled states of higher quality, by using a series of local operations and classical communication. For instance, if we start with two entangled pairs, namely, $A_1 - B_1$ and $A_2 - B_2$, by applying a CNOT gate on co-located memories $A_1 - A_2$ and $B_1 - B_2$, and measuring A_1 and B_1, we will end up with an entangled state of higher quality between A_2 and B_2 provided that the measurement results agree [12]. This suggests that such distillation techniques may end up being probabilistic even if the employed CNOT operation has been deterministic. In addition, given the transmission delay caused by the classical communication element, such techniques may not help with decoherence that happens in memories in the meantime. To remedy this issue, we need one-way distillation techniques, whose basis we cover next.

24.1.5 Quantum Error Correction

Error correction is an established technique to help us approach the channel capacity in a classical communications setting. Its extension to quantum settings has also been a useful means to design fault-tolerant quantum computing operations and also to help with entanglement distillation in quantum communications. In classical error correction codes, a sequence of bits of length M is mapped to a sequence of bits of length $N > M$. These additional bits will then help us with detecting bit-flip errors, and, possibly, where those erroneous bits are. In QEC codes, there are two major differences. First, in the qubit space, in addition to bit-flip errors, we should also correct for phase-flip errors. Secondly, in the design of a QEC code, we only need to map a set of basis states to a larger space. This means that the QEC encoder is agnostic to the superposition coefficients that define the input. For instance, in a simple repetition code, qubit $|0\rangle$ ($|1\rangle$) is mapped to $|000\rangle$ ($|111\rangle$), but $|0\rangle + |1\rangle$ is mapped into $|000\rangle + |111\rangle$, which is a Greenberger-Horne-Zeilinger (GHZ) entangled state [38]. In the latter case, we have effectively mapped a single qubit into a multipartite entangled state. This is another manifestation of how entanglement can help with quantum communications.

24.2 Quantum Repeaters Based on Heralded Entanglement Distribution

In this section, I give an overview of key ideas behind quantum repeater platforms that rely on repeat-until-success, or, equivalently, heralded, techniques for entanglement distribution. Here is the key idea behind the operation of such quantum repeater settings. If we want to entangle two parties A and B at a distance $L = 2^n L_0$, as shown in Figure 24.1(a), we divide the link into 2^n segments, distribute entanglement over each *elementary* link of length L_0, *store* that entangled state into a pair of quantum memories, and then perform entanglement swapping at the middle nodes to distribute entanglement between the far two ends. This is a nested procedure for entanglement distribution, because of which n is called the nesting level. For example, in Figure 24.1(b), for $n = 2$, we have two stages of entanglement swapping. The modules labeled by BSM1 enable us to distribute entanglement over segments with length $2L_0$, whereas BSM2 takes two entangled pairs at this distance and further extends the entanglement to $4L_0$. The same procedure can be repeated, where, in each nesting level, we double the range over which entanglement is distributed.

Depending on how the BSM and the entanglement distillation are performed, I describe three subclasses of quantum repeaters: Probabilistic, where photonic partial BSMs are employed and there is no active distillation in place; semi-probabilistic, where BSMs are deterministic, but the employed entanglement distillation techniques are heralded, and therefore probabilistic; and encoded repeaters, where both BSMs and entanglement distillation techniques are deterministic. Let us start with the simplest scheme, which is of interest today, and then cover the more advanced variants.

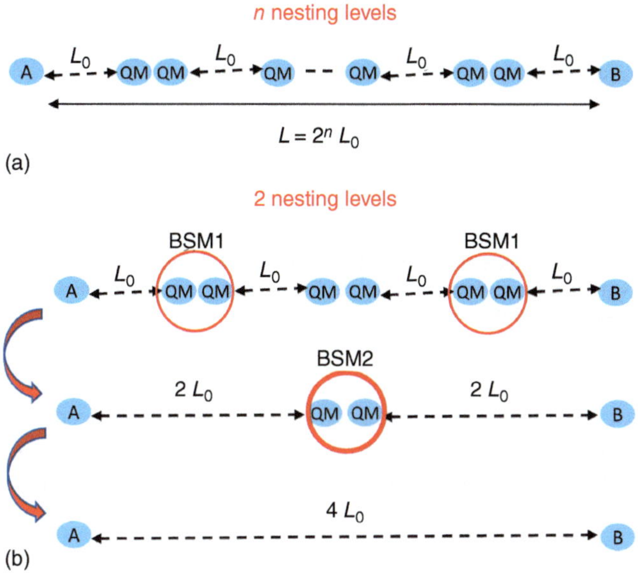

Figure 24.1 (a) A quantum repeater link with nesting level n. (b) An example quantum repeater link with nesting level 2. After distributing entanglement over elementary links with length L_0, we have to do two series of BSMs. BSM1 refers to the first stage of entanglement swapping, where entanglement is shared at distance $2L_0$. BSM2 would then extend the entanglement to the far ends. For probabilistic BSMs, BSM2 cannot be done until we learn about the success of BSM1s. That would require classical communication between the nodes, which causes delay, and possible repetition of some steps until success, which reduces the rate. QM: quantum memory. The picture is taken from Ref. [39].

24.2.1 Probabilistic Quantum Repeaters

Probabilistic quantum repeaters could offer a practical route to building the first operational examples of quantum repeaters. In such repeaters, the BSM operation is probabilistic. In general, the BSM is a joint operation on two logical quantum memories. It is not, however, straightforward for all memory systems to implement the required direct interaction. In lieu of that, the state of memories needs to be read, i.e. transferred into photons on which a partial BSM is performed.

By using probabilistic BSMs, in our nested entanglement distribution, there would be inevitable delays in going from one nesting level to another. For instance, in Figure 24.1(b), we need to know the results of the two BSM1s before applying BSM2 to the corresponding memories [39]. The key advantage of such a wait is the increase in the rate of our end-to-end entanglement distribution. In Figure 24.1(b), we can, in parallel, attempt to entangle the two elementary links, which, takes, roughly, the same order of magnitude that it takes to entangle one elementary link. In that sense, the entanglement generation rate across the remote nodes would scale with the probability of entangling an elementary link, which can be much shorter than the overall link. However, the additional wait implies that we need memories with

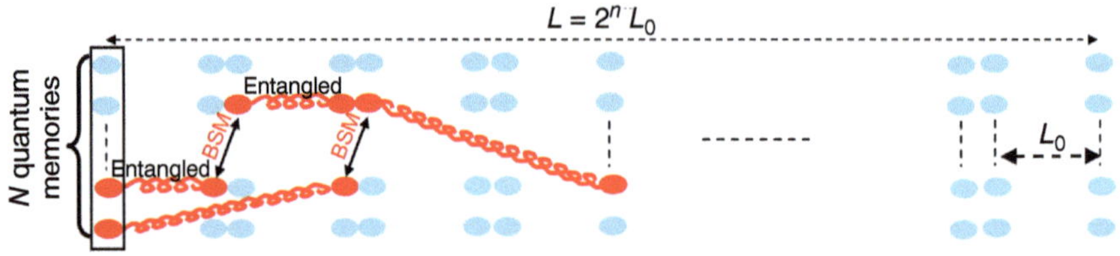

Figure 24.2 A probabilistic quantum repeater with multiple memories per node. In each round, entanglement distribution is attempted on all available elementary links. BSMs, at different nesting levels, will also be performed by matching as many entangled pairs as possible. The picture is taken from Ref. [39].

longer coherence times, or, if we do not have such memories, we may need to do some measurements blindly, i.e. without knowing the results of the previous measurement, at the cost of losing out on some of the rate advantage that quantum repeaters are supposed to offer [40].

There are certain techniques that alleviate the storage time requirement. For instance, in general, if we repeat this scheme in parallel among many banks of memories, see Figure 24.2, then we can use a mix and match technique to do BSMs on already entangled links as soon as they are available [40, 41]. Such a method requires efficient optical switching techniques that enable directing photons from different memories to BSM stations. An alternative way to implement this idea is to use multi-mode memories [42, 43]. Such memories are particularly useful to increase the chance of entanglement distribution across an elementary link, so even if the rest of BSMs need to be done blindly, we may still get some practical advantage [44].

To keep things simple, probabilistic repeaters may not use any distillation techniques. Given that the memories may not allow deterministic CNOT gates, even simple distillation techniques may hardly offer any rate advantage. Depending on the protocols used, probabilistic repeaters may benefit from the inconclusive BSM results in the virtue of post-selection. That is, in some cases, the non-desired terms can end up being selected out by our probabilistic measurement. In the end, the probabilistic repeaters are potentially useful for first breakthroughs in repeater technologies, but may not effectively cover arbitrarily long distances, for which more advanced solutions are required.

24.2.2 Semi-Probabilistic Quantum Repeaters

The original proposal for quantum repeaters relies on *deterministic*, but possibly *erroneous*, gates for BSM and distillation operations [45]. In their seminal work, the authors assume that the initial entanglement distribution and storage over elementary links have already been achieved. In fact, they assume that there are banks of memories, over all elementary links, with many entangled memory pairs in each. The objective is then to create a high quality entangled state between two remote

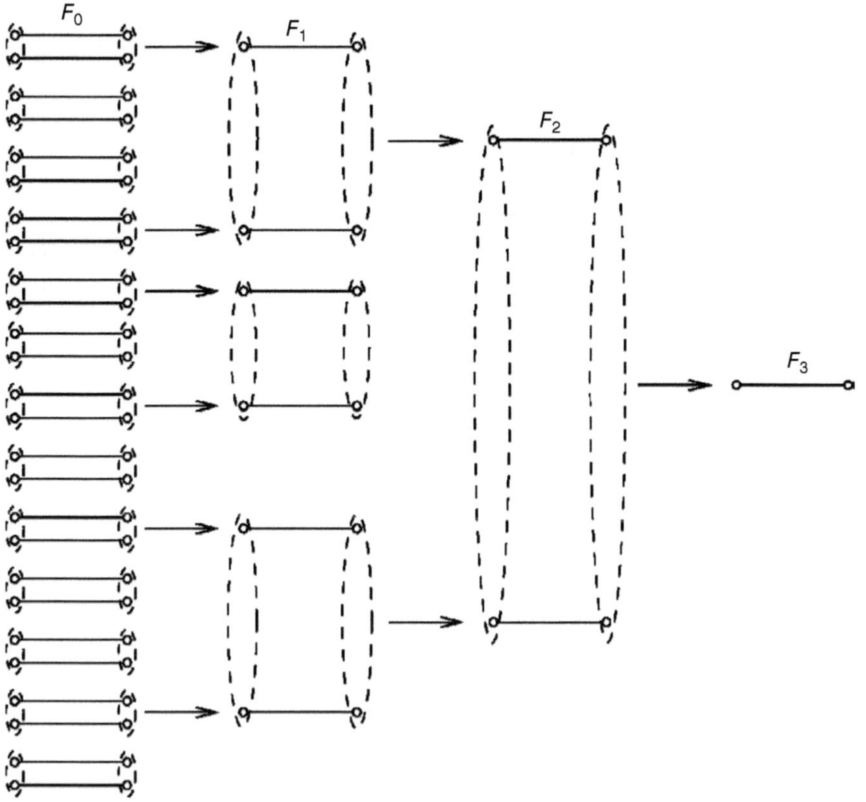

Figure 24.3 The schematic of nested distillation scheme taken from Ref. [11]. Source: Dür et al. [11]/with permission of American Physical Society.

nodes. Assuming that the BSMs can be done in a deterministic way, this, at first instance, may look like an easy job, as it seems that, without any unnecessary waiting, we can perform the BSMs at all nodes simultaneously and resolve the issues of low rate and long storage time in probabilistic schemes. However, in reality, nature might have set other traps for us as errors in such a setting can propagate quite rapidly. In effect, the focus, in semi-probabilistic schemes, is on correcting for errors that may happen because of our employed deterministic gates and/or imperfections in our initial entanglement distribution.

The adopted distillation technique in this class of quantum repeaters is the heralded, but probabilistic, schemes described in Section 24.1.4. That is, after a few rounds of BSMs, we may need to distill the generated entangled state before proceeding to higher nesting levels. For instance, if we need to use distillation techniques after n_d nesting levels, we can perform simultaneous BSMs over all memory banks up to this nesting level. Then, we can choose N pairs of entangled states at a distance $2^{n_d}L_0$ and perform a chosen distillation technique to generate $M < N$ pairs of entangled states of higher quality; see Figure 24.3 for an example. Once this process has succeeded across our repeater chain, we can then do the next set of BSMs and continue this way until we distribute entanglement between the far end nodes. Given that the employed distillation technique is probabilistic, this will in effect turn such schemes into a sort of probabilistic schemes (the reason for the proposed terminology here), albeit with possibly higher performance than fully probabilistic repeaters given that the gates are deterministic and n_d can be larger than one. Nevertheless, such schemes are not resilient against memory decoherence, because they also have to deal with the waiting time, to learn about the success of entanglement distillation, in their operation. At long distances, this would result in the exponential decay of the rate, in such repeaters, with $\sqrt{L}$, although the onset of this effect can be delayed by choosing memories with long coherence times [40]. Moreover, given that the distillation process requires two-qubit gates, the operation error in all gates needs to be kept small (sub percent [11], typically) in order for the whole scheme to properly work.

24.2.3 Encoded Quantum Repeaters

The alternative solution to dealing with errors is to use QEC. In encoded quantum repeaters [46], instead of ideally distributing only Bell states over elementary links, we aim to share *encoded* Bell states. For instance, if via a QEC scheme, $|0\rangle$ ($|1\rangle$) is mapped to $|\mathbf{0}\rangle$ ($|\mathbf{1}\rangle$), instead of sharing $|0\rangle|0\rangle + |1\rangle|1\rangle$, we share $|\mathbf{0}\rangle|\mathbf{0}\rangle + |\mathbf{1}\rangle|\mathbf{1}\rangle$ between two elementary nodes.

This can be done by first sharing multiple bipartite entangled states over the link and then convert them, using local operation and classical communication, to the desired encoded state; see Figure 24.4 for an example. The immediate benefit of this scheme is that errors that occur during the entanglement swapping stage can possibly be corrected by the syndrome information obtained at the middle nodes in a deterministic way and with no need for two-way classical communication. For certain classes of QEC codes, such syndrome measurements are as simple as doing two-qubit BSMs [46]. This means that, after the initial entanglement distribution, there is no additional waiting, associated with repeat-until-success operations, for the distillation part. This is clearly an advantage over semi-probabilistic quantum repeaters. The price to pay though is that, now, we need quantum gates with lower error rates typically on the order of 0.001–0.01 or below [47, 48]. But, if such gates are available, such repeaters can also reduce some of the errors caused by the memory decoherence once we are waiting to learn about the success of the initial entanglement distribution.

Using the above techniques, we can design quantum repeaters with a modestly high key rate. The limitation is mainly from the original requirement for entangling elementary links, which is still probabilistic, and the trade-off between having more nesting levels, and, therefore, higher entanglement generation rates, versus having fewer nesting levels, hence less accumulated error and distillation. One can, however, ask the question that now that we allow for advanced QEC operations to be used for distillation, can we also use them to combat loss? This in effect can make all quantum repeater operations deterministic paving the way for fast and reliable quantum communications.

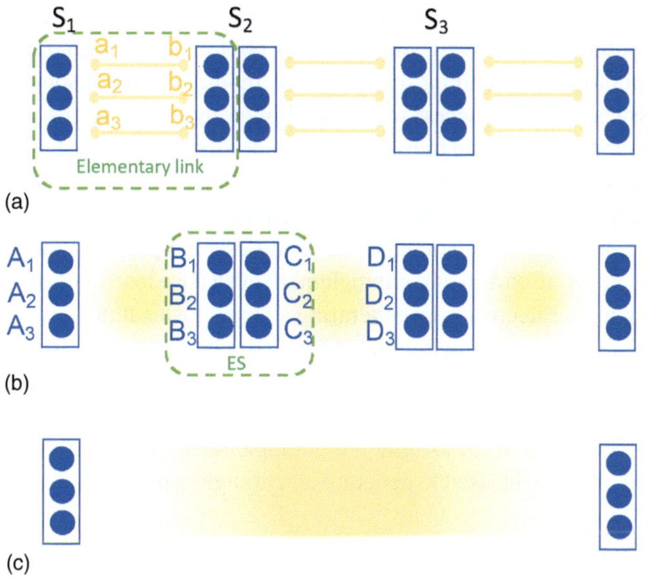

Figure 24.4 The schematic of an encoded quantum repeater taken from Ref. [47]: (a) We first generate auxiliary Bell states over all elementary link, which will be converted to encoded Bell states by local operation and classical communication; (b) we perform BSMs on the encoded entangled pairs. For certain classes of codes, this can be as simple as performing two-qubit BSMs on corresponding memories; and (c) by passing the information obtained during the BSM operations to the end nodes, they can retrieve and share an encoded entangled state. Source: Jing et al. [47]/with permission of American Physical Society.

24.3 Memory-Less Quantum Repeaters

The most advanced protocols for quantum repeaters leave little room for probabilistic operations. In such schemes, we use multipartite entanglement to combat the loss effect. This can be achieved in different ways, two main categories of which I will explain below. The common feature of all these schemes is that we no longer need quantum memories for storage purposes at the core of the network. As there is no free lunch, however, this will impose more stringent requirements on other parts of the system. For instance, we may need more reliable quantum processing capabilities, but, then, the quantum storage may not be the key bottleneck.

24.3.1 All-photonic Quantum Repeaters

Although historically they are not the first memory-less repeaters proposed, all-photonic quantum repeaters [49] use the same concept of entanglement distribution, as we have seen in Section 24.2, for their operation. In the end, what they produce is entanglement between two remote nodes. They, however, achieve this goal without using quantum storage in the middle nodes of the network.

Here is the key idea. Suppose we have a quantum relay structure, where, at the middle of each elementary node, there is an entangled photon source, generating a pair of entangled photons and sending them to BSM modules. Such a solution has critical scaling issues as for its success we need all photons to survive the path loss, hence its entanglement generation rate drops exponentially with distance. Now, imagine that instead of a simple entangled photon pair in the quantum relay structure, we use a more sophisticated photonic source of entanglement that generates a cluster state [50], i.e. a multipartite entangled state. If designed properly, then it may still be possible to generate end-to-end entanglement even if we lose some of the photons in our cluster state. Effectively, by generating a large cluster state, it is as if we are attempting to entangle each link in parallel many times, while we only need some of them to succeed. In the multiple-memory structure of Figure 24.2, we still need to know which entanglling attempts have been successful. In the cluster-based solution, this requirement is removed as now the entanglement embedded in this multipartite entangled state can be used to generate an entangling path between the two remote nodes.

The above idea is certainly a neat way to distribute entanglement at fast rates. The simulation in [49] suggests that for $L = 1{,}000$ km and $L_0 = 4$ km, with a cluster state of tens of photons, we can achieve a key generation rate on the order of MHz. This scheme, however, requires large cluster states for which high-quality single-photon sources and/or quantum processors with highly efficient light interfaces may be needed [51]. One should also account for the resilience of this setup to errors at any of its components.

24.3.2 One-way Quantum Repeaters

Thus far, all introduced quantum repeaters offer a scalable solution for entanglement distribution at, in principle, arbitrarily long distances. So long as QDT is concerned, we can then use teleportation to transfer any unknown states. A valid question to raise is whether we can transfer an unknown qubit without necessarily having entangled states prepared in advance. The answer to this question is positive, and the idea, as was alluded to before, relies on using QEC to protect our qubits against loss and error. Here is an example of such a scheme as proposed in [52].

Suppose the qubit to be transferred is given by $\alpha|0\rangle + \beta|1\rangle$. We first encode this qubit as

$$|\Psi\rangle^{(m,n)} = \alpha\,|+\rangle_1^{(m)} \cdots |+\rangle_n^{(m)} + \beta\,|-\rangle_1^{(m)} \cdots |-\rangle_n^{(m)}, \tag{24.1}$$

where n is the number of logical quantum states, and m is the number of physical qubits in each logical state with $|\pm\rangle^{(m)} = |0\rangle^{\otimes m} \pm |1\rangle^{\otimes m}$. For instance, the above state can be represented by m single photons per mode in n different, e.g. temporal, modes. Qubits $|0\rangle$ and $|1\rangle$ can then represent the polarization of the photons. More resource efficient encoding is possible [52], but we leave it out for the sake of simplicity. Now, suppose, this encoded multipartite entangled state is transferred from one node to its adjacent node. This encoding has the property that the original quantum state can be recovered provided that at least (i) one photon survives in each logical state, and (ii) one logical state, with all its m constituent photons, is fully received. We can then repeat the same encoding at every repeater node, until reaching the final destination. This can, in principle, result in a high quantum state transfer rate on the order of tens of MHz.

These are certain practical issues that need to be managed once we get to implement memory-less repeater schemes. The scheme in [52], for instance, works without explicitly requiring quantum storage in the middle nodes, but, instead, it needs reliable and efficient quantum processing units that can handle QEC encoding and decoding. Additional error analysis on this system [53, 54] suggests that operation errors as low as $10^{-4} - 10^{-3}$ may be needed. There are also alternative schemes for direct transmission of quantum states that combine the merits of encoded and all-photonic repeaters [55, 56]. In such schemes, an encoded state is teleported to the next node by generating a high-quality encoded Bell state at each intermediate node and performing a multipartite BSM on the transmitted state and one half of the locally generated encoded Bell state. The other half will then be sent to the next node. In such schemes, generating high-quality local encoded states could be the main challenge. In any case, for such QEC techniques to counter loss, we typically need the total loss in each elementary link to be less than 3 dB. This effectively requires us to have very short links and therefore, many intermediate nodes to cover long distances.

Finally, note that whether our quantum repeater relies on one-way quantum communications or entanglement distribution, the two functionalities can be interchanged. In one-way quantum repeaters, if the initial qubit is entangled with another qubit at the source, this scheme can also be used for entanglement generation as well. In the other case, teleportation can be used to enable quantum data exchange.

24.4 Summary and Discussion

Quantum repeaters could be one of the most demanding technologies to be developed on our way to fully functional quantum networks. The good news is that this journey has already started. In this chapter, we briefly looked at different classes of quantum repeaters, some simpler, but less efficient, and some more advanced but rather complicated, that have been proposed to achieve reliable QDT. Since early work on quantum repeaters, now over 30 years ago, there has been substantial progress in different related directions. On the experimental side, many different types of quantum memory units, as the essential component of many quantum repeater structures, have been developed and demonstrated. In parallel, quantum computing has seen a boost with the development of the first examples of noisy intermediate-scale quantum computing machines [57]. Examples of memory-based single-node repeater setups have also been successfully demonstrated [27, 58], and progress has been made towards the first entanglement-based quantum network in Netherlands [59]. There are also other related work on the networking, routing, and resource optimisation in quantum networks [60, 61], which I could not cover in this chapter. Overall, while we should not underestimate the challenges we may face along the way, all these efforts, and many others that I did not get to cite or mention, can eventually lead us to our grand vision of implementing the quantum Internet.

References

1 Kimble, H.J. (2008). The quantum internet. *Nature* 453: 1023–1030.
2 Munro, W.J., Azuma, K., Tamaki, K., and Nemoto, K. (2015). Inside quantum repeaters. *IEEE Journal of Selected Topics in Quantum Electronics* 21 (3): 78–90.
3 Pirandola, S., Andersen, U.L., Banchi, L. et al. (2020). Advances in quantum cryptography. *Advances in Optics and Photonics* 12 (4): 1012–1236.
4 Lo Piparo, N. and Razavi, M. (2015). Long-distance trust-free quantum key distribution. *Journal of Selected Topics in Quantum Electronics* 21: 6600508.
5 Barz, S., Kashefi, E., Broadbent, A. et al. (2012). Demonstration of blind quantum computing. *Science* 335 (6066): 303–308.
6 Zhang, Z. and Zhuang, Q. (2021). Distributed quantum sensing. *Quantum Science and Technology* 6 (4): 043001.
7 Wehner, S., Elkouss, D., and Hanson, R. (2018). Quantum internet: a vision for the road ahead. *Science* 362 (6412): eaam9288.
8 Liorni, C., Kampermann, H., and Bruß, D. (2021). Quantum repeaters in space. *New Journal of Physics* 23 (5): 053021.
9 Bennett, C.H., Brassard, G., Crépeau, C. et al. (1993). Teleporting an unknown quantum state via dual classical and Einstein-Podolsky-Rosen channels. *Physical Review Letters* 70: 1895–1899.
10 Sangouard, N., Simon, C., de Riedmatten, H., and Gisin, N. (2011). Quantum repeaters based on atomic ensembles and linear optics. *Reviews of Modern Physics* 83: 33–80.
11 Dür, W., Briegel, H.-J., Cirac, J.I., and Zoller, P. (1999). Quantum repeaters based on entanglement purification. *Physical Review A* 59: 169–181.
12 Bennett, C.H., Brassard, G., Popescu, S. et al. (1996). Purification of noisy entanglement and faithful teleportation via noisy channels. *Physical Review Letters* 76: 722–725.
13 Deutsch, D., Ekert, A., Jozsa, R. et al. (1996). Quantum privacy amplification and the security of quantum cryptography over noisy channels. *Physical Review Letters* 77: 2818; Erratum *Physical Review Letters* 80: 2022 (1998).
14 Muralidharan, S., Li, L., Kim, J. et al. (2016). Optimal architectures for long distance quantum communication. *Scientific Reports* 6 (1): 20463.
15 Duan, L.-M., Lukin, M.D., Cirac, J.I., and Zoller, P. (2001). Long-distance quantum communication with atomic ensembles and linear optics. *Nature* 414: 413–418.
16 Yu, Y., Ma, F., Luo, X.-Y. et al. (2020). Entanglement of two quantum memories via fibres over dozens of kilometres. *Nature* 578 (7794): 240–245.

17 Lloyd, S., Shahriar, M.S., Shapiro, J.H., and Hemmer, P.R. (2001). Long distance, unconditional teleportation of atomic states via complete Bell state measurements. *Physical Review Letters* 87: 167903.

18 Ritter, S., Nölleke, C., Hahn, C. et al. (2012). An elementary quantum network of single atoms in optical cavities. *Nature* 484: 195–200.

19 Duan, L.-M. and Monroe, C. (2010). Quantum networks with trapped ions. *Reviews of Modern Physics* 82: 1209.

20 Stute, A., Casabone, B., Schindler, P. et al. (2012). Tunable ion-photon entanglement in an optical cavity. *Nature* 485: 482.

21 Childress, L. and Hanson, R. (2013). Diamond NV centers for quantum computing and quantum networks. *MRS Bulletin* 38: 134–138.

22 Ishikawa, T., Fu, K.-M.C., Santori, C. et al. (2012). Optical and spin coherence properties of nitrogen-vacancy centers placed in a 100 nm thick isotopically purified diamond layer. *Nano Letters* 12 (4): 2083–2087.

23 Fang, K., Acosta, V.M., Santori, C. et al. (2013). High-sensitivity magnetometry based on quantum beats in diamond nitrogen-vacancy centers. *Physical Review Letters* 110: 130802.

24 Gao, W.B., Imamoglu, A., Bernien, H., and Hanson, R. (2015). Coherent manipulation, measurement and entanglement of individual solid-state spins using optical fields. *Nature Photonics* 9: 363–371.

25 Togan, E., Chu, Y., Trifonov, A.S. et al. (2010). Quantum entanglement between an optical photon and a solid-state qubit. *Nature Letters* 466: 730–735.

26 Bernien, H., Hensen, B., Pfaff, W. et al. (2013). Heralded entanglement between solid-state qubits separated by three meters. *Nature* 497: 86–90.

27 Bhaskar, M.K., Riedinger, R., Machielse, B. et al. (2020). Experimental demonstration of memory-enhanced quantum communication. *Nature* 580 (7801): 60–64.

28 Jelezko, F., Gaebel, T., Popa, I. et al. (2004). Observation of coherent oscillation of a single nuclear spin and realization of a two-qubit conditional quantum gate. *Physical Review Letters* 93 (13): 130501.

29 Waldherr, G., Wang, Y., Zaiser, S. et al. (2014). Quantum error correction in a solid-state hybrid spin register. *Nature* 506 (7487): 204–207.

30 Taminiau, T.H., Cramer, J., van der Sar, T. et al. (2014). Universal control and error correction in multi-qubit spin registers in diamond. *Nature Nanotechnology* 9 (3): 171.

31 Kalb, N., Reiserer, A.A., Humphreys, P.C. et al. (2017). Entanglement distillation between solid-state quantum network nodes. *Science* 356 (6341): 928–932.

32 Calsamiglia, J. and Lütkenhaus, N. (2001). Maximum efficiency of a linear-optical Bell-state analyzer. *Applied Physics B* 72 (1): 67–71.

33 Grice, W.P. (2011). Arbitrarily complete bell-state measurement using only linear optical elements. *Physical Review A* 84: 042331.

34 Ewert, F. and van Loock, P. (2014). 3/4-efficient Bell measurement with passive linear optics and unentangled ancillae. *Physical Review Letters* 113: 140403.

35 Olivo, A. and Grosshans, F. (2018). Ancilla-assisted linear optical Bell measurements and their optimality. *Physical Review A* 98: 042323.

36 Bruschi, D.E., Barlow, T.M., Razavi, M., and Beige, A. (2014). Repeat-until-success quantum repeaters. *Physical Review A* 90: 032306.

37 Zaidi, H.A. and van Loock, P. (2013). Beating the one-half limit of Ancilla-free linear optics bell measurements. *Physical Review Letters* 110: 260501.

38 Pan, J.-W., Bouwmeester, D., Daniell, M. et al. (2000). Experimental test of quantum nonlocality in three-photon Greenberger-Horne-Zeilinger entanglement. *Nature* 403: 515–519.

39 Razavi, M. (2018). *An Introduction to Quantum Communications Networks*, 2053–2571. Morgan & Claypool Publishers. ISBN 978-1-6817-4653-1.

40 Razavi, M., Piani, M., and Lütkenhaus, N. (2009). Quantum repeaters with imperfect memories: cost and scalability. *Physical Review A* 80: 032301.

41 Razavi, M., Thompson, K., Farmanbar, H. et al. (2009). Physical and architectural considerations in quantum repeaters. *Proceedings of the SPIE*, volume 7236, p. 723603, San Jose, CA.

42 Simon, C., de Riedmatten, H., Afzelius, M. et al. (2007). Quantum repeaters with photon pair sources and multimode memories. *Physical Review Letters* 98: 190503.

43 Afzelius, M., Simon, C., de Riedmatten, H., and Gisin, N. (2009). Multimode quantum memory based on atomic frequency combs. *Physical Review A* 79: 052329.

44 Guha, S., Krovi, H., Fuchs, C.A. et al. (2015). Rate-loss analysis of an efficient quantum repeater architecture. *Physical Review A* 92: 022357.

45 Briegel, H.-J., Dür, W., Cirac, J.I., and Zoller, P. (1998). Quantum repeaters: the role of imperfect local operations in quantum communication. *Physical Review Letters* 81 (26): 5932–5935.

46 Jiang, L., Taylor, J.M., Nemoto, K. et al. (2009). Quantum repeater with encoding. *Physical Review A* 79: 032325.

47 Jing, Y., Alsina, D., and Razavi, M. (2020). Quantum key distribution over quantum repeaters with encoding: using error detection as an effective postselection tool. *Physical Review Applied* 14: 064037.

48 Jing, Y. and Razavi, M. (2021). Simple efficient decoders for quantum key distribution over quantum repeaters with encoding. *Physical Review Applied* 15: 044027.

49 Azuma, K., Tamaki, K., and Lo, H.-K. (2015). All-photonic quantum repeaters. *Nature Communications* 6: 6787.

50 Varnava, M., Browne, D.E., and Rudolph, T. (2006). Loss tolerance in one-way quantum computation via counterfactual error correction. *Physical Review Letters* 97: 120501.

51 Buterakos, D., Barnes, E., and Economou, S.E. (2017). Deterministic generation of all-photonic quantum repeaters from solid-state emitters. *Physical Review X* 7: 041023.

52 Munro, W.J., Stephens, A.M., Devitt, S.J. et al. (2012). Quantum communication without the necessity of quantum memories. *Nature Photonics* 6: 771–781. https://doi.org/doi:10.1038/nphoton.2012.243.

53 Muralidharan, S., Kim, J., Lütkenhaus, N. et al. (2014). Ultrafast and fault-tolerant quantum communication across long distances. *Physical Review Letters* 112: 250501.

54 Namiki, R., Jiang, L., Kim, J., and Lütkenhaus, N. (2016). Role of syndrome information on a one-way quantum repeater using teleportation-based error correction. *Physical Review A* 94: 052304.

55 Ewert, F., Bergmann, M., and van Loock, P. (2016). Ultrafast long-distance quantum communication with static linear optics. *Physical Review Letters* 117: 210501.

56 Lee, S.-W., Ralph, T.C., and Jeong, H. (2019). Fundamental building block for all-optical scalable quantum networks. *Physical Review A* 100: 052303.

57 Arute, F., Arya, K., Babbush, R. et al. (2019). Quantum supremacy using a programmable superconducting processor. *Nature* 574 (7779): 505–510.

58 Langenfeld, S., Thomas, P., Morin, O., and Rempe, G. (2021). Quantum repeater node demonstrating unconditionally secure key distribution. *Physical Review Letters* 126: 230506.

59 Pompili, M., Hermans, S.L.N., Baier, S. et al. (2021). Realization of a multinode quantum network of remote solid-state qubits. *Science* 372 (6539): 259–264.

60 da Silva, F.F., Torres-Knoop, A., Coopmans, T. et al. (2021). Optimizing entanglement generation and distribution using genetic algorithms. *Quantum Science and Technology* 6 (3): 035007.

61 Chakraborty, K., Elkouss, D., Rijsman, B., and Wehner, S. (2020). Entanglement distribution in a quantum network: a multicommodity flow-based approach. *IEEE Transactions on Quantum Engineering* 1: 1–21.

25

Long-Distance Satellite-Based Quantum Communication

Yuan Cao[1,2,3] *and Juan Yin*[1,2,3]

[1] *Hefei National Research Center for Physical Sciences at the Microscale and School of Physical Sciences, University of Science and Technology of China, 230026, Anhui, Hefei, China*
[2] *CAS Center for Excellence in Quantum Information and Quantum Physics, University of Science and Technology of China, 230026, Hefei, China*
[3] *Hefei National Laboratory, University of Science and Technology of China, 230088, Hefei, China*

25.1 Introduction

Traditional public-key cryptography usually relies on the perceived computational intractability of certain mathematical functions. However, history shows that advances in cracking defeated many advances in classical cryptography. It might come as a surprise that the very fundamental principle of quantum mechanics was to solve this long-standing problem of information security that mathematicians have struggled with for centuries.

The first idea along these lines was proposed in the 1970s by Stephen Wiesner. He designed quantum banknotes using quantum two-state systems and conjugate encoding, which would be impossible to counterfeit. The principal drawback of Wiesner's idea was that it would require quantum information in superposition states to be held captive and kept coherent for long periods, which appears to be beyond even the technological capability of today, 50 years later. Inspired by Wiesner's idea, in the 1980s, Charles Bennett and Gilles Brassard put forward a feasible protocol of quantum key distribution (QKD) known as BB84 [1]. This protocol permits two distant communicating parties to produce a common, random string of secret bits, called a secret key. This key can then be used alongside the one-time pad encryption, which, as strictly proven by Shannon in 1949 [2], is an unconditionally secure method to encrypt and decrypt a message.

Due to photon loss in the channel, the range of secure QKD via direct transmission of single photons is limited to a few hundred kilometers. Unlike classical bits, the quantum signal of QKD cannot be noiselessly amplified due to the implications of the quantum no-cloning theorem [3], where the security of the QKD is rooted [4].

Photonic Quantum Technologies: Science and Applications, First Edition. Edited by Mohamed Benyoucef.

One potential solution to this problem is via the use of quantum repeaters [5]. However, quantum memories and repeaters' current state remains far from practical applications in realistic long-distance quantum communications [6].

Another more promising solution for global-scale QKD is to utilize satellites to conveniently connect two distant locations on Earth. An essential advantage of satellite-based free-space quantum communications is that the photon loss induced by atmospheric absorption and scattering predominantly occurs only in the lower ~10 km of the atmosphere, with about a 3 dB loss on a clear day. Most photon transmission is across a near-vacuum environment, with almost no absorption and decoherence. The loss caused by beam diffraction is approximately proportional to the square of the distance. By contrast, the losses in fiber channels are predominantly due to the absorption and scattering of the fiber medium, which is proportional to the exponent of the distance. Thus, for long communicating distances (typically hundreds to thousands of kilometers), the satellite-ground free-space channels will have advantages over fiber-based channels in terms of channel losses.

Furthermore, trusted-node topologies are employed for fiber networks to build the QKD network. However, ground-based nodes are at fixed locations, lack flexibility, and are vulnerable to constant surveillance and probes. By placing a satellite above the Earth's atmosphere, direct links can be established between ground stations and the satellite, thus enabling communication between any distant points on the planet. Moreover, more ambitious projects extending the network scale to deeper space, such as to the Moon or planets in the solar system, can also be realized with satellites.

In this chapter, we first briefly review the step-by-step ground-based feasibility studies. During these feasibility studies, the necessary toolbox for satellite-based long-distance quantum communications has been gradually developed, including robust and compact quantum light sources, narrow beam divergence, time synchronization, and rapid acquiring, pointing, and tracking (APT) technologies, which are the key to the optimization of the link efficiency and overcoming the atmospheric turbulence. Then, we introduced a series of satellite-ground quantum communication experiments with Micius and other quantum satellite projects worldwide. Finally, this chapter will end with an outlook on the future work needed to build global-scale practical quantum networks eventually.

25.2 Ground-Based Feasibility Studies

Thorough feasibility studies and simulations on the ground must be systematically implemented to verify the scientific possibilities, evaluate the risks, and develop the technologies before constructing and launching a costly satellite. The ground-based feasibility demonstration and the development of key technologies cover the following five aspects: overcoming the effective atmospheric thickness, testing the feasibility of satellite-ground channels, testing moving objects, time synchronization, and polarization maintenance and compensation.

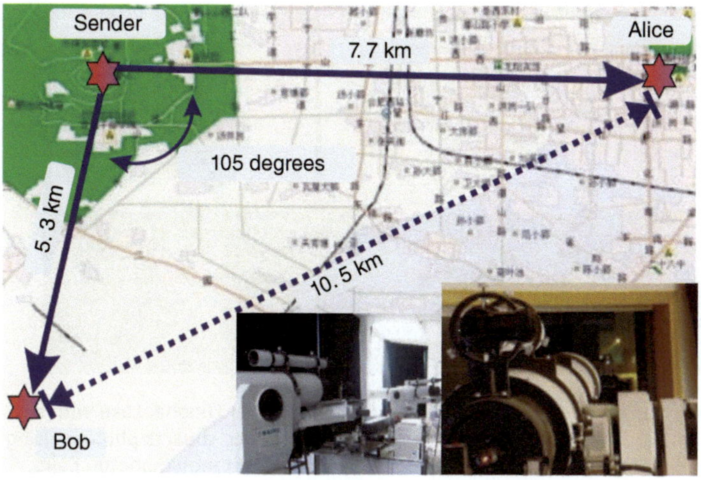

Figure 25.1 Overview of the Hefei 13 km entanglement distribution experiment. Source: Peng et al. [7]/with permission of American Physical Society.

Taking China's quantum science experiment satellite Micius as an example, the research team completed a series of ground-based feasibility verification experiments before the formal satellite project was approved. The first step involves verifying if the effective atmospheric thickness is favorable for the passage of single and entangled photons. In 2005, entangled photon pairs were bidirectionally distributed over Hefei city, one arm across 5.3 km and the other across 7.7 km, conclusively exceeding the effective atmospheric thickness (Figure 25.1) [7]. Narrowband (2.8 nm) filtering and time synchronization (with a precision of 20 ns) were employed to reduce the background counts from the noisy city environment. The two-photon count rates were 10 kHz and 150300 Hz at the sender and the receivers, respectively. The physical separation between the two receivers was 10.5 km, which enabled the performance of a space-like Bell test with a measured S value of 2.45 ± 0.09.

Three years later, single photons were transmitted over the Great Wall of China with an optical free-space distance of 16 km [8]. These two studies demonstrated that entanglement can still survive after both entangled photons have passed through the noisy ground atmosphere at a distance beyond the effective thickness of the aerosphere.

To experimentally verify the feasibility of satellite-based QKD through the one-downlink channel, Wang et al. conducted a complete verification study of the decoy-state QKD over the 97-km free-space link and demonstrated the possibility of achieving a high signal-to-noise ratio and overcoming the obstacle of a high-loss environment [9].

Two-downlink channels are required for the satellite-to-ground entanglement distribution, requiring two independent channels between three different locations. From 2008 to 2010, Pan's group performed a ground-based feasibility study in

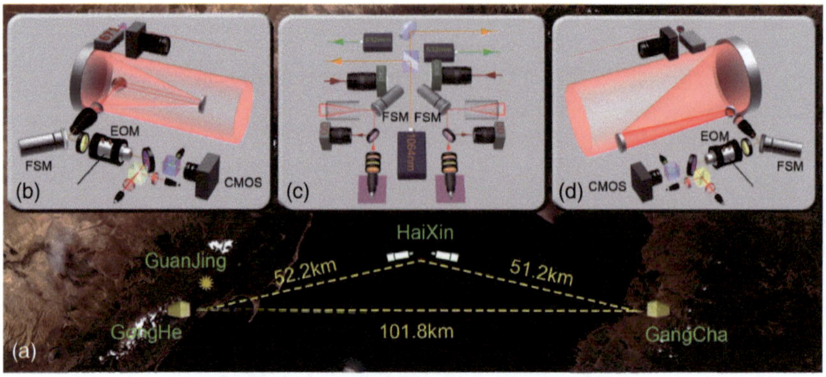

Figure 25.2 Bidirectional two-link entanglement distribution over Qinghai Lake with a distance of 101.8 km. (a) Satellite image of experiment site. (b) Alice collects photons using a 600 mm Cassegrain telescope. (c) Charlie generates and sends entangled photon pairs. (d) Bob collects photons using a 400 mm off-axis reflecting telescope. Source: Yin et al. [10]/with permission of Springer Nature.

Qinghai Lake, China [10]. In the study, entangled photon pairs were distributed over a two-link free-space channel with distances of 51.2 and 52.2 km to two receivers separated by 101.8 km (Figure 25.2). For the study, a crucial enabling technology was developed, i.e. closed-loop tracking, operated with a bandwidth and precision of 150 Hz and 3.5 μrad, respectively. The authors obtained a two-photon rate of 6.5 MHz and measured two-photon correlation functions, violating Bell's inequality by 2.4 standard deviations. The average two-link overall attenuation was 79.5 dB, which is higher than the estimated two-downlink loss based on the low-earth orbit (LEO) satellite.

For eventual satellite-based quantum communications, one should consider that the satellite performs rapid, relative angular motions concerning the ground stations, including unwanted random movement. For a typical LEO satellite at an altitude of 400–800 km, the angular velocity can reach 20 mrad/s, and the angular acceleration can reach 0.23 mrad/s^2. A verification environment that incorporates all possible motion modes and simulations of extreme events, including vibration, random motion, and attitude change, is highly desirable.

To this end, Wang et al. carried out two other experiments, in addition to the 97-km high loss one mentioned above, for the direct and full-scale experimental verifications toward satellite-ground QKD [9]. They implemented the experiments with a turntable and a hot-air balloon to simulate the platform in a rapidly moving orbit and the vibration, random motion, and attitude change related to the LEO satellite. The turntable (see Figure 25.3(a)) provides motion with a maximum angular velocity of 21 mrad/s and a maximum angular acceleration of 8.7 mrad/s^2. With a distance of 40 km between the transmitter and receiver, such a motion regime covers the possible range of motion parameters for a 400–800 km LEO satellite. The floating hot air balloon (see Figure 25.3(b)), with a distance of 20 km from the ground receiver, was employed as a randomly vibrating and floating platform, which afforded an average angular velocity of 10.5 mrad/s and an average angular

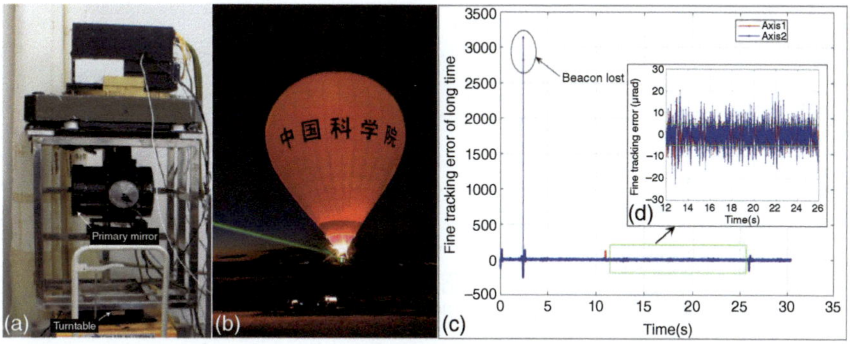

Figure 25.3 (a) The QKD transmitter is mounted on a turntable which has approximate dimensions of 500 mm×450 mm×600 mm (but 500 mm×500 mm×1,000 mm including the supporting metal frame). This is used for simulation of satellite orbiting. (b) Photo of the rising and erupting hot-air balloon in the floating platform experiment. (c) Tracking error in the hot-air balloon simulation. Long-time fine tracking error with the beacon light sometimes out of the field of view. Inset: Fine tracking error in the stabilized time area. When the beacon light is in the tracking field, tracking accuracy is about 5 μrad. Source: Wang et al. [9]/with permission of Springer Nature.

acceleration of 1.7 mrad/s^2 owing to its random motion. The balloon could perform random and dramatic motions, which positioned it out of the view of the field. This motivated the researchers to recapture the target rapidly, typically within 3–5 s (see Figure 25.3(c)). These verification environments incorporate more extreme events, including vibration, random movement, and attitude change, compared to what would suffer from an actual LEO satellite.

Furthermore, by utilizing the cube-corner retrorefector on satellites to simulate the quasi-single-photon source, one can verify the feasibility of establishing a quantum signal link between the satellite and the ground station [11]. Yin et al. performed an experimental simulation of a quasi-single-photon transmitter on the satellite with an average photon number of 0.85 per pulse and a full divergence angle of 38 μrad sending to the ground [12].

In addition to the comprehensive and systematic ground-based verification works dedicated to the Micius satellite, many other experiments have been conducted worldwide, and the scientific conclusions have been consolidated. For instance, in the experiment [13] performed in Vienna, one of the entangled photons was detected locally, while the other one was sent through free space across 7.8 km. A better testbed for free-space quantum communications – the link between the Canary islands of La Palma and Tenerife – was employed by Schmitt-Manderbach et al. [14] for testing decoy-state QKD over a distance of 144 km, as shown in Figure 25.4. Later, Ursin and coworkers adopted the same experimental configuration as in [13] to send triggered single photons across a distance of 144 km (one link) using a free-space link between the Canary Islands La Palma and Tenerife [15]. Due to the various atmospheric influences at such a long distance scale, the apparent bearing of the receiver station varied in tens of seconds to minutes. To maximize and stabilize the link efficiency, an active stabilization of the optical link was

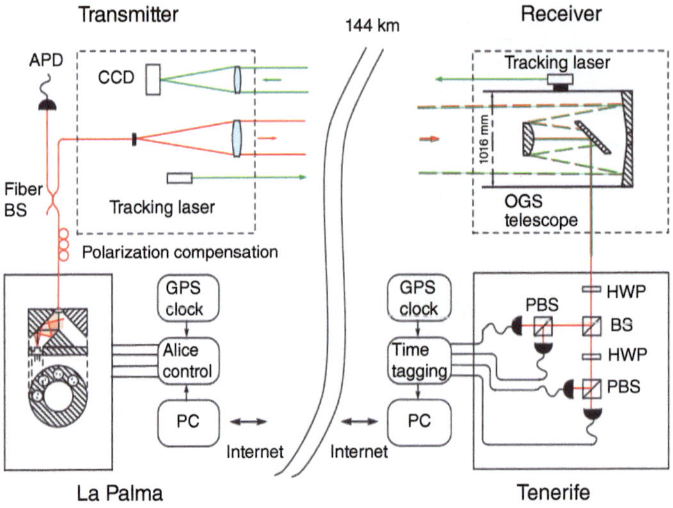

Figure 25.4 Schematics of the 144km free-space decoy-state quantum key distribution. BS, beam splitter; PBS, polarizing beam splitter; HWP, half-wave plate; APD, avalanche photo diode. Source: Schmitt-Manderbach et al. [14], American Physical Society.

implemented via a closed-loop tracking system to correct the beam drifts induced by atmospheric changes, reducing the beam drift from 70 μrad (10 m) to 7 μrad (1 m). Using this one-link free-space channel, quantum teleportation has also been demonstrated [16] with a channel attenuation of approximately 30 dB. A further experiment from the same group upgraded the ß-barium borate (BBO) crystal into a more efficient down-conversion crystal, i.e. periodically poled $KTiOPO_4$, which generated entangled photons at ~1 MHz and sent the two photons through the free-space channel across 144 km [17, 18].

To test the APT with moving objects, Nauerth et al. conducted a QKD experiment from an airplane to the ground [19]. The airplane was moving at a speed of 290 km/h, at a distance of 20 km, which corresponds to an angular velocity of 4 mrad/s. The transmitting beam was narrowed by a divergence of 180 μrad. To establish a stable link with this divergence, fine-pointing assemblies were implemented and optimized on both sides, with a precision upper bound of 150 μrad. With these efforts, the experiment yielded an asymptotically secure key at a rate of 7.9 bit/s.

In addition to employing moving platforms to demonstrate the feasibility of downlink channels, uplink channels were verified by the research team from Canada with a truck and airplane. In 2015, they reported the first demonstration of QKD from a stationary transmitter to a receiver platform located on a moving truck [20]. In this experiment, QKD was implemented with a moving receiver at an angular speed similar to that of a satellite at an altitude of 600 km. Furthermore, they equipped a receiver prototype on an airplane to demonstrate QKD via an uplink channel [21], as shown in Figure 25.5 They specifically designed the receiver prototype to consist of many components that were compatible with the environment and resource constraints of a satellite. Their above ground-based feasibility experiments on uplink

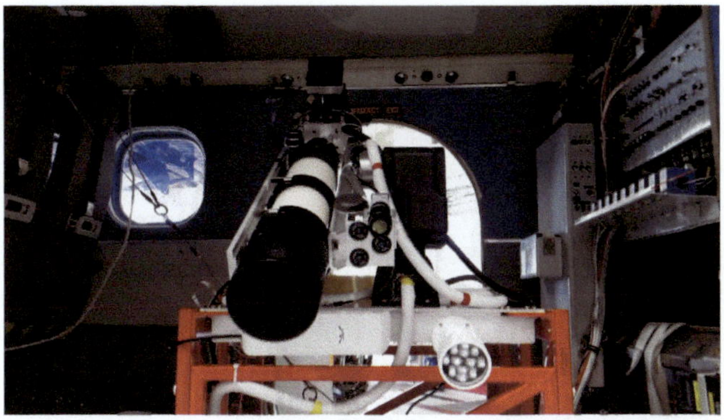

Figure 25.5 The receiver apparatus facing out the port-side door of the NRC Twin Otter research aircraft. Source: Pugh et al. [21]/with permission of IOP Publishing.

channels with moving platforms provided solid technical support for the follow-up satellite project – QEYSSat.

In 2017, scientists from Japan used a classical laser source on the LEO satellite SOCRATES to test the feasibility of the satellite-to-ground quantum-limited link [22]. Additionally, Günthner et al. [23] completed a similar experiment using the classical laser source from a geostationary satellite.

25.3 Satellite-Based Quantum Communication Experiments with Micius

The Micius program was officially approved in 2011. Construction of the prototype satellite started in 2012 and was completed in 2014. After that, the project turned to build the flight model of the satellite, which was completed in November 2015. Finally, on August 16, 2016, the Micius satellite was successfully launched by the Long March-2D rocket from the Jiuquan Satellite Launch Centre, China. The orbit was circular and sun-synchronous, with an altitude of 500 km.

A double-decker design is employed in the Micius satellite (see Figure 25.6). The payloads for the science experiments are composed of two optical transmitters (transmitter 1 & 2), a space-borne entangled photon source (the upper layer of the satellite), an experimental control processor, and two APT control boxes (the lower layer of the satellite), as shown in Figure 25.6(a).

25.3.1 Satellite-to-Ground Quantum Key Distribution

After launching the Micius satellite, the first goal was to establish a space-ground quantum link and perform the satellite-to-ground QKD [24], as shown in Figure 25.7. The downlink has reduced beam spreading compared to the uplink because the beam wandering occurs at the end of the transmission path, which is typically smaller than the effect from the beam diffraction. The 300-mm-aperture

Figure 25.6 The full view of the Micius satellite and main payloads [63]. (a) The photograph of the Micius satellite before launching. (b) The transmitter 1 for QKD, entanglement distribution and teleportation. (c) The transmitter 2 specially for entanglement distribution. (d) The experimental control box. (e) The entangled-photon source.

telescope equipped in the satellite produced a near-diffraction-limited far-field divergence of about 10 μrad. Such a narrow divergence beam from the fast-traveling satellite (with a speed of about 7.6 km/s) requires a fast and precise APT. Tracking accuracy of approximately 1.2 μrad was achieved, which is much smaller than the beam divergence. It should be noted that due to the quiet environment in out space, the tracking accuracy is better than the previous ground tests, which deliberately set a more stringent condition. A diffraction loss of approximately 22 dB was obtained at 1200 km, whereas the loss due to pointing error was below 3 dB. Additionally, the loss due to atmospheric absorption was 3–8 dB.

Apart from being used in the APT, the beacon laser also serves for obtaining the arrival time of the single photons in order to compensate for the space-ground clock drift. The obtained time synchronization jitter is 0.5 ns, which is useful for filtering the background noise. Additionally, a spectral bandpass filter is used in the receiver to reduce the background scattering. Finally, a motorized half-wave plate is used to dynamically compensate for the time-dependent photon polarization rotation during the satellite passage.

In the experiment, the scientists employed the decoy-state BB84 protocol, a form of QKD that uses weak coherent pulses at high channel loss and is immune to photon-number-splitting eavesdropping. Since September 2016, QKD has been performed routinely under good atmospheric conditions. The QKD experiments performed on 23 different days, with different physical distances between the satellite and the ground stations. The shortest satellite-to-station distance depends on the highest altitude angle of the day, and varies from 507.0 km at 85.7° to 1,034.7 km at 25.0°. The sifted key that is obtained has a peak key rate of 40.2 kbit/s at 530 km and decreases for larger distance, for instance, to 1.2 kbit/s at 1,034.7 km. The quantum bit error rates are measured to be 1%-3%. By performing error correction and privacy amplification, the secure final key was 300,939 bits when the statistical failure probability was set to be 10^{-9}, corresponding to a key rate of 1.1 kbits/s. This

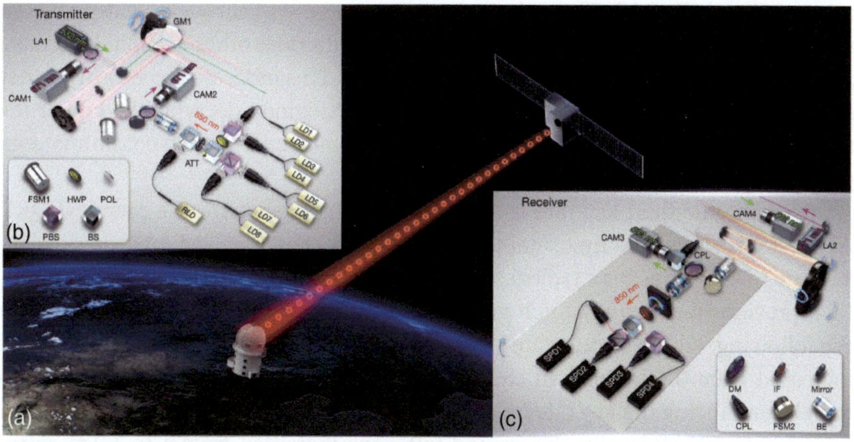

Figure 25.7 Illustration of the experimental set-up. (a) Overview of the satellite-to-ground quantum key distribution. (b) Schematic of the decoy-state QKD transmitter, one of the satellite's payloads. (c) Schematic of the decoy-state QKD decoder at the Xinglong ground station, which is equipped with a 1,000-mm-aperture telescope. Source: Liao et al. [24]/with permission of Springer Nature.

key rate is around 20 orders of magnitudes greater than that expected using an optical fiber of the same length.

Based on these results, the scientists further show the feasibility of an intercontinental quantum communication network by making the Micius the trusted relay [25]. A secret key is created between China and Europe at locations separated by 7600 km on Earth, as shown in Figure 25.8.

Let us denote the random keys shared between Micius and Xinglong as MX, and between Micius and Graz as MG. Micius can simply perform a bitwise exclusive OR operation (⊕) between MX and MG of the same string length, which then yields

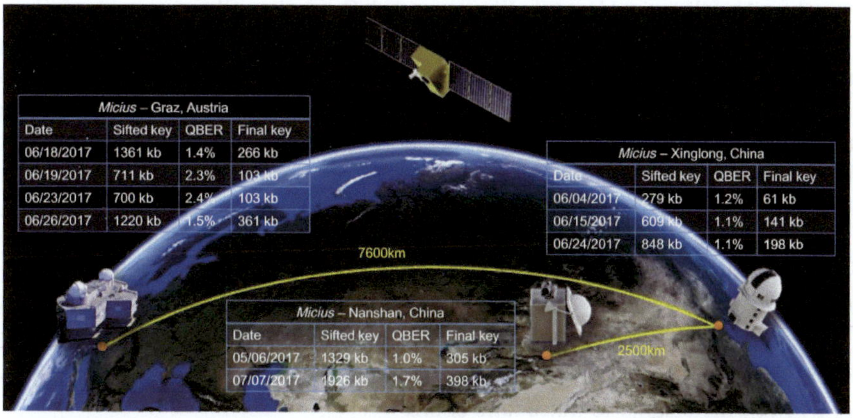

Figure 25.8 Illustration of the three cooperating ground stations (Graz, Nanshan, and Xinglong). Listed are all paths used for key generation and the corresponding final key length. Source: Liao et al. [25]/with permission of American Physical Society.

a new string: MX⊕MG. Then, the new string can be sent through a classical communications channel to Xinglong or Graz, which decodes other original keys using another exclusive OR (i.e. MG=(MX⊕MG)⊕MX). This process can be easily understood since Micius uses MX to encrypt MG and Xinglong decrypts the cipher text to recover MG, shared with Graz. Such a key is known only to both communicating parties and the satellite, but no fourth party.

Improving the final key rate is always one of main goals for the practical QKD. About 3 years after the first satellite-to-ground QKD, it achieved an average secret key rate of 47.8 *K*bps for a typical satellite pass, which is more than 40 times higher than previous results [24]. Such great improvement of the final key rate is due to the following: (i) the signal state ratio, increased from 0.5 to 0.72, the Z-base ratio increased from 0.5 to 0.889 at the satellite and 0.5 to 0.9 on the ground station, thereby enhancing the key rate by 2.34 times; (ii) the repetition frequency increased from 100 to 200 MHz; (iii) the ground telescope increased from 1 to 1.2 m, corresponding to an increment about 1.5 times; (iv) the QBER is reduced and the raw key size increased to about 2 times; (v) the ground coupling efficiency increased from 14%-40%, corresponding to an increment of about three times [26].

25.3.2 Satellite-Based Quantum Entanglement Distribution

Long-distance entanglement distribution is essential for both foundational tests of quantum physics and scalable quantum networks. Owing to channel loss, however, the previously achieved distance was limited to ~300 km [28]. This is mainly due to the photon loss in the channel (optical fibers or terrestrial free space), which generally scales exponentially with the channel length. For example, using a bidirectional distribution of an entangled source of photon pairs with a 10-MHz count rate directly through two 600-km telecommunication fibers with a loss of 0.2 dB/km, we can only obtain 10^{-17}/s two-photon coincidence events.

Three ground stations are cooperating with the satellite, located at Delingha in Qinghai, Nanshan in Urumqi, Xinjiang, and Gaomeigu Observatory in Lijiang, Yunnan, for the mission of entanglement distribution [27]. At Delingha, Lijiang and Nanshan station, the receiving telescopies has diameters of 1200, 1800, and 1200 mm, respectively. The physical distance between Delingha and Lijiang (Nanshan) is 1203 km (1120 km). The separation between the orbiting satellite and these ground stations varies from 500 to 2000 km. The entanglement distribution was achieved both between Delingha and Lijiang and between Delingha and Nanshan. The experiment involving Delingha and Lijiang is described below.

The satellite (Figure 25.6(e)) emits 5.9 million entangled photon pairs per second, which are then sent out using two telescopes. Using a reference laser on the satellite, the overall two-downlink channel attenuation can be measured in real time, which varies from 64 to 82 dB. The experiment observed an average two-photon count rate of 1.1 Hz, with a signal-to-noise ratio of 8:1.

The distributed entangled two photons were then employed for a Bell test. The experimental configuration and Pockels cells used were fast enough to close the locality and freedom-of-choice loopholes. The Bell test ran 1167 trials during an

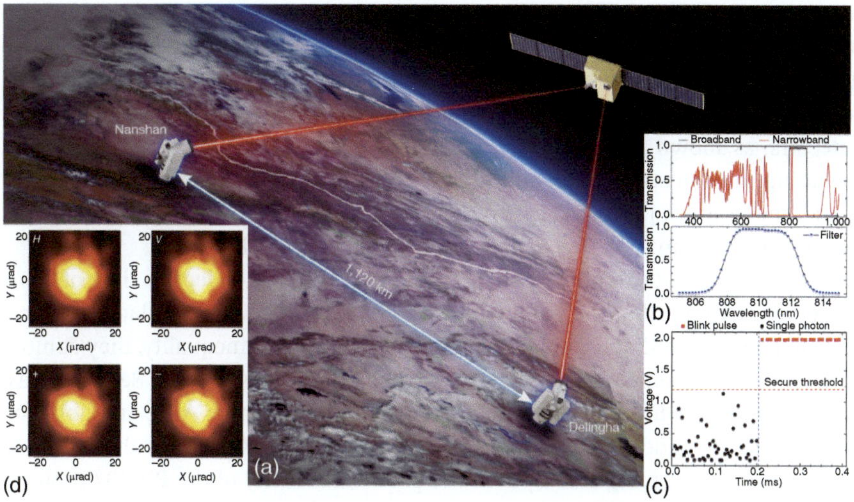

Figure 25.9 Overview of the experimental setup of quantum key distribution based on the entanglement distribution. Source: Yin et al. [30]/with permission of Springer Nature. (a) An illustration of the Micius satellite and two ground stations. Image credit: Fengyun-3C/Visible and Infrared Radiometer, with permission (2020). (b), (c), and (d) are monitoring and filtering against side channels. (b) The transmission of broad-bandwidth and narrow-bandwidth filters. (c) The output of monitoring circuit with/without blinding attack. Without blinding attack, the outputs are random avalanching single-photon-detection signals (black dots). With blinding attack (starting from 0.2 ms), the output signals are at around 2 V, which is clearly above the security threshold, thus triggering the security alarm. (d) The system detection efficiency of the four polarizations in the spatial domain. With the spatial filter, the four efficiencies are identical.

effective time of 1059 s. A violation of the Bell inequality of 2.374 ± 0.093 by 4 standard deviations was obtained. The result again confirms the nonlocal feature of entanglement and excludes the models of reality, which sits on the notions of locality and realism, in a new space scale with thousands of kilometers.

After the first satellite-based entanglement distribution [27, 29], a later experiment of entanglement-based QKD was [30] performed between the ground station of Delingha and Nanshan with a physical separation of 1120 km. The receiving efficiencies were considerably improved using a higher efficiency telescope and follow-up optics.

Entanglement-based QKD is particularly attractive because of its inherent source-independent security [31, 32] where the security can be established without any assumption on trusted relay.

In particular, a special effort of [30] was made to ensure its implementation is practically secure against all known side channels. Due to the source-independent nature of the entanglement-based QKD, the system was immune to any loophole in the source, and all left is to ensure the security on the detection sides in both ground stations. In general, the side channels, known and to be known, on the detection primarily violate the assumption of fair-sampling. Note that the concept of "fair-sampling" here refers to the consistency of different receiving detectors on spatial and spectral freedom of the arrival beams, which differs from the experimental requirement of addressing the "fair-sampling loophole" of the Bell test. Experimentally, Yin et al. ensured the validity of the fair-sampling by filtering in different degrees of freedom, including frequency, spatial and temporal modes. Also, countermeasures were taken for the correct operation of the single-photon detectors, as shown in Figure 25.9. They considered all known detection attacks, including the detector-related attacks [33–35], wavelength-dependent attack [36], spatial-mode attack [37], and other possible side-channels. Consequently, the secret key, generated by this QKD system, is secured under realistic devices.

By running 1021 trials of the Bell test during an effective collection time of 226 s, Yin et al. observed that the parameter S was 2.56 ± 0.07 with a violation of local realism by 8 standard deviations. Having violated the Bell's inequality, they demonstrated the entanglement-based QKD using the protocol presented by Bennett, Brassard, and Mermin and coworkers in 1992 (BBM92), where both Alice and Bob took measurements randomly along the H/V and +/- basis [38]. Due to the efforts to ensure the fair-sampling assumption, the practical security of the BBM92 protocol is compatible with the E91 [39].

Within 3100 s data collection time, 6208 initial coincidences were obtained, which gave 3100 bits of sifted keys with 140 erroneous bits. The quantum bit error rate was $4.5\% \pm 0.4\%$. After error correction and privacy amplification, the secure key rate of 0.43 bits/s in the asymptotic limit of infinite long key and a finite secret-key rate of 0.12 bits/s were obtained. More details on the final key rate are discussed in Ref. [40]. The results increase the secure distance of practical QKD for ground users by 10 times to the order of a thousand kilometers, representing a key step toward the Holy Grail of cryptography. Note that with the newly developed entangled photon source with a 1 GHz generation rate [41] increased the secure key rate by about 2 orders of magnitude directly.

25.3.3 Ground-to-Satellite Quantum Teleportation

Utilizing the Micius satellite, the scientists performed quantum teleportation of a single photon from an observatory ground station in Ngari to the satellite, an uplink (for an overview, see Figure 25.10(a)) [42]. The uplink teleportation experiment has two additional challenges compared with the previous downlink work. First, the teleportation of an independent single photon requires a multi-photon interferometry with a coincidence count rate several orders of magnitude lower than typical single- or two-photon experiments. Second, the atmospheric turbulence in the uplink channel occurs at the beginning of the transmission path, which causes beam wandering and broadening that increases the amount of spreading of the traveling beams.

A very compact design of ultra-bright four-photon sources, which used both collinear and noncollinear spontaneous parametric down-conversion (SPDC) (see Figure 25.10(b)), was employed to meet the extreme condition of the field experiment in Ngari. The four-photon interferometry system was integrated into a compact platform with a dimension of 460 mm×510 mm×100 mm and weighing less than 20 kg. The pump laser was used for two identical multi-photon modules

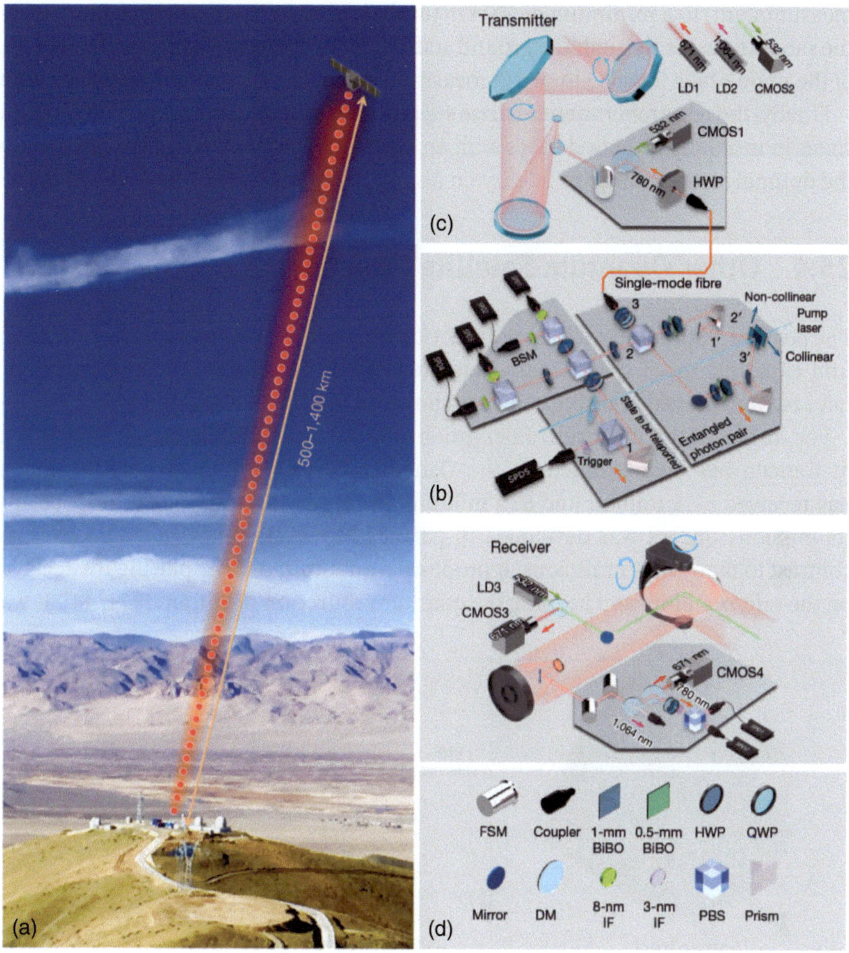

Figure 25.10 Overview of the setup for ground-to-satellite quantum teleportation of a single photon over distances of up to 1,400 km. Source: Ren et al. [42]/with permission of Springer Nature. (a) A schematic of the satellite is overlaid on a photograph of the Ngari ground station in Tibet. The separation between the satellite and the ground station varies from about 500 to 1,400 km during quantum teleportation. (b) The compact multi-photon setup for teleportation at the ground station. (c) The transmitter at the ground station. (d) The receiver on the satellite.

built sequentially, where the multiplexed four-photon count rate was 8200/s. Note that with the newly developed SPDC source [43] improved the four-photon count rate by a factor of ~10.

The teleported single photons from a single-mode fiber were transmitted through a 130-mm-diameter off-axis reflecting telescope (Figure 25.10(c)) and received by a 300-mm-diameter telescope in the satellite (Figure 25.10(d)). Both the transmitter and receiver were equipped with APT systems to optimize the uplink efficiency. The physical distance between the ground station and the satellite varies from a maximum of 1400 km (at an altitude angle of 14.5°, the starting point of our measurement) to a minimum of 500 km (at the highest altitude angle of 76.0°, when the satellite passes through the ground station above the top). Here, the channel loss of the uplink falls from 52 to 41 dB, measured using a high-intensity reference laser.

Finally, the researchers demonstrate successful quantum teleportation of six input states in mutually unbiased bases with an average fidelity of 0.80 ± 0.01, well above the optimal state-estimation fidelity on a single copy of a qubit (the classical limit).

25.4 Other Quantum Satellite Projects

The positive results and exciting prospect kick-started an international race on quantum experiments in space. Many satellite projects for quantum communications have been approved and supported, as shown in Figure 25.11.

For instance, the Quantum Encryption and Science Satellite (QEYSSat) project in Canada has been studied by the Canadian Space Agency since 2010, and it has received \$1.5 million and \$30 million funding in 2017 and 2019, respectively. Its mission concept was developed in partnership with Honeywell Aerospace. In contrast to many other missions, it proposes a quantum uplink, placing the receiver on the microsatellite and keeping the quantum source on the ground [21, 44]. It was

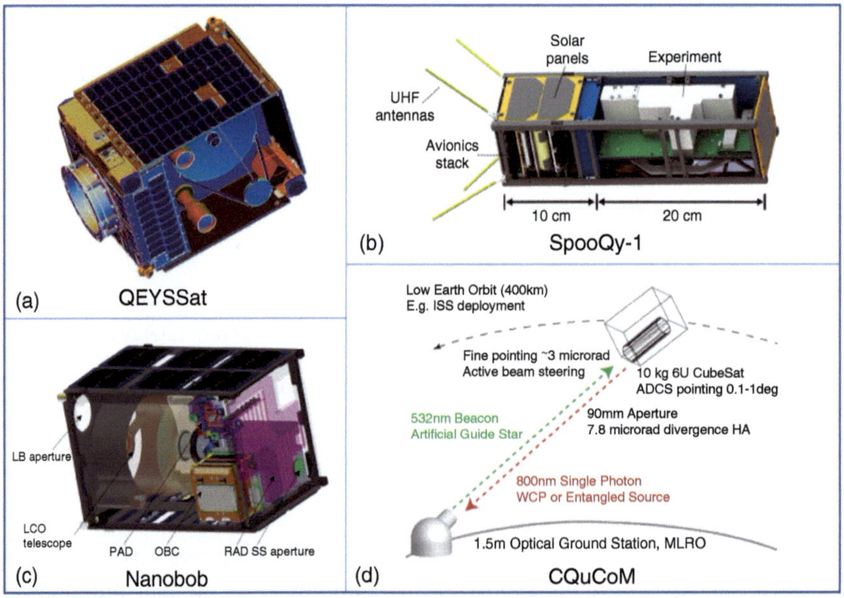

Figure 25.11 Other quantum satellite plans besides Micius. (a) The Quantum Encryption and Science Satellite (QEYSSat) project in Canada. Source: Jennewein et al. [44]/with permission of SPIE. (b) The 3U Cubesat involving an entangled photon source developed by the group in National University of Singapore. Source: Villar et al. [45]/with permission of Optica Publishing Group. (c) The CubeSat-based mission concept Nanobob proposed by researchers of France and Austria. Source: Kerstel et al. [46] Springer Nature/CC BY 4.0. (d) The CubeSat Quantum Communications Mission (CquCoM) jointly undertaking by a joint research team. Source: Oi et al. [47].

reported in [48] that NASA plans to build a quantum satellite link, which was called "Marconi 2.0." The main idea behind Marconi 2.0 is to establish a space-based quantum link between Europe and North America by the mid- to late-2020s.

In addition to the traditional "big-space" paradigm of satellite, many other teams worldwide have started a new paradigm based on nanosatellites, even the CubeSat standard [46, 47].

A group at the National University of Singapore has been committed to designing and developing quantum sources based on nanosatellites and CubeSat platforms. They developed a correlated photon pair source as the pathfinder for their future plan of space applications [49–51]. Recently, they developed an entangled photon-pair source onboard a 3U CubeSat, SpooQy-1, which was launched successfully to the ISS in April 2019. The CubeSat was then deployed into orbit from ISS on June 17, 2019 [45].

Bedington et al. provided a table of notable satellite QKD proposals [52]. In addition to the university consortia and national agencies, the international space race also involves private companies, e.g. QKDSat (ArQit) [53] and QUARTZ [54]. Also, a more ambitious quantum communication infrastructure project is taking shape in Europe. The European Quantum Communication Infrastructure (EuroQCI) initiative aims to build a secure quantum communication infrastructure that will span the entire European Union (EU), including its overseas territories [55].

25.5 Outlook

Although the Micius satellite greatly enhances the scale and capability of quantum experiments in space, Micius only marks the beginning. There is much space for improvement.

One of the main drawbacks of the current satellite-based quantum communication missions is that they can only work at night, significantly limiting their practical application. Therefore, one of the following steps toward the practical satellite-based quantum secure communication network is to demonstrate daylight free-space quantum communication. The main challenge is the intense background noise from the scattered sunlight, typically five orders of magnitude greater than the background noise during nighttime. In general, it is suggested that the signal-to-noise ratio can be by combining detection timing, narrowband filters, and spatial filtering.

A preliminary verification of free-space QKD in daylight under conditions of high channel loss (~48 dB) over 53 km was reported by Liao et al. [56]. To increase the signal-to-noise ratio, first, Liao et al.. chose a working wavelength of 1,550 nm. Compared to 800 nm, the telecom-band wavelength has the transmission slightly higher, and Rayleigh scattering ~14 times smaller. Further, the sunlight intensity at 1,550 nm is ~5 times weaker than that at 800 nm. Second, free-space single-mode fiber-coupling was developed with an efficiency of 30% for the indoor test and 5% in the outdoor. The field of view for the receiving system is reduced below 10 μrad to reduce the background noise. Finally, ultralow-noise up-conversion single-photon detectors were used with a built-in spectral filtering employing volume Bragg grating with a bandwidth of 0.16 nm. Such narrowband filtering reduces noise by a factor of ~100 compared to the 3–10 nm filters used in previous experiments at night. A combination of the three key toolbox enabled a decoy-state QKD with a final key rate of 20–400 bits per second, where the variation was mainly due to the atmospheric environment.

Due to the longer distances and the associated diffraction loss, new techniques need to be developed to increase the link efficiency in the future, including large-size telescopes, better APT systems, and wavefront correction through adaptive optics [57–62].

A LEO satellite alone is not enough to support the construction of the global-scale quantum communication network. In general, attention should be paid to two aspects: increasing the number of satellites and raising the orbital altitude. It is necessary to build a quantum constellation combining LEO satellites and high-earth-orbit (HEO) satellites. Compared to the LEO satellites, high-orbit satellites can provide a much longer service time and broader coverage. Combining a high-orbit satellite and multiple LEO satellites can form a quantum constellation for global services.

Through these further efforts, we could envision a global quantum communication infrastructure with quantum constellation and ground-based fiber networks, as shown in Figure 25.12. Fiber-based network on the ground provides secure communication services for distance cities. Meanwhile, a quantum constellation with LEO

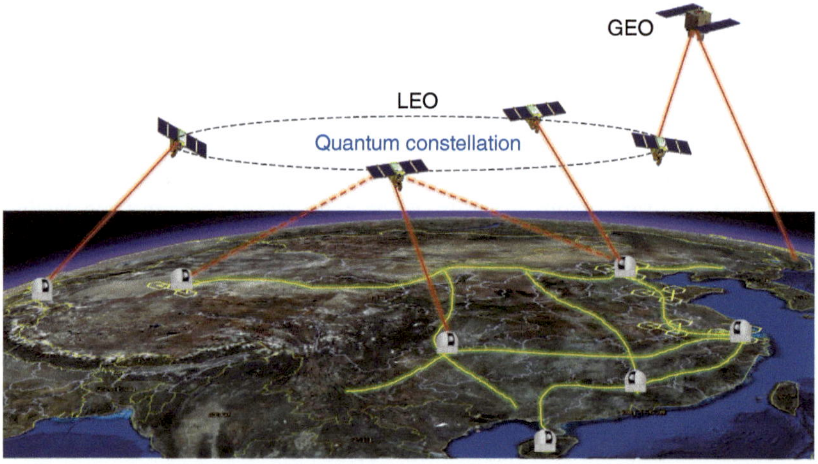

Figure 25.12 The roadmaps toward the global quantum communication network. Intra-city metropolitan networks will be created using fibers. Quantum repeaters can connect the metropolitan networks. Long-distance and intercontinental quantum communication will be realized via satellite-based quantum channels [63].

and the high-orbit satellites connecting key nodes on the fiber networks and movable nodes, even ships in the ocean. Based on the above analysis, we suggest that the simplest quantum constellation should include at least three low-orbit satellites and one high-orbit satellite. In this configuration, assuming that not less than 100 ground stations need to be covered, each ground station needs more than 50 times of QKD links with satellite per year and can obtain about 2 Mbits for each satellite passage. Then, each station can obtain 100 Mbits per year, and the quantum constellation can output about 10 Gbits of keys per year totally, which can support the basic function of voice communication. Also, HEO satellites can provide 24-hr QKD services at a key rate of 1 kbps for some important areas, which can provide the basic needs of text communication.

References

1 Bennett, C.H. and Brassard, G. (1984). Quantum cryptography: public key distribution and coin tossing. *Proceedings of the IEEE International Conference on Computers, Systems and Signal Processing*, pp. 175–179. New York: IEEE Press.

2 Shannon, C.E. (1949). Communication theory of secrecy systems. *Bell System Technical Journal* 28 (4): 656–715.

3 Wootters, W.K. and Zurek, W.H. (1982). A single quantum cannot be cloned. *Nature* 299: 802.

4 Gisin, N., Ribordy, G., Tittel, W., and Zbinden, H. (2002). Quantum cryptography. *Reviews of Modern Physics* 74: 145–195.

5 Briegel, H.-J., Dür, W., Cirac, J.I., and Zoller, P. (1998). Quantum repeaters: the role of imperfect local operations in quantum communication. *Physical Review Letters* 81: 5932–5935.

6 Yang, S.J., Wang, X.-J., Bao, X.-H., and Pan, J.-W. (2016). An efficient quantum light-matter interface with sub-second lifetime. *Nature Photonics* 10: 381.

7 Peng, C.-Z., Yang, T., Bao, X.-H. et al. (2005). Experimental free-space distribution of entangled photon pairs over 13 km: towards satellite-based global quantum communication. *Physical Review Letters* 94: 150501.

8 Jin, X.-M., Ren, J.-G., Yang, B. et al. (2010). Experimental free-space quantum teleportation. *Nature Photonics* 4 (6): 376–381.

9 Wang, J.-Y., Yang, B., Liao, S.-K. et al. (2013). Direct and full-scale experimental verifications towards ground-satellite quantum key distribution. *Nature Photonics* 7: 387.

10 Yin, J., Ren, J.-G., Lu, H. et al. (2012). Quantum teleportation and entanglement distribution over 100-kilometre free-space channels. *Nature* 488: 185.

11 Villoresi, P., Jennewein, T., Tamburini, F. et al. (2008). Experimental verification of the feasibility of a quantum channel between space and earth. *New Journal of Physics* 10 (3): 033038.

12 Yin, J., Cao, Y., Liu, S.-B. et al. (2013). Experimental quasi-single-photon transmission from satellite to earth. *Optics Express* 21 (17): 20032–20040.

13 Resch, K.J., Lindenthal, M., Blauensteiner, B. et al. (2005). Distributing entanglement and single photons through an intra-city, free-space quantum channel. *Optics Express* 13: 202.

14 Schmitt-Manderbach, T., Weier, H., Fürst, M. et al. (2007). Experimental demonstration of free-space decoy-state quantum key distribution over 144 km. *Physical Review Letters* 98: 010504.

15 Ursin, R., Tiefenbacher, F., Schmitt-Manderbach, T. et al. (2007). Entanglement-based quantum communication over 144km. *Nature Physics* 3: 481.

16 Ma, X.S., Herbst, T., Scheidl, T. et al. (2012). Quantum teleportation over 143 kilometres using active feed-forward. *Nature* 489: 269.

17 Fedrizzi, A., Herbst, T., Poppe, A. et al. (2007). A wavelength-tunable fiber-coupled source of narrowband entangled photons. *Optics Express* 15 (23): 15377–15386.

18 Fedrizzi, A., Ursin, R., Herbst, T. et al. (2009). High-fidelity transmission of entanglement over a high-loss free-space channel. *Nature Physics* 5 (6): 389–392.

19 Nauerth, S., Moll, F., Rau, M. et al. (2013). Air-to-ground quantum communication. *Nature Photonics* 7: 382.

20 Bourgoin, J.-P., Higgins, B.L., Gigov, N. et al. (2015). Free-space quantum key distribution to a moving receiver. *Optics Express* 23 (26): 33437–33447.

21 Pugh, C.J., Kaiser, S., Bourgoin, J.-P. et al. (2017). Airborne demonstration of a quantum key distribution receiver payload. *Quantum Science and Technology* 2 (2): 024009.

22 Takenaka, H., Carrascocasado, A., Fujiwara, M. et al. (2017). Satellite-to-ground quantum-limited communication using a 50-kg-class microsatellite. *Nature Photonics* 11 (8): 502–508.

23 Günthner, K., Khan, I., Elser, D. et al. (2017). Quantum-limited measurements of optical signals from a geostationary satellite. *Optica* 4 (6): 611–616.

24 Liao, S.-K., Cai, W.-Q., Liu, W.-Y. et al. (2017). Satellite-to-ground quantum key distribution. *Nature* 549: 43–47.

25 Liao, S.-K., Cai, W.-Q., Handsteiner, J. et al. (2018). Satellite-relayed intercontinental quantum network. *Physical Review Letters* 120: 030501.

26 Chen, Y.-A., Zhang, Q., Chen, T.-Y. et al. (2021). An integrated space-to-ground quantum communication network over 4,600 kilometres. *Nature* 589: 214–219.

27 Yin, J., Cao, Y., Li, Y.H. et al. (2017). Satellite-based entanglement distribution over 1200 kilometers. *Science* 356: 1140.

28 Inagaki, T., Matsuda, N., Tadanaga, O. et al. (2013). Entanglement distribution over 300 km of fiber. *Optics Express* 21 (20): 23241–23249.

29 Yin, J., Cao, Y., Li, Y.-H. et al. (2017). Satellite-to-ground entanglement-based quantum key distribution. *Physical Review Letters* 119: 200501.

30 Yin, J., Li, Y.-H., Liao, S.-K. et al. (2020). Entanglement-based secure quantum cryptography over 1,120 kilometres. *Nature* 582: 501–505.

31 Koashi, M. and Preskill, J. (2003). Secure quantum key distribution with an uncharacterized source. *Physical Review Letters* 90: 057902.

32 Ma, X., Fung, C.-H.F., and Lo, H.-K. (2007). Quantum key distribution with entangled photon sources. *Physical Review A* 76: 012307.

33 Zhao, Y., Fred Fung, C.-H., Qi, B. et al. (2008). Experimental demonstration of time-shift attack against practical quantum key distribution systems. *Physical Review A* 78: 042333.

34 Lydersen, L., Wiechers, C., Wittmann, C. et al. (2010). Hacking commercial quantum cryptography systems by tailored bright illumination. *Nature Photonics* 4 (10): 686–689.

35 Weier, H., Krauss, H., Rau, M. et al. (2011). Quantum eavesdropping without interception: an attack exploiting the dead time of single-photon detectors. *New Journal of Physics* 13 (7): 073024.

36 Li, H.-W., Wang, S., Huang, J.-Z. et al. (2011). Attacking a practical quantum-key-distribution system with wavelength-dependent beam-splitter and multiwavelength sources. *Physical Review A* 84: 062308.

37 Sajeed, S., Chaiwongkhot, P., Bourgoin, J.-P. et al. (2015). Security loophole in free-space quantum key distribution due to spatial-mode detector-efficiency mismatch. *Physical Review A* 91: 062301.

38 Bennett, C.H., Brassard, G., and Mermin, N.D. (1992). Quantum cryptography without bell's theorem. *Physical Review Letters* 68: 557–559.

39 Ekert, A.K. (1991). Quantum cryptography based on bell's theorem. *Physical Review Letters* 67: 661–663.

40 Lim, C.C.-W., Xu, F., Pan, J.-W., and Ekert, A. (2021). Security analysis of quantum key distribution with small block length and its application to quantum space communications. *Physical Review Letters* 126: 100501.

41 Cao, Y., Li, Y.-H., Zou, W.-J. et al. (2018). Bell test over extremely high-loss channels: towards distributing entangled photon pairs between earth and the moon. *Physical Review Letters* 120: 140405.

42 Ren, J.-G., Xu, P., Yong, H.-L. et al. (2017). Ground-to-satellite quantum teleportation. *Nature* 549: 70.

43 Zhong, H.-S., Li, Y., Li, W. et al. (2018). 12-photon entanglement and scalable scattershot boson sampling with optimal entangled-photon pairs from parametric down-conversion. *Physical Review Letters* 121: 250505.

44 Jennewein, T., Bourgoin, J.P., Higgins, B. et al. (2014). QEYSSAT: a mission proposal for a quantum receiver in space. *Advances in Photonics of Quantum Computing, Memory, & Communication VII*.

45 Villar, A., Lohrmann, A., Bai, X. et al. (2020). Entanglement demonstration on board a nano-satellite. *Optica* 7 (7): 734–737.

46 Kerstel, E., Gardelein, A., Barthelemy, M. et al., The CSUG Team (2018). Nanobob: a CubeSat mission concept for quantum communication experiments in an uplink configuration. *EPJ Quantum Technology* 5: 6.

47 Oi, D.K., Ling, A., Vallone, G. et al. (2017). CubeSat quantum communications mission. *EPJ Quantum Technology* 4 (1): 6.

48 Oberhaus, D. (2020). NASA's plan to turn the ISS into a quantum laser lab. (accessed 25 November 2022).

49 Tang, Z., Chandrasekara, R., Tan, Y.C. et al. (2016). The photon pair source that survived a rocket explosion. *Scientific Reports* 6: 25603.

50 Tang, Z., Chandrasekara, R., Tan, Y.C. et al. (2016). Generation and analysis of correlated pairs of photons aboard a nanosatellite. *Physical Review Applied* 5: 054022.

51 Grieve, J.A., Bedington, R., Tang, Z. et al. (2018). SpooQySats: CubeSats to demonstrate quantum key distribution technologies. *Acta Astronautica* 151: 103.

52 Bedington, R., Arrazola, J.M., and Ling, A. (2017). Progress in satellite quantum key distribution. *npj Quantum Information* 3 (1): 30.

53 ESA (2018). D/TIA partners with UK-based ArQit to develop First Quantum Encryption Satellite. (accessed 25 November 2022).

54 ESA (2018). 10 new business and research partners join quartz. (accessed 25 November 2022).

55 Scudo, P.F. and Lewis, A.M. (2021). Quantum communications infrastructure architecture: theoretical background, network structure and technologies. A review of recent studies from a European public infrastructure perspective. *arXiv:quant-ph/2110.06762.*

56 Liao, S.-K., Yong, H.-L., Liu, C. et al. (2017). Long-distance free-space quantum key distribution in daylight towards inter-satellite communication. *Nature Photonics* 11: 509–513.

57 Gruneisen, M.T., Sickmiller, B.A., Flanagan, M.B. et al. (2016). Adaptive spatial filtering of daytime sky noise in a satellite quantum key distribution downlink receiver. *Optical Engineering* 55 (2): 1–11.

58 Gruneisen, M.T., Flanagan, M.B., Sickmiller, B.A. et al. (2015). Modeling daytime sky access for a satellite quantum key distribution downlink. *Optics Express* 23 (18): 23924–23934.

59 Gruneisen, M.T., Flanagan, M.B., and Sickmiller, B.A. (2017). Modeling satellite-Earth quantum channel downlinks with adaptive-optics coupling to single-mode fibers. *Optical Engineering* 56 (12): 1–17.

60 Gruneisen, M.T., Eickhoff, M.L., Newey, S.C. et al. (2021). Adaptive-optics-enabled quantum communication: a technique for daytime space-to-earth links. *Physical Review Applied* 16: 014067.

61 Yang, K.-X., Abulizi, M., Li, Y. et al. (2020). Single-mode fiber coupling with a M-SPGD algorithm for long-range quantum communications. *Optics Express* 28 (24): 36600–36610.

62 Gong, Y.-H., Yang, K.-X., Yong, H.-L. et al. (2018). Free-space quantum key distribution in urban daylight with the SPGD algorithm control of a deformable mirror. *Optics Express* 26 (15): 18897–18905.
63 Lu, C.-Y., Cao, Y., Peng, C.-Z. et al. (2022). Micius quantum experiments in space. *Reviews of Modern Physics* 94(3).

26

Quantum Communication Networks for 6G

Riccardo Bassoli[1,2], Frank H.P. Fitzek[1,2], and Holger Boche[3,4,5]

[1] *Deutsche Telekom Chair of Communication Networks, Technische Universität Dresden, 01187, Dresden, Germany*
[2] *Centre for Tactile Internet with Human-in-the-Loop (CeTI), Cluster of Excellence, 01062, Dresden, Germany*
[3] *Department of Electrical and Computer Engineering, Technische Universität München, 85748, München, Germany*
[4] *Munich Center for Quantum Science and Technology (MCQST), 80799, München, Germany 44780,*
[5] *Cyber Security in the Age of Large-Scale Adversaries (CASA), Cluster of Excellence, Bochum, Germany*

26.1 Introduction

Few years ago, the deployment and commercialization of fifth generation (5G) started. The vision of 5G represented the arrival point of a long scientific and research journey. At the beginning of the last century, the first radar, telegraphs, and telephones were realized. The applications of electromagnetic fields in telecommunications started strongly focused on the physical layer and the hardware. Independently, computing started during the 1930s–1940s. After the II World War, the so-called Age of Information started, with the results by Shannon and the design and realization of the Internet. Next, between 1994 and 2004, wireless cellular communications arose transposing the previous experience obtained with the telephone network. Only with fourth generation (4G) and finally with 5G, the paradigms of computing, the experience obtained with, and the capabilities of the Internet, converged to the idea of wireless and mobility. The paradigm in communication networks shifted from store-and-forward to compute-and-forward, so that software obtained a key role in communications [1].

Currently, the research focus by academia and industry moved to the next generation, the so-called sixth generation (6G), which first aims at addressing the unanswered promises by 5G, which have not been completed yet. This is, for example, widespread network softwarization and programmability, with fully implemented network slicing of virtualized core and edge network. However, 6G has also a wider vision than its predecessor 5G. 6G envisions the ubiquitous integration between human life and communication networks [2]. This sets very demanding and stringent requirements, which have to be mutually satisfied by communication and computing technologies that were not designed to be so tightly integrated. Moreover, the classical technologies experience some intrinsic limitations of the fundamental

Photonic Quantum Technologies: Science and Applications, First Edition. Edited by Mohamed Benyoucef.

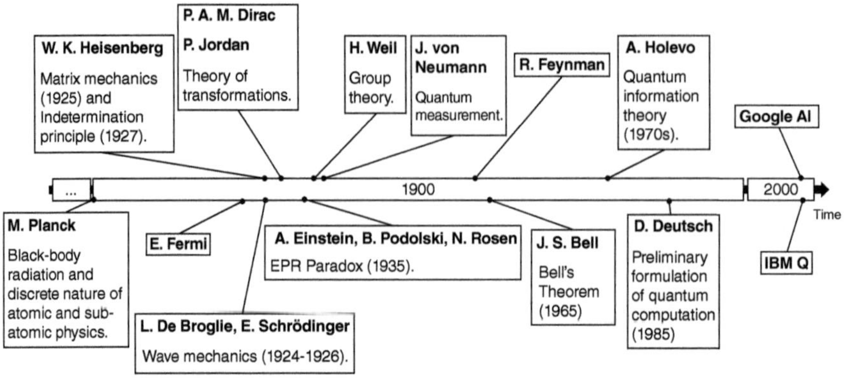

Figure 26.1 General timeline of research in quantum mechanics and its applications.

key performance indicators (KPIs), which arise from their inherent interdependence and correlations. That is why, the scientific and industrial communities are looking for new different resources to be used, which can neglect the correlations happening in the current "classical" domain. The time is now mature for quantum technologies to open new horizons in communication networks and computing. Next, a question becomes prominent, considering the experience gained in the last century: what is the current status of quantum-mechanical applications?

As it is possible to see in Figure 26.1, quantum mechanics started at the beginning of the last century from some experiments that showed the limits of the previous theories to describe physical phenomena at atomic and subatomic level. The formulation of quantum mechanics took some decades and significant research effort by the international scientific community, in order to reach the current mathematical description that we use today, based on the Dirac notation and group theory [3]. With a shift of about 20 years in respect of its classical counterpart, information theory was interconnected to quantum mechanics opening the way to quantum information theory and quantum error correction theory. This also helped the enhancement of quantum computing theory started with Feynman.

In the last decade, we reached the first implementations of quantum computing platforms and quantum communications for security applications, the so-called quantum key distribution (QKD). We are now in the same condition of classical communications in the 1960s and 1970s when the idea of the Internet started. In fact, in 2018, the vision of a Quantum Internet interconnecting quantum computers worldwide oriented the effort of the scientific and the industrial communities. Nevertheless, the Quantum Internet, in respect of its counterpart, does not start from the scratch. In fact, there is the concurrent effort toward 6G that can give significant help to the evolution of quantum communication technologies.

In this panorama, it is important to discuss a topic that is not so well covered in the research community: can we integrate quantum and 6G communication networks? If yes, how? What are the mutual benefits obtained with a quantum-6G? Sections 26.2–26.4 of this chapter try to introduce and set the problem, providing the main conceptual means to orientate is such a complex, heterogeneous, and interdisciplinary research area, where many challenges are still open and under clarification.

26.2 What Is 6G?

The scientific work on the 5G of wireless cellular networks was one of the main research efforts of the last decade. Currently, the standardization is going to reach its Release 18, which will be completed by 2024. Currently, International Telecommunication Union (ITU) has started working on the "International Mobile Telecommunications (IMT) Future Technology Trends Towards 2030 and Beyond." This report will provide the high-level vision and guidelines for the research and design on 6G. The prestandardization effort by the scientific and industrial communities has started worldwide [4]. In Asia, China, Japan, Korea, and India are carrying on relevant national strategies. Next, in North America, Alliance for Telecommunications Industry Solutions (ATIS) has begun the Next G Alliance, while in Europe, various 6G projects have been funded [4]. In particular, in January 2021, the European Union (EU) has seen the kick-off of its 6G Flagship project, called Hexa-X.

The first architectural pillar of 6G will be Artificial Intelligence (AI) [5]. While, in 5G, AI was mainly applied as a service/application for end users, in 6G, it is going to be applied to every network aspect and protocol stack's layer, toward autonomous network operation and management. For example, the 6G network will employ AI for optimizing the usage of the Radio Access Network (RAN) such as managing resources and ensuring the required quality-of-service (QoS). Moreover, the employment of AI will not only be in a centralized but also in a distributed manner. Network elements like functionalities and tasks will be realized as softwarized agents, which will constitute virtual chains to achieve an objective. An intelligent procedure usually consists of four main steps: sense, plan, decide, and act, which are flanked by learning and the final environmental feedback. Sensing is critical for a correct information gathering, including data mining, labeling, and classification. Next, planning refers to the interpretation of mined data according to a model of data analytics so that the results can be used for planning. Deciding consists of identifying the potential actions according to policies and analytics. Finally, acting is the actual realization of the action on the network. The environmental information helps in checking the effects. All the previous states are part of a control loop, which mainly realizes the learning that continuously improves the performance of the AI.

As already started with the advent of 5G, 6G will virtualize or, more specifically, softwarize all the network protocol stack operations and tasks. This will actually result into a modular softwarization so that microservice or intelligent agents will be the building blocks of any network aspects. Then, this is also what will happen to AI. Learning entities will be flexibly instantiated across the network, replacing the monolithic functions by making them more resource efficient. Next, the integration of AI in the network management and operations will also open the way to proactive operations, with the capability to estimate the future network states. This will permit ideally to achieve the so-called "negative" latency. This networking paradigm has been named anticipatory networking, which is a new important advantage obtained in 6G intelligent networks. Nevertheless, the critical trade-off arising is the one combining latency reduction with reliability and resilience reduction, since the network decisions are not deterministic anymore.

6G networks require an efficient employment and integration of intelligence into the architecture. This implies the choice of the optimal implementation paradigm. Centralized learning implies sending all data from each device to a centralized server in the cloud to train an AI model. On the other hand, decentralized incremental learning consists of learning locally at the devices side, while only sending the model parameters to a centralized server in the cloud that stores them. This helps in communication efficiency and privacy assurance. Federated learning (FL) combines the pros of centralized and decentralized incremental AI. Multiple network nodes collaboratively train a centralized AI model in a distributed and iterative way until convergence of the centralized model is achieved. Specifically, the nodes only exchange local model parameters instead of the entire network data mined; next, the centralized model can then be shared to the individual network nodes.

The employment of AI in the future 6G architecture can have multiple benefits. First, it could be used for management and orchestration, traffic classification, and time-series forecasting. In particular, referred to management and orchestration, some examples of use cases could be traffic forecasts, automated Virtual Network Function (VNF) placement and network slicing, network self-healing, and hidden patterns discovery, to proactively assist network planning and sizing. Next, it could be applied to data clustering, information extraction, and detection of network anomalies. Network automation is also a pivotal AI application. Side by side, the characteristics of FL could be exploited to implement collaborative machine learning. The employment of AI also has a "spatial" perspective according to what area of the network is going to be withing the AI domain. First, AI could be locally applied within RAN, core, and edge domains. Second, AI could also be applied in a more cross-domain way to realize self-healing and self-management. This is the enabler of the 6G so-called "continuum orchestration" [5].

From an architectural perspective, 6G can be defined as an "ecosystem" of network, or as a so-called Network of Networks (NoN). By definition, an NoN is an heterogeneous architecture (conceptually similar to the Internet), which can provide global and ideally limitless connectivity to a huge number of heterogeneous devices. This vision has started with the advent of 5G, but it has become mature within the 6G context. Moreover, with 6G, this vision not only remained limited to terrestrial two-dimensional networks, but it also extended to Non-Terrestrial Networks (NTN), realizing a three-dimensional architecture. Other important elements of an NoN are flexible multiconnectivity, subterahertz, visible light, and device-to-device (D2D) communications, also hosting local ad hoc networks within the structured network. Since low latency and then mobile edge computing (MEC) are important pillars of future 6G networks, Local Area Network (LAN) play a fundamental role in the 6G architecture and infrastructure. Non-Public Network (NPN) played an important role in local and metropolitan area networks, providing

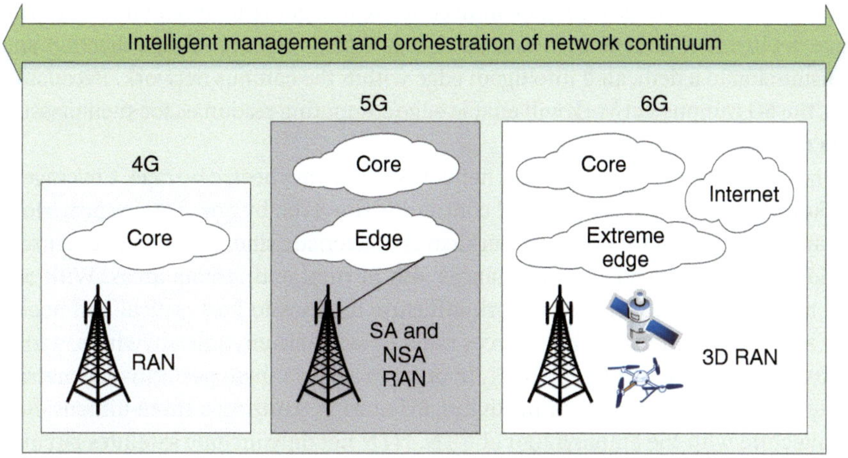

Figure 26.2 6G and the concept of device-edge-cloud continuum orchestration.

specialized infrastructures and architectures to specific users and customers. Their employment was mainly dedicated to industrial, educational, and public-safety scenarios, which required dedicated suitable networks to satisfy their specific KPI. 5G first and now 6G are generalizing NPN, seamlessly integrating them into the whole 6G management and orchestration of the network continuum. That is why, now it is important to talk about 6G campus networks and their role in the architecture and the infrastructure. Before 5G, management and orchestration were mainly performed within the core network. With 5G, the management and orchestration of softwarized network functions has been performed per network area, according to different orchestration domains like core network, transport network, and edge. On the other hand, 6G requires a new architectural design since previous communication networks have not been capable to realize the concept of network continuum and continuum orchestration (see Figure 26.2).

A campus network is a network consisting of an interconnection of LAN within a limited geographical area [6]. The main objectives of 6G campus networks will be cost reduction, security assurance, lower energy usage, and increased flexibility and adaptivity of the RAN and edge. In that sense, open-software solutions are planned to be used in the 6G campus networks like the so-called OpenRAN. OpenRAN is an open interface down to the antenna that can be reconfigured through software. The generalization of campus networks in 6G architecture will not only consist of an open-software implementation of the RAN but also in the inclusion of edge computing resources hosting microservices and local intelligent entities (i.e. agents). 6G campus networks can represent specific and suitable deployments for industrial scenarios, public services, etc. Especially, local and virtual operators can provide campus network resources to a region or a city in order to train robots or to manipulate digital twins from remote. Future industry will massively use collaborative and automated robots, also interacting with humans. For example, future industrial scenarios will consist of hundreds of robots, sensors, machines, etc. The tasks performed by robots and machines are becoming more and more complex, then the number of sensors that humans will need to wear is going to greatly increase. The huge deployment of heterogeneous sensors will generate huge data collection and transmission to a dedicated intelligent edge within the campus network. Because of this, the 6G campus network will enable edge computing resources for such massive data management and processing.

One of the key aspects of wireless networks has always been coverage. Coverage is the key for service provisioning and continuity. However, by now, service provisioning and continuity have not been a critical characteristic since broadband end users could accept the provided performances also in rural and remote areas. With 5G, and maximally with 6G, the network infrastructure has to host critical and necessary services. Then, network resources must be available anytime anywhere, with a probability closer and closer to 100%. In order to achieve these performance metrics ("anywhere" means global connectivity), 6G aims at realizing a three-dimensional architecture with the employment of NTN. NTN not only include satellites but also unmanned aerial vehicle (UAV) and high-altitude platforms (HAPS). The use of these platforms can be twofold: transparent and regenerative. Transparent means that NTN relays of the signal between the end users and the terrestrial base station. Regenerative means that the base station is placed on the third dimension, and then a backhaul/fronthaul is set up with a terrestrial gateway to connect to the terrestrial infrastructure. The former does not need high computing and communication resources. The latter instead requires the adaptation of the NTN equipment to place a fully functional RAN. Furthermore, the third dimension can be used to host computing and communication resources for edge and core functionalities. This implies the realization of constellations, embodying aerial- and satellite-based data centers. NTN can also embrace the campus networks scenarios in 6G network continuum, which opens to the realization of three-dimensional campus networks. If UAVs, HAPS, and satellites host communication interfaces and computing/storage, they can potentially provide any network functionality and operation to meet the 6G KPIs in remote and rural areas. Nevertheless, in case of three-dimensional campus networks, the problem of resource allocation and management becomes more critical due to the variability of the infrastructure characteristics. For example, low earth orbit (LEO) nanosatellites change continuously their positions in the orbits and HAPS can be forced to adjust their altitude and positioning based on environmental and atmospheric conditions.

The definition of KPIs depends on the service they are referred to. First, 6G must satisfy all the indicators of 5G. Next, more stringent KPIs have to be added to host 6G new services and applications. For example, the case of Digital Twins for manufacturing states the following KPI [6]:

- Availability [%] 99.99–99.999999,
- Reliability [%] 99.9–99.999999,
- Maintainability High (low tolerance to downtime in use cases where the digital twin is essential for maximizing the efficiency of the factory),
- Service latency [ms] 0.1–100,
- Data rate (minimum expected, desired, maximum) [Gbit/s] Peak: 10–100, Average: 1–10,

- Agent availability [%] 99.99–99.999999,
- Agent reliability [%] 99.9–99.999999.

26.3 6G Intrinsic Limitations: Why Do We Need Other Technologies?

Section 26.2 has given a brief overview of the complexity and heterogeneity of 6G networks. Big data mining and classification, massive distributed computing, complete network softwarization, unprecedented security, and widespread employment of AI pose critical challenges to classical communication and computing technologies. Next, the following tries to highlight the critical trade-offs that are limiting classical communication technologies for achieving their goals.

Figure 26.3 depicts the impact of software and virtualization on some principal KPIs. The software abstraction introduces additional latency for packet processing. Packet input/output and processing operations in the virtual environments (e.g. virtual machines, containers, etc.) are slower considering the KPIs of 6G. The massive network softwarization is also getting into the layers of the network protocol stack via the paradigm of programmable protocol stack (PPS) so that latency due to network virtualization is expected to constantly increase. When we talk about end-to-end latency in 6G, it is important to keep in mind that it is the sum of various contributions due to the propagation, the transmission on the link (inversely proportional to the available capacity of the communication channel), the queuing operation, and the processing. The price to pay is not only for the latency. Softwarization and paradigms like cloud and edge computing are highly increasing the energy usage of the network. Moreover, the management of Virtual Network Functions VNFs increases the communication load and thus reduces the energy efficiency. Next, software environments are more prone to security threats and privacy issues (if we

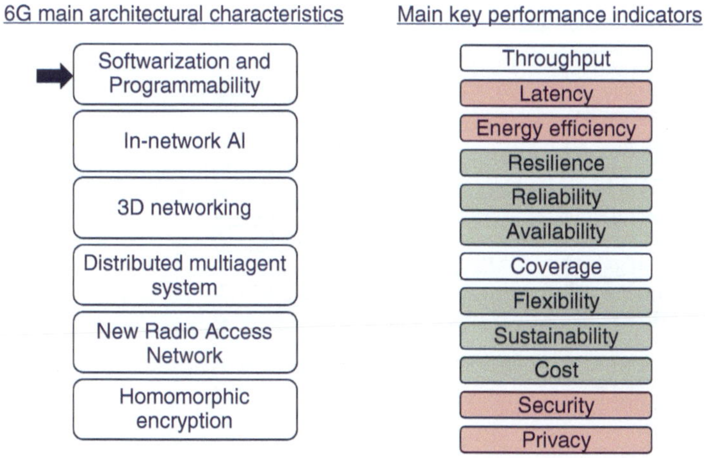

Figure 26.3 Main pros and cons of softwarization and programmability.

consider the offloading of functionalities to third-party data centers for example). Software functions are more prone to failures than hardware ones, but they can recover more flexibly and more rapidly. This can imply a gain in resilience. Finally, the gain in resilience, flexibility, reliability, and sustainability (software is more scalable than hardware) that it is obtained with softwarization and programmability can somehow limit the performances of the network according to other metrics.

The complex and heterogeneous 6G system will also require networks to rely less on human intervention and more on intelligence and AI for network management and operations. In fact, as previously described, in-network intelligence will be a key element of future networks, enabled by softwarization and programmability. A widespread employment of AI in each layer or function of the network will have various potential benefits (see Figure 26.4). First, it will increase resilience, reliability, and availability, since AI can rapidly address network outages or imminent failures of both infrastructure and communications. Next, AI will also have a fundamental impact on end-to-end latency. The employment of AI for proactive algorithms helps for a significant reduction of latency [7]. Moreover, the potential is to open the possibility for a so-called "negative" latency, which will help in achieving the very stringent 6G KPI [8]. The research community in the field calls this paradigm anticipatory networking, in which anticipation means that prediction techniques are used, and networking means that the optimization of routing and network-related functions is performed. However, even if anticipatory networking and in-network intelligence increase the communication resilience and reliability, they can also fail in their decision process, since they are probabilistic in nature. Then, this can be somehow incompatible with the objective of almost no network failure. Furthermore, the cost of AI in terms of computing and communication can be very high [3].

The three-dimensional networking mentioned in Section 26.2 will permit the provisioning of network functions, services, and operations to rural and remote areas, enhancing the network coverage. This paradigm will mainly affect the characteristics of 6G Layers 3 and 4. Moreover, it will reduce the cost of network

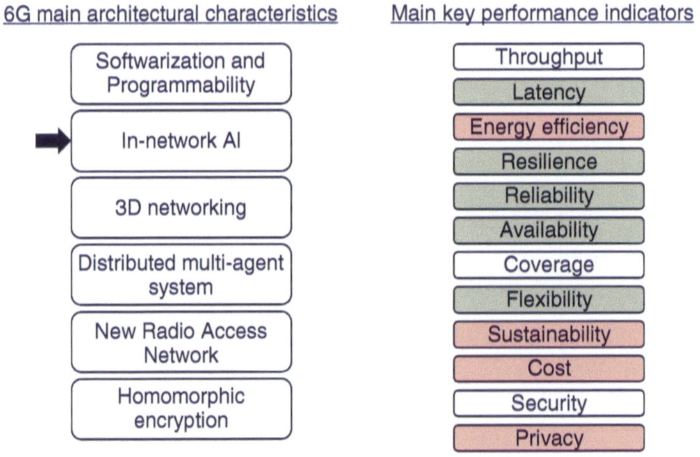

Figure 26.4 Main pros and cons of in-network intelligence.

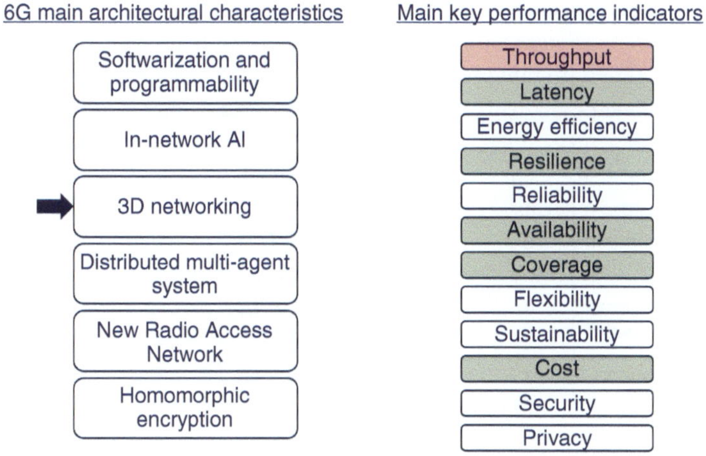

Figure 26.5 Main pros and cons of three-dimensional networking.

deployment in areas where the terrestrial network is impracticable. In case of natural disasters or terrorist attacks [9], three-dimensional networking can highly increase the resilience and the availability of network resources for both communication and computing [10]. Finally, it can also reduce the end-to-end latency where terrestrial routes are too long. Figure 26.5 depicts these positive effects. So, what are the trade-offs in this case? The use of HAPS, UAVs, and nanosatellites increases the impact of the Doppler's effect on the communication link [10]. Especially in the case of nanosatellites, the constellation should be carefully designed to ensure continuous availability of resources. Moreover, these platforms have quite more limited resources for computing and communication since they are battery powered [11]. In parallel, the throughput provided by these devices has quite more limitations that the one provided by a terrestrial infrastructure (e.g. fiber-based or cellular-based) [12].

The placement of computing at a given data center (e.g. big, micro, and femto) has impact on resilience, capacity, and latency. The distributed nature of the in-network computing with related intelligence has negative impact on the energy efficiency, since the control traffic to handle intelligent network agents greatly increases. While positive effects can be in line with the ones obtained with in-network AI and softwarization, the price to be paid is the computing cost (energy usage) and the sustainability once massively widespread (see Figure 26.6). Additionally, while security may be increased in respect of security threats like denial-of-service (DoS), the distributed nature of agent-based systems spreads the processing of data (especially sensitive one) in the network, subsequently reducing privacy.

6G will also consider the change of the RAN to provide the required data rates in the order of hundreds of Gbit/s. This will combine the use of mmWaves as per 5G design and the employment of Terahertz frequencies. While the goals in terms of coverage and throughput will be satisfied, the negative impact will be on the computing latency needed at the baseband unit (BBU) (either hardware-based or softwarized), as shown in Figure 26.7. The scope of the BBU is to perform

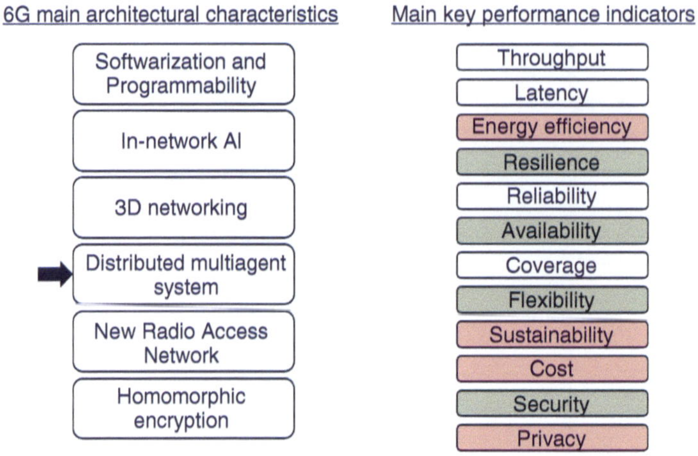

Figure 26.6 Main pros and cons of distributed multiagent systems.

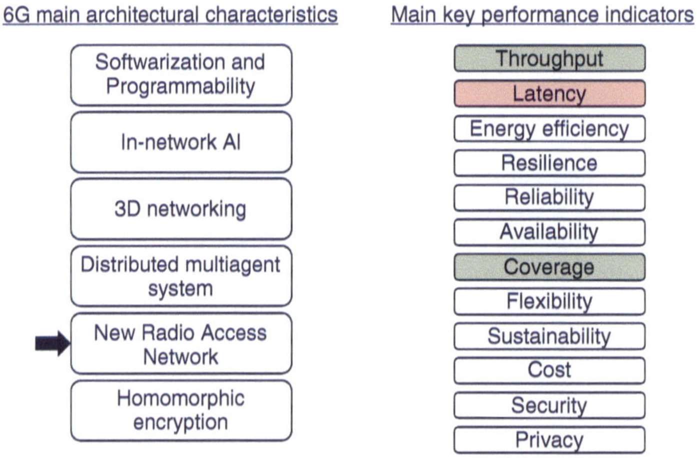

Figure 26.7 Main pros and cons of the new radio access network.

the baseband processing of the signals to/from the wireless link. The operations of a BBU involve physical-layer signal processing, reduction of interference, modulation/demodulation, error-correction coding, radio scheduling, encryption/decryption of data packets both in downlink (network to users) and uplink (users to network). Fast Fourier transfor (FFT) and inverse fast Fourier transfor (IFFT) are also performed at the BBU. The computing required at the BBU has been experimentally demonstrated to be increased by augmenting the throughput of the wireless communication.

Finally, it is worth mentioning the goal of trustworthiness, translated in unprecedented security and privacy. This is something that can be achieved via different advanced schemes like homomorphic encryption [13]. However, these schemes highly increase the processing load and thus the latency and the energy usage (see Figure 26.8).

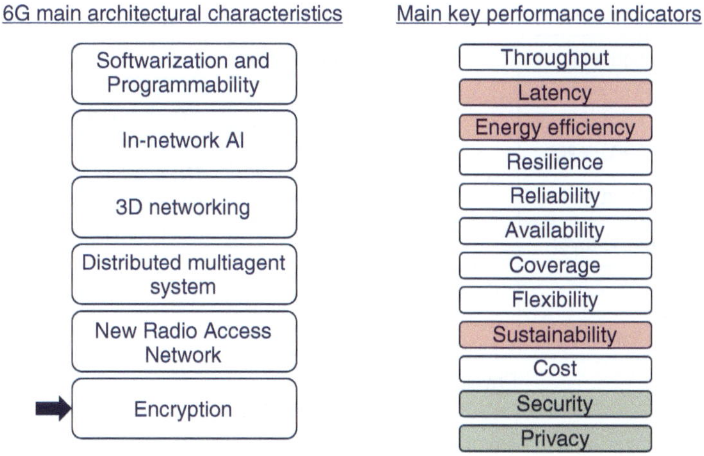

Figure 26.8 Main pros and cons of encryption.

In order to go beyond the trade-offs, in which classical technologies got stuck in 6G, new radical resources are necessary for computing and communication. This means changing the intrinsic nature of the network architecture that the research community is envisioning for 6G. These new resources have been identified in molecular-biological communications and quantum-mechanical communications [2]. In the context of the discussion of this chapter, the focus is on quantum-mechanical resources like quantum superposition, quantum entanglement, and no-cloning theorem. These resources have been identified as the ones providing mutual satisfaction of KPIs in computing and communication avoiding trade-offs.

26.4 The Vision of the Quantum Internet

Before discussing the general characteristics and the definitions that have been provided about the so-called Quantum Internet, first we will briefly recall what the Internet is. The Encyclopædia Britannica defines "[…] Internet, a system architecture that has revolutionized communications and methods of commerce by allowing various computer networks around the world to interconnect. Sometimes referred to as a network of networks […]" [14]. Then, the important achievements were the interoperability of all the computers and the worldwide extension. In fact, the Internet reformed the previous data communication systems, which were connecting few devices and were covering very limited geographic areas. Originally, each communication system was ad hoc, running specific proprietary protocols.

At the end of the 1970s, the process of making the Internet a homogeneous network started. The work achieved a final shape after about ten years, when the Open Systems Interconnection (OSI) reference model was released by the International Organization for Standardization (ISO). Specifically, this represented the Internet standard solution to the so-called "layering problem" in communication systems. In [15], layering is defined as a "[…] structuring technique which permits the

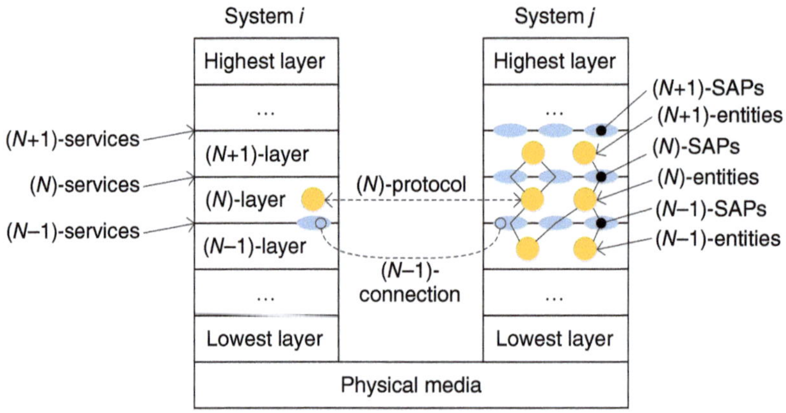

Figure 26.9 Logical layering of a communication network.

network of Open Systems to be viewed as logically composed of a succession of layers, each wrapping the lower layers and isolating them from the higher layers [...]." Figure 26.9 shows the logical individual elements composing the layering procedure. An important assumption of layering in communication networks is the independence of how services work in each layer in respect of the other ones.

The physical media consist of the communication and computing hardware of the network, and the physical communication link. Next, each layer contains various internal elements called entities. Entities can be peer entities if they are in the same layer. The entities are the internal services that a layer can provide to its neighboring layers. The entities of the highest layer work for their own goals, while the entities within the other layers are all interconnected to each other, working collaboratively to provide the service. Next, the entities within layers of different network nodes collaborate via protocols, which establish how the (i)-entities work using the ($i-1$)-services to perform the respective (i)-functions. The Service Access Points (SAPs) are the principal logical interfaces that entities of different layers use to provide their services. It is important to notice that one (i)-SAP can be used by only one (i)-entity and ($i+1$)-entity. On the other hand, one (i)-entity and ($i+1$)-entity can use more than a (i)-SAP (see Figure 26.9). Next, the (i)-layer employs (i)-connections between (i)-SAPs. The end of a connection at SAP is called Connection End-Point (CEP). Various connections can run among the same SAPs. Identifiers and addresses are assigned to Connection End-Points CEPs.

From this more general layering conceptual framework, the specific and well-known ISO-OSI reference model was derived to enable the Internet communications [15, 16]. This layering model specifically consists of six layers above the physical media. The lowest layer is called Link Layer and the highest layer is the Application Layer. In between, there are the Network, Transport, Session, and Presentation Layer. Subsequently, the Transport Control Protocol (TCP)-Internet Protocol (IP) layering arose [17]. Other reference models were provided during the years related to different communication technologies, as for example in the case of Broadband Integrated Services Digital Network (B-ISDN) [18].

In a similar way to what has happened for the classical Internet, the preliminary development of quantum computing platforms has brought the interest to how effectively set up remote communication channels among them. In fact, in the last decade, some big companies worldwide have been implementing preliminary quantum computers. However, the current situation is the same mentioned above for the classical Internet: no standard exists. This means that each quantum computer is realized in an ad hoc manner, without the possibility of direct interoperability. In the moment of writing, a proper standardization of quantum computing platforms and, more generally, of quantum hardware is still missing. Nevertheless, the research interest started to grow for how to interconnect distributed quantum devices via an Internet-like network. This is the so-called Quantum Internet.

The start of the work on the Quantum Internet can be officially located in 2018, when the Quantum Internet Alliance (QIA) began its scientific activities. In parallel, the Internet Research Task Force (IRTF) Quantum Internet Research Group (qirg) started the effort to define the standard characteristics and terminology of the Quantum Internet. In their current draft, the definition of quantum networks says "[...] distributed systems of quantum devices that utilise fundamental quantum mechanical phenomena [...] to achieve capabilities beyond what is possible with nonquantum (classical) networks [...] such devices may range from simple photonic devices [...] to large-scale quantum computers of the future [...]" [19]. From this quote, it is possible to feel the influence of wireless cellular networks and 5G/6G perspectives, since the devices considered are not merely computers but also sensors, etc.

Before dealing with the network-related aspects, it is important to mention what the unit of information is. Quantum systems are characterized by quantum states. These quantum states can represent information. In the current vision of the Quantum Internet, the unit of information is the two-dimensional quantum state, called quantum bit (qubit). Mathematically speaking, the qubit is the superposition of two vector states simultaneously, so that it is described by an infinite number of bits in its superposition state ($\aleph_0$, which is the size of the set of natural numbers).

Next, the IRTF qirg draft also states that "[...] Entanglement as the fundamental resource [...]" [19]. In fact, entanglement is one of the pivotal resources, which make quantum communication networks so attractive. Its communication and computing potentials are still far from being completely unveiled. Generally speaking, entanglement is a strong and inherent correlation that multiparticle quantum systems can have, which also remain valid when such particles are distant from each other. Entanglement is also the fundamental pillar of the basic quantum protocols called teleportation and dense coding [3].

The pillar of the classical Internet is the TCP-IP protocol suite. This implies the grouping of bits of information into packets and frames to be routed through the network. When we read in [19] that "[...] Packets are the fundamental unit in a classical network [...]," it is important to highlight that this is partially correct. In the sense that this claim comes after having solved the network layering problem for a quantum network and, if there is a layer handling information units called packets, then packets are the fundamental units within such a layer. Subsequently, the text says: "[...] In a quantum network, the entangled pairs of qubits are the basic unit of networking. These qubits themselves do not carry any headers. Therefore, quantum networks will have to send all control information via separate classical channels which the repeaters will have to correlate with the qubits stored in their memory [...]" [19]. Even in this case, networking implies the already- solved layering problem, which is still an open challenge for the Quantum Internet. Moreover, entanglement is not a resource for networking since entangled quantum systems should be reliably routed through the network to enhance classical communication. As interestingly said, the classical communication will have an important role in the control plane of the network. As classical communication networks, the Quantum Internet also expects to consist of a control and data plane. This abstraction starts giving some initial design guidelines for the realization of the Quantum Internet architecture. Potentially, the control plane will be highly integrated with the classical networks in order to allow for the reliable communication of quantum information.

A critical aspect of quantum communication networks is the "efficiency" of quantum systems in respect to distance. Entangled quantum systems, sent across a medium, can get coupled to it, and then they can lose the properties that we need to exploit. In fibers, the quantum communication cannot go beyond 100–150 km, which can become 1000 km only in free space. This cannot be solved via amplification since quantum systems cannot be copied as per the no-cloning theorem [3]. The solution to this intrinsic limit is the use of so-called quantum repeaters, which exploit the procedure of entanglement swapping to maintain the "quality" of quantum systems and entanglement for long distance [3]. However, the cost of quantum repeaters is very high and the operations they should perform (like quantum error correction) are still highly complex to be realized for massive and commercial deployment. This currently makes impracticable their placement every 100–150 km worldwide.

Recently, few research works have started solving the layering problem for the Quantum Internet by following the successful past of the ISO OSI solution for the Internet.

Figure 26.10 depicts a solution to the network layering problem for the Quantum Internet, proposed by Dahlberg et al. [20]. The system is abstracted into five layers above the physical media. The Lowest Layer is the so-called Physical Layer, which performs protocols highly integrated with the quantum hardware. The Layer 1 is the Link Layer, which, in the Internet, generally performs link-level operations like medium access control and error-correcting procedures at bit level. In the Quantum Internet perspective it is mostly focused on reliably sending entangled quantum systems on demand across the quantum network. However, while [20] defined and specified the characteristics of entities and protocols, no definition or description of the SAPs and CEPs has been provided.

The entities and protocols designed relate to the Upper-Physical and Link Layer. In the former, the midpoint heralding protocol (MHP) represents a control protocol for handling the generation of entangled quantum systems. The main parameter of this protocol is the cycle, which defines the timing precision and subsequently affects the communication metrics like the data rate. The CEPs identifier is assigned to each

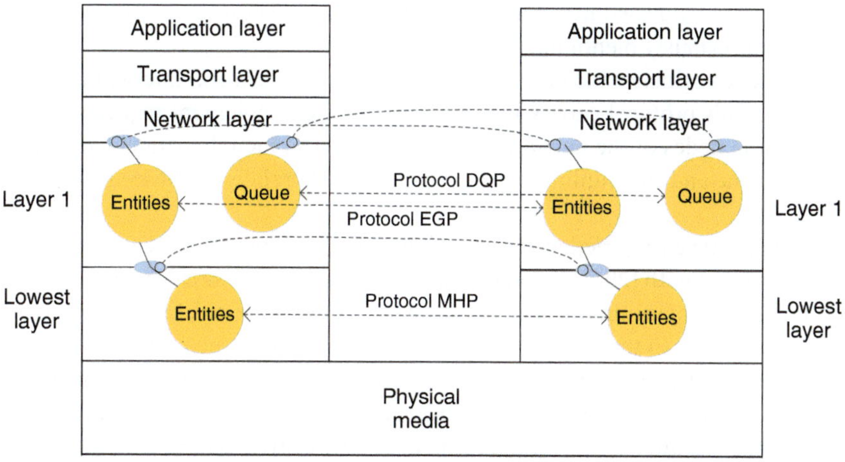

Figure 26.10 Proposed solution to network layering for the Quantum Internet by Dahlberg et al. [20]. Source: Adapted from Dahlberg et al. [20].

photon. The protocol employs a time-division communication paradigm to access to the physical media. The MHP interconnects two kinds of entities: create-and-keep and create-and-measure. The first only deals with unitary operations on quantum systems. Generally speaking, unitary operations are operations that change the characteristics of the quantum system without collapsing it into its probabilistic classical values (i.e. its eigenvalues) [3]. On the other hand, the create-and-measure allows for measurements on the quantum systems. Referred to this, the classical results (bits) obtained with the measurement are made available via the SAP to the entities of link layer using the entanglement generation protocol (EGP).

The Link-Layer entities employ two main protocols: the EGP and the distributed queue protocol (DCP). The former uses distributed queues, Quantum memory management (QMM), fidelity estimation unit (FEU), and a scheduler to generate entangled quantum systems according to the required performances. The management of the distributed queues is necessary to handle the upper-layers requests for entangled quantum systems. The role of the QMM entity selects the quantum systems to be entangled. Next, the FEU is the entity, which estimates the fidelity of the entangled systems, ensuring the required KPIs. Finally, the scheduling entity manages the queuing policies.

The FEU receives the indications by the Network Layer, via the specific SAP, stating how many entangled quantum systems to be generated. Next, the FEU establishes the required values for the fidelity and the time slot for the entanglement generation. Just for clarity, the fidelity of a quantum system is a value in the range $0 \leq f \leq 1$, where 1 means perfect fidelity. Normally, values in the range $f \leq 0.5$ imply that the quantum systems are not usable. The value of fidelity helps estimating how much noise (e.g. environmental) and errors (e.g. gate and channel) have affected the quantum system under consideration. The scheduling entity handles the requests in the distributed queues so that the QMM entity is capable to reserve the requested number of quantum systems. These are tracked using the identifier previously mentioned.

The queue entities, which are distributed across the network, interwork through the DCP. This protocol carries the information of entanglement requests coming from all the nodes of the network. These requests also include parameters like the creation time and the minimum time, which directly relates to the cycle described previously.

Another article [21] tried to address the layering problem for the Quantum Internet. In contrast with [20, 21] tried to apply the logic of the classical ISO OSI layering solution to the Quantum Internet. In that sense, Upper-Physical, Link, and Network Layer are similar to the ones of the Internet. Furthermore, the authors introduced an additional layer, called the Connectivity Layer. The Connectivity Layer deals with maintaining the communication link between network nodes and quantum error correction procedures. This layer also manages the distribution of entangled quantum systems in the network. With this conceptual approach, Pirker and Dür [21] tried to decouple pure Link-Layer operations from the Quantum Internet connectivity procedures.

Next, the Link Layer of [21] deals with the generation and distribution of entanglement, by ensuring the specified level of fidelity. The operations mentioned above performed by the quantum repeaters (e.g. entanglement swapping) are also considered Link-Layer functionalities. Similar to the classical Internet, the Network Layer should manage the routing of entangled quantum systems within the network, via the employment of quantum routers. However, the discussion provided by Pirker and Dür [21] remains very general. Figure 26.11 shows the described approach by Pirker and Dür [21].

Finally, a third attempt to start addressing the layering problem of the Quantum Internet was discussed in [22]. In this work, the approach was different from the previous one. The authors tried to start from the classical architecture seeing how to "procedurally" add the quantum operations for entanglement generation and distribution. This different approach was motivated by the current critical limitations experienced by entanglement storage and distribution technologies.

In particular, Nötzel and DiAdamo [22] envisioned the punctual changes to the Upper-Physical and Link Layer to enable entanglement-assisted classical communications. The proposed protocol is called generate entanglement when idle (GEWI). The protocol starts at the source, which is the one generating and

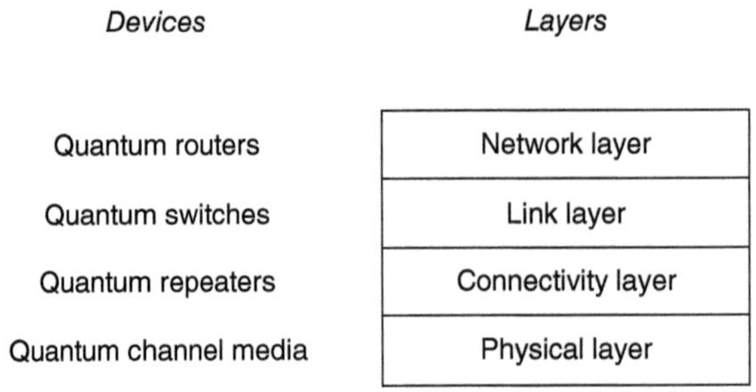

Figure 26.11 Proposed solution to network layering for the Quantum Internet by Pirker and Dür [21]. Source: Adapted from Pirker and Dür [21].

distributing entanglement when there are no data to be sent. When there are data at the source to be sent, no entanglement is stored, then the protocol transmits data without entanglement assistance.

26.5 The Architectural Convergence of Quantum Technologies and 6G

In the above discussion, Section 26.2 described the main architectural aspects and characteristics that 6G envisions. However, Section 26.3 showed that 6G future networks present some intrinsic limitations deriving from trade-offs arising from classical technologies. On the other hand, Section 26.4 showed the high-level characteristics of the Quantum Internet and the first preliminary attempt for solving its layering problem. The initial deployment of 6G is planned for 2030 while the one of the Quantum Internet for 2035. This means that the long-term research and standardization plan are still approaching the two network designs independently. On the other hand, a more suitable approach should be the integration of quantum resources and technologies in 6G to collapse and avoid some intrinsic trade-offs. This should start from the common architectural aspects to which both 6G and the Quantum Internet are converging.

Main architectural aspects that were mentioned in Section 26.2 are softwarization and in-network intelligence. Their combined trade-offs and limitations were highlighted in Section 26.3. Can quantum technologies and thus the Quantum Internet help in this context? The authors in [23] described how the trade-off between big-data mining, latency, and energy efficiency can be collapsed so that it is possible to concurrently gain in all these KPIs. This can be obtained with the design and realization of a quantum control plane.

Figure 26.12 depicts the solution proposed. Classical streams of 2^n bits can be encoded into n qubits. This is a lossy encoding process. However, AI algorithms do not need all collected information to manage the network and the control plane, but their goal is to identify patterns and statistical trends. If the control plane needs to decide and act using the data plane parameters, about 2^{11} parameters are expected to be sent from the various data plane devices to the control plane and hypervisor. This implies an exponential growth of data to be sent and data to be processed. Next, energy efficiency (bit per Joule), transmission latency (the inverse of the available capacity), and computing latency explode. The targeted end-to-end latency and increase of energy efficiency mentioned in Section 26.2 cannot be reached. Nevertheless, 2^{11} parameters can be encoded in a with loss into 11 qubits. Figure 26.12 depicts the proposed solution in a hierarchical network, consisting of quantum-classical switches (qSi), controllers (qCi), and a centralized hypervisor (Hyp). A logarithmic evolution instead of exponential is obtained. It is important to notice that this approach assumes that the devices of the network are equipped with reliable quantum memories (QRAM), which is something not affordable in the short term.

Section 26.4 mentioned the limitation that quantum communications experience with distance. This is something that will push farther the commercialization of

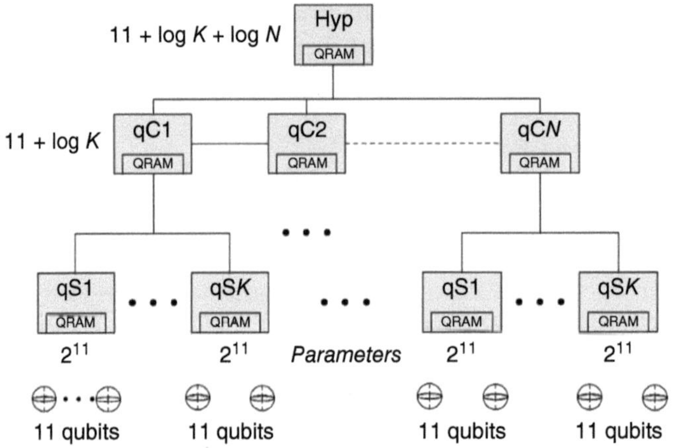

Figure 26.12 Hierarchical 6G virtual architecture showing data plane and control plane, consisting of *K* quantum switches (qS1, etc.), *N* controllers (qC1, etc.), and a hypervisor (Hyp).

a worldwide Quantum Internet and it will discourage companies and operators to approach long-distance quantum communications because of their high costs and investments needed. However, 6G architectural evolution and some recent experiments in quantum space communication can provide the suitable and sustainable solution to this infrastructural and architectural problem.

QKD is the first widespread application of quantum technologies in the field of security. Nowadays, several commercial applications already exist. In 2016, the first important satellite-based QKD was performed by China via their satellite called Micius. The Micius satellite distributed quantum keys to the two ground stations Nanshan and Delingha, distant 1120 km. The Micius satellite was flying in an orbit of altitude 500 km, a LEO orbit [24]. Next, in 2018, the subsequent experiment represented the first intercontinental satellite-based QKD realization. A joint Chinese-Austrian team performed QKD between the quantum satellite Micius and multiple ground stations located in Xinglong, Nanshan, and Graz. Figure 26.13 depicts a high-level representation of these two experiments.

The interest to use satellites for QKD has been growing more and more for few years, and now experiments in LEO employing lower-cost nanosatellites (e.g. CubeSats) are going to be performed. The article [25] describes in detail the design characteristics of the CubeSat, the link, and the OGS. The size of the CubeSat is a six-unit (6U) Cubesat (nominal size 12 cm × 24 cm × 36 cm). The various parts of the satellite can be seen in Figure 26.14. First, 2U are for the platform systems. The on-board computer (OBC) is responsible for routine operations of the satellite. The electrical power system (EPS) the system that store and distribute power to the satellite, supplied by body-mounted solar panels. Next, the communications (COMMS) include several radio systems for classical communications. A ultra high frequency (UHF) dipole array is used for tracking, telemetry, and control, and provides redundancy for low-speed data transmission (in the order of kb/s). An S-band patch antenna is used for high-speed uplink (in the order of Mb/s). For high-speed downlink of mission data, X-band CubeSat transmitters are commercially available and can also

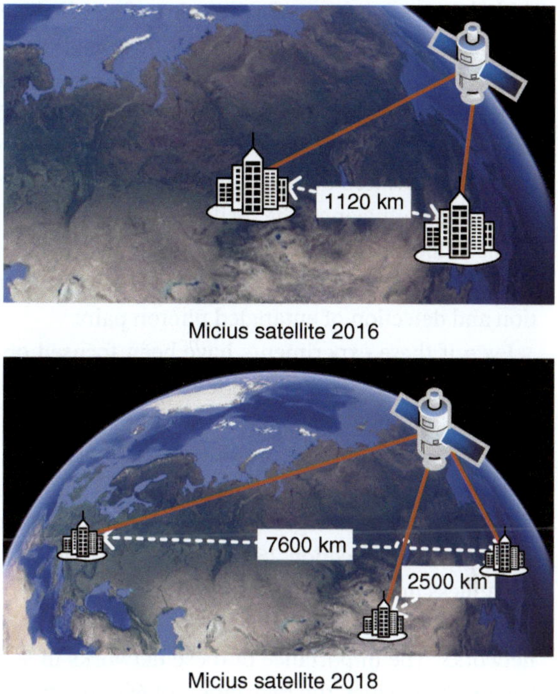

Figure 26.13 Experiments of Quantum Key Distribution with the Chinese Satellite Micius.

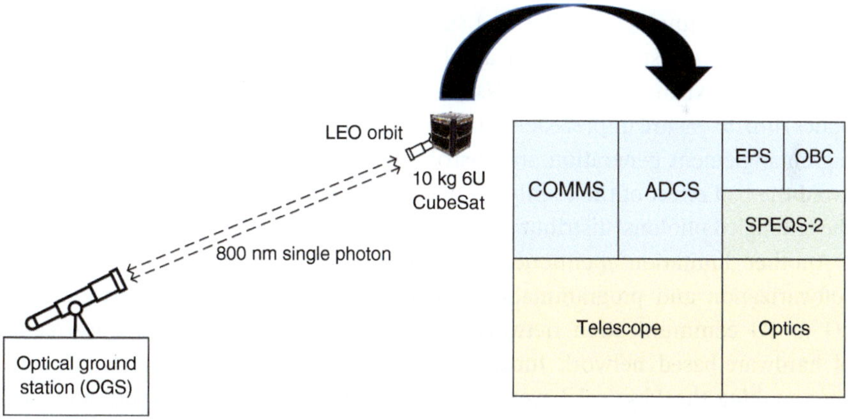

Figure 26.14 Conceptual representation of the CubeSat quantum communications mission, communicating with an Optical Ground Station (OGS). Internal structure of the CubeSat.

provide Mb/s data rates. The attitude determination and control systems (ADCS) are used to provide coarse pointing by rotating the CubeSat to align the transmitting telescope with the optical ground station during quantum transmission. The accuracy of ADCS is fundamental for the performance of the quantum communication. The COMMS module consists of a global positioning system (GPS) patch antenna is also incorporated for tracking the position and speed. Onboard GPS enables precise orbital determination and calibration of two-line element measurements, necessary for the OGS to get the satellite position and also for the ADCS to point to the transmitter telescope toward the OGS. This can allow the optical beacon tracker (OBT) for locking the beacon sent up by the OGS. Figure 26.14 shows the light-red blocks of the CubeSat (3U) that are dedicated to the quantum communication. It is possible to see that research on miniaturizing these devices and the telescope is necessary to increase the efficiency, the resilience, and reducing the costs. The light-green block of the CubeSat is the quantum source, which has the estimated size of 1U. Small photon-entangling quantum system (SPEQS-2) contains the devices for the generation and detection of entangled photon pairs.

Even if these experiments have been focused on QKD and the challenges for a reliable satellite-based entanglement distribution are still widely open, the use of satellites for quantum communications is highly attractive because of their significant lower costs in respect of terrestrial quantum repeaters. This can represent a trend in the quantum technologies and the Quantum Internet that can shorten the time to wait for an integration between future 6G and the Quantum Internet. In fact, as mentioned in Section 26.2, 6G will also have a three-dimensional nature, so that a seamless integration seems more affordable.

Another way to avoid for the moment the need of repeaters is the focus on campus networks. The importance of these networks in 6G has already been mentioned in Section 26.2. Thus, the realization of 6G-quantum campus networks can be highly important to make the commercialization and the employment of quantum communication technologies easier and sooner. Moreover, this can be highly beneficial for future industry, education, and health. For example, quantum routers could interconnect existing classical LANs via wired (fiber-based) or wireless (free-space optics) link to ensure unprecedented security, resilience, and low latency, by exploiting entanglement generation and distribution. However, free-space links have to avoid the bad effect of the highly changing atmospheric conditions that can affect the entangled photons' distribution.

Another limitation mentioned in Section 26.3 is given by the widespread softwarization and programmability of future 6G architecture. The trend from 1G to 6G communication networks has been the more and more abstraction of hardware-based network functionalities into software-based VNFs. This has been making the Physical Layer and the network infrastructure more and more general purpose to focus on computing provisioning. Nevertheless, physical layer service integration (PLSI) can represent a turning point in this trend by having quantum physical media and Physical Layer available and seamlessly integrated with the classical ones. The functionalities that are now expected to run in classical softwarized environments can be translated to be run in the quantum domain and, specifically, at the quantum Physical Layer. This can also imply a cross-layer design of the future 6G-quantum protocol stack. The research of the potential entities that can be moved back to the Physical Layer in the quantum domain is still widely open. However, PLSI can represent an important design approach and paradigm to collapse various existing classical trade-offs related to the load of softwarization and security. This can thus imply the provisioning of concurrently secure, reliable, and low-latency quantum-6G communications.

This chapter has tried to clarify the current status of the research on 6G and the Quantum Internet. As it was possible to see, the independent roadmaps are still in their infancy. In fact, their horizon to provide first deployments is expected to be between 2030 and 2035. Then, their efficient and effective integration can come even later if the current roadmaps remain independent as they are now. Thus, with the discussion provided in this chapter, the objective was to suggest potential contact points for the two roadmaps to mutually find the solutions of their limitations and issues in the other domain, 6G in the Quantum Internet, and vice versa.

References

1 Fitzek, F.H.P., Granelli, F., and Seeling, P. (2020). *Computing in Communication Networks - From Theory to Practice*, vol. 1, 1e. Elsevier.

2 Bassoli, R., Fitzek, F.H.P., and Strinati, E.C. (2021). Why do we need 6G? *ITU Journal on Future and Evolving Technologies* 2 (6).

3 Bassoli, R., Boche, H., Deppe, C. et al. (2021). *Quantum Communication Networks*, 1e. Springer. ISBN 978-3-030-62938-0.

4 Uusitalo, M.A., Rugeland, P., Boldi, M.R. et al. (2021). 6G vision, value, use cases and technologies from European 6G flagship project Hexa-X. *IEEE Access* 9: 160004–160020.

5 Hexa-X (2021). D5.1 – Initial 6G Architectural Components and Enablers. (accessed 26 November 2022).

6 Hexa-X (2022). D1.3 – Targets and requirements for 6G - initial E2E architecture. (accessed 26 November 2022).

7 Seeling, P. and Fitzek, F.H.P. (2021). Anticipatory networking: negative latency for ubiquitous computing. *2021 IEEE 18th Annual Consumer Communications & Networking Conference (CCNC)*, pp. 1–4.

8 Bui, N., Cesana, M., Hosseini, S.A. et al. (2017). A survey of anticipatory mobile networking: context-based classification, prediction methodologies, and optimization techniques. *IEEE Communication Surveys and Tutorials* 19 (3): 1790–1821.

9 Bassoli, R., Sacchi, C., Granelli, F., and Ashkenazi, I. (2019). A virtualized border control system based on UAVs: design and energy efficiency considerations. *2019 IEEE Aerospace Conference*, pp. 1–11.

10 Bonafini, S., Bassoli, R., Granelli, F. et al. (2020). Virtual baseband unit splitting exploiting small satellite platforms. *2020 IEEE Aerospace Conference*, pp. 1–14.

11 Granelli, F., Costa, C., Zhang, J. et al. (2020). Design of an on-demand agile 5G multi-access edge computing platform using aerial vehicles. *IEEE Communications Standards Magazine* 4 (4): 34–41.

12 Bassoli, R., Granelli, F., Sacchi, C. et al. (2020). CubeSat-based 5G cloud radio access networks: a novel paradigm for on-demand anytime/anywhere connectivity. *IEEE Vehicular Technology Magazine* 15 (2): 39–47.

13 Hexa-X (2021). D1.1 – 6G Vision, use cases and key societal values. (accessed 26 November 2022).

14 Dennis, M.A. (2022). Internet. In: *Encyclopædia Britannica*. Encyclopædia Britannica, Inc.

15 Day, J.D. and Zimmermann, H. (1983). The OSI reference model. *Proceedings of the IEEE* 71 (12): 1334–1340.

16 Burg, F.M. and Di Iorio, N. (1989). Networking of networks: interworking according to OSI. *IEEE Journal on Selected Areas in Communications* 7 (7): 1131–1142.

17 Meyer, D. and Zobrist, G. (1990). TCP/IP versus OSI. *IEEE Potentials* 9 (1): 16–19.

18 Staalhagen, L. (1996). A comparison between the OSI reference model and the B-ISDN protocol reference model. *IEEE Network* 10 (1): 24–33.

19 Kozlowski, W., Wehner, S., Van Meter, R. et al. (2022). IETF Quantum Internet Research Group (qirg): Architectural Principles for a Quantum Internet. (accessed 26 November 2022).

20 Dahlberg, A., Skrzypczyk, M., Coopmans, T. et al. (2019). A link layer protocol for quantum networks. *Proceedings of the ACM Special Interest Group on Data Communication*, SIGCOMM '19, pp. 159–173, New York, NY, USA: Association for Computing Machinery. ISBN 9781450359566.

21 Pirker, A. and Dür, W. (2019). A quantum network stack and protocols for reliable entanglement-based networks. *New Journal of Physics* 21 (3): 033003.

22 Nötzel, J. and DiAdamo, S. (2020). Entanglement-assisted data transmission as an enabling technology: a link-layer perspective. *2020 IEEE International Symposium on Information Theory (ISIT)*, pp. 1955–1960.

23 Ferrara, R., Bassoli, R., Deppe, C. et al. (2021). The computational and latency advantage of quantum communication networks. *IEEE Communications Magazine* 59 (6): 132–137.

24 Yin, J., Li, Y.-H., Liao, S.-K. et al. (2020). Entanglement-based secure quantum cryptography over 1,120 kilometres. *Nature* 582: 501–505.

25 Oi, D.K.L., Ling, A., Vallone, G. et al. (2017). CubeSat quantum communications mission. *EPJ Quantum Technology* 4.

Index

a

acousto-optical modulator 191, 242
action-spectroscopic methods 314
action-spectroscopy techniques 312
active components, in quantum photonic circuits 605–608
  E/O modulators 607
  phase shifters 605, 607
  single photon detectors 607, 608
adiabatic rapid passage 67
AFC-based quantum memory 438–440
AFC quantum memory 448, 449
afterpulses 98
alkali atoms 447
all-diamond-based spin-based functional quantum systems 378
angular distribution of radiation (ADR) 154, 161–163
anisotropic hyperfine coupling model 414
arbitrary photo-detector position 255
artificial and real atoms 468–470
artificial atoms 97, 183, 468, 470
artificial intelligence 106, 120, 122, 123, 717
atom-atom entanglement 219, 221–225
atomic clocks 12
  optical 334–342
atomic frequency comb (AFC) 433, 435, 438–440
atomic motion 197–198, 444, 447
atomic qubits
  coherence 219–221
  distributed entanglement 226–227
  long distance atom-atom entanglement 221–225
  rubidium 210
  state preparation, control, and measurement of single atoms 215–219
  trapping single atoms 212–214
attenuated pulsed lasers 37
Autler–Townes splitting (ATS) 433, 442

b

Bell inequalities 10, 15, 24, 226
Bell-state measurement (BSM) 88, 452, 453
Bell state projection measurement 209
Bell's theorem 10, 14–16
bichromatic far-detuned Raman lasers 325
biexciton 66, 69, 71, 72, 474, 483, 493
Big Bell Test 226
blackbody radiation 109, 110, 323, 344
blinking 38, 72, 352, 469, 529
Bloch sphere 42, 390
Bohr–Einstein debate 13–14
Bohr magneton 219, 355, 393, 407, 539
Bose–Einstein condensates (BEC) 19, 109, 114
boson sampling (BS) 53, 525, 581, 622
bottom-up approach 203, 470
broadband enhancement, of extraction efficiency 490

Photonic Quantum Technologies: Science and Applications, First Edition. Edited by Mohamed Benyoucef.

broadband optical-pumping schemes 310
built-in type memory 435
Bulls-eye gratings 380

C

CaH^+ state-detection scheme 325
Carr-Purcell Meiboom-Gill (CPMG) 532
centralized learning 718
Chebyshev polynomial technique 415
chemical vapor deposition (CVD) 471, 512
chip-based QKD 86, 471, 512
chip-to-chip transportation, of entangled quantum states 630
CMOS-compatible nanofabrication methods 379
CNT-$TbPc_2$ device 279
code breaking methods 21
co-flip process 398
coherent one-way QKD 94–95
coherent superposition 20
cold atom clocks 12
colloidal nanocrystals 352, 353
  diluted magnetic semiconductor colloidal NCs 367–368
  neutral and charged excitons spin 353–355
  perovskite QDs 368–369
  photocharging visualized via electron spin coherence 361–363
  photoluminescence in magnetic field 355–359
  spin-flip Raman scattering 363–364
  surface magnetism 364–366
  time-resolved Faraday rotation 359–361
color centers 511–524
  applications 543–544
  diamond
    Group IV Defect Centers 514–516
    material considerations 512–513
    nitrogen-vacancy defect center 513–514
    other defects 516
  host materials 518–519
  optical micro-and nanostructures 519–521
    fabrication in diamond 521–523
    fabrication in SiC 523–524
  quantum emitters 524
    at cryogenic temperature 525–527
    at room temperature 525
    state-of-the-art 528–530
  quantum memories 530
    cycling transitions 532–533
    figures of merit 530–532
    state-of-the-art 534–537
  2-qubit quantum gates 537
    spin-photon entanglement 537–539
  remote spin–spin entanglement 541–542
  silicon carbide
    material properties 516–517
    prominent color centers 517–518
  solid-state multi-qubit registers 539–541
complementarity approach 91
complementary metal-oxide semiconductor (CMOS) 517, 600
conduction band electrons 381
continuously measured Gaussian systems 117, 120
continuous variable (CV) 433
  QKD 95
  quantum systems 608–619
controlled reversible inhomogeneous broadening (CRIB) 433, 442
cooling techniques 114
Copenhagen interpretation 10, 13
correctness criterion 90
Coulomb crystals 311, 325
coupled-mode theory 584
coupled resonator optical waveguides (CROWs) 603
critical point drying (CPD) 590
Cr_7Ni-Cu system 291
cryogenic systems 144
cryptography 21
CV QKD 627

d

dangling bond magnetic polaron (DBMP) 365
DC trap potential 239
deal electric dipole emitter 154–155
Debye–Waller factor (DWF) 512, 524
decoherence
  blackbody radiation 109
  effects 108
  process 107
  rates 109
decoy-state QKD 92–93
Deep Quantum Learning 123
delta-kick cooling 114, 115
demultiplexing 73
  single photon sources 605
density matrix exponentiation 122
detection efficiency 99, 161–163
device-Edge-Cloud Continuum Orchestration concept 719
device-independent (DI)
  protocols 209
  QKD 96
dielectric and electrostatic engineering 568–569
difference-frequency generation (DFG) 517
differential phase shift (DPS) 94
different quantum emitters (QEs) 565
diluted magnetic semiconductor colloidal NCs 367–368
dipole-dipole interaction 248, 392, 398
discrete-time quantum walk 199
discrete variable (DV) quantum systems 608
distributed Bragg reflectors (DBR) 476, 477, 572
distributed multi-agent systems 724
DiVincenzo criteria 275
DLCZ quantum memory 440–442
Doppler cooling 211, 236, 242, 311
Doppler laser cooling 115, 212, 236, 237, 311
dry etching methods 521
Duan-Lukin-Cirac-Zoller (DLCZ) 433
dynamic nuclear polarization (DNP) 396
dynamics circular polarization (DCP) 356

e

EIT-based quantum memory 439, 447
electric dipole emission 158–161
electric dipole moment of electron (eEDM) 305
electromagnetically induced transparency (EIT) 311, 433, 435–438
electromechanical actuation 588
electronic energy level, for TBA[$TbPc_2$] 272
electronic optical modulator (EOM) 435
electron paramagnetic resonance (EPR) pulses 288
electron spin initialization 387
electron spin polarization 394, 397, 399, 401, 404
electron spin qubit relaxation 400–403
electron spin relaxation 398–405, 411, 420, 480, 534
electron Zeeman splitting 389
electrostriction 588
elongated nanorod (NR) 352
emission type memory 435
energy-level structure 305, 306, 312, 315, 320
energy splitting 354, 406, 410
engineering 2D semiconductor quantum emitters 567
entanglement 14, 20, 619
entanglement-based (EB) schemes 87
entanglement swapping 88, 144, 222–223, 452, 678, 728, 730
entropic approach 91
entropy production (EP) 115
entropy production rate (EPR) 115, 118
epitaxial quantum dots 466, 469, 609
epithelial-to-mesenchymal transition 174
error correction protocol 88, 466
error verification 88
evanescently-coupled waveguides 585

evaporative cooling 114
excitation intensity distribution 163–165
exciton magnetic polaron (EMP) 365, 368
excitons, in quantum dots 471–475
extraction efficiency, defined 487

f

Fabry-Perot cavities 71, 187
fast Fourier transform (FFT) spectra 360
feedback cooling 112, 113, 119, 666
Fermi-contact interaction 405
fiber-based QKD experiments 101
fiber based repeater 138–139
fiber optics 466
fiber-to-chip couplers 602, 603
fifth generation (5G) of wireless cellular networks 717
fine structure splitting (FSS) 69, 355, 474, 476, 483, 569, 609
finite difference time domain (FDTD) numerical simulation method 491
first quantum revolution 11–12, 25, 269, 295
5G deployment and commercialisation 715
flexoelectricity 588
fluorescence lifetime 160, 161, 165–167, 171
fluorescence line narrowing (FLN) technique 356
fluorescence quantum yield (QY) 165
fluorescence spectroscopy
  applications
    absolute quantum yield measurements 165–171
    MIET/GIET imaging 171–178
  theory
    angular distribution of radiation 161–163
    electric dipole emission 158–161
    excitation intensity distribution 163–165
    ideal electric dipole emitter 154–155
    plane waves with planar layered structures 155–158
    synthesis 165
focused ion beam (FIB) 521, 570, 611
Forster resonance energy transfer (FRET) 171, 175
Frank-van der Merwe (layer by layer) growth mode 381
free-space optics 581, 600, 734
4f-SMMs 271, 273
full-stack quantum processors 379
full width half maximum (FWHM) 216, 412
fundamental energy-exchange process, thermodynamic interpretation of 106
fundamentals of trapping and laser cooling of ions in a Paul trap 236

g

GaAs/AlGaAs QD, microphotoluminescence spectrum for 473, 474
GaAs-based nanostructures 466, 467
gas atomic ensembles 447–448
gated cooling probe detection (GCPD) scheme 242
Gaussian BS (GBS) 624
geometric phase control and spin echo 389
g-factor 17, 479, 480
global secure quantum communication network
  challenges 132–136
  entanglement 129
gradient echo memory (GEM) 139, 442
graphene-induced energy transfer (GIET) 154, 171
Group IV Defect Centers 514–516
Grover algorithm with $TbPc_2$ spin transistor 291

h

Hanbury Brown-Twiss (HBT)
  experiment 33
  interferometer 1
Harrow, Hassidim, and Lloyd (HHL) algorithm 122
Heisenberg relations 13
Heralded entanglement
  quantum repeaters based 680–685
  via entanglement swapping 222–223
heralded photon pair source 613, 617
heterometallic trinuclear [ErCeEr] complex 291
heterostructure device functionalization 570–571
hexagonal boron nitride (hBN) 563–565
hierarchical 6G virtual architecture 732
high-dimensional (HD) QKD 95
high-pressure high-temperature (HPHT) technique 513
Hofstadter butterfly effect 567
hole spin relaxation dynamics 381, 407
hole spin relaxation in single quantum dots 405
hole spin storage devices 411–413
Hong-Ou-Mandel (HOM)
  effect 56
  experiments 43
  interference 614
Huang-Rhys (HR) theory 524
hyperfine couplings 404
hyperfine-driven QTM (*hf*-QTM), in TBA[$TbPc_2$] 275

i

III-V QDs 378, 466, 479, 480
III-V semiconductor 378
  material system GaAs 381
implementation security 89
InAs/InP QDs onto Si photonic chip 611
InAs QDs on distributed Bragg reflectors 476, 477
Inductively Coupled Plasma (ICP) etching process 601
information reconciliation 89
InGaAs/GaAs quantum dots 408
InGaAs quantum dots 394
InP-based nanostructures 467
InP-based photonic crystal (PhC) 491
InP-based QDs 466, 475–476
InP-based single-photon detectors 4
in-plane hole hyperfine coupling strength 414
integrated circuits 581
integrated optical routers 582
integrated photonic devices for QKD 97
interference-based (IB) strategy 88
internal quantum efficiency 486–491
intrinsic electron spin coherence time 398
inverse flexoelectric effect 588
ionization fragments 218
ion trapping 306, 309
isotropic Fermi-contact hyperfine coupling 408

j

Jaynes-Cummings interaction 119

k

Knight field 392, 397, 402, 415
Kramers ion, degeneracy in 274

l

Lamb–Dicke regime 308, 309, 311, 315
Landau-Zener-Stückelberg (LZS) model 275
lanthanide-based SMMs 271
Larmor frequency 219, 354, 360, 389–391, 394, 540
laser 12
laser cooling 12, 115, 184, 211, 235–238, 326
laser-induced charge transfer (LICT) 312
layering, in communication networks 726
Lévy statistics 19
light emission of uncorrelated ion crystals 238

light-matter interaction 2–3
Lindblad equation 437
linear-optical quantum computing 450–451
linear Paul trap 259, 261, 306, 307
linear-quadrupole RF traps (LQTs) 306, 307
link layer 726, 728–730
local density of optical states (LDOS) 592
locality hypothesis 15
long-distance quantum networks 227
long-range free-space transmission challenges, in terrestrial environment 131

m

machine learning-assisted quantum computational problem 121
magnetic fields 220, 355, 389
magnetic noise resilient systems 283
magnetic traps 184, 187
magneto-optical traps (MOT) 184, 212
  detecting single atoms 187
  single neutral atoms 186–187
Majorana flops 184
Mathieu equations 307
Maxwell-Bloch equation 437
Maxwell's equations 154, 156
measurement-based model of quantum computation (MBQC) 622
measurement-based quantum technologies 378
measurement-device-independent (MDI) 93–94, 629
measurement-device-independent quantum key distribution (MDI-QKD) 453
measurement of the collective light emission of two ion crystal 259
memory based repeater 139
Merkulov-Efros-Rosen (MER) model 392, 393
mesoscopic objects 19, 20
metal-Induced Energy Transfer (MIET) 161, 171
metal-organic chemical vapor deposition (MOCVD) 471
Moiré heterostructures 570
molecular beam epitaxy (MBE) 352, 381, 470, 471
molecular hydrides 309
molecular-ion quantum technologies 306
  cooling of trapped ions 311–312
  ion trapping 306–309
  moleccular ion generation and state initialization 309–311
  quantum-logic experiments on single trapped molecular ions 317
molecular magnetism 271
molecular quantum logic spectroscopy 323
molecular-state-dependent optical-dipole forces 315
motional atomic state control 192
multidimensional path-encoded entanglement on silicon chip 619–621
multiphoton Greenberger-Horne-Zeilinger (GHZ) entangled state 619
multiplexing 63
  for deterministic photon generation 617–618

n

nanocavity-based method 169
nanocrystal (NC) 352
nano-opto-electro-mechanical systems (NOEMS)
  challenges 593–594
  characteristic of 586
  device principles 583
    device speed 586–587
    electromechanical actuation 587–588
    tunable beam splitters 584–586
  dynamical response 586

fabrication procedure 588–590
QPIC applications 590
light-matter interaction control 592–593
routing and switching single photons 590–592
stochastic noise 593
switches 591
nanoplatelets (NPLs) 352
nanowire-QDs 97
narrow-band cavity-based approaches 490
network layering for Quantum Internet 729, 730
network softwarization 721
and programmability 715
neutral excitons 353
nitrogen-vacancy centers (NV) 512
nitrogen-vacancy defect center 513–514
Noisy Intermediate Scale Quantum computers 24
noisy intermediate scale quantum (NISQ) technology 600
nondestructive state detection of MgH^+ ions 315
non-destructive state determination and coherent manipulation of CaH^+ ions 324
non-invasive molecular spectroscopy, of single molecular ions 319
non-Kramers ion, degeneracy in 274
nonlinear optical sources 54–63
nonlinear photon sources
fundamentals of 612–617
multiplexing, for deterministic photon generation 617–618
parasitic effects 618–619
normalization constant 34
novel fabrication techniques 529
N^2-scaling law of Dicke superradiance 247
n-type Schottky photodiode 383
nuclear spin bath 396
nuclear spin dynamics 409
numerical apertures (NAs) 187

o

O-band 466
on-chip Hanbury-Brown and Twiss experiment 611
on-chip silicon photonics 624
one-time pad encryption scheme 85
open quantum system's theory 110
optical coherence, for quantum communications 294
optical conveyor belts 191
orthogonal polarizations 192
for two dimensions 192
optical delay line (ODL) 433–435
optical detection of magnetic resonance (ODMR) 531
optical-dipole force (ODF) 315, 317
optical dipole trap (ODT) 185–187, 211, 213
detecting single atoms 187
laser 219
optical fibers 142, 466
optical injection locking technique 96, 97
optical link technologies 140, 141
optical loop quantum memory 434
optically active III-V QD nanostructures 378
optically active III-V semiconductor quantum dots (QDs) 379
optically active InP-based QDs, heterogeneous or/and monolithic integration of 497
optically active spin systems 378
optically controlled electron spins in semiconductor QDs 496
optically detected magnetic resonance (ODMR) 513
optically generated excitonic spin state 384
optical micro-and nanostructures 519–524
optical nanocavities 380
optical parametric oscillation (OPO) 54
optical phase-locked loop 97
optical-pumping schemes 326

optical spin storage photodiodes 383
optimization 122
optospintronic 293
orbital angular momentum (OAM) degree of freedom 629, 630
orthogonal polarizations 192
OSIRIS terminal 141
Overhauser field 393–398, 401, 402, 407–413, 419

p

parameter estimation 88, 89
passive components, in quantum photonic circuits 602–605
Pauli spin blockade 388
Paul traps 306
perovskite semiconductors 368–369
phasematching 55
phase-sensitive forces, for molecular state identification 320
phonon-sideband (PSB) 512, 524
photocharging visualized via electron spin coherence 361–363
photoluminescence excitation (PLE) spectroscopy 386
photoluminescence in magnetic field 355–359
photoluminescence (PL) spectrum 355
photon 1–2
photon anti-bunching 31, 35, 470
photon coherence 41
photon correlation functions 32
photon extraction efficiency 486–491
photonic cluster states 378
photonic crystal cavities, quantum dots in 491–493
photonic integrated circuits (PICs) 377
photonic quantum entanglement 130
photon-number purity and brightness 60
photon-number splitting (PNS) attack 92
photon pair generation, in silicon waveguides 613
pick-and-place technique 611, 612
piezoelectricity 587
Pockels effect 582, 587, 607
point spread function (PSF) 239
Poissonian photon number distribution 56
polarization beam splitter (PBS) 434
polarization-dependent AC-Stark shift 221
polarization gradient cooling 212
post-quantum cryptography 85
precision measurements 305, 315
prepare-and-measure (PM) strategy 87
privacy amplification 88, 89
probe laser 387
public-key cryptography 85
pump-probe Faraday rotation 480
pump-probe measurement technique 389
Purcell effect 39, 481, 491
Purcell enhancement 39

q

QD-based modular photonic modes 380
Q-factor 493
Q-sensors 284
quadrupolar couplings 404
quantisation 270–273
quantization axis 237
quantum algorithms for optimal classical decision problems 121
quantum bit (qubit) 275, 276, 727
quantum coherence and relaxation 273, 274
quantum communication 89, 292, 294, 451, 465, 470, 599
  defined 624
  generic scheme 624
  mission 733
  quantum key distribution 625–630
quantum computation 11
quantum computational supremacy 622
quantum computer 270
quantum computing 22–24, 288, 292, 599

data intensive tools for 120–123
devices 106
quantum correlations 87
quantum cryptography 21–22, 36
quantum dashes (QDashes) 475
quantum decoherence 20
quantum devices and circuits 465
quantum dots (QDs) 63–75, 352, 465, 511
in binary barriers 481
magneto-optical studies 483–486
optical properties 482–483
photon extraction efficiency 486–491
in photonic crystal cavities 491–493
radiative lifetime 493–494
challenges and future directions 496–497
excitons in 471–475
formation of 470–471
InGaAlAs/InP DBR structure 481
InP-based QDs 475–476
in quaternary barriers
epitaxial growth 477–478
structural properties 478–479
quantum effects, in single-molecule magnets
QTM 274–275
quantisation 270–273
quantum bits and multilevel systems 275–276
quantum coherence 273–274
quantum emitters (QEs) 31, 35, 566, 582, 587, 591, 592
brightness 46
at cryogenic temperature 525–527
deterministic excitation of 44
photon indistinguishability 46
at room temperature 525
single-photon purity 46
state-of-the-art 528–530
wavelength 46
quantum entanglement 16, 130
quantum error correction (QEC) 290, 622
quantum game theory 122
quantum gas microscopes 190
quantum information 16
processing 106, 619
measurement-based model of quantum computation 622
rudimentary quantum computers 622–624
technologies 599
quantum internet 725
quantum jumps 19
quantum key distribution (QKD) 85, 130, 452–454, 469, 510, 600, 625–627, 630, 732, 733
advantage 86
classical post-processing phase 86
coherent one-way method 94–95
components 96, 100
light sources 96–98
linear optical elements 100
single-photon detectors 98–99
continuous-variable 95
decoy-state 92–93
differential phase shift 94
fundamentals 86–89
high-dimensional 95
implementation security problem 95–96
measurement-device-independent 93–94
performance 100–101
quantum communication phase 86
security
definition 89–91
proofs 89
proof techniques 91–92
twin-field 94
quantum leftover hash lemma (LHL) 91
quantum light emitters, resources for 466
quantum-logic experiments, on single trapped molecular ions 317
molecular quantum logic spectroscopy, using resolved-sideband Raman transitions 321–323

quantum-logic experiments, on single trapped molecular ions (*contd.*)
  molecular state detection and spectroscopy via motional qubit 321
  quantum-non-demolition experiment with N_2^+ ions 315–321
quantum logic gate 22
quantum-logic schemes 306
quantum-logic spectroscopy 315, 322
quantum mechanics 9, 13, 25
  application 106
  formulation 716
  learning 122
  research timeline 716
  and single objects 17–21
quantum memory (QM) 139, 142, 144, 211, 433, 453
  applications
    linear-optical quantum computing 450–451
    quantum key distribution 452–454
    quantum repeater 451–452
    single photons detection 454
  atomic frequency comb 438–440
  DLCZ Scheme 440–442
  electromagnetically induced transparency 435–438
  long storage time 444
  optical delay line 434–435
  performance criteria
    bandwidth 445–446
    efficiency 443–444
    fidelity 444–445
    integratability 446
    storage time 444
    working wavelength 443
  physical realization
    gas atomic ensemble 447–448
    solid-state atomic ensemble 448–450
  protocols 511
quantum Monte-Carlo methods 19
Quantum Monte-Carlo Wave Function 10, 19
quantum networks 727
quantum networks in space
  global communication 129
    data protection 129
    information transfer, sources of 130
    secure communication 129 (*see* global secure quantum communication network)
quantum-non-demolition experiment 315–321
quantum-nondemolition state detection 315
quantum optics 509
quantum optimization 123
quantum photonic integrated circuits (QPIC) 581
  light-matter interaction control 592–593
  routing and switching single photons 590–592
quantum photonics 379, 581, 600
  applications 5–6
  communications 600
  experiments 600
  light-matter interaction 2–3
  photon 1–2
  single-photon detectors 4–5
  single-photon source 3–4
Quantum Principal Component Analysis 122
Quantum Random Access Memory (QRAM) 122
quantum repeaters 728
quantum revolutions 9, 11
  first 11–12
  second 21–24
quantum-safe cryptosystems 85
quantum science 105
quantum sensing (Q-sensing) 282–285, 599
quantum sensors, advantages of 283
quantum simulations 285–288
quantum simulators 285
quantum state transport 197
Quantum Support Vector Machines 122

quantum systems 183, 727
quantum technologies 105, 581
quantum thermodynamics (QT) 115
quantum tunneling of the magnetisation (QTM) 274–275, 287
quantum walks 198–201
quantum yield (QY) 165–171
quasi-static approximation (QSA) 394
2-qubit quantum gates, spin-photon entanglement 537–539
qubits 122, 305
qu*d*its *vs.* qubits 276
quenching circuit 98

r

Rabi oscillations 388
Rabi spectroscopy 317
radiative lifetime 493–494
radio access network 724
Raman-scattered single photon 440
Raman sideband cooling 311
rare earth ions doped solids (REIDS) 144, 438, 448
reflection high-energy electron diffraction (RHEED) 481
repeaters 138–140, 142
resonance-enhanced multiphoton dissociation (REMPD) 314
ripening effect 481
rovibrational quantum-state preparation techniques, for molecular ions 310
rubidium 210
rudimentary quantum computers (RQCs) 624
Russell–Saunders (RS) coupling scheme 271
Rydberg dipolar blockade mechanism 233

s

Sagnac-loop configuration 97
satellite-based entanglement distribution 734
satellite-based quantum networks 132
Scattershot BS (SBS) 624
Schmidt number 58
second quantum revolution 11, 21–24, 269
secrecy criterion 90
security, of QKD 85
self-assembled quantum dots 381
self-differencing technique 98
semiconductor colloidal nanocrystals 353
semiconductor nanocrystals (NCs) 352
semiconductor quantum dots 351, 469, 496
semiconductor quantum emitters 567
sideband cooling 112
sifting 88, 89
signal-to-noise ratio (SNR) 447
silicon carbide
  material properties 516–517
  prominent color centers 517–518
silicon-on-insulator (SOI) waveguide platform 581
silicon PIC for QKD application 628
silicon quantum photonics 600
  active components 605–606
  applications 619–630
  fabrication process 601
    electron beam lithography 601
    heater fabrication process 601
    ICP etching process 601
    with thermal tuning 602
    UV lithography 602
  passive components 602–605
  path encoding 619
  quantum communication 624–630
  quantum information processing 619
    measurement-based model of quantum computation 622
    rudimentary quantum computers 622–624
  quantum state of light 608
    nonlinear source on silicon 612–619
    solid-state single photon sources 611

silicon-vacancy (SiV) center 512
single InGaAs QD 382
single-mode Hamiltonian 1
single-molecule magnets (SMMs)
  and quantum effects 270–276
  quantum properties 269
  for quantum technologies 282
    quantum communication 292–294
    quantum computing 288–292
    quantum sensing 282–285
    quantum simulations 285–288
  for single molecule devices 277
    spin transistors 277–279
    spin valve 279–282
single-photon avalanche diode (SPAD) 98
single-photon detectors 4–5
single-photon emission 563, 570
single-photon sources 3–4, 36, 466, 467, 470, 476, 497
  for multi-photon applications
    motivation and applications 53–54
    nonlinear optical sources 54–63
    quantum dots 63–75
single-pixel SNSPD 97
single quantum system 17–18
6G networks
  architectural pillar 717
  Device-Edge-Cloud Continuum Orchestration concept 719
  future architecture 718
  vs. 5G 717
  intrinsic limitations 721–725
  optimal implementation paradigm 718
  softwarization and programmability 734
slow relaxation 415–420
software abstraction 721
softwarization and programmability 721, 722
sold-state atomic ensembles 448–450
solid-state color centers
  color centers 511–524
  photonic quantum-technology applications 510
solid-state nanosystems 377
solid-state quantum emitters in SiC and GaN platforms 609
solid-state spin qubits 377
sources for entangled photons 141
space-based quantum network 133
spatial multiplexing 618
spectrally broadband approach 490
spectral multiplexing schemes 618
spectral purity 57
spectroscopic transitions in molecular systems 305
spin-dependent control technique 192
spin-dependent optical lattice 191
Spin-dependent transport 198
spin dynamics, in InAs/InAlGaAs/InP self-assembled QDs 480
spin flip-flop process 398
spin-flip Raman scattering (SFRS) 363–364
spin-lattice relaxation time 273
Spin Orbit Coupling (SOC) 270
spin physics 354, 479
Spin-polarised Scanning Tunnelling Microscope (SP-STM) 284
spin-qubits for measurement based QIP 378
spin sensing method 388
spin-sensitive reset-pump-probe-read sequence 400
spin-spin relaxation time 273
spin transistors 277–279
spontaneous four-wave mixing (SFWM) 54, 55
spontaneous parametric down-conversion (SPDC) 54, 517
squared normalization constant 252
standard error of mean (SEM) 175
state-of-the-art photonic nanostructures 380
state-selective photoionisation schemes 310, 311
stochastic master equation (SME) 117

strain-tuning 569
Stranski-Krastanow (SK) growth 382, 410, 470
superconducting nanowire single-photon detectors (SNSPDs) 99, 607, 608, 629
superconducting transition-edge sensor 99
superpositions of states 20
Support Vector Machine 122
supramolecular spin valves 281
symmetric Dicke states 243
symmetric InP-based quantum dots, as quantum light sources 494, 496

t

$TbPc_2$ molecule, Q-sensing in 284
$TbPc_2$ quantum spin transistors 278
Tb_2Pc_3 quantum spin transistors 279, 280
technology readiness level (TRL) 135, 140
telecom quantum emitters 496
temporal multiplexing nonlinear sources 605
ternary alloy InGaAs 381
theory of collective Light emission of ion crystals in correlated Dicke states 246
theory of correlated Dicke states 243
theory of measuring collective light emission from linear ion crystals in Dicke-correlated states 254
thermal-tunable silicon photonics chips fabrication 601
thermodynamics, for energetically efficient quantum information processing 106
thermo-optic effect 582
thermo-optic (TO) phase shifters 605
three-dimensional harmonic trap potential 236
three-dimensional networking 722, 723
Time-resolved pump-probe Faraday rotation (FR) 359–361
titanium-based superconducting transition-edge detectors 608
transfer printing technique 612
transform-limited emission 510
transition metal dichalcogenide semiconductors (TMDS) 565–566
trapped ions, cooling of 312
trapping neutral atoms 184
trapping single atoms 212–214
trapping single neutral atoms
  with high fidelity 189
  individual neutral atoms 185–186
  magnetic traps 184
  MOT 186
  motional atomic state 194
  optical cavities 187
  optical dipole traps 185–187
  precision transport of 191–192
  quantum delocalization 198
  quantum state and position 190
  quantum state transport 197
  quantum walks of 198–200
  qubit state of 187–188
  sensors 202
  trapping neutral atoms 184
Trojan-horse attack (THA) 95
tunable beam splitters 584–586, 593
twin-field (TF) QKD 94
2D quantum emitters 569, 572–573
2D quantum photonics
  dielectric and electrostatic engineering 568–569
  heterostructure device functionalization 570–571
  hexagonal boron nitride (hBN) 563
  integrated photonics 572–573
  Moiré heterostructures 570
  semiconductor quantum emitters 567
  site-controlled implantation 569–570
  strain-tuning 569
  transition metal dichalcogenide semiconductors (TMDS) 565–566

2D quantum photonics (*contd.*)
  Van der Waals heterostructures 567
two-level quantum systems 377
two-photon absorption (TPA) 517, 618
two-photon interference 2, 3, 5, 94, 510, 528

u
untrusted nodes 140
  property 136
  in quantum key distribution 137
  stages of 137
    entangled photon sources 138
    mirror systems and coupled systems 137
    repeater 138–140

v
van der Waals 571
  heterostructures 567
vapor etching 590
Voigt configuration 389, 480, 483

w
wafer bonding technique 612
wave-particle duality 11
workplan, quantum technology 116

y
Young-type interferences 238

z
Zeeman-level selective ionization scheme 217
Zeeman-splitting-assisted quantum logic spectroscopy (ZSQLS) scheme 325
zero-phonon line (ZPL) 365, 512, 520, 522